Cyclic Cohomology at 40: Achievements and Future Prospects

Proceedings of Symposia in
PURE MATHEMATICS

Volume 105

Cyclic Cohomology at 40: Achievements and Future Prospects

Virtual Conference
Cyclic Cohomology at 40: Achievements
and Future Prospects
September 27–October 1, 2021
Fields Institute for Research
in Mathematical Sciences
Toronto, ON Canada

A. Connes
C. Consani
B. I. Dundas
M. Khalkhali
H. Moscovici
Editors

AMERICAN
MATHEMATICAL
SOCIETY
Providence, Rhode Island

2020 *Mathematics Subject Classification.* Primary 11M55, 14G40, 16E40, 16T05, 16T10, 18N10, 18M05, 19D55, 46L87, 58B34, 81T75, 81R60.

For additional information and updates on this book, visit
www.ams.org/bookpages/pspum-105

Library of Congress Cataloging-in-Publication Data

Names: Virtual Conference on Cyclic Cohomology at 40 (2021 : Fields Institute for Research in Mathematical Sciences, Toronto (Ont.)), Connes, Alain, editor. | Consani, Caterina, editor. | Dundas, B. I. (Bjørn Ian), editor. | Khalkhali, Masoud, 1963- editor. | Moscovici, Henri, 1944- editor.

Title: Cyclic cohomology at 40 : achievements and future prospects / edited by A. Connes, C. Consani, B. I. Dundas, M. Khalkhali, H. Moscovici.

Description: Providence, Rhode Island : American Mathematical Society, [2023] | Series: Proceedings of symposia in pure mathematics, 0082-0717 ; volume 105 | "Virtual Conference on Cyclic Cohomology at 40: Achievements and Future Prospects, September 27–October 1, 2021, Fields Institute for Research in Mathematical Sciences, Toronto, ON Canada." | Includes bibliographical references.

Identifiers: LCCN 2022045496 | ISBN 9781470469771 (paperback) | ISBN 9781479470472634 (ebook)

Subjects: LCSH: Homology theory–Congresses. | Algebraic cycles–Congresses. | Noncommutative differential geometry–Congresses. | AMS: Number theory – Zeta and L-functions: analytic theory – Relations with noncommutative geometry. | Algebraic geometry – Arithmetic problems. Diophantine geometry – Arithmetic varieties and schemes; Arakelov theory; heights. | Associative rings and algebras – Homological methods – (Co)homology of rings and algebras (e.g. Hochschild, cyclic, dihedral, etc.). | Associative rings and algebras – Hopf algebras, quantum groups and related topics – Hopf algebras and their applications. | Associative rings and algebras – Hopf algebras, quantum groups and related topics – Bialgebras. | K-theory – Higher algebraic K-theory – K-theory and homology; cyclic homology and cohomology. | Functional analysis – Selfadjoint operator algebras (C^*-algebras, von Neumann (W^*-) algebras, etc.) – Noncommutative differential geometry. | Global analysis, analysis on manifolds – Infinite-dimensional manifolds – Noncommutative geometry (à la Connes). | Quantum theory – Quantum field theory; related classical field theories – Noncommutative geometry methods. | Quantum theory – Groups and algebras in quantum theory – Noncommutative geometry.

Classification: LCC QA612.3 .V578 2023 | DDC 514/.23–dc23/eng20221230

LC record available at https://lccn.loc.gov/2022045496

♾ The paper used in this book is acid-free and falls within the guidelines
established to ensure permanence and durability.
Visit the AMS home page at `https://www.ams.org/`

10 9 8 7 6 5 4 3 2 1   28 27 26 25 24 23

Contents

Preface

This volume collects together the Proceedings of a (on-line) Conference at the Fields Institute (September 27-October 1 2021) dedicated to celebrating the 40-th anniversary of the birth announcement of cyclic cohomology at the Oberwolfach meeting in September 1981, where the talk "Spectral sequence and homology of currents for operator algebras" described the theory as well as its main properties. Cyclic cohomology was originally conceived as a tool in noncommutative differential geometry – an area of research, nascent at the time, dedicated to the exploration of geometric spaces whose algebras of coordinates are convolution algebras reflecting not only the points of the spaces but also their interrelations. These spaces arise naturally in pure mathematics and in quantum physics. Aside from the obvious examples of groupoids in differential geometry and phase space in quantum physics, there are some key examples of spaces where this new point of view applies; these are the micro structure of space-time and the Arithmetic Site underlying the geometry of the prime numbers.

Cyclic cohomology plays a fundamental role in an extensive machinery that allows for the formulation and investigation of the geometric properties of noncommutative structures. Given a noncommutative algebra A, one classifies the differential graded algebras (DGA) with A in degree 0 and which are endowed with a closed graded trace. These structures are characterized by the multilinear form τ on A giving the trace of the element $a_0da_1da_2\dots da_n$ of the DGA. This multilinear functional, called the "character", fulfills two key properties: it is a Hochschild cocycle and it is cyclic. The cyclic cochains form a subcomplex of the Hochschild complex, whose cohomology is the cyclic cohomology of A. The key result is that K-homology classes on A define such a DGA with closed graded trace, and that there is a general index formula computing the Fredholm index in terms of the pairing of the cyclic cohomology character of the K-homology class with the straightforward K-theory Chern character.

The forgetful map from cyclic cohomology to Hochschild cohomology, together with the boundary operator B playing the role of the de Rham boundary of currents, are then part of a long exact sequence involving the periodicity operator S. The latter is defined by means of the tensor product with the cyclic cohomology of the scalar field: a polynomial ring in one generator of degree 2. These general facts provide the theoretical framework of numerous index formulas and they play a central role in noncommutative geometry.

The circle of ideas involving cyclic homology turned out to be important well beyond the core field of noncommutative geometry. In particular, it has a strong footing in number theory, algebraic geometry, and in homotopy theory (algebraic topology). Developments in all these areas are interconnected and influence each other. Indeed, the implementation of the purely algebraic definitions of cyclic (co)homology and their functional analytic variants directly inspired the construction of the highly successful topological Hochschild homology (THH) and topological cyclic homology (TC) of Bokstedt, Hsiang and Madsen in the world of ring spectra. In turn these advancements supplied a remarkable computational tool in algebraic K-theory. We refer to the survey of B. Dundas in this volume for a historical overview of this part. With respect to the previously existing Hochschild homology, the novelty implemented by the cyclic aspect is the natural circle action on the topological Hochschild homology. This provides a computational tool of great relevance through the implementation of trace maps providing relations between topological cyclic homology and algebraic K-theory.

From the start, and in parallel to algebraic K-theory, the development of cyclic theory has been highly interdisciplinary and transverse to the traditional classification of mathematical fields. This book of Proceedings gives an overview of the richness of the subject.

In algebraic topology, the cyclotomic structure obtained using the cyclic subgroups of the circle action on topological Hochschild homology gives rise to remarkably significant arithmetic structures intimately related to crystalline cohomology, through the de Rham-Witt complex, and to Fontaine's theory of periods. At the non-archimedean places, the work of L. Hesselholt has shown that using the cyclic structure on topological Hochschild homology (THH) and the additional properties due to replacing the integers by the sphere spectrum as a base, one recovers not only the Witt construction and the de Rham-Witt complex but also elements of Fontaine's theory. In his contribution to this volume, Hesselholt highlights some of the spectacular recent results on topological cyclic homology of Nikolaus-Scholze, Bhatt–Morrow–Scholze, and Antieau-Mathew-Morrow-Nikolaus and explains how the Fargues-Fontaine curve with its decomposition into a punctured curve and the formal neighborhood of the puncture, naturally appears from various forms of topological cyclic homology and maps between them. We refer to the contribution of B. Tsygan in this volume for the construction of the de Rham-Witt complex in the noncommutative framework.

Recently, there has been an extensive body of work done on Hochschild and cyclic homology in the framework of infinity categories. In his contribution to this volume, D. Gepner gives a user friendly introduction to higher algebra generalizing ordinary algebra, that is algebra in the setting of ordinary, discrete, set based category theory. On the other hand, the contribution of R. Meyer and D. Mukherjee is part of a program to define analytic cyclic homology theories for bornological algebras over non-archimedean fields.

This book also reports a number of exciting developments in the complex theory. An active area of research exhibits the role of cyclic homology in higher index theory and the topology of manifolds. We refer to the survey article of P. Piazza and X. Tang on their work on the pairings of cyclic cocycles with secondary K-theory index class and the applications to Atiyah-Patodi Singer index theory in the context of proper Lie group actions. As a by-product, they obtain an interesting analog of the Connes-Moscovici higher index theorem for manifolds with boundaries in the context of proper Lie group actions. The analytic difficulty of capturing the invariance of higher signatures (or the vanishing of indices due to positive scalar curvature) in terms of cyclic cohomology index formulas consists in passing from the framework of topological K-theory of C^*-algebras, where the invariance takes place, to the cyclic cohomology of appropriate subalgebras suitable for differential geometric purposes. In their contributed article J. Wang, Z. Xie, and G. Yu introduce the new ℓ^1-index theory and its pairing with cyclic cohomology for both closed manifolds and compact manifolds with boundary, and prove an ℓ^1-version of the higher Atiyah-Patodi-Singer index theorem for manifolds with boundary. This article opens a new domain of exploration.

Another very active research domain highlighted in this book is the computation of index cocycles of pseudodifferential operators in terms of noncommutative residues and the replacement of abstract K-theory statements in index theory by concrete formulas with an unlimited range of applications. This is illustrated in several contributions, as follows. A. Savin and E. Schrohe extend the residue and trace expansions to the algebra of Fourier integral operators on Euclidean space. A. Gorokhovsky and E. van Erp obtain a formula for the index of a pseudodifferential operator with invertible principal symbol in the extended Heisenberg calculus of Epstein and Melrose. Furthermore, the contribution of J. Block, N. Higson and J. Sanchez presents an overview of the work of Perrot on new index cocycles.

Several articles address the problem of computing the cyclic homology of algebras occurring naturally in noncommutative geometry. R. Ponge presents a survey of the large amount of results on the cyclic homology of cross products by group actions, and M. Puschnigg gives a new method to obtain the periodic cyclic homology of crossed products for actions of discrete groups directly, without passing through Hochschild and cyclic homology. In his contribution, M. Lorentz adresses the Hochschild cohomology of the uniform Roe algebras. Y. Song and X. Tiang compute the variation of orbital integrals for the Cartan motion group under the deformation groupoid which embodies the Mackey principle in representation theory of Lie groups. The work of M. Pfaum applies methods of complex algebraic geometry to obtain general localization results in noncommutative geometry, by sheafification of the algebras under consideration and reduction of the computation to the stalks of the sheaf. Finally A. Baldare M. Benameur and V. Nistor provide a general principle which allows one to compare cyclic homology with K-theory after tensoring by scalars.

In the framework of noncommutative geometry the group symmetries are extended to Hopf algebra actions, while Lie algebra cohomology is greatly extended to the cyclic cohomology of Hopf algebras, which becomes the natural receptacle for characteristic classes. For this fundamental aspect of the general theory we refer to the contribution of H. Moscovici on the van Est analogy in Hopf cyclic cohomology, and to the extensive survey of M. Khalkhali and I. Shapiro. Both these articles open new directions of research.

The development of index theory in the noncommutative context has shown that the most embracing notion of geometric space suitable to develop the general theory is that of a smooth groupoid. We refer to the article of P. Carillo Rouse for the construction of the assembly map in this context.

The quantized calculus associated with a K-homology class was the key construction in the original development of the theory: we refer to the contribution of E. McDonald, F. Sukochev and X. Xiong for their work on the delicate analytic properties involved. The link between the characters of the obtained DGAs generated the periodicity operator S in cyclic cohomology. We refer to the contribution of J. Cuntz for the comparison of S with Bott periodicity.

The impact of cyclic theory on physics is reported in two fundamental contributions. In solid state physics, it is described by the contribution of E. Prodan on cyclic cocycles and quantized pairings in materials science. In quantum field theory, it is explained in the breakthrough contribution of T. van Nuland and W. van Suijlekom, where the perturbations of the spectral action are amazingly encoded by an entire cocycle whose cyclic properties survive at the one loop level and are intimately related to the Ward identities of perturbative expansions of gauge theories.

Finally, as a far-reaching goal, one can envisage that cyclic homology (in some guise of deRham-Witt or crystalline cohomology) should give access to the sought-for cohomology of the Arakelov compactification of the spectrum of the integers. On this topic, the joint contributions of A. Connes and C. Consani describe the way the zeros of the Riemann zeta function appear using the trace map on the noncommutative adèle class space, and the role of cyclic homology in the arithmetic context, after introducing a suitable structure sheaf for the Arakelov compactification of the spectrum of the integers. The enrichment of the cyclic theory due to working over the absolute base (i.e. the sphere spectrum, here understood in its primitive categorical characterization), refines the combinatorial backbone provided by the cyclic category, through the introduction of the new pericyclic category, explicitly involving the cyclotomic structure.

Proceedings of Symposia in Pure Mathematics
Volume **105**, 2023
https://doi.org/10.1090/pspum/105/01893

Chern-Connes-Karoubi character isomorphisms and algebras of symbols of pseudodidifferential operators

Alexandre Baldare, Moulay Benameur, and Victor Nistor

ABSTRACT. We introduce a class of algebras for which the Chern-Connes-Karoubi character is an isomorphism after tensoring with $\mathbb{C}$. We provide several examples of such algebras, such as invariant sections of Azumaya bundles, crossed products with finite groups and certain algebras of pseudodifferential operators.

1. Introduction and statement of main results

1.1. Motivation: invariant operators and their index. Let G be a compact Lie group acting on a smooth manifold M without boundary. Let $E \to M$ be an equivariant bundle and P be a G-invariant pseudodifferential operator acting on the sections of E. One of the *main motivations* for this paper is to *study the equivariant index* of P using methods of non-commutative geometry [**9**–**11**, **14**]. More precisely, we are interested in the index $\mathrm{ind}(\pi_\alpha(P))$ of the restriction $\pi_\alpha(P)$ of the G-invariant operator P to a generic isotypical component $\alpha \in \hat{G}$ [**5**–**7**], because the index provides the main obstruction to the invertibility of these operators. Moreover, the equivariant index $\mathrm{ind}_G(P) \in R(G)$ satisfies

$$\text{(1.1)} \qquad \mathrm{ind}_G(P) = \sum_{\alpha \in \widehat{G}} \frac{\mathrm{ind}(\pi_\alpha(P))}{\dim(\alpha)} [\alpha] \in R(G)\,.$$

Let $\mathcal{C}^\infty(S^*M; \mathrm{End}(E))^G$ be the algebra of G-invariant symbols. We may assume, without loss of generality, that P has order zero. Let us consider the K_1^{top}-class $[\sigma_0(P)] \in K_1^{\mathrm{top}}(\mathcal{C}^\infty(S^*M; \mathrm{End}(E))^G)$ of the principal symbol $\sigma_0(P) \in \mathcal{C}^\infty(S^*M; \mathrm{End}(E))^G$ of P. It is known [**10**, **14**, **15**, **17**, **33**] that the index can be expressed as the pairing $\phi_*([\sigma_0(P)])$ between a cyclic cocycle ϕ on $\mathcal{C}^\infty(S^*M; \mathrm{End}(E))^G$ and the class of $\sigma_0(P)$ in $K_1^{\mathrm{top}}(\mathcal{C}^\infty(S^*M; \mathrm{End}(E))^G)$. (We write K^{top} for the *topological* K-theory functors to distinguish them from their algebraic counterparts, see Remark 2.6.) We are thus lead to study the periodic cyclic homology of the symbol algebra $\mathcal{C}^\infty(S^*M; \mathrm{End}(E))^G$. This turns out to be a textbook application of non-commutative geometry using what we dub "Connes algebras," a class of algebras that we introduce and study in this paper.

2020 *Mathematics Subject Classification.* Primary 58B34, 46L80, 19D55 ; Secondary 16E40, 16S35, 14A22.

Key words and phrases. Cyclic homology, cyclic cohomology, character, K-theory, Representation ring, Lie group, Pseudodidifferential operator, principal symbol, manifold with corners.

For a complex algebra A, let

$$\mathrm{Ch} : K_j^{\mathrm{alg}}(A) \to \mathrm{HP}_j^{\mathrm{top}}(A) , \quad j = 0, 1 , \tag{1.2}$$

be the Chern-Connes-Karoubi character, where $\mathrm{HP}_j^{\mathrm{top}}(A)$ is a periodic cyclic homology of A defined using a suitable completion for the cyclic complex [**10–13, 23**]. Here K_j^{alg} is the algebraic K-theory functor, which is needed since A is not necessarily topological (see Remark 2.6 for its definition and relation to topological K-theory). In order to study the equivariant index of P, we are further led to study whether the map (1.2) induces isomorphisms

$$\mathrm{Ch} : K_j^{\mathrm{top}}(\mathcal{C}^\infty(S^*M; \mathrm{End}(E))^G) \otimes \mathbb{C} \to \mathrm{HP}_j^{\mathrm{top}}(\mathcal{C}^\infty(S^*M; \mathrm{End}(E))^G)$$

($j = 0, 1$). Let $\mathcal{C}(S^*M; \mathrm{End}(E))^G$ be the algebra of continuous, G-invariant sections of $\mathrm{End}(E)$, which is, of course, the C^*-completion of $\mathcal{C}^\infty(S^*M; \mathrm{End}(E))^G$. We are also led to study whether the inclusion $\mathcal{C}^\infty(S^*M; \mathrm{End}(E))^G \subset \mathcal{C}(S^*M; \mathrm{End}(E))^G$ yields isomorphisms in topological K-theory. We prove that the answer to both these questions is affirmative. The techniques for proving these isomorphisms turn out to apply to a more general setting than that of the algebra $\mathcal{C}^\infty(S^*M; \mathrm{End}(E))^G$. We thus introduce and study the class of "Connes algebras," which is, roughly, the class of algebras for which both maps (induced by the character and by inclusion) are isomorphisms. See also [**26, 36**] for earlier papers that used some similar techniques.

1.2. Connes algebras and statement of main results. Let us consider a category of cyclic complexes (this is, roughly, a category of topological algebras for which a definite choice of completion of the cyclic complex has been made, Definition 2.2). We assume that we are given for each $i \in \{0, 1\}$ a suitable K_i-theory functor that is close to topological K_i-theory and is such that the Chern-Connes-Karoubi character (Equation 1.2) extends to this category. A *Connes algebra* $\mathcal{A}$ for the given K_i-functors is a locally convex topological algebra $\mathcal{A}$ together with a continuous Banach algebra norm $\| \cdot \|_0$ on $\mathcal{A}$ and a given completion of the algebraic cyclic complex of $\mathcal{A}$ with the following properties:

(i) Let $\overline{\mathcal{A}}$ be the completion of $\mathcal{A}$ with respect to the norm $\| \cdot \|_0$. Then the inclusion $\mathcal{A} \to \overline{\mathcal{A}}$ induces an isomorphism in K-theory:

$$K_i(\mathcal{A}) \xrightarrow{\sim} K_i(\overline{\mathcal{A}}) , \quad i = 0, 1 . \tag{1.3}$$

(ii) The Chern-Connes-Karoubi character induces an isomorphism "after tensoring with $\mathbb{C}$," that is:

$$\mathrm{Ch} : K_i(\mathcal{A}) \otimes \mathbb{C} \xrightarrow{\sim} \mathrm{HP}_i^{\mathrm{top}}(\mathcal{A}) , \quad i = 0, 1 , \tag{1.4}$$

where the topological periodic cyclic homology is defined with respect to a suitable completion of the algebraic cyclic complex of $\mathcal{A}$.

See Definition 3.5 for more details.

For the applications in this paper, we shall choose $K_i = \mathrm{RK}_i$, where RK_i were introduced by Phillips [**34**] (but see also [**16, 17**] for further insight and important generalizations). The definition of the groups RK_i is recalled below. In this paper, it will be convenient to consider $i \in \mathbb{Z}/2\mathbb{Z} \simeq \{0, 1\}$.

Let X be a manifold with corners (but otherwise $\mathcal{C}^\infty$) and $\mathcal{F} \to X$ be a smooth fiber bundle of finite-dimensional, semi-simple algebras. That is, the fibers of $\mathcal{F}$ are isomorphic to finite direct sums of matrices. We let $\mathcal{C}^\infty(X; \mathcal{F})$ denote the set of smooth sections of $\mathcal{F}$. If $\mathcal{F}$ has simple fibers (i.e. matrix algebras) then this is the prototype of an Azumaya algebra with center $\mathcal{C}^\infty(X)$. We shall need also the following two ideals of $\mathcal{C}^\infty(X;\mathcal{F})$ associated to any closed subset $Y \subset X$. First, $\mathcal{C}_0^\infty(X,Y;\mathcal{F})$ consists of those sections that vanish on Y and, second, $\mathcal{C}_\infty^\infty(X,Y)$ consists of those sections that vanish to *infinite order* on Y. If $\mathcal{F} = \underline{\mathbb{C}}$ (the trivial vector bundle $X \times \mathbb{C} \to X$), then we drop it from the notation.

THEOREM 1.1. *Let X be a compact manifold with corners, $\mathcal{F} \to X$ be a smooth fiber bundle of finite-dimensional, semi-simple algebras, and $Y \subset \partial X$ be a union of closed faces of X. We assume that a compact Lie group G acts smoothly on X and $\mathcal{F}$ such that $GY = Y$. Then the algebras $\mathcal{C}^\infty(X;\mathcal{F})^G$, $\mathcal{C}_0^\infty(X,Y;\mathcal{F})^G$, and $\mathcal{C}_\infty^\infty(X,Y;\mathcal{F})^G$ are Connes algebras.*

Essentially the same method of proof leads to the same result when Y is a more general closed subset of X, but that is not needed for our main result and including a proof would greatly extend the length of the paper.

Let $\mathcal{K}$ be the algebra of compact operators. Thus, in particular, the algebra $\mathcal{C}^\infty(S^*M;\mathrm{End}(E))^G \simeq \Psi^0(M;E)^G/\big(\Psi^0(M;E)^G \cap \mathcal{K}\big)$ is a Connes algebra. To be able to work with the concept of a Connes algebra, we need to make some assumptions that are satisfied in the case of main interest in this paper, that is, that of the algebra $\mathcal{C}^\infty(S^*M;\mathrm{End}(E))^G$. First, we assume that the chosen K-theory functor (the one with respect to which we define the concept of a Connes algebra) is homotopy invariant and satisfies a six term exact sequence for *admissible* short exact sequences of algebras. (The class of admissible exact sequences needs to be specified each time, in our applications, all exact sequences will be admissible as long the ideal and the quotient have the induced topologies. See Sections 2 and 3 for more details.) We also assume that this category satisfies excision in periodic cyclic homology (again for *admissible* short exact sequences) and that the Chern-Connes-Karoubi character is a natural transformation between the exact sequence in K-theory and periodic cyclic homology (associated to, again, an *admissible* short exact sequence of algebras). We then have the following result that will play an important role in our study of the algebras $\mathcal{C}^\infty(S^*M;\mathrm{End}(E))^G$. (In our applications, *all exact sequences will be admissible* if the topologies are compatible.)

THEOREM 1.2. *Let $\mathfrak{C}$ be a category of topological cyclic complexes that satisfies excision in periodic cyclic homology and in K-theory. Let $0 \to \mathcal{I} \to \mathcal{A} \to \mathcal{A}/\mathcal{I} \to 0$ be an admissible short exact sequence in $\mathfrak{C}$ of algebras with a fixed Banach algebra norm such that the corresponding Banach space completion of this sequence is exact as well. Assume that two of the algebras $\mathcal{I}$, $\mathcal{A}$, and $\mathcal{A}/\mathcal{I}$ are Connes algebras. Then the third one is a Connes algebra as well.*

Possible choices of the category $\mathfrak{C}$ and of the K-groups are:

- the category of m-algebras and K-groups of Cuntz with the admissible sequences being the *linear-split* exact sequences [**16, 17**];
- the subcategory of *Fréchet* m-algebras, in which *all exact sequences are admissible* [**30**]. This is the case that is relevant for the applications in this paper, and hence this is the case that we shall consider in detail. Moreover, in this case, the K-theory groups of Cuntz coincide with the representable K-theory groups introduced earlier by Phillips [**34**].

Our result for the algebra $\mathcal{C}^\infty(S^*M; \mathrm{End}(E))^G$ is obtained by iterating Theorem 1.2 for something that we call a C-stratification of ideals (Definition 3.8). More precisely, we have the following result.

THEOREM 1.3. *The algebra $\mathcal{C}^\infty(S^*M; \mathrm{End}(E))^G$ is a Fréchet m-algebra with a composition series*

$$0 = I_N \subset I_{N-1} \subset \ldots \subset I_0 = \mathcal{C}^\infty(S^*M; \mathrm{End}(E))^G$$

consisting of closed nuclear, Fréchet m-algebra ideals with the following property: For each $j = 0, 1, \ldots, N-1$, there exists a compact manifold with corners Y_j, a smooth bundle $\mathcal{E}_j \to Y_j$ of semisimple algebras and an algebra morphism

$$\phi_j : I_j/I_{j+1} \to \mathcal{C}_0^\infty(Y_j, \partial Y_j; \mathcal{E}_j) := \{f : Y_j \to \mathcal{E}_j \mid f \text{ smooth and } f|_{\partial Y_j} = 0\}$$

that is a homeomorphism onto its image $\phi_j(I_j/I_{j+1})$ and $\cap_{n\in\mathbb{N}}\mathcal{C}_0^\infty(Y_j, \partial Y_j; \mathcal{E}_j)^n \subset \phi_j(I_j/I_{j+1})$.

For $k \in \mathbb{Z}/2\mathbb{Z}$, we let $H^k_{\mathrm{per}}(X, Y) := \oplus_{i\in k} H^i(X, Y) \otimes \mathbb{C}$, that is, the direct sum of all singular cohomology groups of the pair (X, Y) of the same parity as k and with complex coefficients. In particular, as in [**26**], we have

COROLLARY 1.4. *Using the notation of Theorem 1.3, the algebra morphisms ϕ_j induce isomorphisms in periodic cyclic homology. Moreover, there is a manifold with corners $\mathfrak{Y}_k$ that is a finite covering $\mathfrak{Y}_k \to Y_k$ such that the center of $\mathcal{C}^\infty(Y_k; \mathcal{E}_k)$ is isomorphic to $\mathcal{C}^\infty(\mathfrak{Y}_k)$ and hence*

$$\mathrm{HP}_k^{\mathrm{top}}(I_j/I_{j+1}) \simeq H^k_{\mathrm{per}}(\mathfrak{Y}_j, \partial\mathfrak{Y}_j), \quad k \in \mathbb{Z}/2\mathbb{Z},$$

and we have a spectral sequence with $E^1_{-p,q} := H^{q-p}_{\mathrm{per}}(\mathfrak{Y}_p, \partial\mathfrak{Y}_p)$ convergent to $\mathrm{HP}^{\mathrm{top}}_{q-p}(\mathcal{C}^\infty(M, \mathrm{End}(E))^G$.

Our algebras are built out of Azumaya algebras, so we include several results proving that they, as well as other algebras related to Azumaya algebras, are Connes algebras. In particular, the commutative algebras modelled by smooth functions are Connes algebras (this is just the initial step in the proof of Theorem 1.1, which is more general).

1.3. Contents of the paper. We start is Section 2 with some preliminaries on periodic cyclic homology and topological K-theory. In Section 3, we briefly review excision from an abstract view point and introduce the concepts of Connes algebras, C-smooth stratifications, and C-smooth algebras. We conclude this section by showing that any C-smooth algebra is a Connes algebra. In Section 4, we show that topologically nilpotent algebras are Connes algebras and show that the space of smooth sections $\mathcal{C}_0^\infty(X, Y\mathcal{F})$ of a finite rank bundle $\mathcal{F} \to X$ of semisimple algebras over a manifold with corners X vanishing on a union of closed faces Y is a Connes algebra. This are the first steps for the proof of the main example of Connes algebra discussed in this paper, that is the set of G-invariant smooth sections $\mathcal{C}^\infty(X, Y, \mathcal{F})^G$ of such a bundle $\mathcal{F}$ with $GY = Y$. This example is treated in Section 5 by exhibiting a C-smooth stratification using blow-up constructions. In Section 6, we then deduce from the previous sections that when the group G is finite then the crossed product $\mathcal{C}_0^\infty(X, Y; \mathcal{F}) \rtimes G$ is a Connes algebra, that the algebras of pseudodifferential operators $\Psi^0(M, E)^G$, $\Psi^{-\infty}(M, E)^G$ and $\Psi^{-1}(M, E)^G$, over a closed G-manifold M, with coefficients in a vector bundle E, are Connes algebras. We also show that the algebras of smooth families of pseudodifferential operators

$\Psi^0(M|B,E)$, $\Psi^{-\infty}(M|B,E)$ and $\Psi^{-1}(M|B,E)$ are Connes algebras. Appendix A is devoted to the proof of a technical result needed in the proof that $\mathcal{C}^\infty(X,Y,\mathcal{F})^G$ is a Connes algebra.

2. Background material and preliminary results

For us, a *locally convex topological algebra* is a complex algebra $\mathcal{A}$ endowed with a compatible locally convex topology such that the multiplication $\mathcal{A}\times\mathcal{A}\to\mathcal{A}$ is continuous. (The multiplication is thus assumed to be continous jointly in both variables.) In other words, for any continuous semi-norm p on $\mathcal{A}$ there exists a continuous semi-norm p' such that $p(ab)\le p'(a)p'(b)$, $\forall a,b\in\mathcal{A}$, see [**11**, Chap 3, Appendix B]. All topological algebras considered in this paper will be assumed to be locally convex.

2.1. Review of periodic cyclic homology. We briefly recall the topological periodic cyclic homology for topological algebras that will be used in this paper, see for instance [**10, 11, 23, 37**] for the standard material used in this section. The topological cyclic homology considered in this paper is such that it is the same for an algebra and for its completion. If $\mathcal{A}$ is a topological algebra, we let $\mathcal{A}^+:=\mathcal{A}\times\mathbb{C}$ denote its unitalization with the product topology and the product $(x,\lambda)(y,\mu)=(xy+\lambda y+\mu x,\lambda\mu)$. In particular, if $\mathcal{A}$ is a locally convex (complete) topological algebra, then $\mathcal{A}^+$ is equipped with the semi-norms given for instance by $p_\alpha(a,\lambda):=p_\alpha(a)+|\lambda|$. An exact sequence

$$0\to\mathcal{I}\to\mathcal{A}\to\mathcal{B}\to0 \tag{2.1}$$

of locally convex, topological algebras is by definition a sequence of continuous algebra morphisms between locally convex, topological algebras such that $\mathcal{A}\to\mathcal{B}$ is surjective, $\mathcal{B}$ has the induced quotient topology, and $\mathcal{I}$ maps homeomorphically onto its kernel. So it is an algebraically exact sequence with a compatibility of the topologies. In the following, $\mathfrak{A}$ will be a category of locally convex, topological algebras.

DEFINITION 2.1. Let $\mathcal{A}$ be a topological algebra. *A topological cyclic complex on* $\mathcal{A}$ is a graded vector space $\mathcal{C}(\mathcal{A}):=\big(\mathcal{C}_n(\mathcal{A})\big)_{n\ge0}$, where, for all n, $\mathcal{C}_n(\mathcal{A})$ is a suitable completion of the (algebraic) tensor product $\mathcal{A}^+\otimes\mathcal{A}^{\otimes n}$. We require that these completions be such that the usual differentials b and B extend by continuity to maps denoted by the same symbols: $b:\mathcal{C}_n(\mathcal{A})\to\mathcal{C}_{n-1}(\mathcal{A})$ and $B:\mathcal{C}_n(\mathcal{A})\to\mathcal{C}_{n+1}(\mathcal{A})$, see for instance [**10, 11**] or [**23**, Chapter II]. The Hochschild, cyclic, and periodic cyclic homologies of a topological cyclic complex $(\mathcal{A},\mathcal{C}(\mathcal{A}))$ are the corresponding groups defined using the mixed complex $(\mathcal{C}(\mathcal{A}),b,B)$. In particular, the *periodic (topological) cyclic homology groups* $\mathrm{HP}_*^{\mathrm{top}}(\mathcal{A})$ of $\mathcal{A}$ are the homology groups of the complex $\big(\prod_{n\in2\mathbb{Z}_+}\mathcal{C}_n(\mathcal{A}),\prod_{n\in2\mathbb{Z}_+}\mathcal{C}_{n+1}(\mathcal{A}),b+B\big)$ ($\mathbb{Z}/2\mathbb{Z}$-graded).

The topological cyclic complexes on (locally convex) topological algebras form a category.

DEFINITION 2.2. Let $\mathfrak{A}$ be a category of locally convex algebras with continuous algebra morphisms, as above. A *category of topological cyclic complexes* (based on the objects of $\mathfrak{A}$) is a category $\mathfrak{C}$ whose objects are pairs $(\mathcal{A},\mathcal{C}(\mathcal{A}))$ consisting of topological algebra $\mathcal{A}$ in $\mathfrak{A}$ together with a topological cyclic complex $\mathcal{C}(\mathcal{A})$ of

$\mathcal{A}$. The morphisms in $\mathfrak{C}$ are continuous, linear maps $f = (f_a, f_c) : (\mathcal{A}, \mathcal{C}(\mathcal{A})) \to (\mathcal{B}, \mathcal{C}(\mathcal{B}))$ such that $f_a : \mathcal{A} \to \mathcal{B}$ is an algebra morphism in $\mathfrak{A}$ and $f_c(a_0 \otimes a_1 \otimes \ldots \otimes a_n) = f_a(a_0) \otimes f_a(a_1) \otimes \ldots \otimes f_a(a_n)$.

2.2. Review of topological K-theory. In order to define a topological K-functor satisfying the properties which are necessary for our study of Connes algebras (such as periodicity) some further conditions need to be imposed on the chosen category of locally convex algebras, see for instance [**11, 16, 34**]. In our applications, we will use the Cuntz category of m-algebras. We have therefore devoted this brief review to the topological K-functor for Fréchet m-algebras, as introduced and studied by Phillips in [**34**]. For general m-algebras, we refer the interested reader to the excellent survey [**17**].

Recall that a given complete topological algebra $\mathcal{A}$ as above is an m-algebra when its topology can be defined by a family of submultiplicative semi-norms [**16, 17, 31**]. A Fréchet m-algebras is an m-algebra which is Fréchet, in other words, it is a complete topological algebra whose topology is defined by a countable family of submultiplicative semi-norms. Notice that each of these two categories is stable under projective tensor products. A first important example of a Fréchet m-algebra is the algebra $\mathcal{R}$ of infinite complex matrices with rapidly decreasing entries, see [**11, 16, 34**]. An element $R = (R_{ij})_{i,j\in\mathbb{N}} = \sum_{i,j\in\mathbb{N}} R_{ij}E_{ij}$ in $\mathcal{R}$ ($R_{ij} \in \mathbb{C}$) satisfies:

$$p_n(R) := \sum_{i,j\in\mathbb{N}} (1+i+j)^n |R_{ij}| < +\infty, \quad \forall \in \mathbb{N}.$$

Given any complete topological algebra $\mathcal{A}$, the completed projective tensor product algebra $\mathcal{R}\hat{\otimes}\mathcal{A}$ will be denoted $\mathcal{R}\mathcal{A}$. It is an m-algebra when $\mathcal{A}$ is an m-algebra [**17**]. It is a Fréchet algebra when $\mathcal{A}$ is a Fréchet algebra [**34**]. Recall that any continuous homomorphism $\varphi : \mathcal{A} \to \mathcal{B}$ of locally convex topological algebras extends to the unitalizations as well as to a homomorphism of locally convex topological algebras $M_N(\mathcal{A}) \to M_N(\mathcal{B})$, $N \in \mathbb{N}$, and $\mathcal{R}\mathcal{A} \to \mathcal{R}\mathcal{B}$.

We also consider the topological algebra $\mathcal{C}^\infty([a,b])\hat{\otimes}\mathcal{A} \simeq \mathcal{C}^\infty([a,b];\mathcal{A})$ of smooth $\mathcal{A}$-valued functions on $[a,b]$. As in the Introduction, the algebra of complex smooth functions on $[0,1]$ which vanish with all their derivatives at 0 and 1 will be denoted $\mathcal{C}^\infty_\infty([0,1],\{0,1\})$. We the define the *smooth suspension* $\mathcal{S}\mathcal{A}$ of the topological algebra $\mathcal{A}$ by $\mathcal{S}\mathcal{A} := \mathcal{C}^\infty_\infty([0,1],\{0,1\})\hat{\otimes}\mathcal{A}$. Again, if $\mathcal{A}$ is an m-algebra (respectively, a Fréchet m-algebra) then so are all the above tensor products.

We recall next the definiton of the groups RK_0 and RK_1 introduced by Phillips [**34**], but we formulate it in the more general setting of locally convex topological algebras.

DEFINITION 2.3. Let $\mathcal{A}$ be a locally convex topological algebra.

(1) Two idempotents $e_0, e_1 \in \mathcal{A}$ will be called *smoothly homotopic* if there exists an idempotent $e \in \mathcal{C}^\infty([0,1],\mathcal{A})$ such that $e(0) = e_0$ and $e(1) = e_1$.
(2) Denote by $\bar{P}(\mathcal{A})$ the set of idempotents $e \in M_2((\mathcal{R}\mathcal{A})^+)$ such that $e - \begin{pmatrix} 1 & 0 \\ 0 & 0 \end{pmatrix} \in M_2(\mathcal{R}\mathcal{A})$. Then the set of smooth homotopy classes in $\bar{P}(\mathcal{A})$ is denoted $\mathrm{RK}_0(\mathcal{A})$.
(3) Denote by $\bar{U}(\mathcal{A})$ the set of invertible elements $u \in (\mathcal{R}\mathcal{A})^+$ such that $u - 1 \in \mathcal{R}\mathcal{A}$. Then the set of smooth homotopy classes in $\bar{U}(\mathcal{A})$ is denoted $\mathrm{RK}_1(\mathcal{A})$.

If $\mathcal{A}$ is a Fréchet m-algebra, Phillips has shown that $\mathrm{RK}_0(\mathcal{A})$ and $\mathrm{RK}_1(\mathcal{A})$ are abelian groups that depend functorially on $\mathcal{A}$ [**34, 35**]. This result obviously remains true in the category of locally convex topological algebras, however, the category of Fréchet m-algebras has several additional nice properties. In the following, the term "excision" is used in the usual sense, which is recalled in Definition 2.4. In the following $\mathfrak{A}$ will be a category of locally convex algebras with continuous algebra morphisms for which we shall assume that there is given a distinguished class of exact sequences of $\mathfrak{A}$ called *admissible exact sequences.*

DEFINITION 2.4. Let $\mathfrak{A}$ be a category whose objects are locally convex algebras and whose morphisms are (suitable) continuous algebra morphisms. We shall say that $\mathfrak{A}$ *satisfies the excision property for* RK if, for any admissible short exact sequence $0 \to \mathcal{I} \to \mathcal{A} \to \mathcal{B} \to 0$ in $\mathfrak{A}$, there exist natural boundary maps $\mathrm{RK}_0(\mathcal{B}) \to \mathrm{RK}_1(\mathcal{I})$ and $\mathrm{RK}_1(\mathcal{B}) \to \mathrm{RK}_0(\mathcal{I})$ that, together with the functorial morphisms, yield a six-term exact sequence:

$$\begin{array}{ccccc} \mathrm{RK}_0(\mathcal{I}) & \longrightarrow & \mathrm{RK}_0(\mathcal{A}) & \longrightarrow & \mathrm{RK}_0(\mathcal{B}) \\ \uparrow & & & & \downarrow \\ \mathrm{RK}_1(\mathcal{B}) & \longleftarrow & \mathrm{RK}_1(\mathcal{A}) & \longleftarrow & \mathrm{RK}_1(\mathcal{I}) . \end{array}$$

THEOREM 2.5 (Phillips). *The representable K-theory functors* RK_0 *and* RK_1 *satisfy stability, Bott periodicity, and excision (or six term short exact sequence) on the category of Fréchet m-algebras (in which all exact sequences are admissible).*

REMARK 2.6. In addition to the RK_i functors, we shall also use the *algebraic* K-theory functors $K_i^{\mathrm{alg}}(A)$, $i = \{0,1\}$ defined as the Grothendieck group of the semi-group of finitely generated, projective modules over the unital algebra A, for $i = 0$, respectively as the limit of the abelianizations of $GL_n(A)$, as $n \to \infty$, for $i = 1$. If A is a topological algebra with unit, we shall also use the *topological* K-theory functors $K_i^{\mathrm{top}}(A)$, which are the quotient of their algebraic counterparts with respect to homotopy. For a non-unital algebra A, $K_n(A)$ is the kernel of $K_n(A^+) \to K_n(\mathbb{C})$, where K denotes any of the K-functors considered. We have natural surjections $K_i^{\mathrm{alg}}(A) \to K_i^{\mathrm{top}}(A)$. If A happens to be a Banach algebra, then $K_0^{\mathrm{alg}}(A) \simeq K_0^{\mathrm{top}}(A)$, but, in general, $K_1^{\mathrm{alg}}(A) \to K_1^{\mathrm{top}}(A)$ is not injective. We also have natural maps $K_i^{\mathrm{top}}(A) \to \mathrm{RK}_i(A)$ if A is a locally convex topological algebra, which are isomorphisms if A is a Banach algebra or if it is a subalgebra of a Banach algebra that is stable under holomorphic functional calculus, see [**11**, Chapter 3, Appendix C, Proposition 3].

REMARK 2.7. Recall the Chern-Connes-Karoubi character $\mathrm{Ch} : K_i^{\mathrm{alg}}(A) \to \mathrm{HP}_i^{\mathrm{alg}}(A)$, defined for any complex algebra [**10, 11, 23**]. This definition extends right away to the RK-functors. Indeed, let $\mathfrak{C}$ be a category of topological cyclic complexes and $(\mathcal{A}, \mathcal{C}(\mathcal{A}))$ be an object of $\mathfrak{C}$. Then the natural map of complexes $\mathcal{C}^{\mathrm{alg}}(\mathcal{A}) := (\mathcal{A}^{\otimes(n+1)})_{n\in\mathbb{Z}_+} \to \mathcal{C}(\mathcal{A})$ will induce a natural group morphism $\mathrm{HP}_*^{\mathrm{alg}}(\mathcal{A}) \to \mathrm{HP}_*^{\mathrm{top}}(\mathcal{A})$ such that Ch descends to a map $\mathrm{Ch} : (K_i^{\mathrm{alg}}(\mathcal{A})/\sim) \to \mathrm{HP}_i^{\mathrm{top}}(\mathcal{A})$, $i = 0,1$ where $\sim$ denotes the smooth homotopy of projections or of invertible elements. As a consequence, we obtain a Chern-Connes-Karoubi character $\mathrm{Ch} : \mathrm{RK}_i(\mathcal{A}) \to \mathrm{HP}_i^{\mathrm{top}}(\mathcal{A})$ extending the classical Chern-Connes-Karoubi character on algebraic K-theory:

$$\mathrm{RK}_i(\mathcal{A}) \to \left(K_i^{\mathrm{alg}}(\mathcal{R}\widehat{\otimes}\mathcal{A})/\sim\right) \xrightarrow{\mathrm{Ch}} \mathrm{HP}_i^{\mathrm{top}}(\mathcal{R}\widehat{\otimes}\mathcal{A}) \xrightarrow{Tr_*} \mathrm{HP}_i^{\mathrm{top}}(\mathcal{A}) .$$

3. Connes' principle and main results

We introduce a class of algebras for which the Chern-Connes-Karoubi character is an isomorphism (after tensoring with $\mathbb{C}$). Several examples of such algebras will be provided in Sections 4, 5, and 6 in relation with algebras of symbols of certain pseudodifferential operators.

3.1. Categories with the excision property for HP and for RK.

We shall treat excision from an abstract view point. In the following, we will have to specify each time which are the "admissible" exact sequences that we consider.

DEFINITION 3.1. Let $\mathfrak{A}$ be a category of locally convex topological algebras in which a class of exact sequences, called *admissible* is given. Let $\mathfrak{C}$ be a category of topological cyclic complexes. We shall say that $0 \to \mathcal{I} \to \mathcal{A} \to \mathcal{B} \to 0$ is *an admissible exact sequence* in $\mathfrak{C}$ if it is an admissible exact sequence of topological algebras in $\mathfrak{A}$, $\mathcal{C}(\mathcal{B})$ is the quotient of $\mathcal{C}(\mathcal{A})$ under the induced morphism and $\mathcal{C}(\mathcal{I}) \to ker\big(\mathcal{C}(\mathcal{A}) \to \mathcal{C}(\mathcal{B})\big)$ is a homeomorphism onto its image.

In particular, the cyclic complexes of $\mathcal{B}$ and $\mathcal{I}$ are determined by that of $\mathcal{A}$.

DEFINITION 3.2. We shall say that a category $\mathfrak{C}$ of topological cyclic complexes *satisfies the excision property for periodic cyclic homology* (or for HP) if for any object $(\mathcal{A}, \mathcal{C}(\mathcal{A}))$ of $\mathfrak{C}$ and for any admissible short exact sequence $0 \to \mathcal{I} \to \mathcal{A} \to \mathcal{B} \to 0$ in $\mathfrak{C}$ (see Definition 3.1) the natural inclusion $\mathcal{C}(\mathcal{I}) \to \ker\big(\mathcal{C}(\mathcal{A}) \to \mathcal{C}(\mathcal{B})\big)$ induces an isomorphism of the corresponding periodic cyclic homology groups.

It is known that the category of Fréchet m-algebras satisfies excision, but, for general m-algebras, the result is known only for *split* exact sequences [**17**–**19**, **30**].

REMARK 3.3. A consequence of the excision property of Definition 3.2 is that there exist boundary maps $\mathrm{HP}_0^{\mathrm{top}}(\mathcal{B}) \to \mathrm{HP}_1^{\mathrm{top}}(\mathcal{I})$ and $\mathrm{HP}_1^{\mathrm{top}}(\mathcal{B}) \to \mathrm{HP}_0^{\mathrm{top}}(\mathcal{I})$ that, together with the functorial morphisms, yield a six-term exact sequence in periodic cyclic homology:

$$\begin{array}{ccccc} \mathrm{HP}_0^{\mathrm{top}}(\mathcal{I}) & \longrightarrow & \mathrm{HP}_0^{\mathrm{top}}(\mathcal{A}) & \longrightarrow & \mathrm{HP}_0^{\mathrm{top}}(\mathcal{B}) \\ \uparrow & & & & \downarrow \\ \mathrm{HP}_1^{\mathrm{top}}(\mathcal{B}) & \longleftarrow & \mathrm{HP}_1^{\mathrm{top}}(\mathcal{A}) & \longleftarrow & \mathrm{HP}_1^{\mathrm{top}}(\mathcal{I}) . \end{array}$$

Thus Phillips' Theorem 2.5 in particular states that the category of *Fréchet* m-algebras satisfies the excision property for RK. See also [**17**]. In the same way, the results of Cuntz state that if $\mathfrak{C}$ is the larger category of m-algebras with $K_* := kk_*(\mathbb{C}, \bullet)$ (where kk is the Cuntz bivariant K-theory functor) then this category also satisfies the K-excision property. We shall need the following extension of a result in [**33**].

PROPOSITION 3.4. *Let $\mathfrak{C}$ be a category of topological cyclic complexes that satisfies excision for* HP *and for* RK *and let $0 \to \mathcal{I} \to \mathcal{A} \to \mathcal{B} \to 0$ be a short exact sequence in $\mathfrak{C}$. Assume that for two of the algebras $\mathcal{I}$, $\mathcal{A}$, and $\mathcal{B}$, the Chern-Connes-Karoubi character* $\mathrm{Ch} : \mathrm{RK}_i \otimes \mathbb{C} \to \mathrm{HP}_i^{\mathrm{top}}$ *is an isomorphism, then* Ch *is an isomorphism also for the third algebra.*

PROOF. The results in [**33**] and Cuntz [**16**] imply that Ch is a natural transformation from the six-term exact sequence of Definition 2.4 to the six-term exact sequence of Remark 3.3. The result then follows from the Five Lemma [**4**]. □

See also [**22, 36**].

3.2. Connes algebras. Recall that a *Banach algebra-norm* $\|\cdot\|_0$ on a topological algebra $\mathcal{A}$ is a continuous norm on $\mathcal{A}$ such that $\|ab\|_0 \leq \|a\|_0\|b\|_0$.

DEFINITION 3.5. Let $(\mathcal{A}, \mathcal{C}(\mathcal{A}))$ be a topological cyclic complex on $\mathcal{A}$, where $\mathcal{A}$ is an algebra endowed with a continuous Banach norm $\|\cdot\|_0$. Let $\overline{\mathcal{A}}$ be the completion of $\mathcal{A}$ in the norm $\|\cdot\|_0$. Then $(\mathcal{A}, \mathcal{C}(\mathcal{A}), \|\cdot\|_0)$ is called a *Connes algebra* if, for all $j \in \mathbb{Z}/2\mathbb{Z}$, the following two maps

(i) $\mathrm{RK}_j(\mathcal{A}) \to \mathrm{RK}_j(\overline{\mathcal{A}})$ and
(ii) $\mathrm{Ch} : \mathrm{RK}_j(\mathcal{A}) \otimes \mathbb{C} \to \mathrm{HP}_j^{\mathrm{top}}(\mathcal{A})$

are isomorphisms.

The isomorphism (i) and the Chern-Connes-Karoubi character

$$\mathrm{Ch} : \mathrm{RK}_n(\mathcal{A}) \longrightarrow \mathrm{HP}_n^{\mathrm{top}}(\mathcal{A}) \tag{3.1}$$

then yield Connes' character

$$\widetilde{\mathrm{Ch}} : K_n(\overline{\mathcal{A}}) \longrightarrow \mathrm{HP}_n^{\mathrm{top}}(\mathcal{A})\,. \tag{3.2}$$

We are ready now to prove Theorem 1.2.

PROOF OF THEOREM 1.2. Item (i) of the definition of Connes algebra (Definition 3.5) is a direct consequence of the six terms exact sequences and a standard diagram chase. Item (ii) of that definition was proved in Proposition 3.4. □

We are interested in Connes algebras because of "Connes' principle", which can be stated as follows:

THEOREM 3.6 (Connes' principle). *If $\mathcal{A}$ is a Connes algebra with $\overline{\mathcal{A}}$ its completion with respect to the given norm $\|\cdot\|_0$, then Connes' character $\widetilde{\mathrm{Ch}} : K_i(\overline{\mathcal{A}}) \otimes \mathbb{C} \to \mathrm{HP}_i^{\mathrm{top}}(\mathcal{A})$ is an isomorphism, for all $i \in \mathbb{Z}/2\mathbb{Z}$.*

Connes' principle follows immediately from the definition of Connes algebras and the preceding discussion. Moreover, it permeates his earlier works on cyclic homology and it is what sets them apart from other related works on cyclic homology. It is, of course, due to Connes. It is a useful principle since the K-groups of C^*-algebras are notoriously difficult to compute.

REMARK 3.7. Let us gather some basic observations about the class of Connes algebras.

(1) The class of Connes algebra is stable with respect to direct sums.
(2) The algebra $\mathcal{C}^\infty(M)$ of smooth functions over a closed manifold M is a Connes algebra with respect to the sup-norm. This follows directly from Connes' result [**10**]. Such algebras are basically the building blocks of the class of Connes algebras.

(3) Moreover, for any (Fréchet) Connes m-algebra $\mathcal{A}$, the (Fréchet) m-algebra $\mathcal{C}^\infty(M,\mathcal{A})$ is a Connes algebra for the sup-norm associated with the given Banach norm on $\mathcal{A}$. See [**34**] for item (i) of the definition of Connes algebras. Item (ii) follows from Künneth formulas in cyclic homology, see [**24, 28**].
(4) In the same way, the (Fréchet) m-algebra $\mathcal{RA}$ is a Connes algebra when $\mathcal{A}$ is a Connes algebra. One can use many Banach completions of $\mathcal{RA}$, see [**17**]. See also [**34**] for the case of Fréchet m-algebras.

In view of definition 3.5, we now introduce the *Banach version* $\mathfrak{C}_0$ of a category of cyclic complexes $\mathfrak{C}$ as the category of triples $(\mathcal{A},\mathcal{C}(\mathcal{A}),\|\cdot\|_0)$ where $(\mathcal{A},\mathcal{C}(\mathcal{A}))$ is an objet in $\mathfrak{C}$ (a locally convex topological algebra $\mathcal{A}$ together with a choice of completion of its cyclic mixed complex $\mathcal{C}(\mathcal{A})$) and a continuous Banach algebra norm on $\mathcal{A}$. The morphisms in $\mathfrak{C}_0$ are the morphisms in $\mathfrak{C}$ that are continuous with respect to the given Banach algebra norms. The main result of this paper states that the category of Connes algebras behaves well with respect to exact sequences. For simplicity, we shall assume from now on that all exact sequences in $\mathfrak{A}$ are admissible. This is the case for Fréchet m-algebras, which is the case that is used in our main result.

DEFINITION 3.8. Let $\mathfrak{C}$ be a category of topological cyclic complexes that satisfies excision for both HP and RK and let $(\mathcal{A},\mathcal{C}(\mathcal{A}),\|\cdot\|_0)$ be an object in its Banach version $\mathfrak{C}_0$.

(1) A *C-smooth stratification* of $\mathcal{A}$ is a sequence of two-sided ideals of $\mathcal{A}$ in $\mathfrak{C}$,
$$0 = \mathcal{I}_N \subset \mathcal{I}_{N-1} \subset \ldots \subset \mathcal{I}_1 \subset \mathcal{I}_0 = \mathcal{A}\,,$$
such that $\mathcal{I}_k/\mathcal{I}_{k+1}$ are Connes algebras for the Banach norms induced from $\mathcal{A}$.
(2) The topological algebra $\mathcal{A}$ is a C-smooth algebra if it admits such a C-smooth stratification.

We shall also write $I_j := \overline{\mathcal{I}}_j$. We are ready now to state the main result of this section.

THEOREM 3.9. *Every C-smooth algebra $\mathcal{A}$ in a category that satisfies excision in K-theory and in periodic cyclic homology is a Connes algebra. In particular, the Connes character* $\widetilde{\mathrm{Ch}} : K_i^{\mathrm{top}}(\overline{\mathcal{A}}) \otimes \mathbb{C} \longrightarrow \mathrm{HP}_i^{\mathrm{top}}(\mathcal{A})$ *is an isomorphism.*

PROOF. We shall proceed by induction on N. If $N = 0$ or 1, there is nothing to prove. Assume then that $N \geq 2$ and that the result is known for $\mathcal{I}_1$ and let us prove it for $\mathcal{A} = \mathcal{I}_0$. The completion of the exact sequence

$$0 \to \mathcal{I}_1 \to \mathcal{I}_0 \to \mathcal{I}_0/\mathcal{I}_1 \to 0 \tag{3.3}$$

with respect to the given Banach space norms is exact since we have assumed that the norms are induced from $\mathcal{A}$. Moreover, this is an exact sequence in $\mathfrak{C}$ since we have assumed that I_1 is an ideal of I_0 in $\mathfrak{C}$ and that $\mathfrak{C}$ satisfies the excision property for HP. We know that $\mathcal{I}_0/\mathcal{I}_1$ is a Connes algebra by the hypothesis. We also know that $\mathcal{I}_1$ is a Connes algebra by the induction hypothesis. Theorem 1.2 then gives that $\mathcal{I}_0$ is also a Connes algebra. □

4. Some examples of Connes algebras

We already mentioned many standard examples of Connes algebras and we now list some others that will be used in the next section in the study of the algebra of G-equivariant symbols on manifolds. The standard Goodwillie argument plays an important part in the applications and computations, see [**21**]. Recall that this argument is based on the vanishing of the periodic cyclic homology for the class of (topologically) nilpotent algebras. These algebras appear naturally when dealing with appropriate quotients of ideals, and form, in some sense, a class of "negligible" Connes algebras, since their K-theory groups as well as their periodic cyclic spaces are trivial. Let us restrict ourselves again to the category of Fréchet m-algebras (and hence, all exact sequences will be admissible). Also, from now on in this paper, the category $\mathfrak{C}$ will be the category of Fréchet m-algebras with the cyclic complex defined using the projective tensor product. The category $\mathfrak{C}_0$ is the subcategory of $\mathfrak{C}$ of algebras equipped with a fixed Banach algebra norm. The morphisms are the morphisms in $\mathfrak{C}$ that are continuous with respect to the fixed norm. The exact sequences are supposed to yield exact sequences of Banach algebras (for the induced norms).

DEFINITION 4.1. [Meyer] A complete locally convex topological algebra $\mathcal{N}$ will be called a *topologically nilpotent algebra,* if, for any semi-norm p on $\mathcal{N}$, there exists $k \geq 1$, such that $p(\mathcal{N}^k) = 0$ (that is, $p(f_1 \dots f_k) = 0$ for any $f_1, \dots, f_k \in \mathcal{N}$).

Notice that such a topologically nilpotent algebra cannot be unital.

PROPOSITION 4.2. *Let $\mathcal{N}$ be a topologically nilpotent Fréchet m-algebra with a Banach norm $\|\cdot\|_0$. Then $\mathcal{N}$ is spectrally invariant in its Banach completion $\overline{\mathcal{N}}$ with respect to $\|\cdot\|_0$. Moreover, all groups $\mathrm{HP}_i(\mathcal{N})$, $K_0^{\mathrm{alg}}(\mathcal{N})$, $K_i^{\mathrm{top}}(\mathcal{N})$, $\mathrm{RK}_i(\mathcal{N})$, $K_0^{\mathrm{alg}}(\overline{\mathcal{N}})$, $K_i^{\mathrm{top}}(\overline{\mathcal{N}})$, and $\mathrm{RK}_i(\overline{\mathcal{N}})$, $i \in \mathbb{Z}/2\mathbb{Z}$, vanish and hence $\mathcal{N}$ is a Connes algebra.*

PROOF. It is well known that the topological periodic cyclic homology of any topologically nilpotent algebra $\mathcal{N}$ is trivial. For the topological periodic cyclic homology, we refer for instance to [**21, 29, 30**]. As for the vanishing of K-theory groups, this is easy to check. First, for any $x \in M_N(\mathcal{N})$, the power series $\sum_{k=0}^{\infty} x^k$ converges, and hence $1 - x$ is invertible. This shows that $K_1^{\mathrm{top}}(\mathcal{N}) = 0$. Let $p_0 \in M_N(\mathbb{C})$ and let $e \in M_N((\mathcal{N})^+)$ be any idempotents which satisfies that $e - p_0 \in M_N(\mathcal{N})$. Then $u := ep_0 + (1-e)(1-p_0)$ satisfies $eu = up_0$ and is an invertible element in $M_N(\mathcal{N}^+)$. Hence e and p_0 are equivalent. This proves that $K_0^{\mathrm{alg}}(\mathcal{N}) = 0$, and hence that $K_0^{\mathrm{top}}(\mathcal{N}) = 0$, again since the later is the quotient of the former, see Remark 2.6.

Finally, let $\overline{\mathcal{N}}$ be some Banach completion of $\mathcal{N}$. Next, both $\mathcal{RN}$ and $\overline{N}$ are also topologically nilpotent. The same arguments gives then $\mathrm{RK}_0(\mathcal{N}) = \mathrm{RK}_1(\mathcal{N}) = 0$. They also show that $K_0^{\mathrm{alg}}(\overline{\mathcal{N}}) = 0 = K_0^{\mathrm{top}}(\overline{\mathcal{N}})$ and $K_1^{\mathrm{top}}(\overline{\mathcal{N}}) = 0$. □

Let us deduce some consequences for the algebras of smooth functions on compact manifolds with corners. Let X be a compact manifold with corners. Recall that, for a closed subspace $Y \subset X$, $\mathcal{C}_0^\infty(X, Y)$ is the ideal in $\mathcal{C}^\infty(X)$ consisting of those complex valued smooth functions that vanish on Y and that $\mathcal{C}_\infty^\infty(X, Y)$ is the set of those smooth complex valued functions that *vanish to infinite order* on Y. It is an ideal of $\mathcal{C}_0^\infty(X, Y)$.

We shall need Azumaya bundles over compact manifolds with corners.

NOTATION 4.3.

(i) $\mathcal{S}, \mathcal{S}_i \to X$ are smooth, bundles of finite dimensional, *simple* algebras.
(ii) $\mathcal{F} \to X$ is a smooth bundle of finite dimensional, *semi-simple* algebras.
(iii) $E \to X$ is a smooth vector bundle on X.
(iv) $\mathcal{C}^\infty(X; E)$ is the vector space of smooth sections of E and $\mathcal{C}(X; \mathcal{F})$ is the vector space of *continuous* sections of $\mathcal{F}$.
(v) Let $Y \subset X$ be a closed subset, then $\mathcal{C}_0(X \smallsetminus Y; E)$ is the vector space of *continuous* sections of $\mathcal{F}$ that *vanish* on Y, $\mathcal{C}_0^\infty(X, Y; E)$ is the vector space of smooth sections of E that *vanish* on Y, and $\mathcal{C}_\infty^\infty(X, Y; E)$ is the vector space of smooth sections of E that *vanish to infinite order* on Y.

So $\mathcal{C}^\infty(X)$, $\mathcal{C}_0^\infty(X, Y)$, and $\mathcal{C}_\infty^\infty(X, Y)$ coincide respectively with $\mathcal{C}^\infty(X; E)$, $\mathcal{C}_0^\infty(X, Y; E)$, and $\mathcal{C}_\infty^\infty(X, Y; E)$ for $E = \underline{\mathbb{C}}$, the one dimensional trivial bundle.

The fibers of $\mathcal{S}$ and $\mathcal{S}_i$ are matrix algebras (one block), so they have natural C^*-norms. The fibers of $\mathcal{F}$ are direct sums of matrix algebras, so they also have natural C^*-norms. In particular, locally we have $\mathcal{F} \simeq \oplus_{i=1}^k \mathcal{S}_i$, but *not* globally. Similarly, we have locally that $\mathcal{S} \simeq \mathrm{End}(E)$, but not globally. In fact, every point x of X has an open neighborhood U such that $\mathcal{S}|_U \simeq U \times M_n(\mathbb{C})$ for some $n \in \mathbb{N}$. In particular, this allows us to define natural C^*-norms on the spaces of smooth sections, which by completion give rise to continuous sections.

The algebra $\mathcal{C}^\infty(X, \mathcal{F})$ is a prototype of an Azumaya algebra (over its center) and in Proposition 4.5 will show that it is a Connes algebra. Recall that a unital Fréchet m-algebra $\mathcal{A}$ is an Azumaya algebra over its center $\mathcal{Z}$ if $\mathcal{A}$ is a finitely generated projective module over Z such that

$$\mathcal{A} \otimes_{\mathcal{Z}} \mathcal{A}^{op} = \mathrm{End}_{\mathcal{Z}}(\mathcal{A}),$$

where $\mathcal{A}^{op}$ is the algebra $\mathcal{A}$ with the opposite product and $\mathrm{End}_{\mathcal{Z}}(\mathcal{A})$ is the algebra of continuous $\mathcal{Z}$-linear maps on $\mathcal{A}$. We shall denote by $\mathrm{Prim}(A)$ the *Primitive ideal spectrum* of a C^*-algebra A, that is, the set of primitive ideals of A with the hull-kernel topology [**20**]. (Recall that an ideal $I \subset A$ is primitive if, by definition, it is the kernel of a non-zero irreducible $*$-representation of A.)

LEMMA 4.4. *Let X be a compact manifold with corners and let $\mathcal{F} \to X$ be a locally trivial bundle of semisimple algebras as above. Let Y be a closed subspace of X. Set $\mathcal{A} := \mathcal{C}^\infty(X, \mathcal{F})$, $\mathcal{I} := \mathcal{C}_0^\infty(X, Y; \mathcal{F})$, and let A and I be their completions in the "sup-norm."*

(1) *Let $\mathfrak{X} := \mathrm{Prim}(A)$. Then $\mathfrak{X}$ is a finite covering of X and hence a manifold with corners.*
(2) *The center $\mathcal{Z}$ of $\mathcal{A}$ is isomorphic to $\mathcal{C}^\infty(\mathfrak{X})$. Moreover, $\mathcal{Z} \simeq \mathcal{C}^\infty(X, Z(\mathcal{F}))$, where $Z(\mathcal{F}) \subset \mathcal{F} \to X$ is the center of $\mathcal{F}$.*
(3) *The algebra $\mathcal{A}$ is an Azumaya algebra over $\mathcal{Z}$, more precisely, $\mathcal{A} \simeq \mathcal{C}^\infty(\mathfrak{X}; \mathcal{F}_A)$, where $\mathcal{F}_A$ is a bundle of* simple *algebras over $\mathfrak{X}$, canonically obtained from $\mathcal{F}$.*
(4) *The algebra $A \simeq \mathcal{C}(\mathfrak{X}; \mathcal{F}_A)$ is an Azumaya algebra over its center $Z \simeq \mathcal{C}(\mathfrak{X})$.*

PROOF. We may assume X to be connected. On each trivialization U of $\mathcal{F}$, we have $\mathcal{F}|_U \simeq U \times \bigoplus_{k=1}^p M_{n_k}(\mathbb{C})$. This implies that the primitive ideal spectrum

$\mathfrak{X} := \operatorname{Prim}(A)$ of the C^*-completion $A := \mathcal{C}(X, \mathcal{F})$ of $\mathcal{A}$ is a covering of X. Hence $\mathfrak{X}$ has the structure of a manifold with corners. Next, the center $\mathcal{Z}$ of $\mathcal{A}$ coincides with the bundle over X with fibers given by the centers of the fibers of $\mathcal{F}$. It is isomorphic to the algebra $\mathcal{C}^\infty(\mathfrak{X})$ as can be checked locally.

For the third item, let $\mathcal{F}_A \to \mathfrak{X}$ be the fiber bundle of simple algebras given by $\mathcal{F}_A|_{U\times\{k\}} \simeq U \times M_{n_k}(\mathbb{C})$, with the transition functions induced by $\mathcal{F}$. Then $\mathcal{A} \simeq \mathcal{C}^\infty(\mathfrak{X}, \mathcal{F}_\mathcal{A})$ and hence the $\mathcal{Z}$-module $\mathcal{A}$ is projective and finitely generated. The proof is completed by a local inspection. Recall that if we denote by $\mathbb{C}^p$ the center of $\bigoplus_{k=1}^p M_{n_k}(\mathbb{C})$, then we have

$$(\oplus_{k=1}^p M_{n_k}(\mathbb{C})) \otimes_{\mathbb{C}^p} (\oplus_{k=1}^p M_{n_k}(\mathbb{C}))^{op} \cong \operatorname{End}_{\mathbb{C}^p} (\oplus_{k=1}^p M_{n_k}(\mathbb{C})) .$$

The last statement is thus a direct consequence of (3). □

The following proposition is well-known for X smooth without boundary (or corners) [**10, 11, 22, 23**]. Its proof generalizes right away to manifolds with corners and to sections of semi-simple bundles.

PROPOSITION 4.5. *Let X be a compact manifold with corners, $\mathcal{F} \to X$ be a fiber bundle of semi-simple algebras over X, and $\mathcal{Z} \simeq \mathcal{C}^\infty(\mathfrak{X})$ be the center of $\mathcal{A} = \mathcal{C}^\infty(X, \mathcal{F})$. Then the inclusion $i : \mathcal{Z} \to \mathcal{A}$ induces isomorphisms $\operatorname{HP}_n(\mathcal{Z}) \to \operatorname{HP}_n(\mathcal{A})$ and $\operatorname{RK}_n(\mathcal{Z}) \otimes \mathbb{Q} \to \operatorname{RK}_n(\mathcal{A}) \otimes \mathbb{Q}$, $n \in \mathbb{Z}/2\mathbb{Z}$. Let Z and A be the norm completions of these algebras. Then, similarly, the inclusion $Z \subset A$ induces isomorphisms $\operatorname{RK}_n(Z) \otimes \mathbb{Q} \to \operatorname{RK}_n(A) \otimes \mathbb{Q}$, $n \in \mathbb{Z}/2\mathbb{Z}$. Consequently, $\mathcal{Z}$ and $\mathcal{A}$ are Connes algebras with*

$$\operatorname{HP}_n(\mathcal{A}) \simeq \operatorname{HP}_n(\mathcal{Z}) \simeq H^n_{\mathrm{per}}(\mathfrak{X}) .$$

PROOF. First, let us prove that if $\mathcal{F} = \underline{\mathbb{C}} := X \times \mathbb{C}$, then $\mathcal{C}^\infty(X; \mathcal{F}) = \mathcal{C}^\infty(X)$ is a Connes algebra as in the case without corners. We follow the classical proof (see [**10, 11, 22, 23**] for the justifications not included here). Let $i \in \mathbb{Z}/2\mathbb{Z}$. since $\mathcal{C}^\infty(X)$ is stable under holomorphic calculus in $\mathcal{C}(X)$, we have that $K_i^{\mathrm{top}}(\mathcal{C}^\infty(X)) \to K_i^{\mathrm{top}}(\mathcal{C}(X))$ is an isomorphism. Since $\mathcal{C}(X)$ is a C^*-algebra, we also have $\operatorname{RK}_i(\mathcal{C}^\infty(X)) \simeq K_i^{\mathrm{top}}(\mathcal{C}^\infty(X)) \simeq K_i^{\mathrm{top}}(\mathcal{C}(X)) \simeq \operatorname{RK}_i(\mathcal{C}(X)) \simeq K^i(X)$. Next, because X has the homotopy type of a finite CW-complex and Ch is a natural transformation of cohomology theories that is an isomorphism for spheres, we also have that $\operatorname{Ch} : \operatorname{RK}_i(\mathcal{C}(X)) \otimes \mathbb{C} \simeq K^i(X) \otimes \mathbb{C} \to H^i_{\mathrm{per}}(X)$ is an isomorphism. Finally, we have that the classical proof of $\operatorname{HP}_i(\mathcal{C}^\infty(X)) \simeq H^i_{\mathrm{per}}(X)$ (using the Hochschild-Kostant-Rosenberg-Connes isomorphism) applies without change. We have thus obtained the sequence of isomorphisms

(4.1)
$$\operatorname{RK}_i(\mathcal{C}^\infty(X)) \otimes \mathbb{C} \simeq K_i^{\mathrm{top}}(\mathcal{C}(X)) \otimes \mathbb{C} \simeq K^i(X) \otimes \mathbb{C} \simeq H^i_{\mathrm{per}}(X) \simeq \operatorname{HP}_i(\mathcal{C}^\infty(X))$$

whose composition is the Chern-Connes-Karoubi map, which is hence an isomorphism. This proves that $\mathcal{C}^\infty(X)$ is a Connes algebra.

The center $\mathcal{Z}$ of $\mathcal{A}$ is $\mathcal{Z} \simeq \mathcal{C}^\infty(\mathfrak{X})$. By Proposition 4.5, by replacing X with $\mathfrak{X}$ and $\mathcal{F}$ with $\mathcal{F}_A$, we may assume that $\mathcal{F}$ consists of simple fibers (i.e. matrix algebras) and hence that $\mathcal{A}$ is Azumaya over X. In particular, we may replace $\mathfrak{X}$ with X in what follows. Let us consider then the natural embeddings:

$$\mathcal{Z} \xhookrightarrow{\ i\ } \mathcal{A} \xhookrightarrow{\ j\ } \operatorname{End}_\mathcal{Z}(\mathcal{A}) \xhookrightarrow{\ k\ } M_N(\mathcal{Z}) \xhookrightarrow{\ l\ } M_N(\mathcal{A}) ,$$

where the morphism k comes from an embedding $\mathcal{F} \subset X \times \mathbb{C}^N$ of vector bundles over X. It is well known that $\mathrm{HP}_n(M_N(\mathcal{Z})) \simeq \mathrm{HP}_n(\mathcal{Z})$ and that $\mathrm{RK}_n(M_N(\mathcal{Z})) \simeq \mathrm{RK}_n(\mathcal{Z})$ and that the same results hold for $\mathcal{A}$. We thus obtain the compositions $k_* \circ j_* \circ i_* : K^n(X) \to K^n(X)$ and $l_* \circ k_* \circ j_* : K_n^{\mathrm{top}}(\mathcal{A}) \to K_n^{\mathrm{top}}(\mathcal{A})$. The first morphism is multiplication with the class $[\mathcal{A}] \in K^0(X)$, which is an invertible element in $K^0(X) \otimes \mathbb{Q}$. Similarly, $K_*^{\mathrm{top}}(\mathcal{A})$ is a module over $K^*(X)$ using the isomorphism $\mathcal{Z} \otimes_{\mathcal{Z}} \mathcal{A} \simeq \mathcal{A}$ and $l \circ k \circ j$ is obtained from $k \circ j \circ i$ by tensoring with $id_{\mathcal{A}}$ over $\mathcal{Z}$. Hence $l_* \circ k_* \circ j_*$ is also multiplication by $[\mathcal{A}]$. It follows that both compositions $k_* \circ j_* \circ i_*$ and $l_* \circ k_* \circ j_*$ become isomorphisms after tensoring with $\mathbb{Q}$ and hence $i_* : \mathrm{RK}_n(\mathcal{Z}) \to \mathrm{RK}_n(\mathcal{A})$ is an isomorphism. The same argument gives that the inclusion $Z \to A$ induces an isomorphism $\mathrm{RK}_n(Z) \otimes \mathbb{Q} \to \mathrm{RK}_n(A) \otimes \mathbb{Q}$.

To obtain the corresponding result in periodic cyclic homology, we first notice that we have also maps $\mathrm{HH}_n(i) : \mathrm{HH}_n(\mathcal{Z}) \to \mathrm{HH}_n(\mathcal{A})$ of Hochschild homology groups that can be localized using the $\mathcal{Z}$-module structure on both groups. Once we localize, the same argument as above (really just Morita equivalence) gives that all localizations to maximal ideals in $\mathcal{Z}$ of $\mathrm{HH}_n(i)$ are isomorphisms, and hence $\mathrm{HH}_n(i)$ is an isomorphism as well. (This is the argument from [**26**].) Hence i induces an isomorphism in periodic cyclic homology as well. Since we have proved that $\mathcal{Z} \simeq \mathcal{C}^\infty(\mathfrak{X}) = \mathcal{C}^\infty(X)$ is a Connes algebra, it follows that $\mathcal{A}$ is a Connes algebra as well (since, up to tensoring with $\mathbb{C}$, it has the same RK and HP groups). □

The following result will be a basic step in the proofs of the results of the next section.

PROPOSITION 4.6. *Let $Y \subset X$ be a closed subspace and $\mathcal{I}$ be a closed subalgebra of $\mathcal{C}_0^\infty(X, Y; \mathcal{F})$ that contains $\mathcal{C}_\infty^\infty(X, Y; \mathcal{F})$. Then, for all $i \in \mathbb{Z}/2\mathbb{Z}$, the natural inclusions induce isomorphisms:*

(1) $\mathrm{RK}_i(\mathcal{C}_0^\infty(X, Y; \mathcal{F})) \simeq \mathrm{RK}_i(\mathcal{C}_0(X \smallsetminus Y; \mathcal{F}))$;
(2) $\mathrm{RK}_i(\mathcal{C}_\infty^\infty(X, Y; \mathcal{F})) \simeq \mathrm{RK}_i(\mathcal{I})$;
(3) $\mathrm{RK}_i(\mathcal{I}) \simeq \mathrm{RK}_i(\mathcal{C}_0^\infty(X, Y; \mathcal{F}))$;
(4) $\mathrm{HP}_i^{\mathrm{top}}(\mathcal{C}_\infty^\infty(X, Y; \mathcal{F})) \simeq \mathrm{HP}_i^{\mathrm{top}}(\mathcal{I})$; *and*
(5) $\mathrm{HP}_i^{\mathrm{top}}(\mathcal{I}) \simeq \mathrm{HP}_i^{\mathrm{top}}(\mathcal{C}_0^\infty(X, Y; \mathcal{F}))$.

PROOF. The algebra $\mathcal{C}_0^\infty(X, Y, \mathcal{F})$ is stable under holomorphic functional calculus in $\mathcal{C}_0(X \smallsetminus Y, \mathcal{F})$, and hence the inclusion induces an isomorphism $\mathrm{RK}_i(\mathcal{C}_0^\infty(X, Y, \mathcal{F})) \to K_i^{\mathrm{top}}(\mathcal{C}_0(X \smallsetminus Y, \mathcal{F}))$. The Fréchet m-algebra $\mathcal{C}_0^\infty(X, Y, \mathcal{F})$ can be endowed with the uniform norm and is an object in the category $\mathfrak{C}_0$ whose completion is precisely the C^*-algebra $\mathcal{C}_0(X \smallsetminus Y, \mathcal{F})$. The same observation holds for the smaller subalgebra $\mathcal{C}_\infty^\infty(X, Y, \mathcal{F})$, which is also a Fréchet m-algebra. In particular, we have an isomorphism $\mathrm{RK}_i(\mathcal{C}_\infty^\infty(X, Y, \mathcal{F})) \to K_i^{\mathrm{top}}(\mathcal{C}_0(X \smallsetminus Y, \mathcal{F}))$.

Now notice that the quotient Fréchet m-algebra $\mathcal{I}/\mathcal{C}_\infty^\infty(X, Y, \mathcal{F})$ is topologically nilpotent when endowed with the quotient topology. Indeed, the semi-norms on the quotient $\mathcal{C}_0^\infty(X, Y, \mathcal{F})/\mathcal{C}_\infty^\infty(X, Y, \mathcal{F})$, can be taken to be a sequence induced by the standard (semi-)norms on $\mathcal{C}_0^\infty(X, Y, \mathcal{F})/\mathcal{C}_\ell^\infty(X, Y, \mathcal{F})$, where $\mathcal{C}_\ell^\infty(X, Y, \mathcal{F})$ is the algebra of smooth functions vanishing of order $\leq \ell$ on Y. Therefore since any such semi-norm on $\mathcal{C}_0^\infty(X, Y, \mathcal{F})/\mathcal{C}_\ell^\infty(X, Y, \mathcal{F})$ vanishes on ℓ-products, we conclude that $\mathcal{C}_0^\infty(X, Y, \mathcal{F})/\mathcal{C}_\infty^\infty(X, Y, \mathcal{F})$ is topologically nilpotent as claimed. This also gives that $\mathcal{I}/\mathcal{C}_\infty^\infty(X, Y, \mathcal{F})$ is topologically nilpotent. Hence $\mathrm{HP}_i(\mathcal{I}/\mathcal{C}_\infty^\infty(X, Y, \mathcal{F})) = 0 = \mathrm{RK}_i(\mathcal{I}/\mathcal{C}_\infty^\infty(X, Y, \mathcal{F}))$. Then, excision in K-theory and in topological periodic cyclic homology for the short exact sequence

$$0 \to \mathcal{C}^\infty_\infty(X, Y, \mathcal{F}) \hookrightarrow \mathcal{I} \longrightarrow \mathcal{I}/\mathcal{C}^\infty_\infty(X, Y, \mathcal{F}) \to 0$$

of Fréchet m-algebras gives that $\mathrm{RK}_i(\mathcal{C}^\infty_\infty(X, Y; \mathcal{F})) \to \mathrm{RK}_i(\mathcal{I})$ and

$$\mathrm{HP}^{\mathrm{top}}_i(\mathcal{C}^\infty_\infty(X, Y; \mathcal{F})) \to \mathrm{HP}^{\mathrm{top}}_i(\mathcal{I})$$

are isomorphisms for all $i \in \mathbb{Z}/2\mathbb{Z}$. By replacing $\mathcal{I}$ with $\mathcal{C}^\infty_0(X, Y; \mathcal{F})$, we obtain that $\mathrm{RK}_i(\mathcal{C}^\infty_\infty(X, Y; \mathcal{F})) \to \mathrm{RK}_i(\mathcal{C}^\infty_0(X, Y; \mathcal{F}))$ and $\mathrm{HP}^{\mathrm{top}}_i(\mathcal{C}^\infty_\infty(X, Y; \mathcal{F})) \to \mathrm{HP}^{\mathrm{top}}_i(\mathcal{C}^\infty_0(X, Y; \mathcal{F}))$ are isomorphisms. Hence the inclusion $\mathcal{I} \to \mathcal{C}^\infty_0(X, Y; \mathcal{F})$ gives that the maps $\mathrm{RK}_i(\mathcal{I}) \to \mathrm{RK}_i(\mathcal{C}^\infty_0(X, Y; \mathcal{F}))$ and $\mathrm{HP}^{\mathrm{top}}_i(\mathcal{I}) \to \mathrm{HP}^{\mathrm{top}}_i(\mathcal{C}^\infty_0(X, Y; \mathcal{F}))$ are also isomorphisms. This completes the proof. □

THEOREM 4.7. *Let X be a compact manifold with corners and $Y \subset \partial X$ be a union of closed faces of X. Let $\mathcal{I}$ be a closed subalgebra of $\mathcal{C}^\infty_0(X, Y; \mathcal{F})$ containing $\mathcal{C}^\infty_\infty(X, Y; \mathcal{F})$. Then the algebras $\mathcal{C}^\infty_\infty(X, Y; \mathcal{F})$, $\mathcal{I}$, and $\mathcal{C}^\infty_0(X, Y; \mathcal{F})$ are all Connes algebras with respect to the uniform norm of $\mathcal{C}_0(X \smallsetminus Y, \mathcal{F})$.*

PROOF. We shall assume for simplicity that $\mathcal{F} = \mathbb{C}$, the general case being the same using that $\mathcal{C}^\infty(X, \mathcal{F})$ is a Connes algebra, see Proposition 4.5. Recall that the boundary depth of a boundary face is the number of vanishing coordinates in a corner chart $\mathbb{R}^n \times [0, \infty)^k$, see [**3**] for the precise definition. We shall proceed by induction on the dimension of X. If $\dim X = 0$ then $\partial X = \emptyset$ and the cardinal of X is finite say k. Thus $\mathcal{C}^\infty(X) \cong \mathbb{C}^k$, which is trivially a Connes algebra. Similarly, if $\dim X = 1$ then ∂X and hence also $Y = \{x_1, \cdots, x_k\}$ are a finite sets. Applying Theorem 1.2 to the exact sequence

$$0 \to \mathcal{C}^\infty_0(X, Y) \to \mathcal{C}^\infty(X) \to \mathbb{C}^k \to 0\,,$$

we get that $\mathcal{C}^\infty_0(X, Y)$ is a Connes algebra.

Now assume that the statment is true for smaller dimensions than that of X, that is, that the algebras $\mathcal{C}^\infty_0(Z, Z_Y)$ are Connes algebras for all compact manifolds with corners Z such that $\dim Z < \dim X$ (and $Z_Y \subset \partial Z$ a union of closed faces of Z). Denote by N the maximal depth of X, i.e. $N = \max \mathrm{depth}(F)$, where F runs over all the boundary faces. Let us introduce the ideals $\mathcal{I}_k := \mathcal{C}^\infty_0(X, Y \cap \cup_{\mathrm{depth}(F) \leq k} F)$ and consider the stratification

$$\mathcal{I}_N = \mathcal{C}^\infty_0(X, Y) \subset \mathcal{I}_{N-1} \subset \cdots \subset \mathcal{I}_1 \subset \mathcal{I}_0 = \mathcal{C}^\infty(X).$$

Then all quotients are given by

$$\mathcal{I}_k/\mathcal{I}_{k+1} \cong \bigoplus_{F \subset Y, \mathrm{depth}(F) = k+1} \mathcal{C}^\infty_0(F, Y \cap \partial F).$$

Indeed, the restriction map $r : \mathcal{I}_k \to \bigoplus_{F \subset Y, \mathrm{depth}(F) = k+1} \mathcal{C}^\infty_0(F, Y \cap \partial F)$ is clearly well defined and has kernel exactly $\mathcal{I}_{k+1}$. Moreover, r is surjective because, if $f \in \mathcal{C}^\infty_0(F, Y \cap \partial F)$, then we can extend f to a smooth function $\tilde{f}$ on X that vanishes on the other faces $\subset Y$ of the same depth as F since the surjectivity is a local statement (using a partition of unity) and, locally, we may assume that F is an embedded boundary face and hence that it has a tubular neighborhood. That is, we can assume that $\tilde{f}$ is equal to 0 on any boundary face contained in Y distinct from F and of depth $k + 1$.

Since $\mathcal{C}^\infty_0(F, Y \cap \partial F)$ are Connes algebras by the induction hypothesis, we obtain that $\mathcal{I}_k/\mathcal{I}_{k+1}$ is a Connes algebra as finite direct sum of Connes algebras.

Now, $\mathcal{C}^\infty(X)$ is a Connes algebra and all quotients $\mathcal{I}_k/\mathcal{I}_{k+1}$ are Connes algebras thus the above stratification is C-smooth and applying Theorem 1.2 several times, we deduce that $\mathcal{I}_k$ is a Connes algebra for all k. In particular, $\mathcal{I}_N = \mathcal{C}_0^\infty(X,Y)$ is a Connes algebra.

The statements about $\mathcal{I} \subset \mathcal{C}_0^\infty(X,Y)$ is a consequence of Proposition 4.2 since $\mathcal{C}_0^\infty(X,Y)/\mathcal{I}$ is topologically nilpotent (it is also a consequence of Proposition 4.6). The statement about $\mathcal{C}_\infty^\infty(X,Y)$ is a particular case. □

5. Group actions on Azumaya bundles

The goal of this section is to prove Theorem 5.9.

5.1. ***G*-manifolds with corners.** We shall use the notation from [**25**] (page 4 and pages 49-50) for transformation groups and the notation from [**3**] for manifolds with corners and blow-ups. See also [**1, 27**]. The reader should also consult these references for the missing definitions or proofs. The definition of the blow-up $[X : Y]$ is recalled in the Appendix, see Equation (A.1).

We let X be a compact manifold with corners with a smooth G-action, We may assume that X is endowed with a smooth Riemannian metric and that the action of G is isometric. We shall consider *submanifolds* of X in the strong sense that they have tubular neighborhoods. Thus, if $Y \subset X$ is a *submanifold with corners* of X, then Y is closed and there is a neighborhood U of Y in X such that U is G-diffeomorphic to the normal vector bundle $N^XY := TX|_Y/TY \to Y$ (the normal vector bundle to Y in X) via a diffeomorphism of manifolds with corners mapping the zero section of N^XY to Y. This is the concept of manifolds with corners considered in [**2**]. In particular, a submanifold with corners of X is also a p-submanifold of X [**3, 27**], but the converse is not true. A submanifold with corners is called *an interior p-submanifold* in [**1**].

If $x \in X$, G_x denotes the isotropy group of x, namely,

$$G_x := \{\gamma \in G \mid \gamma(x) = x\}\,. \tag{5.1}$$

Given a subgroup $H \subset G$, we let (H) denote the set of subgroups of G conjugated to H. If $K \subset G$ is another subgroup, we will write $(H) \le (K)$ if H is conjugated to a subgroup of K. Since X is compact, we know that there exist only finitely many conjugacy classes $C_j = (H_j) = (G_{x_j})$, $j = 1, \ldots, N$, of isotropy groups G_x, $x \in X$. Then $C_j \le C_k$ if there is a subgroup in C_j that is contained in a subgroup in C_k. We may assume the order (numbering) to be such that,

$$C_j \le C_k \Longrightarrow j \ge k.$$

An order with these properties will be called an *admissible* order.

In the following, H and K will denote subgroups of G. We shall use the following standard notations [**25**]

$$\begin{aligned} X_H &:= \{x \in M \mid G_x = H\}\,, \\ X(H) &:= \{x \in X \mid (G_x) = (H)\}\,, \\ X(\ge H) &:= \{x \in X \mid (G_x) \ge (H)\}\,, \quad \text{and} \\ X(\le H) &:= \{x \in X \mid (G_x) \le (H)\}\,. \end{aligned} \tag{5.2}$$

As usual, X^H denotes the set of fixed points by H. It is known then that

$$GX^H = X(\ge H) \ \text{ and } \ GX_H = X(H)\,. \tag{5.3}$$

We also denote by $N(H) := \{g \in G \mid g^{-1}Hg = H\}$ the normalizer of H in G. Moreover, $N(H)$ acts on X^H and we have the following diffeomorphism that will be used later on:

$$X(H) \simeq (G/H) \times_{N(H)/H} X_H\,, \tag{5.4}$$

with the induced action of $N(H)/H$ on X_H (Proposition 1.91 and Corollaries 1.92 and 1.94 of [**25**]). Moreover, the action of $N(H)/H$ on X_H is free. See [**25**] for the following obvious lemma.

LEMMA 5.1. *We have* $\overline{X(H)} \subset X(\geq H)$. *Thus, if* H *is a maximal isotropy group, then* $X(H)$ *is closed.*

As in [**1**], we shall say that the action of G on X is *boundary intersection free* if, given a closed face F of M and $g \in G$, we have either $gF = F$ or $gF \cap F = \emptyset$. Notice that if the action of G on X is boundary intersection free, then so is the action of any subgroup $H \subset G$. The first part of the following statement is a result from [**1**].

LEMMA 5.2. *Assume that the action of* G *on* X *is boundary intersection free. Then* X^G *is a (closed) submanifold with corners of* X *and* $\partial[X^G] = [\partial X]^G = X^G \cap \partial X$. *Similarly,* $X(G)$ *is a manifold with corners and* $\partial[X(G)] = [\partial X](G) = X(G) \cap \partial X$. *If* $H \subset G$ *is a maximal isotropy subgroup, then* $X(H)$ *will be a submanifold with corners of* X. *If* H *is not a maximal isotropy subgroup, then* $X(H)$ *will still be a manifold with corners, but not closed, in general, and hence not a submanifold with corners of* X, *in general.*

PROOF. Let us fix a G-invariant metric on X. Let $x \in X^G$. Let us suppose that x belongs to an open face F of X and let us choose a G-invariant neighborhood U of x in F. By using the exponential map in directions normal to F, we obtain then that x has a G-invariant neighborhood of the form $U \times [0,1)^j$ with the action of G being diagonal and trivial on $[0,1)^j$ since G is compact and its action on X is boundary intersection free. Hence $(U \times [0,1)^j)^G = U^G \times [0,1)^j$ and U^G is a smooth manifold without boundary. This proves that X^G is a manifold with corners. We further have that

$$\begin{aligned}\partial\big[(U \times [0,1)^j)^G\big] &= \partial\big[U^G \times [0,1)^j\big] \\ &= U^G \times \partial\big[[0,1)^j\big] = \Big(U \times \partial\big[[0,1)^j\big]\Big)^G = \big(\partial(U \times [0,1)^j)\big)^G.\end{aligned}$$

This proves that $\partial[X^G] = [\partial X]^G = X^G \cap \partial X$, since this is a local property. By choosing a tubular neighborhood of U^G in U, we obtain a tubular neighborhood of $U^G \times [0,1)^j$ in $U \times [0,1)^j$ and hence that X^G is a submanifold with corners. The other statements are proved in a similar way by using that $U(H)$ is an immersed submanifold of U that is closed if H is a maximal isotropy subgroup (Lemma 5.1). □

Lemma 5.2 allows us to unambiguously write for any closed subgroup of G

$$\partial X^H = \partial[X^H] = [\partial X]^H = X^H \cap \partial X$$

and

$$\partial X(H) = \partial[X(H)] = [\partial X](H) = X(H) \cap \partial X.$$

That is "$\partial(\cdot)$ commutes with $\cdot^H$ and with $\cdot(H)$."

REMARK 5.3. The statement of the last lemma is a local statement, so no assumption of having embedded faces is required.

We formulate the following result as a lemma, for the purpose of further referencing it. Except the statement about V_1 (which follows from Lemma 5.2), it is Lemma 1.81 of [**25**].

LEMMA 5.4. *Assume again that the action of G on the compact manifold with corners X is boundary intersection free, and let $\{H_1, H_2, \ldots, H_N\}$ be a complete set of representatives of conjugacy classes of isotropy groups with an admissible ordering (that is, $(H_i) \geq (H_j)$ implies $i \leq j$). For $1 \leq k \leq N$, we set*

$$V_k := X(H_1) \cup X(H_2) \cup \ldots \cup X(H_k). \tag{5.5}$$

Then V_k is a closed subset of X, but not a submanifold, in general, except V_1, which is a submanifold with corners of X.

The subspaces $X(H_j)$, $j = 1, \ldots, N$ define a stratification of X.

5.2. G-equivariant bundles and algebras. Recall the objects introduced in the notation 4.3, objects that from now on will be endowed with an action of our compact Lie group G. In particular, $\mathcal{S}, \mathcal{S}_i \to X$ will be G-equivariant bundles of finite dimensional *simple* algebras, and $\mathcal{F} \to X$ will be a G-equivariant bundle of finite dimensional, *semi-simple* algebras. Given a closed subset $Y \subset X$ with $GY = Y$, and a G-equivariant vector bundle E, recall also the ideals $\mathcal{C}_0^\infty(X, Y; E)$ and $\mathcal{C}_\infty^\infty(X, Y; E)$ of 4.3, which will now carry a G-action.

Recall that if $E \to X$ is a G-equivariant vector bundle then $E^G \to X^G$ is a vector bundle as image of the projection $p_G : E|_{X^G} \to E^G$ given by $p_G(x, v) = (x, \int_G gv\, dg)$, for any $x \in X^G$ and $v \in E_x$.

PROPOSITION 5.5. *Let $\mathcal{I}$ be as in Proposition 4.6, that is let $\mathcal{I}$ be a subalgebra of $\mathcal{C}_0^\infty(X, Y; \mathcal{F})$ containing $\mathcal{C}_\infty^\infty(X, Y; \mathcal{F})$. We endow $\mathcal{I}$ with the C^*-norm of $\mathcal{C}_0(X \smallsetminus Y; \mathcal{F})$ and we complete the cyclic mixed complexes of $\mathcal{I}$, $\mathcal{C}_0^\infty(X, Y; \mathcal{F})$ and $\mathcal{C}_\infty^\infty(X, Y; \mathcal{F})$ with respect to the projective tensor product. Let us assume that $GY = Y$. Assume also that the action of G on X has a single isotropy type H and that $Y \subset \partial X$ is a union of closed faces of Y. Then $\mathcal{I}^G$ is a Connes algebra. In particular, both*

$$\mathcal{C}_0^\infty(X, Y; \mathcal{F})^G \simeq \mathcal{C}_0^\infty(X^H/N(H), Y^H/N(H); \mathcal{F}^H/N(H)) \quad \textit{and}$$
$$\mathcal{C}_\infty^\infty(X, Y; \mathcal{F})^G \simeq \mathcal{C}_\infty^\infty(X^H/N(H), Y^H/N(H); \mathcal{F}^H/N(H))$$

are Connes algebras.

PROOF. Let $\Gamma := N(H)/H$. The assumptions combined with the diffeomorphism of Equation (5.4) give $X = X(H) \simeq (G/H) \times_\Gamma X_H = (G/H) \times_\Gamma X^H$, and hence

$$\mathcal{C}_0^\infty(X, Y; \mathcal{F})^G \simeq \mathcal{C}_0^\infty(X^H, Y^H; \mathcal{F}^H)^\Gamma \quad \text{and} \quad \mathcal{C}_\infty^\infty(X, Y; \mathcal{F})^G \simeq \mathcal{C}_\infty^\infty(X^H, Y^H; \mathcal{F}^H)^\Gamma.$$

Since the action of $\Gamma := N(H)/H$ on X^H is free, the quotient X^H/Γ is also a manifold with corners and $\mathcal{F}^H$ descends to a bundle of algebras $\mathcal{F}^H/\Gamma \to X^H/G$ such that

$$\mathcal{C}_0^\infty(X^H, Y^H; \mathcal{F}^H)^\Gamma \simeq \mathcal{C}_0^\infty(X^H/\Gamma, Y^H/\Gamma; \mathcal{F}^H/\Gamma) \quad \text{and}$$
$$\mathcal{C}_\infty^\infty(X^H, Y^H; \mathcal{F}^H)^\Gamma \simeq \mathcal{C}_\infty^\infty(X^H/\Gamma, Y^H/\Gamma; \mathcal{F}^H/\Gamma).$$

The result then follows from Theorem 4.7 applied to $(X^H/\Gamma, Y^H/\Gamma, \mathcal{F}^H/\Gamma)$ and $\mathcal{I}^G$. □

Note that, even if $\mathcal{F}$ is a *trivial* bundle of semisimple algebras, $\mathcal{F}^H/\Gamma$ need not be trivial, and hence the extra generality afforded by our setting of sections of algebra bundles is necessary.

5.3. Blow-ups of singular strata. Recall that Y is a closed submanifold of the compact manifold with corners X and that $[X : Y]$ denotes the blow-up manifold. The definition of the blow-up is recalled in the Appendix in Equation (A.1). We refer to the Appendix for more on the blow-up. Let $\{H_1, H_2, \dots, H_N\}$ be a complete set of representatives of conjugacy classes of isotropy groups of G acting on X with admissible ordering. Let

$$V_k := X(H_1) \cup X(H_2) \cup \ldots \cup X(H_k),$$

$k = 1, \dots, N$, which is a closed subset of X for each k, by Lemma 5.4. Let $\mathcal{F}$ be a G-equivariant bundle of finite dimensional semisimple algebras on X.

NOTATION 5.6. We let $I_0 := \mathcal{C}^\infty_\infty(X, \partial X; \mathcal{F})$ and $I_k := \mathcal{C}^\infty_0(X, V_k; \mathcal{F}) \cap I_0$, for $k \geq 1$.

In particular,

$$0 = I_N \subset I_{N-1} \subset \ldots \subset I_1 \subset I_0 := \mathcal{C}^\infty_\infty(X, \partial X; \mathcal{F}) .$$

We assume from now on that the action of G on X is boundary intersection free, as in Lemma 5.2. See Equation (A.1) in the appendix for the definition of the blow-up. We want some sort of "resolution" of the sets V_k. Let H_j be as above (arranged in an admissible order) and let us define by induction X_k and Y_k as follows

$$\begin{cases} X_1 := X\,, & \\ Y_k := X_k(H_k)\,, & \text{if } X_k \text{ was defined,} \\ X_{k+1} := [X_k : Y_k] := [X_k : X_k(H_k)]\,, & \text{if } Y_k \text{ was defined.} \end{cases} \tag{5.6}$$

Then X_k is a compact manifold with corners with isotropy types $\{H_k, H_{k+1}, \dots, H_N\}$, $k \leq N$, as in [**1**]. This procedure works since Y_k is a closed submanifold with corners of X_k by Lemmas 5.1 and 5.2, since H_k is a maximal isotropy group of X_k (see also [**1, 25**]).

We next want to identify the effect of these blow-ups on the sets $X_{k+1}(H_j)$, $j \geq k+1$, which we know are manifolds with corners in view of Lemma 5.2 (which the reader should review now since it will be used again below). First of all, the disjoint union decomposition defining X_{k+1} as the blow-up of X_k along its subset $Y_k := X_k(H_k)$ of points of isotropy of type H_k, namely $X_{k+1} = [X_k \smallsetminus Y_k] \cup SN^{X_k} Y_k$, gives for $j \geq k+1$

$$X_{k+1}(H_j) = [X_k \smallsetminus Y_k](H_j) \cup [SN^{X_k} Y_k](H_j) . \tag{5.7}$$

Since β is a diffeomorphism outside $SN^{X_k} Y_k := SN^{X_k} X_k(H_k)$, we further have

$$[X_{k+1} \smallsetminus SN^{X_k} Y_k](H_j) \;=\; [X_k \smallsetminus Y_k](H_j) \;=\; X_k(H_j)\,, \tag{5.8}$$

since Y_k has isotropy type H_k and $j \geq k+1$. Lemma 5.2 allows us to give an unambigous sense to $\partial X(H) = [\partial X](H) = \partial[X(H)]$ and, together with the fact that $SN^{X_k}Y_k$ is contained in the boundary of X_{k+1} gives that the blow-down map induces for $j \geq k+1$ diffeomorphisms

(5.9)
$$\beta : X_{k+1}(H_j) \smallsetminus \partial X_{k+1}(H_j) \simeq X_k(H_j) \smallsetminus \partial X_k(H_j) \simeq \ldots \simeq X_1(H_j) \smallsetminus \partial X_1(H_j)\,.$$

(Thus, in case $X_1 = X$ does not have a boundary, $Y_k := X_k(H_k)$ will be a compact manifold with corners with interior diffeomorphic to the stratum $X_1(H_k)$, that is, a compactification of $X(H_k)$.)

Recall that $V_k := X(H_1) \cup X(H_2) \cup \ldots \cup X(H_k)$. We obtain then (with $Y_{k+1} := X_{k+1}(H_{k+1})$ and $X_1 = X$, as before) that

$$\beta(Y_{k+1}) \subset V_{k+1} \quad \text{and} \quad \beta(\partial Y_{k+1}) \subset V_k \cup \partial X\,. \tag{5.10}$$

Consequently, given a smooth section f of $\mathcal{F}$ on V_{k+1} that vanishes on $V_k \cup \partial X$, then $\beta^*(f) := f \circ \beta$ will be a smooth section of (the pull-back of) $\mathcal{F}$ on the compact manifold $Y_{k+1} := X_{k+1}(H_{k+1})$ that vanishes on its boundary. This proves the second inclusion of the following proposition, the proof of the first one (by induction on N) being relegated to the next subsection.

Proposition 5.7. *Assume that the action of G on X is boundary intersection free. Let $\beta : X_{k+1} \to X_1 := X$ be the composition of all the blow-down maps and $J_{k+1} := \beta^*(I_k)/\beta^*(I_{k+1})$. Let $0 \leq k \leq N-1$. Then $I_k/I_{k+1} \to J_{k+1}$ is an isomorphism of topological algebras and*

$$\mathcal{C}^{\infty}_{\infty}(Y_{k+1}, \partial Y_{k+1}; \mathcal{F}) \subset J_{k+1} \subset \mathcal{C}^{\infty}_{0}(Y_{k+1}, \partial Y_{k+1}; \mathcal{F})\,.$$

It is also easy to prove that $\mathcal{C}^{\infty}_{c}(Y_{k+1}, \partial Y_{k+1}; \mathcal{F}) \subset J_{k+1}$, but we have not been able to use this observation to prove the first inclusion of Proposition 5.7. In any case, this adds credibility to our statement and justifies the postponement of its proof.

Corollary 5.8. *We use the notation and hypotheses of Proposition* 5.7 *(in particular, the action of G on X is boundary intersection free). Then the algebras J_{k+1} and J_{k+1}^G are Connes algebras.*

Proof. The fact that J_{k+1} is a Connes algebra is a consequence of Theorem 4.7 and Proposition 5.7. Since $Y_{k+1} := X_{k+1}(H_{k+1})$ has a single isotropy type, Proposition 5.5 also yields right away that $\mathcal{C}^{\infty}_{\infty}(Y_{k+1}, \partial Y_{k+1}; \mathcal{F})^G$ is a Connes algebra. Since $J_{k+1}^G/\mathcal{C}^{\infty}_{\infty}(Y_{k+1}, \partial Y_{k+1}; \mathcal{F})^G$ is a topologically nilpotent algebra we obtain that J_{k+1}^G is a Connes algebra as well. □

As a corollary, we are now in position to prove the following result.

Theorem 5.9. *We use the notation and hypotheses of Proposition* 5.7 *(in particular, the action of G on X is boundary intersection free). Then*

(1) *The algebras $\mathcal{C}^{\infty}_{\infty}(X, \partial X; \mathcal{F})^G$ and $\mathcal{C}^{\infty}_{0}(X, \partial X; \mathcal{F})^G$ are Connes algebras.*
(2) *Assume furthermore that X has embedded faces and that $Y \subset \partial X$ is a G-invariant union of closed faces of X, then the algebra $\mathcal{C}^{\infty}_{\infty}(X, Y; \mathcal{F})^G$ and $\mathcal{C}^{\infty}_{0}(X, Y; \mathcal{F})^G$ are Connes algebras.*

Proof. The fact that $\mathcal{C}^{\infty}_{\infty}(X, \partial X; \mathcal{F})^G$ is a Connes algebra follows by applying Theorem 3.9 to the stratification (or composition series) I_k^G which has subquotients J_{k+1}^G that are Connes algebras by Corollary 5.8. The fact that $\mathcal{C}^{\infty}_{0}(X, \partial X; \mathcal{F})^G$ is a Connes algebra follows from the fact that $\mathcal{C}^{\infty}_{0}(X, \partial X; \mathcal{F})^G/\mathcal{C}^{\infty}_{\infty}(X, \partial X; \mathcal{F})^G$ is topologically nilpotent. For the second point, notice that $\mathcal{C}_0^\infty(X,Y;\mathcal{F})^G$ has a composition series with subquotients $\mathcal{C}_0^\infty(F,\partial F;\mathcal{F})^G$, where F ranges through a set of closed faces of X. □

5.4. Proof of Proposition 5.7. The proof of Proposition 5.7 is an induction on N, the number of isotropy types of X and is a consequence of Lemma 5.10. We shall freely use the notation of the previous subsection and Appendix A (but we also recall some of the most important ones from time to time). Since the result is local, we may assume that $\mathcal{F} = \underline{\mathbb{C}}$, and thus drop it from the notation.

If $N = 1$, $J_1 = I_0 = \mathcal{C}_\infty^\infty(X_1, \partial X_1)$, $Y_1 = X = X_1$ and there is nothing to prove.

Let us now turn to the induction step, by assuming the result to be true if there are $N-1$ isotropy types and prove it if there are N isotropy types. Let $\{H_1, H_2, \ldots H_N\}$ be the isotropy types of $X =: X_1$. To simplify notation, let $Y := Y_1$, $\tilde{X} := [X_1 : Y_1] = [X : Y_1]$ where, we recall $Y_1 := X(H_1)$. Let $\beta : \tilde{X} \to X = X_1$ be the blow-down map. Similarly, let $\tilde{Y} := \beta^{-1}(Y) := \beta^{-1}(Y_1)$.

For the induction step, we only need to prove that $\mathcal{C}_\infty^\infty(Y_{k+1}, \partial Y_{k+1}) \subset J_{k+1}$ since the inclusion $J_{k+1} \subset \mathcal{C}_0^\infty(Y_{k+1}, \partial Y_{k+1})$ is obvious (and was already discussed). Let us consider the composition series of 5.6 for $\tilde{X}$, but we shift the indices to account for the fact that $\tilde{X}$ has only $N-1$ isotropy types:

$$\tilde{I}_k := \mathcal{C}_0^\infty(\tilde{X}, \tilde{Y}_k) \cap \mathcal{C}_\infty^\infty(\tilde{X}, \partial \tilde{X}), \quad k \geq 2, \tag{5.11}$$

and $\tilde{I}_1 := \mathcal{C}_\infty^\infty(\tilde{X}, \partial\tilde{X})$. (Thus the definition of I_k is obtained from the definition of $\tilde{I}_k$ by removing all the symbols $\tilde{}$.)

Recall from Lemma A.1 that the pull back $\beta^*(f) := f \circ \beta$ defines an isomorphism $\beta^* : \mathcal{C}_0(X \smallsetminus Y) \to \mathcal{C}_0(\tilde{X} \smallsetminus \tilde{Y})$. Moreover, Lemma A.3 shows that it also defines an isomorphism

$$\beta^* : \mathcal{C}_\infty^\infty(X, Y) \longrightarrow \mathcal{C}_\infty^\infty(\tilde{X}, \tilde{Y}).$$

To complete the proof, we shall need the following result, which is best formulated as a lemma.

Lemma 5.10. *Let $\beta^{*-1} : \mathcal{C}_\infty^\infty(\tilde{X}, \tilde{Y}) \to \mathcal{C}_\infty^\infty(X, Y)$ be the map of Lemma* A.3. *Then $\beta^{*-1}(\tilde{I}_k) \subset I_k$, $k = 1, \ldots, N$.*

Proof. We have that $\partial\tilde{X} = \tilde{Y} \cup \beta^{-1}(\partial X)$. Hence $\tilde{I}_1 := \mathcal{C}_\infty^\infty(\tilde{X}, \partial\tilde{X}) = \mathcal{C}_\infty^\infty(\tilde{X}, \tilde{Y}) \cap \mathcal{C}_\infty^\infty(\tilde{X}, \beta^{-1}(\partial X))$. Equation (5.9) gives

$$\tilde{I}_k := \mathcal{C}_0^\infty(\tilde{X}, \tilde{Y}_k) \cap \mathcal{C}_\infty^\infty(\tilde{X}, \tilde{Y}) \cap \mathcal{C}_\infty^\infty(\tilde{X}, \beta^{-1}(\partial X)), \tag{5.12}$$

for all k (that is, $k = 1, \ldots, N$). Let then $f \in \tilde{I}_k$. By Lemma A.3, $\beta^{*-1}(f) \in \mathcal{C}_\infty^\infty(X, Y)$. It is obvious that $\beta^{*-1}(f)$ also vanishes to infinite order on ∂X since $Y = Y_1$ is an (interior) submanifold with corners (so $\partial X \smallsetminus Y$ is dense in ∂X). It follows that $\beta^{*-1}(f)$ also vanishes on Y_k since f vanishes (even of infinite order) on $\tilde{Y}_1$ and it vanishes on $\tilde{Y}_k \smallsetminus \tilde{Y}_1$, which is mapped bijectively onto its image by β. Hence $\beta^{*-1}(f) \in \mathcal{C}_\infty^\infty(X, Y) \cap \mathcal{C}_0^\infty(X, Y_k) =: I_k$. □

We can now complete the proof. By the induction hypothesis (which gives the first inclusion in the next displayed equation) and since the space Y_{k+1} is the same for both X and $\tilde{X}$ (by construction), we have

$$\mathcal{C}_\infty^\infty(Y_{k+1}, \partial Y_{k+1}; \mathcal{F}) \subset \tilde{I}_k/\tilde{I}_{k+1} \xrightarrow{\beta^{*-1}} I_k/I_{k+1}. \tag{5.13}$$

6. Further applications

We now include a few direct applications of our previous results.

6.1. Crossed product with finite groups. We keep the notations of the previous subsections. Then we obtain the following result (see also [**8, 36**]):

THEOREM 6.1. *Assume that G is a finite group. Let X be a compact, boundary intersection free G-manifold with corners and let $\mathcal{E} \to X$ be a G-algebra bundle over X with simple algebra fibers and let $Y \subset \partial X$ be a G-invariant union of closed faces of X. Then the crossed product algebra $\mathcal{C}^\infty(X, Y; \mathcal{E}) \rtimes G$ is a Connes Fréchet m-algebra that is spectrally invariant in its C^*-completion $\mathcal{C}_0(X \smallsetminus Y; \mathcal{E}) \rtimes G$. In particular, the Chern-Connes-Karoubi character induces an isomorphism*

$$\widetilde{\mathrm{Ch}} : K_n^{\mathrm{top}}(\mathcal{C}_0(X \smallsetminus Y, \mathcal{E}) \rtimes G) \xrightarrow{\simeq} \mathrm{HP}_n^{\mathrm{top}}(\mathcal{C}^\infty(X, Y; \mathcal{E}) \rtimes G)\,, \quad n = 0, 1\,.$$

PROOF. We denote $\mathcal{A} := \mathcal{C}^\infty(X, Y, \mathcal{E})$, $A = \mathcal{C}_0(X \smallsetminus Y; \mathcal{E})$, $\mathcal{B} := \mathcal{A} \rtimes G$ and $B := A \rtimes G$. The algebra $\mathcal{B}$ can be identified with the algebra $(\mathcal{A} \otimes \mathrm{End}(V))^G$ where V is the finite dimensional G-representation on $V = \ell^2 G$ (the regular representation). In the same way, the algebra B can be identified with the C^*-algebra $(A \otimes \mathrm{End}(V))^G$. Indeed, an algebra isomorphism is obtained by identifying any kernel $k : G \times G \to \mathcal{A}$, which is G-equivariant, with a function of a single variable in G, and it is easy to check that this is a topological identification for the smooth functions and a C^*-algebra isomorphism for the completions. More precisely, the isomorphism is given by the map $\Phi : (\mathcal{A} \otimes \mathbb{C}[G \times G])^G \to \mathcal{A} \rtimes G$ defined for $k \in (\mathcal{A} \otimes \mathbb{C}[G \times G])^G$ by $\Phi(k)(g) = k(e, g)$, where $e \in G$ denotes the identity element. Since $k(h, g) = h(k(e, h^{-1}g))$ by G-invariance, we clearly get

$$\begin{aligned}\Phi(k \cdot k')(g) = \sum_{h \in G} k(e, h) k'(h, g) &= \sum_{h \in G} k(e, h) h(k'(e, h^{-1}g)) \\ &= \sum_{h \in G} \Phi(k)(h) h(\Phi(k')(h^{-1}g)) = \Phi(k) * \Phi(k')(g).\end{aligned}$$

The inverse map is given by $\Phi^{-1}(f)(g, h) = g(f(g^{-1}h))$ which is clearly G-invariant because

$$u(\Phi^{-1}(f)(u^{-1}g, u^{-1}h)) = g(f(g^{-1}h)) = \Phi^{-1}(f)(g, h).$$

Similar computations give $\Phi^{-1}(f_1 * f_2) = \Phi^{-1}(f_1) \cdot \Phi^{-1}(f_2)$. Therefore, it remains to show that the Fréchet m-algebra $(\mathcal{A} \otimes \mathrm{End}(V))^G$ is a Connes algebra which is spectrally invariant in its C^*-completion $(A \otimes \mathrm{End}(V))^G$. Now, this is a consequence of Theorem 5.9. □

6.2. Pseudodifferential operators. Assume now that M is a compact, *smooth* manifold (so without corners). Let $B^*M \subset T^*M$ be the set of vector of length ≤ 1. Then there is an identification $S^0_{cl}(T^*M) \simeq \mathcal{C}^\infty(B^*M)$ using a suitable compactification of T^*M. There exists also a "quantization map" $q : S^0_{cl}(T^*M) \to \Psi^0(M)$ and an isomorphism $\mathcal{R} \simeq \Psi^{-\infty}(M)$ such that the induced map $\mathcal{C}^\infty(B^*M) \oplus \mathcal{R} \to \Psi^0(M)$ is onto. We endow $\Psi^0(M)$ with the induced Fréchet topology that makes it an m-algebra. We define similarly the topology on algebras of families of order zero pseudodifferential operators.

6.2.1. *Isotypical components.* We shall denote by $\mathcal{B}(\mathcal{H})$ the C^*-algebra of bounded operators over a Hilbert space $\mathcal{H}$. Let as before $E \to M$ be a G-equivariant hermitian vector bundle. Let α be an irreducible unitary representation of G and denote by $L^2(M,E)_\alpha \cong \alpha \otimes L^2(M, E\otimes\alpha^*)^G$ the isotypical component associated with α. Let us consider the restriction map $\pi_\alpha : \mathcal{B}(L^2(M,E))^G \to \mathcal{B}(L^2(M,E)_\alpha)^G$, see [**5**–**7**]. Recall that we have $\mathcal{K}(L^2(M,E))^G = \bigoplus \mathcal{K}(L^2(M,E)_\alpha)^G$ and $\mathcal{K}(L^2(M,E)_\alpha)^G \cong \mathcal{K}(L^2(M, E\otimes\alpha^*)^G)$, see [**5**]. We get then the following result.

PROPOSITION 6.2. *The algebra $\pi_\alpha(\Psi^{-\infty}(M,E)^G)$ is a Connes algebra. If G is finite, then $\Psi^{-\infty}(M,E)^G$ is also a Connes algebra.*

PROOF. If α does not appear in $L^2(M,E)$ then $\pi_\alpha(\Psi^{-\infty}(M,E)^G) = 0 = \mathcal{K}(L^2(M,E)_\alpha)^G$ and this is clearly a Connes algebra. Assume now that α appears in $L^2(M,E)$. We have $\pi_\alpha(\Psi^{-\infty}(M,E)^G) = \Psi^{-\infty}(M,E)^G \cap \mathcal{K}(L^2(M,E)_\alpha)^G \cong \mathcal{R}$. Thus the result follows. □

A similar proof to Theorem 5.9 (using also the argument in the proof of Proposition 6.3) gives also that $\pi_\alpha(\Psi^{-\infty}(M,E)^G)$ is a Connes algebra as well (this result is one of the main motivation for this paper). The proof will be included in a forthcoming publication, in order not increase the length of this paper too much.

6.2.2. *Families of operators.* Let $p : M \to B$ be a locally trivial fibration of riemannian compact manifolds without boundary. Let $\pi : E \to M$ be a vector bundle. Denote by $\Psi^m(M|B,E)$ the set of smooth families $P = (P_b)_{b\in B}$ of pseudodifferential operators. Denote by $S^*(M|B) \to M$ the cosphere bundle of the vertical tangent bundle $T(M|B) = \ker dp$.

PROPOSITION 6.3. *The algebras $\mathcal{C}^\infty(S^*(M|B), \mathrm{End}(E))$, $\Psi^{-\infty}(M|B,E)$, $\Psi^{-1}(M|B,E)$ and $\Psi^0(M|B,E)$ are Connes algebras.*

PROOF. This is completely similar to the case of one operator, see Proposition 6.2. The result for the first algebra $\mathcal{C}^\infty(S^*(M|B), \mathrm{End}(E))$ is clear. The result for the third algebra $\Psi^{-1}(M|B,E)$ follows from the fact that

$$\Psi^{-1}(M|B,E)/\Psi^{-\infty}(M|B,E)$$

is topologically nilpotent, using Proposition 4.2, and the proof that the second algebra is a Connes algebra given below. The result for the last algebra is then a consequence of Theorem 3.9 using the exact sequence

$$0 \to \Psi^{-1}(M|B,E) \to \Psi^0(M|B,E) \to \mathcal{C}^\infty(S^*(M|B), \mathrm{End}(E)) \to 0.$$

To finish the proof notice that $\Psi^{-\infty}(M|B,E) \cong \mathcal{R}\mathcal{C}^\infty(B)$. Indeed, for each b, the operator P_b is isomorphic to an element $(m_{ij}(b)) \in \mathcal{R}$ and this depends smoothly on b as can be checked on any trivialisation U of $p : M \to B$ because $\Psi^{-\infty}(M|B,E)|_U \cong \mathcal{C}^\infty(U, \Psi^{-\infty}(F,E'))$, where F is the typical fiber of $p : M \to B$ and E' the typical fiber of $p \circ \pi : E \to B$. □

In particular, let M be a compact, *smooth* manifold (so without corners) and $E \to M$ be a vector bundle and Γ be a finite group acting on M and E. Then the algebra $\Psi^0(M;E)^\Gamma$ is a Connes algebra. Indeed, Proposition 6.2 shows that $\Psi^{-\infty}(M,E)^\Gamma$ is a Connes algebra therefore $\Psi^{-1}(M,E)^\Gamma$ is a Connes algebra because the quotient is topologically nilpotent and $\mathcal{C}^\infty(S^*M, \mathrm{End}(E))^\Gamma$ is a Connes algebra by Theorem 5.9 applied with $X = S^*M$, $\partial X = \emptyset$ and $G = \Gamma$. Thus Theorem 1.2 applies to the exact sequence

$$0 \to \Psi^{-1}(M, E)^\Gamma \to \Psi^0(M, E)^\Gamma \to \mathcal{C}^\infty(S^*M, \mathrm{End}(E))^\Gamma \to 0.$$

and gives that $\Psi^0(M, E)^\Gamma$ is a Connes algebra. However, if Γ is infinite, this result is not true anymore, one needs to consider a weaker property, thus leading to "weak Connes algebras," for which we only require the Chern-Connes-Karoubi character to be injective with dense image. Indeed, take $M = G$ a non discrete compact Lie group, (for instance $S^1 = \{z \in \mathbb{C}, |z| = 1\}$). Then $\Psi^{-\infty}(G)^G \cong \mathcal{C}^\infty(G)$ and therefore $RK_j(\Psi^{-\infty}(G)^G) \cong RK_j(\Psi^{-1}(G)^G) \cong K_j(C^*G)$, where $C^*G \cong \mathcal{K}(L^2(G))^G$ is the C^*-algebra of the group G. But $K_0(C^*G) = R(G)$ is the representation ring of G and $K_1(C^*G) = 0$. It is well known that $R(G) = \bigoplus_{\hat{G}} \mathbb{Z}$, where $\hat{G}$ is the set of isomorphism classes of irreducible unitary representations of G. Moreover, $HP_*(C^\infty(G)) = \mathcal{C}^\infty(G)^G$, see [**29, 32**] for instance and thus the Connes' character has only dense image. Now since $\mathcal{C}^\infty(S^*G)^G$ is a Connes algebra, we get that $\Psi^0(G)^G$ can not be a Connes algebra. See also the Appendix of the second (enhanced) version of [**36**], available from the author's home page.

Appendix A. Smooth functions and blow-ups

Recall that if X is a manifold with corners and $Y \subset X$ is a submanifold with corners, then the normal bundle N^XY of Y in X is diffeomorphic to an open neighborhood U of Y in X by a diffeomorphism ϕ that maps the zero section of the normal bundle N^XY to Y. Let SN^XY be the unit sphere bundle of N^XY. (So its fibers are spheres, as the name indicates it.) Then [**3, 27**] the *blow-up of X along Y* (or with respect to Y) is the *disjoint union*

$$[X : Y] := (X \smallsetminus Y) \cup SN^XY\,. \tag{A.1}$$

It comes equipped with the structure of a smooth manifold with corners and a *blow-down* map $\beta : [X : Y] \to X$ that is the identity on $X \smallsetminus Y$ and is the bundle projection $SN^XY \to Y$ on SN^XY. We shall identify without further comment $X \smallsetminus Y$ with $[X : Y] \smallsetminus \beta^{-1}(Y)$ in what follows. Assume that a Lie group Γ acts smoothly on X such that $\Gamma Y = Y$. Then the action of Γ on X lifts to a smooth action of Γ on $[X : Y]$ (this was proved in the case G compact in [**1**] and in general in [**3**]).

Let us denote $[X : Y]$ by $\tilde{X}$ and set $\tilde{Y} := \beta^{-1}(Y)$.

LEMMA A.1. *The pull back $\beta^*(f) := f \circ \beta$ defines an isomorphism $\beta^* : \mathcal{C}_0(X \smallsetminus Y) \to \mathcal{C}_0(\tilde{X} \smallsetminus \tilde{Y})$.*

PROOF. This is because the blow-down map $\beta : \tilde{X} \to X$ is such that X has the quotient topology and $\beta(\tilde{Y}) = Y$. □

Let us denote by $\mathrm{Diff}(X)$, respectively, $\mathrm{Diff}(\tilde{X})$ the *algebra of differential operators* with smooth coefficients on X, respectively, $\tilde{X}$. Let r_Y be a smooth function on X such that, close to Y it is the distance to Y and otherwise it is > 0 outside Y. Then it is known that r_Y lifts to a smooth function $r_Y \circ \beta$ on $\tilde{X}$ whose zero set is exactly $\tilde{Y}$. Notice that

$$\begin{aligned}\mathcal{C}^\infty_\infty(X,Y) &:= \{u \in \mathcal{C}_0(X \smallsetminus Y) \mid Pu \in \mathcal{C}_0(X \smallsetminus Y) \ \text{ for all } P \in \mathrm{Diff}(X)\}\,,\\ \mathcal{C}^\infty_\infty(\tilde{X},\tilde{Y}) &:= \{u \in \mathcal{C}_0(\tilde{X},\tilde{Y}) \mid Pu \in \mathcal{C}_0(\tilde{X} \smallsetminus \tilde{Y}) \ \text{ for all } \tilde{P} \in \mathrm{Diff}(\tilde{X})\}\,,\\ & r_Y^{-1}\mathcal{C}^\infty_\infty(X,Y) \subset \mathcal{C}^\infty_\infty(X,Y)\,, \quad \text{and}\\ & r_Y^{-1}\mathcal{C}^\infty_\infty(\tilde{X},\tilde{Y}) \subset \mathcal{C}^\infty_\infty(\tilde{X},\tilde{Y})\,.\end{aligned} \tag{A.2}$$

LEMMA A.2. *We have the following equalities as operators on* $\mathcal{C}^\infty_c(X \smallsetminus Y) = \mathcal{C}^\infty_c(\tilde{X} \smallsetminus \tilde{Y})$*:*

(1) $\mathrm{Diff}(X) \subset \cup_{k=1}^\infty r_Y^{-k}\,\mathrm{Diff}(\tilde{X})$ *and, similarly,*
(2) $\mathrm{Diff}(\tilde{X}) \subset \cup_{k=1}^\infty (r_Y \circ \beta)^{-k}\mathcal{C}^\infty(\tilde{X})\,\mathrm{Diff}(X)$.

PROOF. This is a standard fact about blow-ups. For the first relation, since r_Y is smooth on $\tilde{X}$, $\cup_{k\in\mathbb{N}} r_Y^{-k}\,\mathrm{Diff}(\tilde{X})$ is an algebra. It is enough then to prove our statement for a system of generators of $\mathrm{Diff}(X)$. This is clear if P is a multiplication operator, since a smooth function on X lifts to a smooth function on $\tilde{X}$. Let v be a vector field on X. Then it is known (see, for instance [**3**]) that there exists a vector field $\tilde{v}$ on $\tilde{X}$ that restricts to v in the interior. Hence $\tilde{v} \in \mathrm{Diff}(\tilde{X})$ and $v = r_Y \circ \beta^{-1}\tilde{v} \in \cup_{k\in\mathbb{N}} r_Y^{-k}\,\mathrm{Diff}(\tilde{X})$, as desired.

It is easy to see in local coordinates that the algebra $\mathrm{Diff}(\tilde{X})$ is generated by $\mathcal{C}^\infty(\tilde{X})$, $(r_Y \circ \beta)^{-1}$, and the lifts of vector fields $r_Y v$, with v a vector field on X. □

We now come to the following lemma which is used in the proof of Proposition 5.7, but which is obviously of independent interest.

LEMMA A.3. *The pull back* $\beta^*(f) := f \circ \beta$ *defines an isomorphism*

$$\beta^* : \mathcal{C}^\infty_\infty(X,Y) \to \mathcal{C}^\infty_\infty(\tilde{X},\tilde{Y})\,.$$

PROOF. This follows from the previous two lemmas and Equation (A.2). Indeed, let $f \in \mathcal{C}^\infty_\infty(X,Y)$ and $\tilde{P} \in \mathrm{Diff}(\tilde{X})$. We want to prove that $\tilde{P}(f \circ \beta) \in \mathcal{C}_0(\tilde{X} \smallsetminus \tilde{Y})$. Then, by the second part of Lemma A.2 we may assume that $\tilde{P} = a r_Y^{-k} P$, with $P \in \mathrm{Diff}(X)$ and $a \in \mathcal{C}^\infty(\tilde{X})$. Then $r_Y^{-k} P f \in \mathcal{C}^\infty_\infty(X,Y)$ and hence $\tilde{P}(f \circ \beta) = a(r_Y^{-k} P f) \circ \beta \in \mathcal{C}_0(\tilde{X} \smallsetminus \tilde{Y})$, by Lemma A.1. This shows that the map $\beta^* : \mathcal{C}^\infty_\infty(X,Y) \to \mathcal{C}^\infty_\infty(\tilde{X},\tilde{Y})$ is well defined. It is obviusly injective since $X \smallsetminus Y$ is dense in both X and $\tilde{X}$. Let us prove that it is onto. Let $g \in \mathcal{C}^\infty_\infty(\tilde{X},\tilde{Y})$. Then $g = f \circ \beta$ with some $f \in \mathcal{C}_0(X \smallsetminus Y)$, again by lemma A.1. Let $P \in \mathrm{Diff}(X)$. We similarly want to prove that $Pf \in \mathcal{C}_0(X \smallsetminus Y)$. We have that $P = r_Y^{-k} Q$ for some $Q \in \mathrm{Diff}(\tilde{X})$. Then $Pf \circ \beta = r_Y^{-k} Q g$. Since $r_Y^{-k} Q g \in \mathcal{C}_0(\tilde{X} \smallsetminus \tilde{Y})$ by the assumption on g, we have that $Pf \in \mathcal{C}_0(X \smallsetminus Y)$, again by Lemma A.1. □

Acknowledgements

We thank Bernd Ammann, Alain Connes, Joachim Cuntz, Matthias Lesch, and Markus Pflaum for useful discussions. We also thank an anonymous referee for carefully reading our paper.

References

[1] Pierre Albin and Richard Melrose, *Resolution of smooth group actions*, Spectral theory and geometric analysis, Contemp. Math., vol. 535, Amer. Math. Soc., Providence, RI, 2011, pp. 1–26, DOI 10.1090/conm/535/10532. MR2560748

[2] Bernd Ammann, Robert Lauter, and Victor Nistor, *On the geometry of Riemannian manifolds with a Lie structure at infinity*, Int. J. Math. Math. Sci. **1-4** (2004), 161–193, DOI 10.1155/S0161171204212108. MR2038804

[3] Bernd Ammann, Jérémy Mougel, and Victor Nistor, *A comparison of the Georgescu and Vasy spaces associated to the N-body problems and applications*, Ann. Henri Poincaré **23** (2022), no. 4, 1141–1203, DOI 10.1007/s00023-021-01109-1. MR4407735

[4] M. F. Atiyah and I. G. Macdonald, *Introduction to commutative algebra*, Addison-Wesley Publishing Co., Reading, Mass.-London-Don Mills, Ont., 1969. MR0242802

[5] Alexandre Baldare, Rémi Côme, Matthias Lesch, and Victor Nistor, *Fredholm conditions for invariant operators: finite abelian groups and boundary value problems*, J. Operator Theory **85** (2021), no. 1, 229–256, DOI 10.7900/jot. MR4198971

[6] A. Baldare, R. Côme, M. Lesch, and V. Nistor, *Fredholm conditions for restrictions of invariant pseudodifferential to isotypical components*, Max Planck Preprint, February 2020.

[7] Alexandre Baldare, Rémi Côme, and Victor Nistor, *Fredholm conditions for operators invariant with respect to compact Lie group actions*, C. R. Math. Acad. Sci. Paris **359** (2021), 1135–1143, DOI 10.5802/crmath.25. MR4334861

[8] Paul Baum and Alain Connes, *Chern character for discrete groups*, A fête of topology, Academic Press, Boston, MA, 1988, pp. 163–232, DOI 10.1016/B978-0-12-480440-1.50015-0. MR928402

[9] Alain Connes, *Sur la théorie non commutative de l'intégration* (French), Algèbres d'opérateurs (Sém., Les Plans-sur-Bex, 1978), Lecture Notes in Math., vol. 725, Springer, Berlin, 1979, pp. 19–143. MR548112

[10] Alain Connes, *Noncommutative differential geometry*, Inst. Hautes Études Sci. Publ. Math. **62** (1985), 257–360. MR823176

[11] Alain Connes, *Noncommutative geometry*, Academic Press, Inc., San Diego, CA, 1994. MR1303779

[12] Alain Connes and Max Karoubi, *Caractère multiplicatif d'un module de Fredholm* (French, with English summary), C. R. Acad. Sci. Paris Sér. I Math. **299** (1984), no. 19, 963–968. MR774679

[13] Alain Connes and Max Karoubi, *Caractère multiplicatif d'un module de Fredholm* (French, with English summary), K-Theory **2** (1988), no. 3, 431–463, DOI 10.1007/BF00533391. MR972606

[14] A. Connes and H. Moscovici, *The local index formula in noncommutative geometry*, Geom. Funct. Anal. **5** (1995), no. 2, 174–243, DOI 10.1007/BF01895667. MR1334867

[15] A. Connes and G. Skandalis, *The longitudinal index theorem for foliations*, Publ. Res. Inst. Math. Sci. **20** (1984), no. 6, 1139–1183, DOI 10.2977/prims/1195180375. MR775126

[16] Joachim Cuntz, *Bivariante K-Theorie für lokalkonvexe Algebren und der Chern-Connes-Charakter* (German, with English summary), Doc. Math. **2** (1997), 139–182. MR1456322

[17] Joachim Cuntz, *Excision in periodic cyclic theory for topological algebras*, Cyclic cohomology and noncommutative geometry (Waterloo, ON, 1995), Fields Inst. Commun., vol. 17, Amer. Math. Soc., Providence, RI, 1997, pp. 43–53, DOI 10.1007/s002220050115. MR1478700

[18] Joachim Cuntz and Daniel Quillen, *On excision in periodic cyclic cohomology* (English, with English and French summaries), C. R. Acad. Sci. Paris Sér. I Math. **317** (1993), no. 10, 917–922, DOI 10.1007/s002220050115. MR1249360

[19] Joachim Cuntz and Daniel Quillen, *On excision in periodic cyclic cohomology. II. The general case* (English, with English and French summaries), C. R. Acad. Sci. Paris Sér. I Math. **318** (1994), no. 1, 11–12. MR1260526

[20] Jacques Dixmier, *Les C^*-algèbres et leurs représentations* (French), Les Grands Classiques Gauthier-Villars. [Gauthier-Villars Great Classics], Éditions Jacques Gabay, Paris, 1996. Reprint of the second (1969) edition. MR1452364

[21] Thomas G. Goodwillie, *Relative algebraic K-theory and cyclic homology*, Ann. of Math. (2) **124** (1986), no. 2, 347–402, DOI 10.2307/1971283. MR855300

[22] Max Karoubi, *K-theory: An introduction*, Grundlehren der Mathematischen Wissenschaften, Band 226, Springer-Verlag, Berlin-New York, 1978. MR0488029

[23] Max Karoubi, *Homologie cyclique et K-théorie* (French, with English summary), Astérisque **149** (1987), 147. MR913964

[24] Christian Kassel, *A Künneth formula for the cyclic cohomology of $\mathbf{Z}/2$-graded algebras*, Math. Ann. **275** (1986), no. 4, 683–699, DOI 10.1007/BF01459145. MR859338

[25] Katsuo Kawakubo, *The theory of transformation groups*, Translated from the 1987 Japanese edition, The Clarendon Press, Oxford University Press, New York, 1991. MR1150492

[26] David Kazhdan, Victor Nistor, and Peter Schneider, *Hochschild and cyclic homology of finite type algebras*, Selecta Math. (N.S.) **4** (1998), no. 2, 321–359, DOI 10.1007/s000290050034. MR1669952

[27] Chris Kottke, *Functorial compactification of linear spaces*, Proc. Amer. Math. Soc. **147** (2019), no. 9, 4067–4081, DOI 10.1090/proc/14452. MR3993798

[28] Jean-Louis Loday, *Cyclic homology*, Grundlehren der mathematischen Wissenschaften [Fundamental Principles of Mathematical Sciences], vol. 301, Springer-Verlag, Berlin, 1992. Appendix E by María O. Ronco, DOI 10.1007/978-3-662-21739-9. MR1217970

[29] Ralf Meyer, *Local and analytic cyclic homology*, EMS Tracts in Mathematics, vol. 3, European Mathematical Society (EMS), Zürich, 2007, DOI 10.4171/039. MR2337277

[30] Ralf Meyer, *Excision in Hochschild and cyclic homology without continuous linear sections*, J. Homotopy Relat. Struct. **5** (2010), no. 1, 269–303. MR2812922

[31] Ernest A. Michael, *Locally multiplicatively-convex topological algebras*, Mem. Amer. Math. Soc. **11** (1952), 79. MR51444

[32] Toshikazu Natsume and Ryszard Nest, *The cyclic cohomology of compact Lie groups and the direct sum formula*, J. Operator Theory **23** (1990), no. 1, 43–50. MR1054814

[33] Victor Nistor, *Higher index theorems and the boundary map in cyclic cohomology*, Doc. Math. **2** (1997), 263–295. MR1480038

[34] N. Christopher Phillips, *K-theory for Fréchet algebras*, Internat. J. Math. **2** (1991), no. 1, 77–129, DOI 10.1142/S0129167X91000077. MR1082838

[35] N. Christopher Phillips, *The Toeplitz operator proof of noncommutative Bott periodicity*, J. Austral. Math. Soc. Ser. A **53** (1992), no. 2, 229–251. MR1175714

[36] Maarten Solleveld, *Some Fréchet algebras for which the Chern character is an isomorphism*, K-Theory **36** (2005), no. 3-4, 275–290 (2006), DOI 10.1007/s10977-006-7106-y. MR2275008

[37] B. L. Tsygan, *Homology of matrix Lie algebras over rings and the Hochschild homology* (Russian), Uspekhi Mat. Nauk **38** (1983), no. 2(230), 217–218. MR695483

Institut fur Analysis, Welfengarten 1, 30167 Hannover, Germany
Email address: alexandre.baldare@math.uni-hannover.de

IMAG, Univ. Montpellier, CNRS, 34090 Montpellier, France
Email address: moulay.benameur@umontpellier.fr

IECL, Université Lorraine, CNRS, 57000 Metz, France
Email address: nistor@univ-lorraine.fr
URL: http://www.iecl.univ-lorraine.fr/~Victor.Nistor

Proceedings of Symposia in Pure Mathematics
Volume **105**, 2023
https://doi.org/10.1090/pspum/105/01894

On Perrot's Index Cocycles

Jonathan Block, Nigel Higson, and Jesus Sanchez Jr

Abstract. We give a simplified version of a construction due to Denis Perrot that recovers the Todd class of the complexified tangent bundle of a smooth manifold from a JLO-type cyclic cocycle. The construction takes place within an algebraic framework, rather than the customary functional-analytic framework of the JLO theory. The series expansion for the exponential function is used in place of the heat kernel from the functional-analytic theory; the Dirac operator chosen is far from elliptic; and a remarkable new trace discovered by Perrot replaces the operator trace. In its full form, Perrot's theory constitutes a wholly new approach to index theory. The account presented here covers most but not all of his approach.

1. Introduction

The purpose of this paper is to give an exposition of some remarkable ideas about index theory in the framework of cyclic cohomology that Denis Perrot introduced about a decade ago [**Per13b**]. Despite the originality of Perrot's work, his ideas seem not to have been studied in much detail, or at any rate not by many. Our aim is to address this circumstance by presenting a streamlined account of many of the main ideas to as wide an audience as possible.

In [**Per13b**], Perrot gives a more or less complete account of the Atiyah-Singer index theorem using his new approach. We shall not go that far. Our aim instead will be to present Perrot's striking construction of the Todd class within the context of cyclic cohomology theory. But in the course of doing so we shall come into contact with many of Perrot's most fascinating discoveries.

In later works Perrot has applied his new approach to index theory to new contexts—involving groupoids [**Per13a, Per16**] and, in joint work with Rudy Rodsphon, foliations [**PR14**]. The latter work settles in almost complete generality a longstanding open problem of Alain Connes and Henri Moscovici [**CM95, CM98**]. These applications give ample evidence of the power of the new approach. But once again our aims will be more limited, and we shall not discuss these developments here.

Let M be a smooth, closed manifold and let ∇ be a torsion-free connection on M. The curvature of ∇ is a 2-form R on M with values in $\mathrm{End}(TM)$, and the Todd

2020 *Mathematics Subject Classification.* Primary 19D55, 19K56; Secondary 58J40.
Key words and phrases. Index theory, cyclic cohomology, pseudodifferential operators.
The second and third authors were partially supported by NSF grant DMS-1952669.

form associated to R is

$$\operatorname{Todd}(R) = \det\left(\frac{R}{\exp(R)-1}\right).$$

This is a closed differential form on M of mixed even degree. Although its origins are in algebraic geometry, the Todd class is probably most famous for the role it plays in the Atiyah-Singer index theorem. For instance, if P is an elliptic $N\times N$ system of pseudodifferential operators on M, then

$$\operatorname{Index}(P) = \int_{S^*M} \operatorname{ch}(\sigma)\operatorname{Todd}\big(R/2\pi i\big),$$

where σ is the symbol of P, which is a smooth map from the cosphere bundle S^*M into invertible $N\times N$ matrices, and $\operatorname{ch}(\sigma)$ is its Chern character, which is a closed differential form on S^*M of mixed odd degree. The (normalized) Todd form is pulled back from M to S^*M.

The purpose of this paper is to present, following Perrot, a new construction of the Todd class using Dirac operators and cyclic cohomology.

Let us recall very briefly that if A is a complex associative algebra, and if

$$C^p(A) = \big\{\text{ complex } (p+1)\text{-multilinear functionals on } A\,\big\},$$

then there are differentials

$$b\colon C^p(A)\to C^{p+1}(A) \quad \text{and} \quad B\colon C^p(A)\to C^{p-1}(A)$$

with $(b+B)^2=0$. A *periodic cyclic cocycle* is a finitely supported family of $\{\varphi_p\}_{p\geq 0}$ with $\varphi_p\in C^p(A)$ and

$$B\varphi_{p+1} + b\varphi_{p-1} = 0 \qquad \forall p\geq 0.$$

See for instance [**Con94**, Ch.III, Sec.1.β]. If A is the algebra of smooth functions on a smooth, closed manifold, and if C is a de Rham current in degree p, then the formula

$$\varphi_C(a^0,\dots,a^p) = \frac{1}{p!}\int_C a^0\,da^1\cdots da^p$$

defines an element $\varphi_C\in C^p(A)$ with

$$b\varphi_C = 0 \quad \text{and} \quad B\varphi_C = \varphi_{d'C},$$

where d' is the de Rham differential on currents. So if C is closed, then φ_C is a periodic cyclic cocycle (concentrated in degree p).

It follows from the above that if we pull the Todd form back from M to S^*M, as in the index theorem, and if we denote by $\operatorname{Todd}^{2q}(R)$ the degree $2q$-component, then for $p+2q = 2\dim(M)-1$ the formula

$$\varphi_p(a^0,\dots,a^p) = \frac{1}{p!}\int_{S^*M} a^0\,da^1\cdots da^p\operatorname{Todd}^{2q}(R) \qquad (p \text{ odd}),$$

defines a (mixed degree) periodic cyclic cocycle.

It is this cocycle that Perrot constructs in a new way. From the point of view of the formalism of cyclic theory, Perrot's construction is a variation on the famous JLO formula

$$\operatorname{STr}\Big(\int_{\Sigma^p} a^0\exp(-s_0 D\!\!\!/^2)[D\!\!\!/,a^1]\exp(-s_1 D\!\!\!/^2)\cdots[D\!\!\!/,a^p]\exp(-s_p D\!\!\!/^2)\,ds\Big).$$

See [**JLO88, Qui88**]. What makes it noteworthy are:

(i) Perrot's definition of the Dirac operator — among other things it is *not* an elliptic operator;

(ii) Perrot's approach to the heat operators $\exp(-s\not{D}^2)$ — they are defined *algebraically*, using the series expansion for the exponential function; and

(iii) Perrot's definition of the (super)trace in the JLO formula — it is not by any means an operator trace, or even a residue trace [**Wod87, Kas89**], although it is related to the latter.

Our goal in this paper is to discuss each of these points in detail. But to conclude our introduction, let us indicate the parts of Perrot's work that we shall *not* cover in the paper.

In [**Per13b**], Perrot works throughout with the algebra of order zero classical pseudodifferential symbols on a smooth, closed manifold M. Our more limited goals in this paper make it possible for us to substantially simplify matters by working instead with the *commutative* associated graded algebra of polyhomogeneous smooth functions on the cotangent bundle. We shall be able to explain Perrot's realization of the Todd form while working exclusively in this context, thereby replicating a major part of [**Per13a**]. But to complete the proof of the index theorem it would be necessary to lift the entire discussion from polyhomogeneous functions to symbols.

In the context of pseudodifferential symbols, Perrot actually constructs *two* periodic cyclic cocycles. One is effectively the cocycle that we shall construct and compute in this paper. The other is constructed in a very similar fashion, but Perrot shows that it is precisely the so-called Radul cocyle [**Rad91**], which produces the analytic index [**Nis97, Per12**]. He also proves that his two cocycles are cohomologous. So, putting everything together, the computations of the two cocycles amount to a proof of the Atiyah-Singer index theorem for elliptic pseudodifferential systems.

A note on the text. Nearly all the arguments presented below are adapted from Perrot's paper [**Per13b**], some of them very closely. In places, for example in Section 2 where we present a simple global construction of the Dirac operator, we have been able to take advantage of our simplified context to streamline some of Perrot's constructions. In other places, for example in our treatment of the coordinate independence of Perrot's trace in Section 4, we have chosen an approach that deviates from [**Per13b**], typically because we felt it was illuminating to do so. But overall we owe a huge debt to [**Per13b**].

2. Perrot's Dirac Operator

Throughout the paper, M will always be a smooth manifold without boundary. We shall always denote by n the dimension of M. In various places (where we need to integrate over M) we shall in addition assume that M is compact. We shall be working with vector fields, differential forms, etc, on M; all these will be assumed to be *smooth*.

DEFINITION 2.1. Throughout the paper, we shall denote by $\dot{T}^*M$ the complement of the set of zero covectors in the cotangent bundle T^*M:

$$\dot{T}^*M = T^*M \setminus M.$$

In this section we shall work with a fixed affine connection ∇ on the tangent bundle of M (eventually we will require ∇ to be torsion-free). Following Perrot (but with some variations, as explained in the introduction) we are going to use ∇ to construct a Dirac operator on the total space $\dot{T}^*M$ that acts on sections of the pullback to $\dot{T}^*M$ of the exterior algebra bundle of M. But we should say at the outset that this Dirac operator will *not* be elliptic, and in particular it will not be a typical Dirac operator from index theory.

The construction requires some facts about horizontal vector fields on $\dot{T}^*M$. In what follows we shall denote by $\pi\colon \dot{T}^*M \to M$ the standard projection mapping.

DEFINITION 2.2. We shall identify vector fields on M with fiberwise linear smooth functions on T^*M via the isomorphism of $C^\infty(M)$-modules

$$\alpha\colon \Big\{ \text{vector fields on } M \Big\} \xrightarrow{\cong} \left\{ \begin{matrix} \text{fiberwise linear smooth} \\ \text{functions on } T^*M \end{matrix} \right\}$$

defined by $\alpha(X)(\omega) = \omega(X)$ for every 1-form ω on M. In local coordinates,

$$\alpha\Big(\frac{\partial}{\partial x^i}\Big) = \xi_i,$$

where $\xi_1, \dots, \xi_n$ are the usual fiberwise (linear) coordinate functions on T^*M dual to $x^1, \dots, x^n$, defined by $\xi_i(\alpha) = \alpha(\partial/\partial x^i)$.

DEFINITION 2.3. Let X be a vector field on M. Its *horizontal lift* to the total space of the cotangent bundle T^*M (with respect to the connection ∇) is the vector field X^H on T^*M that is π-related to X and that is characterized by the further requirement that

$$X^H(\alpha(Y)) = \alpha(\nabla_X Y),$$

for every vector field Y on M.

In local coordinates, if we introduce the usual Christoffel symbols by

$$\nabla_{\partial/\partial x^i} \frac{\partial}{\partial x^j} = \Gamma^k_{ij} \frac{\partial}{\partial x^k}$$

(with the usual summation convention), then

$$\Big(\frac{\partial}{\partial x^i}\Big)^H = \frac{\partial}{\partial x^i} + \Gamma^k_{ij}\, \xi_k \frac{\partial}{\partial \xi_j}, \tag{2.4}$$

where $\xi_1, \dots, \xi_n$ are, as above, the fiberwise linear coordinate functions on T^*M dual to $x^1, \dots, x^n$.

DEFINITION 2.5. We introduce the following additional identification: we define an isomorphism of $C^\infty(M)$-modules

$$\gamma\colon \operatorname{End}(TM) \xrightarrow{\cong} \Big\{ \text{vertical vector fields } a_{ij}\, \xi_i\, \partial/\partial \xi_j \text{ on } T^*M \Big\}$$

by the formula $\gamma(A)(\alpha(X)) = \alpha(A(X))$. Thus in local coordinates, if $A(\partial/\partial x^i) = a_{ij}\partial/\partial x^j$, then

$$\gamma(A) = a_{ij}\, \xi_i \frac{\partial}{\partial \xi_j}.$$

LEMMA 2.6. *If X and Y are vector fields on M, then*

$$[X^H, Y^H] - [X, Y]^H = \gamma(R(X, Y))$$

PROOF. Both sides of the identity are vertical vector fields, so it suffices to check that both sides agree on the functions $\alpha(Z)$ in Definition 2.2. The identity in this case is an immediate consequence of the definitions. $\square$

DEFINITION 2.7. Form the exterior algebra bundle Λ^*T^*M over M. Equip it with the connection induced from ∇, and then pull the bundle and its connection back from M to the manifold $\dot{T}^*M$. For brevity, we shall write

$$S = \pi^*\Lambda^*T^*M$$

for the bundle, and keep the notation ∇ for the connection.

In order to express the connection on S in local coordinates it will be convenient to introduce the following notation:

DEFINITION 2.8. Given local coordinates $x^1, \ldots, x^n$ on an open set $U \subseteq M$, define operators ψ^i and $\overline{\psi}_j$ acting on sections of S by

$$\begin{aligned} \psi^i &= \text{left exterior multiplication by } dx^i \text{ on } \pi^*\Lambda^*T^*M \\ \overline{\psi}_j &= \text{contraction by } \partial/\partial x^j \text{ on } \pi^*\Lambda^*T^*M. \end{aligned}$$

Using the above notation, along with the previously introduced Christoffel symbols Γ^k_{ij}, the induced connection on $\pi^*\Lambda^*T^*M$ is

$$\begin{aligned} \nabla_{(\partial/\partial x^i)^H} &= \frac{\partial}{\partial x^i} + \Gamma^k_{i\ell}\,\xi_k\,\frac{\partial}{\partial \xi_\ell} - \Gamma^k_{i\ell}\,\psi^\ell\,\overline{\psi}_k \\ \nabla_{\partial/\partial\xi_j} &= \frac{\partial}{\partial \xi_j}. \end{aligned} \tag{2.9}$$

DEFINITION 2.10. Given local coordinates on an open set $U \subseteq M$, define operators

$$\not{D}_{\text{horiz}}, \not{D}_{\text{vert}} \colon \Gamma\big(T^*U, \pi^*\Lambda^*T^*M\big) \longrightarrow \Gamma\big(T^*U, \pi^*\Lambda^*T^*M\big)$$

using the formulas

$$\not{D}_{\text{horiz}} = \psi^i \nabla_{(\partial/\partial x^i)^H} \quad \text{and} \quad \not{D}_{\text{vert}} = \overline{\psi}_j\, \nabla_{\partial/\partial\xi_j}.$$

Using the Christoffel symbols for ∇ and the standard local frame for S associated to a coordinate system, we find that

$$\not{D}_{\text{horiz}} = \psi^i \frac{\partial}{\partial x^i} + \Gamma^k_{ij}\psi^i\xi_k\frac{\partial}{\partial\xi_j} - \Gamma^k_{ij}\psi^i\psi^j\overline{\psi}_k \quad \text{and} \quad \not{D}_{\text{vert}} = \overline{\psi}^j\frac{\partial}{\partial\xi_j}.$$

LEMMA 2.11. *The operators $\not{D}_{\text{horiz}}$ and $\not{D}_{\text{vert}}$ are independent of the choice of local coordinate system used to define them.*

PROOF. This follows from the $C^\infty(M)$-linearity of the map α. □

LEMMA 2.12. *If the connection ∇ on TM is torsion-free, then*

$$\not{D}_{\text{horiz}} = \psi^i \frac{\partial}{\partial x^i} + \Gamma^k_{ij}\psi^i\xi_k\frac{\partial}{\partial\xi_j}$$

in any local coordinate system.

PROOF. This is because $\Gamma^k_{ij} = \Gamma^k_{ji}$ for a torsion-free connection, so that the third term in the defining formula

$$\not{D}_{\text{horiz}} = \psi^i \frac{\partial}{\partial x^i} + \Gamma^k_{ij}\psi^i\xi_k\frac{\partial}{\partial\xi_j} - \Gamma^k_{ij}\psi^i\psi^j\overline{\psi}_k$$

is zero in this case. □

LEMMA 2.13. *If the connection ∇ on TM is torsion-free, then*

$$\not{D}^2_{\text{horiz}} = \tfrac{1}{2}\psi^i\psi^j\gamma\big(R(\partial/\partial x^i, \partial/\partial x^j)\big)$$

in any local coordinate system.

PROOF. Fix a local coordinate system on $U \subseteq M$. Give S the local frame associated to these coordinates, and define operators

$$\partial_i^H = \frac{\partial}{\partial x^i} + \Gamma^k_{i\ell}\xi_k\frac{\partial}{\partial \xi_\ell}$$

on smooth sections of S (these are not the operators $\nabla_{(\partial/\partial x^i)^H}$, nor do they have any particularly special meaning; they are merely introduced for the sake of the computation). We now calculate from Lemma 2.12 that

$$\begin{aligned}\not{D}^2_{\text{horiz}} &= \psi^i\partial_i^H \cdot \psi^j\partial_j^H\\ &= \psi^i\psi^j\partial_i^H\partial_j^H\\ &= \sum_{i<j}\psi^i\psi^j\big(\partial_i^H\partial_j^H - \partial_j^H\partial_i^H\big).\end{aligned}$$

But it follows from Lemma 2.6 that

$$\partial_i^H\partial_j^H - \partial_j^H\partial_i^H = \gamma\big(R(\partial/\partial x^i, \partial/\partial x^j)\big),$$

and as a result

$$\begin{aligned}\not{D}^2_{\text{horiz}} &= \sum_{i<j}\psi^i\psi^j\gamma\big(R(\partial/\partial x^i, \partial/\partial x^j)\big)\\ &= \tfrac{1}{2}\psi^i\psi^j\gamma\big(R(\partial/\partial x^i, \partial/\partial x^j)\big),\end{aligned}$$

as required. □

DEFINITION 2.14. We shall write

$$\gamma(R) = \tfrac{1}{2}\psi^i\psi^j\gamma\big(R(\partial/\partial x^i, \partial/\partial x^j)\big).$$

This is a first-order linear differential operator acting on section of S, independent of the choice of coordinates. More explicitly, if we write

$$R\Big(\frac{\partial}{\partial x^i}, \frac{\partial}{\partial x^j}\Big)\frac{\partial}{\partial x^\ell} = R^k_{ij\ell}\frac{\partial}{\partial x^k},$$

then

$$\gamma(R) = \frac{1}{2}\psi^i\psi^j R^k_{ij\ell}\xi_k\frac{\partial}{\partial \xi_\ell}.$$

In this paper, following Perrot we shall be concerned with the "Dirac operator" $D = \not{D}_{\text{horiz}} + \not{D}_{\text{vert}}$ (or actually a small variation of this; see Definition 3.18) and its square. Lemma 2.13 is obviously relevant to the computation of the square, as is the simple formula

$$D^2_{\text{vert}} = 0. \tag{2.15}$$

What remains is to compute the cross-term

$$\{\not{D}_{\text{horiz}}, \not{D}_{\text{vert}}\} = \not{D}_{\text{horiz}}\not{D}_{\text{vert}} + \not{D}_{\text{vert}}\not{D}_{\text{horiz}}.$$

LEMMA 2.16. *If the connection ∇ on TM is torsion-free, then*

$$\{\not{D}_{\text{horiz}}, \not{D}_{\text{vert}}\} = \nabla_{\partial_{\xi_i}}\nabla_{(\partial_{x^i})^H}$$

(*with the summation convention, as always*) *in any local coordinate system.*

PROOF. We shall use the same operators ∂_i^H introduced in the proof of Lemma 2.13. With these, we may write

$$\begin{aligned}\{\not{D}_{\text{horiz}}, \not{D}_{\text{vert}}\} &= (\psi^i \partial_i^H)(\overline{\psi}_j \frac{\partial}{\partial \xi_j}) + (\overline{\psi}_j \frac{\partial}{\partial \xi_j})(\psi^i \partial_i^H) \\ &= \psi^i \overline{\psi}_j \partial_i^H \frac{\partial}{\partial \xi_j} + \overline{\psi}_j \psi^i \frac{\partial}{\partial \xi_j} \partial_i^H.\end{aligned} \tag{2.17}$$

Next, we compute that

$$\partial_i^H \frac{\partial}{\partial \xi_j} = \Big(\frac{\partial}{\partial x^i} + \Gamma_{i\ell}^k \xi_k \frac{\partial}{\partial \xi_\ell}\Big)\frac{\partial}{\partial \xi_j} = \frac{\partial}{\partial \xi_j}\partial_i^H - \Gamma_{i\ell}^j \frac{\partial}{\partial \xi_\ell},$$

and as a result we find from (2.17) that

$$\begin{aligned}\{\not{D}_{\text{horiz}}, \not{D}_{\text{vert}}\} &= (\psi^i \overline{\psi}_j + \overline{\psi}_j \psi^i)\frac{\partial}{\partial \xi_j}\partial_i^H - \Gamma_{i\ell}^j \psi^i \overline{\psi}_j \frac{\partial}{\partial \xi_\ell} \\ &= \frac{\partial}{\partial \xi_i}\Big(\frac{\partial}{\partial x^i} + \Gamma_{i\ell}^k \xi_k \frac{\partial}{\partial \xi_\ell}\Big) - \Gamma_{i\ell}^j \psi^i \overline{\psi}_j \frac{\partial}{\partial \xi_\ell} \\ &= \frac{\partial^2}{\partial \xi_i \partial x^i} + \Gamma_{i\ell}^k \xi_k \frac{\partial^2}{\partial \xi_i \partial \xi_\ell} + \Gamma_{i\ell}^i \frac{\partial}{\partial \xi_\ell} - \Gamma_{i\ell}^j \psi^i \overline{\psi}_j \frac{\partial}{\partial \xi_\ell}.\end{aligned} \tag{2.18}$$

On the other hand,

$$\begin{aligned}\nabla_{\partial_{\xi_i}} \nabla_{(\partial/\partial x^i)^H} &= \frac{\partial}{\partial \xi_i}\left(\frac{\partial}{\partial x^i} + \Gamma_{i\ell}^k \xi_k \frac{\partial}{\partial \xi_\ell} - \Gamma_{i\ell}^k \psi^\ell \overline{\psi}_k\right) \\ &= \frac{\partial^2}{\partial \xi_i \partial x^i} + \Gamma_{i\ell}^k \xi_k \frac{\partial^2}{\partial \xi_i \partial \xi_\ell} + \Gamma_{i\ell}^i \frac{\partial}{\partial \xi_\ell} - \Gamma_{i\ell}^k \psi^\ell \overline{\psi}_k \frac{\partial}{\partial \xi_i}.\end{aligned} \tag{2.19}$$

Since ∇ is torsion-free, $\Gamma_{i\ell}^k = \Gamma_{\ell i}^k$, and so we find that (2.18) and (2.19) are equal, as required. □

If we write

$$\nabla^2 = \nabla_{\partial_{\xi_i}} \nabla_{(\partial_{x^i})^H}$$

(this is coordinate-independent), then the computations in this section may be summarized as follows:

$$\not{D}_{\text{horiz}}^2 = \gamma(R), \quad \not{D}_{\text{vert}}^2 = 0 \quad \text{and} \quad \{\not{D}_{\text{horiz}}, \not{D}_{\text{vert}}\} = \nabla^2. \tag{2.20}$$

These identities will be crucial in the sequel.

REMARK 2.21. Using the connection ∇ we may identify the tangent bundle of $\dot{T}^*M$ with the pullback of $TM \oplus T^*M$ to $\dot{T}^*M$. If we equip the latter with its canonical symmetric bilinear form, which of course is nondegenerate but of indefinite signature, then S carries an irreducible representation of the associated bundle of Clifford algebras, and the operator $\not{D}_{\text{horiz}} + \not{D}_{\text{vert}}$ is a Dirac operator in the usual sense, although since the bilinear form is not definite, the operator is not elliptic.

3. Infinite-Order Polyhomogeneous Operators

In this section we shall describe the infinite-order linear partial differential operators (presented as formal series) with which we shall work in the rest of the paper. We shall begin by considering differential operators that act on scalar functions, but then we shall turn to differential operators acting on spinors, which will be our principal objects of interest.

DEFINITION 3.1. We shall denote by E the Euler vector field on $\dot{T}^*M$, so that

$$(Ef)(\alpha) = \frac{d}{dt}\Big|_{t=0} f(e^t\alpha)$$

for all smooth functions f on $\dot{T}^*M$ and all $\alpha \in \dot{T}^*M$. We shall say that a smooth function on $\dot{T}^*M$ is *homogeneous of degree* $k \in \mathbb{Z}$ if it is a k-eigenfunction for the action of E, or in other words if

$$f(e^t\alpha) = e^{kt}f(\alpha)$$

for all $t \in \mathbb{R}$ and all $\alpha \in \dot{T}^*M$. We shall say that f is *polyhomogenous* if it is a (finite) linear combination of smooth homogeneous functions of various integer degrees. We shall write

$$\mathrm{P}^k(M) = \{\, \text{Smooth degree-}k \text{ homogeneous functions on } \dot{T}^*M \,\}$$

and

$$\mathrm{P}(M) = \{\, \text{Smooth polyhomogeneous functions on } \dot{T}^*M \,\}.$$

DEFINITION 3.2. A linear partial differential operator D on $\dot{T}^*M$, acting on smooth functions on $\dot{T}^*M$, is *homogeneous of degree* $k \in \mathbb{Z}$ if $\mathrm{ad}_E(D) = kD$, where E is the Euler vector field, and *polyhomogeneous* if it is a (finite) linear combination of homogeneous operators of various integer degrees. We shall denote by $\mathrm{PDO}(M)$ the algebra of polyhomogeneous linear partial differential operators on $\dot{T}^*M$.

REMARK 3.3. In local coordinates, the polyhomogeneous operators are linear combinations of operators of the form

$$f_{\alpha\beta}\frac{\partial^\alpha}{\partial x^\alpha}\frac{\partial^\beta}{\partial \xi_\beta},$$

with f a polyhomogeneous smooth function on $\dot{T}^*M$.

DEFINITION 3.4. We shall denote by F the field of formal complex Laurent series $\sum_{-N}^{\infty} a_k\varepsilon^k$ (with only finite singular parts). We shall denote by $\mathrm{FPDO}(M)$ the algebra of formal Laurent series in ε with coefficients from $\mathrm{PDO}(M)$ (again, with only finite singular parts).

We turn now to one of the main constructions of the paper, which is that of a subspace

$$\mathrm{FB}(M) \subseteq \mathrm{FPDO}(M)$$

on which will be defined a crucial trace functional. The subspace is not a subalgebra, let alone an ideal, but it is for instance a bimodule under the left and right actions of $\mathrm{PDO}(M)$ on $\mathrm{FPDO}(M)$ (the name is supposed to suggest this, and Perrot calls his version, on which ours is very closely modelled, the *bimodule of trace class operators*). We shall define the bimodule in this section and construct the trace in the next section.

The space $\mathrm{FB}(M)$ will be defined in stages, and to begin we shall work in a fixed coordinate system on an open subset $U \subseteq M$. We shall use the following operator extensively:

DEFINITION 3.5. Let U be an open subset of M and let $x^1, \ldots, x^n$ be coordinates on U. Given these coordinates, the associated *Laplacian* on $\dot{T}^*U$ is

$$\Delta = \frac{\partial^2}{\partial x^i \partial \xi_i}$$

(summation convention).

REMARKS 3.6. The operator Δ is *not* invariant under changes of coordinates (except for linear changes of coordinates). In addition, the operator Δ is *not* elliptic.

Following Perrot we introduce the following increasing filtration on the algebra $\mathrm{PDO}(U)$.

PROPOSITION/DEFINITION 3.7. *Let U be an open subset of M and let $x^1, \dots, x^n$ be coordinates on U. Define an increasing filtration of the algebra* $\mathrm{PDO}(U)$*, indexed by $\mathbb{Z}$, and an associated notion of order on partial differential operators on U, that we shall call* bimodule order*, as follows. The bimodule filtration is constructed by considering at first all increasing algebra filtrations*

$$\cdots \subseteq \mathrm{PDO}_p(U) \subseteq \mathrm{PDO}_{p+1}(U) \subseteq \cdots \subseteq \mathrm{PDO}(U)$$

indexed by $p \in \mathbb{Z}$ and with $\partial/\partial x^i \in \mathrm{PDO}_3(U)$, $\partial/\partial \xi_i \in \mathrm{PDO}_{-1}(U)$, and $f \in \mathrm{PDO}_{2p}$ if f is homogeneous of degree p. Then define the pth bimodule filtration space to be the intersection of all $\mathrm{PDO}_p(U)$ over all these filtrations. Of course we write $\text{bimodule-order}(T) \leq p$ *if T belongs to this intersection. We have the following identities:*

$$\begin{aligned} \text{bimodule-order}(\partial/\partial x^i) &= 3 \qquad && \forall i \\ \text{bimodule-order}(\partial/\partial \xi_j) &= -1 \qquad && \forall j \\ \text{bimodule-order}(f) &= 2p \qquad && \textit{if } \text{degree}(f) = p. \end{aligned}$$

PROOF. A straightforward computation in local coordinates. □

REMARK 3.8. We shall introduce a second, quite different filtration on differential operators in Section 6, and it should not be confused with the bimodule filtration defined here.

DEFINITION 3.9. We shall denote by $\mathrm{FB}(U) \subseteq \mathrm{FPDO}(U)$ the space of all Laurent series

$$\sum_k \varepsilon^k D_k \exp(\varepsilon\Delta) \qquad (D_k \in \mathrm{PDO}(U))$$

for which there exists $N > 0$ with

$$\text{bimodule-order}(D_k) \leq k + N,$$

for all k.

REMARK 3.10. It is not so easy to explain at the outset the reason for defining the bimodule filtration, and hence $\mathrm{FB}(U)$, in the way that we just have. In fact there is a certain amount of flexibility in how one might define $\mathrm{FB}(U)$, which needs to be small enough that the trace functional to be discussed in the next section is well-defined, and large enough that it is closed under some simple operations, most notably conjugation by $\exp(\varepsilon\Delta)$. The definition given seems to be the simplest one that meets these requirements.[1]

The following is another easy computation:

---

[1]One might perhaps compare this to the situation with traceable Hilbert space operators in index theory, where one might work with smoothing operators, or trace-class operators, or operators whose singular value sequence has rapid decay, and so on.

LEMMA 3.11. *If T is any operator in* PDO(U), *then*

$$\text{bimodule-order}\big([\Delta, T]\big) \leq \text{bimodule-order}(T) + 1. \qquad \square$$

Now suppose that $T \in FPDO(U)$ and that $s \in \mathbb{R}$. The conjugated operator

$$\mathrm{Ad}_{\exp(\varepsilon s\Delta}(T) = \exp(\varepsilon s\Delta)T\exp(-\varepsilon s\Delta) \in \mathrm{FPDO}(U)$$

may be alternatively expressed as

$$\mathrm{Ad}_{\exp(\varepsilon s\Delta)}(T) = \exp(\varepsilon s\,\mathrm{ad}_\Delta)(T) = \sum_{k=0}^{\infty} \frac{s^k\varepsilon^k}{k!}\,\mathrm{ad}_\Delta^k(T)$$

(thanks to the powers ε^k, the sum makes sense in FPDO(U); to use language that will be introduced in the next section, the sum is convergent in the adic topology). Using Lemma 3.11 we find that:

COROLLARY 3.12. *If $T \in$ PDO(U), then for every $s \in \mathbb{R}$ we may write*

$$\mathrm{Ad}_{\exp(\varepsilon s\Delta)}(T) = \sum_{k=0}^{\infty} \varepsilon^k T_k,$$

where $T_k \in$ PDO(U) and order(T_k) $\leq k +$ order(T). In particular the subspace FB(M) $\subseteq$ FPDO(M) *is closed under left and right multiplication by elements of* PDO(M) *in the algebra* FPDO(U). $\square$

COROLLARY 3.13. *If $X \in$ FB(U), then $\mathrm{Ad}_{\exp(\varepsilon s\Delta)}(X) \in$ FB(U) for every $s \in \mathbb{R}$.* $\square$

The bimodule filtration on PDO(U) is independent of the choice of coordinates on U; this can be checked by the usual change-of-coordinates formulas. The following result is a bit more involved:

LEMMA 3.14. *Let U be an open subset of M, let $x^1, \ldots, x^n$ be coordinates on U, and let Δ and* FB(U) *be the associated Laplacian and bimodule. If Δ' is the Laplacian associated to another coordinate system on U, then*

$$\exp(\varepsilon\Delta') \in \mathrm{FB}(U).$$

PROOF. The Laplacian Δ' in a new coordinate system has the form

$$\Delta' = \Delta + a_i\frac{\partial}{\partial\xi_i} + b_{jk\ell}\xi_j\frac{\partial^2}{\partial\xi_k\partial\xi_\ell}$$

in terms of the original coordinate system, where a_i and $b_{jk\ell}$ are smooth functions on U, and so bimodule-order($\Delta - \Delta'$) ≤ 0.

To proceed, within FPDO(U) there is a Duhamel-type formula

$$\exp(\varepsilon\Delta') - \exp(\varepsilon\Delta) = \varepsilon\int_0^1 \exp(\varepsilon s\Delta')(\Delta' - \Delta)\exp(\varepsilon(1-s)\Delta)\,ds.$$

The integrand, viewed as a power series in ε, is a polynomial function of s in each degree, with values in PDO(U). So the definition of the integral presents no problems (and the formula may be proved in the usual way).

By iterating the formula we obtain the perturbation series

$$\exp(\varepsilon\Delta') = \sum_{p=0}^{\infty}\varepsilon^p\int_{\Sigma^p}\exp(\varepsilon s_0\Delta)R\exp(\varepsilon s_1\Delta)R\cdots$$
$$\cdots\exp(\varepsilon s_{p-1}\Delta)R\exp(\varepsilon s_p\Delta)\,ds;$$

in FPDO(U), where $R = \Delta' - \Delta$ and where Σ^p is the standard p-simplex. Writing the above as

$$\exp(\varepsilon\Delta') = \sum_{p=0}^{\infty} \varepsilon^p \int_{\Sigma^p} \mathrm{Ad}_{\exp(\varepsilon s_0 \Delta)}(R)\, \mathrm{Ad}_{\exp(\varepsilon(s_0+s_1)\Delta)}(R) \cdots$$
$$\cdots \mathrm{Ad}_{\exp(\varepsilon(s_0+s_1+\cdots+s_{p-1})\Delta)}(R)\, ds \cdot \exp(\varepsilon\Delta)$$

and using Corollary 3.12 we obtain the result. □

THEOREM 3.15. *The subspace* FB(U) $\subseteq$ FPDO(U) *is independent of the choice of local coordinates.*

PROOF. First, Lemma 3.14 shows that $\exp(\varepsilon\Delta') \in$ FB(U). Second, because the bimodule order is obtained from an algebra filtration, FB(U) is in fact closed under left multiplication by series $\sum_k \varepsilon^k T_k$ with $T_k \in$ PDO(U) and bimodule-order(T_k) $\leq k + N$ for some N and all k. These two observations show that the FB-bimodule defined using Δ' is included in the FB-bimodule defined using Δ. By symmetry, the two are in fact equal. □

We can now define a global version of the bimodule FB(U):

DEFINITION 3.16. We denote by FB(M) $\subseteq$ FPDO(M) the subspace of all elements that restrict to elements in FB(U) whenever U is an open subset of M that supports a coordinate system.

We shall now make the minor changes needed to carry over all of the results above from operators that act on scalar functions to operators that act on sections of the bundle S. The most direct way to do proceed is to adapt the notion of bimodule order to this context:

DEFINITION 3.17. Let U be an open subset of M and let $x^1, \ldots, x^n$ be coordinates on U. Trivialize the bundle S over U using the associated frame for the exterior algebra bundle over U, and define an increasing filtration of the algebra PDO(U, S), indexed by $\mathbb{Z}$, and an associated notion of *bimodule order* on partial differential operators on U, by

$$\begin{aligned}
\text{bimodule-order}(\partial/\partial x^i) &\leq 3 && \forall i\\
\text{bimodule-order}(\partial/\partial \xi_j) &\leq -1 && \forall j\\
\text{bimodule-order}(f) &\leq 2p && \text{if } f \text{ is a scalar function of degree } p\\
\text{bimodule-order}(\psi^i) &\leq 0 && \forall i\\
\text{bimodule-order}(\overline{\psi}_j) &\leq 0 && \forall j
\end{aligned}$$

Thus the bimodule order on PDO(U, S) does not involve the bundle S at all. It is coordinate-independent, and all of the results and definitions given above pass over to the context of operators on S without any change at all (of course one must replace homogeneous scalar functions on $\dot{T}^*U$ with homogeneous endomorphisms of S, noting that since S is pulled back from M, the notion of homogeneity makes sense here, as it does for *any* pull-back bundle, since the operation of scalar multiplication on the fibers of $\dot{T}^*M$ lifts canonically to an action on the bundle).

To conclude this section we shall situate the Dirac operator and its exponential within the bimodule FB(M, S).

DEFINITION 3.18. The *Dirac operator* on $\dot{T}^*M$, associated to a torsion-free connection on M, is the operator

$$\not{D} = \varepsilon \not{D}_{\text{horiz}} + \not{D}_{\text{vert}},$$

whose summands are described in Definition 2.10.

It follows from the formulas in (2.20) that

$$\not{D}^2 = \varepsilon \nabla^2 + \varepsilon^2 \gamma(R),$$

and therefore the exponentials

$$\exp(s\not{D}^2) = \sum_{k=0}^{\infty} \frac{s^k}{k!} \left(\varepsilon \nabla^2 + \varepsilon^2 \gamma(R)\right)^k$$

are well-defined elements of $\operatorname{FPDO}(M, S)$.

We shall use the following two lemmas when we construct the periodic cyclic cocycle associated to the Dirac operator $\not{D}$ in Section 5.

LEMMA 3.19. *If $T \in \operatorname{PDO}(U, S)$, then for every $s \in \mathbb{R}$ we may write*

$$\operatorname{Ad}_{\exp(s\not{D}^2)}(T) = \sum_{k=0}^{\infty} \varepsilon^k T_k$$

where $T_k \in PDO(U, S)$ and $\operatorname{order}(T_k) \leq k + \operatorname{order}(T)$.

PROOF. This is a variation of Corollary 3.13 and it is proved the same way, using the formula

$$\varepsilon \Delta - \not{D}^2 = \varepsilon R_1 + \varepsilon^2 R_2, \tag{3.20}$$

where R_0 and R_1 have bimodule orders 0 and 1, respectively. □

LEMMA 3.21. *If $\not{D}$ is the Dirac operator associated to any torsion-free connection on M, then $\exp(\not{D}^2) \in \operatorname{FB}(M, S)$.*

PROOF. The method use to prove Lemma 3.14 applies here, in view of (3.20). □

4. Perrot's Trace

The most remarkable ingredient in Perrot's work is a supertrace functional defined on the bimodule $\operatorname{FB}(M, S)$ (this space is $\mathbb{Z}/2\mathbb{Z}$-graded in the usual way, using the degree operator on differential forms). We shall construct the supertrace in this section.

As we did in the previous section, we shall begin by ignoring spinors: we shall first construct a trace functional on the space $\operatorname{FB}(M)$ (which is not $\mathbb{Z}/2\mathbb{Z}$-graded). The first ingredient in the definition of the trace here is the usual method of "integrating" homogeneous functions on $\dot{T}^*M$ of degree $-n$, which we shall now review.

DEFINITION 4.1. If W is any smooth manifold, then we shall denote by $\operatorname{Dens}(W)$ the $C^\infty(W)$-module of smooth densities on W (see [**BGV92**, pp.30-31] or [**Lee03**, Ch.14] for instance).

View the cotangent sphere bundle S^*M as the quotient of $\dot{T}^*M$ by the action of positive scalar multiplication on $\dot{T}^*M$ (generated by the Euler vector field) and denote by

$$p\colon \dot{T}^*M \longrightarrow S^*M$$

the projection mapping. In addition, denote by ν the volume form on $\dot{T}^*M$ that in any local coordinate system has the form

$$\nu = dx^1 dx^2 \cdots dx^n d\xi_1 d\xi_2 \cdots d\xi_n.$$

DEFINITION 4.2. Suppose that $f \in \mathrm{P}^{-n}(M)$. The $(2n-1)$-form $f \cdot \iota_E \nu$ is basic for the action of $\mathbb{R}$ on $\dot{T}^*M$, and we shall denote by σ_f the unique $(2n-1)$-form on S^*M such that

$$p^*\sigma_f = f \cdot \iota_E \nu.$$

DEFINITION 4.3. Throughout the paper we shall equip S^*M with the unique orientation such that $\int_{S^*M} \sigma_f \geq 0$ (with σ_f as in Definition 4.2) for every nonnegative $f \in \mathrm{P}^{-n}(M)$.

We obtain from the definitions above an isomorphism of $C^\infty(S^*M)$-modules

$$\mathrm{P}^{-n}(M) \xrightarrow{\cong} \mathrm{Dens}(S^*M). \tag{4.4}$$

We shall use this identification throughout the rest of the paper, and for $f \in \mathrm{P}^{-n}(M)$ we shall write

$$⨍_{S^*M} f = \text{integral of the associated density} = \int_{S^*M} \sigma_f. \tag{4.5}$$

For general $f \in \mathrm{P}(M)$ we shall write

$$⨍_{S^*M} f = ⨍_{S^*M} f^{[-n]}, \tag{4.6}$$

where $f^{[-n]}$ denotes the degree $-n$ component of f.

REMARK 4.7. The above integral is a commutative version of the noncommutative residue on pseudodifferential symbols. Indeed, up to an overall constant, the noncommutative residue of an order $-n$ pseudodifferential symbol is the integral of its principal symbol function, as defined above.

The field F of formal Laurent series, the algebra $\mathrm{FPDO}(U)$ and its subspace $FB(U)$ all carry the usual adic topologies, in which a base for the open neighborhoods of 0 is the collection of spaces

$$\Big\{ \sum_{k=\ell}^{\infty} \varepsilon^k D_k \Big\}$$

for $\ell = 0, 1, 2, \ldots$ (with D_k either a scalar, in the case of F, or a partial differential operator in the other cases).

Now denote by $\mathrm{FB}_c(U)$ the subspace of $\mathrm{FB}(U)$ comprised of series whose coefficients are partial differential operators on $\dot{T}^*U$ that are compactly supported in the U-direction, and define $\mathrm{P}_c(U)$ similarly. The main result of this section will be as follows:

THEOREM 4.8. *Let $U \subseteq M$ be an open set and with a fixed system of local coordinates. There is a unique continuous (for the adic topologies) F-linear functional*

$$\mathrm{Tr}_F\colon \mathrm{FB_c}(U) \longrightarrow F$$

such that

$$\mathrm{Tr}_F\big(f \cdot \exp(\varepsilon\Delta) \cdot g\big) = \varepsilon^{-n} \int_{S^*U} fg$$

for all $f, g \in \mathrm{P}_c(U)$. This functional has the following trace property:

$$\mathrm{Tr}_F(DX) = \mathrm{Tr}_F(XD) \qquad \forall X \in \mathrm{FB}_c(U),\ \forall D \in \mathrm{PDO}(U).$$

The uniquess part of the theorem is easy to prove.

PROOF OF UNIQUENESS IN THEOREM 4.8. Since the *finite* sums

$$\sum \varepsilon^k D_k \exp(\varepsilon\Delta)$$

are dense in $\mathrm{FB_c}(U)$, it suffices to show that Tr_F is determined on these by the formula in the statement of the theorem. We shall show that each such finite sum is a finite Laurent polynomial in ε with coefficients of the form $f\ \exp(\varepsilon\Delta)\, g$. This certainly suffices. To do so, we need only note that the formulas

$$\begin{aligned} f\ \exp(\varepsilon\Delta)\, x^j g - f x^j\ \exp(\varepsilon\Delta)\, g & \\ &= \varepsilon f \frac{\partial}{\partial \xi_j} \exp(\varepsilon\Delta)\, g \\ &= \varepsilon \frac{\partial}{\partial \xi_j} f \cdot \exp(\varepsilon\Delta)\, g - \varepsilon \frac{\partial f}{\partial \xi_j} \exp(\varepsilon\Delta)\, g \end{aligned}$$

and

$$\exp(\varepsilon\Delta)\, \xi^i - \xi^i \exp(\varepsilon\Delta) = \varepsilon \frac{\partial}{\partial x^i} f\ \exp(\varepsilon\Delta)\, g - \varepsilon \frac{\partial f}{\partial x^i} \exp(\varepsilon\Delta)\, g,$$

show that the span of all $f\ \exp(\varepsilon\Delta)\, g$ over the Laurent polynomials in ε is closed under left multiplication by operators in $\mathrm{PDO}(U)$. □

Although the proof of uniqueness suggests an approach to the construction of the trace, it is more convenient to proceed in a different direction. Recall that if A is a positive-definite $2n{\times}2n$ matrix, and if $p\colon \mathbb{R}^{2n} \to \mathbb{C}$ is a polynomial function, then

$$\begin{aligned} \int_{\mathbb{R}^{2n}} p(w) \exp\big(-\tfrac{1}{2}\langle w, Aw\rangle\big)\, dw & \\ &= \frac{(2\pi)^n}{\sqrt{\det(A)}} p(\sqrt{-1}\partial_w) \exp\big(-\tfrac{1}{2}\langle w, A^{-1}w\rangle\big)\Big|_{w=0} \end{aligned}$$

(see for instance [**Hor83**, Sec.7.6]). The following definition in effect extends this formula to the matrix $A = \left[\begin{smallmatrix} 0 & I \\ I & 0 \end{smallmatrix}\right]$, which is not positive-definite.

DEFINITION 4.9. Let $p\colon \mathbb{R}^{2n} \to \mathbb{C}$ be a polynomial function We define

$$\begin{aligned} \Big\langle p(\partial/\partial x, \partial/\partial\xi) \exp(\varepsilon\Delta)\Big\rangle & \\ &= \varepsilon^{-n} p\big(\sqrt{-1}\partial/\partial x, \sqrt{-1}\partial/\partial\xi\big) \exp\big(\varepsilon^{-1} x^i \xi_i\big)\ \Big|_{x=0,\xi=0} \\ &= \varepsilon^{-n} p(\partial/\partial x, \partial/\partial\xi) \exp\Big(-\varepsilon^{-1} x^i \xi_i\Big)\ \Big|_{x=0,\xi=0} \end{aligned}$$

This value lies in $\mathbb{C}[\varepsilon^{-1}, \varepsilon]$.

REMARK 4.10. A comparison of the two integral formulas shows that we have dropped an overall multiplicative constant of $(2\pi i)^n$. For our purposes it is a little tidier to do so.

Perrot calls the construction in the definition above the *contraction* of

$$p(\partial/\partial x, \partial/\partial \xi) \exp(\varepsilon\Delta).$$

It will be convenient to follow Perrot and extend this concept of contraction to cover more cases, as follows:

DEFINITION 4.11. If D is any operator in $\mathrm{PDO}(U)$, say

$$D = \sum_{\alpha,\beta} f_{\alpha\beta} \frac{\partial^\alpha}{\partial x^\alpha} \frac{\partial^\beta}{\partial \xi_\beta},$$

then we define $\langle D \exp(\varepsilon\Delta)\rangle \in \mathrm{P}(U)[\varepsilon^{-1}, \varepsilon]$ by

$$\begin{aligned}\langle D \exp(\varepsilon\Delta)\rangle &= \sum_{\alpha,\beta} f_{\alpha\beta} \cdot \Big\langle \frac{\partial^\alpha}{\partial x^\alpha} \frac{\partial^\beta}{\partial \xi_\beta} \exp(\varepsilon\Delta)\Big\rangle \\ &= \varepsilon^{-n} \sum_{\alpha,\beta} f_{\alpha\beta} \cdot \Big(\frac{\partial^\alpha}{\partial x^\alpha} \frac{\partial^\beta}{\partial \xi_\beta} \exp(-\varepsilon^{-1} x^i \xi_i)\Big)\Big|_{x=0,\xi=0}.\end{aligned}$$

Later on we shall make use of the following alternative formula for the trace. Form the function $\exp(\varepsilon^{-1}(x^i - y^i)(\xi_i - \eta_i))$ on the direct product of two copies of $\dot{T}^*U$, and then for an operator $D \in \mathrm{PDO}(U)$ form the function

$$D \exp(-\varepsilon^{-1}(x^i - y^i)(\xi_i - \eta_i))$$

on the product, with D acting on the first copy (the x- and ξ-variables) alone. If the restriction to the diagonal of this function is regarded as a function on $\dot{T}^*U$ via the projection onto the first factor of the product, then

$$\langle D \exp(\varepsilon\Delta)\rangle = \varepsilon^{-n} \big(D \exp(-\varepsilon^{-1}(x^i - y^i)(\xi_i - \eta_i))\big)\Big|_{x^i=y^i,\ \xi_i=\eta_i}. \tag{4.12}$$

We are now ready to give the formula for Perrot's trace functional.

DEFINITION 4.13. Let $X \in \mathrm{FB}(U)$. Write X as a Laurent series in ε with finite singular part,

$$X = \sum \varepsilon^k D_k \exp(\varepsilon\Delta),$$

as in Definition 3.9. Assuming that each D_k has compact support in the U-direction, we define $\mathrm{Tr}_F(X)$ in the field of complex Laurent series with finite singular parts by

$$\mathrm{Tr}_F(X) = \sum_k \varepsilon^k \int_{S^*M} \langle D_k \exp(\varepsilon\Delta)\rangle.$$

Since an infinite sum is involved, we need to check that the definition actually makes sense:

LEMMA 4.14. *The infinite sum in Definition 4.13 converges in the adic topology on the field F of formal Laurent series with finite singular parts.*

PROOF. The bimodule order plays an important role here: we shall use the fact that if $X \in \mathrm{FB}(U)$, and if

$$X = \sum_{\alpha,\beta,p,k} \varepsilon^k f_{\alpha\beta pk} \frac{\partial^\alpha}{\partial x^\alpha} \frac{\partial^\beta}{\partial \xi_\beta} \exp(\varepsilon\Delta),$$

then according to Definition 3.9, there is some N for which

$$p + 3|\alpha| - |\beta| \leq k + N$$

for all α, β, p and k for which $f_{\alpha\beta pk} \neq 0$.

Recall that

$$\fint_{S^*M} f_{\alpha\beta pk} \neq 0 \quad \Rightarrow \quad p = -n.$$

In addition, it follows from Definition 4.9 that

$$\Big\langle \frac{\partial^\alpha}{\partial x^\alpha} \frac{\partial^\beta}{\partial \xi_\beta} \exp(\varepsilon\Delta) \Big\rangle \neq 0 \quad \Rightarrow \quad \alpha \neq \beta,$$

while if $\alpha = \beta$, then

$$\Big\langle \frac{\partial^\alpha}{\partial x^\alpha} \frac{\partial^\beta}{\partial \xi_\beta} \exp(\varepsilon\Delta) \Big\rangle = \text{scalar} \cdot \varepsilon^{-|\alpha|-n}.$$

Of course if $\alpha = \beta$, then

$$p + 3|\alpha| - |\beta| = p + 2|\alpha|.$$

and so if in addition $p = -n$, then

$$-n + 2|\alpha| \leq k + N$$

and therefore

$$|\alpha| \leq \frac{k}{2} + m,$$

where $m = (N + n)/2$. We find that the series giving the trace in Definition 4.13 has the form

$$\mathrm{Tr}_F(X) = \sum_k \sum_{|\alpha| \leq k/2+m} c_{k,\alpha}\, \varepsilon^{k-|\alpha|-n}.$$

Since $k - |\alpha| - n \geq k/2 + m - n$ under the indicated condition on α, the sum is certainly convergent in the adic topology, as required: the powers of ε are bounded from below each power occurs only finitely many times in the sum. □

We shall now prove that Tr_F has the trace property in Theorem 4.8. To do so, we shall follow the argument of Perrot closely. The proof obviously reduces immediately to consideration of the three cases

$$D = \frac{\partial}{\partial x^i}, \quad D = \frac{\partial}{\partial \xi_j} \quad \text{and} \quad D = f. \tag{4.15}$$

Moreover using the continuity of the trace and its F-linearity, the proof simultaneously reduces to the case where

$$X = g \frac{\partial^\alpha}{\partial x^\alpha} \frac{\partial^\beta}{\partial \xi_\beta} \exp(\varepsilon\Delta).$$

In the first two cases from (4.15) we have

$$[D, X] = \frac{\partial g}{\partial x^i} \frac{\partial^\alpha}{\partial x^\alpha} \frac{\partial^\beta}{\partial \xi_\beta} \exp(\varepsilon\Delta) \quad \text{and} \quad [D, X] = \frac{\partial g}{\partial \xi_j} \frac{\partial^\alpha}{\partial x^\alpha} \frac{\partial^\beta}{\partial \xi_\beta} \exp(\varepsilon\Delta).$$

To handle these two cases of the proof it suffices to show that

$$\fint_{S^*U} \frac{\partial g}{\partial x^i} = 0 \quad \text{and} \quad \fint_{S^*U} \frac{\partial g}{\partial \xi_j} = 0.$$

In the first case we may as well assume that g has degree $-n$ (otherwise the integral is certainly zero, by definition), and then by Stokes' theorem

$$\fint_{S^*U} \frac{\partial g}{\partial x^i} = \int_{S^*U} d\,\iota_{\partial/\partial x_i}\sigma_g = 0,$$

in the notation of Definition 4.2. In the second case we may as well assume that g has degree $1-n$. But then the form $\iota_E \iota_{\partial/\partial\xi_j} g\nu$ on $\dot{T}^*U$ is basic, so it is the pullback of a form τ_g on S^*U, and from

$$d(\iota_E \iota_{\partial/\partial\xi_j} g\nu) = \iota_E \frac{\partial g}{\partial \xi_j}\nu,$$

we find that

$$\fint_{S^*U} \frac{\partial g}{\partial \xi_j} = \int_{S^*U} \sigma_{\partial g/\partial\xi_j} = \int_{S^*U} d\tau_g = 0,$$

again by Stokes' theorem.

It remains to consider the case $D = f$ in (4.15). Consider first the cases where f is a coordinate function: $f = x^i$ or $f = \xi_j$.

LEMMA 4.16. *For any i, and j and any α and β,*

$$\Big\langle \big[x^i, \frac{\partial^\alpha}{\partial x^\alpha}\frac{\partial^\beta}{\partial \xi_\beta}\exp(\varepsilon\Delta)\big]\Big\rangle = 0 \quad \textit{and} \quad \Big\langle \big[\xi_j, \frac{\partial^\alpha}{\partial x^\alpha}\frac{\partial^\beta}{\partial \xi_\beta}\exp(\varepsilon\Delta)\big]\Big\rangle = 0.$$

PROOF. We shall present the argument for x^i (the argument for ξ_j is exactly the same) and we'll describe the case of n=1 dimensions (higher dimensions present additional notational complexities but are otherwise identical).

We calculate that

$$\big[\frac{\partial^a}{\partial x^a}\frac{\partial^b}{\partial \xi^b}\exp(\varepsilon\Delta), x\big] = a\frac{\partial^{a-1}}{\partial x^{a-1}}\frac{\partial^b}{\partial \xi^b}\exp(\varepsilon\Delta) + \varepsilon\frac{\partial^a}{\partial x^a}\frac{\partial^{b+1}}{\partial \xi^{b+1}}\exp(\varepsilon\Delta)$$

and therefore that the contraction is

$$\frac{\partial^b}{\partial \xi^b}\Big(a\frac{\partial^{a-1}}{\partial x^{a-1}} + \varepsilon\frac{\partial^a}{\partial x^a}\frac{\partial}{\partial \xi}\Big)\exp(-\varepsilon^{-1}x\xi)\Big|_{x=0,\xi=0}.$$

But

$$\Big(a\frac{\partial^{a-1}}{\partial x^{a-1}} + \varepsilon\frac{\partial^a}{\partial x^a}\frac{\partial}{\partial \xi}\Big)\exp(-\varepsilon^{-1}x\xi) = (-1)^{a-1}\varepsilon^{-a}\xi^{a-1}x\exp(-\varepsilon^{-1}x\xi).$$

So, thanks to the factor of x, the contraction is 0. ☐

REMARK 4.17. The lemma can be understood more conceptually (but at the expense of a longer argument) using the fact that the contraction is derived from an integral formula on $\mathbb{R}^{2n}$, and the fact that the integral of a partial derivative in that context is zero.

PROOF OF THE TRACE PROPERTY IN THEOREM 4.8. The calculations in the proof of Lemma 4.16 show that if p is any polynomial on $\mathbb{R}^{2n}$, possibly with coefficients that are polynomials in ε, then

$$\big[x^i, p\big(\frac{\partial}{\partial x}, \frac{\partial}{\partial \xi}\big)\exp(\varepsilon\Delta)\big] = q\big(\frac{\partial}{\partial x}, \frac{\partial}{\partial \xi}\big)\exp(\varepsilon\Delta)$$

for some other q. The same is true for commutators with ξ_j, and as a result the lemma in fact shows that if μ and ν are any multi-indices, then

$$\Big\langle \mathrm{ad}_x^\mu\, \mathrm{ad}_\xi^\nu \Big(p\Big(\frac{\partial}{\partial x}, \frac{\partial}{\partial \xi}\Big) \exp(\varepsilon\Delta)\Big)\Big\rangle = 0.$$

But if f is any polyhomogeneous function, then

$$\Big[p\Big(\frac{\partial}{\partial x}, \frac{\partial}{\partial \xi}\Big) \exp(\varepsilon\Delta), f\Big] = \sum_{\mu,\nu\geq 0} \frac{1}{\mu!\nu!}\frac{\partial^\mu\partial^\nu f}{\partial x^\mu \partial \xi^\nu}\cdot \mathrm{ad}_x^\mu\, \mathrm{ad}_\xi^\nu \Big(p\Big(\frac{\partial}{\partial x}, \frac{\partial}{\partial \xi}\Big) \exp(\varepsilon\Delta)\Big),$$

with the sum convergent in the adic topology. So the contraction of the commutator is 0. We have handled the third and final case in (4.15), and therefore the proof is complete. □

The second crucial fact about Tr_F is its coordinate independence:

THEOREM 4.18. *The trace functional* Tr_F *is independent of the choice of coordinate system used to define it.*

PROOF. The theorem isn't an immediate consequence of the uniqueness part of Theorem 4.8 because the Laplacian Δ is not invariant under coordinate changes. What we need to show is that if Tr_F and Δ are constructed using a first coordinate system, and if Δ' is the Laplacian in a second coordinate system, then

$$\mathrm{Tr}_F(f\exp(\varepsilon\Delta)) = \fint_{S^*M} f = \mathrm{Tr}_F(f\exp(\varepsilon\Delta')).$$

As we already noted in the proof of Lemma 3.14, the second Laplacian has the form

$$\Delta' = \Delta + a_i \frac{\partial}{\partial \xi_i} + b^j_{k\ell}\xi_j \frac{\partial^2}{\partial\xi_k\partial\xi_\ell}$$

in the first coordinate system, where a_i and $b^j_{k\ell}$ are smooth functions on M. If $T = \Delta' - \Delta$ then we can write (using the contraction operation associated to the first coordinate system)

$$\big\langle f\exp(\varepsilon\Delta')\big\rangle - \big\langle f\exp(\varepsilon\Delta)\big\rangle = \int_0^1 \big\langle f\exp(s\varepsilon\Delta)\varepsilon T\exp((1{-}s)\varepsilon\Delta')\big\rangle\, ds.$$

If we expand the operators $\exp(s\varepsilon\Delta)\varepsilon T\exp((1{-}s)\varepsilon\Delta')$ as power series in ε, then the coefficient of ε^k, which is an element of $\mathrm{PDO}(U)$, has differential order $k{-}1$ in the x-direction (because T includes no x-derivative) and is homogeneous of degree $-k$ in the vertical direction (since each of Δ, Δ' and T has degree -1). This means that if we write the same operators in the standard form

$$\exp(s\varepsilon\Delta)\varepsilon T\exp((1{-}s)\varepsilon\Delta') = \sum_{k=0}^\infty \varepsilon^k D_k \exp(\varepsilon\Delta)$$

and then write

$$D_k = \sum g_{k,\alpha,\beta}\frac{\partial^\alpha}{\partial x^\alpha}\frac{\partial^\beta}{\partial\xi_\beta},$$

then

$$|\alpha| \geq |\beta| \quad\Rightarrow\quad g_{k,\alpha,\beta} = 0.$$

The contraction is therefore zero and the theorem is proved. □

With Theorems 4.8 and 4.18 available, we can now extend the trace beyond coordinate charts to a continuous and F-linear trace functional

$$\mathrm{Tr}_F\colon \mathrm{FB}(M) \longrightarrow F$$

in the obvious way using partitions of unity. But for the purposes of this paper we only need a part of the trace we have just defined, which is scalar-valued, not F-valued:

DEFINITION 4.19. Let M be a smooth, closed manifold. The *Perrot trace* on $\mathrm{FB}(M)$ is

$$\mathrm{Tr}\colon \mathrm{FB}(M) \longrightarrow \mathbb{C}$$
$$\mathrm{Tr}(X) = \text{Coefficient of } \varepsilon^0 \text{ in } \mathrm{Tr}_F(X).$$

Finally, we need to pass from a (scalar) trace on $\mathrm{FB}(M)$ to a supertrace on $\mathrm{FB}(M,S)$. First denote by $\Gamma(\mathrm{End}(S))$ the smooth polyhomogeneous sections of the endomorphism bundle of S. This is a $\mathbb{Z}/2\mathbb{Z}$-graded algebra over $\mathrm{P}(M)$ and there is a standard $\mathrm{P}(M)$-linear supertrace

$$\mathrm{str}_\Lambda\colon \Gamma(\mathrm{End}(S)) \longrightarrow \mathrm{P}(M)$$

In local coordinates:

$$\mathrm{str}_\Lambda(\overline{\psi}_\nu \psi^\mu) = \begin{cases} 1 & |\mu| = |\nu| = n \\ 0 & \text{otherwise.} \end{cases}$$

Then, in any coordinate chart $U \subseteq M$ we may use the natural (coordinate-dependent) vector space isomorphism

$$\mathrm{FB}(U,S) \cong \Gamma(\mathrm{End}(S|_U)) \otimes_{\mathrm{P}(U)} \mathrm{FB}(U)$$

to define

$$\mathrm{STr}_F\colon \mathrm{FB}(U,S) \longrightarrow F$$
$$\mathrm{STr}_F(A \otimes X) = \mathrm{Tr}(\mathrm{str}_\Lambda(A)X)$$

(strictly speaking this is defined only on elements compactly supported in the U-direction). Thanks to Theorem 4.8, the above is a bimodule supertrace for the left and right actions of the algebra

$$\mathrm{PDO}(U,S) \cong \Gamma(\mathrm{End}(S|_U)) \otimes_{\mathrm{P}(U)} \mathrm{PDO}(U).$$

Thanks to Theorems 4.8 and 4.18 it is independent of the choice of coordinates used to define it.

We can now use partitions of unity to define (independently of all choices) a supertrace

$$\mathrm{STr}_F\colon \mathrm{FB}(M) \longrightarrow F,$$

and finally we make the following definition:

DEFINITION 4.20. Let M be a smooth, closed manifold. The *Perrot supertrace* on $\mathrm{FB}(M,S)$ is

$$\mathrm{STr}\colon \mathrm{FB}(M,S) \longrightarrow \mathbb{C}$$
$$\mathrm{STr}(X) = \text{Coefficient of } \varepsilon^0 \text{ in } \mathrm{STr}_F(X).$$

5. The JLO-Type Cocycle

The purpose of this section is to construct from the Dirac operator $\not{D}$, together with Perrot's supertrace and one more ingredient, a periodic cyclic cocycle for the algebra $C^\infty(S^*M)$.

Smooth functions on S^*M can be viewed as order-one homogeneous functions on $\dot{T}^*M$. Seen this way, the algebra $C^\infty(S^*M)$, together with bimodule $FB(M,S)$, equipped with its supertrace, and the operator $\not{D}$ form something similar to one of Connes' *spectral triples* [**Con94**, Sec.IV.2][2]: the bimodule, with its trace, plays a role similar to the role of the Hilbert space in a spectral triple, with its associated operator trace (or perhaps with a residue trace obtained from this operator trace; see for instance [**Hig04**] for an exposition).

With this in mind we shall attempt to adapt the well-known JLO-cocycle in periodic cyclic theory to Perrot's context. Let Σ^p be the standard p-simplex. If $s \in \Sigma^p$ and $X^0, \ldots, X^p \in \mathrm{PDO}(M,S)$, then let us write

$$\big\langle X^0, \ldots, X^p \big\rangle_s = X^0 \exp(s_0 \not{D}^2) X^1 \exp(s_1 \not{D}^2) \cdots X^p \exp(s_p \not{D}^2). \tag{5.1}$$

The most obvious JLO cocycle that one might try to attach to $\not{D}$ is given by the formula

$$JLO^{\not{D}}(a^0, \ldots, a^p) = \int_{\Delta^p} \mathrm{STr}\Big(\big\langle a^0, [\not{D}, a^1], \ldots, [\not{D}, a^p] \big\rangle_s \Big)\, ds. \tag{5.2}$$

This is the original formula given by Jaffe, Lesniewski and Osterwalder [**JLO88**]; see [**Qui88**, Sec.8] or [**Con94**, Sec.IV.8] for expositions.

However it follows from the supertrace property that the cocycle (5.2) vanishes in odd degrees, whereas the Todd class that we are trying to understand corresponds to a cyclic cocycle that vanishes in even degrees. So the above is not the cocycle we are seeking.[3]

To address this issue we need one further ingredient, as follows. Fix a Riemannian metric on M, and hence a Euclidean structure on the bundle T^*M. Define

$$q\colon \dot{T}^*M \longrightarrow (0,\infty), \qquad q(\alpha) = \|\alpha\|,$$

and then define a derivation

$$\delta\colon \mathrm{PDO}(M,S) \longrightarrow \mathrm{PDO}(M,S)$$

by the formula

$$\delta(X) = \mathrm{ad}_{\log(q)}(X)$$

(as usual we are adapting Perrot's work here to our simplified context). Note that

$$\delta\Big(\frac{\partial}{\partial x^i}\Big) = -\frac{\partial q}{\partial x^i} \cdot \frac{1}{q} \quad \text{and} \quad \delta\Big(\frac{\partial}{\partial \xi_j}\Big) = -\frac{\partial q}{\partial \xi_j} \cdot \frac{1}{q},$$

while δ vanishes on other (local) generators of $\mathrm{PDO}(M,S)$. So δ does indeed map $\mathrm{PDO}(M,S)$ into itself, despite the fact that $\log(q)$ is not a polyhomogeneous function.

---

[2] In this reference the older term *K-cycle* is used, rather than *spectral triple*.

[3] In fact it may be shown that (5.2) vanishes in *all* degrees. In the more elaborate context of Perrot's work on pseudodifferential symbols, this corresponds to the vanishing of the noncommutative residue on pseudodifferential projectors [**Per13b**, Thm.6.5].

Now δ extends to a derivation of the algebra $\mathrm{FPDO}(M, S)$ by applying δ to each coefficient in a formal series. And this extension maps the bimodule $\mathrm{FB}(M, S)$ to itself. This follows from the formula

$$\begin{aligned}\delta(\exp(\varepsilon\Delta)) &= \int_0^1 \exp(s\varepsilon\Delta)\varepsilon\delta(\Delta)\exp((1-s)\varepsilon\Delta)\,ds\\ &= \int_0^1 \mathrm{Ad}_{\exp(s\varepsilon\Delta)}\big(\varepsilon\delta(\Delta)\cdot\exp(\varepsilon\Delta)\big)\,ds\end{aligned}$$

together with Lemma 3.12.

LEMMA 5.3. *If* $X \in \mathrm{FB}(M, S)$, *then* $\mathrm{STr}(\delta(X)) = 0$.

PROOF. Let X be any linear partial differential operator on $\dot{T}^*M$ and let $u \in \mathbb{R}$. Consider the binomial-type formula

$$\mathrm{ad}_{q^u}(X) = \sum_{k=1}^{\infty} \binom{u}{k} \mathrm{ad}_q^k(X) q^{u-k},$$

which actually involves a finite sum on the right-hand side, since if h is any smooth function on $\dot{T}^*M$, then the action of ad_h is locally nilpotent on partial differential operators. Differentiating the formula with respect to u and then setting $u = 0$, we obtain

$$\mathrm{ad}_{\log(q)}(X) = \sum_{k=1}^{\infty} \frac{1}{k}(-1)^{k-1}\,\mathrm{ad}_q^k(X) q^{-k}$$

Again the right-hand side is a finite sum, and it is a combination of commutators, since $\mathrm{ad}_q^k(X)q^{-k} = \mathrm{ad}_q^k(Xq^{-k})$ for all k.

Now we turn to $\mathrm{FPDO}(M, S)$. The above formulas involve infinite sums when $X \in \mathrm{FPDO}(M, S)$, but they are convergent in the adic topology and for that reason remain valid. Restricting to $\mathrm{FB}(M, S)$, it follows that every element of the form $\delta(X) = \mathrm{ad}_{\log(q)}(X)$ is a limit of sums of commutators, and since the trace is continuous the trace of $\delta(X)$ is zero. □

REMARK 5.4. It is worth noting that the trace does *not* extend as a trace if we adjoin $\log(q)$ to the bimodule $\mathrm{FB}(M)$ (so this is not the explanation for the lemma above). For instance, in case where M is a circle, if $q = |\xi|$, and if s is the characteristic function of $\{\,\xi > 0\,\}$ (which is a polyhomogeneous function), then

$$\left[\frac{\partial}{\partial\xi}, s\log(q)\exp(\varepsilon\Delta)\right] = s\xi^{-1}\exp(\varepsilon\Delta),$$

and the trace of the right-hand side is nonzero (despite it being a commutator in the enlarged bimodule).

REMARK 5.5. On a related note, although the definition of δ requires a choice of Riemannian metric, different choices yield derivations that differ only by an inner derivation.

Using Lemma 5.3 it is not difficult to adjust the definition of the JLO cocycle so as to incorporate δ. The following theorem is a generalization of the well-known fact that if δ is any derivation on an algebra, and if τ is a trace with $\tau(\delta(a)) = 0$ for all a, then the formula $\varphi(a^0, a^1) = \tau(a^0\delta(a^1))$ defines a cyclic cocycle. As Rodsphon explains in [**Rod15**], the theorem fits very naturally into Quillen's formalism for the JLO cocycle [**Qui88**], and this is probably the best way of understanding it.

First a simple preliminary calculation:

LEMMA 5.6. *Let ∇ be a torsion-free connection on M and let $\not{D}$ be the associated Dirac operator, as in Definition* 3.18. *If $p \geq 0$, $s \in \Sigma^{p+1}$ and $X_0, \ldots, X_{p+1} \in \operatorname{PDO}(M, S)$, then*

$$\langle X^0, \ldots, X^{p+1} \rangle_s \in \operatorname{FB}(M, S),$$

to use the notation of (5.1).

PROOF. This is a combination of Lemmas 3.19 and 3.21. □

THEOREM 5.7. *The formulas*

$$JLO^{\not{D},\delta}(a^0, \ldots, a^p) = \sum_{k=1}^{p+1} (-1)^k \int_{\Sigma^{p+1}} \operatorname{STr}\Big(\langle a^0, [\not{D}, a^1], \ldots$$
$$\ldots, [\not{D}, a^{k-1}], \delta(\not{D}), [\not{D}, a^k], \ldots, [\not{D}, a^p] \rangle_s \Big)\, ds$$

for p odd and $a^0, \ldots, a^p \in C^\infty(S^*M)$ define a periodic cyclic cocycle for the algebra $C^\infty(S^*M)$.

REMARK 5.8. The integrands in the defining formula for $JLO^{\not{D},\delta}$ are polynomial functions of $s \in \Sigma^{p+1}$, so the integrals are certainly well-defined, and moreover these polynomial functions are identically zero for p sufficiently large (this will be made clear in the next section), so that only finitely many components of $JLO^{\not{D},\delta}$ are nonzero, as required in the definition of periodic cyclic cocycle. This is in contrast to the form of the usual JLO cocycle (involving the Levi-Civita Dirac operator and the operator trace), which is rather an *entire* cyclic cocycle [**Con94**, Ch.4, Sec.7].

REMARK 5.9. In general the Quillen formalism would produce a formula with further terms, involving factors $\delta(a^j)$; see [**Rod15**, Sec.3]. But in the case at hand, $\delta(a^j) = 0$.

We shall not prove Theorem 5.7 in this paper, firstly because it is a general result that is encompased by the Quillen formalism and is explained perfectly well by Rodsphon in [**Rod15**], and secondly because we shall not actually need the result: we shall calculate all the components of $JLO^{\not{D},\delta}$, and it will be evident from these computations that they constitute a periodic cyclic cocycle.

6. Perrot Order of a Differential Operator

The remainder of the paper will be dedicated to the computation of the JLO-type cocycle that was defined in the previous section. We shall introduce a new notion of order on differential operators that we shall call the *Perrot order* (it is distinct from the bimodule order discussed earlier). The Perrot order has the property that operators of negative order have vanishing trace. Roughly speaking, upon analyzing the constituents of the cocycle, we shall find that many parts have negative order, and so they can be removed without altering the value of the cocycle.

The Perrot order resembles somewhat the notion of order that Getzler uses to compute the index of the spinor Dirac operator [**Get83**] (see [**Roe98**] or [**HY19**] for surveys). Both have the property that the curvature operator[4] acquires the same order as the square of the Dirac operator, which leads to the curvature contributing to leading order computations. But the leading order is 2 for Getzler, while it is 0 for Perrot. An important consequence is that the formal exponential $\exp(\not{D}^2)$ also has Perrot order 0. Indeed all of the constituents of the JLO cocycle introduced in the previous section have Perrot order 0.

Despite its role in capturing curvature, the Perrot order does not in fact depend on the choice of connection, whereas Getzler's certainly does (which is a paradox, perhaps, but not a contradiction). But to begin with, to define the Perrot order it is convenient to fix a torsion-free connection on M (and then later we shall show that the Perrot order does not depend on this choice).

DEFINITION 6.1. The *Perrot order* of an operator in $\mathrm{PDO}(M, S)$ (which is an integer, possibly negative) is defined by the following:

(i) If f is any degree p-homogeneous smooth function on $\dot{T}^*M$, then Perrot-order$(f) \leq p$.
(ii) If X is any vector field on M, with horizontal lift X^H on $\dot{T}^*M$, then Perrot-order$(\nabla_{X^H}) \leq 1$.
(iii) If Y is any vertical vector field on $\dot{T}^*M$ (that is, if Y commutes with functions pulled back from M), and if Y is translation-invariant on each cotangent fiber, then Perrot-order$(\nabla_Y) \leq -1$.
(iv) If ω is any 1-form on M, acting on sections of S over $\dot{T}^*M$ by left exterior multiplication, then Perrot-order$(\omega) \leq 0$.
(v) If X is any vector field on M, and if $\iota(X)$ denotes the action on sections of S by contraction, then Perrot-order$(\iota(X)) \leq 1$.

REMARK 6.2. Compare Proposition/Definition 3.7, which explains the way in which an increasing filtration on $\mathrm{PDO}(M, S)$ is constructed from inequalities such as those above. As with the bimodule order considered there, the inequalities in the definition above are actually equalities; this will be clear from the lemma below.

An operator has Perrot order p or less if and only if its restriction to every $\dot{T}^*U$, with $U \subseteq M$ open, has Perrot order p or less. So the Perrot order may be computed locally. The following lemma shows that moreover the Perrot order is easily determined in local coordinates. It also shows that the Perrot order is independent of the choice of affine connection ∇.

LEMMA 6.3. *Let $x_1, \ldots, x_n$ be coordinates on an open set $U \subseteq M$ and let $p \in \mathbb{Z}$. If the bundle S is trivialized over $\dot{T}^*U$ using the standard frame for the exterior algbra bundle associated to these coordinates, then the linear space of all operators of Perrot order p on less on $\dot{T}^*U$ is precisely the linear span of operators of the form*

$$f\psi^\mu \overline{\psi}_\nu \frac{\partial^\alpha}{\partial x^\alpha} \frac{\partial^\beta}{\partial \xi_\beta}$$

with f a homogeneous smooth scalar function on $\dot{T}^*U$ and with

$$\mathrm{degree}(f) + |\alpha| - |\beta| + |\nu| \leq p.$$

[4]In Perrot's case, we are referring here to the operator $\gamma(R)$ that was introduced in Definition 2.14 and appears in the formula for $\not{D}^2$. The actual curvature operators $R(X, Y)$ have order 1; see the proof of Lemma 6.6.

Here α and β are multi-indices with non-negative integer entries, while μ and ν are multi-indices with 0-1 *entries.*

PROOF. It follows from (2.9) that

$$\frac{\partial}{\partial x^i} = \nabla_{(\partial/\partial x^i)^H} - \Gamma^k_{i\ell}\,\xi_k\,\frac{\partial}{\partial \xi_\ell} + \Gamma^k_{i\ell}\,\psi^\ell\,\overline{\psi}_k.$$

All three terms on the right-hand side have Perrot order 1 or less, and therefore $\partial/\partial x^i$ has Perrot order 1 or less. In addition

$$\frac{\partial}{\partial \xi_j} = \nabla_{\partial/\partial \xi_j}.$$

It follows that all the operators in the statement of the lemma have Perrot order p or less.

In the reverse direction, if we define $\mathrm{PDO}_p(U,S)$ to be the linear span of the operators in the statement of the lemma, then we obtain an algebra filtration for which the associated order satisfies the relations in Definition 6.1. It therefore follows from the construction of the Perrot filtration and order (*c.f.* Propositon/Definition 3.7) that every operator of Perrot order p or less belongs to $\mathrm{PDO}_p(U,S)$. □

Let us now extend the Perrot order to the algebra $\mathrm{FPDO}(M,S)$ and to the bimodule $\mathrm{FB}(M,S)$ as follows: a formal series $\sum \varepsilon^k T_k$ has Perrot order p or less if and only if each T_k has Perrot order p or less. (With this definition, there are many elements of infinite Perrot order, but that will not be an issue for us.)

PROPOSITION 6.4. *Let M be a closed manifold. The supertrace vanishes on every element of* $\mathrm{FB}(M,S)$ *of negative Perrot order.*

PROOF. It suffices to consider negative-order generating elements

$$f\psi^\mu\overline{\psi}_\nu\frac{\partial^\alpha}{\partial x^\alpha}\frac{\partial^\beta}{\partial \xi_\beta}\exp(\varepsilon\Delta)$$

of the bimodule $\mathrm{FB}(U,S)$ in a coordinate neighborhood. According to the definition of the supertrace and the formula for the contraction operation in Definition 4.9, if $\alpha \neq \beta$, then the supertrace of this element is zero. In addition, if $|\nu| \neq n$ then the supertrace is zero. If $\alpha = \beta$ and if $|\nu| = n$, then the Perrot order is equal to $\mathrm{degree}(f) - n$, and if this is negative, then by the definition of the integral $⨍ f$ in (4.6) the supertrace is zero. □

Our purpose in the remainder of this section is to develop a technique to compute the supertrace of order zero operators. In the lemma below we shall use the fact that if $\pi\colon W \to M$ is any submersion with compact fibers, then there is an induced morphism of $C^\infty(M)$-modules

$$\pi_*\colon \mathrm{Dens}(W) \longrightarrow \mathrm{Dens}(M)$$

that is characterized by the formula

$$\int_M \pi_*(\alpha) = \int_W \alpha$$

(in the context of oriented smooth manifolds this is the operator of integration over the fiber).

LEMMA 6.5. *There is a unique morphism of $C^\infty(M)$-modules*

$$\mathrm{str}\colon \mathrm{FB}(M) \longrightarrow \mathrm{Dens}(M)$$

such that

$$\mathrm{STr}(X) = \int_M \mathrm{str}(X)$$

for every $X \in \mathrm{FB}^0(M)$.

PROOF. The trace was constructed (locally, at first) as the integral of a density on S^*M, and pushing this density forward along the projection $S^*M \to M$ we obtain from a partition of unity argument the existence of a map as in the statement of the lemma.

To prove uniqueness, if str and str′ are two morphisms of modules as in the statement of the lemma, then from the identities

$$\begin{aligned}\int_M f\cdot \mathrm{str}(X) = \int_M \mathrm{str}(f\cdot X) &= \mathrm{STr}(X)\\ &= \int_M \mathrm{str}'(f\cdot X) = \int_M f\cdot \mathrm{str}'(X)\end{aligned}$$

we conclude that

$$\int_M f\cdot \big(\mathrm{str}(X) - \mathrm{str}'(X)\big) = 0$$

for all $f \in C^\infty(M)$, and hence $\mathrm{str}(X) - \mathrm{str}'(X) = 0$. □

We shall now evaluate the density $\mathrm{str}(X) \in \mathrm{Dens}(M)$ defined above at a single point $m \in M$ under the assumption that $X \in \mathrm{FB}(M,S)$ has order zero.

Form the associated graded algebra

$$\mathrm{gr}\,\mathrm{PDO}(M,S) = \bigoplus_p \mathrm{PDO}_p(M,S)/\,\mathrm{PDO}_{p-1}(M,S)$$

for the filtration by Perrot order. Notice that since smooth functions on M (viewed as operators on sections of S by pointwise multiplication) have Perrot order zero, the associated graded algebra is in fact an algebra over $C^\infty(M)$. We can therefore speak of the fiber

$$\mathrm{gr}\,\mathrm{PDO}(M,S)\big|_m$$

at a given point $m \in M$, which is an algebra in its own right. Our first aim is to compute this fiber.

LEMMA 6.6. *If X and Y are vector fields on M, then the degree 1 classes of ∇_{X^H} and ∇_{Y^H} commute with one another in the associated graded algebra.*

PROOF. Let $\alpha \in \dot{T}^*M$ and let $m = \pi(\alpha)$. Because the connection we are using on $\pi^*\Lambda^*T^*M$ is a pullback, the curvature operator $R(X^H, Y^H)|_\alpha$ is the curvature operator $R(X,Y)|_m$ for the original connection:

$$\nabla_{X^H}\nabla_{Y^H} - \nabla_{Y^H}\nabla_{X^H} = \nabla_{[X^H,Y^H]} + R(X,Y) \tag{6.7}$$

(this is an identity of operators acting on the sections of $S = \pi^*\Lambda^*T^*M$).

If we introduce local coordinates on M and write the curvature operator on TM as

$$R\Big(\frac{\partial}{\partial x^i}, \frac{\partial}{\partial x^j}\Big)\frac{\partial}{\partial x^\ell} = R^k_{ij\ell}\,\frac{\partial}{\partial x^k},$$

as in Definition 2.14, then the curvature operator as it appears in (6.7) is

$$R\Big(\frac{\partial}{\partial x^i},\frac{\partial}{\partial x^j}\Big)=\psi^\ell\overline{\psi}_k R^k_{ij\ell},$$

and in particular the operator $R(X,Y)$ in (6.7) has Perrot order 1.

In addition, according to Lemma 2.6,

$$[X^H,Y^H]=[X,Y]^H+\gamma\big(R(X,Y)\big).$$

The covariant derivatives attached to the two terms on the right have Perrot-orders 1 and 0, respectively. So we find from (6.7) that

$$\begin{aligned}\text{Perrot-order}\big(\nabla_{X^H}\nabla_{Y^H}-\nabla_{Y^H}\nabla_{X^H}\big)&\\ &\le 1\\ &<\text{Perrot-order}\big(\nabla_{X^H}\big)+\text{Perrot-order}\big(\nabla_{Y^H}\big),\end{aligned}$$

which proves the lemma. □

REMARK 6.8. The lemma stands in contrast to the situation with the Getzler's order, in which the commutator covariant derivatives is a curvature operator.

Other, more easily derived, relations in the associated graded algebra are as follows:

(i) The degree -1 classes of the operators ∇_Y associated to vertical vector fields that are fiberwise translation invariant (that is, scalar combinations of the $\nabla_{\partial/\partial\xi_j}$) commute with one another and the classes of the ∇_{X^H}.
(ii) The degree 0 and 1 (respectively) classes of the exterior multiplication and contraction operators anticommute with one another and commute with all the classes of covariant derivatives already mentioned.
(iii) The degree p classes of scalar degree p functions commute with one another and with all the classes mentioned above, except that

$$[\nabla_{\partial/\partial\xi_j}][f]-[f][\nabla_{\partial/\partial\xi_j}]=[\partial f/\partial\xi_j]$$

as in $\mathrm{PDO}(M,S)$.

DEFINITION 6.9. For $m\in M$, denote by A_m the tensor product of the symmetric algebra of T_mM (whose elements we shall write as constant coefficient differential operators on T_mM), the algebra of polyhomogeneous partial differential operators on T^*_mM and the exterior algebra $\Lambda^*(T_mM\oplus T^*_mM)$ (which, given local coordinates on M near m, we shall regard as generated by $dx^i\in T^*_mM$ and $d\xi_j\in T_mM$).

We shall suppress tensor product signs when describing elements of A_m, such as for example

$$dx^i\frac{\partial}{\partial x^i}+d\xi_j\frac{\partial}{\partial\xi_j}\in \mathrm{A}_m \tag{6.10}$$

We equip A_m with an integer grading by assigning degree 1 to each vector in T_mM, the usual degree to each homogeneous operator on T^*_mM, degree 0 to each dx^i and degree 1 to each $d\xi_j$. For instance, the terms in (6.10) have degrees 1 and 0, respectively.

LEMMA 6.11. *Let ∇ be a torsion-free connection on M and let $m \in M$. There is a unique isomorphism of complex $\mathbb{Z}$-graded algebras*

$$\mathrm{Symb}_m^\nabla\colon \operatorname{gr}\mathrm{PDO}(M,S)\big|_m \xrightarrow{\cong} \mathrm{A}_m$$

such that for all local coordinates near m,

(i) $[\nabla_{(\partial/\partial x^i)^H}] \mapsto \partial/\partial x^i$ *and* $[\nabla_{\partial/\partial\xi_j}] \mapsto \partial/\partial\xi_j$.
(ii) *If f is any polyhomogeneous function on $\dot{T}^*M$, then $f \mapsto f|_{\dot{T}^*_m M}$.*
(iii) $[\psi^i] \mapsto dx^i$ *and* $[\overline{\psi}_j] \mapsto d\xi_j$.

PROOF. The reversed correspondences define a morphism of graded algebras in the reverse direction by the universal property inherent in the definition of A_m as a tensor product. It follows from the local coordinate description of the Perrot order in Lemma 6.3 that this morphism is an isomorphism. □

REMARK 6.12. Even though the Perrot order and therefore the associated graded algebra $\operatorname{gr}\mathrm{PDO}(M,S)$, are independent of the choice of connection used in their definitions, the homomorphism in Lemma 6.11 *does* depend on the choice of connection.

We shall now extend the symbol morphism Symb_m^∇ from polyhomogeneous differential operators to elements of the bimodule $\mathrm{FB}(M,S)$. To do so we shall first look to Definition 3.9 to find a suitable target for the symbol homomorphism.

DEFINITION 6.13. Define the *bimodule order* of an element in A_m by

$$\text{bimodule-order}\Big(\frac{\partial}{\partial x^i}\Big) = 3, \quad \text{bimodule-order}\Big(\frac{\partial}{\partial \xi_j}\Big) = -1$$
$$\text{bimodule-order}\big(\psi^i\big) = 0, \quad \text{bimodule-order}\big(\overline{\psi}_j\big) = 0$$

and $\text{bimodule-order}(f) = 2\,\text{degree}(f)$ for a homogeneous function f. Denote by FA_m the space of formal Laurent series

$$\sum \varepsilon^k D_k \exp(\varepsilon\Delta) \in A_m[\varepsilon^{-1},\varepsilon]]$$

with finite singular parts for each of which there exists N with

$$\text{bimodule-order}(D_k) \le k + N$$

for all k.

The space FA_m is a bimodule over A_m, and applying the symbol homomorphism termwise to a Laurent series, we obtain a morphism

$$\mathrm{Symb}_m^\nabla\colon \operatorname{gr}\mathrm{FB}(M,S)|_m \longrightarrow \mathrm{FA}_m$$

that is compatible with bimodule structures. We are going to factor Perrot's supertrace on order zero elements through this symbol morphism, as follows: we shall define a supertrace morphism

$$\mathrm{str}_m\colon \mathrm{FA}_m \longrightarrow \mathrm{Dens}(M)|_m$$

and show that

$$\mathrm{str}(X)|_m = \mathrm{str}_m(\mathrm{Symb}_m^\nabla(X))$$

for every $X \in \mathrm{FB}(M,S)$ of Perrot order 0.

To begin the definition of str_m, suppose we are given an order $-n$ homogeneous smooth function f on $\dot{T}^*_m M$, and let $\alpha \in \Lambda^n T_m M$. Think of α as a (translation-invariant) top-degree form on $\dot{T}^*_m M$. If E is the Euler vector field, as usual, then the contracted form $\iota_E(f \cdot \alpha)$ is basic for the action of the positive real numbers on $\dot{T}^*_m M$ by scalar multiplication, and it is therefore the pullback of a form $\sigma_{f,\alpha}$ on the sphere $S^*_m M$. Consider next the integral

$$\int_{S^*_m M} \sigma_{f,\alpha}, \tag{6.14}$$

formed using the orientation on the sphere for which the integral is positive when f is positive. This orientation depends on α, and so the integral is not bilinear in f and α, but rather

$$\int_{S^*_m M} \sigma_{f,t\alpha} = |t| \int_{S^*_m M} \sigma_{f,\alpha} \qquad \forall t \in \mathbb{R}.$$

The map $\alpha \mapsto \int_{S^*_m M} \sigma_{f,\alpha}$ therefore defines a density at m (see [**BGV92**, pp.30-31] or [**Lee03**, Ch.14] again), which we shall write as

$$⨍_{S^*_m M} f \in \operatorname{Dens}(M)|_m \tag{6.15}$$

(and as before we extend this to polyhomogeneous f by selecting the degree $-n$ component). This construction has the property that if f is a smooth, degree $-n$ function on $\dot{T}^* M$ that is compactly supported in the M-direction, then

$$\int_M \Big(⨍_{S^*_m M} f|_{\dot{T}^*_m M} \Big) = ⨍_{S^* M} f. \tag{6.16}$$

To define a trace[5] functional on FA_m, we use the integral above and essentially the same contraction operation that we used for Perrot's trace: namely for $D \in \mathrm{A}_m$ we define

$$\big\langle D \exp(\varepsilon \Delta) \big\rangle = \varepsilon^{-n} \big(D \exp(-\varepsilon^{-1}(x^i - y^i)(\xi_i - \eta_i)) \big)\Big|_{x^i = y^i = 0,\ \xi_i = \eta_i}.$$

This is a polyhomogeneous function on $\dot{T}^*_m M$ with values in the exterior algebra $\Lambda^*(T_m M \oplus T^*_m M)$. Then we define

$$\operatorname{str}_{m,F}\big(D \exp(\varepsilon\Delta)\big) = \text{coefficient of } dx^1 \cdots dx^n d\xi_1 \cdots d\xi_n \text{ in } ⨍_{S^*_m M} \big\langle D \exp(\varepsilon\Delta) \big\rangle, \tag{6.17}$$

and then extend termwise to the Laurent series in FA_m. The result is a Laurent series in ε and, just as we did when we constructed Perrot's trace, we then define str_m by taking the coefficient of ε^0.

Thanks to our direct mimicry of the constructions in Section 4, and thanks to (6.16), the following result is clear:

[5]The functional is indeed a trace, although we shall not prove this fact because we shall not need it.

LEMMA 6.18. *Let $x_1, \ldots, x_n$ be local coordinates on M, defined near a point $m \in M$. If ∇ is the canonical flat connection on TM associated to these coordinates (in which all the Christoffel symbols Γ^k_{ij} are identically zero), then*

$$\operatorname{str}(X)\big|_m = \operatorname{str}_m(\operatorname{Symb}^\nabla_m(X))$$

for every $X \in \operatorname{FB}(M, S)$ of Perrot order 0. □

In fact, the same result holds for *any* connection:

THEOREM 6.19. *If ∇ is* any *torsion-free connection on TM, if $m \in M$, and if*

$$\operatorname{Symb}^\nabla_m \colon \operatorname{FB}_0(M) \longrightarrow \operatorname{FA}_m$$

is the symbol map at m defined by ∇ and acting on Perrot order zero operators, then

$$\operatorname{str}(X)\big|_m = \operatorname{str}_m(\operatorname{Symb}^\nabla_m(X))$$

for every $X \in \operatorname{FB}^0(M)$.

PROOF. Given ∇ and m, choose local coordinates near m for which all the Christoffel symbols for ∇ vanish at m. Then define ∇^0 to be the canonical flat connection for these coordinates. By examining the two morphisms $\operatorname{Symb}^\nabla_m$ and $\operatorname{Symb}^{\nabla^0}_m$ in this given coordinate system, we see that these two symbol morphisms are in fact equal. So the theorem follows from the previous lemma. □

REMARK 6.20. Needless to say, the theorem relies heavily on the coordinate independence of Perrot's trace (and in particular the theorem is far from trivial).

7. Computation of the JLO cocycle

In this final section we shall use Theorem 6.19 and one final additional computation to identify the cocycle $JLO^{\not{D},\delta}$ from Section 5 with the cocycle associated with the Todd class that was described in Section 1. The starting point is the following computation:

LEMMA 7.1. *If ∇ is any torsion-free connection on M, then*

$$\text{Perrot-order}\big(\not{D}^2_{\text{vert}}\big) \le 0 \quad \textit{and} \quad \text{Perrot-order}\big(\{\not{D}_{\text{horiz}}, \not{D}_{\text{vert}}\}\big) \le 0.$$

In addition if a is any smooth, degree zero function on $\dot{T}^*M$, then

$$\text{Perrot-order}\big([\not{D}, a]\big) \le 0.$$

PROOF. This evident from the formulas in Definition 2.10 and in Lemmas 2.13 and 2.16. □

It follows that the argument of the supertrace in the formula for

$$JLO^{\not{D},\delta}(a^0, \ldots, a^p)$$

has Perrot order 0, and therefore we may use the techniques of Section 6 to compute the supertrace.

With this in mind, we now fix a point $m \in M$. As explained in the Section 6, the quantity $JLO^{\not{D},\delta}(a^0, \ldots, a^p)$ is the integral of a density on M; we shall denote by

$$JLO^{\not{D},\delta}(a^0, \ldots, a^p)|_m \in \operatorname{Dens}(M)|_m$$

the value of this density at m.

If we write

$$R\Big(\frac{\partial}{\partial x^i},\frac{\partial}{\partial x^j}\Big)\frac{\partial}{\partial x^k} = R^k_{ij\ell}\frac{\partial}{\partial x^\ell},$$

as we did earlier, and then define

$$R^k_\ell = \tfrac{1}{2}R^k_{ij\ell}dx^i dx^j\big|_m \in \Lambda^2 T^*_m M,$$

then

$$\operatorname{Symb}^\nabla_m(\exp(s\not{D}^2)) = \exp\big(s(\varepsilon\Delta{+}\varepsilon^2\xi_k R^k_\ell\partial/\partial\xi_\ell)\big)\tag{7.2}$$

Now choose coordinates $x^1,\dots,x^n$ near m for which the Christoffel symbols Γ^k_{ij} vanish at m. Then

$$\begin{aligned}\operatorname{Symb}^\nabla_m([\not{D},a]) &= \varepsilon\frac{\partial a}{\partial x^i}dx^i + \frac{\partial a}{\partial \xi_j}d\xi_j\\ \operatorname{Symb}^\nabla_m(\delta(\not{D})) &= -\varepsilon q^{-1}\frac{\partial q}{\partial x^i}dx^i - q^{-1}\frac{\partial q}{\partial \xi_j}d\xi_j\end{aligned}\tag{7.3}$$

where as in Section 5, $q\colon \dot{T}^*M \to (0,\infty)$ is the norm-function associated to a metric on T^*M.

For brevity, we shall write

$$\begin{aligned}\operatorname{Symb}^\nabla_m([\not{D},a]) &= d_\varepsilon a = \varepsilon d_{\text{horiz}}a + d_{\text{vert}}a\\ \operatorname{Symb}^\nabla_m(\delta(\not{D})) &= -q^{-1}d_\varepsilon q = -\varepsilon q^{-1}d_{\text{horiz}}q - q^{-1}d_{\text{vert}}q.\end{aligned}$$

In addition, to streamline some of the formulas that follow, we shall write

$$\Delta_R = \Delta{+}\varepsilon\xi_k R^k_\ell\partial/\partial\xi_\ell = \Delta{+}\varepsilon\xi\cdot R\cdot\partial/\partial\xi.$$

Then according to the formulas (7.2) and (7.3) above and the formula in Theorem 5.7, the density $JLO^{\not{D},\delta}(a^0,\dots,a^p)|_m$ is the sum from $k=1$ to $k=p{+}1$ of the following integrals:

$$\begin{aligned}(-1)^{k-1}\int_{\Sigma^{p+1}} &\operatorname{str}_m\big(a^0\exp(s_0\varepsilon\Delta_R)d_\varepsilon a^1\exp(s_1\varepsilon\Delta_R)\cdots\\ &\dots d_\varepsilon a^{k-1}\exp(s_{k-1}\varepsilon\Delta_R)q^{-1}dq\exp(s_k\varepsilon\Delta_R)d_\varepsilon a^k\cdots\\ &\qquad\cdots\exp(s_{p-1}\varepsilon\Delta_R)d_\varepsilon a^{p-1}\exp(s_p\varepsilon\Delta_R)\big)\,ds.\end{aligned}\tag{7.4}$$

Using the formula

$$\exp(s\varepsilon\Delta_R)X\exp(-s\varepsilon\Delta_R) = \exp(s\varepsilon\operatorname{ad}_{\Delta_R})(X)$$

we can write the argument of str_m in (7.4) as an infinite linear combination (convergent in the adic topology) of terms

$$\varepsilon^k X_0\operatorname{ad}^{k_0}_{\Delta_R}(X_1)\operatorname{ad}^{k_1}_{\Delta_R}(X_2)\cdots\operatorname{ad}^{k_{p-2}}_{\Delta_R}(X_{p-1})\operatorname{ad}^{k_{p-1}}_{\Delta_R}(X_p)\exp(\varepsilon\Delta_R),\tag{7.5}$$

where each X_j is one of the $d_\varepsilon a^i$ or $q^{-1}d_\varepsilon q$.

Next, as long as any $X\in \mathrm{A}_m$ includes no ξ-derivatives

$$\operatorname{ad}_{\Delta_R}(X) = \big(\partial/\partial x^j + \varepsilon\xi_i R_{ij}\big)[\partial/\partial\xi_j, X],$$

and we note that in this case the result on the right-hand side also includes no ξ-derivatives. Now we appeal to the following result, whose proof we shall defer for a moment:

Proposition 7.6. *If an element $X \in A_m$ includes no ξ-derivatives, then*

$$\operatorname{str}_m\big(\varepsilon^r(\partial_{x^j} + \varepsilon\xi_i R_{ij})X\exp(\varepsilon\Delta_R)\big) = 0$$

for all r and all j.

It follows from the lemma that all the terms (7.4) have vanishing trace, except when all k_j are zero, and hence that (7.4) is equal to

$$(-1)^{k-1}\int_{\Sigma^{p+1}} \operatorname{str}_m\big(a^0 d_\varepsilon a^1 \cdots d_\varepsilon a^{k-1} q^{-1} d_\varepsilon q \cdots d_\varepsilon a^p \exp(\varepsilon\Delta_R)\big)\, ds.$$

The integrand now no longer depends on s, and so the integral is simply

$$\frac{(-1)^{k-1}}{(p+1)!}\operatorname{str}_m\big(a^0 d_\varepsilon a^1 \cdots d_\varepsilon a^{k-1} q^{-1} d_\varepsilon q \cdots d_\varepsilon a^p \exp(\varepsilon\Delta_R)\big).$$

In summary, then:

$$\begin{aligned}(7.7)\quad & JLO^{\not{D},\delta}(a^0,\dots,a^p)|_m \\ &= \sum_{k=1}^{p+1}\frac{(-1)^{k-1}}{(p+1)!}\operatorname{str}_m\big(a^0 d_\varepsilon a^1 \cdots d_\varepsilon a^{k-1} q^{-1} d_\varepsilon q \cdots d_\varepsilon a^p \exp(\varepsilon\Delta_R)\big) \\ &= \sum_{k=1}^{p+1}\frac{1}{(p+1)!}\operatorname{str}_m\big(q^{-1} d_\varepsilon q\, a^0 d_\varepsilon a^1 \cdots d_\varepsilon a^p \exp(\varepsilon\Delta_R)\big) \\ &= \frac{1}{p!}\operatorname{str}_m\big(q^{-1} d_\varepsilon q\, a^0 d_\varepsilon a^1 \cdots d_\varepsilon a^p \exp(\varepsilon\Delta_R)\big).\end{aligned}$$

Now we quote another result, whose proof, like that of Proposition 7.6, we shall defer for a short time:

Proposition 7.8. *If $X \in A_m$ has no x- or ξ-derivatives, then*

$$\big\langle X\exp(\varepsilon\Delta_R)\big\rangle = \varepsilon^{-n} X\,\mathrm{Todd}(\varepsilon^2 R).$$

To continue the computation, let us now write

$$\begin{aligned}&\big[q^{-1} d_\varepsilon q\, a^0 d_\varepsilon a^1 \cdots d_\varepsilon a^p\,\mathrm{Todd}(\varepsilon^2 R)\big]^{\mathrm{top}} \\ &\qquad = \text{top exterior form-degree part of} \\ &\qquad\qquad q^{-1} d_\varepsilon q\, a^0 d_\varepsilon a^1 \cdots d_\varepsilon a^p\,\mathrm{Todd}(\varepsilon^2 R).\end{aligned}$$

The Todd class is a polynomial in ε, for which the coefficient of ε^k lies in $\Lambda^k T^*M$ (with only even k appearing). Each of the d_ε-differentials lies in $\varepsilon\Lambda^1 T^*_m M + \Lambda^1 T_m M$. It therefore follows from an examination of degree in $\Lambda^* T^*_m M$ that

$$\big[q^{-1} d_\varepsilon q\, a^0 d_\varepsilon a^1 \cdots d_\varepsilon a^p\,\mathrm{Todd}(\varepsilon^2 R)\big]^{\mathrm{top}} = \varepsilon^n\,\big[q^{-1} dq\, a^0 da^1 \cdots da^p\,\mathrm{Todd}(R)\big]^{\mathrm{top}}.$$

As a result, it follows from (7.7) and Proposition 7.8, and the definition of str_m in (6.17) that

$$\begin{aligned}(7.9)\quad & \frac{1}{p!}\operatorname{str}_m\big(q^{-1} d_\varepsilon q\, a^0 d_\varepsilon a^1 \cdots d_\varepsilon a^p \exp(\varepsilon\Delta_R)\big) \\ &\qquad = \text{coefficient of } dx^1\cdots dx^n d\xi_1 \cdots d\xi_n \\ &\qquad\qquad \text{in } \frac{1}{p!}\int_{S^*_m M}[q^{-1} dq\, a^0 da^1 \cdots da^p\,\mathrm{Todd}(R)]^{\mathrm{top}}.\end{aligned}$$

In this formula $[q^{-1}dq\, a^0 da^1 \cdots da^p\, \mathrm{Todd}(R)]^{\mathrm{top}}$ is to be regarded as a polyhomogeneous, exterior algebra-valued function on $\dot{T}^*_m M$ and then written as a scalar function times $dx^1 \cdots dx^n d\xi_1 \cdots d\xi_n$; the integral $⨍$ is applied to this scalar function as in (6.15).

The right-hand side in (7.9) involves a contraction by the Euler vector field, so let us now note that

$$\begin{aligned}\iota_E\big(q^{-1}dq\, a^0da^1 \cdots da^p\, \mathrm{Todd}(R)\big) &= \iota_E\big(q^{-1}dq\big)a^0da^1 \cdots da^p\, \mathrm{Todd}(R)\\ &= a^0da^1 \cdots da^p\, \mathrm{Todd}(R),\end{aligned}$$

since $\iota_E\big(q^{-1}dq\big) = 1$ while $\iota_E\big(a^0da^1 \cdots da^p\, \mathrm{Todd}(R)\big) = 0$. It follows that

$$\mathrm{str}_m\big(q^{-1}d_\varepsilon q\, a^0 d_\varepsilon a^1 \cdots d_\varepsilon a^p \exp(\varepsilon\Delta_R)\big) \;=\; \pi_*\big([a^0da^1 \cdots da^p\, \mathrm{Todd}(R)]^{\mathrm{top}}\big)\big|_m,$$

where $\pi\colon S^*M \to M$ and where the differential form is treated as a density using the orientation of S^*M given in Definition 4.3. We have arrived at our goal:

THEOREM 7.10. *Let M be a smooth, closed manifold, let ∇ be a torsion-free connection on TM with curvature R, and let $\not{D}$ be the associated Dirac operator, as in Definition 3.18. If S^*M is equipped with the orientation in Definition 4.3, then*

$$JLO^{\not{D},\delta}(a^0,\ldots,a^p) = \frac{1}{p!}\int_{S^*M} a^0da^1 \cdots da^p\, \mathrm{Todd}(R)$$

for all $p > 0$ and all $a^0,\ldots,a^p \in C^\infty(S^*M)$. □

It remains to prove Propositions 7.6 and 7.8.

PROOF OF PROPOSITIONS 7.6 AND 7.8. The following beautiful argument (along with everything else in this section) is due to Perrot [**Per13b**, Lem.4.4], to which we refer for full details. It is reminiscent of approaches to the Baker-Campbell-Hausdorff formula going back to Schur [**Sch89**].

For $t \in \mathbb{R}$ form the element

$$Y(t) = \exp(\varepsilon\Delta + t\varepsilon^2\xi\cdot R\cdot\partial/\partial\xi)\exp(-\varepsilon\Delta)$$

in $\mathrm{A}_m[\varepsilon]$; it is a polynomial function of t and ε thanks to the nilpotence of the subspace

$$\Lambda^2(T_mM \oplus T^*_mM) \subseteq \Lambda^*(T_mM \oplus T^*_mM).$$

If $X \in \mathrm{A}_m[\varepsilon]$, then according to (4.12),

$$\begin{aligned}&\big\langle X\exp(\varepsilon\Delta + t\varepsilon^2\xi\cdot R\cdot\partial/\partial\xi)\big\rangle\\ &\qquad= \varepsilon^{-n}\Big(X\cdot Y(t)\exp\big(-\varepsilon^{-1}(x^i - y^i)(\xi_i - \eta_i)\big)\Big)\Big|_{x^i=y^i,\ \xi_i=\eta_i}\end{aligned}$$

Consider now the (form-valued) function

$$H(t) = \varepsilon^{-n}Y(t)\exp\big(-\varepsilon^{-1}(x^i - y^i)(\xi_i - \eta_i)\big)$$

on the product of $\dot{T}^*T_mM$ with itself. A direct computation shows that it satisfies a differential equation of the type

$$\frac{dH}{dt}(t) = L(t)H(t),$$

where $L(t)$ is a polynomial in t and ε with coefficients in the algebra A_m, and that the function

$$\mathrm{Todd}(\varepsilon^2 tR) \cdot \exp\Big(-\eta \cdot \varepsilon R \cdot (x-y) - (\xi-\eta) \cdot \frac{t\varepsilon R}{1-\exp(-t\varepsilon^2 R)} \cdot (x-y)\Big)$$

satisfies the same differential equation, with the same initial condition at $t = 0$. The two solutions are necessarily equal, and the two propositions follow by setting $t = 1$ in the explicit solution, applying respectively $(\partial/\partial x^j + \varepsilon\xi_i R_{ij})X$, as in the first proposition, or X alone as in the second, and restricting to the diagonal to obtain an explicit formula for the contractions. □

Acknowledgment

Beyond acknowledging our indebtedness to Perrot, it is a pleasure to thank Rudy Rodsphon for sharing his insights into Perrot's work during a number of stimulating conversations.

References

[BGV92] Nicole Berline, Ezra Getzler, and Michèle Vergne, *Heat kernels and Dirac operators*, Grundlehren der mathematischen Wissenschaften [Fundamental Principles of Mathematical Sciences], vol. 298, Springer-Verlag, Berlin, 1992, DOI 10.1007/978-3-642-58088-8. MR1215720

[CM95] A. Connes and H. Moscovici, *The local index formula in noncommutative geometry*, Geom. Funct. Anal. **5** (1995), no. 2, 174–243, DOI 10.1007/BF01895667. MR1334867

[CM98] A. Connes and H. Moscovici, *Hopf algebras, cyclic cohomology and the transverse index theorem*, Comm. Math. Phys. **198** (1998), no. 1, 199–246, DOI 10.1007/s002200050477. MR1657389

[Con94] Alain Connes, *Noncommutative geometry*, Academic Press, Inc., San Diego, CA, 1994. MR1303779

[Get83] Ezra Getzler, *Pseudodifferential operators on supermanifolds and the Atiyah-Singer index theorem*, Comm. Math. Phys. **92** (1983), no. 2, 163–178. MR728863

[Hig04] Nigel Higson, *The local index formula in noncommutative geometry*, Contemporary developments in algebraic K-theory, ICTP Lect. Notes, XV, Abdus Salam Int. Cent. Theoret. Phys., Trieste, 2004, pp. 443–536, DOI 10.1007/b94118. MR2175637

[Hor83] Lars Hörmander, *The analysis of linear partial differential operators. I*, Grundlehren der mathematischen Wissenschaften [Fundamental Principles of Mathematical Sciences], vol. 256, Springer-Verlag, Berlin, 1983. Distribution theory and Fourier analysis, DOI 10.1007/978-3-642-96750-4. MR717035

[HY19] Nigel Higson and Zelin Yi, *Spinors and the tangent groupoid*, Doc. Math. **24** (2019), 1677–1720. MR4033830

[JLO88] Arthur Jaffe, Andrzej Lesniewski, and Konrad Osterwalder, *Quantum K-theory. I. The Chern character*, Comm. Math. Phys. **118** (1988), no. 1, 1–14. MR954672

[Kas89] Christian Kassel, *Le résidu non commutatif (d'après M. Wodzicki)* (French), Astérisque **177-178** (1989), Exp. No. 708, 199–229. Séminaire Bourbaki, Vol. 1988/89. MR1040574

[Lee03] John M. Lee, *Introduction to smooth manifolds*, Graduate Texts in Mathematics, vol. 218, Springer-Verlag, New York, 2003, DOI 10.1007/978-0-387-21752-9. MR1930091

[Nis97] Victor Nistor, *Higher index theorems and the boundary map in cyclic cohomology*, Doc. Math. **2** (1997), 263–295. MR1480038

[Per12] Denis Perrot, *Extensions and renormalized traces*, Math. Ann. **352** (2012), no. 4, 911–940, DOI 10.1007/s00208-011-0665-0. MR2892456

[Per13a] Denis Perrot. Local index theory for certain Fourier integral operators on Lie groupoids. Preprint, 2013. `arXiv:1401.0225`.

[Per13b] Denis Perrot, *Pseudodifferential extension and Todd class*, Adv. Math. **246** (2013), 265–302, DOI 10.1016/j.aim.2013.07.004. MR3091807

[Per16] Denis Perrot. Index theory for improper actions: localization at units. Preprint, 2016. `arXiv:1612.04090`.

[PR14] Denis Perrot and Rudy Rodsphon. An equivariant index theorem for hypoelliptic operators. Preprint, 2014. arXiv:1412.5042.

[Qui88] Daniel Quillen, *Algebra cochains and cyclic cohomology*, Inst. Hautes Études Sci. Publ. Math. **68** (1988), 139–174 (1989). MR1001452

[Rad91] A. O. Radul, *Lie algebras of differential operators, their central extensions and W-algebras* (Russian), Funktsional. Anal. i Prilozhen. **25** (1991), no. 1, 33–49, DOI 10.1007/BF01090674; English transl., Funct. Anal. Appl. **25** (1991), no. 1, 25–39. MR1113120

[Rod15] Rudy Rodsphon, *Zeta functions, excision in cyclic cohomology and index problems*, J. Funct. Anal. **268** (2015), no. 5, 1167–1204, DOI 10.1016/j.jfa.2014.11.012. MR3304597

[Roe98] John Roe, *Elliptic operators, topology and asymptotic methods*, 2nd ed., Pitman Research Notes in Mathematics Series, vol. 395, Longman, Harlow, 1998. MR1670907

[Sch89] Friedrich Schur, *Neue Begründung der Theorie der endlichen Transformationsgruppen* (German), Math. Ann. **35** (1889), no. 1-2, 161–197, DOI 10.1007/BF01443876. MR1510602

[Wod87] Mariusz Wodzicki, *Noncommutative residue. I. Fundamentals*, K-theory, arithmetic and geometry (Moscow, 1984), Lecture Notes in Math., vol. 1289, Springer, Berlin, 1987, pp. 320–399, DOI 10.1007/BFb0078372. MR923140

Department of Mathematics, University of Pennsylvania, 209 South 33rd Street Philadelphia, Pennsylvania 19104

Email address: blockj@math.upenn.edu

Department of Mathematics, Penn State University, University Park, Pennsylvania 16802

Email address: higson@psu.edu

Department of Mathematics, Penn State University, University Park, Pennsylvania 16802

Email address: jxs1504@psu.edu

Proceedings of Symposia in Pure Mathematics
Volume **105**, 2023
https://doi.org/10.1090/pspum/105/01895

The Chern-Baum-Connes assembly map for Lie groupoids

P. Carrillo Rouse

This paper is dedicated to the 40th *anniversary of cyclic cohomology*

Abstract. For a Lie groupoid $\mathscr{G}$, for which there are Thom isomorphisms in periodic cyclic (co)homology for appropriate $\mathscr{G}$-vector bundles, we construct a morphism from the left hand side of the Baum-Connes asssembly map of $\mathscr{G}$ to the periodic cyclic homology group of the Lie groupoid convolution algebra,

$$K^*_{top}(\mathscr{G}) \xrightarrow{Ch_{BC}} HP_*(C^\infty_c(\mathscr{G})).$$

This morphism, that we call here the Chern-Baum-Connes assembly map (CBC-map), is realized by assembling, via geometric pushforward maps in periodic cyclic homology, Chern-Connes character morphisms associated to $\mathscr{G}$-proper $spin^c$ cocompact manifolds together with an explicit cohomological assembly map in periodic cyclic homology. The main result behind is the wrong way functoriality of the pushforward morphisms in periodic cyclic homology. Next, we show that the CBC-map factors as the Baum-Connes assembly map followed by the Chern-Connes character for $\mathscr{G}$ if this later morphism extends to the C^*-algebra K-theory. More generally we show that if for a given periodic cyclic cocycle on $C^\infty_c(\mathscr{G})$ the associated pairing extends to the $K-$theory of the groupoid C^*-algebra then the composition of the Baum-Connes assembly map followed by the associated extended morphism coincides with the Connes pairing of the given periodic cyclic cycle with the Chern-Baum-Connes assembly map.

Contents

1. Introduction

In [**1**–**3**], Baum and Connes initiated a program on index theory for manifolds under the action of groups (discrete and Lie) or groupoids (at least for holonomy groupoids of regular foliations in their case) that lead to the so-called Baum-Connes assembly map associated to a Lie groupoid $\mathscr{G}$ and to the ensuing Baum-Connes conjecture that states that the assembly map is an isomorphism. The conjecture is at the present time still open for a large class of groups and Lie groupoids, even though it has been proved for many cases (or in some cases just the injectivity or the surjectivity). One of the main reasons this conjecture became an important source of research is that it is directly related to some other important conjectures in geometry, topology and analysis, for example the Novikov conjecture, the stable Gromov-Lawson conjecture, the Connes-Kasparov conjecture and the Kadison-Kaplansky conjecture for mention the most famous ones, see for instance [**13**, **18**, **19**]. The assembly map in question is a group morphism

$$\mu_{BC} : K^*_{top}(\mathscr{G}) \longrightarrow K_*(C^*_r(\mathscr{G})) \tag{1.1}$$

where the left hand side $K^*_{top}(\mathscr{G})$ is a group constructed by assembling (as we will explain below with more details) the topological K-theory groups associated to $\mathscr{G}-$ proper manifolds ($spin^c$ and cocompact) while the right hand side stands for the topological K-theory of the reduced C^*-algebra of the groupoid. This assembly morphism has been also being studied extensively in the recent years in the context of higher secondary invariants, Gromov-Lawson indices, problems related to (existence and classification of) positive scalar curvature metrics and surgery on manifolds, higher orbital integrals, transverse index theory, higher index theory for Lie groupoids, for example see [**5**], [**6**], [**11**], [**9**], [**12**], [**14**], [**15**], [**17**], [**22**], [**23**], [**24**], [**26**] or [**27**] just to mention some recent developments related in some way to our paper.

Now, in the original papers of Baum and Connes, ref.cit., the cycles for the left hand side are composed by classes of couples (M, x), where M is a $\mathscr{G}-$proper, $spin^c$, cocompact manifold and x is an element of the equivariant K-theory group[1] $K^*_{\mathscr{G}}(M)$ and the equivalence is generated by being equal up to a $\mathscr{G}$-family index, to be more precise, if $f : M \to N$ is an appropriate $\mathscr{G}$-equivariant map between manifolds as above then we shall have $(M, x) \sim (N, f_! x)$ where the shriek map $f_!$ is a $\mathscr{G}$-family index morphism, in the original papers of Baum and Connes the assembly was then defined to be

$$[M, x] \mapsto \pi_M!(x) \in K_*(C^*_r(\mathscr{G}))$$

where $\pi_M!(x)$ is $\mathscr{G}$ higher index associated to the momentum map $\pi_M : M \to P$ where $\mathscr{G} \rightrightarrows P$. Now, even if Baum and Connes gave the fundamental ideas and directions to follow, not all proofs were completed, but just sketched. For example, in those papers the well-definedness of the left hand side groups was sketched and not proved. The fact that it is well defined is a consequence of the wrong way functoriality theorem proved in a more general context in [**11**]. This wrong way functoriality also implies that the originally defined Baum-Connes assembly map is well defined.

After the original formulation and mentioned works of Baum and Connes, several and different direcctions occurred, in fact the assembly map for a large class of locally compact groups was formalised some years after by Baum, Connes and Higson in [**4**] using the powerful tool of Kasparov's KK-theory to properly define a "Left hand side" and the assembly map, with the time other models for this left hand side also entered into the game (Baum-Douglas model, Lück model). The

[1]A model for this can be the K-theory group of the C^*-algebra $C^*_r(M \rtimes \mathscr{G})$ associated to the action groupoid $M \rtimes \mathscr{G}$.

emergence of KK-theory have given deep and powerful results on the injectivity and surjectivity of the assembly map and have allowed to establish several properties. However, one of the main motivations of the original ideas of Baum-Connes was to be able to give geometric formulas for higher index formulas, to be able to use classic topological tools for computing higher analytical indices, nevertheless actual index formulae computations of the assembly map are rare, for example in [**1**] Baum and Connes sketched a construction of a Chern character for discrete groups from the left hand side that would eventually lead to an actual pairing with higher currents or higher traces.

In [**9**], together with B.L. Wang and H. Wang, we completed the program on the Chern character for discrete groups as stated in [**1**] and [**2**] by Baum and Connes, in particular by defining an assembled Chern character

$$K^*_{top}(\Gamma) \xrightarrow{ch^{top}_\mu} HP_*(\mathbb{C}\Gamma) \tag{1.2}$$

for Γ a discrete group that allowed us to go further and give an explicit index formula for the Connes pairing in periodic cyclic (co)homology of the group algebra $\mathbb{C}\Gamma$. The main idea in [**9**] was to extend the wrong way functoriality techniques used in [**11**] to the context of periodic cyclic (co)homology. Indeed, these tehcniques are based on the use of Lie deformation groupoids and the description of the pushforward maps using these kind of groupoids. So in particular for the appropriate algebras one can apply very similar constructions in the realm of periodic cyclic (co)homology because one can use then some properties shared with K-theory as Bott periodicity, six term exact sequence, homotopy invariance, Morita invariance (at least Lie groupoid Morita invariance) and compatibility with the Thom isomorphism.

In this paper, for a given Lie groupoid $\mathscr{G} \rightrightarrows P$, we want to initiate the study of the hypothetical existence of a commutative diagram of assembled morphisms

$$\begin{array}{ccc} K^*_{top}(\mathscr{G}) & \xrightarrow{\mu_{BC}} & K_*(C^*_r(\mathscr{G})) \\ {\scriptstyle Ch_{top}}\downarrow & & \downarrow{\scriptstyle Ch_{CO}} \\ H^*_{top}(\mathscr{G}) & \xrightarrow[\mu_{HP}]{} & HP_*(C^\infty_c(\mathscr{G})) \end{array} \tag{1.3}$$

where Ch_{CO} would be the hypothetical extension of the Chern-Connes character

$$Ch_{CO} : K_*(C^\infty_c(\mathscr{G})) \to HP_*(C^\infty_c(\mathscr{G})).$$

In this article we will use geometric deformation groupoids and associated pushforward morphism in periodic cyclic (co)homology to construct two morphisms above Ch_{top} and μ_{HP} for a large class of Lie groupoids[2]. Additionally, the composition will yield a morphism

$$Ch_{BC} : K^*_{top}(\mathscr{G}) \to HP_*(C^\infty_c(\mathscr{G}))$$

that we will call the *Chern-Baum-Connes assembly map* and that will satisfy that the above diagram (1.3) commutes if the Chern-Connes character for $\mathscr{G}$ extends to the C^*-algebra $K-$theory groups as above.

[2]That we conjecture to be all the Lie groupoids, see next remark, but that contains at least all foliation groupoids (Morita equivalent to an étale groupoid), discrete and compact groups for example.

Now, the existence of the extension morphism Ch_{CO} in the right of the diagarm (1.3) is a very hard analytic problem, unknown in general for discrete groups and for many interesting Lie groupoids, and possibly not true for some other examples. For having a more flexible version of the extension problem above we will see in our main theorem below, theorem 1.3, that if for a given periodic cyclic cocycle on $C_c^\infty(\mathscr{G})$ the associated pairing extends to the $K-$theory of the reduced C^*-algebra then the composition of the Baum-Connes assembly map followed by the associated extended morphism coincides with the Connes pairing of the given periodic cyclic cycle with the Chern-Baum-Connes assembly map.

To develop this program, an important and fundamental point to remark is that to be able to apply constructions in periodic cyclic theory similar to those used in K-theory one needs, besides suitable groupoids, appropriate smooth subalgebras of the C^*-algebras of the corresponding groupoids. In particular appropriate enough to have the right exact sequences that lead to classic arguments in algebraic topology (use of homotopy invariance, periodicity, six term exact sequences, etc.). In section 2 we explain which are the main groupoids used in this paper and which algebras we are using in each case when applying periodic cyclic (co)homology to them. In section 3 we construct and study the deformation index type morphisms applied to those algebras/groupoids in periodic cyclic (co)homology.

Next, to give a detailed description of the contents of the paper we start by introducing the class of Lie groupoids that we will consider, which we believe to be all Lie groupoids, see remark below. Let $\mathscr{G} \rightrightarrows P$ be a Lie groupoid, we will say that $\mathscr{G}$ satisfies the Thom isomorphism property in periodic cyclic (co)homology or simply that it is a HP-Thom Lie groupoid if the following two properties are satisfied

(1) For every $E \to M$ $\mathscr{G}$-*spin*^{c} proper vector bundle there is an unique isomorphism

$$\mathcal{T}_E^{\mathscr{G}} : HP_*(C_c^\infty(M \rtimes \mathscr{G})) \longrightarrow HP_{*+r_E}(\mathscr{S}_c(E \rtimes \mathscr{G})) \tag{1.4}$$

where $* + r_E = * + rank(E) \bmod 2$ (see section 2 for the definition of $\mathscr{S}_c(E \rtimes \mathscr{G})$), such that the following diagram commutes

$$\begin{array}{ccc} K_*(C_c^\infty(M \rtimes \mathscr{G})) & \xrightarrow[\cong]{\mathcal{T}_E^{\mathscr{G}}} & K_{*+r_E}(\mathscr{S}_c(E \rtimes \mathscr{G})) \\ \Big\downarrow Ch_{CO} & & \Big\downarrow Ch_{CO} \\ HP_*(C_c^\infty(M \rtimes \mathscr{G})) & \xrightarrow[\mathcal{T}_E^{\mathscr{G}}]{\cong} & HP_{*+r_E}(\mathscr{S}_c(E \rtimes \mathscr{G})) \end{array} \tag{1.5}$$

where the morphism $\mathcal{T}_E^{\mathscr{G}}$ on the top is the Thom isomorphism[3] in K-theory of the correspondent algebras and where the vertical morphisms stand for the correspondent Chern-Connes characters.

(2) For every $E \to M$ $\mathscr{G}$-*spin*^{c} proper vector bundle there is an isomorphism

$$\mathcal{T}_{\mathscr{G}}^E : HP_{*+r_E}(\mathscr{S}_c(E \rtimes \mathscr{G})) \longrightarrow HP_*(C_c^\infty(M \rtimes \mathscr{G})) \tag{1.6}$$

[3]In principle the Thom isomorphism in K-theory is defined for the correspondent C^*-algebras, but in this particular case the smooth algebras are stable under holomorphic calculus since we are assuming proper action.

where $*+r_E = *+rank(E)$ mod 2, such that the following property is satisfied

- For every $\omega \in HP_*(M \rtimes \mathscr{G})$ and every $\delta \in HP_{*+r_E}(\mathscr{S}_c(E \rtimes \mathscr{G}))$ the following equality of Connes pairings holds

$$\langle \omega, \mathcal{T}_{\mathscr{G}}^{E}(\delta) \rangle = \langle \mathcal{T}_{E}^{\mathscr{G}}(\omega), \delta \rangle. \tag{1.7}$$

REMARK 1.1. Some important points to remark about this Thom isomorphism hypothesis in periodic cyclic (co)homology:

- If it exists, the above Thom isomorphism in periodic cyclic homology is more like a twisted Thom isomorphism. Indeed, in the cases where geometrical computations are avalaible (e.g. classic spaces case or discrete groups case for instance), the cohomological Thom isomorphism does not commute with the K-theoretical Thom isomorphism via the Chern character. However, it does commute under the presence of appropriate Todd classes. In this paper we will use for simplicity the terminology Thom isomorphism for the above morphism.
- The Thom isomorphism in K-theory above holds in all generality, it can be obtained by some descent procedure from the classic topological K-theory Thom isomorphism for spaces by using for example the work of Le Gall on groupoid equivariant KK-theory, see [**20**]. For more details the reader can see appendix A.2 in [**11**].
- As far as the author knows, there is no proof of the Periodic Cyclic (co)homology Thom isomorphism for general Lie groupoids. It is certainly highly expected to be true in all generality.
- Some interesting cases in which the above property of being HP-Thom is known to hold are
  1. Foliation groupoids, that is, Lie groupoids which are Morita equivalent to étale groupoids. This can be seen using for example the work of Tu and Xu for proper etale groupoids, [**25**], together with the Lie groupoid Morita invariance of the Periodic Cyclic (co)homology, see for example [**21**]. This case covers the cases of discrete groups and holonomy groupoids of regular foliations.
  2. Compact Lie groups, for example by following the work of Block and Getzler in [**7**].

As for the K-theoretical case and the discrete group case developed in [**11**] and [**9**], the above Thom property, together with the use of appropriate deformation groupoids algebras plays a fundamental role to have:

- A pushforward morphism

$$HP_*(C_c^\infty(M \rtimes \mathscr{G})) \xrightarrow{f!} HP_*(C_c^\infty(N \rtimes \mathscr{G})) \tag{1.8}$$

  for appropriate maps $f: M \to N$ between $\mathscr{G}$–manifolds, definition 4.1;
- the pushforward functoriality for the morphisms above, theorem 4.2;
- an assembled group

$$H_{top}^*(\mathscr{G}) := \varinjlim_{f_!} HP_*(C_c^\infty(M \rtimes \mathscr{G})), \tag{1.9}$$

with manifolds M being $\mathscr{G}$-proper $spin^c$ cocompact, and an assembled Chern character (corollary 4.4)

$$Ch_{top} : K^*_{top}(\mathscr{G}) \longrightarrow H^*_{top}(\mathscr{G}) \tag{1.10}$$

given by

$$Ch_{top}([M,x]) = [M, Ch^M_{CO}(x)]. \tag{1.11}$$

where $Ch^M_{CO} : K_*(C^\infty_c(M \rtimes \mathscr{G})) \to HP_*(C^\infty_c(M \rtimes \mathscr{G}))$ is the Chern-Connes character;

- a cohomological assembly map (theorem 4.5)

$$\mu_{HP} : H^*_{top}(\mathscr{G}) \longrightarrow HP_*(C^\infty_c(\mathscr{G})), \tag{1.12}$$

that is explicitly given by

$$\mu_{HP}([M,\tau]) = \pi_M!(\tau). \tag{1.13}$$

where $\pi_M : M \to P$ is the $\mathscr{G}$−momentum map given by the $\mathscr{G}$-action on M; and

- the commutativity of diagram (1.3) if the Chern-Connes of $\mathscr{G}$ extends to the K-theory of the C^*-algebra.

We summarize the last results in the following theorem.

THEOREM 1.2. *For any HP-Thom Lie groupoid $\mathscr{G}$ there is a well defined morphism*

$$K^*_{top}(\mathscr{G}) \xrightarrow{Ch_{BC}} HP_*(C^\infty_c(\mathscr{G})) \tag{1.14}$$

given by the composition of the Chern character

$$Ch_{top} : K^*_{top}(\mathscr{G}) \longrightarrow H^*_{top}(\mathscr{G}) \tag{1.15}$$

followed by the cohomological assembly

$$\mu_{HP} : H^*_{top}(\mathscr{G}) \longrightarrow HP_*(C^\infty_c(\mathscr{G})), \tag{1.16}$$

inducing a well-defined pairing

$$K^*_{top}(\mathscr{G}) \times HP^*(C^\infty_c(\mathscr{G})) \to \mathbb{C} \tag{1.17}$$

given by

$$\langle [M,x], \tau \rangle := \langle \pi_M!(Ch^M_{CO}(x)), \tau \rangle. \tag{1.18}$$

Moreover, the commutativity of diagram (1.3) *holds if the Chern-Connes character morphism for $\mathscr{G}$ extends to $K_*(C^*_r(\mathscr{G}))$.*

We call the morphism Ch_{BC} above the Chern-Baum-Connes assembly map of the Lie groupoid $\mathscr{G}$.

The construction of the assembly map above includes of course the case of discrete groups. Now, for discrete groups, the assembly map we constructed in [**9**] is in principle different, in fact in that paper we use the fact that an explicit computation for the periodic cyclic (co)homology groups is known in terms of group cohomology groups to give a more geometric description in that case and a formula for the index pairing. We will study elsewhere the relation between these two assembly maps, this would be very interesting since for the assembly map studied in this paper we have an explicit relation with the Baum-Connes map, theorem above and below, while in [**9**] we obtained explicit index formulas for the Connes pairing.

We conclude this introduction with one of the main result of this article, theorem 4.7.

THEOREM 1.3. *Let $\mathscr{G} \rightrightarrows P$ be a Lie groupoid satisfying the Thom isomorphism property in periodic cyclic (co)homology and let $\tau \in HP_*(C_c^\infty(\mathscr{G}))$. Suppose that the morphism*

$$\tilde{\varphi}_\tau : K_*(C_c^\infty(\mathscr{G})) \longrightarrow \mathbb{C} \tag{1.19}$$

induced from the pairing with $HP_*(C_c^\infty(\mathscr{G}))$ (via the Chern-Connes character) extends to a morphism

$$\varphi_\tau : K_*(C_r^*(\mathscr{G})) \longrightarrow \mathbb{C}. \tag{1.20}$$

Then, for every $[M,x] \in K_{top}^*(\mathscr{G})$ we have

$$\varphi_\tau(\mu_{BC}([M,x])) = \langle Ch_{BC}([M,x]), \tau \rangle. \tag{1.21}$$

One of the main interest of the theorem above is that for every (HP-Thom) Lie groupoid, and for every periodic cyclic cocycle, the right hand side of the formula above always makes sense.

2. Semidirect product groupoids and algebras used in this article

In this section we will describe explicitly the main groupoids we use in the present article as well as the associated algebras. For more details on Lie groupoids, generalized morphisms and particularly deformation Lie groupoids and their use in index theory the reader can see [**11**], [**16**] or [**9**].

Let $\mathscr{G} \rightrightarrows P$ be a Lie groupoid and M be a $\mathscr{G}$-manifold with moment map $\pi_M : M \to P$ that we will always assume to be a submersion, we can consider the following groupoids and algebras:

- **The semi-direct groupoid $M \rtimes \mathscr{G}$:** It is the groupoid

$$M \rtimes \mathscr{G} \rightrightarrows M \tag{2.1}$$

with $M \rtimes \mathscr{G} = \{(m,\gamma) \in M \times \mathscr{G} : \pi_M(m) = t(\gamma)\}$ and with structure maps

(1) Source and target maps defined by
$$s(m,g) = m \cdot g, \;\; t(m,g) = m,$$
(2) groupoid product
$$(m,g) \cdot (mg,h) := (m,gh),$$
(3) unit map
$$u(m) = (m, \pi_M(m)),$$
where $\pi_M(m) \in \mathscr{G}$ is the unit element associated to the groupoid structure on $\mathscr{G}$, and
(4) inverse map
$$(m,g)^{-1} := (mg, g^{-1}).$$

For this groupoid we will usually consider the usual groupoid convolution algebra of complex valued compactly supported C^∞-functions $C_c^\infty(M \rtimes \mathscr{G})$.

- **The semi-direct groupoid** $(M \times_P M) \rtimes \mathscr{G}$**:** It is the groupoid associated to the diagonal action on $M \times_P M$, given by

(2.2)
$$(M \times_P M) \rtimes \mathscr{G} \rightrightarrows M$$

with $(M \times_P M) \rtimes \mathscr{G} = \{(m', m, \gamma) \in M \times M \times \mathscr{G} : \pi_M(m) = \pi_M(m') = t(\gamma)\}$ and with structure maps

(1) Source and target maps defined by
$$s(m, n, \gamma) = n \cdot \gamma, \ \ t(m, , n, \gamma) = m,$$
(2) groupoid product
$$(m, n, \gamma) \cdot (n\gamma, l, \eta) := (m, l\mathscr{G}^{-1}, \gamma\eta)$$
(3) unit map
$$u(m) = (m, m, \pi_M(m)),$$
where $\pi_M(m) \in \mathscr{G}$ is the unit element associated to the groupoid structure on $\mathscr{G}$, and
(4) inverse map
$$(m, n, \gamma)^{-1} := (n\gamma, m\gamma, \gamma^{-1}).$$

For this groupoid we will usually consider the usual groupoid convolution algebra of complex valued compactly supported C^∞-functions $C_c^\infty((M \times_P M) \rtimes \mathscr{G})$.

- **The semi-direct groupoid** $T_\pi M \rtimes \mathscr{G}$**:** It is the groupoid associated to the infinitesimal action of $\mathscr{G}$ on the vertical tangent space $T_\pi M := Kerd\pi_M$, it is given by

(2.3)
$$T_\pi M \rtimes \mathscr{G} \rightrightarrows M$$

with $T_\pi M \rtimes \mathscr{G} = \{((m, V), \gamma) \in T_\pi M \times \mathscr{G} : \pi_M(m) = t(\gamma)\}$ and with structure maps

(1) Source and target maps defined by
$$s(m, V, \gamma) = m \cdot \gamma, \ \ t(m, V, \gamma) = m,$$
(2) groupoid product
$$(m, V, \gamma) \cdot (m\gamma, W, \eta) := (m, d_m(R_\gamma)(V) + W, \gamma\eta),$$
where $R_\gamma : \pi_M^{-1}(t(\gamma)) \to \pi_M^{-1}(s(\gamma))$ stands for the right action by γ,
(3) unit map
$$u(m) = (m, 0_m, \pi_M(m)),$$
where $\pi_M(m) \in \mathscr{G}$ is the unit element associated to the groupoid structure on $\mathscr{G}$, $0_m \in (T_\pi M)_m$ is the origin vector, and
(4) inverse map
$$(m, V, \gamma)^{-1} := (m\gamma, -d_m R_\gamma(V), \gamma^{-1}).$$

For this groupoid we will usually consider the Schwartz type groupoid convolution algebra of complex valued C^∞- fiber rapidly decreasing functions with compact support in direction of the base and in the $\mathscr{G}$-direction, $\mathscr{S}_c(T_\pi M \rtimes \mathscr{G})$. To be more precise an element in this algebra is a C^∞, complex valued function f on $T_\pi M \times \mathscr{G}$ such that $f(V, \gamma) = 0$ for γ outside a compact subset of $\mathscr{G}$ and such that the functions $f_\gamma : TM_{t(\gamma)} \to \mathbb{C}$, where

$TM_{t(\gamma)} = \pi^{-1}(t(\gamma))$, defined by $f_\gamma(V) := f(V,\gamma)$ are in the Schwartz type space $\mathscr{S}_c(TM_{t(\gamma)})$ defined in [**8**] Definition 4.6.

The algebra product is given as the convolution product of $C_c^\infty(T_\pi M \rtimes \mathscr{G})$ where the groupoid under consideration is the groupoid $T_\pi M \rtimes \mathscr{G} \rightrightarrows M$ as above, the product is well-defined since we are taking fiberwise Schwartz functions, see [**8**] for more details.

REMARK 2.1. For a $\mathscr{G}-$vector bundle $E \to M$ we can consider the analog algebra $\mathscr{S}_c(E \rtimes \mathscr{G})$.

- **The semi-direct deformation groupoid** $D_f \rtimes \mathscr{G}$**:** Let $f : M \longrightarrow N$ be a $\mathscr{G}$-equivariant C^∞-map between two $\mathscr{G}-$ manifolds M and N, we consider the $\mathscr{G}-$vector bundle $T_f := T_\pi M \oplus f^* T_\pi N$ over M and the deformation groupoid

$$D_f : T_f \bigsqcup (M \times_P M) \times_P N \times (0,1] \rightrightarrows f^*T_\pi N \bigsqcup (M \times_P N) \times (0,1] \tag{2.4}$$

given as the normal groupoid associated to the $\mathscr{G}$-map

$$M \overset{\Delta \times f}{\longrightarrow} (M \times_P M) \times_P N.$$

Here the $\mathscr{G}$-action on $(M \times_P M) \times_P N$ preserves the image of M under the previous map $\Delta \times f$. Hence, by functoriality of the deformation to the normal cone construction, we have an action of $\mathscr{G}$ on D_f with moment map $\pi_f : D_f \to P$ given by the composition of the deformation map of the pair couple $((M \times_P M) \times_P N, M) \to (P, P)$. This is canonically induced from the moment maps of M and N, followed by the first projection $D(P,P) = P \times [0,1] \to P$ (where $D(\cdot,\cdot)$ denotes the deformation to the normal cone). This induces a semi-direct product Lie groupoid

$$D_f \rtimes \mathscr{G} \rightrightarrows f^*T_\pi N \bigsqcup (M \times_P N) \times (0,1] \tag{2.5}$$

with $D_f \rtimes \mathscr{G} = \{(A,\gamma) \in D_f \times \mathscr{G} : \pi_f(A) = t(\gamma)\}$. The groupoid structure is given fiberwisely over $[0,1]$ as the groupoid structure of

$$((M \times_P M) \times_P N) \rtimes \mathscr{G} \rightrightarrows M \times_P N$$

for $t \neq 0$, and at $t = 0$ the groupoid structure is

$$(T_\pi M \oplus f_* T_\pi N) \rtimes \mathscr{G} \rightrightarrows f_* T_\pi N.$$

This groupoid is a LIE groupoid and we could consider its convolution algebra but we will need a slightly larger algebra, indeed for this groupoid we consider the Schwartz type algebra $\mathscr{S}_c(D_f \rtimes \mathscr{G})$ as an immediate generalisation of the Schwartz type algebra for the tangent groupoid defined in [**8**]. More precisely an element in $\mathscr{S}_c(D_f \rtimes \mathscr{G})$ is a C^∞-function $f : D_f \rtimes \mathscr{G} \to \mathbb{C}$ such that $f_\gamma = f(\cdot,\gamma) : (D_f)_\gamma \to \mathbb{C}$ is zero for all γ outside a compact set of $\mathscr{G}$, and all of them belong to the vector space $\mathscr{S}_c((D_f)_\gamma)$ defined in [**8**] Section 4.2, where $(D_f)_\gamma$ is the deformation to the normal cone associated to the immersion induced by f (since it commutes with the moment maps) of $\pi_M^{-1}(t(\gamma))$ in $\pi^{-1}(t(\gamma))$ where π_M is the $\mathscr{G}$-moment map of M and π the $\mathscr{G}$-moment map of $(M \times_P M) \times_P N$. The convolution product in this algebra is induced by the groupoid product in $D_f \rtimes \mathscr{G}$.

The fundamental property of this algebra is to be an intermediate algebra

$$C_c^\infty(D_f \rtimes \mathscr{G}) \subset \mathscr{S}_c(D_f \rtimes \mathscr{G}) \subset C_r^*(D_f \rtimes \mathscr{G}) \tag{2.6}$$

such that the canonical evaluation morphisms yield

$$\mathscr{S}_c(D_f \rtimes \mathscr{G})|_{t=0} = \mathscr{S}_c(T_f \rtimes \mathscr{G}) \tag{2.7}$$

and

$$\mathscr{S}_c(D_f \rtimes \mathscr{G})|_{t=1} = C_c^\infty(((M \times_P M) \times_P N) \rtimes \mathscr{G}). \tag{2.8}$$

Moreover, as seen below, there is a short exact sequence of the form.

$$0 \longrightarrow J \longrightarrow \mathscr{S}_c(D_f \rtimes \mathscr{G}) \xrightarrow{e_0} \mathscr{S}_c(T_f \rtimes \mathscr{G}) \longrightarrow 0. \tag{2.9}$$

As we will see in the next section, the above kernel algebra J has trivial Cyclic periodic (co)homology groups and hence the evaluation at $t = 0$ will induce isomorphisms in these (co)homology groups.

Schwartz type algebras for $\mathscr{G}$-vector bundles and the fiberwise Fourier isomorphism. We already introduced in the last section some algebras that use Schwartz functions in a certain way. In this section we want to explain the reason of using Schwartz algebras in the present article and some important facts related to these algebras.

Given a $\mathscr{G}$-proper equivariant vector bundle $E \to M$ there are two kinds of Schwartz algebras considered in this article. As sets they are the same, the vector space of complex valued C^∞-functions rapidly decreasing in the direction of the vector bundle fibers, with compact support in direction of the base and in the $\mathscr{G}$-direction, denoted by $\mathscr{S}_c(E \rtimes \mathscr{G})$.

Now, there are two different algebra products in these spaces depending on which groupoid one is considering. Indeed, if one considers the semi-direct product groupoid

$$E \rtimes \mathscr{G} \rightrightarrows M$$

associated to the $\mathscr{G}$-vector bundle structure, then there is a convolution product $(\mathscr{S}_c(E \rtimes \mathscr{G}), *)$ using this groupoid structure. On the other hand, one might consider the action groupoid

$$E \rtimes \mathscr{G} \rightrightarrows E$$

associated to the $\mathscr{G}$-space E. In this case there is a product algebra $(\mathscr{S}_c(E \rtimes \mathscr{G}), \cdot)$ that uses the groupoid structure of this action groupoid. These two algebras are not the same at all. In the context of this paper, we use both examples, for the semi-direct product groupoid

$$T_\pi M \rtimes \mathscr{G} \rightrightarrows M,$$

and for the action groupoid

$$T_\pi^* M \rtimes \mathscr{G} \rightrightarrows T_\pi^* M,$$

or for the bundles T_f and T_f^* respectively. As in the classical case when $\mathscr{G}$ is the trivial group we have a Fourier type algebra isomorphism

$$(\mathscr{S}_c(T_\pi M \rtimes \mathscr{G}), *) \cong (\mathscr{S}_c(T_\pi^* M \rtimes \mathscr{G}), \cdot),$$

which indeed is a particular case of the following proposition.

PROPOSITION 2.2. *Given a $\mathscr{G}$-proper equivariant vector bundle $E \to M$, we have the following two properties:*

(1) *The fiberwise Fourier transform gives an algebra isomorphism*

$$(\mathscr{S}_c(E \rtimes \mathscr{G}), *) \cong (\mathscr{S}_c(E^* \rtimes \mathscr{G}), \cdot). \tag{2.10}$$

(2) *The convolution algebra $(\mathscr{S}_c(E \rtimes \mathscr{G}), *)$ is stable under holomorphic calculus as a subalgebra of $(C_r^*(E \rtimes \mathscr{G}), *)$.*

The proof of the above two statements is quite classical, for instance one can follow the same lines as the proof of Proposition 4.5 in [**10**].

Now, one important consequence for us is that, in the above situation, we have the following isomorphisms

$$HP_*(\mathscr{S}_c(E^* \rtimes \mathscr{G}), \cdot) \cong HP_*(\mathscr{S}_c(E \rtimes \mathscr{G}), *) \tag{2.11}$$

and

$$HP^*(\mathscr{S}_c(E^* \rtimes \mathscr{G}), \cdot) \cong HP^*(\mathscr{S}_c(E \rtimes \mathscr{G}), *), \tag{2.12}$$

induced by the Fourier algebra isomorphism.

3. Deformation morphisms in periodic cyclic (co)homology

Let $f : M \longrightarrow N$ be a $\mathscr{G}$-equivariant C^∞-map between two $\mathscr{G}-$manifolds M and N. We consider the $\mathscr{G}-$vector bundle $T_f := T_\pi M \oplus f^*T_\pi N$ over M as the previous section. We are going to use deformation groupoids to construct a morphism

$$I_f : HP_*(\mathscr{S}(T_f \rtimes \mathscr{G})) \longrightarrow HP_*(C_c^\infty((M \times_P M) \times_P N) \rtimes \mathscr{G})). \tag{3.1}$$

Consider the deformation groupoid

$$D_f := T_f \bigsqcup ((M \times_P M) \times_P N) \times (0,1] \tag{3.2}$$

over

$$f^*T_\pi N \bigsqcup (M \times_P N) \times (0,1] \tag{3.3}$$

given as above as the normal groupoid associated to the $\mathscr{G}$-map

$$M \stackrel{\Delta \times f}{\longrightarrow} (M \times_P M) \times_P N.$$

As we already discussed above, the action of $\mathscr{G}$ on $(M \times_P M) \times_P N$ (diagonal on $M \times_P M$) leaves invariant the image of M by $\Delta \times f$ and then we have, by functoriality of the deformation to the normal cone construction, an induced semi-direct product groupoid

$$D_f \rtimes \mathscr{G} \rightrightarrows f^*T_\pi N \bigsqcup (M \times_P N) \times (0,1].$$

The following result follows directly from proposition 4.6 and lemma 5.5 in [**10**].

PROPOSITION 3.1. *For a $\mathscr{G}$-map $f : M \to N$ as above the restriction to zero, or equivalently to the closed subgroupoid $T_f \rtimes \mathscr{G}$ of $D_f \rtimes \mathscr{G}$, induces a short exact sequence*

$$0 \longrightarrow J \longrightarrow \mathscr{S}_c(D_f \rtimes \mathscr{G}) \stackrel{e_0}{\longrightarrow} \mathscr{S}_c(T_f \rtimes \mathscr{G}) \longrightarrow 0 \tag{3.4}$$

where $J = \lambda \cdot \mathscr{S}_c(D_f \rtimes \mathscr{G})$ for $\lambda : D_f \rtimes \mathscr{G} \to [0,1]$ the canonical projection.

As an immediate consequence of the homotopy invariance and of the six term short exact sequence for periodic cyclic homology we have the following corollary.

COROLLARY 3.2. *With the notations of the last proposition we have that*

$$HP_*(J) = 0 \tag{3.5}$$

and

$$(e_0)_* : HP_*(\mathscr{S}_c(D_f \rtimes \mathscr{G})) \to HP_*(\mathscr{S}_c(T_f \rtimes \mathscr{G})) \tag{3.6}$$

is an isomorphism.

DEFINITION 3.3. For a $\mathscr{G}$-map as above we consider the morphism

$$I_f : HP_*(\mathscr{S}_c(T_f \rtimes \mathscr{G})) \longrightarrow HP_*(C_c^\infty((M \times_P M) \times_P N) \rtimes \mathscr{G})). \tag{3.7}$$

given as the composition of $(e_0)_*^{-1}$ followed by $(e_1)_*$. We call it the deformation morphism associated to the the $\mathscr{G}$-map f.

4. Wrong way functoriality in Periodic cyclic homology and the Chern-Baum-Connes assembly map

Wrong way functoriality. In this section we will only work with Lie groupoids satisfying the Thom isomorphism property as defined in the introduction.

DEFINITION 4.1 (Pushward maps in Periodic cyclic homology). Let $\mathscr{G} \rightrightarrows P$ be a HP-Thom Lie groupoid. Let $f : M \longrightarrow N$ be a $\mathscr{G}$-equivariant, K-oriented, C^∞-map between two $\mathscr{G}-$proper cocompact manifolds M and N. Let $r_f := rank(T_f)$. We define the morphism

$$f_! : HP_{*-r_f}(C_c^\infty(M \rtimes \mathscr{G})) \longrightarrow HP_*(C_c^\infty(N \rtimes \mathscr{G})). \tag{4.1}$$

as the composition of the following morphisms

- The Thom isomorphism in Periodic cyclic homology

$$Th : HP_{*-r_f}(C_c^\infty(M \rtimes \mathscr{G})) \longrightarrow HP_*(\mathscr{S}_c(T_f^* \rtimes \mathscr{G})), \tag{4.2}$$

- the isomorphism

$$F : HP_*(\mathscr{S}_c(T_f^* \rtimes \mathscr{G})) \longrightarrow HP_*(\mathscr{S}_c(T_f \rtimes \mathscr{G})) \tag{4.3}$$

  induced from the Fourier algebra isomorphism,
- the deformation morphism

$$I_f : HP_*(\mathscr{S}_c(T_f \rtimes \mathscr{G})) \longrightarrow HP_*(C_c^\infty(((M \times_P M) \times_P N) \rtimes \mathscr{G})), \tag{4.4}$$

  and
- the isomorphism induced from the Morita equivalence (isomorphism by Proposition 3.6 in [**21**]) of groupoids $((M \times_P M) \times_P N) \rtimes \mathscr{G} \sim N \rtimes \mathscr{G}$

$$\mathcal{M} : HP_*(C_c^\infty(((M \times_P M) \times_P N) \rtimes \mathscr{G})) \longrightarrow HP_*(C_c^\infty(N \rtimes \mathscr{G})). \tag{4.5}$$

As in the case of K-theory groups, we can prove that the above-defined push-forward morphisms are functorial. This is one of our main results, stated as follows:

THEOREM 4.2 (Pushforward functoriality in delocalised cohomology). *Let $\mathscr{G} \rightrightarrows P$ be a HP-Thom Lie groupoid. Let $f : M \longrightarrow N$ and $g : N \longrightarrow L$ be $\mathscr{G}$ K-oriented C^∞-maps between $\mathscr{G}-$proper cocompact manifolds. Then*

$$(g \circ f)_! = g_! \circ f_! \tag{4.6}$$

as morphisms from $HP_*(C_c^\infty(M \rtimes \mathscr{G}))$ *to* $HP_{*+r_{g\circ f}}(C_c^\infty(L \rtimes \mathscr{G}))$ $(r_{g\circ f} = r_f + r_g \bmod 2)$.

The proof of the theorem above follows exactly the same steps of the proof of theorem 4.1 in [**9**] or Theorem 4.2 in [**11**] and is based on (co)homological properties of the periodic cyclic homology groups as homotopy invariance, Thom isomorphism (here by hypothesis), Lie groupoid Morita invariance, and naturality.

The above wrong way functoriality theorem allows us to consider the following assembled groups.

DEFINITION 4.3. For a HP-Thom Lie groupoid $\mathscr{G} \rightrightarrows P$ and for $* = 0, 1 \bmod 2$ we can consider the abelian group

$$H_{top}^*(\mathscr{G}) = \varinjlim_{f_!} HP_*(M \rtimes \mathscr{G}). \tag{4.7}$$

where the limit is taken over the $\mathscr{G}$ proper cocompact *spin*^{c} manifolds with submersive moment map and of fiber dimension (the rank its vertical tangent bundle) equal to $*$ modulo 2.

Equivalently it can be defined as the free abelian group with generators (M, τ) with M a $\mathscr{G}$ proper cocompact *spin*^{c} manifold and $\tau \in HP_*(C_c^\infty(M \rtimes \mathscr{G}))$, and the equivalence relation generated by $\tau \sim f_!(\tau)$ for $f : M \to N$ as above.

The following is an immediate corollary of the push-forward functoriality theorems in K-theory (Theorem 4.2 in [**11**]) and in periodic cyclic homology, Theorem 4.2 above.

COROLLARY 4.4. *For any HP-Thom Lie groupoid* $\mathscr{G}$, *there is a well-defined Chern character morphism*

$$Ch_{top} : K_{top}^*(\mathscr{G}) \longrightarrow H_{top}^*(\mathscr{G}) \tag{4.8}$$

given by

$$Ch_{top}([M, x]) = [M, Ch_{CO}^M(x)]. \tag{4.9}$$

Now, the push-forward construction above and its functoriality might be slightly more general. In particular, N does not have to be $\mathscr{G}$-proper to define a similar morphism in cyclic periodic homology. Indeed, if $f : M \to N$ is $\mathscr{G}$-map between two $\mathscr{G}$-manifolds with M proper and T_f carrying a $\mathscr{G}$-proper, $\mathscr{G} - spin^c$ vector bundle structure, then we can consider the pushforward map

$$f_! : HP_*(C_c^\infty(M \rtimes \mathscr{G})) \to HP_*(C_c^\infty(N \rtimes \mathscr{G})) \tag{4.10}$$

defined as in Definition 4.1. The pushforward functoriality theorem can be applied in the same way to prove that if

(4.11)

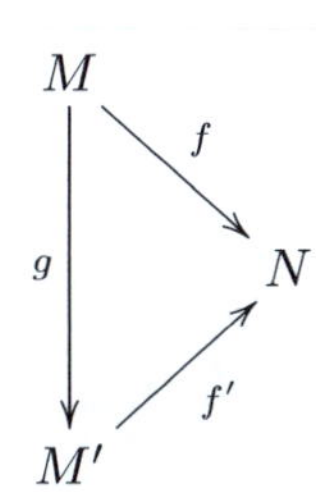

is a commutative diagram of $\mathscr{G}$-equivariant maps between two $\mathscr{G}$-manifolds with M, M' proper and T_f, $T_{f'}$ having $\mathscr{G}-spin^c$ vector bundle structures, then the following diagram is commutative

$$
\begin{array}{ccc}
HP_*(C_c^\infty(M \rtimes \mathscr{G})) & & \\
 & \searrow^{f_!} & \\
\downarrow g_! & & HP_*(C_c^\infty(N \rtimes \mathscr{G})) \\
 & \nearrow_{f'_!} & \\
HP_*(C_c^\infty(M' \rtimes \mathscr{G})) & & .
\end{array}
\tag{4.12}
$$

Besides the case when N is also $\mathscr{G}$-proper there is also the very interesting case when N is a point. In particular we can define $\pi_M!$ for the moment map π_M of M a $\mathscr{G}$-proper $spin^c$ cocompact manifold as above.

We have the following theorem whose proof follows the same steps as Theorem 4.2 together with the observation leading to (4.12).

THEOREM 4.5. *The assembly of the above morphisms μ_M^{top} gives rise to a well defined morphism*

$$\mu_{top} : H^*_{top}(\mathscr{G}) \longrightarrow HP_*(C_c^\infty(\mathscr{G})), \tag{4.13}$$

that is explicitly given by

$$\mu_{top}([M,\tau]) = \pi_M!(\tau). \tag{4.14}$$

where $\pi_M : M \to P$ is the $\mathscr{G}-$momentum map given by the $\mathscr{G}$-action on M.

The Chern-Baum-Connes assembly map. We summarise the last results in the following theorem.

THEOREM 4.6. *For any HP-Thom Lie groupoid $\mathscr{G}$ there is a well defined morphism*

$$K^*_{top}(\mathscr{G}) \xrightarrow{\ Ch_{BC}\ } HP_*(C_c^\infty(\mathscr{G})) \tag{4.15}$$

given by the composition of the Chern character

$$ch^{top} : K^*_{top}(\mathscr{G}) \longrightarrow H^*_{top}(\mathscr{G}) \tag{4.16}$$

followed by the cohomological assembly

$$\mu_{top} : H^*_{top}(\mathscr{G}) \longrightarrow HP_*(C_c^\infty(\mathscr{G})), \tag{4.17}$$

inducing a well-defined pairing

$$K^*_{top}(\mathscr{G}) \times HP^*(C_c^\infty(\mathscr{G})) \to \mathbb{C} \tag{4.18}$$

given by

$$\langle [M,x],\tau\rangle := \langle \pi_M!(Ch^M_{CO}(x)),\tau\rangle. \tag{4.19}$$

Moreover, the commutativity of diagram (1.3) holds if the Chern-Connes character morphism for $\mathscr{G}$ extends to $K_*(C^*_r(\mathscr{G}))$.

We call the morphism Ch_{BC} above the Chern-Baum-Connes assembly map of the Lie groupoid $\mathscr{G}$.

The construction of the assembly map above includes of course the case of discrete groups. Now, for discrete groups, the assembly map we develop in [**9**] is in principle different. Indeed, in that paper we use the fact that an explicit computation for the periodic cyclic (co)homology groups is known in terms of group cohomology groups to give a more geometric description in that case and a formula for the index pairing. We will study elsewhere the relation between these two assemblies. This is interesting because for the Chern-Baum-Connes assembly studied in this paper we can answer the question raised in the introduction of comparing this assembly map with the Baum-Connes assembly map whenever this makes sense. Indeed, we have the following theorem.

THEOREM 4.7. *Let $\mathscr{G} \rightrightarrows P$ be a Chern-Thom Lie groupoid and let $\tau \in HP_*(C_c^\infty(\mathscr{G}))$. Suppose that the morphism*

$$\tilde{\varphi}_\tau : K_*(C_c^\infty(\mathscr{G})) \longrightarrow \mathbb{C} \tag{4.20}$$

induced from the pairing with $HP_*(C_c^\infty(\mathscr{G}))$ (via the Chern-Connes character) extends to a morphism

$$\varphi_\tau : K_*(C_r^*(\mathscr{G})) \longrightarrow \mathbb{C}. \tag{4.21}$$

Then, for every $[M, x] \in K^*_{top}(\mathscr{G})$ we have

$$\varphi_\tau(\mu_{BC}([M, x])) = \langle Ch_{BC}([M, x]), \tau\rangle. \tag{4.22}$$

PROOF. Let $M \xrightarrow{\pi_M} P$ be a $\mathscr{G}$-proper $spin^c$ manifold, $x \in K^*(M \rtimes \mathscr{G})$ and $\tau \in HP_*(C_c^\infty(\mathscr{G}))$ satisfying the extension property in the statement above. We have to show that

$$\varphi_\tau(\pi_M!(x)) = \langle \pi_M!(Ch^M_{CO}(x), \tau)\rangle \tag{4.23}$$

where on the left hand side $\pi_M!$ stands for the K-theoretical pushforward (the pairing makes sense by the extension hypothesis), and where on the right hand side $\pi_M!$ stands for the cyclic periodic homology pushforward and $Ch^M_{CO}(x)$ for the Chern-Connes character of x.

Let us consider the morphism

$$\pi_M^* : HP^*(C_c^\infty(\mathscr{G})) \longrightarrow HP^*(C_c^\infty(M \rtimes \mathscr{G})) \tag{4.24}$$

given as the composition of the following morphisms

- The isomorphism

$$HP^*(C_c^\infty(\mathscr{G})) \xrightarrow[\approx]{\mathcal{M}} HP^*(C_c^\infty((M \times_P M) \rtimes \mathscr{G})) \tag{4.25}$$

  induced from the canonical groupoid Morita equivalence,
- the morphism

$$HP^*(C_c^\infty((M \times_P M) \rtimes \mathscr{G})) \xrightarrow{(e_1)^*} HP^*(\mathsf{S}_c(G_M^{tan} \rtimes \mathscr{G})) \tag{4.26}$$

  induced from the evaluation at $t = 1$ morphism,
- the isomorphism

$$HP^*(\mathsf{S}_c(G_M^{tan} \rtimes \mathscr{G})) \xrightarrow{(e_0^*)^{-1}} HP^*(\mathsf{S}_c(T^\pi M \rtimes \mathscr{G})) \tag{4.27}$$

  induced from the evaluation at $t = 0$ morphism,

- the isomorphism

$$HP^*(\mathsf{S}_c(T^\pi M \rtimes \mathscr{G})) \xrightarrow[\approx]{\mathcal{F}} HP^*(\mathsf{S}_c(T^*_\pi M \rtimes \mathscr{G})) \tag{4.28}$$

induced from the Fourier algebra isomorphism $\mathsf{S}_c(T_\pi M \rtimes \mathscr{G}) \approx \mathsf{S}(T^*_\pi M \rtimes \mathscr{G})$,

- the equivariant Thom isomorphism

$$HP^*(\mathsf{S}_c(T^*_\pi M \rtimes \mathscr{G})) \xrightarrow[\approx]{Th} HP^*(C_c^\infty(M \rtimes \mathscr{G})) \tag{4.29}$$

By a direct computation, for getting the equality (4.23), it is enough to prove the following two points

A. For every $\omega \in HP_*(M \rtimes \mathscr{G})$ and for every $\delta \in HP^*(C_c^\infty(\mathscr{G}))$ the following equality holds

$$\langle \omega, \pi_M^*(\delta) \rangle = \langle \pi_M!(\omega), \delta \rangle. \tag{4.30}$$

B. For every $x \in K^*(M \times \mathscr{G})$ the following equality holds

$$\varphi_\tau(\pi_M!(x)) = \langle x, \pi_M^*(\tau) \rangle \tag{4.31}$$

where $\langle x, \pi_M^*(\tau) \rangle := \langle Ch_{CO}^M(x), \pi_M^*(\tau) \rangle$.

Proof of A. Since both morphisms, $\pi_M!$ and π_M^* are defined by compositions of Thom isomorphisms, Morita isomorphisms and morphisms induced from algebra morphisms, the equality (4.30) follows immediately from the following general properties of Connes' pairing:

Naturality: Given a morphism of algebras $\phi : A \to B$ we have that

$$\langle \phi_*(a), b \rangle = \langle a, \phi^*(b) \rangle \tag{4.32}$$

for every $a \in HP_*(A)$ and every $b \in HP^*(B)$. This property is quite standard and holds in all generality.

Thom invariance: Given a $\mathscr{G}$-*spin*^{c} vector bundle $E \to M$ we have that for every $\omega \in HP_*(M \rtimes \mathscr{G})$ and every $\delta \in HP_{*+r_E}(\mathscr{S}_c(E \rtimes \mathscr{G}))$ the following equality of Connes pairings holds

$$\langle \omega, \mathcal{T}^E_{\mathscr{G}}(\delta) \rangle = \langle \mathcal{T}^{\mathscr{G}}_E(\omega), \delta \rangle. \tag{4.33}$$

This property holds by hypothesis since we are assuming $\mathscr{G}$ to be a HP-Thom Lie groupoid.

Morita invariance: For a $\mathscr{G}$-manifold M as above if we denote by

$$M_{HP_*} : HP_*(C_c^\infty((M \times_P M) \rtimes \mathscr{G})) \longrightarrow HP_*(C_c^\infty(\mathscr{G})) \tag{4.34}$$

and by

$$M_{HP^*} : HP^*(C_c^\infty(\mathscr{G})) \longrightarrow HP^*(C_c^\infty((M \times_P M) \rtimes \mathscr{G})) \tag{4.35}$$

the canonical isomorphisms induced from the Lie groupoid Morita equivalence

$$(M \times_P M) \rtimes \mathscr{G} \longrightarrow \mathscr{G},$$

we have that the following equality holds

$$\langle M_{HP_*}(\mu), \rho \rangle = \langle \mu, M_{HP^*}(\rho) \rangle \tag{4.36}$$

for every $\mu \in HP_*(C_c^\infty((M \times_P M) \rtimes \mathscr{G}))$ and every $\rho \in HP^*(C_c^\infty(\mathscr{G}))$. This property holds, by a direct computation, at least in this case since the morphism induced at the level of cycles consists in integrating the $(M \times_P M)$ part along the diagonal. The fact that these are isomorphism follows from proposition 3.6 in [**21**].

Proof of B. By the properties above, naturality, Morita invariance, and Thom invariance of the Connes pairing, we have the following commutative diagram (4.37)

$$
\begin{array}{ccccc}
K_*(C^*(M \rtimes \mathscr{G})) & \xleftarrow[\cong]{j} & K_*(C_c^\infty(M \rtimes \mathscr{G})) & \xrightarrow{\langle \cdot, \pi_M^*(\tau) \rangle} & \mathbb{C} \\
{\scriptstyle Th}\downarrow{\scriptstyle \cong} & & {\scriptstyle \cong}\downarrow{\scriptstyle Th} & & \downarrow{\scriptstyle =} \\
K_*(C^*(T_\pi M \rtimes \mathscr{G})) & \xleftarrow[\cong]{j} & K_*(\mathscr{S}_c(T_\pi M \rtimes \mathscr{G})) & \xrightarrow{\langle \cdot, (e_0^*)^{-1}(e_1^*(M_{HP^*}(\tau))) \rangle} & \mathbb{C} \\
{\scriptstyle e_0}\uparrow{\scriptstyle \cong} & & \uparrow{\scriptstyle e_0} & & \downarrow{\scriptstyle =} \\
K_*(C^*(D_{\pi_M} \rtimes \mathscr{G})) & \xleftarrow{j} & K_*(\mathscr{S}_c(D_{\pi_M} \rtimes \mathscr{G})) & \xrightarrow{\langle \cdot, e_1^*(M_{HP^*}(\tau)) \rangle} & \mathbb{C} \\
{\scriptstyle e_1}\downarrow & & {\scriptstyle e_1}\downarrow & & \downarrow{\scriptstyle =} \\
K_*(C^*((M \times_P M) \rtimes \mathscr{G})) & \xleftarrow{j} & K_*(C_c^\infty((M \times_P M) \rtimes \mathscr{G})) & \xrightarrow{\langle \cdot, M_{HP^*}(\tau) \rangle} & \mathbb{C} \\
{\scriptstyle M_K}\downarrow{\scriptstyle \cong} & & \downarrow{\scriptstyle M_{K_\infty}} & & \downarrow{\scriptstyle =} \\
K_*(C^*(\mathscr{G})) & \xleftarrow{j} & K_*(C_c^\infty(\mathscr{G})) & \xrightarrow{\langle \cdot, \tau \rangle} & \mathbb{C}
\end{array}
$$

with the curved arrows $\pi_M!: K_*(C^*(M \rtimes \mathscr{G})) \to K_*(C^*(\mathscr{G}))$ and $\varphi_\tau: K_*(C^*(\mathscr{G})) \to \mathbb{C}$,

where the morphisms denoted by j are the morphisms induced by the canonical inclusion of algebras, and where we are using the hypothesis that φ_τ extends the pairing morphism given by τ, *i.e.* the commutativity of the last bottom diagram. Also we recall that the two first morphisms j on the top are isomorphisms because M is a $\mathscr{G}$-proper *spin*^{c} manifold. Now, the proof of **B** follow by direct diagram chasing if the morphisms

$$(e_0)_* : K_*(\mathscr{S}_c(D_{\pi_M} \rtimes \mathscr{G})) \longrightarrow K_*(\mathscr{S}_c(T_\pi M \rtimes \mathscr{G})) \tag{4.38}$$

are surjective. This last statement follows from the fact that we are assuming that $T_\pi M$ is a $\mathscr{G}$-*spin*^{c} proper vector bundle, in particular with an invariant $\mathscr{G}$−metric. Indeed, one can use the local structure of the deformation to the normal cone D_{π_M} around $T_\pi M$ as in appendix C of [**9**]. This completes the proof of B and of the theorem. □

One of the main interests of the theorem above is that the right hand side of the formula above always makes sense, for every (HP-Thom) Lie groupoid and for every periodic cyclic cocycle.

Acknowledgments

I want to thank my collaborators and friends Bai-Ling Wang and Hang Wang for several years of very stimulating and rich mathematical exchanges and collaborations. I want as well take this occasion to deeply thank Alain Connes for his work on Noncommutative geometry and in particular for his main role in the discovery of cyclic cohomology.

References

[1] Paul Baum and Alain Connes, *Chern character for discrete groups*, A fête of topology, Academic Press, Boston, MA, 1988, pp. 163–232, DOI 10.1016/B978-0-12-480440-1.50015-0. MR928402

[2] Paul Baum and Alain Connes, *K-theory for discrete groups*, Operator algebras and applications, Vol. 1, London Math. Soc. Lecture Note Ser., vol. 135, Cambridge Univ. Press, Cambridge, 1988, pp. 1–20, DOI 10.1007/978-1-4612-3762-4_1. MR996437

[3] Paul Baum and Alain Connes, *Geometric K-theory for Lie groups and foliations*, Enseign. Math. (2) **46** (2000), no. 1-2, 3–42. MR1769535

[4] Paul Baum, Alain Connes, and Nigel Higson, *Classifying space for proper actions and K-theory of group C^*-algebras*, C^*-algebras: 1943–1993 (San Antonio, TX, 1993), Contemp. Math., vol. 167, Amer. Math. Soc., Providence, RI, 1994, pp. 240–291, DOI 10.1090/conm/167/1292018. MR1292018

[5] Moulay Benameur, *The gromov-lawson index and the baum-connes assembly map*, arXiv:2103.03495v1, 2021.

[6] Moulay-Tahar Benameur and Indrava Roy, *The Higson-Roe sequence for étale groupoids. I. Dual algebras and compatibility with the BC map*, J. Noncommut. Geom. **14** (2020), no. 1, 25–71, DOI 10.4171/JNCG/358. MR4107510

[7] Jonathan Block and Ezra Getzler, *Equivariant cyclic homology and equivariant differential forms*, Ann. Sci. École Norm. Sup. (4) **27** (1994), no. 4, 493–527. MR1290397

[8] Paulo Carrillo Rouse, *A Schwartz type algebra for the tangent groupoid*, K-theory and noncommutative geometry, EMS Ser. Congr. Rep., Eur. Math. Soc., Zürich, 2008, pp. 181–199, DOI 10.4171/060-1/7. MR2513337

[9] P. Carrillo Rouse, B.L. Wang, and H. Wang, *Topological k-theory for discrete groups and index theory*, arXiv:2012.12359 [math.KT], 2020.

[10] Paulo Carrillo Rouse, *Compactly supported analytic indices for Lie groupoids*, J. K-Theory **4** (2009), no. 2, 223–262, DOI 10.1017/is008003015jkt069. MR2551909

[11] Paulo Carrillo Rouse and Bai-Ling Wang, *Geometric Baum-Connes assembly map for twisted differentiable stacks* (English, with English and French summaries), Ann. Sci. Éc. Norm. Supér. (4) **49** (2016), no. 2, 277–323, DOI 10.24033/asens.2283. MR3481351

[12] X. Chen, J. Wang, Z. Xie, and G. Yu, *Delocalized eta invariants, cyclic cohomology and higher rho invariants*, arXiv:1901.02378v2, 2019.

[13] Alain Connes, *Noncommutative geometry*, Academic Press, Inc., San Diego, CA, 1994. MR1303779

[14] Robin J. Deeley and Magnus Goffeng, *Relative geometric assembly and mapping cones Part II: Chern characters and the Novikov property*, Münster J. Math. **12** (2019), no. 1, 57–92, DOI 10.17879/85169762441. MR3928083

[15] Nigel Higson and John Roe, *K-homology, assembly and rigidity theorems for relative eta invariants*, Pure Appl. Math. Q. **6** (2010), no. 2, Special Issue: In honor of Michael Atiyah and Isadore Singer, 555–601, DOI 10.4310/PAMQ.2010.v6.n2.a11. MR2761858

[16] Michel Hilsum and Georges Skandalis, *Morphismes K-orientés d'espaces de feuilles et fonctorialité en théorie de Kasparov (d'après une conjecture d'A. Connes)* (French, with English summary), Ann. Sci. École Norm. Sup. (4) **20** (1987), no. 3, 325–390. MR925720

[17] Peter Hochs, Yanli Song, and Xiang Tang, *An index theorem for higher orbital integrals*, Math. Ann. **382** (2022), no. 1-2, 169–202, DOI 10.1007/s00208-021-02233-3. MR4377301

[18] Masoud Khalkhali, *Basic noncommutative geometry*, 2nd ed., EMS Series of Lectures in Mathematics, European Mathematical Society (EMS), Zürich, 2013, DOI 10.4171/128. MR3134494

[19] Matthias Kreck and Wolfgang Lück, *The Novikov conjecture*, Oberwolfach Seminars, vol. 33, Birkhäuser Verlag, Basel, 2005. Geometry and algebra, DOI 10.1007/b137100. MR2117411

[20] Pierre-Yves Le Gall, *Théorie de Kasparov équivariante et groupoïdes. I* (French, with English and French summaries), K-Theory **16** (1999), no. 4, 361–390, DOI 10.1023/A:1007707525423. MR1686846

[21] Denis Perrot. Local index thoeory for certain fourier integral operators on lie groupoids. arXiv:1401.0225v2, 2014.

[22] Markus J. Pflaum, Hessel Posthuma, and Xiang Tang, *The transverse index theorem for proper cocompact actions of Lie groupoids*, J. Differential Geom. **99** (2015), no. 3, 443–472. MR3316973

[23] P. Piazza, T. Schick, and V. Zenobi. Higher rho numbers and the mapping of analytic surgery to homology. arXiv:1905.11861v2, 2019.

[24] Y. Song and X. Tang. Higher rho numbers and the mapping of analytic surgery to homology. arXiv:1910.00175, 2019.

[25] Jean-Louis Tu and Ping Xu, *Chern character for twisted K-theory of orbifolds*, Adv. Math. **207** (2006), no. 2, 455–483, DOI 10.1016/j.aim.2005.12.001. MR2271013

[26] Shmuel Weinberger, Zhizhang Xie, and Guoliang Yu, *Additivity of higher rho invariants and nonrigidity of topological manifolds*, Comm. Pure Appl. Math. **74** (2021), no. 1, 3–113, DOI 10.1002/cpa.21962. MR4178180

[27] Vito Felice Zenobi, *Adiabatic groupoid and secondary invariants in K-theory*, Adv. Math. **347** (2019), 940–1001, DOI 10.1016/j.aim.2019.03.003. MR3922452

Institut de Mathématiques de Toulouse, Université Toulouse III, 118, Route de Narbonne, 31400 Toulouse, France.

Email address: paulo.carrillo@math.univ-toulouse.fr

Proceedings of Symposia in Pure Mathematics
Volume **105**, 2023
https://doi.org/10.1090/pspum/105/01896

Hochschild homology, trace map and ζ-cycles

Alain Connes and Caterina Consani

Abstract. In this paper we consider two spectral realizations of the zeros of the Riemann zeta function. The first one involves all non-trivial (*i.e.* non-real) zeros and is expressed in terms of a Laplacian intimately related to the prolate wave operator. The second spectral realization affects only the critical zeros and it is cast in terms of sheaf cohomology. The novelty is that the base space is the Scaling Site playing the role of the parameter space for the ζ-cycles and encoding their stability by coverings.

1. Introduction

In this paper we give a Hochschild homological interpretation of the zeros of the Riemann zeta function. The root of this result is in the recognition that the map

$$(\mathcal{E}f)(u) = u^{1/2} \sum_{n>0} f(nu)$$

which is defined on a suitable subspace of the linear space of complex-valued even Schwartz functions on the real line, is a trace in Hochschild homology, if one brings in the construction the projection $\pi : \mathbb{A}_{\mathbb{Q}} \to \mathbb{Q}^\times \backslash \mathbb{A}_{\mathbb{Q}}$ from the rational adèles to the adèle classes (see Section 3). In this paper, we shall consider two spectral realizations of the zeros of the Riemann zeta function. The first one involves all non-trivial (*i.e.* non-real) zeros and is expressed in terms of a Laplacian intimately related to the prolate wave operator (see Section 4). The second spectral realization is sharper since it affects only the critical zeros. The main players are here the ζ-cycles introduced in [**7**], and the Scaling Site [**6**] as their parameter space, which encodes their stability by coverings. The ζ-cycles give the theoretical geometric explanation for the striking coincidence between the low lying spectrum of a perturbed spectral triple therein introduced (see [**7**]), and the low lying (critical) zeros of the Riemann zeta function. The definition of a ζ-cycle derives, as a by-product, from scale-invariant Riemann sums for complex-valued functions on the real half-line $[0, \infty)$ with vanishing integral. For any $\mu \in \mathbb{R}_{>1}$, one implements the

---

2020 *Mathematics Subject Classification.* Primary 11M55, 11M06, 16E40; Secondary 11F72, 58B34.

Key words and phrases. Riemann zeta function, spectral realization, zeta cycle, Hochschild homology, adèle class space, Scaling Site.

The second author is partially supported by the Simons Foundation collaboration grant n. 691493.

linear (composite) map

$$\Sigma_\mu \mathcal{E} : \mathcal{S}_0^{\text{ev}} \to L^2(C_\mu)$$

from the Schwartz space $\mathcal{S}_0^{\text{ev}}$ of real valued even functions f on the real line, with $f(0) = 0$, and vanishing integral, to the Hilbert space $L^2(C_\mu)$ of square integrable functions on the circle $C_\mu = \mathbb{R}_+^*/\mu^{\mathbb{Z}}$ of length $L = \log \mu$, where

$$(\Sigma_\mu g)(u) := \sum_{k\in\mathbb{Z}} g(\mu^k u).$$

The map Σ_μ commutes with the scaling action $\mathbb{R}_+^* \ni \lambda \mapsto f(\lambda^{-1}x)$ on functions, while $\mathcal{E}$ is invariant under a normalized scaling action on $\mathcal{S}_0^{\text{ev}}$. In this set-up one has

DEFINITION. A **ζ-cycle** is a circle C of length $L = \log \mu$ whose Hilbert space $L^2(C)$ contains $\Sigma_\mu \mathcal{E}(\mathcal{S}_0^{\text{ev}})$ as a non dense subspace.

Next result is known (see [**7**] Theorem 6.4)

THEOREM 1.1. *The following facts hold*

(i) *The spectrum of the scaling action of $\mathbb{R}_+^*$ on the orthogonal space to $\Sigma_\mu \mathcal{E}(\mathcal{S}_0^{\text{ev}})$ in $L^2(C_\mu)$ is contained in the set of the imaginary parts of the zeros of the Riemann zeta function $\zeta(z)$ on the critical line $\Re(z) = \frac{1}{2}$.*

(ii) *Let $s > 0$ be a real number such that $\zeta(\frac{1}{2} + is) = 0$. Then any circle C whose length is an integral multiple of $\frac{2\pi}{s}$ is a ζ-cycle, and the spectrum of the action of $\mathbb{R}_+^*$ on $(\Sigma_\mu \mathcal{E}(\mathcal{S}_0^{\text{ev}}))^\perp$ contains s.*

Theorem 1.1 states that for a countable and dense set of values of $L \in \mathbb{R}_{>0}$, the Hilbert spaces $\mathcal{H}(L) := (\Sigma_\mu \mathcal{E}(\mathcal{S}_0^{\text{ev}}))^\perp$ are non-trivial and, more importantly, that as L varies in that set, the spectrum of the scaling action of $\mathbb{R}_+^*$ on the family of the $\mathcal{H}(L)$'s is the set Z of imaginary parts of critical zeros of the Riemann zeta function. In fact, given the proven stability of ζ-cycles under coverings, the same element of Z occurs infinitely many times in the family of the $\mathcal{H}(L)$'s. This stability under coverings displays the Scaling Site $\mathscr{S} = [0,\infty) \rtimes \mathbb{N}^\times$ as the natural parameter space for the ζ-cycles. In this paper, we show (see Section 5) that after organizing the family $\mathcal{H}(L)$ as a sheaf over $\mathscr{S}$ and using sheaf cohomology, one obtains a spectral realization of critical zeros of the Riemann zeta function. The key operation in the construction of the relevant arithmetic sheaf is given by the action of the multiplicative monoid $\mathbb{N}^\times$ on the sheaf of smooth sections of the bundle L^2 determined by the family of Hilbert spaces $L^2(C_\mu)$, $\mu = \exp L$, as L varies in $(0,\infty)$. For each $n \in \mathbb{N}^\times$ there is a canonical covering map $C_{\mu^n} \to C_\mu$, where the action of n corresponds to the operation of sum on the preimage of a point in C_μ under the covering. This action turns the (sub)sheaf of smooth sections vanishing at $L = 0$ into a sheaf $\mathcal{L}^2$ over $\mathscr{S}$. The family of subspaces $\Sigma_\mu \mathcal{E}(\mathcal{S}_0^{\text{ev}}) \subset L^2(C_\mu)$ generates a closed subsheaf $\overline{\Sigma\mathcal{E}} \subset \mathcal{L}^2$ and one then considers the cohomology of the related quotient sheaf $\mathcal{L}^2/\overline{\Sigma\mathcal{E}}$. Because of the property of $\mathbb{R}_+^*$-equivariance under scaling, this construction determines a spectral realization of critical zeros of the Riemann zeta function, also taking care of eventual multiplicities. Our main result is the following

THEOREM 1.2. *The cohomology $H^0(\mathscr{S}, \mathcal{L}^2/\overline{\Sigma\mathcal{E}})$ endowed with the induced canonical action of $\mathbb{R}_+^*$ is isomorphic to the spectral realization of critical zeros of the Riemann zeta function, given by the action of $\mathbb{R}_+^*$, via multiplication with λ^{is}, on* *the quotient of the Schwartz space* $\mathcal{S}(\mathbb{R})$ *by the closure of the ideal generated by multiples of* $\zeta\left(\frac{1}{2}+is\right)$.

This paper is organized as follows. Section 2 recalls the main role played by the (image of the) map $\mathcal{E}$ in the study of the spectral realization of the critical zeros of the Riemann zeta function. In Section 3 we show the identification of the Hochschild homology HH_0 of the noncommutative space $\mathbb{Q}^\times\backslash\mathbb{A}_\mathbb{Q}$ with the coinvariants for the action of $\mathbb{Q}^\times$ on the Schwartz algebra, using the (so-called) "wrong way" functoriality map $\pi_!$ associated to the projection $\pi:\mathbb{A}_\mathbb{Q}\to\mathbb{Q}^\times\backslash\mathbb{A}_\mathbb{Q}$. We also stress the relevant fact that the Fourier transform on adèles becomes canonical after passing to HH_0 of the adèle class space of the rationals. The key Proposition 3.3 describes the invariant part of such HH_0 as the space of even Schwartz functions on the real line and identifies the trace map with the map $\mathcal{E}$. Section 4 takes care of the two vanishing conditions implemented in the definition of $\mathcal{E}$ and introduces the operator $\Delta=H(1+H)$ (H being the generator of the scaling action of $\mathbb{R}_+^*$ on $\mathcal{S}(\mathbb{R})^{\mathrm{ev}}$) playing the role of the Laplacian and intimately related to the prolate operator. Finally, Section 5 is the main technical section of this paper since it contains the proof of Theorem 1.2.

2. The map $\mathcal{E}$ and the zeros of the zeta function

The adèle class space of the rationals $\mathbb{Q}^\times\backslash\mathbb{A}_\mathbb{Q}$ is the natural geometric framework to understand the Riemann-Weil explicit formulas for L-functions as a trace formula [**3**]. The essence of this result lies mainly in the delicate computation of the principal values involved in the distributions appearing in the geometric (right-hand) side of the semi-local trace formula of *op.cit.* (see Theorem 4 for the notations)

$$\mathrm{Trace}(R_\Lambda U(h)) = h(1)\int_{\substack{\lambda\in C_S\\ |\lambda|\in[\Lambda^{-1},\Lambda]}} d^*\lambda + \sum_{\nu\in S}\int'_{\mathbb{Q}_\nu^*}\frac{h(u^{-1})}{|1-u|}d^*u + o(1) \qquad \text{for } \Lambda\to\infty$$

(later recast in the softer context of [**13**]). There is a rather simple analogy related to the spectral (left-hand) side of the explicit formulas for a global field K (see [**4**] Section 2 for the notations)

$$\hat h(0)+\hat h(1) - \sum_{\chi\in\widehat{C_{K,1}}}\sum_{\rho\in Z_{\tilde\chi}}\hat h(\tilde\chi,\rho) = \sum_\nu\int'_{K_\nu^*}\frac{h(u^{-1})}{|1-u|}d^*u$$

which may help one to realize how the sum over the zeros of the zeta function appears. Here this relation is simply explained. Given a complex-valued polynomial $P(x)\in\mathbb{C}[x]$, one may identify the set of its zeros as the spectrum of the endomorphism T of multiplication by the variable x computed in the quotient algebra $\mathbb{C}[x]/(P(x))$. It is well known that the matrix of T, in the basis of powers of x, is the companion matrix of $P(x)$. Furthermore, the trace of its powers, readily computed from the diagonal terms of powers of the companion matrix in terms of the coefficients of $P(x)$, gives the Newton-Girard formulae.[1] If one transposes this result to the case of the Riemann zeta function $\zeta(s)$, one sees that the multiplication

[1]This is an efficient way to find the power sum of roots of $P(x)$ without actually finding the roots explicitly. Newton's identities supply the calculation via a recurrence relation with known coefficients.

by $P(x)$ is replaced here with the map

$$\mathcal{E}(f)(u) := u^{1/2} \sum_{n=1}^{\infty} f(nu), \tag{2.1}$$

while the role of T (the multiplication by the variable) is played by the scaling operator $u\partial_u$. These statements may become more evident if one brings in the Fourier transform. Indeed, let $f \in \mathcal{S}(\mathbb{R})^{\text{ev}}$ be an even Schwartz function and let $w(f)(u) = u^{1/2}f(u)$ be the unitary identification of f with a function in $L^2(\mathbb{R}_+^*, d^*u)$, where $d^*u := du/u$ denotes the Haar measure. Then, by composing w with the (multiplicative) Fourier transform $\mathbb{F} : L^2(\mathbb{R}_+^*, d^*u) \to L^2(\mathbb{R})$, $\mathbb{F}(h)(s) = \int_{\mathbb{R}_+^*} h(u)u^{-is}d^*u$ one obtains

$$\mathbb{F}(w(f)) = \psi, \qquad \psi(z) = \int_{\mathbb{R}_+^*} f(u)u^{\frac{1}{2}-iz}d^*u.$$

The function $\psi(z)$ is holomorphic in the complex half-plane $\mathfrak{H} = \{z \in \mathbb{C} \mid \Im(z) > -\frac{1}{2}\}$ since $f(u) = O(u^{-N})$ for $u \to \infty$. Moreover, for $n \in \mathbb{N}$, one has

$$\int_{\mathbb{R}_+^*} u^{1/2}f(nu)u^{-iz}d^*u = n^{-1/2+iz}\int_{\mathbb{R}_+^*} v^{1/2}f(v)v^{-iz}d^*v.$$

In the region $\Im(z) > \frac{1}{2}$ one derives, by applying Fubini theorem, the following equality

$$\int_{\mathbb{R}_+^*} \sum_{n=1}^{\infty} u^{1/2}f(nu)u^{-iz}d^*u = \left(\sum_{n=1}^{\infty} n^{-1/2+iz}\right)\int_{\mathbb{R}_+^*} v^{1/2}f(v)v^{-iz}d^*v.$$

Thus, for all $z \in \mathbb{C}$ with $\Im(z) > \frac{1}{2}$ one obtains

$$\int_{\mathbb{R}_+^*} \mathcal{E}(f)(u)u^{-iz}d^*u = \zeta(\frac{1}{2} - iz)\psi(z). \tag{2.2}$$

If one assumes now that the Schwartz function f fulfills $\int_{\mathbb{R}} f(x)dx = 0$, then $\psi(\frac{i}{2}) = 0$. Both sides of (2.2) are holomorphic functions in $\mathfrak{H}$: for the integral on the left-hand side, this can be seen by using the estimate $\mathcal{E}(f)(u) = O(u^{1/2})$ that follows from the Poisson formula. This proves that (2.2) continues to hold also in the complex half-plane $\mathfrak{H}$. Thus one sees that the zeros of $\zeta(\frac{1}{2} - iz)$ in the strip $|\Im(z)| < \frac{1}{2}$ are the common zeros of all functions

$$\mathbb{F}(\mathcal{E}(f))(z), \qquad f \in \mathcal{S}(\mathbb{R})_1^{\text{ev}} := \{f \in \mathcal{S}(\mathbb{R})^{\text{ev}} \mid \int_{\mathbb{R}} f(x)dx = 0\}.$$

One may eventually select the even Schwartz function $f(x) = e^{-\pi x^2}\left(2\pi x^2 - 1\right)$ to produce a specific instance where the zeros of $\mathbb{F}(\mathcal{E}(f))$ are exactly the non-trivial zeros of $\zeta(\frac{1}{2} - iz)$, since in this case $\psi(z) = \frac{1}{4}\pi^{-\frac{1}{4}+\frac{iz}{2}}(-1 - 2iz)\Gamma\left(\frac{1}{4} - \frac{iz}{2}\right)$.

3. Geometric interpretation

In this section we continue the study of the map $\mathcal{E}$ to achieve a geometric understanding of it. This is obtained by bringing in the construction the adèle class space of the rationals, whose role is to grant for the replacement, in (2.1), of the summation over the monoid $\mathbb{N}^\times$ with the summation over the group $\mathbb{Q}^\times$. Then, up to the factor $u^{1/2}$, $\mathcal{E}$ is understood as the composite $\iota^* \circ \pi_!$, where the map $\iota : \mathbb{Q}^\times\backslash\mathbb{A}_{\mathbb{Q}}^*/\hat{\mathbb{Z}}^\times \to \mathbb{Q}^\times\backslash\mathbb{A}_{\mathbb{Q}}/\hat{\mathbb{Z}}^\times$ is the inclusion of idèle classes in adèle classes and $\pi : \mathbb{A}_{\mathbb{Q}}/\hat{\mathbb{Z}}^\times \to \mathbb{Q}^\times\backslash\mathbb{A}_{\mathbb{Q}}/\hat{\mathbb{Z}}^\times$ is induced by the projection $\mathbb{A}_{\mathbb{Q}} \to \mathbb{Q}^\times\backslash\mathbb{A}_{\mathbb{Q}}$. We shall discuss the following diagram

$$\begin{array}{ccc} Y = \mathbb{A}_{\mathbb{Q}}/\hat{\mathbb{Z}}^\times & & \\ \big\downarrow{\scriptstyle \pi} & & \\ X = \mathbb{Q}^\times\backslash\mathbb{A}_{\mathbb{Q}}/\hat{\mathbb{Z}}^\times & \xleftarrow{\ \iota\ } & \mathbb{Q}^\times\backslash\mathbb{A}_{\mathbb{Q}}^\times/\hat{\mathbb{Z}}^\times = \mathbb{R}_+^* \end{array}$$

The conceptual understanding of the map $\pi_!$ uses the Hochschild homology of noncommutative algebras.

We recall that the space of adèle classes *i.e.* the quotient $\mathbb{Q}^\times\backslash\mathbb{A}_{\mathbb{Q}}$ is encoded algebraically by the cross-product algebra

$$\mathcal{A} := \mathcal{S}(\mathbb{A}_{\mathbb{Q}}) \rtimes \mathbb{Q}^\times.$$

The Schwartz space $\mathcal{S}(\mathbb{A}_{\mathbb{Q}})$ is acted upon by (automorphisms of) $\mathbb{Q}^\times$ corresponding to the scaling action of $\mathbb{Q}^\times$ on rational adèles. An element of $\mathcal{A}$ is written symbolically as a finite sum

$$\sum a(q)U(q), \qquad a(q) \in \mathcal{S}(\mathbb{A}_{\mathbb{Q}}).$$

From the inclusion of algebras $\mathcal{S}(\mathbb{A}_{\mathbb{Q}}) \subset \mathcal{S}(\mathbb{A}_{\mathbb{Q}}) \rtimes \mathbb{Q}^\times = \mathcal{A}$ one derives a corresponding morphism of Hochschild homologies

$$\pi_! : \mathrm{HH}(\mathcal{S}(\mathbb{A}_{\mathbb{Q}})) \longrightarrow \mathrm{HH}(\mathcal{A}).$$

Here, we use the shorthand notation $\mathrm{HH}(A) := \mathrm{HH}(A, A)$ for the Hochschild homology of an algebra A with coefficients in the bimodule A. In noncommutative geometry, the vector space of differential forms of degree k is replaced by the Hochschild homology $\mathrm{HH}_k(A)$. If the algebra A is commutative and for $k = 0$, $\mathrm{HH}_0(A) = A$, so that 0-forms are identified with functions. Indeed, the Hochschild boundary map

$$b : A^{\otimes 2} \to A, \qquad b(x \otimes y) = xy - yx$$

is identically zero when the algebra A is commutative. This result does not hold when $A = \mathcal{A}$, since $\mathcal{A} = \mathcal{S}(\mathbb{A}_{\mathbb{Q}}) \rtimes \mathbb{Q}^\times$ is no longer commutative. It is therefore meaningful to bring in the following

PROPOSITION 3.1. *The kernel of $\pi_! : HH_0(\mathcal{S}(\mathbb{A}_{\mathbb{Q}})) \to HH_0(\mathcal{A})$ is the $\mathbb{C}$-linear span E of functions $f - f_q$, with $f \in \mathcal{S}(\mathbb{A}_{\mathbb{Q}})$, $q \in \mathbb{Q}^\times$, and where we set $f_q(x) := f(qx)$.*

PROOF. For any $f, g \in \mathcal{S}(\mathbb{A}_{\mathbb{Q}})$ and $q \in \mathbb{Q}^\times$ one has

$$fg - (fg)_q = fg - U(q)fgU(q^{-1}) = xy - yx, \qquad x := fU(q^{-1}),\ y := U(q)g. \tag{3.1}$$

One knows ([**14**] Lemma 1) that any function $f \in \mathcal{S}(\mathbb{R})$ is a product of two elements of $\mathcal{S}(\mathbb{R})$. Moreover, an element of the Bruhat-Schwartz space $\mathcal{S}(\mathbb{A}_{\mathbb{Q}})$ is a finite linear combination of functions of the form $e \otimes f$, with $e^2 = e$. Thus any $f \in \mathcal{S}(\mathbb{A}_{\mathbb{Q}})$ can be written as a finite sum of products of two elements of $\mathcal{S}(\mathbb{A}_{\mathbb{Q}})$, so that (3.1) entails $f - f_q \in \ker \pi_!$. Conversely, let $f \in \ker \pi_!$. Then there exists a finite number of pairs $x_i, y_i \in \mathcal{A}$ such that $f = \sum[x_i, y_i]$. Let $P : \mathcal{A} \to \mathcal{S}(\mathbb{A}_{\mathbb{Q}})$ be the projection on the coefficient $a(1)$ of $U(1) = 1$ *i.e.*

$$P\left(\sum a(q)U(q)\right) := a(1).$$

Then $f = \sum[x_i, y_i]$ implies $f = \sum P([x_i, y_i])$. We shall prove that for any pair $x, y \in \mathcal{A}$ one has $P([x,y]) \in E$. Indeed, one has

$$x = \sum a(q)U(q),\ y = \sum b(q')U(q'),\ [x,y] = \sum(a(q)b(q')_q - b(q')a(q)_{q'})U(qq')$$

so that

$$P([x,y]) = \sum(a(q)b(q^{-1})_q - b(q^{-1})a(q)_{q^{-1}}).$$

This projection belongs to E because

$$a(q)b(q^{-1})_q - b(q^{-1})a(q)_{q^{-1}} = h_q - h, \qquad h = b(q^{-1})a(q)_{q^{-1}}.$$

This completes the proof. □

Proposition 3.1 shows that the image of $\pi_! : \mathrm{HH}_0(\mathcal{S}(\mathbb{A}_\mathbb{Q})) \to \mathrm{HH}_0(\mathcal{A})$ is the space of coinvariants for the action of $\mathbb{Q}^\times$ on $\mathcal{S}(\mathbb{A}_\mathbb{Q})$, *i.e.* the quotient of $\mathcal{S}(\mathbb{A}_\mathbb{Q})$ by the subspace E.

An important point now to remember is that the Fourier transform becomes canonically defined on the above quotient. Indeed, the definition of the Fourier transform on adèles depends on the choice of a non-trivial character α on the additive, locally compact group $\mathbb{A}_\mathbb{Q}$, which is trivial on the subgroup $\mathbb{Q} \subset \mathbb{A}_\mathbb{Q}$. It is defined as follows

$$\mathbb{F}_\alpha(f)(y) := \int_\mathbb{R} f(x)\alpha(xy)dx.$$

The space of characters of the compact group $G = \mathbb{A}_\mathbb{Q}/\mathbb{Q}$ is one dimensional as a $\mathbb{Q}$-vector space, thus any non-trivial character α as above is of the form $\beta(x) = \alpha(qx)$, so that

$$\mathbb{F}_\beta(f)(y) := \int_\mathbb{R} f(x)\alpha(qxy)dx = \mathbb{F}_\alpha(f)_q(y).$$

Therefore, the difference $\mathbb{F}_\beta - \mathbb{F}_\alpha$ vanishes on the quotient of $\mathcal{S}(\mathbb{A}_\mathbb{Q})$ by E and this latter space is preserved by $\mathbb{F}_\alpha$ since $\mathbb{F}_\alpha(f_q) = \mathbb{F}_\alpha(f)_{q^{-1}}$.

3.1. HH, Morita invariance and the trace map. Let us recall that given an algebra A, the trace map

$$\mathrm{Tr} : M_n(A) \to A, \qquad \mathrm{Tr}((a_{ij}) := \sum a_{jj}$$

induces an isomorphism in degree zero Hochschild homology which extends to higher degrees. If A is a convolution algebra of the étale groupoid of an equivalence relation $\mathcal{R}$ with countable orbits on a space Y, and $\pi : Y \to Y/\mathcal{R}$ is the quotient map, the trace map takes the following form

$$\mathrm{Tr}(f)(x) := \sum_{\pi(j)=x} f(j,j). \tag{3.2}$$

The trace induces a map on HH_0 of the function algebras, provided one takes care of the convergence issue when the size of equivalence classes is infinite. If the relation $\mathcal{R}$ is associated with the orbits of the free action of a discrete group Γ on a locally compact space Y, the convolution algebra is the cross product of the algebra of functions on Y by the discrete group Γ. In this case, the étale groupoid is $Y \rtimes \Gamma$, where the source and range maps are given *resp.* by $s(y,g) = y$ and $r(y,g) = gy$. The elements of the convolution algebra are functions $f(y,g)$ on $Y \rtimes \Gamma$. The diagonal terms in (3.2) correspond to the elements of $Y \rtimes \Gamma$ such that

$s(y,g) = r(y,g)$, meaning that $g = 1$ is the neutral element of Γ, since the action of Γ is assumed to be free. Then, the trace map is

$$\mathrm{Tr}((f)(x) = \sum_{\pi(y)=x} f(y,1).$$

This sum is meaningful in the space of the proper orbits of Γ. For a lift $\rho(x) \in Y$, with $\pi(\rho(x)) = x$ the trace reads as

$$\mathrm{Tr}(f)(x) = \sum_{g\in\Gamma} f(g\rho(x),1). \tag{3.3}$$

In the case of $Y = \mathbb{A}_{\mathbb{Q}}$ acted upon by $\Gamma = \mathbb{Q}^\times$, the proper orbits are parameterized by the idèle classes and this space embeds in the adèle classes using the inclusion

$$\iota : \mathbb{Q}^\times\backslash\mathbb{A}_{\mathbb{Q}}^\times \to \mathbb{Q}^\times\backslash\mathbb{A}_{\mathbb{Q}}.$$

We identify the idèle class group $C_{\mathbb{Q}} = \mathbb{Q}^\times\backslash\mathbb{A}_{\mathbb{Q}}^\times$ with $\hat{\mathbb{Z}}^\times \times \mathbb{R}_+^*$, using the canonical exact sequence affected by the modulus

$$1 \to \hat{\mathbb{Z}}^\times \to C_{\mathbb{Q}} \xrightarrow{\mathrm{Mod}} \mathbb{R}_+^* \to 1.$$

There is a natural section $\rho : C_{\mathbb{Q}} \to \mathbb{A}_{\mathbb{Q}}^\times$ of the quotient map, given by the canonical inclusion

$$\hat{\mathbb{Z}}^\times \times \mathbb{R}_+^* \subset \mathbb{A}_{\mathbb{Q}}^f \times \mathbb{R} = \mathbb{A}_{\mathbb{Q}}.$$

Next, we focus on the $\hat{\mathbb{Z}}^\times$-invariant part of $\mathcal{S}(\mathbb{A}_{\mathbb{Q}})$. Then, with the notations of Proposition 3.1 we have

LEMMA 3.2. *The following facts hold*

(i) *Let* $h \in \mathcal{S}(\mathbb{A}_{\mathbb{Q}})^{\hat{\mathbb{Z}}^\times}$, *then there exists* $f \in \mathcal{S}(\mathbb{R})^{\mathrm{ev}}$ *with* $h - 1_{\hat{\mathbb{Z}}} \otimes f \in E^{\hat{\mathbb{Z}}^\times}$.

(ii) *Let* $f \in \mathcal{S}(\mathbb{R})^{\mathrm{ev}}$ *and* $\tilde{f} = (1_{\hat{\mathbb{Z}}} \otimes f)U(1) \in \mathcal{S}(\mathbb{A}_{\mathbb{Q}})^{\hat{\mathbb{Z}}^\times} \rtimes \mathbb{Q}^\times$, *then one has*

$$\mathrm{Tr}(\tilde{f})(u) = 2\sum_{n\in\mathbb{N}^\times} f(nu) \qquad \forall u \in \mathbb{R}_+^*. \tag{3.4}$$

(iii) *Let* $f \in \mathcal{S}(\mathbb{R})^{\mathrm{ev}}$, *then* $1_{\hat{\mathbb{Z}}} \otimes f \in E^{\hat{\mathbb{Z}}^\times} \iff f = 0$.

PROOF. (i) By definition, the elements of the Bruhat-Schwartz space $\mathcal{S}(\mathbb{A}_{\mathbb{Q}})$ are finite linear combinations of functions on $\mathbb{A}_{\mathbb{Q}}$ of the form ($S \ni \infty$ is a finite set of places)

$$f = \otimes f_v, \qquad f_v = 1_{\mathbb{Z}_v}\ \forall v \notin S, \quad f_\infty \in \mathcal{S}(\mathbb{R}), \quad f_p \in \mathcal{S}(\mathbb{Q}_p) \quad \forall p \in S\backslash\infty,$$

where $\mathcal{S}(\mathbb{Q}_p)$ denotes the space of locally constant functions with compact support. An element of $\mathcal{S}(\mathbb{Q}_p)$ which is $\mathbb{Z}_p^*$-invariant is a finite linear combination of characteristic functions $(1_{\mathbb{Z}_p})_{p^n}(x) := 1_{\mathbb{Z}_p}(p^n x)$. Thus an element $h \in \mathcal{S}(\mathbb{A}_{\mathbb{Q}})^{\hat{\mathbb{Z}}^\times}$ is a finite linear combination of functions of the form

$$f = \otimes f_v, \qquad f_v = 1_{\mathbb{Z}_v} \quad \forall v \notin S, \quad f_\infty \in \mathcal{S}(\mathbb{R}), \quad f_p = (1_{\mathbb{Z}_p})_{p^{n_p}} \quad \forall p \in S\backslash\infty$$

With $q = \prod p^{-n_p}$ one has with $\ell(x) := f(qx)$, $\ell = 1_{\hat{\mathbb{Z}}} \otimes g$, $\ell - f \in E^{\hat{\mathbb{Z}}^\times}$ and the replacement of g with its even part $\frac{1}{2}(g(x) + g(-x))$ does not change the class of f modulo $E^{\hat{\mathbb{Z}}^\times}$.

(ii) Let $f \in \mathcal{S}(\mathbb{R})^{\mathrm{ev}}$ and $\tilde{f} = (1_{\hat{\mathbb{Z}}} \otimes f)U(1) \in S(\mathbb{A}_{\mathbb{Q}})^{\hat{\mathbb{Z}}^\times} \rtimes \mathbb{Q}^\times$. For $u \in \mathbb{R}_+^*$ the lift $\rho(u) \in \mathcal{S}(\mathbb{A}_{\mathbb{Q}})$ is $(1, u) \in \hat{\mathbb{Z}}^\times \times \mathbb{R}_+^* \subset \mathbb{A}_{\mathbb{Q}}^f \times \mathbb{R} = \mathbb{A}_{\mathbb{Q}}$, and by applying (3.3) one has

$$(3.5)\quad \mathrm{Tr}(\tilde{f})(u) = \sum_{q \in \mathbb{Q}^\times} \tilde{f}(q\rho(x), 1) = \sum_{q \in \mathbb{Q}^\times} (1_{\hat{\mathbb{Z}}} \otimes f)(q, qu) = 2 \sum_{n \in \mathbb{N}^\times} f(nu) \quad \forall u \in \mathbb{R}_+^*.$$

(iii) Let $f \in \mathcal{S}(\mathbb{R})^{\mathrm{ev}}$ such that $1_{\hat{\mathbb{Z}}} \otimes f \in E^{\hat{\mathbb{Z}}^\times}$. Let $\tilde{f} = (1_{\hat{\mathbb{Z}}} \otimes f)U(1) \in S(\mathbb{A}_{\mathbb{Q}})^{\hat{\mathbb{Z}}^\times} \rtimes \mathbb{Q}^\times$. By Proposition 3.1 the Hochschild class in $\mathrm{HH}_0(\mathcal{A})$ of $\tilde{f}$ is zero, thus $\mathrm{Tr}(\tilde{f}) = 0$. It follows from (3.4) that $\mathcal{E}(f)(u) = 0 \ \forall u \in \mathbb{R}_+^*$. Then (2.2) implies that the function $\psi(z) = \int_{\mathbb{R}_+^*} f(u)u^{\frac{1}{2}-iz}d^*u$ is well defined in the half-plane $\Im(z) > \frac{1}{2}$ where it vanishes identically, thus $f = 0$. The converse of the statement is obvious. □

The next statement complements Proposition 3.1, with a description of the range of $\pi_! : \mathrm{HH}_0(\mathcal{S}(\mathbb{A}_{\mathbb{Q}})^{\hat{\mathbb{Z}}^\times}) \to \mathrm{HH}_0(\mathcal{A})^{\hat{\mathbb{Z}}^\times}$; it also shows that the map $\mathcal{E}(f)(u) = u^{1/2}\sum_{n=1}^{\infty} f(nu)$ coincides, up to the factor $\frac{u^{1/2}}{2}$, with the trace map (3.5). We keep the notations of Lemma 3.2

PROPOSITION 3.3. *The map* $\mathcal{S}(\mathbb{R})^{\mathrm{ev}} \to \pi_!\left(HH_0(\mathcal{S}(\mathbb{A}_{\mathbb{Q}})^{\hat{\mathbb{Z}}^\times})\right)$, $f \mapsto \tilde{f}$ *is an isomorphism, this means that* $\pi_!\left(HH_0(\mathcal{S}(\mathbb{A}_{\mathbb{Q}})^{\hat{\mathbb{Z}}^\times})\right)$ *is determined by the images of the elements of the subalgebra* $1_{\hat{\mathbb{Z}}} \otimes \mathcal{S}(\mathbb{R})^{\mathrm{ev}} \subset \mathcal{S}(\mathbb{A}_{\mathbb{Q}})^{\hat{\mathbb{Z}}^\times}$. *Furthermore, one has the identity*

$$\mathcal{E}(f)(u) = \frac{u^{1/2}}{2}\ \mathrm{Tr}(\tilde{f})(u) \quad f \in \mathcal{S}(\mathbb{R})^{\mathrm{ev}},\ \forall u \in \mathbb{R}_+^*.$$

PROOF. The first statement follows from Lemma 3.2 (i) and (iii). The second statement from (ii) of the same lemma. □

4. The Laplacian $\Delta = H(1 + H)$

This section describes the spectral interpretation of the squares of non-trivial zeros of the Riemann zeta function in terms of a suitable Laplacian. It also shows the relation between this Laplacian and the prolate wave operator.

4.1. The vanishing conditions. One starts with the exact sequence

$$0 \to \mathcal{S}(\mathbb{A}_{\mathbb{Q}})_1 \to \mathcal{S}(\mathbb{A}_{\mathbb{Q}}) \xrightarrow{\epsilon} \mathbb{C}(1) \to 0$$

associated to the kernel of the $\mathbb{Q}^\times$-invariant linear functional $\epsilon(f) = \int_{\mathbb{A}_{\mathbb{Q}}} f(x)dx \in \mathbb{C}(1)$. By implementing in the above sequence the evaluation $\delta_0(f) := f(0)$, one obtains the exact sequence

$$0 \to \mathcal{S}(\mathbb{A}_{\mathbb{Q}})_0 \to \mathcal{S}(\mathbb{A}_{\mathbb{Q}}) \xrightarrow{(\delta_0, \epsilon)} \mathbb{C}(0) \oplus \mathbb{C}(1) \to 0.$$

The next lemma shows that both $\mathcal{S}(\mathbb{A}_{\mathbb{Q}})_0$ and $\mathcal{S}(\mathbb{A}_{\mathbb{Q}})_1$ have a description in terms of the ranges of two related differential operators. For simplicity of exposition, we restrict our discussion to the $\hat{\mathbb{Z}}^\times$-invariant parts of these function spaces.

LEMMA 4.1. *Let* $H : \mathcal{S}(\mathbb{A}_{\mathbb{Q}}) \to \mathcal{S}(\mathbb{A}_{\mathbb{Q}})$, $H := x\partial_x$ *be the generator of the scaling action of* $1 \times \mathbb{R}_+^* \subset \mathrm{GL}_1(\mathbb{A}_{\mathbb{Q}})$. *Then one has*

(i) H *commutes with the action of* $\mathrm{GL}_1(\mathbb{A}_{\mathbb{Q}})$ *and restricts to* $\mathcal{S}(\mathbb{A}_{\mathbb{Q}})^{\hat{\mathbb{Z}}^\times}$.
(ii) $(1 + H)$ *induces an isomorphism* $\mathcal{S}(\mathbb{R})^{\mathrm{ev}} \to \mathcal{S}(\mathbb{R})_1^{\mathrm{ev}}$.
(iii) $H(1 + H)$ *induces an isomorphism* $\mathcal{S}(\mathbb{R})^{\mathrm{ev}} \to \mathcal{S}(\mathbb{R})_0^{\mathrm{ev}}$.

PROOF. (i) follows since $\mathrm{GL}_1(\mathbb{A}_\mathbb{Q})$ is abelian, thus H commutes with the action of $\mathrm{GL}_1(\mathbb{A}_\mathbb{Q})$.
(ii) The functional $f \mapsto \epsilon(f) = \int_{\mathbb{A}_\mathbb{Q}} f(x)dx$ vanishes on the range of $1+H$ since

$$\epsilon((1+H)f) = \int_{\mathbb{A}_\mathbb{Q}} \partial(xf)dx = 0,$$

thus the range of $(1+H)$ is contained in $\mathcal{S}(\mathbb{R})_1^{\mathrm{ev}}$. The equation $Hf+f=0$ implies that $xf(x)$ is constant, hence $f=0$, for $f \in \mathcal{S}(\mathbb{R})$. Thus $(1+H) : \mathcal{S}(\mathbb{R})^{\mathrm{ev}} \to \mathcal{S}(\mathbb{R})_1^{\mathrm{ev}}$ is injective. Let now $f \in \mathcal{S}(\mathbb{R})_1^{\mathrm{ev}}$ and let $\widehat{f}$ be its Fourier transform. Then $\widehat{f} \in \mathcal{S}(\mathbb{R})^{\mathrm{ev}}$ and $\widehat{f}(0)=0$. The function $g(x) := \widehat{f}(x)/x$, $g(0) := 0$ is smooth, and $g \in \mathcal{S}(\mathbb{R})^{\mathrm{odd}}$, while the function

$$h(x) := \int_{-\infty}^{x} g(y)dy$$

fulfills $h \in \mathcal{S}(\mathbb{R})^{\mathrm{ev}}$ and $\partial_x h = g$. One has $Hh = \widehat{f}$, thus $(-1-H)\widehat{h} = f$. This shows that $(1+H) : \mathcal{S}(\mathbb{R})^{\mathrm{ev}} \to \mathcal{S}(\mathbb{R})_1^{\mathrm{ev}}$ is also surjective.
(iii) Since the evaluation $f \mapsto f(0)$ is invariant under scaling, it vanishes in the range of H. Similarly, one sees that the functional $f \mapsto \int_{\mathbb{A}_\mathbb{Q}} f(x)dx$ vanishes on the range of $1+H$. Thus the range of $H(1+H)$ is contained in $\mathcal{S}(\mathbb{R})_0$. The equation $Hf = 0$ implies that the function f is constant and hence $f = 0$, for $f \in \mathcal{S}(\mathbb{R})$. Similarly $Hf+f=0$ implies that $xf(x)$ is constant and hence $f=0$ for $f \in \mathcal{S}(\mathbb{R})$. Thus $H(1+H) : \mathcal{S}(\mathbb{R}) \to \mathcal{S}(\mathbb{R})_0$ is injective. Let now $f \in \mathcal{S}(\mathbb{R})^{\mathrm{ev}}$ with $f(0) = 0$. Then the function $g(x) := f(x)/x$, $g(0) := 0$, is smooth, $g \in \mathcal{S}(\mathbb{R})^{\mathrm{odd}}$ and there exists a unique $h \in \mathcal{S}(\mathbb{R})^{\mathrm{ev}}$ such that $\partial_x h = g$. One has $Hh = f$ so that $(-1-H)\widehat{h} = \widehat{f}$. Thus if $\widehat{f}(0) = 0$ one has $\widehat{h}(0) = 0$ and there exists $k \in \mathcal{S}(\mathbb{R})^{\mathrm{ev}}$ with $Hk = \widehat{h}$. Then $-(1+H)\widehat{k} = h$ and $H(1+H)\widehat{k} = -f$. This shows that $H(1+H) : \mathcal{S}(\mathbb{R})^{\mathrm{ev}} \to \mathcal{S}(\mathbb{R})_0^{\mathrm{ev}}$ is surjective and an isomorphism. □

4.2. The Laplacian $\Delta = H(1+H)$ and its spectrum. This section is based on the following heuristic dictionary suggesting a parallel between some classical notions in Hodge theory on the left-hand side, and their counterparts in noncommutative geometry, for the adèle class space of the rationals. The notations are inclusive of those of Section 3

Algebra of functions	Cross-product by $\mathbb{Q}^\times$
Differential forms	Hochschild homology
Star operator $\star$	$\iota \times \mathbb{F}$
Differential d	Operator H
$\delta := \star d \star$	Operator $1+H$
Laplacian	$\Delta := H(1+H)$

Next Proposition is a variant of the spectral realization in [**8**, **9**].

PROPOSITION 4.2. *The following facts hold*

(i) *The trace map* Tr *commutes with* $\Delta = H(1+H)$ *and the range of* $\mathrm{Tr} \circ \Delta$ *is contained in the strong Schwartz space* $\mathbf{S}(\mathbb{R}_+^*) := \cap_{\beta \in \mathbb{R}} \mu^\beta \mathcal{S}(\mathbb{R}_+^*)$, *with* μ *denoting the Modulus.*

(ii) *The spectrum of* Δ *on the quotient of* $\mathbf{S}(\mathbb{R}_+^*)$ *by the closure of the range of* $\operatorname{Tr}\circ\Delta$ *is the set (counted with possible multiplicities)*

$$\left\{\left(z-\frac{1}{2}\right)^2-\frac{1}{4}\mid z\in\mathbb{C}\backslash\mathbb{R},\ \zeta(z)=0\right\}.$$

PROOF. (i) The trace map of (3.5) commutes with Δ. By Lemma 4.1 (iii) the range of Δ is $\mathcal{S}(\mathbb{R})_0^{\mathrm{ev}}$ thus the range of $\mathcal{E}\circ(H(1+H))$ is contained in $\mathbf{S}(\mathbb{R}_+^*)$ (see [**9**], Lemma 2.51).
(ii) By construction, $\mathbf{S}(\mathbb{R}_+^*)$ is the intersection, indexed by compact intervals $J\subset\mathbb{R}$, of the spaces $\cap_{\beta\in J}\,\mu^\beta\mathcal{S}(\mathbb{R}_+^*)$. The Fourier transform

$$\mathbb{F}(f)(z):=\int_{\mathbb{R}_+^*}f(u)u^z d^*u$$

defines an isomorphism of these function spaces with the Schwartz spaces $\mathcal{S}(I)$, labeled by vertical strips $I:=\{z\in\mathbb{C}\mid\Re(z)\in J\}$, of holomorphic functions f in I with $p_{k,m}(f)<\infty$ for all $k,m\in\mathbb{N}$ where

$$p_{k,m}(f)=\sum_{n=0}^{m}\frac{1}{n!}\sup_I(1+|z|)^k\cdot|\partial^n f(z)|\,.$$

These norms define the Fréchet topology on $\mathcal{S}(I)$.

It then follows from Lemma 4.125 of [**9**] that, for I sufficiently large, the quotient of $\mathcal{S}(I)$ by the closure of the range of $\operatorname{Tr}\circ\Delta$ decomposes into a direct sum of finite-dimensional spaces associated to the projections $\Pi(N)$, $N\in\mathbb{Z}$ which fulfill the following properties

(1) Each $\Pi(N)$ is idempotent.
(2) The sequence of $\Pi(N)$'s is of tempered growth.
(3) The rank of $\Pi(N)$ is $O(\log|N|)$ for $|N|\to\infty$.
(4) For any $f\in\mathcal{S}(I)$ the sequence $f\,\Pi(N)$ is of rapid decay and

$$\sum_{N\in\mathbb{Z}}f\,\Pi(N)=f\qquad\forall f\in\mathcal{S}(I)\,.$$

This direct sum decomposition commutes with Δ since both $\Pi(N)$ and the conjugate of Δ by the Fourier transform $\mathbb{F}$ are given by multiplication operators. The conjugate of H by $\mathbb{F}$ is the multiplication by $-z$, so that the conjugate of Δ is the multiplication by $-z(1-z)$. The spectrum of Δ is the union of the spectra of the finite-dimensional operators $\Delta_N:=\Pi(N)\Delta=\Delta\Pi(N)$. By [**9**], Corollary 4.118, and the proof of Theorem 4.116, the finite-dimensional range of $\Pi(N)$ is described by the evaluation of $f\in\mathcal{S}(I)$ on the zeros $\rho\in Z(N)$ of the Riemann zeta function which are inside the contour γ_N, *i.e.* by the map

$$\mathcal{S}(I)\ni f\mapsto f|_{Z(N)}=(f^{(n_\rho)}(\rho))_{\rho\in Z(N)}\in\mathbb{C}^{(n_\rho)}$$

where $\mathbb{C}^{(n_\rho)}$ denotes the space of dimension n_ρ of jets of order equal to the order n_ρ of the zero ρ of the zeta function. Moreover, the action of Δ_N is given by the matrix associated with the multiplication of $f\in\mathcal{S}(I)$ by $-z(1-z)$: this gives a triangular matrix whose diagonal is given by n_ρ terms all equal to $-\rho(1-\rho)$. Thus the spectrum of Δ on the quotient of $\mathbf{S}(\mathbb{R}_+^*)$ by the closure of the range of $\operatorname{Tr}\circ\Delta$ is the set (counted with multiplicities)

$$\left\{\left(\rho-\frac{1}{2}\right)^2-\frac{1}{4}\mid\rho\in\mathbb{C}\backslash\mathbb{R},\ \zeta(\rho)=0\right\}.$$

□

COROLLARY 4.3. *The spectrum of* Δ *on the quotient of the strong Schwartz space* $\mathbf{S}(\mathbb{R}_+^*)$ *by the closure of the range of* $\mathrm{Tr}\circ\Delta$ *is negative if and only if the Riemann Hypothesis holds.*

PROOF. This follows from Proposition 4.2 and the fact that for $\rho\in\mathbb{C}$

$$\left(\rho-\frac{1}{2}\right)^2-\frac{1}{4}\leqslant 0\iff\rho\in[0,1]\cup\frac{1}{2}+i\mathbb{R}.$$

□

REMARK 4.4. The main interest of the above reformulation of the spectral realization of [**8, 9**] in terms of the Laplacian Δ is that the latter is intimately related to the prolate wave operator W_λ that is shown in [**10**] to be self-adjoint and have, for $\lambda=\sqrt{2}$ the same UV spectrum as the Riemann zeta function. The relation between Δ and W_λ is that the latter is a perturbation of Δ by a multiple of the Harmonic oscillator.

5. Sheaves on the Scaling Site and $H^0(\mathscr{S},\mathcal{L}^2/\overline{\Sigma\mathcal{E}})$

Let $\mu\in\mathbb{R}_{>1}$ and Σ_μ be the linear map on functions $g:\mathbb{R}_+^*\to\mathbb{C}$ of sufficiently rapid decay at 0 and ∞ defined by

$$(\Sigma_\mu g)(u)=\sum_{k\in\mathbb{Z}}g(\mu^k u). \tag{5.1}$$

We shall denote with $\mathcal{S}_0^{\mathrm{ev}}$ the linear space of real-valued, even Schwartz functions $f\in\mathcal{S}(\mathbb{R})$ fulfilling the two conditions $f(0)=0=\int_{\mathbb{R}}f(x)dx$. The map

$$\mathcal{E}:\mathcal{S}_0^{\mathrm{ev}}\to\mathbb{R},\qquad(\mathcal{E}f)(u)=u^{1/2}\sum_{n>0}f(nu) \tag{5.2}$$

is proportional to a Riemann sum for the integral of f. The following lemma on scale invariant Riemann sums justifies the pointwise "well-behavior" of (5.2) (see [**7**] Lemma 6.1)

LEMMA 5.1. *Let* f *be a complex-valued function of bounded variation on* $(0,\infty)$. *Assume that* f *is of rapid decay for* $u\to\infty$, $O(u^2)$ *when* $u\to 0$, *and that* $\int_0^\infty f(t)dt=0$. *Then the following properties hold*

(i) *The function* $(\mathcal{E}f)(u)$ *in* (5.2) *is well-defined pointwise, is* $O(u^{1/2})$ *when* $u\to 0$, *and of rapid decay for* $u\to\infty$.
(ii) *Let* $g=\mathcal{E}(f)$, *then the series* (5.1) *is geometrically convergent, and defines a bounded and measurable function on* $\mathbb{R}_+^*/\mu^{\mathbb{Z}}$.

We recall that a sheaf over the Scaling Site $\mathscr{S}=[0,\infty)\rtimes\mathbb{N}^\times$ is a sheaf of sets on $[0,\infty)$ (endowed with the euclidean topology) which is equivariant for the action of the multiplicative monoid $\mathbb{N}^\times$ [**6**]. Since we work in characteristic zero we select as structure sheaf of $\mathscr{S}$ the $\mathbb{N}^\times$-equivariant sheaf $\mathcal{O}$ whose sections on an open set $U\subset[0,\infty)$ define the space of smooth, complex-valued functions on U. The next proposition introduces two relevant sheaves of $\mathcal{O}$-modules.

PROPOSITION 5.2. *Let $L \in (0,\infty)$, $\mu = \exp L$, and $C_\mu = \mathbb{R}_+^*/\mu^{\mathbb{Z}}$. The following facts hold*

(i) *As L varies in $(0,\infty)$, the pointwise multiplicative Fourier transform*

$$\mathbb{F} : L^2(C_\mu) \to \ell^2(\mathbb{Z}), \qquad \mathbb{F}(\xi)(n) = L^{-\frac{1}{2}} \int_{C_\mu} \xi(u) u^{-\frac{2\pi i n}{L}} d^*u \tag{5.3}$$

defines an isomorphism between the family of Hilbert spaces $L^2(C_\mu)$ and the restriction to $(0,\infty)$ of the trivial vector bundle $L^2 = [0,\infty) \times \ell^2(\mathbb{Z})$.

(ii) *The smooth sections vanishing at $L = 0$ of the vector bundle L^2 together with the linear maps*

$$\sigma_n : L^2(C_{\mu^n}) \to L^2(C_\mu), \qquad \sigma_n(\xi)(u) = \sum_{j=1}^{n} \xi(\mu^j u) \tag{5.4}$$

define a sheaf $\mathcal{L}^2$ of $\mathcal{O}$-modules on $\mathscr{S}$.

(iii) *For $f \in \mathcal{S}_0^{\mathrm{ev}}$, the maps $\Sigma\mathcal{E}(f)(L) := \Sigma_\mu \mathcal{E}(f)$ as in (5.1) define smooth (global) sections $\Sigma\mathcal{E}(f)$ of $\mathcal{L}^2$.*

(iv) *For any open set $U \subset [0,\infty)$, the submodules closure of $C_0^\infty(U, \Sigma\mathcal{E}(\mathcal{S}_0^{\mathrm{ev}}))$ in $C_0^\infty(U, L^2)$ are $\mathbb{N}^\times$-equivariant and define a subsheaf $\overline{\Sigma\mathcal{E}} \subset \mathcal{L}^2$ of closed $\mathcal{O}$-modules on $\mathscr{S}$.*

(v) *For $U \subset [0,\infty)$ open, a section $\xi \in C_0^\infty(U, L^2)$ belongs to the submodule $C_0^\infty(U, \overline{\Sigma\mathcal{E}})$ if and only if each Fourier component $\mathbb{F}(\xi)(n)$ as in (5.3) belongs to the closure in $C_0^\infty(U, \mathbb{C})$ of the submodule generated by the multiples of the function $\zeta\left(\frac{1}{2} - \frac{2\pi i n}{L}\right)$ ($\zeta(z)$ is the Riemann zeta function).*

PROOF. (i) holds since the Fourier transform is a unitary isomorphism.

(ii) The sheaf $\mathcal{L}^2$ on $[0,\infty)$ is defined by associating to an open subset $U \subset [0,\infty)$ the space $\mathcal{F}(U) = C_0^\infty(U, L^2)$ of smooth sections vanishing at $L = 0$ of the vector bundle L^2. The action of $\mathbb{N}^\times$ on $\mathcal{L}^2$ is given, for $n \in \mathbb{N}^\times$ and for any pair of opens U and U' of $[0,\infty)$, with $nU \subset U'$, by

$$\mathcal{F}(U,n) : C_0^\infty(U', L^2) \to C_0^\infty(U, L^2), \qquad \mathcal{F}(U,n)(\xi)(x) = \sigma_n(\xi(nx)). \tag{5.5}$$

Note that with $\mu = \exp x$ one has $\xi(nx) \in L^2(C_{\mu^n})$ and $\sigma_n(\xi(nx)) \in L^2(C_\mu)$. By construction one has: $\sigma_n \sigma_m = \sigma_{nm}$, thus the above action of $\mathbb{N}^\times$ turns $\mathcal{L}^2$ into a sheaf on $\mathscr{S} = [0,\infty) \rtimes \mathbb{N}^\times$.

(iii) By Lemma 5.1 (i), $\mathcal{E}(f)(u)$ is pointwise well-defined, it is $O(u^{1/2})$ for $u \to 0$, and of rapid decay for $u \to \infty$. By (ii) of the same lemma one has

$$\mathbb{F}(\Sigma_\mu(\mathcal{E}(f)))(n) = L^{-\frac{1}{2}} \int_{C_\mu} \Sigma_\mu(\mathcal{E}(f))(u) u^{-\frac{2\pi i n}{L}} d^*u = L^{-\frac{1}{2}} \int_{\mathbb{R}_+^*} \mathcal{E}(f)(u) u^{-\frac{2\pi i n}{L}} d^*u.$$

It then follows from [**7**] (see (6.4) which is valid for $z = \frac{2\pi n}{L} \in \mathbb{R}$) that

$$\mathbb{F}(\Sigma_\mu(\mathcal{E}(f)))(n) = L^{-\frac{1}{2}} \zeta\left(\frac{1}{2} - \frac{2\pi i n}{L}\right) \int_{\mathbb{R}_+^*} u^{\frac{1}{2}} f(u) u^{-\frac{2\pi i n}{L}} d^*u. \tag{5.6}$$

Since $f \in \mathcal{S}_0^{\mathrm{ev}}$, with $w(f)(u) := u^{1/2} f(u)$, the multiplicative Fourier transform $\mathbb{F}(w(f)) = \psi$, $\psi(z) := \int_{\mathbb{R}_+^*} f(u) u^{\frac{1}{2} - iz} d^*u$ is holomorphic in the complex half-plane defined by $\Im(z) > -5/2$ [**7**]. Moreover, by construction $\mathcal{S}_0^{\mathrm{ev}}$ is stable under the operation $f \mapsto u\partial_u f + \frac{1}{2} f$, hence $w(\mathcal{S}_0^{\mathrm{ev}})$ is stable under $f \mapsto u\partial_u f$. This operation multiplies $\mathbb{F}(w(f))(z) = \psi(z)$ by iz. This argument shows that for any integer

$m > 0$, $z^m\psi(z)$ is bounded in a strip around the real axis and hence that the derivative $\psi^{(k)}(s)$ is $O(|s|^{-m})$ on $\mathbb{R}$, for any $k \geqslant 0$. By applying classical estimates due to Lindelof [**11**], (see [**1**] inequality (56)), the derivatives $\zeta^{(m)}(\frac{1}{2}+iz)$ are $O(|z|^\alpha)$ for any $\alpha > 1/4$. Thus all derivatives ∂_L^m of the function (5.6), now re-written as $h(L,n) := L^{-\frac{1}{2}}\zeta\left(\frac{1}{2} - \frac{2\pi i n}{L}\right)\psi(\frac{2\pi n}{L})$, are sequences of rapid decay as functions of $n \in \mathbb{Z}$. It follows that $\Sigma\mathcal{E}(f)$ is a smooth (global) section of the vector bundle L^2 over $(0,\infty)$. Moreover, when $n \neq 0$ the function $h(L,n)$ tends to 0 when $L \to 0$ and the same holds for all derivatives $\partial_L^m h(L,n)$. In fact, for any $m, k \geqslant 0$, one has

$$\sum_{n\neq 0} |\partial_L^m h(L,n)|^2 = O(L^k) \quad \text{when} \quad L \to 0.$$

This result is a consequence of the rapid decay at ∞ of the derivatives of the function ψ, and the above estimate of $\zeta(z)$ and its derivatives. For $n = 0$ one has $h(L,0) = L^{-\frac{1}{2}}\zeta(\frac{1}{2})\psi(0)$.
(iv) For any open subset $U \subset [0,\infty)$ the vector space $C_0^\infty(U, L^2)$ admits a natural Frechet topology with generating seminorms of the form ($K \subset U$ compact subset)

$$p_K^{(n,m)}(\xi) := \max_K L^{-n}\|\partial_L^m \xi(L)\|_{L^2}. \tag{5.7}$$

One obtains a space of smooth sections $C_0^\infty(U, \Sigma\mathcal{E}) \subset C_0^\infty(U, L^2)$ defined as sums of products $\sum h_j \Sigma\mathcal{E}(f_j)$, with $f_j \in \mathcal{S}_0^{\mathrm{ev}}$ and $h_j \in C_0^\infty(U, L^2)$. The map $\sigma_n : L^2(C_{\mu^n}) \to L^2(C_\mu)$ in (ii) is continuous, and from the equality $\sigma_n \circ \Sigma_{\mu^n} = \Sigma_\mu$ it follows (here we use the notations as in the proof of (ii)) that the sections $\xi \in C^\infty(U', L^2(C'))$ which belong to $C^\infty(U', \overline{\Sigma\mathcal{E}(\mathcal{S}_0^{\mathrm{ev}})})$ are mapped by $\mathcal{F}(U,n)$ inside $C^\infty(U, \overline{\Sigma\mathcal{E}(\mathcal{S}_0^{\mathrm{ev}})})$. In this way one obtains a sheaf $\overline{\Sigma\mathcal{E}} \subset \mathcal{L}^2$ of $\mathcal{O}$-modules over $\mathscr{S}$.
(v) Let $\xi \in H^0(U, \overline{\Sigma\mathcal{E}})$. By hypothesis, ξ is in the closure of $C_0^\infty(U, \Sigma\mathcal{E}(\mathcal{S}_0^{\mathrm{ev}})) \subset C_0^\infty(U, L^2)$ for the Frechet topology. The Fourier components of ξ define continuous maps in the Frechet topology, thus it follows from (5.6) that the functions $f_n = \mathbb{F}(\xi)(n)$ are in the closure, for the Frechet topology on $C_0^\infty(U, \mathbb{C})$, of $C_0^\infty(U, \mathbb{C})g_n$, where $g_n(L) := \zeta\left(\frac{1}{2} - \frac{2\pi i n}{L}\right)$ is a multiplier of $C_0^\infty(U, \mathbb{C})$. This conclusion holds thanks to the moderate growth of the Riemann zeta function and its derivatives on the critical line. Conversely, let $\xi \in C_0^\infty(U, L^2)$ be such that each of its Fourier components $\mathbb{F}(\xi)(n)$ belongs to the closure for the Frechet topology of $C_0^\infty(U, \mathbb{C})$, of $C_0^\infty(U, \mathbb{C})g_n$. Let $\rho \in C_c^\infty([0,\infty), [0,1])$ defined to be identically equal to 1 on $[0,1]$ and with support inside $[0,2]$. The functions $\alpha_k(x) := \rho((kx)^{-1})$ $(k > 1)$ fulfill the following three properties

(1) $\alpha_k(x) = 0, \quad \forall x < (2k)^{-1}, \qquad \alpha_k(x) = 1, \quad \forall x > k^{-1}.$
(2) $\alpha_k \in C_0^\infty([0,\infty))$.
(3) For all $m > 0$ there exists $C_m < \infty$ such that $|x^{2m}\partial_x^m \alpha_k(x)| \leqslant C_m k^{-1}$ $\forall x \in [0,1], k > 1$.

To justify (3), note that $x^2\partial_x f((kx)^{-1}) = -k^{-1}f'((kx)^{-1})$ and that the derivatives of ρ are bounded. Thus one has $|(x^2\partial_x)^m \alpha_k(x)| \leqslant \|\rho^{(m)}\|_\infty k^{-m}$ $\forall x \in [0,\infty)$, $k > 1$ which implies (3) by induction on m.

Thus, when $k \to \infty$ one has $\alpha_k\xi \to \xi$ in the Frechet topology of $C_0^\infty(U, L^2)$. This is clear if $0 \notin U$ since then, on any compact subset $K \subset U$, all α_k are identically equal to 1 for $k > (\min K)^{-1}$. Assume now that $0 \in U$ and let $K = [0,\epsilon] \subset U$. With the notation of (5.7) let us show that $p_K^{(n,m)}((\alpha_k - 1)\xi) \to 0$ when $k \to \infty$.

Since $\alpha_k(x) = 1$, $\forall x > k^{-1}$ one has, using the finiteness of $p_K^{(n+1,m)}(\xi)$

$$\max_K L^{-n}\|((\alpha_k - 1)\partial_L^m \xi)(L)\|_{L^2} \leqslant \max_{(0,1/k]} L^{-n}\|(\partial_L^m \xi)(L)\|_{L^2} \overset{k\to\infty}{\to} 0.$$

Then one obtains

$$L^{-n}\partial_L^m((\alpha_k - 1)\xi)(L) = L^{-n}((\alpha_k - 1)\partial_L^m \xi)(L) + \sum_1^m L^{-n}\binom{m}{j}(\partial_L^j \alpha_k)(L)(\partial_L^{m-j}\xi)(L).$$

Thus using (3) above and the finiteness of the norms $p_K^{(n+2j,m-j)}(\xi)$ one derives: $p_K^{(n,m)}((\alpha_k - 1)\xi) \to 0$ when $k \to \infty$. It remains to show that $\alpha_k \xi$ belongs to the submodule $C_0^\infty(U, \overline{\Sigma\mathcal{E}})$. It is enough to show that for $K \subset (0, \infty)$ a compact subset with $\min K > 0$, one can approximate ξ by elements of $C_0^\infty(U, \Sigma\mathcal{E})$ for the norm $p_K^{(0,m)}$. Let P_N be the orthogonal projection in $L^2(C_\mu)$ on the finite-dimensional subspace determined by the vanishing of all Fourier components $\mathbb{F}(\xi)(\ell)$ for any $\ell, |\ell| > N$. Given $L \in K$ and $\epsilon > 0$ there exists $N(L, \epsilon) < \infty$ such that

$$\|(1 - P_N)\partial_L^j \xi(L)\| < \epsilon \quad \forall j \leqslant m, \ N \geqslant N(L, \epsilon). \tag{5.8}$$

The smoothness of ξ implies that there exists an open neighborhood $V(L, \epsilon)$ of L such that (5.8) holds in $V(L, \epsilon)$. The compactness of K then shows that there exists a finite N_K such that

$$\|(1 - P_N)\partial_L^j \xi(L)\| < \epsilon \quad \forall j \leqslant m, \ L \in K, \ N \geqslant N_K.$$

It now suffices to show that one can approximate $P_N \xi$, for the norm $p_K^{(0,m)}$, by elements of $C_0^\infty(U, \Sigma\mathcal{E})$. To achieve this result, we let $L_0 \in K$ and $\delta_j \in C_c^\infty(\mathbb{R}_+^*)$, $|j| \leqslant N$ be such that

$$\mathbb{F}(\delta_j)\left(\frac{2\pi j'}{L_0}\right) = \begin{cases} 1 & j = j' \\ 0 & j \neq j' \end{cases} \quad \forall j', |j'| \leqslant N, \quad \int_{\mathbb{R}_+^*} u^{1/2}\delta_j(u)d^*u = 0 \quad \forall j, |j| \leqslant N.$$

One construct δ_j starting with a function $h \in C_c^\infty(\mathbb{R}_+^*)$ such that $\mathbb{F}(h)\left(\frac{2\pi j}{L_0}\right) \neq 0$ and acting on h by a differential polynomial whose effect is to multiply $\mathbb{F}(h)$ by a polynomial vanishing on all $\frac{2\pi j'}{L_0}$, $j' \neq j$ and at $i/2$. By hypothesis each Fourier component $\mathbb{F}(\xi)(n)$ belongs to the closure in $C_0^\infty(U, \mathbb{C})$ of the multiples of the function $\zeta\left(\frac{1}{2} - \frac{2\pi i n}{L}\right)$. Thus, given $\epsilon > 0$ one has functions $f_n \in C_0^\infty(U, \mathbb{C})$, $|n| \leqslant N$ such that

$$\max_K |\partial_L^j\left(\mathbb{F}(\xi)(n) - \zeta\left(\frac{1}{2} - \frac{2\pi i n}{L}\right) f_n(L)\right)| \leqslant \epsilon \quad \forall j \leqslant m, \quad |n| \leqslant N.$$

We now can find a small open neighborhood V of L_0 and functions $\phi_j \in C^\infty(V)$, $|j| \leqslant N$ such that

$$\sum \phi_j(L)\mathbb{F}(\delta_j)\left(\frac{2\pi n}{L}\right) = L^{\frac{1}{2}} f_n(L) \quad \forall L \in V. \tag{5.9}$$

This is possible because the determinant of the matrix $M_{n,j}(L) = \mathbb{F}(\delta_j)\left(\frac{2\pi n}{L}\right)$ is non-zero in a neighborhood of L_0 where $M_{n,j}(L_0)$ is the identity matrix. The even functions $d_j(u)$ on $\mathbb{R}$, which agree with $u^{-1/2}\delta_j(u)$ for $u>0$, are all in $\mathcal{S}_0^{\rm ev}$ since $\int_{\mathbb{R}} d_j(x)dx = 2\int_{\mathbb{R}_+^*} u^{1/2}\delta_j(u)d^*u = 0$. One then has

$$\mathbb{F}(\Sigma_\mu(\mathcal{E}(d_j)))(n) = L^{-\frac{1}{2}}\zeta\left(\frac{1}{2}-\frac{2\pi i n}{L}\right)\mathbb{F}_\mu(\delta_j)\left(\frac{2\pi n}{L}\right)$$

by (5.6), and by (5.9) one gets

$$\sum \phi_j(L)\mathbb{F}(\Sigma_\mu(\mathcal{E}(d_j)))(n) = \zeta\left(\frac{1}{2}-\frac{2\pi i n}{L}\right) f_n(L) \quad \forall L \in V.$$

One finally covers K by finitely many such open sets V and use a partition of unity subordinated to this covering to obtain smooth functions $\varphi_\ell \in C_c^\infty(0,\infty)$, $g_\ell \in \mathcal{S}_0^{\rm ev}$ such that the Fourier component of index n, $|n| \leqslant N$, of $\sum \varphi_\ell \Sigma\mathcal{E}(g_\ell)$ is equal to $\zeta\left(\frac{1}{2}-\frac{2\pi i n}{L}\right) f_n(L)$ on K. This shows that ξ belongs to the closure of $C_0^\infty(U, \Sigma\mathcal{E}(\mathcal{S}_0^{\rm ev})) \subset C_0^\infty(U, L^2)$. □

We recall that the space of global sections $H^0(\mathcal{T},\mathcal{F})$ of a sheaf of sets $\mathcal{F}$ in a Grothendieck topos $\mathcal{T}$ is defined to be the set $\mathrm{Hom}_{\mathcal{T}}(1,\mathcal{F})$, where 1 denotes the terminal object of $\mathcal{T}$. For $\mathcal{T} = \mathscr{S}$ and $\mathcal{F}$ a sheaf of sets on $[0,\infty)$, 1 assigns to an open set $U \subset [0,\infty)$ the single element $*$, on which $\mathbb{N}^\times$ acts as the identity. Thus, we understand an element of $\mathrm{Hom}_{\mathscr{S}}(1,\mathcal{F})$ as a global section ξ of $\mathcal{F}$, where $\mathcal{F}$ is viewed as a sheaf on $[0,\infty)$ invariant under the action of $\mathbb{N}^\times$.

With the notations of Proposition 5.2 and for $\xi \in \mathrm{Hom}_{\mathscr{S}}(1,\mathcal{L}^2)$, we write $\widehat{\xi}(L,n) := \mathbb{F}(\xi)(n)$ for the (multiplicative) Fourier components of ξ. Then we have

Lemma 5.3. *The following facts hold*

(i) *The map*

$$\gamma : H^0(\mathscr{S},\mathcal{L}^2) \to C_0^\infty([0,\infty),\mathbb{C}) \times C_0^\infty([0,\infty),\mathbb{C}) \qquad \gamma(\xi) = \widehat{\xi}(\pm 1) \tag{5.10}$$

is an isomorphism of $\mathbb{C}$-vector spaces.

(ii) *The subspace $\gamma(H^0(\mathscr{S},\overline{\Sigma\mathcal{E}}))$ is the closed ideal (for the Frechet topology on $C_0^\infty([0,\infty),\mathbb{C})$), generated by the multiplication with the functions $\zeta\left(\frac{1}{2}\mp\frac{2\pi i}{L}\right)$.*

Proof. (i) Let $\xi \in \mathrm{Hom}_{\mathscr{S}}(1,\mathcal{L}^2)$: this is a global section $\xi \in C_0^\infty([0,\infty),L^2)$ invariant under the action of $\mathbb{N}^\times$, *i.e.* such that $\sigma_n(\xi(nL)) = \xi(L)$ for all pairs (L,n). The Fourier components $\widehat{\xi}(L,n)$ of any such section are smooth functions of $L \in [0,\infty)$ vanishing at $L=0$, for $n \neq 0$, as well as all their derivatives. The equality $\sigma_n(\xi(L)) = \xi(L/n)$ entails, for $n>0$,

$$\widehat{\xi}(L,n) = L^{-\frac{1}{2}}\int_{C_\mu} \xi(L)(u)u^{-\frac{2\pi i n}{L}}d^*u = L^{-\frac{1}{2}}\int_{C_\mu^{1/n}} \xi(L/n)(u)u^{-\frac{2\pi i}{L}}d^*u =$$
$$= n^{-\frac{1}{2}}\widehat{\xi}(L/n,1).$$

This shows that the $\widehat{\xi}(L,n)$ are uniquely determined, for $n>0$ by the function $\widehat{\xi}(L,1)$ and, for $n<0$, by the function $\widehat{\xi}(L,-1)$. With $g(L) = \widehat{\xi}(L,0)$ one has: $g(L) = n^{-\frac{1}{2}}g(L/n)$ for all $n>0$. This implies, since $\mathbb{Q}_+^*$ is dense in $\mathbb{R}_+^*$ and g is assumed to be smooth, that g is proportional to $L^{-\frac{1}{2}}$ and hence identically 0, since it corresponds to a global section smooth at $0 \in [0,\infty)$. This argument proves that γ is injective. Let us show that γ is also surjective. Given a pair of functions $f_\pm \in C_0^\infty([0,\infty),\mathbb{C})$ we construct a global section $\xi \in H^0(\mathscr{S},\mathcal{L}^2)$ such that

$\gamma(\xi) = (f_+, f_-)$. One defines $\xi(L) \in L^2(C_\mu)$ by by means of its Fourier components set to be $\widehat{\xi}(L,0) := 0$, and for $n \neq 0$ by

$$\widehat{\xi}(L,n) := |n|^{-\frac{1}{2}} f_{\mathrm{sign(n)}}(L/n).$$

Since $f_\pm(x)$ are of rapid decay for $x \to 0$, $\sum |\widehat{\xi}(L,n)|^2 < \infty$, thus $\xi(L) \in L^2(C_\mu)$. All derivatives of $f_\pm(x)$ are also of rapid decay for $x \to 0$, thus all derivatives $\partial_L^k(\xi(L))$ belong to $L^2(C_\mu)$ and that the L^2-norms $\|\partial_L^k(\xi(L))\|$ are of rapid decay for $L \to 0$. By construction $\sigma_n(\xi(L)) = \xi(L/n)$, which entails $\xi \in H^0(\mathscr{S}, \mathcal{L}^2)$ with $\gamma(\xi) = (f_+, f_-)$.
(ii) Let $\xi \in H^0(\mathscr{S}, \overline{\Sigma\mathcal{E}})$. By Proposition 5.2 (v), the functions $f_\pm = \widehat{\xi}(L, \pm 1)$ are in the closure, for the Frechet topology on $C_0^\infty([0,\infty), \mathbb{C})$, of the ideal generated by the functions $\zeta\left(\frac{1}{2} \mp \frac{2\pi i}{L}\right)$. Conversely, let $\xi \in H^0(\mathscr{S}, \mathcal{L}^2)$ and assume that $\gamma(\xi)$ is in the closed submodule generated by multiplication with $\zeta\left(\frac{1}{2} \mp \frac{2\pi i}{L}\right)$. The $\mathbb{N}^\times$-invariance of ξ implies $\widehat{\xi}(L,n) = |n|^{-\frac{1}{2}}\widehat{\xi}(L/|n|, \mathrm{sign}(n))$ for $n \neq 0$. Thus the Fourier components $\widehat{\xi}(L,n)$ belong to the closure in $C_0^\infty(U, \mathbb{C})$ of the multiples of the function $\zeta\left(\frac{1}{2} - \frac{2\pi i n}{L}\right)$, then Proposition 5.2 (v) again implies $\xi \in H^0(\mathscr{S}, \overline{\Sigma\mathcal{E}})$. $\square$

The action of $\mathbb{R}_+^*$ on the sheaf $\mathcal{L}^2$ is given by the action ϑ on the Fourier components of its sections ξ. With $\mu = \exp L$, $L \in (0,\infty)$, $n \in \mathbb{N}^*$ and $\lambda \in \mathbb{R}_+^*$, this is

(5.11)
$$\widehat{\vartheta(\lambda)\xi}(L,n) = L^{-\frac{1}{2}} \int_{C_\mu} \xi(\lambda^{-1}u) u^{-\frac{2\pi i n}{L}} d^*u = \lambda^{-\frac{2\pi i n}{L}} L^{-\frac{1}{2}} \int_{C_\mu} \xi(v) v^{-\frac{2\pi i n}{L}} d^*v =$$
$$= \lambda^{-\frac{2\pi i n}{L}} \widehat{\xi}(L,n)$$

The following result explains in particular how the quotient sheaf $\mathcal{L}^2/\overline{\Sigma\mathcal{E}}$ on $\mathscr{S}$ handles eventual multiplicities of critical zeros of the zeta function.

THEOREM 5.4. *The induced action of* $\mathbb{R}_+^*$ *on the global sections* $H^0(\mathscr{S}, \mathcal{L}^2/\overline{\Sigma\mathcal{E}})$ *is canonically isomorphic to the action of* $\mathbb{R}_+^*$*, via multiplication with* λ^{is}*, on the quotient of the Schwartz space* $\mathcal{S}(\mathbb{R})$ *by the closure of the ideal generated by multiples of* $\zeta\left(\frac{1}{2} + is\right)$.

PROOF. We first show that the canonical map $q: H^0(\mathscr{S}, \mathcal{L}^2) \to H^0(\mathscr{S}, \mathcal{L}^2/\overline{\Sigma\mathcal{E}}))$ is surjective. Let $\xi \in H^0(\mathscr{S}, \mathcal{L}^2/\overline{\Sigma\mathcal{E}}))$: as a section of $\mathcal{L}^2/\overline{\Sigma\mathcal{E}}$ on $[0,\infty)$, there exists an open neighborhood $V = [0,\epsilon)$ of $0 \in [0,\infty)$ and a section $\eta \in C_0^\infty(V, L^2)$ such that the class of η in $C_0^\infty(V, L^2/\overline{\Sigma\mathcal{E}})$ is the restriction of ξ to V. The Fourier components $\widehat{\eta}(L,n)$ are meaningful for $L \in V$. Since ξ is $\mathbb{N}^\times$-invariant, for any $n \in \mathbb{N}^\times$ the class of $\mathcal{F}(V/n, n)(\eta)$, with $\mathcal{F}(V/n,n)(\eta)(L) := \sigma_n(\eta(nL))$ (see (5.4)) is equal to the class of the restriction of η in $C_0^\infty(V/n, L^2/\overline{\Sigma\mathcal{E}})$. We thus obtain

$$\eta(L) - \mathcal{F}(V/n,n)(\eta) \in C_0^\infty(V/n, \overline{\Sigma\mathcal{E}})$$

Furthermore, the Fourier components of $\alpha = \mathcal{F}(V/n,n)(\eta)$ are given by

$$\widehat{\alpha}(L,k) = n^{\frac{1}{2}} \widehat{\eta}(nL, nk).$$

Thus, the functions

$$\widehat{\eta}(L,k) - \widehat{\alpha}(L,k) = \widehat{\eta}(L,k) - n^{\frac{1}{2}} \widehat{\eta}(nL, nk)$$

are the Fourier components of an element in $C_0^\infty(V/n, \overline{\Sigma\mathcal{E}})$. By Proposition 5.2 (v), these components are in the closure in $C_0^\infty(V/n, \mathbb{C})$ of the multiples of the function $\zeta\left(\frac{1}{2} - \frac{2\pi i k}{L}\right)$. For $k = 1$ the function $\widehat{\eta}(L, 1) - n^{\frac{1}{2}}\widehat{\eta}(nL, n)$ is in the closure in $C_0^\infty(V/n, \mathbb{C})$ of the multiples of the function $\zeta\left(\frac{1}{2} - \frac{2\pi i}{L}\right)$. This implies that $\widehat{\eta}(L, n) - n^{-\frac{1}{2}}\widehat{\eta}(L/n, 1)$ is in the closure in $C_0^\infty(V, \mathbb{C})$ of the multiples of the function $\zeta\left(\frac{1}{2} - \frac{2\pi i n}{L}\right)$. For $k = -1$ one obtains a similar result for the Fourier components $\widehat{\eta}(L, n)$, $n < 0$. Thus, again by Proposition 5.2 (v), one knows that the class of η in $C_0^\infty(V, L^2/\overline{\Sigma\mathcal{E}})$) s not altered if one replaces η with $\eta_1 \in C_0^\infty(V, L^2)$

$$\widehat{\eta}_1(L, n) := |n|^{-\frac{1}{2}}\eta(L/|n|, \operatorname{sign}(n)).$$

Next step is to extend the functions $\eta(L, \pm 1) \in C_0^\infty(V, \mathbb{C})$ to $f_\pm \in C_0^\infty([0, \infty), \mathbb{C})$ fulfilling the following property. For any open set $U \subset [0, \infty)$ and a section $\beta \in C_0^\infty(U, L^2)$, with the class of β in $C_0^\infty(U, L^2/\overline{\Sigma\mathcal{E}})$) being the restriction of ξ to U, the functions $\widehat{\beta}(L, \pm 1) - f_\pm(L)$ belong to the closure in $C_0^\infty(U, \mathbb{C})$ of the multiples of the function $\zeta\left(\frac{1}{2} \mp \frac{2\pi i}{L}\right)$. To construct $f_\pm$ one considers the sheaf $\mathcal{G}_\pm$ which is the quotient of the sheaf of $C_0^\infty([0, \infty), \mathbb{C})$ functions by the closure of the ideal subsheaf generated by the multiples of the function $\zeta\left(\frac{1}{2} \mp \frac{2\pi i}{L}\right)$. Since the latter is a module over the sheaf of C^∞ functions, it is a fine sheaf, thus a global section of $\mathcal{G}_\pm$ can be lifted to a function. By Proposition 5.2 (v), the Fourier components $\widehat{\xi}_j(L, \pm 1)$ of local sections ξ_j of L^2 representing ξ define a global section of $\mathcal{G}_\pm$. The functions $f_\pm$ are obtained by lifting these sections. By appealing to Lemma 5.3, we let $\phi \in H^0(\mathscr{S}, L^2)$ to be the unique global section such that $\gamma(\phi) = (f_+, f_-)$. Then we show that $q(\phi) = \xi$. We have already proven that the restrictions to $V = [0, \epsilon)$ are the same. Thus it is enough to show that given $L_0 > 0$ and a lift $\xi_0 \in C_0^\infty(U, L^2)$ of ξ in a small open interval U containing L_0, the difference $\delta = \phi - \xi_0$ is a section of $\overline{\Sigma\mathcal{E}}$. Again by Proposition 5.2 (v), it suffices to show that the Fourier components $\widehat{\delta}(L, n)$ are in the closure of the ideal generated by multiples of $\zeta\left(\frac{1}{2} - \frac{2\pi i n}{L}\right)$. The $\mathbb{N}^\times$-invariance of ξ shows that $\mathcal{F}(U/n, n)(\xi_0)$ (see (5.5)) is a lift of ξ in U/n. Thus by the defining properties of the functions $f_\pm$ one has

$$\widehat{\mathcal{F}(U/n, n)}(\xi_0)(\pm 1) - f_\pm \in \overline{C^\infty(U, \mathbb{C})\zeta_\pm}, \qquad \text{for} \quad \zeta_\pm(L) = \zeta\left(\frac{1}{2} \mp \frac{2\pi i}{L}\right).$$

With a similar argument and using the invariance of ϕ under the action of $\mathcal{F}(U/n, n)$, one obtains that $\widehat{\delta}(n)$ is in the closure of the ideal generated by the multiples of $\zeta\left(\frac{1}{2} - \frac{2\pi i n}{L}\right)$.

We have thus proved that $q : H^0(\mathscr{S}, \mathcal{L}^2) \to H^0(\mathscr{S}, \mathcal{L}^2/\overline{\Sigma\mathcal{E}})$) is surjective. One then obtains the exact sequence

$$0 \to H^0(\mathscr{S}, \overline{\Sigma\mathcal{E}})) \to H^0(\mathscr{S}, \mathcal{L}^2) \to H^0(\mathscr{S}, \mathcal{L}^2/\overline{\Sigma\mathcal{E}})) \to 0. \tag{5.12}$$

This sequence is equivariant for the action (5.11) of ϑ of $\mathbb{R}_+^*$ on the bundle L^2. For $h \in L^1(\mathbb{R}_+^*, d^*u)$ one has

$$\widehat{(\vartheta(h)\xi)}(L, n) = \mathbb{F}(h)\left(\frac{2\pi n}{L}\right)\widehat{\xi}(L, n). \tag{5.13}$$

To obtain the required isomorphism between the two spectral realizations, one uses the isomorphism (5.10) of Lemma 5.3. Denote for short $(C_0^\infty)^2 = C_0^\infty([0, \infty)) \times$

$C_0^\infty([0,\infty))$. One maps the Schwartz space $\mathcal{S}(\mathbb{R})$ to $(C_0^\infty)^2$ by

$$\Phi(f) = (\Phi_+(f), \Phi_-(f)), \qquad \Phi_\pm(f)(L) := f\left(\pm\frac{2\pi}{L}\right).$$

Φ is well defined since all derivatives of $\Phi_\pm(f)(L)$ tend to 0 when $L \to 0$ (any function $f \in \mathcal{S}(\mathbb{R})$ is of rapid decay as well as all its derivatives). The exact sequence (5.12), together with Lemma 5.3, then gives an induced isomorphism

$$\gamma : H^0(\mathscr{S}, \mathcal{L}^2/\overline{\Sigma\mathcal{E}})) \simeq (C_0^\infty)^2/\left(\overline{C_0^\infty\zeta_+} \times \overline{C_0^\infty\zeta_-}\right).$$

In turn, the map Φ induces a morphism

$$\tilde{\Phi} : \mathcal{S}(\mathbb{R})/(\mathcal{S}(\mathbb{R})\zeta) \to (C_0^\infty)^2/\left(\overline{C_0^\infty\zeta_+} \times \overline{C_0^\infty\zeta_-}\right).$$

By (5.13) this morphism is equivariant for the action of $\mathbb{R}_+^*$. The map Φ is not an isomorphism since elements of its range have finite limits at ∞. However it is injective and its range contains all elements of $(C_0^\infty)^2$ which have compact support. Since $\zeta_\pm(L) = \zeta\left(\frac{1}{2} \mp \frac{2\pi i}{L}\right)$ tends to a finite non-zero limit when $L \to 0$, $\tilde{\Phi}$ is an isomorphism. □

REMARK 5.5. By a Theorem of Whitney (see [**12**], Corollary 1.7), the closure of the ideal of multiples of $\zeta\left(\frac{1}{2} + is\right)$ in $\mathcal{S}(\mathbb{R})$ is the subspace of those $f \in \mathcal{S}(\mathbb{R})$ which vanish of the same order as ζ at every (critical) zero $s \in Z$. Thus if any such zero is a multiple zero of order $m > 1$, one finds that the action of $\mathbb{R}_+^*$ on the global sections of the quotient sheaf $\mathcal{L}^2/\overline{\Sigma\mathcal{E}}$ admits a non-trivial Jordan decomposition of the form

$$\vartheta(\lambda)\xi = \lambda^{is}(\xi + N(\lambda)\xi),$$

with $N(\lambda)^m = 0$ and $(1 + N(u))(1 + N(v)) = 1 + N(uv)$ for all $u, v \in \mathbb{R}_+^*$.

References

[1] R. J. Backlund, *Über die Nullstellen der Riemannschen Zetafunktion* (German), Acta Math. **41** (1916), no. 1, 345–375, DOI 10.1007/BF02422950. MR1555156

[2] Alain Connes, *Noncommutative geometry*, Academic Press, Inc., San Diego, CA, 1994. MR1303779

[3] Alain Connes, *Trace formula in noncommutative geometry and the zeros of the Riemann zeta function*, Selecta Math. (N.S.) **5** (1999), no. 1, 29–106, DOI 10.1007/s000290050042. MR1694895

[4] Alain Connes, *An essay on the Riemann hypothesis*, Open problems in mathematics, Springer, [Cham], 2016, pp. 225–257. MR3526936

[5] Alain Connes and Caterina Consani, *Geometry of the arithmetic site*, Adv. Math. **291** (2016), 274–329, DOI 10.1016/j.aim.2015.11.045. MR3459019

[6] Alain Connes and Caterina Consani, *Geometry of the scaling site*, Selecta Math. (N.S.) **23** (2017), no. 3, 1803–1850, DOI 10.1007/s00029-017-0313-y. MR3663595

[7] A. Connes, C. Consani, *Spectral triples and ζ-cycles*, Preprint 2021, `arXiv:2106.01715`

[8] Alain Connes, Caterina Consani, and Matilde Marcolli, *The Weil proof and the geometry of the adèles class space*, Algebra, arithmetic, and geometry: in honor of Yu. I. Manin. Vol. I, Progr. Math., vol. 269, Birkhäuser Boston, Boston, MA, 2009, pp. 339–405, DOI 10.1007/978-0-8176-4745-2_8. MR2641176

[9] Alain Connes and Matilde Marcolli, *Noncommutative geometry, quantum fields and motives*, American Mathematical Society Colloquium Publications, vol. 55, American Mathematical Society, Providence, RI; Hindustan Book Agency, New Delhi, 2008, DOI 10.1090/coll/055. MR2371808

[10] Alain Connes and Henri Moscovici, *The UV prolate spectrum matches the zeros of zeta*, Proc. Natl. Acad. Sci. USA **119** (2022), no. 22, Paper No. e2123174119, 7. MR4477577

[11] E. L. Lindelöf, *Quelques remarques sur la croissance de la fonction $\zeta(s)$*, Bulletin des Sciences Mathématiques, Tome 32, (1908).
[12] B. Malgrange, *Ideals of differentiable functions*, Tata Institute of Fundamental Research Studies in Mathematics, vol. 3, Tata Institute of Fundamental Research, Bombay; Oxford University Press, London, 1967. MR0212575
[13] Ralf Meyer, *On a representation of the idele class group related to primes and zeros of L-functions*, Duke Math. J. **127** (2005), no. 3, 519–595, DOI 10.1215/S0012-7094-04-12734-4. MR2132868
[14] Kenichi Miyazaki, *Distinguished elements in a space of distributions*, J. Sci. Hiroshima Univ. Ser. A **24** (1960), 527–533. MR131752

Collège de France and IHES, France
Email address: `alain@connes.org`

Department of Mathematics, Johns Hopkins University, Baltimore, Maryland 21218
Email address: `cconsan1@jhu.edu`

Proceedings of Symposia in Pure Mathematics
Volume **105**, 2023
https://doi.org/10.1090/pspum/105/01897

Cyclic theory and the pericyclic category

Alain Connes and Caterina Consani

Pour moi le "paradis originel" pour l'algèbre topologique, n'est nullement la sempiternelle catégorie Δ semi-simpliciale, si utile soit-elle, et encore moins celle des espaces topologiques (qui l'une et l'autre s'envoient dans la 2-catégorie des topos, qui en est comme une enveloppe commune), mais bien la catégorie **Cat** *des petites catégories, vue avec un oeil de géomètre par l'ensemble d'intuitions, étonnamment riche, provenant des topos. En effet, les topos ayant comme catégories des faisceaux d'ensembles les $\widehat{\mathcal{C}}$, avec $\mathcal{C}$ dans* **Cat**, *sont de loin les plus simples des topos connus, et c'est pour l'avoir senti que j'insiste tant sur ces topos "catégoriques" dans SGA IV.*

Alexander Grothendieck

ABSTRACT. We give a historical perspective on the role of the cyclic category Λ in the development of cyclic theory. This involves a continuous interplay between the extension in characteristic one and in $\mathbb{S}$-algebras, of the traditional development of cyclic theory, and the geometry of the toposes associated with several small categories involved. We clarify the link between various existing presentations of the cyclic and the epicyclic categories and we exemplify the role of the absolute coefficients by presenting the ring of the integers as polynomials in powers of 3, with coefficients in the spherical group ring $\mathbb{S}[C_2]$ of the cyclic group of order 2. Finally, we introduce the pericyclic category which unifies and refines two conflicting notions of epicyclic space existing in the literature.

1. Introduction

The introduction (by the first author of this article) of cyclic cohomology as the analog of de Rham cohomology in noncommutative geometry, was inspired by the development of the quantized calculus [**3**–**6**] where cyclic cohomology of noncommutative algebras is the natural receptacle for the Chern character of K-homology cycles. Both the periodicity operator $S : \mathrm{HC}^n(\mathcal{A}) \to \mathrm{HC}^{n+2}(\mathcal{A})$ and the long exact

---

2020 *Mathematics Subject Classification.* Primary 55U40, 18F10; Secondary 14C40, 18B40.
Key words and phrases. Cyclic categories, Grothendieck toposes, Γ-sets, characterstic one, pericyclic category.
The second author was partially supported by the Simons Foundation collaboration grant n. 691493.

sequence relating cyclic and Hochschild cohomologies of an algebra $\mathcal{A}$ stem naturally from the analytic theory. After many repetitive, algebraic manipulations it became clear [**4**] that laying at the core of the theory there is a small category Λ: the cyclic category, encoding combinatorially the rules of computations for an algebra through a representation of Λ in the vector space $\mathcal{A}^\natural$ (cyclic module) obtained as the direct sum $\mathcal{A}^\natural := \bigoplus \mathcal{A}^{\otimes n}$ of tensor powers of $\mathcal{A}$. This result shows, in particular, that cyclic cohomology and homology are two instances of derived functors in homological algebra. At the conceptual level, it determines an embedding of the category of algebras in the larger abelian category of cyclic modules, thus providing a hint for a motivic development of noncommutative geometry.
The cyclic category Λ is, by definition, an extension of the category Δ of finite ordinals. Its classifying space $B\Lambda$ is the infinite projective space $\mathbb{P}^\infty(\mathbb{C})$ [**4**]. This fact points out the similarity of Λ with the compact group $U(1)$, as it is well known that $BU(1) = \mathbb{P}^\infty(\mathbb{C})$. However, one also ought to keep in mind that while shifting our focus from a small category to the homotopy type of its classifying space we lose some relevant information. For instance, $B\Delta$ is contractible (the category Δ has a final object), while the restriction to Δ of the cyclic representation in $\mathcal{A}^\natural$ is known to govern Hochschild homology.
A better way to store geometrically the relevant categorical information is using the associated presheaf topos, as explained in the quote of A. Grothendieck at the beginning of this article. A small category $\mathcal{C}$ is geometrically represented by the (Grothendieck) topos $\widehat{\mathcal{C}}$ of contravariant functors $\mathcal{C} \longrightarrow \mathfrak{Sets}$ to sets. Both Hochschild and cyclic homologies are described naturally by derived functors on sheaves of abelian groups on the corresponding toposes $\widehat{\Delta}$ and $\widehat{\Lambda}$. The additional structure provided by the λ-operations determines two extensions of Δ and Λ, yielding to the epicyclic refinement $\tilde{\Lambda}$ of the cyclic category [**1, 10, 28**] which is, in essence, the semi-direct product of Λ by an action of the multiplicative monoid $\mathbb{N}^\times$ of non-zero positive integers. This action determines a functor $\mathbf{Mod} : \tilde{\Lambda} \longrightarrow \mathbb{N}^\times$, and at the topos level, a geometric morphism from the epicyclic topos to the topos $\widehat{\mathbb{N}^\times}$ [**11**]. The notion of "point" in a topos leads, even for toposes of presheaf type, to remarkable new spaces. For example, the presheaf topos $\widehat{\mathbb{N}^\times}$ [**12**] gave the first example of a Grothendieck topos whose space of points is noncommutative, *i.e.* a quotient of an ordinary space by an equivalence relation, without any Borel section. The relation between the space of points of $\widehat{\mathbb{N}^\times}$ and the adèle class space of the rationals $\mathbb{Q}^\times\backslash\mathbb{A}_\mathbb{Q}$ [**7**] has culminated with the discovery of the topos $[0,\infty) \rtimes \mathbb{N}^\times$ [**14**], and the result that the space of its points describes the "ζ-sector" of $\mathbb{Q}^\times\backslash\mathbb{A}_\mathbb{Q}$: *i.e.* the quotient $\mathbb{Q}^\times\backslash\mathbb{A}_\mathbb{Q}/\hat{\mathbb{Z}}^\times$ of the adèle class space of $\mathbb{Q}$ by the right action of $\hat{\mathbb{Z}}^\times$. The upgrading of $\widehat{\mathbb{N}^\times}$ to the topos $[0,\infty) \rtimes \mathbb{N}^\times$ was motivated by the search for suitable structure sheaves on these toposes providing a precise geometric structure. As explained in Section 2, the algebras involved here are of "characteristic one" and these structures appear naturally in the study of the cyclic and epicyclic categories. In Section 3 we show how an independent study of these algebras in "characteristic one" and the discovery of the existence for these structures of an analog of the Frobenius endomorphisms, led us to the definition of the geometries of the Arithmetic and Scaling sites.
The search for a universal arithmetic then guided us to the study of algebras in the context of Segal's Γ-rings. This development is reported in Section 4 and illustrates the versatility of the "paradis originel" for the topological algebras mentioned by

Grothendieck. Indeed, the fundamental objects of the relevant tensor category of $\mathbb{S}$-modules (*i.e.* Γ-sets) are defined as functors from a small category $\Gamma^{\rm o}$ (a skeleton of the category of finite pointed sets) to sets. Moreover, the existence of the absolute base $\mathbb{S}$ is dictated by the development of topological cyclic cohomology within the realm of ring spectra. As shown in [**20**], Segal's Γ spaces provide the right framework for topological Hochschild and cyclic theories. By the Dold-Kan correspondence they give rise to the tools apt to develop homological algebra for $\mathbb{S}$-algebras. In Section 4 we explain how this point of view has brought us to the definition of the structure sheaf for the Arakelov compactification $\overline{\operatorname{Spec}\,\mathbb{Z}}$ of the spectrum of the integers and the Riemann-Roch theorem for $\overline{\operatorname{Spec}\,\mathbb{Z}}$. In Section 5 we focus on the resulting presentation of the ring $\mathbb{Z}$ of the integers as polynomials with coefficients in the spherical group ring $\mathbb{S}[C_2] = \mathbb{S}[\pm 1]$, where C_2 is the cyclic group of order 2. This algebra also appears as global sections of the structure sheaf of $\overline{\operatorname{Spec}\,\mathbb{Z}}$.

Finally, Section 7 introduces the main novelty of the present paper namely the *pericyclic category* Π. This category unifies and refines two conflicting notions of epicyclic space existing in the literature. The epicyclic category of Section 6 was originally introduced by T. Goodwillie in a letter to F. Waldhausen and it was implemented in [**1**] to define the corresponding notion of epicyclic space. Later on, Goodwillie introduced in [**24**] a topological category denoted $S\mathcal{RF}$, also called "epicyclic", which is at the root of topological cyclic homology. The pericyclic category Π (see §7.3) is the combinatorial version of a slightly different version $S'\mathcal{RF}$ of $S\mathcal{RF}$. We show that the two notions of epicyclic space mentioned above are encoded by two covariant functors $\Pi \longrightarrow \mathfrak{Sets}$ and we expect that the pericyclic category Π will play the role of a relevant backbone in our future development of a combinatorial version of cyclic homology (hinted at in Section 4).

2. Cyclic categories and projective geometry

Our original motivation for studying exotic algebraic structures of "characteristic one" was to achieve a good algebraic understanding of the cyclic and epicyclic categories, and in particular of the intriguing isomorphism between the cyclic category and its opposite $\Lambda^{\rm o}$. It turns out this isomorphism can be read as the transposition of linear maps in projective geometry, in the natural extension of linear algebra within the context of semifields. The conceptual reason behind the algebraic interpretation of the self-duality of the cyclic category requires some explanation. Originally, Λ has been defined in terms of homotopy classes of degree one, non-decreasing self-maps of the circle S^1, which are compatible with roots of unity [**4**]. A better interpretation of it, (see [**21**]), is obtained by lifting these maps to the universal cover $\mathbb{R}$ of S^1, and then by restricting these maps to $\mathbb{Z}$. In this way, the morphisms between the objects in Λ are described by non-decreasing maps $\phi : \mathbb{Z} \to \mathbb{Z}$ fulfilling the equality, for some $n, m \in \mathbb{N}$

$$\phi(x + n) = \phi(x) + m \qquad \forall x \in \mathbb{Z}. \tag{2.1}$$

An equivalence between two such maps ϕ, ϕ' is expressed by the following relation

$$\phi \sim \phi' \iff \exists k \in \mathbb{Z} \text{ s.t. } \phi'(x) = \phi(x) + km \qquad \forall x \in \mathbb{Z}. \tag{2.2}$$

We shall see that one may interpret the non-decreasing condition together with (2.1) as a linearity property, and (2.2) as a projective equivalence. This is made possible by appealing to the theory of semirings and semifields which is the natural development of the classical theory of rings and fields when one drops the existence of additive inverses. Semirings arise naturally as endomorphisms of any object in a category with a zero object (*i.e.* initial and final) in which the natural maps from direct sums to direct products are isomorphisms. Moving from fields to semifields introduces a few new objects of great interest: the first one is the Boolean semifield $\mathbb{B} = \{0, 1\}$, the unique finite semifield which is not a field. The addition in $\mathbb{B}$ is idempotent ($1+1 := 1$) and a semiring contains $\mathbb{B}$ if and only if it is also idempotent (aka of "characteristic one"). One also sees that there is only one semifield whose multiplicative group is infinite cyclic: this is the tropical semifield $\mathbb{Z}_{\max}$. More precisely, $\mathbb{Z}_{\max}$ is the set $\mathbb{Z} \cup \{-\infty\}$ endowed with the operation

$$n \vee m := \max(n, m) \qquad \forall n, m \in \mathbb{Z} \cup \{-\infty\}$$

replacing the classical addition in $\mathbb{Z}$, while the addition of the integers defines the multiplication in $\mathbb{Z}_{\max}$. In particular, the multiplicative group of $\mathbb{Z}_{\max}$ is the infinite cyclic group $\mathbb{Z}$.
Next, for each integer $n > 0$, one defines a $\mathbb{Z}_{\max}$-module E_n whose underlying set is $\mathbb{Z} \cup \{-\infty\}$ with the same $\vee$ operation, and where the action of $\mathbb{Z}_{\max}$ on E_n is defined by

$$k \bullet m := m + kn \qquad \forall k \in \mathbb{Z}_{\max},\ m \in E_n.$$

One has the following interpretation for a map $\phi : \mathbb{Z} \to \mathbb{Z}$

$$\phi \text{ is non decreasing} \iff \phi(n \vee m) = \phi(n) \vee \phi(m) \qquad \forall n, m \in \mathbb{Z}.$$

Therefore, the non-decreasing maps $\phi : \mathbb{Z} \to \mathbb{Z}$ fulfilling (2.1) can be seen as the non-zero elements of the set $\mathrm{Hom}_{\mathbb{Z}_{\max}}(E_n, E_m)$ which determine, in view of (2.2), projective equivalence classes of ($\mathbb{Z}_{\max}$-)linear maps. Thus, one concludes that the cyclic category Λ coincides with the category of projective morphisms between the modules E_n, and that the isomorphism $\Lambda \simeq \Lambda^o$ is expressed in terms of transpositions of such maps [**9**]. This algebraic description of the cyclic category hints at the extension from Λ to the epicyclic category $\tilde{\Lambda}$ which is known, by the work of J. L. Loday [**28**], to govern the λ-operations in cyclic homology. In classical projective geometry over fields, morphisms connecting two projective spaces over fields K and K' are constructed from linear maps $\phi : E \to E'$ of vector spaces, twisted by a morphism $\sigma : K \to K'$ of the fields, that satisfy the rule

$$\phi(\lambda\xi) = \sigma(\lambda)\phi(\xi) \quad \forall \lambda \in K,\ \xi \in E.$$

This suggests the extension of the cyclic category by implementing the endomorphisms of the semifield $\mathbb{Z}_{\max}$. In this respect, one of the striking features of an algebra in "characteristic one" is that for any idempotent semifield F (*i.e.* the addition of any element with itself gives itself) the map

$$\mathrm{Fr}_a : F \to F, \qquad \mathrm{Fr}_a(x) := x^a \quad \forall x \in F \tag{2.3}$$

is an injective endomorphism, for any positive integer a ([**22**], Propositions 4.43-4.44). These endomorphisms are the analogs of the Frobenius endomorphism in finite characteristic, and for $F = \mathbb{Z}_{\max}$, they are the only non-trivial ones. Using these morphisms to allow for twisted maps between projective spaces amounts to extending (2.1) to

$$\phi(x + n) = \phi(x) + m\,a \quad \forall x \in \mathbb{Z}.$$

The epicyclic category $\tilde{\Lambda}$ is described precisely as the extension of the cyclic category obtained by keeping the same objects $\mathbb{P}(E_n)$ and implementing new morphisms given by general (twisted) morphisms of projective spaces [**9**]. Furthermore, there is a natural "degree" functor **Mod** : $\tilde{\Lambda} \longrightarrow \mathbb{N}^\times$ which associates to a twisted morphism by Fr_a its degree $a \in \mathbb{N}^\times$. The emerging geometric picture involves geometric morphisms of the toposes dual to the categories. To promote, when the algebra $\mathcal{A}$ is commutative, the cyclic module $\mathcal{A}^\natural$ as a sheaf of vector spaces over the topos dual to the epicyclic category one proceeds as follows. The forgetful functor which associates to a map of projective spaces $\mathbb{P}(E_n) \to \mathbb{P}(E_m)$ the underlying morphism $\mathbb{Z}/n\mathbb{Z} \to \mathbb{Z}/m\mathbb{Z}$ of sets of points, connects $\tilde{\Lambda}$ to the category $\mathfrak{Fin}$ of finite sets. When an algebra $\mathcal{A}$ over a field k is commutative, the cyclic module $\mathcal{A}^\natural$ extends to a covariant functor from $\mathfrak{Fin}$ to vector spaces. To understand this functor as a sheaf of vector spaces of a topos, one considers the topos associated with the small category $\tilde{\Lambda}^{\mathrm{o}}$ (the opposite category of $\tilde{\Lambda}$) which maps to $\mathbb{N}^\times$ by the "degree" functor

$$\begin{array}{c} \tilde{\Lambda}^{\mathrm{o}} \\ \mathbf{Mod}\downarrow \\ \mathbb{N}^\times \end{array}$$

This functor extending $\mathcal{A}^\natural$ governs the λ-operations in cyclic homology and is discussed in details in §7.3. In [**8**], these operations are implemented to obtain the cyclic cohomological interpretation of Serre's local L-factors of arithmetic varieties.

3. Geometric developments

The main role played by $\mathbb{Z}_{\max}$ in the understanding of the cyclic and epicyclic categories, and the canonical action (2.3) of $\mathbb{N}^\times$ by Frobenius endomorphisms on this semifield, led us to endow the "base" topos $\widehat{\mathbb{N}^\times}$ with the structure sheaf prescribed exactly by this action. The outcoming geometric structure is the *Arithmetic Site* [**12**] which, despite being defined using only countable data of combinatorial nature, admits a one-parameter family of correspondences whose composition law reflects a fundamental action of the multiplicative group $\mathbb{R}_+^*$ of positive real numbers. This is the group of Galois automorphisms of the semifield $\mathbb{R}_+^{\max}$ of positive real numbers under the two operations of max for addition and usual multiplication, which plays a central role both in the semiclassical limit of quantum mechanics *i.e.* in idempotent analysis, and tropical geometry. The space of points of the Arithmetic Site over $\mathbb{R}_+^{\max}$ is canonically isomorphic to the ζ-sector of the adèle class space $\mathbb{Q}^\times\backslash\mathbb{A}_\mathbb{Q}$ of the rationals [**12**], and the Galois action of $\mathbb{R}_+^*$ on this space reflects geometrically the action of the idèle class group on the ζ-sector, in complete analogy with the geometric development in finite characteristic. The extension of scalars to $\mathbb{R}_+^{\max}$ defines a new geometric object: the *Scaling Site* [**14**]. As a topos, this is the semi-direct product $[0,\infty) \rtimes \mathbb{N}^\times$, where $[0,\infty)$ is the euclidean half line $[0,\infty)$. The points of the topos $[0,\infty) \rtimes \mathbb{N}^\times$ coincide with the "ζ-sector" already selected by the points of the Arithmetic Site over the same scalar extension. The new additional structure inherited by the Scaling Site is the structure sheaf made by piecewise affine, convex functions. A Riemann-Roch theorem analogous to that holding for elliptic curves is verified for the restriction of the Scaling Site to its periodic orbits [**14**]. A new striking feature is that the dimensions of the cohomologies involved are real numbers and they correspond, in the world of "characteristic one", to the real dimensions of Murray and von Neumann. This Riemann-Roch

formula implements only H^0 and H^1 for divisors. One uses Serre duality to define H^1, by-passing in this way, the development of a full cohomological machine. The module H^0 of global sections is naturally filtered using the p-adic valuation for the periodic orbit associated with the prime p, and this filtration gives rise to continuous dimensions.
The development in [**15**] of homological algebra in "characteristic one" has shown that the lack of additive inverses creates considerable difficulties in the process of passing from abelian categories to their counterpart, for idempotent semirings.

4. Universal cyclic theory

Surprisingly, one can circumvent the open problem of the elaboration of homological algebra for idempotent structures by developing cyclic homology in a world parallel to that of noncommutative geometry! This is the world of algebraic K-theory as promoted by D. Quillen and F. Waldhausen. We see it parallel to noncommutative geometry since it is transverse to the standard classification of mathematics into distinct branches. Cyclic homology was dragged into this context at the beginning of the 80's, and then developed for spectra, in the world of "brave new rings". At the Oberwolfach meeting in 1987, which was organized to sponsor interactions between functional analysis and algebraic topology, T. Goodwillie [**23**] emphasized that in the world of "brave new rings", the ring $\mathbb{Z}$ of the integers becomes an algebra over a more fundamental ring, namely the sphere spectrum $\mathbb{S}$. While the homotopical constructions are not particularly appealing to functional analysts, as iterated loop spaces may get quickly out of explicit control, a more concrete and basic version of the theory has gradually emerged and it is accounted for in [**20**]. The framework in question is that of G. Segal Γ-spaces which provides a workable model of connective spectra.
Our independent viewpoint on this subject is that the theory of Γ-spaces supplies the required constructions to produce the basic homological algebra, available through the Dold-Kan correspondence, for a very concrete algebraic set-up solely based on the Segal category Γ^{o}. This category is a skeleton of the category $\mathfrak{Fin}_*$ of pointed finite sets. Its main role is to provide the broadest generalization of a commutative and associative operation. These two properties combine to give a meaning to a "finite sum" and to obtain a covariant functor $\mathcal{F} : \mathfrak{Fin}_* \longrightarrow \mathfrak{Sets}_*$, where the image of a map $f : X \to Y$ of finite pointed sets uses the finite sum over the inverse image $f^{-1}(y)$ to obtain the value at $y \in Y$. This construction generalizes broadly the Eilenberg-MacLane functor which replaces an abelian group A with the covariant functor $\mathrm{HA} : \mathfrak{Fin}_* \longrightarrow \mathfrak{Sets}_*$ that assigns to a finite pointed set X the pointed set of A-valued divisors on X (taking the value 0 at the base point). Divisors push forward by the formula

$$f_*(D)(y) := \sum_{f(x)=y} D(x). \tag{4.1}$$

The Eilenberg-MacLane functor determines a fully faithful inclusion of abelian groups into the category of (covariant) functors $\Gamma^{\mathrm{o}} \longrightarrow \mathfrak{Sets}_*$, where morphisms are natural transformations of functors. Then, by implementing an idea of G. Lydakis, one defines the smash product of such functors starting from the closed structure determined by the smash product in $\mathfrak{Fin}_*$. The identity functor $\mathrm{Id} : \Gamma^{\mathrm{o}} \longrightarrow \mathfrak{Sets}_*$ defines the easiest example of such an algebra in this tensor category and it describes

the sphere spectrum $\mathbb{S}$ in its most elementary form. This is the fundamental brave new ring advocated by Goodwillie in Oberwolfach.
In [**13**] we show that this theory of $\mathbb{S}$-algebras embodies all the previous attempts to extend ordinary algebra to reach a "universal arithmetic".
The main difficulty that one encounters in trying to extend the structure sheaf of Spec $\mathbb{Z}$ by taking into account the point at ∞ (the archimedean place), comes from the lack of an analog, for the archimedean prime, of the subring of $\mathbb{Q}$ which, for a non-archimedean prime p, is the localization of $\mathbb{Z}$ at p. In p-adic terms this corresponds to elements of $\mathbb{Q}$ which belong to the local ring $\mathbb{Z}_p$ of p-adic integers (whose p-adic norm is $\leqslant 1$). At the archimedean place, one meets the difficulty that the condition $|q| \leqslant 1$ does not define a subgroup of $\mathbb{Q}$. However, this problem is successfully solved by the theory of $\mathbb{S}$-algebras. Indeed, the Eilenberg-MacLane functor H replaces faithfully a ring R by the $\mathbb{S}$-algebra HR and this applies in particular to the rings $\mathbb{Z}$ and $\mathbb{Q}$. In this categorical universe the $\mathbb{S}$-algebra H$\mathbb{Q}$ admits the non-trivial subalgebra $\mathcal{O}_\infty : \Gamma^{\mathrm{o}} \longrightarrow \mathfrak{Sets}_*$ defined by the condition on $\mathbb{Q}$-valued divisors:

$$\sum_{x \in X} |D(x)| \leqslant 1. \tag{4.2}$$

This condition is stable by push forward and as such it defines an $\mathbb{S}$-module. Moreover, since the product map

$$\mathcal{O}_\infty(X) \wedge \mathcal{O}_\infty(Y) \to \mathcal{O}_\infty(X \wedge Y)$$

is the restriction of the product map for H$\mathbb{Q}$ which respects the basic condition (4.2), one sees that $\mathcal{O}_\infty$ is an $\mathbb{S}$-subalgebra of H$\mathbb{Q}$. This fact allows one to define the structure sheaf of $\overline{\text{Spec } \mathbb{Z}}$ as a subsheaf of the constant sheaf $\mathbb{Q}$ [**13**].
A relevant application of this construction is the detection of the Gromov norm, which is obtained through a refinement of the ordinary homology with rational coefficients using $\mathcal{O}_\infty$ as $\mathbb{S}$-module [**16**].
With the point at infinity taken into account, $\overline{\text{Spec } \mathbb{Z}}$ behaves like a projective curve, and the global sections of its structure sheaf form a very simple extension of the basic algebra $\mathbb{S}$. This is the spherical group ring $\mathbb{S}[C_2] = \mathbb{S}[\pm 1]$, which associates to a finite pointed set X the smash product of X with the pointed, multiplicative monoid $\{-1, 0, 1\}$. By working over this base, one gains the categorical interpretation of additive inverses! For an abelian group A, for example, the $\mathbb{S}$-module HA is naturally an $\mathbb{S}[\pm 1]$-module using the rule $(-1) \times x = -x$ for all $x \in$ A.
This construction seems to have the potential to go quite far. In the recent article [**19**] for example, we show how to parallel, on $\overline{\text{Spec } \mathbb{Z}}$, the adelic proof of the Riemann-Roch formula for function fields given by A. Weil, including the use of Pontryagin duality. The definition of the cohomologies is based on the universal arithmetic explained above and on an extension for $\mathbb{S}$-modules of the Dold-Kan correspondence. To an Arakelov divisor D on $\overline{\text{Spec } \mathbb{Z}}$ one associates a Γ-space $\mathbf{H}(D)$ using a short complex of $\mathbb{S}[\pm 1]$-modules directly related to the divisor. The two cohomologies $H^0(D)$ and $H^1(D)$ are then defined *independently*, as well as the notion of their dimension over $\mathbb{S}[\pm 1]$. The Riemann-Roch formula also proven in *op.cit.* equates the *integer valued* Euler characteristic of a divisor D with a simple modification of the traditional expression (*i.e.* the degree of the divisor plus log 2). The integer-valued topological side of the formula involves, besides the ceiling function, the division by log 3. This result unveils the special role of the number 3

among the integers and the triadic expansion of the integers as polynomials $P(X)$, with coefficients in $\{-1,0,1\}$ evaluated at $X=3$. In Section 5 we shall explore the extent to which one can view $\mathbb{Z}$ as a polynomial ring over $\mathbb{S}[\pm 1]$.
A long-range goal is the computation of the universal cyclic cohomology for $\overline{\operatorname{Spec}\,\mathbb{Z}}$.

5. $\mathbb{S}[\pm 1][3]$ and the ring of integers

In our recent work [**19**] on the arithmetic Riemann-Roch theorem for $\overline{\operatorname{Spec}\,\mathbb{Z}}$, the number 3 plays the role of a natural generator, allowing one to label the integers $m \in \mathbb{Z}$ as polynomials in powers of 3 with coefficients in $\mathbb{S}[\pm 1]$

$$P(X)=\sum_{j=0}^{k} a_j X^j, \quad a_j \in \{-1,0,1\},\ \forall j. \tag{5.1}$$

The subtlety of the description of the ring structure of the integers $\mathbb{Z}$ in this parametrization derives from the addition of polynomials with coefficients in the multiplicative monoid $\{-1,0,1\}$. Here $\{-1,0,1\}$ is endowed with the obvious multiplication rule, thus one knows how to multiply monomials *i.e.* $a\,X^n \times b\,X^m = (ab)\,X^{n+m}$, while the product of polynomials is obtained uniquely, using the distributive law, provided one knows how to add them. The sum of two monomials $a\,X^n, b\,X^m$ of different degrees is simply the polynomial $a\,X^n + b\,X^m$ and when the degrees are the same the distributivity law $a\,X^n + b\,X^n = (a+b)X^n$ reduces the reconstruction of the sum to polynomials of degree 0.
We shall make sure that addition is defined on polynomials of degree 0 in such a way that the following rule holds:

The sum of 3 polynomials of degree 0 is a polynomial of degree $\leqslant 1$.

Using this hypothesis it follows, by induction on the degree n, that the sum of two polynomials of degree $\leqslant n$ is a polynomial of degree $\leqslant n+1$. The need to require that the sum of three polynomials of degree 0 is of degree $\leqslant 1$ (rather than for just the sum of two), is due to the role of the carry over in adding arbitrary polynomials. Let us now consider the addition of two elements of $\{-1,0,1\}$. If one of them is 0 or if they have opposite signs, then the sum is the obvious one. Thus, using the symmetry $x \mapsto -x$, we can assume that they are both equal to 1. The only new rule that we add is the following

$$1+1=X-1. \tag{5.2}$$

By applying (5.2), one adds two polynomials of degree 0 using the following table

$+$	-1	0	1
-1	$1-X$	-1	0
0	-1	0	1
1	0	1	$X-1$

To specify the sum of three polynomials of degree 0, it is enough to know how to add to the polynomial $X-1$ the elements of $\{-1,0,1\}$. Using the associativity of the sum and (5.2), one has

$$(X-1)+1=X, \qquad (X-1)-1=X-(1+1)=X-(X-1)=1.$$

It follows then that, as required, the sum of 3 polynomials of degree 0 is a polynomial of degree $\leqslant 1$. The following result determines the sum of any pair of polynomials under (5.2)

PROPOSITION 5.1. *Let $P(X)$ and $Q(X)$ be two polynomials as in* (5.1)*, of degree at most n. Then there exists a unique polynomial of degree at most $n+1$ which coincides, using the rule* (5.2)*, with the sum $P(X) + Q(X)$.*

PROOF. Let assume (by induction) that the statement holds for any two polynomials of degree $< n$ and let consider a pair $P(X)$, $Q(X)$ both of degree n. We decompose $P(X) = P_1(X) + \alpha_n X^n$, $Q(X) = Q_1(X) + \beta_n X^n$, with $\alpha_n, \beta_n \in \{-1, 0, 1\}$. Then

$$P(X) + Q(X) = P_1(X) + Q_1(X) + (\alpha_n + \beta_n)X^n,$$

where $P_1(X) + Q_1(X)$ is, by induction, the sum of a polynomial of degree $n-1$ with a term γX^n, $\gamma \in \{-1, 0, 1\}$. Then, from the specification of the sum of three polynomials of degree zero we have $\alpha_n + \beta_n + \gamma = \delta + \epsilon X$, with $\delta, \epsilon \in \{-1, 0, 1\}$. Thus the sum $P(X)+Q(X)$ can be written as a polynomial of degree at most $n+1$, where the terms of degree $\geqslant n$ are $\delta X^n + \epsilon X^{n+1}$. □

We obtain the following intriguing conclusion

PROPOSITION 5.2. *The set of polynomials as in* (5.1)*, under the addition stated in Proposition* 5.1*, and the unique associated product, forms a ring isomorphic to $\mathbb{Z}$.*

PROOF. The map

$$\theta : \sum \alpha_j X^j \mapsto \sum \alpha_j 3^j$$

is a bijection from the set of polynomials as in (5.1) with the integers. This map is compatible with addition and multiplication in view of Proposition 5.1. □

There is a function that allows one to compute effectively sums of polynomials as above. This is a function of three variables α, β, γ in $\{-1, 0, 1\}$ which specifies their sum as a polynomial of degree at most 1 in X. To determine $\alpha + \beta + \gamma$ one can think of it as a sum of 3 real numbers and write the set $\{-1, 0, 1\} + \gamma$ where this sum belongs, to obtain the coefficient of X. At the algebraic level, $\alpha + \beta + \gamma$ can be written explicitly using the finite field $\mathbb{F}_3 = \{-1, 0, 1\}$ as follows

$$\alpha + \beta + \gamma = \overline{\alpha + \beta + \gamma} + \overline{\sigma(\alpha, \beta, \gamma)}\ X$$

where $\overline{m}$ is the residue of m modulo 3 and

$$\sigma(\alpha, \beta, \gamma) := \alpha\beta\gamma - \alpha^2\beta - \alpha^2\gamma - \alpha\beta^2 - \alpha\gamma^2 - \beta^2\gamma - \beta\gamma^2.$$

This determines the algebraic expression for the sum of three polynomials of degree 0 as in the proof of Proposition 5.1. Then, by induction on n, one obtains an algebraic expression for the coefficients of X^n in the sum $\sum_j \alpha_j X^j + \sum_j \beta_j X^j$ of two polynomials. Indeed, one defines by induction polynomials $s_n(\alpha_0, \ldots, \alpha_{n-1}, \beta_0, \ldots, \beta_{n-1})$ with coefficients in $\mathbb{F}_3$ as follows

$$s_0 := 0, \quad s_n = \sigma(\alpha_{n-1}, \beta_{n-1}, s_{n-1}).$$

Then the coefficient of X^n in the sum $\sum_j \alpha_j X^j + \sum_j \beta_j X^j = \sum_j \gamma_j X^j$ is given by

$$\gamma_n = \alpha_n + \beta_n + s_n. \tag{5.3}$$

This construction provides a sequence of polynomials $\gamma_n(\alpha_1, \ldots, \alpha_n, \beta_1, \ldots, \beta_n)$ and finally one has

LEMMA 5.3. *The polynomial $\gamma_n(\alpha_1, \ldots, \alpha_n, \beta_1, \ldots, \beta_n)$ vanishes provided the components α_n, α_{n-1} and β_n, β_{n-1} all vanish.*

PROOF. It is enough to show that $s_n = 0$ if $\alpha_{n-1} = 0$ and $\beta_{n-1} = 0$. This follows from the inductive definition of $s_n = \sigma(\alpha_{n-1}, \beta_{n-1}, s_{n-1})$. □

Since one works over $\mathbb{F}_3$ $x^n = x$ if n is an odd number, while $x^n = x^2$ if n is a non-zero even number. Thus one can reduce the exponents of the variables in the polynomials γ_n to be at most 2. One has $s_1(\alpha_0, \beta_0) = -\alpha_0\beta_0^2 - \alpha_0^2\beta_0$, while s_2 in reduced form is given by

$$s_2(\alpha_0, \alpha_1, \beta_0, \beta_1) = \alpha_1\alpha_0^2\beta_0^2 + \alpha_0^2\beta_0\beta_1^2 + \alpha_1^2\alpha_0^2\beta_0 + \alpha_0^2\beta_0^2\beta_1 - \alpha_1\alpha_0^2\beta_0\beta_1 + \alpha_1^2\alpha_0\beta_0^2 + \alpha_0\beta_0^2\beta_1^2 + \alpha_1\alpha_0\beta_0 - \alpha_1\alpha_0\beta_0^2\beta_1 + \alpha_0\beta_0\beta_1 - \alpha_1\beta_1^2 - \alpha_1^2\beta_1$$

To write in full $s_3(\alpha_0, \alpha_1, \alpha_2, \beta_0, \beta_1, \beta_2)$ would take several lines.

Conceptually, one sees that the above construction is described by the addition rule of two Witt vectors over $\mathbb{F}_3$. The number 3 is the only prime for which the Witt vectors with only finitely many non-zero components form an additive subgroup of the Witt ring. More precisely one can state the following

PROPOSITION 5.4. *Witt vectors with only finitely many non-zero components form a subring of the ring* $\mathbf{W}(\mathbb{F}_3)$. *This subring is isomorphic to* $\mathbb{Z} \subset \mathbb{Z}_3$.

PROOF. For any prime p the map

$$\mathbf{W}(\mathbb{F}_p) \ni \xi = (\xi_j) \mapsto \tilde{\tau}(\xi) := \sum_{j=0}^{\infty} \tau(\xi_j)p^j,$$

where τ is the Teichmüller lift, is known to be a ring isomorphism between the Witt ring $\mathbf{W}(\mathbb{F}_p)$ and the ring $\mathbb{Z}_p$ of p-adic integers. For $p = 3$, $\tau(\mathbb{F}_3) = \{-1, 0, 1\} \subset \mathbb{Z} \subset \mathbb{Z}_3$, thus $\tilde{\tau}$ maps Witt vectors with only finitely many non-zero components into the subring $\mathbb{Z} \subset \mathbb{Z}_3$. This map is an injective ring homomorphism and it surjects onto $\mathbb{Z}$ since by Proposition 5.2 any integer can be written as a finite sum of powers of 3 with coefficients in $\{-1, 0, 1\}$. □

6. The cyclic and epicyclic categories as subcategories of Cat

In Section 2 we reviewed the cyclic and the epicyclic categories using concepts of linear algebra in characteristic one, in particular we have described the objects of these categories as projective spaces over the semifield $\mathbb{Z}_{\max}$. Given these results it seems natural to wonder about the existence of further presentations of the same categories: in particular whether their objects and morphisms may be realized within the "paradis originel" provided by the category **Cat** of small categories. This idea was developed by D. Kaledin in [**27**] and it turns out to be a slight variation of our earlier viewpoint developed in [**10**]. In that paper we described the cyclic and epicyclic categories in terms of oriented groupoids: this result is reviewed here below. This section is dedicated to comparing our viewpoint with that of [**27**] that has the advantage to interpret the collection of the objects as small categories (and as such objects of **Cat**) and the morphisms as functors, thus presenting the two categories as subcategories of **Cat**.

6.1. Oriented groupoids. We recall that a groupoïd G can be seen as a small category in which every morphism is invertible. We denote by $G^{(0)}$ the set of objects of G and by $r, s : G \to G^{(0)}$ *resp.* the range and the source of the morphisms of G. We view $G^{(0)} \subset G$ as the subset of the identity morphisms of G.

The following definition is a direct generalization of the notion of the right ordered group for a groupoïd G. Given $X \subset G$, we let $X^{-1} := \{\gamma^{-1} \mid \gamma \in X\}$.

DEFINITION 6.1. An oriented groupoïd (G, G_+) is a groupoïd G endowed with a subcategory $G_+ \subset G$, fulfilling the following relations

$$G_+ \cap G_+^{-1} = G^{(0)}, \qquad G_+ \cup G_+^{-1} = G. \tag{6.1}$$

Let H be a group acting on a set X. The semi-direct product $G := X \ltimes H$ is a groupoïd with source, range and composition law defined respectively as follows

$$s(x,h) := x, \quad r(x,h) := hx, \quad (x,h) \circ (y,k) := (y, hk).$$

As it happens in any groupoïd, the composite $\gamma \circ \gamma'$ of two elements of G is only defined when $s(\gamma) = r(\gamma')$: here this holds if and only if $x = ky$.
There is a canonical homomorphism of groupoïds $\rho : G \to H$, $\rho(x,h) = h$. We recall from [**10**] (Lemma 2.2) the following

LEMMA 6.2. *Let (H, H_+) be a right ordered group. Assume that H acts on a set X. Then the semi-direct product $G = X \ltimes H$, with $G_+ := \rho^{-1}(H_+)$, is an oriented groupoïd.*

We shall consider, in particular, the right ordered group $(H, H_+) = (\mathbb{Z}, \mathbb{Z}_+)$ acting by translation on the set $X = \mathbb{Z}/(m+1)\mathbb{Z}$ of integers modulo $m+1$. Then, by applying the above lemma one obtains the oriented groupoïd

$$\mathfrak{g}(m) := \big(\mathbb{Z}/(m+1)\mathbb{Z}\big) \ltimes \mathbb{Z}.$$

The relation with the epicyclic category is provided by the following statement (see [**10**] Corollary 3.11)

PROPOSITION 6.3. *The epicyclic category $\tilde{\Lambda}$ is canonically isomorphic to the category whose objects are the oriented groupoïds $\mathfrak{g}(m)$, $m \geqslant 0$, and the morphisms are the non-trivial morphisms of oriented groupoïds.*
The functor which associates to a morphism of oriented groupoïds its class up-to equivalence (of small categories) corresponds, under the canonical isomorphism, with the functor **Mod** $: \tilde{\Lambda} \longrightarrow \mathbb{N}^\times$ *sending a semilinear map of semimodules over $\mathbb{F} = \mathbb{Z}_{\max}$ to the corresponding injective endomorphism $\mathrm{Fr}_n \in \mathrm{End}(\mathbb{F})$ cf.* [**9**].

6.2. Comparison with [27]. We recall the presentation of the cyclic category given in [**27**]: the following notations and the three definitions are taken literally from *op.cit.*

Let $[1]_\Lambda$ be the category with one object and whose endomorphisms are determined by non-negative integers $a \in \mathbb{Z}_+$. The inclusion $\mathbb{Z}_+ \subset \mathbb{Z}$ induces a functor $[1]_\Lambda \longrightarrow \mathbf{pt}_{\mathbb{Z}}$. By the Grothendieck construction, sets with a $\mathbb{Z}$-action correspond to small categories discretely bi-fibered over $\mathbf{pt}_{\mathbb{Z}}$, thus a $\mathbb{Z}$-set S induces, by pull-back, a discrete bi-fibration $[1]_\Lambda^S \to [1]_\Lambda$. For any integer $n \geqslant 1$, let $[n]_\Lambda$ be the category corresponding to the $\mathbb{Z}$-set $\mathbb{Z}/n\mathbb{Z}$. The category $[n]_\Lambda$ is equivalent to the additive monoid $\mathbb{N}$ of non-negative integers. The degree of a functor is defined in terms of the corresponding morphism of monoids

DEFINITION 6.4. A functor $f : [n]_\Lambda \longrightarrow [m]_\Lambda$ is said to be vertical if it is a discrete bi-fibration, horizontal if $\deg(f) = 1$ and non-degenerate if $\deg(f) \neq 0$.

DEFINITION 6.5. The *cyclotomic category* ΛR is the small category whose objects $[n]$ are indexed by positive integers and whose morphisms $[n] \to [m]$ are given by non-degenerate functors $f : [n]_\Lambda \longrightarrow [m]_\Lambda$.

DEFINITION 6.6. The cyclic category Λ is the subcategory of ΛR with the same set of objects and whose morphisms are the horizontal ones.

Next, we relate the above definitions (taken literally from [**27**]) the description of the epicyclic category given in [**10**] and here recalled in §6.1. There is a natural functor $\mathcal{P}$ which associates to an oriented groupoïd G the subcategory $G_+ = \mathcal{P}(G)$, viewed as a small category. For $G = \mathfrak{g}(n)$, one has

$$\mathcal{P}(\mathfrak{g}(n)) = [n+1]_\Lambda \qquad \forall n \geqslant 0. \tag{6.2}$$

The link with the epicyclic category is then provided by the following

PROPOSITION 6.7. *The functor $\mathcal{P}$ identifies the epicyclic category $\tilde{\Lambda}$ (as described in Proposition* 6.3*) with the cyclotomic category ΛR. From the commutativity of the diagram*

$$\begin{array}{ccc} \tilde{\Lambda} & \xrightarrow{\ \mathcal{P}\ } & \Lambda R \\ {\scriptstyle \mathbf{Mod}}\big\downarrow & & \big\downarrow{\scriptstyle \deg} \\ \mathbb{N}^\times & \xrightarrow[\ \mathrm{Id}\ ]{} & \mathbb{N}^\times \end{array}$$

the degree as in [**27**] *corresponds to the module as in Proposition* 6.3.

Since the cyclic category Λ is the subcategory of $\tilde{\Lambda}$ with the same objects, and morphisms fulfilling $\mathbf{Mod}(f) = 1$, it follows from the above identification that Definition 6.6 determines the cyclic category Λ inside $\tilde{\Lambda}$.

7. The pericyclic category Π

In §6.2 it has been shown that the epicyclic category $\tilde{\Lambda}$ can be identified with the subcategory ΛR of **Cat** (the name cyclotomic category for ΛR seems, therefore, to be redundant and possibly also creating confusion). Another source of confusion is that the name *epicyclic* is used in the literature with two different meanings. The epicyclic category $\tilde{\Lambda}$ of Section 6 was originally introduced by T. Goodwillie in a letter to F. Waldhausen, however later on Goodwillie introduced in [**24**] another category, also called "epicyclic" (denoted with $S\mathcal{RF}$ *cf.* [**20**]) and here described in §7.1. In the same §7.1 we also correct a "typo", sometimes reported in the literature, regarding the composition rule for morphisms in the category $S\mathcal{RF}$. In §7.2 we show that using the category $S'\mathcal{RF}$ (with the modified rule implemented), a cyclotomic spectrum gives rise to a functor from $S'\mathcal{RF}$ to spectra. In §7.3 we introduce the *pericyclic category* Π, as the combinatorial version of $S'\mathcal{RF}$ and prove that the two notions of epicyclic space (in relation to the two epicyclic categories mentioned above) are encoded in terms of (covariant) functors $\Pi \longrightarrow \mathfrak{Sets}$. Finally, §7.4 investigates the points of the topos dual to a small category $\mathcal{R}$ constituent of the pericyclic category.

7.1. The category $S\mathcal{RF}$. The small category $\mathcal{RF}$ (see [**20**] Definition 6.2.3.2) has one object a for each integer $a > 0$ and for any pair of objects a, b the morphisms connecting them are

$$\mathrm{Hom}_{\mathcal{RF}}(a,b) := \{f_{r,s} \mid r, s \in \mathbb{N}^\times, a = rbs\}.$$

In particular one has: $\mathrm{Hom}_{\mathcal{RF}}(a,b) = \varnothing$ unless $b|a$. The composition of morphisms is assigned by

$$f_{r,s} \circ f_{p,q} = f_{rp,sq}.$$

Thus one sees that the following maps define functors sending all the objects to the unique object $\bullet$ of $\mathbb{N}^\times$ and acting on morphisms as follows

(7.1)
$$\mathrm{Res}: \mathcal{RF} \longrightarrow \mathbb{N}^\times, \quad \mathrm{Res}(f_{p,q}) := p \in \mathbb{N}^\times, \quad \mathrm{Fr}: \mathcal{RF} \longrightarrow \mathbb{N}^\times, \quad \mathrm{Fr}(f_{p,q}) := q \in \mathbb{N}^\times.$$

The small category $\mathcal{SRF}$ (see *op.cit.* Remark 6.2.3.5) is topological and has the same objects as $\mathcal{RF}$ but implements an angular parameter θ in the morphisms (of $\mathcal{RF}$), viewed as an element of $\mathbb{T} = \mathbb{R}/\mathbb{Z}$ or equivalently understood as the element $e(\theta) := \exp(2\pi i\theta)$ of $U(1) := \{z \in \mathbb{C} \mid |z| = 1\}$. An *epicyclic space* is a functor from $\mathcal{SRF}$ to topological spaces. In this framework the first index r in the morphisms $f_{r,s}$ corresponds to restriction maps (see **[20]**, 6.2.3) while the second index labels the *inclusions of fixed points*.

Let us carefully review the role of such inclusions. Given a $\mathbb{T}$-space X, *i.e.* a space X endowed with an action $\theta \mapsto \rho_\theta \in \mathrm{Aut}(X)$ of $\mathbb{T}$ on X, one associates to an integer $q > 0$ the cyclic subgroup $C_q = \{\theta \in \mathbb{T} \mid q\theta = 0\}$ (equivalently the group of q-th roots of unity in $U(1)$) and the subspace X^{C_q} of the C_q fixed points. The canonical inclusion $i_q : X^{C_q} \to X$ is $\mathbb{T}$-equivariant by construction, however, because of the canonical isomorphism

$$\alpha : \mathbb{T} \to \mathbb{T}/C_q, \quad \alpha(\theta) = \frac{1}{q}\theta, \quad \alpha^{-1} : \mathbb{T}/C_q \to \mathbb{T}, \quad \alpha^{-1}(\theta) = q\theta$$

one modifies the action of $\mathbb{T}$ on X^{C_q} by composing with α. For this new action $\rho' = \rho \circ \alpha$ on X^{C_q} fulfilling $\rho'_{q\theta} = \rho \circ \alpha(q\theta) = \rho_\theta$, one has

$$\rho_\theta i_q(x) = i_q(\rho'_{q\theta}(x)). \tag{7.2}$$

Next, we review in detail the notion of an epicyclic space given in **[20]** (Definition 6.2.3.1). For any integer $k > 0$, the edgewise subdivision is an endofunctor $\mathrm{sd}^k : \Delta \longrightarrow \Delta$ obtained by concatenating k copies of an ordinal: it multiplies the cardinality by k. Given a simplicial set Y one lets $\mathrm{sd}_k(Y) := Y \circ \mathrm{sd}^k$ be the simplicial set obtained by pre-composition with these endofunctors. If moreover Y is a cyclic set then the cyclic structure extends to $\mathrm{sd}_k(Y)$ provided one replaces the cyclic category Λ with the k-cyclic category Λ_k defined by replacing the cyclic group C_n involved as automorphisms of the object of cardinality n, by the group C_{nk} with generator $t_{(n,k)}$, while keeping the same relations of the generator with faces and degeneracies. The fixed points Z^{C_k} in a k-cyclic set Z form a cyclic set. Thus, given a cyclic set Y and an integer $k > 0$ the fixed points $\mathrm{sd}_k(Y)^{C_k}$ in the edgewise subdivision $\mathrm{sd}_k(Y)$ form a cyclic set.

DEFINITION 7.1. An *epicyclic space* (Y, ϕ) is a pointed cyclic space Y equipped with pointed cyclic maps $\phi_q : \mathrm{sd}_q(Y)^{C_q} \to Y$, $\phi_1 = \mathrm{Id}$, so that the following diagrams commute

$$\begin{array}{ccc} \mathrm{sd}_k(\mathrm{sd}_r(Y)^{C_r})^{C_k} & \xrightarrow{id} & \mathrm{sd}_{kr}(Y)^{C_{kr}} \\ \Big\downarrow{\scriptstyle \mathrm{sd}_k(\phi_r)} & & \Big\downarrow{\scriptstyle \phi_{kr}} \\ \mathrm{sd}_k(Y)^{C_k} & \xrightarrow{\phi_k} & Y \end{array}$$

The geometric realization $|Z|$ of a k-cyclic set Z admits a canonical action of the group $\mathbb{R}/k\mathbb{Z}$, and the fixed points of this action under the subgroup $\mathbb{Z}/k\mathbb{Z} \sim C_k$

give rise to the geometric realization of the cyclic set Z^{C_k}. Moreover, one has a canonical homeomorphism of geometric realizations ([**20**] Lemmas 6.2.2.1, 6.2.2.2)

$$D_a : |\mathrm{sd}_a(Y)| \to |Y|, \qquad D_a(s + a\mathbb{Z})(y) = \left(\frac{s}{a} + \mathbb{Z}\right) D_a(y).$$

Given an epicyclic space (Y, ϕ), one introduces for each integer $a > 0$ the $\mathbb{T}$-space

$$Y(a) := |\mathrm{sd}_a(Y)^{C_a}| \simeq |Y|^{C_a}$$

and the maps $i_q : Y(qa) \to Y(a)$ corresponding to the inclusion of the fixed points

$$Y(qa) \simeq |Y|^{C_{qa}} \subset |Y|^{C_a} \simeq Y(a).$$

Then it follows from (7.2) that the compatibility of the maps i_q with the $\mathbb{T}$-action is expressed by the following formula

$$\rho_\theta i_q(x) = i_q(\rho_{q\theta}(x)). \tag{7.3}$$

Now we come to the point of issue. The composition of the morphisms is given in [**24**] by the formula

$$(\theta, f_{r,s}) \circ (\tau, f_{p,q}) = (\theta + s\tau, f_{rp,sq}). \tag{7.4}$$

However, one should associate to the pair $(\theta, f_{r,s})$ the morphism $\phi_r i_s \rho_\theta$, instead of $\phi_r \rho_\theta i_s$, the reason being that the latter labeling is redundant since ρ_θ applied to the range of i_s is unchanged when θ is multiplied by s-torsion. Then, standing (7.3), (7.4) ought to be replaced with

$$(\theta, f_{r,s}) \circ (\tau, f_{p,q}) = (q\theta + \tau, f_{rp,sq}) \tag{7.5}$$

which corresponds to

$$i_s \rho_\theta i_q \rho_\tau = i_s i_q \rho_{q\theta} \rho_\tau.$$

Next, we rely on (7.5) to define the variant $S'\mathcal{RF}$ of the category $S\mathcal{RF}$. Thus by construction $S'\mathcal{RF}$ is an extension of $\mathcal{RF}$ by $U(1)$, and there is a natural forgetful functor $\sigma : S'\mathcal{RF} \longrightarrow \mathcal{RF}$ which is the identity on objects and maps $(z, f_{r,s}) \mapsto f_{r,s}$. We now compare this category with the monoid $U(1) \ltimes \mathbb{N}^\times$ considered in [**1**], whose elements are pairs $(z, s) \in U(1) \ltimes \mathbb{N}^\times$, and the composition law is defined by

$$(z_1, s) \circ (z_2, q) = (z_1 z_2^s, sq).$$

One obtains the following commutative diagram

$$\begin{array}{ccc} (U(1) \ltimes \mathbb{N}^\times)^o & \xleftarrow{\mathrm{Id}\times\mathrm{Fr}} & S'\mathcal{RF} \\ {\scriptstyle p}\downarrow & & \downarrow{\scriptstyle \sigma} \\ \mathbb{N}^\times & \xleftarrow{\mathrm{Fr}} & \mathcal{RF} \end{array} \tag{7.6}$$

7.2. Cyclotomic spectra. The notion for spectra corresponding to the construction reviewed above is that of *cyclotomic spectrum*. The general notion of G-spectrum is here used in the special case G$=\mathbb{T}$, where the compact Lie group $\mathbb{T}$ has the special property that it is canonically isomorphic to its quotient $\mathbb{T}/C$ by any non-trivial closed subgroup $C \subset \mathbb{T}$. To an integer $n \in \mathbb{N}^\times$ corresponds the subgroup C_n of n-th roots of unity; there is an action Ψ of the monoid $\mathbb{N}^\times$ on $\mathbb{T}$-spectra, by restriction to the fixed points of C_n and then viewing the corresponding $\mathbb{T}/C_n$-spectrum as a $\mathbb{T}$-spectrum (see [**25**] Section 2, [**26**] page 9).

DEFINITION 7.2. A cyclotomic spectrum E is a $\mathbb{T}$-spectrum together with morphisms of $\mathbb{T}$-spectra $\epsilon_k : \Psi^k(E) \to E$ such that the following diagrams commute

$$\begin{array}{ccc} \Psi^k(\Psi^r(E)) & \xrightarrow{\simeq} & \Psi^{kr}(E) \\ {\scriptstyle \Psi^k(\epsilon_r)}\downarrow & & \downarrow{\scriptstyle \epsilon_{kr}} \\ \Psi^k(E) & \xrightarrow{\epsilon_k} & E \end{array}$$

By implementing (7.5) to define the variant $S'\mathcal{RF}$ of the category $S\mathcal{RF}$, we obtain

LEMMA 7.3. *A cyclotomic spectrum E defines a (covariant) functor from $S'\mathcal{RF}$ to the category of $RO(\mathbb{T})$-graded topological spaces.*

PROOF. Let consider the inclusion of fixed points, *i.e.* for $a = sb$ the map $(\theta, f_{1,s}) : a \to b$. Let i_s^a be the inclusion $E^{C_a} \subset E^{C_{a/s}}$. The action of $(\theta, f_{1,s})$ is $i_s^a \circ E(\frac{\theta}{a})$. Thus (except for the inclusions) the action of $(\theta, f_{1,s}) : a \to b$ composed with the action of $(\tau, f_{1,q}) : qa \to a$ is

$$E\Big(\frac{\theta}{a}\Big) \circ E\Big(\frac{\tau}{aq}\Big) = E\Big(\frac{\theta}{a} + \frac{\tau}{aq}\Big) = E\Big(\frac{q\theta + \tau}{aq}\Big)$$

which agrees with (7.5). The action of the maps $f_{r,1}$ commutes with the action of $\mathbb{T}$ and corresponds to the composition of the map $\beta : E^{C_r} \to \Phi^{C_r}(E)$ with the structure map $\Phi^{C_r}(E) \to E$ of the cyclotomic spectrum. These are all maps of $\mathbb{T}$-spectra by construction. Thus the parameters r, q do not interfere with the validity of (7.5). □

7.3. The pericyclic category Π. There is a combinatorial version of the topological category $S'\mathcal{RF}$: this is the pericyclic category Π that we shall introduce here below.
In the diagram (7.6) we replace the monoid $(U(1) \ltimes \mathbb{N}^\times)^{\mathrm{o}}$ with the category $\tilde\Lambda^{\mathrm{o}}$. Indeed, on the geometric realization of an epicyclic space in the sense of [**1**] (*i.e.* as a covariant functor from $\tilde\Lambda^{\mathrm{o}}$ to spaces) one has, by applying Theorem A of *op.cit.* a natural *right* action of the monoid $U(1) \ltimes \mathbb{N}^\times$, *i.e.* a left action of $(U(1) \ltimes \mathbb{N}^\times)^{\mathrm{o}}$. This shows that $\tilde\Lambda^{\mathrm{o}}$ is the combinatorial version of the monoid $(U(1) \ltimes \mathbb{N}^\times)^{\mathrm{o}}$. In the following, we describe the missing entries (*i.e.* the ?'s) in the pullback diagram

(7.7)
$$\begin{array}{ccc} \tilde\Lambda^{\mathrm{o}} & \xleftarrow{?} & ? \\ {\scriptstyle \mathbf{Mod}}\downarrow & & \downarrow{\scriptstyle ?} \\ \mathbb{N}^\times & \xleftarrow[\mathrm{Fr}]{} & \mathcal{RF}. \end{array}$$

DEFINITION 7.4. The pericyclic category Π is the subcategory of the product category $\tilde\Lambda^{\mathrm{o}} \times \mathcal{RF}$ sharing the same objects and whose morphisms

$$\mathrm{Hom}_\Pi((n,a),(m,b)) \subset \mathrm{Hom}_{\tilde\Lambda^{\mathrm{o}}}(n,m) \times \mathrm{Hom}_{\mathcal{RF}}(a,b) \tag{7.8}$$

are given by those pairs $(h,f) \in \mathrm{Hom}_{\tilde\Lambda^{\mathrm{o}}}(n,m) \times \mathrm{Hom}_{\mathcal{RF}}(a,b)$ fulfilling the equality

$$\mathrm{Fr}(f) = \mathrm{Mod}(h). \tag{7.9}$$

Thus an object of Π is a pair (n, a) with $n \geqslant 0$, $n \in \mathbb{N}$ and $a \in \mathbb{N}^\times$. Then we complete (7.7) to a pullback diagram as follows

$$\begin{array}{ccc} \tilde{\Lambda}^{\mathrm{o}} & \xleftarrow{\ \lambda\ } & \Pi \\ {\scriptstyle \mathbf{Mod}}\downarrow & & \downarrow{\scriptstyle \pi} \\ \mathbb{N}^\times & \xleftarrow[\mathrm{Fr}]{} & \mathcal{RF} \end{array}$$

by implementing the two projections arising from (7.8), namely the functors $\lambda : \Pi \longrightarrow \tilde{\Lambda}^{\mathrm{o}}$ and $\pi : \Pi \longrightarrow \mathcal{RF}$ defined by

$$\lambda(n, a) := n, \quad \lambda(h, f) := h, \qquad \pi(n, a) := a, \quad \pi(h, f) := f.$$

To obtain a natural section of λ we introduce the following further morphisms in Π. For $n \geqslant 0$, $s > 0$ and $d > 0$ integers, with $m = s(n+1) - 1$, one lets

$$\chi(n, s, d) := (\mathrm{Id}_n^{s,\mathrm{o}}, f_{1,s}) \in \mathrm{Hom}_\Pi((n, sd), (m, d)).$$

This is meaningful since $(\mathrm{Id}_n^{s,\mathrm{op}}, f_{1,s}) \in \mathrm{Hom}_{\tilde{\Lambda}^{\mathrm{o}}}(n, m) \times \mathrm{Hom}_{\mathcal{RF}}(sd, d)$ fulfills (7.9). Let W be the set of morphisms of the form $\mathrm{Id} \times f_{r,1} : (n, a) \to (n, b)$ (for $br = a$). We localize the category Π on W: this is the category $\Pi[W^{-1}]$ obtained by inverting the morphisms in W. One obtains in this way a small category equivalent to $\tilde{\Lambda}^{\mathrm{o}}$

LEMMA 7.5. *The functor* $\theta : \tilde{\Lambda}^{\mathrm{o}} \longrightarrow \Pi[W^{-1}]$

$$\theta(n) := (n, 1), \quad \theta(\ell) = \ell \times f_{1,1}, \ \forall \ell \in \mathrm{Hom}_\Lambda(n, m), \quad \theta(\mathrm{Id}_n^{s,\mathrm{o}}) = \chi(n, s, 1) \circ f_{s,1}^{-1}$$

defines an equivalence of categories.

PROOF. By construction, the restriction of θ to the cyclic category Λ is an isomorphism with a subcategory of Π. Let $p_n^s := \chi(n, s, 1) \circ f_{s,1}^{-1} : (n, 1) \to (n, s) \to (s(n+1) - 1, 1)$. One easily checks that these morphisms of $\Pi[W^{-1}]$ fulfill the presentation of $\tilde{\Lambda}^{\mathrm{o}}$ as an extension of the cyclic category. Indeed, this follows from the construction of Π as a subcategory of $\tilde{\Lambda}^{\mathrm{o}} \times \mathcal{RF}$, so that one can check the relations in $\tilde{\Lambda}^{\mathrm{o}} \times \mathcal{RF}$ using the fact that they hold in $\tilde{\Lambda}^{\mathrm{o}}$. This proves that θ is a functor. Moreover, the composite $\lambda \circ \theta$ is the identity thus θ is faithful. Finally, after inverting the morphisms in W one sees that an object (n, a) of Π is isomorphic to the object $(n, 1)$ which is in the range of θ. Thus θ is an equivalence of categories. $\square$

Lemma 7.5 shows that covariant functors $\Pi[W^{-1}] \longrightarrow \mathfrak{Sets}$ correspond to epicyclic sets in the sense of [**1**]. For an epicyclic space in the sense of [**20**] (Definition 6.2.3.1), *i.e.* of Definition 7.1 we have

PROPOSITION 7.6. *Let* (Y, ϕ) *be an epicyclic space as in Definition* 7.1. *The following equalities define a covariant functor:*

$$\beta : \Pi \longrightarrow \mathfrak{Sets}, \qquad \beta(n, a) = \ sd_a(Y)^{C_a}(n)$$

$$\beta((\ell, \mathrm{Id}_a)) = \ sd_a(Y)^{C_a}(\ell) : \beta(n, a) \to \beta(m, a), \ \forall \ell \in \mathrm{Hom}_\Lambda(n, m) \tag{7.10}$$

$$\beta(\chi(n, u, v)) : \beta(n, uv) \subset \beta(u(n+1) - 1, v) \tag{7.11}$$

$$\beta((\mathrm{Id}, f_{q,1})) = \phi_q : \beta(n, qa) = \ sd_{aq}(Y)^{C_{aq}}(n) \to \beta(n, a) = sd_a(Y)^{C_a}(n).$$

The functor β *encodes faithfully the epicyclic space* (Y, ϕ).

PROOF. Let us first explain the meaning of the inclusion in (7.11). By construction, one has for all $a > 0$ and $n \geqslant 0$

$$\mathrm{sd}_a(Y)(n) = Y(\mathrm{sd}^a(n)) = Y(a(n+1)-1),$$

so that

$$\mathrm{sd}_{uv}(Y)(n) = Y(uv(n+1)-1) = \mathrm{sd}_v(Y)(u(n+1)-1).$$

The fixed points under the cyclic subgroups involve the action of the subgroup C_{uv} for $\beta(n, uv) = \mathrm{sd}_{uv}(Y)^{C_{uv}}(n)$, and its subgroup $C_v \subset C_{uv}$ for $\beta(u(n+1)-1, v) = \mathrm{sd}_v(Y)^{C_v}(u(n+1)-1)$. Thus the inclusion in (7.11) is the inclusion of fixed points. The structure maps ϕ_q are cyclic maps $\mathrm{sd}_{aq}(Y)^{C_{aq}} \to \mathrm{sd}_a(Y)^{C_a}$ and this shows that $\beta((\mathrm{Id}, f_{q,1}))$ commutes with the maps $\beta((\ell, \mathrm{Id}_a))$ of (7.10) defining the cyclic action. Moreover $\beta((\mathrm{Id}, f_{q,1}))$ commutes with the inclusion of fixed points (7.11), *i.e.* the following diagram is commutative for any n, q, u, v

$$\begin{array}{ccc} \beta(n, quv) & \xrightarrow{\beta(\chi(n,u,qv))} & \beta(u(n+1)-1, qv) \\ \downarrow{\scriptstyle \beta((\mathrm{Id}, f_{q,1}))} & & \downarrow{\scriptstyle \beta((\mathrm{Id}, f_{q,1}))} \\ \beta(n, uv) & \xrightarrow[\beta(\chi(n,u,v))]{} & \beta(u(n+1)-1, v). \end{array}$$

It remains to check that the inclusion of fixed points (7.11) fulfills the relation of $\mathrm{Id}_n^{s,\mathrm{o}}$ in the category $\tilde{\Lambda}^{\mathrm{o}}$ with respect to the cyclic action (7.10). This follows at the combinatorial level from the above discussion of the inclusion of fixed points. $\square$

Proposition 7.6 applies, in particular, to the cyclic nerve of a small category $\mathcal{C}$ (*cf.* [**20**] Example 6.2.3.3). This is the cyclic set

$$B_n^{\mathrm{cy}}\mathcal{C} = \{c_n \overset{f_0}{\leftarrow} c_0 \overset{f_1}{\leftarrow} c_1 \leftarrow \cdots \overset{f_n}{\leftarrow} c_n\}$$

formed by $n+1$ composable morphisms of $\mathcal{C}$. The face maps are defined by composition, and the degeneracies by insertion of identities, while the cyclic structure is given by cyclic permutations. In [**20**] (Example 6.2.3.3), one gets an epicyclic space (in the sense of [**20**]), using the isomorphisms

$$\phi_k : (\mathrm{sd}_k B^{\mathrm{cy}}\mathcal{C})^{C_k} \simeq B^{\mathrm{cy}}\mathcal{C}. \tag{7.12}$$

Proposition 7.6 thus defines a covariant functor $B^{\mathrm{epi}}\mathcal{C} : \Pi \longrightarrow \mathfrak{Sets}$. Moreover, since the ϕ_k are invertible, Lemma 7.5 applies and one gets an epicyclic set in the sense of [**1**]. This functor is the same as the epicyclic set associated to the cyclic nerve in *op.cit.* Indeed, the latter is the functor $\tilde{\Lambda}^{\mathrm{o}} \longrightarrow \mathfrak{Sets}$ which extends the canonical cyclic structure of $B^{\mathrm{cy}}\mathcal{C}$ by the additional map $p_n^k : B_n^{\mathrm{cy}} \to B_{k(n+1)-1}^{\mathrm{cy}}$ associated to the morphism $\mathrm{Id}_n^{k,\mathrm{o}} \in \mathrm{Hom}_{\tilde{\Lambda}^{\mathrm{o}}}(n, k(n+1)-1)$. The map p_n^k is the k-fold amplification, mapping the element $(c_n \overset{f_0}{\leftarrow} c_0 \overset{f_1}{\leftarrow} c_1 \leftarrow \cdots \overset{f_n}{\leftarrow} c_n)$ of B_n^{cy} into its repetition k-times

$$c_n \overset{f_0}{\leftarrow} c_0 \overset{f_1}{\leftarrow} c_1 \leftarrow \cdots \overset{f_n}{\leftarrow} c_n \overset{f_0}{\leftarrow} c_0 \overset{f_1}{\leftarrow} c_1 \leftarrow \cdots\cdots \leftarrow c_0 \overset{f_1}{\leftarrow} c_1 \leftarrow \cdots \overset{f_n}{\leftarrow} c_n$$

The map p_n^k is the inverse of ϕ_k in (7.12) and this corresponds exactly to the equality $\theta(\mathrm{Id}_n^{s,\mathrm{op}}) = \chi(n, s, 1) \circ f_{s,1}^{-1}$ of Lemma 7.5.

7.4. The small category $\mathcal{R}$ and points of the topos $\hat{\mathcal{R}}$. To some extent, one may view the kernel of either one of the two functors in (7.1) as the category $\mathcal{R}$ whose objects are the same as for $\mathcal{RF}$, and where there exists a unique morphism $f(a,b) \in \mathrm{Hom}_{\mathcal{R}}(a,b)$ if $b|a$, while otherwise there is no morphism. Since the opposite category $\mathcal{R}^{\mathrm{o}}$ has a morphism $a \to b$ exactly when $a|b$, it would seem that after a process of completion providing the points of the topos $\hat{\mathcal{R}}$ as colimits of objects of $\mathcal{R}^{\mathrm{o}}$, these limits ought to be naturally interpreted in terms of supernatural numbers. It is natural to think of the objects of $\mathcal{RF}$ by associating to the object $n \in \mathbb{N}^\times$ the cyclic subgroup $C_n \subset \mathbb{Q}/\mathbb{Z}$ given by the n-torsion:

$$C_n := \{r \in \mathbb{Q}/\mathbb{Z} \mid nr = 0\} = \frac{1}{n}\mathbb{Z}/\mathbb{Z}.$$

LEMMA 7.7. *The category of points of the topos $\hat{\mathcal{R}}$ is equivalent to the category of subgroups of $\mathbb{Q}$ containing $\mathbb{Z}$, with morphisms given by inclusion.*

PROOF. Let $F : \mathcal{R} \longrightarrow \mathfrak{Sets}$ be a flat functor. Let X_a be the image by F of the object a of $\mathcal{R}$. One has a unique map $F(a,b) : X_a \to X_b$ for any $a, b \in \mathbb{N}^\times$ with $b|a$. The filtering conditions for the category $\int_{\mathcal{R}} F$ read as

(1) $X_a \neq \varnothing$ for some $a \in \mathbb{N}^\times$.
(2) For any $x \in X_a$, $y \in X_b$ $\exists$ $c \in \mathbb{N}^\times$ and $z \in X_c$ such that $a|c$, $b|c$ and $F(c,a)z = x$, $F(c,b)z = y$.

The third filtering condition is automatically fulfilled since for objects i, j of $\int_{\mathcal{R}} F$ there is at most one morphism $i \to j$, thus $\alpha, \beta : i \to j$ implies $\alpha = \beta$. The second condition implies, by taking $a = b$, that there is at most one element in X_a for any $a \in \mathbb{N}^\times$. Let $J := \{a \in \mathbb{N}^\times \mid X_a \neq \varnothing\}$. One then has

$$a \in J, \quad b|a \implies b \in J, \quad a, b \in J \implies \exists c \in J, \ a|c, \ b|c. \tag{7.13}$$

The first implication follows from the existence of the morphism $F(a,b) : X_a \to X_b$ while the second one follows from (2). Thus $1 \in J$ and the flat functors $F : \mathcal{R} \longrightarrow \mathfrak{Sets}$ correspond to subsets $J \subset \mathbb{N}^\times$ fulfilling (7.13). A morphism of functors $F \to F'$ exists if and only if $J \subset J'$, and in that case is unique since both the sets involved have one element. We use the following correspondence between subsets $J \subset \mathbb{N}^\times$ fulfilling (7.13) and subgroups H such that $\mathbb{Z} \subset H \subset \mathbb{Q}$

$$H^J := \bigcup_{n \in J} \frac{1}{n}\mathbb{Z} \subset \mathbb{Q}, \qquad J_H := \{n \in \mathbb{N}^\times \mid \frac{1}{n} \in H\}.$$

The conditions (7.13) show that H^J is a filtering union of subgroups and is hence a subgroup of $\mathbb{Q}$ which contains $\mathbb{Z}$ since $1 \in J$. Conversely, for $\mathbb{Z} \subset H \subset \mathbb{Q}$, the subset $J_H \subset \mathbb{N}^\times$ fulfills the conditions (7.13). Finally, it is easy to see that the maps $J \mapsto H^J$ and $H \mapsto J_H$ are inverse of each other. □

In [**12**] we proved that the non-trivial subgroups of $\mathbb{Q}$ are parameterized by the quotient space $\mathbb{A}^f_{\mathbb{Q}}/\hat{\mathbb{Z}}^*$ (where $\mathbb{A}^f_{\mathbb{Q}}$ are the finite adèles of $\mathbb{Q}$), while the subgroups of $\mathbb{Q}$ containing $\mathbb{Z}$ are parameterized by the subset $\hat{\mathbb{Z}}/\hat{\mathbb{Z}}^* \subset \mathbb{A}^f_{\mathbb{Q}}/\hat{\mathbb{Z}}^*$. This latter subset surjects onto the quotient $\mathbb{Q}^\times_+\backslash\mathbb{A}^f_{\mathbb{Q}}/\hat{\mathbb{Z}}^*$ which parametrizes the points of $\widehat{\mathbb{N}^\times}$. Here, one has a natural geometric morphism of toposes associated with the functor

$$\rho : \mathcal{R} \longrightarrow \mathbb{N}^\times, \qquad \rho(f(a,b)) := a/b$$

The next proposition provides a topos theoretic interpretation of the quotient map $\mathbb{A}^f_{\mathbb{Q}}/\hat{\mathbb{Z}}^* \to \mathbb{Q}^\times_+\backslash\mathbb{A}^f_{\mathbb{Q}}/\hat{\mathbb{Z}}^*$.

PROPOSITION 7.8. *The geometric morphism of toposes* $\hat{\mathcal{R}} \longrightarrow \widehat{\mathbb{N}^\times}$ *associated with the functor* ρ *maps the category of points of* $\hat{\mathcal{R}}$ *to the category of points of* $\widehat{\mathbb{N}^\times}$ *as the quotient map*

$$\hat{\mathbb{Z}}/\hat{\mathbb{Z}}^* \subset \mathbb{A}_{\mathbb{Q}}^f/\hat{\mathbb{Z}}^* \to \mathbb{Q}_+^\times\backslash\mathbb{A}_{\mathbb{Q}}^f/\hat{\mathbb{Z}}^*. \tag{7.14}$$

PROOF. For a point p of $\hat{\mathcal{R}}$, let $J \subset \mathbb{N}^\times$ be the associated subset. The pullback p^* associates to a contravariant functor $Z : \mathcal{R} \longrightarrow \mathfrak{Sets}$ the set

$$\left(\coprod_{a\in J} Z_a\right)/\sim\; = \varinjlim_{a\in J} Z_a \tag{7.15}$$

as the colimit of the filtering diagram of sets $Z_a \to Z_b$ for $a|b$ (Z is contravariant). The contravariant functor $y(\bullet) : \mathbb{N}^\times \to \mathfrak{Sets}$ which associates to the unique object $\bullet$ of $\mathbb{N}^\times$ the set $\mathbb{N}^\times = \mathrm{Hom}_{\mathbb{N}^\times}(\bullet, \bullet)$ on which $\mathbb{N}^\times$ acts by multiplication, once composed with ρ defines the following contravariant functor

$$Z = y(\bullet)\circ\rho : \mathcal{R} \longrightarrow \mathfrak{Sets}, \qquad Z(a) = \mathbb{N}^\times, \qquad Z(f(a,b)) = y(\bullet)(a/b). \tag{7.16}$$

Thus the flat functor $F : \mathbb{N}^\times \longrightarrow \mathfrak{Sets}$ associated with the image of the point p by the geometric morphism, is given by the action of $\mathbb{N}^\times$ on the filtering colimit (7.15) for the functor Z of (7.16). The computation of this colimit gives

$$\varinjlim_{a\in J} \mathbb{N}^\times = \bigcup_{a\in J} \frac{1}{a}\mathbb{Z}_+ = (H_J)_+.$$

This shows that at the level of the subgroups of $\mathbb{Q}$ the map of points associates with H ($\mathbb{Z} \subset H \subset \mathbb{Q}$) the group H viewed as an abstract ordered group. This corresponds precisely to the map (7.14). □

References

[1] D. Burghelea, Z. Fiedorowicz, and W. Gajda, *Power maps and epicyclic spaces*, J. Pure Appl. Algebra **96** (1994), no. 1, 1–14, DOI 10.1016/0022-4049(94)90081-7. MR1297435

[2] Olivia Caramello and Nicholas Wentzlaff, *Cyclic theories*, Appl. Categ. Structures **25** (2017), no. 1, 105–126, DOI 10.1007/s10485-015-9414-y. MR3606497

[3] A. Connes, *Spectral sequence and homology of currents for operator algebras, Math. Forschungsinst. Oberwolfach Tagungsber., 41/81, Funktionalanalysis und C^*-Algebren, 27-9/3-10*, 1981.

[4] Alain Connes, *Cohomologie cyclique et foncteurs* Ext^n (French, with English summary), C. R. Acad. Sci. Paris Sér. I Math. **296** (1983), no. 23, 953–958. MR777584

[5] Alain Connes, *Noncommutative differential geometry*, Inst. Hautes Études Sci. Publ. Math. **62** (1985), 257–360. MR823176

[6] Alain Connes, *Noncommutative geometry*, Academic Press, Inc., San Diego, CA, 1994. MR1303779

[7] Alain Connes, *Trace formula in noncommutative geometry and the zeros of the Riemann zeta function*, Selecta Math. (N.S.) **5** (1999), no. 1, 29–106, DOI 10.1007/s000290050042. MR1694895

[8] Alain Connes and Caterina Consani, *Cyclic homology, Serre's local factors and λ-operations*, J. K-Theory **14** (2014), no. 1, 1–45, DOI 10.1017/is014006012jkt270. MR3238256

[9] Alain Connes and Caterina Consani, *Projective geometry in characteristic one and the epicyclic category*, Nagoya Math. J. **217** (2015), 95–132, DOI 10.1215/00277630-2887960. MR3343840

[10] Alain Connes and Caterina Consani, *The cyclic and epicyclic sites*, Rend. Semin. Mat. Univ. Padova **134** (2015), 197–237, DOI 10.4171/RSMUP/134-5. MR3428418

[11] Alain Connes and Caterina Consani, *Cyclic structures and the topos of simplicial sets*, J. Pure Appl. Algebra **219** (2015), no. 4, 1211–1235, DOI 10.1016/j.jpaa.2014.06.002. MR3282133

[12] Alain Connes and Caterina Consani, *Geometry of the arithmetic site*, Adv. Math. **291** (2016), 274–329, DOI 10.1016/j.aim.2015.11.045. MR3459019

[13] Alain Connes and Caterina Consani, *Absolute algebra and Segal's Γ-rings: au dessous de* $\overline{\mathrm{Spec}(\mathbb{Z})}$, J. Number Theory **162** (2016), 518–551, DOI 10.1016/j.jnt.2015.12.002. MR3448278

[14] Alain Connes and Caterina Consani, *Geometry of the scaling site*, Selecta Math. (N.S.) **23** (2017), no. 3, 1803–1850, DOI 10.1007/s00029-017-0313-y. MR3663595

[15] Alain Connes and Caterina Consani, *Homological algebra in characteristic one*, High. Struct. **3** (2019), no. 1, 155–247. MR3939048

[16] Alain Connes and Caterina Consani, $\overline{\mathrm{Spec}\,\mathbb{Z}}$ *and the Gromov norm*, Theory Appl. Categ. **35** (2020), Paper No. 6, 155–178. MR4066805

[17] Alain Connes and Caterina Consani, *Segal's gamma rings and universal arithmetic*, Q. J. Math. **72** (2021), no. 1-2, 1–29, DOI 10.1093/qmath/haaa042. MR4271378

[18] Alain Connes and Caterina Consani, *On absolute algebraic geometry the affine case*, Adv. Math. **390** (2021), Paper No. 107909, 44, DOI 10.1016/j.aim.2021.107909. MR4291468

[19] A. Connes, C. Consani, *Riemann Roch for* $\overline{\mathrm{Spec}\ \mathbb{Z}}$, Preprint (2022). Available at arXiv:2205.01391

[20] Bjørn Ian Dundas, Thomas G. Goodwillie, and Randy McCarthy, *The local structure of algebraic K-theory*, Algebra and Applications, vol. 18, Springer-Verlag London, Ltd., London, 2013. MR3013261

[21] B. L. Feĭgin and B. L. Tsygan, *Additive K-theory*, K-theory, arithmetic and geometry (Moscow, 1984), Lecture Notes in Math., vol. 1289, Springer, Berlin, 1987, pp. 67–209, DOI 10.1007/BFb0078368. MR923136

[22] Jonathan S. Golan, *Semirings and their applications*, Kluwer Academic Publishers, Dordrecht, 1999. Updated and expanded version of *The theory of semirings, with applications to mathematics and theoretical computer science* [Longman Sci. Tech., Harlow, 1992; MR1163371 (93b:16085)], DOI 10.1007/978-94-015-9333-5. MR1746739

[23] T. Goodwillie, *A Topologist View of Cyclic Homology*, Math. Forschungsinst. Oberwolfach Tagungsber., 26/87, Zyklische Kohomologie und ihre Anwendungen, 14-6/20-6, 1987.

[24] T. Goodwillie, *Notes on the cyclotomic trace* Lecture notes for a series of seminar talks at MSRI, 1990-1991.

[25] Lars Hesselholt and Ib Madsen, *On the K-theory of finite algebras over Witt vectors of perfect fields*, Topology **36** (1997), no. 1, 29–101, DOI 10.1016/0040-9383(96)00003-1. MR1410465

[26] Lars Hesselholt and Ib Madsen, *On the K-theory of local fields*, Ann. of Math. (2) **158** (2003), no. 1, 1–113, DOI 10.4007/annals.2003.158.1. MR1998478

[27] D. Kaledin, *Hochschild-Witt complex*, Adv. Math. **351** (2019), 33–95, DOI 10.1016/j.aim.2019.05.007. MR3949985

[28] Jean-Louis Loday, *Cyclic homology*, 2nd ed., Grundlehren der mathematischen Wissenschaften [Fundamental Principles of Mathematical Sciences], vol. 301, Springer-Verlag, Berlin, 1998. Appendix E by María O. Ronco; Chapter 13 by the author in collaboration with Teimuraz Pirashvili, DOI 10.1007/978-3-662-11389-9. MR1600246

[29] Kenichi Miyazaki, *Distinguished elements in a space of distributions*, J. Sci. Hiroshima Univ. Ser. A **24** (1960), 527–533. MR131752

Collège de France and Institut des Hautes Études Scientifiques, France
Email address: alain@connes.org

Department of Mathematics, Johns Hopkins University, Baltimore, Maryland 21218
Email address: cconsan1@jhu.edu

Proceedings of Symposia in Pure Mathematics
Volume **105**, 2023
https://doi.org/10.1090/pspum/105/01898

The image of Bott periodicity in cyclic homology

Joachim Cuntz

Abstract. We analyze the relationship between Bott periodicity in topological K-theory and the natural periodicity of cyclic homology. This is a basis for understanding the multiplicativity, in odd dimensions, of a bivariant Chern-Connes character from K-theory to cyclic theory.

1. Introduction

There are two principal tools for studying the 'algebraic topology' of an algebra over $\mathbb{C}$: topological K-theory on the one hand and periodic cyclic homology (which generalizes de Rham theory to the not necessarily commutative setting) on the other. One of the first striking observations is the fact that both theories are periodic - they are naturally defined as $\mathbb{Z}/2$-graded theories. In fact, periodic cyclic theory HP_* has a completely intrinsic $\mathbb{Z}/2$-periodicity. On the other hand, on topological K-theory (for algebras over $\mathbb{C}$), the periodicity is due to the non-trivial Bott-periodicity isomorphism. In the case where the ground field is $\mathbb{C}$ and on a natural category of algebras these theories can be compared via a Chern-Connes character (which generalizes the classical Chern character in the commutative case). It is an obvious problem to understand the connection, under the character, between the periodicity operations which are given on the one hand by a 'Bott map' and on the other essentially by Connes' S-operator.

For such a study, we have to fix a suitable category of topological algebras on which both theories have reasonable properties and can be compared - and where the Bott isomorphism holds. There is such a natural category - the category of locally convex algebras. Second, for our study we have to choose the correct setting. This is the bivariant one. Both theories have been developed to bivariant theories with an associative composition product. These give a particularly efficient tool for understanding and computing K-theoretic and cyclic invariants. Bott periodicity is then naturally defined as a product with a bivariant element. As we will see, a Chern-Connes character can be defined in the bivariant setting. It gives a multiplicative transformation from bivariant K-theory, which is denoted on the category of locally convex algebras by $kk_*(A, B)$, to bivariant periodic cyclic theory

2020 *Mathematics Subject Classification.* Primary 19D55, 19E20, 19K35, 58B34.

Gefördert durch Deutsche Forschungsgemeinschaft (DFG) im Rahmen der Exzellenzstrategie des Bundes und der Länder EXC 2044 –390685587, Mathematik Münster: Dynamik–Geometrie–Struktur.

$HP_*(A, B)$. One immediate problem then consists in determining the image of the bivariant Bott element.

The existence of a bivariant Chern-Connes character in the even case as a multiplicative transformation $kk_0(A, B) \to HP_0(A, B)$ follows directly from the fact that kk_0 is the universal functor from the category of locally convex algebras to an additive category subject to three natural properties (half exactness, invariance under smooth homotopies and stability) together with the fact that HP_* shares these properties. There is an obvious candidate for an extension to the odd case. But unfortunately, for kk_* made $\mathbb{Z}/2$-periodic, this candidate will not be compatible with the product of two odd elements, because the product uses Bott periodicity. Therefore the extension of the Chern-Connes character from a multiplicative transformation $kk_0 \to HP_0$ to a multiplicative transformation $kk_* \to HP_*$ is linked to a precise understanding of the image of the Bott periodicity isomorphism in HP_*.

A key observation in comparing Bott periodicity in K-theory to the periodicity in HP_* is the appearance of the numerical factor $2\pi i$. Already in his foundational paper [**2**] Connes (motivated by the example of a complex torus and indirectly also by Bott periodicity) used this factor in his definition of the S-operator. The factor $2\pi i$ then also appeared in discussions of a Chern-Connes character in [**13**], [**12**], [**15**], [**17**], [**5**], [**8**]. As we will see it comes from the connection between the algebra of functions on $[0, 1]$, that vanish at 0 and 1, annd the algebra of functions on S^1 that vanish in 1, see [**15**], [**5**]. The implication for the behaviour of the product of odd elements in the bivariant theory under the bivariant Chern-Connes character has been discussed in [**17**], [**5**], [**7**].

The connection between Bott periodicity in topological K-theory and in periodic cyclic homology has been treated in the literature, notably in [**15**], [**5**]. In [**15**] a character has been established for certain bivariant kk-elements which is partially multiplicative. In [**5**] a fully multiplicative character has been constructed. In both cases the behaviour of Bott periodicity under the character was important. We feel that the issue deserves a more complete presentation. This is the purpose of the present note. An outcome of our discussion will be that a multiplicative character in odd dimensions cannot be entirely canonical, but depends on the choice of a square root of $2\pi i$.

We start by recalling the basics about bivariant K-theory kk_*. For ease of exposition we use the approach from [**8**], but applied to the natural class of complete locally convex algebras with submultiplicative seminorms (m-algebras). As a first step towards defining the Chern-Connes character we define a multiplicative transformation $\chi_* : kk_* \to HP_*$ from $\mathbb{Z}$*-graded* (!) bivariant K-theory. This transformation is also compatible with the boundary maps in long exact sequences but it does not give a multiplicative transformation from kk_* when we make it $\mathbb{Z}/2$-periodic using Bott periodicity.

We then give a treatment, which we think is somewhat more systematic and complete than the one in [**5**], of the behaviour of Bott periodicity and of the product under the character and use it to define the final form of the multiplicative Chern-Connes character ch_*. This character descends to a transformation from $\mathbb{Z}/2$-periodic bivariant K-theory to HP_*. Since the boundary maps in the long exact sequences associated with an extension of algebras are given as products by elements of odd degree, it turns out that ch_* has to multiply the boundary maps by $\sqrt{2\pi i}$. One outcome of our discussion also is that we can alternatively obtain

ch_* in the same natural way as χ_* if we modify the boundary maps in the complex defining HP_* by multiplying them by $\sqrt{2\pi i}$. Denote the homology of the so modified complex by $\overline{HP}_*$. Then ch_* will be compatible with the boundary maps in $\overline{HP}_*$, see Remark 5.5(b).

We mention though that the approach we describe here is fundamentally equivalent to the one in [**5**]. We also note that our arguments carry over without a problem to arbitrary locally convex algebras if we use the framework of [**11**].

2. The setting

For convenience we will work with the simplest natural category suitable for the discussion of the image of the Bott map. This is the category of algebras over $\mathbb{C}$ with a complete locally convex topology defined by a family of submultiplicative seminorms. Following [**5**] we call these algebras m-algebras. The natural tensor product in this category is the completed projective tensor product, denoted by $A \otimes B$. By an extension of m-algebras we mean an exact sequence $0 \to I \xrightarrow{j} E \xrightarrow{q} B \to 0$ of m-algebras where j, q are continuous homomorphisms. Moreover we will always assume that an extension admits a continuous linear split, i.e. a continuous linear map $s : B \to E$ such that $q \circ s = \mathrm{id}$. Using shorthand we write $I \to E \to B$ for such an extension.

The universal extension. Let A be an m-algebra. The non-unital tensor algebra $T_{alg}A = A \oplus A^{\otimes 2} \oplus A^{\otimes 3} \oplus \cdots$ can be completed to an m-algebra TA such that this becomes the universal m-algebra generated by an image of A under a continuous linear map (the linear inclusion of A onto the first direct summand in TA).
There is a natural surjective homomorphism $TA \to A$ mapping a tensor of the form $a_1 \otimes \ldots \otimes a_n$ to the product $a_1 \ldots a_n$ in A. We denote its kernel by JA. The extension $JA \to TA \to A$ is universal among all extension of A admitting a continuous linear splitting. Indeed, if $I \to E \to A$ is any extension of m-algebras with a continuous linear splitting $s : A \to E$, then there is a canonical commutative diagram of the form

$$\begin{array}{ccccccccc} 0 & \to & JA & \to & TA & \to & A & \to & 0 \\ & & \downarrow \gamma_s & & \downarrow & & \downarrow \mathrm{id} & & \\ 0 & \to & I & \to & E & \to & A & \to & 0 \end{array}$$

where the map $TA \to E$ is the unique algebra homomorphism extending the linear map $TA \supset A \xrightarrow{s} E$ and γ_s its restriction to JA. We call the map $\gamma_s : JA \to I$ the classifying map for the extension. It depends on s only up to smooth homotopy, see [**5**] for details.

We also have to specify the class of functions that we allow for homotopies. Since HP_* is only invariant under differentiable homotopies, this will be the class of infinitely differentiable, or smooth, functions on intervals. Given an m-algebra A, we denote by $A[0,1], A(0,1], A(0,1)$ the algebras of smooth A-valued functions on $[0,1]$ with all higher derivatives vanishing at 0 and 1, the functions themselves vanishing at 0 in the second case and at 0 and 1 in the third case. Two homomorphisms $\varphi_0, \varphi_1 : A \to B$ are called diffotopic (or differentiably homotopic), if there is a homomorphism $\Phi : A \to B[0,1]$ such that $ev_0 \circ \Phi = \varphi_0 \; ev_1 \circ \Phi = \varphi_1$.
Finally, we need an appropriate notion of stabilization of an m-algebra A by matrices. The simplest choice for our purposes here is given by the completed projective

tensor product $\mathcal{K}\otimes A$ of A by the algebra $\mathcal{K}$ of $\mathbb{N}\times\mathbb{N}$-matrices with rapidly decreasing coefficients (which is a natural m-algebra completion of the algebra M_∞ of finite matrices of arbitrary size, see [**5**]).

We can now recall the definition, from [**8**], of the bivariant theory kk_* slightly adapted to the setting we are considering here. Denote by $\gamma_A : JA \to A(0,1)$ the classifying map for the cone extension $A(0,1) \to A(0,1] \to A$. We denote by $[A,B]$ the set of diffotopy classes of homomorphisms $A \to B$. For an m-algebra D, we define J^nD recursively by $J^{n+1}D = J(J^nD)$. Using the map $\gamma_A : JA \to A(0,1)$ for each m,n we get a map $[J^nA, B(0,1)^m] \longrightarrow [J^{n+1}A, B(0,1)^{m+1}]$.
Given two m-algebras A,B we can now define

$$kk_n(A,B) = \varinjlim_k [J^{k-n}A, \mathcal{K}\otimes B(0,1)^k]$$

(note that $n<0$ is allowed).
kk_n is an abelian group with the addition

$$[X,\mathcal{K}\otimes Y]\times[X,\mathcal{K}\otimes Y]\to[X,\mathcal{K}\otimes Y]$$

induced for m-algebras X,Y by the inclusion $\mathcal{K}\oplus\mathcal{K}\hookrightarrow\mathcal{K}$

There is a natural product $kk_s(A_1,A_2)\times kk_t(A_2,A_3) \longrightarrow kk_{s+t}(A_1,A_3)$. For simplicity of notation we explain it for $s=t=0$. The general case follows from the identities $kk_s(A,B) = kk_0(A,B(0,1)^s)$, $kk_{-s}(A,B) = kk_0(J^sA,B)$.
Assume that two elements a in $kk_0(A_1,A_2)$ and b in $kk_0(A_2,A_3)$ are represented by $\alpha : J^nA_1 \to \mathcal{K}\otimes A_2(0,1)^n$ and $\beta : J^mA_2 \to \mathcal{K}\otimes A_3(0,1)^m$, respectively. We define their product $a\cdot b$ in $kk_0(A_1,A_3)$ by the following composition of maps

$$\begin{aligned} J^m(J^nA_1) \quad &\overset{J^m(\alpha)}{\longrightarrow} \quad J^m(\mathcal{K}\otimes A_2(0,1)^n) \to \\ &\mathcal{K}\otimes(J^mA_2)(0,1)^n \quad \overset{\beta(0,1)^n}{\longrightarrow} \quad \mathcal{K}\otimes A_3(0,1)^{n+m} \end{aligned}$$

The second arrow is defined as the composition of the natural map $J^m(A_2\otimes E)\to J^m(A_2)\otimes E$ (choosing $E=\mathcal{K}\otimes(0,1)^n$), that exists, by definition of J, for any E, with the isomorphism of $\mathcal{K}\otimes(J^mA_2)(0,1)^n$ that switches the orientation of one of the intervals $(0,1)$ in $(0,1)^n$ according to the parity of mn (i.e. we take $(-1)^{mn}$ times the diffotopy class of this map). Of course we also use the isomorphism $\mathcal{K}\otimes\mathcal{K}\cong\mathcal{K}$. The sign is necessary because the map

$$J^m(\mathcal{K}\otimes A_2(0,1)^n) \to \mathcal{K}\otimes(J^mA_2)(0,1)^n$$

permutes the noncommutative m-fold suspension J^m with the n-fold ordinary suspension $A(0,1)^n$. The sign is explained by the following easy lemma that then shows that the product is well defined (i.e. does not depend on the choice of representatives for a and b in the inductive limits defining kk_0).

LEMMA 2.1. *For an m-algebra D as above we denote by $\gamma_D : JD \to D(0,1)$ the classifying map for the cone extension $D(0,1)\to D(0,1]\to D$.*
(a) Let $I \to E \overset{\pi}{\longrightarrow} A$ be a linearly split extension of locally convex algebras and α its classifying map. Then the two maps $J^2A \to I(0,1)$ defined by the compositions $J^2A \overset{J(\alpha)}{\longrightarrow} JI \overset{\gamma_I}{\longrightarrow} I(0,1)$ and $J^2A \overset{-\gamma_{JA}}{\longrightarrow} (JA)(0,1) \overset{\alpha(0,1)}{\longrightarrow} I(0,1)$ are diffotopic. Here $-\gamma_{JA}$ stands for γ_{JA} followed by a change of orientation on $(0,1)$.
(b) The natural map $\gamma_{JA} : J^2A \to (JA)(0,1)$ is diffotopic to the composition of the natural maps $J^2A \to J(A(0,1))$ and $J(A(0,1)) \to (JA)(0,1)$.

PROOF. (a) The two compositions are classifying maps for the two 2-step extensions in the first and last row of the following commutative diagram

$$\begin{array}{ccccccccc} 0 & \to & I(0,1) & \to & I(0,1] & \to & E & \to & A & \to & 0 \\ & & \| & & \downarrow & & \downarrow & & \downarrow & & \\ 0 & \to & I(0,1) & \to & E(0,1] & \to & Z_\pi & \to & A & \to & 0 \\ & & \| & & \uparrow & & \uparrow & & \uparrow & & \\ 0 & \to & I(0,1) & \to & E(0,1) & \to & A[0,1) & \to & A & \to & 0 \end{array}$$

where $Z_\pi \subset E \oplus A[0,1]$ is the mapping cylinder. They are diffotopic since they also are classifying maps for the extension in the middle row. Here we use the cone $A[0,1)$ in place of $A(0,1]$ and thus reverse the orientation of $[0,1]$.
(b) The first map is classifying for the first row in the commutative diagram

$$\begin{array}{ccccccccc} 0 & \to & (JA)(0,1) & \to & (TA)(0,1) & \to & A(0,1] & \to & A & \to & 0 \\ & & \uparrow\varphi & & \uparrow & & \uparrow & & \uparrow & & \\ 0 & \to & J(A(0,1)) & \to & T(A(0,1)) & \to & A(0,1] & \to & A & \to & 0 \end{array}$$

while the second map is the composition of the classifying map for the second row composed with the first vertical arrow φ in the diagram - thus also classifying for the first row. □

Note that every continuous homomorphism $\alpha : A \to B$ induces an element $kk(\alpha)$ in $kk_0(A,B)$ and that $kk(\beta \circ \alpha) = kk(\alpha) \cdot kk(\beta)$. We denote by 1_A the element $kk(\mathrm{id}_A)$ induced by the identity homomorphism of A. It is a unit in the ring $kk_0(A,A)$.
From its definition one derives the following properties of the functor kk.

(E1) Differentiable homotopy invariance: The evaluation maps $A[0,1] \to A$ induce kk_0-equivalences.
(E2) Stability: The natural inclusion $A \hookrightarrow \mathcal{K} \otimes A$ induces a kk_0-equivalence.
(E3) Half-exactness: Every extension $I \to E \to A$ with a continuous linear split induces short exact sequences in both variables of kk_0.

REMARK 2.2. By property (E3) combined with (E1) every extension $I \to E \to A$ of m-algebras induces long exact sequences in both variables of kk_*. The boundary maps $kk_n \to kk_{n-1}$ are given by product with the element $kk(\gamma) \in kk_0(JE,I) = kk_{-1}(E,I)$ for the classifying map γ of the extension.

We know that kk_0 is best viewed as an additive category with objects m-algebras and morphisms $kk_0(A,B)$. Then $\alpha \to kk(\alpha)$ defines a functor from m-algebras to the category kk_0 which is universal with the properties above:

THEOREM 2.3. *Let E be a functor from m-algebras to an additive category $\mathcal{C}$ which satisfies the properties $E1, E2, E3$ in both variables. Then there is a unique functor $F : kk_0 \to \mathcal{C}$ such that $E = F \circ kk$.*

PROOF. We sketch the simple essence of the argument here and give a detailed construction of F (for the case $E = HP$) in section 4.
The properties of half-exactness and homotopy invariance imply that any extension $I \to A \to B$ of m-algebras with a continuous linear split induces long exact sequences

under E in both variables (in particular it follows that E is automatically additive). We apply the long exact sequences to the standard extensions

$$JA \to TA \to A \qquad A(0,1) \to A(0,1] \to A$$

Write $E(A,B)$ for the image, under E of the m-algebra morphisms $A \to B$ in the morphisms of $\mathcal{C}$. The fact that E applied to the contractible algebras TA and $A(0,1]$ gives 0, shows that there are isomorphisms $E(JA, B(0,1)) \cong E(A,B)$ and by iteration $E(J^nA, B(0,1)^n) \cong E(A,B)$. Using stability of E we get moreover $E(J^nA, \mathcal{K} \otimes B(0,1)^n) \cong E(A,B)$. Let x be an element of $kk_0(A,B)$ represented by a homomorphism $\varphi : J^nA \to \mathcal{K} \otimes B(0,1)^n$. We set $F(x) = E(\varphi) \in E(J^nA, \mathcal{K} \otimes B(0,1)^n) \cong E(A,B)$. The explicit description of F in section 4 will make it obvious that F is well defined and a functor (i.e. multiplicative). □

We also note that $kk_0(\mathbb{C}, B)$ gives the usual topological K-theory (e.g. when applied to a Banach algebra or to a Fréchet algebra), see [**5**], Section 7.

3. Bott periodicity

The well known proofs of Bott periodicity, e.g. [**1**], [**14**], [**4**] use explicitly or implicitly an index map. There is a universal model for the index map that is encoded in the Toeplitz extension. Algebraically, the Toeplitz algebra (for our purposes over the field $\mathbb{C}$) is the universal unital $\mathbb{C}$-algebra $\mathcal{T}$ generated by two elements u, v satisfying the relation $vu = 1$. It is easy to check that the ideal generated by the defect projection $e = 1 - uv$ in $\mathcal{T}$ is isomorphic to the algebra M_∞ of finite matrices of arbitrary size, and the quotient by this ideal is isomorphic to the algebra $\mathbb{C}[z, z^{-1}]$ of Laurent polynomials. There is a well known simple algebraic argument that goes back to [**4**] showing that the natural inclusion $\mathbb{C} \to \mathcal{T}$ induces an isomorphism $E(\mathbb{C}) \to E(\mathcal{T})$ for every homotopy invariant (under polynomial homotopies), split-exact and matrix stable functor E. As a consequence, if we denote by $\mathcal{T}_0$ the kernel of the natural map $\mathcal{T} \to \mathbb{C}$, then $E(\mathcal{T}_0) = 0$. The argument for these facts is so simple and general that it carries over to all reasonable completions of $\mathcal{T}$, in particular to natural C*-algebra completions or m-algebra completions. The purely algebraic form of the argument has been used more recently also in [**3**].

For our purposes we use m-algebra completions $\overline{\mathcal{T}}, \overline{\mathcal{T}}_0$ of $\mathcal{T}, \mathcal{T}_0$, see [**5**], [**7**]. We get extensions $\mathcal{K} \to \overline{\mathcal{T}} \to C^\infty(S^1)$, $\mathcal{K} \to \overline{\mathcal{T}}_0 \to C^\infty(S^1 \setminus 1)$ where $C^\infty(S^1 \setminus 1)$ denotes the algebra of C^∞-functions on the circle S^1 that vanish at the point 1. The general argument mentioned above then shows that $\overline{\mathcal{T}}_0$ is equivalent to 0 in kk_0 see [**5**]. This means in particular that $kk_n(\overline{\mathcal{T}}_0 \otimes A, B)$ and $kk_n(A, \overline{\mathcal{T}}_0 \otimes B)$ are 0 for all m-algebras A, B and all n.

Bott periodicity for kk_* means that

$$kk_{n+2}(B,A) = kk_n(B, A(0,1)^2) \cong kk_n(B,A)$$

or, equivalently, that $kk_{n-2}(A,B) = kk_n(J^2A, B) \cong kk_n(A,B)$ for all A, B, n. These periodicity isomorphisms are implemented by an invertible element in $kk_{-2}(A,A)$ or by its inverse in $kk_2(A,A)$. This element has a natural representation. We note that any extension $I \to D \to Q$ of m-algebras with a continuous linear split where D is contractible, or more generally equivalent to 0 in kk_0, defines an invertible element in $kk_{-1}(Q,I)$ (this follows from the long exact sequences in both variables). Now, given an m-algebra A, we have two extensions of that type

$$\mathcal{K} \otimes A \to \overline{\mathcal{T}}_0 \otimes A \to C^\infty(S^1 \setminus 1) \otimes A \qquad A(0,1) \to A(0,1] \to A$$

They determine invertible elements $\tau_A \in kk_{-1}(C^\infty(S^1 \setminus 1) \otimes A, \mathcal{K} \otimes A)$ and $\rho_A \in kk_{-1}(A, A(0,1))$, respectively. Moreover the natural inclusion $\mathbb{C}(0,1) \to C^\infty(S^1 \setminus 1)$ as functions on S^1 vanishing with all derivatives at 1, determines an element σ_A in $kk_0(A(0,1), C^\infty(S^1 \setminus 1) \otimes A)$. It is invertible by differentiable homotopy invariance of kk_0 (by an argument as e.g. in [**5**], 1.1, p.143 there is a homotopy inverse to σ). Finally denote by κ_A the element in $kk_0(A, \mathcal{K} \otimes A)$ induced by the natural inclusion $A \to \mathcal{K} \otimes A$. The product $\rho_A \sigma_A \tau_A \kappa_A^{-1}$ defines an invertible element in $kk_{-2}(A, A)$. This fundamental element has been used explicitly or implicitly in many places, e.g. [**4**], [**14**], [**15**], [**5**]. We record it in the following definition

DEFINITION 3.1. For an m-algebra A we denote by b_A the element $\rho_A \sigma_A \tau_A \kappa_A^{-1}$ of $kk_{-2}(A, A)$ as above.

4. A multiplicative transformation $kk_* \to HP_*$

We now come to the connection of kk_* to bivariant periodic cyclic homology HP_*. Of course we have to use the version of HP_* for topological algebras. This means that for m-algebras, in the algebraic definition [**9**] of HP_* we have to replace algebraic tensor products by completed projective tensor products. Recall then that any extension $I \to D \to A$ of m-algebras with a continuous linear splitting induces long exact sequences in both variables of kk_*. For such an extension the classifying map $\gamma : JA \to I$ gives an element in $kk_{-1}(A, I)$. It is important for our purposes to note that the boundary maps $kk_n \to kk_{n-1}$ in both long exact sequences under kk_* are given by the product by this element.

By excision [**10**],[**6**] HP satisfies (E3). It therefore has long exact sequences in both variables. Given an m-algebra A and another m-algebra B we consider the long exact sequences in $HP_*(B, A)$ induced in the second variable by the universal extension $JA \to TA \to A$ on the one hand and by the cone extension $A(0,1) \to A(0,1] \to A$ on the other hand. We denote by δ the boundary map for the first sequence, by $\bar\delta$ the boundary maps for the second sequence and obtain a commutative diagram

$$
\begin{array}{ccccc}
\to & HP_0(B,A) & \xrightarrow{\delta} & HP_1(B,JA) & \to \\
& \downarrow = & & \downarrow \cdot\gamma_A & \\
\to & HP_0(B,A) & \xrightarrow{\bar\delta} & HP_1(B,A(0,1)) & \to
\end{array}
\tag{1}
$$

Here $\gamma_A : JA \to A(0,1)$ denotes the natural map and $\cdot\gamma_A$ denotes product by γ_A on the right (note that here and elsewhere we denote the product in kk_* and in HP_* in the order opposite to the composition of homomorphisms). Since $HP_*(B, TA) = HP_*(B, A(0,1]) = 0$ by contractibility of TA and $A(0,1]$, we see that $\delta, \bar\delta$ are isomorphisms.

Thus there are invertible elements $\alpha_A \in HP_1(A, JA)$ and $\beta_A \in HP_1(A(0,1), A)$ such that $\delta = \cdot\alpha_A$ and $\bar\delta = \cdot\beta_A^{-1}$ (explicitly they are determined by $\alpha_A = \delta(1_A)$, $\beta_A^{-1} = \bar\delta(1_A)$, see [**10**] Theorem 5.5). The commutative diagram (1) shows that $\alpha_A\gamma_A\beta_A = 1_A$ where 1_A denotes the identity element of $HP_0(A, A)$ and also $\gamma_A \cdot \beta_A \cdot \alpha_A = 1_{JA}$, $\beta_A \cdot \alpha_A \cdot \gamma_A = 1_{A(0,1)}$.

By iteration we get invertible elements $\alpha_A^n = \alpha_A \alpha_{JA} \cdots \alpha_{J^n A} \in HP_n(A, J^n A)$ and $\beta_A^n = \beta_{A(0,1)^n} \cdots \beta_{A(0,1)} \beta_A \in HP_n(A(0,1)^n, A)$. Let moreover $\gamma^n \in HP_0(J^n A, A(0,1)^n)$

be the element induced by the canonical homomorphism $J^nA \to A(0,1)^n$. The following relations are derived by iteration from diagram (1)

$$(2)\qquad \alpha_A^n \cdot \gamma^n \cdot \beta_A^n = 1_A \qquad \gamma^n \cdot \beta_A^n \cdot \alpha_A^n = 1_{J^nA} \qquad \beta_A^n \cdot \alpha_A^n \cdot \gamma^n = 1_{A(0,1)}$$

As a first step towards constructing a multiplicative Chern-Connes character we can now define a multiplicative transformation χ from the $\mathbb{Z}$-graded theory kk_* to HP_*.

THEOREM 4.1. *There are natural maps $\chi_s : kk_s(A,B) \to HP_s(A,B)$ such that*

$$\chi_{s+t}(\varphi \cdot \psi) = \chi_s(\varphi) \cdot \chi_t(\psi)$$

in $HP_{s+t}(A,C)$ for $\varphi \in kk_s(A,B)$ and $\psi \in kk_t(B,C)$ and such that $\chi_0(1_A) = 1_A$ for each A. Here s and t are in $\mathbb{Z}$ and HP_s is 2-periodic in s.

PROOF. Ignoring first the stabilization by $\mathcal{K}$ assume that $\varphi \in kk_s(A,B)$ is represented by a homomorphism $\eta : J^{n-s}A \to B(0,1)^n$. We then define

$$\chi_s(\varphi) = \alpha_A^{n-s} HP_0(\eta) \beta_B^n$$

The relation $\alpha_A^n \cdot \gamma^n \cdot \beta_A^n = 1_A$ generalizes to $\alpha_A^n \cdot \theta^n \cdot \beta_B^n = \theta$ if θ is an element of $HP_0(A,B)$ determined by a homomorphism $A \to B$ and θ^n is the element of $HP_0(J^nA, B(0,1)^n)$ determined by the homomorphism $J^nA \to B(0,1)^n$ induced by θ. Since the inductive limit in the definition of kk_s is taken exactly with respect to the passage from homomorphisms $A \to B$ to homomorphisms $J^nA \to B(0,1)^n$, this shows that χ is well defined. The compatibility with the product follows from the relations (2) and Lemma 2.1.
Finally, the natural inclusion $B \to \mathcal{K} \otimes B$ induces an invertible element ι in $HP_0(B, \mathcal{K} \otimes B)$. The formula for χ then becomes

$$(3)\qquad \chi_s(\varphi) = \alpha_A^{n-s} HP_0(\eta) \beta_B^n \iota^{-1}$$

if φ is represented by the map $\eta : J^{n-s}A \to \mathcal{K} \otimes B(0,1)^n$. The stabilization by $\mathcal{K}$ does not lead to any essential change in the arguments for well-definedness and multiplicativity. □

REMARK 4.2. Note that the formula for χ_0 makes the universal functor $kk_0 \to HP_0$ from Theorem 2.3 explicit.

5. The image of Bott periodicity in HP_* and the final form of the Chern-Connes character

In section 4 we have seen that, essentially due to the universal property of the kk-functor, there is a completely canonical multiplicative transformation χ from the $\mathbb{Z}$-graded(!) theory kk_* to the a priori also $\mathbb{Z}$-graded theory HP_*. But HP_* is actually $\mathbb{Z}/2$-periodic in a canonical way (due to Connes' S-operator). On the other hand kk_* is periodic too - not by definition but using the product with the Bott element b_A from Definition 3.1. The following computation has been made first in [**15**].

THEOREM 5.1. *(a) One has $\chi_{-2}(b_{\mathbb{C}}) = (2\pi i)^{-1} 1_{\mathbb{C}}$.*
(b) Given an m-algebra A one has $\chi_{-2}(b_A) = (2\pi i)^{-1} 1_A$.

PROOF. (a) The element $b_{\mathbb{C}}$ is defined as the product $\rho_{\mathbb{C}}\sigma_{\mathbb{C}}\tau_{\mathbb{C}}\kappa_{\mathbb{C}}^{-1}$ where $\tau_{\mathbb{C}}$, $\rho_{\mathbb{C}}$ are the elements of kk_{-1} determined by the extensions

$$(4)\qquad \mathcal{K}\to\overline{\mathcal{T}}_0\to C^\infty(S^1\smallsetminus 1)\qquad \mathbb{C}(0,1)\to\mathbb{C}(0,1]\to\mathbb{C}$$

and $\sigma_{\mathbb{C}},\kappa_{\mathbb{C}}$ are the isomorphisms in kk_0 induced by the maps $\mathbb{C}(0,1)\to C^\infty(S^1\smallsetminus 1)$ and $\mathbb{C}\to\mathcal{K}$. By naturality of the boundary maps in kk_*, $b_{\mathbb{C}}$ can also be written as the product $\rho_{\mathbb{C}}\overline{\sigma}_{\mathbb{C}}\overline{\tau}_{\mathbb{C}}\kappa_{\mathbb{C}}^{-1}$ where $\overline{\tau}_{\mathbb{C}}$ denotes the element of kk_{-1} determined by the extension $\mathcal{K}\to\overline{\mathcal{T}}\to C^\infty(S^1)$ and $\overline{\sigma}_{\mathbb{C}}$ the element of kk_0 induced by the inclusion $\mathbb{C}(0,1)\to C^\infty(S^1)$.

Given any extension $I\to E\to Q$ with classifying map $\gamma: JQ\to I$, the product by $\chi_{-1}(\gamma)$ gives the boundary maps in the long exact sequences associated by the extension in the first or second variable of HP, respectively, see [**10**], Theorem 5.5. So the product by $\chi_{-1}(\rho_{\mathbb{C}})$ and $\chi_{-1}(\overline{\tau}_{\mathbb{C}})$ on the right gives the boundary maps $HP_*(\mathbb{C},\mathbb{C})\to HP_{*-1}(\mathbb{C},\mathbb{C}(0,1))$ and $HP_*(\mathbb{C},C^\infty(S^1)\to HP_{*-1}(\mathbb{C},\mathcal{K})$ for the extensions $\mathbb{C}(0,1)\to\mathbb{C}(0,1]\to\mathbb{C}$ and $\mathcal{K}\to\overline{\mathcal{T}}\to C^\infty(S^1)$, respectively. We are going to determine these boundary maps using that $HP_*(\mathbb{C},A)=HP_*(A)$.
In fact, since all the algebras involved have small homological dimension, it turns out that we can do the computation in HX_* rather than in HP_*. Recall that $HX_*(A)$ denotes the homology of the $\mathbb{Z}/2$-graded X-complex

$$X(A):\qquad A\quad \underset{b}{\overset{\natural d}{\rightleftarrows}}\quad \Omega^1(A)_\natural$$

which is the quotient of the periodic $B-b$-complex $\widehat{\Omega}_*(A)$ by the first subcomplex F_1 in the Hodge filtration. Here $\natural$ denotes quotient by commutators. One has

$$(5)\qquad HX_0(A)=SHC_2(A)\qquad HX_1(A)=HC_1(A)$$

where HC_* denotes ordinary cyclic homology, see [**9**], p.393.
By the argument in [**2**], p.127 and formula (4) above one has $HC_1(\mathbb{C}(0,1))=HX_1(\mathbb{C}(0,1))=\mathbb{C}$, $SHC_0(\mathbb{C}(0,1))=HX_0(\mathbb{C}(0,1))=0$. Using the fact that $\mathbb{C}(0,1)$ is h-unital and excision for HC_* one gets that $HX_*(\mathbb{C}(0,1])=0$. For the extension $\mathbb{C}(0,1)\to\mathbb{C}(0,1]\to\mathbb{C}$ we then get the following commutative diagram

$$(6)\qquad \begin{array}{ccccccc} \longrightarrow & HP_*(\mathbb{C}(0,1)) & \longrightarrow & HP_*(\mathbb{C}(0,1]) & \longrightarrow & HP_*(\mathbb{C}) & \longrightarrow\\ & \downarrow\cong & & \downarrow\cong & & \downarrow\cong & \\ \longrightarrow & HX_*(\mathbb{C}(0,1):\mathbb{C}(0,1]) & \longrightarrow & HX_*(\mathbb{C}(0,1]) & \longrightarrow & HX_*(\mathbb{C}) & \longrightarrow \end{array}$$

where $HX_*(\mathbb{C}(0,1):\mathbb{C}(0,1]$ denotes the homology of the relative complex i.e. the homology of the kernel of $X_*(\mathbb{C}(0,1])\to X_*(\mathbb{C})$. The rows are exact and the first downward arrow is an isomorphism by the five-lemma. Since $\mathbb{C}(0,1)$ is h-unital $HX_*(\mathbb{C}(0,1):\mathbb{C}(0,1])$ is isomorphic to $HX_*(\mathbb{C}(0,1))$.
Similarly, we get a commutative diagram for the Toeplitz extension

$$(7)\qquad \begin{array}{cccccccc} \longrightarrow & HP_*(\mathcal{K}) & \longrightarrow & HP_*(\overline{\mathcal{T}}) & \longrightarrow & HP_*(C^\infty(S^1)) & \longrightarrow\\ & \downarrow\cong & & \downarrow\cong & & \downarrow\cong & \\ \longrightarrow & HX_*(\mathcal{K}) & \longrightarrow & HX_*(\overline{\mathcal{T}}) & \longrightarrow & HX_*(C^\infty(S^1)) & \longrightarrow \end{array}$$

The second row is exact because $\mathcal{K}$ is h-unital. The first and third downward arrows are isomorphisms because $\mathcal{K}$ and $C^\infty(S^1)$ have homological dimension 1, and finally the middle downward arrow is an isomorphism by the five-lemma.
We now have to identify the boundary maps δ_1,δ_2 for the second rows in (6) and

(7). They will represent $\chi_{-1}(\rho_{\mathbb{C}})$ and $\chi_{-1}(\overline{\tau}_{\mathbb{C}})$. By definition of the boundary map in the long exact sequence for an extension of complexes, in (6) we have to lift the element 1 in $X_0(\mathbb{C})$ to an element in $X_0(\mathbb{C}(0,1])$. For such a lift we choose a monotone function $h \in \mathbb{C}(0,1]$ such that $h(1) = 1$. We then apply the boundary operator $\natural d$ in the X-complex. This gives $\delta_1(1) = \natural\, dh$.
Under the map $X(\mathbb{C}(0,1)) \to X(C^\infty(S^1))$ induced by the inclusion $\overline{\sigma} : \mathbb{C}(0,1) \to C^\infty(S^1)$, $\natural\, dh$ is mapped to $(2\pi i)^{-1} w^{-1} dw$ where $w = e^{2\pi i h}$ (this is because modulo commutators $w^{-1}dw = e^{-2\pi i h}\, 2\pi i\, dh\, e^{2\pi i h}$). By homotopy invariance of HP_* the element $w^{-1}dw$ becomes in $HP_*(C^\infty(S^1)) = HX_*(C^\infty(S^1))$ equal to $z^{-1}dz$ with $z = e^{2\pi i t}$ the standard generator of $C^\infty(S^1)$. Now, to compute $\delta_2(z^{-1}dz)$, we lift z, z^{-1} to u, v and $z^{-1}dz$ to vdu in $X_1(\overline{\mathcal{T}})$. When we apply the boundary b in the X-complex to vdu, we obtain $vu - uv = e$. In conclusion $\delta_2 \overline{\sigma}_* \delta_1(1) = (2\pi i)^{-1} e$ with e the standard minimal idempotent in $\mathcal{K}$ (here $\overline{\sigma}_*$ denotes the map induced on HX_0 by $\overline{\sigma}$).
(b) follows from (a) by exterior product, see e.g. [**10**], p.86. In fact b_A is the exterior product $b_{\mathbb{C}} \otimes \mathrm{id}_A$. □

Since the ideals $\mathcal{K}$ and $\mathbb{C}(0,1)$ are h-unital, the computation above can in fact be done in non-periodic cyclic homology and Nistor has found in [**15**] that b_A corresponds in the non-periodic case to $2\pi i\, S$.

REMARK 5.2. In [**5**] we had given a proof for the formulas in Theorem 5.1 that reduces the computation of the boundary maps to the purely algebraic case of finitely generated algebras over $\mathbb{C}$. For the present paper we have chosen the proof given above in order to emphasize the role of the inclusion map $\overline{\sigma}$ from $\mathbb{C}(0,1)$ into $C^\infty(S^1)$.

Recall the invertible elements $\alpha_A^n = \alpha_A \alpha_{JA} \cdots \alpha_{J^n A} \in HP_n(A, J^n A)$ and $\beta_A^n = \beta_{A(0,1)^n} \cdots \beta_{A(0,1)} \beta_A \in HP_n(A(0,1)^n, A)$ from section 4. Left and right multiplication by these elements gives the transformation $\chi_n : kk_n \to HP_n$. In particular $\chi_0 : kk_0(A,B) \to HP_0(A,B)$ is the universal functor described in section 2. By the discussion in section 4 it sends a homomorphism $\varphi : J^n A \to B(0,1)^n$ representing an element of $kk_0(A,B)$ to $\alpha_A^n \varphi \beta_B^n$. This part of χ therefore is the canonical choice for the Chern-Connes character $kk_0 \to HP_0$.
Now what happens to $\chi_{2n} : kk_{2n} \to HP_{2n}$. The group $kk_n(A,B)$ is isomorphic to $kk_{n-2}(A,B)$ under left multiplication $x \mapsto b_A \cdot x$ by the Bott element and the two groups are identified under this isomorphism in the periodic $\mathbb{Z}/2$-graded K-theory. We have to check how this identification behaves under χ. This means that we have to compute $\chi_{n-2}(b_A \cdot x)$ in terms of $\chi_n(x)$.
But this has been achieved in Theorem 5.1 and the answer is $\chi_{n-2}(b_A \cdot x) = (2\pi i)^{-1} \chi_n(x)$. Therefore, if we define

$$ch_{-2n}(x) = (2\pi i)^n \chi_{-2n}(x) \tag{8}$$

for an element $x \in kk_{-2n}(A,B)$ then we get a multiplicative transformation ch from the even part $kk_{2*}(A,B)$ of kk to $HP_0(A,B)$. Moreover

$$ch_{-2n-2}(b_A \cdot x) = (2\pi i)^{n+1} \chi_{-2n-2}(b_A \cdot x) = (2\pi i)^{n+1} (2\pi i)^{-1} \chi_{-2n}(x) = ch_{-2n}(x).$$

Thus ch_* is invariant under $b_A \cdot$ and descends to a transformation $kk_{ev} \to HP_0$ from $kk_{2\star}$ made $\mathbb{Z}/2$-periodic. This is the natural choice for the Chern-Connes character in the even case.

The situation is more complicated for the odd part of the character. We have to extend ch from the even part to a transformation $ch_* : kk_* \to HP_*$ that is multiplicative and descends to kk_* made periodic via the Bott map. Multiplicativity means in particular that $ch_{-1}(x)ch_{-1}(y) = ch_{-2}(xy) = 2\pi i\chi_{-2}(xy)$ for $x, y \in kk_{-1}(A, A)$. This equation leads more or less inevitably to the following definition.

DEFINITION 5.3. We define, given $x \in kk_n(A, B)$, the bivariant Chern-Connes character by $ch_n(x) = (2\pi i)^{-n/2}\chi_n(x)$.

Thus, more explicitly, if x in $kk_n(A, B)$ is represented by $\varphi : J^{k-n}A \to B(0,1)^k$, then we set

$$ch_n(x) = (2\pi i)^{-n/2}\alpha_A^{k-n} HP_0(\varphi)\beta_B^k \tag{9}$$

For more clarity, in this formula we have suppressed the identification of B with $\mathcal{K} \otimes B$ via ι^{-1} as in (3). Observe that we cannot avoid the non-canonical choice of a square root for $2\pi i$ that we have to make once and for all.

THEOREM 5.4. *The transformation ch_* defined in* (5.3) *extends the definition* (8) *to the case of odd n. It is multiplicative, i.e. it satisfies $ch_s(x)ch_t(y) = ch_{s+t}(xy)$ and it descends to a multiplicative transformation $kk_{ev/od} \to HP_{ev/od}$ from kk_* made periodic via the Bott map.*

PROOF. Multiplicativity follows from the multiplicativity of χ. The compatibility with the Bott map follows as before from Theorem 5.1. □

REMARK 5.5. (a) Given an extension $I \to A \to B$ of m-algebras denote by x in $kk_{-1}(B, I)$ the element defined by the classifying map $JB \to I$. The boundary maps in the long exact sequences for kk_* and HP_* are given as the product by x and $\chi(x)$ respectively. Since χ is multiplicative it therefore respects the boundary maps (see also [**16**] for a more precise statement). On the other hand $ch_{-1}(x) = \sqrt{2\pi i}\chi_{-1}(x)$. Therefore ch_* multiplies the boundary maps by $\sqrt{2\pi i}$.
(b) Consider the theory $\overline{HP}_*$ which is obtained if we multiply the boundary map $B - b$ in the complex defining HP_* by $\sqrt{2\pi i}$. Since the kernels and images of $B - b$ remain unchanged we have $\overline{HP}_* \cong HP_*$. This isomorphism is in fact an equality of homology groups. The only difference between $\overline{HP}$ and HP is in the boundary maps for long exact sequences. The boundary maps in the long exact sequences in $\overline{HP}_*$ are defined using the boundary map in the complex and therefore are also multiplied by $\sqrt{2\pi i}$. We can now have a look at what happens to the transformation $\overline{\chi}_* : kk_* \to \overline{HP}_*$ defined in analogy to χ_*. Since the maps α^n and β^n used in Theorem 4.1 for the definition of χ_* are defined using the boundary maps in the exact sequences in HP_* for the universal and cone extensions, to obtain their analogues $\overline{\alpha}^n$, $\overline{\beta}^n$ in $\overline{HP}_*$ we have to multiply α^n, β^n by $(2\pi i)^{n/2}$ and $(2\pi i)^{-n/2}$, respectively. Thus, if we define $\overline{\chi}_* : kk_* \to \overline{HP}_*$ in analogy to (9) using $\overline{\alpha}^n$, $\overline{\beta}^n$, then $\overline{\chi}$ has all the required properties - it is multiplicative, compatible with boundary maps and descends to a multiplicative transformation from kk_*, made $\mathbb{Z}/2$-periodic, to $\overline{HP}_*$. In fact, under the natural identification $\overline{HP}_* = HP_*$ it is equal to the character ch_* constructed above.

References

[1] M. F. Atiyah, *K-theory*, W. A. Benjamin, Inc., New York-Amsterdam, 1967. Lecture notes by D. W. Anderson. MR0224083

[2] Alain Connes, *Noncommutative differential geometry*, Inst. Hautes Études Sci. Publ. Math. **62** (1985), 257–360. MR823176

[3] Guillermo Cortiñas and Andreas Thom, *Bivariant algebraic K-theory*, J. Reine Angew. Math. **610** (2007), 71–123. MR2359851

[4] Joachim Cuntz, *K-theory and C^*-algebras*, Algebraic K-theory, number theory, geometry and analysis (Bielefeld, 1982), Lecture Notes in Math., vol. 1046, Springer, Berlin, 1984, pp. 55–79, DOI 10.1007/BFb0072018. MR750677

[5] Joachim Cuntz, *Bivariante K-Theorie für lokalkonvexe Algebren und der Chern-Connes-Charakter* (German, with English summary), Doc. Math. **2** (1997), 139–182. MR1456322

[6] Joachim Cuntz, *Excision in periodic cyclic theory for topological algebras*, Cyclic cohomology and noncommutative geometry (Waterloo, ON, 1995), Fields Inst. Commun., vol. 17, Amer. Math. Soc., Providence, RI, 1997, pp. 43–53, DOI 10.1007/s002220050115. MR1478700

[7] Joachim Cuntz, *Cyclic theory, bivariant K-theory and the bivariant Chern-Connes character*, Cyclic homology in non-commutative geometry, Encyclopaedia Math. Sci., vol. 121, Springer, Berlin, 2004, pp. 1–71, DOI 10.1007/978-3-662-06444-3_1. MR2052771

[8] Joachim Cuntz, *Bivariant K-theory and the Weyl algebra*, K-Theory **35** (2005), no. 1-2, 93–137. MR2240217

[9] Joachim Cuntz and Daniel Quillen, *Cyclic homology and nonsingularity*, J. Amer. Math. Soc. **8** (1995), no. 2, 373–442. MR1303030 (96e:19004)

[10] Joachim Cuntz and Daniel Quillen, *Excision in bivariant periodic cyclic cohomology*, Invent. Math. **127** (1997), no. 1, 67–98, DOI 10.1007/s002220050115. MR1423026

[11] Joachim Cuntz and Andreas Thom, *Algebraic K-theory and locally convex algebras*, Math. Ann. **334** (2006), no. 2, 339–371, DOI 10.1007/s00208-005-0722-7. MR2207702

[12] G. A. Elliott, T. Natsume, and R. Nest, *Cyclic cohomology for one-parameter smooth crossed products*, Acta Math. **160** (1988), no. 3-4, 285–305. MR945014

[13] Max Karoubi, *Homologie cyclique des groupes et des algèbres* (French, with English summary), C. R. Acad. Sci. Paris Sér. I Math. **297** (1983), no. 7, 381–384. MR732839

[14] G. G. Kasparov, *Equivariant KK-theory and the Novikov conjecture*, Invent. Math. **91** (1988), no. 1, 147–201, DOI 10.1007/BF01404917. MR918241

[15] Victor Nistor, *A bivariant Chern-Connes character*, Ann. of Math. (2) **138** (1993), no. 3, 555–590, DOI 10.2307/2946556. MR1247993

[16] Victor Nistor, *Higher index theorems and the boundary map in cyclic cohomology*, Doc. Math. **2** (1997), 263–295. MR1480038

[17] Michael Puschnigg, *Asymptotic cyclic cohomology*, Lecture Notes in Mathematics, vol. 1642, Springer-Verlag, Berlin, 1996, DOI 10.1007/BFb0094458. MR1482804

Mathematisches Institut, Einsteinstr. 62, 48149 Münster, Germany
Email address: cuntz@uni-muenster.de

Proceedings of Symposia in Pure Mathematics
Volume **105**, 2023
https://doi.org/10.1090/pspum/105/01899

Applications of topological cyclic homology to algebraic K-theory

Bjørn Ian Dundas

ABSTRACT. Algebraic K-theory has applications in a broad range of mathematical subjects, from number theory to functional analysis. It is also fiendishly hard to calculate. Presently there are two main inroads: motivic homotopy theory and cyclic homology. I've been asked to present an overview of the applications of topological cyclic homology to algebraic K-theory "from a historical perspective". The timeline spans from the very early days of algebraic K-theory to the present, starting with ideas in the seventies around the "tangent space" of algebraic K-theory all the way to the current state of affair where we see a resurgence in structural theorems, calculations and a realization that variants of cyclic homology have important things to say beyond the moorings to K-theory.

In what follows I try to give a quasi-historical overview of topological cyclic homology, TC, from the perspective of algebraic K-theory by selecting some papers I think are good illustrations. These papers may or may not be the "first" or the "best" in any sense and since I've chosen them there is some bias, but regardless, by starting from these and see who cites them and who they cite I hope it should be possible to get a fairly broad grasp of the subject. I have largely ignored development that is not about the **connection** between the two theories. To cite C. Weibel [**Wei99**] "Like all papers, this one has a finiteness obstruction [...] We apologize in advance to the large number of people whose important contributions we have skipped over."

Before we start on the historical development, there are some ICM/ECM addresses and books that should be mentioned for reference: M. Karoubi [**Kar87**] (which was my main source as a student), T. G. Goodwillie [**Goo91**], I. Madsen [**Mad94a**], [**Mad94b**], J.-L. Loday [**Lod98**] (the readers who wish to check the definitions of more algebraic matters like Kähler differentials, Hochschild homology and so on should have this handy since it would break the flow to do so here), L. Hesselholt [**Hes02**], B. I. Dundas, Goodwillie and R. McCarthy [**DGM13**] and J. Rognes [**Rog14**].

Copyright conundrums rob this paper for the most enjoyable part of the original talk [from which this paper was much too hastily assembled]: the happy faces of many of my heroes and fellow mathematicians. Please picture them yourself or do a web search for the talk.

I'll use the following shorthand: K-theory means algebraic K-theory, ring spectrum and $\mathbb{S}$-algebra means E_1-ring spectrum and commutative ring spectrum means E_∞-ring spectrum

1. Prehistory

The history of homological invariants of K-theory goes far back, but for our purposes a natural place to start is with S. M. Gersten's d log map and S. Bloch's interpretation [**Blo73**] of this map as providing information about the "tangent space" of algebraic K-theory. This can be seen as a beginning of the story for many reasons and it actually appears in the 1972 Battelle proceedings in which D. Quillen first develops fully his algebraic K-theory.

If A is a commutative ring and Ω^*_A its ring of Kähler differentials, Gersten defines a graded ring homomorphism

$$\mathrm{d}\log\colon K_*(A) \to \Omega^*_A$$

via Grothendieck's theory of Chern classes, extending the map $\mathrm{GL}_1(A) \to \Omega^1_A$ sending a unit g to $\frac{1}{g}\,dg$. Bloch defines the *tangent space* $TK_n(A)$ of K-theory at A to be the kernel of the map $K_n(A[\epsilon]/\epsilon^2) \to K_n(A)$ induced by $\epsilon \mapsto 0$ and proves that Ω^{n-1}_A splits off $TK_n(A)$ if A is a local ring with 2 invertible. When $n = 2$ this was known to van der Kallen, and one can think of this as a first step towards identifying relative K-theory in terms of cyclic homology-like invariants.

Another observation in the seventies with implications to TC is K. Dennis' trace map to Hochschild homology

$$K_n(A) \to \mathrm{HH}_n(A),$$

related to the Hattori-Stallings trace [**Hat65**], [**Sta65**]. Simultaneously, there were deep investigations by many authors of the connections to crystalline cohomology, Chow groups, algebraic cycles, Chern classes and so on.

A quick word about how the maps of Gersten and Dennis fit in the picture to emerge may be in order. Thinking of K-theory as having something to do with vector bundles and Kähler differentials with differential forms, Gersten's map seems very geometric. Upon staring at a concrete formula, the Dennis trace map may seem more algebraic. This is an illusory distinction. In the smooth case the Hochschild-Kostant-Rosenberg theorem asserts that the "antisymmetrization map" $\epsilon\colon \Omega^n_A \to \mathrm{HH}_n(A)$ is an isomorphism, and – getting a bit ahead of ourselves – for general commutative A the de Rham differential $d\colon \Omega^n_A \to \Omega^{n+1}_A$ corresponds under ϵ to Connes' operator $B\colon \mathrm{HH}_n(A) \to \mathrm{HH}_{n+1}(A)$ coming from the circle action on HH which will be central when we get to cyclic homology. Connes' idea is that with Hochschild homology we gain access also to "non-commutative geometry". For more background and definitions the reader may consult [**Lod98**] and [**Con94**].

In the second half of the seventies F. Waldhausen enters the fray in full force. He is motivated by the study of manifolds and the prospect that algebraic K-theory may have much more to offer in this direction if one allows a generalization of rings he envisions as "Brave New Rings". In modern parlance, he is considering algebras over the sphere spectrum $\mathbb{S}$, and in particular spherical group rings $\mathbb{S}[G]$ whose algebraic K-theory gives deep information about the classifying space BG of the (simplicial or topological) group G. Through work of for instance J. H. C. Whitehead and C. T. C. Wall and the famous s-cobordism theorem of Barder-Mazur-Stallings it was known that important information about a pointed connected space X could be obtained by looking at the K-theory of the integral group ring $\mathbb{Z}[\pi_1 X]$. However, it was clear that much information was lost. It was primarily the lower K-groups that carried valuable information and even that picture was marred by fringe effects and missing torsion information. Replacing the fundamental group by the loop group is a good idea, but Waldhausen's daring leap is to realize that the relevant questions involve not linear algebra over the integers, but in a stable context. Hence the focus shifts to spherical group rings. The interested reader may consult the book by Waldhausen, Jahren and Rognes [**WJR13**] and its references for the geometric consequences, in particular [**Wal87b**] and [**Wal87a**]. We will have a few more details to offer when we revisit this theme in Section 2.1 (rational results) and Section 8.3 (integral results) where the relation with cyclic homology is at the forefront.

In [**Wal79**] Waldhausen describes how the Dennis trace for a ring A factors through a universal homology theory under K-theory he dubs *stable K-theory*

$$K(A) \to K^S(A) \to \mathrm{HH}(A)$$

(here and later $K(A)$ is a naturally chosen spectrum with $\pi_n K(A) = K_n(A)$ – this turns out to be an important shift of paradigm: the groups are no longer our main target). The first map makes sense also for A an $\mathbb{S}$-algebra, and for spherical group rings it splits (with a geometrically important complement), and a natural question is what HH should mean for $\mathbb{S}$-algebras. Eventually Waldhausen and Goodwillie conclude that there ought to exist a Hochschild homology over the sphere spectrum with tensor replaced by the (at the time still conjectural) smash product of spectra. They call this theory *topological Hochschild homology* THH and conjecture that stable K-theory and topological Hochschild homology are equivalent.

2. Cyclic homology in the 80's

In the early eighties A. Connes defined *cyclic homology*, HC [**Con83**], [**Con85**], a de Rham-like theory for associative rings which can also be seen as the homotopy orbits $\mathrm{HH}(A)_{h\mathbb{T}}$ by the cyclic action on the standard representation of Hochschild homology, see [**Lod98**]. Although not Connes' main focus at the time, to us the central questions at this juncture might be

- What is the relation between algebraic K-theory and cyclic homology?
- In particular, is there a "d log" like Gersten's or more generally some sort of Chern theory for associative rings and how does it compare with Bloch's "tangent space" insight?
- How does cyclic homology fit in Waldhausen's program? Is there a version for $\mathbb{S}$-algebras and how much information about K-theory can be retained?

The first years of cyclic homology see an explosion of ideas appearing simultaneously. Independently, B. Tsygan [**Tsy83**] had discovered the theory and dubbed it *additive K-theory* since (at least rationally) it is "like K-theory but with the Lie algebra $\mathfrak{gl}(A)$ playing the rôle of the general linear group $\mathrm{GL}(A)$"; a point of view investigated both in Feigin-Tsygan [**FT87**] and Loday-Quillen [**LQ84**]. This offers a tantalizing geometric perspective: if we pretend that $\mathrm{GL}_n(A)$ is a Lie group G, then the Lie algebra $\mathfrak{gl}_n(A)$ should be the tangent space $T_1 G$ at the unit $1 \in G$ and the inverse of the exponential map $T_1G \to G$ should be a local isomorphism connecting K-theory (made from $\mathrm{GL_n}$) to cyclic homology (made from $\mathfrak{gl}_n$). A naïve look at the formula $\log(1-x) = \sum_{j>0} \frac{x^j}{j}$ gives the impression that this idea only works over the rationals (we need to divide out by the denominators) and somehow $(x^j)_{j>0}$ should converge rather rapidly to zero. This fits hand in glove with Bloch's approach: he considered the situation $x^2 = 0$ (to interpret the tangent space).

2.1. Rational results. In hindsight it is clear that the some of the ideas developed independently in [**HS82**] by W. C. Hsiang and R. Staffeldt are tangent to Connes and Tsygan's, but the goal is different: Hsiang and Staffeldt prove that rationally Waldhausen's algebraic K-theory of simply connected spaces can be described by the homology of Lie algebras and the K-theory of $\mathbb{Z}$. Since we're working rationally, the K-theory of $\mathbb{Z}$ is known from A. Borel's calculation [**Bor74**].

Another indication that things were under control over the rationals came in the form of Goodwillie's confirmation [**Goo85b**] of Waldhausen's conjecture that stable K-theory and Hochschild homology are rationally equivalent. This was followed up by [**Goo86**] showing that if $A \to B$ is a map of simplicial rings such that $\pi_0 A \to \pi_0 B$ is a nilpotent extension (i.e., a surjective ring homomorphism with nilpotent kernel), then the *relative* K-theory $K(A \to B) = \mathrm{fiber}\{K(A) \to K(B)\}$ agrees with the (shifted) relative cyclic homology

$$K(A \to B) \simeq_{\mathbb{Q}} \Sigma \mathrm{HH}(A \to B)_{h\mathbb{T}}. \tag{2.1.0}$$

Goodwillie's result subsumes that of Hsiang and Staffeldt in the sense that they study the rational K-theory of the 1-connected map of spherical group rings $\mathbb{S}[\Omega X] \to \mathbb{S}$ for X a simply connected space. However, since K-theory commutes with rationalization this amounts to studying the map $\mathbb{Q}[\Omega X] \to \mathbb{Q}$ which fits in Goodwillie's framework (phrased in terms of simplicial rings after strictifying the loop group into a simplicial group e.g., via the Kan loop group).

A crucial ingredient in Goodwillie's analysis is that the Dennis trace factors over the homotopy *fixed* points $\mathrm{HH}(A)^{h\mathbb{T}}$ of the circle action on Hochschild homology

$$K(A) \to \mathrm{HH}(A)^{h\mathbb{T}} \to \mathrm{HH}(A),$$

(see [**Goo85a**] and C. H. Hood and D. S. D. Jones [**HJ87**]) and the equivalence (2.1.0) is induced by this factorization and by the so-called "norm map" $\Sigma\mathrm{HH}(A)_{h\mathbb{T}} \to \mathrm{HH}(A)^{h\mathbb{T}}$. The homotopy groups of the homotopy fixed points are often referred to as "negative cyclic homology". The positive/negative point of view comes from the grading on the bicomplexes used to calculate (negative) cyclic homology, mirroring the skeleton filtration of the classifying space $B\mathbb{T} = \mathbb{C}P^\infty$ of the circle.

The cofiber of the norm map $\Sigma\mathrm{HH}(A)_{h\mathbb{T}} \to \mathrm{HH}(A)^{h\mathbb{T}}$ (denoted $\mathrm{HH}(A)^{t\mathbb{T}}$ and called the *Tate construction* or *periodic cyclic homology*) plays an important rôle in many situations, and its siblings will reappear below. For now it suffices to notice that Goodwillie proved that evaluating at $t = 0$ gives an equivalence $\mathrm{HH}(A[t])^{t\mathbb{T}} \xrightarrow{\sim} \mathrm{HH}(A)^{t\mathbb{T}}$ for rational A.

With this development, calculations of rational algebraic K-theory all of a sudden seem feasible, spawning an interest in relevant calculations of cyclic homology; a somewhat arbitrary list could contain papers by D. Burghelea, G. Cortiñas, Z. Fiedorowicz, S. Geller, L. Reid, M. Vigué-Poirrier, Villamayor and Weibel

[**VPB85**], [**BF86**], [**CnV87**], [**GRW89**] and certainly a lot of others, some of which will resurface later.

2.2. Interpretations of the trace map. Connes and Karoubi [**CK84**] and Weibel [**Wei87**] provide a good way of expressing the relation between various forms of K-theory and of cyclic homology. In motivic theory one of the important axioms is $\mathbb{A}^1$-invariance: the affine line should be contractible. In the affine situation this means that the theory shouldn't see the difference between a ring A and its polynomial ring $A[t]$. However, unless you assume that A is particularly nice (for instance regular), $K(A[t])$ and $K(A)$ will be very different, and the difference is given by the (relative) K-theory of the category of nilpotent endomorphisms of finitely generated projective modules.

There is a universal way of imposing $\mathbb{A}^1$-invariance on a functor F from simplicial rings to spectra: let $HF(A)$ be what you get if you apply F degreewise to the simplicial ring $\Delta A = \{[q] \mapsto A[t_0, t_1, \ldots, t_q]/\sum t_j = 1\}$. Then $HF(A[t]) \overset{\sim}{\to} HF(A)$ and in the case where already $F(A[t]) \overset{\sim}{\to} F(A)$ the canonical map $F(A) \to HF(A)$ is an equivalence. Applying this construction to K-theory you get $\mathbb{A}^1$-invariant K-theory with history going back to Karoubi and O. Villamayor [**KV71**] and we may define nil-K-theory $nilK(A)$ to be the fiber of the canonical map $K(A) \to HK(A)$. When the degreewise extension of F is a homotopy functor, the contractibility of ΔA forces $HF(A)$ to be contractible too; this is for instance the case with cyclic homology $\mathrm{HH}(A)_{h\mathbb{T}}$. Hence, for A rational the Tate-theory $\mathrm{HH}(A)^{t\mathbb{T}}$ is the $\mathbb{A}^1$ invariant theory associated with the homotopy fixed points $\mathrm{HH}(A)^{h\mathbb{T}}$ and the trace gives a map of fiber sequences

$$\begin{array}{ccccc} nilK(A) & \longrightarrow & K(A) & \longrightarrow & HK(A) \\ \downarrow & & \downarrow & & \downarrow \\ \Sigma\mathrm{HH}(A)_{h\mathbb{T}} & \longrightarrow & \mathrm{HH}(A)^{h\mathbb{T}} & \longrightarrow & \mathrm{HH}(A)^{t\mathbb{T}}. \end{array}$$

In the relative nilpotent situation of Goodwillie's theorem (2.1.0) the spectra to the right both vanish and in the absolute situation the perspective of the trace as the logarithm from a Lie group to its tangent space (in the guise of cyclic homology) propagandized for above should be viewed as a statement about the left vertical map.

3. 'Rationally' is *not* enough: THH, TC

While all of this was happening, M. Bökstedt managed to give a convincing definition of topological Hochschild homology, THH, conforming with Goodwillie and Waldhausen's expectations of a Hochschild homology over the sphere spectrum. Even though his definition predates the modern view on ring spectra, what Bökstedt does is to write down an explicit description (in terms of objects later recognized as good building blocks for (highly structured) ring spectra) closely mimicking the algebraic construction of Hochschild homology and show that the result has the correct homotopical properties. Perhaps most impressively, he was even able to calculate the pivotal cases $\mathrm{THH}(\mathbb{F}_p)$ and $\mathrm{THH}(\mathbb{Z})$. Bökstedt's calculations showed that the homotopy ring $\mathrm{THH}_*(\mathbb{F}_p)$ is polynomial on a generator in degree two, starkly contrasting with (derived) Hochschild homology $\mathrm{HH}(\mathbb{F}_p)$ (a. k. a. Shukla homology) which is a divided power algebra on a generator in degree two. So, although $\mathrm{THH}(\mathbb{F}_p)$ and $\mathrm{HH}(\mathbb{F}_p)$ have the same homotopy *groups*, the *ring structures* are vastly different, a phenomenon that is at the heart of many of the ensuing calculations of K-theory through trace methods.

M. Jibladze, T. Pirashvili and Waldhausen soon realized that THH of ordinary rings coincides with an old homology theory of Mac Lane's and also with the homology of the category of projective modules and of certain functor categories [**JP91**], [**PW92**]. This correspondence led to some extremely efficient calculations by V. Franjou, J. Lannes, L. Schwartz and Pirashvili [**FLS94**], [**FP98**]. In hindsight, some of these groups were already known to L. Breen [**Bre78**] through similar considerations.

3.1. The algebraic K-theory Novikov conjecture. The potential of Bökstedt's construction was immediately demonstrated through Bökstedt, Hsiang and Madsen's successful attack [**BHM93**] in the late 80's on the algebraic K-theory version of the Novikov conjecture. They prove that the assembly map

$$K(\mathbb{Z})\wedge BG_+ \to K(\mathbb{Z}[G])$$

to the K-theory of the integral group ring is rationally injective whenever the group G has finitely generated homology in each dimension. To achieve this they define "topological cyclic homology" TC. Connes discovered that the Hochschild complex is "cyclic" and so (its realization) comes with a natural action by the circle $\mathbb{T}$. The same goes for topological Hochschild homology, but TC is not just a homotopy orbit/fixed point-type construction applied to THH. With the explicit construction Bökstedt gave, there are fairly apparent maps given in terms of restricting equivariant maps to fixed points. These restriction maps give a structure on the various fixed point spectra akin to Witt-ring constructions in algebra, a feature of which is captured by saying the THH is a "cyclotomic spectrum". The important ushot is a map $\widehat{\Gamma}_1\colon \mathrm{THH} \to \mathrm{THH}^{tC_p}$. The actual construction of TC is somewhat involved, and we cheat by citing the much later interpretation [**NS18**] of T. Nikolaus and P. Scholze, where TC is the homotopy equalizer

$$\mathrm{TC}(A) \longrightarrow \mathrm{THH}(A)^{h\mathbb{T}} \underset{\varphi}{\overset{\mathrm{can}}{\rightrightarrows}} \mathrm{THH}(A)^{t\mathbb{T}^\wedge},$$

where $\mathrm{can}\colon \mathrm{THH}(A)^{h\mathbb{T}} \to \mathrm{THH}(A)^{t\mathbb{T}^\wedge}$ is the profinite completion of the cofiber of the above mentioned norm map $\Sigma\mathrm{THH}(A)_{h\mathbb{T}} \to \mathrm{THH}(A)^{hT}$ and the Frobenius φ comes from the cyclotomic structure (or more generally, from the fact that smash powers support "geometric diagonals"). Note that if we ignore the profinite completion and replace φ by the trivial map, $\mathrm{TC}(A)$ is replaced by (a suspension of) the orbit spectrum $\mathrm{THH}(A)_{h\mathbb{T}}$ – the most naïve definition imaginable of "topological cyclic homology".

The Dennis trace factors over TC giving rise to the *(cyclotomic) trace*

$$K(A) \to \mathrm{TC}(A)$$

(this is amazing and it wouldn't have worked with the "naïve TC": the composite $K(A) \to \mathrm{TC}(A) \to \mathrm{THH}(A)^{h\mathbb{T}}$ does not factor over $\Sigma\mathrm{THH}(A)_{h\mathbb{T}}$).

The projection $\mathbb{S} \to \mathbb{Z}$ induces a rational equivalence $K(\mathbb{S}[G]) \xrightarrow{\sim}_{\mathbb{Q}} K(\mathbb{Z}[G])$, so for Bökstedt, Hsiang and Madsen's purposes it is just as good to work over the sphere spectrum, which is advantageous since they understand $\mathrm{TC}(\mathbb{S}[G])$ in terms of free loop spaces and power maps (when applied to spherical group rings the above mentioned restriction maps split. See also the precursor [**CCGH87**] and the later [**BCC$^+$96**]). This means that they understand the assembly map for TC

$$\mathrm{TC}(\mathbb{S})\wedge BG_+ \to \mathrm{TC}(\mathbb{S}[G])$$

and eventually the K-theory version of the Novikov conjecture is "reduced" to understanding the difference between K-theory and TC of the (p-adic) integers. In hindsight we know (see Section 6) that after profinite completion the homotopy groups of these spectra agree in non-negative dimensions, but in [**BHM93**] they get through knowing somewhat less.

3.2. Stable K-theory is THH. Once Bökstedt had defined THH, Waldhausen proposed a program to prove that stable K-theory and topological Hochschild homology agree for all $\mathbb{S}$-algebras, see [**SSW96**]. Indeed, he had already established the "vanishing of the mystery homology theory" by geometric means [**Wal87a**], which amounts to the equivalence $\mathbb{S} \xrightarrow{\sim} K^S(\mathbb{S})$. Together with the obvious $\mathbb{S} \xrightarrow{\sim} \mathrm{THH}(\mathbb{S})$ one has that stable K-theory and THH coincide in the initial case.

The first full proof was almost orthogonal to this program: Dundas and McCarthy proved $K^S \simeq \mathrm{THH}$ for simplicial rings [**DM94**] by comparing both theories to the homology of the category of finitely generated projective modules through McCarthy's setup for amalgamating (topological) Hochschild homology and Waldhausen's K-theory. The equivalence was later extended to all connective $\mathbb{S}$-algebra through a denseness argument [**Dun97**].

4. The first calculation of TC

With TC established as a success for obtaining information about K-theory, Bökstedt and Madsen attack the most prestigious calculation: What is $\mathrm{TC}(\mathbb{Z})$? Their (successful) attack was until recently the standard model for next to all calculations of TC (and – together with motivic methods – for K-theory). The papers [**BM94**] and [**BM95**] are hard to read, but a treasure trove of ideas. Somewhat streamlined, the procedure is as follows:

(1) Calculate THH (Bökstedt had already done that for $\mathbb{Z}$ and $\mathbb{F}_p$)
(2) Consider cartesian squares comparing categorical and homotopical fixed points of the cyclic groups C_{p^n} of orders powers of the prime p

$$\begin{array}{ccc} \mathrm{THH}^{C_{p^n}} & \xrightarrow{R} & \mathrm{THH}^{C_{p^{n-1}}} \\ \downarrow{\scriptstyle \Gamma_n} & & \downarrow{\scriptstyle \widehat{\Gamma}_n} \\ \mathrm{THH}^{hC_{p^n}} & \longrightarrow & \mathrm{THH}^{tC_{p^n}}. \end{array}$$

(3) Prove that $\widehat{\Gamma}_1$ is an equivalence in high degrees.
(4) Calculate the homotopy fixed points $\mathrm{THH}^{hC_{p^n}}$ and $\mathrm{THH}^{h\mathbb{T}}$ keeping track of known elements coming from K-theory using fixed point/Tate spectral sequences
(5) Assemble the pieces keeping in mind possible Witt-like extensions to obtain TC.

Bökstedt and Madsen do this at odd primes for the integers, and Rognes does $p = 2$ [**Rog99a**]. These calculations expose periodic phenomena as predicted by the Lichtenbaum-Quillen conjecture and later expanded to a "red-shift" paradigm, see Section 8.4.

Item 2 in the procedure above is not phrased like this in the earliest papers. Nikolaus and Scholze use the diagram to "define" the categorical fixed points by means of homotopy invariant constructions, using only the existence of a map $\widehat{\Gamma}_1$ which arises from the cyclotomic structure. As a matter of fact, their repackaging of the theory has led to new ways of interpreting this attack which seem very fruitful, and entirely new and elegant methods for handling TC have been developed. For a striking example see e.g., R. Liu and G. Wang's preprint [?https://doi.org/10.48550/arxiv.2012.15014].

The reason that proving that $\widehat{\Gamma}_1$ is an equivalence in high degrees suffices is Tsalidis' result [**Tsa97**] saying that then this is also true for all the Γ_n's via an argument akin to Carlsson's proof of the Segal conjecture.

5. How far apart are K-theory and TC?

Inspired by his calculus of functor and his rational results, Goodwillie conjectured at the 1990 ICM in Kyoto [**Goo91**] that:

> Let K^{inf} be the fiber of the trace $K \to \text{TC}$. If $A \to B$ is a map of connective ring spectra such that $\pi_0 A \to \pi_0 B$ is a nilpotent extension, then
>
> $$K^{\text{inf}}A \to K^{\text{inf}}B$$
>
> is an equivalence.

The conjecture was stated only for 1-connected maps, and Goodwillie's vision was that one should prove that

(1) Both K-theory and TC are analytic functors
(2) The trace induces an equivalence of differentials

$$D_1 K(A) \xrightarrow{\sim} D_1 \text{TC}(A).$$

According to Goodwillie's calculus of functors one would then have that

Within the "radius of convergence" the fiber $K^{\text{inf}}(A)$ of the trace $K(A) \to \text{TC}(A)$ is constant.

In the guise of stable K-theory we've seen that $D_1 K(A)$ is equivalent to topological Hochschild homology and Hesselholt [**Hes94**] proves that so is $D_1 \text{TC}(A)$. From this McCarthy [**McC97**] proves Goodwillie's conjecture after profinite completion for simplicial rings, and this was extended to connective $\mathbb{S}$-algebras in [**Dun97**]. The integral result stated above as Goodwillie's conjecture was established around 1996 and documented in [**DGM13**].

6. Calculations of TC amount to calculations in K-theory

Algebraic K-theory was already fairly well understood "away from the prime p" by results like Gabber rigidity [**Gab92**] which, for instance, implies that the projection $\mathbb{Z}_p \to \mathbb{F}_p$ from the p-adics to the prime field induces an equivalence in K-theory after completion at any prime different from p. What topological cyclic homology has to offer is that we get hold of information "at the prime".

It all starts with the prime field $\mathbb{F}_p$, which Quillen tells us is fairly dull at p:

$$K(\mathbb{F}_p) \simeq_p H\mathbb{Z},$$

where $\simeq_p$ designates an equivalence after p-completion and $H\mathbb{Z}$ is the integral Eilenberg-Mac-Lane spectrum. Madsen [**Mad94b**] shows that the cyclotomic trace is an equivalence except for noise in negative degrees: $\mathrm{TC}(\mathbb{F}_p) \simeq_p \Sigma^{-1}H\mathbb{Z} \vee H\mathbb{Z}$. Goodwillie's conjecture then implies that for all $n > 0$

$$K(\mathbb{Z}/p^n\mathbb{Z}) \simeq_p \mathrm{TC}(\mathbb{Z}/p^n\mathbb{Z})[0, \infty),$$

where the spectrum on the right hand side is the connective cover. Now, investigations by, for instance, A. Suslin and I. Panin [**Pan86**] show that after profinite completion, $K(\mathbb{Z}_p)$ is the limit of the $K(\mathbb{Z}/p^n\mathbb{Z})$s, and when Hesselholt and Madsen [**HM97b**] establish the same result for TC they are in position to conclude that Bökstedt and Madsen's calculation of $\mathrm{TC}(\mathbb{Z})$ amounts to a calculation of the p-completion of $K(\mathbb{Z}_p)$!

On the TC side, [**HM97b**] cements and enhances the innovations of [**BM94**] and [**BM95**]. The paper also provides a lot of valuable perspectives (plus that the end result is more general than I've stated) and should be seen in tandem with [**Hes96**] which realizes Bloch's program connecting K-theory and crystalline cohomology [**Blo77**].

At the end of the millennium, the best structural result may be summarized by

> Let K^{inf} be the fiber of the trace $K \to \mathrm{TC}$. If k is a perfect field of characteristic $p > 0$, A a connective ring spectrum whose ring of path components $\pi_0(A)$ is a $W(k)$-algebra which is finitely generated as a module, then
>
> $$K^{\mathrm{inf}}(A) \simeq_p \Sigma^{-2}H\mathrm{coker}\{\pi_0 A \xrightarrow{1-F} \pi_0 A\},$$
>
> where F is the Frobenius.

7. Where do we go from here? TC in the 21st century.

The pioneering age of topological cyclic homology ends around the turn of the millennium. In the 21st century TC consolidates its rôle as a work horse in the study K-theory. Firstly, TC provides calculations of K-theory that seemed totally out of reach in 1990. We give some examples of this in Section 8.

Secondly, after 2000 the structural understanding of TC expands immensely, shedding light on the nature of TC as well as on K-theory and we give a brief account in Section 9. Many of the structural results were pivotal to the calculations; for instance, understanding how TC behaves under localization was needed for calculations of the K-theory of local number fields. This structural understanding has even spilled over to neighboring fields – a technical part of the calculation of K-theory of the integers via motivic cohomology as presented by Rognes and Weibel [**RW00**] relied on understanding the multiplicative structure of TC.

Last but not least, new frameworks and perspectives on TC and the trace have materialized; a much too brief discussion is concentrated in Section 9.2.

8. An explosion of calculations

8.1. Local fields. In [**HM03**] Hesselholt and Madsen muster the full force of the theory to calculate the K-theory of local fields. To achieve this they use an amalgam of

- the methods for calculation of TC initiated by Bökstedt, Madsen and Hesselholt, enhanced through setting up a framework for handling the "de Rham-Witt complex"
- Goodwillie's conjecture
- a setup for handling localization through log-formalism combined with TC of categories – a precursor of log-methods and localization for ring spectra.

The last point is crucial and we will return to a fuller discussion of log-methods and localization for ring spectra in Section 9.4. The calculation of the K-theory of local fields [**HM03**] accentuates how structurally different K-theory and TC are, in particular how differently they behave with respect to localization. Whereas TC of a field of characteristic 0 is pretty uninformative in positive degrees, the K-theory of a local field F sits in a fiber sequence $K(k) \to K(\mathcal{O}_F) \to K(F)$ where $\mathcal{O}_F$ is the ring of integers and k is the residue field. The failure is not so much TC not respecting Waldhausen's localization sequence as TC not allowing a reformulation of the base spectrum as $\mathrm{TC}(F)$. Taking the consequence of this, Hesselholt and Madsen simply keep the category Waldhausen provides and take TC of that using [**DM96**] to obtain a map from the localization sequence in K-theory. See also [**Dun98**] for a structural comparison of K-theory and TC.

From an algebraic and equivariant point of view, the treatment of the de Rham-Witt complex may be even more interesting. In the paper this mixture of the ring of Witt vectors and the de Rham complex (with log-poles) is important because it provides a framework for handling the interplay between the various fixed point spectra of THH.

The setup has been revisited many times, for instance in Hesselholt, M. Larsen and A. Lindenstrauss' [**HLL20**].

A reason the methods work for local but not global fields is exemplified by the observation that $K(\mathbb{Z})$ and $K(\mathbb{Z}_p)$ are vastly (and interestingly) different, while $\mathrm{TC}(\mathbb{Z}) \tilde{\to}_p \mathrm{TC}(\mathbb{Z}_p)$.

8.2. Truncated polynomials, nilterms, endomorphisms. Calculations of K-theory of truncated polynomial rings go far back to the infancy of K-theory, and with Goodwillie's conjecture an established fact such calculations became open for attack; the first published account being [**HM97a**] where the answer is given in terms of big Witt vectors. This topic have been followed up by several authors, for instance V. Angeltveit's [**Ang17**]. Also, there is an interesting interplay between $\mathbb{A}^1$-invariance and nil-invariance, making calculations of the K-theory of nilpotent endomorphism somewhat more accessible.

The identification between stable K-theory and THH factors over the K-theory of the category of endomorphisms, which for a very long time had been known to be closely related to the theory of Witt vectors. Before the existence of TC, Goodwillie had written a widely circulated letter to Waldhausen indicating a theory in this direction. McCarthy soon realized that there was a close connection between the Taylor tower of K-theory and such a Witt-like theory and saw it as an extension of the TR-construction used to build TC, see e.g., Lindenstrauss, McCarthy [**LM12**], A. Blumberg, D. Gepner, G. Tabuada [**BGT16**] and the recent [**MP22**] by C. Malkiewich and K. Ponto.

The vanishing of negative K-groups is a somewhat related question that has been addressed through trace methods. For the localization square

$$\begin{array}{ccc} K(\mathbb{P}_A) & \longrightarrow & K(A[t]) \\ \downarrow & & \downarrow \\ K(A[t^{-1}]) & \longrightarrow & K(A[t,t^{-1}]) \end{array}$$

to be cartesian (where $\mathbb{P}_A$ is the projective line in Quillen's sense; $K(\mathbb{P}_A) \simeq K(A) \vee K(A)$ [**Qui73**]), Bass observed that you need to extend the definition of K-theory to also have negative homotopy groups (see e.g., [**TT90**]) and Weibel formulated a question about whether the negative K-groups vanish below the dimension of the input. Using (topological) cyclic homology this was shown in characteristic zero by in [**CnHSW08**] by G. Cortiñas, C. Haesemeyer, M. Schlichting and Weibel and in positive characteristic (over infinite perfect fields, assuming strong resolution of singularities) by T. Geisser and Hesselholt in [**GH10**].

8.3. Algebraic K-theory of spaces, ring spectra and assembly. Before Grothendieck, J. H. C. Whitehead developed K-theoretical methods for group rings, and to this day (spherical) group rings are among the most important inputs for K-theory via Waldhausen's algebraic K-theory of spaces as discussed briefly in Section 1. Goodwillie's conjecture opened a new line of investigation. For an overview of the situation and some more background, see Rognes' very readable lecture [**Rog99b**] where he also outlines the link between automorphisms of manifolds and algebraic K-theory in more detail.

As an example let's look at the initial case: $\mathbb{S} \to \mathbb{Z}$. As promised, $K(\mathbb{S})$ contains much geometric information: indeed, according to Waldhausen there is an isomorphism

$$K_i(\mathbb{S}) \cong \pi_i\mathbb{S} \oplus \operatorname{colim}_n \pi_{i-2}\mathrm{Diff}(D^{n+1} \operatorname{rel} D^n),$$

where $\mathrm{Diff}(D^{n+1} \operatorname{rel} D^n)$ is the space of diffeomorphisms of the closed unit $(n+1)$-dimensional disc that are fixed on the Northern hemisphere. The colimit is actually attained in the sense that each homotopy class of such diffeomorphisms of discs of sufficiently high dimension is present in the algebraic K-theory of the sphere spectrum. On the other hand, K-theory detects a lot of number theory, starting with class field theory in dimension zero and continuing in all dimensions as codified for instance in the Lichtenbaum–Quillen conjecture.

So, the induced map $K(\mathbb{S}) \to K(\mathbb{Z})$ spans all the way from information about diffeomorphisms of discs to number theory, but the chasm is not wider than what TC can handle. Goodwillie's conjecture tells us that the square

$$\begin{array}{ccc} K(\mathbb{S}) & \longrightarrow & \mathrm{TC}(\mathbb{S}) \\ \downarrow & & \downarrow \\ K(\mathbb{Z}) & \longrightarrow & \mathrm{TC}(\mathbb{Z}) \end{array}$$

is cartesian. Add to that the calculation of $\mathrm{TC}(\mathbb{Z})$ and the homotopy theoretic control of $\mathrm{TC}(\mathbb{S})$ we see that we're in an extremely interesting situation. J. Klein and Rognes attacked the problem in [**KR97**] and the issue has been revisited by many since, see e.g., Rognes' identification [**Rog03**] of the homotopy type of $K(\mathbb{S})$ and Blumberg and M. Mandell's complement on the homotopy groups in [**BM19**].

The next space one wants to understand the algebraic K-theory of is the circle S^1. As is commented by Madsen in [**Mad94b**], thanks to the insight provided by Waldhausen, K. Igusa [**Igu88**], F. T. Farrell and L. E. Jones [**FJ91**] and M. Weiss and R. Williams [**WW88**], knowing the cofiber of the assembly map $K(\mathbb{S})\wedge S^1_+ \to K(\mathbb{S}[\Omega S^1]) = K(\mathbb{S}[t,t^{-1}])$ (which is called the topological Whitehead spectrum $Wh^{\mathrm{TOP}}(S^1)$ of the circle) goes a long way towards understanding the homeomorphisms of manifolds with negative sectional curvature. For further details, see e.g., the introduction to [**Hes09**] where Hesselholt provides some calculations which should be seen in conjunction with J. Grunewald, Klein and T. Macko's [**GKM08**].

As we saw, TC was created to solve the Novikov conjecture. There is a host of refined conjectures and results in this direction – essentially about how $K(A[G])$ can be decomposed into information about $K(A)$ and information about the subgroup lattice of G – but just some of the progress can be attributed TC. This progress can be followed by consulting the paper [**LRRV19**] by W. Lück, H. Reich, Rognes and M. Varisco.

Ausoni and Rognes' calculations of the algebraic K-theory of topological K-theory [**AR02**] (to be commented further on in the next section) is another prime example of calculations of K-theory of ring spectra through trace methods. Later this calculation was given a geometric interpretation through two-vector bundles [**BDRR11**], [**BDRR13**], but the concrete calculation needed to show that the Dirac monopole splits as a virtual two vector bundle [**ADR08**] was only accessible through a comparison using the trace.

Many important ring spectra are Thom spectra. Blumberg, C. Schlichtkrull and R. Cohen [**BCS10**] show that Thom spectra are very manageable on the THH-side – a far-reaching extension of the equivalence between $\mathrm{THH}(\mathbb{S}[G])$ and the free loop space $\mathrm{Map}(S^1, BG)$. Most provocatively, $H\mathbb{F}_p$ is a Thom spectrum (but admittedly not through an E_∞-map). Following the references to [**BCS10**] will lead to many recent results where familiar spectra are viewed as Thom spectra, and, among other things, to reformulations of Bökstedt's calculation, see e.g., the multiplicative version in A. Krause and Nikolaus [**KN19**].

Another amazing equivariant result about THH of Thom spectra comes in the form of S. Lunøe-Nielsen and Rognes' "Segal conjecture" for complex cobordism MU [**LNR11**]. This makes it attractive to try to understand algebraic K-theory of spaces not via descent over $\mathbb{S} \to \mathbb{Z}$, but over $\mathbb{S} \to \mathrm{MU}$, see Section 9.4.

See also Malkiewich [**Mal17**] for a dual picture.

8.4. Red-shift/chromatic behavior beyond LQC. Chromatic convergence (from now on: complete everything at a prime p and ignore fine points wrt. what localization to use) tells us that the sphere spectrum is the limit of the tower

$$\cdots \to L_n\mathbb{S} \to L_{n-1}\mathbb{S} \to \cdots \to L_1\mathbb{S} \to L_0\mathbb{S} = H\mathbb{Q}$$

of localizations at the Johnson-Wilson theories $E(n)$ with coefficients $E(n)_* = \mathbb{Z}_{(p)}[v_1, \ldots, v_n, v_n^{-1}]$. Waldhausen [**Wal84**] speculated that applying K-theory would yield a tower starting at $K(\mathbb{Q})$ and converging to $K(\mathbb{S})$ – giving a highly structured ascent from number theory to manifolds! See J. McClure and Staffeldt [**MS93**] for more details.

C. Ausoni and Rognes take the first step up this ladder beyond the Lichtenbaum-Quillen conjecture in [**AR02**] where they calculate the K-theory of the so-called Adams summand; followed up by Ausoni's calculation [**Aus10**] of the K-theory of the connective complex K-theory, ku. Part of the pattern that emerges can be seen from the following table (think of $\mathbb{Z}_p$ as "$\mathbb{F}_p[[v_0]]$"):

the homotopy of	$\mathrm{TC}(\mathbb{F}_p)$	is a finitely generated free $\mathbb{Z}_p$-module
the mod-p homotopy of	$\mathrm{TC}(\mathbb{Z})$	is a finitely generated free $\mathbb{F}_p[v_1]$-module
the mod-(p, v_1) homotopy of	$\mathrm{TC}(\mathrm{ku})$	is a finitely generated free $\mathbb{F}_p[v_2]$-module

Ausoni and Rognes also point to conjectural calculations showing a clear pattern, but depending on properties of the Brown-Petersen spectra that must be somewhat adjusted. They synthesize the picture into a "red-shift conjecture" in [**MR208**]. See also [**Rog14**] which among many other things points towards a motivic filtration for commutative ring spectra, see e.g., [**HRW22**].

The chromatic properties of K-theory has been revisited several times, e.g., B. Bhatt, D. Clausen, A. Mathew [**BCM20**], Ausoni, B. Richter and Rognes [**AR20**], [**AR12**]. Personally, my favorite approach to red-shift is through iterations, noting that (for a commutative ring spectrum) iterating THH is simply tensoring with a torus whose rich group of symmetries should structure the chromatic behavior. A first start can be found in T. Veen's thesis [**Vee18**] which shows that the periodic self-map v_{n-1} can be detected in the n-fold iterated K-theory $K^{(n)}(\mathbb{F}_p)$, at least for $n \leq p > 3$.

Crucial for this way of thinking is the extension of the "cyclotomic" concept for THH alluded to in Section 3.1 to the realization that the geometric fixed points of $X \otimes A$, where X is a G-space and A a commutative ring spectrum, is given by the homotopy G-orbits $X_{hG} \otimes A$ through the "geometric diagonal" [**BCD10**], [**BDS16**] ($X \otimes A$ is the categorical tensor in commutative ring spectra – a notion that already appears in J. McClure, R. Schwänzl and R. Vogt's interpretation of topological Hochschild homology [**MSV97**]). Put another way, we really don't have a good picture for why there should be red-shift for K-theory, but the very tight connection between K-theory and TC offers us the alternative – and to me reasonable – explanation that it is is really about the symmetries in stable homotopy theory to which K-theory is just an innocent victim.[1]

[1]**Stop the press!** Since this talk was held, the history of the red-shift conjecture has been seriously altered and most likely even come to a satisfactory conclusion for commutative ring spectra with the preprint on the chromatic Nullstellensatz by R. Burklund, T. Schlank and A. Yuan [**BSY22**]. Precisely, their claim is (Theorem 9.11)

> Let A be a nontrivial commutative ring spectrum of nonnegative height.
> Then the height of $K(A)$ is exactly one greater than the height of A.

This statement relies on J. Hahn's result [**Hah16**] on the chromatic support of commutative ring spectra to give a good characterization of the height. Burklund, Schlank and Yuan need to show that the height of $K(A)$ is strictly greater that the height of A since the fact that the difference between the heights of $K(A)$ and A is not more than one follows from M. Land, Mathew, L. Meier, G. Tamme [**LMMT20**] and Clausen, Mathew, Nauman, Noel [**CMNN20a**]. They achieve this through their Nullstellensatz which guarantees a map to a Lubin-Tate theory where Yuan [Yua21] already has shown the result.

9. Structural results

9.1. Structure on TC. Although "we" are awesome at calculating TC we know very little about the deeper structure of TC. For instance, even though TC preserves much structure, we don't understand the commutative structure of $\mathrm{TC}(A)$ – or even $\mathrm{THH}(A)$ – of a commutative ring spectrum A. My favourite example is the following: the commutator $S^1 \to S^1 \vee S^1$ (attaching the 2-cell in the torus) induces a map of augmented commutative $H\mathbb{F}_p$-algebras

$$\mathrm{THH}(\mathbb{F}_p) \to \mathrm{THH}(\mathbb{F}_p) \wedge_{H\mathbb{F}_p} \mathrm{THH}(\mathbb{F}_p).$$

Is it trivial? On homotopy it is the trivial map $\mathbb{F}_p[x] \to \mathbb{F}_p[x_1, x_2]$, $x \mapsto 0$ and furthermore as an *associative* map it is trivial and all primary E_∞-obstructions to triviality vanish. If it *is* trivial as an E_∞-map, we have a slew of results relevant to red-shift.

That said, we know some things, for instance from H. Bergsaker and Rognes [**BR10**] and also Angeltveit and Rognes [**AR05**]. Another point of view relevant for the multiplicative properties is the "norm" or "smash power" point of view exemplified by [**CDD11**] and [**ABG$^+$18a**] and which we touched upon in Section 8.4.

9.2. New frameworks. We've already mentioned Nikolaus and Scholze's new take on TC. This perspective is so useful and has proven to be so influential that I've allowed myself to anachronistically refer to it several times already – even when introducing TC in Section 3. See Hesselholt and Nikolaus' survey [**HN20**] for examples of this framework used on old and new problems.

There is another very important point of view that puts things in a different perspective. From the early days the universal properties of K-theory was quite apparent: roughly stated as "algebraic K-theory is the universal theory under the nerve that satisfies Waldhausen's additivity theorem". Things were finally firmly organized around this principle with Tabuada's additive and localizing invariants, see [**Tab14**] and together with Blumberg and Gepner [**BGT13**], [**BGT14**] and also by C. Barwick [**Bar16**]. In general, the ∞-categorical perspective has permeated much of the more recent literature. An example where the Goodwillie conjecture is significantly extended can be found in E. Elden and V. Sosnilo's [**ES20**].

9.3. Back to algebraic geometry. Grothendieck designed K_0 to answer questions in algebraic geometry and the trace methods were from the start built around Chern characters. As we've seen, TC doesn't always behave well with respect to localizations, but in the smooth case many things are fine, and Geisser and Hesselholt set up a successful theory for TC of schemes in [**GH99**] via hypercohomology as an alternative to TC of the exact category of algebraic vector bundles. Along the way they also provide a lot of structural insight and important results. In subsequent papers Geisser and Hesselholt follow up with a string of impressive

---

Note that full commutativity is essential to this setup; what happens in the E_n-case for $n < \infty$ is not understood. Forerunners to this phenomenon were noticed already by Ausoni and Rognes in their conjectural calculations for BP-theories in the introduction to [**AR02**]. For recent development in this direction the reader may enjoy Hahn and D. Wilson's preprint [**HW20**].

When the increase in height is established, it may be the time to look back at the (admittedly rather weak) calculational evidence that lead to the conjecture and see whether there is something more detailed to be said about the structure of algebraic K-theory (or perhaps rather: about the interplay between group actions and commutativity).

calculation, much, but not all, in positive characteristic. This line of investigation continues to bear fruit, with calculations in algebraic geometry and feeding back to the theory, see for instance [**GH06b**].

Some more recent results in this direction are connected with p-adic Hodge theory, see e.g., Bhatt, Morrow and Scholze's [**BMS18**] and [**BMS19**]. Scholze's theory of perfectoid rings plays a central rôle here, as does Hesselholt's much earlier paper [**Hes06**]. A commutative $\mathbb{Z}_p$-algebra R is perfectoid if there exists a nonzero-divisor $\pi \in R$ with $p \in \pi^p R$ such that R is complete and separated in the π-adic topology and such that the Frobenius $\phi\colon R/\pi \to R/\pi$ is a bijection.

As an appetizer, let's mention that Bhatt, Morrow and Scholze extend Bökstedt periodicity to all perfectoid rings R: $\pi_*\mathrm{THH}(R)_p \cong R[u]$ and identify Nikolaus and Scholze's maps $\phi, \mathrm{can}\colon \mathrm{THH}(R)_p^{h\mathbb{T}} \to \mathrm{THH}(R)_p^{t\mathbb{T}}$ on homotopy groups as

$$A_{\mathrm{inf}}(R)[u,v]/(uv-\xi) \xrightarrow[\mathrm{can}(u)=\xi\sigma,\, \mathrm{can}(v)=\sigma^{-1}]{\phi(u)=\sigma,\, \phi(v)=\phi(\xi)\sigma^{-1}} A_{\mathrm{inf}}(R)[\sigma,\sigma^{-1}].$$

Here $|u| = -|v| = 2$ and $A_{\mathrm{inf}}(R)$ is Fontaine's universal p-complete pro-infinitesimal thickening given as the ring of Witt vectors of the limit of R/p along the Frobenius map and ξ is a generator of the kernel of the associated map $\theta\colon A_{\mathrm{inf}}(R) \to R$. A fascinating point in their calculation is where they work not over $\mathbb{S}$ but the polynomial ring $\mathbb{S}[z]$ showing that the cyclotomic structure fits equally well here.

This is a point of departure where the interest in the homological side becomes clear, even at the expense of the connections to K-theory, a theme that we unfortunately don't have the occasion to explore here and where I hope someone can write an accessible survey. Even so, there is payback to K-theory, for instance in the form of S. Kelly and Morrow's proof of Gersten injectivity for valuation rings over fields [**KM21**].

9.4. Localization and Galois theory for ring spectra. A central question in K-theory is the following: if a group G acts on a ring A through ring homomorphisms (of particular interest is the case of a Galois extension $A^G \subseteq A$), to what extent is the natural map from the K-theory of the fixed ring to the homotopy fixed points of the K-theory

$$K(A^G) \to K(A)^{hG}$$

an equivalence? At the beginning of the millennium Rognes embarked on the quest to extend Galois theories to ring spectra (part of the motivation can be found in the introduction of [**AR02**]) and much of this program is documented in [**Rog08**]. This tour de force has seen applications in many direction and is a guiding principle also for the trace invariant.

Ever since the 1980s (see e.g., R. Thomason's perspective on K-theory [**TT90**]) descent has been a central tool. K-theory satisfies Nisnevich descent, but not necessarily étale descent – unless one localizes. Clausen, Mathew, Nauman and Noel address and partially answer the question about to what extent a Galois action passes through K-theory in [**CMNN20b**]. Successful applications that use both trace methods and descent have often proceeded by proving descent in parallel in K-theory and TC, see for instance [**Mor16**] and [**CMNN20b**].

Indeed, many applications of trace methods to descent have taken the following form: assume given a cosimplicial ring spectrum $A_\bullet$ (or some other diagram), such that for each $[n] \in \Delta$ you have control over the trace $K(A_{[n]}) \to \mathrm{TC}(A_{[n]})$, to what extent does that give you control over the trace on $A = \mathrm{holim}_\Delta A_\bullet$? In **[Dun97]** this was used to transport theorems from simplicial rings to connective ring spectra by considering the Amitzur complex $[n] \mapsto H\mathbb{Z}^{\wedge n+1}$ converging to $\mathbb{S}$. While of theoretical importance, the spectra $H\mathbb{Z}^{\wedge n+1}$ are computationally useless, and in [**DR18**, 4] the prospect of using descent along $\mathbb{S} \to MU$ is suggested. This has the advantage that $MU^{\wedge n+1}$ is more accessible than $H\mathbb{Z}^{\wedge n+1}$ since MU is a nice Thom spectrum, $MU^{\wedge n+1} \simeq MU \wedge BU_+^{\times n}$ and the Segal-type control of the equivariant structure offered by [**LNR11**].

The reasons that prompted the combination of trace and of log-methods in [**HM03**] continue to be relevant when considering ring spectra. This need was first addressed by Rognes in [**Rog09**] and has since been refined and followed up, see e.g., Rognes, S. Sagave and Schlichtkrull [**SS12**], [**RSS15**].

Another need that became clear with [**AR02**] was a good theory for localizations for ring spectra, and the first test was whether there is a map of localization sequences

$$\begin{array}{ccccc} K(\mathbb{Z}) & \longrightarrow & K(ku) & \longrightarrow & K(KU) \\ \downarrow & & \downarrow & & \downarrow \\ \mathrm{TC}(\mathbb{Z}) & \longrightarrow & \mathrm{TC}(ku) & \longrightarrow & \mathrm{TC}(ku \mid KU), \end{array}$$

where KU is the localization of ku you get by inverting the Bott element $u \in ku_2$. This challenge was taken up by Blumberg and Mandell who establish this in [**BM08**] and [**BM12**]. Unfortunately, this picture does not extend beyond ku to higher Brown-Peterson spectra, see B. Antieau, T. Barthel and Gepner [**ABG18b**] who use the trace to show that the fiber of localization can't be as hoped. That said, according to [**ABG18b**, Theorem 0.3], if A is a ring spectrum and $a \in \pi_t A$ is an element such that $\{1, a, a^2, \dots\}$ satisfies the Ore condition, then the localization $A \to A[a^{-1}]$ gives rise to a fiber sequence

$$K(End_A(A/a)^{\mathrm{op}}) \to K(A) \to K(A[a^{-1}]).$$

9.5. Singularities/closed excision. Except for the applications of TC to nilpotent extensions, the initial development focused on the smooth case (nicely complemented by motivic methods). However, for K-theory "closed excision" was important from the start. In the affine case it can be stated as follows: consider a "Milnor square", a cartesian square of rings

$$\begin{array}{ccc} A & \longrightarrow & B \\ \downarrow & & \downarrow \\ C & \longrightarrow & D; \end{array}$$

where $B \twoheadrightarrow D$ is surjective. What can you say about $K(A)$? In the rational case this received quite some attention in the eighties, but it was not before Cortiñas [**Cn06**], Geisser and Hesselholt [**GH06a**], Dundas and H. Kittang [**DK08**], [**DK13**] and M. Land and G. Tamme [**LT19**] that the full result was established

> Let K^{inf} be the fiber of the trace $K \to \mathrm{TC}$. If
>
> $$\begin{array}{ccc} A & \longrightarrow & B \\ \downarrow & & \downarrow \\ C & \longrightarrow & D \end{array}$$
>
> is a cartesian square of connective ring spectra where $\pi_0 B \twoheadrightarrow \pi_0 D$ is surjective, then the associated square
>
> $$\begin{array}{ccc} K^{\text{inf}}A & \longrightarrow & K^{\text{inf}}B \\ \downarrow & & \downarrow \\ K^{\text{inf}}C & \longrightarrow & K^{\text{inf}}D \end{array}$$
>
> is also cartesian.

Note that Land and Tamme prove more than closed excision.

As with Goodwillie's conjecture, closed excision opens up the possibility for calculations of K-theory, see for instance Hesselholt, Angeltveit T. Gerhardt and Nikolaus [**Hes07**], [**AG11**] and [**Hes14**].

9.6. Rigidity. I end the talk with a personal note and a cool result with an amazing proof. In 1996 I applied for a grant to show "rigidity" in the following form: If $A \to B$ is a map of connective ring spectra so that $\pi_0 A \to \pi_0 B$ is a complete extension (a surjection with a complete kernel), then

$$K^{\text{inf}}A \to K^{\text{inf}}B$$

is an equivalence after profinite completion. I didn't get the grant (which is my only excuse for not delivering – though I shared some beers with Randy McCarthy the one evening we both believed that we knew a proof).

In the case when $\pi_0 A$ and $\pi_0 B$ are commutative there is a generalization of the completeness assumption. A surjection of commutative rings $f\colon R \to S$ is a henselian extension if for all polynomials $q \in R[x]$ with reduction $q_S \in S[x]$ the induced function

$$\{\text{roots } \alpha \text{ of } q \text{ such that } q'(\alpha) \text{ is a unit}\} \to \{\text{roots } \beta \text{ of } q_S \text{ such that } q_S'(\beta) \text{ is a unit}\}$$

is surjective. The proof of rigidity by Clausen, Mathew and Morrow [**CMM21**] is truly wonderful. A crucial ingredient is the fact (which I never would have guessed) that TC/p commutes with filtered colimits and also that the trace mod p^n is split injective on homotopy groups when evaluated at local $\mathbb{F}_p$-algebras.

> Let K^{inf} be the fiber of the trace $K \to \mathrm{TC}$. If $A \to B$ is a map of connective ring spectra such that $\pi_0 A \to \pi_0 B$ is a henselian extension, then
>
> $$K^{\text{inf}}A \to K^{\text{inf}}B$$
>
> is an equivalence after profinite completion.

Exercise. Rigidity shouldn't depend on commutativity assumptions. State and prove the correct theorem.

Redshift should depend on some commutativity. Explore the borderline cases and explain the fine-structure.

Merging closed excision with Goodwillie's conjecture allows for more general statements. Find and prove a satisfactory statement for higher dimensional cartesian cubes.

Acknowledgements

I thank the referee for many good suggestions. At certain points the suggestions have forced me to abandon the idea of largely leaving out material that had not yet made it to publication by the time I gave the talk. This risks being partial to the emerging papers that I know about at the expense of some potentially breathtaking ones that have not yet caught my attention. For this I apologize.

References

[ABG+18a] Vigleik Angeltveit, Andrew J. Blumberg, Teena Gerhardt, Michael A. Hill, Tyler Lawson, and Michael A. Mandell, *Topological cyclic homology via the norm*, Doc. Math. **23** (2018), 2101–2163. MR3933034

[ABG18b] Benjamin Antieau, Tobias Barthel, and David Gepner, *On localization sequences in the algebraic K-theory of ring spectra*, J. Eur. Math. Soc. (JEMS) **20** (2018), no. 2, 459–487, DOI 10.4171/JEMS/771. MR3760300

[ADR08] Christian Ausoni, Bjørn Ian Dundas, and John Rognes, *Divisibility of the Dirac magnetic monopole as a two-vector bundle over the three-sphere*, Doc. Math. **13** (2008), 795–801. MR2466184

[AG11] Vigleik Angeltveit and Teena Gerhardt, *On the algebraic K-theory of the coordinate axes over the integers*, Homology Homotopy Appl. **13** (2011), no. 2, 103–111, DOI 10.4310/HHA.2011.v13.n2.a7. MR2846160

[Ang17] Vigleik Angeltveit, *Witt vectors and truncation posets*, Theory Appl. Categ. **32** (2017), Paper No. 8, 258–285. MR3607214

[AR02] Christian Ausoni and John Rognes, *Algebraic K-theory of topological K-theory*, Acta Math. **188** (2002), no. 1, 1–39, DOI 10.1007/BF02392794. MR1947457

[AR05] Vigleik Angeltveit and John Rognes, *Hopf algebra structure on topological Hochschild homology*, Algebr. Geom. Topol. **5** (2005), 1223–1290, DOI 10.2140/agt.2005.5.1223. MR2171809

[AR12] Christian Ausoni and John Rognes, *Algebraic K-theory of the first Morava K-theory*, J. Eur. Math. Soc. (JEMS) **14** (2012), no. 4, 1041–1079, DOI 10.4171/JEMS/326. MR2928844

[AR20] Christian Ausoni and Birgit Richter, *Towards topological Hochschild homology of Johnson-Wilson spectra*, Algebr. Geom. Topol. **20** (2020), no. 1, 375–393, DOI 10.2140/agt.2020.20.375. MR4071375

[Aus10] Christian Ausoni, *On the algebraic K-theory of the complex K-theory spectrum*, Invent. Math. **180** (2010), no. 3, 611–668, DOI 10.1007/s00222-010-0239-x. MR2609252

[Bar16] Clark Barwick, *On the algebraic K-theory of higher categories*, J. Topol. **9** (2016), no. 1, 245–347, DOI 10.1112/jtopol/jtv042. MR3465850

[BCC+96] M. Bökstedt, G. Carlsson, R. Cohen, T. Goodwillie, W. C. Hsiang, and I. Madsen, *On the algebraic K-theory of simply connected spaces*, Duke Math. J. **84** (1996), no. 3, 541–563, DOI 10.1215/S0012-7094-96-08417-3. MR1408537

[BCD10] Morten Brun, Gunnar Carlsson, and Bjørn Ian Dundas, *Covering homology*, Adv. Math. **225** (2010), no. 6, 3166–3213, DOI 10.1016/j.aim.2010.05.018. MR2729005

[BCM20] Bhargav Bhatt, Dustin Clausen, and Akhil Mathew, *Remarks on $K(1)$-local K-theory*, Selecta Math. (N.S.) **26** (2020), no. 3, Paper No. 39, 16, DOI 10.1007/s00029-020-00566-6. MR4110725

[BCS10] Andrew J. Blumberg, Ralph L. Cohen, and Christian Schlichtkrull, *Topological Hochschild homology of Thom spectra and the free loop space*, Geom. Topol. **14** (2010), no. 2, 1165–1242, DOI 10.2140/gt.2010.14.1165. MR2651551

[BDRR11] Nils A. Baas, Bjørn Ian Dundas, Birgit Richter, and John Rognes, *Stable bundles over rig categories*, J. Topol. **4** (2011), no. 3, 623–640, DOI 10.1112/jtopol/jtr016. MR2832571

[BDRR13] Nils A. Baas, Bjørn Ian Dundas, Birgit Richter, and John Rognes, *Ring completion of rig categories*, J. Reine Angew. Math. **674** (2013), 43–80, DOI 10.1515/crelle.2012.024. MR3010546

[BDS16] M. Brun, B. I. Dundas, and M. Stolz, *Equivariant structure on smash powers*, e-prints 2016, `arXiv:1605.05939`.

[BF86] D. Burghelea and Z. Fiedorowicz, *Cyclic homology and algebraic K-theory of spaces. II*, Topology **25** (1986), no. 3, 303–317, DOI 10.1016/0040-9383(86)90046-7. MR842427

[BGT13] Andrew J. Blumberg, David Gepner, and Gonçalo Tabuada, *A universal characterization of higher algebraic K-theory*, Geom. Topol. **17** (2013), no. 2, 733–838, DOI 10.2140/gt.2013.17.733. MR3070515

[BGT14] Andrew J. Blumberg, David Gepner, and Gonçalo Tabuada, *Uniqueness of the multiplicative cyclotomic trace*, Adv. Math. **260** (2014), 191–232, DOI 10.1016/j.aim.2014.02.004. MR3209352

[BGT16] Andrew J. Blumberg, David Gepner, and Gonçalo Tabuada, *K-theory of endomorphisms via noncommutative motives*, Trans. Amer. Math. Soc. **368** (2016), no. 2, 1435–1465, DOI 10.1090/tran/6507. MR3430369

[BHM93] M. Bökstedt, W. C. Hsiang, and I. Madsen, *The cyclotomic trace and algebraic K-theory of spaces*, Invent. Math. **111** (1993), no. 3, 465–539, DOI 10.1007/BF01231296. MR1202133

[Blo73] S. Bloch, *On the tangent space to Quillen K-theory*, Algebraic K-theory, I: Higher K-theories (Proc. Conf., Battelle Memorial Inst., Seattle, Wash., 1972), Lecture Notes in Math., Vol. 341, Springer, Berlin, 1973, pp. 205–210. MR0466264

[Blo77] Spencer Bloch, *Algebraic K-theory and crystalline cohomology*, Inst. Hautes Études Sci. Publ. Math. **47** (1977), 187–268 (1978). MR488288

[BM94] M. Bökstedt and I. Madsen, *Topological cyclic homology of the integers*, Astérisque **226** (1994), 7–8, 57–143. K-theory (Strasbourg, 1992). MR1317117

[BM95] M. Bökstedt and I. Madsen, *Algebraic K-theory of local number fields: the unramified case*, Prospects in topology (Princeton, NJ, 1994), Ann. of Math. Stud., vol. 138, Princeton Univ. Press, Princeton, NJ, 1995, pp. 28–57. MR1368652

[BM08] Andrew J. Blumberg and Michael A. Mandell, *The localization sequence for the algebraic K-theory of topological K-theory*, Acta Math. **200** (2008), no. 2, 155–179, DOI 10.1007/s11511-008-0025-4. MR2413133

[BM12] Andrew J. Blumberg and Michael A. Mandell, *Localization theorems in topological Hochschild homology and topological cyclic homology*, Geom. Topol. **16** (2012), no. 2, 1053–1120, DOI 10.2140/gt.2012.16.1053. MR2928988

[BM19] Andrew J. Blumberg and Michael A. Mandell, *The homotopy groups of the algebraic K-theory of the sphere spectrum*, Geom. Topol. **23** (2019), no. 1, 101–134, DOI 10.2140/gt.2019.23.101. MR3921317

[BMS18] Bhargav Bhatt, Matthew Morrow, and Peter Scholze, *Integral p-adic Hodge theory*, Publ. Math. Inst. Hautes Études Sci. **128** (2018), 219–397, DOI 10.1007/s10240-019-00102-z. MR3905467

[BMS19] Bhargav Bhatt, Matthew Morrow, and Peter Scholze, *Topological Hochschild homology and integral p-adic Hodge theory*, Publ. Math. Inst. Hautes Études Sci. **129** (2019), 199–310, DOI 10.1007/s10240-019-00106-9. MR3949030

[Bor74] Armand Borel, *Stable real cohomology of arithmetic groups*, Ann. Sci. École Norm. Sup. (4) **7** (1974), 235–272 (1975). MR387496

[BR10] Håkon Schad Bergsaker and John Rognes, *Homology operations in the topological cyclic homology of a point*, Geom. Topol. **14** (2010), no. 2, 735–772, DOI 10.2140/gt.2010.14.755. MR2602850

[Bre78] Lawrence Breen, *Extensions du groupe additif* (French), Inst. Hautes Études Sci. Publ. Math. **48** (1978), 39–125. MR516914

[BSY22] Robert Burklund, Tomer M. Schlank, and Allen Yuan, *The Chromatic Nullstellensatz*, arXiv:2207.09929, 2022.

[CCGH87] G. E. Carlsson, R. L. Cohen, T. Goodwillie, and W. C. Hsiang, *The free loop space and the algebraic K-theory of spaces*, K-Theory **1** (1987), no. 1, 53–82, DOI 10.1007/BF00533987. MR899917

[CDD11] Gunnar Carlsson, Christopher L. Douglas, and Bjørn Ian Dundas, *Higher topological cyclic homology and the Segal conjecture for tori*, Adv. Math. **226** (2011), no. 2, 1823–1874, DOI 10.1016/j.aim.2010.08.016. MR2737802

[CK84] Alain Connes and Max Karoubi, *Caractère multiplicatif d'un module de Fredholm* (French, with English summary), C. R. Acad. Sci. Paris Sér. I Math. **299** (1984), no. 19, 963–968. MR774679

[CMM21] Dustin Clausen, Akhil Mathew, and Matthew Morrow, *K-theory and topological cyclic homology of henselian pairs*, J. Amer. Math. Soc. **34** (2021), no. 2, 411–473, DOI 10.1090/jams/961. MR4280864

[CMNN20a] Dustin Clausen, Akhil Mathew, Niko Naumann, and Justin Noel, *Descent and vanishing in chromatic algebraic K-theory via group actions*, arXiv:2011.08233, 2020.

[CMNN20b] Dustin Clausen, Akhil Mathew, Niko Naumann, and Justin Noel, *Descent in algebraic K-theory and a conjecture of Ausoni-Rognes*, J. Eur. Math. Soc. (JEMS) **22** (2020), no. 4, 1149–1200, DOI 10.4171/JEMS/942. MR4071324

[Cn06] Guillermo Cortiñas, *The obstruction to excision in K-theory and in cyclic homology*, Invent. Math. **164** (2006), no. 1, 143–173, DOI 10.1007/s00222-005-0473-9. MR2207785

[CnHSW08] G. Cortiñas, C. Haesemeyer, M. Schlichting, and C. Weibel, *Cyclic homology, cdh-cohomology and negative K-theory*, Ann. of Math. (2) **167** (2008), no. 2, 549–573, DOI 10.4007/annals.2008.167.549. MR2415380

[CnV87] Guillermo H. Cortiñas and Orlando E. Villamayor, *Cyclic homology of $K[\mathbf{Z}/2\mathbf{Z}]$*, Rev. Un. Mat. Argentina **33** (1987), no. 1-2, 55–61 (1990). MR1101119

[Con83] Alain Connes, *Cohomologie cyclique et foncteurs* Ext^n (French, with English summary), C. R. Acad. Sci. Paris Sér. I Math. **296** (1983), no. 23, 953–958. MR777584

[Con85] Alain Connes, *Noncommutative differential geometry*, Inst. Hautes Études Sci. Publ. Math. (1985), no. 62, 257–360. MR823176 (87i:58162)

[Con94] Alain Connes, *Noncommutative geometry*, Academic Press, Inc., San Diego, CA, 1994. MR1303779

[DGM13] Bjørn Ian Dundas, Thomas G. Goodwillie, and Randy McCarthy, *The local structure of algebraic K-theory*, Algebra and Applications, vol. 18, Springer-Verlag London, Ltd., London, 2013. MR3013261

[DK08] Bjørn Ian Dundas and Harald Øyen Kittang, *Excision for K-theory of connective ring spectra*, Homology Homotopy Appl. **10** (2008), no. 1, 29–39. MR2369021

[DK13] Bjørn Ian Dundas and Harald Øyen Kittang, *Integral excision for K-theory*, Homology Homotopy Appl. **15** (2013), no. 1, 1–25, DOI 10.4310/HHA.2013.v15.n1.a1. MR3031812

[DM94] Bjørn Ian Dundas and Randy McCarthy, *Stable K-theory and topological Hochschild homology*, Ann. of Math. (2) **140** (1994), no. 3, 685–701, DOI 10.2307/2118621. MR1307900

[DM96] Bjørn Ian Dundas and Randy McCarthy, *Topological Hochschild homology of ring functors and exact categories*, J. Pure Appl. Algebra **109** (1996), no. 3, 231–294, DOI 10.1016/0022-4049(95)00089-5. MR1388700

[DR18] Bjørn Ian Dundas and John Rognes, *Cubical and cosimplicial descent*, J. Lond. Math. Soc. (2) **98** (2018), no. 2, 439–460, DOI 10.1112/jlms.12141. MR3873116

[Dun97] Bjørn Ian Dundas, *Relative K-theory and topological cyclic homology*, Acta Math. **179** (1997), no. 2, 223–242, DOI 10.1007/BF02392744. MR1607556

[Dun98] Bjørn Ian Dundas, *K-theory theorems in topological cyclic homology*, J. Pure Appl. Algebra **129** (1998), no. 1, 23–33, DOI 10.1016/S0022-4049(97)00083-2. MR1626647

[ES20] Elden Elmanto and Vladimir Sosnilo, *On nilpotent extensions of ∞-categories and the cyclotomic trace*, arXiv:2010.09155, 2020.

[FJ91] F. Thomas Farrell and Lowell E. Jones, *Rigidity in geometry and topology*, Proceedings of the International Congress of Mathematicians, Vol. I, II (Kyoto, 1990), Math. Soc. Japan, Tokyo, 1991, pp. 653–663. MR1159252

[FLS94] Vincent Franjou, Jean Lannes, and Lionel Schwartz, *Autour de la cohomologie de Mac Lane des corps finis* (French, with English and French summaries), Invent. Math. **115** (1994), no. 3, 513–538, DOI 10.1007/BF01231771. MR1262942

[FP98] Vincent Franjou and Teimuraz Pirashvili, *On the Mac Lane cohomology for the ring of integers*, Topology **37** (1998), no. 1, 109–114, DOI 10.1016/S0040-9383(97)00005-0. MR1480880

[FT87] B. L. Feĭgin and B. L. Tsygan, *Additive K-theory*, K-theory, arithmetic and geometry (Moscow, 1984), Lecture Notes in Math., vol. 1289, Springer, Berlin, 1987, pp. 67–209, DOI 10.1007/BFb0078368. MR923136

[Gab92] Ofer Gabber, *K-theory of Henselian local rings and Henselian pairs*, Algebraic K-theory, commutative algebra, and algebraic geometry (Santa Margherita Ligure, 1989), Contemp. Math., vol. 126, Amer. Math. Soc., Providence, RI, 1992, pp. 59–70, DOI 10.1090/conm/126/00509. MR1156502

[GH99] Thomas Geisser and Lars Hesselholt, *Topological cyclic homology of schemes*, Algebraic K-theory (Seattle, WA, 1997), Proc. Sympos. Pure Math., vol. 67, Amer. Math. Soc., Providence, RI, 1999, pp. 41–87, DOI 10.1090/pspum/067/1743237. MR1743237

[GH06a] Thomas Geisser and Lars Hesselholt, *Bi-relative algebraic K-theory and topological cyclic homology*, Invent. Math. **166** (2006), no. 2, 359–395, DOI 10.1007/s00222-006-0515-y. MR2249803

[GH06b] Thomas Geisser and Lars Hesselholt, *On the K-theory and topological cyclic homology of smooth schemes over a discrete valuation ring*, Trans. Amer. Math. Soc. **358** (2006), no. 1, 131–145, DOI 10.1090/S0002-9947-04-03599-8. MR2171226

[GH10] Thomas Geisser and Lars Hesselholt, *On the vanishing of negative K-groups*, Math. Ann. **348** (2010), no. 3, 707–736, DOI 10.1007/s00208-010-0500-z. MR2677901

[GKM08] Joachim Grunewald, John R. Klein, and Tibor Macko, *Operations on the A-theoretic nil-terms*, J. Topol. **1** (2008), no. 2, 317–341, DOI 10.1112/jtopol/jtm012. MR2399133

[Goo85a] Thomas G. Goodwillie, *Cyclic homology, derivations, and the free loopspace*, Topology **24** (1985), no. 2, 187–215, DOI 10.1016/0040-9383(85)90055-2. MR793184

[Goo85b] Thomas G. Goodwillie, *On the general linear group and Hochschild homology*, Ann. of Math. (2) **121** (1985), no. 2, 383–407, DOI 10.2307/1971179. MR786354

[Goo86] Thomas G. Goodwillie, *Relative algebraic K-theory and cyclic homology*, Ann. of Math. (2) **124** (1986), no. 2, 347–402, DOI 10.2307/1971283. MR855300

[Goo91] Thomas G. Goodwillie, *The differential calculus of homotopy functors*, Proceedings of the International Congress of Mathematicians, Vol. I, II (Kyoto, 1990) (Tokyo), Math. Soc. Japan, 1991, pp. 621–630. MR1159249 (93g:55015)

[GRW89] S. Geller, L. Reid, and C. Weibel, *The cyclic homology and K-theory of curves*, J. Reine Angew. Math. **393** (1989), 39–90, DOI 10.1515/crll.1989.393.39. MR972360

[Hah16] Jeremy Hahn, *On the Bousfield classes of H_∞-ring spectra*, `arXiv:1612.04386`, 2016.

[Hat65] Akira Hattori, *Rank element of a projective module*, Nagoya Math. J. **25** (1965), 113–120. MR175950

[Hes94] Lars Hesselholt, *Stable topological cyclic homology is topological Hochschild homology*, Astérisque **226** (1994), 8–9, 175–192. K-theory (Strasbourg, 1992). MR1317119

[Hes96] Lars Hesselholt, *On the p-typical curves in Quillen's K-theory*, Acta Math. **177** (1996), no. 1, 1–53, DOI 10.1007/BF02392597. MR1417085

[Hes02] Lars Hesselholt, *Algebraic K-theory and trace invariants*, Proceedings of the International Congress of Mathematicians, Vol. II (Beijing, 2002) (Beijing), Higher Ed. Press, 2002, pp. 415–425. MR1957052 (2004d:19005)

[Hes06] Lars Hesselholt, *On the topological cyclic homology of the algebraic closure of a local field*, An alpine anthology of homotopy theory, Contemp. Math., vol. 399, Amer. Math. Soc., Providence, RI, 2006, pp. 133–162, DOI 10.1090/conm/399/07517. MR2222509

[Hes07] Lars Hesselholt, *On the K-theory of the coordinate axes in the plane*, Nagoya Math. J. **185** (2007), 93–109, DOI 10.1017/S0027763000025757. MR2301459

[Hes09] Lars Hesselholt, *On the Whitehead spectrum of the circle*, Algebraic topology, Abel Symp., vol. 4, Springer, Berlin, 2009, pp. 131–184, DOI 10.1007/978-3-642-01200-6_7. MR2597738

[Hes14] Lars Hesselholt, *On the K-theory of planar cuspical curves and a new family of polytopes*, Algebraic topology: applications and new directions, Contemp. Math., vol. 620, Amer. Math. Soc., Providence, RI, 2014, pp. 145–182, DOI 10.1090/conm/620/12368. MR3290091

[HJ87] Christine E. Hood and John D. S. Jones, *Some algebraic properties of cyclic homology groups*, K-Theory **1** (1987), no. 4, 361–384, DOI 10.1007/BF00539623. MR920950

[HLL20] Lars Hesselholt, Michael Larsen, and Ayelet Lindenstrauss, *On the K-theory of division algebras over local fields*, Invent. Math. **219** (2020), no. 1, 281–329, DOI 10.1007/s00222-019-00909-x. MR4050106

[HM97a] Lars Hesselholt and Ib Madsen, *Cyclic polytopes and the K-theory of truncated polynomial algebras*, Invent. Math. **130** (1997), no. 1, 73–97, DOI 10.1007/s002220050178. MR1471886

[HM97b] Lars Hesselholt and Ib Madsen, *On the K-theory of finite algebras over Witt vectors of perfect fields*, Topology **36** (1997), no. 1, 29–101, DOI 10.1016/0040-9383(96)00003-1. MR1410465

[HM03] Lars Hesselholt and Ib Madsen, *On the K-theory of local fields*, Ann. of Math. (2) **158** (2003), no. 1, 1–113, DOI 10.4007/annals.2003.158.1. MR1998478

[HN20] Lars Hesselholt and Thomas Nikolaus, *Topological cyclic homology*, Handbook of homotopy theory, CRC Press/Chapman Hall Handb. Math. Ser., CRC Press, Boca Raton, FL, [2020] ©2020, pp. 619–656. MR4197995

[HRW22] Jeremy Hahn, Arpon Raksit, and Dylan Wilson, *A motivic filtration on the topological cyclic homology of commutative ring spectra*, arXiv:2206.11208, 2022.

[HS82] W. C. Hsiang and R. E. Staffeldt, *A model for computing rational algebraic K-theory of simply connected spaces*, Invent. Math. **68** (1982), no. 2, 227–239, DOI 10.1007/BF01394056. MR666160

[HW20] Jeremy Hahn and Dylan Wilson, *Redshift and multiplication for truncated Brown-Peterson spectra*, 2020.

[Igu88] Kiyoshi Igusa, *The stability theorem for smooth pseudoisotopies*, K-Theory **2** (1988), no. 1-2, vi+355, DOI 10.1007/BF00533643. MR972368

[JP91] Mamuka Jibladze and Teimuraz Pirashvili, *Cohomology of algebraic theories*, J. Algebra **137** (1991), no. 2, 253–296, DOI 10.1016/0021-8693(91)90093-N. MR1094244

[Kar87] Max Karoubi, *Homologie cyclique et K-théorie* (French, with English summary), Astérisque **149** (1987), 147. MR913964

[KM21] Shane Kelly and Matthew Morrow, *K-theory of valuation rings*, Compos. Math. **157** (2021), no. 6, 1121–1142, DOI 10.1112/s0010437x21007119. MR4264079

[KN19] Achim Krause and Thomas Nikolaus, *Bökstedt periodicity and quotients of DVRs*, Compos. Math. **158** (2022), no. 8, 1683–1712, DOI 10.1112/s0010437x22007655. MR4490929

[KR97] John R. Klein and John Rognes, *The fiber of the linearization map $A(*) \to K(\mathbf{Z})$*, Topology **36** (1997), no. 4, 829–848, DOI 10.1016/S0040-9383(96)00040-7. MR1432423

[KV71] Max Karoubi and Orlando Villamayor, *K-théorie algébrique et K-théorie topologique. I* (French), Math. Scand. **28** (1971), 265–307 (1972), DOI 10.7146/math.scand.a-11024. MR313360

[LM12] Ayelet Lindenstrauss and Randy McCarthy, *On the Taylor tower of relative K-theory*, Geom. Topol. **16** (2012), no. 2, 685–750, DOI 10.2140/gt.2012.16.685. MR2928981

[LMMT20] Markus Land, Akhil Mathew, Lennart Meier, and Georg Tamme, *Purity in chromatically localized algebraic K-theory*, 2020.

[LNR11] Sverre Lunøe-Nielsen and John Rognes, *The Segal conjecture for topological Hochschild homology of complex cobordism*, J. Topol. **4** (2011), no. 3, 591–622, DOI 10.1112/jtopol/jtr015. MR2832570

[Lod98] Jean-Louis Loday, *Cyclic homology*, 2nd ed., Grundlehren der mathematischen Wissenschaften [Fundamental Principles of Mathematical Sciences], vol. 301, Springer-Verlag, Berlin, 1998. Appendix E by María O. Ronco; Chapter 13 by the author in collaboration with Teimuraz Pirashvili, DOI 10.1007/978-3-662-11389-9. MR1600246

[LQ84] Jean-Louis Loday and Daniel Quillen, *Cyclic homology and the Lie algebra homology of matrices*, Comment. Math. Helv. **59** (1984), no. 4, 569–591, DOI 10.1007/BF02566367. MR780077

[LRRV19] Wolfgang Lück, Holger Reich, John Rognes, and Marco Varisco, *Assembly maps for topological cyclic homology of group algebras*, J. Reine Angew. Math. **755** (2019), 247–277, DOI 10.1515/crelle-2017-0023. MR4015233

[LT19] Markus Land and Georg Tamme, *On the K-theory of pullbacks*, Ann. of Math. (2) **190** (2019), no. 3, 877–930, DOI 10.4007/annals.2019.190.3.4. MR4024564

[LW22] Ruochuan Liu and Guozhen Wang, *Topological cyclic homology of local fields*, Invent. Math. **230** (2022), no. 2, 851–932, DOI 10.1007/s00222-022-01134-9. MR4493328

[Mad94a] Ib Madsen, *Algebraic K-theory and traces*, Current developments in mathematics, 1995 (Cambridge, MA), Int. Press, Cambridge, MA, 1994, pp. 191–321. MR1474979

[Mad94b] Ib Madsen, *The cyclotomic trace in algebraic K-theory*, First European Congress of Mathematics, Vol. II (Paris, 1992), Progr. Math., vol. 120, Birkhäuser, Basel, 1994, pp. 213–241. MR1341845

[Mal17] Cary Malkiewich, *Cyclotomic structure in the topological Hochschild homology of DX*, Algebr. Geom. Topol. **17** (2017), no. 4, 2307–2356, DOI 10.2140/agt.2017.17.2307. MR3686399

[McC97] Randy McCarthy, *Relative algebraic K-theory and topological cyclic homology*, Acta Math. **179** (1997), no. 2, 197–222, DOI 10.1007/BF02392743. MR1607555

[Mor16] Matthew Morrow, *Pro CDH-descent for cyclic homology and K-theory*, J. Inst. Math. Jussieu **15** (2016), no. 3, 539–567, DOI 10.1017/S1474748014000413. MR3505658

[MP22] Cary Malkiewich and Kate Ponto, *Periodic points and topological restriction homology*, Int. Math. Res. Not. IMRN **4** (2022), 2401–2459, DOI 10.1093/imrn/rnaa174. MR4381922

[MR208] *Guido's book of conjectures*, Monographies de L'Enseignement Mathématique [Monographs of L'Enseignement Mathématique], vol. 40, L'Enseignement Mathématique, Geneva, 2008. A gift to Guido Mislin on the occasion of his retirement from ETHZ June 2006; Collected by Indira Chatterji. MR2499538

[MS93] J. E. McClure and R. E. Staffeldt, *The chromatic convergence theorem and a tower in algebraic K-theory*, Proc. Amer. Math. Soc. **118** (1993), no. 3, 1005–1012, DOI 10.2307/2160154. MR1164148

[MSV97] J. McClure, R. Schwänzl, and R. Vogt, *$THH(R) \cong R \otimes S^1$ for E_∞ ring spectra*, J. Pure Appl. Algebra **121** (1997), no. 2, 137–159, DOI 10.1016/S0022-4049(97)00118-7. MR1473888

[NS18] Thomas Nikolaus and Peter Scholze, *On topological cyclic homology*, Acta Math. **221** (2018), no. 2, 203–409, DOI 10.4310/ACTA.2018.v221.n2.a1. MR3904731

[Pan86] I. A. Panin, *The Hurewicz theorem and K-theory of complete discrete valuation rings* (Russian), Izv. Akad. Nauk SSSR Ser. Mat. **50** (1986), no. 4, 763–775, 878. MR864175

[PW92] Teimuraz Pirashvili and Friedhelm Waldhausen, *Mac Lane homology and topological Hochschild homology*, J. Pure Appl. Algebra **82** (1992), no. 1, 81–98, DOI 10.1016/0022-4049(92)90012-5. MR1181095

[Qui73] Daniel Quillen, *Higher algebraic K-theory. I*, Algebraic K-theory, I: Higher K-theories (Proc. Conf., Battelle Memorial Inst., Seattle, Wash., 1972), Lecture Notes in Math., Vol. 341, Springer, Berlin, 1973, pp. 85–147. MR0338129

[Rog99a] John Rognes, *Topological cyclic homology of the integers at two*, J. Pure Appl. Algebra **134** (1999), no. 3, 219–286, DOI 10.1016/S0022-4049(97)00155-2. MR1663390

[Rog99b] John Rognes, *Two-primary algebraic K-theory of spaces and related spaces of symmetries of manifolds*, Algebraic K-theory (Seattle, WA, 1997), Proc. Sympos. Pure Math., vol. 67, Amer. Math. Soc., Providence, RI, 1999, pp. 213–229, DOI 10.1090/pspum/067/1743242. MR1743242

[Rog03] John Rognes, *The smooth Whitehead spectrum of a point at odd regular primes*, Geom. Topol. **7** (2003), 155–184, DOI 10.2140/gt.2003.7.155. MR1988283

[Rog08] John Rognes, *Galois extensions of structured ring spectra. Stably dualizable groups*, Mem. Amer. Math. Soc. **192** (2008), no. 898, viii+137. MR2387923

[Rog09] John Rognes, *Topological logarithmic structures*, New topological contexts for Galois theory and algebraic geometry (BIRS 2008), Geom. Topol. Monogr., vol. 16,

Geom. Topol. Publ., Coventry, 2009, pp. 401–544, DOI 10.2140/gtm.2009.16.401. MR2544395

[Rog14] John Rognes, *Algebraic K-theory of strict ring spectra*, Proceedings of the International Congress of Mathematicians—Seoul 2014. Vol. II, Kyung Moon Sa, Seoul, 2014, pp. 1259–1283. MR3728661

[RSS15] John Rognes, Steffen Sagave, and Christian Schlichtkrull, *Localization sequences for logarithmic topological Hochschild homology*, Math. Ann. **363** (2015), no. 3-4, 1349–1398, DOI 10.1007/s00208-015-1202-3. MR3412362

[RW00] J. Rognes and C. Weibel, *Two-primary algebraic K-theory of rings of integers in number fields*, J. Amer. Math. Soc. **13** (2000), no. 1, 1–54, DOI 10.1090/S0894-0347-99-00317-3. Appendix A by Manfred Kolster. MR1697095

[SS12] Steffen Sagave and Christian Schlichtkrull, *Diagram spaces and symmetric spectra*, Adv. Math. **231** (2012), no. 3-4, 2116–2193, DOI 10.1016/j.aim.2012.07.013. MR2964635

[SSW96] Roland Schwänzl, Ross Staffeldt, and Friedhelm Waldhausen, *Stable K-theory and topological Hochschild homology of A_∞ rings*, Algebraic K-theory (Poznań, 1995), Contemp. Math., vol. 199, Amer. Math. Soc., Providence, RI, 1996, pp. 161–173, DOI 10.1090/conm/199/02479. MR1409624

[Sta65] John Stallings, *Centerless groups—an algebraic formulation of Gottlieb's theorem*, Topology **4** (1965), 129–134, DOI 10.1016/0040-9383(65)90060-1. MR202807

[Tab14] Gonçalo Tabuada, *Galois descent of additive invariants*, Bull. Lond. Math. Soc. **46** (2014), no. 2, 385–395, DOI 10.1112/blms/bdt108. MR3194756

[Tsa97] Stavros Tsalidis, *On the topological cyclic homology of the integers*, Amer. J. Math. **119** (1997), no. 1, 103–125. MR1428060

[Tsy83] B. L. Tsygan, *Homology of matrix Lie algebras over rings and the Hochschild homology* (Russian), Uspekhi Mat. Nauk **38** (1983), no. 2(230), 217–218. MR695483

[TT90] R. W. Thomason and Thomas Trobaugh, *Higher algebraic K-theory of schemes and of derived categories*, The Grothendieck Festschrift, Vol. III, Progr. Math., vol. 88, Birkhäuser Boston, Boston, MA, 1990, pp. 247–435, DOI 10.1007/978-0-8176-4576-2_10. MR1106918

[Vee18] Torleif Veen, *Detecting periodic elements in higher topological Hochschild homology*, Geom. Topol. **22** (2018), no. 2, 693–756, DOI 10.2140/gt.2018.22.693. MR3748678

[VPB85] Micheline Vigué-Poirrier and Dan Burghelea, *A model for cyclic homology and algebraic K-theory of 1-connected topological spaces*, J. Differential Geom. **22** (1985), no. 2, 243–253. MR834279

[Wal79] Friedhelm Waldhausen, *Algebraic K-theory of topological spaces. II*, Algebraic topology, Aarhus 1978 (Proc. Sympos., Univ. Aarhus, Aarhus, 1978), Lecture Notes in Math., vol. 763, Springer, Berlin, 1979, pp. 356–394. MR561230

[Wal84] Friedhelm Waldhausen, *Algebraic K-theory of spaces, localization, and the chromatic filtration of stable homotopy*, Algebraic topology, Aarhus 1982 (Aarhus, 1982), Lecture Notes in Math., vol. 1051, Springer, Berlin, 1984, pp. 173–195, DOI 10.1007/BFb0075567. MR764579

[Wal87a] Friedhelm Waldhausen, *Algebraic K-theory of spaces, concordance, and stable homotopy theory*, Algebraic topology and algebraic K-theory (Princeton, N.J., 1983), Ann. of Math. Stud., vol. 113, Princeton Univ. Press, Princeton, NJ, 1987, pp. 392–417. MR921486

[Wal87b] Friedhelm Waldhausen, *An outline of how manifolds relate to algebraic K-theory*, Homotopy theory (Durham, 1985), London Math. Soc. Lecture Note Ser., vol. 117, Cambridge Univ. Press, Cambridge, 1987, pp. 239–247. MR932266

[Wei87] Charles A. Weibel, *Nil K-theory maps to cyclic homology*, Trans. Amer. Math. Soc. **303** (1987), no. 2, 541–558, DOI 10.2307/2000683. MR902784

[Wei99] Charles A. Weibel, *The development of algebraic K-theory before 1980*, Algebra, K-theory, groups, and education (New York, 1997), Contemp. Math., vol. 243, Amer. Math. Soc., Providence, RI, 1999, pp. 211–238, DOI 10.1090/conm/243/03695. MR1732049

[WJR13] Friedhelm Waldhausen, Bjørn Jahren, and John Rognes, *Spaces of PL manifolds and categories of simple maps*, Annals of Mathematics Studies, vol. 186, Princeton University Press, Princeton, NJ, 2013, DOI 10.1515/9781400846528. MR3202834

[WW88] Michael Weiss and Bruce Williams, *Automorphisms of manifolds and algebraic K-theory. I*, K-Theory **1** (1988), no. 6, 575–626, DOI 10.1007/BF00533787. MR953917

DEPARTMENT OF MATHEMATICS, UNIVERSITY OF BERGEN, NORWAY

Proceedings of Symposia in Pure Mathematics
Volume **105**, 2023
https://doi.org/10.1090/pspum/105/01900

Algebraic K-theory and generalized stable homotopy theory

David Gepner

1. Introduction

1.1. Homotopy theory and higher category theory. In order to place algebraic K-theory in the higher categorical context and employ higher categorical tools in its study, it might be helpful to begin with some background and motivation from homotopy theory and higher category theory. The overarching idea of higher category theory is to systematically replace set-based mathematics with space-based mathematics. The relevant notion of space, or *homotopy type*, should, according to Grothendieck's *homotopy hypothesis*, coincide with that of ∞-groupoid, or *higher set*. This allows us to prescribe precise meaning to these concepts: namely, we can model homotopy types as cofibrant and fibrant topological spaces (cell complexes), higher sets as cofibrant and fibrant simplicial sets (that is, Kan complexes), and we obtain the same thing, as these model categories are equivalent in the sense of Quillen [**17**]. The main reference for higher set theory is Lurie's book [**13**]; topoi generalize the category of sets, and ∞-topoi generalize the higher sets, or ∞-groupoids. We use these terms interchangeably, as we are free to replace any representing topological space by a weakly equivalent cell complex, or equivalently a Kan complex.

Higher categories are endowed with homotopy types as the spaces of morphisms between objects, called *mapping spaces*, which are not typically discrete. It doesn't really make sense to ask whether two elements (points) of a homotopy type are *equal*; rather, they are *equivalent* if they correspond to points which can be connected by a path, in some (hence any) model for the homotopy type. But then any two such paths might form a nontrivial loop, leading to higher automorphisms, and so on. The notion of equality only makes sense after passing to discrete invariants such as homotopy groups.

There are a number of ways of making this perspective precise. The most obvious, and the easiest to get off the ground, is to use topologically (or better, simplicially) enriched categories. Following the notation of [**13**], we will write Cat_Δ for the category of simplically enriched categories. A weak equivalence of simplicially (or topologically) enriched categories is an enriched functor $f\colon \mathcal{C} \to \mathcal{D}$ which is fully faithful and essentially surjective *up to homotopy*. Homotopically fully faithful means that, for every pair of objects s and t of $\mathcal{C}$, the map $\mathrm{Map}_{\mathcal{C}}(s,t) \to \mathrm{Map}_{\mathcal{D}}(f(s), f(t))$ is a weak equivalence of simplicial sets (or topological spaces), while homotopically essentially surjective means that the induced functor $f\colon h\mathcal{C} \to h\mathcal{D}$ is essentially surjective on the lever of *homotopy categories*, namely the (ordinary) category obtained by applying the functor $\pi_0\colon \mathrm{Set}_\Delta \to \mathrm{Set}$ to each mapping simplicial set (or topological space).

The chief issue which arises when working with simplicial (or topological) categories is that the simplicial category of simplicial functors from $\mathcal{D}$ to $\mathcal{C}$ is not in general invariant under weak equivalence. To obtain the homotopically correct simplicial category of functors we must replace $\mathcal{D}$ with a sufficiently "free" version $\mathcal{D}' \to \mathcal{D}$ of itself, and this is an intractable procedure in general. The inability to control the category of functors and natural transformations is the primary reason we are forced to look for less rigid models, such as Joyal's theory of quasicategories [**12**], originally discovered by Boardmann–Vogt [**9**] and developed extensively by Lurie in [**13**], [**14**], and elsewhere. Indeed, in practice, it is usually easier to apply the *homotopy coherent nerve* functor

$$\mathrm{N} : \mathrm{Cat}_\Delta \to \mathrm{Fun}(\Delta^{\mathrm{op}}, \mathrm{Set})$$

and work in simplicial sets, which are able to encode ∞-categories very efficiently, though at a cost — the loss of predetermined models for the mapping spaces between objects, for instance. Here N is the right adjoint of the colimit preserving functor $\mathfrak{C}$: $\mathrm{Fun}(\Delta^{\mathrm{op}}, \mathrm{Set}) \to \mathrm{Cat}_\Delta$ determined by defining $\mathrm{Map}_{\mathfrak{C}[\Delta^n]}(i,j)$ to be the simplicial set of partially ordered subsets of $\{i, i+1, \ldots, j-1, j\}$. This is a thickening of the category $\{0 \to 1 \to \cdot \to n-1 \to n\}$ of which Δ^n is the (usual) nerve, obtained by keeping track of the all possible paths from i to j.

The theory of quasicategories has the distinct advantage that it allows for an easy construction of the ∞-category $\mathrm{Fun}(\mathcal{D}, \mathcal{C})$ of functors from an ∞-category $\mathcal{D}$ to an ∞-category $\mathcal{C}$, namely as the exponential

$$\mathrm{Fun}(\mathcal{D}, \mathcal{C}) := \mathcal{C}^{\mathcal{D}}$$

in the (cartesian closed) category of simplicial sets. This is a completely combinatorial object: if X and Y are simplicial sets, an m-simplex of X^Y is natural transformation $\Delta^m \times Y \to X$ of functors $\Delta^{\mathrm{op}} \to \mathrm{Set}$, so it is completely determined by a compatible family of *functions* $\Delta^m_n \times Y_n \to X_n$, $n \in \mathbf{N}$.

THEOREM 1.1.1. [**13**, Theorem 2.2.5.1] *The homotopy coherent nerve functor*

$$\mathrm{N} : \mathrm{Cat}_\Delta \to \mathrm{Fun}(\Delta^{\mathrm{op}}, \mathrm{Set}),$$

from the Bergner–Dwyer–Kan model structure on simplically enriched categories to the Joyal model structure on simplicial sets, is a right Quillen equivalence. In particular, the nerve $\mathrm{N}(\mathcal{C})$ *of a fibrant simplicially enriched category* $\mathcal{C}$ *(one in which each mapping space* $\mathrm{Map}_{\mathcal{C}}(A, B)$ *is a Kan complex) is a quasicategory.*

Higher algebra generalizes ordinary algebra, by which we mean algebra in the setting of ordinary, discrete, set based category theory. While spectra are classical objects in algebraic topology, the main reference for the ∞-category of spectra, and higher algebra in general, is Lurie's book [**14**]; abelian categories generalize the category of abelian groups, and their higher categorical analogues are stable ∞-categories, which generalize the ∞-category of spectra. Spectra are the higher categorical version of the category of abelian groups, and one recovers abelian groups from spectra as the heart of the standard t-structure (a generalization of ordinary truncation in the stable setting, where the usual notion of truncation is no longer well behaved). We can do algebra in the ∞-category of spectra because it carries a symmetric monoidal structure which refines the ordinary tensor product of abelian groups.

Higher algebra is truly *homotopical* in nature, meaning that many of its most important objects (such as the sphere spectrum, or the algebraic K-theory spectrum) simply do not exist within the *homological* world of chain complexes or derived categories. The part of the theory which can be formulated in these terms is differential graded algebra. We will assume that the reader is familiar with differential graded algebra and differential graded categories. The main points to keep in mind are that a differential graded category determines a simplicially enriched category via the Dold-Kan correspondence, which asserts that there is a suitably monoidal equivalece of categories between connective chain complexes and simplicial abelian groups.

Specifically, the functor $\mathrm{Ch}_{\geq 0}(\mathrm{Ab}) \to \mathrm{Ab}_\Delta$ admits a lax monoidal structure [**19**], and lax monoidal functors between enriching monoidal categories allow us to functorially change enrichment. Since the truncation functor $\tau_{\geq 0} : \mathrm{Ch}(\mathrm{Ab}) \to \mathrm{Ch}_{\geq 0}(\mathrm{Ab})$ and underlying set functor $\mathrm{Ab} \to \mathrm{Set}$ are also lax monoidal, we obtain a composite lax monoidal functor

$$\mathrm{Ch}(\mathrm{Ab}) \to \mathrm{Ch}_{\geq 0}(\mathrm{Ab}) \to \mathrm{Fun}(\Delta^{\mathrm{op}}, \mathrm{Ab}) \to \mathrm{Fun}(\Delta^{\mathrm{op}}, \mathrm{Set}),$$

and hence a functor from differential graded to simplicially enriched categories. We write $(-)_\Delta : \mathrm{Cat}_{\mathrm{dg}} \to \mathrm{Cat}_\Delta$ for this functor and $\mathrm{N}(\mathcal{C}) := \mathrm{N}(\mathcal{C}_\Delta)$ for the homotopy coherent nerve of the differential graded category $\mathcal{C}$. Composing with the homotopy coherent nerve functor, we obtain a functorial way of passing from homological algebra to homotopical algebra.

We conclude by recalling the definition of the ∞-category of ∞-categories. The primary objects of study in this survey — stable ∞-categories and their algebraic K-theory — are nothing more than ∞-categories which have certain nice properties, such as the existence of finite limits and colimits. It is a fact that if $\mathcal{C}$ and $\mathcal{D}$ are quasicategories, then the exponential object $\mathcal{D}^{\mathcal{C}}$ is again a quasicategory, typically denoted $\mathrm{Fun}(\mathcal{C}, \mathcal{D})$ and called the ∞-category of functors from $\mathcal{C}$ to $\mathcal{D}$. There is a maximal Kan complex $\mathrm{Map}(\mathcal{C}, \mathcal{D})$ inside of $\mathrm{Fun}(\mathcal{C}, \mathcal{D})$, which basically forgets the noninvertible natural transformations (they are actually not forgotten, they can be reconstructed from as functors $\Delta^1 \times \mathcal{C} \to \mathcal{D}$), resulting in a fibrant (i.e. Kan complex enriched) simpicial category of quasicategories, functors, (invertible) natural transformations, and so on. We write Cat_∞ for the nerve of this Kan-enriched category of quasicategories. Similarly, the ∞-category Gpd_∞ of ∞-groupoid can be constructed as the nerve of the full simplicial subcategory of quasicategories spanned by the Kan complexes. This is in keeping with Grothendieck's *homotopy hypothesis*, that the ∞-category of ∞-groupoids should be modeled by Kan complexes. One can equivalently define $\mathrm{Gpd}_\infty \subset \mathrm{Cat}_\infty$ as the full subcategory consisting of those ∞-categories $\mathcal{C}$ for which any morphism $\Delta^1 \to \mathcal{C}$ is an equivalence.

1.2. The sphere spectrum. The story of the sphere spectrum begins with one of the most basic mathematical objects, the set of natural numbers $\mathbf{N}$, the numbers which express magnitudes of finite discrete quantities. We ought to include zero as a natural number, and mathematically this endows $\mathbf{N}$ with the structure of a commutative semiring with additive and multiplicative identity. Since subtraction is an essential mathematical operation with real world applications, we ought to include additive inverses, and in doing so we arrive at the ring of integers $\mathbf{Z}$, arguably the fundamental object of modern algebra and number theory.

Certain key advances in algebraic K-theory, homotopy theory, and higher category theory suggest that $\mathbf{Z}$ is perhaps not the most initial ring which is relevant in arithmetic geometry. Fortunately, there are a plethora of rings in the higher categorical sense which are more initial than the integers. For the purposes of this paper, a higher categorical ring is a ring spectrum in the sense of stable homotopy theory. This is a good choice since the symmetric monoidal ∞-category of spectra is the universal stable ∞-category, in a suitable sense which we will describe in the next section. In particular, it carries a symmetric monoidal structure, which we will also describe in the next section, and a (commutative) ring spectrum is a (commutative) algebra object in this symmetric monoidal ∞-category.

In any case, one cannot easily escape having to deal with spectra in the study of algebraic K-theory. To start, it is *necessarily* a spectrum valued functor from some input ∞-category, which one might take to be rings, schemes, stable ∞-categories, etc. (we will take stable ∞-categories throughout, unless otherwise noted). Further related evidence of this is the spectacular theorem of Dundas–Goodwillie–McCarthy, which asserts that the fiber of the cyclotomic trace map $\mathrm{K}(R) \to \mathrm{TC}(R)$, from the algebraic K-theory to the topological Hochschild homology of a connective ring spectrum R (e.g. R a discrete ring), is invariant under nilimmersions.[1] Moreover, the topological cyclic homology is calculated from the circle action on the topological Hochschily homology of R, which means its Hochschild homology relative to the sphere. This is important, because the topological (i.e. higher categorical) theory is a stronger and in some sense more well-behaved invariant than Hochschild homology relative to $\mathbf{Z}$ — the basic case of the prime field $\mathbf{F}_p$ already demonstrates this, where the former is a polynomial algebra over $\mathbf{F}_p$ on a degree two generator, whereas the latter is a divided power algebra over $\mathbf{F}_p$ on a degree two generator. Finally, the sphere spectrum itself is a K-theory spectrum in a certain sense, as it can be identified with the group completion of the commutative semiring ∞-category $\mathrm{Fin}^{\simeq}$ of finite sets and bijections. We can use this as a definition.

We will define spectra and their homotopy groups in the next section. For now, suffice it to say that the ∞-category *connective* spectra, the full subcategory of the ∞-category of spectra spanned by those objects whose negative homotopy groups all vanish, is equivalent — via the functor which sends a spectrum to its infinite loop space — to the ∞-category of grouplike symmetric monoidal ∞-groupoids. We can avail ourselves of this equivalence.

Definition 1.2.1. The *sphere spectrum* $\mathbf{S}$ is the connective spectrum whose infinite loop space is the group completion of the commutative monoid $\mathrm{Fin}^{\simeq}$ of finite sets and bijections.

The spectrum structure encodes the addition, but the multiplication is induced from the symmetric monoidal structure. This is the higher categorical analogue of the ring of integers, $\mathbf{Z}$. Part of the utility of this definition is due to the following easy facts.

[1]That is, maps of connective ring spectra $f\colon R' \to R$ such that f is a surjection on π_0 with nilpotent kernel

LEMMA 1.2.2. *The ∞-groupoid* $\mathrm{Fin}^{\simeq} \simeq \coprod_{n\in\mathbf{N}} B\Sigma_n$ *is the free symmetric monoidal ∞-groupoid on one generator. The (infinite loop space of) sphere spectrum* $\mathbf{S}$ *is the free grouplike symmetric monoidal ∞-groupoid on one generator.*

Why is this object referred to as the sphere? In homotopy theory, hence also in higher category theory, commutativity has a geometric interpretation in terms of configuration spaces of points in Euclidean space. These geometric considerations govern the $\mathbf{E}_n$-operads, which interpolate between associative and commutative structures. Loop space theory identifies the ∞-category of $\mathbf{E}_n$-groups with the ∞-category of pointed spaces with trivial homotopy groups in dimensions less than n. Moreover, we know from algebraic topology that the quintessential example of an $\mathbf{E}_n$-group is an n-fold loop space; in fact, the free $\mathbf{E}_n$-group on a pointed space X is the space $\Omega^n\Sigma^n X$. Hence the free grouplike $\mathbf{E}_\infty$-monoid on one generator is

$$\mathbf{S} \simeq \operatorname{colim}_{n\to\infty} \operatorname{End}(S^n) \simeq \operatorname{colim}_{n\to\infty} \Omega^n\Sigma^n S^0 \simeq \Omega^\infty\Sigma^\infty S^0.$$

Here Ω^∞ denotes the underlying pointed space of a spectrum, and Σ^∞ its left adjoint. It is the (homotopy) colimit of the spaces of maps from the sphere to itself. Note that $\pi_0\mathbf{S} \cong \operatorname{colim}_{n\to\infty} \pi_n S^n \cong \mathbf{Z}$, while the higher homotopy groups of $\mathbf{S}$ are not known in general — this is the "stable homotopy groups of spheres" problem, one of the central questions in modern algebraic topology.

2. Abstract stable homotopy theory

2.1. Excisive functors and spectrum objects. We typically write $\mathcal{S}$ denotes the ∞-category of spaces, meaning (in keeping with Grothendieck's homotopy hypothesis) the ∞-category Gpd_∞ of ∞-groupoids, or the nerve N(Kan) of the simplicial category of Kan complexes. We write $\mathcal{S}_*$ for the ∞-category of pointed spaces, and $\mathcal{S}_*^{\mathrm{fin}}$ for the finite pointed spaces.

DEFINITION 2.1.1. A *cohomology theory* $\{F^n\}_{n\in\mathbf{Z}}$ is a $\mathbf{Z}$-graded family of functors $F^n : \mathrm{Ho}(\mathcal{S}_*^{\mathrm{op}}) \to \mathrm{Ab}$ and natural isomorphisms $\sigma^n : F^n \to F^{n+1}\circ\Sigma$ satisfying the following exactness conditions:

(1) For any cofiber sequence $X \to Y \to Z$ and integer n the sequence of abelian groups $F^n(Z) \to F^n(Y) \to F^n(X)$ is exact.
(2) For any (possibly infinite) wedge decomposition $X \simeq \bigvee_{i\in I} X_i$ and integer n, the homomorphism $F^n(X) \to \prod_{i\in I} F^n(X_i)$ is an isomorphism.

The *Brown representability theorem* [**11**] implies that any cohomology theory is represented by a spectrum which is unique up to homotopy. The Brown representability theorem itself asserts that any suitably left exact functor $F : \mathcal{S}^{\mathrm{op}} \to \mathrm{Set}_*$ is representable by a pointed space. In particular, the cohomology theory $F = \{F^n\}_{n\in\mathbf{Z}}$ is such that

$$F^n \cong \pi_0 \operatorname{Map}_{\mathcal{S}_*}(-, A_n) \colon \mathcal{S}^{\mathrm{op}} \longrightarrow \mathrm{Ab}$$

for some pointed space A_n, and the resulting sequence of pointed spaces $\{A_n\}$ can be chosen so that the suspension isomorphisms $\sigma^n : F^n \to F^{n+1}\circ\Sigma$ are induced by equivalences $\eta_n : A_n \to \Omega A_{n+1}$ in $\mathcal{S}_*$. Moreover, the (iterated) loop space structure induces the (abelian) group structure on the represented functor, and any such group structure necessarily arises in this way. This is how spectra arose historically.

A spectrum A therefore, is the data of an $\mathbf{N}$-indexed sequence of pointed spaces $\{A_n\}_{n\in\mathbf{N}}$ together with equivalences $A_n \simeq \Omega A_{n+1}$. By applying the loop functor Ω, we see that it is actually enough just to specify the value A_n on any strictly increasing sequence of natural numbers, and that this can be used to specify a unique (up to contractible space of choices) value A_n for every integer n. We think of a spectrum as a choice of *delooping* of its zero space A_0, which is also its so-called *infinite loop space*, $A_0 \simeq \Omega^\infty A$. The functor Ω^∞ admits a left adjoint Σ^∞ from point spaces to spectra, called the *suspension spectrum.* We can also form the suspension spectrum of an unpointed space X by first adding a disjoint basepoint.

There is a more canonical presentation of spectra in terms of excisive functors, so we will use this as our official definition. Excisive functors also play an important role in algebraic K-theory proper, such as the proof of the Dundas–Goodwillie–McCarthy theorem.

DEFINITION 2.1.2. Let $\mathcal{C}$ be an ∞-category with finite limits. A functor

$$F : \mathcal{S}_*^{\mathrm{fin}} \to \mathcal{C}$$

is said to be *excisive* if F sends cocartesian squares to cartesian squares. Similarly, a functor

$$F : \mathcal{S}_*^{\mathrm{fin}} \to \mathcal{C}$$

is said to be *reduced* if F sends initial objects to final objects.

We write

$$\mathrm{Sp}(\mathcal{C}) := \mathrm{Exc}_*(\mathcal{S}_*^{\mathrm{fin}}, \mathcal{C}) \subset \mathrm{Fun}_*(\mathcal{S}_*^{\mathrm{fin}}, \mathcal{C})$$

for the full subcategory of *spectrum objects* in $\mathcal{C}$, i.e. the reduced excisive functors $\mathcal{S}_*^{\mathrm{fin}} \to \mathcal{C}$. When $\mathcal{C} = \mathcal{S}$ we write simply

$$\mathrm{Sp} := \mathrm{Sp}(\mathcal{S})$$

for the ∞-category of spectra. This turns out to a stable ∞-category, basically because a reduced functor $F\colon \mathcal{S}_*^{\mathrm{fin}} \to \mathcal{C}$ admits an *excisive approximation*

$$\partial F \simeq \operatorname{colim} \Omega^n \circ F \circ \Sigma^n.$$

One shows that the excisive approximation is left adjoint to the inclusion of the full subcategory, and it is easy to see from the formula for the excisive approximation — the first derivative in the Goodwillie calculus — that Σ is actually an *inverse* to Ω in $\mathrm{Exc}_*(\mathcal{S}_*^{\mathrm{fin}}, \mathcal{C})$.

Let $\mathcal{C}$ be an ∞-category with finite limits, and let $\mathcal{D}$ denote an inverse limit

$$\mathcal{D} \simeq \lim\{\cdots \xrightarrow{\Omega} \mathcal{C}_* \xrightarrow{\Omega} \mathcal{C}_* \xrightarrow{\Omega} \mathcal{C}_*\}$$

in Cat_∞. Then $\mathcal{D} \simeq \mathrm{Sp}(\mathcal{C})$, though the term *spectrum object* in $\mathcal{C}$ is meant to describe an object of the limit, as it amounts to specifying a sequence ("spectrum") $A = \{A_n\}_{n\in\mathbf{N}}$ of pointed objects $A_n \in \mathcal{C}_*$ equipped with specified equivalences $A_n \simeq \Omega A_{n+1}$. We regard the A_n as chosen *deloopings* of A_0, the *underlying object* of the spectrum. When $\mathcal{C} = \mathcal{S}$, each of the A_n are canonically grouplike $\mathbf{E}_\infty$-monoid spaces.

The reason for the terminology "spectrum object" is that a reduced excisive functor determines, and is determined by, its value on any infinite sequence $\{S^{n_0}, S^{n_1}, S^{n_2}, \ldots\}$ of spheres of strictly increasing dimension. The most canonical choice is the sequence of all spheres $\{S^0, S^1, S^2, \ldots\}$, so that evaluation on this family of spheres induces a map

$$\mathrm{Exc}_*(\mathcal{S}_*^{\mathrm{fin}}, \mathcal{C}) \longrightarrow \lim\{\cdots \xrightarrow{\Omega} \mathcal{C}_* \xrightarrow{\Omega} \mathcal{C}_* \xrightarrow{\Omega} \mathcal{C}_*\}$$

which sends the excisive functor F to the sequence of pointed objects $\{F(S^n)\}_{n\in\mathbf{N}}$ and equivalences $F(S^n) \xrightarrow{\simeq} \Omega F(\Sigma S^n)$. In other words, writing $\mathrm{Sph}_* \subset \mathcal{S}_*^{\mathrm{fin}}$ for the smallest full pointed subcategory spanned by the spheres and the zero object, we see that a spectrum object in $\mathcal{C}$ is, equivalently, a reduced functor $F\colon \mathrm{Sph}_* \to \mathcal{C}$ such that, for all $n \in \mathbf{N}$, the canonical map $F(S^n) \to \Omega F(\Sigma S^n)$ is an equivalence. This is an explicit model for the limit $\mathrm{Sp}(\mathcal{C}) \simeq \lim\{\cdots \xrightarrow{\Omega} \mathcal{C}_* \xrightarrow{\Omega} \mathcal{C}_* \xrightarrow{\Omega} \mathcal{C}_*\}$ along the lines of the classical notion of spectrum object in algebraic topology.

2.2. Stable ∞-categories. Spectra are a special case of the stabilization of an ∞-category — specifically, the case in which the ∞-category being stabilized is the ∞-category $\mathcal{S}$ of spaces (a.k.a. ∞-groupoids, Kan complexes, homotopy types, etc.).

DEFINITION 2.2.1. A *zero object* of an ∞-category $\mathcal{C}$ is an object which is both initial and final. An ∞-category $\mathcal{C}$ is *pointed* if $\mathcal{C}$ admits a zero object.

DEFINITION 2.2.2. Let $\mathcal{C}$ be a pointed ∞-category. A *triangle* in $\mathcal{C}$ is a commutative square $\Delta^1 \times \Delta^1 \to \mathcal{C}$ of the form

$$\begin{array}{ccc} A & \xrightarrow{f} & B \\ \downarrow & & \downarrow{\scriptstyle g} \\ 0 & \longrightarrow & C, \end{array}$$

where 0 denotes a zero object of $\mathcal{C}$.

We say that a triangle in $\mathcal{C}$ is *left exact* if it is cartesian (a pullback), *right exact* if it is cocartesian (a pushout), and *exact* if it is cartesian and cocartesian. What do these exactness properties mean in some basic cases? Given a triangle of the form

$$\begin{array}{ccc} A & \longrightarrow & 0 \\ \downarrow & & \downarrow \\ 0 & \longrightarrow & B, \end{array}$$

then $A \simeq \Omega B$ if the triangle is left exact and $\Sigma A \simeq B$ if the triangle is right exact. If the triangle is exact, we have equivalences $A \simeq \Omega\Sigma A$ and $\Sigma\Omega B \simeq B$.

One often writes $A \xrightarrow{f} B \xrightarrow{g} C$ for a triangle in $\mathcal{C}$, though it is important to remember that a choice of composite $h = g \circ f$ and nullhomotopy $h \simeq 0$ in $\mathrm{Map}(A, C)$ are also part of the data. The notion of a bicartesian square seems to be a useful notion in ∞-category theory and algebraic K-theory, providing us with a notion of exact sequence *in* a stable ∞-category, as well as a notion of exact sequence *of* stable ∞-categories (a Verdier sequence, see below).

DEFINITION 2.2.3. A *stable ∞-category* is an ∞-category $\mathcal{C}$ with a zero object, finite limits and colimits, and which satisfies the following condition: a commutative square $\Delta^1 \times \Delta^1 \to \mathcal{C}$ in $\mathcal{C}$ is cartesian if and only if it is cocartesian (in other words, a pullback if and only if it is a pushout).

PROPOSITION 2.2.4. [**14**, Corollary 1.4.2.11] *Let $\mathcal{C}$ be a pointed ∞-category. The following conditions are equivalent:*

(1) $\mathcal{C}$ is stable.
(2) $\mathcal{C}$ admits finite colimits and $\Sigma : \mathcal{C} \to \mathcal{C}$ is an equivalence.
(3) $\mathcal{C}$ admits finite limits and $\Omega : \mathcal{C} \to \mathcal{C}$ is an equivalence.

If $\mathcal{C}$ is stable, we may regard the suspension and loops endofunctors as mutually inverse shift functors, as we tend to do in the theory of derived categories.

DEFINITION 2.2.5. Let $\mathcal{C}$ and $\mathcal{D}$ be ∞-categories which admit finite limits and finite colimits. A functor $f : \mathcal{D} \to \mathcal{C}$ is *left exact* if f preserves finite limits, *right exact* if f preserves finite colimits, and *exact* if f is both left and right exact (i.e., f preserves finite limits and finite colimits).

We write $\mathrm{Cat}_\infty^{\mathrm{st}} \subset \mathrm{Cat}_\infty$ for the (not full) subcategory consisting of the stable ∞-categories and the exact functors. This is the higher categorical analogue of the category of triangulated categories, with the distinct advantage that $\mathrm{Cat}_\infty^{\mathrm{st}}$ is a *subcategory* of Cat, meaning that stability is a property and not extra structure.

THEOREM 2.2.6. [**14**, Lemma 1.1.2.13] *The homotopy category $h\mathcal{C}$ of a stable ∞-category $\mathcal{C}$ is canonically a triangulated category: the shift functor is induced from the suspension functor $\Sigma : \mathcal{C} \to \mathcal{C}$, and a triangle $A \xrightarrow{f} B \xrightarrow{g} C \xrightarrow{h} \Sigma A$ in $h\mathcal{C}$ is exact if and only if there exist exact triangles*

$$\begin{array}{ccccc} A & \xrightarrow{f'} & B & \longrightarrow & 0 \\ \downarrow & & {\scriptstyle g'}\downarrow & & \downarrow \\ 0 & \longrightarrow & C & \xrightarrow{h'} & D \end{array}$$

in $\mathcal{C}$ such that f', g', and h' are representatives of f, g and h composed with the equivalence $\Sigma A \to D$, respectively.

REMARK 2.2.7. A cohomological functor $F : \mathrm{Sp}^{\mathrm{op}} \to \mathrm{Sp}$ induces a functor $\pi_0 F : \mathrm{Sp}^{\mathrm{op}} \to \mathrm{Ab}$ which necessarily factors through the triangulated homotopy category of spectra. There is a version of Brown representability for triangulated categories which are compactly generated in the appropriate sense, due to Neeman [**15**]. This is not a corollary of the corresponding formal result for compactly generated stable ∞-categories, namely that $\mathrm{Sp}(\mathcal{C}) \subset \mathrm{Fun}(\mathcal{C}^{\mathrm{op}}, \mathrm{Sp})$ is the full subcategory consisting of the cohomological functors. Indeed, triangulated categories aren't always homotopy categories of stable ∞-categories and don't necessarily even admit finite limits or colimits; rather, the requisite exactness properties are encoded by the triangulated structure.

2.3. Bounded t-structures. Let $\mathcal{C}$ is a stable ∞-category. We recall some notions which are well-known from homological algebra and import them into the world of homotopical algebra. Given an object A of $\mathcal{C}$, we sometimes write $A[n]$ for the n-fold suspension $\Sigma^n A$ of A, $n \in \mathbf{Z}$. Similarly, if $\mathcal{D} \subset \mathcal{C}$ is a full subcategory of $\mathcal{C}$, we write $\mathcal{D}[n] \subset \mathcal{C}$ for the full subcategory consisting the objects of the form $B[n]$, where B is an object of the full subcategory $\mathcal{D}$. We shall see later that algebraic K-theory is the analogue of a homology theory on the ∞-category of stable ∞-categories, so we use homological, as opposed to cohomological, indexing conventions.

The data of a t-structure on a stable ∞-category $\mathcal{C}$ is equivalent to the data of a t-structure on its triangulated homotopy category $h\mathcal{C}$ in the sense originally defined by Beilinson-Bernstein-Deligne [**4**] (save for the homological indexing).

DEFINITION 2.3.1. A *t-structure* on a stable ∞-category $\mathcal{C}$ consists of a pair of full subcategories $\mathcal{C}_{\geq 0} \subset \mathcal{C}$ and $\mathcal{C}_{\leq 0} \subset \mathcal{C}$, satisfying the following conditions:

(1) $\mathcal{C}_{\geq 0}[1] \subset \mathcal{C}_{\geq 0}$ and $\mathcal{C}_{\leq 0} \subset \mathcal{C}_{\leq 0}[1]$;
(2) If $A \in \mathcal{C}_{\geq 0}$ and $B \in \mathcal{C}_{\leq 0}$, then $\mathrm{Map}_{\mathcal{C}}(A, B[-1]) = 0$;
(3) Every $A \in \mathcal{C}$ fits into an exact triangle of the form $\tau_{\geq 0}A \to A \to \tau_{\leq -1}A$ with $\tau_{\geq 0}A \in \mathcal{C}_{\geq 0}$ and $\tau_{\leq -1}A \in \mathcal{C}_{\leq 0}[-1]$.

REMARK 2.3.2. The truncations $\tau_{\geq n}$ and $\tau_{\leq n}$ are in fact functorial in the sense that the inclusions $\mathcal{C}_{\geq n} \to \mathcal{C}$ and $\mathcal{C}_{\leq n} \to \mathcal{C}$ admit right and left adjoints, respectively.

DEFINITION 2.3.3. An exact functor $f\colon \mathcal{C} \to \mathcal{D}$ between stable ∞-categories equipped with t-structures is *left t-exact* if $f(\mathcal{C}_{\leq 0}) \subset \mathcal{D}_{\leq 0}$ and *right t-exact* if $f(\mathcal{C}_{\geq 0}) \subset \mathcal{D}_{\geq 0}$. An exact functor is *t-exact* if is both left and right t-exact. We set $\mathcal{C}_{\geq n} = \mathcal{C}_{\geq 0}[n]$ and $\mathcal{C}_{\leq n} = \mathcal{C}_{\leq 0}[n]$.

PROPOSITION 2.3.4. [**14**, Remark 1.2.1.12] *The intersection $\mathcal{C}_{\geq 0} \cap \mathcal{C}_{\leq 0}$ is an abelian category (that is, it is equivalent to the nerve of an abelian* 1*-category).*

DEFINITION 2.3.5. The *heart* of a t-structure $(\mathcal{C}_{\geq 0}, \mathcal{C}_{\leq 0})$ on a stable ∞-category $\mathcal{C}$ is the abelian category $\mathcal{C}_{\geq 0} \cap \mathcal{C}_{\leq 0}$, and is denoted $\mathcal{C}^{\heartsuit}$.

DEFINITION 2.3.6. Let $\mathcal{C}$ be a stable ∞-category equipped with a t-structure $(\mathcal{C}_{\geq 0}, \mathcal{C}_{\leq 0})$ and let A be an object of $\mathcal{C}$. There is an equivalence of functors $\tau_{\geq 0}\tau_{\leq 0} \simeq \tau_{\leq 0}\tau_{\geq 0} : \mathcal{C} \to \mathcal{C}^{\heartsuit}$ [**14**, Proposition 1.2.1.10]. The *homotopy groups* $\pi_n A$, $n \in \mathbf{Z}$, are defined via the formula $\pi_n A = \tau_{\geq 0}\tau_{\leq 0}(A[-n]) \in \mathcal{C}^{\heartsuit}$.

REMARK 2.3.7. In the stable setting, passing to homotopy groups is a homological functor in the sense that there are long exact sequences

$$\cdots \to \pi_{n+1}C \to \pi_n A \to \pi_n B \to \pi_n C \to \pi_{n-1}A \to \cdots$$

in $\mathcal{C}^{\heartsuit}$ associated to an exact triangle $A \to B \to C$ in $\mathcal{C}$.

The standard example of a stable ∞-category with a t-structure comes from homological algebra. If $\mathcal{A}$ is a small abelian category, then the bounded derived ∞-category $\mathcal{D}^b(\mathcal{A})$ admits a canonical t-structure in which $\mathcal{D}^b(\mathcal{A})_{\geq n}$ consists of the complexes A such that $\mathrm{H}_i(A) = 0$ for $i < n$, and similarly for $\mathcal{D}^b(\mathcal{A})_{\leq n}$. If $\mathcal{A}$ is a Grothendieck abelian category then the *unbounded* derived ∞-category $\mathcal{D}(\mathcal{A})$ admits a t-structure with the same description as above. The heart of these t-structure is the abelian category $\mathcal{A}$.

More generally, if R is a connective ring spectrum, the stable presentable ∞-category LMod_R of left R-module spectra admits a t-structure with $\mathrm{LMod}_{R\geq 0} \simeq \mathrm{LMod}_R^{\mathrm{cn}}$, the ∞-category of connective left R-module spectra, called the *Postnikov t-structure.* The heart of this t-structures is the abelian category of left $\pi_0 R$-modules.

DEFINITION 2.3.8. A t-structure $(\mathcal{C}_{\geq 0}, \mathcal{C}_{\leq 0})$ on a stable ∞-category $\mathcal{C}$ is *bounded* if the inclusion of the full subcategory of bounded objects

$$\mathcal{C}^b = \bigcup_{n \in \mathbf{N}} \mathcal{C}_{\geq -n} \cap \mathcal{C}_{\leq n} \subset \mathcal{C}$$

is an equivalence. There are analogous notions of right and left bounded.

Many of the most important examples of stable ∞-categories arise in algebraic geometry via the formation of the derived category of quasicoherent sheaves. Whether or not this induces a t-structure on the subcategory of coherent sheaves is an important question in general, and has to do with whether or not the scheme is regular or singular. See [**18**] for the classification of bounded t-structures on the derived category of a Notherian commutative ring with finite Krull dimension.

The ∞-category Sp of spectra admits a left and right complete t-structure with heart $\mathrm{Sp}^{\heartsuit} \simeq \mathrm{Ab}$ the category of abelian groups. The identity functor $\mathrm{Ab} \to \mathrm{Ab}$ is right exact and so determines a right t-exact functor $\mathcal{D}^{-}(\mathbf{Z}) \simeq \mathcal{D}^{-}(\mathrm{Ab}) \to \mathrm{Sp}$ with image the *Eilenberg-MacLane spectra.* Strictly speaking, these are the *generalized* Eilenberg-MacLane spectra: an Eilenberg-MacLane spectrum is a generalized Eilenberg-MacLane spectrum which has homotopy concentrated in a single degree, or equivalently is the image of a chain complex with homology concentrated in a single degree.

REMARK 2.3.9. Serre showed that higher homotopy groups of spheres are always finitely generated, and in fact almost always finite: since S^0 is discrete, we may assume $n > 0$, in which case $(\pi_m S^n) \otimes_{\mathbf{Z}} \mathbf{Q} \cong 0$ unless $m = n$ or $m = 2n-1$. The stable homotopy groups of spheres are the colimit of the unstable homotopy groups, and since they satisfy the suspension axiom, can be deduced from the stable homotopy groups of the 0-sphere, i.e. the homotopy groups of the sphere spectrum $\mathbf{S}$. Part of the moral of this story is that the algebraic K-groups are a categorification of stable homotopy groups, though finite generation even in basic cases is mostly unknown (this is content of the Bass conjecture).

2.4. Symmetric monoidal structures. In practice, most of the ∞-categories which we'd like to stabilize are *presentable* ∞-categories, meaning they arise as localizations of ∞-categories of presheaves. More precisely, they are presented by *generators and relations* in the sense that they are generated under small colimits by a small ∞-category, modulo the relation of localization, which is required to preserve small objects in a certain precise sense.

Presentable ∞-categories admit all small colimits, and morphism of presentable ∞-categories are functors which preserve small colimits; by the adjoint functor theorem, such functors are automatically left adjoints. We write LPr for the ∞-category of presentable ∞-categories and left adjoint functors, and $\mathrm{LPr}_{\mathrm{st}} \subset \mathrm{LPr}$ for the full subcategory spanned by the stable presentable ∞-categories. There are obvious dual notions, denoted RPr and $\mathrm{RPr}_{\mathrm{st}}$, involving right adjoint functors.

One of the most remarkable features of LPr is that, like Cat_∞, it is symmetric monoidal. But the tensor product functor $\otimes \colon \mathrm{LPr} \times \mathrm{LPr} \to \mathrm{LPr}$ behaves more like the tensor product of abelian groups than the cartesian product of ∞-categories, and for the basically the same reason, namely that presentable ∞-categories have sums (small colimits). Nevertheless, the universal property of the tensor product is the same: writing $\mathrm{LFun}(\mathcal{A}, \mathcal{B})$ for the (again presentable) ∞-category of colimit preserving (hence left adjoint) functors from $\mathcal{A}$ to $\mathcal{B}$, we have an equivalence of (presentable) ∞-categories

$$\mathrm{LFun}(\mathcal{A} \otimes \mathcal{B}, \mathcal{C}) \simeq \mathrm{LFun}(\mathcal{A}, \mathrm{LFun}(\mathcal{B}, \mathcal{C}))$$

for any three presentable ∞-categories $\mathcal{A}$, $\mathcal{B}$, and $\mathcal{C}$.

PROPOSITION 2.4.1. [**14**, Proposition 4.8.2.18] *The inclusion of the full subcategory* $\mathrm{LPr}_{\mathrm{st}} \subset \mathrm{LPr}$ *admits a left adjoint* $\mathrm{LPr} \to \mathrm{LPr}_{\mathrm{st}}$.

REMARK 2.4.2. The left adjoint "stabilization" functor is given by tensoring with the ∞-category of spectra. By virtue of the equivalence $\mathcal{C} \otimes \mathrm{Sp} \simeq \mathrm{Sp}(\mathcal{C})$, the ∞-category of spectrum objects in $\mathcal{C}$ is a description of its stabilization.

COROLLARY 2.4.3. *Let $\mathcal{A}$ and $\mathcal{B}$ be presentable ∞-categories such that $\mathcal{A}$ is stable. Then $\mathcal{A} \otimes \mathcal{B}$ is stable.*

We now turn to certain symmetric monoidal presentable ∞-categories of interest, namely the ∞-categories of spaces and spectra. Given spectra A and B, their smash product is usually written $A \wedge B$ is the literature. We follow the convention of [**14**] and write $A \otimes B$ instead, emphasizing the analogy with the tensor product of abelian groups (or more precisely the derived tensor product of chain complexes). By construction, the symmetric monoidal structure on the ∞-category of spectra is compatible with the cartesian symmetric monoidal structure on the ∞-category of spaces via the suspension spectrum functor

$$\Sigma^\infty_+ : \mathcal{S} \longrightarrow \mathrm{Sp}\,.$$

This means that, for spaces $X_1, \ldots, X_n$, there is a canonical equivalence

$$(\Sigma^\infty_+ X_1) \otimes \cdots \otimes (\Sigma^\infty_+ X_n) \simeq \Sigma^\infty_+(X_1 \times \cdots \times X_n).$$

REMARK 2.4.4. The colimit preserving symmetric monoidal structure on the ∞-category $\mathcal{S}_*$ of pointed spaces is the *smash product*, namely $X \wedge Y \simeq X \times Y / X \wedge Y$, where $X \wedge Y$ (the *wedge product*) is the coproduct of X and Y in $\mathcal{S}_*$ (i.e., the disjoint union of X and Y with their respective basepoints identified). The corresponding functor $\Sigma^\infty : \mathcal{S}_* \to \mathrm{Sp}$ is again the unique symmetric monoidal left adjoint functor, and we deduce that

$$(\Sigma^\infty X_1) \otimes \cdots \otimes (\Sigma^\infty X_n) \simeq \Sigma^\infty(X_1 \wedge \cdots \wedge X_n)$$

for any finite collection of pointed spaces $X_1, \ldots, X_n$.

Using the description of spectra as the limit of the tower associated to the endofunctor $\Omega : \mathcal{S}_* \to \mathcal{S}_*$, we obtain maps

$$\Omega^{\infty - n} : \mathrm{Sp} \to \mathcal{S}_*$$

by projection to the nth factor. The reason for this notation is that we have equivalences $\Omega^\infty \simeq \Omega^n \Omega^{\infty-n}$. Indeed, for any spectrum A, $\Omega^\infty A \simeq A_0 \simeq \Omega^n A_n \simeq \Omega^n \Omega^{\infty-n} A$. The family of functors $\{\Omega^{\infty-n}\}_{n \in \mathbf{N}}$, or any infinite subset thereof, form a conservative family of functors $\mathrm{Sp} \to \mathcal{S}_*$ in RPr. They admit left adjoints, denoted $\Sigma^{\infty-n} : \mathcal{S}_* \to \mathrm{Sp}$.

REMARK 2.4.5. Any spectrum A admits a canonical presentation by desuspended suspension spectra

$$A \simeq \mathrm{colim}_{n \to \infty} \Sigma^{\infty-n} \Omega^{\infty-n} A \simeq \mathrm{colim}_{n \to \infty} \Sigma^{-n} \Sigma^\infty A_n$$

in which the maps $\Sigma^{\infty-n} A_n \to \Sigma^{\infty-n-1} A_{n+1}$ correspond to the composites

$$A_n \to \Omega^\infty \Sigma^n \Omega^n \Sigma^\infty A_n \to \Omega^\infty \Sigma^n \Omega^{n+1} \Sigma^\infty A_{n+1} \simeq \Omega^{\infty-n} \Sigma^{\infty-n-1} A_{n+1}.$$

Using the fact that $\Sigma^{\infty-m} X \otimes \Sigma^{\infty-n} Y \simeq \Sigma^{\infty-m-n} X \wedge Y$, we can easily write down explicit formulas for the spaces in the smash product of any finite sequence of spectra.

If B is a fixed spectrum, the functor $\mathrm{Sp} \xrightarrow{(\mathrm{id},B)} \mathrm{Sp}\times\mathrm{Sp} \xrightarrow{\otimes} \mathrm{Sp}$, which sends A to $A\otimes B$, preserves colimits. It follows from the adjoint functor theorem that this functor admits a right adjoint. This right adjoint is the *mapping spectrum* functor $\mathcal{F}(B,-) : \mathrm{Sp} \to \mathrm{Sp}$, which admits the following description: if A is a spectrum, then $\mathcal{F}(B,A)$ is the spectrum given by the formula

$$\mathcal{F}(B,A)_n \simeq \mathrm{Map}_{\mathrm{Sp}}(B,\Sigma^n A),$$

with structure maps

$$\mathrm{Map}_{\mathrm{Sp}}(B,\Sigma^n A) \simeq \mathrm{Map}_{\mathrm{Sp}}(B,\Omega\Sigma^{n+1} A) \simeq \Omega\,\mathrm{Map}_{\mathrm{Sp}}(B,\Sigma^{n+1} A).$$

Here the first equivalence uses the fact that Sp is a stable ∞-category and therefore the composite functor $\Omega\Sigma$ is equivalent to the identity.

3. Algebraic K-theory

3.1. The Grothendieck group. The Grothendieck group $G(M)$ of a monoid M is the initial group $G(M)$ equipped with a monoid map $M \to G(M)$. More precisely, we can characterize the Grothendieck group functor

$$G\colon \mathrm{Mon} \to \mathrm{Gp}$$

as a left adjoint of the forgetful functor $F\colon \mathrm{Gp} \to \mathrm{Mon}$. The Grothendieck group is in some sense the first step in K-theory: if R is a ring, we can consider the (commutative) monoid of finitely generated projective R-modules, in which the monoidal product is given by the sum. The Grothendieck group of this monoid is $\mathrm{K}_0(R)$, the zeroth algebraic K-group of R.

Sometimes the Grothendieck group refers to the zeroth algebraic K-group of an algebro-geometric object. If X is a (possibly non affine) algebraic variety or scheme, then vector bundles on X are typically not projective $\mathcal{O}_X$-modules, though this is still true locally on X. In this case, K_0 is more than the Grothendieck group completion; we also want to quotient by a relation of the form $[V] = [U] + [W]$, whenever $0 \to U \to V \to W \to 0$ is a short exact sequence of vector bundles on X. This exact sequence relation is actually more fundamental than the group completion, as the latter is automatic provided we pass to the derived category of X and use the notion of exactness which is relevant there. The reason this forces group completion is that, for any vector bundle V, the shift ΣV of V into (homological) degree one (a perfectly good object of the derived category of X) participates in an exact sequence of the form $V \to 0 \to \Sigma V$, so that $[\Sigma V] = -[V]$ in $\mathrm{K}_0(X)$.

The derived category is canonically triangulated, and the Grothendieck group functor K_0 is naturally a functor from triangulated categories and exact functors to abelian groups.[2] If $\mathcal{C}$ is a small triangulated category, $\mathrm{K}_0(\mathcal{C})$ can be described as the free abelian group on the objectd of $\mathcal{C}$, modulo the relation $[B] = [A] + [C]$ whenever there exists a distinguished triangle $A \to B \to C$ in $\mathcal{C}$. So K-theory takes the free abelian group on a set, and imposes the relations witnessed by exact sequences in $\mathcal{C}$. We will categorify this construction, and it the process manage to construct algebraic K-theory itself, without relying on any pre-existing definition of algebraic K-theory.

[2]But be careful: the algebraic K-theory functor itself does not descend to a functor of triangulated categories!

3.2. Exact sequences of stable ∞-categories. The ∞-category $\mathrm{Cat}_\infty^{\mathrm{st}}$ is not stable. It is pointed, with the trivial stable ∞-category 0 acting as a zero object, but the loops functor $\Omega\colon \mathrm{Cat}_\infty^{\mathrm{st}} \to \mathrm{Cat}_\infty^{\mathrm{st}}$ is far from an endoequivalence; rather, $\Omega\mathcal{C} \simeq 0$ for any stable ∞-category $\mathcal{C}$.

While stable ∞-categories admit canonical notions of exact sequence, obtained by taking successive (co)fibers, this is a special case of a more general notion of exact sequence in any pointed ∞-category.

DEFINITION 3.2.1. A *localization sequence* in $\mathrm{Cat}_\infty^{\mathrm{st}}$ is a cartesian and cocartesian square of the form

$$\begin{array}{ccc} \mathcal{A} & \longrightarrow & \mathcal{B} \\ \downarrow & & \downarrow \\ 0 & \longrightarrow & \mathcal{C} \end{array}$$

(that is, one of the two intermediate vertices must be a zero object).

Localization sequences are sometimes also called Verdier sequences, Verdier localizations, and Verdier quotients (the latter terminology tends to refer to the essentially surjective map $\mathcal{B} \to \mathcal{C}$, from which the fully faithful map $\mathcal{A} \to \mathcal{B}$ arises as the fiber). It is common to simply write $\mathcal{A} \to \mathcal{B} \to \mathcal{C}$ for a localization sequence, leaving implicit the requirement that the composition $\mathcal{A} \to \mathcal{B} \to \mathcal{C}$ is null, and that the resulting square is bicartesian (in other words, $\mathcal{A}$ is the fiber of $\mathcal{B} \to \mathcal{C}$, and $\mathcal{C}$ is the cofiber of $\mathcal{A} \to \mathcal{B}$). The reason this is possible is that these are properties of the composable sequence of maps $\mathcal{A} \to \mathcal{B} \to \mathcal{C}$, and not extra structure.

We also have the notion of *additivity sequence*, or *semi-orthogonal decomposition*. This is roughly analogous to a split-exact sequence, as it is an exact sequence of stable ∞-categories such that the functors $\mathcal{A} \to \mathcal{B}$ and $\mathcal{B} \to \mathcal{C}$ admit right adjoints. For instance, any localization sequences of presentable stable ∞-categories, where the maps are required to preserve colimits, is automatically split.

REMARK 3.2.2. Given a localization sequence $\mathcal{A} \xrightarrow{i} \mathcal{B} \xrightarrow{j} \mathcal{C}$, the fact that i is fully faithful (because of the exactness condition) implies that the left adjoint functor $i_! : \mathrm{Ind}(\mathcal{A}) \to \mathrm{Ind}(\mathcal{B})$ is fully faithful with right adjoint i^*, and the right adjoint $j^* : \mathrm{Ind}(\mathcal{C}) \to \mathrm{Ind}(\mathcal{B})$ of the functor $j_! : \mathrm{Ind}(\mathcal{B}) \to \mathrm{Ind}(\mathcal{C})$ is also fully faithful. Hence the exact sequence $\mathrm{Ind}(\mathcal{A}) \to \mathrm{Ind}(\mathcal{B}) \to \mathrm{Ind}(\mathcal{C})$ is a semi-orthogonal decomposition of $\mathrm{Ind}(\mathcal{B})$ by the full subcategories $\mathrm{Ind}(\mathcal{A})$ and $\mathrm{Ind}(\mathcal{C})$.

DEFINITION 3.2.3. A *localizing invariant* is a functor $F\colon \mathrm{Cat}_\infty^{\mathrm{st}} \to \mathrm{Sp}$ which preserves localization sequences, in the sense that it preserves zero objects and bicartesian squares of the form

$$\begin{array}{ccc} \mathcal{A} & \longrightarrow & \mathcal{B} \\ \downarrow & & \downarrow \\ 0 & \longrightarrow & \mathcal{C} \end{array}.$$

REMARK 3.2.4. Let X be a scheme (which we suppose quasicompact and quasiseparated, for simplicity), let $j\colon U \to X$ be a quasicompact open immersion, and let $i\colon Z \to X$ be the complementary closed inclusion (we only care about the underlying closed subspace of X, the scheme structure is not relevant here). Then $j^*\colon \mathrm{Perf}(X) \to \mathrm{Perf}(U)$ determines a localization sequence of small stable

∞-categories, with kernel $\mathrm{Perf}_Z(X)$, the full subcategory of perfect complexes $\mathcal{F}$ on X which are supported on Z (in the sense that $j^*\mathcal{F}$ is zero in $\mathrm{Perf}(U)$).

It is not typically a semi-orthogonal decomposition: for instance, if $X = \operatorname{Spec} \mathbf{Z}_{(p)}$ and $U = \operatorname{Spec} \mathbf{Q}_{(p)}$, then the right adjoint $j_*\colon \mathrm{Mod}_{\mathbf{Q}_{(p)}} \to \mathrm{Mod}_{\mathbf{Z}_{(p)}}$ does not preserve compact objects, and so does not restrict to a functor $j_*\colon \mathrm{Perf}_{\mathbf{Q}_{(p)}} \to \mathrm{Perf}_{\mathbf{Z}_{(p)}}$. On the other hand, if $X = \operatorname{Spec} A \times B$ is a product of rings, and $U = \operatorname{Spec} A$, then $j_*\colon \mathrm{Mod}_U \to \mathrm{Mod}_X$ does preserve compact objects.

REMARK 3.2.5. A localizing invariant $F\colon \mathrm{Cat}_\infty^{\mathrm{st}} \to \mathrm{Sp}$ preserves Mayer–Vietoris squares, in the following sense. If X is a scheme (again, we assume quasicompact and quasiseparated, for simplicity) which is the union $X = U \cup V$ of two quasicompact open subschemes U and V of X, then

$$\begin{array}{ccc} U \cap V & \longrightarrow & U \\ \downarrow & & \downarrow \\ V & \longrightarrow & X \end{array}$$

is a bicartesian square of schemes. We obtain a cartesian square

$$\begin{array}{ccc} \mathrm{Perf}(X) & \longrightarrow & \mathrm{Perf}(V) \\ \downarrow & & \downarrow \\ \mathrm{Perf}(U) & \longrightarrow & \mathrm{Perf}(U \cap V) \end{array}$$

of small stable ∞-categories in which the maps are all Verdier localizations. In particular, the fibers $\mathrm{Perf}(X, V) \simeq \mathrm{Perf}(U, U \cap V)$ are equivalent, and we get a commutative diagram of spectra

$$\begin{array}{ccccc} F(X, V) & \longrightarrow & F(X) & \longrightarrow & F(V) \\ \downarrow & & \downarrow & & \downarrow \\ F(U, U \cap V) & \longrightarrow & F(U) & \longrightarrow & F(U \cap V) \end{array}$$

in which the rows are exact and the left vertical map is an equivalence. It follows that the square

$$\begin{array}{ccc} F(X) & \longrightarrow & F(V) \\ \downarrow & & \downarrow \\ F(U) & \longrightarrow & F(U \cap V) \end{array}$$

is bicartesian, and that F preserves Mayer–Vietoris squares.

3.3. Waldhausen's $S_\bullet$-construction. The most general forms of algebraic K-theory are obtained via Waldhausen's $S_\bullet$-construction, a means of "universally splitting short exact sequences". This is precisely what is needed to turn a category $\mathcal{C}$, equipped with a suitable notion of exact sequence, into a spectrum, which we might then hope to compute something about. For instance, its homotopy groups are the algebraic K-groups of the category.

Since we may regard exact (or Waldhausen) categories as exact ∞-categories, we lose nothing by supposing that $\mathcal{C}$ is an ∞-category.[3] Then, for each natural number n, writing $\mathrm{Ar}(\Delta^n) = \mathrm{Fun}(\Delta^1, \Delta^n)$ for the arrow category, one defines

$$S_n\mathcal{C} \subset \mathrm{Fun}(\mathrm{Ar}(\Delta^n), \mathcal{C})$$

to be the full subcategory consisting of those functors which are reduced, in the sense that they send identity arrows to zero objects, and coexcisive, in the sense that certain squares are sent to pushouts. Had we taken the dual approach and encoded homotopy pullbacks using fibrations (instead of homotopy pushouts via cofibrations), we would have considered reduced excisive functors. The correct approach depends on the context, and whether or not the relevant objects are approximated cofibrantly or fibrantly (the homotopical version of projective or injective resolution in homological algebra). Frequently, however, either would work equally well.

Part of the reason for this is that many of the most natural and interesting examples come from still more elaborate structures in which both cofibrations and fibrations are encoded. For instance, they might be extracted from Quillen model categories, or exact ∞-categories (often exact categories, or even abelian categories), or stable ∞-categories. We will always assume the latter, and simply take the viewpoint that algebraic K-theory is an invariant of stable ∞-categories and exact functors. This is a good choice because $\mathrm{Cat}_\infty^{\mathrm{st}}$ receives compatible maps both from algebra and geometry, by associating to an algebro-geometric object its stable ∞-category of *perfect modules*, and we do not need to specify auxiliary notions of cofibration or fibration.

Let $\mathcal{C}$ be a stable ∞-category. Following Waldhausen, we define a simplicial stable ∞-category $S_\bullet\mathcal{C}$ in which the objects are iterated exact sequences in $\mathcal{C}$ of length n. We encode the data of an iterated exact sequence as a functor $F\colon \mathrm{Ar}(\Delta^n) \simeq \mathrm{Fun}(\Delta^1, \Delta^n) \to \mathcal{C}$ which is reduced (it sends degenerate 1-simplices to zero) and excisive (for $i \leq j \leq k \in [n]$, the square spanned by $(i,j) \to (i,k) \to (j,k)$ in $\mathrm{Ar}(\Delta^n) \simeq \mathrm{Fun}(\Delta^1, \Delta^n)$ is sent to a bicartesian square in $\mathcal{C}$). In particular, $S_0\mathcal{C} \simeq 0$, the trivial stable ∞-category, while $S_1\mathcal{C} \simeq \mathcal{C}$, and $S_2\mathcal{C} \simeq \mathrm{Fun}(\Delta^1, \mathcal{C})$, since an exact sequence in a stable ∞-category $\mathcal{C}$ determined, and is determined by, an arrow in $\mathcal{C}$. In general, $S_n\mathcal{C} \simeq \mathrm{Fun}(\Delta^{n-1}, \mathcal{C})$, though this obscures the simplicial structure, which uses the fact that $\mathcal{C}$ admits certain (co)limits.

DEFINITION 3.3.1. Let $\mathcal{C}$ be a stable ∞-category. The connective algebraic K-theory of $\mathcal{C}$ is the space $\Omega^\infty\mathrm{K}(\mathcal{C})$ given by the formula $\Omega^\infty\mathrm{K}(\mathcal{C}) := \Omega|\iota S_\bullet\mathcal{C}|$.

The (pre)spectrum structure on the infinite loop space $\Omega^\infty\mathrm{K}(\mathcal{C})$ is given by the formula

$$\mathrm{K}(\mathcal{C})_n \simeq \Omega^n|\iota S_\bullet^{(n)}\mathcal{C}|,$$

where $S_\bullet^{(n)}$ denotes the n-fold simplicial stable infinity category obtained by iterating the $S_\bullet$-construction, and the realization is understood to be the colimit of the multisimplicial space $\iota S_\bullet^{(n)}\mathcal{C}$. It follows from Waldhausen's additivity theorem that $\{\mathrm{K}(\mathcal{C})_n\}_{n\in\mathbf{N}_{>0}}$ is a spectrum, as for $n > 0$ the map

$$|\iota S_\bullet^{(n)}\mathcal{C}| \to \Omega|\iota S_\bullet^{(n+1)}\mathcal{C}|$$

is an equivalence.

[3]The fact that Waldhausen categories have a class of weak equivalences is not an issue here, as we can use that remember them in the ∞-category as well, or attempt to invert them.

Proposition 3.3.2 (Waldhausen Additivity). *Let $\mathcal{A} \to \mathcal{B} \to \mathcal{C}$ be an additive sequence of stable ∞-categories. Then* $\mathrm{K}(\mathcal{A}) \to \mathrm{K}(\mathcal{B}) \to \mathrm{K}(\mathcal{C})$ *is a (split) fiber sequence of pointed spaces. In particular,* $\Omega^\infty\mathrm{K}(\mathcal{A}) \to \Omega^\infty\mathrm{K}(\mathcal{B}) \to \Omega^\infty\mathrm{K}(\mathcal{C})$ *is a (split) fiber sequence of pointed spaces.*

Remark 3.3.3. It is probably better to define $\mathrm{K}(\mathcal{C})$ as $\mathrm{K}(\mathcal{C}^{\mathrm{idem}})$, where $\mathcal{C}^{\mathrm{idem}}$ denotes the idempotent completion of $\mathcal{C}$. In practice, $\mathcal{C} = \mathrm{Perf}(X)$ is perfect complexes on some geometric object X, meaning compact (or possibly dualizable, in commutative contexts, though in practice these tend to agree) objects in the derived ∞-category of X, and such objects are already stable under retracts. But the long exact localization sequence in K-theory is best suited to the idempotent complete case, and the difference only occurs in $\mathrm{K}_0(\mathcal{C})$, where $\mathrm{K}_0(\mathcal{C}^{\mathrm{idem}})$ contains $\mathrm{K}_0(\mathcal{C})$ as a summand. The distinction is essential for negative K-groups of $\mathrm{K}(\mathcal{C})$, as nonzero elements in negative degree precisely correspond to missing idempotents in Verdier quotients of stable ∞-categories derived from $\mathcal{C}$. We ignore this technical distinction for simplicity.

3.4. Noncommutative motives as a stabilization of $\mathrm{Cat}_\infty^{\mathrm{st}}$. If $\mathcal{C}$ is a presentable ∞-category, its stabilization $\mathrm{Sp}(\mathcal{C})$ is the initial stable presentable ∞-category equipped with a colimit preserving functor $\Sigma^\infty_+\mathcal{C} \to \mathrm{Sp}(\mathcal{C})$. As observed, in the case of interest, where $\mathcal{C} = \mathrm{Cat}_\infty^{\mathrm{st}}$, we have that $\mathrm{Sp}(\mathcal{C}) \simeq 0$, the trivial stable ∞-category. Consequently, it might be more useful to consider presentable stable ∞-categories $\mathcal{D}$ equipped with a functor $\mathcal{C} \to \mathcal{D}$ which needn't commute with all colimits.

Since $\mathrm{Cat}_\infty^{\mathrm{st}}$ is compactly generated, it is completely determined by its full subcategory of compact objects, $\mathrm{Cat}_\infty^{\mathrm{st},\omega}$. There is a canonical stable ∞-category generated by the small ∞-category $\mathrm{Cat}_\infty^{\mathrm{st},\omega}$, namely the ∞-category of presheaves of spectra

$$\mathcal{P}(\mathrm{Cat}_\infty^{\mathrm{st},\omega}; \mathrm{Sp}) = \mathrm{Fun}(\mathrm{Cat}_\infty^{\mathrm{st},\omega,\mathrm{op}}, \mathrm{Sp})$$

on $\mathrm{Cat}_\infty^{\mathrm{st},\omega}$. The (restricted) Yoneda embedding gives a functor

$$\Sigma^\infty\colon \mathrm{Cat}_\infty^{\mathrm{st}} \to \mathcal{P}(\mathrm{Cat}_\infty^{\mathrm{st},\omega}) \to \mathcal{P}_{\mathrm{Sp}}(\mathrm{Cat}_\infty^{\mathrm{st},\omega}) \simeq \mathrm{Fun}(\mathrm{Cat}_\infty^{\mathrm{st},\omega,\mathrm{op}}, \mathrm{Sp})$$

which we will also (abusively) refer to as Σ^∞, as we are regarding passing to the associated noncommutative motive as the correct stabilization of $\mathrm{Cat}_\infty^{\mathrm{st}}$. We want to consider only the "excisive" presheaves, namely those functors $F : \mathrm{Cat}_\infty^{\mathrm{st},\omega,\mathrm{op}} \to \mathrm{Sp}$ which satisfy $F(0) \simeq 0$ and $F(\mathcal{C}) \to F(\mathcal{B}) \to F(\mathcal{A})$ is exact whenever $\mathcal{A} \to \mathcal{B} \to \mathcal{C}$ is exact.

Remark 3.4.1. The precise condition we should require is locality for the map $\Sigma^\infty\mathcal{B}/\Sigma^\infty\mathcal{A} \to \Sigma^\infty\mathcal{C}$, which is to say that $F(\mathcal{C}) \to F(\mathcal{B}) \to F(\mathcal{A})$ is a fiber sequence, or a left exact sequence. Since spectra is stable, any left exact sequence is also right exact, hence exact.

Definition 3.4.2. The stable ∞-category of localizing (respectively, additive) motives is the full subcategory of $\mathcal{P}_{\mathrm{Sp}}(\mathrm{Cat}_\infty^{\mathrm{st},\omega})$ consisting of those functors F : $\mathrm{Cat}_\infty^{\mathrm{st},\omega,\mathrm{op}} \to \mathrm{Sp}$ which satisfy $F(0) \simeq 0$ and such that $F(\mathcal{C}) \to F(\mathcal{B}) \to F(\mathcal{S})$ is exact whenever $\mathcal{A} \to \mathcal{B} \to \mathcal{C}$ is exact (respectively, split exact).

Theorem 3.4.3. *The algebraic K-theory functor* $\mathrm{K}\colon \mathrm{Cat}_\infty^\omega \to \mathrm{Sp}$ *is stably corepresented by the unit; that is,* $\mathrm{K}(\mathcal{A}) \simeq \mathrm{Map}(\Sigma^\infty \mathrm{Sp}^\omega, \Sigma^\infty\mathcal{A})$.

Here $\Sigma^\infty \mathcal{A}$ denotes the noncommutative motive of $\mathcal{A}$, its Yoneda image in the localization of $\mathcal{P}_{\mathrm{Sp}}(\mathrm{Cat}_\infty^{\mathrm{st},\omega})$. This suggests that, by analogy, algebraic K-theory might be regarded as a categorification of stable (co)homotopy.

Recall that (ordinary) spectra determine both a contravariant cohmology theory, as well as a covariant homology theory, on the ∞-category $\mathcal{S}$ of spaces.

There is also the dual homology theory, $G(\mathcal{A}) \simeq \mathrm{Map}(\Sigma^\infty \mathcal{A}, \Sigma^\infty \mathrm{Sp}^\omega)$. In particular we obtain *bivariant K-theory* spectra

$$\mathrm{K}(\mathcal{A}, \mathcal{B}) = \mathrm{Map}(\Sigma^\infty \mathcal{A}, \Sigma^\infty \mathcal{B}),$$

making algebraic K-theory into a bifunctor

$$\mathrm{K}(-,-)\colon \mathrm{Cat}_\infty^{\mathrm{st},\mathrm{op}} \times \mathrm{Cat}_\infty^{\mathrm{st}} \to \mathrm{Sp}\,.$$

REMARK 3.4.4. The G-theory spectrum of a ring R is given by the formula $G(R) \simeq \mathrm{Map}(\Sigma^\infty \mathrm{Perf}(R), \Sigma^\infty \mathrm{Sp}^\omega)$. Note that this is the formal dual of K-theory, as it is computed as maps to, instead of from, the unit $\Sigma^\infty \mathrm{Sp}^\omega$.

3.5. The theorem of the heart. There is a beautiful relationship between the K-theory of stable ∞-categories and the K-theory of abelian categories via a higher categorical version of Quillen's dévissage. The classical dévissage theorem states that the K-theory of an abelian category agrees with that of any Serre subcategory with the property that every object of the former has a finite filtration whose quotients lie in the latter. The basic K-theoretic relations allow one to show that the K-theory of a filtered object agrees with the alternating sum of the K-theory of the associated graded.

To illustrate, let's consider the (ordinary) category of $\mathbf{Z}_{(p)}$-modules. Inverting p, we obtain $\mathbf{Q}_{(p)}$-modules, and a localization sequence

$$\mathrm{Mod}(\mathbf{Z}_{(p)}, \mathbf{F}_p) \to \mathrm{Mod}(\mathbf{Z}_{(p)}) \to \mathrm{Mod}(\mathbf{Q}_{(p)})$$

in which the kernel, the $\mathbf{Z}_{(p)}$-modules supported on $\mathbf{F}_p$, are precisely the p-torsion $\mathbf{Z}_{(p)}$-modules. Using dévissage, the K-theory of the kernel is equivalent to $\mathbf{F}_p$, and Quillen's famous K-theoretic localization sequence

$$\mathrm{K}(\mathbf{F}_p) \to \mathrm{K}(\mathbf{Z}_{(p)}) \to \mathrm{K}(\mathbf{Q}_{(p)})$$

falls out.

The higher categorical version of dévissage is best encapsulated by the Barwick–Neeman *theorem of the heart* (see **[3]**), which identifies the (connective) algebraic K-theory of a stable ∞-category $\mathcal{C}$, equipped with a bounded t-structure, with the (connective) algebraic K-theory of its heart. This latter is easier to compute, at least in principle.

THEOREM 3.5.1. *Let $\mathcal{C}$ be a stable ∞-category equipped with a bounded t-structure and let $\mathcal{C}^\heartsuit \to \mathcal{C}$ denote the inclusion of the heart. The induced map*

$$\tau_{\geq 0}\mathrm{K}(\mathcal{C}^\heartsuit) \to \tau_{\geq 0}\mathrm{K}(\mathcal{C})$$

is an equivalence of spectra. In particular, for $n \geq 0$, $\mathrm{K}_n(\mathcal{C}^\heartsuit) \to \mathrm{K}_n(\mathcal{C})$ is an isomorphism.

Oftentimes in cases of interest, the negative K-theory of $\mathcal{C}$ vanishes. This is the case, for instance, if the heart of the bounded t-structure is a noetherian abelian category **[2]**.

THEOREM 3.5.2. *Suppose that $\mathcal{C}$ is a stable ∞-category equipped with a bounded t-structure. Then* $\mathrm{K}_{-1}(\mathcal{C}) = 0$. *If additionally* $\mathcal{C}^{\heartsuit}$ *is noetherian, then* $\mathrm{K}_{-n}(\mathcal{C}) = 0$ *for all* $n > 0$.

4. Hochschild and Cyclic Homology

4.1. The cyclic bar construction. Algebraic K-theory (or at least its connective cover, the infinite loop space) can be obtained via Waldhausen's $S_\bullet$-construction:

$$\Omega^\infty \mathrm{K}(\mathcal{C}) \simeq \Omega|\iota S_\bullet \mathcal{C}|.$$

There is a similar formula for the Hochschild homology of $\mathcal{C}$, in terms of the cyclic bar construction. Since we are working in the ∞-category of stable ∞-categories, we are essentially working **S**-linearly, and this is essential. There is a version of Hochschild homology relative to any commutative ring (or ring spectrum) R, which historically might have been a field, a slightly easier case as any module over a field is flat, so the tensor product doesn't need to be derived. Hence the absolute (that is, relative to the sphere) version of Hochschild homology is usually referred to as *topological Hochschild homology*, and denoted THH.

While there are a number of quite intrinsic ways to define the Hochschild homology of a stable ∞-category $\mathcal{C}$, the most elementary first forgets that $\mathcal{C}$ is a stable ∞-category, and only remembers that $\mathcal{C}$ is a spectrally-enriched ∞-category. This is not a radical operation, as there is a fully faithful inclusion

$$\mathrm{Cat}_\infty^{\mathrm{st}} \to \mathrm{Cat}(\mathrm{Sp})$$

of stable ∞-categories and exact functors into spectral ∞-categories and spectral and functors. The left adjoint functor is the stable envelope (a.k.a. perfect complexes, at least if we additionally require our stable ∞-categories to be idempotent complete), and is a symmetric monoidal left adjoint.

Topological Hochschild homology is Morita invariant, meaning that it inverts those maps of the form

$$\mathcal{A} \to \mathrm{Perf}(\mathcal{A}) \simeq \mathrm{Fun}^{\mathrm{st}}(\mathcal{A}^{\mathrm{op}}, \mathrm{Sp})^\omega.$$

This implies that it does not depend on the choice of spectral representative for the stable ∞-category. The advantage of this approach is that, as a small spectrally enriched ∞-category, $\mathcal{C}$ can be chose to have a set of objects, as opposed to a space of objects, and this set can often be quite small, as it only needs to generate the stable ∞-category under spectral colimits. This can make it much easier to index the cyclic bar construction, as we often only have to sum over a small set of objects.

DEFINITION 4.1.1. Let $\mathcal{A}$ and $\mathcal{B}$ be small spectrally enriched ∞-categories. The $\mathcal{A}$ and $\mathcal{B}$ are Morita equivalent if there is an equivalence of spectral ∞-categories $\mathrm{Perf}(\mathcal{A}) \simeq \mathrm{Perf}(\mathcal{B})$.

We could equally well have asked for an equivalence $\mathrm{Mod}(\mathcal{A}) \simeq \mathrm{Mod}(\mathcal{B})$. By definition, $\mathrm{Perf}(\mathcal{A}) \simeq \mathrm{Mod}(\mathcal{A})^\omega$ is the full subcategory of compact objects, and $\mathrm{Mod}(\mathcal{A}) \simeq \mathrm{Ind}(\mathrm{Perf}(\mathcal{A})$ is a compactly generated stable ∞-category. In other words, perfect (compact) modules generate the entire module ∞-category under filtered colimits, but $\mathrm{Perf}(\mathcal{A})$ has the advantage that it is a small stable ∞-category. In fact we have the following precise relationship between spectral ∞-categories and stable ∞-categories (see [**5**]), which identifies the latter as the Morita localization of the former.

THEOREM 4.1.2. *Let* Cat(Sp) *denote the* ∞*-category of spectrally enriched* ∞*-categories, and let* $\mathrm{Mor} \subset \mathrm{Fun}(\Delta^1, \mathrm{Cat}(\mathrm{Sp}))$ *denote the class of Morita equivalences. Then* $\mathrm{Cat}_\infty^{\mathrm{st}} \simeq \mathrm{Cat}(\mathrm{Sp})[\mathrm{Mor}^{-1}]$.

DEFINITION 4.1.3. Let $\mathcal{C}$ be a small spectrally enriched ∞-category. The topological Hochschild homology $\mathrm{THH}(\mathcal{C})$ of $\mathcal{C}$ is the spectrum obtained as the realization of the simplicial spectrum which in degree n is

$$\coprod_{A_0,\ldots,A_n \in \mathcal{C}} \underline{\mathrm{Map}}_{\mathcal{C}}(A_0, A_1) \otimes \underline{\mathrm{Map}}_{\mathcal{C}}(A_1, A_2) \otimes \cdots \otimes \underline{\mathrm{Map}}_{\mathcal{C}}(A_{n-1}, A_n) \otimes \underline{\mathrm{Map}}_{\mathcal{C}}(A_n, A_0).$$

The face map

$$\coprod_{A_0,\ldots,A_n} \underline{\mathrm{Map}}(A_0, A_1) \otimes \cdots \otimes \underline{\mathrm{Map}}(A_{i-1}, A_i) \otimes \underline{\mathrm{Map}}(A_i, A_{i+1}) \otimes \cdots \otimes \underline{\mathrm{Map}}(A_n, A_0)$$

$$\longrightarrow \coprod_{A_0\ldots,A_{n-1}} \underline{\mathrm{Map}}(A_0, A_1) \otimes \cdots \otimes \underline{\mathrm{Map}}(A_{i-1}, A_i) \otimes \cdots \otimes \underline{\mathrm{Map}}(A_{n-1}, A_0)$$

composes the maps $A_{i-1} \to A_i \to A_{i+1}$ and reindexes, setting $i+1$ to i.

REMARK 4.1.4. It is important to note that we index modulo n, revealing the cyclic nature of this particular simplicial spectrum.

REMARK 4.1.5. Hochschild homology is Morita invariant, meaning that it only depends on the (idempotent completion of) the (derived) ∞-category of modules over the ring or scheme of interest. Consequently, if $\mathcal{C}$ is the stable ∞-category of perfect modules for a ring spectrum R, then there is an equivalence of spectra $\mathrm{THH}(R) \simeq \mathrm{THH}(\mathcal{C})$.

In the case of a ring spectrum R, one can compute the topological Hochschild homology as the tensor product

$$R \bigotimes_{R^{\mathrm{op}} \otimes R} R \simeq \mathrm{THH}(R),$$

where R^{op} denotes the opposite algebra. If R is a commutative ring spectrum, the canonical equivalence $R^{\mathrm{op}} \simeq R$ reduces this formula to the even more straightforward expression

$$S^1 \otimes R \simeq \mathrm{THH}(R).$$

Here S^1 denotes the circle, which appears because it is the pushout $\mathrm{pt} \coprod_{S^0} \mathrm{pt}$.

Since the ∞-category of commutative ring spectra admits all colimits, it is canonically tensored over the ∞-category of spaces. This makes the circle action of Hochschild homology especially clear, though it is present, as a relic of the cyclic bar construction, even in the associative case.

REMARK 4.1.6. As $\mathbf{S}$ is the initial ring spectrum, commutative or otherwise, we see that $\mathrm{THH}(\mathbf{S}) \simeq \mathbf{S}$. In particular, $\pi_0\mathrm{THH}(\mathbf{S}) \cong \pi_0\mathbf{S} \cong \mathbf{Z}$.

LEMMA 4.1.7. *The functor*

$$\mathrm{THH}\colon \mathrm{Cat}(\mathrm{Sp}) \to \mathrm{Sp}$$

is Morita invariant. In particular, it descends to a functor of ∞*-categories*

$$\mathrm{THH}\colon \mathrm{Cat}_\infty^{\mathrm{st}} \simeq \mathrm{Cat}(\mathrm{Sp})[\mathrm{Mor}^{-1}] \to \mathrm{Sp}\,.$$

THEOREM 4.1.8. *The topological Hochschild homology functor* $\mathrm{THH}\colon \mathrm{Cat}_\infty^{\mathrm{st}} \to \mathrm{Sp}$ *is a localizing invariant, and there is an isomorphism* $\pi_0 \,\mathrm{Map}(\mathrm{K}, \mathrm{THH}) \cong \mathbf{Z}$.

The fact that algebraic K-theory is corepresentable in the ∞-category of noncommutative motives yields the following identification of the spectrum of natural transformations of localizing functors $\mathrm{K} \to E$.

THEOREM 4.1.9. *Given a localizing invariant* $F\colon \mathrm{Cat}_\infty^{\mathrm{st}} \to \mathrm{Sp}$ *with values in the stable* ∞*-category of spectra, we have a natural equivalence*

$$\mathrm{Nat}(\mathrm{K}, F) \simeq F(\mathrm{Sp}^\omega) \simeq F(\mathbf{S}),$$

where $\mathrm{Nat}(\mathrm{K}, F)$ *denotes the spectrum of natural transformations from* K *to* F *as localizing invariants from small stable* ∞*-categories to spectra.*

Applying the Theorem to the localizing invariant THH, we conclude that there is an equivalence of spectra

$$\mathrm{Nat}(\mathrm{K}(-), \mathrm{THH}(-)) \cong \mathrm{THH}(\mathrm{Sp}^\omega) \simeq \mathrm{THH}(\mathbf{S}) \simeq \mathbf{S}.$$

It follows that homotopy classes of natural transformations from K to THH are in bijective correspondence with elements of $\pi_0(\mathbf{S})$. The standard natural transformation $\mathrm{K} \to \mathrm{THH}$ is the Dennis trace map, and is related to taking the trace of a matrix.

THEOREM 4.1.10. *The topological Dennis trace is (up to homotopy) the natural transformation given by the identity element* $1 \in \pi_0\mathrm{THH}(\mathbf{S}) \cong \pi_0(\mathbf{S}) \cong \mathbf{Z}$.

The results of [**6**] imply that the Dennis trace is the unique multiplicative natural transformation from K to THH.

4.2. Topological cyclic homology. We have seen that the algebraic K-groups of a stable ∞-category $\mathcal{C}$ arise as homotopy groups of the associated noncommutative motive. Similarly, the topological cyclic homology groups of $\mathcal{C}$ are encoded as the homotopy groups of the cyclotomic spectrum associated to $\mathcal{C}$, namely the topological Hochschild homology of $\mathcal{C}$.

As Connes observed, the Hochschild homology construction has more functoriality than a simplicial object: it refines to a cyclic object. Very roughly, Connes' cyclic category Λ^{op} is an amalgamation of the simplicial indexing category Δ^{op} and the finite cyclic groups; instead of objects corresponding to total orders $[0 \to 1 \to \cdots n-1 \to n]$, they correspond to cycles by identifying initial and final objects. The realization of a cyclic object is the colimit of the underlying simplicial object, which is obtained by restricting along the canonical functor $\Delta^{\mathrm{op}} \to \Lambda^{\mathrm{op}}$. This functor factors as a composition $\Delta^{\mathrm{op}} \to \Lambda_\infty^{\mathrm{op}} \to \Lambda^{\mathrm{op}}$, where the functor to $\Lambda_\infty^{\mathrm{op}}$, the *paracyclic category*, is cofinal, and Λ^{op} the quotient of $\Lambda_\infty^{\mathrm{op}}$ by an action of the integers $\mathbf{Z}$. It follows that realizations of cyclic objects come equipped with a natural circle action, as the circle is equivalent to the classifying space of the integers.

The classical approach to cyclic homology uses the full apparatus of equivariant homotopy theory. For finite subgroups $G \subset S^1$, the circle action results in a set of compatible maps

$$\phi^G\mathrm{THH} \to \mathrm{THH},$$

where $\phi^G\mathrm{THH}$ is viewed as an $S^1/G \cong S^1$-spectrum via the isogeny identification of S^1/G with itself.

For each natural number n, the functor $\mathrm{TR}^n\colon \mathrm{Cat}_\infty^{\mathrm{st}} \to \mathrm{Sp}$ is given by

$$\mathrm{TR}^n(\mathcal{C}) = \mathrm{THH}(\mathcal{C})^{C_{p^{n-1}}};$$

that is, formation fixed points with respect to the induced $C_{p^{n-1}}$ action. The inclusion of fixed points and the cyclotomic structure give rise to Frobenius and restriction maps

$$F, R\colon \mathrm{TR}^n \to \mathrm{TR}^{n-1}.$$

We define $\mathrm{TC}^n(\mathcal{C})$ to be the equalizer of F and R. Equivalently, we have a pullback square of spectra

$$\begin{array}{ccc} \mathrm{TC}^n(\mathcal{C}) & \longrightarrow & \mathrm{TR}^n(\mathcal{C}) \\ \big\downarrow & & \big\downarrow {\scriptstyle F \mid R} \\ \mathrm{TR}^{n-1}(\mathcal{C}) & \longrightarrow & \mathrm{TR}^{n-1}(\mathcal{C}) \times \mathrm{TR}^{n-1}(\mathcal{C}) \end{array}$$

in which the bottom map is the diagonal, or an exact sequence of spectra

$$\mathrm{TC}^n(\mathcal{C}) \to \mathrm{TR}^n(\mathcal{C}) \overset{F-R}{\to} \mathrm{TR}^{n-1}(\mathcal{C}).$$

Finally, we define

$$\mathrm{TC}(\mathcal{C}) = \lim_n \mathrm{TC}^n(\mathcal{C}),$$

where we form the limit over the maps induced by the restriction R. This definition is equivalent to the one originally given in [**10**]. The functor TC is Morita invariant, and consequentlyvinduces a functor of ∞-categories

$$\mathrm{TC}\colon\ \mathrm{Cat}_\infty^{\mathrm{st}} \simeq \mathrm{Cat}(\mathrm{Sp})[\mathrm{Mor}^{-1}] \to \mathrm{Sp}\,.$$

This is the *topological cyclic homology* functor.

The corepresentability result for THH morally ought to apply for TC, except that TC itself is not a localizing invariant. It is Morita invariant and satisfies localization, as shown in [**8**], but it does not preserve filtered colimits. For this reason, the filtered colimit preservation is often dropped from the definition of a localizing invariant. We can instead regard TC as a kind of pro-object, and observe that its constituent pieces TC^n are in fact localizing invariants, to obtain the desired corepresentability result.

REMARK 4.2.1. Groundbreaking work of Nikolaus and Scholze on topological cyclic homology [**16**] reformulates the notion of cyclotomic spectra, in a considerably simpler way, on the full subcategory of spectra consisting of the bounded below objects. They introduce a stable ∞-category of *cyclotomic spectra* in which topological cyclic homology is corepresented by the cyclotomic sphere $\mathbf{S}^{\mathrm{cyc}}$, which is the sphere $\mathbf{S}$ together with the cyclotomic structure coming from the equivalence $\mathbf{S} \simeq \mathrm{THH}(\mathbf{S})$.

4.3. The Dundas–Goodwillie–McCarthy Theorem. The fact that algebraic K-theory is corepresented by the sphere spectrum, together with the well-known calculations of $\mathrm{THH}_0(\mathbf{S})$ and $\mathrm{TC}_0(\mathbf{S})$, implies that there are canonical trace maps $\mathrm{K} \to \mathrm{THH}$ (the Dennis trace) and $\mathrm{K} \to TC$ (the cyclotomic trace). In fact, the cyclotomic trace factors the Dennis trace map

$$\mathrm{K} \to \mathrm{TC} \to \mathrm{THH};$$

the map $\mathrm{TC} \to \mathrm{THH}$ is the forgetful functor evident from the construction of topological cyclic homology from topological Hochschild homology.

We have seen that homotopical algebra is a nonlinear version of homological algebra. The precise degree of nonlinearity varies: if we work unstably, we have no linearity whatsoever, whereas if we work stably, we have $\mathbf{S}$-linearity, a higher categorical version of $\mathbf{Z}$-linearity. As we saw earlier, these rings are related via the Barratt–Priddy theorem, as the group completions of the semirings $\mathrm{Fin}^{\simeq}$ and $\mathbf{N}$. The symmetric monoidal functor $\mathrm{Fin}^{\simeq} \to \mathbf{N}$ induces a map of commutative rings $\mathbf{S} \to \mathbf{Z}$ which is the 0-truncation, as $\pi_0\mathbf{S} \cong \mathbf{Z}$. The Nishida nilpotence theorem asserts that the positive degree elements of the homotopy ring $\pi_*\mathbf{S}$ (which are torsion, by work of Serre) are nilpotent, which suggests that we might regard $\mathbf{S}$ as a higher categorical deformation of $\mathbf{Z}$.

There is a heuristic in derived geometry that higher homotopy groups of connective ring spectra R ought to be regarded as nilpotent elements (regardless of whether or not this is actually the case in π_*R), so we might define a deformation of a connective ring spectrum R to be any map $\tilde{R} \to R$ which induces a surjection $\pi_0\tilde{R} \to \pi_0 R$ with nilpotent kernel. The powerful Dundas–Goodwillie–McCarthy theorem asserts that for deformations, the difference in K-theory, the fiber of the map $\mathrm{K}(\tilde{R}) \to \mathrm{K}(R)$, is the same as the difference in cyclic homology (over the sphere), the fiber of the map $\mathrm{TC}(\tilde{R}) \to \mathrm{TC}(R)$. Since TC is computable at least in principle, and increasingly in practice, this allows us to calculate the K-theory of deformations of R in terms of $\mathrm{K}(R)$ and $\mathrm{TC}(\tilde{R})$ over $\mathrm{TC}(R)$.

THEOREM 4.3.1 (Dundas–Goodwillie–McCarthy). *Let $\tilde{R} \to R$ be a deformation of ring spectra, in the sense that $\pi_0\tilde{R} \to \pi_0 R$ is surjective with nilpotent kernel. Then*

$$\begin{array}{ccc} \mathrm{K}(\tilde{R}) & \longrightarrow & \mathrm{K}(R) \\ \downarrow & & \downarrow \\ \mathrm{TC}(\tilde{R}) & \longrightarrow & \mathrm{TC}(R) \end{array}$$

is bicartesian square of spectra.

REMARK 4.3.2. It is important to observe that this theorem *requires* the use of the sphere spectrum. It fails for cyclic homology over other bases, though if one is only interested in rational information then Goodwillie's original treatment, using rational cyclic homology, suffices.

4.4. Applications. The ring of p-local integers $\mathbf{Z}_{(p)}$ captures the behavior of the ring of integers $\mathbf{Z}$ near the prime p. We can also localize the sphere spectrum at a prime p to capture its p-local behavior. The ∞-category of modules for the p-local sphere is the ∞-category of p-local spectra. In fact we can localize at any of the primes of the sphere spectrum, the *chromatic primes.* These are classified, and there is a $\mathbf{N}$-indexed family of primes of the sphere which lie over a fixed (nonzero) prime $p \in \mathrm{Spec}(\mathbf{Z})$, denoted $K(n)$, and called the nth Morava K-theory.[4]

The (presentable) stable ∞-category of spectra is the Ind-completion of the stable ∞-category Sp^{ω} of finite (or compact, they are equivalent) spectra, and the latter decomposes in the usual way — using the arithmetic square which splices together the p-primary information with the rational information — into the stable ∞-categories of finite p-local spectra $\mathrm{Sp}^{\omega}_{(p)}$. At a fixed prime p, the ∞-category of finite p-local spectra is filtered by chromatic height n.

[4]There is a relationship to topological K-theory; namely, $K(1) \simeq KU/p$ is complex topological K-theory modulo p. For $n > 1$, there is really no relationship with K-theory, the Morava K-theories are really the spectra which arise as residue fields of primes in $\mathbf{S}$.

The height is the same as the height from the theory of (commutative, one dimensional) formal group laws; this is because, in practice, and say at (the ring spectrum R corresponding to the residue field of) the chromatic prime in question, the formal group law appears as the ring spectrum evaluated at the infinite complex projective space $\mathbf{CP}^\infty$, which turns out to be a (commutative, one dimensional) formal group. The ring spectrum MU, representing complex cobordism theory, has homotopy groups which form a graded ring isomorphic to the Lazard ring classifying these formal groups, and once an ordinary prime p is fixed, the p-localization $MU_{(p)}$ splits into a number of copies of a spectrum called BP (named for Brown–Peterson, who discovered it), whose graded homotopy ring classifies p-typical formal group laws. But p-typical formal groups still have a height n, and there are truncations $BP\langle n\rangle$ of BP, with

$$\pi_* BP\langle n\rangle \cong \mathbf{Z}_{(p)}[v_1, \ldots, v_n],$$

which see only the height less than or equal to n phenomena. We can kill v_n to obtain $BP\langle n-1\rangle$ or invert v_n to obtain a ring spectrum known as $E(n)$, the nth Johnson–Wilson theory.

Generalizing Quillen's famous localization sequence $\mathrm{K}(\mathbf{F}_p) \to \mathrm{K}(\mathbf{Z}_{(p)}) \to \mathrm{K}(\mathbf{Q})$, or its integral version $\bigoplus_p \mathrm{K}(\mathbf{F}_p) \to \mathrm{K}(\mathbf{Z}) \to \mathrm{K}(\mathbf{Q})$, is the following result of Blumberg–Mandell [**7**].

THEOREM 4.4.1. *Inverting and killing the Bott element in ku determines a localization sequence*

$$\mathrm{K}(\mathbf{Z}) \to \mathrm{K}(ku) \to \mathrm{K}(KU)$$

in algebraic K-theory. Similarly, for each prime p, there is a localization sequence

$$\mathrm{K}(\mathbf{Z}_{(p)}) \to \mathrm{K}(BP\langle 1\rangle) \to \mathrm{K}(E(1)).$$

Here $BP\langle 1\rangle$ and $E(1)$ are summands of the p-local complex K-theory spectra $ku_{(p)}$ and $KU_{(p)}$, respectively. This gives us a fundamental localization sequence for each chromatic prime of height less than two, and it is natural to ask if this persists to higher heights.

For higher chromatic heights, we have the p-local theories $BP\langle n\rangle$ and its v_n-periodic variant $E(n)$, but not integral theories which localize to the $BP\langle n\rangle$ at a prime p (with the possible exception of elliptic cohomology, at chromatic height 2). Nevertheless, we can invert v_n in $BP\langle n\rangle$ to form $E(n)$, and kill v_n to obtain $BP\langle n-1\rangle$, so it is reasonable to ask whether or not this determines a localization sequence

$$\mathrm{K}(BP\langle n-1\rangle) \to \mathrm{K}(BP\langle n\rangle) \to \mathrm{K}(E(n))$$

in algebraic K-theory. One can show that the fiber of the map $\mathrm{K}(BP\langle n\rangle) \to \mathrm{K}(E(n))$ is the algebraic K-theory of the a certain endomorphism spectrum $A\langle n-1\rangle$ related to $BP\langle n-1\rangle$, but is not actually connective. It turns out that the trace map to (rational, even) Hochschild homology suffices to show that these two rings have different K-theories as soon as $n > 1$, which implies that the sequence $\mathrm{K}(BP\langle n-1\rangle) \to \mathrm{K}(BP\langle n\rangle) \to \mathrm{K}(E(n))$ is not exact. See [**1**] for details.

Acknowledgments

It is a pleasure to thank the organizers of this conference — namely Alain Connes, Katia Consani, Masoud Khalkhali, and Henri Moscovici — for their efforts. I would also like to thank Bjørn Dundas for his efforts in editing the lecture notes from the talks, as well as an anonymous referee for a number of helpful comments and suggestions.

References

[1] Benjamin Antieau, Tobias Barthel, and David Gepner, *On localization sequences in the algebraic K-theory of ring spectra*, J. Eur. Math. Soc. (JEMS) **20** (2018), no. 2, 459–487, DOI 10.4171/JEMS/771. MR3760300

[2] Benjamin Antieau, David Gepner, and Jeremiah Heller, *K-theoretic obstructions to bounded t-structures*, Invent. Math. **216** (2019), no. 1, 241–300, DOI 10.1007/s00222-018-00847-0. MR3935042

[3] Clark Barwick, *On exact ∞-categories and the theorem of the heart*, Compos. Math. **151** (2015), no. 11, 2160–2186, DOI 10.1112/S0010437X15007447. MR3427577

[4] A. A. Beĭlinson, J. Bernstein, and P. Deligne, *Faisceaux pervers* (French), Analysis and topology on singular spaces, I (Luminy, 1981), Astérisque, vol. 100, Soc. Math. France, Paris, 1982, pp. 5–171. MR751966

[5] Andrew J. Blumberg, David Gepner, and Gonçalo Tabuada, *A universal characterization of higher algebraic K-theory*, Geom. Topol. **17** (2013), no. 2, 733–838, DOI 10.2140/gt.2013.17.733. MR3070515

[6] Andrew J. Blumberg, David Gepner, and Gonçalo Tabuada, *Uniqueness of the multiplicative cyclotomic trace*, Adv. Math. **260** (2014), 191–232, DOI 10.1016/j.aim.2014.02.004. MR3209352

[7] Andrew J. Blumberg and Michael A. Mandell, *The localization sequence for the algebraic K-theory of topological K-theory*, Acta Math. **200** (2008), no. 2, 155–179, DOI 10.1007/s11511-008-0025-4. MR2413133

[8] Andrew J. Blumberg and Michael A. Mandell, *Localization theorems in topological Hochschild homology and topological cyclic homology*, Geom. Topol. **16** (2012), no. 2, 1053–1120, DOI 10.2140/gt.2012.16.1053. MR2928988

[9] J. M. Boardman and R. M. Vogt, *Homotopy invariant algebraic structures on topological spaces*, Lecture Notes in Mathematics, Vol. 347, Springer-Verlag, Berlin-New York, 1973. MR0420609

[10] M. Bökstedt, W. C. Hsiang, and I. Madsen, *The cyclotomic trace and algebraic K-theory of spaces*, Invent. Math. **111** (1993), no. 3, 465–539, DOI 10.1007/BF01231296. MR1202133

[11] Edgar H. Brown Jr., *Cohomology theories*, Ann. of Math. (2) **75** (1962), 467–484, DOI 10.2307/1970209. MR138104

[12] A. Joyal, *Quasi-categories and Kan complexes*, J. Pure Appl. Algebra **175** (2002), no. 1-3, 207–222, DOI 10.1016/S0022-4049(02)00135-4. Special volume celebrating the 70th birthday of Professor Max Kelly. MR1935979

[13] Jacob Lurie, *Higher topos theory*, Annals of Mathematics Studies, vol. 170, Princeton University Press, Princeton, NJ, 2009, DOI 10.1515/9781400830558. MR2522659

[14] Jacob Lurie. *Higher Algebra.* Available at `http://www.math.harvard.edu/~lurie/papers/HA.pdf`, version dated 18 September 2017.

[15] Amnon Neeman, *The Grothendieck duality theorem via Bousfield's techniques and Brown representability*, J. Amer. Math. Soc. **9** (1996), no. 1, 205–236, DOI 10.1090/S0894-0347-96-00174-9. MR1308405

[16] Thomas Nikolaus and Peter Scholze, *On topological cyclic homology*, Acta Math. **221** (2018), no. 2, 203–409, DOI 10.4310/ACTA.2018.v221.n2.a1. MR3904731

[17] Daniel G. Quillen, *Homotopical algebra*, Lecture Notes in Mathematics, No. 43, Springer-Verlag, Berlin-New York, 1967. MR0223432

[18] Harry Smith, *Bounded t-structures on the category of perfect complexes over a Noetherian ring of finite Krull dimension*, Adv. Math. **399** (2022), Paper No. 108241, 21, DOI 10.1016/j.aim.2022.108241. MR4383013

[19] Charles A. Weibel, *An introduction to homological algebra*, Cambridge Studies in Advanced Mathematics, vol. 38, Cambridge University Press, Cambridge, 1994, DOI 10.1017/CBO9781139644136. MR1269324

Department of Mathematics, John Hopkins University, Baltimore, Maryland 21218

Proceedings of Symposia in Pure Mathematics
Volume **105**, 2023
https://doi.org/10.1090/pspum/105/01901

Cyclic cohomology and the extended Heisenberg calculus of Epstein and Melrose

Alexander Gorokhovsky and Erik van Erp

ABSTRACT. In this paper we present a formula for the index of a pseudodifferential operator with invertible principal symbol in the *extended* Heisenberg calculus of Epstein and Melrose. The results of this paper build on our previous work, where we restricted attention to the Heisenberg calculus proper.

CONTENTS

1. Introduction

In this paper we present a formula for the index of a pseudodifferential operator with invertible principal symbol in the *extended* Heisenberg calculus of Epstein and Melrose [**3**, **4**, **9**].

The extended Heisenberg calculus includes, as subalgebras, the algebra of Heisenberg pseudodifferential operators *as well as* the algebra of classical pseudodifferential operators. Thus, an index formula for the extended Heisenberg calculus is a common generalization of the Atiyah-Singer index formula for elliptic operators on the one hand, and the index theorem for Heisenberg elliptic operators on the other hand. The monograph [**3**] of Epstein and Melrose aimed to find such a formula, but the project was unfinished. Our results here build on the work we did in [**7**], where we restricted attention to the Heisenberg calculus proper. Note that it is a non-trivial exercise to derive the Atiyah-Singer formula in its standard form from our formula.

2020 *Mathematics Subject Classification.* Primay 46L80, 58J20, 58J40.

A hypoelliptic operator of order zero in the extended Heisenberg calculus on a compact contact manifold M is a bounded Fredholm operator. Its symbol determines an element in the K-theory of the noncommutative algebra of Heisenberg symbols $\mathscr{S}^{0,0}_{eH}$. We construct a character homomorphism

$$\chi \circ s\colon K_1(\mathscr{S}^{0,0}_{eH}) \to H^{odd}(M)$$

which assigns a closed differential form to an invertible principal symbol (of order zero). As in our previous paper [**7**], the construction is heavily based on ideas and constructions from cyclic theory.

We show that if $\mathscr{L}$ is a pseudodifferential operator of order zero in the extended Heisenberg calculus with invertible principal symbol $\sigma = \sigma_{eH}(\mathscr{L})$, then

$$\operatorname{Index} \mathscr{L} = \int_M \chi(s(\sigma)) \wedge \hat{A}(M).$$

(See Theorem 8.1.)

The paper is organized as follows. In section 2 we describe the algebra of principal symbols in the extended Heisenberg calculus and relate them to sections of bundles of Weyl algebras. In section 3 we describe connections on a bundle of Weyl algebras, and section 4 describes a regularized trace on the fibers of such bundles. Section 5 gives a review of cyclic theory and the Chern character in this context. In section 6 we describe an extension of the algebra of principal symbols in the extended Heisenberg calculus which is well suited for cyclic theory. Section 7 contains the description of our characteristic map χ. Finally in the section 8 we state and prove the main result of the paper – the index formula for order zero hypoelliptic operators in the extended Heisenberg calculus.

2. Symbols in the extended Heisenberg calculus

In this section we briefly review the structure of the algebra of Heisenberg and extended Heisenberg principal symbols. We assume the reader has some familiarity with the Heisenberg calculus [**1, 10**]. We follow the treatment of Epstein and Melrose [**3, 9**]. This section serves to fix our notation. See also [**7**] for a more detailed discussion of some of the material in this section.

2.1. Contact manifolds. Throughout this paper, M is a smooth closed orientable manifold of dimension $2n+1$. We assume that M is equipped with a contact 1-form, i.e. a 1-form α such that $\alpha(d\alpha)^n$ is a nowhere vanishing volume form. Such an M will be called a *contact manifold*. Note that M is oriented by the form $\alpha(d\alpha)^n$.

The Reeb field T is the vector field on M determined by $d\alpha(T, -) = 0$ and $\alpha(T) = 1$. We denote by $H \subset TM$ the bundle of tangent vectors that are annihilated by α. Then $TM \cong H \oplus \underline{\mathbb{R}}$, where $\underline{\mathbb{R}} = M \times \mathbb{R}$ is the trivial line bundle.

The restriction of the 2-form $\omega := -d\alpha$ to each fiber $H_p, p \in M$, is a symplectic form, $\omega_p := -d\alpha|H_p$. We assume that a compatible complex structure $J \in \operatorname{End}(H)$ has been chosen, i.e. an endomorphism of H with $J^2 = -\mathrm{Id}$, $\omega_p(Jv, Jw) = \omega_p(v, w)$, $\omega_p(Jv, v) \geq 0$. The combination of the symplectic and complex structures make H into a complex Hermitian vector bundle, which we denote by $H^{1,0}$. We shall reserve the notation H for the real vector bundle. We denote by P_H the $U(n)$ principal bundle of orthonormal frames in $H^{1,0}$. P_H is equipped with a right action of $U(n)$. The map $v \mapsto \omega(v, \cdot)$ establishes a canonical isomorphism of bundles H and H^*, allowing to transfer to H^* complex and Hermitian structures. In particular, we will use it to identify the bundle of orthonormal frames in $(H^*)^{1,0}$ with P_H.

Examples of contact manifolds are: the Heisenberg group, boundaries of strictly pseudoconvex domains (e.g. the odd-dimensional unit sphere in $\mathbb{C}^n$), the cosphere bundle of a manifold.

2.2. The Weyl algebra. Let V be a $2n$-dimensional real vector space with symplectic form ω. The Weyl algebra $\mathcal{W}(V,\omega)$ consists of smooth complex-valued functions $a \in C^\infty(V,\mathbb{C})$ with properties:

- There is an integer $m \in \mathbb{Z}$ such that for every multi-index α there is $C_\alpha > 0$ with

$$|(\partial_v^\alpha a)(v)| \le C_\alpha(1 + \|v\|^2)^{\frac{1}{2}(m-|\alpha|)} \qquad \text{for all } v \in V. \tag{2.1}$$

  The integer m is the Weyl order of a.
- a has a 1-step polyhomogeneous asymptotic expansion

$$a \sim \sum_{j=-m}^{\infty} a_j \qquad a_j(sv) = s^{-j} a_j(v) \quad s > 0, v \neq 0. \tag{2.2}$$

The product of two symbols $a, b \in \mathcal{W}(V,\omega)$ is

$$(a\#b)(v) = \frac{1}{(2\pi)^{2n}} \iint e^{2i\omega(x,y)} a(v+x) b(v+y)\, dxdy,$$

where $dx = dy = |\omega^n|/n!$. The Weyl algebra is $\mathbb{Z}$-filtered by the Weyl order. We denote by $\mathcal{W}^m$ the space of elements of order m,

$$\cdots \subset \mathcal{W}^{-2} \subset \mathcal{W}^{-1} \subset \mathcal{W}^0 \subset \mathcal{W}^1 \subset \mathcal{W}^2 \subset \cdots$$

Schwartz class functions $\mathcal{S}(V,\omega)$ form a two-sided ideal in $\mathcal{W}(V,\omega)$. Denote the quotient algebra by $\mathcal{B}(V,\omega) = \mathcal{W}(V,\omega)/\mathcal{S}(V,\omega)$, with quotient map

$$\lambda : \mathcal{W}(V,\omega) \to \mathcal{B}(V,\omega).$$

Elements of $\mathcal{B}(V,\omega)$ can be identified with formal series $\sum_{l=-m}^{\infty} a_l$. The product $a \star b = c$ of formal series $a = \sum a_l$ and $b = \sum b_m$ in $\mathcal{B}$ is given by the formal series

$$a \star b = \sum c_p,$$

where c_p is homogeneous of degree $-p$, and given by the finite sum

$$c_p := \sum_{2k+l+m=p} B_k(a_l, b_m),$$

$$B_k(f,g) = \left(\frac{i}{2}\right)^k \sum_{|\alpha|+|\beta|=k} \frac{1}{\alpha!\beta!}(-1)^{|\beta|}(\partial_x^\alpha \partial_\xi^\beta f)(\partial_x^\beta \partial_\xi^\alpha g),$$

where x, ξ are symplectic (Darboux) coordinates on V. We denote by $\mathcal{B}_q(V,\omega)$ the subset of $\mathcal{B}(V,\omega)$ consisting of formal series $a = \sum_{l=-m}^{\infty} a_{2l}$ for which all the terms a_{2l} are homogeneous of even degree $-2l$. $\mathcal{B}_q(V,\omega)$ is a subalgebra of $\mathcal{B}(V,\omega)$.

DEFINITION 2.1. Let $\iota : \mathcal{B}_q(V,\omega) \to \mathcal{B}_q(V,\omega)$ be the linear map

$$a = \sum_{j=-m}^{\infty} a_{2j} \in \mathcal{B}_q(V,\omega) \qquad \iota(a) := \sum_{j=-m}^{\infty} (-1)^j a_{2j} \in \mathcal{B}_q(V,\omega),$$

where $a_{2j} = a_{2j}(v)$ is homogeneous of degree $-2j$.

The map ι is an involution,

$$\iota(a \star b) = \iota(b) \star \iota(a).$$

We let

$$\mathcal{W}_q(V,\omega) := \{w \in \mathcal{W}(V,\omega) \mid \lambda(w) \in \mathcal{B}_q(V,\omega)\},$$

$$\mathcal{W}_q^{2m}(V,\omega) := \{w \in \mathcal{W}^{2m}(V,\omega) \mid \lambda(w) \in \mathcal{B}_q(V,\omega)\}.$$

2.3. Action of the symplectic and unitary groups. As can be seen from the definition of the $\#$ product, a linear symplectic transformation $\phi: V \to V$ determines an automorphism of the algebra $\mathcal{W}(V,\omega)$. For $\phi \in \mathrm{Sp}(V)$, $a \in \mathcal{W}(V,\omega)$ let $\phi(a) := a \circ \phi^{-1}$. Then

$$\phi(a)\#\phi(b) = \phi(a\#b).$$

This action preserves the ideal $\mathcal{S}(V,\omega)$, and so descends to an action on $\mathcal{B}(V,\omega)$. It preserves $\mathcal{B}_q(V,\omega) \subset \mathcal{B}(V,\omega)$ and hence $\mathcal{W}_q(V,\omega) \subset \mathcal{W}(V,\omega)$.

Let $\mathfrak{g} \subset \mathcal{W}(V,\omega)$ be the subspace of homogeneous polynomials of degree 2 that are purely imaginary (i.e. with values in $i\mathbb{R}$). $\mathfrak{g}$ is a (real) Lie algebra endowed with the bracket

$$[X,Y] = X\#Y - Y\#X = i\{X,Y\}.$$

Let $V^* := \mathrm{Hom}(V,\mathbb{R}) \subset \mathcal{W}$ be the (real) dual space of V. If $X \in \mathfrak{g}$ and $f \in V^*$ then $[X,f] = \{iX,f\}$ is real-valued and homogeneous of degree 1, and so $[X,f] \in V^*$. We obtain a morphism of Lie algebras,

$$\mu^*: \mathfrak{g} \to \mathrm{End}\, V^* \qquad \mu^*(X) := [X,\cdot\,].$$

The map $V \ni v \mapsto \omega(v,\cdot) \in V^*$ establishes an isomorphism $V \to V^*$ which can be used to define a symplectic form ω^* on V^*. For $f,g \in V^*$ we have $\{f,g\} = \omega^*(f,g)\cdot 1$, where $1 \in \mathcal{W}$ is a constant function on V.

If $X \in \mathfrak{g}$ then $\mu^*(X) \in \mathfrak{sp}(V^*)$. The map $\mu^*: \mathfrak{g} \to \mathfrak{sp}(V^*)$ is a Lie algebra isomorphism. We will identify the groups of symplectic transformations $\mathrm{Sp}(V) \cong \mathrm{Sp}(V^*)$ as well as the corresponding Lie algebras $\mathfrak{sp}(V) \cong \mathfrak{sp}(V^*)$ via the isomorphism $V \to V^*$: $v \mapsto \omega(v,\cdot)$. We therefore obtain an isomorphism of Lie algebras

$$\mu: \mathfrak{g} \to \mathfrak{sp}(V). \tag{2.3}$$

The action of $Sp(V)$ on $\mathcal{W}(V,\omega)$ induces an action of the Lie algebra $\mathfrak{sp}(V)$ by derivations,

$$(\phi.w)(v) := -w(\phi(v)) \qquad v \in V,\ w \in \mathcal{W}(V,\omega),\ \phi \in \mathfrak{sp}(V).$$

This is an action by inner derivations:

$$(\phi.w)(v) = [\mu^{-1}(\phi), w] \qquad w \in \mathcal{W}(V,\omega),\ \phi \in \mathfrak{sp}(V).$$

Similarly $Sp(V)$ and $\mathfrak{sp}(V)$ act on $\mathcal{W}(V,\omega)^{op}$ by automorphisms and derivations respectively. The action of $\mathfrak{sp}(V)$ is again by inner derivations:

$$(\phi.w)(v) = [-\mu^{-1}(\phi), w]_{op} \qquad w \in \mathcal{W}(V,\omega)^{op},\ \phi \in \mathfrak{sp}(V),$$

where $[\cdot,\cdot]_{op}$ is the commutator in $\mathcal{W}(V,\omega)^{op}$.

2.4. Bundle of Weyl algebras. Fix Darboux coordinates (x,ξ), $x = (x_1,\dots,x_n)$, $\xi = (\xi_1,\dots,\xi_n)$ on V, thus identifying $V \cong \mathbb{R}^{2n}$. In this situation, we will omit the vector space from the notations, writing simply $\mathcal{W}$ for $\mathcal{W}(\mathbb{R}^{2n})$, etc.

Recall that on a compact contact manifold M with contact form α, $H = \operatorname{Ker}\alpha \subset TM$ is a symplectic bundle with symplectic form $\omega = -d\alpha$. As before, we denote by $J \in \operatorname{End}(H)$ is a compatible almost complex structure, and by P_H is the principal $U(n)$-bundle of frames of $(H^*)^{1,0}$.

Since $U(n) \subset \operatorname{Sp}(2n)$ acts on $\mathcal{W}_q \subset \mathcal{W}$ and $\mathcal{W}_q^{op} \subset \mathcal{W}^{op}$ by automorphisms, we can form associated bundles:

$$W_H^+ := P_H \times_{U(n)} \mathcal{W}_q \quad W_H^- := P_H \times_{U(n)} \mathcal{W}_q^{op}.$$

The elements of the algebra of smooth sections of W_H^+ can be described as smooth $U(n)$-invariant functions from P_H to $\mathcal{W}_q$,

$$\mathcal{W}_H := C^\infty(M; W_H^+) = C^\infty(P_H; \mathcal{W}_q)^{U(n)}.$$

where the action of $\phi \in U(n)$ on $s\colon P_H \to \mathcal{W}_q$ is given by $(\phi \cdot s)(p) := \phi(s(p\phi))$. In other words, s is invariant if $s(p\phi) = \phi^{-1}(s(p))$. Similarly, the algebra can be described as

$$\mathcal{W}_H^{op} = C^\infty(M; W_H^-) = C^\infty(P_H; \mathcal{W}_q)^{U(n)}.$$

Finally, we can construct the bundle

$$B := P_H \times_{U(n)} \mathcal{B}_q$$

and the algebra

$$\mathcal{B}_H := C^\infty(M; B) = C^\infty(P_H; \mathcal{B}_q)^{U(n)}.$$

The map λ then defines a homomorphism $\lambda\colon \mathcal{B}_H \to \mathcal{W}_H$.

2.5. Heisenberg principal symbols. Let Ψ_H^m be the set of Heisenberg pseudodifferential operators of order $m \in \mathbb{Z}$. The algebra of order zero principal symbols,

$$\mathscr{S}_H^0 := \Psi_H^0/\Psi_H^{-1}$$

is a subalgebra of $\mathcal{W}_H \oplus \mathcal{W}_H^{op}$,

$$\mathscr{S}_H^0 = \{(w_+, w_-) \in \mathcal{W}_H^0 \oplus (\mathcal{W}_H^0)^{op} \mid \lambda(w_+) = \iota \circ \lambda(w_-)\},$$

where $\mathcal{W}_H^0$, $(\mathcal{W}_H^0)^{op}$ denote the subalgebras of $\mathcal{W}_H$ and $\mathcal{W}_H^{op}$ respectively consisting of sections of order 0 in the Weyl algebra.

2.6. Principal symbols in the extended Heisenberg calculus. Let $\Psi_{eH}^{m,k}$ be the set of pseudodifferential operators in the extended Heisenberg calculus that have classical order $m \in \mathbb{Z}$, and Heisenberg order $k \in \mathbb{Z}$ (see [**3**]). The algebra of order $(0,0)$ principal symbols,

$$\mathscr{S}_{eH}^{0,0} := \Psi_{eH}^{0,0}/\Psi_{eH}^{-1,-1}$$

is

$$\mathscr{S}_{eH}^{0,0} = \{(w_+, w_-, f)\},$$

where

- $w_+ \in \mathcal{W}_H^0$, $w_- \in (\mathcal{W}_H^0)^{op}$
- $f : [0,1] \to C^\infty(S(H^*))$ is a smooth function; here $S(H^*)$ is the sphere bundle of H^*.

- $f(0) \in C^\infty(S(H^*))$ is the leading term in the formal series $\lambda(w_+)$, while $f(1) \in C^\infty(S(H^*))$ is the leading term in the formal series $\lambda(w_-)$.

The product is

$$(w_+, w_-, f)(u_+, u_-, g) = (w_+ \# u_+, u_- \# w_-, fg),$$

where fg is pointwise multiplication of the functions f, g.

3. Connection and curvature for a bundle of Weyl algebras

Let $\Omega^\bullet(P_H; \mathcal{W})_{basic}$ be the space of basic $\mathcal{W}$-valued forms. Recall that an $\mathcal{W}$-valued differential form η on P_H is called basic if:

- η is horizontal, i.e. $\iota_X \eta = 0$ for every vertical vector field X on P_H;
- η is $U(n)$ invariant, i.e. $\phi^* \eta = \phi^{-1}(\eta)$ for every $\phi \in U(n)$.

k-forms with values in the bundle $\mathcal{W}_H$ are, by definition, basic $\mathcal{W}$-valued k-forms. We denote

$$\Omega^k(\mathcal{W}_H) = \Omega^k(M; W_H^+) = \Omega^k(P_H; \mathcal{W}_q)_{basic}.$$

A unitary connection ∇ on $H \cong H^*$ can be represented by a connection 1-form

$$\beta \in \Omega^1(P_H; \mathfrak{u}(n)).$$

The curvature of ∇ is

$$\theta := d\beta + \frac{1}{2}[\beta, \beta] \in \Omega^2(P_H, \mathfrak{u}(n))_{basic}.$$

The 1-form β defines a covariant derivative

$$\nabla^+ : \mathcal{W}_H \to \Omega^1(\mathcal{W}_H)$$

by

$$\nabla^+(a) := da + \beta \cdot a \in \Omega^1(\mathcal{W}_H), \qquad a \in \mathcal{W}_H = C^\infty(P_H; \mathcal{W}_q)^{U(n)}.$$

This covariant derivative extends to a derivation

$$\nabla^+ : \Omega^k(\mathcal{W}_H) \to \Omega^{k+1}(\mathcal{W}_H)$$

defined by the same formula:

$$\nabla^+(\eta) := d\eta + \beta \cdot \eta.$$

The curvature of ∇^+ is defined by

$$\boldsymbol{\theta}^+ := \mu^{-1}(\theta) \in \Omega^2(\mathcal{W}_H) \tag{3.1}$$

Then

$$(\nabla^+)^2(\eta) = [\boldsymbol{\theta}^+, \eta] \qquad \eta \in \Omega^\bullet(\mathcal{W}_H).$$

LEMMA 3.1. *With the definitions above we have*

$$\nabla^+(\boldsymbol{\theta}^+) = 0.$$

In an entirely analogous manner, connection form β yields a covariant derivative

$$\nabla^- : \Omega^k(\mathcal{W}_H^{op}) \to \Omega^{k+1}(\mathcal{W}_H^{op}),$$

where

$$\Omega^k(\mathcal{W}_H^{op}) = \Omega^k(M; W_H^-) = \Omega^k(P_H; \mathcal{W}_q^{op})_{basic}.$$

It satisfies

$$(\nabla^-)^2(\eta) = [\boldsymbol{\theta}^-, \eta] \qquad \eta \in \Omega^\bullet(\mathcal{W}_H^{op}),$$

where $\boldsymbol{\theta}^- = -\boldsymbol{\theta}^+ \in \Omega^2(\mathcal{W}_H^{op})$

Finally, the same construction defines a covariant derivative

$$\nabla^{+}:\Omega^{k}(\mathcal{B}_H)\to\Omega^{k+1}(\mathcal{B}_H),$$

where

$$\Omega^{k}(\mathcal{B}_H)=\Omega^{k}(M;B)=\Omega^{k}(P_H;\mathcal{B}_q)_{basic}$$

satisfying

$$(\nabla^{+})^{2}(\eta)=[\boldsymbol{\theta}^{+},\eta]\qquad \eta\in\Omega^{\bullet}(\mathcal{B}_H)$$

and

$$\lambda(\nabla^{+}(\eta)=\nabla^{+}(\lambda(\eta))\qquad \eta\in\Omega^{\bullet}(\mathcal{W}_H).$$

4. A regularized trace for the bundle of Weyl algebras

Weyl quantization associates with $a\in\mathcal{W}=\mathcal{W}(\mathbb{R}^{2n})$ a pseduidifferential operator $A=\mathrm{Op}^{w}(a)$ defined (formally) on functions $u\in\mathcal{S}(\mathbb{R}^{n})$ as

$$(4.1)\qquad (Au)(x)=\frac{1}{(2\pi)^{n}}\iint e^{i(x-y)\cdot\xi}a\Big(\frac{x+y}{2},\xi\Big)u(y)\,dy\,d\xi.$$

We denote by $\mathcal{H}$ the harmonic oscillator

$$\mathcal{H}=\sum_{j=1}^{n}\Big(-\frac{\partial^{2}}{\partial x_j^{2}}+x_j^{2}\Big)=\mathrm{Op}^{w}\Big(\sum_{j=1}^{n}(\xi_j^{2}+x_j^{2})\Big).$$

$\mathcal{H}$ is a strictly positive unbounded selfadjoint operator on $L^{2}(\mathbb{R}^{n})$.

For $a\in\mathcal{W}_q^{2m}$ of even order $2m$, $\mathrm{Op}^{w}(a)\mathcal{H}^{-z}$ is of trace class if $\operatorname{Re}z>n+m$. Define the zeta-function

$$\zeta_a(z):=\mathrm{Tr}(\mathrm{Op}^{w}(a)\mathcal{H}^{-z}),\qquad \operatorname{Re}z>n+m.$$

The zeta function is holomorphic for $\operatorname{Re}z>n+m$, and it extends to a meromorphic function with at most simple poles at $m+n,m+n-1,m+n-2,\ldots$. The residue at $z=0$ of the zeta-function gives a residue trace on $\mathcal{W}_q$,

$$\mathrm{Res}:\mathcal{W}_q\to\mathbb{C}\qquad \mathrm{Res}(a)=\lim_{z\to0}z\zeta_a(z).$$

Res is a trace on $\mathcal{W}_q$ that vanishes on the Schwartz class ideal $\mathcal{S}$. It follows that residue induces a trace on the quotient $\mathcal{B}_q=\mathcal{W}_q/\mathcal{S}$ which we also denote Res. An explicit formula for $\mathrm{Res}\,a$ in terms of the asymptotic expansion $a=\sum_j a_{2j}\in\mathcal{B}_q$ is

$$\mathrm{Res}\,a=-\frac{1}{2(2\pi)^{n}}\int_{S^{2n-1}}a_{2n}(\theta)d\theta,$$

where S^{2n-1} is the unit sphere in V.

We denote by $\overline{\mathrm{Tr}}(a)$ the constant term at $z=0$ of the zeta-function,

$$\overline{\mathrm{Tr}}(a)=\lim_{z\to0}\Big(\zeta_a(z)-\frac{1}{z}\mathrm{Res}(a)\Big).$$

If $a\in\mathcal{S}$ then $\mathrm{Tr}(\mathrm{Op}^{w}(a)\mathcal{H}^{-z})$ is an entire function, and we see that

$$\overline{\mathrm{Tr}}(a)=\mathrm{Tr}\,\mathrm{Op}^{w}(a)=\frac{1}{(2\pi)^{n}}\int a(x,\xi)dx\,d\xi\qquad \forall a\in\mathcal{S}(\mathbb{R}^{2n}).$$

However the functional $\overline{\mathrm{Tr}}$ is not a trace on $\mathcal{W}_q$. More exlicitely, we have the following result. First note that even though $\log(x^{2}+\xi^{2})\notin\mathcal{B}_q$, the formula

$$(4.2)\qquad \delta(a)=a\star\log(x^{2}+\xi^{2})-\log(x^{2}+\xi^{2})\star a$$

defines a derivation δ of $\mathcal{B}_q$. Direct calculation shows the following

LEMMA 4.1. $\overline{\mathrm{Tr}}([a,b]) = \operatorname{Res}\lambda(a)\delta(\lambda(b))$.

The regularized trace $\overline{\mathrm{Tr}}$ is *not* invariant under the action of the symplectic group $\mathrm{Sp}(2n) := \mathrm{Sp}(\mathbb{R}^{2n})$. However it is invariant under the action of unitary subgroup $U(n) \subset \mathrm{Sp}(2n)$:

$$\overline{\mathrm{Tr}}(\phi(a)) = \overline{\mathrm{Tr}}(a) \qquad a \in \mathcal{W},\ \phi \in U(n).$$

Clearly the same invariance holds for $\overline{\mathrm{Tr}}$ on $\mathcal{W}^{op}$. Therefore we have regularized traces defined on sections in the bundles of Weyl algebras,

$$\overline{\mathrm{Tr}} : \mathcal{W}_H \to C^\infty(M) \qquad \overline{\mathrm{Tr}} : \mathcal{W}_H^{op} \to C^\infty(M).$$

5. The Chern character in cyclic homology

In this section we give a very brief overview of the periodic cyclic homological complex, mostly to fix the notation. We also recall the Chern character map into the cyclic homology. The standard reference for this material is [**8**].

For a complex unital algebra A set $C_l(A) := A \otimes (A/(\mathbb{C}\cdot 1))^{\otimes l}$, $l \geq 0$. One defines differentials $b\colon C_l(A) \to C_{l-1}(A)$ and $B\colon C_l(A) \to C_{l+1}(A)$ by

$$b(a_0 \otimes a_1 \otimes \dots a_l) := \sum_{i=0}^{l-1} (-1)^i a_0 \otimes \dots a_i a_{i+1} \otimes \dots a_l + (-1)^l a_l a_0 \otimes a_1 \otimes \dots a_{l-1},$$

$$B(a_0 \otimes a_1 \otimes \dots a_l) := \sum_{i=0}^{l} (-1)^{li} 1 \otimes a_i \otimes a_{i+1} \otimes \dots a_{i-1} \ (\text{with } a_{-1} := a_l).$$

One verifies directly that b, B are well defined and satisfy $b^2 = 0$, $B^2 = 0$, $Bb + bB = 0$. Let u be a formal variable of degree -2. The space of periodic cyclic chains of degree $i \in \mathbb{Z}$ is defined by

$$CC_i^{per}(A) = \left(C_\bullet(A)[u^{-1}, u]]\right)_i = \prod_{-2n+l=i} u^n C_l(A).$$

Note that $CC_i^{per}(A) = uCC_{i+2}^{per}(A)$. We will write a chain in $CC_i^{per}(A)$ as $\alpha = \sum_{i+2m\geq 0} u^m \alpha_{i+2m}$ where $\alpha_l \in C_l(A)$. The boundary is given by $b + uB$ where b and B are the Hochschild and Connes boundaries of the cyclic complex. The homology of this complex is periodic cyclic homology, denoted $HC_\bullet^{per}(A)$.

If $r \in M_n(A)$ is invertible the following formula defines a cycle in the periodic cyclic complex:

$$\mathrm{Ch}(r) := -\frac{1}{2\pi i} \sum_{l=0}^{\infty} (-1)^l\, l!\, u^l \operatorname{tr}(r^{-1} \otimes r)^{\otimes(l+1)} \in CC_1^{per}(A), \tag{5.1}$$

where $\operatorname{tr}\colon (A \otimes M_n(\mathbb{C}))^{\otimes k} \to A^{\otimes k}$ is the map given by

$$\operatorname{tr}(a_0 \otimes m_0) \otimes (a_1 \otimes m_1) \otimes \dots (a_k \otimes m_k) = (\operatorname{tr} m_0 m_1 \dots m_k) a_0 \otimes a_1 \otimes \dots a_k.$$

In the case of interest to us, A will be a Fréchet algebra. In that case we will use a projective tensor product to define $C_l(A) = A \otimes (A/(\mathbb{C}\cdot 1))^{\otimes l}$. One can define the topological K-theory of a Fréchet algebra as $K_1(A) := \pi_0(GL(A))$. With these definitions, (5.1) defines the (odd) Chern character homomorphism

$$\mathrm{Ch}\colon K_1(A) \to HC_1^{per}(A) \tag{5.2}$$

from topological K-theory to periodic cyclic homology. If $t \mapsto r_t$, $t \in [0,1]$ is a (piecewise) smooth map from $[0,1]$ to $GL_n(A)$ then

$$\mathrm{Ch}(r_1) - \mathrm{Ch}(r_0) = (b + uB)\,\mathrm{Tch}(r_t),$$

where

$$\operatorname{Tch}(r_t) := -\frac{1}{2\pi i}\int_0^1 \xi(t)dt,$$

$$\xi(t) = \operatorname{tr}(r_t^{-1}\frac{dr_t}{dt}) + \sum_{l=0}^{\infty}(-1)^{l+1}l!u^{l+1}\sum_{j=0}^{l}\operatorname{tr}\left((r_t^{-1}\otimes r_t)^{\otimes j+1}\otimes r_t^{-1}\frac{dr_t}{dt}\otimes(r_t^{-1}\otimes r_t)^{\otimes l-j}\right),$$

which shows in particular that Chern character is well-defined.

6. An extension of the algebra of symbols

In this section we define an extension of the algebra of extended Heisenberg symbols, algebra $\mathscr{E}_H$, which is a convenient source of the characteristic map in the next section.

Let $\mathcal{B}_H$ be the algebra of smooth sections in the bundle over M whose fiber at $p \in M$ is $\mathcal{B}_q(H_p^*, \omega_p^*)$. We let $\mathscr{E}_H$ be the algebra

$$\mathscr{E}_H := \{(w_+, w_-, r)\},$$

where

- $w_+ \in \mathcal{W}_H$, $w_- \in \mathcal{W}_H^{op}$.
- $r : [0,1] \to \mathcal{B}_H$ is a smooth function.
- $r(1) = \lambda(w_+)$, $r(0) = \iota \circ \lambda(w_-)$.

There is the obvious algebra homomorphism,

$$\mathscr{E}_H \to \mathscr{S}_{eH}^{0,0} \quad (w_+, w_-, r) \mapsto (w_+, w_-, f),$$

where $f(t)$ is the leading term of the series $r(t) \in \mathcal{B}_H$. This quotient map has a canonical section,

$$s\colon \mathscr{S}_{eH}^{0,0} \to \mathscr{E}_H \quad (w_+, w_-, f) \mapsto (w_+, w_-, r(t))$$

where $r(t) = r_0 + tr_1$ is linear in t, $r_0 = \lambda(w_-)$, $r_1 = \lambda(w_+) - \iota \circ \lambda(w_-)$. While this section s is not an algebra homomorphism, it maps invertible elements to invertible elements and gives a well-defined map of homotopy classes of invertible elements. Thus, this section s determines a map in K-theory,

$$K_1(\mathscr{S}_{eH}^{0,0}) \to K_1(\mathscr{E}_H).$$

A pseudodifferential operator $\mathscr{L} \in \Psi_{eH}^{0,0}$ with invertible principal symbol $\sigma_{eH}(\mathscr{L}) \in \mathscr{S}_{eH}^{0,0}$ determines a class

$$[\sigma_{eH}(\mathscr{L})] \in K_1(\mathscr{S}_{eH}^{0,0})$$

and, via s, also a class

$$[s(\sigma_{eH}(\mathscr{L}))] \in K_1(\mathscr{E}_H).$$

We shall express the index of $\mathscr{L}$ as a function of this symbol class in $K_1(\mathscr{E}_H)$.

7. The characteristic map

In this section, we associate with the algebra $\mathscr{E}_H$ the complex $\mathcal{C}_\bullet(\mathscr{E}_H)$ and define a characteristic map from this complex to de Rham complex.

As a vector space,

$$\mathcal{C}_\bullet(\mathscr{E}_H) := CC_\bullet^{per}(\mathcal{W}_H) \oplus CC_\bullet^{per}(\mathcal{W}_H^{op}) \oplus CC_{\bullet+1}^{per}(\mathcal{B}_H).$$

The differential is given by

(7.1) $$\partial(\zeta^+, \zeta^-, \gamma) = ((b+uB)\zeta^+, (b+uB)\zeta^-, \lambda_*(\zeta^+) - (\iota\circ\lambda)_*(\zeta_-)) - (b+uB)\gamma),$$

where $(\zeta^+,\zeta^-,\gamma) \in \mathcal{C}_\bullet(\mathscr{E}_H)$. The homology of the complex $\mathcal{C}_\bullet(\mathscr{E}_H)$ is denoted by $\mathcal{H}_\bullet(\mathscr{E}_H)$, and the homology of the cycle $(\zeta^+,\zeta^-,\gamma) \in \mathcal{C}_\bullet(\mathscr{E}_H)$ is denoted by $[\zeta^+,\zeta^-,\gamma] \in \mathcal{H}_\bullet(\mathscr{E}_H)$. We have a group homomorphism

$$\mathrm{Ch}\colon K_1(\mathscr{E}_H) \to \mathcal{H}_1(\mathscr{E}_H)$$

given by

$$\mathrm{Ch}\colon [(w_+,w_-,r_t)] \mapsto [\mathrm{Ch}(w_+),\mathrm{Ch}(w_-),\mathrm{Tch}(r_t)]. \tag{7.2}$$

We will now define a morphism of complexes

$$\mathcal{C}_\bullet(\mathscr{E}_H) \to \Omega^\bullet(M)[u^{-1},u].$$

We first construct maps

$$\Phi^+\colon CC_\bullet(\mathcal{W}_H) \to \Omega^\bullet(M), \quad \Phi^-\colon CC_\bullet(\mathcal{W}_H^{op}) \to \Omega^\bullet(M)[u]$$

and

$$\phi\colon CC_\bullet(\mathcal{B}_H) \to \Omega^{\bullet-1}(M)[u]$$

described as follows. Choose a unitary connection on H. As in the Section 3 we can define connections ∇^+,∇^-, with curvatures $\boldsymbol{\theta}^+,\boldsymbol{\theta}^- = -\boldsymbol{\theta}^+$. The definition of the maps $\Phi^\pm$ is inspired by [**2, 5, 6**] and is given by

$$\Phi^\pm(a_0\otimes a_1\otimes a_k) =$$
$$\sum_{m=0}^{k}(-1)^{m(k-m)}\sum_{i_0,i_1,\dots i_{k+1}}\frac{(-1)^{i_0+\dots+i_{k+1}}}{(i_0+i_1+\dots+i_{k+1}+k+1)!}\overline{\mathrm{Tr}}(\boldsymbol{\theta}^\pm)^{i_0}\nabla^\pm(a_m)(u\boldsymbol{\theta}^\pm)^{i_1}\nabla^\pm(a_{m+1})$$
$$\dots(u\boldsymbol{\theta}^\pm)^{i_{k+1-m}}a_0(u\boldsymbol{\theta}^\pm)^{i_{k+2-m}}\dots\nabla^\pm(a_{m-1})(u\boldsymbol{\theta}^\pm)^{i_{k+1}},$$

or equivalently

$$\Phi^\pm(a_0\otimes a_1\otimes a_k) =$$
$$\sum_{m=0}^{k}(-1)^{m(k-m)}\int_{\Delta^{k+1}}\overline{\mathrm{Tr}}\, e^{-t_0u\boldsymbol{\theta}^\pm}\nabla^\pm(a_m)e^{-t_1u\boldsymbol{\theta}^\pm}\nabla^\pm(a_{m+1})\dots$$
$$e^{-t_{k+1-m}u\boldsymbol{\theta}^\pm}a_0e^{-t_{k+2-m}u\boldsymbol{\theta}^\pm}\dots\nabla^\pm(a_{m-1})e^{-t_{k+1}u\boldsymbol{\theta}^\pm})dt_1\dots dt_{k+1}.$$

The map ϕ is given

$$\phi(b_0\otimes b_1\otimes b_k) = \sum_{m=1}^{k}\sum_{i_0,i_1,\dots i_k}\frac{(-1)^{i_0+\dots+i_k+m-1}}{(i_0+i_1+\dots i_k+k)!}$$
$$\mathrm{Res}\, b_0(u\boldsymbol{\theta}^+)^{i_0}\nabla^+(b_1)(u\boldsymbol{\theta}^+)^{i_1}\dots\delta(b_m)\dots\nabla^+(b_k)(u\boldsymbol{\theta}^+)^{i_k},$$

where δ is the derivation defined in (4.2).

PROPOSITION 7.1. *The maps $\Phi^\pm$, ϕ defined above satisfy the following relations:*

$$\Phi^+\circ(b+uB)-(ud)\circ\Phi^+ = \phi\circ\lambda_*,$$
$$\Phi^-\circ(b+uB)-(ud)\circ\Phi^- = (-1)^n\phi\circ(\iota\circ\lambda)_*,$$
$$\phi\circ(b+uB)-(ud)\circ\phi = 0.$$

PROOF. These equalities follow from straightforward calculations.
Since the connection is unitary and $\overline{\mathrm{Tr}}$ is unitarily invariant, we have

$$\overline{\mathrm{Tr}}(\nabla(\cdot)) = d\,\overline{\mathrm{Tr}}(\cdot) \qquad \overline{\mathrm{Tr}}[\boldsymbol{\theta},\cdot] = 0.$$

This is used in the first calculation, as well as the equality $\overline{\mathrm{Tr}}[A,B] = \mathrm{Res}\,A\delta(B)$.

The second identity follows from the first, using

$$\nabla^-(\iota(b)) = \iota(\nabla^+(b)), \quad \iota(\boldsymbol{\theta}^-) = \boldsymbol{\theta}^+, \quad \delta(\iota(b)) = -\iota(\delta(b)), \quad \iota(a) \star_- \iota(b) = \iota(a \star b).$$

The third equality also follows from the first, since λ is surjective.

□

THEOREM 7.2. *The map* $\Phi: \mathcal{C}_\bullet(\mathscr{E}_H) \to (\Omega^\bullet(M)[u], ud)$ *given by*

$$\Phi(\xi^+, \xi^-, \gamma) := \Phi^+(\xi) - (-1)^n \Phi(\xi^-) - \phi(\gamma)$$

is a morphism of complexes.

PROOF. This is an immediate consequence of the identities in the preceding proposition, as well as the definition of the boundary (7.1) in the complex $\mathcal{C}_\bullet(\mathscr{E}_H)$ □

DEFINITION 7.3. We define the characteristic map

$$\chi: K_1(\mathcal{E}_H) \to H^{odd}(M)$$

as the composition

$$K_1(\mathcal{E}_H) \xrightarrow{\text{Ch}} \mathcal{H}_1(\mathcal{E}_H) \xrightarrow{\Phi} H^{odd}(M)[u^{-1}, u] \xrightarrow{R} H^{odd}(M).$$

Here the first map is given in equation (7.2), the second is gven in Theorem 7.2, and $R(\sum u^q \alpha_q) = \sum (2\pi i)^{-q} \alpha_q$, $\alpha_q \in H^\bullet(M)$.

8. The index formula

Let $\mathscr{L} \in \Psi^{0,0}_{eH}$ be an operator in the extended calculus with invertible principal symbol

$$\sigma_{eH}(\mathscr{L}) \in \mathscr{S}^{0,0}_{eH} := \Psi^{0,0}_{eH} / \Psi^{-1,-1}_{eH}.$$

THEOREM 8.1. *Let M be a closed smooth manifold of dimension $2n+1$ with contact form α. We orient M by the volume form $\alpha(d\alpha)^n$. Let*

$$\mathscr{L}: C^\infty(M, \mathbb{C}^r) \to C^\infty(M, \mathbb{C}^r)$$

be a pseudodifferential operator of order zero in the extended Heisenberg calculus that acts on sections in a trivial bundle $M \times \mathbb{C}^r$. Assume that $\sigma_{eH}(\mathscr{L}) \in M_r(\mathscr{S}^{0,0}_{eH})$ is invertible. Then the index of $\mathscr{L}$ is

$$\operatorname{Index} \mathscr{L} = \int_M \chi(s(\sigma_{eH}(\mathscr{L}))) \wedge \hat{A}(M).$$

Here $s : \mathscr{S}^{0,0}_{eH} \to \mathscr{E}_H$ is the canonical section, and $\chi : K_1(\mathscr{E}_H) \to H^{odd}(M)$ is the character homomorphism of Definition 7.3.

PROOF. Both $\operatorname{Index} \mathscr{L}$ and $\int_M \chi(s(\sigma_{eH}(\mathscr{L}))) \wedge \hat{A}(M)$ depend only on the class $[\sigma_{eH}(\mathscr{L})] \in K_1(\mathscr{S}^{0,0}_{eH})$. Thus we need to verify that the two resulting group homomorphisms $K_1(\mathscr{S}^{0,0}_{eH}) \to \mathbb{C}$ coincide.

The embedding $\mathscr{S}^0_H \to \mathscr{S}^{0,0}_{eH}$ induces an isomorphism in K-theory. This can be seen by considering the diagram in K-theory that corresponds to the following commutative diagram,

$$\begin{array}{ccccccccc} 0 & \longrightarrow & \mathcal{S}(H^*,\omega)\oplus\mathcal{S}(H^*,-\omega) & \longrightarrow & \mathscr{S}^0_H & \longrightarrow & C^\infty(S(H^*)) & \longrightarrow & 0 \\ & & \Big\downarrow{\scriptstyle =} & & \Big\downarrow{\scriptstyle \subset} & & \Big\downarrow & & \\ 0 & \longrightarrow & \mathcal{S}(H^*,\omega)\oplus\mathcal{S}(H^*,-\omega) & \longrightarrow & \mathscr{S}^{0,0}_{eH} & \longrightarrow & C^\infty(S(H^*)\times[0,1]) & \longrightarrow & 0 \end{array}$$

Here $\mathcal{S}(H^*,\omega)$ is the algebra of smooth sections in the bundle whose fibers are the Schwartz class ideals $\mathcal{S}(H^*_p,\omega^*_p)$, and $S(H^*)$ is the unit sphere bundle of H^*. The map $C^\infty(S(H^*))\to C^\infty(S(H^*)\times[0,1])$ is determined by the homotopy equivalence $S(H^*)\times[0,1]\to S(H^*)$. The rows in the diagram are short exact sequences.

We see that it suffices to verify that the homomorphism

$$K_1(\mathscr{S}^{0,0}_{eH})\to\mathbb{C}\qquad \sigma\mapsto\int_M \chi(s(\sigma))\wedge\hat{A}(M)$$

computes the index of operators in the Heisenberg calculus with invertible principal symbol in $\mathscr{S}^0_H\subset\mathscr{S}^{0,0}_{eH}$. But in this case our formula reduces to Theorem 9.1 of [**7**]. □

References

[1] R. Beals and P. Greiner, *Calculus on Heisenberg manifolds*, Annals of Mathematics Studies, vol. 119, Princeton University Press, Princeton, NJ, 1988, DOI 10.1515/9781400882397. MR953082

[2] A. Connes, *Noncommutative geometry*, Academic Press, Inc., San Diego, CA, 1994. MR1303779

[3] C. Epstein and R. Melrose, *The Heisenberg algebra, index theory and homology*. Unpublished manuscript, available at `https://math.mit.edu/~rbm/book.html`.

[4] C. L. Epstein, *Lectures on indices and relative indices on contact and CR-manifolds*, Woods Hole mathematics, Ser. Knots Everything, vol. 34, World Sci. Publ., Hackensack, NJ, 2004, pp. 27–93, DOI 10.1142/9789812701398_0002. MR2123367

[5] A. Gorokhovsky, *Characters of cycles, equivariant characteristic classes and Fredholm modules*, Comm. Math. Phys. **208** (1999), no. 1, 1–23, DOI 10.1007/s002200050745. MR1729875

[6] A. Gorokhovsky, *Explicit formulae for characteristic classes in noncummutative geometry*, ProQuest LLC, Ann Arbor, MI, 1999. Thesis (Ph.D.)–The Ohio State University. MR2699561

[7] A. Gorokhovsky and E. van Erp, *The Heisenberg calculus, index theory and cyclic cohomology*, Adv. Math. **399** (2022), Paper No. 108229, 62, DOI 10.1016/j.aim.2022.108229. MR4383012

[8] J.-L. Loday, *Cyclic homology*, Grundlehren der mathematischen Wissenschaften [Fundamental Principles of Mathematical Sciences], vol. 301, Springer-Verlag, Berlin, 1992. Appendix E by María O. Ronco, DOI 10.1007/978-3-662-21739-9. MR1217970

[9] R. Melrose, *Homology and the Heisenberg algebra*, Séminaire sur les Équations aux Dérivées Partielles, 1996–1997, École Polytech., Palaiseau, 1997, pp. Exp. No. XII, 11. Joint work with C. Epstein and G. Mendoza. MR1482818

[10] M. E. Taylor, *Noncommutative microlocal analysis. I*, Mem. Amer. Math. Soc. **52** (1984), no. 313, iv+182, DOI 10.1090/memo/0313. MR764508

University of Colorado, Boulder, Campus Box 395, Boulder, Colorado, 80309
Email address: `gorokhov@colorado.edu`

Dartmouth College, 6188, Kemeny Hall, Hanover, New Hampshire, 03755
Email address: `jhamvanerp@gmail.com`

Proceedings of Symposia in Pure Mathematics
Volume **105**, 2023
https://doi.org/10.1090/pspum/105/01902

Topological cyclic homology and the Fargues–Fontaine curve

Lars Hesselholt

Introduction

This paper is an elaboration of my lecture at the conference. The purpose is to explain how the Fargues–Fontaine curve and its decomposition into a punctured curve and the formal neighborhood of the puncture naturally appear from various forms of topological cyclic homology and maps between them. I make no claim of originality. My purpose here is to highlight some of the spectacular material contained in the papers of Nikolaus–Scholze [**16**], Bhatt–Morrow–Scholze [**3**], and Antieau–Mathew–Morrow–Nikolaus [**1**] on topological cyclic homology and in the book by Fargues–Fontaine [**7**] on their revolutionary curve.

1. The Fargues–Fontaine curve

We give a brief introduction to the Fargues–Fontaine curve and refer to their book [**7**] for details. The lecture notes by Lurie [**12**] are also helpful.

We define a completely valued field to be a pair $(C, \mathcal{O}_C)$ of a field C and a proper subring $\mathcal{O}_C \subset C$ that is a complete valuation ring of rank 1. The field C is the quotient field of $\mathcal{O}_C$ and the requirement that the inclusion $\mathcal{O}_C \subset C$ be proper is equivalent to the requirement that the value group of $\mathcal{O}_C$ be non-trivial. An isomorphism of completely valued fields $\psi\colon (C, \mathcal{O}_C) \to (C', \mathcal{O}_{C'})$ is an isomorphism of fields $\psi\colon C \to C'$ that restricts to an isomorphism of rings $\psi\colon \mathcal{O}_C \to \mathcal{O}_{C'}$.

We will be interested in the case, where C is algebraically closed and $\mathcal{O}_C$ is of residue characteristic $p > 0$. In this case, "tilting" is a correspondence that to an algebraically closed completely valued field $(C, \mathcal{O}_C)$ of residue characteristic $p > 0$ assigns the algebraically closed completely valued field $(C^\flat, \mathcal{O}_{C^\flat})$ of characteristic $p > 0$, where $\mathcal{O}_{C^\flat}$ is the limit of the diagram of $\mathbb{F}_p$-algebras

$$\cdots \xrightarrow{\varphi} \mathcal{O}_C/p \xrightarrow{\varphi} \mathcal{O}_C/p \xrightarrow{\varphi} \mathcal{O}_C/p$$

with φ the Frobenius. The non-trivial fact that $C^\flat$ is algebraically closed is proved in [**7**, Proposition 2.1.11]. The tilting correspondence is many-to-one. More precisely, given an algebraically closed completely valued field $(F, \mathcal{O}_F)$ of characteristic $p > 0$,

2020 *Mathematics Subject Classification.* Primary 11S70, 19D55; Secondary 14F30.

The author was partially supported by the Danish National Research Foundation through the Copenhagen Center for Geometry and Topology (DNRF151) and by JSPS Grant-in-Aid for Scientific Research number 21K03161.

an untilt of $(F, \mathcal{O}_F)$ is a pair $((C, \mathcal{O}_C), \iota)$ of an algebraically closed completely valued field $(C, \mathcal{O}_C)$ and an isomorphism of completely valued fields

$$(F, \mathcal{O}_F) \xrightarrow{\iota} (C^\flat, \mathcal{O}_{C^\flat}).$$

The isomorphism ι gives rise to an isomorphism of value groups

$$F^\times/\mathcal{O}_F^\times \longrightarrow C^\times/\mathcal{O}_C^\times$$

and the "perfectoid" nature of the respective valuation rings implies that for every pair of elements $\varpi_F \in \mathcal{O}_F$ and $\varpi_C \in \mathcal{O}_C$ of equal and sufficiently small but nonzero valuation, the isomorphism ι gives rise to an isomorphism of rings

$$\mathcal{O}_F/\varpi_F \longrightarrow \mathcal{O}_C/\varpi_C.$$

We note that the common ring is not a field, but rather is a highly non-noetherian nilpotent thickening of its (algebraically closed) residue field.

An isomorphism of untilts $\psi\colon ((C, \mathcal{O}_C), \iota) \to ((C', \mathcal{O}_{C'}, \iota')$ is an isomorphism of completely valued fields $\psi\colon (C, \mathcal{O}_C) \to (C', \mathcal{O}_{C'})$ with the property that the diagram

$$\begin{array}{ccc} & \xrightarrow{\iota} & (C^\flat, \mathcal{O}_{C^\flat}) \\ (F, \mathcal{O}_F) & & \downarrow \psi^\flat \\ & \xrightarrow{\iota'} & (C'^\flat, \mathcal{O}_{C'^\flat}) \end{array}$$

commutes. Up to isomorphism, there is one untilt of $(F, \mathcal{O}_F)$ of characteristic p. But there are many non-isomorphic untilts of $(F, \mathcal{O}_F)$ of characteristic 0. We note that if $((C, \mathcal{O}_C), \iota)$ is an untilt of $(F, \mathcal{O}_F)$, then so is $((C, \mathcal{O}_C), \iota \circ \varphi)$, and that if C is of characteristic 0, then these two untilts are non-isomorphic. So the group $\varphi^{\mathbb{Z}}$ acts freely on the set of isomorphism classes of characteristic 0 untilts of F. Fargues and Fontaine show in [**7**, Théorème 6.5.2] that their curve

$$X = X_F \longrightarrow \operatorname{Spec}(\mathbb{Q}_p)$$

parametrizes the orbits of this action in the sense that there is a canonical bijection from the set $|X|$ of closed points onto the set of $\varphi^{\mathbb{Z}}$-orbits of isomorphism classes of characteristic 0 untilts of $(F, \mathcal{O}_F)$. Moreover, if $x \in |X|$ corresponds to the class of $((C, \mathcal{O}_C), \iota)$ via this bijection, then the residue field $k(x)$ is canonically isomorphic to C. The additional structure of the valuation ring $\mathcal{O}_C \subset C$ and the isomorphism $\iota\colon (F, \mathcal{O}_F) \to (C^\flat, \mathcal{O}_{C^\flat})$ arises from the "perfectoid" nature of the situation, a fact that is hidden away in the proof of [**7**, Théorème 6.5.2].

One may wonder if all untilts of $(F, \mathcal{O}_F)$ have abstractly isomorphic underlying completely valued fields. More precisely, given an untilt $((C, \mathcal{O}_C), \iota)$, we let

$$\operatorname{Aut}(F, \mathcal{O}_F)/(\operatorname{Aut}(C, \mathcal{O}_C) \times \varphi^{\mathbb{Z}}) \longrightarrow |X|$$

be the map that to ψ assigns the class of $((C, \mathcal{O}_C), \iota \circ \psi)$ and ask if this map is a bijection. It follows from [**7**, Corollaire 2.2.23] and [**17**, Corollary 6] that the answer to this question is affirmative if and only if the valued field $(F, \mathcal{O}_F)$ is spherically complete (a.k.a. maximally complete) in the sense that every decreasing sequence of discs in F has non-empty intersection. This is a stronger condition than completeness, which is the condition that every decreasing sequence of discs, whose radii tend to 0, has non-empty intersection. For example, the completion $\mathbb{C}_p$ of an algebraic closure of $\mathbb{Q}_p$ with respect to the unique extention of the p-adic absolute value is not spherically complete, and neither is its tilt $\mathbb{C}_p^\flat$.

The Fargues–Fontaine curve behaves as a proper, regular, connected curve of genus 0 over $\mathbb{Q}_p$, except that it is not of finite type. Strikingly, the map

$$\mathbb{Q}_p \longrightarrow H^0(X, \mathcal{O}_X)$$

induced by the structure map is an isomorphism. Hence, there is a great discrepancy between the field $\mathbb{Q}_p$ of global sections and the residue fields $k(x)$ at closed points $x \in |X|$, all of which all are algebraically closed completely valued extensions of $\mathbb{Q}_p$ by [**7**, Théorème 6.5.2]. This phenomenon is one that the Fargues–Fontaine curve shares with the twistor projective line. The complex projective line $\mathbb{P}^1_{\mathbb{C}}$ has two real forms, namely, the real projective line $\mathbb{P}^1_{\mathbb{R}}$ and the twistor projective line

$$X = \widetilde{\mathbb{P}}^1_{\mathbb{R}} \longrightarrow \operatorname{Spec}(\mathbb{R}),$$

which is the Brauer–Severi variety of the quaternions $\mathbb{H}$. Its ring of global sections is $\mathbb{R}$, whereas the residue field at every closed point $x \in |X|$ is an algebraically closed field that contains $\mathbb{R}$. These residue fields are of course all isomorphic to $\mathbb{C}$. In particular, the twistor projective line has no real points.

2. Curves

In general, a connected curve over a field

$$X \longrightarrow \operatorname{Spec}(k)$$

can be understood as follows. If we choose a regular closed point $\infty \in |X|$, then the open complement $X \smallsetminus \{\infty\} \subset X$ is affine, and the quasi-coherent ideal

$$0 \longrightarrow \mathcal{O}_X(-1) \longrightarrow \mathcal{O}_X \longrightarrow i_{\infty *}k(\infty) \longrightarrow 0$$

is invertible. Moreover, writing $\mathcal{O}_X(n)$ for its $(-n)$-fold tensor power, we have

$$X \simeq \operatorname{Proj}(P) \longrightarrow \operatorname{Spec}(k),$$

where P is the graded k-algebra

$$P = \textstyle\bigoplus_{n\in\mathbb{Z}} H^0(X, \mathcal{O}_X(n)).$$

Associated with the closed point $\infty \in |X|$, we have the maps

$$U \xrightarrow{\ i\ } X \xleftarrow{\ j\ } Y$$

of formal schemes over k, where i and j are the canonical maps from the open complement of $\{\infty\} \subset X$ and the formal completion along $\{\infty\} \subset X$, respectively. In Grothendieck's philosophy, the map j should be viewed as the open and affine immersion of a tubular neighborhood of $\{\infty\} \subset X$, whereas the map i should be viewed as the closed immersion of the closed complement of said neighborhood. This point of view is substantiated by the stable recollement

$$\operatorname{QCoh}(U) \underset{i^!}{\overset{i^*}{\leftrightarrows}}\!\!\xrightarrow{i_* \simeq i_!} \operatorname{QCoh}(X) \underset{j_*}{\overset{j_!}{\leftrightarrows}}\!\!\xrightarrow{j^! \simeq j^*} \operatorname{QCoh}(Y)$$

among the corresponding stable ∞-categories of quasi-coherent modules.[1] We give a proof of this is in [**11**, Theorem 1.7], but we also refer to [**5**, Lecture V] for an explanation of the open-closed reversal in this situation. So we have a cartesian diagram in the ∞-category $\mathrm{QCoh}(X)$ of quasi-coherent $\mathcal{O}_X$-modules

$$\begin{array}{ccc} \mathcal{O}_X(n) & \longrightarrow & i_*i^*\mathcal{O}_X(n) \\ \downarrow & & \downarrow \\ j_*j^*\mathcal{O}_X(n) & \longrightarrow & i_*i^*j_*j^*\mathcal{O}_X(n). \end{array}$$

We interpret $\mathcal{O}_X(n)$ as the sheaf of meromorphic functions on X that are regular away from ∞ and whose pole order at ∞ is at most n, and $i_*i^*\mathcal{O}_X(n)$ as the sheaf of meromorphic functions on X that are regular away from ∞. Similarly, we interpret $j_*j^*\mathcal{O}_X(n)$ as the sheaf of formal meromorphic functions near ∞ that are regular away from ∞ and whose pole order at ∞ is at most n, and $i_*i^*j_*j^*\mathcal{O}_X(n)$ as the sheaf of formal meromorphic functions near ∞ that are regular away from ∞. It follows that the k-vector spaces $H^0(X,\mathcal{O}_X(n))$ and $H^1(X,\mathcal{O}_X(n))$ are canonically identified with the limit and cokernel, respectively, of the diagram

$$\begin{array}{ccc} & & H^0(X,i_*i^*\mathcal{O}_X(n)) \\ & & \downarrow \\ H^0(X,j_*j^*\mathcal{O}_X(n)) & \longrightarrow & H^0(X,i_*i^*j_*j^*\mathcal{O}_X(n)). \end{array} \tag{2.1}$$

As a simple example, let us first consider the complex projective line

$$X = \mathbb{P}^1_{\mathbb{C}} \longrightarrow \mathrm{Spec}(\mathbb{C}).$$

If we choose a coordinate t^{-1} at $\infty \in |X|$, then the diagram (2.1) takes the form

$$\begin{array}{ccc} & & \mathbb{C}[t] \\ & & \downarrow \\ t^n\mathbb{C}[[t^{-1}]] & \longrightarrow & \mathbb{C}((t^{-1})) \end{array} \tag{2.2}$$

with the two maps given by the canonical inclusions. So for $n \geq 0$, we have

$$H^0(X,\mathcal{O}_X(n)) = \mathbb{C}\cdot\{1,t,\dots,t^n\}$$
$$H^1(X,\mathcal{O}_X(n)) = 0,$$

where $\mathbb{C}\cdot S$ is the $\mathbb{C}$-vector space generated by S, whereas for $n<0$, we have

$$H^0(X,\mathcal{O}_X(n)) = 0$$
$$H^1(X,\mathcal{O}_X(n)) = \mathbb{C}\cdot\{t^{-1},\dots,t^{n+1}\}.$$

In particular, we find that the graded $\mathbb{C}$-algebra P is given by

$$P \simeq \bigoplus\nolimits_{n\geq 0}\mathbb{C}\cdot\{1,t,\dots,t^n\} \longrightarrow \mathbb{C}[x,y],$$

where the nth component of the right-hand isomorphism takes t^i to x^iy^{n-i}.

[1] Let $Y^{(n)} \subset X$ be the nth infinitesimal neighborhood of $\{\infty\} \subset X$. We define $\mathrm{QCoh}(Y)$ to be the colimit $\mathrm{colim}_n\,\mathrm{QCoh}(Y^{(n)})$ in the ∞-category of presentable ∞-categories and right adjoint functors, or equivalently, as the limit $\lim_n \mathrm{QCoh}(Y^{(n)})$ in the ∞-category of presentable ∞-categories and left adjoint functors. The functor $j^! \simeq j^*$ is Grothendieck's local cohomology.

In the case of the twistor projective line

$$X = \widetilde{\mathbb{P}}^1_{\mathbb{R}} \longrightarrow \operatorname{Spec}(\mathbb{R}),$$

the diagram (2.1) takes the form

$$\begin{array}{ccc} & & \mathbb{R}[u,v]/(u^2+v^2+1) \\ & & \big\downarrow \\ t^n\mathbb{C}[[t^{-1}]] & \longrightarrow & \mathbb{C}((t^{-1})) \end{array} \tag{2.3}$$

with the horizontal map given by the canonical inclusion and with the vertical map obtained by solving the complex linear system of equations

$$\begin{aligned} u + iv &= t \\ u - iv &= -t^{-1}. \end{aligned}$$

We remark that now the identification of the terms in the bottom row uses the fact that a complete discrete valuation ring of residue characteristic 0 admits a unique coefficient field, which, in general, relies on the axiom of choice. Thus, the $\mathbb{C}$-algebra structure on the two rings is somewhat of a mirage, as opposed to the $\mathbb{C}$-algebra structure on the associated graded rings for the t-adic filtrations. This filtration, in turn, induces a $\mathbb{Z}$-graded ascending filtration

$$\cdots \subset \operatorname{Fil}_{m-1} H^i(X, \mathcal{O}_X(n)) \subset \operatorname{Fil}_m H^i(X, \mathcal{O}_X(n)) \subset \cdots$$

of the cohomology groups in question, and we find that for $n \geq 0$,

$$\operatorname{gr}_m H^0(X, \mathcal{O}_X(n)) \simeq \begin{cases} \mathbb{R} \cdot 1 & \text{if } m = 0 \\ \mathbb{C} \cdot t^m & \text{if } 0 < m \leq n \end{cases}$$

are the only nonzero graded pieces, whereas for $n < 0$,

$$\operatorname{gr}_m H^1(X, \mathcal{O}_X(n)) \simeq \begin{cases} \mathbb{C} \cdot t^m & \text{if } n+1 \leq m \leq -1 \\ \mathbb{C}/\mathbb{R} \cdot 1 & \text{if } m = 0 \end{cases}$$

are the only nonzero graded pieces. In particular, the quotient $\mathbb{C}/\mathbb{R}$ of the residue field at ∞ by the subfield of global sections appears as $H^1(X, \mathcal{O}_X(-1))$. We find that there is an isomorphism of graded $\mathbb{R}$-algebras

$$P = \bigoplus\nolimits_{n \geq 0} H^0(X, \mathcal{O}_X(n)) \longrightarrow \mathbb{R}[x,y,z]/(x^2+y^2+z^2)$$

whose nth component takes the class of $u^i v^j$ to $x^i y^j z^{n-(i+j)}$.

In the case of the Fargues–Fontaine curve

$$X = X_F \longrightarrow \operatorname{Spec}(\mathbb{Q}_p),$$

the diagram (2.1) is expressed in terms of Fontaine's period rings[2] as

$$\begin{array}{ccc} & & B_e \\ & & \big\downarrow \\ \operatorname{Fil}_n B_{\mathrm{dR}} & \longrightarrow & B_{\mathrm{dR}} \end{array} \tag{2.4}$$

[2] The notation B_e stems from [**4**, (3.7.2)]. The subscript e indicates the exponential case.

The ring B_{dR} is a complete discrete valuation field with residue field $C = k(\infty)$,[3] and the horizontal map is the canonical inclusion of the $(-n)$th power of the maximal ideal of its valuation ring $B_{\mathrm{dR}}^+ = \mathrm{Fil}_0 B_{\mathrm{dR}}$. The ring B_e is a principal ideal domain, the nonzero ideals of which are in canonical one-to-one correspondence with the closed points $x \in |X \smallsetminus \{\infty\}|$. This non-trivial fact, instigated by Berger's discovery that every finitely generated ideal in B_e is principal [**2**, Proposition 1.1.9], was the discovery which led Fargues and Fontaine to realize that they had a curve in their hands. The proof is given in [**7**, Théorème 6.5.2], and Colmez' recounting of this discovery process in [**7**, Préface] is quite illuminating. The discrete valuation on B_{dR} again gives rise to an ascending $\mathbb{Z}$-graded filtration

$$\cdots \subset \mathrm{Fil}_{m-1} H^i(X, \mathcal{O}_X(n)) \subset \mathrm{Fil}_m H^i(X, \mathcal{O}_X(n)) \subset \cdots$$

of the cohomology groups, and for $n \geq 0$,

$$\mathrm{gr}_m H^0(X, \mathcal{O}_X(n)) = \begin{cases} \mathbb{Q}_p \cdot 1 & \text{if } m = 0 \\ C \cdot t^m & \text{if } 0 < m \leq n \end{cases}$$

are the only nonzero graded pieces, whereas for $n < 0$,

$$\mathrm{gr}_m H^1(X, \mathcal{O}_X(n)) = \begin{cases} C \cdot t^m & \text{if } n+1 \leq m \leq -1 \\ C/\mathbb{Q}_p \cdot 1 & \text{if } m = 0 \end{cases}$$

are the only nonzero graded pieces. Here $t^{-1} \in \mathrm{Fil}_{-1} B_{\mathrm{dR}}$ is a local parameter at ∞, that is, a generator of this B_{dR}^+-module, which is free of rank 1. Again, the quotient $C/\mathbb{Q}_p$ of the residue field at ∞ and the subfield of global sections appears as $H^1(X, \mathcal{O}_X(-1))$, but this now an infinite dimensional $\mathbb{Q}_p$-vector space, reflecting the fact that the Fargues–Fontaine curve is not of finite type over $\mathbb{Q}_p$. Nevertheless, Fargues and Fontaine show in [**7**, Théorème 6.2.1] that the graded $\mathbb{Q}_p$-algebra

$$P = \bigoplus\nolimits_{n \geq 0} H^0(X, \mathcal{O}_X(n))$$

is a graded unique factorization domain, all of whose irreducible elements are of degree 1. Thus, closed points $x \in |X|$ are in canonical one-to-one correspondence with $\mathbb{Q}_p$-lines in $H^0(X, \mathcal{O}_X(1))$, or equivalently, extensions of the form

$$0 \longrightarrow \mathbb{Q}_p \longrightarrow H^0(X, \mathcal{O}_X(1)) \longrightarrow k(x) \longrightarrow 0;$$

see [**7**, Théorème 6.5.2].

3. The formal neighborhood of $\infty \in |X|$

We proceed to explain how to obtain the diagram (2.4) for the Fargues–Fontaine curve from a diagram of spectra. We begin with the lower line in the diagram, and recall some general theory following Nikolaus–Scholze [**16**, Theorem I.4.1].

We will use the language of ∞-categories following Joyal and Lurie [**14**], but we will use the term "anima" or "animated set" for what Lurie calls a space to emphasize that these should be viewed as sets with internal symmetries rather than anything resembling a topological space.

If $f \colon T \to S$ is any map of anima, then we have the restriction along f, $f^! \simeq f^*$, and the left and right Kan extensions along f, $f_!$ and f_*, between the corresponding ∞-categories of spectrum-valued presheaves:

[3] While there does exist an isomorphism $B_{\mathrm{dR}} \simeq C((t^{-1}))$ of complete discrete valuation fields, such an isomorphism is highly non-canonical and not useful in practice.

$$\mathrm{Sp}^T \; \begin{matrix} \xrightarrow{\quad f_! \quad} \\ \xleftarrow{\; f^! \simeq f^* \;} \\ \xrightarrow{\quad f_* \quad} \end{matrix} \; \mathrm{Sp}^S$$

In fact, these functors are part of a six-functor formalism on the ∞-category of anima in the sense of Mann [**15**, Definition A.5.7]. Indeed, this claim is a trivial instance of [**15**, Proposition A.5.10], where every map of anima $f\colon T \to S$ is declared to be a local isomorphism, and where a map of anima $f\colon T \to S$ is declared to be proper if its fibers are compact projective anima, or equivalently, anima that are equivalent to finite sets, including the empty set. In this six-functor formalism, we have $f^! \simeq f^*$ for every map $f\colon T \to S$, which reflects the fact that every map of anima is a local isomorphism. Now, the ∞-categories Sp^S and Sp^T are compactly generated presentable stable ∞-categories, and the "homology" functor $f_!$ preserves compact objects, because its right adjoint $f^!$ preserves all colimits. However, the "cohomology" functor f_* does not preserve compact objects, and by [**16**, Theorem I.4.1], there is a unique a map $f_* \to f_*^T$ to a "Tate cohomology" functor that takes all compact objects to zero and that is initial with this property. The fiber of this map preserves colimits, which implies that it necessarily is of the form $X \mapsto f_!(X \otimes D_f)$ for a unique $D_f \in \mathrm{Sp}^T$.

We recall from [**13**, Theorem 5.6.2.10] that every group in anima G is the loop group ΩBG of a pointed connected anima

$$1 \xrightarrow{\;s\;} BG,$$

and we first apply the general theory above to the unique map

$$BG \xrightarrow{\;f\;} 1.$$

If G is the group in anima underlying a compact Lie group G, then $D_f \simeq S^{\mathfrak{g}}$ is the suspension spectrum of the one-point compactification of its Lie algebra with the adjoint G-action. So in this situation, the defining fiber sequences takes the form

$$f_!(S^{\mathfrak{g}} \otimes X) \longrightarrow f_*(X) \longrightarrow f_*^T(X),$$

where $X \in \mathrm{Sp}^{BG}$. This sequence is commonly written as

$$(S^{\mathfrak{g}} \otimes X)_{hG} \longrightarrow X^{hG} \longrightarrow X^{tG}.$$

The Postnikov filtration of X gives rise to the spectral sequences

$$\begin{aligned} E_{i,j}^2 &= H^{-i}(BG, \pi_j(s^*(X))) \Rightarrow \pi_{i+j}(X^{hG}) \\ E_{i,j}^2 &= \hat{H}^{-i}(BG, \pi_j(s^*(X))) \Rightarrow \pi_{i+j}(X^{tG}), \end{aligned}$$

and the edge homomorphism of the former,

$$\pi_j(X^{hG}) \xrightarrow{\;\theta\;} \pi_j(s^*(X)),$$

is induced by the unit map

$$f_*(X) \xrightarrow{\;\theta\;} f_*s_*s^*(X) \simeq s^*(X).$$

The groups in the two E^2-terms can be interpreted as the cohomology and Tate cohomology of BG with coefficients in the local system $\pi_j(s^*(X))$. If G is finite, then these can be identified with the group cohomology and Tate cohomology of the group G with coefficients in the G-module $\pi_j(s^*(X))$.

It follows from [**16**, Theorem I.4.1] that if $E \in \mathrm{CAlg}(\mathrm{Sp}^{BG})$ is a commutative algebra in spectra with G-action, then the map $f_*(E) \to f_*^T(E)$ promotes to a map of commutative algebras in spectra. This implies that the spectral sequences become spectral sequences of bigraded anticommutative rings and that the edge homomorphism θ becomes an map of graded anticommutative rings; see [**9**]. If the G-action on E is trivial in the sense that $E \simeq f^*s^*(E)$, then a choice of trivialization and the unit map combine to give a section

$$s^*(E) \xrightarrow{\ \sigma\ } f_*f^*s^*(E) \simeq f_*(E)$$

of the edge homomorphism θ. So in this situation, the map $f_*(E) \to f_*^T(E)$ becomes a map of commutative algebras in $s^*(E)$-modules in spectra. So the induced map on homotopy groups becomes a map of anticommutative graded $\pi_*(s^*(E))$-algebras, and the spectral sequences above become spectral sequences of anticommutative bigraded $\pi_*(s^*(E))$-algebras.

We are interested in the group $G \simeq U(1)$, and since it is abelian, the G-action on its Lie algebra is trivial. So in this case, the fiber sequence above becomes

$$\Sigma E_{hG} \longrightarrow E^{hG} \longrightarrow E^{tG}.$$

For instance, in the case of $E \simeq \mathrm{HH}(A/R)$, this gives Connes' sequence

$$\Sigma\,\mathrm{HC}(A/R) \longrightarrow \mathrm{HC}^-(A/R) \longrightarrow \mathrm{HP}(A/R)$$

relating cyclic homology, negative cyclic homology, and periodic cyclic homology. Choosing a generator $\bar{v} \in H^2(BU(1), \mathbb{Z})$, the spectral sequences take the form[4]

$$E^2 = \pi_*(s^*(E))[\bar{v}] \Rightarrow \pi_*(E^{hU(1)})$$
$$E^2 = \pi_*(s^*(E))[\bar{v}^{\pm 1}] \Rightarrow \pi_*(E^{tU(1)}).$$

We refer to the filtrations of the respective abutments induced by the two spectral sequences as the "Nygaard" filtrations.

We will consider $E \in \mathrm{CAlg}(\mathrm{Sp}^{BU(1)})$ equipped with a "Bott" element

$$\beta \in \pi_2(E^{hU(1)}) \longrightarrow \pi_2(E^{tU(1)})$$

and obtain the desired filtered ring from the filtered graded ring $\pi_*(E^{tU(1)})$ by the following procedure:

(1) Invert the Bott element.
(2) Complete with respect to the Nygaard filtration.
(3) Extract the filtered subring consisting of homogeneous elements of degree 0.

The image of the induced ring homomorphism

$$(\pi_*(E^{hU(1)})[\beta^{-1}]^\wedge)_0 \longrightarrow (\pi_*(E^{tU(1)})[\beta^{-1}]^\wedge)_0$$

agrees with the 0th stage of the Nygaard filtration. If $\pi_*(s^*(E))$ is concentrated in even degrees, then so are $\pi_*(E^{hU(1)})$ and $\pi_*(E^{tU(1)})$ and all differentials in the respective spectral sequences are zero. In this case, the map above is injective.

[4] Here $\pi_*(-)$ indicates homotopy groups and not pushforward along some map π.

Suppose first that the $U(1)$-action on E is trivial. If the homotopy groups $\pi_*(s^*(E))$ are concentrated in even degrees, then there are isomorphisms

$$\pi_*(E^{hU(1)}) \simeq \pi_*(s^*(E))[[v]] \simeq \lim_n \pi_*(s^*(E))[v]/(v^{n+1})$$
$$\pi_*(E^{tU(1)}) \simeq \pi_*(s^*(E))((v)) \simeq \pi_*(s^*(E))[[v]][v^{-1}]$$

of anticommutative graded $\pi_*(s^*(E))$-algebras. Here the limit is calculated in the category of graded $\pi_*(s^*(E))$-algebras and $v \in \pi_{-2}(E^{hU(1)})$ is a choice of lift of $\bar{v}$, that is, a "complex orientation" of E. So with $t^{-1} = \beta v$, we have

$$(\pi_*(E^{hU(1)})[\beta^{-1}])_0 \simeq (\pi_*(s^*(E))[\beta^{-1}])_0[[t^{-1}]]$$
$$(\pi_*(E^{tU(1)})[\beta^{-1}])_0 \simeq (\pi_*(s^*(E))[\beta^{-1}])_0((t^{-1}))$$

Hence, writing $R = (\pi_*(s^*(E))[\beta^{-1}])_0$, we see that the graded $\pi_*(s^*(E))$-algebra $\pi_*(E^{tU(1)})$ with the Nygaard filtration gives rise to the R-algebra

$$R((t^{-1}))$$

with the t-adic filtration. This is the filtered R-algebra that we would obtain from the formal neighborhood of an R-valued point $\infty \in |X|$ of a curve $X \to \operatorname{Spec}(R)$.

Let us now consider the Fargues–Fontaine curve

$$X = X_F \longrightarrow \operatorname{Spec}(\mathbb{Q}_p)$$

associated to an algebraically closed completely valued field $(F, \mathcal{O}_F)$ of characteristic $p > 0$, and let $\infty \in |X|$ be a closed point corresponding to an untilt $((C, \mathcal{O}_C), \iota)$ of $(F, \mathcal{O}_F)$. We consider the commutative algebra in spectra with $U(1)$-action

$$E = \mathrm{THH}(\mathcal{O}_C, \mathbb{Z}_p) \in \mathrm{CAlg}(\mathrm{Sp}^{BU(1)})$$

given by the p-adic completion of the topological Hochschild homology of $\mathcal{O}_C$. In this case, Bhatt–Morrow–Scholze show in [**3**, Proposition 6.2] that

$$\pi_*(E^{hU(1)}) \simeq \mathrm{TC}^-_*(\mathcal{O}_C, \mathbb{Z}_p) \simeq A_{\mathrm{inf}}(\mathcal{O}_F)[u,v]/(uv - \xi)$$
$$\pi_*(E^{tU(1)}) \simeq \mathrm{TP}_*(\mathcal{O}_C, \mathbb{Z}_p) \simeq A_{\mathrm{inf}}(\mathcal{O}_F)[v^{\pm 1}],$$

where $u \in \mathrm{TC}^-_2(\mathcal{O}_C, \mathbb{Z}_p)$ and $v \in \mathrm{TC}^-_{-2}(\mathcal{O}_C, \mathbb{Z}_p)$, and where ξ is a generator of the kernel of the edge homomorphism

$$\mathrm{TC}^-_0(\mathcal{O}_C, \mathbb{Z}_p) \simeq A_{\mathrm{inf}}(\mathcal{O}_F) \xrightarrow{\ \theta\ } \mathcal{O}_C.$$

The element u is called a Bökstedt element. It is not a Bott element, but rather a divided Bott element. More precisely, the element

$$\beta = \varphi^{-1}(\mu)u$$

is a Bott element, but $\varphi^{-1}(\mu) \in A_{\mathrm{inf}}(\mathcal{O}_F)$ is not a unit; see also [**10**, Section 3.2]. So by inverting β, we also invert $\varphi^{-1}(\mu)$, which, in particular, inverts p, since

$$\theta(\varphi^{-1}(\mu)) = \zeta_p - 1 \in \mathcal{O}_C$$

is a pseudo-uniformizer. It follows that

$$(\pi_*(E^{tU(1)})[\beta^{-1}]^\wedge)_0 \simeq (\mathrm{TP}_*(\mathcal{O}_C, \mathbb{Z}_p)[\beta^{-1}]^\wedge)_0 \simeq B_{\mathrm{dR}}$$

as a filtered $\mathbb{Q}_p$-algebra. It is a discrete valuation field with residue field C. We stress that, after inverting β, we do not subsequently p-complete, which would leave us with zero. So this procedure is distinct from Morava $K(1)$-localization.

4. The affine complement of $\infty \in |X|$

We will next explain how to obtain the right-hand vertical map in (2.4) from the "crystalline Chern character" of Antieau–Mathew–Morrow–Nikolaus [**1**]. The definition of this map relies on the modern approach to topological cyclic homology and cyclotomic spectra due to Nikolaus–Scholze [**16**].

To briefly recall the definition, let $p\colon BU(1) \to BU(1)$ be the map of pointed anima induced by the p-power map of the abelian group $U(1)$, and let

$$\mathrm{Sp}_p^{BU(1)} \xrightarrow{p_*^T} \mathrm{Sp}_p^{BU(1)}$$

be the associated Tate cohomology functor. A p-complete cyclotomic spectrum is a pair (X, φ) of a p-complete spectrum[5] with $U(1)$-action and a "Frobenius" map

$$X \xrightarrow{\varphi} p_*^T(X)$$

of spectra with $U(1)$-action. Nikolaus and Scholze show that p-complete cyclotomic spectra can be organized into a stable symmetric monoidal ∞-category equipped with a conservative symmetric monoidal forgetful functor

$$\mathrm{CycSp}_p \longrightarrow \mathrm{Sp}_p^{BU(1)}$$

to the symmetric monoidal stable ∞-category of p-complete spectra with $U(1)$-action. The tensor unit is given by the pair $\mathbf{1} \simeq (f^*(\mathbb{S}_p), \varphi)$ of the p-complete sphere spectrum with trivial $U(1)$-action and the composition

$$f^*(\mathbb{S}_p) \longrightarrow p_*f^*(\mathbb{S}_p) \longrightarrow p_*^T f^*(\mathbb{S}_p)$$

of the adjunct of the equivalence $p^*f^* \simeq (fp)^* \simeq f^*$ and the canonical map. Now, p-complete topological cyclic homology is the symmetric monoidal functor

$$\mathrm{CycSp}_p \xrightarrow{\mathrm{TC}} \mathrm{Sp}_p$$

corepresented by $\mathbf{1} \in \mathrm{CycSp}_p$. If (X, φ) is a p-complete cyclotomic spectrum with X bounded below, then $\mathrm{TC}(X, \varphi)$ is given by the equalizer

$$\mathrm{TC}(X,\varphi) \longrightarrow \mathrm{TC}^-(X) \simeq f_*(X) \underset{\mathrm{can}}{\overset{\varphi}{\rightrightarrows}} \mathrm{TP}(X) \simeq f_*^T(X)$$

of the canonical map "can" and the cyclotomic Frobenius "φ" defined as follows. Since $fp \simeq f$, we have a diagram of p-complete spectra

$$\begin{array}{ccccc}
\Sigma f_! p_!(X) \simeq \Sigma f_!(X) & \longrightarrow & f_*(X) & \longrightarrow & f_*^T(X) \\
\downarrow & & \| & & \downarrow \\
f_* p_!(X) & \longrightarrow & f_* p_*(X) & \longrightarrow & f_* p_*^T(X)
\end{array}$$

[5] The inclusion $j_*\colon \mathrm{Sp}_p \to \mathrm{Sp}$ of the full subcategory of p-complete spectra admits a left adjoint $j^*\colon \mathrm{Sp} \to \mathrm{Sp}_p$, and the unit map $\eta\colon X \to j_*j^*(X)$ is p-completion. The functor j^* admits an essentially unique promotion to a symmetric monoidal functor, which, in turn, promotes j_* to a lax symmetric monoidal functor. Hence, we obtain an induced adjunction of the associated ∞-categories of commutative algebras.

in which the two rows are fiber sequences. Since X is bounded below, the Tate orbit lemma [**16**, Lemma I.2.1] shows that the left-hand vertical map is an equivalence, and hence, so is the right-hand vertical map. So in this case, we have the map

$$X^{hU(1)} \simeq f_*(X) \xrightarrow{f_*(\varphi)} f_*p_*^T(X) \simeq f_*^T(X) \simeq X^{tU(1)},$$

which, by abuse of notation, we also denote by $\varphi\colon \mathrm{TC}^-(X) \to \mathrm{TP}(X)$.

We recall the most economical way of associating to a commutative algebra in p-complete spectra R a commutative algebra in p-complete cyclotomic spectra

$$(\mathrm{THH}(R, \mathbb{Z}_p), \varphi)$$

following [**16**, Section IV.2].

In general, let G be a group in anima, and let $s\colon 1 \to BG$ be the corresponding connected pointed anima. If $\mathcal{C}$ is any presentable ∞-category, then we have

$$\mathcal{C} \underset{s^*}{\overset{s_!}{\rightleftarrows}} \mathcal{C}^{BG},$$

where the right adjoint s^* takes an object of $\mathcal{C}$ with G-action to the underlying object of $\mathcal{C}$, and where the left adjoint $s_!$ takes an object of $\mathcal{C}$ to the free object of $\mathcal{C}$ with G-action. We have a cartesian square of pointed anima

$$\begin{array}{ccc} G & \xrightarrow{s'} & 1 \\ \big\downarrow{\scriptstyle s'} & & \big\downarrow{\scriptstyle s} \\ 1 & \xrightarrow{s} & BG \end{array}$$

and the base-change formula applied to this square shows that

$$s^*s_! \simeq s'_!s'^*$$

as endofunctors of $\mathcal{C}$. Informally, if X is an object of $\mathcal{C}$, then $s^*s_!(X)$ is the object of $\mathcal{C}$ underlying the free object of $\mathcal{C}$ with G-action associated with X, whereas $s'_!s'^*(X)$ is the colimit in $\mathcal{C}$ of the constant diagram with value X indexed by G.

We apply this with $G \simeq U(1)$ and $\mathcal{C} \simeq \mathrm{CAlg}(\mathrm{Sp}_p)$. Given $R \in \mathrm{CAlg}(\mathrm{Sp}_p)$, we define its p-complete topological Hochschild homology to be the commutative algebra in p-complete spectra with $U(1)$-action given by

$$\mathrm{THH}(R, \mathbb{Z}_p) \simeq s_!(R) \in \mathrm{CAlg}(\mathrm{Sp}_p)^{BU(1)}$$

and the base-change formula shows that its underlying commutative algebra in p-complete spectra is given by

$$s^*\,\mathrm{THH}(R, \mathbb{Z}_p) \simeq s^*s_!(R) \simeq s'_!s'^*(R) \simeq R^{\otimes U(1)},$$

where the right-hand term is suggestive notation for the colimit in $\mathrm{CAlg}(\mathrm{Sp}_p)$ of the constant diagram with value R indexed by the anima underlying $U(1)$.

We next recall the definition of the Frobenius map

$$\mathrm{THH}(R, \mathbb{Z}_p) \xrightarrow{\varphi} p_*^T(\mathrm{THH}(R, \mathbb{Z}_p)),$$

where we abuse notation and also write p_*^T for the endofunctor of

$$\mathrm{CAlg}(\mathrm{Sp}_p)^{BU(1)} \simeq \mathrm{CAlg}(\mathrm{Sp}_p^{BU(1)})$$

induced by the lax symmetric monoidal endofunctor p_*^T of $\mathrm{Sp}_p^{BU(1)}$. We consider the following diagram of pointed anima

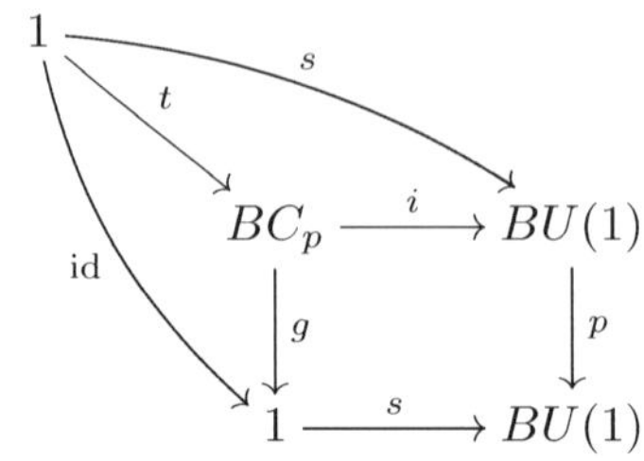

with the square cartesian. As part of the structure of any symmetric monoidal ∞-category $\mathcal{C}$, there is a symmetric monoidal functor

$$\mathcal{C} \xrightarrow{t_!^{\otimes}} \mathcal{C}^{BC_p}$$

which, informally, takes X to $X^{\otimes p}$ together with an equivalence between the induced functor of commutative algebras and

$$\mathrm{CAlg}(\mathcal{C}) \xrightarrow{t_!} \mathrm{CAlg}(\mathcal{C})^{BC_p} \simeq \mathrm{CAlg}(\mathcal{C}^{BC_p}).$$

We now recall from [**16**, Proposition III.3.1] that there is a unique lax symmetric monoidal transformation called the Tate diagonal[6]

$$X \xrightarrow{\delta} g_*^T t_!^{\otimes}(X).$$

So for $R \in \mathrm{CAlg}(\mathrm{Sp}_p)$, we get the composite map

$$R \xrightarrow{\delta} g_*^T t_!(R) \xrightarrow{\eta} g_*^T i^* i_! t_!(R) \simeq g_*^T i^* s_!(R) \simeq s^* p_*^T s_!(R)$$

of commutative algebras in p-complete spectra, whose adjunct

$$s_!(R) \xrightarrow{\varphi} p_*^T s_!(R)$$

is the desired map of commutative algebras in p-complete spectra with $U(1)$-action.

The crystalline Chern character of [**1**] is obtained as follows. Given a p-complete cyclotomic spectrum $X = (X, \varphi)$, we may form the tensor product

$$X \otimes \mathrm{THH}(\mathbb{F}_p, \mathbb{Z}_p)$$

in the symmetric monoidal ∞-category CycSp_p of p-complete cyclotomic spectra and consider the canonical map

$$\mathrm{TC}(X \otimes \mathrm{THH}(\mathbb{F}_p, \mathbb{Z}_p)) \longrightarrow \mathrm{TC}^-(X \otimes \mathrm{THH}(\mathbb{F}_p, \mathbb{Z}_p))$$

from its topological cyclic homology to its negative topological cyclic homology. We consider this map for $X \simeq \mathrm{THH}(\mathcal{O}_C, \mathbb{Z}_p)$, take homotopy groups, invert the Bott element, Nygaard complete the target, and extract the subrings of homogeneous elements of degree 0. By [**1**, Theorem 2.12], the resulting map takes the form

$$(\mathrm{TC}_*(\mathcal{O}_C/p, \mathbb{Z}_p)[\beta^{-1}])_0 \longrightarrow (\mathrm{TP}_*(\mathcal{O}_C, \mathbb{Z}_p)[\beta^{-1}]^{\wedge})_0,$$

[6] The generalized Segal conjecture for C_p proved in [**16**, Theorem III.1.7] states that if X is bounded below, then the Tate diagonal is an equivalence.

and this map is the desired crystalline Chern character. Finally, the identification of this map with the map $B_e \to B_{\mathrm{dR}}$ in (2.4) is a consequence of [**3**, Corollary 8.23] and [**3**, Proposition 6.2]. We conclude that the diagram

$$\begin{array}{ccc} & & (\mathrm{TC}_*(\mathcal{O}_C/p,\mathbb{Z}_p)[\beta^{-1}])_0 \\ & & \downarrow \\ \mathrm{Fil}_n(\mathrm{TP}_*(\mathcal{O}_C,\mathbb{Z}_p)[\beta^{-1}]^\wedge)_0 & \longrightarrow & (\mathrm{TP}_*(\mathcal{O}_C,\mathbb{Z}_p)[\beta^{-1}]^\wedge)_0 \end{array} \tag{4.1}$$

obtained from the various flavors of topological cyclic homology and maps between them is canonically identified with the diagram of period rings (2.4).

5. Dreams

We would like to similarly obtain the diagram (2.3) for the twistor projective line from a diagram analogous to (4.1). Given a closed point $\infty \in |X|$ with residue field $k(\infty) \simeq \mathbb{C}$, one might hope to obtain the bottom line in (2.3) from the periodic cyclic homology $\mathrm{HP}_*(\mathbb{C}/\mathbb{R})$. This will not work, however, due to the lack of a Bott element $\beta \in \mathrm{HC}_2^-(\mathbb{C}/\mathbb{R})$. The liquid theory of Clausen–Scholze [**6**] points to an alternative. Let $0 < r < 1$ be a real number. The Harbater ring is the subring

$$\mathbb{Z}((T))_{>r} \subset \mathbb{Z}((T))$$

consisting of the Laurent series that for some $r' > r$ converge absolutely on a disc of radius r' centered at the origin. Harbater shows in [**8**] that this ring is a principal ideal domain; see also [**6**, Theorem 7.1]. In particular, if $x \in \mathbb{C}$ and $|x| \leq r$, then the kernel of the continuous ring homomorphism

$$\mathbb{Z}((T))_{>r} \longrightarrow \mathbb{C}$$

that to T assigns x is a principal ideal, and hence, the Hochschild homology

$$\mathrm{HH}_*(\mathbb{C}/\mathbb{Z}((T))_{>r}) = \mathbb{C}\langle \bar{u}\rangle = \mathbb{C}[\bar{u}]$$

is a (divided) polynomial algebra on a generator $\bar{u}$ of degree 2. So with

$$E = \mathrm{HH}(\mathbb{C}/\mathbb{Z}((T))_{>r}) \in \mathrm{CAlg}(\mathrm{Sp}^{BU(1)}),$$

we find that

$$\begin{aligned} \pi_*(E^{hU(1)}) &\simeq \mathrm{HC}_*^-(\mathbb{C}/\mathbb{Z}((T))_{>r}) \simeq \mathbb{C}[[\xi]][u,v]/(uv-\xi) \\ \pi_*(E^{tU(1)}) &\simeq \mathrm{HC}_*^-(\mathbb{C}/\mathbb{Z}((T))_{>r}) \simeq \mathbb{C}[[\xi]][v^{\pm 1}] \end{aligned}$$

with $u \in \mathrm{HC}_2^-(\mathbb{C}/\mathbb{Z}((T))_{>r})$ and $v \in \mathrm{HC}_{-2}^-(\mathbb{C}/\mathbb{Z}((T))_{>r})$ and with ξ is a generator of the kernel of the edge homomorphism

$$\mathrm{HC}_0^-(\mathbb{C}/\mathbb{Z}((T))_{>r}) \simeq \mathbb{C}[[\xi]] \xrightarrow{\ \theta\ } \mathbb{C},$$

analogously to $E \simeq \mathrm{THH}(\mathcal{O}_C,\mathbb{Z}_p)$. Hence, for a "Bott element" of the form $\beta = fu$ with $f \in \mathbb{C}[[\xi]]^\times$ a unit, we obtain the bottom row in (2.3) with $t^{-1} = \beta v$.

Acknowledgments

The author would like to express his gratitude to an anonymous referee for a number of helpful comments.

References

[1] Benjamin Antieau, Akhil Mathew, Matthew Morrow, and Thomas Nikolaus, *On the Beilinson fiber square*, Duke Math. J. **171** (2022), no. 18, 3707–3806, DOI 10.1215/00127094-2022-0037. MR4516307

[2] Laurent Berger, *Construction de* (ϕ, Γ)*-modules: représentations* p*-adiques et* B*-paires* (French, with English and French summaries), Algebra Number Theory **2** (2008), no. 1, 91–120, DOI 10.2140/ant.2008.2.91. MR2377364

[3] Bhargav Bhatt, Matthew Morrow, and Peter Scholze, *Topological Hochschild homology and integral* p*-adic Hodge theory*, Publ. Math. Inst. Hautes Études Sci. **129** (2019), 199–310, DOI 10.1007/s10240-019-00106-9. MR3949030

[4] Spencer Bloch and Kazuya Kato, *L-functions and Tamagawa numbers of motives*, The Grothendieck Festschrift, Vol. I, Progr. Math., vol. 86, Birkhäuser Boston, Boston, MA, 1990, pp. 333–400. MR1086888

[5] D. Clausen and P. Scholze, *Condensed mathematics and complex geometry*, Lecture notes, Bonn–Copenhagen, 2022.

[6] D. Clausen and P. Scholze, *Lectures on analytic geometry*, Lecture notes, Bonn, 2020.

[7] Laurent Fargues and Jean-Marc Fontaine, *Courbes et fibrés vectoriels en théorie de Hodge p-adique* (French, with English and French summaries), Astérisque **406** (2018), xiii+382. With a preface by Pierre Colmez. MR3917141

[8] David Harbater, *Convergent arithmetic power series*, Amer. J. Math. **106** (1984), no. 4, 801–846, DOI 10.2307/2374325. MR749258

[9] A. Hedenlund and J. Rognes, *A multiplicative Tate spectral sequence for compact Lie group actions*, `arXiv:2008.09095`, 2020.

[10] Lars Hesselholt and Thomas Nikolaus, *Topological cyclic homology*, Handbook of homotopy theory, CRC Press/Chapman Hall Handb. Math. Ser., CRC Press, Boca Raton, FL, [2020] ©2020, pp. 619–656. MR4197995

[11] L. Hesselholt and P. Pstragowski, *Dirac geometry II: Coherent cohomology*, in preparation.

[12] J. Lurie, *The Fargues–Fontaine curve*, available at math.ias.edu/~lurie, 2018.

[13] J. Lurie, *Higher algebra*, available at math.ias.edu/~lurie, 2017.

[14] Jacob Lurie, *Higher topos theory*, Annals of Mathematics Studies, vol. 170, Princeton University Press, Princeton, NJ, 2009, DOI 10.1515/9781400830558. MR2522659

[15] L. Mann, *A p-adic 6-functor formalism in rigid-analytic geometry*, `arXiv:2206.02022`, 2022.

[16] Thomas Nikolaus and Peter Scholze, *On topological cyclic homology*, Acta Math. **221** (2018), no. 2, 203–409, DOI 10.4310/ACTA.2018.v221.n2.a1. MR3904731

[17] Bjorn Poonen, *Maximally complete fields*, Enseign. Math. (2) **39** (1993), no. 1-2, 87–106. MR1225257

Nagoya University, Japan, and University of Copenhagen, Denmark

Email address: `larsh@math.nagoya-u.ac.jp`

Proceedings of Symposia in Pure Mathematics
Volume **105**, 2023
https://doi.org/10.1090/pspum/105/01903

Hopf Cyclic Cohomology and Beyond

Masoud Khalkhali and Ilya Shapiro

Abstract. This paper is an introduction to Hopf cyclic cohomology with an emphasis on its most recent developments. We cover three major areas: the original definition of Hopf cyclic cohomology by Connes and Moscovici as an outgrowth of their study of transverse index theory on foliated manifolds, the introduction of Hopf cyclic cohomology with coefficients by Hajac-Khalkhali-Rangipour-Sommerhäuser, and finally the latest episode on unifying the coefficients as well as extending the notion to more general settings beyond Hopf algebras. In particular, the last section discusses the relative Hopf cyclic theory that arises in the braided monoidal category settings.

Contents

1. Introduction

Cyclic homology can be understood as a noncommutative analogue of de Rham cohomology of smooth manifolds. In the same vein, Hopf cyclic cohomology should be understood as the noncommutative analogue of group homology and Lie algebra homology. Having said so, it must be understood that there was no "royal road" to either theories and their discovery came indirectly through questions related to index theory of elliptic operators and its abstract formulation called K-homology theory. Once cyclic cohomology was discovered, it turned out that it is possible to

2020 *Mathematics Subject Classification.* Primary 58B34, 19D55, 16E40, 16T05, 16T10, 18D05, 18D10, 18D15.

Key words and phrases. Hopf algebras, bialgebras, cyclic homology, Hopf cyclic cohomology, noncommutative geometry, monoidal categories, braided categories.

The first author was supported in part by an NSERC Grant. The second author was supported in part by NSERC Grant #820455.

define it in a purely algebraic setting and in fact several different, but equivalent, variations are possible. The original approach by Connes, based on quantized calculus, and his route to the discovery of cyclic cohomology is well documented in his 1981 Oberwolfach talk and his written report, where cyclic cohomology was first announced [**10**] (for quantized calculus see [**12**]):

"The transverse elliptic theory for foliations requires as a preliminary step a purely algebraic work, of computing for a noncommutative algebra $\mathcal{A}$ the cohomology of the following complex: n-cochains are multilinear functions $\varphi(f^0,\dots,f^n)$ of $f^0,\dots,f^n \in \mathcal{A}$ where

$$\varphi(f^1,\dots,f^0) = (-1)^n\varphi(f^0,\dots,f^n)$$

and the boundary is

$$\begin{aligned} b\varphi(f^0,\dots,f^{n+1}) &= \varphi(f^0f^1,\dots,f^{n+1}) - \varphi(f^0,f^1f^2,\dots,f^{n+1}) + \dots \\ &\quad +(-1)^{n+1}\varphi(f^{n+1}f^0,\dots,f^n). \end{aligned}$$

The basic class associated to a transversally elliptic operator, for $\mathcal{A}$ = the algebra of the foliation, is given by:

$$\varphi(f^0,\dots,f^n) = Trace\,(\varepsilon F[F,f^0][F,f^1]\dots[F,f^n]), \quad f^i \in \mathcal{A}$$

where

$$F = \begin{pmatrix} 0 & Q \\ P & 0 \end{pmatrix}, \quad \varepsilon = \begin{pmatrix} 1 & 0 \\ 0 & -1 \end{pmatrix},$$

and Q is a parametrix of P. An operation

$$S : H^n(\mathcal{A}) \to H^{n+2}(\mathcal{A})$$

is constructed as well as a pairing

$$K(\mathcal{A}) \times H(\mathcal{A}) \to \mathbb{C}$$

where $K(\mathcal{A})$ is the algebraic K-theory of A. It gives the index of the operator from its associated class φ. Moreover $\langle e,\varphi\rangle = \langle e, S\varphi\rangle$ so that the important group to determine is the inductive limit $H_p = \underrightarrow{Lim} H^n(\mathcal{A})$ for the map S. Using the tools of homological algebra the groups $H^n(\mathcal{A},\mathcal{A}^*)$ of Hochschild cohomology with coefficients in the bimodule $\mathcal{A}^*$ are easier to determine and the solution of the problem is obtained in two steps:
1) the construction of a map

$$B : H^n(\mathcal{A},\mathcal{A}^*) \to H^{n-1}(\mathcal{A})$$

and the proof of a long exact sequence

$$\cdots \to H^n(\mathcal{A},\mathcal{A}^*) \xrightarrow{B} H^{n-1}(\mathcal{A}) \xrightarrow{S} H^{n+1}(\mathcal{A}) \xrightarrow{I} H^{n+1}(\mathcal{A},\mathcal{A}^*) \to \cdots$$

where I is the obvious map from the cohomology of the above complex to the Hochschild cohomology.
2) The construction of a spectral sequence with E_2 term given by the cohomology of the degree -1 differential $I \circ B$ on the Hochschild groups $H^n(\mathcal{A},\mathcal{A}^*)$ and which converges strongly to a graded group associated to the inductive limit.

This purely algebraic theory is then used. For $\mathcal{A} = C^\infty(V)$ one gets the de Rham homology of currents, and for the pseudo-torus, i.e. the algebra of the Kronecker foliation, one finds that the Hochschild cohomology depends on the Diophantine *nature of the rotation number while the above theory gives H_p^0 of dimension 2 and H_p^1 of dimension 2, as expected, but from some remarkable cancelations."*

Our goal in this paper is to give a short survey of aspects of Hopf cyclic cohomology where we have been personally active in. Hopf cyclic cohomology was defined by Connes and Moscovici in their seminal article [**15**] (see also [**16, 17**]) which deals with index theory for transversally elliptic operators. The resulting theory can be regarded as a noncommutative analogue of construction of characteristic classes in Chern-Weil theory through group cohomology. The role of group cohomology is now played by Hopf cyclic cohomology. Due to noncommutativity and non-cocommutativity of the Hopf algebras involved, defining this cohomology theory was not straightforward at all. In fact one of the main steps in [**15**] was to obtain a *noncommutative characteristic map*

$$\chi_\tau : HC^\bullet_{(\delta,\sigma)}(H) \longrightarrow HC^\bullet(A),$$

for an action of a Hopf algebra H on an algebra A endowed with an invariant trace $\tau : A \to \mathbb{C}$. Here, the pair (δ, σ) consists of a grouplike element $\sigma \in H$ and a character $\delta : H \to \mathbb{C}$ satisfying certain compatibility conditions to be explained later in this paper. This intriguing index theoretic construction naturally begged for its purely algebraic underpinnings to be abstracted in order to build the theory independently of its original index theory connections.

In [**43**] a new approach was indeed found by defining a cyclic cohomology theory for triples (C, H, M), where C is a coalgebra endowed with an action of a Hopf algebra H and M is an H-module and an H-comodule satisfying some extra compatibility conditions. It was observed that the theory of Connes and Moscovici corresponds to $C = H$ equipped with the regular action of H and M a one dimensional H-module with an extra structure.

One of the main ideas of [**43**] was to view the Hopf-cyclic cohomology as the cohomology of the *invariant* part of certain natural complexes attached to (C, H, M). This should remind one of the cohomology of the Lie algebra of a Lie group as the invariant part of the de Rham cohomology of the group. Another important idea in [**43**] was to introduce *coefficients* into the theory. This also explained the important role played by the so called modular pair in involution (δ, σ) in [**15**].

The module M is a noncommutative analogue of coefficients for Lie algebra cohomology and group cohomology. And a natural question was to identify the most general type of modules. Now the periodicity condition $\tau_n^{n+1} = id$ for the cyclic operator and the fact that all simplicial and cyclic operators have to descend to the invariant complexes, puts restrictions on the type of the H-modules M one can work with. This problem that was partly solved in [**43**] was then completely solved in the twin papers of Hajac-Khalkhali-Rangipour-Sommerhäuser [**25, 26**] by introducing the class of *stable anti-Yetter-Drinfeld modules* over a Hopf algebra. It was shown that the category of anti-Yetter-Drinfeld modules over a Hopf algebra H is obtained from the category of Yetter-Drinfeld H-modules by twisting the latter by a modular pair in involution.

Starting with Section 5 we focus on developing a conceptual understanding of the coefficients, i.e., (stable) anti-Yetter-Drinfeld modules and their generalizations. The main benefit of this approach, besides a general method of extending the definition to other Hopf-like or bialgebra-like settings, is that it suggests that stable anti-Yetter-Drinfeld modules, or rather their mixed generalizations, are exactly the

D-modules on the noncommutative "space" obtained as a quotient of a point by the symmetries given by a Hopf algebra. We see this by observing that stable anti-Yetter-Drinfeld modules arise as the naive cyclic homology category of the category of representations of a Hopf algebra. On the other hand [**5**] shows that $HH(QCohX)^{S^1}$, the cyclic homology category, in the ∞-setting, of the category of quasi-coherent sheaves is, roughly speaking, the category of D-modules.

Once the correct abelian category of coefficients is found, and the algebras are mapped to it, the cohomology is obtained as an Ext, just as in [**11**]. More precisely, given an $A \in \mathcal{C}$ a unital associative algebra, and $M \in HH(\mathcal{C})^{S^1}$ a coefficient, the definition of the analogue of Hopf-cyclic cohomology of A with coefficients in M can be obtained as [**63**]:

$$HC^{\bullet}(A, M) = Ext^{\bullet}_{HH(\mathcal{C})^{S^1}}(ch(A), M),$$

where $ch(A)$ is defined in (7.8). The analogue of the deRham cohomology of M would then be $Ext^{\bullet}_{HH(\mathcal{C})^{S^1}}(ch(1), M)$. We will not actually deal with such cohomology groups in this paper outside of Theorem 7.13. Our focus instead will be on understanding the coefficients, namely the categories $HH(\mathcal{C})$, $HC(\mathcal{C})$, and a bit of $HH(\mathcal{C})^{S^1}$; the latter does not appear outside of Sections 7.6 and 7.7. We note that the naive treatment of Section 5 can be moved to the ∞-setting for a more direct comparison with [**5**].

All algebras, coalgebras and Hopf algebras in this paper are over a fixed field k of characteristic zero, and are unital or counital. Tensors and homs are denoted $\otimes, Hom$ and will be over k, unless specified otherwise. We use Sweedler's notation for comultiplication Δ, with summation understood, and write

$$\Delta(c) = c^{(1)} \otimes c^{(2)}.$$

The *iterated comultiplication* maps

$$\Delta^n = (\Delta \otimes I) \circ \Delta^{n-1} : C \longrightarrow C^{\otimes(n+1)}, \quad \Delta^1 = \Delta,$$

will be written as

$$\Delta^n(c) = c^{(1)} \otimes \cdots \otimes c^{(n+1)},$$

where summation is understood.

2. Hopf algebras in cyclic homology

Even before the introduction of Hopf cyclic cohomology, Hopf algebras played a role in cyclic homology theory, acting as symmetries of associative algebras. In this section we shall briefly look at two such examples.

2.1. Deformation complex and Deligne's conjecture. In 1960's Gerstenhaber understood that deformations of an associative algebra A is controlled by Hochschild cohomology of A with coefficients in A, usually denoted as $H^{\bullet}(A, A)$ [**23**]. In particular he showed that $H^{\bullet}(A, A)$ is a graded Poisson algebra, i.e., a commutative differential graded algebra equipped with a compatible differential graded Lie algebra structure with a shifted degree. A structure that is usually called a Gerstenhaber algebra. He had a similar result for deformations of Lie algebras. The product and the Lie bracket operations are defined on the level of cochains and we have a cochain complex $(C^{\bullet}(A, A), \delta, \cup, [\ ,\ ])$, where

$$C^n(A, A) = Hom_k(A^{\otimes n}, A).$$

The Jacobi identity holds for cochains, but the cup product is only homotopy associative. Compatibility of the cup product with Lie algebra structure is also only up to homotopy. This structure can be rather loosely called a homotopy Gerstenhaber algebra. And a major question is what is the full algebraic structure hidden here? Deligne's conjecture gives a precise answer to this question: The Hochschild cochain complex is an algebra over the singular chain operad of the little squares operad E_2 [**20**].

In [**38**] an algebraic approach to this question is developed. It uses a differential graded Hopf algebra that acts as symmetries of the bar complex and a bar-cobar duality. A central tool here is the notion of $X-$complex for differential graded coalgebras as in [**60**] (see also [3] for the case of algebras). Let us explain this construction briefly. Recall that the bar construction is a functor from the category of differential graded algebras to the category of differential graded coalgebras

$$\mathcal{B} : \mathcal{DGA} \to \mathcal{DGC},$$

where $\mathcal{B}(A) = \oplus_n A^{\otimes n}$. A crucial property of the bar constructions is that its coderivations is isomorphic to the deformation complex:

$$\text{Coder}(\mathcal{B}(A), \mathcal{B}(A)) = C^\bullet(A, A).$$

It is shown in [**38**] that If A is a homotopy Gerstenhaber algebra, then $\mathcal{B}(A)$ is a Hopf algebra.This is how Hopf algebras emerge as relevant to operations on cyclic homology.

The X-complex of a differential graded coalgebra C, denoted $\hat{X}(C)$ is the $\mathbb{Z}/2\mathbb{Z}$-graded complex

$$C \to \Omega^1 C_\natural$$

where $\Omega^1 C_\natural$ is the cocommutator subspace of the C-bicomodule of universal differential 1-forms on C. By a result of Quillen [**60**], the cyclic chain complex of an algebra A is isomorphic to the X-complex of the bar construction

$$\mathcal{B}(A) \to \Omega^1(\mathcal{B}(A))_\natural.$$

From this point of view operations on cyclic and Hochschild complex of homotopy Gerstenhaber algebras are predicted by Kunneth formula for the X-complex of differential graded coalgebras.

More precisely, let V be a homotopy Gerstenhaber algebra. Then it is shown in [**38**] that there are natural maps of supercomplexes

$$\hat{X}(BV) \otimes \hat{X}(BV) \longrightarrow \hat{X}(BV), \quad \hat{X}(BV) \otimes \hat{X}(BV_0) \longrightarrow \hat{X}(BV_0).$$

Applied to $V = C^\bullet(A, A)$, one can get many of the operations.

The main point here is that Hopf algebra symmetries lead to operations in cyclic homology and to a host of intriguing relations among them. We should also mention that the bar construction and Quillen's theorem above should be useful to define cyclic homology of A_∞-algebras and algebras over operads in general. The point is that the bar construction of an A_∞-algebra is a differential graded coalgebra and hence one can consider its X-complex as its cyclic complex. For a recent survey of the status of Deligne's conjecture the reader can consult the paper of Kaufmann and Zhang [**34**].

2.2. Cyclic homology of Hopf crossed products. An important question in cyclic cohomology theory is to compute the cyclic cohomology of a crossed product algebra $A\sharp G$ where the group G acts on the algebra A by automorphisms. This question has been studied by many since the beginning of the theory in 1980's. If G is a discrete group, there is a spectral sequence, due to Feigin and Tsygan [**22**], which converges to the cyclic homology of the crossed product algebra. This result generalizes Burghelea's calculation of the cyclic homology of a group algebra [**8**]. For a recent survey we refer to [**58**] in this volume and references therein.

In [**1**] a spectral sequence is obtained which computes the cyclic cohomology of a Hopf crossed product algebra. The Hopf algebra need not be commutative or cocommutative. This result was later extended to compute the Hopf cyclic cohomology of bicrossed products of two Hopf algebras in [**56, 57**]. This gave a uniform method to compute the Hopf cyclic cohomology of Connes-Moscovici Hopf algebras in all dimensions since the latter are known to be Hopf bicrossed product algebras [**15**].

Let a Hopf algebra $\mathcal{H}$ act on an algebra A such that the coalgebra structure of H is compatible with the algebra structure of A and A is an $\mathcal{H}$-module. We say A is a Hopf module algebra. We can then form the Hopf crossed product algebra $A\sharp\mathcal{H}$. In [**1**] it is shown that there exists an isomorphism between $C_{\bullet}(A^{op}\sharp\mathcal{H}^{cop})$ the cyclic module of the crossed product algebra $A^{op}\sharp\mathcal{H}^{cop}$, and $\Delta(A\natural\mathcal{H})$, the cyclic module defined as the diagonal of a so called cylindrical module $A\natural\mathcal{H}$. When the antipode of the Hopf algebra is invertible, it was shown that the cyclic homology $HC_{\bullet}(A\sharp\mathcal{H})$ can be approximated by a spectral sequence and the of E^0, E^1 and E^2 of this spectral sequence was explicitly computed.

3. The work of Connes and Moscovici

In this section we recall the approach by Connes and Moscovici towards their definition of a cyclic cohomology theory for Hopf algebras. The local index formula [**14**] gives the Connes-Chern character $Ch(A, h, D)$ of a regular spectral triple (A, h, D) as a cyclic cocycle in the (b, B)-bicomplex of the algebra A. For spectral triples of interest in transverse geometry [**15**], this cocycle is in the image of the Connes-Moscovici characteristic map χ_τ defined below. To identify this class in terms of characteristic classes of foliations, it turned out that it would be helpful to show that it is the image of a cocycle for a cohomology theory for Hopf algebras. This is rather similar to the situation for classical characteristic classes which are pull backs of group cohomology classes via a classifying map.

We can formulate this problem abstractly as follows. Let H be a Hopf algebra, $\delta : H \to k$ a character and $\sigma \in H$ a grouplike element. Following [**15–17**], we say that (δ, σ) is a *modular pair* if $\delta(\sigma) = 1$, and a *modular pair in involution* if

$$\widetilde{S}_\delta^2(h) = \sigma h \sigma^{-1},$$

for all h in H. Here the *δ-twisted antipode* $\widetilde{S}_\delta : H \to H$ is defined as

$$\widetilde{S}_\delta(h) = \delta(h^{(1)})S(h^{(2)}).$$

Now let A be an H-module algebra and $\tau : A \to k$ be a δ- invariant σ-trace on A. The Connes-Moscovici characteristic map is defined as

$$\chi_\tau : H^{\otimes n} \longrightarrow Hom(A^{\otimes(n+1)},\ k),$$

$$\chi_\tau(h_1 \otimes \cdots \otimes h_n)(a_0 \otimes \cdots \otimes a_n) = \tau(a_0 h_1(a_1) \cdots h_n(a_n)).$$

This is a generalization of a map defined for an action of a Lie algebra on an algebra in [**9**] which was then fully developed in [**12**]. The question is if one can define a cocyclic module structure on the collection of spaces $\{H^{\otimes n}\}_{n\geq 0}$ such that the characteristic map χ_τ turns into a morphism of cocyclic modules? The face, degeneracy, and cyclic operators for $Hom(A^{\otimes(n+1)}, k)$ are defined as

$$\begin{aligned}
\delta_i^n \varphi(a_0, \cdots, a_{n+1}) &= \varphi(a_0, \cdots, a_i a_{i+1}, \cdots, a_{n+1}), \quad i = 0, \cdots, n,\\
\delta_{n+1}^n \varphi(a_0, \cdots, a_{n+1}) &= \varphi(a_{n+1} a_0, a_1, \cdots, a_n),\\
\sigma_i^n \varphi(a_0, \cdots, a_n) &= \varphi(a_0, \cdots, a_i, 1, \cdots, a_n), \quad i = 0, \cdots, n,\\
\tau_n \varphi(a_0, \cdots, a_n) &= \varphi(a_n, a_0, \cdots, a_{n-1}).
\end{aligned}$$

The relation

$$h(ab) = h^{(1)}(a) h^{(2)}(b)$$

shows that the face operators on $H^{\otimes n}$ must involve the coproduct of H:

$$\begin{aligned}
\delta_0^n(h_1 \otimes \cdots \otimes h_n) &= 1 \otimes h_1 \otimes \cdots \otimes h_n,\\
\delta_i^n(h_1 \otimes \cdots \otimes h_n) &= h_1 \otimes \cdots \otimes h_i^{(1)} \otimes h_i^{(2)} \otimes \cdots \otimes h_n,\\
\delta_{n+1}^n(h_1 \otimes \cdots \otimes h_n) &= h_1 \otimes \cdots \otimes h_n \otimes \sigma,
\end{aligned}$$

With this definition, we have

$$\chi_\tau \delta_i^n = \delta_i^n \chi_\tau.$$

Similarly, the degeneracy operators on $H^{\otimes n}$ should be defined as

$$\sigma_i^n(h_1 \otimes \cdots \otimes h_n) = h_1 \otimes \cdots \otimes \varepsilon(h_i) \otimes \cdots \otimes h_n.$$

A challenging part is to define the *cyclic operator* $\tau_n : H^{\otimes n} \to H^{\otimes n}$. Compatibility of the latter with χ_τ implies the relation

$$\tau(a_0 \tau_n(h_1 \otimes \cdots \otimes h_n)(a_1 \otimes \cdots \otimes a_n)) = \tau(a_n h_1(a_0) h_2(a_1) \cdots h_n(a_{n-1})).$$

Now the (δ, σ)-invariance property of τ shows that

$$\tau(a_1 h(a_0)) = \tau(h(a_0)\sigma(a_1)) = \tau(a_0 \tilde{S}_\delta(h)(\sigma(a_1)).$$

This suggests a definition for $\tau_1 : H \to H$ as

$$\tau_1(h) = \tilde{S}_\delta(h)\sigma.$$

Note that the condition $\tau_1^2 = I$ is equivalent to the involutive condition $\tilde{S}_\delta^2(h) = \sigma h \sigma^{-1}$. In general one can show that for any n:

$$\begin{aligned}
\tau(a_n h_1(a_0) \cdots h_n(a_{n-1})) &= \tau(h_1(a_0) \cdots h_n(a_{n-1})\sigma(a_n))\\
&= \tau(a_0 \tilde{S}_\delta(h_1)(h_2(a_1) \cdots h_n(a_{n-1})\sigma(a_n)))\\
&= \tau(a_0 \tilde{S}_\delta(h_1) \cdot (h_2 \otimes \cdots \otimes h_n \otimes \sigma)(a_1 \otimes \cdots \otimes a_n).
\end{aligned}$$

This suggests that the *Hopf-cyclic operator* $\tau_n : H^{\otimes n} \to H^{\otimes n}$ should be defined as

$$\tau_n(h_1 \otimes \cdots \otimes h_n) = \tilde{S}_\delta(h_1) \cdot (h_2 \otimes \cdots \otimes h_n \otimes \sigma),$$

where $\cdot$ denotes the diagonal action defined by

$$h \cdot (h_1 \otimes \cdots \otimes h_n) := h^{(1)} h_1 \otimes h^{(2)} h_2 \otimes \cdots \otimes h^{(n)} h_n.$$

We let $\tau_0 = I : H^{\otimes 0} = k \to H^{\otimes 0}$, be the identity map.

Connes and Moscovici proved the remarkable result that the above face, degeneracy, and cyclic operators on $\{H^{\otimes n}\}_{n\geq 0}$ define a cocyclic module [**15–17**]. The resulting cyclic cohomology groups are denoted by $HC^n_{(\delta,\sigma)}(H)$, $n = 0, 1, \cdots$ and we have the desired characteristic map

$$\chi_\tau : HC^n_{(\delta,\sigma)}(H) \to HC^n(A).$$

The Hopf cyclic cohomology groups $HC^n_{(\delta,\sigma)}(H)$ are computed for several key examples in [**15**–**17**, **56**, **57**]. For a recent survey and references we refer to the article of Moscovici in this volume [**55**]. In particular for applications to transverse index theory, the (periodic) cyclic cohomology of the Connes-Moscovici Hopf algebra $\mathcal{H}_1$ plays an important role. It is shown in [**15**] that the periodic groups $HP^n_{(\delta,1)}(\mathcal{H}_1)$ are canonically isomorphic to the Gelfand-Fuchs cohomology of the Lie algebra of formal vector fields $\mathfrak{a}_1$ on the line:

$$H^\bullet(\mathfrak{a}_1, \mathbb{C}) = HP^\bullet_{(\delta,1)}(\mathcal{H}_1).$$

The following interesting elements appear. It can be directly checked that the elements $\delta_2' := \delta_2 - \frac{1}{2}\delta_1^2$ and δ_1 are cyclic 1-cocycles on $\mathcal{H}_1$, and

$$F := X \otimes Y - Y \otimes X - \delta_1 Y \otimes Y$$

is a cyclic 2-cocycle. See [**18**] for a survey and relations between these cocycles and the Schwarzian derivative, Godbillon-Vey cocycle, and the transverse fundamental class of Connes [**13**], respectively.

4. Hopf cyclic cohomology with coefficients

In [**43**] (see also [**2**] for an equivariant version) the Hopf cyclic cohomology was extended to a so called *invariant cyclic cohomology* theory for triples (C, H, M) where C is a H-module coalgebra and M is an H-module. The three pieces C, H, M had to satisfy some compatibility conditions so that the resulting complex turns out to be a cyclic module. An important question which was left open in [**43**] was to identify the most general class of coefficients M for this Hopf cyclic cohomology. This problem was completely solved in the twin papers of Hajac-Khalkhali-Rangipour-Sommerhauser [**25**, **26**]. It was shown that the most general coefficients are the class of so called *stable anti-Yetter-Drinfeld modules* (AYD modules). It was found that anti-Yetter-Drinfeld modules are twistings by modular pairs in involution of Yetter-Drinfeld modules. This means that the category of anti-Yetter-Drinfeld modules is a 'mirror' of the category of Yetter-Drinfeld modules. It is quite surprising that when the general formalism of cyclic cohomology theory, namely the theory of (co)cyclic modules [**11**], is applied to Hopf algebras, variations of notions like Yetter-Drinfeld modules which are of importance in low dimensional topology and invariants of knots and ribbons appear naturally.

4.1. Stable anti Yetter-Drinfeld modules. Let H be a Hopf algebra, M a left H-module and left H-comodule. Then M is called a left-left *Yetter-Drinfeld H-module* if

$$\rho(hm) = h^{(1)}m^{(-1)}S(h^{(3)}) \otimes h^{(2)}m^{(0)},$$

for all $h \in H$ and $m \in M$ [**53**, **61**, **68**]. We denote the category of left-left Yetter-Drinfeld modules over H by ${}^H_H\mathcal{YD}$. This notion is closely related to the Drinfeld double of finite dimensional Hopf algebras. In fact if H is finite dimensional, then one can show that the category ${}^H_H\mathcal{YD}$ is isomorphic to the category of left modules over the Drinfeld double $D(H)$ of H [**53**].

The category ${}^H_H\mathcal{YD}$ is a braided monoidal category under the braiding

$$c_{M,N} : M \otimes N \to N \otimes M, \quad m \otimes n \mapsto m^{(-1)} \cdot n \otimes m^{(0)},$$

and the tensor product $M \otimes N$ of two Yetter-Drinfeld modules defined by structure maps

$$h(m \otimes n) = h^{(1)}m \otimes h^{(1)}n, \quad (m \otimes n) \mapsto m^{(-1)}n^{(-1)} \otimes m^{(0)} \otimes n^{(0)}.$$

The category ${}^H_H\mathcal{YD}$ is the *center* of the monoidal category ${}_H\mathcal{M}$ of left H-modules. Recall that the (left) center $\mathcal{ZC}$ of a monoidal category [**33**] is a category whose objects are pairs $(X, \sigma_{X,-})$, where X is an object of $\mathcal{C}$ and $\sigma_{X,-} : X \otimes - \to - \otimes X$ is a natural isomorphism satisfying certain compatibility axioms with associativity and unit constraints. It can be shown that the center of a monoidal category is a braided monoidal category and $\mathcal{Z}({}_H\mathcal{M}) =^H_H \mathcal{YD}$ [**33**].

Let us give an example of a class of Yetter-Drinfeld modules. Let $H = kG$ be the group algebra of a discrete group G. Let $\mathcal{G}$ be a groupoid whose objects are G and its morphisms are defined by

$$Hom(g, h) = \{k \in G;\ kgk^{-1} = h\}.$$

Recall that an *action* of a groupoid $\mathcal{G}$ on the category Vec of vector spaces is simply a functor $F : \mathcal{G} \to$ Vec. Then it is easily seen that we have a one-one correspondence between YD modules for kG and actions of $\mathcal{G}$ on Vec. This example clearly shows that one can think of an YD module over kG as an *equivariant sheaf* on G and of YD modules as noncommutative analogues of equivariant sheaves on a topological group. See Section 7.4 for the dual case of $H = \mathcal{O}_G$, i.e, functions on a monoid G.

Unlike Yetter-Drinfeld modules, the definition of anti-Yetter-Drinfeld modules and stable anti-Yetter-Drinfeld modules was entirely motivated by cyclic cohomology theory. In fact the anti-Yetter-Drinfeld condition guarantees that the simplicial and cyclic operators are well defined on invariant complexes and the stability condition implies that the crucial periodicity condition for cyclic modules are satisfied. A center-like description is possible, see Section 7.5.

DEFINITION 4.1. ([**25**]) A left-left anti-Yetter-Drinfeld H-module is a left H-module and left H-comodule such that

$$\rho(hm) = h^{(1)}m^{(-1)}S^{-1}(h^{(3)}) \otimes h^{(2)}m^{(0)},$$

for all $h \in H$ and $m \in M$. M is called stable if in addition we have

$$m^{(-1)}m^{(0)} = m,$$

for all $m \in M$.

The following lemma from [**25**] shows that 1-dimensional SAYD modules correspond to Connes-Moscovici's modular pairs in involution:

LEMMA 4.2. *There is a one-one correspondence between modular pairs in involution (δ, σ) on H and SAYD module structure on $M = k$, defined by*

$$h.r = \delta(h)r, \quad r \mapsto \sigma \otimes r,$$

for all $h \in H$ and $r \in k$. We denote this module by $M = {}^\sigma k_\delta$.

Let ${}^H_H\mathcal{AYD}$ denote the category of left-left anti-Yetter-Drinfeld H-modules, where morphisms are H-linear and H-colinear maps. Unlike YD modules, anti-Yetter-Drinfeld modules do not form a monoidal category under the standard tensor product. This can be checked easily on 1-dimensional modules given by modular pairs in involution. The following result of [**25**], however, shows that the tensor product of an anti-Yetter-Drinfeld module with a Yetter-Drinfeld module is again anti-Yetter-Drinfeld.

LEMMA 4.3. *Let M be a Yetter-Drinfeld module and N be an anti-Yetter-Drinfeld module (both left-left). Then $M \otimes N$ is an anti-Yetter-Drinfeld module under the diagonal action and coaction:*

$$h(m \otimes n) = h^{(1)}m \otimes h^{(1)}n, \quad (m \otimes n) \mapsto m^{(-1)}n^{(-1)} \otimes m^{(0)} \otimes n^{(0)}.$$

In particular, using a modular pair in involution (δ, σ), we obtain a full and faithful functor

$${}^H_H\mathcal{YD} \to^H_H \mathcal{AYD}, \quad M \mapsto \bar{M} =^{\sigma} k_\delta \otimes M.$$

This result clearly shows that AYD modules can be regarded as the *twisted analogue* or *mirror image* of YD modules, with twistings provided by modular pairs in involution.

Noncommutative principal bundles, also known as Hopf-Galois extensions, give rise to large classes of examples of SAYD modules [**25**]. Let P be a right H-comodule algebra, and let

$$B := P^H = \{p \in P;\ \rho(p) = p \otimes 1\}$$

be the space of coinvariants of P. It is easy to see that B is a subalgebra of P. The extension $B \subset P$ is called a Hopf-Galois extension if the map

$$can : P \otimes_B P \to B \otimes H, \quad p \otimes p' \mapsto p\rho(p'),$$

is bijective. The bijectivity assumption allows us to define the translation map $T : H \to P \otimes_B P$,

$$T(h) = can^{-1}(1 \otimes h) = h^{(\bar{1})} \otimes h^{(\bar{2})}.$$

It can be checked that the centralizer $Z_B(P) = \{p \mid bp = pb\ \forall b \in B\}$ of B in P is a subcomodule of P. There is an action of H on $Z_B(P)$ defined by $ph = h^{(1)}ph^{(2)}$ called the Miyashita-Ulbrich action. It is shown that this action and coaction satisfy the Yetter-Drinfeld compatibility condition. On the other hand if B is central, then by defining the new action $ph = (S^{-1}(h))^{(2)}p(S^{-1}(h))^{(1)}$and the right coaction of P we have a SAYD module.

4.2. Hopf cyclic cohomology with coefficients. In the work of Hajac-Khalkhali-Rangipour-Sommerhaeuser [**26**] 3 types of Hopf-cyclic cohomology theory are defined. In this section we give a brief account of two of them.

Let A be a left H-module algebra and M be a left-right SAYD H-module. Then the spaces $M \otimes A^{\otimes(n+1)}$ are right H-modules via the diagonal action

$$(m \otimes \widetilde{a})h := mh^{(1)} \otimes S(h^{(2)})\widetilde{a},$$

where the left H-action on $\widetilde{a} \in A^{\otimes(n+1)}$ is via the left diagonal action of H.

We define the space of *equivariant cochains on A with coefficients in M* by

$$C^n_H(A, M) := Hom_H(M \otimes A^{\otimes(n+1)}, k).$$

More explicitly, $f : M \otimes A^{\otimes(n+1)} \to k$ is in $C^n_H(A, M)$, if and only if

$$f((m \otimes a_0 \otimes \cdots \otimes a_n)h) = \varepsilon(h)f(m \otimes a_0 \otimes \cdots \otimes a_n),$$

for all $h \in H, m \in M$, and $a_i \in A$. It is shown in [**26**] that the following operators define a cocyclic module structure on $\{\mathcal{C}^n_H(A,M)\}_{n\in\mathbb{N}}$:

$$(\delta_i f)(m \otimes a_0 \otimes \cdots \otimes a_n) = f(m \otimes a_0 \otimes \cdots \otimes a_i a_{i+1} \otimes \cdots \otimes a_n),\ 0 \leq i < n,$$

$$(\delta_n f)(m \otimes a_0 \otimes \cdots \otimes a_n) = f(m^{(0)} \otimes (S^{-1}(m^{(-1)})a_n)a_0 \otimes a_1 \otimes \cdots \otimes a_{n-1}),$$

$$(\sigma_i f)(m \otimes a_0 \otimes \cdots \otimes a_n) = f(m \otimes a_0 \otimes \cdots \otimes a_i \otimes 1 \otimes \cdots \otimes a_n),\ 0 \leq i \leq n,$$

$$(\tau_n f)(m \otimes a_0 \otimes \cdots \otimes a_n) = f(m^{(0)} \otimes S^{-1}(m^{(-1)})a_n \otimes a_0 \otimes \cdots \otimes a_{n-1}).$$

We denote the resulting cyclic cohomology theory by $HC^n_H(A,M), n = 0,1,\cdots$. For $M = H$ and H acting on M by conjugation and coacting via coproduct (Example 2.2.3.), we obtain the equivariant cyclic cohomology theory of Akbarpour and Khalkhali for H-module algebras [**1,2**].

Here is another variation. Connes-Moscovici's original example of Hopf-cyclic cohomology belongs to this class of theories. Let C be a left H-module coalgebra, and M be a right-left SAYD H-module. Let

$$\mathcal{C}^n(C,M) := M \otimes_H C^{\otimes(n+1)} \quad n \in \mathbb{N}.$$

It can be checked that, thanks to the SAYD condition on M, the following operators are well defined and define a cocyclic module, denoted $\{\mathcal{C}^n(C,M)\}_{n\in\mathbb{N}}$. In particular the crucial periodicity conditions $\tau_n^{n+1} = id$, $n = 0,1,2\cdots$, are satisfied [**26**]:

$$\begin{aligned}
\delta_i(m \otimes c_0 \otimes \cdots \otimes c_{n-1}) &= m \otimes c_0 \otimes \cdots \otimes c_i^{(1)} \otimes c_i^{(2)} \otimes c_{n-1},\ 0 \leq i < n,\\
\delta_n(m \otimes c_0 \otimes \cdots \otimes c_{n-1}) &= m^{(0)} \otimes c_0^{(2)} \otimes c_1 \otimes \cdots \otimes c_{n-1} \otimes m^{(-1)} c_0^{(1)},\\
\sigma_i(m \otimes c_0 \otimes \cdots \otimes c_{n+1}) &= m \otimes c_0 \otimes \cdots \otimes \varepsilon(c_{i+1}) \otimes \cdots \otimes c_{n+1},\ 0 \leq i \leq n,\\
\tau_n(m \otimes c_0 \otimes \cdots \otimes c_n) &= m^{(0)} \otimes c_1 \otimes \cdots \otimes c_n \otimes m^{(-1)} c_0.
\end{aligned}$$

For $C = H$ and $M = {}^{\sigma}k_{\delta}$, the cocyclic module $\{\mathcal{C}^n_H(C,M)\}_{n\in\mathbb{N}}$ is isomorphic to the cocyclic module of Connes-Moscovici [**15**], attached to a Hopf algebra endowed with a modular pair in involution. This example is truly fundamental and started the whole theory.

It is possible to interpret the stable anti-Yetter-Drinfeld H-modules as a local system over H, that is a module equipped with a noncommutative flat connection [**36**]. Excision in Hopf cyclic cohomology is studied in [**37**]. Other references include [**24, 40, 44–47**].

5. Beyond Hopf algebras, a brief detour into limits

Let I be a small category ([**52**] should suffice for this section). Recall that an I-indexed diagram in a category $\mathcal{T}$ is a functor $D_\bullet : I \to \mathcal{T}$. These naturally form the category $Fun(I,\mathcal{T})$. We have an obvious functor $c : \mathcal{T} \to Fun(I,\mathcal{T})$ that assigns to an object $T \in \mathcal{T}$ the constant diagram with $c(T)_i = T$ for all $i \in I$ and for any $f \in Hom_I(i,j)$, the induced map $f_* \in Hom_{\mathcal{T}}(c(T)_i, c(T)_j)$ is Id_T. Should they exist, we denote by $\varinjlim_I$ the left adjoint of c and by $\varprojlim_I$ the right adjoint of c to denote the colimit and limit functors; the use of the latter will be delayed until Section 7.

Since we are interested in the I-indexed diagrams in the 2-category $\mathcal{C}at$ of small categories enriched over Vec, i.e., vector spaces, let us briefly spell out the subtleties that this categorification entails. A diagram $\mathcal{C}_\bullet : I \to \mathcal{C}at$ consists of categories $\mathcal{C}_i$, an assignment of $f_* \in Fun(\mathcal{C}_i, \mathcal{C}_j)$ to every $f \in Hom_I(i,j)$, and natural isomorphisms $\iota_{f,g} \in NatIso(f_*g_*, (fg)_*)$. The latter satisfy

$$\iota_{f,gh}\iota_{g,h} = \iota_{fg,h}\iota_{f,g}.$$

We can define the weak version of the above by relaxing the condition that $\iota_{f,g}$'s be isomorphisms.

Given two diagrams $\mathcal{C}_\bullet$ and $\mathcal{D}_\bullet$ a morphism of diagrams is the data $(F_\bullet, \beta_f)$, where $F_i \in Fun(\mathcal{C}_i, \mathcal{D}_i)$ and $\beta_f \in NatIso(f_*^{\mathcal{D}}F_i, F_j f_*^{\mathcal{C}})$ for $f \in Hom_I(i,j)$. The β_f's satisfy

$$\iota_{f,g}^{\mathcal{C}}\beta_f\beta_g = \beta_{fg}\iota_{f,g}^{\mathcal{D}}.$$

A weak morphism is obtained by again relaxing the condition that β_f's be isomorphisms.

A constant diagram $c(\mathcal{T})_\bullet$ is obtained from a category $\mathcal{T}$ in an obvious manner; namely, $c(\mathcal{T})_i = \mathcal{T}$, $f_* = Id_{\mathcal{T}}$, and $\iota_{f,g} = Id$. By a (weak) section of $\mathcal{C}_\bullet$ we mean a (weak) morphism from $c(\mathrm{Vec})_\bullet$ to $\mathcal{C}$. Explicitly, it is a choice of $C_i \in \mathcal{C}_i$ together with $\beta_f \in Hom_{\mathcal{C}_j}(f_*C_i, C_j)$, where the β_f is an isomorphism if the section is not weak. The compatibility condition becomes:

$$\beta_f f_*(\beta_g) = \beta_{fg}\iota_{f,g}.$$

It is immediate that (weak) morphisms map (weak) sections to (weak) sections, furthermore a weak section of $c(\mathcal{T})_\bullet$ is just an object in $Fun(I, \mathcal{T})$, while a strong section is much less reasonable, requiring all maps in I to be mapped to isomorphisms in $\mathcal{T}$. Thus, as tempting as it is to ignore the weak versions of the above concepts, one does need them as we shall see below. Of course a strong version of any of the above may freely be used in place of a weak one.

Let us again define $\varinjlim_I$ to be the left adjoint to c, the inclusion of constant diagrams into (weak) diagrams. According to the above discussion, we see that if our goal (and it is, to first order) is to obtain objects in $Fun(I, \mathcal{T})$ from weak sections in $\mathcal{C}_\bullet$ we need only obtain an object in $Fun(\varinjlim_I \mathcal{C}_\bullet, \mathcal{T})$. In particular, given an $M \in \varinjlim_I \mathcal{C}_\bullet$ we obtain an object in $Fun(I^{op}, \mathrm{Vec})$ via

$$Hom_{\varinjlim_I \mathcal{C}_\bullet}(-, M)$$

and similarly, an object in $Fun(I, \mathrm{Vec})$ via

$$Hom_{\varinjlim_I \mathcal{C}_\bullet}(M, -).$$

On the other hand, taking $\mathcal{T} = \varinjlim_I \mathcal{C}_\bullet$ and the functor to be Id, we see that weak sections of $\mathcal{C}_\bullet$ yield functors from I to $\varinjlim_I \mathcal{C}_\bullet$.

We will next focus on trying to understand $\varinjlim_I \mathcal{C}_\bullet$ in a particular setting of direct relevance to cyclic homology.

6. Algebras and monoidal categories

Let us briefly recall the definition of a cyclic object associated to an algebra. More precisely, we mean an associative algebra with 1. We observe that if A is an algebra then this structure is encoded into a functor $A_\bullet : \Delta \to \mathrm{Vec}$ where Δ is the simplex category of finite linearly ordered sets, with non-decreasing maps (made small). More precisely, let $n \in \Delta$ denote the set consisting of $0, 1, \cdots, n$ and set $A_n = A^{\otimes(n+1)}$. For $f \in Hom_\Delta(i, j)$ set

$$f_*(a_0 \otimes \cdots \otimes a_i) = \otimes_{s=0}^{j} f_*^s(a_0 \otimes \cdots \otimes a_i)$$
$$f_*^s(a_0 \otimes \cdots \otimes a_i) = \prod_{l \in f^{-1}(s)} a_l$$

where the induced ordering on $f^{-1}(s)$ is used to define the product, and $\prod_\emptyset = 1$.

The above $A_\bullet$ extends to the Connes' cyclic category Λ. One then applies cyclic duality [**51**], i.e., the identification of Λ^{op} with Λ by essentially exchanging the roles of faces and degeneracies to obtain

$$A_\bullet : \Lambda^{op} \to \text{Vec}.$$

The latter is the classical cyclic object associated to the unital associative algebra A. One obtains the simplicial object by considering the restriction of the digram $A_\bullet$ to the simplex category Δ^{op}. The cyclic and Hochschild cohomologies of A are then expressed as an *Ext* in the abelian categories of cyclic and simplicial objects respectively [**11**]; see also [**32**] for the mixed complexes version.

6.1. Monoidal categories. Roughly speaking, a monoidal category $\mathcal{C}$ [**29**] is a category equipped with a bifunctor $\otimes : \mathcal{C} \times \mathcal{C} \to \mathcal{C}$ written as $(X, Y) \mapsto X \otimes Y$ and an associativity constraint $\alpha_{X,Y,Z} : X \otimes (Y \otimes Z) \simeq (X \otimes Y) \otimes Z$ that satisfies a coherence (pentagon) axiom. There is also a monoidal unit $1 \in \mathcal{C}$ with natural requirements. We will not be precise here since we will reinterpret this structure below.

For the sake of clarity of exposition we begin with a construction that will be imitated below, but will not itself be immediately useful to us. The definitions are themselves essentially a copy of those of $A_\bullet$ above. We define a strong Δ-indexed diagram in $\mathcal{C}at$ by setting $\mathcal{C}_n = \mathcal{C}^{\times(n+1)}$. To every $f \in Hom_\Delta(i, j)$ we assign

$$f_*(X_0, \cdots, X_i)_s = \bigotimes_{l \in f^{-1}(s)} X_l, \quad \text{with} \quad \bigotimes_\emptyset = 1.$$

Note that the associator eliminates the ambiguity of bracketing and defines the required structure of $\iota_{f,g} : f_* g_* \simeq (fg)_*$.

REMARK 6.1. There are settings, such as that of multiplier Hopf algebras [**67**] when the $\iota_{f,g}$ above are not isomorphisms, so that we do not really have a monoidal category. We nevertheless do have a weak diagram as above.

Let A be a unital associative algebra in $\mathcal{C}$. The definition is exactly what one would expect, namely a monoid object in $\mathcal{C}$. Note that $(A, \cdots, A) \in \mathcal{C}^{\times(n+1)}$ together with $f_*(A, \cdots, A)_s = \otimes_{l \in f^{-1}(s)} A \to A$ given by multiplication when $f^{-1}(s) \neq \emptyset$ and the unit, when it is empty, gives a *weak* section of $\mathcal{C}_\bullet$ as defined above.

Everything extends to diagrams and their sections over Λ. One then applies cyclic duality to obtain a diagram and its weak section over Λ^{op}. Recall that the input to this procedure is an algebra A in a monoidal category $\mathcal{C}$. We will take this no further here, though in principle we see that given the pair $(\mathcal{C}, A)$ we obtain a cyclic object in $\varinjlim_{\Lambda^{op}} \mathcal{C}_\bullet$. The issue is that we don't know how to compute the limit category. We begin to address this deficiency in the next section, by changing the limit.

6.2. A variation on the notion of a monoidal category. To address the difficulty pointed out in the previous section, we need [**63**] to replace the category $\mathcal{C}^{\times k}$ with a more sophisticated category $\mathcal{C}^{\boxtimes k}$. The definition of the latter is more subtle than the former, but it is worth it for us for two reasons: it is actually not that sophisticated in our examples and it makes the limit category fairly computable.

Let $\mathcal{C} = {}_H\mathcal{M}$, i.e., as an abelian category it is given as the category of left H-modules for some algebra H. Then $\mathcal{C}^{\boxtimes 2} = {}_{H^2}\mathcal{M}$, i.e., the category of modules with two commuting left H-actions; similarly $\mathcal{C}^{\boxtimes n}$ is defined as the category of modules with n commuting left H-actions. We have a functor $\mathcal{C}^{\times n} \to \mathcal{C}^{\boxtimes n}$ given in our example by $(X_1, \cdots, X_n) \mapsto X_1 \otimes_k \cdots \otimes_k X_n$ where k is the field over which H is an algebra. We will assume that the monoidal product on $\mathcal{C}$ extends to an appropriately associative functor

$$\Delta^* : \mathcal{C}^{\boxtimes 2} \to \mathcal{C},$$

i.e., the associator extends to $\Delta^*(Id \boxtimes \Delta^*) \simeq \Delta^*(\Delta^* \boxtimes Id)$ that satisfies the pentagon axiom. The notation is motivated by the case of H a Hopf algebra (or more generally, a bialgebra) where the product is given by the pullback along the algebra map $\Delta : H \to H \otimes H$ and its right adjoint $\Delta_* : \mathcal{C} \to \mathcal{C}^{\boxtimes 2}$, to be defined shortly.

REMARK 6.2. Going back to Remark 6.1 we point out that in the case of multiplier Hopf algebras the appropriate "monoidal structure" on $\mathcal{C}$ actually depends on the desired type of the Hopf-cyclic theory. One of the appropriate structures on $\mathcal{C}$ is monoidal and extends, the other is either weakly monoidal as in the previous remark, or strongly monoidal in the sense of this section. Thus, a weakly monoidal structure can extend to a strongly monoidal one. We further remark that in the multiplier case the extension has a left, not right, adjoint.

We can repeat the constructions of Section 6.1 to obtain a Λ^{op}-indexed diagram $\mathcal{C}^{\boxtimes(\bullet+1)}$ with a diagram morphism from the $\mathcal{C}^{\times(\bullet+1)}$ constructed therein. Thus, the new diagram inherits a weak section (arising from $A \in \mathcal{C}$) from the old one. The problem now reduces to the computation of

$$HC(\mathcal{C}) = \varinjlim_{\Lambda^{op}} \mathcal{C}^{\boxtimes(\bullet+1)}.$$

To solve it we will assume the existence of right adjoints. This is a reasonable assumption, though sometimes we would need the left adjoints instead, the discussion would be similar.

It will be useful to us later to also consider the restriction of the diagram $\mathcal{C}^{\boxtimes(\bullet+1)}$ to the index category Δ^{op}. Let us denote the resulting colimit as follows:

$$HH(\mathcal{C}) = \varinjlim_{\Delta^{op}} \mathcal{C}^{\boxtimes(\bullet+1)}.$$

Observe that the colimit over the smaller diagram results in a category that has an extra structure when compared to the original colimit, namely an $\varsigma \in Aut(Id_{HH(\mathcal{C})})$ that controls stability (see Definition 4.1) in the sense that an object X of $HH(\mathcal{C})$ is stable if $\varsigma_X = Id$.

7. Towards understanding the limit category

Let us assume that the monoidal product of $\mathcal{C}$, originally given as a bifunctor $\otimes : \mathcal{C} \times \mathcal{C} \to \mathcal{C}$, not only extends to a functor $\Delta^* : \mathcal{C} \boxtimes \mathcal{C} \to \mathcal{C}$, but that the latter has a right adjoint Δ_*. Under the additional assumption that we are looking for the direct limit (colimit) defining $HC(\mathcal{C})$ in $\mathcal{C}at_*$, i.e., categories with functors possessing right adjoints, we have that

$$HC(\mathcal{C}) = \varprojlim_{\Lambda} \mathcal{C}_*^{\boxtimes(\bullet+1)} \tag{7.1}$$

where the diagram $\mathcal{C}_*^{\boxtimes(\bullet+1)}$ is obtained from $\mathcal{C}^{\boxtimes(\bullet+1)}$ by replacing all functors with their right adjoints. Note that the inverse limit (limit) is computed in $\mathcal{C}at^*$.

We observe that the limit (7.1) can be explicitly described as the category of strong sections of $\mathcal{C}_*^{\boxtimes(\bullet+1)}$. This holds in general, but in the case of our indexing category Λ we have an even simpler description. In the following we use the faces δ's and degeneracies s's notation. For example $\delta_0 = \Delta_*$ and $\delta_1 = \sigma\Delta_*$ for $\delta_i : \mathcal{C} \to \mathcal{C}^{\boxtimes 2}$. Note that $\sigma : \mathcal{C}^{\boxtimes 2} \to \mathcal{C}^{\boxtimes 2}$ just interchanges the two H-actions.

A strong section, in the case of the Connes' cyclic category as index, reduces to an $M \in \mathcal{C}$ equipped with an isomorphism τ in $\mathcal{C}^{\boxtimes 2}$ subject to conditions, namely:

$$\begin{aligned} \tau : \delta_0(M) \simeq \delta_1(M) &\quad \text{structure in } \mathcal{C}^{\boxtimes 2}, \\ s_0(\tau) = Id_M &\quad \text{unit condition in } \mathcal{C}, \\ \delta_2(\tau)\delta_0(\tau) = \delta_1(\tau) &\quad \text{associativity condition in } \mathcal{C}^{\boxtimes 3}, \\ s_1(\tau) = Id_M &\quad \text{stability condition in } \mathcal{C}. \end{aligned} \tag{7.2}$$

The last, stability condition, is omitted for the description of $HH(\mathcal{C})$, instead, it equips the latter with a

$$\varsigma = s_1(\tau) \in Aut(Id_{HH(\mathcal{C})}).$$

In light of this description we see that $HH(\mathcal{C})$ and $HC(\mathcal{C})$ consist of objects of $\mathcal{C}$ with extra structure with the distinction between $HH(\mathcal{C})$ and $HC(\mathcal{C})$ being an extra property on the latter. This is very much like the definition of the center of a monoidal category $\mathcal{Z}(\mathcal{C})$; we will come back to this in Section 7.5.

REMARK 7.1. The same arguments apply if we assume that the monoidal product $\otimes : \mathcal{C}\times\mathcal{C} \to \mathcal{C}$ extends to $\Delta^! : \mathcal{C}\boxtimes\mathcal{C} \to \mathcal{C}$ that has a left adjoint $\Delta_!$. The difference is illustrated in the subsequent section.

7.1. Categorical traces. We observe that structure/properties (7.2) imply that, for $M \in \mathcal{C}$ that satisfies them, the functor

$$T(-) = Hom_{\mathcal{C}}(-, M)$$

is a (contravariant) trace from $\mathcal{C}$ to Vec [**6**, **28**, **30**, **31**, **59**]. Namely, we can define natural transformations $\iota_{X,Y} : T(X \otimes Y) \simeq T(Y \otimes X)$, for $X, Y \in \mathcal{C}$ via [**63**]:

$$\begin{aligned} Hom_{\mathcal{C}}(X \otimes Y, M) = Hom_{\mathcal{C}}(\Delta^*(X \boxtimes Y), M) &= Hom_{\mathcal{C}^{\boxtimes 2}}(X \boxtimes Y, \Delta_* M) \\ &\overset{\tau}{\to} Hom_{\mathcal{C}^{\boxtimes 2}}(X \boxtimes Y, \sigma\Delta_* M) = Hom_{\mathcal{C}}(\Delta^*(Y \boxtimes X), M) \\ &= Hom_{\mathcal{C}}(Y \otimes X, M). \end{aligned}$$

These isomorphisms satisfy: $\iota_{1,X} = Id = \iota_{X,1}$, and for $X, Y, Z \in \mathcal{C}$ we have $\iota_{X\otimes Y,Z} = \iota_{Y,Z\otimes X}\iota_{X,Y\otimes Z}$. Note that $\iota_{1,X} = Id$ is automatic and $\iota_{X,1} = Id$ will be relaxed in Section 7.6.

THEOREM 7.2 ([**28**]). *Any $T : \mathcal{C} \to Vec$ with this structure makes $T(A^{\otimes(\bullet+1)})$, for $A \in \mathcal{C}$ an algebra, into a cocyclic object. Similarly, for $C \in \mathcal{C}$ a coalgebra, we have that $T(C^{\otimes(\bullet+1)})$ is a cyclic object.*

Here, we were operating under the assumption that we are using the right adjoint to the product Δ^*. Similarly, we see that $L(-) = Hom_{\mathcal{C}}(N, -)$ is a (covariant) trace from $\mathcal{C}$ to Vec if the left adjoint $\Delta_!$ to the product $\Delta^!$ is used instead. The trace L produces (co)-cyclic objects from (co)-algebras in $\mathcal{C}$.

7.2. Monadicity. Let $U : HC(\mathcal{C}) \to \mathcal{C}$ denote the functor that forgets the extra structure. Under our assumptions above U has a left adjoint F; the pair (F, U) is monadic. Namely, there exists a monad on $\mathcal{C}$, i.e., an algebra in $Fun(\mathcal{C}, \mathcal{C})$, that we will denote $\mathcal{A}$, such that $HC(\mathcal{C})$ consists of modules in $\mathcal{C}$ over $\mathcal{A}$. Normally, monadicity is decided based on some general theorems, while the candidate monad is simply UF, but in our case, we are interested in an explicit description [**63**] that we provide below. Note that if we consider $(\Delta_!, \Delta^!)$ instead of (Δ^*, Δ_*) then the pair (U, F) is comonadic.

Observe that the only extra structure on $M \in \mathcal{C}$ that turns it into an object in $HC(\mathcal{C})$ is the isomorphism $\tau : \Delta_* M \to \sigma\Delta_*$, while the rest are properties of τ. It is immediate that if we set

$$\mathcal{A}' = \Delta^* \sigma \Delta_*$$

then τ is encoded into $act : \mathcal{A}'(M) \to M$ via adjunction. We now need to encode the properties of τ into those of act. A reasonable guess is that $\mathcal{A}'$ is a monad such that modules over it satisfy the unit and associativity conditions and furthermore, $\mathcal{A}'$ has a central element ς such that $s_1(\tau)$ is given by its action. Thus, $\mathcal{A}$ should be $\mathcal{A}'/(\varsigma - 1)$; this is not always true.

The problem is not the unit and stability, those are immediate from the definition of $\mathcal{A}'$. The problem is associativity, as in general the product on $\mathcal{A}'$ is only partially defined.

7.3. The role of property A. If $\mathcal{C}$ is a monoidal category with the extended product that admits a right adjoint, consider the associator: $\Delta^*(Id \boxtimes \Delta^*) \simeq \Delta^*(\Delta^* \boxtimes Id)$. Via adjunction it yields a map

$$(Id \boxtimes \Delta^*)(\Delta_* \boxtimes Id) \to \Delta_* \Delta^*. \tag{7.3}$$

DEFINITION 7.3. Roughly speaking, we say [**63**] that $\mathcal{C}$ has property A if (7.3) is an isomorphism. In [**63**] there are other technical conditions listed as well, but this is the key one.

Recall that $\mathcal{C}$ is rigid if every object has both right and left duals, i.e., we have objects X^* and ${}^*X$ such that the functor pairs $(-\otimes X, -\otimes X^*)$ and $(X \otimes -, {}^*X \otimes -)$ are adjoint pairs; also see Remark 7.9.

Having property A is strongly related to being rigid, in the following sense (rigid does imply it). It is not hard to show that ${}_H\mathcal{M}$ has property A if H is a Hopf algebra, but need not have it if H is only a bialgebra. The former yields a rigid category only if one considers finite dimensional H-modules, yet the full module category has property A as well; the bialgebra yields a category of modules that does not have property A even if one restricts to finite dimensional modules.

Let us consider the example of G, a finite monoid, i.e., a finite group without inverses. Associate to it a monoidal category $\mathcal{C} = \mathrm{Vec}_G$, i.e., the category of G-graded vector spaces with the product:

$$(V \otimes W)_g = \bigoplus_{xy=g} V_x \otimes W_y. \tag{7.4}$$

Then (7.3) for $\mathcal{C}$ is equivalent to the commutative diagram that expresses the associativity of the binary operation on G being Cartesian. The latter is equivalent to G being a group. More generally, it should be a simple exercise to show that ${}_H\mathcal{M}$ has property A for a bialgebra H if and only if H has an invertible antipode (so that it is, in particular, a Hopf algebra).

It turns out that if $\mathcal{C}$ has property A, then $\mathcal{A}'$ is our monad:

THEOREM 7.4 ([**63**]). *If $\mathcal{C}$ has property A then*

$$HH(\mathcal{C}) = \mathcal{A}'_{\mathcal{C}}\text{-}mod.$$

If $\mathcal{C}$ does not have property A, then $\mathcal{A}'$ merely generates the required monad subject to certain relations (work-in-progress of the second author). We will see an example of this phenomenon in (7.7).

The difference between $\Delta^*\sigma\Delta_*$ itself being a monad and it merely generating one is striking. The classical descriptions of $HH(\mathcal{C})$ and, thus, $HC(\mathcal{C})$ (or rather what they coincide with in Hopf-like settings) relies on $\Delta^*\sigma\Delta_*$ being a monad. Since it fails to be one, in bialgebra-like settings, attempts to copy the classical definition [**35**] fail as well. One should instead consider the monad that $\Delta^*\sigma\Delta_*$ generates; see (7.7).

REMARK 7.5. If the extended product admits a left adjoint instead then we talk about comonadicity, i.e., we are looking for a comonad, whose comodules in $\mathcal{C}$ yield $HC(\mathcal{C})$. Everything works in the same manner.

7.4. Examples. Let H be a Hopf algebra. Then $\mathcal{C} = {}_H\mathcal{M}$ has property A. The extended product in this case admits both a right and a left adjoint. By examining the resulting monad and comonad we deduce:

THEOREM 7.6 ([**63**]). *The category $HC(\mathcal{C})$ consists of stable anti-Yetter-Drinfeld contramodules (see [**7**] for the definition) in the monad case. And $HC(\mathcal{C})$ consists of stable anti-Yetter-Drinfeld modules in the comonad case.*

Though the categories of stable anti-Yetter-Drinfeld contramodules and stable anti-Yetter-Drinfeld modules are related by an adjoint pair of functors [**62**] that are often equivalences, they are probably not the same in general. Yet we abuse notation by speaking of the category $HC(\mathcal{C})$ without reference to the context (assumed existence of right vs left adjoints) in which the colimits were taken.

One should not get too attached to stable elements of $HH(\mathcal{C})$, as we will argue in Section 7.6 that it is only the first approximation to the correct setting for cyclic homology; we will replace stable objects by the so-called ς-mixed objects.

From the general discussion it follows that for $A \in {}_H\mathcal{M}$ an algebra:

$$C^n = Hom_H(A^{\otimes(n+1)}, M) \tag{7.5}$$

is a cocyclic module if M is a stable anti-Yetter-Drinfeld contramodule. On the other hand,

$$C_n = Hom_H(N, A^{\otimes(n+1)}) \tag{7.6}$$

is a cyclic module if N is a stable anti-Yetter-Drinfeld module.

REMARK 7.7. Note that (7.5) and (7.6) do not cover all the possible Hopf-cyclic theories. While we did address coalgebras in Section 7.1, we omitted a discussion of certain other traces, namely ones not obtained via Hom functors. They still fit into the general framework of Section 5, and are still obtained from objects in $HC(\mathcal{C})$.

Returning to the example of $\mathcal{C} = \mathrm{Vec}_G$ with G a finite monoid, let us begin by assuming that G is a group. It is well known that in the group case, $HC(\mathcal{C})$ consists of the full subcategory of G-graded G-equivariant vector spaces, with respect to the adjoint action of G on itself; the latter is $HH(\mathcal{C})$. The stability condition (specifying the full subcategory) is that the action of g on V_g is Id. The description of $HH(\mathcal{C})$ in the group case is via the representation category of a groupoid, namely the adjoint action groupoid of G on G.

REMARK 7.8. Above we see a common phenomenon. Namely, as does happen in the Hopf case, outside of hard to find pathological examples [**27**]: $HH(\mathrm{Vec}_G) = \mathcal{Z}(\mathrm{Vec}_G)$. On the other hand, for bialgebras this almost never happens.

To demonstrate the non property A case, consider the case of a monoid G. It is not hard to show (work-in-progress of the second author), that $HH(\mathcal{C})$ is again given as the representation category of a groupoid, but it is not an action groupoid, nor will we have $HH(\mathcal{C})$ coincide with $\mathcal{Z}(\mathcal{C})$. More precisely, the groupoid is defined as follows. The objects consist of $x \in G$ as before, but the arrows are given by pairs (x, y) that have source yx and target xy. The composition is freely generated by these arrows subject to the relations:

$$\begin{aligned} (1, x) &= Id \quad \text{redundant}, \\ (x, yz)(y, zx) &= (xy, z), \quad \text{for every triple } (x, y, z), \\ (x, 1) &= \varsigma. \end{aligned} \tag{7.7}$$

This of course reduces to the action groupoid in the group case, but for a monoid, the behaviour of the groupoid (7.7) is very chaotic, unlike the different groupoid whose representations give the monoidal center.

The construction of the groupoid (7.7) by the free generation with imposed relations, is exactly the meaning behind the statement that our monad $\mathcal{A}'$ in general is not given by $\Delta^* \sigma \Delta_*$, but is rather generated by it. Unlike the group case, where the relations allow us to replace any sequence of generators by a single generator, the monoid case produces a rich variety of new arrows from the generating set.

7.5. Biclosed categories and centers. In this section we provide a perspective that is less general than the preceding discussion, but makes the definition of the Hochschild homology category $HH(\mathcal{C})$ as close as possible to that of the monoidal center $\mathcal{Z}(\mathcal{C})$. This is used in [**48, 49**] to explore the coefficients in the Hopf algebroid, weak Hopf, and quasi-Hopf setting settings.

A monoidal category $\mathcal{C}$ is biclosed if for all $X \in \mathcal{C}$, the functors $- \otimes X$ and $X \otimes -$ have right adjoints that we denote by $X \triangleright -$ and $- \triangleleft X$ respectively. Namely,

$$Hom_{\mathcal{C}}(X \otimes Y, Z) = Hom_{\mathcal{C}}(X, Y \triangleright Z) = Hom_{\mathcal{C}}(Y, Z \triangleleft X).$$

REMARK 7.9. When $\mathcal{C}$ is rigid, it is biclosed with $X \triangleright Y = Y \otimes X^*$ and $Y \triangleleft X = {}^*X \otimes Y$. Being rigid is much stronger though, as $\mathcal{C} = {}_H\mathcal{M}$ is biclosed if H is a bialgebra, but only rigid if replaced by $\mathcal{C} = {}_H\mathcal{M}_{fd}$ and H has an invertible antipode.

We can consider the center of the bimodule category [**21**] $\mathcal{C}^{op}$ over $\mathcal{C}$ that results from this adjoint action; let us call it $\mathcal{Z}(\mathcal{C}^{op})$. More precisely, if $M \in \mathcal{Z}(\mathcal{C}^{op})$ then for every $X \in \mathcal{C}$ we have isomorphisms:

$$\iota_X : X \triangleright M \to M \triangleleft X,$$

subject to $\iota_1 = Id$ and $\iota_{X\otimes Y} = \iota_X\iota_Y$. We observe that

$$HH(\mathcal{C}) = \mathcal{Z}(\mathcal{C}^{op}).$$

REMARK 7.10. Suppose that $\mathcal{C}$ is pivotal, i.e., we have a monoidal natural isomorphism $Id \simeq (-)^{**}$. Then $HH(\mathcal{C}) = \mathcal{Z}(\mathcal{C})$.

Denote by $\mathcal{Z}'(\mathcal{C}^{op})$ the full subcategory of $\mathcal{Z}(\mathcal{C}^{op})$ that consists of objects such that the identity map $Id \in Hom_{\mathcal{C}}(M, M)$ is mapped to same via

$$Hom_{\mathcal{C}}(M, M) \simeq Hom_{\mathcal{C}}(1, M \triangleright M) \to Hom_{\mathcal{C}}(1, M \triangleleft M) \simeq Hom_{\mathcal{C}}(M, M),$$

where the map in the middle is post-composition with ι_M and the isomorphisms come from the definition of the adjoint action. This condition is stability and so

$$HC(\mathcal{C}) = \mathcal{Z}'(\mathcal{C}^{op}).$$

7.6. The category of ς-cyclic objects. The goal pursued so far in this text has been to associate to an algebra A in a monoidal category $\mathcal{C}$ a cyclic object in a universal manner. This goal is achieved above by constructing just such a cyclic object in what we called $HC(\mathcal{C})$. Note that we did not give an explicit construction, though it can be obtained from the following discussion.

The modification presented in this section is motivated by the observation that the naive definition of $HC(\mathcal{C})$ should be replaced by one in the setting of ∞-categories. Since we are not going to pursue it at this time, we provide the discussion below as a second approximation, sufficient for the Hopf-like settings, to the truly correct setting for cyclic homology. Namely, the colimit of the diagram $\mathcal{C}^{\boxtimes(\bullet+1)}$ should have been computed in the ∞-setting, instead of the more comprehensible setting of the usual categories. We propose **[63]** that the cyclic objects in $HC(\mathcal{C})$ should be replaced with ς-cyclic objects in $HH(\mathcal{C})$.

DEFINITION 7.11. A ς-cyclic object is a paracyclic object with

$$\tau_n^{n+1} = \varsigma,$$

instead of Id as would be in the cyclic case. As this is a worthy replacement for $HC(\mathcal{C})$, we will denote the category of ς-cyclic objects in $HH(\mathcal{C})$ by $HH(\mathcal{C})^{S^1}$.

Since $HC(\mathcal{C})$ consists of stable objects in $HH(\mathcal{C})$ we retain all the cyclic objects in $HC(\mathcal{C})$. We can restate this from the point of view of the mixed complexes **[32]** formulation of cyclic objects. Namely, a ς-mixed object (M, d, h) in $HH(\mathcal{C})$ is a non-positively graded cochain complex (M, d) in $HH(\mathcal{C})$ with a specified homotopy h between ς and Id, i.e.,

$$dh + hd = 1 - \varsigma.$$

Thus, a ς-mixed complex has cohomology in $HC(\mathcal{C})$; the homotopy ensuring this is a very important, perhaps the most important, part of the data. We do not distinguish between ς-cyclic and ς-mixed objects and denote both by $HH(\mathcal{C})^{S^1}$.

Given $A \in \mathcal{C}$ we can give a quick definition of the ς-cyclic object in $HH(\mathcal{C})$ associated to it. Recall that we have an adjoint pair of functors:

$$F : \mathcal{C} \leftrightarrows HH(\mathcal{C}) : U.$$

We note that the functor F is a ς-trace, i.e., it is a modification of the notion of trace from Section 7.1 obtained by replacing $\iota_{X,1} = Id$ with $\iota_{X,1} = \varsigma_{F(X)}$. Set

$$ch_n(A) = F(A^{\otimes(n+1)}) \tag{7.8}$$

to obtain the desired ς-cyclic object. Placing $F(A^{\otimes(n+1)})$ in degree $-n$ with $d = b$ and $h = B$ (see [**51**]) we get the ς-mixed object.

REMARK 7.12. In the Hopf algebra case we obtain a strict [**65**] generalization of stable anti-Yetter-Drinfeld modules, which have been previously used as coefficients. Namely, we get mixed anti-Yetter-Drinfeld modules, i.e., cochain complexes of anti-Yetter-Drinfeld modules equipped with specified homotopies between Id and the stability automorphism.

7.7. On the higher Morita invariance. Classical cyclic homology of algebras is Morita invariant, i.e., it depends not so much on algebras as on their categories of modules. What we discuss here is the categorified version of this observation, and its generalization. More precisely, it is well known that the monoidal center $\mathcal{Z}(\mathcal{C})$ is invariant under duality (under suitable assumptions [**21**]), so that it depends only on the 2-category of $\mathcal{C}$-modules. Drastically different looking monoidal categories can be dual to each other, for example if H is a finite dimensional Hopf algebra, then ${}_H\mathcal{M}$ and ${}_{H^*}\mathcal{M}$ are dual. If $H = kG$ for G a finite group then the latter is G-graded vector spaces, and the former is G-representations.

We mentioned in Section 7.4 that $HH(\mathcal{C})$ yields anti-Yetter-Drinfeld contramodules for $\mathcal{C} = {}_H\mathcal{M}$, where H is a Hopf algebra. This holds in the (Δ^*, Δ_*) setting that we fix for the purposes of this section. It is also true that $HH(\mathcal{M}^H)$ consists of anti-Yetter-Drinfeld modules, where $\mathcal{M}^H$ is the monoidal category of right H-comodules. In general we have an adjoint pair of functors [**62**]:

$$(-)_c : HH({}_H\mathcal{M}) \leftrightarrows HH(\mathcal{M}^H) : \widehat{(-)}, \tag{7.9}$$

and they are equivalences, in particular, if H is finite dimensional; these functors are compatible with stability, thus, relating the cyclic homology categories as well.

The general picture is as follows: let $\mathcal{C}$ and $\mathcal{D}$ be monoidal categories, and $\mathcal{N}$ an admissible [**63**] $(\mathcal{C}, \mathcal{D})$-bimodule category. The admissibility criteria is asymmetrical, an example would be the $({}_H\mathcal{M}, \mathcal{M}^H)$-bimodule Vec that results in the (7.9) above. Another, crucial, example is the $(\text{Vec}, \mathcal{C})$-bimodule $A_{\mathcal{C}}$-mod for $A \in \mathcal{C}$ algebra. It also covers the case of a monoidal functor (with a right adjoint) $\mathcal{C} \to \mathcal{D}$ that corresponds to the $(\mathcal{C}, \mathcal{D})$-bimodule $\mathcal{D}$. The bimodule category $\mathcal{N}$ gives a stability compatible adjoint pair of functors:

$$\mathcal{N}^* : HH(\mathcal{C}) \leftrightarrows HH(\mathcal{D}) : \mathcal{N}_* \tag{7.10}$$

that can also be used to redefine (7.8) as

$$ch(A) = A_{\mathcal{C}}\text{-mod}^*(k) \in HH(\mathcal{C})^{S^1}$$

where k is the trivial mixed complex. The pair $(\mathcal{N}^*, \mathcal{N}_*)$ and (7.10) can be considered as a generalization of Morita invariance that includes cases such as (7.9) that are not always equivalences.

Furthermore, for $A \in \mathcal{C}$ algebra, $\mathcal{N}^*(A) \in \mathcal{D}$ algebra can be defined, and $ch(\mathcal{N}^*(A)) = \mathcal{N}^*(ch(A))$, thus, we obtain a kind of Eckmann - Shapiro lemma:

THEOREM 7.13 ([**64**]). *For $A \in \mathcal{C}$ and $M \in HH(\mathcal{D})^{S^1}$ we have*

$$HC^\bullet_{\mathcal{D}}(\mathcal{N}^* A, M) = HC^\bullet_{\mathcal{C}}(A, \mathcal{N}_* M).$$

As an example, consider an algebra $A \in {}_H\mathcal{M}$ and let M be a stable anti-Yetter-Drinfeld module, then

$$HC^\bullet(A, \widehat{M}) = HC^\bullet(A \sharp H, M).$$

8. Rigid braided categories

In this section (it and its subsection are based on a work-in-progress of the second author) we discuss the existence of the so-called annuli stacking monoidal product [4], as it is known in factorization homology [3], on $HH(\mathcal{B})$, for $\mathcal{B}$ rigid braided [29]. Recall that a monoidal category is braided if there is a braiding isomorphism

$$\tau_{X,Y} : X \otimes Y \simeq Y \otimes X.$$

The braiding is to satisfy certain coherence conditions that ensure that braid diagrams can be used to perform calculations; thus, this structure gives rise to braid group actions on powers $X^{\otimes n}$. The braided category $\mathcal{B}$ is symmetric if $\tau^2_{X,Y} = \tau_{Y,X}\tau_{X,Y} = Id_{X\otimes Y}$; the braid action then factors through the symmetric group. On the opposite side of the spectrum, we say that $\mathcal{B}$ is non-degenerate if $\tau^2_{X,Y} = Id_{X\otimes Y}$ for all Y, implies that $X = 1$.

The annuli stacking product structure depends on the braiding, thus, if the same monoidal category is endowed with a different braiding, the product will change. Contrast that with the observation that the underlying category $HH(\mathcal{B})$ needs no braiding for its definition. When $HH(\mathcal{B})$ is identified with $\mathcal{Z}(\mathcal{B})$, as often happens for a rigid monoidal category (this does not need a braiding), the former inherits a non-degenerate braided (pair of pants product) from the latter; this product is unsatisfactory since it is not compatible with stability, namely ς is not monoidal. The annuli stacking product on the other hand is compatible with stability, but it is not braided, unless the original braiding on $\mathcal{B}$ was symmetric.

Consider the concrete case of a Hopf algebra H. The category of anti-Yetter-Drinfeld modules for H, unlike that of Yetter-Drinfeld modules, does not have a natural monoidal structure. It is, nevertheless, a bimodule category over the braided category of Yetter-Drinfeld modules; see Lemma 4.3, [25]. It is known that often, in fact, in the presence of the original modular pair in involution, one has an identification between the two categories. It is even possible to describe the action of ς in terms of a certain ribbon element of the category of Yetter-Drinfeld modules. This identification endows, however non-canonically, the anti-Yetter-Drinfeld modules with a product. Unfortunately ς is not monoidal with respect to this product. In fact, any ribbon element is not monoidal (it is a twist, see (8.2)) unless the braided category is symmetric, which the category of Yetter-Drinfeld modules never is. Thus, the identification produced by a modular pair in involution does not produce a product on the stable anti-Yetter-Drinfeld modules; also, it is rather arbitrary.

We can demonstrate the contrast between the two types of product on $HH(\mathcal{B})$ in our example (7.4). We assume that G is a group here. It is true that Vec_G is not braided (as G need not be commutative) but the category $Rep(G)$ of representations of G is (in fact it is symmetric), and they share the same Hochschild homology category, as in Section 7.7. Recall that $\mathcal{Z}(\mathrm{Vec}_G) = HH(\mathrm{Vec}_G) = Sh_G G^{ad}$, where the latter is G-graded, G-equivariant (with respect to the $x(-)x^{-1}$ action) vector spaces. The product on $\mathcal{Z}(\mathrm{Vec}_G)$ is the convolution product, i.e, it is given by the same formula is in (7.4). On the other hand the correct product on $HH(\mathrm{Vec}_G)$ is given by

$$(V \otimes W)_x = V_x \otimes W_x, \tag{8.1}$$

which is clearly compatible with stability, i.e., $\varsigma(v_x) = xv_x$ is monoidal with respect to (8.1), but is a twist (8.2) with respect to (7.4).

We describe, roughly, a construction of the product on $HH(\mathcal{B})$. Let $\mathcal{B}$ be rigid braided, the category $HH(\mathcal{B})$ is given not just by a monad on $\mathcal{B}$ as was discussed in Section 7.3, but (using the braiding) by a commutative algebra $\mathcal{H}$ in $\mathcal{Z}(\mathcal{B})$. By a generalization of the usual monoidal product on modules over a commutative ring, we obtain the desired product on $\mathcal{H}_{\mathcal{C}}$-mod $= HH(\mathcal{B})$.

It is possible to describe the product more concretely in our case of interest (see Section 8.2). We will require an additional assumption that $\mathcal{B}$ is balanced, i.e, it is equipped with a twist $\theta \in Aut(Id_{\mathcal{B}})$ satisfying:

$$\theta_{X\otimes Y} = \tau^2_{X,Y}\theta_X \otimes \theta_Y. \tag{8.2}$$

Now, if the braiding on $\mathcal{B}$ is non-degenerate (the opposite of symmetric) then

$$HH(\mathcal{B}) = End(\mathcal{B}), \tag{8.3}$$

as monoidal categories, in contrast with our example above where $\mathcal{B}$ was symmetric. We can extend (8.3) in a way that elucidates $HC(\mathcal{B})$. Namely, under the equivalence in (8.3) we have the correspondence

$$\varsigma = \theta(-)\theta^{-1}.$$

Note that $\mathcal{B}$ may have more than one twist (usually many as they form a torsor over $Aut^{\otimes}(Id_{\mathcal{B}})$), yet ς is ς. The role of different twists will become evident below (see Remark 8.2).

It may seem, especially in light of (8.3), that this canonical product on $HH(\mathcal{B})$ is not very interesting. This is not the case; recall that this product is compatible with stability and, thus, restricts to $HC(\mathcal{B})$. The latter is more interesting. Furthermore, the product plays a crucial role in the next section.

8.1. Relative coefficients and relative centers. Consider a Hopf algebra H in a rigid braided category $\mathcal{B}$ [**54**]. It is immediate that

$$\mathcal{C} = H_{\mathcal{B}}\text{-mod},$$

the category of modules over H in $\mathcal{B}$, is a monoidal category. We can, thus, apply the machinery considered in this paper to $\mathcal{C}$, and obtain $HH(\mathcal{C})$ and $HC(\mathcal{C})$, but it would not remember $\mathcal{B}$. In fact, $\mathcal{C}$ can often be expressed as modules of a certain "product" Hopf algebra T (see (8.8) below), and the machinery does not distinguish between $H_{\mathcal{B}}$-mod and T-mod.

What one wants instead is a relative (with respect to $\mathcal{B}$) version of $HH(\mathcal{C})$ and $HC(\mathcal{C})$. In fact, one may want a relative version of the more classical $\mathcal{Z}(\mathcal{C})$ as well (this is considered in [**50**]). The latter problem is actually simpler, so we will briefly address it as well. It is tempting, as was attempted in [**41**], to try and generalize the old constructions to yield new (anti)-Yetter-Drinfeld modules, now starting with a braided category other than the Vec of before. This works for Yetter-Drinfeld modules, and, in principle, for anti-Yetter-Drinfeld modules in a symmetric category, but it cannot work in general. One needs "coefficients" for this to work.

We begin by observing that if H is a Hopf algebra in $\mathcal{B}$ then we have a braided functor

$$\mathcal{B} \to \mathcal{Z}(\mathcal{C}).$$

This induces a monoidal, with respect to the annuli stacking product, functor

$$HH(\mathcal{B}) \to HH(\mathcal{Z}(\mathcal{C})).$$

The latter monoidal category acts on the right on the category $HH(\mathcal{C})$. We see that $HH(\mathcal{C})$ is a right $HH(\mathcal{B})$-module; the same applies to $HC(-)$. In particular, if $\mathcal{M}$ is a left $HH(\mathcal{B})$-module (with a compatible ς-action) then one may consider the category

$$HH(\mathcal{C}) \boxtimes_{HH(\mathcal{B})} \mathcal{M} \tag{8.4}$$

or its cyclic version: $HC(\mathcal{C}) \boxtimes_{HC(\mathcal{B})} \mathcal{M}^{\varsigma}$. This right module structure of $HH(\mathcal{C})$ over $HH(\mathcal{B})$ is the "memory" of $\mathcal{B}$. As we see in (8.9), it localizes $HC(\mathcal{C})$.

When considering the center version of the above we note that the braided functor $\mathcal{B} \to \mathcal{Z}(\mathcal{C})$ also yields a monoidal, with respect to the annuli stacking product (not the usual product on centers), functor $\mathcal{Z}(\mathcal{B}) \to \mathcal{Z}(\mathcal{Z}(\mathcal{C}))$, and the latter acts on the right on $\mathcal{Z}(\mathcal{C})$. In this case $\mathcal{Z}(\mathcal{B})$ (with the annuli stacking product) has a canonical left module category $\mathcal{B}$, so that the relative center can be defined as

$$\mathcal{Z}_{\mathcal{B}}(\mathcal{C}) = \mathcal{Z}(\mathcal{C}) \boxtimes_{\mathcal{Z}(\mathcal{B})} \mathcal{B}. \tag{8.5}$$

Going back to our Hochschild homology case, the action of $HH(\mathcal{B})$ on $\mathcal{B}$ is equivalent to specifying a twist θ. If $\mathcal{B}$ is symmetric then the trivial twist is available and so the relative Hochschild and cyclic homology categories can be defined as in (8.5). In the non-symmetric case, there is no canonical twist (or none may be available). However, with a specified twist in hand we can define the relative Hochschild and cyclic homology categories:

$$HH_{(\mathcal{B},\theta)}(\mathcal{C}) = HH(\mathcal{C}) \boxtimes_{HH(\mathcal{B})} \mathcal{B}_{\theta} \quad \text{and} \quad HC_{(\mathcal{B},\theta)}(\mathcal{C}) = HC(\mathcal{C}) \boxtimes_{HC(\mathcal{B})} \mathcal{B}^{\theta}_{\theta}. \tag{8.6}$$

REMARK 8.1. The definition (8.4) is rather abstract, but it is possible to describe these categories monadically. The resulting definition (8.6) of relative anti-Yetter-Drinfeld modules (dependent on the θ chosen) looks very similar to the old one. More precisely, they are described as objects in $\mathcal{B}$ that are modules over H and ${}^*H$ such that the actions satisfy a compatibility condition that involves θ. The definition of ς also involves θ. Unlike the Vec case where anti-Yetter-Drinfeld modules are modules over the twisted Drinfeld double, here we have that they are modules over a suitable monad on $\mathcal{B}$, not an algebra in $\mathcal{B}$.

8.2. Examples. Let G be an abelian group and χ a bicharacter such that $\omega(x,y) = \chi(x,y)\chi(y,x)$ is non-degenerate. Let $\mathcal{B} = (\mathrm{Vec}_G, \chi)$, i.e.,

$$\tau : v_x \otimes w_y \mapsto \chi(x,y) w_y \otimes v_x$$

and $\theta(g) := \chi(g,g)$ its canonical ribbon element (special twist). Let

$$I = \theta(G) \subset k^{\times} \quad \text{and} \quad n_i = \#\{x | \theta(x) = i\}$$

then

$$HH(\mathcal{B}) \simeq M_{|G|}(\mathrm{Vec}) \quad \text{and} \quad HC(\mathcal{B}) = \bigoplus_{i \in I} M_{n_i}(\mathrm{Vec}), \tag{8.7}$$

where $M_n(\mathrm{Vec})$ is the monoidal category of matrices with vector space entries. Any other twist is of the form $\theta_x = \theta\omega(x,-)$ for $x \in G$. The twists θ_x and θ_y yield the same theory if $\theta(x) = \theta(y)$. Thus, for an $H \in \mathcal{B}$ there is a unique category of "relative" anti-Yetter-Drinfeld modules, but $\#I$ categories of "relative" stable anti-Yetter-Drinfeld modules in general.

REMARK 8.2. In (8.7) we see that $HC(\mathcal{B})$ has irreducible module categories Vec^{n_i} indexed by $i \in I$. These correspond to twists θ_x with $\theta(x) = i$.

We can go further in a subexample of the above by considering Taft algebras [**66**]. We recall the relevant definitions: suppose that p is an odd prime and ξ is a primitive pth root of 1. Let $\mathcal{B} = (\mathrm{Vec}_{\mathbb{Z}/p}, \xi^{ij})$, and enumerate the twists by $\theta_t = \xi^{i^2+ti}$. Define $H = k[x]/x^p$ with $deg(x) = 1$, $\Delta(x) = 1 \otimes x + x \otimes 1$, $S(x) = -x$, and $\epsilon(x) = 0$. Note that

$$H_{\mathcal{B}}\text{-mod} = T_p(\xi)\text{-mod}, \tag{8.8}$$

where $T_p(\xi)$ is the Taft Hopf algebra. Let $q = \xi^{-1/2}$ then independently of t:

$$HH_{(\mathcal{B},\theta_t)}(T_p(\xi)\text{-mod}) \simeq u_q(sl_2)\text{-mod},$$

where $u_q(sl_2)$ denotes the small quantum group, while

$$HC_{(\mathcal{B},\theta_0)}(T_p(\xi)\text{-mod}) \simeq \mathrm{Vec}.$$

For $t \neq 0$, $HC_{(\mathcal{B},\theta_t)}(T_p(\xi)\text{-mod})$ depends only on $|t|$ but is otherwise non-trivial and sits over a distinct (for $|t| \neq |t'|$) point of $Spec\mathcal{Z}(u_q(sl_2))$.

In light of the above, the question of what is a Hopf-cyclic theory for a Hopf algebra in a braided category has two answers. First, balanced (not braided) categories yield Hopf-cyclic theories. Different twist = Different theory. Second, remembering $\mathcal{B}$ "localizes" $HC(\mathcal{C})$. In the Taft algebra example (which is just a usual Hopf algebra) we have that any stable anti-Yetter-Drinfeld module (in the classical sense) breaks down

$$M = M_0 \oplus \bigoplus_{i=1}^{(p-1)/2} (M_i^1 \oplus M_i^2)$$

with M_0 in a category equivalent to Vec and M_i^j of type i. The index $0, \cdots, (p-1)/2$ runs over the spectrum of $\mathcal{Z}(u_q(sl_2))$; the type $i \neq 0$ categories are not Vec.

Returning to the non-degenerate case of $\mathcal{B} = (\mathrm{Vec}_G, \chi)$ with $\mathcal{C} = H_{\mathcal{B}}$-mod, there exists a category X with $\varsigma \in Aut(Id_X)$ such that

$$HC(\mathcal{C}) = \bigoplus_{i \in I} X^{i\varsigma} \boxtimes \mathrm{Vec}^{n_i}. \tag{8.9}$$

References

[1] R. Akbarpour and M. Khalkhali, *Hopf algebra equivariant cyclic homology and cyclic homology of crossed product algebras*, J. Reine Angew. Math. **559** (2003), 137–152, DOI 10.1515/crll.2003.046. MR1989648

[2] R. Akbarpour and M. Khalkhali, *Equivariant cyclic cohomology of $\mathcal{H}$-algebras*, K-Theory **29** (2003), no. 4, 231–252, DOI 10.1023/B:KTHE.0000009977.24960.5a. MR2029079

[3] David Ayala and John Francis, *Factorization homology of topological manifolds*, J. Topol. **8** (2015), no. 4, 1045–1084, DOI 10.1112/jtopol/jtv028. MR3431668

[4] David Ben-Zvi, Adrien Brochier, and David Jordan, *Quantum character varieties and braided module categories*, Selecta Math. (N.S.) **24** (2018), no. 5, 4711–4748, DOI 10.1007/s00029-018-0426-y. MR3874702

[5] David Ben-Zvi, John Francis, and David Nadler, *Integral transforms and Drinfeld centers in derived algebraic geometry*, J. Amer. Math. Soc. **23** (2010), no. 4, 909–966, DOI 10.1090/S0894-0347-10-00669-7. MR2669705

[6] Roman Bezrukavnikov, Michael Finkelberg, and Victor Ostrik, *On tensor categories attached to cells in affine Weyl groups. III*, Israel J. Math. **170** (2009), 207–234, DOI 10.1007/s11856-009-0026-9. MR2506324

[7] Tomasz Brzeziński, *Hopf-cyclic homology with contramodule coefficients*, Quantum groups and noncommutative spaces, Aspects Math., E41, Vieweg + Teubner, Wiesbaden, 2011, pp. 1–8, DOI 10.1007/978-3-8348-9831-9_1. MR2798432

[8] Dan Burghelea, *The cyclic homology of the group rings*, Comment. Math. Helv. **60** (1985), no. 3, 354–365, DOI 10.1007/BF02567420. MR814144

[9] Alain Connes, *C^* algèbres et géométrie différentielle* (French, with English summary), C. R. Acad. Sci. Paris Sér. A-B **290** (1980), no. 13, A599–A604. MR572645

[10] A. Connes. Spectral sequence and homology of currents for operator algebras. Oberwolfach Tagungsber., 41/81, Funktionalanalysis und C^*-Algebren, 27-9/3-10, 1981.

[11] Alain Connes, *Cohomologie cyclique et foncteurs* Ext^n (French, with English summary), C. R. Acad. Sci. Paris Sér. I Math. **296** (1983), no. 23, 953–958. MR777584

[12] Alain Connes, *Noncommutative differential geometry*, Inst. Hautes Études Sci. Publ. Math. **62** (1985), 257–360. MR823176

[13] A. Connes, *Cyclic cohomology and the transverse fundamental class of a foliation*, Geometric methods in operator algebras (Kyoto, 1983), Pitman Res. Notes Math. Ser., vol. 123, Longman Sci. Tech., Harlow, 1986, pp. 52–144. MR866491

[14] A. Connes and H. Moscovici, *The local index formula in noncommutative geometry*, Geom. Funct. Anal. **5** (1995), no. 2, 174–243, DOI 10.1007/BF01895667. MR1334867

[15] A. Connes and H. Moscovici, *Hopf algebras, cyclic cohomology and the transverse index theorem*, Comm. Math. Phys. **198** (1998), no. 1, 199–246, DOI 10.1007/s002200050477. MR1657389

[16] Alain Connes and Henri Moscovici, *Cyclic cohomology and Hopf algebras*, Lett. Math. Phys. **48** (1999), no. 1, 97–108, DOI 10.1023/A:1007527510226. Moshé Flato (1937–1998). MR1718047

[17] Alain Connes and Henri Moscovici, *Cyclic cohomology and Hopf algebra symmetry*, Lett. Math. Phys. **52** (2000), no. 1, 1–28, DOI 10.1023/A:1007698216597. Conference Moshé Flato 1999 (Dijon). MR1800488

[18] Alain Connes and Henri Moscovici, *Background independent geometry and Hopf cyclic cohomology*, `arXiv:math/0505475`, 2005.

[19] Joachim Cuntz and Daniel Quillen, *Cyclic homology and nonsingularity*, J. Amer. Math. Soc. **8** (1995), no. 2, 373–442, DOI 10.2307/2152822. MR1303030

[20] P. Deligne. Letter to stasheff, Gerstenhaber, May, Schechtman and Drinfeld, May 17 (1993).

[21] Pavel Etingof, Shlomo Gelaki, Dmitri Nikshych, and Victor Ostrik, *Tensor categories*, Mathematical Surveys and Monographs, vol. 205, American Mathematical Society, Providence, RI, 2015, DOI 10.1090/surv/205. MR3242743

[22] B. L. Feĭgin and B. L. Tsygan, *Additive K-theory*, K-theory, arithmetic and geometry (Moscow, 1984), Lecture Notes in Math., vol. 1289, Springer, Berlin, 1987, pp. 67–209, DOI 10.1007/BFb0078368. MR923136

[23] Murray Gerstenhaber, *The cohomology structure of an associative ring*, Ann. of Math. (2) **78** (1963), 267–288, DOI 10.2307/1970343. MR161898

[24] Alexander Gorokhovsky, *Secondary characteristic classes and cyclic cohomology of Hopf algebras*, Topology **41** (2002), no. 5, 993–1016, DOI 10.1016/S0040-9383(01)00015-5. MR1923996

[25] Piotr M. Hajac, Masoud Khalkhali, Bahram Rangipour, and Yorck Sommerhäuser, *Stable anti-Yetter-Drinfeld modules* (English, with English and French summaries), C. R. Math. Acad. Sci. Paris **338** (2004), no. 8, 587–590, DOI 10.1016/j.crma.2003.11.037. MR2056464

[26] Piotr M. Hajac, Masoud Khalkhali, Bahram Rangipour, and Yorck Sommerhäuser, *Hopf-cyclic homology and cohomology with coefficients* (English, with English and French summaries), C. R. Math. Acad. Sci. Paris **338** (2004), no. 9, 667–672, DOI 10.1016/j.crma.2003.11.036. MR2065371

[27] Sebastian Halbig, *Generalized Taft algebras and pairs in involution*, Comm. Algebra **49** (2021), no. 12, 5181–5195, DOI 10.1080/00927872.2021.1939043. MR4328530

[28] Mohammad Hassanzadeh, Masoud Khalkhali, and Ilya Shapiro, *Monoidal categories, 2-traces, and cyclic cohomology*, Canad. Math. Bull. **62** (2019), no. 2, 293–312, DOI 10.4153/cmb-2018-016-4. MR3952519

[29] André Joyal and Ross Street, *Braided tensor categories*, Adv. Math. **102** (1993), no. 1, 20–78, DOI 10.1006/aima.1993.1055. MR1250465

[30] D. Kaledin, *Cyclic homology with coefficients*, Algebra, arithmetic, and geometry: in honor of Yu. I. Manin. Vol. II, Progr. Math., vol. 270, Birkhäuser Boston, Boston, MA, 2009, pp. 23–47, DOI 10.1007/978-0-8176-4747-6_2. MR2641186

[31] D. Kaledin, *Trace theories and localization*, Stacks and categories in geometry, topology, and algebra, Contemp. Math., vol. 643, Amer. Math. Soc., Providence, RI, 2015, pp. 227–262, DOI 10.1090/conm/643/12900. MR3381474

[32] Christian Kassel, *Cyclic homology, comodules, and mixed complexes*, J. Algebra **107** (1987), no. 1, 195–216, DOI 10.1016/0021-8693(87)90086-X. MR883882

[33] Christian Kassel, *Quantum groups*, Graduate Texts in Mathematics, vol. 155, Springer-Verlag, New York, 1995, DOI 10.1007/978-1-4612-0783-2. MR1321145

[34] Ralph M. Kaufmann and Yongheng Zhang, *Permutohedral structures on E_2-operads*, Forum Math. **29** (2017), no. 6, 1371–1411, DOI 10.1515/forum-2016-0052. MR3719307

[35] Atabey Kaygun, *Bialgebra cyclic homology with coefficients*, K-Theory **34** (2005), no. 2, 151–194, DOI 10.1007/s10977-005-1501-7. MR2180111

[36] A. Kaygun and M. Khalkhali, *Excision in Hopf cyclic homology*, K-Theory **37** (2006), no. 1-2, 105–128, DOI 10.1007/s10977-006-0002-7. MR2274671

[37] Atabey Kaygun and Masoud Khalkhali, *Hopf modules and noncommutative differential geometry*, Lett. Math. Phys. **76** (2006), no. 1, 77–91, DOI 10.1007/s11005-006-0062-x. MR2223765

[38] Masoud Khalkhali, *Operations on cyclic homology, the X complex, and a conjecture of Deligne*, Comm. Math. Phys. **202** (1999), no. 2, 309–323, DOI 10.1007/s002200050584. MR1689975

[39] Masoud Khalkhali, *Basic noncommutative geometry*, EMS Series of Lectures in Mathematics, European Mathematical Society (EMS), Zürich, 2009, DOI 10.4171/061. MR2567651

[40] Masoud Khalkhali, *A short survey of cyclic cohomology*, Quanta of maths, Clay Math. Proc., vol. 11, Amer. Math. Soc., Providence, RI, 2010, pp. 283–311. MR2732055

[41] Masoud Khalkhali and Arash Pourkia, *Hopf cyclic cohomology in braided monoidal categories*, Homology Homotopy Appl. **12** (2010), no. 1, 111–155. MR2607413

[42] M. Khalkhali and B. Rangipour, *A new cyclic module for Hopf algebras*, K-Theory **27** (2002), no. 2, 111–131, DOI 10.1023/A:1021193729183. MR1942182

[43] M. Khalkhali and B. Rangipour, *Invariant cyclic homology*, K-Theory **28** (2003), no. 2, 183–205, DOI 10.1023/A:1024552111457. MR1995876

[44] M. Khalkhali and B. Rangipour, *A note on cyclic duality and Hopf algebras*, Comm. Algebra **33** (2005), no. 3, 763–773, DOI 10.1081/AGB-200051130. MR2128410

[45] Masoud Khalkhali and Bahram Rangipour, *Cup products in Hopf-cyclic cohomology* (English, with English and French summaries), C. R. Math. Acad. Sci. Paris **340** (2005), no. 1, 9–14, DOI 10.1016/j.crma.2004.10.025. MR2112033

[46] M. Khalkhali and B. Rangipour, *Para-Hopf algebroids and their cyclic cohomology*, Lett. Math. Phys. **70** (2004), no. 3, 259–272, DOI 10.1007/s11005-004-4303-6. MR2128954

[47] M. Khalkhali and B. Rangipour. A survey of Hopf cyclic cohomology, Noncommutative geometry and number theory, 155-178, Aspects Math., E37, Friedr. Vieweg, Wiesbaden, 2006.

[48] Ivan Kobyzev and Ilya Shapiro, *Anti-Yetter-Drinfeld modules for quasi-Hopf algebras*, SIGMA Symmetry Integrability Geom. Methods Appl. **14** (2018), Paper No. 098, 10, DOI 10.3842/SIGMA.2018.098. MR3853619

[49] Ivan Kobyzev and Ilya Shapiro, *A categorical approach to cyclic cohomology of quasi-Hopf algebras and Hopf algebroids*, Appl. Categ. Structures **27** (2019), no. 1, 85–109, DOI 10.1007/s10485-018-9544-0. MR3901951

[50] Robert Laugwitz, *The relative monoidal center and tensor products of monoidal categories*, Commun. Contemp. Math. **22** (2020), no. 8, 1950068, 53, DOI 10.1142/S0219199719500688. MR4142329

[51] Jean-Louis Loday, *Cyclic homology*, 2nd ed., Grundlehren der mathematischen Wissenschaften [Fundamental Principles of Mathematical Sciences], vol. 301, Springer-Verlag, Berlin, 1998. Appendix E by María O. Ronco; Chapter 13 by the author in collaboration with Teimuraz Pirashvili, DOI 10.1007/978-3-662-11389-9. MR1600246

[52] Saunders Mac Lane, *Categories for the working mathematician*, 2nd ed., Graduate Texts in Mathematics, vol. 5, Springer-Verlag, New York, 1998. MR1712872

[53] Shahn Majid, *Foundations of quantum group theory*, Cambridge University Press, Cambridge, 1995, DOI 10.1017/CBO9780511613104. MR1381692

[54] Shahn Majid, *Algebras and Hopf algebras in braided categories*, Advances in Hopf algebras (Chicago, IL, 1992), Lecture Notes in Pure and Appl. Math., vol. 158, Dekker, New York, 1994, pp. 55–105. MR1289422

[55] H. Moscovici, *On the van Est analogy in Hopf cyclic cohomology*, in Cyclic Cohomology at 40: Achievements and Future Prospects (A. Connes, C. Consani, B. I. Dundas, M. Khalkhali, and H. Moscovici, editors), pp. 297–310, 2023.

[56] Henri Moscovici and Bahram Rangipour, *Cyclic cohomology of Hopf algebras of transverse symmetries in codimension 1*, Adv. Math. **210** (2007), no. 1, 323–374, DOI 10.1016/j.aim.2006.06.008. MR2298827

[57] Henri Moscovici and Bahram Rangipour, *Hopf algebras of primitive Lie pseudogroups and Hopf cyclic cohomology*, Adv. Math. **220** (2009), no. 3, 706–790, DOI 10.1016/j.aim.2008.09.017. MR2483227

[58] R. Ponge, *Cyclic Homology and Group Actions*, in Cyclic Cohomology at 40: Achievements and Future Prospects (A. Connes, C. Consani, B. I. Dundas, M. Khalkhali, and H. Moscovici, editors), pp. 371–394, 2023.

[59] Kate Ponto and Michael Shulman, *Shadows and traces in bicategories*, J. Homotopy Relat. Struct. **8** (2013), no. 2, 151–200, DOI 10.1007/s40062-012-0017-0. MR3095324

[60] Daniel Quillen, *Algebra cochains and cyclic cohomology*, Inst. Hautes Études Sci. Publ. Math. **68** (1988), 139–174 (1989). MR1001452

[61] David E. Radford and Jacob Towber, *Yetter-Drinfel′d categories associated to an arbitrary bialgebra*, J. Pure Appl. Algebra **87** (1993), no. 3, 259–279, DOI 10.1016/0022-4049(93)90114-9. MR1228157

[62] Ilya Shapiro, *On the anti-Yetter-Drinfeld module-contramodule correspondence*, J. Noncommut. Geom. **13** (2019), no. 2, 473–497, DOI 10.4171/JNCG/327. MR3988752

[63] Ilya Shapiro, *Categorified Chern character and cyclic cohomology*, `arXiv:1904.04230`, 2019.

[64] Ilya Shapiro, *Invariance properties of cyclic cohomology with coefficients*, J. Algebra **546** (2020), 484–517, DOI 10.1016/j.jalgebra.2019.10.034. MR4035000

[65] Ilya Shapiro, *Mixed vs stable anti-Yetter-Drinfeld contramodules*, SIGMA Symmetry Integrability Geom. Methods Appl. **17** (2021), Paper No. 026, 10, DOI 10.3842/SIGMA.2021.026. MR4231449

[66] Earl J. Taft, *The order of the antipode of finite-dimensional Hopf algebra*, Proc. Nat. Acad. Sci. U.S.A. **68** (1971), 2631–2633, DOI 10.1073/pnas.68.11.2631. MR286868

[67] A. Van Daele, *Multiplier Hopf algebras*, Trans. Amer. Math. Soc. **342** (1994), no. 2, 917–932, DOI 10.2307/2154659. MR1220906

[68] David N. Yetter, *Quantum groups and representations of monoidal categories*, Math. Proc. Cambridge Philos. Soc. **108** (1990), no. 2, 261–290, DOI 10.1017/S0305004100069139. MR1074714

Department of Mathematics, University of Western Ontario, London, Ontario, Canada, N6A 5B7

Email address: `masoud@uwo.ca`

Department of Mathetics and Statistics, University of Windsor, Windsor, Ontario, Canada, N9B 3P4

Email address: `ishapiro@uwindsor.ca`

Proceedings of Symposia in Pure Mathematics
Volume **105**, 2023
https://doi.org/10.1090/pspum/105/01904

The Hochschild Cohomology of Uniform Roe Algebras

Matthew Lorentz

ABSTRACT. In Rufus Willett's and the authors paper "Bounded Derivations on Uniform Roe Algebras" we showed that all bounded derivations on a uniform Roe algebra $C_u^*(X)$ associated to a bounded geometry metric space X are inner. This naturally leads to the question of whether or not the higher dimensional Hochschild cohomology groups of the uniform Roe algebra vanish also. While we cannot answer this question completely, we are able to give necessary and sufficient conditions for the vanishing of $H_c^n(C_u^*(X), C_u^*(X))$.

1. Introduction

In Rufus Willett's and the authors paper "Bounded Derivations on Uniform Roe Algebras" [**5**] we showed that all bounded derivations on a uniform Roe algebra $C_u^*(X)$ associated to a bounded geometry metric space X are inner. That all bounded derivations are inner is equivalent to the first norm continuous Hochschild cohomology group $H_c^1(C_u^*(X), C_u^*(X))$ vanishing. Indeed, the Hochschild coboundary operator from a C*-algebra $\mathcal{A}$ to the linear maps from $\mathcal{A}$ to itself is given by

$$\partial a(b) = ab - ba, \ \ a, b \in \mathcal{A}.$$

Thus, ∂a is an inner derivation. Next, the coboundary operator from a linear map ϕ to bilinear map from $\mathcal{A}$ to itself is given by

$$\partial \phi(a, b) = a\phi(b) - \phi(ab) + \phi(a)b.$$

Hence, the kernel of this coboundary operator is the set of derivations on $\mathcal{A}$. So, taking this kernel and modding out by the image of the previous coboundary, if zero, means that all derivations on $\mathcal{A}$ are inner. Thus, the first Hochschild cohomology of uniform Roe algebras associated to bounded geometry metric spaces vanishes. It is then natural to ask if the higher groups $H_c^n(C_u^*(X), C_u^*(X))$ also vanish.

The question of whether or not the Hochschild cohomology vanishes in all dimensions for a hyperfinite von Neumann algebra has been answered completely by Kadison and Ringrose.

THEOREM 1.1 ([**4**] Theorem 3.1). *The Hochschild cohomology of a hyperfinite von Neumann algebra vanishes in all dimensions.*

2020 *Mathematics Subject Classification.* Primary 46L89.
Research supported by NSF grants DSM-1564281 and DSM-1901522.

Additionally, there have been many advancements for von Neumann algebras in general. For examples see Sinclair and Smith's book "Hochschild cohomology of von Neumann algebras" [**6**].

While we are not able to answer the question of whether or not the Hochschild cohomology vanishes in all dimensions for uniform Roe algebras, in Section 4 we are able to give conditions for the vanishing of the higher dimensional Hochschild cohomology of a uniform Roe algebra. Specifically:

THEOREM 1.2 (cf. Theorem 4.1). *If every element of $H_c^n(C_u^*(X))$ admits a weakly continuous representation, then $H_c^n(C_u^*(X)) = 0$.*

Note that, since all derivations are automatically weakly continuous by [**3**] Lemma 3, the previous theorem contains the derivations theorem as a special case.

2. Preliminaries

Inner products are linear in the first variable. For a Hilbert space $\mathcal{H}$ we denote the space of bounded operators on $\mathcal{H}$ by $\mathscr{B}(\mathcal{H})$, and the space of compact operators by $\mathscr{K}(\mathcal{H})$.

The Hilbert space of square-summable sequences on a set X is denoted $\ell^2(X)$, and the canonical basis of $\ell^2(X)$ will be denoted $(\delta_x)_{x\in X}$. For $a \in \mathscr{B}(\ell^2(X))$ we define its matrix entries by

$$a_{xy} := \langle \delta_x, a\delta_y \rangle .$$

DEFINITION 2.1 (propagation, uniform Roe algebra). Let X be a metric space and $r \geq 0$. An operator $a \in \mathscr{B}(\ell^2(X))$ *has propagation at most* r if $a_{xy} = 0$ whenever $d(x, y) > r$ for all $(x, y) \in X \times X$. In this case, we write $\text{prop}(a) \leq r$. The set of all operators with propagation at most r is denoted $\mathbb{C}_u^r[X]$. We define

$$\mathbb{C}_u[X] := \{a \in \mathscr{B}(\ell^2(X)) : \text{prop}(a) < \infty\};$$

it is not difficult to see that this is a $*$-algebra. The *uniform Roe algebra*, denoted $C_u^*(X)$, is defined to be the norm closure of $\mathbb{C}_u[X]$ under the norm inherited from $\mathscr{B}(\ell^2(X))$.

DEFINITION 2.2 (*ϵ-r-approximated*). Let X be a metric space. Given $\epsilon > 0$ and $r > 0$, an operator $a \in \mathscr{B}(\ell^2(X))$ can be *ϵ-r-approximated* if there exists $b \in \mathbb{C}_u^r[X]$ such that $\|a - b\| \leq \epsilon$. Note that an operator $a \in \mathscr{B}(\ell^2(X))$ is in the uniform Roe algebra if and only if for all $\epsilon > 0$ there exists an r such that a can be ϵ-r-approximated.

We will be exclusively interested in uniform Roe algebras associated to bounded geometry metric spaces as in the next definition.

DEFINITION 2.3 (bounded geometry). A metric space X is said to have *bounded geometry* if for every $r \geq 0$ there exists an $N_r \in \mathbb{N}$ such that for all $x \in X$, the ball of radius r about x has at most N_r elements.

Lastly, we introduce an averaging operator corresponding to an amenable group. For a full construction, and the proofs of the properties of this operator, see [**5**, Section 3].

THEOREM 2.4. *Let G be a discrete (possibly uncountable) amenable group and let B be a complex Banach space with dual B^*. Then there is a map*

$$\ell^\infty(G, B^*) \to B^* \quad \textit{defined by} \quad a \mapsto \int_G a(g)d\mu(g)$$

which is uniquely determined by the condition

$$\left\langle b, \int_G a(g)d\mu(g)\right\rangle = \int_G (\langle b, a(g)\rangle)d\mu(g) \tag{2.1}$$

for $b \in B$ and $a \in \ell^\infty(G, B^*)$. It is contractive, linear, invariant, and acts as the identity on constant functions. □

We will apply this machinery in the case that $B = \mathcal{L}^1(\ell^2(X))$ is the trace class operators on $\ell^2(X)$. In this case, the dual B^* canonically identifies with $\mathscr{B}(\ell^2(X))$: indeed, if Tr is the canonical trace on $\mathcal{L}^1(\ell^2(X))$, $b \in \mathcal{L}^1(\ell^2(X))$, and $a \in \mathscr{B}(\ell^2(X))$, then the pairing inducing this duality isomorphism is defined by

$$\langle b, a\rangle := \mathrm{Tr}(ba). \tag{2.2}$$

The next lemma says that our averaging process behaves well with respect to propagation. The main point of the lemma is that the collection of operators in $\mathscr{B}(\ell^2(X))$ that have propagation at most r is weak-$*$ closed for the weak-$*$ topology inherited from the pairing with $\mathcal{L}^1(\ell^2(X))$. We refer the reader to [**5**, Lemma 3.3] for a proof.

LEMMA 2.5. *With notation as above, if $r \geq 0$ and $f \in \ell^\infty(G, \mathscr{B}(\ell^2(X)))$ is such that the propagation of each $f(g)$ is at most r for all $g \in G$, then the propagation of $\int_G f(g)d\mu(g)$ is also at most r.* □

2.1. A Result of Braga and Farah. In this subsection, we present a result of Braga and Farah from [**2**, Lemma 4.9]. This theorem will allow us to uniformly ϵ-r-approximate $f \in \ell^\infty(\mathcal{U}, \mathscr{B}(\ell^2(X)))$ where $\mathcal{U}$ is the unitary group of $\ell^\infty(X)$. That is, given $\epsilon > 0$, there exists a *single* $r > 0$ such that for all $u \in \mathcal{U}$ such that $f(u) \in \mathscr{B}(\ell^2(X))$ can be ϵ-r-approximated.

To state the result, let $\mathbb{D} := \{z \in \mathbb{C} \mid |z| \leq 1\}$ denote the closed unit disk in the complex plane. Let I be a countably infinite set, and let $\mathbb{D}^I$ denote as usual the space of all I-indexed tuples $\lambda := (\lambda_i)_{i\in I}$ with each $\lambda_i \in \mathbb{D}$. We fix this notation throughout this section.

DEFINITION 2.6 (symmetrically summable). A sequence $(a_i)_{i\in I}$ is *symmetrically summable* if for all $\lambda \in \mathbb{D}^I$, the sum $\sum_{i\in I} \lambda_i a_i$ converges in the weak operator topology to an element of $C_u^*(X)$. If (a_i) is symmetrically summable and $\lambda = (\lambda_i)$ is in $\mathbb{D}^I$, we write a_λ for the operator $\sum_{i\in I} \lambda_i a_i$.

THEOREM 2.7 (Lemma 4.9 [**2**]). *Let (a_i) be a symmetrically summable collection of operators in $C_u^*(X)$. Then for any $\epsilon > 0$ there exists $r > 0$ such that for all $\lambda \in \mathbb{D}^I$, the operator a_λ is ϵ-r-approximated.*

Next, we upgrade Braga and Farah's lemma to a multilinear version.

DEFINITION 2.8 (separately symmetrically summable). For a finite sequence of countable index sets $\{I_n\}_{n=1}^N$, $N < \infty$, a uniformly bounded family of operators $(a_{(i_1,\ldots,i_N)})_{(\bar{i}\in\prod_{n=1}^N I_n)} \subseteq C_u^*(X)$ is N *separately symmetrically summable* if the following condition holds.

For any $(1 \leq k \leq N)$, and for each fixed

$$\left\{\lambda^{(1)}, \ldots, \lambda^{(k-1)}, \lambda^{(k+1)}, \ldots, \lambda^{(N)}\right\} \in \prod_{\substack{n=1\\ n\neq k}}^{N} \mathbb{D}^{I_n}$$

the sum

$$\sum_{i_k \in I_k} \lambda_{i_k}^{(k)} a_{(\lambda^{(1)},\ldots,\lambda^{(k-1)},i_k,\lambda^{(k+1)},\ldots,\lambda^{(N)})}$$

converges in the weak operator topology to an element

$$a_{(\lambda^{(1)},\ldots,\lambda^{(k)},\ldots,\lambda^{(N)})} \in C_u^*(X).$$

Additionally,

$$\text{for all } \left\{\lambda^{(1)},\ldots,\lambda^{(N)}\right\} \in \prod_{n=1}^{N} \mathbb{D}^{I_n}, \qquad \sup_{(\lambda^{(1)},\ldots,\lambda^{(N)})} \left\| a_{(\lambda^{(1)},\ldots,\lambda^{(N)})} \right\| < \infty.$$

Note that, if $(a_{(i_1,\ldots,i_{N+1})})_{(\bar{i}\in\prod_{n=1}^{N+1} I_n)}$ is $(N+1)$ separately symmetrically summable, then for any fixed $\eta \in \mathbb{D}^{I_{N+1}}$, $(a_{(i_1,\ldots,i_N,\eta)})_{(\bar{i}\in\prod_{n=1}^{N} I_n)}$ is N separately symmetrically summable.

THEOREM 2.9. *Suppose that*

$$(a_{(i_1,\ldots,i_N)})_{(\bar{i}\in\prod_{n=1}^{N} I_n)} \subseteq C_u^*(X)$$

is N separately symmetrically summable. Then for any $\epsilon > 0$ there exists an $r > 0$ such that for all $(\lambda^{(1)},\ldots,\lambda^{(N)}) \in \prod_{n=1}^{N} \mathbb{D}^{I_n}$, the operator $a_{(\lambda^{(1)},\ldots,\lambda^{(N)})}$ is ϵ-r-approximated.

To prove this theorem we induct on N. However, we will need a few lemmas and a definition first. Note that the base case is handled by Theorem 2.7. Proofs for the next two lemmas can be found in [**2**, Lemma 4.6] and [**5**, Lemma 4.5] respectively.

LEMMA 2.10. (1) *If a is a bounded operator on $\ell^2(X)$ such that for all finite rank projections p in $\ell^\infty(X)$ the product pap can be ϵ-r-approximated, then a itself can be ϵ-r-approximated.*
(2) *Say a is a bounded operator on $\ell^2(X)$ and $\epsilon, r > 0$ are such that for all $\delta > 0$, a can be $(\epsilon+\delta)$-r-approximated. Then a can be ϵ-r-approximated.* □

LEMMA 2.11. *Say $(x_i)_{i\in I}$ is a collection in a Banach space such that $\sum_i \lambda_i x_i$ converges in norm for all $(\lambda_i) \in \mathbb{D}^I$. Then for any $\delta > 0$ there exists a finite subset F of I such that for all $(\lambda_i) \in \mathbb{D}^I$*

$$\left\| \sum_{i\in I\setminus F} \lambda_i x_i \right\| < \delta.$$

□

DEFINITION 2.12. Suppose that $(a_{(i_1,\ldots,i_N)})_{(\bar{i}\in\prod_{n=1}^{N} I_n)} \subseteq C_u^*(X)$ is N separately symmetrically summable. Let $\lambda = (\lambda^{(1)},\ldots,\lambda^{(N-1)})$. Then for $\eta \in \mathbb{D}^{I_N}$ we let

$$a_{\lambda,\eta} = \sum_{i_N \in I_N} \eta_{i_N} a_{\lambda,i_N}$$

Then for $\epsilon, r > 0$ define

$$U_{\epsilon,r} := \left\{ \eta \in \mathbb{D}^{I_N} \mid a_{\lambda,\eta} \text{ is } \epsilon\text{-}r\text{-approximated for all } \lambda \in \prod_{n=1}^{N-1} \mathbb{D}^{I_n} \right\}.$$

REMARK 2.13. On the first read it may provide intuition to just consider the $N = 2$ case since the proof of the inductive step is only notationally different.

Suppose that $\epsilon > 0$ is given. If we are considering the $N = 2$ case, and $\{a_{i,j}\}_{i\in I, j\in J}$ is 2 separately symmetrically summable. Then, for each fixed $\eta \in \mathbb{D}^J$, $\{a_{i,\eta}\}_{i\in I}$ is symmetrically summable so by Theorem 2.7 we may write $\mathbb{D}^J$ as the union in line (2.3).

For the inductive step, suppose that $(a_{(i_1,\dots,i_N)})_{(\bar{i}\in\prod_{n=1}^N I_n)}$ is N separately symmetrically summable. Then, for each fixed $\eta \in \mathbb{D}^{I_N}$ we have that $(a_{(i_1,\dots,i_{N-1},\eta)})_{(\bar{i}\in\prod_{n=1}^{N-1} I_n)}$ is $(N-1)$ separately symmetrically summable. Thus, by inductive hypothesis we may write $\mathbb{D}^{I_N}$ as the union

$$\mathbb{D}^{I_N} = \bigcup_{r=1}^{\infty} U_{\epsilon,r}. \tag{2.3}$$

We will first show that the sets in Definition 2.12 are closed for any N separately symmetrically summable $(a_{(i_1,\dots,i_N)})_{(\bar{i}\in\prod_{n=1}^N I_n)}$. Then we will show that if $(a_{(i_1,\dots,i_N)})_{(\bar{i}\in\prod_{n=1}^N I_n)}$ does not satisfy the conclusion of Theorem 2.9, there is $\epsilon > 0$ such that for all $r > 0$, $U_{r,\epsilon}$ is nowhere dense in $\mathbb{D}^{I_N}$. As we have the union in line (2.3), this contradicts the Baire category theorem and we will be done.

LEMMA 2.14. *Suppose that $(a_{(i_1,\dots,i_N)})_{(\bar{i}\in\prod_{n=1}^N I_n)}$ is separately symmetrically summable. Let $\lambda = (\lambda^{(1)}, \dots, \lambda^{(N-1)})$. Then for any $\epsilon, r > 0$ the set $U_{\epsilon,r}$ of Definition 2.12 is closed.*

PROOF. Assume for contradiction that for some $\epsilon, r > 0$, $U_{\epsilon,r}$ is not closed. Then there exists some $\eta \in \overline{U_{\epsilon,r}} \setminus U_{\epsilon,r}$. As $\eta \notin U_{\epsilon,r}$, there exists a $\lambda \in \prod_{n=1}^{N-1} \mathbb{D}^{I_n}$ such that $a_{\lambda,\eta}$ cannot be ϵ-r-approximated. Fix this λ.

Using (the contrapositive of) Lemma 2.10, part (i), there exists a finite rank projection $p \in \ell^\infty(X)$ such that $pa_{\lambda,\eta}p$ cannot be ϵ-r-approximated.

Now, for any $\mu \in \mathbb{D}^{I_N}$, the sum $\sum_{i\in I_N} \mu_i a_{\lambda,i}$ defining $a_{\lambda,\mu}$ is weakly convergent. As p is finite rank, this implies that the sum $\sum_{i\in I_N} p\mu_i a_{\lambda,i}p$ is norm convergent. Hence, using Lemma 2.11, for any $\delta > 0$ there exists a finite subset F of I_N such that

$$\left\| \sum_{i\in I_N\setminus F} p\mu_i a_{\lambda,i} p \right\| < \delta \tag{2.4}$$

for all $\mu \in \mathbb{D}^{I_N}$ (and in particular for $\mu = \eta$).

As F is finite, the set

$$\left\{ \mu \in \mathbb{D}^{I_N} \;\middle|\; |F| \max_{i\in F} \|a_{\lambda,i}\| |\mu_i - \eta_i| < \delta \text{ for all } i \in F \right\} \tag{2.5}$$

is an open neighborhood of η for the product topology. As η is in the closure of $U_{\epsilon,r}$, the set in line 2.5 thus contains some $\theta \in U_{\epsilon,r}$. In particular $pa_{\lambda,\theta}p$ is ϵ-r-approximated, so there is $b \in \mathbb{C}_u^r[X]$ such that $\|pa_{\lambda,\theta}p - b\| \le \epsilon$.

Note that

$$\|pa_{\lambda,\eta}p - b\| \le \|pa_{\lambda,\theta}p - b\| + \|pa_{\lambda,\eta}p - pa_{\lambda,\theta}p\|$$

$$\leq \|pa_{\lambda,\theta}p - b\| + \left\|\sum_{i\in F}(\eta_i - \theta_i)pa_{\lambda,i}p\right\| + \left\|\sum_{i\in I_N\setminus F}\theta_i pa_{\lambda,i}p\right\| + \left\|\sum_{i\in I_N\setminus F}\eta_i pa_{\lambda,i}p\right\|.$$

The first term on the bottom line is bounded above by ϵ by choice of b, the second is bounded above by δ using that θ is in the set in line 2.5, and the third and fourth terms are bounded above by δ using the estimate in line 2.4 (which is valid for all elements η of $\mathbb{D}^{I_N}$).

Now, we have shown that for arbitrary $\delta > 0$, we have found $b \in \mathbb{C}_u^r[X]$ such that $\|pa_{\lambda,\eta}p - b\| \leq \epsilon + 3\delta$. Using Lemma 2.10, part (ii), this implies that $pa_{\lambda,\eta}p$ can be ϵ-r-approximated. This contradicts our assumption in the first paragraph, so we are done. □

LEMMA 2.15. *Suppose that $(a_{(i_1,\ldots,i_N)})_{(\bar{i}\in\prod_{n=1}^N I_n)} \subseteq C_u^*(X)$ is separately symmetrically summable. Let $\lambda = (\lambda^{(1)},\ldots,\lambda^{(N-1)})$. Then, for all $\epsilon > 0$, for any $\theta \in \mathbb{D}^{I_N}$, and any finite $F \subseteq I_N$ there exists an $r > 0$ such that the sum $\sum_{i\in F}\theta_i a_{\lambda,i}$ is ϵ-r-approximated.*

PROOF. Let F be a finite subset of I_N and $\epsilon > 0$ be given. By supposition, for each i, we may write

$$a_{\lambda,i} = b_{\lambda,i} + c_{\lambda,i} \text{ where } b_{\lambda,i} \in \mathbb{C}_u^{r_i}[X] \text{ and } \|c_{\lambda,i}\| < \tfrac{\epsilon}{|F|}.$$

Let $r = \max_{i\in F}\{r_i\}$ and note that $\sum_{i\in F}\theta_i b_{\lambda,i} \in \mathbb{C}_u^r[X]$ for all λ. Additionally,

$$\left\|\sum_{i\in F}\theta_i c_{\lambda,i}\right\| \leq \sum_{i\in F}|\theta_i|\,\|c_{\lambda,i}\| < \epsilon.$$

Hence, $\sum_{i\in F}\theta_i a_{\lambda,i}$ is ϵ-r-approximated for all $\lambda \in \prod_{n=1}^{N-1}\mathbb{D}^{I_n}$. □

LEMMA 2.16. *Suppose that $(a_{(i_1,\ldots,i_N)})_{(\bar{i}\in\prod_{n=1}^N I_n)}$ is a separately symmetrically summable collection of operators in $C_u^*(X)$ that does not satisfy the conclusion of Lemma 2.9. Additionally, let $\lambda = (\lambda^{(1)},\ldots,\lambda^{(N-1)})$. Then there is an $\epsilon > 0$ so that for all $r > 0$ and all finite subsets $F \subseteq I_N$ there exists $\eta \in \mathbb{D}^{I_N}$ such that $\sum_{i\in I_N\setminus F}\eta_i a_{\lambda,i}$ cannot be ϵ-r-approximated.*

PROOF. Let $(a_{(i_1,\ldots,i_N)})_{(\bar{i}\in\prod_{n=1}^N I_n)}$ be as in the statement. Then there exists $\delta > 0$ such that for all $r > 0$ there exists $(\lambda,\eta) \in \left(\prod_{n=1}^{N-1}\mathbb{D}^{I_n}\right) \times \mathbb{D}^{I_N}$ such that $a_{\lambda,\eta}$ is not δ-r-approximable. Fix this λ. Assume for contradiction that the conclusion of the lemma fails. Then there exists $s > 0$ and a finite subset F of I_N such that for all $\xi \in \mathbb{D}^{I_N}$ we have that $\sum_{i\in I_N\setminus F}\xi_i a_{\lambda,i}$ is $\delta/2$-s-approximated. As F is finite, by Lemma 2.15 there is a $t > 0$ such that every element of

$$\left\{\sum_{i\in F}\xi_i a_{\lambda,i} \;\middle|\; \xi \in \mathbb{D}^{I_N}\right\}$$

can be $\delta/2$-t-approximated. Now, for arbitrary $\xi \in \mathbb{D}^{I_N}$,

$$a_{\lambda,\xi} = \sum_{i\in F}\xi_i a_{\lambda,i} + \sum_{i\in I_N\setminus F}\xi_i a_{\lambda,i};$$

as the first term above can be $\delta/2$-s-approximated, and as the second can be $\delta/2$-t-approximated, this implies that $a_{\lambda,\xi}$ can be δ-$\max\{s,t\}$-approximated. As ξ was arbitrary, this contradicts the second sentence in the proof, and we are done. □

As stated at the end of Remark 2.13, the following lemma completes the proof of Corollary 2.9.

LEMMA 2.17. *Suppose that* $(a_{(i_1,\dots,i_N)})_{(\bar{i}\in\prod_{n=1}^N I_n)}$ *is a separately symmetrically summable collection of operators in* $C_u^*(X)$ *that does not satisfy the conclusion of Lemma 2.9. Let* $\lambda = (\lambda^{(1)},\dots,\lambda^{(N-1)})$. *Then there is* $\epsilon > 0$ *such that for each* $r > 0$ *the set* $U_{\epsilon,r}$ *of Definition 2.12 is nowhere dense in* $\mathbb{D}^{I_N}$.

PROOF. Let $(a_{(i_1,\dots,i_N)})_{(\bar{i}\in\prod_{n=1}^N I_n)}$ be as in the statement. Then there exists $\delta > 0$ such that for all $r > 0$ there exists $(\lambda,\eta) \in \left(\prod_{n=1}^{N-1}\mathbb{D}^{I_n}\right)\times\mathbb{D}^{I_N}$ such that $a_{\lambda,\eta}$ is not δ-r-approximable. Fix this λ. Let $\epsilon' > 0$ have the property from Lemma 2.16. We claim that $\epsilon := \epsilon'/2$ has the property required for this lemma. Assume for contradiction that for some $r > 0$, $U_{\epsilon,r}$ is not nowhere dense. Lemma 2.14 implies that $U_{\epsilon,r}$ is closed, and so it contains a point ξ in its interior. Then by definition of the product topology there exists a finite set $F \subseteq I_N$ and $\delta > 0$ such that the set

$$V := \{\nu \in \mathbb{D}^{I_N} \mid |\xi_i - \nu_i| < \delta \text{ for all } i \in F\} \quad \text{is contained in} \quad U_{\epsilon,r}. \tag{2.6}$$

Note that the element $\sum_{i\in F}\xi_i a_{\lambda,i}$ is in $C_u^*(X)$ by assumption, so can be ϵ-s-approximated for some s. Let $b_{\xi,\lambda} \in \mathbb{C}_u^s[X]$ be such that $\|\sum_{i\in F}\xi_i a_{\lambda,i} - b_{\lambda,\xi}\| \le \epsilon$. On the other hand, Lemma 2.16 gives us $\mu \in \mathbb{D}^{I_N}$ so that $\sum_{i\in I\setminus F}\mu_i a_{\lambda,i}$ cannot be ϵ'-$\max\{r,s\}$-approximated. We may further assume that $\mu_i = 0$ for $i \in F$. Define $\theta \in \mathbb{D}^I$ by

$$\theta_i := \begin{cases} \xi_i & i \in F \\ \mu_i & i \notin F \end{cases}$$

Then θ is clearly in the set V of line 2.6, and so $a_{\lambda,\theta}$ is ϵ-r-approximated. Let then $b_{\lambda,\theta} \in \mathbb{C}_u^r[X]$ be such that $\|a_{\lambda,\theta} - b_{\lambda,\theta}\| \le \epsilon$. We then see that

$$\begin{aligned}\|a_{\lambda,\mu} - (b_{\lambda,\theta} - b_{\lambda,\xi})\| &\le \|a_{\lambda,\mu} - a_{\lambda,\theta} + b_{\lambda,\xi}\| + \|a_{\lambda,\theta} - b_{\lambda,\theta}\| \\ &\le \left\| b_{\lambda,\xi} - \sum_{i\in F}\xi_i a_{\lambda,i}\right\| + \|a_{\lambda,\theta} - b_{\lambda,\theta}\|\end{aligned}$$

The terms on the bottom row are each less than ϵ by choice of $b_{\lambda,\xi}$ and $b_{\lambda,\theta}$, and so $\|a_{\lambda,\mu} - (b_{\lambda,\theta} - b_{\lambda,\xi})\| \le 2\epsilon = \epsilon'$. As $b_{\lambda,\xi} + b_{\lambda,\theta}$ has propagation at most $\max\{r,s\}$, this contradicts the assumption that $a_{\lambda,\mu}$ cannot be ϵ'-$\max\{r,s\}$-approximated, so we are done. □

3. Hochschild Cohomology

In this section we introduce Hochschild cohomology, its construction, and several of its properties.

DEFINITION 3.1 (Subdual). Let $\mathcal{A}$ be a C*-algebra and let $\mathcal{V}$ be an $\mathcal{A}$-submodule of a dual module $\mathcal{W}$ (under the same action). We will call such a module $\mathcal{V}$ a *subdual* of $\mathcal{W}$. Note that we are not requiring $\mathcal{V}$ to be a dual space, just that it is a submodule of a specified dual space. Moreover, if $\mathcal{A}$ is a C*-subalgebra of a von Neumann algebra $\mathcal{M}$ where $\mathcal{W}$ is a dual normal $\mathcal{M}$-module and the action of $\mathcal{A}$ on $\mathcal{V}$ is inherited from the $\mathcal{M}$-action on $\mathcal{W}$ then we say that $\mathcal{V}$ is a *subdual normal* $\mathcal{A}$-module of $\mathcal{W}$.

An example of a subdual normal module is the uniform Roe algebra acting on itself by multiplication. $C_u^*(X)$ acts on $\mathscr{B}(\ell^2(X))$ by multiplication making $\mathscr{B}(\ell^2(X))$ a $C_u^*(X)$-module. $\mathscr{B}(\ell^2(X))$ is a dual space with predual $\mathcal{L}^1(\ell^2(X))$, the trace class operators. So $C_u^*(X)$ is a submodule of the dual space $\mathscr{B}(\ell^2(X))$. However, $C_u^*(X)$ is not usually a dual space. This additional structure on the submodule allows us to use the relative weak* topology inherited from the parent module.

By $\mathscr{L}_c^n(\mathcal{A},\mathcal{V})$ we mean the vector space of separately norm continuous multilinear maps from the n-fold Cartesian product of $\mathcal{A}$ to the $\mathcal{A}$-bimodule $\mathcal{V}$ when $n \geq 1$ and $\mathscr{L}_c^0(\mathcal{A},\mathcal{V}) := \mathcal{V}$.

Let $\mathcal{A}$ be a concrete C*-algebra. If $\mathcal{W}$ is a dual normal $\mathcal{A}$-bimodule with subdual $\mathcal{V}$, we use the notation $\mathscr{L}_w^n(\mathcal{A},\mathcal{V})$ to indicate the vector space of multilinear maps that are separately ultraweak-weak* continuous; that is, for $\phi \in \mathscr{L}_w^n(\mathcal{A},\mathcal{V})$

$$\text{if} \quad \{a_\alpha\} \subset \mathcal{A} \quad \text{is a net such that} \quad a_\alpha \to a \in \mathcal{A} \quad \text{ultraweakly in} \quad \mathscr{B}(\mathcal{H})$$

$$\text{then} \quad \phi(\dots, a_\alpha, \dots) \to \phi(\dots, a, \dots) \in \mathcal{V} \quad \text{weak* in } \mathcal{W}.$$

When we write $\mathscr{L}^n(\mathcal{A},\mathcal{V})$ then either subscript may be attached. Considering $\mathcal{A}$ as a module over itself we will simply write $\mathscr{L}(\mathcal{A})$. Additionally, we equip both $\mathscr{L}_c^n(\mathcal{A},\mathcal{V})$ and $\mathscr{L}_w^n(\mathcal{A},\mathcal{V})$ with the operator norm.

REMARK 3.2. Note that while $\mathscr{L}_c^n(\mathcal{A},\mathcal{V})$ is complete in norm, we are not assuming, nor do we require these vector spaces to be complete in norm.

To define the Hochschild cohomology we first construct the cochain complex

$$0 \to \mathscr{L}^0(\mathcal{A},\mathcal{V}) \overset{\partial}{\to} \mathscr{L}^1(\mathcal{A},\mathcal{V}) \overset{\partial}{\to} \cdots \overset{\partial}{\to} \mathscr{L}^n(\mathcal{A},\mathcal{V}) \overset{\partial}{\to} \mathscr{L}^{n+1}(\mathcal{A},\mathcal{V}) \overset{\partial}{\to} \cdots$$

for both the norm continuous and ultraweak-weak* continuous cases where the coboundary operator $\partial : \mathscr{L}^n(\mathcal{A},\mathcal{V}) \to \mathscr{L}^{n+1}(\mathcal{A},\mathcal{V})$ is defined by

$$\begin{aligned}(\partial\phi)(a_1,\dots,a_{n+1}) &= a_1\phi(a_2,\dots,a_{n+1})\\ &+\sum_{j=1}^{n}(-1)^j\phi(a_1,\dots,a_ja_{j+1},\dots,a_{n+1})\\ &+(-1)^{n+1}\phi(a_1,\dots,a_n)a_{n+1} \quad (n\geq 1)\end{aligned}$$

and for $n = 0$

$$(\partial v)(a) = av - va \quad (v \in \mathcal{V}, a \in \mathcal{A}).$$

A straightforward calculation shows that ∂^2 is always zero. The nth Hochschild cohomology group $H_c^n(\mathcal{A},\mathcal{V})$ (resp. $H_w^n(\mathcal{A},\mathcal{V})$ in the ultraweak-weak* case) is the cohomology at the nth step

If we consider $\mathcal{A}$ as a module over itself we simply write $H^n(\mathcal{A})$. The cohomology obtained from this construction is the *Hochschild* cohomology.

DEFINITION 3.3 (multimodular maps). Let $\mathcal{A}$ be a C*-algebra and let $\phi : \mathcal{A}^n \to \mathcal{V}$ be a bounded multilinear map to the Banach $\mathcal{A}$-bimodule $\mathcal{V}$. If $\mathcal{B}$ is a C*-subalgebra of $\mathcal{A}$ we say that ϕ is *$\mathcal{B}$-multimodular* if for any $b \in \mathcal{B}$ the following hold.

(1) $b\phi(a_1,\dots,a_n) = \phi(ba_1,\dots,a_n)$,
(2) $\phi(a_1,\dots,a_{j-1}b,a_j,\dots,a_n) = \phi(a_1,\dots,a_{j-1},ba_j,\dots,a_n)$ and
(3) $\phi(a_1,\dots,a_nb) = \phi(a_1,\dots,a_n)b$

If $\mathcal{B}$ is a C*-subalgebra of $\mathcal{A}$ we use the notation $\mathscr{L}^n(\mathcal{A},\mathcal{V}:\mathcal{B})$ to indicate that the maps are $\mathcal{B}$-multimodular where we may use either subscript,"c" or "w". As before we may construct the Hochschild cohomology of $\mathcal{B}$-multimodular maps which we denote by $H^n(\mathcal{A},\mathcal{V}:\mathcal{B})$ where either subscript c or w may be attached. Additionally, if we are considering $\mathcal{A}$ as a module over itself we simply write $H^n(\mathcal{A}:\mathcal{B})$.

3.1. Sinclair and Smith's 'Reduction of Cocycles'. In this subsection we introduce a method to modify a cocycle, say $\phi \in \mathscr{L}_c^n(\mathcal{A},\mathcal{W})$, by a coboundary to obtain a map in $\mathscr{L}_c^n(\mathcal{A},\mathcal{W}:\mathcal{B})$ where $\mathcal{A}\subseteq \mathscr{B}(\mathcal{H})$ is a C*-algebra, $\mathcal{B}$ is a C*-subalgebra of $\mathcal{A}$, and $\mathcal{W}$ is a dual normal $\mathscr{B}(\mathcal{H})$-bimodule.

LEMMA 3.4 ([**6**] Lemma 3.2.1). *Let $\mathcal{B}$ be a unital subalgebra of a unital C*-algebra $\mathcal{A}$. Let $\mathcal{W}$ be a Banach $\mathcal{A}$-bimodule, and let $\phi \in \mathscr{L}^n(\mathcal{A},\mathcal{W})$ with $\partial\phi = 0$.*
Then for all $b\in\mathcal{B}$ and $x_1,\dots,x_n\in\mathcal{A}$ we have:

(1) $\phi(b,x_2,\dots,x_n)=0$ *if and only if*
$\phi(1,x_2,\dots,x_n)=0$ *and* $\phi(bx_1,x_2,\dots,x_n)=b\phi(x_1,\dots,x_n)$.
(2) *Fix $k\le n$. Then for all $j\in\{2,\dots,k\}$,*
$\phi(x_1,\dots,x_{j-1},b,x_{j+1},\dots,x_n)=0$ *if and only if*
$\phi(x_1,\dots,x_{j-1},1,x_{j+1},\dots,x_n)=0$ *and*
$\phi(x_1,\dots,x_{j-1}b,x_j,\dots,x_n)=\phi(x_1,\dots,x_{j-1},bx_j,\dots,x_n)$
(3) *Additionally,*
$\phi(x_1,\dots,x_{n-1},b)=0$ *if and only if*
$\phi(x_1,\dots,x_{n-1},1)=0$ *and* $\phi(x_1,\dots,x_nb)=\phi(x_1,\dots,x_n)b$ □

LEMMA 3.5 ([**6**] Lemma 3.2.4). *Let $\mathcal{B}$ be a C*-subalgebra spanned by an amenable group $\mathcal{U}$ (with respect to the discrete topology) of unitaries in a unital C*-algebra $\mathcal{A}$, and let $\mathcal{W}$ be a dual Banach $\mathcal{A}$-bimodule. There is a continuous linear map*

$$K_n:\mathscr{L}_c^n(\mathcal{A},\mathcal{W})\to\mathscr{L}_c^{n-1}(\mathcal{A},\mathcal{W})$$

(depending on a choice of invariant mean on $\mathcal{U}$) such that if $\phi\in\mathscr{L}_c^n(\mathcal{A},\mathcal{W})$ satisfies $\partial\phi=0$ then $\phi-\partial(K_n\phi)$ is $\mathcal{B}$-multimodular. Moreover, we have that

$$\|K_n\|\le\tfrac{(n+2)^n-1}{n+1}.$$

□

REMARK 3.6. As we will need it later, let us recall that the map K_n is constructed recursively via

$$J_1:\mathscr{L}_c^n(\mathcal{A},\mathcal{W})\to\mathscr{L}_c^{n-1}(\mathcal{A},\mathcal{W})\quad\text{defined by}$$

$$(J_1\phi)(a_1,\dots,a_{n-1})=\int_{\mathcal{U}}u^*\phi(u,a_1,\dots,a_{n-1})d\mu(u),\tag{3.1}$$

$$G_k:\mathscr{L}_c^n(\mathcal{A},\mathcal{W})\to\mathscr{L}_c^{n-1}(\mathcal{A},\mathcal{W})\quad\text{defined by}$$

$$(G_k\phi)(a_1,\dots,a_{n-1})=\int_{\mathcal{U}}\phi(a_1,\dots,a_ku^*,u,a_{k+1},\dots,a_{n-1})d\mu(u),\tag{3.2}$$

$$J_{k+1}:\mathscr{L}_c^n(\mathcal{A},\mathcal{W})\to\mathscr{L}_c^{n-1}(\mathcal{A},\mathcal{W})\quad\text{defined by}\quad J_{k+1}=J_k+(-1)^kG_k(I-\partial J_k),$$

$$\text{and } K_n=J_n.\tag{3.3}$$

LEMMA 3.7 ([**6**] Lemma 3.2.6). *Let $\mathcal{A}$ be a unital C^*-algebra and let $\mathcal{W}$ be a dual $\mathcal{A}$-bimodule. Suppose that $\mathcal{B}$ is a C^*-subalgebra of $\mathcal{A}$ generated by an amenable group $\mathcal{U}$ of unitaries. Then there is a continuous surjective linear projection*

$$Q_n : \mathscr{L}_c^n(\mathcal{A},\mathcal{W}) \to \mathscr{L}_c^n(\mathcal{A},\mathcal{W}:\mathcal{B})$$

such that $\partial Q_{n-1} = Q_n\partial$ and $\|Q_n\| = 1$. □

We conclude this section with a theorem that will be useful in the next section.

THEOREM 3.8 ([**6**] Theorem 3.2.7). *Let $\mathcal{B}$ be the C^*-algebra generated by an amenable group $\mathcal{U}$ of unitaries in a unital C^*-algebra $\mathcal{A}$, and let $\mathcal{W}$ be a dual $\mathcal{A}$-bimodule. Then*

$$H_c^n(\mathcal{A},\mathcal{W}) \cong H_c^n(\mathcal{A},\mathcal{W}:\mathcal{B})$$

for all $n \in \mathbb{N}$ with isomorphism induced by the natural embedding

$$\mathscr{L}_c^n(\mathcal{A},\mathcal{W}:\mathcal{B}) \hookrightarrow \mathscr{L}_c^n(\mathcal{A},\mathcal{W}).$$

PROOF. Clearly, the natural embedding $\mathscr{L}_c^n(\mathcal{A},\mathcal{W}:\mathcal{B}) \hookrightarrow \mathscr{L}_c^n(\mathcal{A},\mathcal{W})$ induces a homomorphism $H_c^n(\mathcal{A},\mathcal{W}:\mathcal{B}) \to H_c^n(\mathcal{A},\mathcal{W})$. By Lemma 3.5 this map is surjective. Furthermore, if $\phi \in \mathscr{L}_c^n(\mathcal{A},\mathcal{W}:\mathcal{B})$ and $\psi \in \mathscr{L}_c^n(\mathcal{A},\mathcal{W})$ is such that $\phi = \partial\psi$, then with Q_n as in Lemma 3.7,

$$\phi = Q_n\phi = Q_n\partial\psi = \partial Q_{n-1}\psi \quad \text{where} \quad Q_{n-1}\psi \in \mathscr{L}_c^{n-1}(\mathcal{A},\mathcal{W}:\mathcal{B})$$

and so our map is injective. □

REMARK 3.9. If our averaging operator, i.e. the "integral" over the unitary group $\mathcal{U}$, converges in the weak* topology of the dual normal $\mathcal{A}$-bimodule $\mathcal{W}$ to an element in the subdual $\mathcal{V}$ of $\mathcal{W}$ for all $\phi \in \mathscr{L}^n(\mathcal{A},\mathcal{W})$ then we may replace $\mathcal{W}$ with $\mathcal{V}$ everywhere above.

4. A Relation Between Cohomologies

The goal of this section is to prove the following theorem.

THEOREM 4.1. *The natural map $H_w^n(C_u^*(X)) \to H_c^n(C_u^*(X))$ is surjective if and only if*

$$H_c^n(C_u^*(X)) = 0.$$

Note that, by [**3**, Lemma 3], all bounded derivations on any C*-algebra are weakly continuous. Thus, the natural map $H_w^1(\mathcal{A}) \to H_c^1(\mathcal{A})$ is automatically surjective. However, this does not seem to be known for $n \geq 2$. Additionally, our proof of Theorem 4.1 depends on the underlying C*-algebra being a uniform Roe algebra. That being said, Theorem 4.1 strictly generalizes Rufus Willett and the author's work in [**5**].

For notational convenience throughout we let: $A = C_u^*(X)$, $\mathscr{B} = \mathscr{B}(\ell^2(X))$, and $\ell = \ell^\infty(X)$. As a first step towards showing Theorem 4.1 we show that $H_c^n(A,\mathscr{B}:\ell) \cong H_c^n(A:\ell)$ in the following lemma.

LEMMA 4.2. *Let $\phi \in \mathscr{L}^n(A,\mathscr{B}:\ell)$. Then ϕ takes image in the uniform Roe algebra; that is, $\mathscr{L}^n(A,\mathscr{B}:\ell) = \mathscr{L}^n(A:\ell)$.*

PROOF. Let $\phi \in \mathscr{L}^n(A, \mathscr{B} : \ell)$, $(x_1, \ldots, x_n) \in A^n$, and $0 < \epsilon \leq 1$ be given. Set $M = \max\{\|x_i\|\} + 1$ and note that since each $x_i \in A$ we may write each x_i as

$$x_i = a_i + b_i \quad \text{where} \quad a_i \in \mathbb{C}_u^{r_i}[X] \quad \text{and} \quad \|b_i\| < \min\left\{\frac{\epsilon}{n\|\phi\|M^n}, \epsilon\right\}.$$

Moreover, we have that $\|a_i\| < M$. Next, since ϕ is multilinear we may write

$$\phi(x_1, \ldots, x_n) = \phi(a_1, \ldots, a_n) + \phi(a_1, \ldots, a_{n-1}, b_n) + \phi(a_1, \ldots, a_{n-2}, b_{n-1}, x_n) + \cdots + \phi(a_1, b_2, x_3, \ldots, x_n) + \phi(b_1, x_2, \ldots, x_n).$$

Observe that every term but the first in this expansion has a b_i in a single coordinate and either a_i's or x_i's in the remaining coordinates. Thus, the norm for each of the terms with a b_k in the kth coordinate is bounded by

$$\|\phi\| \left(\prod_{i=1}^{n} M\right) \|b_k\| < \frac{\epsilon}{n}$$

Hence, it is enough to show that $\phi(a_1, \ldots, a_n) \in \mathbb{C}_u^{n \cdot r}[X]$ where $r = \max\{r_i\}$.

To show this let p_x be the projection onto the span of the Dirac mass at x, and let $B_x(r)$ denote the closed ball of radius r centered at x. We then define

$$p_{B_x(r)} := \sum_{k \in B_x(r)} p_k.$$

Note that, the sum defining $p_{B_x(r)}$ is finite for any given $r \in \mathbb{N}$ since X has bounded geometry. Next, for any fixed $x \in X$,

$$p_x a_1 = p_x a_1 p_{B_x(r)} \quad \text{and} \quad p_{B_x((i-1)\cdot r)} a_i = p_{B_x((i-1)\cdot r)} a_i p_{B_x(i \cdot r)} \tag{4.1}$$

since each a_i has propagation less than r. Next, fix $x, y \in X$ such that $d(x, y) > n \cdot r$ and observe that

$$p_x \phi(a_1, \ldots, a_n) p_y = \phi(p_x a_1, \ldots, a_n p_y)$$
$$= \phi(p_x a_1 p_{B_x(r)}, \ldots, a_n p_y) = \phi(p_x a_1 p_{B_x(r)}, p_{B_x(r)} a_2, \ldots, a_n p_y)$$

where on the left hand we have used line (4.1) and on the right hand side we use that ϕ is $\ell^\infty(X)$-multimodular.

Continuing this process $n-1$ times we arrive at

$$p_x \phi(a_1, \ldots, a_n) p_y$$
$$= \phi(p_x a_1 p_{B_x(r)}, \ldots, p_{B_x((i-1)\cdot r)} a_i p_{B_x(i \cdot r)} \ldots, p_{B_x((n-1)\cdot r)} a_n p_y).$$

Observe that for any $k \in B_x((n-1) \cdot r)$,

$$d(k, y) \geq d(x, y) - d(x, k) \geq d(x, y) - (n-1) \cdot r > n \cdot r - (n-1) \cdot r = r,$$

and so

$$p_{B_x((n-1)\cdot r)} a_n p_y = 0 \quad \text{since} \quad a_n \in \mathbb{C}_u^r[X].$$

Thus,

$$p_x \phi(a_1, \ldots, a_n) p_y = 0$$

and since $x, y \in X$ were an arbitrary pair satisfying $d(x, y) > n \cdot r$, we have that $\phi(a_1, \ldots, a_n) \in \mathbb{C}_u^{n \cdot r}[X]$ as was to be shown. □

REMARK 4.3. By Lemma 4.2 and Theorem 3.8 we know that

$$H^n_c(A:\ell) \cong H^n_c(A,\mathscr{B}:\ell) \cong H^n_c(A,\mathscr{B}).$$

In Sinclair and Smith [**6**] Theorem 3.3.1 they show that $H^n_c(A,\mathscr{B}) \cong H^n_c(\mathscr{B})$. Hence, by Theorem 1.1 $H^n_c(A:\ell)=0$. Thus, we need only show that the homomorphism

$$H^n_c(A:\ell) \to H^n_c(A) \text{ induced by the inclusion } \mathscr{L}^n_c(A:\ell) \to \mathscr{L}^n_c(A)$$

is a surjection. By Lemma 3.5, averaging over the unitary group of $\ell^\infty(X)$, we know that for a cocycle $\phi \in \mathscr{L}^n_c(A)$, $(\phi - \partial K_n\phi) \in \mathscr{L}^n_c(A:\ell)$. Thus, to show that $H^n_c(A:\ell) \to H^n_c(A)$ is a surjection it suffices to show that $K_n\phi \in \mathscr{L}^{n-1}_c(A)$ so that $\partial K_n\phi$ is a coboundary in $\mathscr{L}^n_c(A)$, for then

$$H^n_c(A:\ell) \ni [\phi - \partial K_n\phi] = [\phi] \text{ in } H^n_c(A).$$

Furthermore, since $H^n_w(A) \to H^n_c(A)$ is a surjection by the hypothesis of Theorem 4.1, we may assume that $\phi \in \mathscr{L}^n_w(A)$.

Before we embark on the proof that $H^n_c(A)=0$ if the map $H^n_w(A) \to H^n_c(A)$ is a surjection for a general n, we show some properties of the map K_n arising from its construction and set some notation.

LEMMA 4.4. *K_n is the sum of $\sum_{k=1}^n 2^{k-1}$ terms (before applying the boundary operator), where the first term is J_1, the next terms are the n-alternating sum of the maps G_k, and the remaining terms for $n \geq 2$ are of the form*

$$G_{j_i}\partial \dots G_{j_1}\partial J_1 \quad \textit{or} \quad G_{j_i}\partial \dots G_{j_2}\partial G_{j_1} \quad \textit{for} \quad j_i > j_{i-1} > \dots > j_1. \tag{4.2}$$

PROOF. Since K_n is defined by $K_n = J_n$ where $J_{k+1} = J_k + (-1)^k(G_k - G_k\partial J_k)$ we will induct on k.

Let $D_k = (G_k - G_k\partial J_k)$, then

$$\begin{aligned} J_{k+1} &= J_k + (-1)^k D_k \\ &= J_{k-1} + (-1)^{k-1}D_{k-1} + (-1)^k D_k \\ &= J_1 + \sum_{j=1}^{k}(-1)^j D_j \\ &= J_1 + \sum_{i=1}^{k}(-1)^i G_i + \sum_{j=1}^{k}(-1)^{j+1}G_j\partial J_j \end{aligned}$$

Note that, since $j \leq k$ for all j in the last summation, by inductive hypothesis our terms are of the form of line (4.2).

Lastly, using the recursive definition of J_{k+1} and letting $|J_{k+1}|$ be the number of terms of J_{k+1},we have

$$|J_{k+1}| = |J_k| + |G_k| + |G_k\partial J_k| = 2\,|J_k| + 1 = 2\sum_{j=1}^{k} 2^{j-1} + 1 = \sum_{j=1}^{k+1} 2^{j-1}$$

as was to be shown. □

LEMMA 4.5. *Let $\phi \in \mathscr{L}^n(A)$ and let $(a_1,\dots,a_n)$, $a_i \in A$ be given. Then*

$$(G_{j_i}\partial \dots G_{j_1}\partial J_1\phi)(a_1,\dots,a_{n-1}) \quad \text{and} \quad (G_{j_i}\partial \dots G_{j_2}\partial G_{j_1}\phi)(a_1,\dots,a_{n-1})$$

are both finite sums of terms of the form

$$\int_{\mathcal{U}} \cdots \int_{\mathcal{U}} \prod_{k=1}^{N}(c_{1,k}v_{1,k})\phi\Big(\prod_{k=1}^{N} c_{2,k}v_{2,k},\dots,\prod_{k=1}^{N} c_{n,k}v_{n,k}\Big)\prod_{k=1}^{N} c_{n+1,k}v_{n+1,k}\; d\mu(u_{j_i})\dots d\mu(u_{j_1})$$

where each $c_{\ell,k}$ is fixed as one of the a_j's or 1, and $v_{\ell,k} \in \mathcal{U}$ the unitary group of $\ell^\infty(X)$. Additionally $N < \infty$.

PROOF. Consider

$$(G_{j_i}\partial \dots G_{j_1}\partial J_1\phi)(a_1,\dots,a_{n-1}).$$

Observe that, after applying G_{j_i}, in the l'th coordinate we will have: a_l, $a_l u_{j_i}^*$, or u_{j_i}. Note that we may write this coordinate as $c_l v_l$ where c_l is fixed as 1 or a_l and $v_l = 1$, u_{j_i}, or $u_{j_i}^*$. Also note that $v_l \in \mathcal{U}$. Thus we may write,

$$(G_{j_i}\partial \dots G_{j_1}\partial J_1\phi)(a_1,\dots,a_{n-1})$$

$$= \int_{\mathcal{U}} (\partial G_{j_{i-1}} \dots \partial G_{j_1}\partial J_1\phi)(c_1v_1,\dots,c_nv_n)\; d\mu(u_{j_i}). \tag{4.3}$$

Next, since our averaging operator is finitely additive and the boundary operator introduces a finite number of terms, we may 'bring in' the averaging operator to each term. Additionally, since the boundary operator just moves one of the arguments to the coordinate to the left, in front of, or behind the map, we may write (after reindexing) a typical term obtained from applying the boundary map in line (4.3) as

$$\int_{\mathcal{U}} c_0v_0(G_{j_{i-1}}\partial \dots G_{j_1}\partial J_1\phi)(c_1v_1c_2v_2,\dots,c_{2n}v_{2n}c_{2n+1}v_{2n+1})c_{2n+2}v_{2n+2}\; d\mu(u_{j_i})$$

where $c_k \in \{1, a_1,\dots,a_{n-1}\}$ and $v_k \in \mathcal{U}$ (note that the c_k's will be fixed differently for each term). Applying this process again it is not hard to see that after applying $G_{j_i}\partial G_{i-1}\partial$ we will have a finite sum of terms of the form

$$\int_{\mathcal{U}}\int_{\mathcal{U}} \prod_{k=1}^{4}(c_{0,k}v_{0,k})(G_{j_{i-2}}\dots\partial J_1\phi)\Big(\prod_{k=1}^{4} c_{1,k}v_{1,k},\dots,\prod_{k=1}^{4} c_{n,k}v_{n,k}\Big)\prod_{k=1}^{4} c_{n+1,k}v_{n+1,k}\; d\mu(u_{j_i})d\mu(u_{j_{i-1}}).$$

Note that the application of the J_1 map does not change our technique and eventually this process must end. Thus, the conclusion holds and we are done. □

DEFINITION 4.6. For each term obtained in the previous lemma the set of $\{c_{l,k}\}$ is fixed for that term. We shall call this a *partial coordinate fixing* of ϕ.

LEMMA 4.7. *Let $(a_1,\dots,a_n)$, $a_i \in A$ and $\phi \in \mathscr{L}^n_w(A)$ be given. Consider*

$$y_0\phi(y_1,\dots,y_n)y_{n+1}, \quad y_i = \prod_{j=1}^{N_i} c_j f_{(i,j)}, \quad \text{where } N_i < \infty \tag{4.4}$$

where $f_{(i,j)}$ is any element in $(\ell^\infty(X))_1$, and each $c_j = a_k$ or 1 is fixed. Then for all $\epsilon > 0$ there exists an $r > 0$ (depending on the partial coordinate fixing of ϕ) such that $y_0\phi(y_1,\dots,y_n)y_{n+1}$ can be ϵ-r-approximated.

PROOF. Let $p_x \in \mathscr{B}(\ell^2(X))$ be the rank one projection onto the span of the Dirac mass at x. For any element f in the unit ball of $\ell^\infty(X)$, we may write f as a strongly (and so weakly) convergent sum

$$f = \sum_{x \in X} f(x)p_x. \tag{4.5}$$

Then, for an arbitrary i, j where $1 \leq i \leq N$ and $0 \leq j \leq n+1$ and $f_{(\ell,k)} \in (\ell^\infty(X))_1$ fixed whenever $\ell, k \neq i, j$, we have that

$$\sum_{x_j \in X} \lambda_{x_j}^{(j)} y_0 \phi \left(y_1, \ldots, \left(\prod_{k=1}^{j-1} c_k f_{(i,k)} \right) c_j p_{x_j} \left(\prod_{k=j+1}^{N_i} c_k f_{(i,k)} \right), \ldots, y_n \right) y_{n+1}$$

weakly converges to

$$y_0 \phi(y_1, \ldots, \prod_{k=1}^{N_i} c_k f_{(i,k)}, \ldots, y_n) y_{n+1}$$

Moreover, (4.4) is bounded above by $\|\phi\| \prod_{k=1}^n \|a_k\|$ for all $f_{(\ell,k)} \in (\ell^\infty(X))_1$. Hence, since the weak and ultraweak topologies coincide on norm bounded sets and $\phi \in \mathscr{L}_w^n(A)$, we have that, for each partial coordinate fixing of ϕ,

$$\prod_{k=1}^{N_0} (c_{0,k} p_{x_{(0,k)}}) \phi \Big(\prod_{k=1}^{N_1} c_{1,k} p_{x_{(1,k)}}, \ldots, \prod_{k=1}^{N_n} c_{n,k} p_{x_{(n,k)}} \Big) \prod_{k=1}^{N_{n+1}} c_{n+1,k} p_{x_{(n+1,k)}}$$

is separately symmetrically summable. Thus, by Corollary 2.9, for all $\epsilon > 0$ there exists an $r > 0$ (depending on the partial coordinate fixing of ϕ) such that $y_0 \phi(y_1, \ldots, y_n) y_{n+1}$ can be ϵ-r-approximated. □

LEMMA 4.8. *$K_n \phi \in \mathscr{L}_c^{n-1}(A)$ whenever $\phi \in \mathscr{L}_w^n(A)$, where K_n is constructed by averaging over the unitary group of $\ell^\infty(X)$.*

PROOF. Let $\epsilon > 0$ be given.

By Lemmas 2.15 and 4.7, $K_n \phi(a_1, \ldots, a_{n-1})$ is the finite sum of finite sums of terms of the form

$$\int_{\mathcal{U}_{j_i}} \cdots \int_{\mathcal{U}_{j_1}} y_0 \phi(y_1, \ldots, y_n) y_{n+1} d\mu(u_{j_1}) \ldots d\mu(u_{j_i})$$

where each term is a different partial coordinate fixing of ϕ (in the sense of Definition 4.6). Using Lemma 4.7 we may write each of these terms as

$$= \int_{\mathcal{U}_{j_i}} \cdots \int_{\mathcal{U}_{j_1}} a(\overline{u}) + b(\overline{u}) d\mu(u_{j_1}) \ldots d\mu(u_{j_i}) \tag{4.6}$$

where each $a(\overline{u}) \in \mathbb{C}_u^r[X]$ and $\|b(\overline{u})\| < \epsilon/M$ for a given $M > 0$ and all $\overline{u} \in \mathcal{U}_{j_i} \times \cdots \times \mathcal{U}_{j_1}$. Thus, taking M and R sufficiently large, since $K_n \phi(a_1, \ldots, a_{n-1})$ is the finite sum of terms as in line (4.6), $K_n \phi(a_1, \ldots, a_{n-1})$ is ϵ-R-approximated. Since ϵ was arbitrary, we are done. □

PROOF OF THEOREM 4.1. By Remark 4.3, to show that $H_c^n(C_u^*(X)) = 0$ it suffices to show that $K_n \phi \in \mathscr{L}_c^{n-1}(C_u^*(X))$ whenever $\phi \in \mathscr{L}_w^n(C_u^*(X))$, which we have done in the previous lemma. □

5. Ultraweak-Weak* Continuous Cohomology

In this section we discuss methods for relating norm continuous and ultraweak-weak* continuous cohomologies which will allow us to obtain the following result.

THEOREM 5.1. *If $H^n_c(C^*_u(X)) = 0$ for all $n \in \mathbb{N}$ then $H^n_w(C^*_u(X)) = 0$ for all $n \in \mathbb{N}$.*

5.1. A Bridge Between Ultraweak-Weak* Continuous and Norm Continuous Cohomology. In this subsection we discuss a method to extend separately continuous multilinear maps to separately ultraweakly continuous multilinear maps. Many of the proofs can be found in Sinclair and Smith [**6**].

LEMMA 5.2 ([**6**] Lemma 3.3.3). *Let $\mathcal{A}$ be a C^*-algebra acting nondegenerately on a Hilbert space $\mathcal{H}$ and let $\mathcal{V}$ be the dual of a Banach space $\mathcal{V}_*$. If ϕ is a bounded n-linear map from $\mathcal{A}^n$ to $\mathcal{V}$ that is separately ultraweak-weak* continuous, then ϕ extends uniquely without change in norm to a bounded n-linear map $\overline{\phi}$ from $(\overline{\mathcal{A}})^n$ to $\mathcal{V}$ that is seperately ultraweak-weak* continuous.* □

The following lemma can be found in Blackadar [**1**] III.5.2.11 or Takesaki [**7**] III.2.4, III.2.14.

LEMMA 5.3. *Let $\mathcal{A}$ be a C^*-algebra acting on a Hilbert space $\mathcal{H}$ with weak closure $\overline{\mathcal{A}}$. If π is the universal representation of $\mathcal{A}$, then there is a projection p in the center of the weak closure $\overline{\pi(\mathcal{A})}$ of $\pi(\mathcal{A})$ and a $*$-isomorphism*

$$\theta : p\overline{\pi(\mathcal{A})} \to \overline{\mathcal{A}} \qquad \text{such that}$$

$$\theta(p\pi(a)) = a \qquad \text{and} \qquad \theta(px) = \pi^{-1}(x) \text{ for all } a \in \mathcal{A} \text{ and } x \in \pi(\mathcal{A}). \tag{5.1}$$

Moreover, θ is a homeomorphism from $p\overline{\pi(\mathcal{A})}$ onto $\overline{\mathcal{A}}$ if both have their ultraweak topologies, since $*$-isomorphisms between von Neumann algebras are ultraweak homeomorphisms. □

LEMMA 5.4 ([**6**] Lemma 3.3.4). *Let $\mathcal{A}$ be a C^*-algebra acting on a Hilbert space $\mathcal{H}$ with weak closure $\overline{\mathcal{A}}$. Let π be the universal representation of $\mathcal{A}$, and let p, θ be as in Lemma 5.3. Additionally, let $\mathcal{V}$ be a dual normal $\overline{\mathcal{A}}$-module. Then $\mathcal{V}$ may be regarded as a dual normal $\overline{\pi(\mathcal{A})}$-module via*

$$x \cdot v = \theta(px)v \qquad \text{and} \qquad v \cdot x = v\theta(px) \tag{5.2}$$

and there are continuous linear maps

$$T_n : \mathscr{L}^n_c(\mathcal{A}, \mathcal{V}) \to \mathscr{L}^n_w(\overline{\pi(\mathcal{A})}, \mathcal{V}),$$
$$S_n : \mathscr{L}^n_w(\overline{\pi(\mathcal{A})}, \mathcal{V}) \to \mathscr{L}^n_w(\overline{\mathcal{A}}, \mathcal{V})$$
$$W_n : \mathscr{L}^n_w(\overline{\pi(\mathcal{A})}, \mathcal{V}) \to \mathscr{L}^n_c(\mathcal{A}, \mathcal{V})$$

such that:

(1) *$\partial T_n = T_{n+1}\partial$, $\partial W_n = W_{n+1}\partial$, and $\partial S_n = S_{n+1}\partial$,*
(2) *$\|T_n\|, \|S_n\|, \|W_n\| \leq 1$,*
(3) *if $\mathcal{B}$ is a C^*-subalgebra of $\mathcal{A}$, T_n maps $\mathcal{B}$-multimodular maps to $\overline{\pi(\mathcal{B})}$-multimodular maps, and S_n and W_n map $\overline{\pi(\mathcal{B})}$-multimodular maps to $\mathcal{B}$-multimodular maps,*
(4) *S_nT_n is a projection from $\mathscr{L}^n_c(\mathcal{A}, \mathcal{V})$ onto $\mathscr{L}^n_w(\mathcal{A}, \mathcal{V})$,*

(5) *if $\mathcal{C}$ is the C^*-subalgebra of $\overline{\pi(\mathcal{A})}$ generated by 1 and p, and if*

$$\psi \in \mathscr{L}_w^n(\overline{\pi(\mathcal{A})}, \mathcal{V} : \mathcal{C}),$$

then $W_n\psi = S_n\psi \in \mathscr{L}_w^n(\overline{\mathcal{A}}, \mathcal{V})$,

(6) *W_nT_n is the identity map on $\mathscr{L}_c^n(\mathcal{A}, \mathcal{V})$.* □

Note that for $\psi \in \mathscr{L}_w^n(\overline{\pi(\mathcal{A})}, \mathcal{V} : \mathcal{C})$ being $\mathcal{C}$-multimodular is equivalent to having the property that $\psi(a_1, \dots, a_n) = 0$ if any of the arguments $a_j \in (1-p)\overline{\pi(\mathcal{A})}$. As we shall need it later, we review the construction of the maps T_n, S_n, and W_n.

REMARK 5.5 (The map T_n). The equation

$$\phi_1(x_1, \dots, x_n) = \phi(\theta(px_1), \dots, \theta(px_n)) \qquad \text{for all} \qquad x_1, \dots, x_n \in \pi(\mathcal{A}) \tag{5.3}$$

defines $\phi_1 \in \mathscr{L}_c^n(\pi(\mathcal{A}), \mathcal{V})$. This map is separately ultraweakly-weak* continuous in each of its arguments, because π is the universal representation of $\mathcal{A}$, so by [**7**] III.2.4 each continuous linear functional on $\pi(\mathcal{A})$ is ultraweakly continuous; that is $\phi_1 \in \mathscr{L}_w^n(\pi(\mathcal{A}), \mathcal{V})$. Moreover, by [**6**] Lemma 3.3.3, we may extend ϕ_1 to $\overline{\phi_1} \in \mathscr{L}_w^n(\overline{\pi(\mathcal{A})}, \mathcal{V})$ without change of norm. The map T_n is then defined by $T_n\phi = \overline{\phi_1}$.

REMARK 5.6. The map $S_n : \mathscr{L}_w^n(\overline{\pi(\mathcal{A})}, \mathcal{V}) \to \mathscr{L}_w^n(\overline{\mathcal{A}}, \mathcal{V})$ is defined by

$$(S_n\psi)(a_1, \dots, a_n) = \psi(\theta^{-1}(a_1), \dots, \theta^{-1}(a_n)).$$

REMARK 5.7. The map $W_n : \mathscr{L}_w^n(\overline{\pi(\mathcal{A})}, \mathcal{V}) \to \mathscr{L}_c^n(\mathcal{A}, \mathcal{V})$ is defined by

$$W_n\psi(a_1, \dots, a_n) = \psi(\pi(a_1), \dots, \pi(a_n)).$$

Note that, if $\mathcal{A} \subseteq \mathcal{V} = \mathscr{B}(\mathcal{H})$ then W_n maps $\mathscr{L}_w^n(\pi(\mathcal{A}), \mathcal{A})$ to $\mathscr{L}_c^n(\mathcal{A})$.

LEMMA 5.8 ([**6**] Lemma 3.3.5). *Let $\mathcal{A}$ be a C^*-algebra and let $\mathcal{V}$ be a dual normal $\mathcal{A}$-bimodule. Then the homomorphism*

$$H_w^n(\mathcal{A}, \mathcal{V}) \to H_c^n(\mathcal{A}, \mathcal{V}) \qquad \textit{induced by} \qquad \mathscr{L}_w^n(\mathcal{A}, \mathcal{V}) \to \mathscr{L}_c^n(\mathcal{A}, \mathcal{V})$$

is surjective. □

THEOREM 5.9 ([**6**] Theorem 3.3.1). *Let $\mathcal{A}$ be a C^*-algebra acting on a Hilbert space $\mathcal{H}$ with weak closure $\overline{\mathcal{A}}$. Additionally, let $\mathcal{V}$ be a dual normal $\overline{\mathcal{A}}$-module. Then,*

$$H_c^n(\mathcal{A}, \mathcal{V}) \cong H_w^n(\mathcal{A}, \mathcal{V}) \cong H_w^n(\overline{\mathcal{A}}, \mathcal{V})$$

PROOF. By the previous lemma we have that the map $H_w^n(\mathcal{A}, \mathcal{V}) \to H_c^n(\mathcal{A}, \mathcal{V})$ is a surjection. To see that this map is injective first note that for $\psi \in \mathscr{L}_c^{n-1}(\mathcal{A}, \mathcal{V})$ we have that $S_{n-1}T_{n-1}\psi \in \mathscr{L}_w^{n-1}(\mathcal{A}, \mathcal{V})$ by Lemma 5.4 (iv). Next, if $\phi \in \mathscr{L}_w^n(\mathcal{A}, \mathcal{V})$ with $\phi = \partial\psi$ where $\psi \in \mathscr{L}_c^{n-1}(\mathcal{A}, \mathcal{V})$, then

$$\phi = \partial\psi = S_nT_n\partial\psi = \partial S_{n-1}T_{n-1}\psi.$$

Thus, the map $H_w^n(\mathcal{A}, \mathcal{V}) \to H_c^n(\mathcal{A}, \mathcal{V})$ induced by the inclusion is also injective and so an isomorphism.

Lastly, by Lemma 5.2, the restriction map $\mathscr{L}_w^n(\overline{\mathcal{A}}, \mathcal{V}) \to \mathscr{L}_w^n(\mathcal{A}, \mathcal{V})$ is an isomorphism and so we are done. □

REMARK 5.10. Note that the ultraweak closure of $C_u^*(X)$ is $\mathscr{B}(\ell^2(X))$. Hence, by the previous theorem we have that

$$H_c^n(C_u^*(X), \mathscr{B}(\ell^2(X))) \cong H_c^n(\mathscr{B}(\ell^2(X))).$$

5.2. On the Vanishing of the Ultraweak-Weak* Continuous Cohomology of Uniform Roe Algebras. Before we prove Theorem 5.1 we will need a few lemmas. Once more for notational convenience throughout we let: $A = C_u^*(X)$, $\mathscr{B} = \mathscr{B}(\ell^2(X))$, and $\ell = \ell^\infty(X)$.

LEMMA 5.11. *Let π be the universal representation of A, and let p be the projection from Lemma 5.3. If $\{q_\alpha\}$ is the net of finite rank projections in ℓ with its usual ordering then*

$$\pi(q_\alpha) \xrightarrow{\text{ultraweakly}} p \qquad \text{in} \qquad \overline{\pi(\ell)}.$$

PROOF. Recall that the double dual of the compact operators $\mathscr{K}(\mathcal{H})^{**}$ is naturally identified with $\mathscr{B}$ (cf. [7] II.1.8). Moreover, since $\mathscr{K}(\mathcal{H})$ is an ideal in $C_u^*(X)$ and $\{q_\alpha\}$ is an approximate unit for $\mathscr{K}(\mathcal{H})$, by Blackadar [1] III.5.2.11, there exists a central projection $q \in A^{**}$ such that

$$\hat{q}_\alpha \to q \quad \text{in the} \quad \sigma(A^{**}, A^*) \quad \text{topology and}$$

$$qA^{**} = \mathscr{K}(\mathcal{H})^{**} \cong p\overline{\pi(A)}$$

Thus, if $\tilde{\pi}$ is the unique linear map from the double dual $\mathcal{A}^{**}$ onto $\overline{\pi(\mathcal{A})}$, the weak closure of $\pi(A)$, we have that $\tilde{\pi}(q) = p$. Moreover, since $\{q_\alpha\} \subseteq \ell$ and $\tilde{\pi}$ is a $\sigma(A^{**}, A^*)$-ultraweak homeomorphism, we have that

$$\pi(q_\alpha) \xrightarrow{\text{ultraweakly}} p \qquad \text{and} \qquad p \in \overline{\pi(\ell)}.$$

□

LEMMA 5.12. *If $\phi \in \mathscr{L}_c^n(A:\ell)$ then $\phi \in \mathscr{L}_w^n(A)$. That is, $\mathscr{L}_c^n(A:\ell) = \mathscr{L}_w^n(A:\ell) \subseteq \mathscr{L}_w^n(A)$.*

PROOF. Since T_n takes ℓ-multimodular maps to $\overline{\pi(\ell)}$-multimodular maps we have

$$\begin{aligned}
S_nT_n\phi(a_1,\dots,a_n) &= T_n\phi(\theta^{-1}(a_1),\dots,\theta^{-1}(a_n)) && \text{by the definition of } S_n \\
&= T_n\phi(p\pi(a_1),\dots,p\pi(a_n)) && \text{by the properties of } \theta \\
&= p\cdot T_n\phi(\pi(a_1),\dots,\pi(a_n)) && \text{since } p \text{ is central and} \\
& && T_n\phi \text{ is } \overline{\pi(\ell)}\text{-multimodular.} \\
&= p\cdot\phi(a_1,\dots,a_n) && \text{by the definition of } T_n. \\
&= \phi(a_1,\dots,a_n) && \text{by line (5.2).}
\end{aligned}$$

Then, since S_nT_n is a projection from $\mathscr{L}_c^n(A)$ onto $\mathscr{L}_w^n(A,\mathscr{B})$, we are done. □

LEMMA 5.13. *If $H_c^{n-1}(A) = 0$ then the map $H_w^n(A) \to H_c^n(A)$ is an injection.*

PROOF. Suppose that $\phi \in \mathscr{L}_w^n(A)$ with $\phi = \partial\psi$ for some $\psi \in \mathscr{L}_c^{n-1}(\mathcal{A})$. So if $H_c^{n-1}(A) = 0$, we have $H_c^{n-1}(A:\ell) \cong H_c^{n-1}(A)$, since by Remark 4.3 $H_c^{n-1}(A:\ell) = 0$. Hence, without loss of generality, we may assume $\psi \in \mathscr{L}_c^{n-1}(A:\ell) \subseteq \mathscr{L}_w^{n-1}(\mathcal{A})$. Thus, $[\phi] = 0$ in $H_w^n(A)$ and we are done. □

We are now ready to prove Theorem 5.1.

PROOF OF THEOREM 5.1. Since derivations are automatically weakly continuous ([3] Lemma 3), $H_w^1(A) = H_c^1(A) = 0$. Next, given any $n > 1$ we have that $H_c^{n-1}(A) = 0$, so by Lemma 5.13, $H_w^n(A) \to H_c^n(A)$ is an injection. Moreover, $H_c^n(A) = 0$, so we must have that $H_w^n(A) = 0$. Since n was arbitrary we are done. □

References

[1] B. Blackadar, *Operator algebras*, Encyclopaedia of Mathematical Sciences, vol. 122, Springer-Verlag, Berlin, 2006. Theory of C^*-algebras and von Neumann algebras; Operator Algebras and Non-commutative Geometry, III, DOI 10.1007/3-540-28517-2. MR2188261

[2] B. M. Braga and I. Farah, *On the rigidity of uniform Roe algebras over uniformly locally finite coarse spaces*, Trans. Amer. Math. Soc. **374** (2021), no. 2, 1007–1040, DOI 10.1090/tran/8180. MR4196385

[3] Richard V. Kadison, *Derivations of operator algebras*, Ann. of Math. (2) **83** (1966), 280–293, DOI 10.2307/1970433. MR193527

[4] Richard V. Kadison and John R. Ringrose, *Cohomology of operator algebras. II. Extended cobounding and the hyperfinite case*, Ark. Mat. **9** (1971), 55–63, DOI 10.1007/BF02383637. MR318907

[5] Matthew Lorentz and Rufus Willett, *Bounded derivations on uniform Roe algebras*, Rocky Mountain J. Math. **50** (2020), no. 5, 1747–1758, DOI 10.1216/rmj.2020.50.1747. MR4170683

[6] Allan M. Sinclair and Roger R. Smith, *Hochschild cohomology of von Neumann algebras*, London Mathematical Society Lecture Note Series, vol. 203, Cambridge University Press, Cambridge, 1995, DOI 10.1017/CBO9780511526190. MR1336825

[7] M. Takesaki, *Theory of operator algebras. I*, Encyclopaedia of Mathematical Sciences, vol. 124, Springer-Verlag, Berlin, 2002. Reprint of the first (1979) edition; Operator Algebras and Non-commutative Geometry, 5. MR1873025

619 Red Cedar Road C318 Wells Hall East Lansing, Michigan 48824
Email address: lorentzm@msu.edu

Proceedings of Symposia in Pure Mathematics
Volume **105**, 2023
https://doi.org/10.1090/pspum/105/01905

Quantum differentiability – the analytical perspective

E. McDonald, F. Sukochev, and X. Xiong

Abstract. The study of quantum differentiability (in the sense of Connes) concerns the characterisation of the singular value sequences of commutators of pointwise multipliers with signs of Dirac operators. This subject ties together several themes in operator theory and harmonic analysis, and is inspired by analytic issues arising from cyclic cohomology and noncommutative differential geometry. We give a unified treatment of some recent results in this area, and a proof of the novel result that the necessary and sufficient conditions for quantum differentiability in noncommutative Euclidean spaces are formally identical to those of commutative Euclidean spaces.

1. Introduction

The notion of quantised calculus originates with Connes' seminal work on noncommutative differential geometry [**3**]. The central feature of quantised calculus is the replacement of the differential df of a smooth function f on a closed Riemannian spin n-manifold X with the operator $đf$, defined as the commutator $\mathrm{i}[\mathrm{sgn}(D), M_f]$ on the Hilbert space of square integrable sections of the spinor bundle of X, where D is a Dirac operator on X. Here, M_f denotes the operator of pointwise multiplication by f on sections of the spinor bundle of X.

Inspired by applications to cyclic cohomology, the central analytic problem in this theory is to relate the spectral properties of $đf$ with the nature of the function f. Of particular interest is the membership of $đf$ in some ideal of compact operators.

Already it was pointed out by Connes in [**4**, Theorem 3] that if $f \in C^\infty(X)$, then $|đf|^n$ belongs to the Macaev-Dixmier ideal $\mathcal{M}_{1,\infty}$. That is, the singular number sequence $\mu(đf) = \{\mu(k, đf)\}_{k=0}^\infty$ obeys

$$\sup_{N\geq 0} \frac{1}{\log(N+2)} \sum_{k=0}^N \mu(k, đf)^n < \infty.$$

For particular manifolds, far more detailed information is available. The best developed case is where $n = 1$ and $X = \mathbb{T}$ is the unit circle. In this case the precise relationship between the singular numbers of $đf = \mathrm{i}[\mathrm{sgn}(D), M_f] \in \mathcal{B}(L_2(\mathbb{T}))$ was

2020 *Mathematics Subject Classification.* Primary 58B34; Secondary 46L87, 47B10.
Key words and phrases. Quantised derivative, quantum tori, quantum Euclidean spaces, trace formula, Sobolev space.
The third author was supported by the National Natural Science Foundation of China (Grant No. 12001136).
The second author is the corresponding author.

determined by Peller. A combination of the results in Chapter 6 of [**26**] asserts that for all $0 < p < \infty$ we have

$$đf \in \mathcal{L}_p \Longleftrightarrow f \in B^{\frac{1}{p}}_{p,p}(\mathbb{T}).$$

The original proof of this equivalence goes back to Peller's famous work [**24**] for $p \geq 1$, and to Peller [**25**] and Semmes [**30**] independently for $0 < p < 1$. Here, $\mathcal{L}_p$ is the Schatten ideal of compact operator T such that $\sum_{k=0}^{\infty} \mu(k,T)^p < \infty$, and $B^{\frac{1}{p}}_{p,p}(\mathbb{T})$ is a Besov space on the unit circle $\mathbb{T}$. It is further possible to characterise those f such that $đf$ belongs to an arbitrary ideal of compact operators having the monotone Riesz-Fischer property [**10**].

Higher dimensional cases have been studied in the harmonic analysis literature. A theorem of Janson and Wolff [**13**] implies that if f is a function on the d-torus $\mathbb{T}^d$ where $d \geq 2$, then the quantised differential $đf$ belongs to $\mathcal{L}_p$ if and only if f is constant for $p \leq d$ and $f \in B^{\frac{d}{p}}_{p,p}(\mathbb{T}^d)$ for $p > d$. When $p = d$, based on the result of Rochberg and Semmes [**29**] describing the weak Schatten $\mathcal{L}_{d,\infty}$ property of $đf$ in terms of the oscillation of f, Connes, Sullivan and Teleman [**6**] sketched an argument which can be adapted to show that $đf \in \mathcal{L}_{d,\infty} \Longleftrightarrow f \in \dot{W}^1_d(\mathbb{T})$, where $\dot{W}^1_d(\mathbb{T})$ is a homogeneous Sobolev space on $\mathbb{T}^d$.

Peller [**25**] and Semmes [**30**] illustrate that, in the one dimensional case, $đf$ is an infinitesimal of infinite order for smooth f. But when $d > 2$, Rochberg-Semmes [**29**] show that $đf$ is an infinitesimal exactly of order $\frac{1}{d}$ in the sense that if $đf$ has order greater than $\frac{1}{d}$ then f is constant.

A general setting for quantised calculus is a spectral triple. A spectral triple (A, H, D) consists of a Hilbert space H, a pre-C^*-algebra A, represented faithfully on H and a self-adjoint operator D acting on H such that every $a \in A$ maps the domain of D into itself and the commutator $[D, a] = Da - aD$ extends from the domain of D to a bounded linear endomorphism of H. Here, the quantised differential $đa$ of $a \in A$ is defined to be the bounded operator $\mathrm{i}[\mathrm{sgn}(D), a]$. This is related to the construction of a Fredholm module from a spectral triple, as in e.g. [**5**].

An intrinsic feature of the quantised calculus is that its definition is indifferent to whether the algebra A is commutative. Over the past few decades since its invention, the quantised calculi for several strictly noncommutative examples have been investigated. In this survey we focus mainly on some of the oldest and best-developed noncommutative spaces: the *noncommutative tori* and *noncommutative Euclidean spaces*. We review below the well-established constructions of spectral triples for these examples, which serve as noncommutative deformations of the standard flat geometry of tori and Euclidean spaces respectively. The analytic background for quantised calculus in these examples was developed by the authors in [**19, 20**] based on the commutative arguments in [**17**].

Let us describe the main results and some crucial methods used in this paper. The first part of our results concerns the estimate $đa \in \mathcal{L}_{p,\infty}$ for a in p dimensional spectral triple (A, H, D). This is done by writing

$$đa = \mathrm{i}[\frac{D}{(1+D^2)^{\frac{1}{2}}}, a] + \mathrm{i}[\mathrm{sgn}(D) - \frac{D}{(1+D^2)^{\frac{1}{2}}}, a].$$

The first term on the right hand side is treated by the theory of double operator integrals (briefly summarised in Section 2), which allows us to write $[\frac{D}{(1+D^2)^{\frac{1}{2}}}, a]$ as the double operator integrals of $[D, a] = -\mathrm{i}\partial a$. The theory of double operator integrals goes back to Birman and Solomyak [**1**]; see [**23, 28**] for important progress. The second term on the right hand side is treated by the Cwikel type estimates, which are described in Section 4 for concrete examples. Cwikel type estimates were first obtained in [**7**], see also [**31**]; recent progress on some noncommutative examples has been made in [**9, 15, 19**]. The novelty of our approach here is that we use a technical instrument derived from [**28**], stated as Proposition 3.6 below. This technique gives a simple unified presentation to all of the examples we consider.

The ingredient in proving necessary conditions for $đa \in \mathcal{L}_{p,\infty}$ is an adaptation of the Connes' trace formula and the use of C^*-algebraic principal symbol calculus in noncommutative geometry. [**21**, Theorem 6.15] states that for every continuous normalised trace φ on $\mathcal{L}_{1,\infty}$, smooth enough element x in $\mathbb{R}^d_\theta$ or $\mathbb{T}^d_\theta$, and every order 0 pseudo differential operator T, we have

$$\varphi(Tx(1+D^2)^{-d/2}) = C_d\Big(\tau_\theta \otimes \int_{\mathbb{S}^{d-1}}\Big)(\mathrm{sym}(T)(x \otimes 1))$$

where C_d is an absolute constant and $\mathrm{sym}(T)$ is the principal symbol of T. With this formula, we are reduced to finding operators T and x such that $\varphi(Tx(1+D^2)^{-d/2}) = \varphi(|đa|^d)$, which is then achieved by principal symbol calculus in [**21**] . In the same spirit, we give a slight strengthening to the computation of $\varphi(|đa|^d)$ by "localising" the trace formulas (see Theorem 5.1 below):

$$\varphi(y|đx|^d) = C_d \int_{\mathbb{S}^{d-1}} \tau_\theta\left(y\Big(\sum_{j=1}^d |\partial_j x - s_j \sum_{k=1}^d s_k \partial_k x|^2\Big)^{\frac{d}{2}}\right) ds,$$

where $x \in \dot{W}^1_d(\mathbb{R}^d_\theta)$ and $y \in C_0(\mathbb{R}^d_\theta)+\mathbb{C}$. It is worth emphasizing that the conceptual advantage of this abstract calculus (together with that in [**20**, Section 5]) is that it can be carried out in a continuous rather than a smooth setting, as well as in the setting where pseudo differential operator theory is not well developed.

The most substantial novelty of this paper is a strengthening of the main result of [**20**]. In that paper we were unable to prove a necessary and sufficient condition for quantum differentiability in the setting of noncommutative Euclidean spaces. The "sufficient" condition in [**20**] made an unnecessary integrability assumption which we were not able to remove at the time. Now, on the basis of new developments in the theory, we can present necessary and sufficient conditions on $x \in L_\infty(\mathbb{R}^d_\theta)$ such that $đx \in \mathcal{L}_{d,\infty}$. The missing ingredient of [**20**] was an assertion of the density of the noncommutative Schwartz class in noncommutative homogeneous Sobolev spaces, which we are now able to prove (see Proposition 4.9 below). With this addition, the necessary and sufficient conditions for quantum differentiability in the noncommutative Euclidean case are formally identical to the commutative Euclidean case. The proof is interesting in its own right, using a noncommutative Mikhlin multiplier theorem and the noncommutative Khintchine inequalities of Lust-Piquard and Pisier [**18**]. A particular result which is of independent interest is the noncommutative Littlewood-Paley inequalities proof proved as Theorem 4.15 below. For additional developments in the theory of Fourier multipliers in noncommutative Euclidean space, see [**11, 14**].

2. Notation and preliminaries

Throughout H is a complex, separable and infinite dimensional Hilbert space. We denote the space of all bounded linear endomorphisms of H by $\mathcal{B}(H)$, and the ideal of compact operators on H is denoted $\mathcal{K}(H)$. The operator norm is denoted $\|\cdot\|_\infty$. For a compact operator T, the singular value function $t \mapsto \mu(t,T)$ is defined by

$$\mu(t,T) = \inf\{\|T-R\|_\infty \ : \ \mathrm{rank}(R) \leq t\}.$$

The sequence $\mu(T) = \{\mu(n,T)\}_{n=0}^\infty$ is called the singular value sequence of T.

Given two compact operators $T, S \in \mathcal{K}(H)$, we say that T *submajorises* S (written $S \prec\prec T$) if for all $t > 0$ we have

$$\int_0^t \mu(s,S)\,ds \leq \int_0^t \mu(s,T)\,ds.$$

An ideal $\mathcal{E} \subset \mathcal{K}(H)$ of compact operators equipped with a Banach norm $\|\cdot\|_{\mathcal{E}}$ is said to be *fully symmetric* if for all $T \in \mathcal{E}$ and $S \in \mathcal{K}(H)$ such that $S \prec\prec T$ we have $S \in \mathcal{E}$ and $\|S\|_{\mathcal{E}} \leq \|T\|_{\mathcal{E}}$.

For $p, q > 0$, the Schatten-Lorentz ideal $\mathcal{L}_{p,q}(H)$ is defined as the space of compact linear operators T such that $\mu(T)$ belongs to the Lorentz sequence space $\ell_{p,q}$. That is, if and only if

$$\|T\|_{p,q} := \left(\sum_{n=0}^\infty (n+1)^{\frac{q}{p}-1}\mu(n,T)^q\right)^{\frac{1}{q}} < \infty.$$

We denote $\mathcal{L}_{p,\infty}(H)$ for the space of compact operators T such that $\|T\|_{p,\infty} := \sup_{n\geq 0}(n+1)^{\frac{1}{p}}\mu(n,T)$. For $p \geq 1$, $\mathcal{L}_p(H)$ is fully symmetric and for $p > 1$, $\mathcal{L}_{p,\infty}(H)$ is fully symmetric. The Schatten ideals $\mathcal{L}_p(H)$ are defined as $\mathcal{L}_{p,q}(H)$. We will often omit H and simply write $\mathcal{K}$, $\mathcal{L}_{p,q}$, etc. for $\mathcal{K}(H)$, $\mathcal{L}_{p,q}(H)$, etc. whenever the Hilbert space is clear from context.

A trace on an ideal $\mathcal{E}$ of $\mathcal{B}(H)$ is a linear functional φ on $\mathcal{E}$ which is unitarily invariant, i.e. $\varphi(UTU^*) = \varphi(T)$ for all $T \in \mathcal{E}$ and unitary operators U. The ideal $\mathcal{L}_{1,\infty}$ has non-vanishing traces, and they are all *singular*, meaning that they vanish on the subspace of finite rank operators. A trace φ on $\mathcal{L}_{1,\infty}$ is said to be continuous if $|\varphi(T)| \lesssim \|T\|_{1,\infty}$ for all $T \in \mathcal{L}_{1,\infty}$ and is said to be normalised if $\varphi(\mathrm{diag}\{\frac{1}{n+1}\}_{n=0}^\infty) = 1$. The most well-known continuous normalised traces on $\mathcal{L}_{1,\infty}$ are the Dixmier traces tr_ω. For further details, see [**16**].

We also briefly refer to the notion of double operator integral, essentially following the conventions of [**8**]. If (Ω, Σ, μ) is a probability space and α, β are bounded measurable functions on $\mathbb{R} \times \Omega$ such that

$$\int_\Omega \sup_{t\in\mathbb{R}} |\alpha(t,\omega)| \sup_{s\in\mathbb{R}} |\beta(s,\omega)|\, d\mu(\omega) < \infty$$

then the double operator integral with symbol

$$\psi(t,s) = \int_\Omega \alpha(t,\omega)\beta(s,\omega)\,d\mu(\omega), \quad t,s \in \mathbb{R}$$

is defined for a given pair of self-adjoint operators A and B as a linear map

$$T_\psi^{A,B} : \mathcal{B}(H) \to \mathcal{B}(H)$$

by the formula

$$T_\psi^{A,B}(X) := \int_\Omega \alpha(A,\omega)X\beta(B,\omega)\,d\mu(\omega), \quad X \in \mathcal{B}(H).$$

It is proved in [8] that $T_\psi^{A,B}$ is well-defined and there is a constant C_ψ such that

$$T_\psi^{A,B}(X) \prec\prec C_\psi X, \quad X \in \mathcal{B}(H).$$

Therefore, $T_\psi^{A,B}$ acts boundedly on any fully symmetric ideal.

We will frequently use the symbol $\lesssim$ to denote inequality with a constant. That is, $x \lesssim y$ means that there is a constant C such that $x \le Cy$. Subscripts will denote parameters, such as $x \lesssim_{p,q,r} y$ means that $x \le C_{p,q,r}y$ where $C_{p,q,r}$ depends on the parameters p, q and r.

3. Quantum differentiability in general

In the paradigm of noncommutative geometry (in the sense of Connes), the basic notion of a geometric space is taken by a spectral triple (A, H, D). Roughly speaking, a spectral triple is an algebraic substitute for the notion of a Riemannian spin manifold in ordinary geometry, however the idea of a spectral triple is slightly different. Indeed, there are many examples of commutative spectral triples that are totally unrelated to geometry in the traditional sense.

A spectral triple consists of a Hilbert space H, a potentially unbounded self-adjoint operator $D : \mathrm{dom}(D) \subset H \to H$, and a $*$-algebra A admitting a representation of bounded operators on H. For notational simplicity, we assume that $A \subset \mathcal{B}(H)$. One says that (A, H, D) is a spectral triple if

(1) every $a \in A$ maps $\mathrm{dom}(D)$ into itself and for every $a \in A$, the commutator $[D, a] : \mathrm{dom}(D) \to H$ admits a bounded extension.
(2) For every $a \in A$, the operators $a(\mathrm{i}+D)^{-1}$ and $[D, a](\mathrm{i}+D)^{-1}$ are compact.

We abbreviate $\partial(a)$ for the bounded extension of $\mathrm{i}[D, a]$

The latter condition could be equivalently be the assertion that $a(1 + D^2)^{-\frac{1}{2}}$ and $\partial(a)(1 + D^2)^{-\frac{1}{2}}$ are compact.

If A contains the identity operator of H, then the spectral triple (A, H, D) is said to be unital. Often one assumes that a spectral triple has additional structure, such as a $\mathbb{Z}/2\mathbb{Z}$-grading γ of H, such that every $a \in A$ is even and D is odd with respect to H. The motivating commutative example of a spectral triple is to take $A = C^\infty(X)$ as an algebra of smooth functions on a compact manifold X, $H = L_2(X, E)$ as the space of square-integrable sections of a Hermitian vector bundle E, and D as a first order differential operator acting on sections of E.

We will denote by $\overline{A}$ the norm closure of A in $\mathcal{B}(H)$. In the classical setting $A = C^\infty(X)$, the C^*-algebra $\overline{A}$ is the algebra of all continuous functions on X.

Quantised calculus can be defined in the setting of spectral triples. Let $F = \mathrm{sgn}(D)$ be the sign of D. It is helpful for us that F be unitary, and so we adopt the convention that $\mathrm{sgn}(0) = 1$. Given $a \in \overline{A}$, we define the *quantised differential* of a as

$$đa := \mathrm{i}[F, a].$$

The factor of i is in place to ensure that $(đa)^* = đ(a^*)$, but is not strictly necessary.

DEFINITION 3.1. A spectral triple (A, H, D) is said to be QC^1 if for every $a \in A$, the commutator

$$\delta(a) = \mathrm{i}[|D|, a] : \mathrm{dom}(D) \to H$$

admits a bounded extension from H to H, and if

$$\delta(a)(\mathrm{i} + D)^{-1}$$

is compact for every $a \in A$. Equivalently, $\delta(a)(1 + D^2)^{-\frac{1}{2}}$ is compact.

Recall that for (H, F) to be a Fredholm module for the C^*-algebra $\overline{A}$, we require at least that $[F, a]$ be compact for every $a \in \overline{A}$. The next lemma demonstrates that this is indeed the case if (A, H, D) is QC^1 in the above sense.

LEMMA 3.2. *Let (A, H, D) be a QC^1-spectral triple. Then $đa$ is compact for every $a \in \overline{A}$.*

PROOF. Initially assume that $a \in A$. Observe that the operator

$$(1 + D^2)^{\frac{1}{2}} - |D| : \mathrm{dom}(D) \to H$$

admits a unique bounded extension. Indeed, by the functional calculus we have

$$(1 + D^2)^{\frac{1}{2}} - |D| = ((1 + D^2)^{\frac{1}{2}} + |D|)^{-1}$$

which is a bounded function of D. It follows that the operator

$$\mathrm{i}[(1 + D^2)^{\frac{1}{2}}, a](1 + D^2)^{-\frac{1}{2}} = \mathrm{i}[(1 + D^2)^{\frac{1}{2}} - |D|, a](\mathrm{i} + D)^{-1} + \delta(a)(1 + D^2)^{-\frac{1}{2}}$$

is compact. Similarly, we have

$$F(1 + D^2)^{\frac{1}{2}} - D = F((1 + D^2)^{\frac{1}{2}} - |D|) \in \mathcal{B}(H).$$

It follows that

$$\mathrm{i}[F(1 + D^2)^{\frac{1}{2}}, a](1 + D^2)^{-\frac{1}{2}}$$

is compact. By the Leibniz rule

$$\begin{aligned} đa = \mathrm{i}[F, a] &= \mathrm{i}[F(1 + D^2)^{\frac{1}{2}}(1 + D^2)^{-\frac{1}{2}}, a] \\ &= \mathrm{i}[F(1 + D^2)^{\frac{1}{2}}, a](1 + D^2)^{-\frac{1}{2}} - \mathrm{i}F[(1 + D^2)^{\frac{1}{2}}, a](1 + D^2)^{-\frac{1}{2}}. \end{aligned}$$

The latter two terms are compact for every $a \in A$. Thus, $đa$ is compact. Due to the inequality

$$\|đa\|_\infty \leq \|\mathrm{sgn}(D)a\|_\infty + \|a\mathrm{sgn}(D)\|_\infty \leq 2\|a\|_\infty$$

this immediately extends to all $a \in \overline{A}$. □

The *spectral dimension* of a spectral triple (A, H, D) is defined in various slightly different ways in the literature. Here, we will adopt the following definition.

DEFINITION 3.3. Let $p > 0$. A QC^1 spectral triple (A, H, D) is said to be have spectral dimension p if for all $a \in A$ we have

$$a(1 + D^2)^{-\frac{1}{2}},\ \partial(a)(1 + D^2)^{-\frac{1}{2}},\ \delta(a)(1 + D^2)^{-\frac{1}{2}} \in \mathcal{L}_{p,\infty}.$$

This definition is different to that of [**2**, Definition 2.15], wherein the spectral dimension p was defined as the infimum over all $s > 0$ such that $|a|(1+D^2)^{-\frac{s}{2}} \in \mathcal{L}_1$ for all $a \in A$.

The next proposition strengthens Lemma 3.2 when (A, H, D) has spectral dimension p, and $a \in A$.

PROPOSITION 3.4. *Let (A, H, D) be a QC^1 spectral triple having spectral dimension p. Then for every $a \in A$ we have*

$$đa \in \mathcal{L}_{p,\infty}.$$

PROOF. We may follow identically the arguments in Lemma 3.2. By the same reasoning given in that proof, we obtain that the operators

$$[(1+D^2)^{\frac{1}{2}}, a](1+D^2)^{-\frac{1}{2}},\ [F(1+D^2)^{\frac{1}{2}}, a](1+D^2)^{-\frac{1}{2}}$$

belong to $\mathcal{L}_{p,\infty}$. Thus,

$$[F,a] = [F(1+D^2)^{\frac{1}{2}}, a](1+D^2)^{-\frac{1}{2}} - F[(1+D^2)^{\frac{1}{2}}, a](1+D^2)^{-\frac{1}{2}}$$

also belongs to $\mathcal{L}_{p,\infty}$. □

Note that the above proposition is not true for all $a \in \overline{A}$. In the specific examples developed below, we will determine necessary conditions on a such that $đa \in \mathcal{L}_{p,\infty}$.

Fundamentally, the proof of Proposition 3.4 relies solely on the algebraic identity

$$đa = \partial(a)(1+D^2)^{-\frac{1}{2}} - F\delta(a)(1+D^2)^{-\frac{1}{2}} + đa(1+D^2)^{-\frac{1}{2}}(|D| + (1+D^2)^{\frac{1}{2}})^{-1}.$$

which gives the inequality

$$\|đa\|_{p,\infty} \lesssim_p \|\partial(a)(1+D^2)^{-\frac{1}{2}}\|_{p,\infty} + \|\delta(a)(1+D^2)^{-\frac{1}{2}}\|_{p,\infty} + \|đa\|_\infty \|a(1+D^2)^{-1}\|_{p,\infty}.$$

The use of $\mathcal{L}_{p,\infty}$ here is not really important. Indeed, if $\mathcal{E}$ is any ideal of $\mathcal{B}(H)$, then if (A, H, D) is a QC^1 spectral triple and $a \in A$ is such that

$$\delta(a)(1+D^2)^{-\frac{1}{2}},\ \partial(a)(1+D^2)^{-\frac{1}{2}},\ a(1+D^2)^{-\frac{1}{2}} \in \mathcal{E}$$

then $đa \in \mathcal{E}$.

A more subtle argument, essentially proved in [**28**], supplies a better inequality for so-called fully symmetric operator ideals. These are the ideals that interpolate between $\mathcal{L}_1(H)$ and $\mathcal{B}(H)$, and this class includes $\mathcal{L}_{p,\infty}$ for $1 < p < \infty$.

The following computation is essentially in [**28**], but see also [**17**, Lemma 9].

LEMMA 3.5. *Let*

$$g(t) = \frac{t}{(1+t^2)^{\frac{1}{2}}}, \quad t \in \mathbb{R}.$$

We have

$$\frac{g(\lambda) - g(\mu)}{\lambda - \mu} = \psi_1(\lambda,\mu)\psi_2(\lambda,\mu)\psi_3(\lambda,\mu)$$

where

$$\psi_1(\lambda,\mu) = 1 + \frac{1-\lambda\mu}{(1+\lambda^2)^{\frac{1}{2}}(1+\mu^2)^{\frac{1}{2}}},\ \psi_2(\lambda,\mu) = \frac{(1+\lambda^2)^{\frac{1}{4}}(1+\mu^2)^{\frac{1}{4}}}{(1+\lambda^2)^{\frac{1}{2}} + (1+\mu^2)^{\frac{1}{2}}}$$

and

$$\psi_3(\lambda,\mu) = (1+\lambda^2)^{-\frac{1}{4}}(1+\mu^2)^{-\frac{1}{4}}.$$

It is immediate that the functions ψ_1 is of a suitable form so that we may define the double operator integrals $T^{D,D}_{\psi_1}$, and $T^{D,D}_{\psi_1}$ is bounded on any fully symmetric operator space.

It was proved in [**17**], as a consequence of [**28**, Lemma 8] that $T^{D,D}_{\psi_2}$ is also bounded on $\mathcal{L}_1(H)$ and $\mathcal{B}(H)$, and consequently on every fully symmetric operator space.

Note that the following lemma does not need any QC^1 assumption.

PROPOSITION 3.6. *Let (A, H, D) be a spectral triple. For any fully symmetric operator space $\mathcal{E}$ we have*

$$\|\text{đ}a\|_{\mathcal{E}} \lesssim_{\mathcal{E}} \|(1+D^2)^{-\frac{1}{4}}\partial(a)(1+D^2)^{-\frac{1}{4}}\|_{\mathcal{E}} + \|a(1+D^2)^{-1}\|_{\mathcal{E}} + \|a^*(1+D^2)^{-1}\|_{\mathcal{E}}.$$

for all $a \in A$ such that the right hand side is defined.

PROOF. We write

$$\text{đ}a = \mathrm{i}[F - \frac{D}{(1+D^2)^{\frac{1}{2}}}, a] + \mathrm{i}[\frac{D}{(1+D^2)^{\frac{1}{2}}}, a].$$

By the functional calculus, we have

$$F - \frac{D}{(1+D^2)^{\frac{1}{2}}} = F(1+D^2)^{-\frac{1}{2}}((1+D^2)^{\frac{1}{2}} + |D|)^{-1}$$

and therefore

$$\|[F - \frac{D}{(1+D^2)^{\frac{1}{2}}}, a]\|_{\mathcal{E}} \lesssim \|a(1+D^2)^{-1}\|_{\mathcal{E}} + \|(1+D^2)^{-1}a\|_{\mathcal{E}}. \tag{3.1}$$

Now we write

$$[\frac{D}{(1+D^2)^{\frac{1}{2}}}, a] = T_{\phi}^{D,D}((1+D^2)^{-\frac{1}{4}}[D,a](1+D^2)^{-\frac{1}{4}})$$

where

$$\phi(\lambda,\mu) = \frac{(1+\lambda^2)^{\frac{1}{4}}(1+\mu^2)^{\frac{1}{4}}(\frac{\lambda}{(1+\lambda^2)^{\frac{1}{2}}} - \frac{\mu}{(1+\mu^2)^{\frac{1}{2}}})}{\lambda - \mu}, \quad \lambda, \mu \in \mathbb{R}.$$

By Lemma 3.5,

$$\phi(\lambda,\mu) = \psi_1(\lambda,\mu)\psi_2(\lambda,\mu)$$

Hence, $T_{\phi}^{D,D} = T_{\psi_1}^{D,D} T_{\psi_2}^{D,D}$ is bounded from $\mathcal{E}$ to $\mathcal{E}$. Therefore,

$$\left\|\left[\frac{D}{(1+D^2)^{\frac{1}{2}}}, a\right]\right\|_{\mathcal{E}} \lesssim_{\mathcal{E}} \|(1+D^2)^{-\frac{1}{4}}\partial(a)(1+D^2)^{-\frac{1}{4}}\|_{\mathcal{E}}.$$

Combined with (3.1), this completes the proof. □

4. Quantum differentiability in examples

Proposition 3.6 implies that in order to estimate the norm of đa in some fully symmetric ideal $\mathcal{E}$, it suffices to look at the three operators

$$(1+D^2)^{-\frac{1}{4}}\partial(a)(1+D^2)^{-\frac{1}{4}},\ a(1+D^2)^{-\frac{1}{2}}.$$

and $a^*(1+D^2)^{-\frac{1}{2}}$. In mathematical physics, the norms of operators of this form in Schatten and weak Schatten ideals are the subject of so-called Cwikel-type estimates. We first explain the commutative case of $\mathbb{R}^d$, for which the important estimate is the following.

THEOREM 4.1 (Cwikel's inequality). [**7**] *Let $d > 1$, and $f \in L_d(\mathbb{R}^d)$. There is a constant C_d such that*

$$\|(1-\Delta)^{-\frac{1}{4}}M_f(1-\Delta)^{-\frac{1}{4}}\|_{d,\infty} \le C_d\|f\|_d.$$

The standard spectral triple for $\mathbb{R}^d$ takes $A = C_c^\infty(\mathbb{R}^d)$, acting by pointwise multiplication on the Hilbert space $L_2(\mathbb{R}^d, \mathbb{C}^{N_d})$, where $N_d = 2^{\lfloor \frac{d}{2} \rfloor}$ and D is the Euclidean Dirac operator

$$D = \sum_{j=1}^{d} -\mathrm{i}\gamma_j \otimes \partial_j.$$

Here, ∂_j is the operator of differentiation in the jth coordinate direction of $\mathbb{R}^d$ and $\{\gamma_j\}_{j=1}^d$ is a family of Clifford matrices. That is, each γ_j is a self-adjoint $N_d \times N_d$ complex matrix such that $\gamma_j\gamma_k + \gamma_k\gamma_j = 2\delta_{j,k}$. Observe that $D^2 = -1 \otimes \Delta$, and that

$$\partial(M_f) = \mathrm{i}[D, 1 \otimes M_f] = \sum_{j=1}^{d} \gamma_j \otimes M_{\partial_j f}.$$

From Cwikel's inequality (Theorem 4.1), Proposition 3.6 implies the following.

PROPOSITION 4.2. *Let $d > 1$ and let $f \in C_c^\infty(\mathbb{R}^d)$. We have*

$$đf := \mathrm{i}[\mathrm{sgn}(D), 1 \otimes M_f] \in \mathcal{L}_{d,\infty}(L_2(\mathbb{R}^d, \mathbb{C}^{N_d}))$$

and

$$\|đf\|_{d,\infty} \lesssim_d \sum_{j=1}^{d} \|\partial_j f\|_{L_d(\mathbb{R}^d)} + \|f\|_{L_d(\mathbb{R}^d)}.$$

PROOF. By Proposition 3.6 and the fact that $\mathcal{L}_{d,\infty}$ is fully symmetric, we have

$$\|đf\|_{d,\infty} \lesssim_d \|(1+D^2)^{-\frac{1}{4}}\partial(M_f)(1+D^2)^{-\frac{1}{4}}\|_{d,\infty} + \|(1\otimes M_f)(1+D^2)^{-1}\|_{d,\infty} + \|(1\otimes M_{\overline{f}})(1+D^2)^{-1}\|_{d,\infty}.$$

Since $\partial(M_f) = \sum_{j=1}^d \gamma_j \otimes M_{\partial_j f}$ and $D^2 = -1 \otimes \Delta$, it follows by the triangle inequality that

$$\|đf\|_{d,\infty} \lesssim_d \sum_{j=1}^{d} \|(1-\Delta)^{-\frac{1}{4}} M_{\partial_j f}(1-\Delta)^{-\frac{1}{4}}\|_{d,\infty} + \|M_f(1-\Delta)^{-1}\|_{d,\infty} + \|M_{\overline{f}}(1-\Delta)^{-1}\|_{d,\infty}.$$

By Cwikel's inequality, we have for every $1 \le j \le d$ that

$$\|(1-\Delta)^{-\frac{1}{4}} M_{\partial_j f}(1-\Delta)^{-\frac{1}{4}}\|_{d,\infty} \lesssim_d \|\partial_j f\|_{L_d(\mathbb{R}^d)}$$

and similarly $\|M_f(1-\Delta)^{-1}\|_{d,\infty} \lesssim_d \|f\|_{L_d(\mathbb{R}^d)}$. Thus,

$$\|đf\|_{d,\infty} \lesssim \sum_{j=1}^{d} \|\partial_j f\|_{L_d(\mathbb{R}^d)} + \|f\|_{L_d(\mathbb{R}^d)}$$

and this completes the proof. □

4.1. Quantum Euclidean spaces and Quantum tori. We recall the definitions of quantum Euclidean spaces and quantum tori, following the expositions of [**20**] and [**19**] respectively.

DEFINITION 4.3. Let $d \geq 2$, and let θ be a $d \times d$ real antisymmetric matrix. For $t \in \mathbb{R}^d$, let $U_\theta(t)$ be the operator on $L_2(\mathbb{R}^d)$ defined by

$$U_\theta(t)g(s) = e^{\mathrm{i}\langle t, \theta s\rangle} g(s-t), \quad s \in \mathbb{R}^d \; g \in L_2(\mathbb{R}^d).$$

The von Neumann algebra generated by the family $\{U_\theta(t)\}_{t\in\mathbb{R}^d}$ is called the *noncommutative Euclidean d-space*, and denoted $L_\infty(\mathbb{R}^d_\theta)$. That is,

$$L_\infty(\mathbb{R}^d_\theta) := \{U_\theta(t)\}''_{t\in\mathbb{R}^d} \subset \mathcal{B}(L_2(\mathbb{R}^d)).$$

where $''$ denotes the double commutant.

The subalgebra of $L_\infty(\mathbb{R}^d_\theta)$ generated by $\{U_\theta(n)\}_{n\in\mathbb{Z}^d}$ is called the *noncommutative d-torus* and denoted $L_\infty(\mathbb{T}^d_\theta)$. That is,

$$L_\infty(\mathbb{T}^d_\theta) := \{U_\theta(n)\}''_{n\in\mathbb{Z}^d} \subset \mathcal{B}(L_2(\mathbb{R}^d)).$$

Observe that when $\theta = 0$, these definitions reproduce $L_\infty(\mathbb{R}^d)$ and $L_\infty(\mathbb{T}^d)$ in their representation on $L_2(\mathbb{R}^d)$ as Fourier multipliers.

The von Neumann algebras $L_\infty(\mathbb{R}^d_\theta)$ and $L_\infty(\mathbb{T}^d_\theta)$ are semifinite. Their traces (both denoted τ_θ) may be defined as follows. Given a Schwartz class function $f \in \mathcal{S}(\mathbb{R}^d)$, denote by $U_\theta(f)$ the norm convergent Bochner integral

$$U(f) = \int_{\mathbb{R}^d} f(t)U(t)\,dt.$$

We define

$$\tau_\theta(U(f)) := f(0).$$

That this is well-defined is not obvious, we refer the reader to [**20**] for further details. The functional τ_θ extends to a normal semifinite trace on $L_\infty(\mathbb{R}^d_\theta)$, and we define the L_p-space $L_p(\mathbb{R}^d_\theta)$ as being the noncommutative L_p-spaces with respect to this trace. That is, for $1 \leq p < \infty$, we define $L_p(\mathbb{R}^d_\theta)$ as being the completion of

$$N_p := \{x \in L_\infty(\mathbb{R}^d_\theta) \; : \; \tau_\theta(|x|^p) < \infty\}$$

with respect to the norm $\|x\|_p := (\tau_\theta(|x|^p))^{\frac{1}{p}}$.

A normal semifinite trace, also conventionally denoted τ_θ, is defined in a similar manner on $L_\infty(\mathbb{T}^d_\theta)$. We define τ_θ initially on the span of $\{U_\theta(n)\}_{n\in\mathbb{Z}^d}$ by the formula

$$\tau_\theta\left(\sum_{n\in\mathbb{Z}^d} c_n U_\theta(n)\right) := c_0.$$

The use of the notation τ_θ in both cases should not cause confusion. It can be proved that τ_θ extends to a normal semifinite trace on $L_\infty(\mathbb{T}^d_\theta)$ and the L_p-spaces $L_p(\mathbb{T}^d_\theta)$ are defined as being the noncommutative L_p spaces associated to the semifinite trace τ_θ.

For $s \in \mathbb{R}^d$, the translation automorphism of $L_\infty(\mathbb{R}^d_\theta)$, denoted T_s, is the linear map defined by

$$T_s(x) = M_{e_s} x M_{e_{-s}}$$

where e_{-s} is the trigonometric function $e_{-s}(t) = \exp(\mathrm{i}\langle t, s\rangle)$. It follows readily from the definition that

$$T_s(U_\theta(t)) = e^{\mathrm{i}\langle t,s\rangle} U_\theta(t),$$

and hence T_s is an automorphism of $L_\infty(\mathbb{R}^d_\theta)$.

The partial derivatives $\{\partial_1, \ldots, \partial_d\}$ are defined as unbounded derivations of $L_\infty(\mathbb{R}^d_\theta)$ in terms of T_s by

$$\partial_j x := \frac{d}{ds_j} T_s(x).$$

Given a multi-index $\alpha = (\alpha_1, \ldots, \alpha_d) \in \mathbb{N}^d$, we denote $\partial^\alpha := \partial_1^{\alpha_1} \cdots \partial_d^{\alpha_d}$. The Laplace operator Δ is defined as

$$\Delta = \sum_{j=1}^d \partial_j^2.$$

The Schwartz space $\mathcal{S}(\mathbb{R}^d_\theta)$ is defined as being the image of the usual (commutative) Schwartz space $\mathcal{S}(\mathbb{R}^d)$ under U_θ. The standard Fréchet topology of $\mathcal{S}(\mathbb{R}^d)$ is transferred to $\mathcal{S}(\mathbb{R}^d_\theta)$ by the isomorphism U_θ. The space $\mathcal{S}'(\mathbb{R}^d_\theta)$ of tempered distributions is then defined as being the topological dual of $\mathcal{S}(\mathbb{R}^d_\theta)$. Every L_p space $L_p(\mathbb{R}^d_\theta)$ embeds continuously into $\mathcal{S}'(\mathbb{R}^d_\theta)$).

If $m \in L_\infty(\mathbb{R}^d)$, we denote $m(D) = m(-\mathrm{i}\partial_1, -\mathrm{i}\partial_2, \ldots, -\mathrm{i}\partial_d)$, defined by functional calculus. This is termed a Fourier multiplier with symbol m.

The space $C^\infty(\mathbb{T}^d_\theta)$ of smooth functions can be defined equivalently as either the subspace of $x \in L_\infty(\mathbb{T}^d_\theta)$ belonging to the domain of ∂^α for all $\alpha \in \mathbb{N}^d$, or equivalently as

$$C^\infty(\mathbb{T}^d_\theta) := \{\sum_{n\in\mathbb{Z}^d} c(n)U_\theta(n) \ : \ c \in \mathcal{S}(\mathbb{Z}^d)\}$$

where $\mathcal{S}(\mathbb{Z}^d)$ is the space of sequences of rapid decay. The Fréchet topology of $C^\infty(\mathbb{T}^d_\theta)$ is equivalent to the Fréchet topology of $\mathcal{S}(\mathbb{Z}^d)$. The topological dual of $C^\infty(\mathbb{T}^d_\theta)$ is the space of distributions on $\mathbb{T}^d_\theta$, and is denoted $\mathcal{D}'(\mathbb{T}^d_\theta)$. For further details, please see [**34**].

Definition 4.4. For $k \geq 0$ and $1 \leq p \leq \infty$, the Sobolev spaces $W^k_p(\mathbb{R}^d_\theta)$ and $W^k_p(\mathbb{T}^d_\theta)$ are defined as the subspaces of $x \in L_p(\mathbb{R}^d_\theta)$ and $y \in L_p(\mathbb{T}^d_\theta)$ such that $\partial^\alpha x \in L_p(\mathbb{R}^d_\theta)$ and $\partial^\alpha y \in L_p(\mathbb{T}^d_\theta)$ for all $|\alpha| \leq k$ respectively. We define

$$\|x\|_{W^k_p(\mathbb{R}^d_\theta)} := \sum_{|\alpha|\leq k} \|\partial^\alpha x\|_p$$

and

$$\|y\|_{W^k_p(\mathbb{T}^d_\theta)} := \sum_{|\alpha|\leq k} \|\partial^\alpha y\|_p$$

The *homogeneous* Sobolev spaces $\dot{W}^k_p(\mathbb{R}^d_\theta)$ and $\dot{W}^k_p(\mathbb{T}^d_\theta)$ are defined as the space of all $x \in \mathcal{S}'(\mathbb{R}^d_\theta)$ (and $y \in \mathcal{D}'(\mathbb{T}^d_\theta)$ respectively) such that the Sobolev seminorm

$$\|x\|_{\dot{W}^k_p(\mathbb{R}^d_\theta)} := \sum_{|\alpha|=k} \|\partial^\alpha x\|_{L_p(\mathbb{R}^d_\theta)}$$

is finite, and respectively

$$\|y\|_{\dot{W}^k_p(\mathbb{T}^d_\theta)} := \sum_{|\alpha|=k} \|\partial^\alpha y\|_{L_p(\mathbb{T}^d_\theta)} < \infty.$$

Extensions and variants of Cwikel's inequality are known for the noncommutative spaces.

THEOREM 4.5. [**19, 20**] *Let $d>1$ and $x\in L_d(\mathbb{T}^d_\theta)$. then $(1-\Delta)^{-\frac14}x(1-\Delta)^{-\frac14}\in\mathcal{L}_{d,\infty}$, and*

$$\|(1-\Delta)^{-\frac14}x(1-\Delta)^{-\frac14}\|_{d,\infty}\le C_d\|x\|_{L_d(\mathbb{T}^d_\theta)}.$$

Similarly, if $x\in L_d(\mathbb{R}^d_\theta)$, then

$$\|(1-\Delta)^{-\frac14}x(1-\Delta)^{-\frac14}\|_{d,\infty}\le C_d\|x\|_{L_d(\mathbb{R}^d_\theta)}.$$

These estimates, combined with Proposition 3.6, yield the following by an identical argument to Proposition 4.2.

THEOREM 4.6. *Let $x\in\mathcal{S}(\mathbb{R}^d_\theta)$. Then*

$$\|đx\|_{d,\infty}\lesssim_d \|x\|_{L_d(\mathbb{R}^d_\theta)}+\|x\|_{\dot W^1_d(\mathbb{R}^d_\theta)}. \tag{4.1}$$

If $x\in C^\infty(\mathbb{T}^d_\theta)$, then

$$\|đx\|_{d,\infty}\lesssim_d \|x\|_{L_d(\mathbb{T}^d_\theta)}+\|x\|_{\dot W^1_d(\mathbb{T}^d_\theta)} \tag{4.2}$$

In these three cases ($\mathbb{R}^d$, $\mathbb{R}^d_\theta$ and $\mathbb{T}^d_\theta$), the inhomogeneous term (that is $\|x\|_d$) can be removed. The argument is quite different for $\mathbb{T}^d_\theta$ compared to $\mathbb{R}^d_\theta$.

4.2. Quantum differentiability on quantum tori. The simplest case to consider is $\mathbb{T}^d_\theta$, and in this section we explain how to remove the $\|x\|_{L_d(\mathbb{T}^d_\theta)}$ term in (4.2). This will lead to the inequality

$$\|đx\|_{d,\infty}\lesssim_d \|x\|_{\dot W^1_d(\mathbb{T}^d_\theta)},\quad x\in C^\infty(\mathbb{T}^d_\theta).$$

This inequality is key to the implication that $x\in\dot W^1_d(\mathbb{T}^d_\theta)\Rightarrow đx\in\mathcal{L}_{d,\infty}$.

Since $đx=\mathrm{i}[\mathrm{sgn}(D),x]=\mathrm{i}[\mathrm{sgn}(D),x-\widehat{x}(0)]$, where $\widehat{x}(0)$ is the constant term of $x\in\mathbb{T}^d_\theta$, it is quite natural to remove the inhomogeneous term $\|x\|_d$ on the right hand side of (4.2). On quantum tori, this is done by the aide of the following Poincaré inequality [**34**, Theorem 2.12]

$$\|x-\widehat{x}(0)\|_d\lesssim_d\|x\|_{\dot W^1_d}. \tag{4.3}$$

Plugging (4.3) into (4.2), we deduce

$$\begin{aligned}\|đx\|_{d,\infty}&=\|\mathrm{i}[\mathrm{sgn}(D),x-\widehat{x}(0)]\|_{d,\infty}\\&\lesssim_d\|x-\widehat{x}(0)\|_d+\|x-\widehat{x}(0)\|_{\dot W^1_d}\\&\lesssim_d\|x\|_{\dot W^1_d}.\end{aligned}$$

Next, we can remove the assumption that $x\in C^\infty(\mathbb{T}^d_\theta)$; in other words, the only condition we need for $đx\in\mathcal{L}_{d,\infty}$ is that $x\in\dot W^1_d(\mathbb{T}^d_\theta)$. Indeed, by the density of $C^\infty(\mathbb{T}^d_\theta)\subset\dot W^1_d(\mathbb{T}^d_\theta)$ (see [**34**, Proposition 2.7(ii)]), we may select a sequence $\{x_n\}_{n\ge1}\subset C^\infty(\mathbb{T}^d_\theta)$ converging to x in the $\dot W^1_d$ norm, which ensures the convergence of $đx_n$ to some operator $T\in\mathcal{L}_{d,\infty}$. On the other hand, it is straightforward to verify that

$$\lim_{n\to\infty}\|đx_n(\eta)-\mathrm{i}[\mathrm{sgn}(D),x](\eta)\|_2=0$$

for any $\eta\in C^\infty(\mathbb{T}^d_\theta)\otimes\mathbb{C}^{N_d}$, the dense subspace of $L_2(\mathbb{T}^d_\theta)\otimes\mathbb{C}^{N_d}$. Consequently , we see that $đx_n$ converges to $\mathrm{i}[\mathrm{sgn}(D),x]$ in the $\mathcal{L}_{d,\infty}$ topology. Moreover,

$$\|\mathrm{i}[\mathrm{sgn}(D),x]\|_{d,\infty}=\lim_{n\to\infty}\|đx_n\|_{d,\infty}\lesssim_d\lim_{n\to\infty}\|x_n\|_{\dot W^1_d}=\|x\|_{\dot W^1_d}.$$

All together, we have proved the following sufficiency of quantum differentiability on quantum tori.

THEOREM 4.7. *If* $x \in \dot{W}^1_d(\mathbb{T}^d_\theta)$, *then* $\mathrm{i}[\mathrm{sgn}(D), x]$, *also denoted as* $đx$, *is a well defined operator in* $\mathcal{L}_{d,\infty}$ *satisfying*

$$\|đx\|_{d,\infty} \lesssim_d \|x\|_{\dot{W}^1_d}.$$

4.3. Quantum differentiability on quantum Euclidean spaces. Now we explain how to remove the $\|x\|_{L_d(\mathbb{R}^d_\theta)}$ term in (4.1). This argument was first given in [**20**], and uses an interesting "dilation" trick. The idea is that in the commutative case, Proposition 4.2 delivers

$$\|đf\|_{d,\infty} \lesssim_d \|f\|_{L_d(\mathbb{R}^d)} + \|\nabla f\|_{L_d(\mathbb{R}^d,\mathbb{C}^d)}.$$

The quantities $\|đf\|_{d,\infty}$ and $\|\nabla f\|_{L_d(\mathbb{R}^d,\mathbb{C}^d)}$ are invariant under a rescaling of the coordinates, while $\|f\|_{L_d(\mathbb{R}^d)}$ is not. By choosing a particular rescaling, the $\|f\|_{L_d(\mathbb{R}^d)}$ term may be removed.

There is an immediate obstacle to using this argument for $\mathbb{R}^d_\theta$, which is that there is no natural "rescaling" action of $\mathbb{R}_+$ on $L_\infty(\mathbb{R}^d_\theta)$ by automorphisms. However, borrowing an idea from [**11**], in [**20**] we defined a "rescaling" by $\lambda \in \mathbb{R}_+$ which is a $*$-homomorphism.

$$\Psi_\lambda : L_\infty(\mathbb{R}^d_\theta) \to L_\infty(\mathbb{R}^d_{\lambda^2\theta})$$

Exploiting the fact that the constants in (4.1) are independent of θ, we are able to use Ψ_λ in place of a rescaling action. We explain the argument briefly here.

Let $\lambda > 0$ and $\Psi_\lambda : L_\infty(\mathbb{R}^d_\theta) \to L_\infty(\mathbb{R}^d_{\lambda^2\theta})$ be the $*$-isomorphism sending $U_\theta(s)$ to $U_{\lambda^2\theta}(s/\lambda)$; unlike the usual dilation map on $\mathbb{R}^d$, the $*$-isomorphism Ψ_λ is defined from the von Neumann algebra $L_\infty(\mathbb{R}^d_\theta)$ onto another von Neumann algebra $L_\infty(\mathbb{R}^d_{\lambda^2\theta})$ defined with respect to the matrix $\lambda^2\theta$. Denoting $\sigma_\lambda : \xi(t) \mapsto \lambda^{d/2}\xi(\lambda t)$ the usual L_2-norm preserving dilation on $\mathbb{R}^d$, one has $\Psi_\lambda(x) = \sigma_\lambda x \sigma_\lambda^*$. Since the operator $\mathrm{sgn}(D)$, viewed as a Fourier multiplier on $\mathbb{R}^d$, commutes with σ_λ (and σ_λ^*), we have

$$\begin{aligned} đ(\Psi_\lambda(x)) &= \mathrm{i}[\mathrm{sgn}(D), \Psi_\lambda(x)] = \mathrm{i}[\mathrm{sgn}(D), \sigma_\lambda x \sigma_\lambda^*] \\ &= \mathrm{i}\sigma_\lambda[\mathrm{sgn}(D), x]\sigma_\lambda^* = \sigma_\lambda đx\, \sigma_\lambda^*, \end{aligned}$$

whence, $\|đ(\Psi_\lambda(x))\|_{d,\infty} = \|đx\|_{d,\infty}$. Applying (4.1) to $\Psi_\lambda(x) \in \mathcal{S}(\mathbb{R}^d_{\lambda^2\theta})$, we obtain

$$\|đ(\Psi_\lambda(x))\|_{d,\infty} \lesssim_d \|\Psi_\lambda(x)\|_d + \|\Psi_\lambda(x)\|_{\dot{W}^1_d}.$$

Since $\|\Psi_\lambda(x)\|_d = \lambda\|x\|_d$ and $\|\Psi_\lambda(x)\|_{\dot{W}^1_d} = \|x\|_{\dot{W}^1_d}$ (see [**20**, Proposition 3.7]), we substitute Ψ_λ into (4.1) which yields

$$\|đx\|_{d,\infty} = \|đ(\Psi_\lambda(x))\|_{d,\infty} \lesssim_d \lambda\|x\|_d + \|x\|_{\dot{W}^1_d}.$$

Letting $\lambda \to 0$ yields for $x \in \mathcal{S}(\mathbb{R}^d_\theta)$

$$\|đx\|_{d,\infty} \lesssim_d \|x\|_{\dot{W}^1_d}.$$

Similar to the quantum torus case, we have the following theorem which characterises the quantum differentiability on $\mathbb{R}^d_\theta$ in full generality.

THEOREM 4.8. *If* $x \in \dot{W}^1_d(\mathbb{R}^d_\theta)$, *then* $\mathrm{i}[\mathrm{sgn}(D), x]$, *also denoted as* $đx$, *is a well defined operator in* $\mathcal{L}_{d,\infty}$ *satisfying*

$$\|đx\|_{d,\infty} \lesssim_d \|x\|_{\dot{W}^1_d}.$$

Compared to [**20**, Theorem 1.1], the assumption that $x \in L_p(\mathbb{R}^d_\theta)$ for some $d < p < \infty$ is not needed any more in Theorem 4.8, which can be proved using the following density result.

PROPOSITION 4.9. *Let $k \geq 0$ and $1 < p < \infty$. Then $\mathcal{S}(\mathbb{R}^d_\theta)$ is dense in $\dot{W}^k_p(\mathbb{R}^d_\theta)$.*

This result is proved in the classical case (that is, $\theta = 0$) in [**12**]. The condition that $p > 1$ cannot be weakened in general. For example, it can be shown that $\mathcal{S}(\mathbb{R})$ is not dense in $\dot{W}^1_1(\mathbb{R})$.

The proof that we supply here is very different to that of [**12**], and is based on a Littlewood-Paley decomposition of $L_\infty(\mathbb{R}^d_\theta)$ and a Mikhlin multiplier theorem.

We begin with a Mikhlin multiplier theorem in the operator-valued case; it is proved in a standard way, and should be known to experts. Let ψ be a smooth function on $\mathbb{R}^d$ supported in the annulus

$$\frac{1}{2} \leq |\xi| \leq 2$$

such that

$$\psi(\xi) + \psi(\frac{\xi}{2}) = 1, \quad 1 \leq |\xi| \leq 2.$$

Denote

$$\psi_j(\xi) = \psi(2^{-j}\xi), \quad j \in \mathbb{Z}.$$

The sequence $\{\psi_j\}_{j\in\mathbb{Z}}$ forms a homogeneous Littlewood-Paley decomposition of $\mathbb{R}^d$. Let ϕ be a smooth function on $\mathbb{R}^d \setminus \{0\}$. Assume that for some $\sigma > \frac{d}{2}$,

$$\sup_{k\in\mathbb{Z}} \|\phi(2^k\cdot)\psi\|_{H^\sigma_2(\mathbb{R}^d)} < \infty, \tag{4.4}$$

where $H^\sigma_2(\mathbb{R}^d)$ denotes the fractional Sobolev space on $\mathbb{R}^d$. (4.4) is known as the Sobolev condition on the symbols of Fourier multipliers, and is slightly stronger than the usual Hörmander condition in the following way: Let k be the (inverse) Fourier transform of ϕ, then k is a Calderón-Zygmund kernel satisfying

$$\begin{cases} \|\widehat{\mathsf{k}}\|_{L_\infty(\mathbb{R}^d)} \lesssim \sup_{k\in\mathbb{Z}} \|\phi(2^k\cdot)\psi\|_{H^\sigma_2(\mathbb{R}^d)}, \\ \sup_{t\in\mathbb{R}^d} \int_{|s|>2|t|} |\mathsf{k}(s-t) - \mathsf{k}(s)|ds \lesssim \sup_{k\in\mathbb{Z}} \|\phi(2^k\cdot)\psi\|_{H^\sigma_2(\mathbb{R}^d)}. \end{cases} \tag{4.5}$$

Let us recall the operator-valued BMO spaces introduced in [**22**]. Let Q be a cube in $\mathbb{R}^d$ with sides parallel to the axes, and $|Q|$ its volume. For a function f with values in a semifinite von Neumann algebra $\mathcal{M}$, the column BMO norm of f is

$$\|f\|_{\mathrm{BMO}^c} = \sup_{Q\subset\mathbb{R}^d} \left\| \frac{1}{|Q|} \int_Q |f(t) - f_Q|^2 dt \right\|_{\mathcal{M}}^{\frac{1}{2}},$$

where $f_Q = \frac{1}{|Q|}\int_Q f(t)dt$. Then

$$\mathrm{BMO}^c(\mathbb{R}^d, \mathcal{M}) = \{f : \int_{\mathbb{R}^d} |f(t)|^2 \frac{dt}{1+|t|^{d+1}} < \infty, \|f\|_{\mathrm{BMO}^c} < \infty\}.$$

The row space $\mathrm{BMO}^r(\mathbb{R}^d, \mathcal{M})$ is defined as the space of f such that $f^* \in \mathrm{BMO}^c(\mathbb{R}^d, \mathcal{M})$, and $\mathrm{BMO}(\mathbb{R}^d, \mathcal{M}) = \mathrm{BMO}^c(\mathbb{R}^d, \mathcal{M}) \cap \mathrm{BMO}^r(\mathbb{R}^d, \mathcal{M})$ with the intersection norm.

The argument below is standard, see for example [**33**, Lemma 2.1].

PROPOSITION 4.10. *Let* k *be a Calderón-Zygmund kernel satisfying* (4.5). *Then the convolution operator* $\mathsf{k}(f)(s) = \int_{\mathbb{R}^d} \mathsf{k}(s-t)f(t)dt$ *is bounded from* $L_\infty(\mathbb{R}^d, \mathcal{M})$ *into* $\mathrm{BMO}(\mathbb{R}^d, \mathcal{M})$.

PROOF. Since $\|f\|_{L_\infty(\mathbb{R}^d,\mathcal{M})} = \|f^*\|_{L_\infty(\mathbb{R}^d,\mathcal{M})}$, it suffices to show that k is bounded from $L_\infty(\mathbb{R}^d, \mathcal{M})$ into $\mathrm{BMO}^c(\mathbb{R}^d, \mathcal{M})$. Denote the center of the cube Q by c, and $\widetilde{Q} = 2Q$ the cube with center c and twice the side length of Q. Decompose $f = g + h$ with $g = f\chi_{\widetilde{Q}}$, where χ stands for the characteristic function. Setting

$$a = \int_{\mathbb{R}^d \setminus \widetilde{Q}} \mathsf{k}(c-t)f(t)dt,$$

we have

$$\mathsf{k}(f)(s) - a = \mathsf{k}(g)(s) + \int_{\mathbb{R}^d} (\mathsf{k}(s-t) - \mathsf{k}(c-t))h(t)dt.$$

Thus

$$\frac{1}{|Q|}\int_Q |\mathsf{k}(f)(s) - a|^2 ds \leq 2(A+B),$$

where

$$A = \frac{1}{|Q|}\int_Q |\mathsf{k}(g)(s)|^2 ds,$$
$$B = \frac{1}{|Q|}\int_Q \Big|\int_{\mathbb{R}^d} (\mathsf{k}(s-t) - \mathsf{k}(c-t))h(t)dt\Big|^2 ds.$$

The term A is estimated by the Plancherel formula and the first condition in (4.5):

$$\begin{aligned}|Q|A &\lesssim \sup_{k\in\mathbb{Z}} \|\phi(2^k\cdot)\psi\|^2_{H_2^\sigma(\mathbb{R}^d)} \int_{\mathbb{R}^d} |\widehat{g}(s)|^2 ds \\ &= \sup_{k\in\mathbb{Z}} \|\phi(2^k\cdot)\psi\|^2_{H_2^\sigma(\mathbb{R}^d)} \int_{\mathbb{R}^d} |g(s)|^2 ds \\ &\lesssim \sup_{k\in\mathbb{Z}} \|\phi(2^k\cdot)\psi\|^2_{H_2^\sigma(\mathbb{R}^d)} |\widetilde{Q}| \|f\|^2_{L_\infty(\mathbb{R}^d,\mathcal{M})}.\end{aligned}$$

The term B is estimated by the convexity of the operator valued function $x \mapsto |x|^2$ and the second condition in (4.5):

$$\begin{aligned}&\Big|\int_{\mathbb{R}^d} (\mathsf{k}(s-t) - \mathsf{k}(c-t))h(t)dt\Big|^2 \\ &\leq \int_{\mathbb{R}^d\setminus\widetilde{Q}} |\mathsf{k}(s-t) - \mathsf{k}(c-t)|dt \cdot \int_{\mathbb{R}^d\setminus\widetilde{Q}} \Big||\mathsf{k}(s-t) - \mathsf{k}(c-t)|^{\frac{1}{2}} h(t)\Big|^2 dt \\ &\lesssim \sup_{k\in\mathbb{Z}} \|\phi(2^k\cdot)\psi\|_{H_2^\sigma(\mathbb{R}^d)} \int_{\mathbb{R}^d\setminus\widetilde{Q}} \Big||\mathsf{k}(s-t) - \mathsf{k}(c-t)|^{\frac{1}{2}} h(t)\Big|^2 dt \\ &\lesssim \sup_{k\in\mathbb{Z}} \|\phi(2^k\cdot)\psi\|^2_{H_2^\sigma(\mathbb{R}^d)} \|f\|^2_{L_\infty(\mathbb{R}^d,\mathcal{M})}.\end{aligned}$$

The boundedness of k from $L_\infty(\mathbb{R}^d, \mathcal{M})$ into $\mathrm{BMO}^c(\mathbb{R}^d, \mathcal{M})$ is therefore proved. □

In order to apply the above multiplier theorem to $\mathbb{R}^d_\theta$, we need a transference method which contains a $*$-isomorphism from $L_\infty(\mathbb{R}^d_\theta)$ into $L_\infty(\mathbb{R}^d) \otimes L_\infty(\mathbb{R}^d_\theta)$ determined by

$$\Phi_\theta : U_\theta(t) \mapsto e^{\mathrm{i}\langle t,\cdot\rangle} U_\theta(t),$$

(see [**11**, Section 1.1.2]). Evidently, for smooth $U_\theta(f) \in \mathcal{S}(\mathbb{R}^d_\theta)$,

$$\Phi_\theta(U_\theta(f))(\xi) = \int_{\mathbb{R}^d} e^{\mathrm{i}\langle t,\xi\rangle} f(t) U_\theta(t) dt$$

is a function on ξ taking values in $L_\infty(\mathbb{R}^d_\theta)$. Thus, for ϕ satisfying (4.4) with Fourier transform k satisfying (4.5), we have

$$\Phi_\theta\big(\phi(D)(U_\theta(f))\big) = \mathsf{k}\Big(\int_{\mathbb{R}^d} e^{\mathrm{i}\langle t,\xi\rangle} f(t) U_\theta(t) dt\Big)$$

Following [**11**, Appendix B], we define

$$\mathrm{BMO}(\mathbb{R}^d_\theta) := \{x \in \mathcal{S}'(\mathbb{R}^d_\theta) \ : \ \Phi_\theta(x) \in \mathrm{BMO}(\mathbb{R}^d, L_\infty(\mathbb{R}^d_\theta))\}$$

and $\|x\|_{\mathrm{BMO}} := \|\Phi_\theta(x)\|_{\mathrm{BMO}(\mathbb{R}^d, L_\infty(\mathbb{R}^d_\theta))}$.

Since Φ_θ is by definition a contraction from $\mathrm{BMO}(\mathbb{R}^d_\theta)$ into $\mathrm{BMO}(\mathbb{R}^d, L_\infty(\mathbb{R}^d_\theta))$, Proposition 4.10 leads to the boundedness of $\phi(D)$ from (the Schwartz class subspace of) $L_\infty(\mathbb{R}^d_\theta)$ into $\mathrm{BMO}(\mathbb{R}^d_\theta)$.

On the other hand, the first condition in (4.5) and the Plancherel formula on $\mathbb{R}^d_\theta$ yield immediately the $L_2(\mathbb{R}^d_\theta)$ boundedness of $\phi(D)$. A theorem of González-Pérez, Junge and Parcet [**11**, Theorem B.8] together with the reiteration theorem of complex interpolation implies that

$$L_p(\mathbb{R}^d_\theta) = [L_2(\mathbb{R}^d_\theta), \mathrm{BMO}(\mathbb{R}^d_\theta)]_{\frac{2}{p}}, \quad 2 < p < \infty.$$

where $[\cdot,\cdot]_{\frac{2}{p}}$ is the functor of complex interpolation. Hence, the L_2-boundedness and $L_\infty - \mathrm{BMO}$ boundedness of $\phi(D)$ implies that $\phi(D)$ is bounded on $L_p(\mathbb{R}^d_\theta)$ for $2 < p < \infty$. Since the adjoint function $\overline{\phi}$ satisfies the same estimates as ϕ, a standard duality argument proves that $\phi(D)$ is also bounded on $L_p(\mathbb{R}^d_\theta)$ for $1 < p < 2$. In summary, we arrive at the following Mikhlin theorem for $\mathbb{R}^d_\theta$.

PROPOSITION 4.11. *Let m be a smooth function on $\mathbb{R}^d \setminus \{0\}$ satisfying* (4.4). *Then for all $1 < p < \infty$ we have*

$$m(D) : L_p(\mathbb{R}^d_\theta) \to L_p(\mathbb{R}^d_\theta).$$

The same is true if m is a smooth function on $\mathbb{R}^d \setminus \{0\}$ satisfying the Mikhlin condition:

$$\sup\{|\xi|^{|\alpha|}|\partial^\alpha_\xi m(\xi)| : \xi \in \mathbb{R}^d \setminus \{0\}, \alpha \in \mathbb{N}^d, |\alpha| \le \lfloor \frac{d}{2} \rfloor + 1\} < \infty. \tag{4.6}$$

Using Proposition 4.11, we can get Littlewood-Paley inequalities for $L_p(\mathbb{R}^d_\theta)$. Denote

$$\dot{\Delta}_j = \psi_j(D), \quad j \in \mathbb{Z}.$$

This is the homogeneous Littlewood-Paley decomposition for $\mathbb{R}^d_\theta$. The inhomogeneous Littlewood-Paley decomposition can be defined by

$$\Delta_j = \begin{cases} \dot{\Delta}_j, & j \ge 1, \\ 1 - \sum_{j\ge 1} \dot{\Delta}_j, & j = 0. \end{cases}$$

LEMMA 4.12. *Let $\varepsilon = \{\varepsilon_j\}_{j\in\mathbb{Z}} \in \{-1,1\}^{\mathbb{Z}}$. Define a multiplier*

$$m_\varepsilon(\xi) = \sum_{j\in\mathbb{Z}} \varepsilon_j \psi_j(\xi).$$

Then m_ε obeys condition (4.6), *with constants uniform in ε.*

PROOF. Let $\xi \in \mathbb{R}^d \setminus \{0\}$. There exists some $j \in \mathbb{Z}$ such that $2^{j-1} \leq |\xi| \leq 2^{j+1}$. Then

$$m_\varepsilon(\xi) = \sum_{k=j-1}^{j+1} \varepsilon_k \psi_k(\xi).$$

For $\alpha \in \mathbb{N}^d$ we have

$$|\partial^\alpha m_\varepsilon(\xi)| = \left| \sum_{k=j-1}^{j+1} \varepsilon_k 2^{-k|\alpha|} \partial^\alpha \psi(2^{-k}\xi) \right| \leq C_\alpha 2^{-j|\alpha|} \leq C_\alpha |\xi|^{-|\alpha|}.$$

□

Now applying Proposition 4.11, we get Littlewood-Paley inequalities for $\mathbb{R}^d_\theta$.

THEOREM 4.13. *Let $1 < p < \infty$. Then there exist constants $0 < c_p < C_p < \infty$ such that for all $u \in L_p(\mathbb{R}^d_\theta)$ we have*

$$c_p \|u\|_p \leq (\mathbb{E}(\|\sum_{j\in\mathbb{Z}} \varepsilon_j \dot{\Delta}_j u\|_p^p))^{\frac{1}{p}} \leq C_p \|u\|_p.$$

Here, $\{\varepsilon_j\}_{j\in\mathbb{Z}}$ are independent Rademacher variables, taking the values ± 1 with probabilities $\frac{1}{2}$. The expectation is over all ε.

PROOF. Lemma 4.12 implies that for all $\varepsilon \in \{-1,1\}^{\mathbb{Z}}$ we have

$$\left\| \sum_{j\in\mathbb{Z}} \varepsilon_j \dot{\Delta}_j u \right\|_p \leq C_p \|u\|_p \tag{4.7}$$

where C_p is independent of ε. Taking the expectation over all ε yields the upper bound.

The lower bound follows from a duality argument. Note that if $|j-k| \geq 2$ then

$$\tau_\theta(\dot{\Delta}_j u \dot{\Delta}_k v) = 0$$

Therefore,

$$\tau_\theta(uv) = \sum_{|j-k|\leq 1} \tau_\theta(\dot{\Delta}_j u \dot{\Delta}_k v).$$

This splits into a sum of three terms,

$$\sum_{j\in\mathbb{Z}} \tau_\theta(\dot{\Delta}_{j-1} u \dot{\Delta}_j v) + \tau_\theta(\dot{\Delta}_j u \dot{\Delta}_j v) + \tau_\theta(\dot{\Delta}_{j+1} u \dot{\Delta}_j v).$$

Examining the middle sum,

$$\sum_{j\in\mathbb{Z}} \tau_\theta(\dot{\Delta}_j u \dot{\Delta}_j u) = \mathbb{E}(\tau(m_\varepsilon(D)u \cdot m_\varepsilon(D)v)).$$

By Hölder's inequality, if $q = \frac{p}{p-1}$ we have

$$|\mathbb{E}(\tau_\theta((m_\varepsilon(D)u) \cdot (m_\varepsilon(D)v)))| \leq \mathbb{E}\|m_\varepsilon(D)u\|_p \sup_\varepsilon \|m_\varepsilon(D)u\|_q.$$

Using a similar argument for the other two summands and applying (4.7) yields

$$|\tau_\theta(uv)| \leq C_q \|v\|_q \mathbb{E}\|m_\varepsilon(D)u\|_p.$$

Therefore,

$$\|u\|_p \leq C_q \mathbb{E}\|m_\varepsilon(D)u\|_p.$$

□

With identical reasoning, we can also get the inhomogeneous Littlewood-Paley inequalities:

$$c_p\|u\|_p \leq (\mathbb{E}\|\sum_{j=0}^{\infty}\varepsilon_j\Delta_j u\|_p^p)^{1/p} \leq C_p\|u\|_p, \quad 1<p<\infty.$$

Another application of the Mikhlin theorem yields Littlewood-Paley characterisations of Sobolev spaces:

COROLLARY 4.14. *For all $s\in\mathbb{R}$ and $1<p<\infty$, we have*

$$(\mathbb{E}\|\sum_{j\in\mathbb{Z}}2^{js}\varepsilon_j\dot{\Delta}_j u\|_p^p)^{1/p} \cong \|u\|_{\dot{W}_p^s},$$

and

$$(\mathbb{E}\|\sum_{j=0}^{\infty}2^{js}\varepsilon_j\Delta_j u\|_p^p)^{1/p} \cong \|u\|_{W_p^s}.$$

The noncommutative Khintchine inequalities (see [**27**, Section 6] and references therein) give us the following Littlewood-Paley description of noncommutative Sobolev spaces.

THEOREM 4.15. *Let $1<p<\infty$ and $s\in\mathbb{R}$. The inhomogeneous Sobolev norm $\|u\|_{W_p^s(\mathbb{R}^d)}$ is equivalent to*

$$\max\{\left\|\left(\sum_{j=0}^{\infty}2^{2js}|\Delta_j u|^2\right)^{1/2}\right\|_p, \left\|\left(\sum_{j=0}^{\infty}2^{2js}|(\Delta_j u)^*|^2\right)^{1/2}\right\|_p\}$$

for $2\leq p<\infty$, and equivalent to

$$\inf\{\left\|\left(\sum_{j=0}^{\infty}2^{2js}|\Delta_j a|^2\right)^{1/2}\right\|_p + \left\|\left(\sum_{j=0}^{\infty}2^{2js}|(\Delta_j b)^*|^2\right)^{1/2}\right\|_p \;:\; u=a+b\}$$

for $1<p\leq 2$. For the homogeneous Sobolev spaces, the seminorm $\|u\|_{\dot{W}_p^s(\mathbb{R}_\theta^d)}$ is equivalent to

$$\max\{\left\|\left(\sum_{j\in\mathbb{Z}}2^{2js}|\dot{\Delta}_j u|^2\right)^{1/2}\right\|_p, \left\|\left(\sum_{j\in\mathbb{Z}}2^{2js}|(\dot{\Delta}_j u)^*|^2\right)^{1/2}\right\|_p\}, \quad 2\leq p<\infty$$

for $2\leq p<\infty$ and equivalent to

$$\inf\{\left\|\left(\sum_{j\in\mathbb{Z}}2^{2js}|\dot{\Delta}_j a|^2\right)^{1/2}\right\|_p + \left\|\left(\sum_{j\in\mathbb{Z}}2^{2js}|(\dot{\Delta}_j b)^*|^2\right)^{1/2}\right\|_p \;:\; u=a+b\}$$

for $1<p\leq 2$.

The need to consider $p<2$ and $p>2$ separately is a purely noncommutative phenomenon, see [**34**, Proposition 4.23] for similar conclusion on quantum tori.

PROOF OF PROPOSITION 4.9. Since it has been shown in [**20**] that $\mathcal{S}(\mathbb{R}^d_\theta)$ is dense in $W^k_p(\mathbb{R}^d_\theta)$, it suffices to show that $W^k_p(\mathbb{R}^d_\theta)$ is dense in $\dot{W}^k_p(\mathbb{R}^d_\theta)$.

Hence, we prove the density of the inclusion $W^k_p(\mathbb{R}^d_\theta) \subset \dot{W}^k_p(\mathbb{R}^d_\theta)$ for the case $p \geq 2$, the case $1 < p \leq 2$ being treated similarly. Let $u \in \dot{W}^k_p(\mathbb{R}^d_\theta)$, and define

$$u_n := \sum_{j \geq -n} \dot{\Delta}_j u.$$

From Theorem 4.15, $u_n \in W^k_p(\mathbb{R}^d_\theta)$, with norm $\|u_n\|_{W^k_p} \lesssim 2^{nk}$. We compute $\|u - u_n\|_{\dot{W}^k_p}$ using the above Littlewood-Paley theorem:

$$\|u - u_n\|_{\dot{W}^k_p} \cong \max\{\|(\sum_{j \leq -n} 2^{2jk}|\dot{\Delta}_j u|^2)^{1/2}\|_p, \|(\sum_{j \leq -n} 2^{2jk}|(\dot{\Delta}_j u)^*|^2)^{1/2}\|_p\}.$$

Noting that $u \in \dot{W}^k_p(\mathbb{R}^d_\theta)$, we have as $n \to \infty$,

$$\sum_{j \geq -n} 2^{2jk}|\dot{\Delta}_j u|^2 \nearrow \sum_{j \in \mathbb{Z}} 2^{2jk}|\dot{\Delta}_j u|^2 \in L_{p/2}(\mathbb{R}^d_\theta),$$

whence

$$(\sum_{j \leq -n} 2^{2jk}|\dot{\Delta}_j u|^2)^{1/2} \searrow 0,$$

in the operator sense. By the order continuity of the L_p-norm, it follows that

$$\lim_{n \to \infty} \|(\sum_{j \leq -n} 2^{2jk}|\dot{\Delta}_j u|^2)^{1/2}\|_p = 0.$$

Similarly,

$$\lim_{n \to \infty} \|(\sum_{j \leq -n} 2^{2jk}|(\dot{\Delta}_j u)^*|^2)^{1/2}\|_p = 0.$$

Therefore, Theorem 4.15 implies that $\|u_n - u\|_{\dot{W}^k_p} \to 0$ and this completes the proof. □

With the above proposition, one repeats the approximation argument in the proof of Theorem 4.7 and concludes Theorem 4.8.

5. Trace formulae

In this section we state the trace formula in localised form for the three examples. We will work on the noncommutative plane $\mathbb{R}^d_\theta$, since it does not matter whether $\theta = 0$ or not, and precisely the same reasoning applies to $\mathbb{T}^d_\theta$.

Before stating and proving the local trace formula, let us introduce some subalgebras of $\mathcal{B}(L_2(\mathbb{R}^d))$. The first one is a dense $*$-subalgebra $\mathcal{A}(\mathbb{R}^d_\theta)$ of $\mathcal{S}(\mathbb{R}^d_\theta)$. Rather than define $\mathcal{A}(\mathbb{R}^d_\theta)$ here, we will only mention that $\mathcal{A}(\mathbb{R}^d_\theta)$ has the property that every $x \in \mathcal{A}(\mathbb{R}^d_\theta)$ can be expressed as a sum of products of elements of $\mathcal{A}(\mathbb{R}^d_\theta)$. For further details see [**20**, Proposition 2.5]. This algebra was introduced for technical reasons. In the extreme cases where $\theta = 0$ or $\det(\theta) \neq 0$, we have $\mathcal{A}(\mathbb{R}^d_\theta) = \mathcal{S}(\mathbb{R}^d_\theta)$. The assertion of [**20**, Theorem 6.1] is that for all $x \in \mathcal{A}(\mathbb{R}^d_\theta)$ we have

$$|đx|^d - |A|^d(1 + D^2)^{-\frac{d}{2}} \in \mathcal{L}_1 \qquad (5.1)$$

where

$$A := \sum_{j=1}^{d} \gamma_j \otimes A_j := \sum_{j=1}^{d} \gamma_j \otimes (\partial_j x + \sum_{k=1}^{d} \frac{\partial_j \partial_k}{-\Delta} \partial_k x).$$

The proof of the main trace formula in [**20**] was on the basis of Connes' trace theorem in the form of [**21**]. This version of Connes' trace theorem was based on a pseudodifferential calculus we briefly define here. Let $C_0(\mathbb{R}^d_\theta)$ denote the norm closure of $\mathcal{S}(\mathbb{R}^d_\theta)$ in $\mathcal{B}(L_2(\mathbb{R}^d))$. For $g \in C(\mathbb{S}^{d-1})$, denote $g(\frac{\mathrm{i}\nabla}{(-\Delta)^{1/2}})$ for the multiplication operator $\xi(t) \mapsto g(\frac{t}{|t|})\xi(t)$ in $\mathcal{B}(L_2(\mathbb{R}^d))$. The set of all $g(\frac{\mathrm{i}\nabla}{(-\Delta)^{1/2}})$ with $g \in C(\mathbb{S}^{d-1})$ form a commutative C^*-subalgebra of $\mathcal{B}(L_2(\mathbb{R}^d))$.

We define the C^*-algebra $\Pi(C_0(\mathbb{R}^d_\theta) + \mathbb{C}, C(\mathbb{S}^{d-1}))$ to be the C^*-subalgebra of $\mathcal{B}(L_2(\mathbb{R}^d))$ generated by $C_0(\mathbb{R}^d_\theta)+\mathbb{C}$ and all $g(\frac{\mathrm{i}\nabla}{(-\Delta)^{1/2}})$ where $g \in C(\mathbb{S}^{d-1})$. Theorem 3.3 of [**21**] asserts that there exists a unique norm-continuous $*$-homomorphism

$$\mathrm{sym} : \Pi(C_0(\mathbb{R}^d_\theta) + \mathbb{C}, C(\mathbb{S}^{d-1})) \longrightarrow \big(C_0(\mathbb{R}^d_\theta) + \mathbb{C}\big) \otimes_{\min} C(\mathbb{S}^{d-1})$$

which maps $x \in C_0(\mathbb{R}^d_\theta)$ to $x \otimes 1$ and $g(\frac{\mathrm{i}\nabla}{(-\Delta)^{1/2}})$ to $1 \otimes g$. Then [**21**, Theorem 6.15] says that for every continuous normalised trace φ on $\mathcal{L}_{1,\infty}$, every $x \in W^d_1(\mathbb{R}^d_\theta)$, and every $T \in \Pi(C_0(\mathbb{R}^d_\theta) + \mathbb{C}, C(\mathbb{S}^{d-1}))$, we have

$$\varphi(Tx(1+D^2)^{-d/2}) = C_d\Big(\tau_\theta \otimes \int_{\mathbb{S}^{d-1}}\Big)(\mathrm{sym}(T)(x \otimes 1)) \tag{5.2}$$

where C_d is a certain constant depending only on the dimension d.

Applying (5.2) to (5.1) gives the following trace formula, which slightly generalises the trace formula of [**20**] by including a "localisation" parameter $y \in C_0(\mathbb{R}^d_\theta) + \mathbb{C}$.

THEOREM 5.1. *Let $x \in \dot{W}^1_d(\mathbb{R}^d_\theta)$ and $y \in C_0(\mathbb{R}^d_\theta) + \mathbb{C}$. Then there is a constant C_d depending only on the dimension d such that for any continuous normalised trace φ on $\mathcal{L}_{1,\infty}$ we have:*

$$\varphi(y|đx|^d) = C_d \int_{\mathbb{S}^{d-1}} \tau_\theta\left(y\Big(\sum_{j=1}^{d} |\partial_j x - s_j \sum_{k=1}^{d} s_k \partial_k x|^2\Big)^{\frac{d}{2}}\right) ds.$$

Here, the integral over $\mathbb{S}^{d-1}$ is taken with respect to the rotation-invariant measure ds on $\mathbb{S}^{d-1}$, and $s = (s_1, \ldots, s_d)$.

PROOF. Assume initially that $x \in \mathcal{A}(\mathbb{R}^d_\theta)$. Theorem 4.8 guarantees $đx \in \mathcal{L}_{d,\infty}$, and thus $y|đx|^d \in \mathcal{L}_{1,\infty}$. By (5.1), for $y \in C_0(\mathbb{R}^d_\theta) + \mathbb{C}$ we have

$$y|đx|^d - y|A|^d(1+D^2)^{-\frac{d}{2}} \in \mathcal{L}_1.$$

For every trace φ on $\mathcal{L}_{1,\infty}$, we therefore have

$$\varphi(y|đx|^d) = \varphi(y|A|^d(1+D^2)^{-\frac{d}{2}}).$$

We have proved in [**20**] that A belongs to $\Pi(C_0(\mathbb{R}^d_\theta)+\mathbb{C}, C(\mathbb{S}^{d-1}))$. Hence, the same is true of $y|A|^d$. Since the principal symbol mapping sym is a C^*-homomorphism and $\mathrm{sym}(y) = y$, we have

$$\mathrm{sym}(y|A|^d) = y \cdot \mathrm{sym}(|A|^d)$$

But $\mathrm{sym}(|A|^d)$ is already computed in the proof of Theorem 1.2 of [**20**]:

$$\mathrm{sym}(|A|^d)(s) = \Big(\sum_{j=1}^{d} |\partial_j x - s_j \sum_{k=1}^{d} s_k \partial_k x|^2\Big)^{\frac{d}{2}}, \quad s \in \mathbb{S}^{d-1}.$$

Therefore

$$\mathrm{sym}(y|A|^d)(s) = y\Big(\sum_{j=1}^{d} |\partial_j x - s_j \sum_{k=1}^{d} s_k \partial_k x|^2\Big)^{\frac{d}{2}}, \quad s \in \mathbb{S}^{d-1}.$$

Applying the Connes' trace formula in the form of (5.2), it follows that

$$\varphi(y|A|^d(1+D^2)^{-\frac{d}{2}}) = C_d \int_{\mathbb{S}^{d-1}} \tau_\theta\Big(y(\sum_{j=1}^{d} |\partial_j x - s_j \sum_{k=1}^{d} s_k \partial_k x|^2\Big)^{\frac{d}{2}}) \, ds.$$

Therefore

$$\varphi(y|{\rm đ}x|^d) = C_d \int_{\mathbb{S}^{d-1}} \tau_\theta\Big(y(\sum_{j=1}^{d} |\partial_j x - s_j \sum_{k=1}^{d} s_k \partial_k x|^2)^{\frac{d}{2}}\Big) \, ds.$$

By virtue of Proposition 4.9, we may remove the assumption $x \in \mathcal{A}(\mathbb{R}^d_\theta)$ by the same approximation argument as in the proof of [**19**, Theorem 1.2], so we omit the details. □

5.1. Necessity condition for quantum differentiability. In this last part of the paper, we state the necessity of the condition $x \in \dot{W}^1_d(\mathbb{R}^d_\theta)$ for ${\rm đ}x \in \mathcal{L}_{d,\infty}$ as a converse to Theorem 4.8. Let us first state the following corollary of Theorem 5.1; the proof is exactly the same as that of [**19**, Corollary 1.3], so we omit the details.

COROLLARY 5.2. *Let $x \in \dot{W}^1_d(\mathbb{R}^d_\theta)$. Then there are constants c_d and C_d depending only on d such that for any continuous normalised trace φ on $\mathcal{L}_{1,\infty}$ we have*

$$c_d \|x\|^d_{\dot{W}^1_d} \leq \varphi(|{\rm đ}x|^d) \leq C_d \|x\|^d_{\dot{W}^1_d}.$$

Next, we show that for $x \in L_\infty(\mathbb{R}^d_\theta)$, if ${\rm đ}x \in \mathcal{L}_{d,\infty}$, then $x \in \dot{W}^1_d(\mathbb{R}^d_\theta)$ and $\|x\|_{\dot{W}^1_d} \lesssim \|{\rm đ}x\|_{d,\infty}$. Our strategy is approximating x by smooth elements $x_n \in L_\infty(\mathbb{R}^d_\theta)$ and then applying Theorem 5.1 and Corollary 5.2. So we begin with the introduction of the required smooth elements in $L_\infty(\mathbb{R}^d_\theta)$.

Taking $\psi \in \mathcal{S}(\mathbb{R}^d)$ such that $\int_{\mathbb{R}^d} \psi(s)\, ds = 1$ and denoting for $\varepsilon > 0$,

$$\psi_\varepsilon(t) = \varepsilon^{-d}\psi(\frac{t}{\varepsilon}),$$

we get a family of smooth elements $U(\psi_\varepsilon) \in \mathcal{S}(\mathbb{R}^d_\theta)$. By [**20**, Theorem 3.8], $U(\psi_\varepsilon)x \to x$ in the $L_p(\mathbb{R}^d_\theta)$ norm as $\varepsilon \to 0$, and thus $U(\psi_\varepsilon) \to 1$ in the strong operator topology of $L_\infty(\mathbb{R}^d_\theta)$.

Moreover, we need the following notion of convolution: Let $x \in L_p(\mathbb{R}^d_\theta)$ for $1 \leq p < \infty$. For $\psi \in L_1(\mathbb{R}^d)$ define:

$$\psi * x := \int_{\mathbb{R}^d} \psi(s) T_{-s}(x)\, ds$$

as an absolutely convergent Bochner integral. When $\varepsilon \to 0$, we have also $\psi_\varepsilon * x \to x$ in the L_p-norm. If $x \in \dot{W}_d^1(\mathbb{R}_\theta^d)$, by the triangle inequality and the fact that T_s commutes with $\operatorname{sgn}(D)$, we have

$$\|\bar{d}(\psi * x)\|_{d,\infty} \lesssim_d \|\psi\|_1 \|\bar{d}x\|_{d,\infty}. \tag{5.3}$$

THEOREM 5.3. *Suppose that $x \in L_\infty(\mathbb{R}_\theta^d)$. If $\bar{d}x$ has bounded extension in $\mathcal{L}_{d,\infty}$, then $x \in \dot{W}_d^1(\mathbb{R}_\theta^d)$, and there is a constant $c_d > 0$ depending only on d such that*

$$\|x\|_{\dot{W}_d^1} \leq c_d \|\bar{d}x\|_{\mathcal{L}_{d,\infty}}.$$

Note that when $d > 2$, the assumption $x \in L_\infty(\mathbb{R}_\theta^d)$ may be replaced by the weaker one $x \in L_d(\mathbb{R}_\theta^d) + L_\infty(\mathbb{R}_\theta^d)$ ([**20**]); but in the strictly noncommutative $\det(\theta) \neq 0$ case, $L_d(\mathbb{R}_\theta^d) + L_\infty(\mathbb{R}_\theta^d) = L_\infty(\mathbb{R}_\theta^d)$. Also note that on quantum tori, one only has to assume additionally $x \in L_2(\mathbb{T}_\theta^d)$ ([**19**]).

PROOF. Suppose that $\bar{d}x \in \mathcal{L}_{d,\infty}$. If we show that $x \in \dot{W}_d^1(\mathbb{R}_\theta^d)$, then Corollary 5.2 will ensure that

$$\|x\|_{\dot{W}_d^1} \lesssim_d \varphi(|\bar{d}x|^d) \lesssim_d \|\varphi\|_{(\mathcal{L}_{1,\infty})^*} \|\bar{d}x\|_{d,\infty}.$$

So we are reduced to proving $x \in \dot{W}_d^1(\mathbb{R}_\theta^d)$.

From [**20**, Lemma 3.12], we may select $\{\psi_\varepsilon\}_{\varepsilon>0}$ and $\{\phi_\varepsilon\}_{\varepsilon>0}$ such that $\psi_\varepsilon * (U(\phi_\varepsilon)x) \in \mathcal{S}(\mathbb{R}_\theta^d)$. The upper bound (5.3) implies:

$$\|\bar{d}(\psi_\varepsilon * (U(\phi_\varepsilon)x))\|_{d,\infty} \lesssim_d \|\psi_\varepsilon\|_1 \|\bar{d}(U(\phi_\varepsilon)x)\|_{d,\infty}.$$

Expanding the commutator using the Leibniz rule, the quasi-triangle inequality and Theorem 4.8:

$$\|\bar{d}(\psi_\varepsilon * (U(\phi_\varepsilon)x))\|_{d,\infty} \lesssim_d \|\psi_\varepsilon\|_1 \big(\|(\bar{d}U(\phi_\varepsilon))x\|_{d,\infty} + \|U(\phi_\varepsilon)\|_\infty \|\bar{d}x\|_{d,\infty}\big). \tag{5.4}$$

By construction $\|\psi_\varepsilon\|_1$ is constant, and $\|U(\phi_\varepsilon)\|_\infty$ and $\|U(\phi_\varepsilon)\|_{\dot{W}_d^1}$ are uniformly bounded as $\varepsilon \to 0$. Then Theorem 4.8 yields the bound $\|(\bar{d}U(\phi_\varepsilon))x\|_{d,\infty} \lesssim_d \|U(\phi_\varepsilon)\|_{\dot{W}_d^1} \|x\|_\infty \lesssim \|x\|_\infty$. So all the terms on the right hand side of (5.4) are bounded.

It follows that $\{\bar{d}(\psi_\varepsilon * (U(\phi_\varepsilon)x))\}_{\varepsilon>0}$ is uniformly bounded in $\mathcal{L}_{d,\infty}$ as $\varepsilon \to 0$. Now applying Corollary 5.2 to $\bar{d}(\psi_\varepsilon * (U(\phi_\varepsilon)x))$, we see that $\{\psi_\varepsilon * (U(\phi_\varepsilon)x)\}_{\varepsilon>0}$ is uniformly bounded in $\dot{W}_d^1(\mathbb{R}_\theta^d)$, so for every $1 \leq j \leq d$, $\{\partial_j(\psi_\varepsilon * (U(\phi_\varepsilon)x))\}_{\varepsilon>0}$ is uniformly bounded in $L_d(\mathbb{R}_\theta^d)$. Since $d \geq 2$, the space $L_d(\mathbb{R}_\theta^d)$ is reflexive and therefore $\{\partial_j(\psi_\varepsilon * (U(\phi_\varepsilon)x))\}_{\varepsilon>0}$ has a weak limit point in $L_d(\mathbb{R}_\theta^d)$. But we know from [**20**, Corollary 3.11] that if $y \in L_{d/(d-1)}(\mathbb{R}_\theta^d)$ or $y \in L_1(\mathbb{R}_\theta^d)$, then $U(\phi_\varepsilon)(\psi_\varepsilon * y) \to y$ in the $L_{d/(d-1)}(\mathbb{R}_\theta^d)$ sense or in the $L_1(\mathbb{R}_\theta^d)$ sense respectively; hence that $\psi_\varepsilon * (U(\phi_\varepsilon)x) \to x$ in the the distributional sense. It follows that the weak limit point of $\{\partial_j(\psi_\varepsilon * (U(\phi_\varepsilon)x))\}_{\varepsilon>0}$ in $L_d(\mathbb{R}_\theta^d)$ must also be $\partial_j x$.

Therefore, $\partial_j x \in L_d(\mathbb{R}_\theta^d)$ for every $1 \leq j \leq d$. That is, $x \in \dot{W}_d^1(\mathbb{R}_\theta^d)$. □

References

[1] M. Sh. Birman and M. Z. Solomyak, *Operator integration, perturbations and commutators* (Russian, with English summary), Zap. Nauchn. Sem. Leningrad. Otdel. Mat. Inst. Steklov. (LOMI) **170** (1989), no. Issled. Line ĭn. Oper. Teorii Funktsiĭ. 17, 34–66, 321, DOI 10.1007/BF01099305; English transl., J. Soviet Math. **63** (1993), no. 2, 129–148. MR1039572

[2] A. L. Carey, V. Gayral, A. Rennie, and F. A. Sukochev, *Index theory for locally compact noncommutative geometries*, Mem. Amer. Math. Soc. **231** (2014), no. 1085, vi+130. MR3221983

[3] Alain Connes, *Noncommutative differential geometry*, Inst. Hautes Études Sci. Publ. Math. **62** (1985), 257–360. MR823176

[4] A. Connes, *The action functional in noncommutative geometry*, Comm. Math. Phys. **117** (1988), no. 4, 673–683. MR953826

[5] Alain Connes, *Noncommutative geometry and reality*, J. Math. Phys. **36** (1995), no. 11, 6194–6231, DOI 10.1063/1.531241. MR1355905

[6] Alain Connes, Dennis Sullivan, and Nicolas Teleman, *Quasiconformal mappings, operators on Hilbert space, and local formulae for characteristic classes*, Topology **33** (1994), no. 4, 663–681, DOI 10.1016/0040-9383(94)90003-5. MR1293305

[7] Michael Cwikel, *Weak type estimates for singular values and the number of bound states of Schrödinger operators*, Ann. of Math. (2) **106** (1977), no. 1, 93–100, DOI 10.2307/1971160. MR473576

[8] P. G. Dodds, T. K. Dodds, F. A. Sukochev, and D. Zanin, *Arithmetic-geometric mean and related submajorisation and norm inequalities for τ-measurable operators: Part II*, Integral Equations Operator Theory **92** (2020), no. 4, Paper No. 32, 60, DOI 10.1007/s00020-020-02586-5. MR4124453

[9] V. Gayral, J. M. Gracia-Bondía, B. Iochum, T. Schücker, and J. C. Várilly, *Moyal planes are spectral triples*, Comm. Math. Phys. **246** (2004), no. 3, 569–623, DOI 10.1007/s00220-004-1057-z. MR2053945

[10] Liliana Gabriela Gheorghe, *Hankel operators in Schatten ideals*, Ann. Mat. Pura Appl. (4) **180** (2001), no. 2, 203–210, DOI 10.1007/s10231-001-8202-5. MR1847404

[11] Adrián Manuel González-Pérez, Marius Junge, and Javier Parcet, *Singular integrals in quantum Euclidean spaces*, Mem. Amer. Math. Soc. **272** (2021), no. 1334, xiii+90, DOI 10.1090/memo/1334. MR4320770

[12] Piotr Hajłasz and Agnieszka Kałamajska, *Polynomial asymptotics and approximation of Sobolev functions*, Studia Math. **113** (1995), no. 1, 55–64. MR1315521

[13] Svante Janson and Thomas H. Wolff, *Schatten classes and commutators of singular integral operators*, Ark. Mat. **20** (1982), no. 2, 301–310, DOI 10.1007/BF02390515. MR686178

[14] Marius Junge, Tao Mei, Javier Parcet, and Runlian Xia, *Algebraic Calderón-Zygmund theory*, Adv. Math. **376** (2021), Paper No. 107443, 72, DOI 10.1016/j.aim.2020.107443. MR4178915

[15] Galina Levitina, Fedor Sukochev, and Dmitriy Zanin, *Cwikel estimates revisited*, Proc. Lond. Math. Soc. (3) **120** (2020), no. 2, 265–304, DOI 10.1112/plms.12301. MR4008371

[16] Steven Lord, Fedor Sukochev, and Dmitriy Zanin, *Singular traces. Vol. 1. Theory*, De Gruyter Studies in Mathematics, vol. 46/1, De Gruyter, Berlin, [2021] ©2021. 2nd edition [of 3099777], DOI 10.1515/9783110378054-201. MR4304074

[17] Steven Lord, Edward McDonald, Fedor Sukochev, and Dmitry Zanin, *Quantum differentiability of essentially bounded functions on Euclidean space*, J. Funct. Anal. **273** (2017), no. 7, 2353–2387, DOI 10.1016/j.jfa.2017.06.020. MR3677828

[18] Françoise Lust-Piquard and Gilles Pisier, *Noncommutative Khintchine and Paley inequalities*, Ark. Mat. **29** (1991), no. 2, 241–260, DOI 10.1007/BF02384340. MR1150376

[19] Edward McDonald, Fedor Sukochev, and Xiao Xiong, *Quantum differentiability on quantum tori*, Comm. Math. Phys. **371** (2019), no. 3, 1231–1260, DOI 10.1007/s00220-019-03384-w. MR4029831

[20] Edward McDonald, Fedor Sukochev, and Xiao Xiong, *Quantum differentiability on noncommutative Euclidean spaces*, Comm. Math. Phys. **379** (2020), no. 2, 491–542, DOI 10.1007/s00220-019-03605-2. MR4156216

[21] Edward McDonald, Fedor Sukochev, and Dmitriy Zanin, *A C^*-algebraic approach to the principal symbol II*, Math. Ann. **374** (2019), no. 1-2, 273–322, DOI 10.1007/s00208-019-01822-7. MR3961311

[22] Tao Mei, *Operator valued Hardy spaces*, Mem. Amer. Math. Soc. **188** (2007), no. 881, vi+64, DOI 10.1090/memo/0881. MR2327840

[23] B. de Pagter, H. Witvliet, and F. A. Sukochev, *Double operator integrals*, J. Funct. Anal. **192** (2002), no. 1, 52–111, DOI 10.1006/jfan.2001.3898. MR1918492

[24] V. V. Peller, *Hankel operators of class $\mathfrak{S}_p$ and their applications (rational approximation, Gaussian processes, the problem of majorization of operators)* (Russian), Mat. Sb. (N.S.) **113(155)** (1980), no. 4(12), 538–581, 637. MR602274

[25] V. V. Peller, *Description of Hankel operators of the class $\mathfrak{S}_p$ for $p > 0$, investigation of the rate of rational approximation and other applications* (Russian), Mat. Sb. (N.S.) **122(164)** (1983), no. 4, 481–510. MR725454

[26] Vladimir V. Peller, *Hankel operators and their applications*, Springer Monographs in Mathematics, Springer-Verlag, New York, 2003, DOI 10.1007/978-0-387-21681-2. MR1949210

[27] Gilles Pisier and Quanhua Xu, *Non-commutative L^p-spaces*, Handbook of the geometry of Banach spaces, Vol. 2, North-Holland, Amsterdam, 2003, pp. 1459–1517, DOI 10.1016/S1874-5849(03)80041-4. MR1999201

[28] D. Potapov and F. Sukochev, *Unbounded Fredholm modules and double operator integrals*, J. Reine Angew. Math. **626** (2009), 159–185, DOI 10.1515/CRELLE.2009.006. MR2492993

[29] Richard Rochberg and Stephen Semmes, *Nearly weakly orthonormal sequences, singular value estimates, and Calderon-Zygmund operators*, J. Funct. Anal. **86** (1989), no. 2, 237–306, DOI 10.1016/0022-1236(89)90054-2. MR1021138

[30] Stephen Semmes, *Trace ideal criteria for Hankel operators, and applications to Besov spaces*, Integral Equations Operator Theory **7** (1984), no. 2, 241–281, DOI 10.1007/BF01200377. MR750221

[31] Barry Simon, *Trace ideals and their applications*, London Mathematical Society Lecture Note Series, vol. 35, Cambridge University Press, Cambridge-New York, 1979. MR541149

[32] F. A. Sukochev, *On a conjecture of A. Bikchentaev*, Spectral analysis, differential equations and mathematical physics: a festschrift in honor of Fritz Gesztesy's 60th birthday, Proc. Sympos. Pure Math., vol. 87, Amer. Math. Soc., Providence, RI, 2013, pp. 327–339, DOI 10.1090/pspum/087/01428. MR3087913

[33] Runlian Xia, Xiao Xiong, and Quanhua Xu, *Characterizations of operator-valued Hardy spaces and applications to harmonic analysis on quantum tori*, Adv. Math. **291** (2016), 183–227, DOI 10.1016/j.aim.2015.12.023. MR3459017

[34] Xiao Xiong, Quanhua Xu, and Zhi Yin, *Sobolev, Besov and Triebel-Lizorkin spaces on quantum tori*, Mem. Amer. Math. Soc. **252** (2018), no. 1203, vi+118, DOI 10.1090/memo/1203. MR3778570

Department of Mathematics, Penn State University, University Park, State College, Pennsylvania 16802
Email address: `eamcd92@gmail.com`

School of Mathematics and Statistics, University New South Wales, Australia
Email address: `f.sukochev@unsw.edu.au`

Institute for Advanced Study in Mathematics, Harbin Institute of Technology, People's Republic of China
Email address: `xxiong@hit.edu.cn`

Proceedings of Symposia in Pure Mathematics
Volume **105**, 2023
https://doi.org/10.1090/pspum/105/01906

Local cyclic homology for nonarchimedean Banach algebras

Ralf Meyer and Devarshi Mukherjee

ABSTRACT. Let V be a complete discrete valuation ring with uniformiser π. We introduce an invariant of Banach V-algebras called local cyclic homology. This invariant is related to analytic cyclic homology for complete, bornologically torsionfree V-algebras. It is shown that local cyclic homology only depends on the reduction mod π of a Banach V-algebra and that it is homotopy invariant, matricially stable, and excisive.

1. Introduction

This article is part of a programme to define analytic cyclic homology theories for bornological algebras over nonarchimedean fields, with the goal of defining well behaved homology theories for algebras over their residue fields, which may have finite characteristic (see [**1, 2, 7, 8**]). In this article, we define a version of cyclic homology that gives good results for Banach algebras over nonarchimedean local fields such as $\mathbb{Q}_p$. Our prototype here is the local cyclic homology theory in the archimedean case, defined first by Puschnigg for inductive systems of "nice" Fréchet algebras over $\mathbb{C}$ (see [**10**]) and then simplified by the first author in the setting of complete bornological algebras over $\mathbb{C}$ (see [**6**]). The remarkable property of local cyclic homology for algebras over $\mathbb{C}$ is that it remains well behaved for C*-algebras. For instance, it is invariant under homotopies that are merely continuous, and it is stable under the C*-algebraic tensor product with the compact operators.

The definition of archimedean local cyclic homology has two key ingredients. The first one is analytic cyclic homology for complete bornological algebras over $\mathbb{C}$. The first author arrived at its definition by rewriting Connes' entire cyclic cohomology in terms of bornologies and taking the most natural "predual" of that cohomology theory. The second key ingredient is to turn a Banach algebra into a bornological algebra using the bornology of precompact subsets. This allows to use locally defined linear maps, which map a continuous function to a nearby smooth function.

Before we can talk about the nonarchimedean version of this, we fix some notation. Let V be a complete discrete valuation ring. Let π be its uniformiser, $\mathbb{F}$ its residue field, and F its fraction field. We assume throughout that F has characteristic zero. The nonarchimedean version of analytic cyclic homology has already been defined in [**2**]. It is a functor HA_* from the category of complete, torsionfree bornological V-algebras to the category of F-vector spaces. We usually

2020 *Mathematics Subject Classification.* 19D55.

work with the enriched version $\mathbb{HA}$ of HA_*, which takes values in the derived category of the quasi-Abelian category of countable projective systems of inductive systems of Banach F-vector spaces; to define $\mathrm{HA}_*(A)$, we first take the homotopy limit of $\mathbb{HA}(A)$, then a colimit. This gives a chain complex of F-vector spaces, whose homology is $\mathrm{HA}_*(A)$.

In the archimedean case, the functors $\mathbb{HA}$ and HA_* are only homotopy invariant for smooth homotopies, and they are not expected to behave well for C*-algebras. The nonarchimedean versions of $\mathbb{HA}$ and HA_* are shown in [**2**] to be invariant under dagger homotopies, which take values in the completed tensor product with the algebra $V[t]^\dagger$. The latter algebra is the Monsky–Washnitzer algebra of the affine plane. It consists of those power series $\sum c_n t^n$ with $c_n \in V$ where the valuations of the coefficients c_n grow at least linearly. It is unclear whether $\mathbb{HA}$ and HA_* behave well for larger algebras such as the π-adic completion $\widehat{V[t]}$, which is defined by asking only for $\lim c_n = 0$.

In this article, we define a nonarchimedean analogue of the precompact bornology, namely, the compactoid bornology. A subset S of a bornological V-module M is called *compactoid* if there is a bounded V-submodule $T \subseteq M$ such that, for every $n \in \mathbb{N}$, there is a finite set $F_n \subseteq T$ with $S \subseteq V F_n + \pi^n \cdot T$. Let B be a Banach algebra over F with a submultiplicative norm. Then its unit ball $D \subseteq B$ is an algebra over V with some nice extra properties, which make it a dagger algebra. For any dagger algebra D, let D' be D with the compactoid bornology. We define the local cyclic homology $\mathbb{HL}(D)$ as $\mathbb{HA}(D')$. We are going to prove that this theory has good homological properties and gives good results for many interesting Banach V-algebras. Namely, local cyclic homology is invariant under continuous homotopy, tensoring with the π-adic completion of finite matrices, and satisfies excision for all extensions of dagger algebras. And we compute it for Banach algebra versions of Leavitt path algebras, Laurent polynomials in several variables, and the Tate algebras of curves over V.

The hardest part of the work was already done in our previous paper [**8**], where we built an analytic cyclic homology theory for algebras over the residue field $\mathbb{F}$. A key result in [**8**] says that $\mathbb{HA}(A)$ for an algebra A over $\mathbb{F}$ is naturally isomorphic to $\mathbb{HA}(D)$ if D is a dagger algebra with $D/\pi D \cong A$ and such that the quotient bornology on $D/\pi D$ is the fine one; we briefly call D fine mod π. Roughly speaking, any dagger algebra lifting of A that is also fine mod π may be used to compute $\mathbb{HA}(A)$. The compactoid bornology is always fine mod π, and we prove that a dagger algebra with the compactoid bornology remains a dagger algebra. Therefore, $\mathbb{HA}(A) \cong \mathbb{HL}(D)$ for any dagger algebra D with $D/\pi D \cong A$. Here D may be π-adically complete or, in other words, a Banach V-algebra.

We could also have defined the local cyclic homology for dagger algebras by $\mathbb{HL}(D) := \mathbb{HA}(D/\pi D)$. This is equivalent to our definition using the precompact bornology. This definition looks very quick, and we use it implicitly to prove the formal properties of local cyclic homology and compute some examples, based on the results in [**8**] showing that analytic cyclic homology for algebras over $\mathbb{F}$ is polynomially homotopy invariant, stable with respect to the $\mathbb{F}$-algebra of finite matrices, and excisive, and based on examples computed there.

While the definition of $\mathbb{HL}(D)$ as $\mathbb{HA}(D/\pi D)$ looks rather quick, it just shifts all the difficulties to the analytic cyclic homology for $\mathbb{F}$-algebras. This is defined by lifting an $\mathbb{F}$-algebra to a projective system of inductive systems of V-algebras and then taking an analytic cyclic homology complex for the latter object. Results in [**8**] allow to replace the lifting that is used to define $\mathbb{HA}(D/\pi D)$ by a pro-dagger algebra. The choices of such liftings that the general theory in [**8**] provides are, however, still rather unwieldy. The main observation in this article is that we may choose the given Banach algebra D with its precompact bornology to compute $\mathbb{HA}(D/\pi D)$. Thus our definition of $\mathbb{HL}(D)$ turns out to be more direct. In addition, it clarifies the similarity with local cyclic homology for bornological $\mathbb{C}$-algebras.

We may also prove the formal properties of $\mathbb{HL}$ using that $\mathbb{HL}(D_1) \cong \mathbb{HL}(D_2)$ if $D_1 \subseteq D_2$ is a dagger subalgebra with $D_1/\pi D_1 = D_2/\pi D_2$. For instance, the homotopy invariance of $\mathbb{HL}$ reduces to the homotopy invariance of analytic cyclic homology for dagger homotopies that is proven in [**2**]. This proof would be very close to the proof in [**6**] that local cyclic homology for bornological $\mathbb{C}$-algebras is homotopy invariant. In the end, we switched to proofs that reduce to $\mathbb{F}$-algebras because these are shorter than proofs that remain within the realm of dagger algebras.

We end this article with an illuminating counterexample. A key open question in the study of analytic cyclic homology for dagger algebras is when the analytic and periodic cyclic homology of a nice dagger algebra agree. This is important because periodic cyclic homology is shown in [**1**] to specialise to rigid cohomology for Monsky–Washnitzer algebras, whereas analytic cyclic homology is shown in [**8**] to depend only on the reduction mod π. In the archimedean case, Khalkhali [**5**] proved that entire and periodic cyclic cohomology are isomorphic for Banach $\mathbb{C}$-algebras of finite projective dimension. This may lead to the hope that finite projective dimension may suffice to show that nonarchimedean analytic and periodic cyclic homology are isomorphic. We prove that this is not the case, by showing that the Tate algebra $\widehat{V[t]}$ with the compactoid bornology is quasi-free – which means of projective dimension 1. But its periodic and analytic cyclic homology differ.

The paper is structured as follows. We begin by recalling some basic definitions and results about bornologies and dagger algebras in Section 2. We define the compactoid bornology and nuclearity in Section 3. Section 4 deals with inheritance properties of the compactoid bornology. We define local cyclic homology in Section 5. We prove that it has the expected homological properties in Section 6. We compute the local cyclic homology of some examples in Section 7.

2. Preliminaries

Let V be a complete discrete valuation ring. Let π be its uniformiser, $\mathbb{F}$ its residue field, and F its fraction field. We assume throughout that F has characteristic zero.

As in [**1, 2, 7, 8**], we use the framework of bornologies to do nonarchimedean analysis. A *bornology* on a set X is a collection of its subsets, which are called *bounded subsets*, such that finite subsets are bounded and finite unions and subsets of bounded subsets remain bounded.

A *bornological V-module* is a V-module M with a bornology such that every bounded subset is contained in a bounded V-submodule. We call a V-module map $f\colon M \to N$ *bounded* if it maps bounded subsets of M to bounded subsets of N. A bornological V-algebra is a bornological V-module with a bounded multiplication map. A *complete* bornological V-module is a bornological V-module in which every bounded subset is contained in a bounded, π-adically complete V-submodule. Every bornological V-module M has a completion $\overline{M}$ (see [**1**, Proposition 2.14]).

EXAMPLE 2.1. The most basic example of a bornology on a V-module is the *fine bornology*, which consists of those subsets that are contained in a finitely generated V-submodule. Any fine bornological V-module is complete. By default, we equip modules over the residue field $\mathbb{F}$ with the fine bornology.

DEFINITION 2.2 ([**7**, Definition 4.1]). We call a bornological V-module M (bornologically) *torsionfree* if multiplication by π is a bornological embedding, that is, M is algebraically torsionfree and $\pi^{-1} \cdot S := \{x \in M : \pi x \in S\}$ is bounded for every bounded subset $S \subseteq M$. A V-module with the fine bornology is bornologically torsionfree if and only if it is torsionfree in the purely algebraic sense. For the rest of this article, we briefly write "torsionfree" instead of "bornologically torsionfree".

DEFINITION 2.3 ([**1**,**7**]). We call a bornological V-algebra D *semidagger* if, for every bounded subset $S \subseteq D$, the V-submodule $\sum_{i=0}^{\infty} \pi^i S^{i+1}$ is bounded in D. A complete, torsionfree, semidagger bornological V-algebra is called a *dagger algebra*.

EXAMPLE 2.4. Any $\mathbb{F}$-algebra with the fine bornology is semidagger and complete.

EXAMPLE 2.5. Let B be a Banach F-algebra. We assume the norm of B to be submultiplicative. Let $D \subseteq B$ be the unit ball. Then $D \cdot D \subseteq D$, and D becomes a π-adically complete, torsionfree V-algebra. Conversely, if such an algebra D is given, then $D \hookrightarrow D \otimes F$ and there is a unique norm on $D \otimes F$ with unit ball D.

Let D be the unit ball of a Banach F-algebra as above. Then we call D with the bornology where all subsets are bounded a *Banach V-algebra*. This bornology makes D a dagger algebra.

DEFINITION 2.6. Any bornology on a V-algebra D is contained in a smallest semidagger bornology, namely, the bornology generated by the V-submodules of the form $\sum_{i=0}^{\infty} \pi^i S^{i+1}$, where $S \subseteq D$ is bounded in the original bornology. This is called the *linear growth bornology*. We denote D with the linear growth bornology by D_{lg}.

If D is torsionfree, then the completion $D^{\dagger} := \overline{D_{\mathrm{lg}}}$ is a dagger algebra (see [**7**, Proposition 3.8] or, in slightly different notation, [**1**, Lemma 3.1.12]).

DEFINITION 2.7. A bornological V-module M is called *fine mod π* if the quotient bornology on $M/\pi M$ is the fine one. Equivalently, any bounded subset is contained in $F + \pi M$ for a finitely generated V-submodule $F \subseteq M$.

3. Compactoid bornologies

Let A be an algebra over the residue field and let D be a dagger algebra with $D/\pi D \cong A$. If, in addition, D is fine mod π, then the main result of [**8**] gives a natural quasi-isomorphism $\mathbb{HA}(D) \cong \mathbb{HA}(A)$. It seems likely, however, that this fails when D is a Banach V-algebra as in Example 2.5. We are going to introduce a nonarchimedean analogue of the precompact bornology on a topological $\mathbb{C}$-vector space, which makes any dagger algebra fine mod π. This is the basis for our definition of local cyclic homology, exactly as in the archimedean case in [**6**].

DEFINITION 3.1. Let M be a bornological V-module. A subset $S \subseteq M$ is *compactoid* if it is contained in a bounded V-submodule $T \subseteq M$ such that, for every $n \in \mathbb{N}$, there is a finite set $F_n \subseteq T$ with $S \subseteq V F_n + \pi^n \cdot T$.

We call M *nuclear* if any bounded subset of M is already compactoid.

REMARK 3.2. The analogous concept of a compactoid V-submodule of a locally convex F-vector space has already been studied by Peter Schneider (see [**11**, Section 12]).

Compactoid subsets are bounded, and the compactoid subsets in M form another bornology. This bornology is always fine mod π.

EXAMPLE 3.3. Let M be a Banach V-module as in Example 2.5. Then $S \subseteq M$ is compactoid if and only if for every $n \in \mathbb{N}$, the image of S in $M/\pi^n M$ is finitely generated. This is the largest torsionfree bornology on M that is fine mod π.

EXAMPLE 3.4. For any V-module M, its fine bornology is nuclear.

The following proposition characterises the compactoid bornology in a different way, which is analogous to a very useful description of the precompact bornology on a Fréchet space over $\mathbb{C}$.

PROPOSITION 3.5. *Let M be a torsionfree bornological V-module and let S be a V-submodule. The following are equivalent:*

(i) *$S \subseteq M$ is compactoid;*
(ii) *there is a bounded V-submodule T containing S such that for each n, the V-submodule of $T/\pi^n T$ generated by the image of S is finitely generated;*
(iii) *there are a bounded V-submodule $T \subseteq M$ and a null sequence (t_n) in T with*

$$S = \left\{ s = \sum_{n=0}^{\infty} c_n t_n : (c_n) \in \ell^\infty(\mathbb{N}, V) \text{ and } s \text{ converges in } T \right\};$$

(iv) *there are a bounded V-submodule $T \subseteq M$ with $S \subseteq T$ and a null sequence (t_k) in T such that $S \subseteq \sum_{k=0}^{\infty} V t_k + \pi^n T$ for each $n \in \mathbb{N}$.*

PROOF. The equivalence between (i) and (ii) is trivial. It is easy to prove that (iii) implies (iv). It remains to show that (ii) implies (iii) and that (iv) implies (ii). Assume (iv) and let T and (t_k) be as in (iv). For fixed $n \in \mathbb{N}$, there is $k_0 \in \mathbb{N}$ with $t_k \in \pi^n T$ for $k > k_0$. Then $S \subseteq \sum_{k=0}^{k_0} V t_k + \pi^n T$. Thus (iv) implies (ii). Conversely, assume (ii), and let S and T be as in (ii). We construct a null sequence (t_k) as in (iii) inductively, by choosing an increasing sequence $(k_n)_{n\in\mathbb{N}}$ and $t_{k_n+1}, \dots, t_{k_{n+1}} \in S \cap \pi^n T$ for all n, such that if $s \in S$ is written as $s \equiv \sum_{i=0}^{k_n} c_i t_i \bmod \pi^n T$, then there are c_i for $i = k_n + 1, \dots, k_{n+1}$ with $s \equiv \sum_{i=0}^{k_{n+1}} c_i t_i \bmod \pi^{n+1} T$. Since $S \subseteq T$, we may start the induction with $k_0 = -1$. Let $\mathsf{Im}_n(S)$ be the image of S in $T/\pi^n T$. In the induction step from n to $n+1$, we use that $\mathsf{Im}_{n+1}(S)$ is finitely generated. Then so is $\left(\sum_{i=0}^{k_n} V t_i + \mathsf{Im}_{n+1}(S)\right) \cap \pi^n T/\pi^{n+1} T$ because V is Noetherian. Let $t_{k_n+1}, \dots, t_{k_{n+1}} \in \pi^n T$ be representatives of generators for this submodule. Let $s \in S$. Then there are coefficients c_i for $0 \le i \le k_n$ with $s \equiv \sum_{i=0}^{k_n} c_i t_i \bmod \pi^n T$. Then $s - \sum_{i=0}^{k_n} c_i t_i \in \left(\sum_{i=0}^{k_n} V t_i + \mathsf{Im}_{n+1}(S)\right) \cap \pi^n T/\pi^{n+1} T$ may be written as $\sum_{i=k_n+1}^{k_{n+1}} c_i t_i$. Then $s \equiv \sum_{i=0}^{k_{n+1}} c_i t_i \bmod \pi^{n+1} T$. The construction ensures that the infinite series $\sum_{i=0}^{\infty} c_i t_i$ converges π-adically in T towards s. □

DEFINITION 3.6. Let M' denote M with the compactoid bornology.

PROPOSITION 3.7. *Let M be a torsionfree bornological V-module. Then the bornological V-module M' is nuclear.*

PROOF. Let $S \subseteq M'$ be bounded. That is, S is compactoid in the bornology of M. Proposition 3.5 provides a bounded V-submodule $T \subseteq M$ and a null sequence (x_n) in T such that

$$S = \left\{ s = \sum_{n=0}^{\infty} c_n x_n : (c_n) \in \ell^\infty(\mathbb{N}, V) \text{ and } s \text{ converges in } T \right\}.$$

Since (x_n) is a null sequence in T, there are $a_n \in \mathbb{N}$ and $y_n \in T$ with $x_n = \pi^{a_n} y_n$, $\lim a_n = \infty$ and $\lim y_n = 0$ in T. The subset

$$S' = \left\{ s = \sum_{n=0}^{\infty} c_n y_n : (c_n) \in \ell^\infty(\mathbb{N}, V), s \text{ converges in } T \right\}$$

is still compactoid in T by Proposition 3.5. We claim that S is compactoid as a subset of S'. For the proof, we show that any π-adically convergent series $S \ni s = \sum_{n=0}^{\infty} c_n x_n$ is π-adically convergent in S'. That s converges in T means that for each $e \geq 1$, there is an n_0 such that $s - \sum_{n=0}^{n_1} c_n x_n \in \pi^e T$ for all $n_1 \geq n_0$. Therefore, there are $\beta_{e,n_1} \in T$ with

$$\pi^e \beta_{e,n_1} = s - \sum_{n=0}^{n_1} c_n x_n = \sum_{n=n_1+1}^{\infty} c_n x_n = \sum_{n=n_1+1}^{\infty} c_n \pi^{a_n} y_n.$$

If n_1 is big enough, then $a_n \geq e$ for all $n > n_1$. Since M is torsionfree, we get

$$\beta_{e,n_1} = \sum_{n=n_1+1}^{\infty} c_n \pi^{a_n - e} y_n.$$

This series converges in T because β_{f,n_2} exist for $n_2 \gg n_1$. Then its limit β_{e,n_1} belongs to S'. Therefore, the series $\sum_{n=0}^{\infty} c_n x_n$ converges in S' as required. □

4. Inheritance properties of the compactoid bornology

Recall that M' denotes M with the compactoid bornology. We are going to prove that M' inherits many properties from M and that the property of being nuclear is preserved by several constructions with bornologies. We will use many of these results in our study of local cyclic homology. Let M be a bornological V-module.

LEMMA 4.1. *If the bornological V-module M is complete, then so is M'.*

PROOF. Let S be a compactoid subset of M. Then there is a bounded V-submodule $T \subseteq M$ such that for any $m \in \mathbb{N}$, there is a finite subset $F_m \subseteq T$ with $S \subseteq V F_m + \pi^m T$. Since M is complete, we may choose T to be a bounded π-adically complete V-submodule. Let $T' := \bigcap_{m \in \mathbb{N}} (V F_m + \pi^m T)$. Then $S \subseteq T'$ and T' is π-adically complete because it is π-adically closed in T. It is compactoid by construction. Thus M' is complete. □

LEMMA 4.2. *If the bornological V-module M is torsionfree, then so is M'.*

PROOF. Let $S \subseteq M$ be a compactoid subset. Let T and F_n be as in the definition of being compactoid and let $n \geq 1$. Since M is torsionfree, $\pi^{-1} T := \{x \in M : \pi x \in T\}$ is bounded. By construction,

$$\pi^{-1} S \subseteq \pi^{-1} V F_{n+1} + \pi^n T \subseteq \pi^{-1} V F_{n+1} + \pi^n \pi^{-1} T. \tag{1}$$

Multiplication by π on M is injective because M is torsionfree. It maps $\pi^{-1} V F_{n+1}$ onto a submodule of $V F_{n+1}$. Since V is Noetherian, it follows that $\pi^{-1} V F_{n+1}$ is again finitely generated. Therefore, (1) witnesses that $\pi^{-1} S$ is compactoid. □

LEMMA 4.3. *On torsionfree bornological F-modules, the assignment $M \mapsto M'$ is the right adjoint functor of the inclusion of the subcategory of nuclear, torsionfree bornological V-modules.*

PROOF. We assume N to be torsionfree to ensure that N' is nuclear by Proposition 3.7. Lemma 4.2 shows that N' is again torsionfree. Let M and N be torsionfree bornological V-modules. Assume that the bornology on M is nuclear. We claim that a V-linear map $\varphi: M \to N$ is bounded if and only if it is bounded as a map to N'. One implication is trivial because compactoid subsets are bounded. Let $S \subseteq M$ be a bounded subset. By assumption, S is compactoid. That is, there is a bounded V-submodule $T \subseteq M$ containing S, such that for each n, there is a finite subset $F_n \subseteq T$ with $S \subseteq VF_n + \pi^n T$. The inclusions $\varphi(S) \subseteq V\varphi(F_n) + \pi^n \varphi(T)$ witness that $\varphi(S) \subseteq N$ is compactoid. That is, φ is bounded as a map $M \to N'$. □

LEMMA 4.4. *If M is a nuclear bornological V-module, then so is $\overline{M}$.*

PROOF. Let $S \subseteq \overline{M}$ be a bounded subset. The inductive limit description of $\overline{M}$ shows that S is contained in the image of the π-adic completion $\widehat{U}$ of some bounded submodule $U \subseteq M$. Since M is compactoid, there is a bounded submodule $U' \subseteq M$ such that the image of U in $U'/\pi^k U'$ is finitely generated for each $k \in \mathbb{N}$. This is equal to the image of $\widehat{U}$ in $\widehat{U'}/\pi^k \widehat{U'} \cong U'/\pi^k U'$. This witnesses that S is compactoid. □

We now investigate the behaviour of the compactoid bornology under tensor products and extensions.

PROPOSITION 4.5. *Let M and N be complete, torsionfree bornological V-modules. Then there is an isomorphism of complete bornological V-modules*

$$M' \overline{\otimes} N' \cong (M \overline{\otimes} N)'.$$

If M and N are nuclear, then so is $M \overline{\otimes} N$.

PROOF. Lemma 4.1 shows that M' and N' are complete. A submodule in $M' \overline{\otimes} N'$ is bounded if it is in the image of $S \widehat{\otimes} T$, where S and T are π-adically complete, compactoid submodules of M and N. By definition, this means that there are bounded, π-adically complete V-submodules S' and T' of M and N such that the images of S in $S' \twoheadrightarrow S'/\pi^n S'$ and of T in $T' \twoheadrightarrow T'/\pi^n T'$ are finitely generated for each n. Then the image of $S \widehat{\otimes} T$ in $S' \widehat{\otimes} T'/\pi^n(S' \widehat{\otimes} T')$ is finitely generated for each n. This says that $S \widehat{\otimes} T$ is compactoid in $M \otimes N$.

Conversely, let $X \subseteq M \overline{\otimes} N$ be a compactoid subset of $M \overline{\otimes} N$. Then there is a bounded π-adically complete submodule $U \subseteq M \overline{\otimes} N$, such that the image of X under the quotient map to $U/\pi^n U$ is finitely generated for each n. There are bounded π-adically complete submodules $S' \subseteq M$ and $T' \subseteq N$ such that U is contained in the image of $S' \widehat{\otimes} T'$ in $M \overline{\otimes} N$. We have $S' \widehat{\otimes} T'/\pi^n(S' \widehat{\otimes} T') \cong (S' \otimes T')/\pi^n(S' \otimes T')$, and the image of X remains finitely generated in $(S' \otimes T')/\pi^n(S' \otimes T')$ for all $n \in \mathbb{N}$. By Proposition 3.5, there is a null sequence $(x_k)_{k \in \mathbb{N}}$ in $S' \widehat{\otimes} T'$ such that X is contained in the image of $\sum_{k=0}^{\infty} V x_k + \pi^n(S' \widehat{\otimes} T')$ in $M \overline{\otimes} N$. Since $\lim x_k = 0$, there are $l_k \to \infty$ with $x_k \in \pi^{2l_k} S' \widehat{\otimes} T'$. Since $S' \widehat{\otimes} T'/\pi^n(S' \widehat{\otimes} T') \cong (S' \otimes T')/\pi^n(S' \otimes T')$, the proof of Proposition 3.5 shows that we may arrange $x_k \in \pi^{2l_k} S' \otimes T'$. Then $x_k = \pi^{2l_k} \sum_{i=1}^{m_i} a_i \otimes b_i$ for some $m_i \in \mathbb{N}$, $a_i \in S'$ and $b_i \in T'$ for all i. Let S and T be the π-adically closed V-submodules in S' and T' generated by the bounded subsets

$\{\pi^{l_k} a_i\}$ and $\{\pi^{l_k} b_i\}$, respectively. By Proposition 3.5, S and T are compactoid submodules of S' and T', and $X \subseteq S \hat{\otimes} T$. That is, X is bounded in $M' \overline{\otimes} N'$.

By definition, a bornological V-module M is nuclear if and only if the canonical map $M' \to M$ is a bornological isomorphism. If $M' = M$ and $N' = N$, then $(M \overline{\otimes} N)' = M' \overline{\otimes} N' = M \overline{\otimes} N$ as well. □

PROPOSITION 4.6. *Let $K \rightarrowtail L \twoheadrightarrow P$ be an extension of complete, torsionfree bornological V-modules and let P be algebraically torsionfree. Then $K' \rightarrowtail L' \twoheadrightarrow P'$ is an extension of complete bornological V-modules as well. In addition, L is nuclear if and only if both K and P are nuclear.*

PROOF. The functoriality of the compactoid bornology (Lemma 4.3) shows that a subset $S \subseteq K$ that is compactoid in K remains compactoid in L. Conversely, we claim that if $S \subseteq K$ is compactoid as a subset of L, it is compactoid as a subset of K. Let $T \supseteq S$ be a bounded subset of L such that the image of S in $T/\pi^n T$ is finitely generated for all $n \in \mathbb{N}$. Since $S \subseteq K$, it follows that $S \subseteq T \cap K$. Since P is torsionfree, $\pi^n x \in K$ for $x \in L$ can only happen if $x \in K$. This says that the canonical map $(T \cap K)/\pi^n(T \cap K) \to T/\pi^n T$ is injective. Therefore, the image of S in $(T \cap K)/\pi^n(T \cap K)$ is finitely generated for all $n \in \mathbb{N}$. This finishes the proof of the claim that the inclusion $K' \to L'$ is a bornological embedding.

The quotient map clearly remains bounded as a map $q: L' \to P'$. It remains to prove that it is a bornological quotient map, that is, any compactoid subset of P is the image of a compactoid subset of L.

Let $S \subseteq P$ be compactoid. Since P is complete, Proposition 3.5 implies that there is a bounded, π-adically complete V-submodule $T \subseteq P$ containing S and a null sequence (x_n) in T such that $S \subseteq \{s = \sum_{n\in\mathbb{N}} c_n x_n : (c_n) \in \ell^\infty(\mathbb{N}, V)\}$; these sequences converge automatically because $\lim x_n = 0$ and T is complete. Since L is complete, there is a bounded, π-adically complete V-submodule $W \subseteq L$ that lifts T. We may lift (x_n) to a null sequence $(\tilde{x}_n)$ in W. Then $\tilde{S} = \{\sum_{n=0}^\infty c_n \tilde{x}_n : (c_n) \in \ell^\infty(\mathbb{N}, V)\}$ is a compactoid submodule of L that lifts S.

By definition, a bornological V-module M is nuclear if and only if the canonical map $M' \to M$ is a bornological isomorphism. The category of bornological V-modules is quasi-Abelian. This implies a version of the Five Lemma. For our two extensions $K' \rightarrowtail L' \twoheadrightarrow P'$ and $K \rightarrowtail L \twoheadrightarrow P$, it says that the map $L' \to L$ is a bornological isomorphism if the maps $K' \to K$ and $P' \to P$ are bornological isomorphisms. Conversely, it is easy to see that subspaces and quotients inherit nuclearity. □

LEMMA 4.7. *Let D be a bornological algebra and let D' be D with the compactoid bornology. This is again a bornological algebra. If D is semidagger, then so is D'.*

PROOF. It is easy to see that the product of compactoid subsets is again compactoid, so that D' is a bornological algebra. Now assume D to be semidagger. We are going to prove that D' is semidagger as well. Let S be a compactoid V-submodule of D. We must prove that $S_1 := \sum_{n=0}^\infty \pi^n S^{n+1}$ is compactoid as well. Since S is compactoid, there are a bounded subset T and a null sequence $(x_n)_{n\in\mathbb{N}}$ in T such that $S \subseteq \sum_{n\in\mathbb{N}} V x_n + \pi^k T$ for each $k \in \mathbb{N}$. Since D is semidagger, the subset $T_{1/2} := \sum_{i=0}^\infty \pi^{\lceil i/2 \rceil} T^{i+1} \supseteq T$ is still bounded (see [**1**, Lemma 3.1.10]). For $n \in \mathbb{N}$,

$i = (i_0, \dots, i_n) \in \mathbb{N}^{n+1}$, let $x_i := x_{i_0} \cdots x_{i_n} \in T^n$. Then

$$S_1 = \sum_{n=0}^{\infty} \pi^n S^{n+1} \subseteq \sum_{n=0}^{\infty} \sum_{i \in \mathbb{N}^{n+1}} V \pi^n x_i + \pi^k \pi^n T^{n+1} \subseteq \left(\sum_{n=0}^{\infty} \sum_{i \in \mathbb{N}^{n+1}} V \pi^n x_i \right) + \pi^k T_{1/2}.$$

We may arrange the countable set of elements $(\pi^n x_i)$ for $n \in \mathbb{N}$, $i \in \mathbb{N}^{n+1}$ into a sequence that converges to zero in $T_{1/2}$. Therefore, the inclusion above witnesses that S_1 is compactoid. □

LEMMA 4.8. *If D is a dagger algebra, then so is D'.*

PROOF. This follows from Lemmas 4.1, 4.2 and 4.7. □

PROPOSITION 4.9. *Let R be a nuclear bornological V-algebra. Then R_{lg} and $R^\dagger = \overline{R_{\mathrm{lg}}}$ are nuclear.*

PROOF. By Lemma 4.4, we only need to prove that R_{lg} is nuclear. Let S be a subset of linear growth in R. Then there is a bounded submodule $T \subseteq R$ with $S \subseteq T_1 := \sum_{j=0}^{\infty} \pi^j T^{j+1}$. Since R is compactoid, Proposition 3.5 gives a bounded submodule $\tilde{T}$ of R and a null sequence $(x_n) \in \tilde{T}$ such that $T \subseteq \sum_{n \in \mathbb{N}} V x_n + \pi^k \tilde{T}$ for each $k \geq 1$. Now we consider the countable set of all products $\pi^j x_{i_1} \cdots x_{i_{j+1}}$ for $j, i_1, \dots, i_{j+1} \in \mathbb{N}$. For each k, there are only finitely many such terms that do not belong to $\pi^k \tilde{T}$. Therefore, we may order these products into a sequence $(y_m)_{m \in \mathbb{N}}$ that converges to 0 in the π-adic topology on $\tilde{T}$. Now it follows from the construction that $T_1 \subseteq \sum_{m \in \mathbb{N}} V y_m + \pi^k \tilde{T}$ for each $k \in \mathbb{N}$. Thus T_1 is compactoid. Hence so is S. □

COROLLARY 4.10. *Let R be a torsionfree bornological algebra with the fine bornology. Then $R^\dagger = \overline{R_{\mathrm{lg}}}$ is nuclear.*

PROOF. The fine bornology on R is nuclear by Example 3.4. Then apply Proposition 4.9. □

Let R be any torsionfree bornological V-algebra. There is a canonical bounded algebra homomorphism $R \to R^\dagger$. Since the compactoid bornology is functorial, it induces a bounded algebra homomorphism $R' \to (R^\dagger)'$. The target is a dagger algebra by Lemma 4.8. So this homomorphism extends uniquely to a bounded homomorphism

$$(R')^\dagger \to (R^\dagger)'.$$

In general, the homomorphism $(R')^\dagger \to (R^\dagger)'$ need not be invertible. The following proposition describes the two situations where we know this:

PROPOSITION 4.11. *Let R be a torsionfree bornological algebra. If R is nuclear or dagger, then the canonical map $(R')^\dagger \to (R^\dagger)'$ is an isomorphism.*

PROOF. Suppose first that R is nuclear. Then $R \cong R'$, so that $(R')^\dagger = R^\dagger$. The canonical map $(R^\dagger)' \to R^\dagger$ generates the required inverse map $(R^\dagger)' \to (R')^\dagger$. Next assume that R is a dagger algebra. Then $R \cong R^\dagger$ and so $(R^\dagger)' \cong R'$. The canonical map $R' \to (R')^\dagger$ provides the required inverse map $(R^\dagger)' \to (R')^\dagger$. □

5. Definition of local cyclic homology

Our definition of local cyclic homology is based on the *analytic* cyclic homology theories introduced in [**2**, **8**]. We do not recall how they are defined. Suffice it to say the following. Let $\overleftarrow{\mathsf{Ind}}(\mathsf{Ban}_F)$ be the quasi-Abelian category of countable projective systems of inductive systems of Banach F-vector spaces. There are functors

$$\mathbb{HA}\colon\{\text{complete, torsionfree bornological } V\text{-algebras}\} \to \mathsf{Der}\Big(\overleftarrow{\mathsf{Ind}}(\mathsf{Ban}_F)\Big),$$

$$\mathbb{HA}\colon\{\mathbb{F}\text{-algebras}\} \to \mathsf{Der}\Big(\overleftarrow{\mathsf{Ind}}(\mathsf{Ban}_F)\Big).$$

Their definitions are based on the Cuntz–Quillen approach to cyclic homology. The first functor is invariant under dagger homotopies and Morita equivalences and satisfies excision for extensions with a bounded V-linear section. The second functor is invariant under polynomial homotopies, stable for matrices and satisfies excision for all extensions of $\mathbb{F}$-algebras. To be more precise, the first functor above is the composite of the functor defined in [**2**] with the dissection functor from the category of complete bornological F-vector spaces to the category of inductive systems of Banach F-vector spaces. This composite is already used in [**8**]. The main result in [**8**] is the following:

THEOREM 5.1 ([**8**, Theorem 5.9]). *Let A be an $\mathbb{F}$-algebra. Let $D = (D_n)_{n\in\mathbb{N}}$ be a projective system of dagger algebras that is fine mod π. Let $\varrho\colon D \to A$ be a pro-homomorphism, represented by a coherent family of surjective homomorphisms $\varrho_n\colon D_n \to A$ for $n \in \mathbb{N}$. Assume that* $\ker \varrho = (\ker \varrho_n)_{n\in\mathbb{N}}$ *is analytically nilpotent. Then there is a canonical quasi-isomorphism $\mathbb{HA}(A) \cong \mathbb{HA}(D)$.*

In particular, if D is a dagger algebra and fine mod π, then $\mathbb{HA}(D)\cong\mathbb{HA}(D/\pi D)$. This result will be the basis for our study of local cyclic homology. Here it is crucial to restrict to dagger algebras that are fine mod π. We do not know how to compute $\mathbb{HA}(D)$ when D is a Banach V-algebra as in Example 2.5.

DEFINITION 5.2. Let D be a dagger algebra and let D' be the algebra D with the compactoid bornology. The *local cyclic homology complex* of D is the chain complex $\mathbb{HL}(D) := \mathbb{HA}(D')$, viewed as an object in the derived category of $\overleftarrow{\mathsf{Ind}}(\mathsf{Ban}_F)$.

LEMMA 5.3. $\mathbb{HL}$ *is a functor on the category of dagger algebras.*

PROOF. The compactoid bornology is functorial by Lemma 4.3, and so is $\mathbb{HA}$. □

THEOREM 5.4. *Let D_1 and D_2 be dagger algebras. If $D_1/\pi D_1 \cong D_2/\pi D_2$, then $\mathbb{HL}(D_1) \cong \mathbb{HL}(D_2)$.*

PROOF. By Lemma 4.8, the algebras D_1' and D_2' are again dagger algebras. They are fine mod π by construction of the compactoid bornology, and $A := D_1'/\pi D_1' \cong D_2'/\pi D_2'$. Theorem 5.1 implies

$$\mathbb{HL}(D_1) \cong \mathbb{HA}(D_1') \cong \mathbb{HA}(A) \cong \mathbb{HA}(D_2') \cong \mathbb{HL}(D_2). \qquad \square$$

Theorem 5.4 explains our choice of the compactoid bornology in the definition of local cyclic homology. Its crucial features are that it is fine mod π and that it still gives a dagger algebra. This ensures that $\mathbb{HL}(D)$ for a dagger algebra D depends only on $D/\pi D$. There are, however, other bornologies that may work just as well. For instance, we could give a dagger algebra the linear growth bornology generated by the fine bornology. Some work would, however, be required to check whether this bornology is complete. Another advantage of the compactoid bornology is that it does not use the multiplication and that it is analogous to the precompact bornology in the archimedean case, which is used in that setting to define local cyclic homology (see [**6**]).

COROLLARY 5.5. *Let R be a torsionfree V-algebra and let $\widehat{R}$ be its π-adic completion. Give R the fine bornology and equip $\widehat{R}$ with the bornology where all subsets are bounded. There are natural quasi-isomorphisms*

$$\mathbb{HA}(R^\dagger) \cong \mathbb{HL}(R^\dagger) \cong \mathbb{HL}(\widehat{R}).$$

PROOF. The algebra $\widehat{R}$ is a Banach V-algebra as in Example 2.5, and $R^\dagger$ is a dagger algebra by construction. It is nuclear by Corollary 4.10. So $\mathbb{HA}(R^\dagger) = \mathbb{HL}(R^\dagger)$. We compute

$$\widehat{R}/\pi\widehat{R} \cong R/\pi R = R_{\mathrm{lg}}/\pi R_{\mathrm{lg}} \cong \overline{R_{\mathrm{lg}}}/\pi \overline{R_{\mathrm{lg}}} = R^\dagger/\pi R^\dagger;$$

the isomorphism $R_{\mathrm{lg}}/\pi R_{\mathrm{lg}} \cong \overline{R_{\mathrm{lg}}}/\pi \overline{R_{\mathrm{lg}}}$ follows because $R_{\mathrm{lg}}/\pi R_{\mathrm{lg}}$ as an $\mathbb{F}$-algebra is bornologically complete as a V-algebra (compare [**2**, Proposition 2.3.3]). Now Theorem 5.4 implies $\mathbb{HL}(R^\dagger) \cong \mathbb{HL}(\widehat{R})$. □

6. Homotopy invariance, excision and matricial stability

We are going to prove that local cyclic homology satisfies some nice formal properties, namely, homotopy invariance, matricial stability, excision, and an analogue of Bass' Fundamental Theorem. These will be used in Section 7 to compute the local cyclic homology of certain Banach V-algebras.

THEOREM 6.1 (Homotopy invariance). *Let D be a dagger algebra. Give $\widehat{V[t]}$ the bornology where all subsets are bounded. There is a quasi-isomorphism*

$$\mathbb{HL}(D) \cong \mathbb{HL}\big(D \mathbin{\overline{\otimes}} \widehat{V[t]}\big).$$

PROOF. The bornological algebras D' and $(D \mathbin{\overline{\otimes}} \widehat{V[t]})'$ are both dagger by Proposition 4.8 and fine mod π. Then Theorem 5.1 and [**8**, Theorem 8.1] imply

$$\mathbb{HL}(D) \cong \mathbb{HA}(D/\pi D) \cong \mathbb{HA}(D/\pi D \otimes \mathbb{F}[t]) \cong \mathbb{HA}\big((D \mathbin{\overline{\otimes}} \widehat{V[t]})'\big) = \mathbb{HL}\big(D \mathbin{\overline{\otimes}} \widehat{V[t]}\big). \square$$

An *elementary homotopy* between two bounded homomorphisms $f, g\colon D_1 \rightrightarrows D_2$ is a bounded homomorphism $F\colon D_1 \to D_2 \mathbin{\overline{\otimes}} \widehat{V[t]}$ with $\mathrm{ev}_0 \circ F = f$ and $\mathrm{ev}_1 \circ F = g$. Let *homotopy* be the equivalence relation generated by elementary homotopy. If $f, g\colon D_1 \rightrightarrows D_2$ are homotopic, then $\mathbb{HL}(f) = \mathbb{HL}(g)$ by Theorem 6.1.

COROLLARY 6.2. *For any Banach V-algebra D, there is a natural quasi-isomorphism $\mathbb{HL}(D) \cong \mathbb{HL}\big(\widehat{D[t]}\big)$. Here $\widehat{D[t]}$ is the Banach V-algebra of all polynomials $\sum_{n=0}^{\infty} d_n t^n$ with $d_n \in D$ for all $n \in \mathbb{N}$ and $\lim d_n = 0$ π-adically in D.*

THEOREM 6.3 (Matrix stability). *Let D be a dagger algebra. For a set Λ, let $M_\Lambda(V)$ be the V-algebra of finitely supported matrices indexed by $\Lambda \times \Lambda$. Let $\lambda \in \Lambda$. The canonical map $i_\lambda\colon D \to D \mathbin{\overline{\otimes}} \widehat{M_\Lambda(V)}$, $x \mapsto e_{\lambda,\lambda} \otimes x$, induces a quasi-isomorphism*

$$\mathbb{HL}(D) \cong \mathbb{HL}\big(D \mathbin{\overline{\otimes}} \widehat{M_\Lambda(V)}\big).$$

Proof. Let $D/\pi D = A$. We compute

$$\mathbb{HL}(D) \cong \mathbb{HA}(A) \cong \mathbb{HA}(M_\Lambda(A)) \cong \mathbb{HA}\big((D \overline{\otimes} \widehat{M_\Lambda(V)})'\big) = \mathbb{HL}\big(D \overline{\otimes} \widehat{M_\Lambda(V)}\big);$$

here the second quasi-isomorphism uses [**8**, Proposition 8.2], and the third quasi-isomorphism uses Theorem 5.1 and $D \overline{\otimes} \widehat{M_\Lambda(V)}/\pi \cdot \big(D \overline{\otimes} \widehat{M_\Lambda(V)}\big) \cong M_\Lambda(A)$. □

Corollary 6.4. *Let D be a Banach V-algebra. There is quasi-isomorphism $\mathbb{HL}(D) \cong \mathbb{HL}\big(\widehat{M_\Lambda(D)}\big)$. Here $\widehat{M_\Lambda(D)}$ is the Banach V-algebra of all matrices $(d_{\lambda,\mu})_{\lambda,\mu\in\Lambda}$ with $d_{\lambda,\mu} \in D$ for all $\lambda, \mu \in \Lambda$ and such that, for each $k \in \mathbb{N}$, there are only finitely many $\lambda, \mu \in \Lambda$ with $d_{\lambda,\mu} \notin \pi^k D$.*

Next, we prove that local cyclic homology satisfies excision for *any* extension of dagger algebras; no bounded linear section is needed.

Theorem 6.5 (Excision Theorem). *An extension of dagger algebras $K \overset{i}{\rightarrowtail} E \overset{p}{\twoheadrightarrow} Q$ induces an exact triangle*

$$\mathbb{HL}(K) \overset{i_*}{\to} \mathbb{HL}(E) \overset{p_*}{\to} \mathbb{HL}(Q) \overset{\delta}{\to} \mathbb{HL}(K)[-1]$$

in the derived category of the quasi-abelian category $\overleftarrow{\mathsf{Ind}(\mathsf{Ban}_F)}$.

Proof. The given extension induces an extension of $\mathbb{F}$-algebras

$$K \otimes_V \mathbb{F} \rightarrowtail E \otimes_V \mathbb{F} \twoheadrightarrow Q \otimes_V \mathbb{F}$$

because Q is torsionfree. This induces a natural exact triangle

$$\mathbb{HA}(K \otimes_V \mathbb{F}) \xrightarrow{i_*} \mathbb{HA}(E \otimes_V \mathbb{F}) \xrightarrow{p_*} \mathbb{HA}(Q \otimes_V \mathbb{F}) \overset{\delta}{\to} \mathbb{HA}(K \otimes_V \mathbb{F})[-1]$$

in the derived category by [**8**, Theorem 8.3]. By Theorem 5.4, this exact triangle is isomorphic to an exact triangle as in the statement of the theorem. □

An exact triangle in a derived category implies a long exact sequence in homology (compare [**2**, Theorem 5.1]).

Theorem 6.6 (Bass Fundamental Theorem). *Let D be a dagger algebra. Then $\mathbb{HL}\big(D \overline{\otimes} \widehat{V[t,t^{-1}]}\big) \cong \mathbb{HL}(D) \oplus \mathbb{HL}(D)[1]$; here $[1]$ means a degree shift.*

Proof. Let $A = D/\pi D$. Then we compute

$$\begin{aligned}\mathbb{HL}\big(D \overline{\otimes} \widehat{V[t,t^{-1}]}\big) &\cong \mathbb{HA}\big((D \overline{\otimes} \widehat{V[t,t^{-1}]})'\big) \cong \mathbb{HA}(A \otimes \mathbb{F}[t,t^{-1}]) \\ &\cong \mathbb{HA}(A) \oplus \mathbb{HA}(A)[1] \cong \mathbb{HA}(D') \oplus \mathbb{HA}(D')[1] \cong \mathbb{HL}(D) \oplus \mathbb{HL}(D)[1],\end{aligned}$$

where $\mathbb{HA}(A \otimes \mathbb{F}[t,t^{-1}]) \cong \mathbb{HA}(A) \oplus \mathbb{HA}(A)[1]$ follows from [**8**, Corollary 8.5]. □

7. Some computations of local cyclic homology

In this section, we compute local cyclic homology for some Banach V-algebras.

Example 7.1. Let E be a directed graph. Let $L(\mathbb{F}, E)$ and $C(\mathbb{F}, E)$ be its Leavitt and Cohn path algebras over $\mathbb{F}$, respectively. Their lifts $L(V, E)$ and $C(V, E)$ are torsionfree V-algebras, which we equip with the fine bornology. Their π-adic completions $\widehat{L(V,E)}$ and $\widehat{C(V,E)}$ are Banach V-algebras. The analytic cyclic homology of the dagger completions $L(V,E)^\dagger$ and $C(V,E)^\dagger$ is computed in [**2**, Theorem 8.1]. With the compactoid bornology, these dagger completions are still dagger algebras by Lemma 4.8, and still fine mod π. Corollary 4.10 shows directly that $L(V,E)^\dagger$ is nuclear.

Theorem 5.4 implies that the local cyclic homology of $\widehat{L(V,E)}$ and $\widehat{C(V,E)}$ is naturally isomorphic to the analytic cyclic homology computed in [**2**], namely,

$$\mathbb{HL}\big(\widehat{L(V,E)}\big) \cong \mathbb{HA}(L(\mathbb{F},E)) \cong \operatorname{coker}(N_E) \oplus \ker(N_E)[1],$$
$$\mathbb{HL}\big(\widehat{C(V,E)}\big) \cong \mathbb{HA}(C(\mathbb{F},E)) \cong F^{(E^0)}.$$

EXAMPLE 7.2. Consider the V-algebra of Laurent polynomials in n variables $L_n(V) = V[t_1,t_1^{-1},\dots,t_n,t_n^{-1}]$. We may write this as $L_{n-1}(V)\otimes V[t_n,t_n^{-1}]$. Taking π-adic completions gives

$$\widehat{L_n(V)} \cong L_{n-1}(V)\mathbin{\widehat{\otimes}} \widehat{V[t,t^{-1}]}.$$

Then Theorem 6.6 implies

$$\mathbb{HL}\Big(\widehat{L_n(V)}\Big) \cong \mathbb{HL}\Big(\widehat{L_{n-1}(V)}\Big) \oplus \mathbb{HL}\Big(\widehat{L_{n-1}(V)}\Big)[1] \cong \mathbb{HL}\Big(\widehat{L_{n-1}(V)}\Big)\otimes (F\oplus F[1]).$$

Iterating this and using $\mathbb{HL}(\widehat{L_0(\mathbb{F})}) = \mathbb{HA}(V) = F$, we identify $\mathbb{HL}(\widehat{L_n(V)})$ with the F-vector space $\Lambda^*(F^n)$, the exterior algebra, with the usual $\mathbb{Z}/2$-grading and the zero boundary map. This is also isomorphic to Berthelot's rigid cohomology of $\mathbb{F}[t,t^{-1}]$. A closer inspection shows that the canonical chain maps

$$\mathbb{HL}(\widehat{L_n(V)}) \leftarrow \mathbb{HL}(L_n(V)^\dagger) \to \mathbb{HP}(L_n(V)^\dagger \otimes F)$$

are quasi-isomorphisms; here $\mathbb{HP}$ denotes the chain complex that computes periodic cyclic homology (see [**1**] for its definition in the current setting).

EXAMPLE 7.3. Let X be a smooth affine variety over $\mathbb{F}$ and let $A = \mathcal{O}(X)$ be its coordinate ring. Elkik [**4**] shows that there is a smooth V-algebra R with $R/\pi R \cong A$. Its π-adic completion $\widehat{R}$ still satisfies $\widehat{R}/\pi\widehat{R} \cong A$. We call $\widehat{R}$ the *Tate algebra of* X. This is a Banach V-algebra. Theorem 5.1 implies

$$\mathbb{HL}(\widehat{R}) := \mathbb{HA}(\widehat{R}') \cong \mathbb{HA}(A).$$

In the situation above, we also know that $\mathbb{HL}(\widehat{R}) = \mathbb{HL}(R^\dagger) = \mathbb{HA}(R^\dagger) \cong \mathbb{HA}(A)$. If X has dimension 1, then $\mathrm{HA}_*(A)$ is the rigid cohomology of X with coefficients in F by [**8**, Corollary 5.6].

THEOREM 7.4. *Let B be a Banach V-algebra and let R be a V-subalgebra of B such that $R/\pi R \cong B/\pi B$. Equip R with the fine bornology. Then*

$$\mathbb{HA}(R^\dagger) = \mathbb{HL}(R^\dagger) \cong \mathbb{HL}(B).$$

In addition, there is a canonical chain map $\mathbb{HA}(R^\dagger) \to \mathbb{HP}(R^\dagger \otimes F)$.

PROOF. By Lemma 4.8, both B' and $(R^\dagger)'$ are dagger algebras. Since they are also fine mod π, Theorem 5.4 gives a quasi-isomorphism $\mathbb{HL}(R^\dagger) \cong \mathbb{HL}(B)$. The periodic cyclic homology complex $\mathbb{HP}(R^\dagger \otimes F)$ for bornological F-algebras like $R^\dagger \otimes F$ is defined in [**1**]. There is a canonical chain map $\mathbb{HA}(R^\dagger) \to \mathbb{HP}(R^\dagger\otimes F)$ because the chain complex $\mathbb{HA}(R^\dagger)$ is defined as a subcomplex of $\mathbb{HP}(R^\dagger \otimes F)$. □

Let us return to the situation of Example 7.3. The periodic cyclic homology $\mathrm{HP}_*(R^\dagger \otimes F)$ is identified in [**1**] with the rigid cohomology of the underlying affine variety over $\mathbb{F}$. The computation of $\mathbb{HA}(A)$ in [**8**, Corollary 5.6] shows that the canonical chain map $\mathbb{HA}(A^\dagger) \to \mathbb{HP}(A^\dagger \otimes F)$ in Theorem 7.4 is a quasi-isomorphism for 1-dimensional varieties. Example 7.2 shows that the same happens for some varieties of higher dimension. It is unclear, however, how to generalise these results to other smooth commutative $\mathbb{F}$-algebras.

One could hope that the canonical chain map $\mathbb{HA}(A^\dagger) \to \mathbb{HP}(A^\dagger \otimes F)$ is an isomorphism when the algebra $A^\dagger$ is "smooth" in the sense that $A^\dagger$ has a resolution by projective $A^\dagger$-bimodules of finite length. The counterexample in the following proposition shows that this cannot work without extra assumptions.

PROPOSITION 7.5. *Let $D := V[t]$ be the coordinate ring of the affine plane. Then $\widehat{D}'$, the π-adic completion of D equipped with the compactoid bornology, is quasi-free, but* $\mathrm{HA}_*(\widehat{D}') \ncong \mathrm{HP}_*(\widehat{D}' \otimes F)$.

PROOF. Define $\Omega^1(D)$ as in [**3**] and let $\Omega^1(\widehat{D}') = \ker\big(\widehat{D}' \overline{\otimes} \widehat{D}' \to \widehat{D}'\big)$. It is easy to see that the map

$$D \otimes D \to \Omega^1(D), \qquad f_1 \otimes f_2 \mapsto f_1 \cdot \mathrm{d}t \cdot f_2$$

is an isomorphism of D-bimodules. Since D is torsionfree, the π-adic completions form an extension $\widehat{\Omega^1(D)} \rightarrowtail \widehat{D \otimes D} \twoheadrightarrow \widehat{D}$. Therefore,

$$\Omega^1\big(\widehat{D}\big) \cong \widehat{\Omega^1(D)} \cong \widehat{D \otimes D} \cong \widehat{D} \mathbin{\widehat{\otimes}} \widehat{D}.$$

Then $\widehat{\Omega^1(D)}' \rightarrowtail \widehat{D \otimes D}' \twoheadrightarrow \widehat{D}'$ is an extension by Proposition 4.6. Proposition 4.5 shows that $\widehat{D \otimes D}' \cong \widehat{D}' \overline{\otimes} \widehat{D}'$. This implies an isomorphism

$$\Omega^1(\widehat{D}') \cong \widehat{D}' \overline{\otimes} \widehat{D}'$$

of $\widehat{D}'$-bimodules. So $\Omega^1(\widehat{D}')$ is a projective $\widehat{D}'$-bimodule. Then $\widehat{D}'$ is quasi-free.

Now let $\underline{\widehat{D}} := \widehat{D}' \otimes F$. This is again quasi-free. Since F has characteristic zero, $\mathbb{HP}(\underline{\widehat{D}}) \cong X(\underline{\widehat{D}})$. The homology of this chain complex is the algebraic de Rham cohomology of $\underline{\widehat{D}}$. It is well known that this differs from the Monsky–Washnitzer or rigid cohomology of the plane (see [**9**]). At the same time,

$$\mathbb{HA}\big(\widehat{D}'\big) = \mathbb{HL}\big(\widehat{D}\big) \cong \mathbb{HL}(D^\dagger) = \mathbb{HA}(D^\dagger) \cong \mathbb{HA}(\mathbb{F}[t])$$

by Theorem 5.4. This agrees with the rigid cohomology of the affine plane. As a result, $\mathrm{HA}_*(\widehat{D}') \ncong \mathrm{HP}_*(\widehat{D}' \otimes F)$. □

Acknowledgments

This article got started by a remark by Joachim Cuntz, and we thank him for several useful discussions.

References

[1] Guillermo Cortiñas, Joachim Cuntz, Ralf Meyer, and Georg Tamme, *Nonarchimedean bornologies, cyclic homology and rigid cohomology*, Doc. Math. **23** (2018), 1197–1245. MR **3874948**

[2] Guillermo Cortiñas, Ralf Meyer, and Devarshi Mukherjee, *Nonarchimedean analytic cyclic homology*, Doc. Math. **25** (2020), 1353–1419. MR **4164727**

[3] Joachim Cuntz and Daniel Quillen, *Algebra extensions and nonsingularity*, J. Amer. Math. Soc. **8** (1995), no. 2, 251–289, doi: 10.2307/2152819. MR **1303029**

[4] Renée Elkik, *Solutions d'équations à coefficients dans un anneau hensélien* (French), Ann. Sci. École Norm. Sup. (4) **6** (1973), 553–603 (1974). MR **345966**

[5] Masoud Khalkhali, *Algebraic connections, universal bimodules and entire cyclic cohomology*, Comm. Math. Phys. **161** (1994), no. 3, 433–446. MR **1269386**

[6] Ralf Meyer, *Local and analytic cyclic homology*, EMS Tracts in Mathematics, vol. 3, European Mathematical Society (EMS), Zürich, 2007. doi: 10.4171/039 MR **2337277**
[7] Ralf Meyer and Devarshi Mukherjee, *Dagger completions and bornological torsion-freeness*, Q. J. Math. **70** (2019), no. 3, 1135–1156, doi: 10.1093/qmath/haz012. MR **4009486**
[8] Ralf Meyer and Devarshi Mukherjee, *Analytic cyclic homology in positive characteristic* (2021), eprint. arXiv:2109.01470.
[9] P. Monsky and G. Washnitzer, *Formal cohomology. I*, Ann. of Math. (2) **88** (1968), 181–217, doi: 10.2307/1970571. MR **248141**
[10] Michael Puschnigg, *Diffeotopy functors of ind-algebras and local cyclic cohomology*, Doc. Math. **8** (2003), 143–245. MR **2029166**
[11] Peter Schneider, *Nonarchimedean functional analysis*, Springer Monographs in Mathematics, Springer-Verlag, Berlin, 2002. doi: 10.1007/978-3-662-04728-6 MR **1869547**

Mathematisches Institut, Universität Göttingen, Bunsenstrasse 3–5, 37073 Göttingen, Germany
Email address: rmeyer2@uni-goettingen.de

Mathematisches Institut, Universität Göttingen, Bunsenstrasse 3–5, 37073 Göttingen, Germany
Email address: devarshi.mukherjee@mathematik.uni-goettingen.de

Proceedings of Symposia in Pure Mathematics
Volume **105**, 2023
https://doi.org/10.1090/pspum/105/01907

On the van Est analogy in Hopf cyclic cohomology

Henri Moscovici

Abstract. We present results relating the Hopf cyclic cohomology of the Hopf algebra $\mathcal{H}_n$ of moving frames with that of the DG Hopf algebra of forms on $\mathrm{GL}(n,\mathbb{R})$, analogous to the van Est isomorphism between Lie algebra cohomology and continuous group cohomology. This analogy allows to find the appropriate Hopf algebras of forms whose DG Hopf cyclic cohomology groups store the geometric characteristic classes not only of standard foliations but also those of framed foliations, as well as those corresponding to the Hopf cyclic cohomology of $\mathcal{H}_n$ relative to $\mathrm{GL}(n,\mathbb{R})$.

Introduction

Hopf cyclic cohomology has emerged from the attempt to unravel the local formula [**7**] for the cyclic cohomological Chern character [**5**] of the hypoelliptic spectral triple [**8**] modeling the space of leaves of a foliation. A close inspection of that formula brought to light a Hopf algebra of 'moving frames' $\mathcal{H}(n,\mathbb{R})$ (abbreviated as $\mathcal{H}_n$), which appeared to assume in the transverse geometry of foliations a role similar to that of $\mathrm{GL}(n,\mathbb{R})$ as structure group of the frame bundle of a manifold. Further examination revealed that the the index formula in question was in fact the expression of a characteristic cocycle in the range of a certain cohomology specific to Hopf algebras. In the particular case of $\mathcal{H}_n$, this Hopf cyclic cohomology turned out to be isomorphic to the Gelfand-Fuchs cohomology of the Lie algebra $\mathfrak{a}_n$ of formal vector fields on $\mathbb{R}^n$ which, together with its relative versions, is well-known to encode the geometric characteristic classes of foliations. The benefit of having the same classes recaptured in Hopf cyclic cohomology is that the latter affords a direct characteristic map to the cyclic cohomology of étale foliations groupoids. Ultimately, the main outcome of the aforementioned investigation was the realization that the Hopf cyclic cohomology $HP^\bullet(\mathcal{H}_n;\mathbb{C}_\delta)$ of $\mathcal{H}(n,\mathbb{R})$ together with its relative versions $HP^\bullet(\mathcal{H}_n,\mathrm{O}_n;\mathbb{C}_\delta)$ and $HP^\bullet(\mathcal{H}_n,\mathrm{GL}(n,\mathbb{R});\mathbb{C}_\delta)$ provide the appropriate repository for all geometric characteristic classes of foliations in the framework of noncommutative geometry.

On the other hand in the classical theory it was known that the characteristic classes can be constructed not just in terms of the usual connection and curvature procedure, but also by simplicial cohomological methods involving $\mathrm{GL}(n,\mathbb{R})$

2020 *Mathematics Subject Classification.* Primary 16E40, 57R32, 58B34.

Key words and phrases. Hopf algebras, Hopf cyclic cohomology, van Est isomorphism, characteristic classes.

as structure group (*cf.* [**1**, **3**, **12**, **20**, **28**]). A. Gorokhovsky figured out how to convert that kind of representation into the formalism of Hopf cyclic cohomology by extending it to differential graded (DG) Hopf algebras. He showed [**17**] that the suitably truncated Hopf cyclic cohomology $HP^\bullet((\Omega^*\,\mathrm{GL}(n,\mathbb{R}))_m)$ of the DG Hopf algebra of differential forms on $\mathrm{GL}(n,\mathbb{R})$ captures the characteristic classes of flat Γ-equivariant vector bundles of rank n, where Γ is a discrete group of diffeomorphisms of the base manifold of dimension m. In particular his results yield an isomorphism

$$HP^\bullet(\mathcal{H}_n, \mathrm{O}_n; \mathbb{C}_\delta) \xrightarrow{\cong} HP^\bullet((\Omega^*\,\mathrm{GL}(n,\mathbb{R}))_n), \tag{0.1}$$

which bears a conspicuous resemblance with the van Est isomorphism for the continuous cohomology of $\mathrm{GL}(n,\mathbb{R})$.

The aim of this note is to show that this analogy is not coincidental, and it in fact extends to other avatars of the van Est isomorphism, thus providing the connection with the absolute cohomology $HP^\bullet(\mathcal{H}_n; \mathbb{C}_\delta)$, which accounts for the characteristic classes of framed foliations, and with the relative cohomology $HP^\bullet(\mathcal{H}_n, \mathrm{GL}(n,\mathbb{R}); \mathbb{C}_\delta)$, which only stores the Chern classes. More specifically, we prove that in addition to (0.1) there are canonical isomorphisms

$$HP^\bullet(\mathcal{H}_n, \mathrm{GL}(n,\mathbb{R}); \mathbb{C}_\delta) \xrightarrow{\cong} HP^\bullet((\Omega^*\,\mathrm{GL}_{\mathrm{alg}}(n,\mathbb{R}))_n), \tag{0.2}$$

where $\Omega^*\,\mathrm{GL}_{\mathrm{alg}}(n,\mathbb{R})$ is the Hopf algebra of forms on $\mathrm{GL}(n,\mathbb{R})$ viewed as an algebraic group, and

$$HP^\bullet(\mathcal{H}_n; \mathbb{C}_\delta) \xrightarrow{\cong} HP^\bullet((\bar{\Omega}^*\,\mathrm{GL}_{\mathrm{germ}}(n,\mathbb{R}))_n), \tag{0.3}$$

where $\bar{\Omega}^*\,\mathrm{GL}_{\mathrm{germ}}(n,\mathbb{R})$ is the Hopf algebra of germs of forms on the group germ $\mathrm{GL}_{\mathrm{germ}}(n,\mathbb{R})$ defined by $\mathrm{GL}(n,\mathbb{R})$.

The material is organized as follows. After recalling the basic definitions related to $\mathcal{H}_n$ and Hopf cyclic cohomology in §1, we outline Gorokhovsky's computation of the cyclic cohomology of $\Omega^*\,\mathrm{GL}(n,\mathbb{R})$ in §2.1, explain the 'behind-the-scenes' role of the van Est isomorphism in §2.2, and then prove the above mentioned results in §2.3 (Theorem 2.4) and §2.4 (Theorem 2.5). Finally, in §2.5 we illustrate the convenience of using $HP^\bullet((\Omega^*\,\mathrm{GL}(n,\mathbb{R}))_n)$ as a repository of characteristic classes by giving examples of representative cocycles.

1. Hopf algebra of moving frames and its cyclic cohomology

We recall in this section the definition of the Hopf algebra of moving frames in $\mathbb{R}^n$, its Hopf cyclic cohomology, and how the relationship with the Gelfand-Fuchs cohomology of the Lie algebra of formal vector fields on $\mathbb{R}^n$.

1.1. Hopf algebra $\mathcal{H}_n$. Consider the groupoid $\Gamma_n = \mathrm{Diff}_n \ltimes \mathbb{R}^n$, where Diff_n denotes the group of diffeomorphisms of $\mathbb{R}^n$ equipped with the discrete topology. Let $F\Gamma_n = \mathrm{Diff}_n \ltimes F\mathbb{R}^n$, where $F\mathbb{R}^n = \mathbb{R}^n \times \mathrm{GL}(n,\mathbb{R})$ denote the 'moving frames' groupoid, with Diff_n acting by prolongation:

$$\varphi(x, \mathbf{y}) := (\varphi(x), \varphi'(x)\mathbf{y}), \text{ where } \varphi'(x) = \big(\partial_j \varphi^i(x)\big) \in \mathrm{GL}(n,\mathbb{R}).$$

The convolution algebra $C_c^\infty(F\Gamma_n)$ is spanned by monomials fU_φ with $f \in C_c^\infty(F\mathbb{R}^n)$ and $\varphi \in \mathrm{Diff}_n$, with the product given by the rule

$$f_1 U_{\varphi_1} \cdot f_2 U_{\varphi_2} = f_1 \left(f_2 \circ \varphi_1{}^{-1}\right) U_{\varphi_1 \varphi_2}, \quad \text{where} \quad U_\varphi(f) = f \circ \varphi^{-1}.$$

The vertical vector fields $Y_i^j = \sum_\mu y_i^\mu \frac{\partial}{\partial y_j^\mu}$ and the horizontal vector fields $X_k = \sum_\mu y_k^\mu \frac{\partial}{\partial x^\mu}$, are made to act on $C_c^\infty(F\Gamma_n)$ as follows: if Z is one of these, then $Z(fU_\varphi) := Z(f)U_\varphi$. Since the right action of $\mathrm{GL}(n,\mathbb{R})$ on $F\mathbb{R}^n$ commutes with the action of Diff_n, the vertical operators so extended remain derivations:

$$Y_i^j(a\,b) \;=\; Y_i^j(a)\,b \,+\, a\,Y_i^j(b)\,, \quad a,b\in C_c^\infty(F\Gamma_n)\,.$$

By contrast, the horizontal vector fields are not Diff_n-invariant, and satisfy instead

$$\varphi_*(X_k) \equiv U_\varphi\,X_k\,U_\varphi^{-1} \;=\; X_k \,-\, \gamma_{jk}^i(\varphi^{-1})\,Y_i^j\,, \qquad \text{where}$$
$$\gamma_{jk}^i(\varphi)(x,\mathbf{y}) = \left(\mathbf{y}^{-1}\cdot\varphi'(x)^{-1}\cdot\partial_\mu\varphi'(x)\cdot\mathbf{y}\right)_j^i\,\mathbf{y}_k^\mu\,.$$

Consequently the extended operators X_k are no longer derivations but satisfy the generalized Leibniz rule

$$X_k(a\,b) \;=\; X_k(a)\,b \,+\, a\,X_k(b) \,+\, \delta_{jk}^i(a)\,Y_i^j(b)\,, \quad a,b\in C_c^\infty(F\Gamma_n).$$

Here δ_{jk}^i are multiplication operators given by

$$\delta_{jk}^i(fU_\varphi) = \gamma_{jk}^i(\varphi^{-1})\,fU_\varphi = -\gamma_{jk}^i(\varphi)\circ\tilde{\varphi}^{-1}\,fU_\varphi$$

and they satisfy the usual Leibniz rule

$$\delta_{jk}^i(a\,b) \;=\; \delta_{jk}^i(a)\,b \,+\, a\,\delta_{jk}^i(b).$$

The operators Id, $\{X_k,\, Y_j^i,\, \delta_{jk}^i\}$ generate a unital subalgebra of the linear operators on $C_c^\infty(F\Gamma_n)$, which defines $\mathcal{H}_n$ as an algebra. Automatically, $\mathcal{H}_n$ is also a Lie algebra with respect to the usual commutator, and the commutator relations between its generators give more insight into its nature. The vector fields satisfy the standard commutation relations of the affine group, and the iterated commutators of the operators δ_{jk}^i's with the X_ℓ's yield further operators

$$\delta_{jk\,\ell_1\ldots\ell_r}^i \;:=\; [X_{\ell_r},\ldots[X_{\ell_1},\delta_{jk}^i]\ldots],$$

which act by multiplication

$$\begin{aligned}\delta_{jk\,\ell_1\ldots\ell_r}^i(f\,U_\varphi) \;&:=\; \gamma_{jk\,\ell_1\ldots\ell_r}^i(\varphi^{-1})\,f\,U_\varphi\,, \qquad &&\text{where}\\ \gamma_{jk\,\ell_1\ldots\ell_r}^i(\varphi) \;&:=\; X_{\ell_r}\cdots X_{\ell_1}\big(\gamma_{jk}^i(\varphi)\big), &&\varphi\in\bar{\mathbf{G}}_n,\end{aligned}$$

and therefore commute with each other. They are no longer derivations but do satisfy (generalized) Leibniz rules.

The Leibniz rules obeyed by the generators extend by multiplicativity to all $h\in\mathcal{H}_n$. In turn, these rules uniquely determine a coproduct $\Delta:\mathcal{H}_n\to\mathcal{H}_n\otimes\mathcal{H}_n$ by the stipulation expressed in the Sweedler notation as follows:

$$\Delta(h) = \sum h_{(1)}\otimes h_{(2)} \quad \text{iff} \quad h(ab) = \sum h_{(1)}(a)h_{(2)}(b). \tag{1.1}$$

The counit $\epsilon:\mathcal{H}_n\to\mathbb{C}$, $\epsilon(h) := h(1)$ and the anti-automorphism $S:\mathcal{H}_n\to\mathcal{H}_n$ defined on generators by

$$S(Y_i^j) = -Y_i^j, \quad S(X_k) = -X_k + \delta_{jk}^i Y_i^j, \quad S(\delta_{jk}^i) = -\delta_{jk}^i,$$

complete the properties which make $\mathcal{H}_n$ a Hopf algebra.

The antipode S is not involutive but it can be twisted into an involution by means of the character $\delta:\mathcal{H}_n\to\mathbb{C}$ which trivially extends the trace map on $\mathfrak{gl}_n(\mathbb{C})$, *i.e.* defined by

$$\delta(Y_i^j) = \delta_i^j, \quad \delta(X_k) = 0, \quad \text{and} \quad \delta(\delta_{jk}^i) = 0.$$

Then $S_\delta(h) := \sum \delta(h_{(1)})S(h_{(2)})$, $h \in \mathcal{H}_n$ satisfies $S_\delta^2 = \mathrm{Id}$.

1.2. Hopf cyclic cohomology. By its very construction, the Hopf algebra $\mathcal{H}_n$ acts on the algebra $C_c^\infty(F\Gamma_n)$, and (1.1) ensures that $C_c^\infty(F\Gamma_n)$ is actually a left $\mathcal{H}_n$-module algebra. $C_c^\infty(F\Gamma_n)$ has a canonical trace, namely

$$\tau(fU_\varphi) = \int_{F\mathbb{R}^n} f\varpi, \quad \text{if } \varphi = \mathrm{Id}, \qquad \tau(fU_\varphi) = 0, \quad \text{if } \varphi \neq \mathrm{Id}, \tag{1.2}$$

where ϖ is the volume form dual to $X_1 \wedge \cdots \wedge X_n \wedge Y_1^1 \wedge \cdots \wedge Y_n^n$ (lexicographically ordered); τ is tracial since ϖ is Diff_n-invariant, and in addition it is δ-invariant with respect to the action of $\mathcal{H}_n$, *i.e.* satisfies

$$\tau(h(a)) = \delta(h)\,\tau(a), \qquad \forall\, h \in \mathcal{H},\ a \in C_c^\infty(F\Gamma_n).$$

The cyclic cohomology of the Hopf algebra $\mathcal{H}_n$ was introduced in [**8**] by importing the standard cyclic structure of the algebra $C_c^\infty(F\Gamma_n)$. via the morphism $\chi_* = \{\chi_q : \mathcal{H}_n^{\otimes q} \to CC^q(C_c^\infty(F\Gamma_n))\}_{q\geq 0}$, where

$$\chi_q(h^1 \otimes \ldots \otimes h^q)(a_0, a_1, \ldots, a_q) := \tau\left(a_0 h^1(a_1)\cdots h^q(a_q)\right). \tag{1.3}$$

The stipulation that χ_* is a morphism of ΔC-objects, where ΔC is Connes' cyclic category, endows $\mathcal{H}_n^\natural := \sum_{q\geq 0}^{\oplus} \mathcal{H}_n^{\otimes q}$ with the basic ΔC-morphisms

$$\begin{aligned}
\partial_0(h^1 \otimes \ldots \otimes h^{q-1}) &= 1 \otimes h^1 \otimes \ldots \otimes h^{q-1}, \\
\partial_j(h^1 \otimes \ldots \otimes h^{q-1}) &= h^1 \otimes \ldots \otimes \Delta h^j \otimes \ldots \otimes h^{q-1} \\
\partial_q(h^1 \otimes \ldots \otimes h^{q-1}) &= h^1 \otimes \ldots \otimes h^{q-1} \otimes 1 \\
\sigma_i(h^1 \otimes \ldots \otimes h^{q+1}) &= h^1 \otimes \ldots \otimes \varepsilon(h^{i+1}) \otimes \ldots \otimes h^{q+1} \\
\tau_q(h^1 \otimes \ldots \otimes h^q) &= S_\delta(h^1)\cdot(h^2 \otimes \ldots \otimes h^q \otimes 1).
\end{aligned} \tag{1.4}$$

The involutive property $S_\delta^2 = \mathrm{Id}$ evidently implies $\tau_1^2 = \mathrm{Id}$, and in fact ensures that $\tau_q^{q+1} = \mathrm{Id}$ for any $q \in \mathbb{N}$ (*cf.* [**9**]). The corresponding cyclic cohomology groups $HC^*(\mathcal{H};\mathbb{C}_\delta)$ are computed from the associated bicomplex $CC^{*,*}(\mathcal{H}_n;\mathbb{C}_\delta)$ with boundary operators

$$b = \sum_{i=0}^{q+1}(-1)^i\partial_i, \quad B = \Big(\sum_{i=0}^{q}(-1)^{(q-1)i}\tau_q^i\Big)\,\sigma_{q-1}\,(1-(-1)^q\tau_q), \tag{1.5}$$

and the periodic Hopf cyclic cohomology $HP^\bullet(\mathcal{H}_n;\mathbb{C}_\delta)$ is the $\mathbb{Z}_2$-graded cohomology of the corresponding total complex.

More generally (*cf.* [**9**, Theorem 1]), to any Hopf algebra $\mathcal{H}$ endowed with a *modular pair in involution* (σ,δ), *i.e.* σ is a group element, δ a character and $\delta(\sigma) = 1$, one associates a (b,B) bicomplex in a similar way, after modifying the operators in (1.4) as follows:

$$\begin{aligned}
\partial_q(h^1 \otimes \ldots \otimes h^{q-1}) &= h^1 \otimes \ldots \otimes h^{q-1} \otimes \sigma \\
\tau_q(h^1 \otimes \ldots \otimes h^q) &= S_\delta(h^1)\cdot(h^2 \otimes \ldots \otimes h^q \otimes \sigma).
\end{aligned} \tag{1.6}$$

The definition of Hopf cyclic cohomology was subsequently extended in [**18**] to a large class of coefficients.

For applications to characteristic classes of foliations, the relative versions of the Hopf cyclic cohomology of $\mathcal{H}_n$ of particular interest are that relative to O_n, resp. to SO_n in the orientable case, and to a lesser extent that relative to $\mathrm{GL}(n,\mathbb{R})$.

To simplify the notation, in the remainder of this section we abbreviate $\mathrm{GL}(n,\mathbb{R})$ by GL_n and $\mathfrak{gl}_n(n,\mathbb{R})$ by $\mathfrak{g}_n$. Letting H denote one of closed subgroups of GL_n enumerated above, we recall that the cohomology of $\mathcal{H}_n$ relative to H is defined as follows (*cf.* [**10**]).

First note that the adjoint action of GL_n on the Lie algebra $\mathfrak{gl}_n \ltimes \mathbb{R}^n$ of the affine group $\mathrm{GL}_n \ltimes \mathbb{R}^n$, with which we identify the frame bundle $F\mathbb{R}^n$, extends to a linear action of GL_n on $\mathcal{H}_n$. With $\mathfrak{h}$ denoting the Lie algebra of H, its universal enveloping algebra $\mathcal{U}(\mathfrak{h})$ is a sub-Hopf algebra of $\mathcal{H}_n$. The quotient $\mathcal{H}_n \otimes_{\mathcal{U}(\mathfrak{h})} \mathbb{C} \equiv \mathcal{H}_n/\mathcal{H}_n\mathcal{U}^+(\mathfrak{h})$ (where $\mathcal{U}^+(\mathfrak{h})$ denotes the ideal of $\mathcal{U}(\mathfrak{h})$ generated by $\mathfrak{h}$) is an $\mathcal{H}_n$-module coalgebra with respect to the coproduct and counit induced from $\mathcal{H}_n$. We denote by $\mathcal{Q}_H$ its subspace of invariants under the action of H on $\mathcal{H}_n$. The cyclic object defining the relative cohomology consists of $\sum_{q\geq 0}^{\oplus} \mathcal{Q}_H^{\otimes q}$ endowed with the basic ΔC-morphisms defined by

$$
\begin{aligned}
\partial_0(c^1\otimes\ldots\otimes c^{q-1}) &= \dot{1}\otimes c^1\otimes\ldots\otimes\ldots\otimes c^{q-1},\\
\partial_i(c^1\otimes\ldots\otimes c^{q-1}) &= c^1\otimes\ldots\otimes\Delta c^i\otimes\ldots\otimes c^{q-1},\\
\partial_n(c^1\otimes\ldots\otimes c^{q-1}) &= c^1\otimes\ldots\otimes c^{q-1}\otimes\dot{1};\\
\sigma_i(c^1\otimes\ldots\otimes c^{q+1}) &= c^1\otimes\ldots\otimes\varepsilon(c^{i+1})\otimes\ldots\otimes c^{q+1},\\
\tau_q(\dot{h}^1\otimes c^2\otimes\ldots\otimes c^q) &= S_\delta(h^1)\cdot(c^2\otimes\ldots\otimes c^q\otimes\dot{1}).
\end{aligned}
\tag{1.7}
$$

The corresponding bicomplex is denoted $CC^*(\mathcal{H}_n, H;\mathbb{C}_\delta)$ and the $\mathbb{Z}_2$-graded cohomology group of the total complex is $HP^\bullet(\mathcal{H}_n, H;\mathbb{C}_\delta)$.

The above cohomology groups have been computed in [**7**] (and by alternative means in [**25, 26**]). They are isomorphic to the Gelfand-Fuchs cohomology of the Lie algebra $\mathfrak{a}_n$ of formal vector fields on $\mathbb{R}^n$, resp. to its relative versions, which in turn are isomorphic to the cohomology of truncated Weil algebras (*cf.* [**15, 16**]), as follows:

$$HP^\bullet(\mathcal{H}_n;\mathbb{C}_\delta) \cong H^\bullet_{\mathrm{cont}}(\mathfrak{a}_n;\mathbb{C}) \cong H^\bullet(W(\mathfrak{gl}_n)_n), \tag{1.8}$$

and relative to $H = \mathrm{O_n}, \mathrm{SO}_n$ or $\mathrm{GL}(n,\mathbb{R})$

$$HP^\bullet(\mathcal{H}_n, H;\mathbb{C}_\delta) \cong H^\bullet_{\mathrm{cont}}\big(\mathfrak{a}_n, H;\mathbb{C}\big) \cong H^\bullet\big(W(\mathfrak{gl}_n, H)_n\big). \tag{1.9}$$

Here $W(\mathfrak{gl}_n) = \wedge\mathfrak{gl}_n^* \otimes S(\mathfrak{gl}_n^*)$ is the Weil algebra, $W(\mathfrak{gl}_n)_m$ denotes the quotient by the ideal generated by $\sum_{k>m} S^k(\mathfrak{gl}_n^*)$, and $W(\mathfrak{gl}_n, H)_m$ stands for the DG subalgebra of H-basic elements in $W(\mathfrak{gl}_n)_m$. Explicit bases for the cohomology groups in the right hand side of (1.8) and (1.9) are given in [**16**], and their geometric realizations via the above isomorphisms can be found in [**23, 24**].

2. Hopf algebras of forms and their cyclic cohomology

2.1. DG Hopf algebra of a Lie group. Let G be an almost connected Lie group. Its algebra of differential forms $\Omega^*(G)$ has an natural structure of a topological (Fréchet) DG Hopf algebra, extending that of the function algebra $C^\infty(G) = \Omega^0(G)$. Indeed $C^\infty(G)$, equipped with its standard topological vector space structure of uniform convergence on compacta of functions and their derivatives, is a Hopf algebra with the coproduct $\Delta : C^\infty(G) \to C^\infty(G)\hat{\otimes} C^\infty(G) = C^\infty(G\times G)$ defined by $\Delta(f)(x,y) = f(xy)$, the counit given by evaluation at the unit $\mathbf{1}\in G$ and the antipode S induced by inversion. As topological vector spaces $\Omega^*(G) \cong C^\infty(G)\otimes\bigwedge\mathfrak{g}^*$, where $\mathfrak{g}$ is the Lie algebra of G, and the Hopf structure of $C^\infty(G) = \Omega^0(G)$ extends naturally to $\Omega^*(G)$. Together with the de Rham differential, the couple $(\Omega^*(G), d)$ is a DG (Fréchet) Hopf algebra.

In [**17**, §3] Gorokhovsky defines a cyclic object associated to an arbitrary DG Hopf algebra $(\mathcal{H}^*, d)$ endowed with a modular pair, by a natural graded extension of the cyclic object associated to $\mathcal{H}^0$. In particular this applies to $(\Omega^*(G), d)$ which, having involutive antipode, carries a trivial modular pair. The cyclic object consists of

$$(\Omega^* G)^\natural := \sum_{q\geq 0} \Omega^*(G)^{\hat{\otimes} q} \equiv \sum_{q\geq 0} \Omega^*(G^{\times q})$$

equipped with the ΔC-morphisms similar to those in (1.4), with the difference that the summands (implicitly present) in the expression of the cyclic operator τ_q acquire appropriate signs (*cf.* [**17**, §3, (3.3)]). The corresponding (b, B)-bicomplex $C^{*,*}(\Omega^*(\mathrm{GL}(n, \mathbb{R})))$ is then upgraded to a tricomplex $C^{*,*,*}(\Omega^*(G), d)$ by the addition of the differential

$$d(\alpha_1 \otimes \cdots \otimes \alpha_q) = \sum_{i=1}^{q} (-1)^{\deg \alpha_1 + \cdots + \deg \alpha_q} \alpha_1 \otimes \cdots d\alpha_i \cdots \otimes \alpha_q),$$

The latter is filtered by the subcomplexes

$$F^m C^\bullet(\Omega^*(G), d) = \{\sum \alpha_1 \otimes \cdots \otimes \alpha_q \mid \deg \alpha_1 + \cdots + \deg \alpha_q > m\},$$

and gives rise, for each $m \in \mathbb{Z}^+$, to a truncated complex

$$C^{*,*,*}(\Omega^*(G)_m, d) = C^{*,*,*}(\Omega^*(G), d)/F^m C^\bullet(\Omega^*(G), d);$$

Its Hochschild, cyclic and periodic cyclic cohomology, formed using only finite cochains, will be respectively denoted by $HH^*\left((\Omega^* G)_m\right)$, $HC^*\left((\Omega^* G)_m\right)$ and $HP^\bullet\left((\Omega^* G)_m\right)$. Gorokhovsky computed them in terms of the cohomology of truncated Weil algebras as follows.

THEOREM 2.1 ([**17**], §6). *With K denoting the maximal subgroup of G, there are canonical isomorphisms*

$$HH^*\left((\Omega^* G)_m\right) \cong H^*\big(W(\mathfrak{g}, K)_m\big), \quad \forall\, m \geq 0, \tag{2.1}$$

$$HC^q\left((\Omega^* G)_m\right) \cong \bigoplus_{i\geq 0} H^{q-2i}\big(W(\mathfrak{g}, K)_m\big). \tag{2.2}$$

For the clarity of the ensuing discussion we outline Gorokhovsky's proof. The crucial observation is that the Hochschild bicomplex $C^{*,*}(\Omega^*(G), b, d)$ coincides with the Bott [**1**] simplicial de Rham bicomplex $\{\Omega^{*,*}(NG), \delta, d\}$ of the nerve NG of G. Recall that $NG := \bigsqcup N_p G$, where $N_p G = G^p$, and $\Omega^{p,q}(NG) = \Omega^q(G^p)$.

As was shown in [**3**] for a general simplicial manifold $X = \{X_p\}$, the cohomology of $\{\Omega^{*,*}(X), \delta, d\}$ is isomorphic to the singular cohomology of the geometric realization of X. On the other hand, Dupont [**12**]) introduced a related de Rham complex, $\{\Omega^*(\|X\|), d\}$, where $\|X\|$ is the "thick" geometric realization of X, which has the advantage of being graded commutative. A q-form $\omega \in \Omega^q(\|X\|)$ is a collection $\{\omega_p\}$ of de Rham q-forms on $\bigsqcup_p X_p$, satisfying the compatibility condition

$$(\varepsilon^i \times \mathrm{Id})^* \omega_p = (\mathrm{Id} \times \varepsilon_i)^* \omega_{p-1},$$

where $\varepsilon_i : X_p \to X_{p-1}$ are the face operators corresponding to the inclusions of the faces $\varepsilon^i : \Delta^{p-1} \to \Delta^p$, $i = 0, 1, \ldots, p$. Dupont has shown that integration over simplices $\mathcal{I}_\Delta : \Omega^*(\|X\|) \to \Omega^*(X)$,

$$\mathcal{I}_\Delta(\alpha) = \int_{\Delta^p} \alpha \mid \Delta^p \times X_p$$

defines a quasi-isomorphism of $\{\Omega^*(\|X\|), d\}$ with with the total complex $\{\Omega^*(X), \delta \pm d\}$.

We next recall that the geometric realization EG of the simplicial manifold $\bar{N}G$, with $\bar{N}_p G = G^{p+1}$ on which G acts diagonally, gives the total space of the universal (right) G-bundle $\pi : EG \to BG$. This bundle has a canonical connection θ induced by the Maurer-Cartan form on G; for a linear group G the expression of $\theta \mid \Delta^p \times \bar{N}_p G$ is

$$\theta(t_0, \ldots, t_p; g_o, \ldots, g_p) = t_0 g_o^{-1} dg_o + \cdots + t_p g_p^{-1} dg_p, \tag{2.3}$$

and the curvature form is $\Omega = d\theta + \theta \wedge \theta$.

The Dupont complex being a $\mathfrak{g}$-DG algebra, by the universal property of the Weil algebra (see [**4**]) this connection determines a morphism $w : W(\mathfrak{g}) \to \Omega^*(EG)$. In turn, the latter descends to K-invariants yielding a morphism $w_K : W(\mathfrak{g}, K) \to \Omega^*(EG/K)$. Using the contractibility of G/K and the canonical construction of geodesic simplices (*cf.* [**12**, §5] for metrics of nonpositive curvature, one defines a simplicial cross-section $s : BG \to EG/K$ (*cf.* [**17**, Eqs. (6.10),(6.11)]), which moreover is equivariant with respect to the cyclic group action (*cf.* [**17**, (6.8)]). The composition

$$\mu = \mathcal{I}_\Delta \circ s^* \circ w_K : W(\mathfrak{g}, K) \to \Omega^*(NG) \tag{2.4}$$

preserves the canonical filtrations and descends to the truncated complexes of any level $m \geq 0$, inducing in cohomology a morphism

$$\mu^*_{HH} : H^*\left(W(\mathfrak{g}, K)_m\right) \to H^q\left((\Omega^*(NG))_m\right) = HH^q\left((\Omega^* G)_m\right). \tag{2.5}$$

In view of results proved in [**20**] and [**28**], it follows that μ^*_{HH} is actually an isomorphism. Finally, since the connection (2.3) and the cross-section s are invariant under the natural cyclic action, the Hochschild cocycles in the image of μ are actually cyclic. This implies that Connes' periodicity exact sequence splits into short exact sequences

$$0 \to HC^{q-2}\left((\Omega^* G)_m\right) \xrightarrow{S} HC^q\left((\Omega^* G)_m\right) \xrightarrow{I} HH^q\left((\Omega^* G)_m\right) \to 0,$$

which in turn implies the isomorphism (2.2) for cyclic cohomology.

2.2. Role of van Est isomorphism. We now revisit the proof of the isomorphism (2.5) and give an alternative argument emphasizing the role of the van Est isomorphism, which will serve as a template for dealing with the counterparts of the isomorphism (2.2) corresponding to $H = \{\mathbf{1}\}$ in (1.8) and and $H = \mathrm{GL}(n, \mathbb{R})$ in (1.9).

A key element of the approach we are about to describe is the linkage made by Bott [**1**] with the continuous group cohomology. Relying on the Dold-Puppe homology theory for nonadditive functors, Bott proved a basic Decomposition Lemma [**1**, §2] for the simplicial de Rham complex $\Omega^*(NG)$ and use it show that for any Lie group G the spectral sequence of the bicomplex $\{\Omega^*(NG), d, \delta\}$ filtered by the degree of the forms converges to the cohomology of BG. In particular (*cf.* [**1**, Theorem 1]) he expressed the E_1-term as continuous group cohomology with coefficients:

$$E_1^{pq} = H_\delta^p\big(\Omega^q(NG)\big) \cong H_{\mathrm{cont}}^{p-q}\big(G; S^q(\mathfrak{g}^*)\big). \tag{2.6}$$

By the van Est isomorphism [**14**],

$$H_{\mathrm{cont}}^*\big(G; S^*(\mathfrak{g}^*)\big) \cong H^*(\mathfrak{g}, K; \mathbb{R}) \otimes S^*(\mathfrak{g}^*)^G, \tag{2.7}$$

where K is the maximal compact subgroup. In the case when G itself is compact one has

$$H_{\mathrm{cont}}^0\big(G; S^*(\mathfrak{g}^*)\big) \cong S^*(\mathfrak{g}^*)^G \quad \text{and} \quad H_{\mathrm{cont}}^k\big(G; S^*(\mathfrak{g}^*)\big), \quad \forall k > 0,$$

and so (2.6)and (2.7) imply that there is no cohomology above the diagonal. This allows Bott to conclude (*cf.* [**1**, Remark (a)]) that the edge homomorphism to the cohomology of the total complex

$$S^*(\mathfrak{g}^*) \to H^*(\Omega^*(NG)),$$

which under the identification $H^*(\Omega^*(NG)) \cong H^*(BG)$ coincides with the Chern-Weil homomorphism, is an isomorphism.

For the remainder of this section, we specialize to the group $\mathrm{GL}(n, \mathbb{R})$ in order to remain consistent with the context of §1. To keep the notation convenient though we will simply denote it by G (except for the rare occasion when G is allowed to be an arbitrary Lie group, in which case this will be explicitly stated).

Returning to the Bott spectral sequence, where now for $G = \mathrm{GL}(n, \mathbb{R})$ and $K = \mathrm{O_n}$, it is well known that

$$\begin{aligned} H^*(\mathfrak{g}, K; \mathbb{R}) &\cong E(h_1, h_3, \ldots, h_{2[\frac{n}{2}]+1}), \\ \text{and} \quad S^*(\mathfrak{g}^*)^G &\equiv P[c_1, c_2, \ldots, c_n]; \end{aligned} \tag{2.8}$$

here E stands for the exterior algebra in generators h_i of degree $\deg h_i = 2i - 1$, and $P[c_1, c_2, \ldots, c_n]$ denotes the polynomial algebra in generators c_i of degree $\deg c_i = 2i$. Since the sequence converges to $H^*(BG) = P[c_2, c_4, \ldots, c_{2[\frac{n}{2}]}]$, it can be seen, successively, that d_{2i-1} sends h_{2i-1} to a non-zero multiple of c_{2i-1}, and that c_{2i}'s survive; also, $d_r = 0$ for $r > 2[\frac{n+1}{2}]$.

A similar pattern occurs for the spectral sequence of the truncated complex $(\Omega^*(NG))_n, d, \delta)$, the main difference being that the polynomial algebra

$$P[c_1, c_2, \ldots, c_n]$$

is replaced by its truncation modulo the ideal of elements of degree $> 2n$, denoted $S^*(\mathfrak{g}*)_n^{\mathrm{GL}_n} \equiv P_n[c_1, c_2, \ldots, c_n]$. In particular (2.6) and (2.7) become

$$H_\delta^p\big(\Omega^q(NG)_n\big) \cong H_{\mathrm{cont}}^{p-q}\big(G; S^q(\mathfrak{g}^*)_n\big), \tag{2.9}$$

respectively

$$H_{\mathrm{cont}}^*\big(G; S^*(\mathfrak{g}^*)_n\big) \cong H^*(\mathfrak{g}, K; \mathbb{R}) \otimes S^*(\mathfrak{g}^*)_n^G. \tag{2.10}$$

Thus, in view of (2.8), the corresponding E_1-term is

$$E_1 \cong WO_n := E(u_1, u_3, \ldots, u_{2[\frac{n}{2}]+1}) \otimes P_n[c_1, c_2, \ldots, c_n]. \tag{2.11}$$

There is a natural inclusion of the DG algebra WO_n into the quotient $W(\mathfrak{g}, K)_n$ of $W(\mathfrak{g}, K)$ by the ideal of elements of degree $> 2n$, and the map $WO_n \hookrightarrow W(\mathfrak{g}, K)_n$ is a quasi-isomorphism. Indeed, this follows from the comparison theorem for the spectral sequences associated to the quotient of the standard filtration (by the ideal generated by the elements of degree $> 2n$), as the E_1 term of both spectral sequences is the same as in (2.11). Although the spectral sequences of $W(\mathfrak{g}, K)_n$ and $\Omega^*(NG)_n$ cannot be directly compared, by chasing their differentials one can still infer that $H^*(W(\mathfrak{g}, K)_n)$ and $H^*(\Omega^*(NG)_n)$ are formally isomorphic.

The elegant argument though, due to Shulman and Stasheff [**28**], involves the semi-simplicial Weil algebra of Kamber and Tondeur [**20**]. Relying on a semi-simplicial generalization of the van Est isomorphism, Shulman and Stasheff arrive to the isomorphism $H^*(W(\mathfrak{g}, K)_n) \cong H^*(\Omega^*(NG)_n)$ not by ad hoc calculations but by exploiting the standard comparison theorem for spectral sequences. To explain their line of argument, we recall that the semi-simplicial Weil algebra $W_1(\mathfrak{g})$ (see [**20, 21**]) is a $\mathfrak{g}$-DG algebra with underlying vector space $\bigoplus_{p\geq 0} W(\mathfrak{g}^{p+1})$ and faces induced by the projections $\mathrm{pr}_k : \mathfrak{g}^{p+1} \to \mathfrak{g}^p$, $0 \leq k \leq p$, that omit the $(k+1)$th factor. The projection $\pi : W_1(\mathfrak{g}) \to W(\mathfrak{g})$ on the first summand ($p = 0$) is a map of $\mathfrak{g}$-DG algebras, and is shown to induce a chain equivalence. This is but a special case of a core result of Kamber and Tondeur ([**20**, Theorem 8.12], also [**21**, §6]), which for $W_1(\mathfrak{g})$ can be stated as follows.

THEOREM 2.2 ([**20**], Ch.8). *Let G be an almost connected Lie group and H a closed subgroup. There is a suitable filtration $\mathcal{F}_1$ of $W_1(\mathfrak{g})$ such that the canonical projection at the level of truncated algebras*

$$\pi_m : W_1(\mathfrak{g}, H)_m := W_1(\mathfrak{g}, H)/\mathcal{F}_1^{m+1} W_1(\mathfrak{g}, H) \to W(\mathfrak{g}, H)_m$$

induces an isomorphism of the associated spectral sequence and hence on cohomology:

$$\pi_m^\natural : H^*(W_1(\mathfrak{g}, H)_m) \xrightarrow{\cong} H^*(W(\mathfrak{g}, H)_m). \tag{2.12}$$

On the other hand, by the universal property of Weil algebras, the canonical connection on the principal G-bundle $\bar{N}G \to NG$ gives rise to a map of simplicial $\mathfrak{g}$-DG algebras

$$\varphi : W_1(\mathfrak{g}) = \bigoplus_{p\geq 0} W(\mathfrak{g}^{p+1}) \to \bigoplus_{p\geq 0} \Omega^*(G^{p+1}) = \Omega^*(\bar{N}G). \tag{2.13}$$

Shulman and Stasheff [**28**] define a compatible filtration $\mathcal{F}$ on $\Omega^*(\bar{N}G)$ which renders φ fitration-preserving, as follows. First they endow $\Omega^*(NG)$ with the standard filtration F by the degree of forms. Then they define $\mathcal{F}$ as the filtration of $\Omega^*(\bar{N}G)$ induced by $\psi : \Omega^*(NG) \to \bar{N}G)$, where $\psi = \{\psi_p\}$ with ψ_p corresponding to the projection $G^{p+1} \to G^p$ on the first p coordinates. In the homogeneous picture $\Omega^*(NG)$ is identified with the subcomplex $\Omega^*(\bar{N}G)_G$ of G-basic forms on $\bar{N}G$. Note that $\mathcal{F}$ also restricts to a filtration on the K-basic forms $\Omega^*(\bar{N}G)_K$.

Both φ and ψ descend to maps φ_m, resp. ψ_m, at the level of the quotient algebras by the $(m+1)$th power of the corresponding ideal, which are shown to give rise to equivalences of spectral sequences.

THEOREM 2.3 ([**28**], pp. 68-70). *Each of the maps*

$$\varphi_m : W_1(\mathfrak{g}, K)/\mathcal{F}_1^{m+1} \to \Omega^*(\bar{N}G)_K/\mathcal{F}^{m+1}, \quad \textit{resp.} \tag{2.14}$$

$$\psi_m : \Omega^*(NG)_m \equiv \Omega^*(\bar{N}G)_G/F^{m+1} \to \Omega^*(\bar{N}G)_K/\mathcal{F}^{m+1} \tag{2.15}$$

is filtration preserving and induces an isomorphism of spectral sequences.

One thus obtains in homology the isomorphisms

$$\varphi^\natural_m : H^*\left(W_1(\mathfrak{g},K)_m\right) \xrightarrow{\cong} H^*\left(\Omega^*(\bar{N}/K)_m\right), \quad \forall\, m \geq -1 \tag{2.16}$$

$$\psi^\natural_m : H^*\left(\Omega^*(NG)_m\right) \xrightarrow{\cong} H^*\left(\Omega^*(\bar{N}/K)_m\right), \tag{2.17}$$

which combined with (2.12) give rise to a canonical isomorphism

$$H^*\left((\Omega^*(NG)_m\right) \cong H^*\big(W(\mathfrak{g},K)_m\big). \tag{2.18}$$

The above isomorphism coincides with that of (2.5), as both are constructed via the universal property of the Weil algebra in Chern-Weil theory. The difference is that the cross-section s in (2.4) is replaced by the equivalence (2.15) of spectral sequences, and the passage through the Dupont complex is altogether bypassed.

In order to see that Theorem 2.3 represents a semi-simplicial generalization of the van Est isomorphism $H^*(\mathfrak{g},K;\mathbb{R}) \cong H^*_{\mathrm{cont}}(G;\mathbb{R})$, note that at the level of 0-th filtration, $W_1(\mathfrak{g},K)/\mathcal{F}_1 = \bigoplus_{p\geq 0}\left(\wedge^*\mathfrak{g}^{p+1}\right)_{K-\mathrm{basic}}$ while $\Omega^0(NG) = \bigoplus_{p\geq 0} C^\infty(G^p)$ is the complex of differentiable group cohomology, and φ_0 and ψ_0 are the very same maps used by van Est in his proof [**14**].

2.3. Hopf algebra of algebraic forms. Regarding $G = \mathrm{GL}(n,\mathbb{R})$ as a real algebraic group, we denote by $\Omega^*_{\mathrm{alg}}(G)$ its graded Hopf algebra of algebraic differential forms. As vector spaces $\Omega^*_{\mathrm{alg}}(G) \cong C_{\mathrm{alg}}(G) \otimes \bigwedge \mathfrak{g}^*$, where $C_{\mathrm{alg}}(G) = \mathbb{R}[g_{ij}; \det g^{-1}]$ is the ring of regular rational functions on G, which is the classical Hopf algebra of a linear algebraic group. The associated Hochschild complex of the DG Hopf algebra $\Omega^*_{\mathrm{alg}}(G)$ coincides with the Bott simplicial de Rham $(\Omega^{*;*}_{\mathrm{alg}}(NG),\delta,d)$, where $\Omega^{p,q}_{\mathrm{alg}}(NG) := \Omega^q_{\mathrm{alg}}(G^p)$.

THEOREM 2.4. *There are canonical isomorphisms*

$$HH^q\left((\Omega^*_{\mathrm{alg}}G)_n\right) \xrightarrow{\cong} H^*\big(W(\mathfrak{g},G)_n\big), \tag{2.19}$$

$$HC^q\left((\Omega^*_{\mathrm{alg}}G)_n\right) \xrightarrow{\cong} \bigoplus_{i\geq 0} H^{q-2i}\big(W(\mathfrak{g},G)_n\big). \tag{2.20}$$

PROOF. The short proof is parallel to Bott's argument in the case of a compact Lie group (*cf.* [**1**, Remark (a)]). Indeed, by the analogue of Bott's Decomposition Lemma, the E_1-term of the spectral sequence for the column filtration of the truncated complex $(\Omega^{*;*}_{\mathrm{alg}}(NG)_n,\delta,d)$ is seen to be

$$E_1^{p,q} = H^p_\delta\big(\Omega^q_{\mathrm{alg}}(G^p)_n\big) \cong H^{p-q}_{\mathrm{alg}}\big(G; S^q(\mathfrak{g}^*)_n\big). \tag{2.21}$$

The algebraic counterpart of the van Est isomorphism, which is a consequence of Hochschild's isomorphism [**19**, Theorem 5.2], then implies:

$$\begin{aligned} &H^0_{\mathrm{alg}}\big(G; S^q(\mathfrak{g}^*)_n\big) \cong S^q(\mathfrak{g}^*)^G_n, \quad \text{and} \\ &H^i_{\mathrm{alg}}\big(G; S^q(\mathfrak{g}^*)_n\big) = 0, \qquad \text{if} \quad i>0. \end{aligned} \tag{2.22}$$

From (2.21) and (2.22) it follows that

$$H^*(\Omega^*_{\mathrm{alg}}(NG)_n) \cong S^*(\mathfrak{g}^*)_n = P_n[c_1,c_2,\dots,c_n].$$

The left hand side is the same as $HH^*\left(\left(\Omega^*_{\mathrm{alg}}G)_n\right)\right)$ while the right hand is the same as $H^*\big(W(\mathfrak{g},G)_n\big)$.

The Shulman-Stasheff argument in §2.2 can also be adapted. The algebraic counterpart of ψ_m in Theorem 2.3 is $\psi_m^{\mathrm{alg}} = \mathrm{Id}$, and the proof that

$$\varphi_m^{\mathrm{alg}} : W_1(\mathfrak{g}, G)/\mathcal{F}_1^{m+1} \to \Omega^*_{\mathrm{alg}}(\bar{N}G)_G/\mathcal{F}^{m+1} \equiv \Omega^*_{\mathrm{alg}}(NG)_m$$

induces an isomorphism of spectral sequences follows along the same lines as in [**28**], with the difference that the semi-simplicial generalization of the van Est bicomplex [**13**, §10] is replaced by its algebraic counterpart, which can be handled as in the proof of [**22**, §2] for the algebraic version of the van Est isomorphism.

We finally recall that (2.20) automatically follows from (2.19). □

2.4. Hopf algebra of germs of forms. We now consider a Hopf algebra of forms on the *group germ* determined by G, which is defined as follows. First, a differential form $\alpha \in \Omega^q(G)$ will be called *locally trivial* if it vanishes identically in a neighborhood of $\mathbf{1} := \mathrm{Id}$ in G. We denote the set of such forms by $\Omega_0^*(G)$. More generally, for any $p \in \mathbb{N}$, we define in a similar way the set $\Omega_0^*(G^p)$ of locally trivial forms for the group G^p. Obviously, $\Omega_0^*(G^p)$ is an ideal in $\Omega^*(G^p)$, and we form the quotient algebra

$$\bar{\Omega}^*(G^p) := \Omega^*(G^p)/\Omega_0^*(G^p), \quad \forall p \in \mathbb{N}.$$

We then define the tensor product $\bar{\otimes}$ of two such algebras by stipulating

$$\bar{\Omega}^*(G^p)\bar{\otimes}\bar{\Omega}^*(G^q) := \bar{\Omega}^*(G^{p+q}),$$

and equip $\bar{\Omega}^*(G)$ with the coproduct

$$\bar{\Delta} : \bar{\Omega}^*(G) \to \bar{\Omega}^*(G)\bar{\otimes}\bar{\Omega}^*(G)$$

induced by $\Delta : \Omega^*(G) \to \Omega^*(G) \otimes \Omega^*(G)$. The latter is well-defined since if $\alpha \in \Omega_0^*(G)$ then $\Delta\alpha \in \Omega_0^*(G^2)$. With this coproduct and the obvious antipode, unit and counit, it is straightforward to verify that $\bar{\Omega}^*(G)$ satisfies the axioms of a DG Hopf algebra. Its Hochschild complex coincides with the simplicial de Rham bicomplex $\{\bar{\Omega}^{*,*}(NG), \delta, d\}$, resp. $\{\bar{\Omega}^{*,*}(\bar{N}G), \delta, d\}$, where $\bar{\Omega}^{p,q}(NG) := \bar{\Omega}^q(G^p)$ and $\bar{\Omega}^{p,q}(\bar{N}G) := \bar{\Omega}^q(G^{p+1})$.

THEOREM 2.5. *There are canonical isomorphisms*

$$HH^q\left((\bar{\Omega}^*G)_n\right) \xrightarrow{\cong} H^*\big(W(\mathfrak{g})_n\big), \tag{2.23}$$

$$HC^q\left((\bar{\Omega}^*G)_n\right) \xrightarrow{\cong} \bigoplus_{i\geq 0} H^{q-2i}\big(W(\mathfrak{g})_n\big). \tag{2.24}$$

PROOF. The analogue of Bott's Decomposition Lemma yields

$$E_1^{p,q} = H_\delta^p\left(\bar{\Omega}^q(G^p)_n\right) \cong H_\square^{p-q}\left(G; S^q(\mathfrak{g}^*)_n\right),$$

where $H_\square^*$ refers to the cohomology of group germs, which by its very definition (see [**27**, §4]) in the case at hand is the cohomology of the complex

$$\{\bar{\Omega}^0\left(G^p; S^q(\mathfrak{g}^*)_n\right), \delta\}$$

. Furthermore, the counterpart of van Est's isomorphism for this case was proved by Świerczkowski [**27**, Theorem 2] and it gives the isomomorphism

$$H_\square^i\left(G; S^q(\mathfrak{g}^*)_n\right) \cong H^i\left(\mathfrak{g}; S^q(\mathfrak{g}^*)_n\right). \tag{2.25}$$

Since $H^*\left(\mathfrak{g}; S^q(\mathfrak{g}^*)_n\right) \cong H^*(\mathfrak{g}; \mathbb{R}) \otimes S^q(\mathfrak{g}^*)_n^G$, it follows that

$$H^*(\mathfrak{g}; \mathbb{R}) \otimes S(\mathfrak{g}^*)_n^G \cong E[h_1, h_2, \ldots, h_n] \otimes P_n[c_1, c_2, \ldots, c_n]$$

After identifying the first differential as being $d_1 u_i = c_i$ for $1 \leq i \leq n$, *i.e.* showing that $E_1 \cong W_n$, one can continues as in the first proof in §2.2.

Again, the more satisfying argument follows the pattern of the second proof in §2.2, with the maps φ_m and ψ_m replaced by $\varphi_m^{\rm germ}: W_1(\mathfrak{g})/\mathcal{F}_1^{m+1} \to \bar{\Omega}^*(\bar{N}G)/\mathcal{F}^{m+1}$ and $\psi_m^{\rm germ} : \bar{\Omega}^*(NG)_m \to \bar{\Omega}^*(\bar{N}G)/\mathcal{F}^{m+1}$. The proof that these maps induce isomorphisms of spectral sequences goes along the same lines as in [**28**], with the necessary modifications similar to those used by Świerczkowski to adapt van Est's proof for the bicomplex in [**13**, §10] to the bicomplex in [**27**, §7]. □

2.5. Examples of representative cocycles. First we point out that, similarly to the realization of $\mathcal{H}_n = \mathcal{H}(n, \mathbb{R})$ as a Hopf algebra of 'moving frames', $\Omega^*((\mathrm{GL}(n, \mathbb{R}))$ has a natural representation as a Hopf algebra of 'moving coframes'. This representation is given by the coaction of $G = \mathrm{GL}(n, \mathbb{R})$ on the étale groupoid $\Gamma_n = \mathrm{Diff}_n \ltimes \mathbb{R}^n$ determined by the first jet map (analogous to the local coaction of $\mathrm{GL}(n, \mathbb{R})$ on the C^*-algebra of a codimension n foliation, mentioned by Connes in [**6**, III 7 α]). Specifically, *cf.* [**17**, §5], the formula

$$\rho(\alpha)(\omega U_\varphi) := J^*(\alpha)\omega U_\varphi, \quad \alpha \in \Omega^*(G), \quad \omega \in \Omega_c^*(\Gamma_n), \tag{2.26}$$

where $J : \Gamma_n \to \mathrm{GL}(n, \mathbb{R})$ is the Jacobian map

$$J(\varphi, x) = \varphi'(x) = \big(\partial_j \varphi^i(x)\big) \in \mathrm{GL}(n, \mathbb{R}),$$

defines an action $\rho : \Omega^*(G) \times \Omega_c^*(\Gamma_n) \to \Omega_c^*(\Gamma_n)$.

Parallel to the case of the action of $\mathcal{H}_n$ on Γ_n, there is a closed graded trace $⨍ : \Omega_c^*(\Gamma_n) \to \mathbb{C}$, given by

$$⨍ \omega U_\varphi = \int \omega, \quad \text{if } \varphi = \mathbf{1}, \qquad ⨍ f U_\varphi = 0, \quad \text{if } \varphi \neq \mathbf{1},$$

which is $\Omega^*(G)$-invariant, *i.e.* satisfies

$$⨍ (d\omega) U_\varphi = 0, \quad \text{and} \qquad ⨍ \rho(\alpha)(\omega U_\varphi) = \epsilon(\alpha) ⨍ \omega U_\varphi.$$

With these at hand, one can define (see [**17**, (3.11)]) a characteristic map of cyclic modules $\kappa_* : \Omega^*(G)^\natural \to \Omega_c^*(\Gamma_n)^\natural$ by setting

$$\kappa_q(\alpha^1 \otimes \ldots \otimes \alpha^q)(a_0, a_1, \ldots, a_q) := (-1)^{\mathrm{d}} ⨍ a_0 \rho(\alpha^1)(a_1) \cdots \rho(\alpha^q)(a_q), \tag{2.27}$$

where $\alpha^i \in \Omega^*(G)$, $a_j \in \Omega_c^*(\Gamma_n)$, and $\mathrm{d} := \sum_{i>j} \deg \alpha^i \deg a_j$.

In order to write down examples of Hopf cyclic classes in $HC^*((\Omega^*(G))_n)$, let $\theta \in \Omega^*(G) \otimes \mathfrak{g}^*$ denote the Maurer-Cartan form $\theta_g = g^{-1}dg$, and define $\log|\det| \in \Omega^0(G)$ by $\log|\det|(g) := \log|\det g|$. Then

$$gv_{1,q} := \log|\det| \otimes \mathrm{Tr}\left(\theta^{\otimes q}\right) \in \Omega^q(G^{q+1})$$

is a cocycle representing in $HC^*((\Omega^*(G))_n)$ the Godbillon-Vey class corresponding to $u_1 c_1^q \in H^{2q+1}(W(\mathfrak{g}, G)_n)$. More generally, cocycles representing generalized Godbillon-Vey classes in $HC^*((\Omega^*(G))_n)$ can be obtained as suitable linear combinations of terms of the form

$$\log|\det| \otimes \mathrm{Tr}\left(\theta^{\otimes \ell_1}\right) \otimes \mathrm{Tr}\left(\theta^{\otimes \ell_2}\right) \otimes \cdots \otimes \mathrm{Tr}\left(\theta^{\otimes \ell_k}\right) \in \Omega^q(G^{q+1})$$

where $\lambda = (\ell_1, \ldots, \ell_k)$ runs over the partitions of q. Carried through the characteristic map (2.27), these cocycles acquire expressions reminiscent of those obtained for the same classes by Bott [**2**] in group cohomology, and by Crainic and Moerdijk [**11**, §5.1] in their Čech-de Rham theory for leaf spaces.

The Chern classes in both $HC^*((\Omega^*(G))_n)$ and $HC^*((\Omega^*_{\rm alg}(G))_n)$ can also be represented by cocycles of a similar form only without the transcendental factor $\log|\det|$.

References

[1] R. Bott, *On the Chern-Weil homomorphism and the continuous cohomology of Lie-groups*, Advances in Math. **11** (1973), 289–303, DOI 10.1016/0001-8708(73)90012-1. MR345115

[2] Raoul Bott, *On the characteristic classes of groups of diffeomorphisms*, Enseign. Math. (2) **23** (1977), no. 3-4, 209–220. MR488080

[3] R. Bott, H. Shulman, and J. Stasheff, *On the de Rham theory of certain classifying spaces*, Advances in Math. **20** (1976), no. 1, 43–56, DOI 10.1016/0001-8708(76)90169-9. MR402769

[4] Henri Cartan, *Notions d'algèbre différentielle; application aux groupes de Lie et aux variétés où opère un groupe de Lie* (French), Colloque de topologie (espaces fibrés), Bruxelles, 1950, Georges Thone, Liège; Masson & Cie, Paris, 1951, pp. 15–27. MR0042426

[5] Alain Connes, *Noncommutative differential geometry*, Inst. Hautes Études Sci. Publ. Math. **62** (1985), 257–360. MR823176

[6] Alain Connes, *Noncommutative geometry*, Academic Press, Inc., San Diego, CA, 1994. MR1303779

[7] A. Connes and H. Moscovici, *The local index formula in noncommutative geometry*, Geom. Funct. Anal. **5** (1995), no. 2, 174–243, DOI 10.1007/BF01895667. MR1334867

[8] A. Connes and H. Moscovici, *Hopf algebras, cyclic cohomology and the transverse index theorem*, Comm. Math. Phys. **198** (1998), no. 1, 199–246, DOI 10.1007/s002200050477. MR1657389

[9] Alain Connes and Henri Moscovici, *Cyclic cohomology and Hopf algebras*, Lett. Math. Phys. **48** (1999), no. 1, 97–108, DOI 10.1023/A:1007527510226. Moshé Flato (1937–1998). MR1718047

[10] Connes, A. and Moscovici, H., *Background independent geometry and Hopf cyclic cohomology*, `arXiv:math/0505475`, 2005.

[11] Marius Crainic and Ieke Moerdijk, *Čech-De Rham theory for leaf spaces of foliations*, Math. Ann. **328** (2004), no. 1-2, 59–85, DOI 10.1007/s00208-003-0473-2. MR2030370

[12] Johan L. Dupont, *Simplicial de Rham cohomology and characteristic classes of flat bundles*, Topology **15** (1976), no. 3, 233–245, DOI 10.1016/0040-9383(76)90038-0. MR413122

[13] W. T. van Est, *Group cohomology and Lie algebra cohomology in Lie groups. I, II*, Nederl. Akad. Wetensch. Proc. Ser. A. **56** Indagationes Math. **15** (1953), 484–492, 493–504. MR0059285

[14] W. T. van Est, *Une application d'une méthode de Cartan-Leray* (French), Nederl. Akad. Wetensch. Proc. Ser. A. **58** Indag. Math. **17** (1955), 542–544. MR0073108

[15] I. M. Gel′fand and D. B. Fuks, *Cohomologies of the Lie algebra of formal vector fields* (Russian), Izv. Akad. Nauk SSSR Ser. Mat. **34** (1970), 322–337. MR0266195

[16] Claude Godbillon, *Cohomologies d'algèbres de Lie de champs de vecteurs formels* (French), Séminaire Bourbaki, 25ème année (1972/1973), Exp. No. 421, Lecture Notes in Math., Vol. 383, Springer, Berlin, 1974, pp. 69–87. MR0418112

[17] Alexander Gorokhovsky, *Secondary characteristic classes and cyclic cohomology of Hopf algebras*, Topology **41** (2002), no. 5, 993–1016, DOI 10.1016/S0040-9383(01)00015-5. MR1923996

[18] Hajac, P. M., Khalkhali, M., Rangipour, B. and Sommerhäuser Y., *Stable anti-Yetter-Drinfeld modules, Hopf-cyclic homology and cohomology with coefficients,* C. R. Math. Acad. Sci. Paris **338** (2004), 587–590 and 667–672.

[19] G. Hochschild, *Cohomology of algebraic linear groups*, Illinois J. Math. **5** (1961), 492–519. MR130901

[20] Franz W. Kamber and Philippe Tondeur, *Foliated bundles and characteristic classes*, Lecture Notes in Mathematics, Vol. 493, Springer-Verlag, Berlin-New York, 1975. MR0402773

[21] Franz W. Kamber and Philippe Tondeur, *Semi-simplicial Weil algebras and characteristic classes*, Tohoku Math. J. (2) **30** (1978), no. 3, 373–422, DOI 10.2748/tmj/1178229977. MR509023

[22] Shrawan Kumar and Karl-Hermann Neeb, *Extensions of algebraic groups*, Studies in Lie theory, Progr. Math., vol. 243, Birkhäuser Boston, Boston, MA, 2006, pp. 365–376, DOI 10.1007/0-8176-4478-4_13. MR2214254

[23] Henri Moscovici, *Geometric construction of Hopf cyclic characteristic classes*, Adv. Math. **274** (2015), 651–680, DOI 10.1016/j.aim.2015.02.003. MR3318164

[24] Henri Moscovici, *Equivariant Chern classes in Hopf cyclic cohomology*, Bull. Math. Soc. Sci. Math. Roumanie (N.S.) **58(106)** (2015), no. 3, 317–330. MR3410260

[25] Henri Moscovici and Bahram Rangipour, *Hopf algebras of primitive Lie pseudogroups and Hopf cyclic cohomology*, Adv. Math. **220** (2009), no. 3, 706–790, DOI 10.1016/j.aim.2008.09.017. MR2483227

[26] Henri Moscovici and Bahram Rangipour, *Hopf cyclic cohomology and transverse characteristic classes*, Adv. Math. **227** (2011), no. 1, 654–729, DOI 10.1016/j.aim.2011.02.010. MR2782206

[27] S. Świerczkowski, *Cohomology of group germs and Lie algebras*, Pacific J. Math. **39** (1971), 471–482. MR311746

[28] Herbert Shulman and James Stasheff, *de Rham theory for* $B\Gamma$, Differential topology, foliations and Gelfand-Fuks cohomology (Proc. Sympos., Pontifícia Univ. Católica, Rio de Janeiro, 1976), Lecture Notes in Mathematics, Vol. 652, Springer, Berlin, 1978, pp. 62–74. MR505650

Department of mathematics, The Ohio State University, Columbus, Ohio 43210
Email address: `moscovici.1@osu.edu`

Proceedings of Symposia in Pure Mathematics
Volume **105**, 2023
https://doi.org/10.1090/pspum/105/01908

Primary and secondary invariants of Dirac operators on G-proper manifolds

Paolo Piazza and Xiang Tang

For the 40th birthday of cyclic cohomology

Abstract. In this article, we survey the recent constructions of cyclic cocycles on the Harish-Chandra Schwartz algebra of a connected real reductive Lie group G and their applications to higher index theory for proper cocompact G-actions.

Contents

1. Introduction

Let G be a connected reductive linear Lie group. The study of index theory on proper cocompact G-manifolds goes back to the 70s, in work of Atiyah [**2**], Atiyah-Schmid [**3**], and Connes-Moscovici [**13**]. It is deeply connected to geometric analysis, noncommutative geometry, and representation theory. In September 2021, in the Conference "Cyclic Cohomology at 40", organized by the Fields Institute, the authors reported on the recent studies of cyclic cocycles on the Harish-Chandra Schwartz algebra $\mathcal{C}(G)$ and their applications to higher index theory. In this paper,

2020 *Mathematics Subject Classification.* Primary 58J20; Secondary 58B34, 58J22, 58J42, 19K56.

Key words and phrases. Lie groups, proper actions, higher orbital integrals, delocalized cyclic cocycles, index classes, relative pairing, excision, Atiyah-Patodi-Singer higher index theory, delocalized eta invariants, higher delocalized eta invariants.

The first author was supported in part by *Ministero Università e Ricerca*, through the PRIN *Spazi di moduli e teoria di Lie*. The second author was supported in part by NSF Grants DMS-1800666 and DMS-1952551.

we expand our two talks and provide a more complete account of related topics. The article is organized as follows.

In Section 2, we introduce the geometric set up for proper cocompact G-manifolds and define the C^*-algebraic index of an invariant elliptic operator on a proper cocompact G-manifold. We also explain the connection to the Connes-Kasparov isomorphism conjecture/theorem.

In Section 3, after briefly introducing the definition of cyclic cohomology, we present two approaches to construct cyclic cocycles on the Harish-Chandra Schwartz algebra $\mathcal{C}(G)$. One source of cyclic cocycles is the differentiable group cohomology of G; the other source is the orbital integral and its generalization to higher cyclic cocycles.

In Section 4, we study higher index theory for proper cocompact G-manifolds that have no boundaries. More precisely, we discuss two higher index theorems corresponding to the above two types of cyclic cocycles on $\mathcal{C}(G)$. We also explain these higher index theorems in the special case of $X = G/K$, with their connections to representation theory.

In Section 5, we extend the study of higher index theory to proper cocompact G-manifolds with boundary. We introduce the geometric set up for proper cocompact G-manifold with boundary and adapt the Melrose b-calculus to investigate the higher index on these manifolds.

In Section 6, we introduce the framework of relative cyclic cohomology and apply it to present two higher Atiyah-Patodi-Singer index theorems associated to the cyclic cocycles introduced in Section 3. Higher rho invariants are introduced as spectral invariants for proper cocompact G-manifolds without boundary.

In Section 7, we discuss applications of these results to interesting problems in topology and geometry; in particular we introduce higher genera for G-proper manifolds without boundary and explain their stability properties and their cut-and-paste behaviour; we also discuss bordism invariance of the rho numbers we have introduced.

2. Invariant elliptic operators on proper cocompact manifolds

Let G be a connected reductive linear Lie group with a maximal compact subgroup K. We introduce in this section the index of a G-equivariant elliptic operator on a proper cocompact G-manifold.

2.1. Geometry of Proper Cocompact Manifolds.

DEFINITION 2.1. A smooth manifold X is called a G-proper manifold if X is equipped with a proper G-action, that is, the associated map

$$G \times X \to X \times X, \quad (g, x) \mapsto (x, gx), \quad g \in G, x \in X,$$

is a proper map. This implies that the stabilizer groups G_x of all points $x \in X$ are compact and that the quotient space X/G is Hausdorff. The action is said to be *cocompact* if the quotient X/G is compact. Finally, a cocompact G-proper manifold can always be endowed with a cut-off function c_X, a smooth compactly supported function on X satisfying

$$\int_G c_X(g^{-1}x)dg = 1, \quad \text{for all } x \in X$$

The following theorem was proved by Abels [**1**] for proper cocompact G-manifolds, and will be used crucially in our study of index theory.

THEOREM 2.2. *Let X be a proper cocompact G-manifold (with or without boundary). There is a compact submanifold Z (with or without boundary) which is equipped with a K-action such that*

$$X \cong G \times_K Z.$$

ASSUMPTION 2.3. *We assume in this paper that the manifolds G/K, X and Z are even dimensional.*

Choose a K-invariant inner product on the Lie algebra $\mathfrak{g}$ of G. We then have an orthogonal decomposition $\mathfrak{g} = \mathfrak{k} \oplus \mathfrak{p}$ where $\mathfrak{k}$ is the Lie algebra of K and $\mathfrak{p}$ its orthogonal complement. Accordingly, we have an isomorphism

$$TY \cong G \times_K (\mathfrak{p} \oplus TZ), \tag{2.4}$$

where in the above decomposition we have abused notation and employed $\mathfrak{p}$ for the trivial vector bundle $\mathfrak{p} \times Z \to Z$.

DEFINITION 2.5. We say that the G-invariant metric $\mathbf{h}$ on X is slice-compatible, if it is obtained by a K-invariant metric on Z and a K-invariant metric on $\mathfrak{p}$ via Equation (2.4).

We assume that that adjoint representation $\mathrm{Ad} : K \to SO(\mathfrak{p})$ admits a lift $\widetilde{\mathrm{Ad}} : K \to \mathrm{Spin}(\mathfrak{p})$. By the following exact sequence of vector bundles and the two out of three lemma for Spin^c-structures,

$$0 \to G \times_K \mathfrak{p} \to TX \to G \times_K TZ \to 0,$$

we see that a K-invariant Spin^c-structure on Z induces a G-invariant Spin^c-structure on X.

DEFINITION 2.6. We shall say that a G-invariant Spin^c-structure on X is slice-compatible if it is associated to a K-invariant metric and a K-invariant Spin^c-structure on the slice in the above way.

We consider a G-equivariant twisted spinor bundle E on X, i.e. $E = \mathcal{S} \otimes W$ (for the spinor bundle $\mathcal{S}$ on X and a G-equivariant Hermitian vector bundle W). We shall consider E as arising from a K-invariant twisted spinor bundle E_Z on Z, defined by the slice-compatible Spin^c-structure on Z and by an auxiliary K-equivariant vector bundle on Z. Then E is equipped with a G-equivariant $\mathrm{Cliff}(TX)$-module structure. A *Clifford connection* ∇^E on E is a connection on E satisfying

$$[\nabla^E_V, \mathbf{c}(V')] = \mathbf{c}(\nabla^{TX}_V V'), \quad V, V' \in C^\infty(X, TX)$$

where $\mathbf{c}$ is the Clifford action and ∇^{TX} is the Levi-Civita connection on X. The Dirac operator associated to the Clifford connection is given by the following composition

$$D\colon C^\infty(X,E) \xrightarrow{\nabla^E} C^\infty(X, T^*X \otimes E) \cong C^\infty(X, TX \otimes E) \xrightarrow{\mathbf{c}} C^\infty(X,E).$$

The vector bundle E on X is defined by a K-equivariant vector bundle E_Z on Z, i.e.

$$E \cong G \times_K (S_{\mathfrak{p}} \otimes E_Z),$$

and E_Z admits a K-equivariant Cliff(TZ)-module structure. Accordingly, we can decompose

$$L^2(Y,E) \cong \left[L^2(G)\otimes S_{\mathfrak{p}}\otimes L^2(Z,E_Z)\right]^K.$$

By the assumption that the metric is slice-compatible, the Dirac operator D decomposes

$$D = D_{G,K}\hat{\otimes}1 + 1\hat{\otimes}D_Z, \tag{2.7}$$

where $D_{G,K}$ is the Spin^c-Dirac operator on $(L^2(G)\otimes S_{\mathfrak{p}})^K$, D_Z is a K-equivariant Dirac operator on E_Z, and $\hat{\otimes}$ means the graded tensor product. We make the following observation,

$$[D_{G,K}\hat{\otimes}1, 1\hat{\otimes}D_Z] = 0, \quad D^2 = D^2_{G,K} + D^2_Z.$$

2.2. Roe's C^*-algebra. Roe's C^*-algebra [**52**] for a complete proper metric space X is a powerful tool to study higher index theory.

More precisely, for $r>0$, let $C_{fp}(X\times X)$ be the space of bounded continuous functions on $X\times X$ satisfying $f(x,y)=0$ for $d(x,y)>r$ for some $r>0$. Elements of $C_{fp}(X\times X)$ naturally define kernels of bounded linear operators on $L^2(X)$. Define $C^*(X)$ to be the completion of $C_{fp}(X\times X)$ with respect to the operator norm on $L^2(X)$. Composition of kernels makes $C^*(X)$ into a C^*-algebra. For a proper cocompact G-manifold X equipped with a G-invariant proper complete metric, we consider $C^G_{fp}(X\times X)$ which is a subspace of $C_{fp}(X\times X)$ consisting of G-invariant bounded continuous functions. And we denote the completion of $C^G_{fp}(X\times X)$ with respect to the operator norm by $C^*_G(X)$, which is a C^*-algebra with respect to operator composition.

In this paper, we will work with smooth subalgebra of $C^*_G(X)$ using the slice theorem (Theorem 2.2), i.e. $X\cong G\times_K Z$. Under the above diffeomorphism, $C^G_{fp}(X\times X)$ can be identified with $\mathcal{A}_G(X)$ which is defined as

$$\mathcal{A}_G(X) := \left(C_c(G)\hat{\otimes}C(S\times S)\right)^{K\times K}.$$

Let $\Psi^{-\infty}(S)$ be the space of smoothing operators on X. Then we consider the subspace $\mathcal{A}^c_G(X)\subset\mathcal{A}_G(X)$ with

$$\mathcal{A}^c_G(X) := \left(C^\infty_c(G)\hat{\otimes}\Psi^{-\infty}(S)\right)^{K\times K}\subset\mathcal{A}_G(X).$$

This is an algebra and it corresponds, under Abels' identification, to the algebra of smoothing G-equivariant operators on X of G-compact support. Notice that the latter or, equivalently, $\mathcal{A}^c_G(X)$ is a subalgebra of the Roe algebra $C^*_G(X)$ in a natural way. More generally, if E is a G-equivariant vector bundle on X, define

$$\mathcal{A}^c_G(X,E) := \left(C^\infty_c(G)\hat{\otimes}\Psi^{-\infty}(S,E|_S)\right)^{K\times K},$$

where $\Psi^{-\infty}(S,E|_S)$ is the space of smoothing operators on $E|_S$. We remark that $C^\infty_c(G)$ and $\Psi^{-\infty}(S)$ are Fréchet algebras, and the tensor products in the above definitions of $\mathcal{A}^c_G(X)$ and $\mathcal{A}^c_G(X,E)$ are projective tensor products. The following description of $\mathcal{A}^c_G(X,E)$ is proved in [**47**, Prop. 1.7]

$$\begin{aligned}&\mathcal{A}^c_G(X,E)\\ &\cong\left\{\Phi: G\to\Psi^{-\infty}(S,E|_S),\ \text{smooth, compactly supported and } K\times K \text{ invariant}\right\}.\end{aligned}\tag{2.8}$$

2.3. The Harish-Chandra Schwartz algebra $\mathcal{C}(G)$. Let $\widehat{G}$ be the tempered dual of isomorphism classes of unitary irreducible representations of $C^*_r(G)$. In noncommutative geometry, $C^*_r(G)$ can be viewed as the algebra of "continuous functions" on $\widehat{G}$. In the following, we introduce the noncommutative version of smooth functions on $\widehat{G}$ which is the Harish-Chandra Schwartz algebra $\mathcal{C}(G)$.

Let v_0 be a unit vector in the spherical representation π of G, and Θ be the Harish-Chandra spherical function on G defined by

$$\Theta(g) := \langle v_0, \pi(g)v_0 \rangle.$$

Let $\mathfrak{g}$ be the Lie algebra of G and $U(\mathfrak{g})$ be the universal enveloping algebra associated to $\mathfrak{g}$. For $V, W \in U(\mathfrak{g})$, let L_V be the differential operator on G associated to the left G-action on G, and R_W be the differential operator on G associated to the right G-action on G.

Consider the Cartan decomposition $G = K \exp \mathfrak{p}$ where $\mathfrak{p}$ is the Lie algebra of the associated parabolic subgroup P. For $X \in \mathfrak{p}$, let $\|X\|$ be a K-invariant Euclidean norm on $\mathfrak{p}$. Let $L : G \to \mathbb{R}^+$ be the function on G defined by,

$$L(g) := \|X\|$$

for $g = k\exp(X)$ with $k \in K$ and $X \in \mathfrak{p}$.

DEFINITION 2.9. Define the seminorm $\nu_{V,W,m}(-)$ by

$$\nu_{V,W,m}(f) := \sup_{g\in G} \| |1 + L(g)|^m \Theta(g)^{-1} L_V R_W(f)(g)\|, \; f \in C^\infty_c(G).$$

The Harish-Chandra Schwartz algebra $\mathcal{C}(G)$ for G is the space of smooth functions f on G such that

$$\nu_{V,W,m}(f) < \infty, \quad \text{for } V, W \in U(\mathfrak{g}), m \geq 0.$$

By [**30**], the Harish-Chandra Schwartz algebra $\mathcal{C}(G)$ is a subalgebra of $C^*_r(G)$ stable under holomorphic functional calculus. Therefore, we have

$$K_i(\mathcal{C}(G)) \cong K_i(C^*_r(G)), \; i = 0, 1.$$

Starting from the Harish-Chandra Schwartz algebra $\mathcal{C}(G)$ we can also define an algebra of smoothing G-equivariant operators on X by considering

$$\mathcal{A}^\infty_G(X, E) := \left(\mathcal{C}(G)\hat{\otimes}\Psi^{-\infty}(S, E|_S)\right)^{K\times K}.$$

One can extend (2.8) and prove the following description of $\mathcal{A}^\infty_G(X, E)$.

$$\mathcal{A}^\infty_G(X, E) \cong \left\{ \begin{array}{l} \Phi : G \to \Psi^{-\infty}(S), \; K \times K \text{ invariant and} \\ g \mapsto \nu_{V,W,m}(\|\Phi(g)\|_\alpha) \text{ bounded } \forall \alpha, m, V, W \end{array} \right\} \tag{2.10}$$

The algebra $\mathcal{A}^\infty_G(X, E)$ is a subalgebra of the Roe C^*-algebra $C^*_G(X, E)$ and it is not difficult to prove, see [**46**, Proposition 3.9], that it is dense and holomorphically closed. Thus we have

$$K_i(\mathcal{A}^\infty_G(X, E)) \cong K_i(C^*_G(X, E)), \; i = 0, 1.$$

We end this subsection by pointing out that there exists a Morita isomorphism $\mathcal{M}$ between $C^*_G(M, E)$ and $C^*_r(G)$; this can be justified by general principles, since $C^*_G(M, E)$ is isomorphic to the C^*-algebra of compact operators $\mathbb{K}(\mathcal{E})$, with $\mathcal{E}$ denoting the $C^*_r(G)$-Hilbert module obtained by closing the space of compactly supported sections of E on X, $C^\infty_c(X, E)$, endowed with the $C^*_r G$-valued inner product $(e, e')_{C^*_r G}(g) := (e, g \cdot e')_{L^2(X,E)}$, $\quad e, e' \in C^\infty_c(X, E)$, $g \in G$ (and it is well known that then $K_i(\mathbb{K}(\mathcal{E})) = K_*(C^*_r(G))$.) Following Hochs-Wang [**28**], we prefer to implement explicitly this isomorphism as follows: we consider $\mathcal{A}^\infty_G(M,E)$ and $\mathcal{C}(G)$ and consider a partial trace map $\mathrm{Tr}_S : \mathcal{A}^\infty_G(X,E) \to \mathcal{C}(G)$ associated to the slice S: if $f \otimes k \in (\mathcal{C}(G))\hat{\otimes}\Psi^{-\infty}(S, E|_S))^{K\times K}$ then

$$\mathrm{Tr}_S(f \otimes k) := f\,\mathrm{Tr}(T_k) = f \int_S \mathrm{tr}\, k(s,s)ds,$$

with T_k denoting the smoothing operator on S defined by k and $\mathrm{Tr}(T_k)$ its functional analytic trace on $L^2(S, E|_S)$. It is proved in [**28**] that this map induces the isomorphism $\mathcal{M}$ between $K_*(C^*_G(X,E))$ and $K_*(C^*_r(G))$.

2.4. The index class of a G-invariant elliptic operator. Recall that X is even-dimensional. Let D be an odd $\mathbb{Z}_2$-graded Dirac operator, equivariant with respect to the G-action. Recall, first of all, the classical Connes-Skandalis idempotent. Let Q be a G-equivariant parametrix of G-compact support with remainders $S_\pm \in \mathcal{A}^c(X,E)$; consider the 2×2 idempotent

$$P_Q := \begin{pmatrix} S_+^2 & S_+(I+S_+)Q \\ S_-D^+ & I - S_-^2 \end{pmatrix}. \tag{2.11}$$

This produces a well-defined class

$$\mathrm{Ind}_c(D) := [P_Q] - [e_1] \in K_0(\mathcal{A}^c_G(X,E)) \text{ with } e_1 := \begin{pmatrix} 0 & 0 \\ 0 & 1 \end{pmatrix}. \tag{2.12}$$

DEFINITION 2.13. The C^*-index associated to D is the class $\mathrm{Ind}_{C^*_G(M,E)}(D) \in K_0(C^*_G(X,E))$ obtained by considering $[P_Q] - [e_1]$ as a formal difference of idempotents with entries in $C^*_G(X,E)$, under the continuous inclusion $\iota : \mathcal{A}^c_G(X,E) \hookrightarrow C^*_G(X,E)$.

One can also give a definition of $\mathrm{Ind}_{C^*_G(X,E)}(D) \in K_0(C^*(X,E)^G)$ using Coarse Index Theory, as in the book of Higson and Roe, see [**21**]; the compatibility of the two definitions is proved in [**49**, Proposition 2.1].

We denote the image through the Morita isomorphism $\mathcal{M}$ of the index class $\mathrm{Ind}_{C^*_G(X,E)}(D) \in K_0(C^*_G(X,E))$ in the group $K_0(C^*_r(G))$ by $\mathrm{Ind}_{C^*_r(G)}(D)$. There are other, well-known descriptions of the latter index class: one, following Kasparov, see [**29**], describes the $C^*_r(G)$-index class as the difference of two finitely generated projective $C^*_r(G)$-modules, using the invertibility modulo $C^*_r(G)$-compact operators of (the bounded-tranform of) D; the other description is via assembly and KK-theory, as in the classic article by Baum, Connes and Higson [**4**]. All these descriptions of the class $\mathrm{Ind}_{C^*_r(G)}(D) \in K_0(C^*_r(G))$ are equivalent. See [**53**] and [**49**, Proposition 2.1].

There is another way of expressing the index class $\mathrm{Ind}_{C^*_G(X,E)}(D) \in K_0(C^*_G(X,E))$; this is due to Connes–Moscovici [**12**] and employs the parametrix

$$Q := \frac{I - \exp(-\frac{1}{2}D^-D^+)}{D^-D^+}D^+ \tag{2.14}$$

with $I - QD^+ = \exp(-\frac{1}{2}D^-D^+)$, $I - D^+Q = \exp(-\frac{1}{2}D^+D^-)$. This particular choice of parametrix produces the idempotent

$$V_{\mathrm{CM}}(D) = \begin{pmatrix} e^{-D^-D^+} & e^{-\frac{1}{2}D^-D^+}\left(\frac{I-e^{-D^-D^+}}{D^-D^+}\right)D^- \\ e^{-\frac{1}{2}D^+D^-}D^+ & I - e^{-D^+D^-} \end{pmatrix} \tag{2.15}$$

where CM stands for Connes and Moscovici. We certainly have $\mathrm{Ind}_{C^*_G(M,E)}(D) = [V_{\mathrm{CM}}(D)] - [e_1]$.
The following result is proved in Hochs-Song-Tang [**24**] and in Piazza-Posthuma [**47**]:

PROPOSITION 2.16. *The idempotent* $V_{\mathrm{CM}}(D)$ *is an element in* $M_{2\times 2}(\mathcal{A}^\infty_G(X,E))$ *(with the identity adjoined).*

As $\mathcal{A}^\infty(M,E)$ is holomorphically closed in $C^*_G(X,E)$ we have

$$\begin{aligned} &\mathrm{Ind}_{C^*_G(X,E)}(D) \equiv \mathrm{Ind}_{\mathcal{A}^\infty(M,E)}(D) \\ &=[V_{\mathrm{CM}}(D)] - [e_1] \in K_0(\mathcal{A}^\infty(M,E)) = K_0(C^*(M,E)^G). \end{aligned} \tag{2.17}$$

We call $[V_{\mathrm{CM}}(D)] - [e_1] \in K_0(\mathcal{A}^\infty(M,E))$ the *smooth* index class associated to D.

As in [**39**], [**16**] we shall also consider the adjoint $V^*_{\mathrm{CM}}(D)$, which is again an element with entries in $\mathcal{A}^\infty_G(X,E)$. It is easy to prove that $[V_{\mathrm{CM}}(D)] - [e_1] = [V^*_{\mathrm{CM}}(D)] - [e_1]$ in $K_0(\mathcal{A}^\infty_G(X,E)) = K_0(C^*_G(X,E))$ through an explicit homotopy.

DEFINITION 2.18. We set $V(D) := V_{\mathrm{CM}}(D) \oplus V^*_{\mathrm{CM}}(D)$.

Let $f_1 = e_1 \oplus e_1$. Then

$$[V(D)] - [f_1] = 2([V_{\mathrm{CM}}(D)] - [e_1]) \equiv 2\,\mathrm{Ind}_{C^*(M,E)^G}(D) \quad \text{in} \quad K_0(C^*_G(X,E)). \tag{2.19}$$

We call $V(D)$ the symmetrized Connes-Moscovici projector; as we shall see, it plays an important role in the proof of explicit higher index formulae.

Let $X = G/K$ with the proper left G action. For a highest weight μ of K, let V_μ be the corresponding unitary irreducible K-representation associated to μ, and $\widetilde{V}_\mu$ be the associated vector bundle on G/K defined by $E_\mu := G \times V_\mu/K$. We consider the Dirac operator D_μ associated to the vector bundle E_μ on X. Applying the construction in Definition 2.18, we obtain an element $\mathrm{Ind}_{C^*_G(X,E_\mu)}(D_\mu) \in K_0(C^*_G(X,E_\mu)^G) = K_0(C^*_r(G))$. Varying μ, we obtain the following morphism of abelian groups

$$\mathrm{Ind} : \mathfrak{Rep}(K) \to K_0(C^*_r(G)),$$

where $\mathfrak{Rep}(K)$ is the representation ring of the compact group K.

Connes and Kasparov conjectured in the early 80s that the above index map Ind is an isomorphism, [**4**], providing a geometric approach to compute the K-theory groups of $C^*_r(G)$. The study of the index map has been one of the driving forces in the study of higher index theory on proper cocompact G-manifolds. The isomorphism theorem is now established in great generality: it is satisfied by all almost connected topological groups [**8**].

3. Cyclic cocycles on $\mathcal{C}(G)$

In this section, we present two methods to construct cyclic cocycles on $\mathcal{C}(G)$.

3.1. Cyclic cohomology. Cyclic cohomology was introduced by Alain Connes [**10**] as the noncommutative version of de Rham cohomology. We briefly recall it below.

DEFINITION 3.1. Let A be a Fréchet algebra over $\mathbb{C}$. The space of Hochschild cochains of degree k of A is defined to be the space

$$C^k(A) \colon = \mathrm{Hom}_{\mathbb{C}}\big(A^{\otimes(k+1)}, \mathbb{C}\big)$$

of all bounded $(k+1)$-linear functionals on A. The Hochschild codifferential $b \colon C^k(A) \to C^{k+1}(A)$ is defined by

$$\begin{aligned} &b\Phi(a_0 \otimes \cdots \otimes a_{k+1}) \\ &= \sum_{i=0}^{k} (-1)^i \Phi(a_0 \otimes \cdots \otimes a_i a_{i+1} \otimes \cdots \otimes a_{k+1}) + (-1)^{k+1} \Phi(a_{k+1} a_0 \otimes a_1 \otimes \cdots \otimes a_k). \end{aligned}$$

The Hochschild cohomology of A is the cohomology of the complex $(C^\bullet(A), b)$.

DEFINITION 3.2. A Hochschild k-cochain $\Phi \in C^k(A)$ is called *cyclic* if

$$\Phi(a_k, a_0, \ldots, a_{k-1}) = (-1)^k \Phi(a_0, a_1, \ldots, a_k), \quad \forall a_0, \ldots, a_k \in A.$$

The subspace $C^k_\lambda(A)$ of cyclic cochains is closed under the Hochschild codifferential. The cyclic cohomology $HC^k_\lambda(A)$ is defined to be the cohomology of the subcomplex of cyclic cochains.

There exists a natural operator $S : HC^k_\lambda(A) \to HC^{k+2}_\lambda(A)$, the periodicity operator, and the periodic cyclic cohomology is defined as follows

$$HP^k(A) := \lim_{n\to\infty} HC^{k+2n}_\lambda(A), \ k = 0, 1.$$

To understand the definition of Hochschild and cyclic cohomology, we look at the definition for $k = 0$. A degree 0 Hochschild cochain Φ on A is a linear functional on A. The coboundary of Φ is a 1-cochain defined as follows

$$b\Phi(a_0, a_1) = \Phi(a_0 a_1) - \Phi(a_1 a_0).$$

The above formula for $b\Phi$ suggests that Φ is a Hochschild cocycle if $\Phi(a_0 a_1) = \Phi(a_1 a_0)$, $\forall a_0, a_1 \in A$, i.e. Φ is a trace on A. For $k = 0$, the cyclic property trivially holds. Hence, the degree 0 cyclic cohomology of A consists of traces on A. In general, cyclic cohomology is a natural generalization of traces.

The cyclic cohomology can also be computed by the following b-B bicomplex on $C^\bullet(A)$. Consider the operator $B : C^k(A) \to C^{k-1}(A)$ by the formula

$$\begin{aligned} B\Phi(a_0 \otimes \ldots \otimes a_{k-1}) := &\sum_{i=0}^{k-1} (-1)^{i(k-1)} \Phi(1, a_i, \cdots, a_{k-1}, a_0, \ldots, a_{i-1}) \\ &- (-1)^{i(k-1)} \Phi(a_i, 1, a_{i+1}, \cdots, a_{k-1}, a_0, \cdots, a_{i-1}). \end{aligned}$$

This defines a differential, i.e., $B^2 = 0$, and we have $[B, b] = 0$, so we can form the (b, B)-bicomplex.

$$
\begin{array}{ccccc}
\cdots & & \cdots & & \cdots \\
{\scriptstyle b}\uparrow & & {\scriptstyle b}\uparrow & & {\scriptstyle b}\uparrow \\
C^2(A) & \xrightarrow{B} & C^1(A) & \xrightarrow{B} & C^0(A) \\
{\scriptstyle b}\uparrow & & {\scriptstyle b}\uparrow & & \\
C^1(A) & \xrightarrow{B} & C^0(A) & & \\
{\scriptstyle b}\uparrow & & & & \\
C^0(A) & & & &
\end{array}
$$

The cyclic cohomology $HC^k_\lambda(A)$ is isomorphic to the degree k total cohomology of the above b-B bicomplex, c.f. [**36**, Sec. 2.4].

For the application to index theory, we are interested in the pairing between cyclic cohomology and K-theory of A. Let $P = (p_{ij}), i, j = 1, \ldots, m$ be an idempotent in $M_m(A)$, the space of $m \times m$ matrices with entries in A. The following formula

$$\langle [\Phi], [P] \rangle : = \frac{1}{k!} \sum_{i_0, \cdots, i_{2k}=1}^{m} \Phi(p_{i_0 i_1}, p_{i_1 i_2}, \ldots, p_{i_{2k} i_0})$$

defines a natural pairing between $[\Phi] \in HP^{\text{even}}(A)$ and $K_0(A)$, i.e.

$$\langle \cdot\, , \cdot \rangle : HP^{\text{even}}(A) \otimes K_0(A) \to \mathbb{C}.$$

In this article, we are interested in constructing cyclic cocycles of the Harish-Chandra Schwartz algebra $\mathcal{C}(G)$ and applying them on the one hand to study the structure of $K_\bullet(\mathcal{C}(G)) \cong K_\bullet(C^*_r(G))$ via the above pairing and, on the other hand, to obtain and study numeric invariants of a Dirac operator through the pairing of the associated K-theory index class with such cyclic cocycles.

Our study is inspired by the computation of the periodic cyclic cohomology of the group algebra of a discrete group, [**7**, **11**]. Let Γ be a discrete group and $\mathbb{C}\Gamma$ be the group algebra, i.e. $\mathbb{C}\Gamma := \bigoplus_{\gamma \in \Gamma} \mathbb{C}\gamma$ such that $\gamma_1 * \gamma_2 := \gamma_1 \gamma_2$.

$$HP^\bullet(\mathbb{C}\Gamma) = \prod_{\hat{\gamma} \in \langle \Gamma \rangle'} \left(H^\bullet(N_\gamma, \mathbb{C}) \otimes HP^\bullet(\mathbb{C}) \right) \times \prod_{\hat{\gamma} \in \langle \Gamma \rangle''} H^\bullet(N_\gamma, \mathbb{C}), \tag{3.3}$$

where $\langle \Gamma \rangle$ is the set of conjugacy classes of Γ, $\langle \Gamma \rangle' \subset \langle \Gamma \rangle$ is the subset of classes of elements $\gamma \in \Gamma$ of finite order, $\langle \Gamma \rangle''$ its complement, and N_γ is the normalizer of γ defined by $C_\gamma / (\gamma)$, the quotient of the centralizer C_γ of γ by the subgroup (γ) generated by γ. Let O_γ be the conjugacy class associated to γ. On $\mathbb{C}\Gamma$, we have the following trace tr_γ associated to O_γ:

$$\mathrm{tr}_\gamma(f) = \sum_{\alpha \in O_\gamma} f(\alpha) = \sum_{\beta \in G / C_\gamma} f(\beta \gamma \beta^{-1}), \tag{3.4}$$

where we recall that C_γ is the centralizer of γ. The trace tr_γ is a degree 0 cyclic cocycle on $\mathbb{C}\Gamma$ associated to the conjugacy class $O_\gamma \in \langle \Gamma \rangle$. The computation (3.3) shows that there are many interesting higher degree cyclic cocycles on the group algebra $\mathbb{C}\Gamma$.

A complete generalization of the computation (3.3) for a general Lie group is still under search. In (3.3), the cyclic cohomology of $\mathbb{C}\Gamma$ is decomposed into conjugacy classes of the group Γ. As a step toward understanding the cyclic cohomology of $\mathcal{C}(G)$, we will discuss below recent developments in the following special cases.

(1) We shall study cyclic cocycles associated to the identity conjugacy class of G. In [**44, 46**] a chain map was introduced from differentiable group cohomology of G to the cyclic cohomology of $\mathcal{C}(G)$ generalizing the trace tr_e associated to the identity element e.
(2) We shall study the orbital integral associated to a conjugacy class of G, a direct generalization of (3.4). Next, following [**54**], we shall define *higher* orbital integrals on $\mathcal{C}(G)$ and study their properties.

3.2. Case of $\mathbb{R}^n$. In this subsection, we look at the abelian group $\mathbb{R}^n$. The Harish-Chandra Schwartz algebra $\mathcal{C}(\mathbb{R}^n)$ consists of Schwartz functions on $\mathbb{R}^n$ with the convolution product

$$f * g(x) = \int_{\mathbb{R}^n} f(y)g(x-y)dy.$$

Under the Fourier transform,

$$\mathcal{F}(f)(\xi) = \int_{\mathbb{R}^n} f(x)\exp^{-2\pi i\xi\cdot x}dx,$$

the image of $\mathcal{F}(f)$ is a Schwartz function on $\widehat{\mathbb{R}}^n$ for $f \in \mathcal{C}(\mathbb{R}^n)$, and the convolution product is transformed to the pointwise multiplication

$$f\cdot g(\xi) = f(\xi)g(\xi),\ \mathcal{F}(f*g) = \mathcal{F}(f)\cdot\mathcal{F}(g).$$

Using the Fourier transform, we conclude that the cyclic cohomology of $\mathcal{C}(\mathbb{R}^n)$ is isomorphic to the one of $\mathcal{S}(\widehat{\mathbb{R}}^n)$, which by [**10**, Theorem 46] is computed as follows

$$HP^i(\mathcal{C}(\mathbb{R}^n)) = \begin{cases} 0, & i \not\equiv n\ (mod\ 2), \\ \mathbb{C}, & i \equiv n\ (mod\ 2). \end{cases}$$

On $\mathcal{S}(\widehat{\mathbb{R}}^n)$, the n-th cyclic cohomology group is generated by the following cocycle

$$\Psi(\hat{f}_0, \cdots, \hat{f}_n) = \int_{\mathbb{R}^n} \hat{f}_0 d\hat{f}_1\cdots d\hat{f}_n.$$

On $\mathcal{C}(\mathbb{R}^n)$, the Fourier transform of Ψ has the following form. Define a function $C : \underbrace{\mathbb{R}^n \times \cdots \times \mathbb{R}^n}_{n} \to \mathbb{R}$ by

$$C(x_1, \cdots, x_n) := \begin{vmatrix} x_1^1 & \cdots & x_1^n \\ & \cdots & \\ x_n^1 & \cdots & x_n^n \end{vmatrix}. \tag{3.5}$$

Define Φ to be a cocycle on $\mathcal{C}(\mathbb{R}^n)$ by

$$\Phi(f_0, \cdots, f_n) := \int_{\mathbb{R}^n} \cdots \int_{\mathbb{R}^n} dx_1\cdots dx_n \\ C(x_1, \cdots, x_n) f_0(-x_1 - \cdots - x_n) f_1(x_1)\cdots f_n(x_n).$$

It is by direct computation that one can show that up to a constant Φ is the Fourier transform of the cyclic cocycle Ψ on $\mathcal{S}(\widehat{\mathbb{R}}^n)$.

PROPOSITION 3.6. *The cyclic cocycle* Φ *is the generator of the cyclic cohomology of* $\mathcal{C}(\mathbb{R}^n)$.

The generalization of Φ to general groups plays the central role in the following constructions.

3.3. Differentiable group cohomology. We observe that the function C defined in Eq. (3.5) is a group cocycle on $\mathbb{R}^n$. And the correspondence, $C \mapsto \Phi$, can be viewed as the analog of $H^\bullet(N_\gamma)$ in the cyclic cohomology of $\mathcal{C}(G)$ for $\gamma = e$ and $G = \mathbb{R}^n$ with $N_e = G$. In [**44, 46**], this construction was generalized to general Lie group(oid)s. In literature, two different but equivalent models have been used in the definition of differentiable group cohomology. We recall them below.

DEFINITION 3.7. Let $C^\infty(G^{\times k})$ be the space of smooth functions on $\underbrace{G \times \cdots \times G}_{k}$.

Define the differential $\delta : C^\infty(G^{\times k}) \to C^\infty(G^{\times k})$ by

$$\begin{aligned} & \delta(\varphi)(g_1, \cdots, g_{k+1}) \\ = \; & \varphi(g_2, \cdots, g_{k+1}) - \varphi(g_1 g_2, \cdots, g_{k+1}) \\ + \; & \cdots + (-1)^k \varphi(g_1, \cdots, g_k g_{k+1}) + (-1)^{k+1} \varphi(g_1, \cdots, g_k). \end{aligned}$$

The differentiable group cohomology $H^\bullet_{\text{diff}}(G)$ is defined to be the cohomology of $(C^\infty(G^{\times \bullet}), \delta)$, which is called the normalized differentiable group cohomology complex.

DEFINITION 3.8.

$$C^k_{\text{diff}}(G) := \left\{ \begin{array}{l} c : G^{\times(k+1)} \to \mathbb{C} \text{ smooth}, \\ c(gg_0, \ldots, gg_k) = c(g_0, \ldots, g_k), \ \forall g, g_0, \ldots, g_k \in G \end{array} \right\}.$$

Define the differential $\partial : C^k_{\text{diff}}(G) \to C^{k+1}_{\text{diff}}(G)$ as

$$(\partial c)(g_0, \ldots, g_{k+1}) := \sum_{i=0}^{k+1} (-1)^i c(g_1, \ldots, \hat{g}_i, \ldots, g_{k+1}).$$

The differentiable group cohomology $H^\bullet_{\text{diff}}(G)$ can also be computed by the cohomology of the chain complex $(C^\bullet_{\text{diff}}(G), \partial)$, which is called the homogenous differentiable group cohomology complex.

In this paper, following the literature, we will use both chain complexes introduced in Definition 3.7 and 3.8. And we will explicitly point out the models used in the respective formulas below.

Fix a Haar measure dg on G.

DEFINITION 3.9. Define a pairing between $C^\infty(G^{\times k})$ and $C^\infty_c(G)^{\hat{\otimes}(k+1)}$ by

$$\langle \hat{\varphi}, f_0 \otimes \cdots \otimes f_k \rangle := \int f_0(g_k^{-1} \cdots g_1^{-1}) f_1(g_1) \cdots f_k(g_k) \varphi(g_1, \cdots, g_k) dg_1 \cdots dg_k.$$

In the above formula, we have used the normalized differentiable cohomology complex in Definition 3.7.

THEOREM 3.10. *Assume that* G *is unimodular, i.e. the Haar measure is bi-invariant. The above pairing descends to a character morphism* χ,

$$\chi : H^\bullet_{\text{diff}}(G) \to HP^\bullet(\mathcal{C}(G)).$$

REMARK 3.11. In [**44**], the character morphism was introduced as a map from $H^\bullet_{\rm diff}(G)$ to $HP^\bullet(C_c^\infty(G))$, where $C_c^\infty(G)$ is the convolution algebra of compactly supported functions. When G has property RD and the homogeneous space G/K has nonpositive sectional curvature, Piazza and Posthuma [**46**] improved this character morphism as a map from $H^\bullet_{\rm diff}(G)$ to the cyclic cohomology of $H_L^\infty(G)$, which for a length function L on G is defined as follows,

$$H_L^\infty(G) = \left\{ f \in L^2(G) | \int_G (1+L(g))^{2k} |f(g)|^2 dg < \infty, \ \forall k \in \mathbb{N} \right\}.$$

For a connected reductive linear Lie group G, G/K is a Hermitian symmetric space of noncompact type [**20**], i.e. G/K has nonpositive sectional curvature. We observe that the same argument for $H_L^\infty(G)$ can be applied to the Harish-Chandra Schwarz algebra $\mathcal{C}(G)$ via the techniques in [**54**, Appendix A.]. This is how we reached the final statement for Theorem 3.10.

EXAMPLE 3.12. For $G = SL_2(\mathbb{R})$, by the van Est isomorphism, $H^2_{\rm diff}(SL_2(\mathbb{R}))$ is generated by the area cocycle, which has the following geometric description. Let $\mathbb{H}$ be the upper half space identified with the homogenous space $SL_2(\mathbb{R})/SO(2)$. Let $[e]$ be the point in $\mathbb{H}$ corresponding to the coset $eSO(2)$ in $SL_2(\mathbb{R})/SO(2)$. As $\mathbb{H}$ is equipped with a metric of constant negative curvature, any two points in $\mathbb{H}$ are connected by a unique geodesic. Given $g_1, g_2, g_3 \in SL_2(\mathbb{R})$, we consider $g_i[e]$, $i = 1, 2, 3$, in $\mathbb{H}$, and the corresponding geodesic triangle $\Delta(g_1[e], g_2[e], g_3[e])$ with vertices $g_1[e], g_2[e], g_3[e]$. Define a smooth function A on $SL_2(\mathbb{R}) \times SL_2(\mathbb{R}) \times SL_2(\mathbb{R})$ as follows,

$$A(g_0, g_1, g_2) := \mathrm{Area}_{\mathbb{H}}(\Delta(g_0[e], g_1[e], g_2[e])),$$

where $g_i \in SL(2,\mathbb{R})$ acts on $\mathbb{H}$ as usual by Möbius transformations. It is straightforward to check that A is a smooth 2-cocycle on $SL_2(\mathbb{R})$ (in the homogeneous differentiable group cohomology chain complex introduced in Definition 3.8), c.f. [**46**, Remark 2.20, (ii)], generalizing the volume cocycle (3.5) for $\mathbb{R}^n$ in Sec. 3.2. Furthermore, its image $\chi(A)$ under the character morphism can be identified with the Connes-Chern character of the fundamental K-cycle of $SL_2(\mathbb{R})$ in [**10**, I.9, Lemma 5]. For $f_0, f_1, f_2 \in \mathcal{C}(SL_2(\mathbb{R}))$,

$$\chi(A)(f_0, f_1, f_2) = \int_{SL_2(\mathbb{R})} \int_{SL_2(\mathbb{R})} f_0((g_1g_2)^{-1}) f_1(g_1) f_2(g_2) \\ \mathrm{Area}_{\mathbb{H}}\left(\Delta([e], g_1[e], g_1g_2[e])\right) dg_1 dg_2.$$

3.4. Higher orbital integrals. The trace tr_γ introduced in Equation (3.4) on $\mathbb{C}\Gamma$ has a natural generalization tr_g for G, i.e.

$$\mathrm{tr}_g(f) := \int_{G/Z_g} f(hgh^{-1}) d(hZ_g). \tag{3.13}$$

For a discrete group Γ, it is not clear whether the trace tr_γ on $\mathbb{C}\Gamma$, for a general element γ, pairs with $K_0(C_r^*(\Gamma))$. Still, although a general statement is missing, the pairing is well-defined for groups of polynomial growths and, more importantly, for Gromov hyperbolic groups. The latter example has been a challenge for some time and has been settled in the affirmative way by Puschnigg, who defined a dense holomorphically closed subalgebra of $C_r^*(\Gamma)$ to which the trace tr_γ extends. See [**51**]. Higher cyclic cocycles associated to the other elements in the Burghelea's decomposition were studied in detail by Chen-Wang-Xie-Yu in [**9**] and by Piazza-Schick-Zenobi in [**50**]; in the above articles (and references therein) many purely geometric applications were also given. Notice that there is an analogy between the Puschnigg algebra and the extendability of delocalized (higher) cyclic cocycles for discrete Gromov hyperbolic groups and the Harish-Chandra Schwartz algebra and the extendability of delocalized (higher) cyclic cocycles for connected real reductive Lie groups.

We go back to the orbital integral (3.13). Hochs and Wang [**28**], building on results of Harish-Chandra [**19**], proved that for a semisimple element[1] g the trace functional tr_g is well defined on the Harish-Chandra Schwartz algebra $\mathcal{C}(G)$ and therefore pairs with $K_0(\mathcal{C}(G)) \cong K_0(C_r^*(G))$. Hochs and Wang [**28**] computed the pairing between tr_g and the index element $\mathrm{Ind}_{C_r^*(G)}(D)$ for a G-equivariant Dirac operator on a proper cocompact G-manifold X; this explicit formula will be recalled in Theorem 4.11 below. As we shall see in Theorem 4.11, when G does not have a compact Cartan subgroup the trace tr_g is a trivial cyclic cohomology class. This raises a natural question to find a generalization of tr_g that pairs nontrivially with $K_0(C_r^*(G))$. The problem was raised and solved in [**54**], as we shall now explain .

Let $K < G$ be a maximal compact subgroup and let $P < G$, $P = MAN$, be a cuspidal parabolic subgroup of G. Using the Iwasawa decomposition $G = KMAN$ we write an element $g \in G$ as

$$g = \kappa(g)\mu(g)e^{H(g)}n \in KMAN = G.$$

We observe that the function $H(g)$ is well defined on G though the components $\kappa(g)$ and $\mu(g)$ in the Iwasawa decomposition may not be unique. Let $\dim(A) = m$. Choosing coordinates of the Lie algebra $\mathfrak{a}$ of A, we define the function

$$H = (H_1, \ldots, H_m) : G \to \mathfrak{a}.$$

Song and Tang defined in [**54**] the following cyclic cocycle on $\mathcal{C}(G)$, the Harish-Chandra Schwartz algebra of G.

DEFINITION 3.14. For $f_0, \ldots, f_m \in \mathcal{C}(G)$ and a semi-simple element $g \in M$, define Φ_g^P by the following integral,

(3.15)

$$\begin{aligned}\Phi_g^P&(f_0, f_1, \ldots, f_m)\\ &:= \int_{h\in M/Z_M(x)} \int_{KN} \int_{G^{\times m}} C\big(H(g_1 \ldots g_m k), H(g_2 \ldots g_m k), \cdots, H(g_m k)\big)\\ &\qquad f_0\big(khgh^{-1}nk^{-1}(g_1 \ldots g_m)^{-1}\big) f_1(g_1) \ldots f_m(g_m) dg_1 \cdots dg_m dk dn dh,\end{aligned}$$

where $Z_M(x)$ is the centralizer of x in M.

It is easy to check that the integral in Equation (3.15) is convergent for $f_0, \ldots, f_m \in C_c^\infty(G)$. Song and Tang proved [**54**, Theorem 3.5] the following property for Φ_g^P.

THEOREM 3.16. *For a cuspidal parabolic subgroup $P = MAN$ of G, and a semisimple element $g \in M$, the cochain Φ_g^P is a cyclic cocycle on $\mathcal{C}(G)$.*

[1] that is, the corresponding operator $\mathrm{Ad}_g : \mathfrak{g} \to \mathfrak{g}$ is diagonalizable.

3.5. From cocycles on $\mathcal{C}(G)$ to cocycles on $\mathcal{A}_G^\infty(X,E)$. In Section 2.3, we introduced $\mathcal{A}_G^\infty(X,E)$, a dense subalgebra of the Roe algebra $C_G^*(X,E)$ that has the same K-theory groups as $\mathcal{C}(G)$. In this subsection, we explain how to lift the cocycles on $\mathcal{C}(G)$ to $\mathcal{A}_G^\infty(X,E)$.

Recall, see (2.10), that an element in $\mathcal{A}_G^\infty(X,E)$ is a K bi-invariant smooth function on G with values in $\Psi^{-\infty}(S)$. The partial trace map

$$\mathrm{Tr}_S : \mathcal{A}_G^\infty(X,E) \to \mathcal{C}(G),\ \mathrm{Tr}_S(A)(g) := \mathrm{Tr}(A(g)) = \int_S \mathrm{tr}\big(A(g)(s,s)\big)ds.$$

Generalizing this partial trace map, we define the map

$$\tau : C^k(\mathcal{C}(G)) \to C^k(\mathcal{A}_G^\infty(X,E))$$

as follows,

$$\tau_\varphi(A_0,\cdots,A_k) := \varphi\big(\mathrm{Tr}\big(A_0(g_0)\circ A_1(g_1)\circ\cdots\circ A_k(g_k)\big)\big).$$

It is easy to check that τ induces a chain map on the Hochschild and cyclic complexes. Thus, see [**46**], we have:

PROPOSITION 3.17. *The chain map τ induces a morphism $\tau : HP^\bullet(\mathcal{C}(G)) \to HP^\bullet(\mathcal{A}_G^\infty(X,E))$.*

We shall work with the following cyclic cocycles on $\mathcal{A}_G^\infty(X,E)$:

- for a differentiable group cocycle φ of G, we shall consider $\tau\big(\chi(\varphi)\big)$, denoted in the sequel by τ_φ^X;
- for $g\in G$, we shall consider $\tau(\mathrm{tr}_g)$, denoted by tr_g^X in the sequel;
- for $g\in G$, we shall consider $\tau\big(\Phi_g^P\big)$, denoted by $\Phi_{X,g}^P$ in the sequel.

4. Higher indices for proper cocompact manifolds without boundaries

In this section, we present higher index theorems on proper cocompact G-manifolds associated to the cyclic cocycles on $\mathcal{C}(G)$ introduced in Section 3.

4.1. L^2-index theorem. We start by reviewing Atiyah's L^2-index theorem on Galois coverings. Let M be a compact smooth manifold without boundary, and E a twisted spinor bundle on M, and D an odd $\mathbb{Z}_2$-graded Dirac operator acting on the sections of E.

Let Γ be the fundamental group of M and X be the universal covering space of M. X is equipped with a proper, free, and cocompact Γ action such that the quotient is M. As X is a covering space of M, the operator D, as a differential operator on M, lifts to an odd $\mathbb{Z}_2$-graded Γ-equivariant operator $\widetilde{D}$ on $\widetilde{E}$, the pullback of E to X.

Let $C_r^*(\Gamma)$ be the reduced group C^*-algebra of Γ. Following Equation (2.17), we consider the index $\mathrm{Ind}_{C_r^*(\Gamma)}(\widetilde{D})$ of the operator $\widetilde{D}$, which is an element of $K_0(C_r^*(\Gamma))$.

The trace tr_e on $\mathbb{C}\Gamma$ naturally extends naturally to a trace on $C_r^*(\Gamma)$. The pairing between tr_e and the index element $\mathrm{Ind}_{C_r^*(\Gamma)}(D)$ was computed by Atiyah [**2**] in the following theorem.

THEOREM 4.1.

$$\langle \mathrm{tr}_e, \mathrm{Ind}_{C_r^*(\Gamma)}(\widetilde{D})\rangle = \mathrm{ind}(D).$$

where on the right hand side we have the Fredholm index of D.

Generalizing Atiyah's Theorem 4.1 to proper Lie group actions, Connes and Moscovici [13] proved the following index formula on homogeneous spaces. Let K be a maximal compact subgroup of G, and $X := G/K$ be the associated homogeneous space, which is equipped with a proper left G action with the quotient being a point. We assume that X is equipped with a G-equivariant Spin^c-structure and consider the G-equivariant Dirac operator D on X obtained by twisting the Spin^c-Dirac operator with the bundle $E_\mu := G \times V_\mu / K$, with V_μ an irreducible unitary K representation associated with highest weight μ. We obtain a well defined index class for D, an element $\mathrm{Ind}_{C_r^*(G)}(D) \in K_0(C_r^*(G))$.

THEOREM 4.2. *Assume that G is unimodular. Let $\mathfrak{g}$ and $\mathfrak{k}$ be the Lie algebras of G and K.*

$$\langle \mathrm{tr}_e, \mathrm{Ind}(D_\mu) \rangle = \langle \widehat{A}(\mathfrak{g}, K) \wedge \mathrm{ch}(V_\mu)_{\mathfrak{p}^*}, [V] \rangle,$$

where $\mathfrak{p}^* \subset \mathfrak{g}^*$ is the conormal space of $\mathfrak{k}$ in $\mathfrak{g}$, and $[V]$ is the fundamental class of $\mathfrak{p}^*$.

Hang Wang [56] generalized the Connes-Moscovici theorem, Theorem 4.2, to general proper cocompact G-manifolds as follows.

THEOREM 4.3. *Let X be a proper cocompact G-manifold which is equipped with a G-equivariant Spin^c-structure. Suppose that D is a G-equivariant Dirac operator on X associated to a twisted G-equivariant twisted spinor bundle $E = \mathcal{S} \otimes F$. The pairing between tr_e and $\mathrm{Ind}_{C_r^*(G)}(D)$ is computed as follows.*

$$\langle \mathrm{tr}_e, \mathrm{Ind}_{C_r^*(G)}(D) \rangle = \int_X c_X \, \mathrm{AS}(D),$$

with c_X a cut-off function for the proper action of G on X and

$$\mathrm{AS}(D) := \widehat{A}\Big(\frac{R_X}{2\pi i}\Big) \, \mathrm{tr}\Big(\exp\big(\frac{R_F}{2\pi i}\big)\Big).$$

Here R_X is the curvature form for the Levi-Civita connection on X and R_F is the curvature form associated to a G-invariant Hermitian metric on G.

4.2. Higher indices associated to differentiable group cohomology. In [12], Connes and Moscovici established a higher version of Atiyah's L^2-index Theorem 4.1 and computed a geometric formula for the pairing between the group cohomology of the fundamental group Γ and the index class $\mathrm{Ind}_{\mathcal{A}_G^c(X,E)}(D) \in K_\bullet(\mathcal{A}_G^c(X,E))$. Inspired by this result, Pflaum, Posthuma, and Tang [44, 45] proved a higher generalization of Theorem 4.3, computing the pairing between the cyclic cocycles in the image of the character morphism $\chi : H^\bullet_{\mathrm{diff}}(G) \to HP^\bullet(\mathcal{C}(G))$ in Theorem 3.10 and the smooth index class

$$\mathrm{Ind}_{\mathcal{C}(G)}(D) \in K_0(\mathcal{C}(G)) \equiv \mathrm{Ind}_{C_r^*(G)}(D) \in K_0(C^*(G)).$$

Let X be a manifold equipped with proper G action, and $\Omega^\bullet_{\mathrm{inv}}(X)$ be the space of G-invariant differential forms on X. The de Rham differential restricts to $\Omega^\bullet_{\mathrm{inv}}(X)$ and the associated cohomology is denoted by $H^\bullet_{\mathrm{inv}}(X)$.

DEFINITION 4.4. The van Est morphism $\Phi : C^k_{\mathrm{diff}}(G) \to \Omega^k_{\mathrm{inv}}(X)$ is defined as follows.

$$f_\varphi(x_0, \cdots, x_k) := \int_{G^{\times(k+1)}} c_X(g_0^{-1}x_0) \cdots c_X(g_k^{-1}x_k) \varphi(g_0, g_1, \cdots, g_k) \mathrm{d}g_0 \cdots \mathrm{d}g_k,$$

$$\Phi(\varphi) := (d_{x_1} d_{x_2} \cdots d_{x_k} f_\varphi)(x, \cdots, x).$$

In the above formulas, we have used the homogeneous differentiable group cohomology chain complex in Definition 3.8, and f_φ is a smooth function on $X^{\times(k+1)}$, and $d_i f_\varphi$ is the differential of f_φ along the x_i component. As the cut-off function c_X is compactly supported along each G orbit, the above integral is convergent.

The following property is proved in [**46**, Prop. 2.5].

PROPOSITION 4.5. *The map* Φ *induces a morphism* $\Phi : H^\bullet_{diff}(G) \to H^\bullet_{Inv}(X)$.

The following theorem computes the explicit formula about the pairing between $\chi(\varphi)$ and $\mathrm{Ind}_{C_r^*(G)}(D)$.

THEOREM 4.6. *Suppose that* G *is unimodular. Let* G *be a connected reductive linear Lie group acting properly and cocompactly on a manifold* X*. For any* $[\varphi] \in H^{2k}_{\mathrm{diff}}(G)$*, the index pairing is given by*

$$\langle \chi(\varphi), \mathrm{Ind}_{C_r^*(G)}(D)\rangle = \frac{1}{(2\pi\sqrt{-1})^k (2k)!}\int_{T^*X} c_X\, \Phi([\varphi]) \wedge \widehat{A}(T^*X) \wedge \mathrm{ch}(F),$$

where $c_X \in C_c^\infty(X)$ *is a cut-off function,* $\widehat{A}(T^*X) := \widehat{A}\left(\frac{R_X}{2\pi i}\right)$ *and* $\mathrm{ch}(F) := \mathrm{tr}\left(\exp\left(\frac{R_F}{2\pi i}\right)\right)$*, as in Theorem* 4.3

REMARK 4.7. Pflaum, Posthuma, and Tang [**45**] proved Theorem 4.6 through the algebraic index theorem method developed by Fedosov [**14**] and [**42**] for general G-invariant elliptic operators on X on a proper cocompact G-manifold for a Lie groupoid G. Recently, Piazza and Posthuma [**47**] presented a new proof of this theorem for the case of Dirac operators using the heat kernel and Getzler's rescaling techniques.

REMARK 4.8. The assumption of unimodularity in Theorem 4.6 can be dropped by working with smooth group cohomology of G with coefficients. This is developed in [**44**].

EXAMPLE 4.9. We consider the case in which G is $\mathbb{R}^2$ and K is trivial. In this case, $X = G/K$ is identified with $\mathbb{R}^2$. Via the isomorphism with $\mathbb{C}$, $\mathbb{R}^2$ is equipped with the $\mathbb{R}^2$ invariant Dolbeault operator D. As there is no L^2-harmonic form on $\mathbb{R}^2$, it was observed by Connes and Moscovici [**13**] that the L^2-index $\langle \mathrm{tr}_e, \mathrm{Ind}_{C_r^*(\mathbb{R}^2)}(D)\rangle$ vanishes. However, the determinant C function in Section 3.2 is a 2-cocycle on $\mathbb{R}^2$, and its image $\chi(C)$ in $HP^{even}(\mathcal{C}(\mathbb{R}^2))$ coincides with the cyclic cocycle Φ introduced in Section 3.2. We can apply Theorem 4.6 to compute

$$\langle \Phi, \mathrm{Ind}_{C_r^*(\mathbb{R}^2)}(D)\rangle = \langle \chi(C), \mathrm{Ind}_{C_r^*(\mathbb{R}^2)}(D)\rangle = \frac{1}{2}.$$

This example shows that higher indices contain interesting information of the operator D beyond the L^2-index. The computation also extends naturally to $\mathbb{R}^{2n} = \mathbb{C}^n$.

4.3. Delocalized indices. Hochs and Wang [**28**] computed the pairing between tr_g and $\mathrm{Ind}_{C_r^*(G)}(D)$ for a twisted Spin^c Dirac operator on a proper cocompact G-manifold on X. To introduce their result, we fix the following set up.

- X^g is the g-fixed point submanifold;
- $\mathcal{N}_{X^g}$ is the normal bundle of X^g in X and $R^{\mathcal{N}}$ is the curvature form associated to the Hermitian connection on $\mathcal{N}_{X^g} \otimes \mathbb{C}$;

- L_{det} is the determinant line bundle of the Spin^c-structure on X and $L_{\text{det}}|_{X^g}$ is its restriction to X^g and R^L is the curvature form associated to the Hermitian connection on $L_{\text{det}}|_{X^g}$;
- R_{X^g} is the Riemannian curvature form associated to the Levi-Civita connection on the tangent bundle of X^g;
- The $\mathrm{AS}_g(D)$ has the following expression for a twisted Dirac operator on $E = \mathcal{S} \otimes F$:

$$\mathrm{AS}_g(D) := \frac{\widehat{A}(\frac{R_{X^g}}{2\pi i})\,\mathrm{tr}\left(g\exp(\frac{R^F}{2\pi i})\right)\exp(\mathrm{tr}(\frac{R^L}{2\pi i}))}{\det\left(1 - g\exp(-\frac{R^{\mathcal{N}}}{2\pi i})\right)^{\frac{1}{2}}}. \tag{4.10}$$

This will also be denoted by $\mathrm{AS}_g(X, E)$ or, if there is no confusion on the vector bundle E, simply by $\mathrm{AS}_g(X)$.

Hochs and Wang proved the following result using heat kernel and Getzler's rescaling techniques [2]

THEOREM 4.11. (1) *If G has a compact Cartan subgroup, then*

$$\langle \mathrm{tr}_g, \mathrm{Ind}_{C_r^*(G)}(D)\rangle = \int_{X^g} c_{X^g}\,\mathrm{AS}_g(D),$$

for a cutoff function $c_{X^g} \in C_c^\infty(X^g)$ for the Z_g-action on X^g, with Z_g equal to the centralizer subgroup of g in G;

(2) *If G does not have a compact Cartan subgroup,*

$$\langle \mathrm{tr}_g, \mathrm{Ind}_{C_r^*(G)}(D)\rangle = 0.$$

4.4. Delocalized higher indices. To improve Theorem 4.11 to allow G to be nonequal rank, in [**24**], Hochs, Song, and Tang computed the pairing between the cocycle Φ_g^P and the index element $\mathrm{Ind}_{C_r^*(G)}(D)$. Let X/AN be the quotient of X with respect to the $AN < G$ action. The group M acts properly and cocompactly on the quotient X/AN; for g a semisimple element of G, $(X/AN)^g$ is the fixed submanifold of the g action on the quotient X/AN; $c^g_{X/AN}$ is a smooth compactly supported cut-off function on $(X/AN)^g$.

Let $\mathfrak{m}$ be the Lie algebra of the group M, and $\mathfrak{a}$ the Lie algebra of the group A. Using the K-invariant metric on $\mathfrak{p}$, we obtain a $K \cap M$-invariant decomposition

$$\mathfrak{p} = (\mathfrak{p} \cap \mathfrak{m}) \oplus \mathfrak{a} \oplus (\mathfrak{k}/(\mathfrak{k} \cap \mathfrak{m})),$$

which induces the $K \cap M$ decomposition of spinors,

$$S_{\mathfrak{p}} \cong S_{\mathfrak{p}\cap\mathfrak{m}} \otimes S_{\mathfrak{a}} \otimes S_{\mathfrak{k}(\mathfrak{k}\cap\mathfrak{m})}.$$

Using the slice theorem, Theorem. 2.2, X/AN can be identified as $X/AN = M \times_{K\cap M} S$. On X/AN, we consider the bundle $E_{X/AN}$,

$$E_{X/AN} = M \times_{M\cap K} \left(S_{\mathfrak{p}\cap\mathfrak{m}} \otimes S_{\mathfrak{k}/(\mathfrak{k}\cap\mathfrak{m})} \otimes W|_S\right) \cong M \times_{M\cap K} \left(S_{\mathfrak{p}\cap\mathfrak{m}} \otimes \widetilde{W}\right),$$

where we denote $S_{\mathfrak{k}/(\mathfrak{k}\cap\mathfrak{m})} \otimes W|_S$ by $\widetilde{W}$ and $M \times_{M\cap K} \widetilde{W}$ by $W_{X/AN}$. We observe that $W_{X/AN}$ is an M-equivariant Hermitian vector bundle on X/AN. On $E_{X/AN}$, we consider the associated Dirac operator $D_{X/AN}$. The index of $D_{X/AN}$ is an element $\mathrm{Ind}_{C_r^*(M)}(D_{X/AN})$ in $K_\bullet(\mathcal{C}(M))$.

Hochs, Song, and Tang [**24**] proved a reduction theorem relating the index pairing for D and the index pairing for $D_{X/AN}$.

[2] We refer also to the recent article [**48**] for a related, detailed discussion of this result.

THEOREM 4.12.

$$\langle \Phi_g^P, \mathrm{Ind}_{C_r^*(G)}(D)\rangle = \langle \mathrm{tr}_g^M, \mathrm{Ind}_{C_r^*(M)}(D_{X/AN})\rangle.$$

The following theorem follows directly from Theorem 4.11 and 4.12.

THEOREM 4.13.

$$\langle \Phi_g^P, \mathrm{Ind}_{C_r^*(G)}(D)\rangle = \int_{(X/AN)^g} c_{(X/AN)^g} \mathrm{AS}(X/AN)_g.$$

4.5. The example of G/K. We look at the example of $X = G/K$ with the left G action. Song and Tang [**54**] prove the following theorem using Harish-Chandra's theory of orbital integrals; it can also be derived from Theorem 4.13 as a corollary.

THEOREM 4.14. *Let H be a Cartan subgroup of G, and $T := K \cap H$ with $t \in T$. Assume that $T < M$ is a Cartan subgroup of M, and P is a maximal cuspidal parabolic subgroup, i.e. T is maximal among all possible P. Let Δ_T^M be the corresponding Weyl denominator.*

(1)

$$\langle \Phi_e^P, \mathrm{Ind}_{C_r^*(G)}(D)\rangle = \frac{1}{|W_{M\cap K}|} \cdot \sum_{w\in W_K} m\left(\sigma^M(w\cdot\mu)\right), \tag{4.15}$$

where $\sigma^M(w\cdot \mu)$ is the discrete series representation of M with Harish-Chandra parameter $w \cdot \mu$, and $m\left(\sigma^M(w\cdot\mu)\right)$ is its Plancherel measure, and $W_{M\cap K}$ is the Weyl group of $M \cap K$;

(2)

$$\langle \Phi_t^P, \mathrm{Ind}(D)\rangle = \frac{\sum_{w\in W_K}(-1)^w e^{w\cdot\mu}(t)}{\Delta_T^M(t)}. \tag{4.16}$$

REMARK 4.17. When G is equal rank, Eq. (4.15) was proved by Connes-Moscovici [**13**], i.e. the L^2-trace of the operator D is the formal degree of the associated discrete series representation of G.

REMARK 4.18. When the kernel of D gives a discrete series representation, i.e. G has equal rank, Equation (4.16) in Theorem 4.14 may be derived from Theorem 4.13 and the computation in Hochs and Wang [**27**]. Here, we do not assume G to have equal rank in Theorem 4.14 and allow the kernel of D to be a limit of discrete series representation. Equation (4.16) suggests that the index pairing can be used to detect some of the limits of discrete series representations of G.

REMARK 4.19. In Theorem 4.15, we have assumed P to be maximal. When P is not maximal, it can be shown that the pairing in Theorem 4.14 vanishes.

As a special case of Theorem 4.6, we have the following theorem generalizing the Connes-Moscovici L^2-index theorem, Theorem 4.2, for homogeneous spaces.

THEOREM 4.20. *Suppose that G is unimodular. For a G-invariant Dirac operator on $X = G/K$ and $[\varphi] \in H^{2k}_{\mathrm{diff}}(G)$, we have*

$$\chi(\varphi)(\mathrm{Ind}_{C_r^*(G)}(D_\mu)) = \frac{1}{(2\pi\sqrt{-1})^k}\left\langle \widehat{A}(\mathfrak{g}, K) \wedge \mathrm{ch}(V_\mu)_{\mathfrak{m}^*} \wedge \Phi([\varphi]), [V]\right\rangle,$$

where $\Phi([\varphi])$ is a class in $H^{2k}(\mathfrak{g}, K)$ defined by a G-invariant closed differential form on G/K via the van Est isomorphism in Proposition 4.5.

4.6. Summary of results for manifolds without boundary. We have introduced 0-cyclic cocycles tr_e and tr_g on the Harish-Chandra algebra $\mathcal{C}(G)$ and we have stated index theorems computing the pairing of these 0-cyclic cocycles with the index class $\mathrm{Ind}_{\mathcal{C}(G)}(D) \in K_0(\mathcal{C}(G)) \equiv \mathrm{Ind}_{C^*_r(G)}(D) \in K_0(C^*(G))$:

$$\mathrm{tr}_e(\mathrm{Ind}_{\mathcal{C}(G)}(D)), \quad \mathrm{tr}_g(\mathrm{Ind}_{\mathcal{C}(G)}(D))$$

These results are due to Wang, c.f. Theorem 4.3, and Hochs-Wang, c.f. Theorem 4.11. We have then stated generalizations of these two theorems: the first theorem, c.f. Theorem 4.6, by Pflaum-Posthuma-Tang, computes the pairing of the index class with the cyclic cocycles associated to cocycles in the differentiable group cohomology introduced in Theorem 3.10; the second theorem, c.f. Theorem 4.13 by Hochs-Song-Tang, computes the pairing between the index class and the higher orbital integrals introduced in Theorem 3.16.

5. Proper cocompact G-manifolds with boundaries

In this section, we introduce the index class for a Dirac operator on a proper cocompact G-manifold.

5.1. Geometry on G proper manifolds with boundary. We start with some generalities. Let Y_0 be a manifold with boundary, G a finitely connected Lie group acting properly and cocompactly on Y_0. We denote by X the boundary of Y_0. There exists a collar neighbourhood U of the boundary ∂Y_0, $U \cong [0,2] \times \partial Y_0$, which is G-invariant and such that the action of G on U is of product type. We assume that Y_0 is endowed with a G-invariant metric $\mathbf{h}_0$ which is of product type near the boundary. We let $(Y_0, \mathbf{h}_0)$ be the resulting Riemannian manifold with boundary; in the collar neighborhood $U \cong [0,2] \times \partial Y_0$ the metric $\mathbf{h}_0$ can be written, through the above isomorphism, as $dt^2 + \mathbf{h}_X$, with $\mathbf{h}_X$ a G-invariant Riemannian metric on $X = \partial Y_0$. We denote by c_{Y_0} a cut-off function for the action of G on Y_0; since the action is cocompact, this is a compactly supported smooth function. We consider the associated manifold with cylindrical ends $\widehat{Y} := Y_0 \cup_{\partial Y_0} ((-\infty, 0] \times \partial Y_0)$, endowed with the extended metric $\widehat{\mathbf{h}}$ and the extended G-action. We denote by $(Y, \mathbf{h})$ the b-manifold associated to $(\widehat{Y}, \widehat{\mathbf{h}})$. See [Melrose-book]. We shall often treat $(\widehat{Y}, \widehat{\mathbf{h}})$ and $(Y, \mathbf{h})$ as the same object. We denote by c_Y the obvious extension of the cut-off function c_{Y_0} for the action of G on Y_0 (constant along the cylindrical end); this is a cut-off function of the extended action of G on Y. If x is a boundary defining function for the cocompact G-manifold Y_0, then the b-metric $\mathbf{h}$ has the following product-structure near the boundary X:

$$\frac{dx^2}{x^2} + \mathbf{h}_X$$

Let us fix a slice Z_0 for the G action on Y_0; thus

$$Y_0 \cong G \times_K Z_0$$

with K a maximal compact subgroup of G and Z_0 a smooth compact manifold with boundary endowed with a K-action. We denote by S the boundary ∂Z_0. Consequently, $Y \cong G \times_K Z$ with Z the b-manifold associated to Z_0 and the boundary

$$X = \partial Y_0 \cong G \times_K S.$$

We shall concentrate on the case where the symmetric space G/K, the G-manifold Y_0 and the K-slice Z_0 are all even dimensional.

EXAMPLE 5.1. Start with an inclusion $K \subset G$ of Lie groups with K compact, and let Z_0 be a compact K-manifold with boundary $\partial Z_0 =: S_0$. Then $Y_0 := G \times_K Z_0$ is an example of manifold with boundary $\partial Y_0 = G \times_K \partial S_0$, equipped with a proper, cocompact action of G. We also have the associated b-manifolds Z and $Y = G \times_K Z$. If we choose a K- invariant inner product on the Lie algebra $\mathfrak{g}$ of G and a K-invariant b-metric g_Z on Z of product-type near the boundary then, as in the closed case, we obtain a G-invariant metric on Y; we call such a metric *slice compatible.*

As an explicit example, consider $G = SL(2,\mathbb{R})$ and $K = SO(2)$ acting on the unit disk $\mathbb{D}^2$ in the complex plane by rotations around the origin. The resulting manifold $Y := SL(2,\mathbb{R}) \times_{SO(2)} \mathbb{D}^2$ is a 4-dimensional fiber bundle over hyperbolic 2-space $SL(2,\mathbb{R})/SO(2) \cong \mathbb{H}^2$ with fiber $\mathbb{D}^2$. The boundary ∂Y of this manifold is isomorphic to $SL(2,\mathbb{R})$.

5.2. Dirac operators.

We assume the existence of a G-equivariant bundle of Clifford modules E_0 on Y_0, endowed with a Hermitian metric, product-type near the boundary, for which the Clifford action is unitary, and equipped with a Clifford connection also of product type near the boundary. Associated to these structures there is a generalized G-invariant Dirac operator D on Y_0 with product structure near the boundary acting on the sections of E_0. We denote by D_∂ the operator induced on the boundary. We employ the same symbol, D, for the associated b-Dirac operator on M, acting on the extended Clifford module E. We also have $D_{\rm cyl}$ on $\mathbb{R} \times \partial Y_0 \equiv {\rm cyl}(\partial Y_0)$. We shall make the following fundamental assumption.

ASSUMPTION 5.2. *There exists $\alpha > 0$ such that*

$$\operatorname{spec}_{L^2}(D_\partial) \cap [-\alpha, \alpha] = \emptyset. \tag{5.3}$$

We also have $D_{\rm cyl}$ on $\mathbb{R} \times \partial Y_0 \equiv {\rm cyl}(\partial Y_0)$. It should be noticed that because of the self-adjointness of D_∂, assumption (5.3) implies the L^2-invertibility of $D_{\rm cyl}$.

EXAMPLE 5.4. As an example where this condition is satisfied we can consider a G-proper manifold with boundary with a G-invariant riemannian metric and a G-invariant Spin structure with the property that the metric on the boundary is of positive scalar curvature. We would then consider the Spin-Dirac operator D; because of the psc assumption on the boundary we do have that D_∂ is L^2-invertible. These manifolds arise as in [**17**] from the slice theorem and a K-Spin-manifold with boundary S endowed with a K-invariant metric which is of psc on ∂S. Such compact manifolds S arise, for example, as follows. Consider a compact K-manifold without boundary N endowed with a K-invariant metric of positive scalar curvature. For the existence of such manifolds see for example [**18, 31, 58**]. We can now perform on this manifold K-equivariant surgeries and produce along the process a K-manifold with boundary W. Under suitable conditions (for example, equivariant surgeries only of codimension at least equal to 3) this manifold with boundary W will have a K-invariant metric of positive scalar curvature. We can now take the connected sum of W with a closed K-manifold not admitting a K-invariant metric of psc. See [**18, 58**]. The result will be a manifold S with a K-invariant metric which is of psc (only) on ∂S.

5.3. The index class and b-calculus. Let Y_0 be a G-proper manifold with boundary, with compact quotient, and let Y be the associated manifold with cylindrical ends, or, equivalently, the associated b-manifold. In the b-picture we consider $\mathcal{E}_b$, the $C^*_r G$-Hilbert module obtained by (double) completion of $\dot{C}^\infty_c(Y, E)$ endowed with the $C^\infty_c(G)$-valued inner product

$$(e, e')_{C^*_c(G)}(x) := (e, x \cdot e')_{L^2_b(Y,E)}, \quad e, e' \in \dot{C}^\infty_c(Y, E),\ x \in G\,.$$

Here the dot means vanishing of infinite order at the boundary; notice that on the right hand side we employ $L^2_b(Y, E)$, the L^2 space associated to the volume form associated to the b metric $\mathbf{h}$. One can prove that $\mathbb{K}(\mathcal{E}_b)$ is isomorphic to the relative Roe algebra $C^*(Y_0 \subset Y, E)^G$, the latter defined by completion of operators with propagation at a finite distance from Y_0. See [**49**] for precise definitions and proofs. We then have the following canonical isomorphisms

$$K_*(\mathbb{K}(\mathcal{E}_b)) \cong K_*(C^*(Y_0 \subset Y, E)^G) \cong K_*(C^*(Y_0, E_0)^G) \cong K_*(C^*_r(G)) \tag{5.5}$$

with the second isomorphism also explained in [**49**] and the third one the Morita isomorphism already explained in these notes.

We now want to prove the existence of an index class $\mathrm{Ind}(D) \in K_0(C^*(Y \subset Y, E)^G)$ under Assumption 5.2,; next, we shall want to find a smooth, i.e. dense and holomorphically closed, subalgebra $\mathcal{A}^\infty(Y, E)$ of $C^*(Y_0 \subset Y, E)^G$ and a smooth representative $\mathrm{Ind}_\infty(D) \in K_0(\mathcal{A}^\infty(Y, E))$ corresponding to $\mathrm{Ind}(D) \in K_0(C^*(Y \subset Y, E)^G)$ under the natural isomorphism between $K_0(\mathcal{A}^\infty(Y, E))$ and $K_0(C^*(Y \subset Y, E)^G)$.

We use the b-pseudodifferential calculus as in [**47**], even though this is not strictly necessary for the first task. For a detailed account of the b-calculus we refer the reader to Melrose book, [**37**].

Let us recall very briefly the basics of the b-calculus. For simplicity we assume that the boundary of our cocompact G-proper manifold Y_0 is connected. As above, we denote by Y the associated b-manifold. We sometimes denote the boundary of Y_0 by X; for notational convenience we set $\partial Y = X$. Finally, we often expunge the vector bundle E from the notation.

We can define the b-stretched product $Y \times_b Y$ which inherits in a natural way an action of $G \times G$ and a diagonal action of G, see [**32**]. Proceeding as in these references, we can define the algebra of G-equivariant b-pseudodifferential operators on Y with G-compact support, denoted ${}^b\Psi^*_{G,c}(Y)$, which is a $\mathbb{Z}$-graded algebra.

Let us fix $\epsilon > 0$. Then, as in the compact case, we can extend the algebra ${}^b\Psi^*_{G,c}(Y)$ and consider the G-equivariant b-calculus of G-compact support with ϵ-bounds, denoted ${}^b\Psi^{*,\epsilon}_{G,c}(Y)$. Thus, by definition,

$${}^b\Psi^{*,\epsilon}_{G,c}(Y) = {}^b\Psi^*_{G,c}(Y) + {}^b\Psi^{-\infty,\epsilon}_{G,c}(Y) + \Psi^{-\infty,\epsilon}_{G,c}(Y).$$

The second term on the right hand side corresponds to the Schwartz kernels on $Y \times_b Y$, smooth in the interior, conormal of order ϵ on the left and right boundaries of $Y \times_b Y$ and smooth up to the front face (and of course G-equivariant). The third term on the right hand side corresponds instead to Schwartz kernels on $Y \times Y$, smooth in the interior and conormal of order ϵ on the boundary hypersurfaces of $Y \times Y$. See [**37**] and for a quick treatment the Appendix in [**38**]. Elements in $\Psi^{-\infty,\epsilon}_{G,c}(Y)$ are called *residual.* Composition formulae for the elements in the calculus with bounds are as in [**38**, Theorem 4].

Restriction to the front face of $Y \times_b Y$ defines the indicial operator $I : {}^b\Psi^{-\infty}_{G,c}(Y) \to {}^b\Psi^{-\infty}_{G,c,\mathbb{R}^+}(\overline{N^+\partial Y})$ and, more generally,

$$I : {}^b\Psi^{-\infty,\epsilon}_{G,c}(Y) \to {}^b\Psi^{-\infty,\epsilon}_{G,c,\mathbb{R}^+}(\overline{N^+\partial Y}) \tag{5.6}$$

with $\overline{N^+\partial Y}$ denoting the compactified inward-pointing normal bundle to the boundary and the subscript $\mathbb{R}^+$ denoting equivariance with respect to the natural $\mathbb{R}^+$-action. See [**37**, Section 4.15], where $\overline{N^+\partial Y}$ is denoted $\widetilde{Y}$. Recall that there is a diffeomorphism of b-manifolds: $\overline{N^+\partial Y} \simeq \partial Y \times [-1,1]$. Metrically we can think of $\overline{N^+\partial Y}$ as the cylinder $\mathrm{cyl}(\partial Y) := \mathbb{R} \times \partial Y$ and with a small abuse of notation we shall adopt this notation henceforth, thus writing (5.6) as

$$I : {}^b\Psi^{-\infty,\epsilon}_{G,c}(Y) \to {}^b\Psi^{-\infty,\epsilon}_{G,c,\mathbb{R}^+}(\mathrm{cyl}(\partial Y)) \tag{5.7}$$

By fixing a cut-off function χ on the collar neighborhood of the boundary, equal to 1 on the boundary, we can define a section s to the indicial homomorphism I:

$$s : {}^b\Psi^{-\infty,\epsilon}_{c,G,\mathbb{R}^+}(\mathrm{cyl}(\partial Y)) \to {}^b\Psi^{-\infty,\epsilon}_{G,c}(Y)\,; \tag{5.8}$$

The linear map s associates to an $\mathbb{R}^+$-invariant operator G on $\mathrm{cyl}(\partial Y)$ an operator on the b-manifold Y; the latter is obtained by pre-multiplying and post-multiplying by the cut-off function χ.
The Mellin transform in the s-variable for an $\mathbb{R}^+$-invariant kernel $\kappa(s,y,y') \in {}^b\Psi^{-\infty,\epsilon}_{G,c,\mathbb{R}^+}(\mathrm{cyl}(\partial Y))$ defines an isomorphism $\mathcal{M}$ between ${}^b\Psi^{-\infty,\epsilon}_{G,c,\mathbb{R}^+}(\mathrm{cyl}(\partial Y))$ and holomorphic families of operators

$$\{\mathbb{R} \times i(-\epsilon,\epsilon) \ni \lambda \to \Psi^{-\infty}_{G,c}(\partial Y)\}$$

which rapidly decrease as $|\mathrm{Re}\lambda| \to \infty$ as functions with values in the Fréchet algebras $\Psi^{-\infty}_{c,G}(\partial Y)$. We denote by $I(R,\lambda)$ or sometimes by $\widehat{R}(\lambda)$, the Mellin transform of $R \in {}^b\Psi^{-\infty,\epsilon}_{c,G,\mathbb{R}^+}(\mathrm{cyl}(\partial Y))$; if $P \in {}^b\Psi^{-\infty,\epsilon}_{G,c}(Y)$ then its indicial family is by definition the indicial family associated to its indicial operator $I(P)$. The inverse $\mathcal{M}^{-1}$ is obtained by associating to a holomorphic family $\{S(\lambda), |\mathrm{Im}\lambda| < \epsilon\}$, rapidly decreasing in $\mathrm{Re}\lambda$, the $\mathbb{R}^+$-invariant Schwartz kernel κ_S that in projective coordinates is given by

$$\kappa_G(s,y,y') = \int_{\mathrm{Im}\lambda = r} s^{i\lambda} G(\lambda)(y,y')d\lambda \tag{5.9}$$

with $r \in (-\epsilon,\epsilon)$.

Let D be an equivariant $\mathbb{Z}_2$-graded odd Dirac operator, of product type near the boundary and let us make Assumption 5.2. We sketch the proof of the existence of a C^*-index class associated to D^+. One begins by finding a symbolic parametrix Q_σ to D^+, with remainders $R^\pm_\sigma$:

$$Q_\sigma D^+ = \mathrm{Id} - R^+_\sigma\,, \quad D^+ Q_\sigma = \mathrm{Id} - R^-_\sigma\,. \tag{5.10}$$

We can choose $Q_\sigma \in {}^b\Psi^{-1}_{G,c}(Y)$, i.e. of G-compact support and then get $R^\pm_\sigma \in {}^b\Psi^{-\infty}_{G,c}(Y)$.

Consider now the indicial operator $I(D^+) \equiv D^+_{\mathrm{cyl}} \equiv \partial/\partial t + D_\partial$; by Assumption 5.2 we know that this operator is L^2-invertible on the b-cylinder $\mathrm{cyl}(\partial Y)$. Consider the smooth kernel $K(Q')$ on $Y \times_b Y$ which is zero outside a neighbourhood U_{ff} of the front face and such that in $U_{\rm ff}$ with (projective) coordinates (x, s, y, y') is equal to

$$K(Q')(s,x,y,y') := \chi(x)\int_{-\infty}^{\infty} s^{i\lambda} K((D_\partial + i\lambda)^{-1} \circ I(R_\sigma^-, \lambda))(y,y')d\lambda\,.$$

Put it differently, $Q' = s(I(D)^{-1}I(R_\sigma)$, with s as in (5.8) and where we have used the inverse Mellin transform for the expression of $I(D)^{-1}I(R_\sigma)$. Notice that because of the presence of $(D_\partial + i\lambda)^{-1}$ this is not compactly supported. Still, by proceeding *exactly* as in [**34**, Lemma 4 and 5] we can establish the following fundamental result:

THEOREM 5.11. *The kernel $K(Q')$ defines a bounded operator on the $C^*_r G$-module $\mathcal{E}_b$: $Q' \in \mathbb{B}(\mathcal{E}_b)$. If $Q^b := Q_\sigma - Q'$, then Q^b provides an inverse of D^+ modulo elements in $\mathbb{K}(\mathcal{E}_b)$. Thus, for a Dirac operator D satisfying the Assumption 5.2 there is a well defined index class*

$$\mathrm{Ind}_{C^*}(D) \in K_0(\mathbb{K}(\mathcal{E}_b)) \equiv K_0(C^*(Y_0 \subset Y, E)^G)\,.$$

We sometime refer to the passage from Q_σ to $Q^b = Q_\sigma - Q'$ as the passage from a symbolic parametrix to a *true parametrix* or an *improved parametrix.*

The b-calculus approach to the C^* index class will prove to be useful in establishing the existence of a *smooth* index class, see below, and for proving a *higher* APS index formula. However, there are other approaches to the C^*-index class. Indeed, in the recent preprint [**26**] a coarse approach is explained. It should not be difficult to show that the two index classes, the one defined here and the one defined in [**26**] are in fact the same. (The proof should proceed as for Galois coverings, where the analogous result is established in [**49**, Proposition 2.4].)
As in the case of Galois coverings, and always under the invertibility Assumption 5.2, it should also be possible to establish the existence of an APS C^*-index class through a boundary value problem defined through the spectral projection $\chi_{(0,\infty)}(D_\partial)$ and show that it is equal to the index class as defined above. See [**33**] for the case of Galois coverings.

Assume now that G is connected, or more generally that $|\pi_0(G)| < \infty$; we can apply to Y the slice theorem and obtain a diffeomorphism

$$G \times_K Z \xrightarrow{\alpha} Y\,.$$

with Z a compact manifold with boundary. One can prove, as in the closed case, that

$${}^b\Psi^{-\infty,\epsilon}_{G,c}(Y) = (C^\infty_c(G)\hat{\otimes}\,{}^b\Psi^{-\infty,\epsilon}(Z))^{K\times K}.$$

A similar description can be given for $\Psi^{-\infty,\epsilon}_{G,c}(Y)$.
We set

$${}^b\mathcal{A}^{c,\epsilon}_G(Y) := {}^b\Psi^{-\infty,\epsilon}_{G,c}(Y) + \Psi^{-\infty,\epsilon}_{G,c}(Y)$$

Similarly, we consider

$$(5.12)\qquad {}^b\mathcal{A}^{c,\epsilon}_{G,\mathbb{R}^+}(\mathrm{cyl}(\partial Y)) := C^\infty_c(G)\hat{\otimes}\,{}^b\Psi^{-\infty,\epsilon}_{\mathbb{R}^+}(\mathrm{cyl}(\partial Z))^{K\times K}$$

and observe that the indicial operator induces a surjective algebra homomorphism

$${}^b\mathcal{A}^{c,\epsilon}_G(Y) \xrightarrow{I} {}^b\mathcal{A}^{c,\epsilon}_{G,\mathbb{R}^+}(\mathrm{cyl}(\partial Y))$$

where I is equal to the indicial operator on ${}^b\Psi^{-\infty,\epsilon}_{G,c}(Y) = (C^\infty_c G)\hat{\otimes}\, {}^b\Psi^{-\infty,\epsilon}(Z))^{K\times K}$ and it is defined as zero on $\Psi^{-\infty,\epsilon}_{G,c}(Y) = (C^\infty_c G)\hat{\otimes}\, \Psi^{-\infty,\epsilon}(Z))^{K\times K}$.

Notation: from now on we shall omit the ϵ from the notations for ${}^b\mathcal{A}^{c,\epsilon}_G(Y)$ and other related algebras.

We obtain in this way algebras of operators on Y fitting into a short exact sequence

$$0 \to \mathcal{A}^c_G(Y) \to {}^b\mathcal{A}^c_G(Y) \xrightarrow{I} {}^b\mathcal{A}^c_{G,\mathbb{R}}(\mathrm{cyl}(\partial Y)) \to 0. \tag{5.13}$$

Here

$$\mathcal{A}^c_G(Y) := \mathrm{Ker}({}^b\mathcal{A}^c_G(Y) \xrightarrow{I} {}^b\mathcal{A}^c_{G,\mathbb{R}}(\mathrm{cyl}(\partial Y))). \tag{5.14}$$

5.4. Harish-Chandra smoothing operators. Recall that

$$\begin{gathered}{}^b\mathcal{A}^c_G(Y) = (C^\infty_c(G)\hat{\otimes}\, {}^b\Psi^{-\infty,\epsilon}(Z))^{K\times K} + (C^\infty_c(G)\hat{\otimes}\, \Psi^{-\infty,\epsilon}(Z))^{K\times K},\\ {}^b\mathcal{A}^c_{G,\mathbb{R}}(\mathrm{cyl}(\partial Y)) = (C^\infty_c(G)\hat{\otimes}\, {}^b\Psi^{-\infty,\epsilon}_{\mathbb{R}}(\mathrm{cyl}(\partial Z)))^{K\times K}.\end{gathered}$$

Employing the Harish-Chandra algebra $\mathcal{C}(G)$ instead of $C^\infty_c(G)$ on the right hand side we can define the algebras ${}^b\mathcal{A}^\infty_G(Y)$ and ${}^b\mathcal{A}^\infty_{G,\mathbb{R}}(\mathrm{cyl}(\partial Y))$ fitting into the short exact sequence

$$0 \to \mathcal{A}^\infty_G(Y) \to {}^b\mathcal{A}^\infty_G(Y) \xrightarrow{I} {}^b\mathcal{A}^\infty_{G,\mathbb{R}}(\mathrm{cyl}(\partial Y)) \to 0 \tag{5.15}$$

with $\mathcal{A}^\infty_G(Y) = \mathrm{Ker}({}^b\mathcal{A}^\infty_G(Y) \xrightarrow{I} {}^b\mathcal{A}^\infty_{G,\mathbb{R}}(\mathrm{cyl}(\partial Y)))$. This extends (5.13) and will play a crucial role in our analysis. Similar short exact sequences can be defined when the operators act on sections of a vector bundle E. One can prove, see [**47**], the following result:

PROPOSITION 5.16. *The algebra $\mathcal{A}^\infty_G(Y,E)$ is a dense and holomorphically closed subalgebra of the Roe algebra $C^*(Y_0 \subset Y, E)^G$. Consequently, $K_*(\mathcal{A}^\infty_G(Y,E)) \simeq K_*(C^*(Y_0 \subset Y, E)^G)$.*

5.5. The smooth index class. Bearing in mind the last proposition and analyzing in great detail the Schwartz kernels of the operator involved, one can prove the following crucial results, see [**47**].

THEOREM 5.17. *Let D be as above and let Q^σ be a symbolic b-parametrix for D. The Connes-Skandalis projector*

$$P^b_Q := \begin{pmatrix} {}^bS_+^2 & {}^bS_+(I + {}^bS_+)Q^b \\ {}^bS_-D^+ & I - {}^bS_-^2 \end{pmatrix} \tag{5.18}$$

associated to a true parametrix $Q^b = Q^\sigma - Q'$ with remainders ${}^bS^\pm$ in $\mathcal{A}^\infty_G(Y_0)$ is a 2×2 matrix with entries[3] *in $\mathcal{A}^\infty_G(Y)$. We thus have a well-defined smooth index class*

$$\mathrm{Ind}_\infty(D) := [P^b_Q] - [e_1] \in K_0(\mathcal{A}^\infty_G(Y)) \equiv K_0(C^*(Y_0 \subset Y)^G) \quad \text{with} \quad e_1 := \begin{pmatrix} 0 & 0 \\ 0 & 1 \end{pmatrix}. \tag{5.19}$$

[3]As usual, the entries really live in a slightly extended algebra because of the identity appearing in the right lower corner of the matrix.

We can consider in particular the parametrix

$$Q_{\exp} := \frac{I - \exp(-\frac{1}{2}D^- D^+)}{D^- D^+} D^+$$

and its associated true parametrix $Q^b_{\exp}$. Using the first, we can define the Connes-Moscovici projector $V_{\rm CM}(D)$; using the latter we define the *improved* Connes-Moscovici projector $V^b_{\rm CM}(D)$. We then have:

THEOREM 5.20. *The projector $V^b_{\rm CM}(D)$ obtained from the true parametrix $Q^b_{\exp}$ associated to $Q_{\exp}$ is a 2×2 matrix with entries in $\mathcal{A}^\infty_G(Y)$ and defines the same smooth index class in $K_0(\mathcal{A}^\infty_G(Y))$ as the Connes-Skandalis projector of Theorem 5.17.*

We can of course extend these results to the associated symmetrized projectors.

5.6. Higher APS-indices. We now want to extract numbers out of our index class $\operatorname{Ind}_\infty(D) \in K_0(\mathcal{A}^\infty_G(Y)) = K_0(C^*(Y_0 \subset Y)^G)$ and we shall do so by pairing the index class with the cyclic cocycles already encountered in the closed case, that is

- the 0-cocycle tr^Y_g, $g \in G$, with $g = e$ or $g \neq e$ a semisimple element,
- the higher cocycles τ^Y_φ, with $\varphi \in H^*_{\rm diff}(G)$,
- the higher delocalized cocycles $\Phi^P_{Y,g}$, with P a cuspidal parabolic subgroup and g a semisimple element.

These cocycles do define elements in $HP^*(\mathcal{A}^c_G(Y,E))$, because the Schwartz kernels of the operators in $\mathcal{A}^c_G(Y,E)$ vanish on all of the boundary hypersurfaces of $Y \times_b Y$. The following proposition is proved as in the closed case, using the information that the corresponding cyclic cocycles on the group G, that is

$$\operatorname{tr}_g\,, \quad \chi(\varphi)\,, \quad \Phi^P_g\,,$$

do extend from $C^\infty_c(G)$ to $\mathcal{C}(G)$.

PROPOSITION 5.21. *The cyclic cocycles*

$$\operatorname{tr}^Y_g\,, \quad \tau^Y_\varphi\,, \quad \Phi^P_{Y,g} \tag{5.22}$$

extend continuously from $\mathcal{A}^c_G(Y,E)$ to $\mathcal{A}^\infty_G(Y,E)$, thus defining elements in the periodic cyclic cohomology $HP^{\rm even}(\mathcal{A}^\infty_G(Y,E))$.

Using the pairing between $HP^{\rm even}(\mathcal{A}^\infty_G(Y,E))$ and $K_0(\mathcal{A}^\infty_G(Y,E))$ we obtain the numerical indices

$$\langle \operatorname{tr}^Y_g, \operatorname{Ind}_\infty(D)\rangle\,, \quad \langle \tau^Y_\varphi, \operatorname{Ind}_\infty(D)\rangle\,, \quad \langle \Phi^P_{Y,g}, \operatorname{Ind}_\infty(D)\rangle\,. \tag{5.23}$$

In the next section we shall present formulae à la Atiyah-Patodi-Singer for these (higher) indices.

6. Higher APS index theorems

In this section we shall finally present index formulae of Atiyah-Patodi-Singer type for the numerical indices appearing in (5.23). We shall employ in a crucial way relative K-theory and relative cyclic cohomology, as in previous work of Moriyoshi-Piazza, Lesch-Moscovici-Pflaum and Gorokhovsky-Moriyoshi-Piazza, see [**16, 35,**

40]. To be more precise, and forgetting the bundle E in the notation, we shall proceed as follows. Consider the surjective homomorphism

$$ {}^b\mathcal{A}^\infty_G(Y) \xrightarrow{I} {}^b\mathcal{A}^\infty_{G,\mathbb{R}^+}(\mathrm{cyl}(\partial Y)). \tag{6.1}$$

Then, as a first step, associated to (6.1), we shall define a relative index class

$$ \mathrm{Ind}_\infty(D, D_\partial) \in K_0({}^b\mathcal{A}^\infty_G(Y), {}^b\mathcal{A}^\infty_{G,\mathbb{R}^+}(\mathrm{cyl}(\partial Y)) \tag{6.2}$$

where, with a small abuse of notation, we do not write the indicial homomorphism. Next, let ω^Y one of the cyclic cocycles appearing in (5.22); then

- to the cyclic cocycle ω^Y on $\mathcal{A}^\infty_G(Y)$ we associate a regularised cyclic cochain on ${}^b\mathcal{A}^\infty_G(Y)$ and a cyclic cocycle $\xi^{\partial Y}$ on ${}^b\mathcal{A}^\infty_{G,\mathbb{R}^+}(\mathrm{cyl}(\partial Y))$ so that $(\omega^Y, \xi^{\partial Y})$ is a *relative* cyclic cocycle for the homomorphism (6.1);
- we call $\xi^{\partial Y}$ the eta cocycle associated to ω^Y and we call $(\omega^Y, \xi^{\partial Y})$ the relative cyclic cocycle associated to ω^Y.
- we prove that

$$\langle \omega^Y, \mathrm{Ind}_\infty(D)\rangle = \langle (\omega^{Y,r}, \xi^{\partial Y}), \mathrm{Ind}_\infty(D, D_\partial)\rangle;$$

Let us see some of details involved in this program.

6.1. Relative index classes. Recall that if $0 \to J \to A \xrightarrow{\pi} B \to 0$ is a short exact sequence of Fréchet algebras then we set $K_0(J) := K_0(J^+, J) \cong \mathrm{Ker}(K_0(J^+) \to \mathbb{Z})$ and $K_0(A^+, B^+) = K_0(A, B)$ with $()^+$ denoting unitalization. See [**6, 21**]. Recall that a relative K_0-element for $A \xrightarrow{\pi} B$ with unital algebras A, B is represented by a triple (P, Q, p_t) with P and Q idempotents in $M_{n\times n}(A)$ and $p_t \in M_{n\times n}(B)$ a path of idempotents connecting $\pi(P)$ to $\pi(Q)$. The excision isomorphism

$$ \alpha_{\mathrm{ex}} : K_0(J) \longrightarrow K_0(A, B) \tag{6.3}$$

is given by $\alpha_{\mathrm{ex}}([(P, Q)]) = [(P, Q, \mathfrak{c})]$ with $\mathfrak{c}$ denoting the constant path.

Let us go back to the parametrix

$$Q_{\exp} := \frac{I - \exp(-\frac{1}{2}D^- D^+)}{D^- D^+} D^+$$

and its associated true parametrix $Q^b_{\exp}$. Using the latter we have defined the *improved* Connes-Moscovici projector $V^b_{\mathrm{CM}}(D)$, with entries in $\mathcal{A}^\infty(Y, E)$. Using the former we can now define the usual Connes-Moscovici projector $V_{\mathrm{CM}}(D)$. We have the following important result:

THEOREM 6.4.

1) *The Connes-Moscovici projector* $V_{\mathrm{CM}}(D)$,

$$V_{\mathrm{CM}}(D) := \begin{pmatrix} e^{-D^- D^+} & e^{-\frac{1}{2}D^- D^+}\left(\frac{I - e^{-D^- D^+}}{D^- D^+}\right) D^- \\ e^{-\frac{1}{2}D^+ D^-} D^+ & I - e^{-D^+ D^-} \end{pmatrix},$$

is a 2×2 *matrix with entries in* ${}^b\mathcal{A}^\infty_G(Y)$;

2) *the Connes-Moscovici projector* $V_{\mathrm{CM}}(D^{\mathrm{cyl}})$ *is a* 2×2 *matrix with entries in* ${}^b\mathcal{A}^\infty_{G,\mathbb{R}}(\mathrm{cyl}(\partial Y))$.

Consider the Connes-Moscovici projections $V_{\mathrm{CM}}(D)$ and $V_{\mathrm{CM}}(D_{\mathrm{cyl}})$ associated to D and D_{cyl}. With $e_1 := \begin{pmatrix} 0 & 0 \\ 0 & 1 \end{pmatrix}$, we have the triple,

(6.5)
$$(V_{\mathrm{CM}}(D), e_1, q_t), \quad t \in [1, +\infty], \quad \text{with } q_t := \begin{cases} V_{\mathrm{CM}}(tD_{\mathrm{cyl}}), & \text{if } \ t \in [1, +\infty), \\ e_1, & \text{if } \ t = \infty. \end{cases}$$

PROPOSITION 6.6. *Under the invertibility Assumption* (5.3), *the Connes-Moscovici idempotents* $V_{\mathrm{CM}}(D)$ *and* $V_{\mathrm{CM}}(D_{\mathrm{cyl}})$ *define through formula* (6.5) *a relative class in* $K_0({}^b\mathcal{A}^\infty_G(Y), {}^b\mathcal{A}^\infty_{G,\mathbb{R}^+}(\mathrm{cyl}(\partial Y))$, *the relative K-theory group associated to the surjective homomorphism*

$${}^b\mathcal{A}^\infty_G(Y) \xrightarrow{I} {}^b\mathcal{A}^\infty_{G,\mathbb{R}^+}(\mathrm{cyl}(\partial Y)).$$

With a small abuse of notation we denote this class by $[V_{\mathrm{CM}}(D), e_1, V_{\mathrm{CM}}(tD_{\mathrm{cyl}})]$.

DEFINITION 6.7. We define the relative (smooth) index class as

$$\mathrm{Ind}_\infty(D, D_\partial) := [V_{\mathrm{CM}}(D), e_1, V_{\mathrm{CM}}(D_{\mathrm{cyl}})] \ \in \ K_0({}^b\mathcal{A}^\infty_G(Y), {}^b\mathcal{A}^\infty_{G,\mathbb{R}^+}(\mathrm{cyl}(\partial Y))\,.$$

Recall now the smooth index class $\mathrm{Ind}_\infty(D) \in K_0(\mathcal{A}^\infty_G(Y))$ defined through Theorem 5.20. The following result plays a crucial role:

THEOREM 6.8. *Let* α_{ex} *be the excision isomorphism for the short exact sequence* $0 \to \mathcal{A}^\infty_G(Y) \to {}^b\mathcal{A}^\infty_G(Y) \xrightarrow{I} {}^b\mathcal{A}^\infty_{G,\mathbb{R}^+}(\mathrm{cyl}(\partial Y)) \to 0$. *Then*

$$\alpha_{\mathrm{ex}}(\mathrm{Ind}_\infty(D)) = \mathrm{Ind}_\infty(D, D_\partial)\,. \tag{6.9}$$

We summarize the content of subsection 5.5 and the present subsection 6.1 as follows:
using the Connes-Moscovici projector(s) we have proved the existence of smooth index classes

$$\mathrm{Ind}_\infty(D) \in K_0(\mathcal{A}^\infty_G(Y)) \ \text{and} \ \mathrm{Ind}_\infty(D, D_\partial) \in K_0({}^b\mathcal{A}^\infty_G(Y), {}^b\mathcal{A}^\infty_{G,\mathbb{R}}(\mathrm{cyl}(\partial Y))),$$

with the first one sent to the second one by the excision isomorphism α_{ex}.

6.2. Relative cyclic cocycles. Recall that the relative cyclic complex associated to a short exact sequence $0 \to J \to A \xrightarrow{\pi} B \to 0$ of algebras is given by

$$CC^k(A, B) := CC^k(A) \oplus CC^{k+1}(B),$$

equipped with the differential

$$\begin{pmatrix} b + B & -\pi^* \\ 0 & -(b + B) \end{pmatrix},$$

where b, B are the usual Hochschild and cyclic differential and π^* denotes the pullback of functionals through the surjective morphism $\pi : A \to B$. In our case, the relevant extension is given, first of all, by

$$0 \to \mathcal{A}^c_G(Y) \to {}^b\mathcal{A}^c_G(Y) \xrightarrow{I} {}^b\mathcal{A}^c_{G,\mathbb{R}^+}(\mathrm{cyl}(Y)) \to 0\,, \quad \mathcal{A}^c_G(Y) := \mathrm{Ker}\, I,$$

where this is now, for simplicity, the short exact sequence for the small b-calculus.

As in the closed case, given a global slice $Z \subset Y$, we can view $A \in {}^b\mathcal{A}^c_G(Y)$ as a map $\Phi_A : G \to {}^b\Psi^{-\infty}(Z)$ by setting $\Phi_A(g, s_1, s_2) := A(s_1, gs_2)$. Likewise, an element $B \in {}^b\mathcal{A}^c_{G,\mathbb{R}^+}(\mathrm{cyl}(\partial Y))$ gives rise, by Mellin transform, to a map Φ_B :

$G \times \mathbb{R} \to \Psi^{-\infty}(\partial Z)$, which is compactly supported on G, and rapidly decreasing on $\mathbb{R}$. In the following, we shall denote by

$$\Phi_A \mapsto \hat{\Phi}_A \text{ the morphism } I : {}^b\mathcal{A}^c_G(Y) \to {}^b\mathcal{A}^c_{G,\mathbb{R}^+}(\text{cyl}(\partial Y)) \quad \text{followed by the Mellin transform.} \tag{6.10}$$

6.3. The b-Trace on G-proper b-manifolds. Let us construct the correct analogue of the b-trace of Melrose in this setting where we have a proper group action. In our geometric setting, a choice of cut-off function c_{Y_0} for the action of G on Y_0 restricts to give a cut-off function $c_{\partial Y_0} := c_{Y_0}|_{\partial Y_0}$ for the G-action on ∂Y_0. We shall also write briefly c_∂. We consider as usual the associated b-manifold Y, endowed with a product-type b-metric $\mathbf{h}$, so that, metrically, Y is a manifold with cylindrical ends, and we shall, by a small abuse of notation, write c_Y for the extension of c_{Y_0} on Y_0 which is constant in the cylindrical coordinate.

Using the b-integral of Melrose, see [**37**], we now define

DEFINITION 6.11. For $A \in {}^b\mathcal{A}^c_G(Y)$ its ${}^bG$-trace is defined as

$${}^b\text{tr}_{c_Y}(A) := {}^b\!\!\int_Y K_A(x,x) c_Y(x) d\text{vol}(x).$$

Remark that the cut-off function on Y_0 has compact support, so the b-regularized integral is indeed well defined. The argument in [**45**, Prop. 2.3] shows that ${}^b\text{tr}_{c_Y}$ is independent of the choice of cut-off function c_Y. Next, using a simple trick with a family $\{c_{Y,\epsilon}\}_{\epsilon>0}$ of cut-off functions converging to the characteristic function on Z, we can rewrite

$${}^b\text{tr}_{c_Y}(A) = {}^b\text{tr}_Y\left(\Phi_A(e)\right). \tag{6.12}$$

As in the usual b-calculus, ${}^b\text{tr}_{c_Y}$ is not a trace, but we have a precise formula for its defect on commutators, directly inspired by Melrose' b-trace formula :

LEMMA 6.13. *For $A_1, A_2 \in {}^b\mathcal{A}^c_G(Y)$, we have*

$${}^b\text{tr}_{c_Y}([\Phi_{A_1}, \Phi_{A_2}]) = \frac{i}{2\pi}\int_{\mathbb{R}}\int_G \text{tr}_{\partial Z}\left(\frac{\partial I(\Phi_{A_1}, h^{-1}, \lambda)}{\partial \lambda} \circ I(\Phi_{A_2}, h, \lambda)\right) dh d\lambda.$$

with $\text{tr}_{\partial Z}$ denoting the usual functional analytic trace on smoothing operators on closed compact manifolds.

6.4. From absolute to relative cyclic cocycles. Let ω^Y be any one of the cyclic cocycles for $\mathcal{A}^\infty_G(Y)$ appearing in (5.22). We can write

$$\omega^Y = \omega \circ \text{tr}_Z$$

where ω is the corresponding cyclic cocycle on the group G and where $\text{tr}_Z : \mathcal{A}^\infty_G(Y, E) \to \mathcal{C}(G)$ is the homomorphism of integration along the slice. The integrals along the slice are well defined because the operators in $\mathcal{A}^\infty_G(Y)$ vanish to order ϵ at all the boundary hypersurfaces of $Y \times_b Y$. If we pass to ${}^b\mathcal{A}^\infty_G(Y)$ we must replace ordinary integration with b-integration , as operators in ${}^b\mathcal{A}^\infty_G(Y)$ do *not* vanish on the front face of $Y \times_b Y$; we obtain in this way a *regularized* multilinear functional $\omega^{Y,r}$; this will not give us a cyclic cocycle anymore, precisely because of Lemma 6.13; however, using the exact form of the Lemma 6.13 we will be able to produce a cyclic cocycle on ${}^b\mathcal{A}^c_{G,\mathbb{R}^+}(\text{cyl}(\partial Y))$, call it $\xi^{\partial Y}$, in such a way that the pair $(\omega^{Y,r}, \xi^{\partial Y})$ is in fact a *relative* cyclic cocycle for the surjective homomorphism ${}^b\mathcal{A}^c_G(Y) \xrightarrow{I} {}^b\mathcal{A}^c_{G,\mathbb{R}^+}(\text{cyl}(\partial Y))$.

6.5. Relative cyclic cocycles associated to orbital integrals. Let us see how the general principle put forward in the previous subsection works in the case $\omega^Y = \mathrm{tr}^Y_g$. If Y_0 is a cocompact G-proper manifold with boundary and Y is the associated b-manifold, then associated to the orbital integral tr_g we have the trace-homomorphism

$$\mathrm{tr}^Y_g : \mathcal{A}^\infty_G(Y) \to \mathbb{C}, \tag{6.14}$$

given explicitly by

$$\mathrm{tr}^Y_g(\kappa) := \int_{G/Z_g} \int_Y c_Y(hgh^{-1}y)\mathrm{tr}(hgh^{-1}\kappa(hg^{-1}h^{-1}y, y))dx\, d(hZ). \tag{6.15}$$

Here dy denotes the b-density associated to the b-metric $\mathbf{h}$ and the cut-off function c_{Y_0} on Y_0 is extended constantly along the cylinder to define c_Y. As already explained, the Y integral converges, given that κ vanishes of order $\epsilon > 0$ on all the boundary hypersurfaces of $Y \times_b Y$.

Now, let X be a closed manifold equipped with a proper, cocompact action of G. We consider $\mathrm{cyl}(X) = X \times \mathbb{R}$ the cylinder over X, equipped with the action of $G \times \mathbb{R}$. Using the Mellin transforms we have an algebra homomorphism

$${}^b\mathcal{A}^c_{G,\mathbb{R}}(\mathrm{cyl}(Y)) \longrightarrow \mathscr{S}(\mathbb{R}, \mathcal{A}^c_G(Y)), \quad A \mapsto \hat{A}. \tag{6.16}$$

PROPOSITION 6.17. *Let X be a cocompact G-proper manifold without boundary; for example $X = \partial Y_0$. Define the following 1-cochain on ${}^b\mathcal{A}^c_{G,\mathbb{R}}(\mathrm{cyl}(X))$*

$$\sigma^X_g(A_0, A_1) = \frac{i}{2\pi} \int_{\mathbb{R}} \mathrm{tr}^X_g(\partial_\lambda \widehat{A}_0(\lambda) \circ \widehat{A}_1(\lambda))d\lambda\,, \tag{6.18}$$

Then $\sigma^X_g(\,,\,)$ is well-defined and is a cyclic 1-cocycle.

Let Y be now a b-manifold and let ∂Y be its boundary. Let $\mathrm{tr}^{Y,r}_g$ be the functional on ${}^b\mathcal{A}^c_G(Y)$:

$$\mathrm{tr}^{Y,r}_g(T) := \int_{G/Z_g} \int^b_Y c_Y(hgh^{-1}y)\mathrm{tr}(hgh^{-1}\kappa(hg^{-1}h^{-1}y, y))dy\, d(hZ)\,.$$

Here κ is the kernel of the operator G, and Melrose's b-integral has been used. This is the regularization of tr^Y_g that one needs to consider when one passes from $\mathcal{A}^c_G(Y)$ to ${}^b\mathcal{A}^c_G(Y)$ (for the time being on kernels of G-compact support). Observe that

$$\mathrm{tr}^{Y,r}_g = \mathrm{tr}_g \circ {}^b\mathrm{tr}_Z$$

with ${}^b\mathrm{tr}_Z : {}^b\mathcal{A}^c_G(Y) \to C^\infty_c(G)$ denoting b-integration along the slice Z. More precisely, as in the closed case, we have an isomorphism

$$\begin{aligned} &{}^b\mathcal{A}^c_G(Y) \\ \cong\ &\left\{\Phi : G \to {}^b\Psi^{-\infty,\epsilon}(Z), \text{ smooth, compactly supported and } K \times K \text{ invariant}\right\} \end{aligned}$$

and ${}^b\,\mathrm{Tr}_Z$ associates to Φ the function $G \ni \gamma \to {}^b\,\mathrm{Tr}(\Phi(\gamma))$.

One can prove the following

PROPOSITION 6.19. *The pair $(\mathrm{tr}^{Y,r}_g, \sigma^{\partial Y}_g)$ defines a relative 0-cocycle for*

$${}^b\mathcal{A}^c_G(Y) \xrightarrow{I} {}^b\mathcal{A}^c_{G,\mathbb{R}}(\mathrm{cyl}(\partial Y)).$$

Moreover, the 0-degree cyclic cocycle $(\mathrm{tr}_g^{Y,r}, \sigma_g^{\partial Y})$ *extends continuously to a relative 0-cocycle for*

$$ {}^b\mathcal{A}^\infty_G(Y) \xrightarrow{I} {}^b\mathcal{A}^\infty_{G,\mathbb{R}}(\mathrm{cyl}(\partial Y)). $$

Finally, the following formula holds:

$$ \langle \mathrm{tr}_g^Y, \mathrm{Ind}_\infty(D)\rangle = \langle (\mathrm{tr}_g^{Y,r}, \sigma_g^{\partial Y}), \mathrm{Ind}_\infty(D, D_\partial)\rangle \,. \tag{6.20} $$

6.6. The delocalized APS index formula on G-proper manifolds. We shall prove the delocalized APS index formula using crucially (6.20). On the right hand-side we have the pairing of the relative cocycle $(\tau_g^{Y,r}, \sigma_g^{\partial Y})$ with the relative index class defined by

(6.21)

$$ (V_{\mathrm{CM}}(D), e_1, q_t)\,, \quad t \in [1, +\infty]\,, \quad \text{with } q_t := \begin{cases} V_{\mathrm{CM}}(tD_{\mathrm{cyl}}) & \text{if } \ t \in [1, +\infty) \\ e_1 & \text{if } \ t = \infty \end{cases} $$

and with $e_1 := \begin{pmatrix} 0 & 0 \\ 0 & 1 \end{pmatrix}$.

By definition of relative pairing we have:

$$ \begin{aligned} &\langle (\mathrm{tr}_g^{Y,r}, \sigma_g^{\partial Y}), (V_{\mathrm{CM}}(D), e_1, q_t)\rangle \\ &= \mathrm{tr}_g^{Y,r}(e^{-D^- D^+}) - \mathrm{tr}_g^{Y,r}(e^{-D^+ D^-}) + \int_1^\infty \sigma_g^{\partial Y}([\dot q_t, q_t], q_t) dt \,. \end{aligned} \tag{6.22} $$

A complicated but totally elementary computation shows that the following Proposition holds:

Proposition 6.23. *The term* $\int_1^\infty \sigma_g^{\partial Y}([\dot q_t, q_t], q_t)dt$*, with* $q_t := V_{\mathrm{CM}}(tD_{\mathrm{cyl}})$ *is equal to*

$$ -\frac{1}{2}\left(\frac{1}{\sqrt{\pi}}\int_1^\infty \mathrm{tr}_g^{\partial Y}(D_\partial \exp(-tD_\partial^2))\frac{dt}{\sqrt{t}}\right) . $$

Thanks to this Proposition we have that

(6.24)

$$ \begin{aligned} &\langle (\mathrm{tr}_g^{Y,r}, \sigma_g^{\partial Y}), (V(D), e_1, q_t)\rangle \\ &= \mathrm{tr}_g^{Y,r}(e^{-D^- D^+}) - \mathrm{tr}_g^{Y,r}(e^{-D^+ D^-}) - \frac{1}{2}\int_1^\infty \frac{1}{\sqrt{\pi}} \tau_g^{\partial M}(D_{\partial Y} \exp(-tD_{\partial Y}^2))\frac{dt}{\sqrt{t}} \,. \end{aligned} $$

As a last step we replace D by sD; in the equality

$$ \langle \mathrm{tr}_g^Y, \mathrm{Ind}_\infty(D)\rangle = \langle (\mathrm{tr}_g^{Y,r}, \sigma_g^{\partial Y}), \mathrm{Ind}_\infty(D, D_\partial)\rangle $$

the left hand side $\langle \tau_g^Y, \mathrm{Ind}_\infty(D)\rangle$ remains unchanged whereas the right hand side becomes

$$ \mathrm{tr}_g^{Y,r}(e^{-s^2 D^- D^+}) - \mathrm{tr}_g^{Y,r}(e^{-s^2 D^+ D^-}) - \frac{1}{2}\int_s^\infty \frac{1}{\sqrt{\pi}} \mathrm{tr}_g^{\partial Y}(D_{\partial Y} \exp(-tD_{\partial Y}^2)\frac{dt}{\sqrt{t}} \,. $$

Summarizing, for each $s > 0$ we have

$$ \begin{aligned} &\mathrm{tr}_g^{Y,r}(e^{-s^2 D^- D^+}) - \mathrm{tr}_g^{Y,r}(e^{-s^2 D^+ D^-}) \\ = \;& \langle \mathrm{tr}_g^Y, \mathrm{Ind}_\infty(D)\rangle + \frac{1}{2}\int_s^\infty \frac{1}{\sqrt{\pi}} \mathrm{tr}_g^{\partial Y}(D_{\partial Y} \exp(-tD_{\partial Y}^2)\frac{dt}{\sqrt{t}} \,. \end{aligned} $$

Now we take the limit as $s \downarrow 0$; using Getzler' rescaling in the b-context one can prove that

$$\lim_{s\downarrow 0} \mathrm{tr}_g^{Y,r}(e^{-s^2 D^- D^+}) - \mathrm{tr}_g^{Y,r}(e^{-s^2 D^+ D^-}) = \int_{Y_0^g} c_{Y_0}^g \mathrm{AS}_g(D_0)$$

with $c_{Y_0}^g \mathrm{AS}_g(D_0)$ defined in Equation (4.10).
Assuming this last result, we can infer that the limit

$$\lim_{s\downarrow 0} \frac{1}{2}\int_s^\infty \frac{1}{\sqrt{\pi}} \mathrm{tr}_g^{\partial Y}(D_{\partial Y}\exp(-tD_{\partial Y}^2))\frac{dt}{\sqrt{t}}$$

exists and equals

$$\int_{Y_0^g} c_{Y_0^g}^g \mathrm{AS}_g(D_0) - \langle \mathrm{tr}_g^Y, \mathrm{Ind}_\infty(D)\rangle\,.$$

We conclude that the following Theorem holds:

THEOREM 6.25. *(0-degree delocalized APS)*
Let G be a connected, linear real reductive group. Let g be a semisimple element. Let Y_0, Y, D, $D_{\partial Y}$ as above. Assume that $D_{\partial Y}$ is L^2-invertible. Then

$$\eta_g(D_{\partial Y}) := \frac{1}{\sqrt{\pi}}\int_0^\infty \mathrm{tr}_g^{\partial Y}(D_{\partial Y}\exp(-tD_{\partial Y}^2))\frac{dt}{\sqrt{t}}$$

exists and for the pairing of the index class $\mathrm{Ind}(D_\infty)\in K_0(\mathcal{A}_G^\infty(Y))\equiv K_0(C^*(Y_0\subset Y)^G)$ with the 0-cocycle $\mathrm{tr}_g^Y \in HC^0((\mathcal{A}_G^\infty(Y)))$ the following delocalized 0-degree APS index formula holds:

$$\langle \mathrm{tr}_g^Y, \mathrm{Ind}_\infty(D)\rangle = \int_{Y_0^g} c_{Y_0^g}\mathrm{AS}_g(D_0) - \frac{1}{2}\eta_g(D_{\partial Y})\,, \tag{6.26}$$

where the integrand $c_{Y_0^g}\mathrm{AS}_g(D_0)$ is defined in the same way as the one in Equation (4.10).

REMARK 6.27. This result was first discussed in the work of Hochs-Wang-Wang [**26**]. Our treatment, centred around the interplay between absolute and relative cyclic cohomology and the b-calculus, is completely different; moreover our treatment allows us to get sharper results compared to [**26**] in the case of a connected linear real reductive group G. More precisely, in Theorem 6.25 we only assume that g is a semisimple element of G to obtain the index formula (6.26), while in [**26**, Theorem 2.1], the authors require that G/Z_g is compact.

REMARK 6.28. If we take $g = e$ then we get

$$\langle \mathrm{tr}_e^Y, \mathrm{Ind}_{C^*}(D)\rangle = \int_Y c_Y \mathrm{AS}(Y) - \frac{1}{2}\eta_G(D_\partial) = \int_{Y_0} c_{Y_0}\mathrm{AS}(Y_0) - \frac{1}{2}\eta_G(D_\partial)$$

with

$$\eta_G(D_\partial) = \frac{2}{\sqrt{\pi}}\int_0^\infty \mathrm{tr}_e^{\partial Y}\, D_\partial e^{-(tD_\partial)^2}dt.$$

Notice however that this particular result holds under much more general assumptions on G than the ones we are currently imposing (G connected reductive linear Lie group). Indeed, the pairing of the index class with tr_e^Y is equal to the pairing of the Morita equivalent $C_r^*(G)$-index class $\mathrm{Ind}_{C_r^*(G)}(D)$ with the canonical trace tr_e on C^*G:

$$\langle\, \mathrm{tr}_e^Y, \mathrm{Ind}_{C^*}(D)\rangle = \langle \mathrm{tr}_e, \mathrm{Ind}_{C^*G}(D)\rangle\,.$$

Proceeding as in [**56**], one checks that $\langle \mathrm{tr}_e^Y, \mathrm{Ind}_{C^*G}(D)\rangle$ is equal to the von Neumann G-index of D^+. A formula for this von Neumann index can be proved in the von Neumann framework by mimicking the proof of Vaillant for Galois coverings of manifolds with cylindrical ends [**55**] (in turn inspired by Melrose' proof on manifolds with cylindrical ends). Thus, assuming only that G is a Lie group but keeping the L^2-invertibility of the boundary operator D_∂, we obtain that $\langle \mathrm{tr}_e^Y, \mathrm{Ind}_{C^*}(D)\rangle$ and $\langle \mathrm{tr}_e, \mathrm{Ind}_{C^*G}(D)\rangle$ are well defined and that

$$\langle \mathrm{tr}_e^Y, \mathrm{Ind}_{C^*}(D), \rangle = \langle \mathrm{tr}_e, \mathrm{Ind}_{C^*G}(D)\rangle = \mathrm{Ind}_{\mathrm{vN}}(D^+) = \int_{Y_0} c_{Y_0} \mathrm{AS}(Y_0) - \frac{1}{2}\eta_G(D_\partial)$$

with $\eta_G(D_\partial) = \frac{2}{\sqrt{\pi}} \int_0^\infty \mathrm{tr}_e^{\partial Y} D_\partial e^{-(tD_\partial)^2} dt$.
This formula is the same as the one appearing in the work of Hochs-Wang-Wang, see [**25**].

6.7. More on delocalized eta invariants. In the previous section we have obtained the well-definedness of $\eta_g(D_\partial)$, with D_∂ being L^2-invertible, as a byproduct of the proof of the delocalized APS index theorem for 0-degree cocycles. In fact, one can show that $\eta_g(D)$ is well defined on a cocompact G-proper manifold even if D does not arise as a boundary operator and without the invertibility assumption. This is the content of the next theorems, partially discussed also in [**25, 26**]. We assume as always when short time limits of the heat kernels are involved, that the metric is slice-compatible.

THEOREM 6.29. *Let $(X, \mathbf{h})$ be a cocompact G-proper manifold without boundary endowed with a slice compatible G-equivariant metric and let D be a Dirac-type operator of the form (2.7). Let g be a semisimple element. The integral*

$$\frac{1}{\sqrt{\pi}} \int_0^\infty \mathrm{tr}_g^X (D \exp(-tD^2)) \frac{dt}{\sqrt{t}} \tag{6.30}$$

converges and defines the delocalized eta invariant associated to D, $\eta_g(D)$.

Notice that the proof of this result is rather delicate, both at $t = 0$ and $t = +\infty$, and employs results of Moscovici-Stanton [**41**], Bismut [**5**], and Zhang [**60**].

6.8. Relative cyclic cocycles associated to smooth group cocycles. In this subsection we want to see how, given a smooth group cocycle $\varphi \in Z^k_{\mathrm{diff}}(G)$, we can pass from the cyclic cocycle τ^Y_φ on $\mathcal{A}^c_G(Y)$ to a relative cyclic cocycle for the surjective homomorphism ${}^b\mathcal{A}^c_G(Y) \xrightarrow{I} {}^b\mathcal{A}^c_{G,\mathbb{R}^+}(\mathrm{cyl}(\partial Y))$. As we have already explained, the first step is to pass from τ^Y_φ to $\tau^{Y,r}_\varphi$, and this is achieved by replacing integrals by b-integrals or, equivalently, traces by b-traces. This is what we do in the next definition.

DEFINITION 6.31. Let Y be a proper G-manifold with boundary, and $\varphi \in Z^k_{\mathrm{diff}}(G)$ be a smooth group cocycle. For $A_0, \ldots, A_k \in {}^b\mathcal{A}^c_G(Y)$, define

$$\tau^{Y,r}_\varphi(A_0, \ldots, A_k) := \int_{G^{\times k}} {}^b\mathrm{tr}_Z \left(\Phi_{A_0}((g_1 \cdots g_k)^{-1}) \circ \Phi_{A_1}(g_1) \circ \ldots \circ \Phi_{A_k}(g_k)\right) \varphi(e, g_1, g_1 g_2, \ldots, g_1 \cdots g_k) dg_1 \cdots dg_k.$$

in the above equation we have used the homogeneous differentiable group cohomology complex introduced in Definition 3.8.

Next, following the general strategy explained at the beginning of this section, we want to define the eta cocycle associated to φ.

DEFINITION 6.32. Let X be a closed manifold equipped with a proper, cocompact action of G, and let $\varphi \in C^k_{\text{diff}}(G)$ be a smooth group cochain. The *eta cochain* on ${}^b\mathcal{A}^c_{G,\mathbb{R}}(\text{cyl}(X))$ associated to φ is defined as

$$\begin{aligned}
&\sigma^X_\varphi(B_0,\ldots,B_{k+1})\\
&:=\frac{(-1)^{k+1}}{2\pi}\int_{G^{k+1}}\int_{\mathbb{R}}\\
&\qquad \operatorname{tr}\left(\hat{B}_0((g_1\cdots g_{k+1})^{-1},\lambda)\circ\hat{B}_1(g_1,\lambda)\circ\cdots\circ\hat{B}_k(g_k,\lambda)\circ\frac{\partial\hat{B}_{k+1}(g_{k+1},\lambda)}{\partial\lambda}\right)\\
&\qquad\qquad \varphi(e,g_1,g_1g_2,\ldots,g_1\cdots g_k)d\lambda dg_1\cdots dg_{k+1}.
\end{aligned}$$

where the notation in (6.10) has been used, and where we have used the homogeneous differentiable group cohomology complex introduced in Definition 3.8.

Using Lemma 6.13 one can prove the following result [**47**]:

PROPOSITION 6.33. *If* $\varphi \in Z^k_{\text{diff}}(G)$ *is a smooth group* **cocycle** *then* $(\tau^{Y,r}_\varphi, \sigma^{\partial Y}_\varphi)$ *is a relative cyclic cocycle for* ${}^b\mathcal{A}^c_G(Y) \xrightarrow{I} {}^b\mathcal{A}^c_{G,\mathbb{R}^+}(\text{cyl}(\partial Y))$

In addition, one can prove the following:

PROPOSITION 6.34. *The pair* $(\tau^{Y,r}_\varphi, \sigma^{\partial Y}_\varphi)$ *extends continuously to a relative k-cocycle for*

$${}^b\mathcal{A}^\infty_G(Y) \xrightarrow{I} {}^b\mathcal{A}^\infty_{G,\mathbb{R}}(\text{cyl}(\partial Y)).$$

Moreover the following formula holds:

$$\langle \tau^Y_\varphi, \text{Ind}_\infty(D)\rangle = \langle(\tau^{Y,r}_\varphi, \sigma^{\partial Y}_\varphi), \text{Ind}_\infty(D, D_\partial)\rangle. \tag{6.35}$$

EXAMPLE 6.36. As explained in Example 3.12, besides the trivial group cocycle, there is an interesting degree 2 cocycle A given by the area of a hyperbolic triangle in $\mathbb{H}$. The corresponding eta 3-cocycle on $\mathscr{S}(\mathbb{R}, C^\infty_c(G))$ is given by

$$\begin{aligned}
&\sigma_A(B_0,B_1,B_2,B_3)\\
&:=-\frac{1}{2\pi}\int_{G^3}\int_{\mathbb{R}}\hat{B}_0((g_1g_2g_3)^{-1},\lambda)*\hat{B}_1(g_1,\lambda)*\hat{B}_2(g_2,\lambda)*\frac{\partial\hat{B}(g_3,\lambda)}{\partial\lambda}\\
&\qquad\qquad \text{Area}(\Delta_{\mathbb{H}}([e],g_1[e],g_1g_2[e]))d\lambda dg_1dg_2dg_3.
\end{aligned}$$

6.9. Higher APS index theorem associated to a group cocycle $\varphi \in Z^k_{\text{diff}}(G)$. Using (6.35), proceeding as we did for the delocalized trace tr^Y_g (and for the corresponding relative cocycle $(\text{tr}^{Y,r}_g, \sigma^{\partial Y}_g)$), using the heat kernel approach to the Pflaum-Posthuma-Tang index formula developed in [**47**], one can establish the following higher Atiyah-Patodi-Singer index theorem

THEOREM 6.37. *Let Y_0, Y and D as above. Assume that the boundary operator D_∂ is L^2-invertible. We consider $[\varphi] \in H^{2p}_{\text{diff}}(G)$ and*

$$\text{Ind}_\varphi(D) := (-1)^p\frac{2p!}{p!}\langle\tau^Y_\varphi, \text{Ind}_\infty(D)\rangle.$$

Then
(6.38)

$$\operatorname{Ind}_\varphi(D) = \int_Y c_Y \operatorname{AS}(Y) \wedge \Phi([\varphi]) - \frac{1}{2}\eta_\varphi(D_\partial) \equiv \int_{Y_0} c_{Y_0} \operatorname{AS}(Y_0) \wedge \Phi([\varphi]) - \frac{1}{2}\eta_\varphi(D_\partial),$$

where

$$\eta_\varphi(D_\partial) := c_p \left[\sum_{i=0}^{2p} \int_0^\infty \sigma_\varphi^{\partial Y}(p_t, \dots, [\dot{p}_t, p_t], \dots, p_t) dt \right] \tag{6.39}$$

with $p_t = V(tD_{\text{cyl}})$ *and* $c_p = (-1)^p \frac{2p!}{p!}$.

For more details we refer to [**47**].

6.10. Higher delocalized APS index theorem. We finally come to the higher delocalized cyclic cocycles $\Phi^P_{Y,g}$, with P a cuspidal parabolic subgroup with Langlands decomposition MAN and g a semisimple element in M. The cyclic cocycle $\Phi^P_{Y,g}$ on $\mathcal{A}^c_G(Y)$ can be written explicitly as

$$\begin{aligned}
&\Phi^P_{Y,g}(A_0, A_1, \dots, A_m)\\
&:= \int_{h\in M/Z_M(g)} \int_{KN} \int_{G^{\times m}} C\big(H(g_1 \cdots g_m k), H(g_2 \cdots g_m k), \dots, H(g_m k)\big)\\
&\quad \operatorname{Tr}\Big(A_0\big(khgh^{-1}nk^{-1}(g_1 \cdots g_m)^{-1}\big) \circ A_1(g_1) \cdots \circ A_m(g_m)\Big) dg_1 \cdots dg_m dk dn dh.
\end{aligned}$$

Substituting the trace over Y with the b-trace we define the cochain $\Phi^{P,r}_{Y,g}$ over ${}^b\mathcal{A}^c_G(Y)$. We also have the corresponding eta cochain $\sigma^P_{\partial Y,g}$ on ${}^b\mathcal{A}^c_{G,\mathbb{R}}(\text{cyl}(Y))$: for $B_0, \dots, B_{m+1} \in {}^b\mathcal{A}^c_{G,\mathbb{R}}(\text{cyl}(Y))$

$$\begin{aligned}
&\sigma^P_{\partial Y,g}(B_0, \dots, B_{m+1})\\
&:= \int_{h\in M/Z_M(g)} \int_{KN} \int_{G^{\times m+1}} \int_{\mathbb{R}} C\big(H(g_1 \cdots g_m k), H(g_2 \cdots g_m k), \dots, H(g_m k)\big)\\
&\qquad \operatorname{Tr}\Big(\hat{B}_0\big(khgh^{-1}nk^{-1}(g_1 \cdots g_m g_{m+1})^{-1}, \lambda\big) \circ \hat{B}_1(g_1, \lambda) \circ \cdots \circ \hat{B}_m(g_m, \lambda)\\
&\qquad\qquad \circ \frac{\partial \hat{B}_{m+1}(g_{m+1}, \lambda)}{\partial \lambda}\Big) dg_1 \cdots dg_{m+1} dk dn dh d\lambda,
\end{aligned}$$

where the notation in (6.10) has been used. One proves that the pair $(\Phi^{r,P}_{Y,g}, \sigma^P_{\partial Y,g})$ defines a relative cyclic cocycle for the homomorphism ${}^b\mathcal{A}^c_G(Y) \xrightarrow{I} {}^b\mathcal{A}^c_{G,\mathbb{R}}(\text{cyl}(\partial Y))$. Moreover:

- the relative cyclic cocycle $(\Phi^{r,P}_{Y,g}, \sigma_{\partial Y,g})$ extends continuously from the pair ${}^b\mathcal{A}^c_G(Y) \xrightarrow{I} {}^b\mathcal{A}^c_{G,\mathbb{R}}(\text{cyl}(\partial Y))$ to the pair ${}^b\mathcal{A}^\infty_G(Y) \xrightarrow{I} {}^b\mathcal{A}^\infty_{G,\mathbb{R}}(\text{cyl}(\partial Y))$.
- the crucial formula $\langle \Phi^P_{Y,g}, \operatorname{Ind}_\infty(D)\rangle = \langle (\Phi^{P,r}_{Y,g}, \sigma^P_{\partial Y,g}), \operatorname{Ind}_\infty(D, D_\partial)\rangle$ holds.

Using the last formula and proceeding as in the previous cases we arrive at the following result:
for each $s \in (0,1]$

$$c_m \langle \operatorname{Ind}_\infty(D), [\Phi^P_{Y,g}]\rangle = \Phi^{P,r}_{Y,g}(V(sD), \dots, V(sD)) - \frac{1}{2}\int_s^\infty \eta^P_g(t) dt$$

with $c_m = (-1)^{\frac{m}{2}} \frac{m!}{(\frac{m}{2})!}$ and $\eta_g^P(t) := 2c_m \sum_{i=0}^m \sigma_{\partial Y,g}^P(p_t, \ldots, [\dot{p}_t, p_t], \ldots, p_t)$. Part of the statement is of course that the t-integral converges at $+\infty$.

We would like to take the limit as $s \downarrow 0$; unfortunately we do not know how to compute the limit of the first term on the right hand side (this should produce the local term in the index formula); in fact we cannot compute this limit even in the closed case.

Open problem. Let $P < G$ a cuspidal parabolic subgroup with Langland 's decomposition $P = MAN$ and let $g \in M$ be a semisimple element. Let D be a G-equivariant Dirac operator on a closed cocompact G-proper manifold X; let $V(D)$ be the symmetrized Connes-Moscovici projector. Can one prove that

$$\lim_{s \downarrow 0} \Phi_{X,g}^P(V(sD), \ldots, V(sD))$$

exists? If so, can one give an explicit formula for it?

Because of these difficulties we go back to the proof of the higher delocalized index formula in the closed case by reduction. Thus we consider $Y_M := Y/AN$, an M-proper manifold, which has a slice decomposition given by $Y_M := M \times_{K \cap M} Z$. The arguments of Hochs, Song and Tang in Theorem 4.12 can be extended (with some efforts) to the case of manifolds with boundary yielding the following theorem (which is one of the main results in [**48**]):

THEOREM 6.40. *Assume that $D_{\partial Y}$ is L^2-invertible and consider the higher index $\langle \Phi_{Y,g}^P, \mathrm{Ind}_\infty(D) \rangle$. The following formula holds:*

$$\langle \Phi_{Y,g}^P, \mathrm{Ind}_\infty(D_Y) \rangle = \int_{(Y_0/AN)^g} c^g_{(Y_0/AN)^g} \mathrm{AS}(Y_0/AN)_g - \frac{1}{2}\eta_g(D_{\partial Y_M})$$

with

$$\eta_g(D_{\partial Y_M}) = \frac{1}{\sqrt{\pi}} \int_0^\infty \mathrm{tr}_g^{\partial Y_M}(D_{\partial Y_M} \exp(-tD^2_{\partial Y_M}))\frac{dt}{\sqrt{t}}.$$

Here $c^g_{(Y_0/AN)^g}$ is a compactly supported smooth cutoff function on $(Y_0/AN)^g$ associated to the $Z_{M,g}$ action on $(Y_0/AN)^g$.

7. Geometric applications

Now that we have explained various index theorems associated to a G-equivariant Dirac operator on a G-proper manifold, we discuss some geometric applications.

7.1. Higher genera. We begin by observing that the homogeneous space G/K is a smooth model for $\underline{E}G$, the classifying space for proper actions of G, see [**4**]: for any smooth proper action of G on a manifold X, there exists a smooth G-equivariant classifying map $\psi_X : X \to G/K$, unique up to G-equivariant homotopy. For any proper action of G on manifold X we consider $\Omega^\bullet_{\mathrm{inv}}(X)$, the complex of G-invariant differential forms on X and its cohomology denoted by $H^\bullet_{\mathrm{inv}}(X)$. For a connected real reductive linear group G, we have the Van Est isomorphism: $H^*_{\mathrm{diff}}(G) \simeq H^\bullet_{\mathrm{inv}}(G/K)$. Consider now $[\alpha] \in H^\bullet_{\mathrm{inv}}(G/K)$ and let $\alpha \in \Omega^\bullet_{\mathrm{inv}}(G/K)$ a representative; consider its pull-back $\psi_X^*\alpha \in \Omega^\bullet_{\mathrm{inv}}(X)$ such that $[\psi_X^*\alpha] = \Phi([\alpha])$. The higher signature associated to $[\alpha]$ is the real number

$$\sigma(X, [\alpha]) := \int_X c_X L(X) \wedge \psi_X^*(\alpha), \tag{7.1}$$

where $L(X)$ is the invariant de Rham form representing the L-class of X. The insertion of the cut-off function c_X, which as we know has compact support, ensures that the integral is well-defined (and it can be shown that it only depends on the class $[L(X) \wedge \psi_X^*(\alpha)] \in H^\bullet_{\rm inv}(X)$). The collection

$$\{\sigma(X, [\alpha]), [\alpha] \in H^\bullet_{\rm inv}(G/K)\} \tag{7.2}$$

are called the higher signatures of X; by the Van Est isomorphism they are labelled by the elements in $H^*_{\rm diff}(G)$. Similarly, the higher $\widehat{A}$ genus associated to X and to $[\alpha] \in H^\bullet_{\rm inv}(G/K)$ is the real number

$$\widehat{A}(X, [\alpha]) := \int_X c_X \widehat{A}(X) \wedge \psi_X^*(\alpha) \tag{7.3}$$

with $\widehat{A}(X)$ the de Rham class associated to the $\widehat{A}$-differential form for a G-invariant metric. The collection

$$\{\widehat{A}(X, [\alpha]), \alpha \in H^\bullet_{\rm inv}(G/K)\} \tag{7.4}$$

are called the higher $\widehat{A}$-genera of Y.
We have:

THEOREM 7.5. *Let G be a connected real reductive Lie group. Let X be an orientable manifold with a proper, cocompact action of G. Then the following holds true:*

(*i*) *each higher signatures $\sigma(X, [\alpha])$, $[\alpha] \in H^\bullet_{\rm inv}(G/K)$, is a G-homotopy invariant of X.*
(*ii*) *if X admits a G-invariant Spin structure and a G-invariant metric of positive scalar curvature*[4] *then each higher $\widehat{A}$-genus $\widehat{A}(X, [\alpha])$, $[\alpha] \in H^\bullet_{\rm inv}(G/K)$, vanishes.*

The theorem follows easily from the higher index formula for differentiable group cocycles explained in the first part of this article and the usual stability properties of the C^* index class of the signature operator and of the Spin-Dirac operator, established in this context by Fukumoto [**15**] and Guo-Mathai-Wang [**17**]. As we have already explained , the theorem in fact holds more generally for G a Lie group with finitely many connected components satisfying property RD, and such that G/K is of non-positive sectional curvature for a maximal compact subgroup K. See [**46**] for more details.

The corresponding APS index theorems can be used to introduce relative higher genera $\sigma(X, \partial X, [\alpha])$ and $\widehat{A}(X, \partial X, [\alpha])$ and prove, for example, additivity results for closed manifolds that are obtained by gluing manifolds with boundary along G-diffeomorphic boundaries. In addition, the relative higher $\widehat{A}$-genera can be used to produce obstructions to the existence of an isotopy from a G-invariant PSC metric on ∂X to a G-invariant metric on ∂X which extends to a G-invariant metric which is PSC on all of X.

7.2. Rho invariants. In this subsection we shall briefly introduce (higher) rho numbers associated to positive scalar curvature metrics and G-equivariant homotopy equivalences. All our Dirac operators will be L^2-invertible; indeed, if we want to consider bordism properties of these numbers we do need L^2-invertibility so

[4] from now on we shall briefly write PSC

as to be able to define an APS index class on the manifold with boundary realising the bordism.

Rho numbers associated to delocalized 0-cocycles. We consider a closed G-proper manifold X **without** boundary, G connected, linear real reductive, $g \in G$ a semisimple element, D_X a G-equivariant L^2-invertible Dirac operator of the form (2.7). We consider

$$\eta_g(D_X) := \frac{1}{\sqrt{\pi}} \int_0^\infty \mathrm{tr}_g^X (D_X \exp(-tD_X^2)) \frac{dt}{\sqrt{t}}.$$

Let X be G-equivariantly Spin and $D_X \equiv D_{\mathbf{h}}$, the Spin Dirac operator associated to a G-equivariant PSC metric $\mathbf{h}$. Then we know that $D_{\mathbf{h}}$ is L^2-invertible. We define

$$\rho_g(\mathbf{h}) := \eta_g(D_{\mathbf{h}})\,.$$

If, on the other hand, $\mathbf{f} : X_1 \to X_2$ is a G-homotopy equivalence, then using [**15**] we know that there exists a bounded perturbation $B_{\mathbf{f}}$ of the signature operator on $X := X_1 \sqcup (-X_2)$, where $-X_2$ is the same manifold as X_2 with the opposite orientation, that makes it invertible. Moreover, one can prove that this perturbation $B_{\mathbf{f}}$ is in $C^*(X, \Lambda^*)^G$. Hence, by density, we conclude that there exists a perturbation $B_{\mathbf{f}}^\infty \in \mathcal{A}_G^\infty(X, \Lambda^*)$ such that $D_X^{\mathrm{sign}} + B_{\mathbf{f}}^\infty$ is L^2-invertible. It is possible to extend Theorem 6.29 to this perturbed situation and define the rho-number of the homotopy equivalence $\mathbf{f}$ as

$$\rho_g(\mathbf{f}) := \eta_g(D_X^{\mathrm{sign}} + B_{\mathbf{f}}^\infty)\,.$$

We refer to [**48**] for the details.

Rho numbers associated to higher delocalized cocycles We can generalize the above definitions and define rho numbers associated the higher cocycles Φ_g^P. More precisely, let $P = MNA$, $g \in M$ as above and consider Φ_g^P and $\Phi_{X,g}^P$ (we recall that X is without boundary). Assume that $\mathbf{h}$ is a slice-compatible G-invariant PSC metric on X. Then

$$\rho_g^P(\mathbf{h}) := \eta_g(D_{X_M}) \tag{7.6}$$

(with X_M the reduced manifold associated to X) is well defined. Notice that it is proved [**48**] that if D_X invertible, then D_{X_M} is also invertible.

Bordism properties. The APS index theorems introduced in this article can be used to study the bordism properties of these rho invariants. We concentrate on the case of psc metrics. We assume, unless otherwise stated, that we are on a G-proper manifold which is endowed with a slice-compatible G-invariant metric and a slice-compatible G-invariant Spin structure.

Let $(Y, \mathbf{h}_0)$ and $(Y, \mathbf{h}_1)$ be two slice-compatible psc metrics. We say that they are G-concordant if there exists a G-invariant metric $\mathbf{h}$ on $Y \times [0,1]$, also slice compatible, which is of PSC, product-type near the boundary and restricts to $\mathbf{h}_0$ at $Y \times \{0\}$ and to $\mathbf{h}_1$ at $Y \times \{1\}$.

The following Proposition is an example of the applications one can envisage for these secondary invariants. Before stating it, we remark that if g is non-elliptic, that is, does not conjugate to a compact element, then every element of the conjugacy class $C(g) := \{hgh^{-1} | h \in G\}$ in G does not have any fixed point on a G-proper manifold.

PROPOSITION 7.7.
1] Assume that the slice-compatible psc metrics $\mathbf{h}_0$ *and* $\mathbf{h}_1$ *on* Y *are* G*-concordant. Assume that* g *is non-elliptic on* Y*. Then* $\rho_g(\mathbf{h}_0) = \rho_g(\mathbf{h}_1)$.
2] Let $P = MAN < G$ *be a cuspidal parabolic subgroup and let* $x \in M$ *be a semisimple element. Assume that the slice-compatible psc metrics* $\mathbf{h}_0$ *and* $\mathbf{h}_1$ *on* Y *are* G*-concordant. Assume that* g *is non-elliptic on* Y_M*. Then* $\rho_g^P(\mathbf{h}_0) = \rho_g^P(\mathbf{h}_1)$*.*
Put it differently, for such g *our (higher) rho invariants are in fact concordance invariants.*

PROOF. In both cases the proof is an immediate consequence of the relevant delocalized APS index theorems. □

REMARK 7.8. In the case of free proper actions of discrete groups, higher rho invariants have been employed very successfully in studying the moduli space of concordant metrics of positive scalar curvature. See [**59**] and [**50**]. It is a challenge to understand whether the results we have just explained in the context of proper actions of Lie groups can be employed in studying the space $\mathcal{R}^+_{G,\text{slice}}(Y)$ of slice-compatible metrics of PSC (if non-empty) or the space $\mathcal{R}^+_G(Y)$ (arbitrary G-equivariant metrics of PSC). For these questions it would be interesting to develop a G-equivariant Stolz' sequence and investigate its basic properties. We leave this task to future research.

Acknowledgments

We take this opportunity to thank Alain Connes for his foundational role in the development of cyclic cohomology. Ideas and questions by Connes and his collaborators have inspired many developments that we have surveyed in this article. We also thank the organizers of the conference, "Cyclic Cohomology at 40", Alain Connes, Katia Consani, Masoud Khalkali and Henri Moscovici, for organizing such an interesting online event, providing the participants great opportunities to meet and exchange ideas.

We are very happy to thank Markus Pflaum, Hessel Posthuma and Yanli Song for inspiring discussions.

References

[1] Herbert Abels, *Parallelizability of proper actions, global K-slices and maximal compact subgroups*, Math. Ann. **212** (1974/75), 1–19, DOI 10.1007/BF01343976. MR375264 ↑

[2] M. F. Atiyah, *Elliptic operators, discrete groups and von Neumann algebras*, Colloque "Analyse et Topologie" en l'Honneur de Henri Cartan (Orsay, 1974), Astérisque, No. 32-33, Soc. Math. France, Paris, 1976, pp. 43–72. MR0420729 ↑

[3] Michael Atiyah and Wilfried Schmid, *A geometric construction of the discrete series for semisimple Lie groups*, Invent. Math. **42** (1977), 1–62, DOI 10.1007/BF01389783. MR463358 ↑

[4] Paul Baum, Alain Connes, and Nigel Higson, *Classifying space for proper actions and K-theory of group C^*-algebras*, C^*-algebras: 1943–1993 (San Antonio, TX, 1993), Contemp. Math., vol. 167, Amer. Math. Soc., Providence, RI, 1994, pp. 240–291, DOI 10.1090/conm/167/1292018. MR1292018 ↑

[5] Jean-Michel Bismut, *Eta invariants and the hypoelliptic Laplacian*, J. Eur. Math. Soc. (JEMS) **21** (2019), no. 8, 2355–2515, DOI 10.4171/jems/887. MR4035848 ↑

[6] Bruce Blackadar, *K-theory for operator algebras*, 2nd ed., Mathematical Sciences Research Institute Publications, vol. 5, Cambridge University Press, Cambridge, 1998. MR1656031 ↑

[7] Dan Burghelea, *The cyclic homology of the group rings*, Comment. Math. Helv. **60** (1985), no. 3, 354–365, DOI 10.1007/BF02567420. MR814144 ↑

[8] Jérôme Chabert, Siegfried Echterhoff, and Ryszard Nest, *The Connes-Kasparov conjecture for almost connected groups and for linear p-adic groups*, Publ. Math. Inst. Hautes Études Sci. **97** (2003), 239–278, DOI 10.1007/s10240-003-0014-2. ↑
[9] Xiaoman Chen, Jinmin Wang, Zhizhang Xie, and Guoliang Yu, *Delocalized eta invariants, cyclic cohomology and higher rho invariants.*arXiv:1901.02378 ↑
[10] Alain Connes, *Noncommutative differential geometry*, Inst. Hautes Études Sci. Publ. Math. **62** (1985), 257–360. MR823176 ↑
[11] Alain Connes, *Noncommutative geometry*, Academic Press, Inc., San Diego, CA, 1994. MR1303779 ↑
[12] Alain Connes and Henri Moscovici, *Cyclic cohomology, the Novikov conjecture and hyperbolic groups*, Topology **29** (1990), no. 3, 345–388, DOI 10.1016/0040-9383(90)90003-3. MR1066176 ↑
[13] Alain Connes and Henri Moscovici, *The L^2-index theorem for homogeneous spaces of Lie groups*, Ann. of Math. (2) **115** (1982), no. 2, 291–330, DOI 10.2307/1971393. MR647808 ↑
[14] Boris Fedosov, *Deformation quantization and index theory*, Mathematical Topics, vol. 9, Akademie Verlag, Berlin, 1996. MR1376365 ↑
[15] Yoshiyasu Fukumoto, *G-homotopy invariance of the analytic signature of proper co-compact G-manifolds and equivariant Novikov conjecture*, J. Noncommut. Geom. **15** (2021), no. 3, 761–795, DOI 10.4171/jncg/420. MR4345201 ↑
[16] Alexander Gorokhovsky, Hitoshi Moriyoshi, and Paolo Piazza, *A note on the higher Atiyah-Patodi-Singer index theorem on Galois coverings*, J. Noncommut. Geom. **10** (2016), no. 1, 265–306, DOI 10.4171/JNCG/234. MR3500822 ↑
[17] Hao Guo, Varghese Mathai, and Hang Wang, *Positive scalar curvature and Poincaré duality for proper actions*, J. Noncommut. Geom. **13** (2019), no. 4, 1381–1433, DOI 10.4171/jncg/321. MR4059824 ↑
[18] Bernhard Hanke, *Positive scalar curvature with symmetry*, J. Reine Angew. Math. **614** (2008), 73–115, DOI 10.1515/CRELLE.2008.003. MR2376283 ↑
[19] Harish-Chandra, *Discrete series for semisimple Lie groups. II. Explicit determination of the characters*, Acta Math. **116** (1966), 1–111, DOI 10.1007/BF02392813. MR219666 ↑
[20] Sigurdur Helgason, *Differential geometry, Lie groups, and symmetric spaces*, Pure and Applied Mathematics, vol. 80, Academic Press, Inc. [Harcourt Brace Jovanovich, Publishers], New York-London, 1978. MR514561 ↑
[21] Nigel Higson and John Roe, *Analytic K-homology*, Oxford Mathematical Monographs, Oxford University Press, Oxford, 2000. Oxford Science Publications. MR1817560 ↑
[22] Peter Hochs, *Quantisation commutes with reduction at discrete series representations of semisimple groups*, Adv. Math. **222** (2009), no. 3, 862–919, DOI 10.1016/j.aim.2009.05.011. MR2553372 ↑
[23] Peter Hochs and Varghese Mathai, *Quantising proper actions on Spin^c-manifolds*, Asian J. Math. **21** (2017), no. 4, 631–685, DOI 10.4310/AJM.2017.v21.n4.a2. MR3691850 ↑
[24] Peter Hochs, Yanli Song, and Xiang Tang, *An index theorem for higher orbital integrals*, Math. Ann. **382** (2022), no. 1-2, 169–202, DOI 10.1007/s00208-021-02233-3. MR4377301 ↑
[25] Peter Hochs, Bai-Ling Wang, and Hang Wang, *An equivariant Atiyah-Patodi-Singer index theorem for proper actions I: the index formula.*arXiv:1904.11146 ↑
[26] Peter Hochs, Bai-Ling Wang, and Hang Wang, *An equivariant Atiyah-Patodi-Singer index theorem for proper actions II: the K-theoretic index.*arXiv:2006.08086 ↑
[27] Peter Hochs and Hang Wang, *Orbital integrals and K-theory classes*, Ann. K-Theory **4** (2019), no. 2, 185–209, DOI 10.2140/akt.2019.4.185. MR3990784 ↑
[28] Peter Hochs and Hang Wang, *A fixed point formula and Harish-Chandra's character formula*, Proc. Lond. Math. Soc. (3) **116** (2018), no. 1, 1–32, DOI 10.1112/plms.12066. MR3747042 ↑
[29] G. G. Kasparov, *The operator K-functor and extensions of C^*-algebras* (Russian), Izv. Akad. Nauk SSSR Ser. Mat. **44** (1980), no. 3, 571–636, 719. MR582160 ↑
[30] V. Lafforgue, *Banach KK-theory and the Baum-Connes conjecture*, Proceedings of the International Congress of Mathematicians, Vol. II (Beijing, 2002), Higher Ed. Press, Beijing, 2002, pp. 795–812. MR1957086 ↑
[31] H. Blaine Lawson Jr. and Shing Tung Yau, *Scalar curvature, non-abelian group actions, and the degree of symmetry of exotic spheres*, Comment. Math. Helv. **49** (1974), 232–244, DOI 10.1007/BF02566731. MR358841 ↑

[32] Eric Leichtnam and Paolo Piazza, *The b-pseudodifferential calculus on Galois coverings and a higher Atiyah-Patodi-Singer index theorem* (English, with English and French summaries), Mém. Soc. Math. Fr. (N.S.) **68** (1997), iv+121. MR1488084 ↑

[33] Eric Leichtnam and Paolo Piazza, *Dirac index classes and the noncommutative spectral flow*, J. Funct. Anal. **200** (2003), no. 2, 348–400, DOI 10.1016/S0022-1236(02)00044-7. MR1979016 ↑

[34] Eric Leichtnam and Paolo Piazza, *Étale groupoids, eta invariants and index theory*, J. Reine Angew. Math. **587** (2005), 169–233, DOI 10.1515/crll.2005.2005.587.169. MR2186978 ↑

[35] Matthias Lesch, Henri Moscovici, and Markus J. Pflaum, *Connes-Chern character for manifolds with boundary and eta cochains*, Mem. Amer. Math. Soc. **220** (2012), no. 1036, viii+92, DOI 10.1090/S0065-9266-2012-00656-3. MR3025890 ↑

[36] Jean-Louis Loday, *Cyclic homology*, Grundlehren der mathematischen Wissenschaften [Fundamental Principles of Mathematical Sciences], vol. 301, Springer-Verlag, Berlin, 1992. Appendix E by María O. Ronco, DOI 10.1007/978-3-662-21739-9. MR1217970 ↑

[37] Richard B. Melrose, *The Atiyah-Patodi-Singer index theorem*, Research Notes in Mathematics, vol. 4, A K Peters, Ltd., Wellesley, MA, 1993, DOI 10.1016/0377-0257(93)80040-i. MR1348401 ↑

[38] Richard B. Melrose and Paolo Piazza, *Families of Dirac operators, boundaries and the b-calculus*, J. Differential Geom. **46** (1997), no. 1, 99–180. MR1472895 ↑

[39] H. Moscovici and F.-B. Wu, *Localization of topological Pontryagin classes via finite propagation speed*, Geom. Funct. Anal. **4** (1994), no. 1, 52–92, DOI 10.1007/BF01898361. MR1254310 ↑

[40] Hitoshi Moriyoshi and Paolo Piazza, *Eta cocycles, relative pairings and the Godbillon-Vey index theorem*, Geom. Funct. Anal. **22** (2012), no. 6, 1708–1813, DOI 10.1007/s00039-012-0197-0. MR3000501 ↑

[41] Henri Moscovici and Robert J. Stanton, *Eta invariants of Dirac operators on locally symmetric manifolds*, Invent. Math. **95** (1989), no. 3, 629–666, DOI 10.1007/BF01393895. MR979370 ↑

[42] Ryszard Nest and Boris Tsygan, *Algebraic index theorem*, Comm. Math. Phys. **172** (1995), no. 2, 223–262. MR1350407 ↑

[43] Ryszard Nest and Boris Tsygan, *Algebraic index theorem for families*, Adv. Math. **113** (1995), no. 2, 151–205, DOI 10.1006/aima.1995.1037. MR1337107 ↑

[44] M. J. Pflaum, H. Posthuma, and X. Tang, *The localized longitudinal index theorem for Lie groupoids and the van Est map*, Adv. Math. **270** (2015), 223–262, DOI 10.1016/j.aim.2014.11.007. MR3286536 ↑

[45] Markus J. Pflaum, Hessel Posthuma, and Xiang Tang, *The transverse index theorem for proper cocompact actions of Lie groupoids*, J. Differential Geom. **99** (2015), no. 3, 443–472. MR3316973 ↑

[46] Paolo Piazza and Hessel B. Posthuma, *Higher genera for proper actions of Lie groups*, Ann. K-Theory **4** (2019), no. 3, 473–504, DOI 10.2140/akt.2019.4.473. MR4043466 ↑

[47] Paolo Piazza and Hessel B. Posthuma, *Higher genera for proper actions of Lie groups, II: The case of manifolds with boundary*, Ann. K-Theory **6** (2021), no. 4, 713–782, DOI 10.2140/akt.2021.6.713. MR4382801 ↑

[48] Paolo Piazza, Hessel B. Posthuma, Yanli Song, and Xiang Tang, *Higher Orbital Integrals, Rho Numbers and Index Theory* (2021).`arXiv:2108.00982` ↑

[49] Paolo Piazza and Thomas Schick, *Rho-classes, index theory and Stolz' positive scalar curvature sequence*, J. Topol. **7** (2014), no. 4, 965–1004, DOI 10.1112/jtopol/jtt048. MR3286895 ↑

[50] Paolo Piazza, Thomas Schick, and Vito Felice Zenobi, *Mapping analytic surgery to homology, higher rho numbers and metrics of positive scalar curvature*.`arXiv:1905.11861` ↑

[51] Michael Puschnigg, *New holomorphically closed subalgebras of C^*-algebras of hyperbolic groups*, Geom. Funct. Anal. **20** (2010), no. 1, 243–259, DOI 10.1007/s00039-010-0062-y. MR2647141 ↑

[52] John Roe, *Index theory, coarse geometry, and topology of manifolds*, CBMS Regional Conference Series in Mathematics, vol. 90, Published for the Conference Board of the Mathematical Sciences, Washington, DC; by the American Mathematical Society, Providence, RI, 1996, DOI 10.1090/cbms/090. MR1399087 ↑

[53] John Roe, *Comparing analytic assembly maps*, Q. J. Math. **53** (2002), no. 2, 241–248, DOI 10.1093/qjmath/53.2.241. MR1909514 ↑

[54] Yanli Song and Xiang Tang, *Higher Orbit Integrals, Cyclic Cocycles, and K-theory of Reduced Group C^*-algebra,* `arXiv:1910.00175`. ↑

[55] Boris Vaillant, *Index Theory for Coverings.* Preprint `arXiv:0806.4043`. ↑

[56] Hang Wang, *L^2-index formula for proper cocompact group actions*, J. Noncommut. Geom. **8** (2014), no. 2, 393–432, DOI 10.4171/JNCG/160. MR3275037 ↑

[57] Antony Wassermann, *Une démonstration de la conjecture de Connes-Kasparov pour les groupes de Lie linéaires connexes réductifs* (French, with English summary), C. R. Acad. Sci. Paris Sér. I Math. **304** (1987), no. 18, 559–562. MR894996 ↑

[58] Michael Wiemeler, *Circle actions and scalar curvature*, Trans. Amer. Math. Soc. **368** (2016), no. 4, 2939–2966, DOI 10.1090/tran/6666. MR3449263 ↑

[59] Zhizhang Xie and Guoliang Yu, *Higher rho invariants and the moduli space of positive scalar curvature metrics*, Adv. Math. **307** (2017), 1046–1069, DOI 10.1016/j.aim.2016.11.030. MR3590536 ↑

[60] Wei Ping Zhang, *A note on equivariant eta invariants*, Proc. Amer. Math. Soc. **108** (1990), no. 4, 1121–1129, DOI 10.2307/2047979. MR1004426 ↑

Dipartimento di Matematica, Sapienza Università di Roma, I-00185 Roma, Italy
Email address: piazza@mat.uniroma1.it

Department of Mathematics and Statistics, Washington University, St. Louis, Missouri, 63130
Email address: xtang@math.wustl.edu

Proceedings of Symposia in Pure Mathematics
Volume **105**, 2023
https://doi.org/10.1090/pspum/105/01909

Localization in Hochschild homology

Markus J. Pflaum

Dedicated to the 40th birthday of cyclic cohomology

ABSTRACT. Localization methods are ubiquitous in cyclic homology theory, but vary in detail and are used in different scenarios. In this paper we will elaborate on a common feature of localization methods in noncommutative geometry, namely sheafification of the algebra under consideration and reduction of the computation to the stalks of the sheaf. The novelty of our approach lies in the methods we use which mainly stem from real instead of complex algebraic geometry. We will then indicate how this method can be used to determine the Hochschild homology theory of more complicated algebras out of simpler ones.

Introduction

Since the invention of cyclic homology theory 40 years ago, powerful methods have been developed to compute the Hochschild and cyclic homology of a given algebra. They range from passing to more tractable resolutions of the algebra like the Connes–Koszul resolution [**7**] over to the localization methods by Loday [**9**], Burghelea [**6**], Brylinski [**5**], Weibel [**24**], Connes–Moscovici [**8**], Teleman [**20, 21**], Nistor [**14**], Brasselet-Pflaum [**2**] and others. Localization essentially amounts to the sheafification of the Hochschild chain complex of the algebra under consideration. The idea is to reduce the computation of the "global" Hochschild or cyclic homology to "local" ones at the stalks. In many cases, the homologies of the stalks are easier to compute. One then has to glue the resulting homologies to a global one and thus obtains under reasonable assumptions the homology of the originally given algebra. In the case of algebraic varieties, a sheafification approach has been originally suggested by Loday [**9**, 3.4] and worked out by Weibel–Geller [**26**] and Weibel [**24**] to define the Hochschild homology of algebraic varieties and schemes. Note that in the case of schemes using a sheaf theoretic version of the Hochschild chain complex is indispensible since the algebra of global sections of the structure sheaf might be trivial, hence so would be its Hochschild homology if it were defined solely through the algebra of global sections. In the case of smooth manifolds or stratified spaces endowed with a structure sheaf of smooth functions on it, Teleman [**20, 21**] used smooth partitions of unity and cutoff functions around the diagonal

2020 *Mathematics Subject Classification.* Primary 16E40, 14P99, 18G99, 55N30.

Key words and phrases. Hochschild homology, localization, cyclic homology.

The author gratefully acknowledges support for this work by the National Science Foundation under Grant No. DMS-1105670 and by the Simons Foundation under Grant No. 359389.

to construct a chain complex quasi-isomorphic to the Hochschild chain complex. Referring to its construction the resulting complex has been called the *diagonal complex*. In many situations it is significantly easier to handle than the original Hochschild chain complex.

Here we propose a sheafification method for the Hochschild chain complex of the global section algebra of a differentiable space or more generally of a sheaf of bornological algebras over a differentiable space. In a certain sense it can be understood as a combination of the approaches by Weibel and Teleman. In the first section of this paper we provide a rather detailed introduction to differentiable spaces from the point of view of real algebraic geometry. We mostly follow [**13**] and supplement it by material from [**16**]. In the second section we prove a localization result related to but stronger than the ones from [**2, 20, 21**] and [**18**]. More precisely, we construct a fine sheaf complex which we call the diagonal sheaf complex and which in degree k consists of relative Whitney fields concentrated around the diagonal of the k-the power of the differentiable space. The chain complex of global sections of the diagonal sheaf complex turns out to be quasi-isomorphic to the original Hochschild chain complex. This allows to reduce the computation of the Hochschild homology of the original algebra to the computation of the Hochschild homologies of the stalks which are often much easier to determine.

Applications of our localization method are given in the last section and an outlook is provided how it can help unravelling the singularity structure of differentiable stratified spaces. Finally, localization in cyclic homology is briefly addressed.

1. Preliminaries from real algebraic geometry

In this section we introduce a category of commutative locally $\mathbb{R}$-ringed spaces which we call differentiable spaces and for which we want to provide localization methods to examine their Hochschild homology. In addition, we recall some notions from the theory of Whitney fields needed in the sequel to construct localized chain complexes.

Locally, the reduced versions of the ringed spaces we study look like a closed subset $X \subset \mathbb{R}^n$ of some Euclidean space with structure sheaf $\mathcal{C}^\infty_{|X}$ given by the sheaf of smooth functions restricted to X. More precisely, for every relatively open subset $U \subset X$ the section space $\mathcal{C}^\infty_{|X}(U)$ coincides with the quotient algebra $\mathcal{C}^\infty(\widetilde{U})/\mathcal{J}_X(\widetilde{U})$ where $\widetilde{U} \subset \mathbb{R}^n$ is an open subset such that $\widetilde{U} \cap X = U$ and where $\mathcal{J}_X(\widetilde{U})$ denotes the closed ideal of all smooth functions on $\widetilde{U}$ vanishing on X. For any unital commutative $\mathbb{R}$-algebra A let $\operatorname{Spec}_{\mathbb{R}}(A)$ be its *real spectrum* that is the space $\operatorname{Hom}_{\mathbb{R}\text{-alg}}(A, \mathbb{R})$ of algebra homomorphisms from A to $\mathbb{R}$ endowed with the Gelfand topology. The canonical projection $p : \mathcal{C}^\infty(\mathbb{R}^n) \to \mathcal{C}^\infty(X)$ then induces a closed embedding of real spectra $p^* : \operatorname{Spec}_{\mathbb{R}}(\mathcal{C}^\infty(X)) \to \operatorname{Spec}_{\mathbb{R}}\left(\mathcal{C}^\infty(\mathbb{R}^n)\right)$ with image being the subset X under the identification $\operatorname{Spec}_{\mathbb{R}}(\mathcal{C}^\infty(\mathbb{R}^n) \cong \mathbb{R}^n$.

To allow for non-reduced versions of the local models we follow [**13**] and call a commutative locally $\mathbb{R}$-ringed space $(X, \mathcal{O})$ an *affine differentiable space* if the following conditions hold true.

(DS1) There exists a homeomorphism $\varphi : X \to \mathsf{X} \subset \mathbb{R}^n$ onto a closed subspace $\mathsf{X} \subset \mathbb{R}^n$.

(DS2) There exists an epimorphism of sheaves $\varphi^\sharp : \mathcal{C}^\infty_{\mathbb{R}^n} \to \varphi_* \mathcal{O}$.

(DS3) The embedding of real spectra

$$(\varphi^\sharp_{\mathbb{R}^n})^* : \operatorname{Spec}_{\mathbb{R}}(\mathcal{O}(X)) \to \operatorname{Spec}_{\mathbb{R}}\big(\mathcal{C}^\infty(\mathbb{R}^n)\big) \cong \mathbb{R}^n$$

induced by the projection $\varphi^\sharp_{\mathbb{R}^n} : \mathcal{C}^\infty(\mathbb{R}^n) \to \mathcal{O}(X)$ has image $\mathsf{X} = \varphi(X)$.

(DS4) For each open $U \subset X$, the section algebra $\mathcal{O}(U)$ coincides with the localization of $\mathcal{O}(X)$ by $S_U := \{g \in \mathcal{O}(X) \mid (\varphi^\sharp_{\mathbb{R}^n})^*(v) \neq 0 \text{ for all } v \in \mathsf{X}\}$.

Conditions (DS1) and (DS2) entail that the pair $(\varphi, \varphi^\sharp) : (X, \mathcal{O}) \to \big(\mathbb{R}^n, \mathcal{C}^\infty\big)$ is a morphism, more precisely even a closed embedding of locally $\mathbb{R}$-ringed spaces. Following [**16**, Sec. 1.3], we call a closed embedding $(\varphi, \varphi^\sharp) : (X, \mathcal{O}) \to \big(\mathbb{R}^n, \mathcal{C}^\infty\big)$ such that the conditions (DS3) and (DS4) hold true a *singular chart* of the affine differentiable space $(X, \mathcal{O})$. We will usually write $\varphi : X \to \mathsf{X} \subset \mathbb{R}^n$ for the underlying map to denote that its image is X. In case the affine differentiable space $(X, \mathcal{O})$ is *reduced* meaning that $\mathcal{O}$ is a sheaf of real valued continuous functions on X, the sheaf morphism $\varphi^\sharp$ is uniquely determined by φ and given by pullback with φ. More precisely, over $V \subset \mathbb{R}^n$ open it has the form

$$\varphi^\sharp_V : \mathcal{C}^\infty(V) \to \mathcal{O}(\varphi^{-1}(V)), \quad f \mapsto f \circ \varphi|_{\varphi^{-1}(V)} \,.$$

A *reduced affine differentiable space* is therefore exactly a commutative locally $\mathbb{R}$-ringed space $(X, \mathcal{O})$ isomorphic to $\big(\mathsf{X}, \mathcal{C}^\infty_{|\mathsf{X}}\big)$ for some closed $\mathsf{X} \subset \mathbb{R}^n$. In the reduced case we often just call the embedding $\varphi : X \to \mathsf{X} \subset \mathbb{R}^n$ inducing the isomorphism a *singular chart* for $(X, \mathcal{C}^\infty)$. Moreover, we denote the structure sheaf of a reduced (affine) differentiable space by $\mathcal{C}^\infty_X$ instead of $\mathcal{O}$ and call the elements of a section space $\mathcal{C}^\infty_X(U)$ the *smooth functions* on the open set $U \subset X$.

Gluing together affine differentiable spaces gives rise to the category we are interested in. More precisely, a commutative locally $\mathbb{R}$-ringed space $(X, \mathcal{O})$ is called a *differentiable space* if it possesses an *atlas* consisting of (local) singular charts that is if there exists a family $(\varphi_i, \varphi_i^\sharp)_{i \in I}$ of continuous maps $\varphi_i : U_i \to \mathbb{R}^{n_i}$ and sheaf morphisms $\varphi_i^\sharp : \mathcal{C}^\infty_{\mathbb{R}^{n_i}} \to (\varphi_i)_* \mathcal{O}_{|U_i}$ with the following properties.

(DS5) The family $(U_i)_{i \in I}$ is an open cover of X.

(DS6) Each of the maps $\varphi_i : U_i \to \mathbb{R}^{n_i}$ is a homeomorphism onto a closed subset of $\mathbb{R}^{n_i}$.

(DS7) The commutative locally $\mathbb{R}$-ringed space $(U_i, \mathcal{O}_{|U_i})$ is an affine differentiable space with singular chart $(\varphi_i, \varphi_i^\sharp) : (U_i, \mathcal{O}_{|U_i}) \to \big(\mathbb{R}^{n_i}, \mathcal{C}^\infty_{\mathbb{R}^{n_i}}\big)$ for every $i \in I$.

REMARK 1.1. In the definition of a differentiable space, the underlying topological space need not necessarily be Hausdorff. We do not need that generality here, so for reasons of convenience we assume for the rest of the paper that X is Hausdorff. As in the affine case, a differentiable space will be called *reduced* if its structure sheaf consists of continuous functions on the underlying topological space. For more details on differentiable spaces see the monograph [**13**].

Next we turn differentiable spaces into a category. To this end we define a morphism between differentiable spaces $(X, \mathcal{O}_X)$ and $(Y, \mathcal{O}_Y)$ as a morphism $(\psi, \psi^\sharp) : (X, \mathcal{O}_X) \to (Y, \mathcal{O}_Y)$ in the category of locally $\mathbb{R}$-ringed spaces. In case both $(X, \mathcal{C}^\infty_X)$ and $(Y, \mathcal{C}^\infty_Y)$ are reduced, $\psi^\sharp$ is given over $V \subset Y$ open by the pullback $\psi^\sharp_V = (\psi|_V)^*$. We say in this situation that ψ is *smooth*. In other words this means that pullback by ψ maps smooth functions to smooth functions.

A particularly interesting class of non-reduced affine differentiable spaces is given by the Whitney fields over a closed subset of some Euclidean space. The interpretation of Whitney fields as forming the structure sheaf of a non-reduced affine differentiable space appears to be new, but quite helpful for our purposes. Originally, Whitney fields played an important role in the extension and composition theory of smooth functions [**1**, **10**, **22**]. They can be defined for any affine differentiable space X with a fixed singular chart. For simplicity, we assume that X is embedded as a closed subspace in some Euclidean space $\mathbb{R}^n$ which in other words means that the singular chart is the identical embedding. By a Whitney field on a relatively open subset $U \subset X$ we then understand a family $F = (F_\alpha)_{\alpha\in\mathbb{N}^n}$ of continuous functions $F_\alpha : U \to \mathbb{R}$ for which there exists a smooth $f : \widetilde{U} \to \mathbb{R}$ defined on some open set $\widetilde{U} \subset \mathbb{R}^n$ such that $U = \widetilde{U} \cap X$ and such that for all $\alpha \in \mathbb{N}^n$ the restriction of the partial derivative $\partial^\alpha f = \frac{\partial^{|\alpha|} f}{\partial^\alpha x}$ coincides with F_α that is if $F_\alpha = \partial^\alpha f|_U$. We denote the space of Whitney fields over U by $\mathcal{E}^\infty(U)$ and the canonical map $\mathcal{C}^\infty(U) \to \mathcal{E}^\infty(U)$, $f \mapsto (\partial^\alpha f)_{\alpha\in\mathbb{N}^n}$ by J^∞. The latter is sometimes called the *jet map*. By Whitney's extension theorem there is an intrinsic definition of the concept of a Whitney field, but we do not need that here and therefore refer the interested reader to [**10**, **22**]. The space of Whitney fields $\mathcal{E}^\infty(U)$ carries a unique structure of an algebra over $\mathbb{R}$ such that the jet map $\mathsf{J}^\infty : \mathcal{C}^\infty(\widetilde{U}) \to \mathcal{E}^\infty(U)$ becomes a morphism of algebras. Its kernel then is an ideal in $\mathcal{C}^\infty(\widetilde{U})$ which we denote by $\mathcal{J}^\infty_X(\widetilde{U})$ or $\mathcal{J}^\infty(X;\widetilde{U})$ and call it the ideal of smooth functions on $\widetilde{U}$ which are *flat* on X. Note that $\mathcal{J}^\infty_X(\widetilde{U})$ is contained in the vanishing ideal $\mathcal{J}_X(\widetilde{U})$. When U runs through the open sets of X, the assignment $U \mapsto \mathcal{E}^\infty(U)$ becomes a fine sheaf on X, the sheaf of *Whitney fields* on X with respect to the embedding $X \hookrightarrow \mathbb{R}^n$. The pair $(X, \mathcal{E}^\infty)$ then becomes an affine differentiable space which in general is not reduced.

There is a relative version of Whitney fields which allows to get rid of the existence of a global singular chart. To this end let $(X, \mathcal{O}_X)$ be a differentiable space and $(Y, \mathcal{C}^\infty_Y)$ a closed differentiable subspace which means that $Y \subset X$ is closed and that $\mathcal{C}^\infty_Y$ is the restriction of the sheaf $\mathcal{C}^\infty_X$ to X. Let $\mathcal{C}^\infty(X;Y)$ be the ideal in the algebra of smooth functions on X for which there exists an atlas $(\varphi_i)_{i\in I}$ of singular charts $\varphi_i : U_i \to A_i \subset \mathbb{R}^{n_i}$ of X such that a smooth function $f : X \to \mathbb{R}$ lies in $\mathcal{C}^\infty(X;Y)$ if and only if for each i there is a smooth function $f_i : \widetilde{U}_i \to \mathbb{R}$ defined on an open neighborhood $\widetilde{U}_i$ of A_i in $\mathbb{R}^{n_i}$ such that f_i is flat on $\varphi_i(Y \cap U_i)$ and such that $f|_{U_i} = f_i \circ \varphi_i$. One can check that if this condition is fulfilled in one atlas of singular charts for X, then it is so too in any other, hence the definition of $\mathcal{C}^\infty(X;Y)$ does not depend on the choice of an atlas. Following [**1**], we call $\mathcal{C}^\infty(X;Y)$ the algebra of smooth functions on X *flat* on Y. It is straightforward to check that associating to each open subset $U \subset X$ the algebra $\mathcal{C}^\infty(U;Y) := \mathcal{C}^\infty(U;Y \cap U)$ gives rise to a fine sheaf on X which actually even is an ideal sheaf in $\mathcal{C}^\infty_X$. Let us describe the section space $\mathcal{C}^\infty(U;Y)$ when U is the domain of some singular chart $\varphi : U \to \mathbb{R}^n$ and $U \cap Y \neq \emptyset$. Choose $\widetilde{U} \subset \mathbb{R}^n$ open such that $U = \widetilde{U} \cap A$ where $A = \varphi(X)$. Put $B = \varphi(Y)$. The intersection $\mathcal{J}^\infty_B(\widetilde{U}) \cap \mathcal{J}_A(\widetilde{U})$ then is an ideal in $\mathcal{J}^\infty_B(\widetilde{U})$ giving rise to a natural isomorphism of algebras

$$\mathcal{C}^\infty(U;Y) \cong \mathcal{J}_B^\infty(\widetilde{U}) \,/\, \mathcal{J}_B^\infty(\widetilde{U}) \cap \mathcal{J}_A(\widetilde{U}) \ .$$

The corresponding algebra of *Whitney fields* on Y *relative* X can now be defined and is given as the quotient algebra

$$\mathcal{E}^\infty(Y;X) := \mathcal{C}^\infty(X)/\mathcal{C}^\infty(X;Y) \ .$$

Finally in this section we state a topological result whose proof is immediate from the fact that the algebra of smooth functions on an open set of some Euclidean space is a nuclear Fréchet space, see e.g. [**23**].

PROPOSITION 1.2. *For $X \subset \mathbb{R}^n$ locally closed, and $Y \subset X$ a relatively closed subset the spaces $\mathcal{C}^\infty(X)$, $\mathcal{E}^\infty(X)$, $\mathcal{C}^\infty(X,Y)$, and $\mathcal{E}^\infty(Y,X)$ all carry in a natural way the structure of a Fréchet space which they inherit as subquotients from the Fréchet space $\mathcal{C}^\infty(U)$, where $U \subset \mathbb{R}^n$ is open such that $X \subset U$ is closed.*

2. The diagonal complex

Let $(X, \mathcal{C}^\infty)$ be a reduced differentiable space as defined in Section 1, and $A = \mathcal{C}^\infty(X)$ be the algebra of smooth functions on X. Since A carries the structure of a nuclear Fréchet algebra, all tensor product topologies on the tensor powers $A^{\otimes k+1}$ coincide, hence the completed tensor product

$$C_k(A) := A^{\hat{\otimes} k+1} = A \hat{\otimes} A \hat{\otimes} \ldots \hat{\otimes} A, \quad k \in \mathbb{N}$$

is uniquely determined. Moreover, $C_k(A)$ coincides with $\mathcal{C}^\infty(X^k)$ since for open $U \subset \mathbb{R}^n$ and $V \subset \mathbb{R}^m$ there is a natural identification $\mathcal{C}^\infty(U \times V) \cong \mathcal{C}^\infty(U) \hat{\otimes} \mathcal{C}^\infty(V)$, see [**23**]. The Hochschild boundary

$$b: \ C_k(A) \to C_{k-1}(A), \ a_0 \otimes \ldots \otimes a_k \mapsto \sum_{0 \le i \le k} (-1)^i \, b_i\big(a_0 \otimes \ldots \otimes a_k\big),$$

where

$$b_i\big(a_0 \otimes \ldots \otimes a_k\big) = \begin{cases} a_0 \otimes \ldots \otimes a_i a_{i+1} \otimes \ldots \otimes a_k & \text{for } 0 \le i < k, \\ a_k a_0 \otimes \ldots \otimes a_{k-1} & \text{for } i = k, \end{cases} \tag{2.1}$$

then is continuous for each $k \in \mathbb{N}^*$, and gives rise to the (continuous) Hochschild complex $\big(C_\bullet(A), b\big)$; see [**19**] for details on the Hochschild homology of Fréchet algebras. The homology of the complex $\big(C_\bullet(A), b\big)$ is the (continuous) Hochschild homology of A and is to be denoted by $HH_\bullet(A)$. In the following, we will construct a chain complex $\big(E_\bullet(A), \beta\big)$, called the diagonal complex, which is quasi-isomorphic to $\big(C_\bullet(A), b\big)$ and which can be interpreted as the complex of global sections of a fine sheaf complex $\big(\mathscr{E}_\bullet, \beta\big)$ over the space X. The diagonal complex originally goes back to the work of Teleman [**20**] on an alternative method to determine the Hochschild homology of the algebra of smooth functions on a smooth manifold. It has been used by Brasselet–Pflaum [**2**] to compute the Hochschild homology of algebras of Whitney fields over subanalytic sets.

To define the diagonal complex, consider for every $k \in \mathbb{N}$ the closed ideal

$$J_k(X) := \mathcal{C}^\infty(X^{k+1}, \Delta_X^{k+1}) \subset \mathcal{C}^\infty(X^{k+1})$$

of smooth functions on X^{k+1} flat on the diagonal

$$\Delta_X^{k+1} := \big\{(x_0, \ldots, x_k) \in X^{k+1} \mid x_0 = \ldots = x_k\big\} \ .$$

In other words, $J_k(X)$ is the set of all $f \in \mathcal{C}^\infty(X^{k+1})$ such that for every singular chart $\varphi : U \to \mathsf{X} \subset \mathbb{R}^n$ over an open $U \subset X$ the function

$$\varphi_* f : \mathsf{X}^{k+1} \to \mathbb{R},\ (v_0, \ldots, v_k) \mapsto f\big(\varphi^{-1}(v_0), \ldots, \varphi^{-1}(v_k)\big)$$

coincides with the restriction $g|_{\mathsf{X}^{k+1}}$ of a function $g \in \mathcal{J}^\infty(\Delta_{\mathsf{X}}^{k+1}; \widetilde{U}^{k+1})$, where $\widetilde{U} \subset \mathbb{R}^n$ is an open neighborhood of X. Define $\mathscr{E}_k(X)$ as the algebra of Whitney functions on the diagonal Δ_X^{k+1} relative to the space X^{k+1} that is put

$$\mathscr{E}_k(X) := \mathcal{E}^\infty(\Delta_X^{k+1}; X^{k+1}) = \mathcal{C}^\infty(X^{k+1})/J_k(X)\,. \tag{2.2}$$

LEMMA 2.1. *The Hochschild boundary b maps $J_k(X)$ into $J_{k-1}(X)$.*

PROOF. Without loss of generality it suffices to prove that for X a closed subset of some open set U in $\mathbb{R}^n$ and $a \in \mathcal{J}^\infty(\Delta_X^{k+1}, U^{k+1})$ the image $b_i(a)$ is in $\mathcal{J}^\infty(\Delta_X^k, U^k)$. To this end denote by ∂_j^α the differential operator

$$a_0 \otimes \ldots \otimes a_l \mapsto a_0 \otimes \ldots \otimes (\partial^\alpha a_j) \otimes \ldots \otimes a_l,$$

where $a_0, \ldots, a_l \in \mathcal{C}^\infty(U)$ and $j \leq l$. The higher Leibniz formula entails for $\alpha \in \mathbb{N}^n$ and $i, j \in \mathbb{N}$ with $0 \leq i \leq k$ and $0 \leq j \leq k-1$ that

$$\partial_j^\alpha\big(b_i(a)\big) = b_i(\widetilde{a}), \tag{2.3}$$

where

$$\widetilde{a} = \begin{cases} \partial_j^\alpha a & \text{for } j < i,\ (j,i) \neq (0,k)\ , \\ \partial_{j+1}^\alpha a & \text{for } j > i\ , \\ \sum\limits_{\beta+\gamma=\alpha} \binom{\alpha}{\beta}\, \partial_k^\beta\, \partial_0^\gamma a & \text{for } (j,i) = (0,k)\ , \\ \sum\limits_{\beta+\gamma=\alpha} \binom{\alpha}{\beta}\, \partial_i^\beta\, \partial_{i+1}^\gamma a & \text{for } j = i\ . \end{cases} \tag{2.4}$$

Eqs. (2.3) and (2.4) imply that for a diagonal element $\Delta^k(v) := (v, \ldots, v) \in U^k$ with $v \in X$ the equality

$$\big(\partial_j^\alpha b_i(a)\big)\big(\Delta^k(v)\big) = b_i \widetilde{a}\big(\Delta^k(v)\big) = \widetilde{a}\big(\Delta^{k+1}(v)\big) = 0$$

holds true, since $\widetilde{a} \in \mathcal{J}^\infty(\Delta_X^{k+1}, U^{k+1})$. This entails the claim. □

As a consequence of the lemma we obtain a short exact sequence of chain complexes:

$$0 \longrightarrow \big(J_\bullet(X), b\big) \longrightarrow \big(C_\bullet(A), b\big) \longrightarrow \big(E_\bullet(A), \beta\big) \longrightarrow 0\,. \tag{2.5}$$

We call the quotient complex $\big(E_\bullet(A), \beta\big)$ the *diagonal complex* of A, and denote the canonical projection by $p_X : C_\bullet(A) \to E_\bullet(A)$.

Letting U run through the open subsets of X we put for each $k \in \mathbb{N}$

$$\mathscr{E}_k(U) := \mathcal{C}^\infty(U^{k+1})/J_k(U). \tag{2.6}$$

By construction, the $\mathscr{E}_k(U)$ form the section spaces of a presheaf on X denoted by $\mathscr{E}_k$. Since the differential β commutes with the restriction morphisms, we even obtain a presheaf of chain complexes $\big(\mathscr{E}_\bullet, \beta\big)$. Its global section space coincides with the diagonal complex of A:

$$E_\bullet(A) = \mathscr{E}_\bullet(X)\ .$$

Even more holds true, but before that we need to introduce the following notion based on the concepts of regularity by Malgrange [**10**] and Tougeron [**22**, Defs. 1.30 & 4.4], cf. also [**16**, Defs. 1.6.6 & 1.6.18]

DEFINITION 2.2. A differentiable space $(X, \mathcal{C}^\infty)$ is called *homologically regular* if for each point $x \in X$ there exists a singular chart $\varphi : U \to \mathbb{R}^n$ defined on an open neighborhood of x such that $\varphi(U) \subset \mathbb{R}^n$ is regular in the sense of Tougeron and if for every $k \in \mathbb{N}^*$ the sets $\varphi(U)^k$ and $\Delta^k_{\mathbb{R}^n}$ are regularly situated in $\mathbb{R}^{kn}$.

THEOREM 2.3. *Let $(X, \mathcal{C}^\infty)$ be a homologically regular reduced differentiable space and let $A = \mathcal{C}^\infty(X)$ be its global section algebra. Then $(\mathscr{E}_\bullet, \beta)$ is a complex of fine sheaves over X. If in addition $(X, \mathcal{C}^\infty)$ is homologically regular, then the chain map $p_X : C_\bullet(A) \to \mathscr{E}_\bullet(X) = E_\bullet(A)$ onto the diagonal complex is a quasi-isomorphism.*

PROOF. We proceed in several steps, and follow and extend the argument given in the proof of [**2**, Prop. 4.2].

Step 1. First we show that $\mathscr{E}_k$ is a complex of fine sheaves. Let $\mathcal{C}^\infty_{|\Delta^{k+1}_X}$ be the restriction of the structure sheaf of X^{k+1} to the diagonal Δ^{k+1}_X. Since the structure sheaf $\mathcal{C}^\infty_{X^{k+1}}$ is fine, $\mathcal{C}^\infty_{|\Delta^{k+1}_X}$ is a a fine sheaf, too. By construction, $\mathscr{E}_k$ is a module over the sheaf $\mathcal{C}^\infty_{|\Delta^{k+1}_X}$, hence $\mathscr{E}_k$ is even a sheaf and has to be fine as well.

Step 2. Next we will define for each $t > 0$ a localization chain map $\Psi_{\bullet,t} : C_\bullet(A) \to C_\bullet(A)$ such that for every $c \in C_k(A)$ the chain $\Psi_{\bullet,t}c$ has support in a t-neighborhood of the diagonal $\Delta^{k+1}(X)$. To this end choose a proper embedding φ of $(X, \mathcal{C}^\infty)$ into the colimit of ringed spaces $(\mathbb{R}^\infty, \mathcal{C}^\infty_{\mathbb{R}^\infty}) := \varinjlim_{n\in\mathbb{N}}(\mathbb{R}^n, \mathcal{C}^\infty_{\mathbb{R}^n})$. Such an embedding exists by [**16**, 1.3.17 Lemma] and [**18**, Prop. A.3]. Denote by $d : X \times X \to \mathbb{R}$ the restriction of the Euclidean metric d_∞ on $\mathbb{R}^\infty$ to X that is put $d(x, y) = d_\infty(\varphi(x), \varphi(y))$ for all $x, y \in X$. By [**18**, Cor. A.4], the function d^2 lies in $\mathcal{C}^\infty(X \times X)$. Now fix a smooth function $\varrho : \mathbb{R} \to [0, 1]$ having support in $(-\infty, \frac{1}{2}]$ and which satisfies $\varrho(r) = 1$ for $r \leq \frac{1}{3}$. For $t > 0$ denote by ϱ_t the rescaled function $r \mapsto \varrho(\frac{r}{t})$. Next define functions $\Psi_{k,i,t} \in \mathcal{C}^\infty(X^{k+1})$ for $k \in \mathbb{N}$ and $i = 0, \ldots, k+1$ by

(2.7)

$$\Psi_{k,i,t}(x_0, \ldots, x_k) = \prod_{j=0}^{i-1} \varrho_t\big(d^2(x_j, x_{j+1})\big) \quad \text{where } x_0, \ldots, x_k \in X,\ x_{k+1} := x_0 \,.$$

Then put $\Psi_{k,t} := \Psi_{k,k+1,t}$. Since $\mathcal{C}^\infty(X^{k+1})$ acts on $C_k(A)$ in a natural way, one obtains for each $t > 0$ a graded map of degree 0

$$\Psi_{\bullet,t} : C_\bullet(A) \to C_\bullet(A),\ C_k(A) \ni c \mapsto \Psi_{k,t}c \,.$$

One immediately checks the commutation relations

$$b_j \Psi_{k,i,t} = \begin{cases} \Psi_{k,i-1,t} b_j, & \text{if } j < i, \\ \Psi_{k,i,t} b_j, & \text{if } j \geq i, \end{cases}$$

which implies that $\Psi_{\bullet,t}$ commutes with the face maps b_i. Hence, $\Psi_{\bullet,t}$ is a chain map. By construction, the support of $\Psi_{k,t}c$ is contained in the neighborhood $U_{k+1,\frac{t}{2}}$ of the diagonal $\Delta^{k+1}(X)$, where for $\varepsilon > 0$

$$U_{k,\varepsilon} := \left\{(x_1, \ldots, x_k) \in X^k \mid \sqrt{d^2(x_1, x_2) + d^2(x_2, x_3) + \ldots + d^2(x_k, x_1)} < \varepsilon\right\}$$

is the so-called *ε-neighborhood* of the diagonal $\Delta^k(X) \subset X^k$.

For later purposes let us mention at this point that the functions $\Psi_{k,i,t}$ and $\Psi_{k,t}$ are restrictions of smooth functions $\Psi^{\infty}_{k,i,t}$ and $\Psi^{\infty}_{k,t}$ on $\left(\mathbb{R}^{\infty}\right)^{k+1}$ which are defined analogously, just with d_{∞} instead of d. In the following, we will denote the restrictions of the functions $\Psi^{\infty}_{k,i,t}$ and $\Psi^{\infty}_{k,t}$ and to the cartesian product Y^{k+1} of a locally closed subset $Y \subset \mathbb{R}^{\infty}$ of $\mathbb{R}^{\infty}$, by the symbols $\Psi^{Y}_{k,i,t}$ and $\Psi^{Y}_{k,t}$, respectively.

Step 3. The chain map $\Psi_{\bullet,t} : C_{\bullet}(A) \to C_{\bullet}(A)$ is homotopic to the identity morphism on $C_{\bullet}(A)$ for every $t > 0$.

To prove this recall first that the Hochschild chain complex of A has the following degeneracy maps for $k \in N$ and $i = 0, \dots, k$:

$$s_{k,i} : C_k(A) \to C_{k+1}(A),\ a_0 \otimes \ldots \otimes a_k \mapsto a_0 \otimes \ldots \otimes a_i \otimes 1 \otimes a_{i+1} \otimes \ldots \otimes a_k . \tag{2.8}$$

Therefore, we can define $\mathcal{C}^{\infty}(X)$-module maps $\eta_{k,i,t} : C_k(A) \to C_{k+1}(A)$ for every integer $k \geq -1$, $i = 1, \cdots, k+2$ and $t > 0$ by

$$\eta_{k,i,t}(c) := \begin{cases} \Psi_{k+1,i,t} \cdot (s_{k,i-1}c) & \text{for } i \leq k+1, \\ 0 & \text{for } i = k+2. \end{cases} \tag{2.9}$$

Here we have assumed $c \in C_k(A)$ and have put $C_{-1}(A) := \{0\}$. For $i = 2, \cdots, k$ one then computes using the above commutation relations for b_j and $\Psi_{k,i,t}$

(2.10)

$$\begin{aligned}(b\eta_{k,i,t} + \eta_{k-1,i,t}b)c = {} & (-1)^{i-1}\Psi_{k,i-1,t}c + \Psi_{k,i-1,t}\sum_{j=0}^{i-2}(-1)^j\, s_{k-1,i-2}b_{k,j}c + \\ & + (-1)^i\Psi_{k,i,t}c + \Psi_{k,i,t}\sum_{j=0}^{i-1}(-1)^j\, s_{k-1,i-1}b_{k,j}c,\end{aligned}$$

and for the two remaining cases $i = 1$ and $i = k+1$

$$(b\eta_{k,1,t} + \eta_{k-1,1,t}b)c = c - \Psi_{k,1,t}c + \Psi_{k,1,t}s_{k-1,0}b_{k,0}c, \tag{2.11}$$

$$\begin{aligned}(b\eta_{k,k+1,t} + \eta_{k-1,k+1,t}b)c = {} & \\ = \Psi_{k,k,t}(-1)^k c + \Psi_{k,k,t} & \sum_{j=0}^{k-1}(-1)^j\, s_{k-1,k-1}b_{k,j}c + (-1)^{k+1}\Psi_{k,t}\, c.\end{aligned} \tag{2.12}$$

These formulas immediately entail that the collection $\left(H_{k,t}\right)_{k\in\mathbb{N}}$ of maps

$$H_{k,t} = \sum_{i=1}^{k+1}(-1)^{i+1}\,\eta_{k,i,t} : C_k(A) \to C_{k+1}(A)$$

forms a homotopy between the identity and the localization morphism $\Psi_{\bullet,t}$. More precisely,

$$\left(bH_{k,t} + H_{k-1,t}b\right)c = c - \Psi_{k,t}\, c \quad \text{for all } c \in C_k(A) . \tag{2.13}$$

This finishes the third step.

Note that by the considerations at the end of the previous step, the functions $\eta_{k,i,t}$ can be also defined on the spaces $C_k(\mathcal{C}^{\infty}(Y))$, where $Y \subset \mathbb{R}^{\infty}$ is again a locally closed subset. We write $\eta^{Y}_{k,i,t}$ for these maps. Observe that for each locally closed

$Z \subset Y$ the diagrams

$$\begin{array}{ccc} C_k(\mathcal{C}^\infty(Y)) & \xrightarrow{\Psi^Y_{k,i,t}} & C_{k+1}(\mathcal{C}^\infty(Y)) \\ \downarrow & & \downarrow \\ C_k(\mathcal{C}^\infty(Z)) & \xrightarrow[\Psi^Z_{k,i,t}]{} & C_k(\mathcal{C}^\infty(Z)) \end{array} \tag{2.14}$$

and

$$\begin{array}{ccc} C_k(\mathcal{C}^\infty(Y)) & \xrightarrow{\eta^Y_{k,i,t}} & C_{k+1}(\mathcal{C}^\infty(Y)) \\ \downarrow & & \downarrow \\ C_k(\mathcal{C}^\infty(Z)) & \xrightarrow[\eta^Z_{k,i,t}]{} & C_k(\mathcal{C}^\infty(Z)) \end{array} \tag{2.15}$$

then commute, where the vertical arrows denote the natural restriction maps.

Step 4. Next we study the operators $K_{k,t} : C_k(A) \to C_{k+1}(A)$ defined for $t \in (0,1]$ by

$$K_{k,t}c := \int_t^1 H_{k,\frac{s}{2}}\big(\partial_s \Psi_{k,s}c\big)ds\,. \tag{2.16}$$

Since Ψ_s is a chain map, the algebraic homotopy relation stated in Eq. (2.13) entails (as in Eq. (4.9) of [**2**]) that for $c \in C_k(A)$

$$\begin{aligned} \big(bK_{k,t} + K_{k-1,t}b\big)c &= \int_t^1 bH_{k,\frac{s}{2}}\big(\partial_s\Psi_{k,s}c\big) + H_{k-1,\frac{s}{2}}b\big(\partial_s\Psi_{k,s}c\big)ds = \\ &= \int_t^1 \partial_s\Psi_{k,s}c - \Psi_{k,\frac{s}{2}}(\partial_s\Psi_{k,s}c)ds = \Psi_{k,1}c - \Psi_{k,t}c\,, \end{aligned} \tag{2.17}$$

where we have used that $\operatorname{supp}\Psi_{k,\frac{s}{2}} \subset U_{k+1,\frac{s}{4}}$ and $(\partial_s\Psi_{k,s})_{|U_{k+1,\frac{s}{4}}} = 0$.

Step 5. In the last step we show that $J_\bullet(X)$ is a contractible complex which will entail the claim that $p_X : C_\bullet(A) \to \mathscr{E}_\bullet(X)$ is a quasi-isomorphism. To this end it suffices to show by Eq. (2.17) that for $c \in J_k(X)$ the relation

$$\lim_{t\searrow 0} \Psi_{k,t}c = 0 \tag{2.18}$$

holds true, and that the operator

$$K_k : J_k(X) \to J_{k+1}(X),\ c \mapsto \lim_{t\searrow 0} K_{k,t}c \tag{2.19}$$

exists and is continuous. The limits in Equations (2.18) and (2.19) are taken with respect to the natural Fréchet topology which the space $J_k(X)$ inherits as a closed subspace of $C_k(A) = \mathcal{C}^\infty(X^{k+1})$; see Prop. 1.2. Describing convergence of a sequence $(f_l)_{l\in\mathbb{N}} \subset J_k(X)$ is somewhat tricky, but it essentially boils down to the criterion that for every compact $K \subset X$ there is an ambient Euclidean space $\mathbb{R}^n \supset K$ (provided by an appropriate singular chart) and a sequence of representatives $F_l \in \mathcal{C}^\infty(\mathbb{R}^{(k+1)n})$ of f_l over Δ_K^{k+1} such that $(F_l)_{l\in\mathbb{N}}$ and all sequences of higher derivatives $(\partial^\alpha F_l)_{l\in\mathbb{N}}$ converge uniformly on Δ_K^{k+1}.

Now denote by B_n the closed ball in $\mathbb{R}^n$ of radius n, and put $C_n := B_n \cap X$. Then $(C_n)_{n\in\mathbb{N}}$ is a compact exhaustion of X. Consider for locally closed $Y \subset \mathbb{R}^\infty$

the maps

$$\mu_k^Y : J_k(Y) \times [0,1] \to J_k(Y), \quad (c,t) \mapsto \begin{cases} \Psi_{k,t}^Y, & \text{if } t > 0, \\ 0, & \text{if } t = 0, \end{cases}$$

$$\mu_{k,i}^Y : J_k(Y) \times [0,1] \to J_k(Y), \quad (c,t) \mapsto \begin{cases} \Psi_{k,i,t}^Y s_{k-1,i}(\partial_t \Psi_{k-1,t}^Y) c, & \text{if } t > 0, \\ 0, & \text{if } t = 0, \end{cases}$$

The maps $\mu_k^{B_n} : J_k(B_n) \times [0,1] \to J_k(B_k)$ and $\mu_{k,i}^{B_n} : J_k(B_n) \times [0,1] \to J_k(B_n)$ are continuous by [**2**, Lem. 4.4]. Note that in the proof of that lemma the assumption is needed that X is homologically regular, meaning that X^{k+1} and Δ_X^{k+1} are regularly situated. Since $J_k(B_n) \to J_k(C_n)$ is a topological identification by the open mapping theorem and Proposition 1.2, it follows by commutativity of the diagrams (2.14) and (2.15) that $\mu_k^{C_n}$ and $\mu_{k,i}^{C_n}$ are continuous as well. But this means that for all $n \in \mathbb{N}$ and $c \in J_k(X)$

$$\lim_{t \searrow 0} \left(\Psi_{k,t} c \right)_{|C_n} = \lim_{t \searrow 0} \mu_k^{C_n}(c_{|C_n}, t) = 0$$

and

$$\begin{aligned} \lim_{t \searrow 0} \left(K_{k,t} c \right)_{|C_n} &= \lim_{t \searrow 0} \sum_{i=1}^{k+1} (-1)^{i+1} \int_t^1 \mu_{k,i}^{C_n}(c_{|C_n}, s) ds = \\ &= \sum_{i=1}^{k+1} (-1)^{i+1} \int_0^1 \mu_{k,i}^{C_n}(c_{|C_n}, s) ds \,, \end{aligned}$$

where the last integral exists and is continuous in $c \in J_k(X)$ by continuity of $\mu_{k,i}^{C_n}$. The proof is finished. □

Before we can apply this result to reduce the computation of the Hochschild homology of the global section algebra to local ones we need to introduce hypercohomology for complexes of sheaves not bounded below. We follow [**24, 26**] and refer to these papers for further details.

Let $(\mathscr{D}^\bullet, \delta)$ be a cochain complex of sheaves on a paracompact topological space X. Since the category of sheaves on X has enough injectives, there exists an injective Cartan–Eilenberg resolution $\mathscr{D}^\bullet \longrightarrow \mathscr{I}^{\bullet,\bullet}$ as described in [**25**, Sec. 5.7]. Up to chain homotopy equivalence, the product total complex $\mathrm{Tot}^{\Pi}(\mathscr{I}^{\bullet,\bullet})$ does not depend on the choice of the injective resolution of $\mathscr{D}^\bullet$. Applying the global section functor results in the total complex $\Gamma\left(X, \mathrm{Tot}^{\Pi} \mathscr{I}^{\bullet,\bullet}\right) \cong \mathrm{Tot}^{\Pi} \Gamma\left(X, \mathscr{I}^{\bullet,\bullet}\right)$ which again up to chain homotopy equivalence is independant of the particular choice of the resolution $\mathscr{D}^\bullet \longrightarrow \mathscr{I}^{\bullet,\bullet}$. The *hypercohomology* of $(\mathscr{D}^\bullet, \delta)$ is then defined by

$$\mathbb{H}^k(X, \mathscr{D}^\bullet) = H^k\left(\Gamma\left(X, \mathrm{Tot}^{\Pi} \mathscr{I}^{\bullet,\bullet}\right)\right) \,.$$

Since a fine sheaf on X is acyclic, one can take fine resolutions instead of injective ones which results in the observation that for a cochain complex $(\mathscr{D}^\bullet, \delta)$ of sheaves which are fine already, the hypercohomology is given by the cohomology of the global sections. More precisely, the following folklore result holds true.

THEOREM 2.4. *Let $(\mathscr{D}^\bullet, \delta)$ be a cochain complex of fine sheaves on a paracompact topological space X. Let $\mathscr{H}^\bullet(\mathscr{D}^\bullet)$ be its cohomology sheaf that is let*

$$\mathscr{H}^k(\mathscr{D}^\bullet) = \ker\left(\delta : \mathscr{D}^k \to \mathscr{D}^{k+1}\right) / \operatorname{im}\left(\delta : \mathscr{D}^{k-1} \to \mathscr{D}^k\right) \,.$$

Then the global section functor commutes with cohomology and the hypercohomology of $(\mathscr{D}^\bullet, \delta)$ is given by

$$\mathbb{H}^k(X, \mathscr{D}^\bullet) = H_k(\Gamma(X, \mathscr{D}^\bullet)) = \Gamma\left(X, \mathscr{H}^k(\mathscr{D}^\bullet)\right) \quad \textit{for all } k \in \mathbb{N}\ . \tag{2.20}$$

PROOF. It only remains to show the second equality in Eq. (2.20). Since $(\mathscr{D}^\bullet, \delta)$ and $(\mathscr{H}^\bullet(\mathscr{D}^\bullet), 0)$ are quasi-isomorphic sheaf complexes, their hypercohomologies coincide. Both cochain complexes consist of fine sheaves, hence their hypercohomology is given by the cohomology of global sections and the claim follows. □

The virtue of Weibel's approach is that it allows to define hypercohomology for cochain complexes not bounded below. After reindexing $\mathscr{C}^k := \mathscr{C}_{-k}$ the components of a chain complex of sheaves $(\mathscr{C}_\bullet, d)$ one therefore obtains a cochain complex of sheaves $(\mathscr{C}^\bullet, d^*)$ for which the hypercohomology groups $\mathbb{H}^k(X, \mathscr{C}^\bullet)$ are defined. We put $\mathbb{H}_k(X, \mathscr{C}_\bullet) = \mathbb{H}^{-k}(X, \mathscr{C}^\bullet)$ and call these the *hyperhomology* groups of $(\mathscr{C}_\bullet, d)$. Let us apply this idea to the Hochschild homology theory of a broad class of sheaves of algebras which we call *sheaves of bornological algebras over a differentiable space* $(X, \mathcal{C}^\infty_X)$. By definition, a sheaf $\mathscr{A}$ defined on X lies in that category if it satisfies the following three conditions.

- (BA1) $\mathscr{A}$ is a sheaf of algebras over $\mathcal{C}^\infty_X$ that is for each open $U \subset X$ the section space $\mathscr{A}(U)$ is an algebra over the commutative ring $\mathcal{C}^\infty_X(U)$ and the corresponding restriction morphism $\mathscr{A}(U) \to \mathscr{A}(V)$ for $V \subset U$ open is an algebra morphism over the restriction homomorphism $\mathcal{C}^\infty_X(U) \to \mathcal{C}^\infty_X(V)$.
- (BA2) The section space $\mathscr{A}(U)$ for $U \subset X$ open is a nuclear LF-space and a bornological algebra with respect to the bornology of von Neumann bounded sets. In particular this means that multiplication is separately continuous.
- (BA3) For each open $U \subset X$ the action $\mathcal{C}^\infty_X(U) \times \mathscr{A}(U) \times \mathscr{A}(U)$ is continuous where $\mathcal{C}^\infty_X(U)$ carries its natural structure of a Fréchet algebra.

Given a sheaf of bornological algebras $\mathscr{A}$ over $(X, \mathcal{C}^\infty_X)$ we call it

- (BA4) *unital* if the section spaces $\mathscr{A}(U)$ together with their restriction morphisms $\mathscr{A}(U) \to \mathscr{A}(V)$ are unital, and
- (BA5) *H-unital* if each of the section spaces $\mathscr{A}(U)$ is an H-unital algebra that is if the Bar complex corresponding to $\mathscr{A}(U)$ is acyclic.

Note that every unital sheaf of bornological algebras is H-unital. Below we will give in Section 3.2 an example of an H-unital but not unital sheaf of bornological algebras.

Assume that $\mathscr{A}$ is a sheaf of bornological algebras over X and let $A = \mathscr{A}(X)$ be its algebra of global sections. Then use the completed bornological tensor product $\hat{\otimes}$ as defined in [**11**, Sec. 2.2] to define for $U \subset X$ open the the space of Hochschild k-chains of the bornological algebra $\mathscr{A}(U)$:

$$\mathscr{C}_k(\mathscr{A})(U) := \mathscr{A}(U)^{\hat{\otimes} k+1} := \underbrace{\mathscr{A}(U) \hat{\otimes} \ldots \hat{\otimes} \mathscr{A}(U)}_{k+1\text{-times}} \ .$$

Associating to every open $U \subset X$ the space $\mathscr{C}_k(\mathscr{A})(U)$ gives rise to a presheaf on X. The Hochschild boundary operators $b : \mathscr{C}_k(\mathscr{A})(U) \to \mathscr{C}_{k-1}(\mathscr{A})(U)$ combine to a morphism of presheaves $b : \mathscr{C}_k(\mathscr{A}) \to \mathscr{C}_{k-1}(\mathscr{A})$. After passing to the sheafifications $\hat{\mathscr{C}}_k(\mathscr{A})$ we therefore obtain a chain complex of sheaves $(\hat{\mathscr{C}}_\bullet(\mathscr{A}), b)$ which we call the *Hochschild chain complex* of $\mathscr{A}$. For the case where $\mathscr{A}$ coincides with the structure sheaf $\mathcal{O}_X$ of a differentiable space X one can interpret $\big(\hat{\mathscr{C}}_\bullet(\mathcal{O}_X), b\big)$ as a "differentiable version" of Weibel's sheaf complex of Hochschild chains of a scheme [**24**]. The *Hochschild homology* $HH_\bullet(\mathscr{A})$ of the sheaf $\mathscr{A}$ on X is now defined as the hyperhomology $\mathbb{H}_\bullet(X, \hat{\mathscr{C}}_\bullet(\mathscr{A}))$. The question arises how $HH_\bullet(\mathscr{A})$ relates to the Hochschild homology $HH_\bullet(A)$ of the global section algebra $A := \mathscr{A}(X)$ and to the global section space of the homology sheaf $\mathscr{H}\mathscr{H}_\bullet(\mathscr{A}) := \mathscr{H}_\bullet\big(\hat{\mathscr{C}}_\bullet(\mathscr{A})\big)$. The answer is given by the next result.

THEOREM 2.5. *Let $\mathscr{A}$ be a sheaf of H-unital bornological algebras over the homologically regular differentiable space $(X, \mathcal{O})$. Then the Hochschild chain complex $\big(\hat{\mathscr{C}}_\bullet(\mathscr{A}), b\big)$ is a complex of fine sheaves on X. The global section space of its homology sheaf coincides with the Hochschild homology of the global section algebra $A = \Gamma(X, \mathscr{A})$. More precisely, for all $k \in \mathbb{N}$*

$$HH_k(A) = HH_k(\mathscr{A}) = HH_k\Big(\Gamma(X, \hat{\mathscr{C}}_\bullet(\mathscr{A}))\Big) = \Gamma\left(X, \mathscr{H}\mathscr{H}_k(\mathscr{A})\right) \ . \tag{2.21}$$

Before we can prove the theorem, we need to extend the diagonal complex to the situation where a sheaf of bornological algebras is given. To define the diagonal complex associated to such a sheaf $\mathscr{A}$ choose a singular chart $(\varphi, \varphi^\sharp) : (U, \mathcal{O}|_U) \to (\mathbb{R}^n, \mathcal{C}^\infty_{\mathbb{R}^n})$ of the differentiable space $(X, \mathcal{O})$ and put for each $k \in \mathbb{N}$

$$\mathscr{J}_k(U) := \mathscr{J}_k^\varphi(U) := \varphi^\sharp_{\mathbb{R}^{kn}}\big(\mathcal{J}^\infty_{\Delta^{k+1}_{\mathbb{R}^n}}(\mathbb{R}^{kn})\big) \cdot \hat{\mathscr{C}}_k(\mathscr{A})(U) \subset \hat{\mathscr{C}}_k(\mathscr{A})(U) \ . \tag{2.22}$$

One immediately checks that assigning to an open $V \subset U$ the section space $\mathscr{J}_k^\varphi(V)$ gives rise to a sheaf $\mathscr{J}_k^\varphi$ on U. Moreover, using transition maps between singular charts one shows that for a second singular chart $(\psi, \psi^\sharp) : (U, \mathcal{O}|_U) \to (\mathbb{R}^m, \mathcal{C}^\infty_{\mathbb{R}^m})$ of X over U the sheaves $\mathscr{J}_k^\varphi$ and $\mathscr{J}_k^\psi$ over U coincide. Hence they glue together to a global sheaf over X which we denote by $\mathscr{J}_k$ and which by construction is a subsheaf of $\hat{\mathscr{C}}_k(\mathscr{A})$. As in the proof of Lemma 2.1 one shows that the Hochschild boundary b maps $\mathscr{J}_k$ to $\mathscr{J}_{k-1}$ hence we obtain a monomorphism of sheaf complexes $\big(\mathscr{J}_\bullet, b\big) \hookrightarrow \big(\hat{\mathscr{C}}_\bullet(\mathscr{A}), b\big)$.

PROPOSITION 2.6. *Assume to be given a sheaf of H-unital bornological algebras $\mathscr{A}$ over the homologically regular differentiable space $(X, \mathcal{O})$. Let $\big(\mathscr{E}_\bullet, \beta\big)$ be its* diagonal sheaf complex *that means let it be the unique quotient making the following sequence of chain complexes of sheaves exact:*

$$0 \longrightarrow \big(\mathscr{J}_\bullet, b\big) \longrightarrow \big(\hat{\mathscr{C}}_\bullet(\mathscr{A}), b\big) \stackrel{p}{\longrightarrow} \big(\mathscr{E}_\bullet, \beta\big) \longrightarrow 0\,. \tag{2.23}$$

Denote by A the global section algebra $\Gamma(X, \mathscr{A})$. Then the canonical chain map $C_\bullet(A) \to \hat{\mathscr{C}}_\bullet(\mathscr{A})(X)$ and the projection $p_X : \hat{\mathscr{C}}_\bullet(\mathscr{A})(X) \to E_\bullet(A) = \mathscr{E}_\bullet(X)$ are quasi-isomorphisms.

PROOF. Proposition 2.7 in [**18**] says that the chain map $C_\bullet(A) \to \hat{\mathscr{C}}_\bullet(\mathscr{A})(X)$ is a quasi-isomorphism, so it remains to show that the projection $p_X : \hat{\mathscr{C}}_\bullet(\mathscr{A})(X) \to E_\bullet(A)$ is a quasi-isomorphism. We only sketch the argument because it is analogous to the proof of Theorem 2.3.

Case 1. Let us first consider the local case that is the case where X is assumed to be a closed subspace of some Euclidean space $\mathbb{R}^n$. The global section algebra $\mathcal{O}(X)$ then coincides with the quotient of $\mathcal{C}^\infty(\mathbb{R}^n)$ by a closed ideal I, and the space X can be identified with the real spectrum of the Fréchet algebra $\mathcal{O}(X) = \mathcal{C}^\infty(\mathbb{R}^n)/I$. Recall the definition of the cutoff functions $\Psi_{k,i,t} \in \mathcal{C}^\infty(\mathbb{R}^{n(k+1)})$ for

$k \in \mathbb{N}$, $i = 0, \dots, k$ and $t > 0$ by Eq. 2.7. Again put $\Psi_{k,t} := \Psi_{k,k,t}$. We will regard the functions $\Psi_{k,i,t}$ and $\Psi_{k,t}$ both as smooth functions on $\mathbb{R}^{n(k+1)}$ and as elements of $\mathcal{O}(X)^{\hat{\otimes}(k+1)} = \mathcal{O}(X^{k+1})$.

Case 1a. After these preparations, we assume that $\mathscr{A}$ is a sheaf of unital bornological algebras over $(X, \mathcal{O})$. Then the degeneracy maps $s_{k,i} : C_k(A) \to C_{k+1}(A)$ defined by Eq. 2.8 exist for the global section algebra A and extend to sheaf morphisms $s_{k,i} : \hat{\mathscr{C}}_k(\mathscr{A}) \to \hat{\mathscr{C}}_{k+1}(\mathscr{A})$. By the natural action of $\mathcal{O}(X^{k+1})$ on $C_k(A)$ we obtain for each $t > 0$ a chain map

$$\Psi_{\bullet,t} : C_\bullet(A) \to C_\bullet(A),\ C_k(A) \ni c \mapsto \Psi_{k,t} \cdot c$$

as in *Step 2* of the proof of Theorem 2.3. *Steps 3* to *5* in that proof also extend. In particular we have $\mathcal{O}(X)$-module maps $\eta_{k,i,t} : C_k(A) \to C_{k+1}(A)$ defined by Eq. 2.9 and a graded map $H_{\bullet,t} : C_\bullet(A) \to C_\bullet(A)$ of degree 1 defined by $H_{k,t} = \sum_{i=1}^{k+1}(-1)^{i+1}\eta_{k,i,t}$. Since Eq. 2.13 then holds as well, $H_{\bullet,t}$ is a homotopy between the identity and the localization morphism $\Psi_{\bullet,t}$. Next define $K_{k,t} : C_k(A) \to C_{k+1}(A)$ as in Eq. 2.16 and observe that like in *Step 4* one proves that $K_{\bullet,t}$ provides a homotopy between the localizations $\Psi_{\bullet,1}$ and $\Psi_{\bullet,t}$. Finally we want to show that the limit $\lim_{t\searrow 0} K_{\bullet,t}$ exists and provides a homotopy between the localization $\Psi_{\bullet,1}$ and 0. To this end recall from *Step 5* in the proof of Theorem 2.3 that for $f \in \mathscr{J}^\infty(\Delta^{k+1}_{\mathbb{R}^n}, \mathbb{R}^{(k+1)n})$ the limit $\lim_{t\searrow 0} \Psi_{k,t} f$ exists and vanishes. Now assume to be given $c \in \mathscr{J}_k(X)$ of the form $c = \sum_{i=1}^{l} f_i c_i$ with $f_i \in \mathscr{J}^\infty(\Delta^{k+1}_{\mathbb{R}^n}, \mathbb{R}^{(k+1)n})$ and $c_i \in \hat{\mathscr{C}}_k(\mathscr{A})(X)$. Then $\lim_{t\searrow 0} \Psi_{k,t} c = 0$ and the limit $K_k c = \lim_{t\searrow 0} K_{k,t} c$ exists. The latter is even continuous in c. Since the space of $c \in \mathscr{J}_k(X)$ of the given form is dense in $\mathscr{J}_k(X)$ by Eq. 2.22, the operator

$$K_k : \mathscr{J}_k(X) \to \mathscr{J}_{k+1}(X),\ c \mapsto \lim_{t\searrow 0} K_{k,t} c$$

therefore is well-defined and continuous. By construction, it is a homotopy between the localization $\Psi_{\bullet,1}$ and 0. Hence the chain complex $\mathscr{J}_\bullet(X)$ is acyclic, and p_X is a quasi-isomorphism as claimed.

Case 1b. Now we come to the case where $\mathscr{A}$ is not necessarily unital, but still assumed to be an H-unital bornological sheaf of algebras. Consider the direct sum $\mathscr{A} \oplus \mathcal{O}$ and denote it by $\mathscr{A}^+$. Since $\mathscr{A}$ is a $\mathcal{O}$-module sheaf and $\mathcal{O}$ is a sheaf of unital algebras, $\mathscr{A}^+$ inherits in antural way the structure of a sheaf of unital algebras such that the following sequence of sheaves is split exact.

$$0 \longrightarrow \mathscr{A} \longrightarrow \mathscr{A}^+ \longrightarrow \mathcal{O} \longrightarrow 0$$

By *Case 1a*, the chain complex $\hat{\mathscr{C}}_\bullet(\mathscr{A}^+)(X)$ is quasi-isomorphic to the diagonal complex $E_\bullet(A^+) = \hat{\mathscr{C}}_\bullet(\mathscr{A}^+)(X)/\mathscr{J}^+_\bullet(X)$, where $A^+ = \mathscr{A}^+(X)$ and where the sheaf complex $\mathscr{J}^+_\bullet$ is defined by Eq. 2.22 with $\mathscr{A}$ replaced by $\mathscr{A}^+$. Since $A = \mathscr{A}(X)$ is H-unital and $E_\bullet(A^+) \cong E_\bullet(A) := \hat{\mathscr{C}}_\bullet(\mathscr{A})(X)/\mathscr{J}_\bullet(X)$, the chain complex $\hat{\mathscr{C}}_\bullet(\mathscr{A})(X)$ has to be quasi-isomorphic to the diagonal complex $E_\bullet(A)$ which finalizes *Case 1*. See [**18**, Lemma 2.4.] for further details in the last argument.

Case 2. It remains to prove the claim when $(X, \mathcal{O})$ is not necessarily affine. We only consider the unital case, the H-unital one can be treated is as in *Case 1b*. Choose a family $(\varphi_i, \varphi_i^\sharp)_{i\in I}$ of singular charts of $(X, \mathcal{O})$ such that the family of its domains $(U_i)_{\in I}$ is a locally finite cover of X and such that each U_i is relatively compact. Let $\mathbb{R}^{n_i}$ be the range of φ_i. Choose a partition of unity $(\psi_i)_{i\in I}$ subordinate partition subordinate to the cover $(U_i)_{i\in I}$ and such that the support of each cutoff function ψ_i is compact. Recall from *Step 2* in the proof of Theorem 2.3 that the cutoff functions $\Psi_{k,t}$ can be globally defined on X by embedding X into the colimit $(\mathbb{R}^\infty, \mathcal{C}^\infty_{\mathbb{R}^\infty})$. Now fix a degree k and choose $\varepsilon_i > 0$ such that $\operatorname{supp} \Psi_{k,\varepsilon_i} \cap \operatorname{supp} \varphi \times X^k$ is relatively compact in U_i^{k+1}. By *Case 1*, there exists for each i a contracting homotopy $H^i_\bullet : \mathscr{J}^{\varphi_i}_\bullet(U_i) \to \mathscr{J}^{\varphi_i}_\bullet(U_i)$. Put for $l \leq k$

$$H_l = \sum_{i \in I} H^i_l \varphi_i \Psi_{l,\varepsilon_i} : \mathscr{J}_k(X) \to \mathscr{J}_{k+1}(X) \, .$$

The equality

$$bH_l c + H_l bc = c$$

then holds for all $c \in \mathscr{J}_l(X)$ and $l \leq k$, hence $\mathscr{J}_\bullet(X)$ is contractible and p_X a quasi-isomorphism. □

3. Applications

In this section we provide several examples of differentiable spaces and sheaves of bornological algebras where our method of localization can be applied and used to determine the corresponding Hochschild homology groups.

3.1. Whitney fields. Let $X \subset \mathbb{R}^n$ be a regular locally closed subset with regularly situated diagonals that is X^k and $\Delta^k_{\mathbb{R}^n}$ are regularly situated for all positive k. Denote for every relatively open $U \subset X$ by $\mathcal{E}^\infty(U)$ the algebra of Whitney fields over U. Then $(X, \mathcal{E}^\infty)$ is a differentiable space with global sections being the Whitney fields over X. For each $x \in X$, the stalk of the sheaf $\hat{\mathscr{C}}_\bullet(\mathcal{E})$ at x coincides with the tensor product of the stalk $\mathcal{E}^\infty_x$ with the Hochschild chain complex $C_\bullet(\mathcal{F}_n)$ of the algebra $\mathcal{F}_n$ of formal power series in n variables. The Hochschild homology of $\mathcal{F}_n$ coincides in degree k with $\mathcal{F}_n \otimes \Lambda^k \mathbb{R}^n$ as one shows for example by using the Connes–Koszul resolution for $\mathcal{F}_n$. By the localization result Theorem 2.5 this entails that the Hochschild homology of $\mathcal{E}^\infty(X)$ is given by

$$HH_k\big(\mathcal{E}^\infty(X)\big) \cong \Omega^k_{\mathcal{E}^\infty}(X) := \mathcal{E}^\infty(X) \otimes \Lambda^k \mathbb{R}^n \, .$$

We thus have recovered [**2**, Thm. 4.8].

3.2. Convolution algebras. Let $\mathsf{G} \rightrightarrows M$ be a proper Lie groupoid. Denote by s, t its source and target map, respectively, and by $\pi : M \to X := M/\mathsf{G}$ the canonical projection onto the space of orbits. Note that X is a differentiable space by [**17**]. Its structure sheaf $\mathcal{C}^\infty_X$ has section spaces $\mathcal{C}^\infty_X(U)$ over $U \subset X$ open given by all continuous functions $f : U \to \mathbb{R}$ such that the pullback $\pi^* f$ is smooth on M. Given a subset $C \subset X$ put $C^0 := \pi^{-1}(C)$ and $C^1 := s^{-1}(C^0)$ and call them the saturations of C in M and G, respectively. A subset $K \subset \mathsf{G}$ then is said to be *longitudinally compact* if for every compact $C \subset X$ the intersection K with the saturation C_1 is compact in G. For each open $U \subset X$ we now define the *convolution algebra* of G over U by

$$\mathscr{A}(U) = \{ f \in \mathcal{C}^\infty(U_1) \mid \operatorname{supp}(f) \text{ is longitudinally compact} \} \, .$$

The product on $\mathscr{A}(U)$ is the *convolution product* given by

$$f_1 * f_2(g) = \int_{\mathsf{G}(t(g),-)} f_1(h) f_2(h^{-1}g) d\lambda^{t(g)}, \, f_1, f_2 \in \mathscr{A}(U), \; g \in U^1,$$

where $(\lambda^x)_{x \in M}$ is a smooth left Haar system on G. One verifies that one thus obtains a sheaf $\mathscr{A}$ of H-unital bornological algebras over the orbit space $X = M/\mathsf{G}$ [**18**].

The sheaf $\mathscr{A}$ is called the *convolution sheaf* of the groupoid G and is sometimes denoted more descriptively by $\mathscr{A}_{\mathsf{G}}$. Its algebra of global sections $A = \Gamma(X, \mathscr{A}_{\mathsf{G}})$ is the *smooth convolution algebra* of G. The localization result Theorem 2.5 applies and gives rise to an isomorphism

$$HH_\bullet(A) = \Gamma\left(X, \mathscr{H}\mathscr{H}_\bullet(\mathscr{A}_{\mathsf{G}})\right) .$$

The stalks of the sheaf $\hat{\mathscr{C}}_\bullet(\mathscr{A})$ of Hochschild chains have been identified in [**18**, Sec. 3]. More precisely, it has been shown in [**18**, Prop. 3.5] that for every orbit $x = \mathsf{G}p \in X$ through a point $p \in M$ there exists a quasi-isomorphism

$$L_{\bullet,x} : \hat{\mathscr{C}}\big(\mathcal{A}_{\mathsf{G}}\big)_{\bullet,x} \to \hat{\mathscr{C}}\big(\mathcal{A}_{G_p \ltimes N_p}\big)_{\bullet,0} , \tag{3.1}$$

where G_p denotes the isotropy group of G at p, N_p its normal space, and $\mathcal{A}_{G_p \ltimes N_p}$ the convolution sheaf of the action groupoid $G_p \ltimes N_p$. Brylinski conjectured in [**3**,**4**] that the Hochschild homology of the convolution algebra $\mathcal{A}_{G \ltimes M}$ of a compact Lie group action on M coincides with the so-called *basic relative forms* on the inertia space $\Lambda_0(G \ltimes M)$. The conjecture has been proved for S^1-actions in [**18**, Thm. 6.10], but even though the conjecture is believed to be correct a complete proof is still lacking. Together with the quasi-isomorphisms 3.1, verification of Brylinski's conjecture would allow to determine the Hochschild and possibly even the cyclic homology of proper groupoids.

3.3. Differentiable stratified spaces. Let $(X, \mathcal{C}^\infty)$ be a homologically regular differentiable stratified space and assume that X is endowed with a (minimal) Whitney stratification compatible with the differentiable structure $\mathcal{C}^\infty_X$ which means that for every stratum $S \subset X$ the canonical embedding $S \hookrightarrow X$ or in other words that $(S, \mathcal{C}^\infty_S) \hookrightarrow (X, \mathcal{C}^\infty_X)$ is an embedding of differentiable spaces. Examples of such spaces are real or complex algebraic varieties, real analytic varieties, semialgebraic or subanalytic sets, orbit spaces of compact Lie group actions and orbit spaces of proper Lie groupoids. Unless X is smooth, the Hochschild homology theory of the global section algebra $A = \mathcal{C}^\infty(X)$ is in general not known. Localization methods as those in Section 2 appear to be helpful for making progress in understanding the cyclic homology theory of such space and for obtaining at least a qualitative picture. By Theorem 2.3, the Hochschild homology of A can be computed by the homology of the diagonal complex $E_\bullet(A)$. The homology of $E_\bullet(A)$ coincides with the global section space of the homology sheaf $\mathscr{H}_\bullet\big(\mathscr{E}_\bullet(\mathcal{C}^\infty_X)\big)$. Denote by $X^\circ \subset X$ the subspace consisting of the union of all strata of depth 0. Then X° is open, dense and smooth, hence its Hochschild homology is known to be isomorphic to the graded space of differential forms on X° that is

$$HH_\bullet(\mathcal{C}^\infty(X^\circ)) \cong \mathscr{H}_\bullet\big(\mathscr{E}_\bullet(\mathcal{C}^\infty_{X^\circ})\big) \cong \Omega^\bullet(X^\circ) .$$

The question arises, how the Hochschild homology sheaf $\mathscr{H}\mathscr{H}_\bullet(\mathcal{C}^\infty_X) := \mathscr{H}_\bullet\big(\hat{\mathscr{C}}(\mathcal{C}^\infty_X)\big)$ looks over a stratum S in the singular locus X_{sing} of X. The following appears to be a reasonable conjecture.

CONJECTURE 3.1. *Assume to be given a homologically regular reduced differentiable stratified space $(X, \mathcal{C}^\infty)$, possibly from an allowable class like locally semialgebraic differentiable spaces or orbit spaces of proper Lie groupoids. Then the Hochschild homology sheaf $\mathscr{H}\mathscr{H}_\bullet(\mathcal{C}^\infty_X)|_S$ restricted to a stratum $S \subset X$ is locally free, and the strata are maximal subsets of X with that property.*

If the conjecture holds true, the stratification theory of a differentiable space from an allowable class could be detected by its Hochschild homology. In particular, Hochschild homology would serve as an invariant of singularity types. To the authors knowledge this observation has not been made in cyclic homology theory yet and could lead to significant progress in the understanding of singularities from the point of view of noncommutative geometry. Strong indication for that philosophy is the work by Michler [**12**], where it has been shown that for complex isolated singularities in quasi-homogeneous complex hypersurfaces the Milnor number is recovered by its cyclic homology. We have another related conjecture.

CONJECTURE 3.2. *The diagonal sheaf complex $\mathscr{E}_\bullet(\mathcal{C}^\infty_X)$ and the Hochschild homology sheaf $\mathscr{H}\mathscr{H}_\bullet(\mathcal{C}^\infty_X)$ associated to a homologically regular reduced differentiable space $(X, \mathcal{C}^\infty)$ are quasi-coherent.*

For complex analytic spaces, Palamodov has shown in [**15**] that the the corresponding analytic Hochschild homology sheaf is coherent which seems to substantiate this conjecture.

3.4. Localization in cyclic homology. Given a differentiable space $(X, \mathcal{O})$ with global section algebra $A = \mathcal{O}(X)$, Connes' boundary operator $B : C_k(A) \to C_{k+1}(A)$ does not commute with the localization maps $\Psi_{\bullet,t}$, $t > 0$. Hence, localization in Hochschild homology cannot be directly transferred to Connes' (b, B)-complex which determines cyclic homology. But the situation is not hopeless since Connes' spectral sequence might still degenerate. For example, in the case where A is the algebra $\mathcal{E}^\infty(X)$ of Whitney fields as in Section 3.1, the spectral sequence degenerates at the E^1-term which allows for the computation of the cyclic homology of $\mathcal{E}^\infty(X)$ as in Connes' original paper [**7**]. This approach has been pursued in [**2**], and the cyclic homology of Whitney fields could be identified with

$$HC_k\big(\mathcal{E}^\infty(X)\big) \cong \Omega^k_{\mathcal{E}^\infty}(X)/d\Omega^{k-1}_{\mathcal{E}^\infty}(X) \oplus H^{k-2}_{\mathrm{WdR}}(X) \oplus H^{k-4}_{\mathrm{WdR}}(X) \oplus \dots ,$$

where $H^k_{\mathrm{WdR}}(X)$ is the k-th *Whitney-de Rham cohomology group* of X that is the k-th cohomology group of the cochain complex $\big(\Omega^\bullet_{\mathcal{E}^\infty}(X), d\big) := \big(\mathcal{E}^\infty(X) \otimes \Lambda^\bullet \mathbb{R}^n, d\big)$. Note, however, that Connes' spectral sequence need not always degenerate. This has been pointed out e.g. in [**3**, Rem. 3.8] referring to an example by Block.

Another approach to localization in cyclic homology comes from the observation that the localization maps $\Psi_{\bullet,t}$ commute with the cyclic operator λ. The homotopy operators $H_{\bullet,t}$ do not have this property, though. Hence the homotopies $H_{\bullet,t}$ do not descend to Connes' cyclic chain complex, and it is not clear whether $\Psi_{\bullet,t}$ defines a quasi-isomorphism on the cyclic chain complex. To circumvent this obstacle it appears to be promising to consider localization within the cyclic bicomplex directly. Progress in this approach is expected to eventually lead to unravel the cyclic homology theory of differentiable spaces and Lie groupoids.

References

[1] Edward Bierstone, Pierre D. Milman, and Wiesław Pawłucki, *Composite differentiable functions*, Duke Math. J. **83** (1996), no. 3, 607–620, DOI 10.1215/S0012-7094-96-08318-0. MR1390657

[2] Jean-Paul Brasselet and Markus J. Pflaum, *On the homology of algebras of Whitney functions over subanalytic sets*, Ann. of Math. (2) **167** (2008), no. 1, 1–52, DOI 10.4007/annals.2008.167.1. MR2373151

[3] Jean-Luc Brylinski, *Algebras associated with group actions and their homology*, unpublished, Brown University preprint, 1987.

[4] Jean-Luc Brylinski, *Cyclic homology and equivariant theories* (English, with French summary), Ann. Inst. Fourier (Grenoble) **37** (1987), no. 4, 15–28. MR927388

[5] Jean-Luc Brylinski, *Central localization in Hochschild homology*, J. Pure Appl. Algebra **57** (1989), no. 1, 1–4, DOI 10.1016/0022-4049(89)90022-4. MR984041

[6] Dan Burghelea, *The cyclic homology of the group rings*, Comment. Math. Helv. **60** (1985), no. 3, 354–365, DOI 10.1007/BF02567420. MR814144

[7] Alain Connes, *Noncommutative differential geometry*, Inst. Hautes Études Sci. Publ. Math. **62** (1985), 257–360. MR823176

[8] Alain Connes and Henri Moscovici, *Cyclic cohomology, the Novikov conjecture and hyperbolic groups*, Topology **29** (1990), no. 3, 345–388, DOI 10.1016/0040-9383(90)90003-3. MR1066176

[9] Jean-Louis Loday, *Cyclic homology, a survey*, Geometric and algebraic topology, Banach Center Publ., vol. 18, PWN, Warsaw, 1986, pp. 281–303. MR925871

[10] B. Malgrange, *Ideals of differentiable functions*, Tata Institute of Fundamental Research Studies in Mathematics, vol. 3, Tata Institute of Fundamental Research, Bombay; Oxford University Press, London, 1967. MR0212575

[11] Ralf Meyer, *Analytic cyclic homology*, Ph.D. thesis, Universität Münster, 1999, `arXiv:9906205`.

[12] Ruth I. Michler, *Torsion of differentials of affine quasi-homogeneous hypersurfaces*, Rocky Mountain J. Math. **26** (1996), no. 1, 229–236, DOI 10.1216/rmjm/1181072113. MR1386162

[13] Juan A. Navarro González and Juan B. Sancho de Salas, *C^∞-differentiable spaces*, Lecture Notes in Mathematics, vol. 1824, Springer-Verlag, Berlin, 2003, DOI 10.1007/b13465. MR2030583

[14] Victor Nistor, *Cyclic cohomology of crossed products by algebraic groups*, Invent. Math. **112** (1993), no. 3, 615–638, DOI 10.1007/BF01232449. MR1218326

[15] Victor P. Palamodov, *Infinitesimal deformation quantization of complex analytic spaces*, Lett. Math. Phys. **79** (2007), no. 2, 131–142, DOI 10.1007/s11005-006-0139-6. MR2301392

[16] Markus J. Pflaum, *Analytic and geometric study of stratified spaces*, Lecture Notes in Mathematics, vol. 1768, Springer-Verlag, Berlin, 2001. MR1869601

[17] Markus J. Pflaum, Hessel Posthuma, and Xiang Tang, *Geometry of orbit spaces of proper Lie groupoids*, J. Reine Angew. Math. **694** (2014), 49–84, DOI 10.1515/crelle-2012-0092. MR3259039

[18] M.J. Pflaum, H.B. Posthuma, and X. Tang, *On the Hochschild homology of convolution algebras of proper Lie groupoids*, J. of Noncommutative Geometry, to appear, `arXiv:2009.03216` [math.KT], 2020.

[19] Joseph L. Taylor, *Homology and cohomology for topological algebras*, Advances in Math. **9** (1972), 137–182, DOI 10.1016/0001-8708(72)90016-3. MR328624

[20] Nicolae Teleman, *Microlocalisation de l'homologie de Hochschild* (French, with English and French summaries), C. R. Acad. Sci. Paris Sér. I Math. **326** (1998), no. 11, 1261–1264, DOI 10.1016/S0764-4442(98)80175-4. MR1649133

[21] Nicolae Teleman, *Localization of the Hochschild homology complex for fine algebras*, Proceedings of "BOLYAI 200" International Conference on Geometry and Topology, Cluj Univ. Press, Cluj-Napoca, 2003, pp. 169–184. MR2112623

[22] Jean-Claude Tougeron, *Idéaux de fonctions différentiables*, Ergebnisse der Mathematik und ihrer Grenzgebiete, Band 71, Springer-Verlag, Berlin-New York, 1972. MR0440598

[23] François Trèves, *Topological vector spaces, distributions and kernels*, Academic Press, New York-London, 1967.

[24] Charles Weibel, *Cyclic homology for schemes*, Proc. Amer. Math. Soc. **124** (1996), no. 6, 1655–1662, DOI 10.1090/S0002-9939-96-02913-9. MR1277141

[25] Charles A. Weibel, *An introduction to homological algebra*, Cambridge Studies in Advanced Mathematics, vol. 38, Cambridge University Press, Cambridge, 1994, DOI 10.1017/CBO9781139644136. MR1269324

[26] Charles A. Weibel and Susan C. Geller, *Étale descent for Hochschild and cyclic homology*, Comment. Math. Helv. **66** (1991), no. 3, 368–388, DOI 10.1007/BF02566656. MR1120653

Department of Mathematics, University of Colorado, Boulder Colorado 80309
Email address: `markus.pflaum@colorado.edu`

Proceedings of Symposia in Pure Mathematics
Volume **105**, 2023
https://doi.org/10.1090/pspum/105/01910

Cyclic Homology and Group Actions

Raphaël Ponge

Abstract. In this article we present the construction of explicit quasi-isomorphisms for crossed products associated with actions of discrete groups. Along the way we recover and clarify various earlier results (in the sense that we obtain explicit chain maps that yield quasi-isomorphisms).

In the framework of noncommutative geometry, cyclic homology [**10–12**] (see also [**46**]) is the relevant extension of de Rham cohomology to the noncommutative. Its dual version, cyclic cohomology, is the natural receptacle for the Chern character in K-homology (see [**12, 15**]). For instance, this is an essential ingredient in the proof of the Novikov conjecture for hyperbolic groups by Connes-Moscovici [**14**] and their local index formula in diffeomorphism-invariant geometry [**15–17**]. Moreover, the version of cyclic cohomology for Hopf algebras is the natural receptacle for the construction of characteristic classes in the noncommutative setting (see, e.g., [**16, 34, 35**]).

Given a group Γ acting on a manifold M, the quotient M/Γ need not be Hausdorff. In the framework of noncommutative geometry, the badly behaved quotient M/Γ is traded for the crossed-product algebra $C^\infty(M) \rtimes \Gamma$. Thus, the questions that arise are the computation of the cyclic homology of $C^\infty(M) \rtimes \Gamma$ and what geometric information can be extracted from it. There is a large amount of research on the cyclic homology of crossed-product algebras and groupoids (see, e.g., [**1, 2, 5, 7, 13, 18, 22, 24, 36–39**]). However, at the notable exception of the characteristic map of Connes [**13**] for the localization at identity, we don't have explicit quasi-isomorphisms.

In this article, we present the explicit construction of such quasi-isomorphisms. The results are expressed in terms of some version of equivariant (co)homology. On the way we simplify and clarify various approaches to the cyclic homology of crossed-product algebras. These results were announced in the notes [**40, 41**] and the proceedings paper [**42**]. There is some overlap between the material of this paper and the material of those articles. The full details should appear elsewhere.

The remainder of this paper is organized as follows. In Section 1, we review some basic facts on cyclic homology. In Section 2, we present our main results on the cyclic homology of $C^\infty(M) \rtimes \Gamma$, where M is a closed manifold acted on by

2020 *Mathematics Subject Classification.* Primary 19D55, 20J06, 55N91.
Key words and phrases. Cyclic homology, group homology, equivariant cohomology.
The research for this paper was partially supported by NFSC grant No. 11971328 (China).

a group Γ of diffeomorphisms. The next three sections are devoted to presenting various homological algebra tools that are needed in the rest of the paper. In Section 3, we review the main results of [**43**] on para S-modules. In Section 4, we present a version of the Eilenberg-Zilber theorem for biparacyclic modules via the construction of a paracyclic version of the Alexander-Whitney map. In Section 5, we present the basic facts on the triangular S-modules of [**40, 42**]. In Section 6, we present general constructions of quasi-isomorphisms for crossed-product algebras $\mathcal{A} \rtimes \Gamma$, where $\mathcal{A}$ is either an algebra over a ring $k \supset \mathbb{Q}$ or a locally convex algebras which is acted on by a (discrete) group Γ. In Section 7, we explain how these results enable us to get the main results of Section 2 in the case of group actions on manifolds.

1. Mixed Complexes and Cyclic Homology

Throughout this section we be k a ring with a unit.

DEFINITION 1.1 ([**9, 29**]). A mixed complex $C = (C_\bullet, b, B)$ is given by

- k-modules C_m, $m \geqslant 0$.
- k-module maps $b : C_\bullet \to C_{\bullet-1}$ and $B : C_\bullet \to C_{\bullet+1}$ such that

$$b^2 = B^2 = bB + Bb = 0.$$

EXAMPLE 1.2. Suppose that M is a closed manifold and $(\Omega^\bullet, d)$ is its de Rham complex of differential forms. Then $(\Omega^\bullet(M), 0, d)$ is a mixed complex.

DEFINITION 1.3. Let $C = (C_\bullet, b, B)$ be a mixed complex.

(1) The *cyclic homology* $\mathrm{HC}_\bullet(C)$ is the homology of the chain complex,

$$C^\natural = (C^\natural, b + Bu^{-1}), \qquad C^\natural_\bullet = k[u] \otimes C_\bullet, \quad \deg(u) = 2, \tag{1.1}$$

$$C^\natural_m = C_m u^0 \oplus C_{m-2} u \oplus \cdots. \tag{1.2}$$

(2) The *periodic cyclic homology* $\mathrm{HP}_\bullet(C)$ is the homology of the chain complex,

$$C^\sharp = (C^\sharp_\bullet, b + B), \qquad C^\sharp_i = \prod_{q \geqslant 0} C_{2q+i}, \; i = 0, 1.$$

REMARK 1.4. The operator $S := u^{-1}$ is called the *periodicity operator.*

DEFINITION 1.5 (Connes [**11**]). A cyclic k-module $C = (C_\bullet, d, s, t)$ is given by

- k-modules C_m, $m \geqslant 0$.
- k-module maps $d : C_\bullet \to C_{\bullet-1}$, $s : C_\bullet \to C_{\bullet+1}$ and $t : C_\bullet \to C_\bullet$ such that
  - C is a simplicial k-module with faces $d_j = t^j d t^{-(j+1)}$ and degeneracies $s_j = t^{j+1} s t^{-(j+1)}$ in degree m.
  - t is a cyclic operator, i.e., $t^{m+1} = 1$ on C_m.

PROPOSITION 1.6 (Connes [**11**]). *Any cyclic k-module $C = (C_\bullet, d, s, t)$ gives rise to a mixed complex $(C_\bullet, b, B)$, with*

$$b = \sum_{0 \leqslant j \leqslant m} (-1)^j d_j, \qquad B = (1 - \tau) s N,$$

where $\tau = (-1)^m t$ and $N = 1 + \tau + \cdots + \tau^m$ on C_m.

DEFINITION 1.7 (Connes [**11, 12**]). Let $\mathcal{A}$ be a unital k-algebra.

(1) The *cyclic k-module* of $\mathcal{A}$ is

$$C(\mathcal{A}) = (C_\bullet(\mathcal{A}), d, s, t), \qquad C_m(\mathcal{A}) = \mathcal{A}^{\otimes(m+1)}, \tag{1.3}$$

$$d(a^0 \otimes \cdots \otimes a^m) = (a^m a^0) \otimes a^1 \otimes \cdots \otimes a^{m-1}, \tag{1.4}$$

$$s(a^0 \otimes \cdots \otimes a^m) = 1 \otimes a^0 \otimes \cdots \otimes a^m, \tag{1.5}$$

$$t(a^0 \otimes \cdots \otimes a^m) = a^m \otimes a^0 \otimes \cdots \otimes a^{m-1}. \tag{1.6}$$

(2) The *cyclic homology* and the *periodic cyclic homology* of $\mathcal{A}$, denoted $\mathrm{HC}_\bullet(\mathcal{A})$ and $\mathrm{HP}_\bullet(\mathcal{A})$, respectively, are the cyclic homology and periodic cyclic homology of $C(\mathcal{A})$.

REMARK 1.8. The ordinary chain complex $(C_\bullet(\mathcal{A}), b)$ is the Hochschild complex of $\mathcal{A}$.

DEFINITION 1.9 (Connes [**11, 12**]). Suppose that $\mathcal{A}$ is a unital locally convex $\mathbb{C}$-algebra.

(1) The *cyclic space* of $\mathcal{A}$ is

$$C(\mathcal{A}) = (C_\bullet(\mathcal{A}), d, s, t), \qquad C_m(\mathcal{A}) = \mathcal{A}^{\hat{\otimes}_\pi (m+1)},$$

where $\hat{\otimes}_\pi$ is the projective topological tensor product and the operators (d, s, t) are defined as in (1.4)–(1.6).

(2) The *cyclic homology* and the *periodic cyclic homology* of $\mathcal{A}$, denoted $\mathrm{HC}_\bullet(\mathcal{A})$ and $\mathrm{HP}_\bullet(\mathcal{A})$, respectively, are the cyclic homology and periodic cyclic homology of $C(\mathcal{A})$.

REMARK 1.10. If $\mathcal{A}$ is a locally convex algebra, we have two definitions for its cyclic space depending on whether we use the algebraic tensor product or the projective tensor product to define the spaces of chains. Throughout this paper, in the case of a locally convex algebra, we shall always use the the projective tensor product to define the spaces of chains $C_m(\mathcal{A})$. Therefore, in this case, the cyclic space $C(\mathcal{A})$ and the cyclic homologies $\mathrm{HC}_\bullet(\mathcal{A})$ and $\mathrm{HP}_\bullet(\mathcal{A})$ will always be meant in the sense of Definition 1.9.

EXAMPLE 1.11 (Connes-Hochschild-Rosenberg Theorem [**12**]). Let M be a closed manifold, and denote by $\mathcal{A}$ the algebra $C^\infty(M)$ of smooth functions on M equipped with its usual locally convex topology. The Hochschild-Kostant-Rosenberg map $\alpha : C_\bullet(\mathcal{A}) \to \Omega^\bullet(M)$ is defined by

$$\alpha\left(f^0 \otimes \cdots \otimes f^m\right) = \frac{1}{m!} f^0 df^1 \wedge \cdots \wedge df^m, \qquad f^j \in C^\infty(M).$$

This is a map of mixed complexes from the mixed complex $(C_\bullet(\mathcal{A}), b, B)$ of $\mathcal{A}$ to the "de Rham mixed complex" $(\Omega^\bullet(M), 0, d)$. It was proved by Connes [**12**] this is a quasi-isomorphism. Thus,

$$\mathrm{HC}_m(\mathcal{A}) = (\ker d \cap \Omega^m(M)) \oplus H^{m-2}(M, \mathbb{C}) \oplus \cdots,$$
$$\mathrm{HP}_i(\mathcal{A}) = \bigoplus_{p \geqslant 0} H^{2p+i}(M, \mathbb{C}), \qquad i = 0, 1.$$

Here $H^\bullet(M, \mathbb{C})$ is the de Rham cohomology of M with coefficients in $\mathbb{C}$.

2. The Cyclic Homology of $C^\infty(M) \rtimes \Gamma$

Throughout this section we let M be a closed manifold and Γ a group acting on M by diffeomorphisms. We denote by $\mathcal{A}_\Gamma$ the crossed-product algebra $C^\infty(M) \rtimes \Gamma$. It is generated by elements fu_ϕ with $f \in C^\infty(M)$ and $\phi \in \Gamma$ subject to the relations,

$$u_\phi^{-1} = u_{\phi^{-1}}, \qquad f^0 u_{\phi_0} f^1 u_{\phi_1} = f^0(f^1 \circ \phi) u_{\phi_0\phi_1}.$$

We equip $\mathcal{A}_\Gamma$ with a topology as follows:

- Given any finite subset $S \subset \Gamma$, set $\mathcal{A}_S := \{au_\phi;\ \phi \in S\}$. As a vector space this is the direct sum of $|S|$-copies of $C^\infty(M)$. We equip it with the corresponding direct-sum topology.
- The topology of $\mathcal{A}_\Gamma$ then is the weakest locally convex topology with respect to which the inclusion $\mathcal{A}_S \hookrightarrow \mathcal{A}_\Gamma$ is continuous for every finite subset $S \subset \Gamma$.

With respect to this topology $\mathcal{A}_\Gamma$ is a locally convex algebra.

2.1. Splitting along conjugacy classes. In what follows, given any $\phi \in \Gamma$ we denote by $[\phi]$ its conjugacy class in Γ.

DEFINITION 2.1. $C(\mathcal{A}_\Gamma)_{[\phi]}$ is the cyclic subspace of $C(\mathcal{A}_\Gamma)$ whose space of degree m is generated by tensor products of the form,

$$f^0 u_{\phi^0} \otimes \cdots \otimes f^m u_{\phi^m}, \qquad f^j \in C^\infty(M), \quad \phi_0 \cdots \phi_m \in [\phi].$$

PROPOSITION 2.2. *We have a direct sum of cyclic spaces,*

$$C(\mathcal{A}_\Gamma) = \bigoplus_{[\phi]} C(\mathcal{A}_\Gamma)_{[\phi]},$$

where the summation runs over all conjugacy classes.

In other words, computing the cyclic homology of $\mathcal{A}_\Gamma$ reduces to the computation the cyclic homology of the delocalized cyclic modules $C(\mathcal{A}_\Gamma)_{[\phi]}$.

2.2. Mixed equivariant homology. In what follows, given a commutative ring k with unit and $k\Gamma$-modules $\mathscr{M}$ and $\mathscr{M}'$, we denote by $\mathscr{M}' \otimes_\Gamma \mathscr{M}$ the quotient of $\mathscr{M} \otimes_k \mathscr{M}$ by the diagonal action of Γ.

DEFINITION 2.3. Let $\mathscr{M}$ be a (left) $k\Gamma$-module.

(1) The *group homology* $H_\bullet(\Gamma, \mathscr{M})$ is the homology of the chain complex,

$$(C_\bullet(\Gamma, \mathscr{M}), \partial \otimes 1), \qquad C_m(\Gamma, \mathscr{M}) = k\Gamma^{m+1} \otimes_\Gamma \mathscr{M}, \tag{2.1}$$

$$\partial(\psi_0, \ldots, \psi_m) = \sum_{0 \leqslant j \leqslant m} (-1)^j (\psi_0, \ldots, \hat{\psi}_j, \ldots, \psi_m), \quad \psi_j \in \Gamma. \tag{2.2}$$

(2) The *group cohomology* $H^\bullet(\Gamma, \mathscr{M})$ is the homology of the cochain complex,

$$(C^\bullet(\Gamma, \mathscr{M}), \partial), \qquad C^m(\Gamma, \mathscr{M}) = \{\Gamma\text{-equivariant maps } u : \Gamma^{m+1} \to \mathscr{M}\},$$

$$\partial u(\psi_0, \ldots, \psi_{m+1}) = \sum_{0 \leqslant j \leqslant m} (-1)^j u(\psi_0, \ldots, \hat{\psi}_j, \ldots, \psi_{m+1}), \quad \psi_j \in \Gamma,\ u \in C^m(\Gamma, \mathscr{M}).$$

REMARK 2.4. For $\mathscr{M} = k = \mathbb{Z}$ we recover the homology and cohomology of the *classifying space* $B\Gamma$.

Suppose that $k = \mathbb{C}$. Borel's model for equivariant cohomology is defined as follows.

DEFINITION 2.5 (Borel). The *equivariant cohomology* $H^\bullet_\Gamma(M)$ of the Γ-manifold M is the cohomology of the cochain complex,

$$(\mathrm{Tot}^\bullet(C(\Gamma,\Omega(M))),\partial+(-1)^p d),\quad \mathrm{Tot}^m(C(\Gamma,\Omega(M))):=\bigoplus_{p+q=m} C^p(\Gamma,\Omega^q(M)).$$

REMARK 2.6. The equivariant cohomology $H^\bullet_\Gamma(M)$ is isomorphic to the cohomology of the homotopy quotient $E\Gamma\times_\Gamma M=(E\Gamma\times M)/\Gamma$, where $E\Gamma$ is the universal contractible Γ-bundle.

DEFINITION 2.7 (Borel). The *even/odd equivariant cohomology* $H^{\mathrm{ev/odd}}_\Gamma(M)$ is the cohomology of the cochain complex,

$$\Big(\bigoplus_{p+q\ \mathrm{even/odd}} C^p(\Gamma,\Omega^q(M)),\partial+(-1)^p d\Big).$$

The even/odd equivariant cohomology is the natural receptacle for the construction of equivariant characteristic classes (see, e.g., [**4, 20**]). Let E be a Γ-equivariant vector bundle over M. As explained in [**4**] (see also [**23**]), given any connection ∇^E on E, the equivariant Chern character $\mathrm{Ch}_\Gamma(E)\in H^{\mathrm{ev}}_\Gamma(M)$ is represented by the equivariant cocycle $\mathrm{Ch}_\Gamma(\nabla^E)=(\mathrm{Ch}^p_\Gamma(\nabla^E))_{p\geqslant 0}$ in $C^{\mathrm{ev}}_\Gamma(M)$, whose components $\mathrm{Ch}^p_\Gamma(\nabla^E)\in C^p(\Gamma,\Omega^\bullet(M))$ are given by

$$\mathrm{Ch}^0_\Gamma(\nabla^E)(\phi_0)=\mathrm{Ch}\left((\phi_0)_*\nabla^E\right), \tag{2.3}$$

$$\mathrm{Ch}^p_\Gamma(\nabla^E)(\phi_0,\ldots,\phi_p)=(-1)^p\,\mathrm{CS}\left((\phi_0)_*\nabla^E,\ldots,(\phi_p)_*\nabla^E\right),\quad \phi_j\in\Gamma. \tag{2.4}$$

Here Ch is the (total) Chern form of the connection $(\phi_0)_*\nabla^E$ and $\mathrm{CS}((\psi_0)_*\nabla^E,\ldots,(\psi_p)_*\nabla^E)$ is the Chern-Simons form of the connections $(\psi_0)_*\nabla^E,\ldots,(\psi_p)_*\nabla^E$ as defined in [**23**].

For our purpose we need to form the tensor product of the chain complex $C_\bullet(\Gamma,\mathbb{C})$ with the de Rham complex $(\Omega^\bullet(M),d)$. Although, it is not usual to form the tensor product of a chain complex and a cochain complex, we still can regard both of them as mixed complexes and then take the tensor product of these mixed complexes. This leads to the following definition.

DEFINITION 2.8. The *equivariant mixed complex* $C^\Gamma(M)^\natural$ is the mixed complex,

$$(\mathrm{Tot}_\bullet(C(\Gamma,\Omega(M))),\partial,(-1)^p d),\quad \mathrm{Tot}_m(C(\Gamma,\Omega(M))):=\bigoplus_{p+q=m} C_p(\Gamma,\Omega^q(M)).$$

The cyclic homology of $C^\Gamma(M)^\natural$ is called the *mixed equivariant homology* of the Γ-manifold M and is denoted $H^\Gamma_\bullet(M)^\natural$. Its periodic periodic cyclic homology is called *even/odd mixed equivariant homology*) and is denoted $H^\Gamma_\bullet(M)^\natural$.

The mixed equivariant homology is a natural receptacle for the cap product between group homology and equivariant cohomology. Namely, we have the following result.

PROPOSITION 2.9. *We have a cap product,*

$$-\frown-: H_\bullet(\Gamma,\mathbb{C})\times H^{ev/odd}_\Gamma(M)\longrightarrow H^\Gamma_{ev/odd}(M)^\natural.$$

2.3. The cyclic homology of $C(\mathcal{A}_\Gamma)_{[\phi]}$ (finite order case). In what follows, given any $\phi\in\Gamma$, we denote by Γ_ϕ its centralizer in Γ and by M^ϕ its fixed-point set in M. The action of Γ on M descends to an action of Γ_ϕ on M^ϕ.

If ϕ has finite order, then M^{ϕ} is a disjoint union of submanifolds of M that are acted on by Γ_{ϕ}. This allows to define a Γ_{ϕ}-equivariant complex for M^{ϕ} as the direct-sum of the equivariant complexes of its submanifold components.

THEOREM 2.10. *Assume that $\phi \in \Gamma$ has finite order. Then the following hold:*

(1) *There are explicit quasi-isomorphisms,*

$$\eta^{\phi} : H_{\bullet}^{\Gamma_{\phi}}(M^{\phi})^{\natural} \xrightarrow{\sim} \mathrm{HC}_{\bullet}(\mathcal{A}_{\Gamma})_{[\phi]}, \tag{2.5}$$

$$\eta^{\phi} : H_{ev/odd}^{\Gamma_{\phi}}(M^{\phi})^{\natural} \xrightarrow{\sim} \mathrm{HP}_{\bullet}(\mathcal{A}_{\Gamma})_{[\phi]}. \tag{2.6}$$

(2) *We have an explicit bilinear map,*

$$\eta^{\phi} \circ (- \frown -) : H_{ev/odd}(\Gamma_{\phi}, \mathbb{C}) \times H_{\Gamma_{\phi}}^{ev/odd}(M^{\phi}) \longrightarrow \mathrm{HP}_{\bullet}(\mathcal{A}_{\Gamma})_{[\phi]}.$$

REMARK 2.11. By using Poincaré duality we can recover the description in terms of equivariant homology of $\mathrm{HC}_{\bullet}(\mathcal{A}_{\Gamma})_{[\phi]}$ and $\mathrm{HP}_{\bullet}(\mathcal{A}_{\Gamma})_{[\phi]}$ by Brylinski-Nistor [**7**], as well the description of $\mathrm{HP}_{\bullet}(\mathcal{A}_{\Gamma})_{[\phi]}$ for $\phi = 1$ by Connes [**13**]. Connes actually exhibited an explicit quasi-isomorphism for $\mathrm{HP}_{\bullet}(\mathcal{A}_{\Gamma})_{[1]}$.

REMARK 2.12. In case Γ_{ϕ} is finite we recover the description of $\mathrm{HC}_{\bullet}(\mathcal{A}_{\Gamma})$ and $\mathrm{HP}_{\bullet}(\mathcal{A}_{\Gamma})$ by Baum-Connes [**2**]. Namely,

$$\mathrm{HC}_m(\mathcal{A}_{\Gamma})_{[\phi]} \simeq (\ker d \cap \Omega^m(M^{\phi})^{\Gamma_{\phi}}) \oplus H^{m-2}(M^{\phi})^{\Gamma_{\phi}} \oplus \cdots,$$
$$\mathrm{HP}_{\bullet}(\mathcal{A}_{\Gamma})_{[\phi]} \simeq H^{\mathrm{ev/odd}}(M^{\phi})^{\Gamma_{\phi}}.$$

Here $H^{\bullet}(M^{\phi})^{\Gamma_{\phi}}$ is the Γ_{ϕ}-invariant cohomology of M^{ϕ}.

REMARK 2.13. In case $M = \{\mathrm{pt}\}$ the crossed product algebra $\mathcal{A}_{\Gamma}$ is just the group ring $\mathbb{C}\Gamma$, and the mixed equivariant complex at $[\phi]$ is just the group homology complex $(C_{\bullet}(\Gamma_{\phi}), \partial, 0)$ seen as a mixed complex. Thus, in this case we recover the results of Burghelea [**8**] (see also [**33**]) on the cyclic homology of $C(\mathbb{C}\Gamma)_{[\phi]}$. Namely,

$$\mathrm{HC}_m(\mathbb{C}\Gamma)_{[\phi]} \simeq H_m(\Gamma_{\phi}, \mathbb{C}) \oplus H_{m-2}(\Gamma_{\phi}, \mathbb{C}) \oplus \cdots . \tag{2.7}$$

Let us briefly explain how the quasi-isomorphisms (2.5)–(2.6) enable us to get extract geometric invariants. Let $(\Omega_{\bullet}^{\Gamma}(M), d)$ be the cochain complex of equivariant de Rham currents. Here $\Omega^{\Gamma}(M)$ consists of Γ-equivariant maps $C : \Gamma \to \Omega_m(M)$, where $\Omega_m(M)$ is the space of m-dimensional de Rhan currents. Any $C \in \Omega^{\Gamma}(M)$ defines a cochain $\varphi_C \in C^m(\mathcal{A}_{\Gamma})$ by

$$\varphi_C(f^0 u_{\phi_0}, \ldots, f^m u_{\phi_m}) = \frac{1}{m!}\langle C(\phi), f^0 d\hat{f}^1 \wedge \cdots \wedge d\hat{f}^m \rangle, \qquad f^j \in \mathcal{A},\ \phi_j \in \Gamma,$$

where we have set $\phi = \phi_0 \cdots \phi_m$ and $\hat{f}^j = f^j \circ (\phi_0 \cdots \phi_{j-1})^{-1}$. This provides us with a map of mixed complexes from $(\Omega_{\bullet}^{\Gamma}(M), d, 0)$ to $(C^{\bullet}(\mathcal{A}_{\Gamma}), B, b)$. The transverse fundamental class of Connes [**13**] and the CM cocycle of an equivariant Dirac spectral triple [**45**] are instances of cocycles arising from equivariant currents.

Let $\phi \in \Gamma$ have finite order, and let $C \in \Omega_{\mathrm{ev/odd}}^{\Gamma}(M)$ be a closed even/odd equivariant current. If $\omega = (\omega_{p,q})$, $\omega_{p,q} \in C_p(\Gamma^{\phi}, \Omega^q(M))$, $p+q$ even/odd, is an even/odd equivariant cycle, then

$$\langle \varphi_C, \eta^{\phi}(\omega) \rangle = \langle C(\phi), \tilde{\omega}_{0,\bullet} \rangle,$$

where $\tilde{\omega}_{0,\bullet} \in \Omega^{\mathrm{ev/odd}}(M)$ is such that $(\tilde{\omega}_{0,\bullet})_{|M^{\phi}} = \omega_{0,\bullet}$.

Let E be a Γ_ϕ-equivariant vector bundle over M^ϕ. Given any connection ∇^E on E, the equivariant Chern character of E is represented by the equivariant cocycle $\mathrm{Ch}_{\Gamma_\phi}(\nabla^E) \in C^{\mathrm{ev}}_{\Gamma_\phi}(M^\phi_a)$ given by (2.3)–(2.4). If $\operatorname{supp} C(\phi) \subset M^\phi$, then for any cycle $\xi \in C_p(\Gamma_\phi, \mathbb{C})$, we have

$$\langle \varphi_C, \eta^\phi(\mathrm{Ch}_\Gamma(\nabla^E) \frown \xi)\rangle = \langle C(\phi), (\mathrm{Ch}_\Gamma(\nabla^E) \frown \xi)_0\rangle.$$

If we write $\xi = \sum_\ell \lambda_\ell(\psi_0^\ell, \ldots, \psi_p^\ell) \otimes_{\Gamma_\phi} 1$, $\psi_j^\ell \in \Gamma_\phi$, $\lambda_\ell \in \mathbb{C}$, then

$$\left(\mathrm{Ch}_\Gamma(\nabla^E) \frown \xi\right)_0 = \sum_\ell (-1)^p \lambda_\ell \,\mathrm{CS}\left((\psi_0^\ell)_* \nabla^E, \ldots, (\psi_p^\ell)_* \nabla^E\right).$$

In particular, for the generator $1 := 1 \otimes_{\Gamma_\phi} 1$ of $C_0(\Gamma_\phi, \mathbb{C})$, we get

$$\langle \varphi_C, \eta^\phi(\mathrm{Ch}_\Gamma(\nabla^E) \frown 1)\rangle = \langle C(\phi), \mathrm{Ch}(\nabla^E)\rangle.$$

2.4. The cyclic homology of $C(\mathcal{A}_\Gamma)_{[\phi]}$ (infinite order case). Throughout this subsection we let $\phi \in \Gamma$ have infinite order. That is, the subgroup $\langle\phi\rangle$ it generates is isomorphic to $\mathbb{Z}$. Thus, if we let $\overline{\Gamma}_\phi = \Gamma_\phi/\langle\phi\rangle$ be the normalizer of ϕ, then we have a group extension,

$$1 \longrightarrow \mathbb{Z} \longrightarrow \Gamma_\phi \longrightarrow \overline{\Gamma}_\phi \longrightarrow 1.$$

This group extension uniquely defines a group cohomology class $e_\phi \in H^2(\overline{\Gamma}_\phi, \mathbb{Z})$, which is called the *Euler class*.

DEFINITION 2.14. Let $u_\phi \in C^2(\overline{\Gamma}_\phi, \mathbb{Z})$ be any 2-cocycle representing e_ϕ. Then $H^\sigma_\bullet(\overline{\Gamma}_\phi, M^\phi)$ is the homology of the chain complex,

$$\left(C^\sigma_\bullet(\overline{\Gamma}_\phi, M^\phi), \partial + (-1)^p(u_\phi \frown -)d\right), \quad \text{where} \tag{2.8}$$

$$C^\sigma_m(\overline{\Gamma}_\phi, M^\phi) := \bigoplus_{p+q=m} \mathbb{C}\overline{\Gamma}_\phi^{p+1} \otimes_{\overline{\Gamma}_\phi} \Omega^q(M^\phi). \tag{2.9}$$

THEOREM 2.15. *Assume that $\phi \in \Gamma$ acts cleanly on M (see Definition 7.1).*

(1) *We have an explicit quasi-isomorphism,*

$$H^\sigma_\bullet(\overline{\Gamma}_\phi, M^\phi) \simeq \mathrm{HC}_\bullet(\mathcal{A}_\Gamma)_{[\phi]}. \tag{2.10}$$

(2) *Under this quasi-isomorphism, the S-operator agrees with the cap product,*

$$e_\phi \frown - : H^\sigma_\bullet(\overline{\Gamma}_\phi, M^\phi) \longrightarrow H^\sigma_{\bullet-2}(\overline{\Gamma}_\phi, M^\phi).$$

REMARK 2.16. The action of ϕ is clean whenever ϕ preserves a metric or an affine connection.

REMARK 2.17. The isomorphism (2.10) clarifies the misidentification of $\mathrm{HC}_\bullet(\mathcal{A}_\Gamma)_{[\phi]}$ in [**18**] and provides an explicit quasi-isomorphism at the level of chains.

REMARK 2.18. In case $M = \{\mathrm{pt}\}$ we recover the identification of $\mathrm{HC}(\mathbb{C}\Gamma)_{[\phi]}$ by Burgelea [**8**],

$$\mathrm{HC}_\bullet(\mathbb{C}\Gamma)_{[\phi]} \simeq H_\bullet(\overline{\Gamma}_\phi, \mathbb{C}). \tag{2.11}$$

In general, we have the following result.

THEOREM 2.19. *The following hold.*

(1) $\mathrm{HC}_\bullet(\mathcal{A}_\Gamma)_{[\phi]}$ *is a module over* $H_\bullet(\overline{\Gamma}_\phi, \mathbb{C})$.
(2) *The S-operator is given by the action of the Euler class e_ϕ.*
(3) $\mathrm{HP}_\bullet(\mathcal{A}_\Gamma)_{[\phi]} = 0$ *whenever e_ϕ is nilpotent in $H^\bullet(\overline{\Gamma}_\phi, \mathbb{C})$.*

REMARK 2.20. The existence of a $H_\bullet(\overline{\Gamma}_\phi, \mathbb{C})$-module structure on $\mathrm{HC}_\bullet(\mathcal{A}_\Gamma)_{[\phi]}$ and the identification of the periodicity operator with the action of the Euler class e_ϕ is originally due to Nistor [**37**]. Nistor's arguments rely on deep homological arguments. Here the $H_\bullet(\overline{\Gamma}_\phi, \mathbb{C})$-module structure arises from an explicit construction of an action at the level of chains (see Section 6).

REMARK 2.21. The nilpotency of e_ϕ is closely related to the Bass and idempotent conjectures (see, e.g., [**21, 27**]). In particular, e_ϕ is rationally nilpotent for every infinite order element $\phi \in \Gamma$ if Γ is in any of the following classes of groups: free products of Abelian groups, hyperbolic groups, and arithmetic groups.

3. Periodicity and Cyclic Homology

From now on we let k be a unital ring. By a k-module we shall always means a *left* k-module.

3.1. S-modules. Following Jones-Kassel [**28, 30**] an *S-module* is any data $(C_\bullet, d, S)$, where $(C_\bullet, d)$ is a chain complex of k-modules and $S : C_\bullet \to C_{\bullet-2}$ is a chain map, i.e., $[S, d] = 0$. This notion, which has its roots in the note of Connes [**11**], naturally comes into play in bivariant cyclic homology [**19, 28**]. Moreover, it encapsulates various approaches to cyclic homology:

- Connes' cyclic complex $(C^\lambda_\bullet, b, S)$ of a cyclic k-module, where $C^\lambda_\bullet = C_\bullet/\operatorname{ran}(1-t)$ with $\tau = (-1)^m t$ and S is Connes' periodicity operator (see [**10–12**]).
- The cyclic complex $(C^\natural_\bullet, b + Bu^{-1}, u^{-1})$ of a mixed complex $(C_\bullet, b, B)$.
- The total complex of the CC-bicomplex of a cyclic k-module with the operators $(b, -b', 1-\tau, N)$ of Connes [**11**, Section 4] and Tsygan [**46**, Proposition 1].

Any S-module $C = (C_\bullet, d, S)$ gives rise to a $\mathbb{Z}_2$-graded chain complex $(C^\sharp_\bullet, d)$, called *periodic complex*, where

(3.1)
$$C^\sharp_i = \varprojlim_S C_{2q+i} = \left\{(x_{2q+i})_{q\geqslant 0};\ x_{2q+i} \in C_{2q+i},\ Sx_{2q+i+2} = x_{2q+i}\right\}, \quad i = 0, 1.$$

An S-map $f : C_\bullet \to C'_\bullet$ between S-modules, is any chain map that is compatible with the S-operators, i.e., $fS = Sf$. Any S-map extends to a chain map between the corresponding periodic complexes. If in addition this is a quasi-isomorphism, then we also have a quasi-isomorphism between the corresponding periodic complexes.

Recall that two chain maps $f, g : C_\bullet \to C'_\bullet$ are chain homotopic if there is a k-module map $\varphi : C_\bullet \to C'_{\bullet+1}$ such that $f - g = d\varphi + \varphi d$. We have chain homotopy equivalence $C_\bullet \underset{g}{\overset{f}{\rightleftarrows}} C'_\bullet$ if f and g are chain maps such that fg and gf are chain homotopic to the identity maps. This is a deformation retract if we further have $fg = 1$. If $C_\bullet$ and $C'_\bullet$ are S-modules and the chain homotopies are compatible with the S-operators, then we further get a chain homotopy equivalence at the level of periodic chains.

3.2. Paracyclic. A *paracyclic k-module* $(C_\bullet, d, s, t)$ satisfies the same axioms as those of a cyclic k-module, except that the cyclic condition $t^{m+1} = 1$ is relaxed into requiring $t = ds$ to be invertible. The paracyclic category was formally introduced by Getzler-Jones [**24**], but there were instances of paracyclic k-modules already in the work of Feigin-Tsygan [**22**].

The para version of mixed complexes is provided by the *parachain complexes* of Getzler-Jones [**24**]. Namely, a parachain complex $(C_\bullet, b, B)$ is like a mixed complex except that the anticommutation relation $bB + Bb = 0$ is relaxed into $bB + Bb = 1 - T$, where $T : C_\bullet \to C_\bullet$ is a k-module isomorphism. Note that T automatically commutes with the operators (b, B), and so this is a parachain complex automorphism.

3.3. Para S-modules. For our purpose we need a para version of S-modules. This is provided by the para-S-modules defined below.

DEFINITION 3.1 ([**43**]). A *para S-module* A *para-S-module* is any data $(C_\bullet, d, S, T)$, where C_m, $m \geqslant 0$ are k-modules, $d : C_\bullet \to C_{\bullet-1}$ and $S : C_\bullet \to C_{\bullet-2}$ are k-module maps and $T : C_\bullet \to C_\bullet$ is a k-module isomorphisms such that

$$d^2 = S(1-T) \qquad \text{and} \qquad [d, S] = [d, T] = [S, T] = 0.$$

In the same way as mixed complexes gives rise to an S-module, any parachain complex $(C_\bullet, b, B)$ gives rise to a para S-module $(C_\bullet^\natural, b + Bu^{-1}, S, T)$, where $C_\bullet^\natural$ and $S = u^{-1}$ are defined as in (1.1)–(1.2) and $T := 1 - (bB + Bb)$. We refer to [**43**] for further examples of para S-modules.

The drawback of the condition $d^2 = S(1-T)$ is that $(C_\bullet, d)$ is not a chain complex anymore, and so we don't have have a direct way to define quasi-isomorphisms for para S-modules. However, we still can make sense of S-maps and chain homotopies much like with S-modules.

An S-map $f : C_\bullet \to C'_\bullet$ is a k-module map which is compatible with the operators (d, S, T). When $T = 1$ we recover the usual notion of S-map between cyclic complexes of mixed complexes in the sense of Kassel [**29**]. We may also define S-homotopies between S-maps as being implemented by k-module maps $\varphi : C_\bullet \to C'_{\bullet+1}$ that are compatible with the (S, T)-operators. This allows us speak about S-homotopy equivalences and S-deformation retracts of para S-modules.

At the periodic level, we still can form the k-modules $C_i^\sharp$, $i = 0, 1$, as in (3.1). The operators (d, T) are compatible with the S-operators, and so they induce operators on $C_\bullet^\sharp$. What we get is a *quasi-complex* $(C_\bullet^\sharp, d, T)$ in the sense of [**47**]. Any S-homotopy equivalence of para S-modules induces a T-compatible homotopy equivalence at the level of periodic chains (see [**43**]).

3.4. Perturbation lemma. If $(C_\bullet, b, B)$ and $(\bar{C}_\bullet, b, B)$ are parachain complexes, then any S-map $f : C_\bullet^\natural \to \bar{C}_\bullet^\natural$ takes the form,

$$f = f^{(0)} + f^{(1)}u^{-1} + f^{(2)}u^{-2} + \cdots,$$

where$f^{(j)} : C_\bullet \to \bar{C}_{\bullet+2j}$, $j \geqslant 0$, are T-compatible k-module maps such that

$$bf^{(0)} = f^{(0)}b, \qquad bf^{(j)} + Bf^{(j-1)} = f^{(j)}b + f^{(j-1)}B, \quad j \geqslant 1. \tag{3.2}$$

In particular, $f^{(0)} : C_\bullet \to \bar{C}_\bullet$ is a chain map with respect to the b-differential. Note that any map of parachain complexes $f : C_\bullet \to \bar{C}_\bullet$ is an S-map with $f^{(0)} = f$ and $f^{(j)} = 0$ for $j \geqslant 1$.

Given a deformation retract $(C_\bullet, b) \xrightarrow{f} (\bar{C}_\bullet, b) \xrightarrow{g} (C_\bullet, b)$, it is natural to ask if we could find an S-deformation retract $C_\bullet^\natural \xrightarrow{f^\natural} \bar{C}_\bullet^\natural \xrightarrow{g^\natural} C_\bullet^\natural$ such that $(f^\natural)^{(0)} = f$ and $(g^\natural)^{(0)} = g$. In this case we say that $f^\natural$ and $g^\natural$ are *coextensions* of f and g, respectively. The answer to this is provided by the following perturbation lemma.

LEMMA 3.2 ([**43**]). *Let* $(C_\bullet, b) \xrightarrow{f} (\bar{C}_\bullet, b) \xrightarrow{g} (C_\bullet, b)$ *a* T*-compatible deformation retract. Then there is a* k*-module map* $\tilde{B} : \bar{C}_\bullet \to \bar{C}_{\bullet+1}$ *such that*

(i) $\tilde{C} := (\bar{C}_\bullet, b, \tilde{B})$ *is a parachain complex .*
(ii) *We have an* S*-deformation retract* $C^\natural_\bullet \xrightarrow{f^\natural} \bar{C}^\natural_\bullet \xrightarrow{g^\natural} C^\natural_\bullet$*, where* $f^\natural$ *and* $g^\natural$ *are coextensions of* f *and* g*, respectively.*
(iii) *If* f *is a map of parachain complexes, then* $\tilde{B} = B$ *and* $\alpha^\natural = \alpha$*.*

REMARK 3.3. For mixed complexes Lemma 3.2 is obtained by Kassel [**30**] as an application of the basic perturbation lemma in homological algebra (see, e.g., [**6**]). A para version of the basic perturbation lemma enables us to get Lemma 3.2 (see [**43**]).

REMARK 3.4. Kassel [**30**] further used the basic perturbation lemma for mixed complexes to compare the various cyclic complexes associated with cyclic modules and their normalizations. We refer to [**43**] for analogues of those results for parachain complexes.

4. Paracyclic Eilenberg-Zilberg Theorem

4.1. Biparacyclic. The following notions were introduced by Getzler-Jones [**24**].

A *biparacyclic* k-module is any data $(C_{\bullet,\bullet}, \bar{d}, \bar{s}, \bar{t}, d, s, t)$, where $C_{p,q}$, $p, q \geqslant 0$, are k-modules and $(C_{\bullet,q}, \bar{d}, \bar{s}, \bar{t})$ and $(C_{p,\bullet}, d, s, t)$ are paracyclic k-modules for all $p, q \geqslant 0$ in such a way that the horizontal operators $(\bar{d}, \bar{s}, \bar{t})$ commute with each of the vertical operators (d, s, t). It is called a *cylindrical* k*-module* if we further have the relation,

$$\bar{t}^{p+1} t^{q+1} = 1 \qquad \text{on } C_{p,q}. \tag{4.1}$$

Any biparacyclic k-module $C = (C_{\bullet,\bullet}, \bar{d}, \bar{s}, \bar{t}, d, s, t)$ gives rise to a *diagonal paracyclic* k*-module* $\mathrm{Diag}(C) := (\mathrm{Diag}_\bullet(C), \bar{d}d, \bar{s}s, \bar{t}t)$, where $\mathrm{Diag}_m(C) = C_{m,m}$. Thanks to (4.1) we obtain a cyclic complex complex if C is cyclindrical.

A *parachain bicomplex* is any data of $(C_{\bullet,\bullet}, \bar{b}, \bar{B}, b, B)$, where $C_{p,q}$, $p, q \geqslant 0$, are k-modules and $(C_{\bullet,q}, \bar{b}, \bar{B})$ and $(C_{p,\bullet}, b, B)$ are parachain complexes for all $p, q \geqslant 0$ in such a way that the horizontal operators $(\bar{b}, \bar{B})$ commute with each of the vertical operators (b, B). In this case $\overline{T} := 1 - (\bar{b}\bar{B} + \bar{B}\bar{b})$ and $T := 1 - (bB + Bb)$ are k-modules automorphisms. It is called a *cylindrical complex* if $\overline{T}T = 1$.

Any parachain bicomplex $C = (C_{\bullet,\bullet}, \bar{b}, \bar{B}, b, B)$ gives rise to a *total parachain complex* $\mathrm{Tot}(C) = (\mathrm{Tot}_\bullet, \mathfrak{b}, \mathfrak{B})$, where

$$\mathrm{Tot}_m(C) = \bigoplus_{p+q=m} C_{p,q}, \qquad \mathfrak{b} = \bar{b} + (-1)^p b, \qquad \mathfrak{B} = \bar{B} + (-1)^p \overline{T} B.$$

We obtain a mixed complex if C is cyclindrical.

Any biparacylic k-module $C = (C_{\bullet,\bullet}, \bar{d}, \bar{s}, \bar{t}, d, s, t)$ gives rise to a parachain bicomplex $C = (C_{\bullet,\bullet}, \bar{b}, \bar{B}, b, B)$. This is a cylindrical complex if C is cylindrical. Therefore, out of a biparacylic k-module we get the following two para S-modules:

- The para S-module $\mathrm{Diag}(C)^\natural$ associated with the diagonal paracyclic k-module $\mathrm{Diag}(C)$.
- The para S-module $\mathrm{Tot}(C)^\natural$ associated with the total parachain complex $\mathrm{Tot}(C)$ of the corresponding parachain bicomplex.

4.2. The Paracyclic Eilenberg-Zilber theorem. Let $C=(C_{\bullet,\bullet},\bar{d},\bar{s},\bar{t},d,s,t)$ be a biparacyclic k-module. As mentioned above we get two para S-modules $\mathrm{Tot}_\bullet(C)^\natural$ and $\mathrm{Diag}(C)^\natural$. At the ordinary chain level, the Eilenberg-Zilber theorem asserts that we have quasi-isomorphisms,

$$\mathrm{Tot}_\bullet(C) \xrightarrow{\sqcup\!\sqcup} \mathrm{Diag}_\bullet(C) \xrightarrow{\mathrm{AW}} \mathrm{Tot}_\bullet(C),$$

where $\sqcup\!\sqcup$ and AW are the shuffle and Alexander-Whitney maps, respectively. We actually obtain a deformation retract at the level of normalized chains. Using the perturbation lemma for parachain complexes (Lemma 3.2 above) enables us to construct explicit coextensions of the shuffle and Alexander-Whitney maps which provide us with a deformation retract of the normalization of $\mathrm{Tot}_\bullet(C)^\natural$ to that of $\mathrm{Diag}(C)^\natural$. Lifting them to unnormalized chains yields the following result.

THEOREM 4.1 (Paracyclic Eilenberg-Zilberg Theorem [**44**]). *Let* $C = (C_{\bullet,\bullet},\bar{d},\bar{s},\bar{t},d,s,t)$ *be a biparacyclic k-module.*

(1) *We have an S-homotopy equivalence,*

$$\mathrm{Tot}_\bullet(C)^\natural \xrightarrow{\sqcup\!\sqcup^\natural} \mathrm{Diag}_\bullet(C)^\natural \xrightarrow{\mathrm{AW}^\natural} \mathrm{Tot}_\bullet(C)^\natural,$$

where $\sqcup\!\sqcup^\natural$ and $\mathrm{AW}^\natural$ are explicit co-extensions of the shuffle and Alexander-Whitney maps, respectively.

(2) *On normalized chains this induces a deformation retract.*

(3) *The S-map $\sqcup\!\sqcup^\natural$ takes the form,*

$$\sqcup\!\sqcup^\natural = \sqcup\!\sqcup + \sqcup\!\sqcup' u^{-1}, \qquad \sqcup\!\sqcup' : \mathrm{Tot}_\bullet(C) \longrightarrow \mathrm{Diag}_{\bullet+2}(C). \tag{4.2}$$

REMARK 4.2. For bicyclic k-modules the above result is due to Bauval [**3**]. She also showed that on tensor product of cyclic modules her map $\sqcup\!\sqcup'$ agrees with the cyclic shuffle map of [**32**] up to conjugation by flip maps.

REMARK 4.3. For cyclindrical k-modules Getzler-Jones [**24**] established the existence of a co-extension of the shuffle map of the form (4.2) without giving an explicit formula for the (para)cylic shuffle map $\sqcup\!\sqcup'$. This implies that $\mathrm{Tot}(C)^\natural$ and $\mathrm{Diag}(C)^\natural$ are quasi-isomorphic.

REMARK 4.4. An attempt to give a formula for the map $\sqcup\!\sqcup'$ along the lines of [**3**] for cylindrical k-modules is presented in [**31**].

5. Triangular S-Modules and Spectral Sequences

5.1. Horizontal triangular S-modules. A (horizontal) *para-triangular S-module* is given by any data $(C_{\bullet,\bullet},\bar{d},\bar{S},\overline{T},b,B)$, where $C_{p,q}$, $p,q\geqslant 0$, are k-modules, $(C_{\bullet,q},\bar{d},\bar{S},\overline{T})$ is a para S-module and $(C_{p,\bullet},b,B)$ is a parachain complex for all $p,q\geqslant 0$, in such a way that the horizontal operators $(\bar{d},\bar{S},\overline{T})$ commute with each of the vertical operators (b,B). We say that have a *triangular S-module* if we further have the relation $\overline{T}T=1$, with $T:=1-(bB+Bb)$.

Any para-triangular S-module $C=(C_{\bullet,\bullet},\bar{d},\bar{S},b,B)$ gives rise to a (total) para S-module $\mathrm{Tot}(C)=(\mathrm{Tot}_\bullet(C),\mathbf{d},\bar{S},\overline{T}T)$, where $\mathrm{Tot}_m(C)=\bigoplus_{p+q=m}C_{p,q}$ and

$$\mathbf{d}=\bar{d}+(-1)^p(b+\overline{T}B\bar{S}).$$

Indeed,

$$\mathbf{d}^2=\bar{d}^2+\overline{T}(bB+Bb)\bar{S}=(1-\overline{T})\bar{S}+\overline{T}(1-T)\bar{S}=(1-\overline{T}T)\bar{S}.$$

In particular, if C is a triangular S-module, then $\mathbf{d}^2 = 0$, and so $\operatorname{Tot}(C)$ is an S-module.

Any parachain bicomplex $C = (C_{\bullet,\bullet}, \bar{b}, \bar{B}, b, B)$ gives rise to a para-triangular S-module,

$$C^{\bar{\sigma}} := \left(C^{\bar{\sigma}}_{\bullet,\bullet}, \bar{b} + \bar{B}u^{-1}, u^{-1}, \overline{T}, b, B\right), \quad \text{where } C^{\bar{\sigma}}_{p,q} = C_{p,q} \oplus C_{p-2,q}u \oplus \cdots .$$

The corresponding total para S-module $\operatorname{Tot}(C^{\bar{\sigma}})$ agrees with the para S-module $\operatorname{Tot}(C)^{\natural}$ defined earlier. If C is a cylindrical complex, then $C^{\bar{\sigma}}$ is a triangular S-module.

Para-triangular S-modules provide us with a natural framework to define the tensor products of para S-modules with parachain complexes. Assume we are given a tensor product on the category $\operatorname{Mod}(k)$ of left k-modules, in the sense of we have a bifunctor $\otimes_k : \operatorname{Mod}(k) \times \operatorname{Mod}(k) \rightarrow \operatorname{Mod}(k')$, where k' is a sub-ring of k containing the unit. If k is commutative, then we shall take this tensor product to be the usual tensor product of k-modules, in which case $k' = k$.

If $\bar{C} = (\bar{C}_{\bullet}, \bar{d}, \bar{S}, \overline{T})$ is a para S-module and $C = (C_{\bullet}, b, B)$ is a parachain complex, then we can form their tensor product as the triangular para S-module,

$$\bar{C} \otimes_k C = \left(\bar{C}_{\bullet} \otimes_k C_{\bullet}, \bar{d}, \bar{S}, \overline{T}, b, B\right). \tag{5.1}$$

With this convention if $\bar{C} = (\bar{C}_{\bullet}, \bar{b}, \bar{B})$ and $C = (C_{\bullet}, b, B)$ are both parachain complexes, then we observe that $\bar{C}^{\natural} \otimes_k C = (\bar{C} \otimes_k C)^{\bar{\sigma}}$, and so we get

$$\operatorname{Tot}\left(\bar{C} \otimes_k C\right)^{\natural} = \operatorname{Tot}\left(\left(\bar{C} \otimes_k C\right)^{\bar{\sigma}}\right) = \operatorname{Tot}\left(\bar{C}^{\natural} \otimes_k C\right). \tag{5.2}$$

This realizes the para S-module $\operatorname{Tot}\left(\bar{C} \otimes_k C\right)^{\natural}$ as a total para S-module. In particular, any left module structure on $\bar{C}^{\natural}$ automatically extends to $\operatorname{Tot}\left(\bar{C} \otimes_k C\right)^{\natural}$ (see Section 6).

5.2. Vertical triangular S-modules. In the definition of horizontal para-triangular S-modules, the horizontal and vertical directions do not play symmetric roles. Reversing their roles leads us to define *vertical para-triangular S-modules* as any data $C = (C_{\bullet,\bullet}, \bar{b}, \bar{B}, d, S, T)$, where $C_{p,q}$, $p, q \geqslant 0$, are k-modules, and $(C_{p,\bullet}, \bar{b}, \bar{B})$ is a parachain complex and $(C_{p,\bullet}, d, \overline{T}S, T)$ is a para S-module for all $p, q \geqslant 0$ with $\overline{T} = 1 - (\bar{b}\bar{B} + \bar{B}\bar{b})$ in such a way that the horizontal operators $(\bar{b}, \bar{B})$ commute with each of the vertical operators (d, S, T). We call it a *vertical triangular S-module* if $\overline{T}T = 1$.

As above, any vertical para-triangular S-module $C = (C_{\bullet,\bullet}, \bar{b}, \bar{B}, d, S, T)$ gives rise to a para S-module $\operatorname{Tot}(C) = (\operatorname{Tot}_{\bullet}(C), \mathbf{d}, S, T)$, with $\mathbf{d} = \bar{b} + \bar{B}S + (-1)^p d$. Here we have

$$\mathbf{d}^2 = (\bar{b}\bar{B} + \bar{B}\bar{b})S + d^2 = (1 - \overline{T})S + (1 - T)\overline{T}S = (1 - \overline{T}T)\bar{S}.$$

Thus, we get an S-module precisely when C is a (vertical) triangular S-module.

Any parachain bicomplex $C = (C_{\bullet,\bullet}, \bar{b}, \bar{B}, b, B)$ gives rise to a para-triangular S-module,

$$C^{\sigma} := \left(C^{\sigma}_{\bullet,\bullet}, \bar{b}, \bar{B}, b + \overline{T}Bu^{-1}, u^{-1}, T\right), \quad \text{where } C^{\sigma}_{p,q} = C_{p,q} \oplus C_{p,q-2}u \oplus \cdots .$$

As above, the total para S-module $\operatorname{Tot}(C^{\sigma})$ agrees with the para S-module $\operatorname{Tot}(C)^{\natural}$. If C is a cylindrical complex, then C^{σ} is a triangular S-module.

If $C = (\bar{C}_\bullet, \bar{b}, \bar{B})$ is a mixed complex and $C = (C_\bullet, d, S, T)$ is a para S-module, then we can form their tensor product as the vertical triangular para S-module,

$$\bar{C} \otimes_k C = \left(\bar{C}_\bullet \otimes_k C_\bullet, \bar{b}, \bar{B}, d, S, T\right).$$

This yields a triangular S-module if C is an S-module. In particular, if $C = (C_\bullet, b, B)$ is a parachain complex, then $\bar{C} \otimes_k C^\natural = (\bar{C} \otimes_k C)^\sigma$, and hence we get

$$\text{(5.3)} \qquad \operatorname{Tot}\left(\bar{C} \otimes_k C\right)^\natural = \operatorname{Tot}\left(\left(\bar{C} \otimes_k C\right)^\sigma\right) = \operatorname{Tot}\left(\bar{C} \otimes_k C^\natural\right).$$

5.3. Spectral sequences. If $C = (C_{\bullet,\bullet}, \bar{d}, \bar{T}, \bar{S}, b, B)$ is a horizontal triangular S-module, we further observe that the filtration of $\operatorname{Tot}(C)$ by columns is a filtration of chain complexes, and so we get for free a spectral sequence computing the homology of $\operatorname{Tot}(C)$. The E^1-term is $E^1_{p,q} = H^v_{p,q}(C)$, where $H^v_{p,q}(C)$ is the degree q homology of the chain complex $(C_{p,\bullet}, b)$, $p \geqslant 0$. The fact that $0 = \mathbf{d}^2 = \bar{d}^2 + \bar{T}(bB + Bb)\bar{S}$ ensures that $\bar{d}$ induces a differential on $H^v_{\bullet,q}(C)$. Denote by $H^h_\bullet H^v_q(C)$ the homology of $(H^v_{\bullet,q}(C), \bar{d})$. This gives the E^2-term, i.e., $E^2_{p,q} = H^h_p H^v_q(C)$.

Likewise, if $C = (C_{\bullet,\bullet}, \bar{b}, \bar{B}, d, S, T)$ is a vertical triangular S-module, the filtration of $\operatorname{Tot}(C)$ by rows is a filtration of chain complexes, and so we also get a spectral sequence. The E^1-term is $E^1_{p,q} = H^h_{p,q}(C)$, where $H^h_{p,q}(C)$ is the degree p homology module of the chain complex $(C_{\bullet,q}, \bar{b})$, $q \geqslant 0$. Here d induces a differential on $H^h_{p,\bullet}(C)$. Let $H^v_\bullet H^h_q(C)$ be the homology of $(H^h_{p,\bullet}(C), d)$. The E^2-term then is $E^2_{p,q} = H^h_p H^v_q(C)$.

Let $C = (C_{\bullet,\bullet}, \bar{b}, \bar{B}, b, B)$ be a cyclindrical module. As mentioned above it gives rise to a horizontal triangular S-module $C^{\bar{\sigma}}$ and a vertical triangular S-modules C^σ whose total S-modules $\operatorname{Tot}(C^{\bar{\sigma}})$ and $\operatorname{Tot}(C)$ both agree with the S-module associated with the cyclic complex $\operatorname{Tot}(C)^\natural$, and hence $H_\bullet(\operatorname{Tot}(C^{\bar{\sigma}})) = H_\bullet(\operatorname{Tot}(C^\sigma)) = \operatorname{HC}_\bullet(\operatorname{Tot}(C))$.

Denote by $E(C^{\bar{\sigma}})$ and $E(C^\sigma)$ the spectral sequences of $C^{\bar{\sigma}}$ and C^σ. For the spectral sequence of $C^{\bar{\sigma}}$, the horizontal differentials $(\bar{b}, \bar{B})$ descend to the vertical homology $H^v_q(C^{\bar{\sigma}})$, so that we get mixed complexes $(H^v_q(C^{\bar{\sigma}}), \bar{b}, \bar{B})$. The E^2-term $E^2_{p,q}(C^{\bar{\sigma}})$ then is given by the p-th degree cyclic homology of this mixed complex. For the spectral sequence $E(C^\sigma)$, the vertical differentials (b, B) descend to $H^h_p(C^\sigma)$ and the E^2-term $E^2_{p,q}(C^\sigma)$ is given by the cyclic homology of the mixed complex $(H^h_p(C^\sigma), b, B)$. Therefore, we have obtained the following result.

PROPOSITION 5.1. *Let $C = (C_{\bullet,\bullet}, \bar{b}, \bar{B}, b, B)$ be a cyclindrical complex. Then the filtration by columns of* $\operatorname{Tot}(C^{\bar{\sigma}})$ *and the filtration by rows of* $\operatorname{Tot}(C)$ *give rise to spectral sequences,*

$$\text{(5.4)} \qquad E^2_{p,q}(C^{\bar{\sigma}}) = \operatorname{HC}_p(H^v_q(C)) \Longrightarrow \operatorname{HC}_{p+q}(\operatorname{Tot}(C)),$$

$$\text{(5.5)} \qquad E^2_{p,q}(C^\sigma) = \operatorname{HC}_q(H^h_p(C)) \Longrightarrow \operatorname{HC}_{p+q}(\operatorname{Tot}(C)).$$

6. The Cyclic Homology of Crossed-Product Algebras

In what follows we let Γ be a (discrete) group on a unital algebra $\mathcal{A}$, which we assume to be either an algebra over a commutative ring $k \supset \mathbb{Q}$ or a local convex algebra, in which case $k = \mathbb{C}$. We denote by $\mathcal{A}_\Gamma$ the crossed-product algebra $\mathcal{A} \rtimes \Gamma$. If $\mathcal{A}$ is a locally convex algebra, we equip $\mathcal{A}_\Gamma$ with the same kind of locally convex topology as described in Section 6. In this case the spaces of chains $C_m(\mathcal{A}_\Gamma)$ and

$C_m(\mathcal{A})$ are defined in terms of the projective topological tensor product, and so the cyclic and periodic cyclic homologies are meant in the sense of Definition 1.9.

6.1. Splitting along conjugacy classes. As in Section 2, given any $\phi \in \Gamma$ we denote by $[\phi]$ its conjugacy class in Γ. We define the cyclic k-submodule $C(\mathcal{A}_\Gamma)_{[\phi]}$ as in Definition 2.1. As in Proposition 2.2 we have a direct sum of cyclic k-modules,

$$C(\mathcal{A}_\Gamma) = \bigoplus_{[\phi]} C(\mathcal{A}_\Gamma)_{[\phi]},$$

where the summation runs over all conjugacy classes. Therefore, the study of the cyclic k-module of $\mathcal{A}_\Gamma$ reduces to that of each of the cyclic submodules $C(\mathcal{A}_\Gamma)_{[\phi]}$.

6.2. The cylindrical complexes $C^\phi(\Gamma, \mathscr{C})$. In what follows we let ϕ be a central element of Γ. The twisted cyclic k-module of Γ is then defined as follows.

DEFINITION 6.1. $C^\phi(\Gamma)$ is the paracyclic $k\Gamma$-module,

$$(C_\bullet(\Gamma), d, s_\phi, t_\phi), \qquad C_m(\Gamma) = k\Gamma^{m+1},$$

where

$$d(\psi_0, \ldots, \psi_m) = (\psi_0, \ldots, \psi_{m-1}), \tag{6.1}$$

$$s_\phi(\psi_0, \ldots, \psi_m) = (\phi^{-1}\psi_m, \psi_0, \ldots, \psi_m), \tag{6.2}$$

$$t_\phi(\psi_0, \ldots, \psi_m) = (\phi^{-1}\psi_m, \psi_0, \ldots, \psi_{m-1}). \tag{6.3}$$

REMARK 6.2. For $\phi = 1$ we recover the usual cyclic k-module of Γ.

REMARK 6.3. On $C_m(\Gamma)$ the operator $T_\phi = t_\phi^{m+1}$ is given by

$$T_\phi(\psi_0, \ldots, \psi_m) = (\phi^{-1}\psi_0, \ldots, \phi^{-1}\psi_m) = \phi^{-1}(\psi_0, \ldots, \psi_m). \tag{6.4}$$

In other words, T_ϕ is given by the action of ϕ^{-1} on $C_\bullet(\Gamma)$.

As in Section 2, given $k\Gamma$-modules $\mathscr{M}$ and $\mathscr{M}'$, we denote by $\mathscr{M}' \otimes_\Gamma \mathscr{M}$ the quotient of $\mathscr{M} \otimes_k \mathscr{M}$ by the diagonal action of Γ. In this case $\mathscr{M}' \otimes_\Gamma \mathscr{M}$ is a k-module.

DEFINITION 6.4. If $\mathscr{C} = (\mathscr{C}_\bullet, b, B)$ is a parachain complex, then $C^\phi(\Gamma, \mathscr{C})$ the parachain bicomplex,

$$\left(C_\bullet(\Gamma) \otimes_\Gamma C_\bullet(\mathcal{A}), \bar{d}, \bar{s}_\phi, \bar{t}_\phi, d_\phi, s, t_\phi\right),$$

where $(\bar{d}, \bar{s}_\phi, \bar{t}_\phi)$ are given by (6.1)–(6.3).

In what follows we shall call *ϕ-parachain complex* any parachain complex $\mathscr{C} = (\mathscr{C}_\bullet, b, B)$ of $k\Gamma$-modules such that the operator $T := 1 - (bB + Bb)$ is given by the action of ϕ^{-1}.

Let $\mathscr{C} = (\mathscr{C}_\bullet, b, B)$ be a ϕ-parachain complex. By definition $C_p(\Gamma) \otimes_\Gamma \mathscr{C}_q$ is the quotient of $C_p(\Gamma) \otimes_k \mathscr{C}_q$ by the diagonal action of Γ. Therefore, in view (6.4) on $C^\phi_{p,q}(\Gamma_\phi, \mathscr{C}) = C_p(\Gamma) \otimes_\Gamma \mathscr{C}_q$ the operator $\overline{T}_\phi T = \bar{t}_\phi^{p+1} T$ is given by

$$\overline{T}_\phi T\left[(\psi_0, \ldots, \psi_p) \otimes_\Gamma \xi\right] = \left[\phi^{-1}(\psi_0, \ldots, \psi_p)\right] \otimes_\Gamma (\phi^{-1}\xi) = (\psi_0, \ldots, \psi_p) \otimes_\Gamma \xi.$$

Thus, $\overline{T}_\phi T = 1$, and so $C^\phi(\Gamma_\phi, \mathscr{C})$ is a cylindrical complex.

The twisted cyclic k-module of $\mathcal{A}$ is defined as follows.

DEFINITION 6.5. $C^{\phi}(\mathcal{A})$ is the paracyclic $k\Gamma$-module,

$$(C_{\bullet}(\mathcal{A}), d_{\phi}, s, t_{\phi}), \qquad C_m(\mathcal{A}) = \mathcal{A}^{\otimes(m+1)},$$

where

$$d_{\phi}(a^0 \otimes \cdots \otimes a^m) = [(\phi^{-1}a^m)a^0] \otimes a^1 \otimes \cdots \otimes a^m, \tag{6.5}$$

$$s(a^0 \otimes \cdots \otimes a^m) = 1 \otimes a^0 \otimes \cdots \otimes a^m, \tag{6.6}$$

$$t_{\phi}(a^0 \otimes \cdots \otimes a^m) = (\phi^{-1}a^m) \otimes a^0 \otimes \cdots \otimes a^{m-1}. \tag{6.7}$$

REMARK 6.6. On $C_m(\mathcal{A})$ the operator $T_{\phi} = t_{\phi}^{m+1}$ is given by

$$T_{\phi}(a^0 \otimes \cdots \otimes a^m) = (\phi^{-1}a^0) \otimes \cdots \otimes (\phi^{-1}a^m) = \phi^{-1}((a^0 \otimes \cdots \otimes a^m).$$

That is, T_{ϕ} is given by the action of ϕ^{-1} on $C_{\bullet}(\mathcal{A})$, and so the parachain complex associated with $C^{\phi}(\mathcal{A})$ is a ϕ-parachain complex.

Specializing the construction of $C^{\phi}(\Gamma_{\phi}, \mathscr{C})$ to $\mathscr{C} = C^{\phi}(\mathcal{A})$ yields a cylindrical complex. In fact, in this case we actually can get a cylindrical module.

DEFINITION 6.7. $C^{\phi}(\Gamma, \mathcal{A})$ is the cylindrical k-module,

$$\left(C_{\bullet}(\Gamma) \otimes_{\Gamma} C_{\bullet}(\mathcal{A}), \bar{d}, \bar{s}_{\phi}, \bar{t}_{\phi}, d_{\phi}, s, t_{\phi}\right),$$

where $(\bar{d}, \bar{s}_{\phi}, \bar{t}_{\phi})$ are given by (6.1)–(6.3) and (d_{ϕ}, s, t_{ϕ}) are given by (6.5)–(6.7).

REMARK 6.8. The cylindrical k-module $C^{\phi}(\Gamma, \mathcal{A})$ is isomorphic to the cylindrical k-module $\mathcal{A}_{\phi}^{\natural}$ of Getzler-Jones [**24**, §4]. The main advantage of using $C^{\phi}(\Gamma, \mathcal{A})$ lies on its construction as an actual tensor product. As we shall see this allows us to break down the study $C^{\phi}(\Gamma, \mathcal{A})$ into separate studies of each of the the paracyclic modules $C^{\phi}(\Gamma)$ and $C^{\phi}(\mathcal{A})$.

REMARK 6.9. The diagonal cyclic k-module $\operatorname{Diag}(C^{\phi}(\Gamma, \mathcal{A}))$ is isomorphic to the cyclic k-module $\mathcal{A}_{\phi}[\Gamma]^{\natural\natural}$ of Feigin-Tsygan [**22**, §§4.1]. It is also isomorphic to the cyclic k-module $\tilde{L}(A, \Gamma, \phi)$ of Nistor [**37**, §§2.5].

Finally, it can be shown that the tensor products $C_m(\Gamma) \otimes_{\Gamma} -$, $m \geqslant 0$, are exact functors from $\operatorname{Mod}(k\Gamma)$ to $\operatorname{Mod}(k)$. Therefore, we obtain the following result.

PROPOSITION 6.10. *If $\alpha : \mathscr{C}_{\bullet} \to \mathscr{C}'_{\bullet}$ is a quasi-isomorphism of ϕ-parachain complexes, then the mixed complex map $\alpha \otimes 1 : \operatorname{Tot}_{\bullet}(C^{\phi}(\Gamma, \mathscr{C})) \to \operatorname{Tot}_{\bullet}(C^{\phi}(\Gamma, \mathscr{C}'))$ is a quasi-isomorphism.*

6.3. From $C(\mathcal{A})_{[\phi]}$ to $C^{\phi}(\Gamma, \mathcal{A})$. As in Section 2, given any $\phi \in \Gamma$, we denote by Γ_{ϕ} the centralizer of ϕ in Γ. By definition ϕ is a central element of Γ_{ϕ}. This allows us to form the biparacyclic k-module $C^{\phi}(\Gamma_{\phi}, \mathcal{A})$. We have a natural embedding of cyclic k-modules $\mu_{\phi} : \operatorname{Diag}_{\bullet}(C^{\phi}(\Gamma_{\phi}, \mathcal{A})) \to C_{\bullet}(\mathcal{A}_{\Gamma})_{[\phi]}$ given by

$$\mu_{\phi}\left((\psi_0, \ldots, \psi_m) \otimes_{\Gamma_{\phi}} (a^0 \otimes \cdots \otimes a^m)\right) = [(\psi_m^{-1}\phi) \cdot a^0]u_{\phi\psi_m^{-1}\psi_0} \otimes (\psi_0^{-1} \cdot a^1)u_{\psi_0^{-1}\psi_1} \otimes \cdots \otimes (\psi_{m-1}^{-1} \cdot a^m)u_{\psi_{m-1}^{-1}\psi_m}. \tag{6.8}$$

This embedding can be shown to be a quasi-isomorphism by using Shapiro's lemma. If $\phi = 1$ this is actually an isomorphism. We refer to [**42**] for an explicit formula for μ_1^{-1}.

By combining this with Theorem 4.1 we then arrive at the following statement.

PROPOSITION 6.11. *Let $\phi \in \Gamma$. The following S-maps are quasi-isomorphisms,*

$$\text{Tot}_\bullet(C^\phi(\Gamma_\phi, \mathcal{A}))^\natural \underset{\text{AW}^\natural}{\overset{\sqcup\!\sqcup^\natural}{\rightleftarrows}} \text{Diag}_\bullet(C^\phi(\Gamma_\phi, \mathcal{A}))^\natural \overset{\mu}{\hookrightarrow} C_\bullet(\mathcal{A}_\Gamma)^\natural_{[\phi]}. \tag{6.9}$$

This reduces the study of the S-module $C_\bullet(\mathcal{A}_\Gamma)^\natural_{[\phi]}$ that of $\text{Tot}_\bullet(C^\phi(\Gamma_\phi, \mathcal{A}))^\natural$. Once again, the advantage of using $C^\phi(\Gamma_\phi, \mathcal{A})$ lies on its construction as an actual tensor product. This allows us to break down its study into separate studies of the paracyclic k-modules $C^\phi(\Gamma)$ and $C^\phi(\mathcal{A})$.

6.4. Finite order case. Let $\phi \in \Gamma$ be central and have finite order. In what follows we denote by $C^\flat(\Gamma)$ the mixed complex $(C(\Gamma), \partial, 0)$, where ∂ is the group differential (2.2). In addition, given any ϕ-invariant mixed complex of $k\Gamma$-modules $\mathscr{C} = (\mathscr{C}_\bullet, b, B)$, we denote by $C^\flat(\Gamma, \mathscr{C})$ the mixed bicomplex of k-modules $C^\flat(\Gamma) \otimes_\Gamma \mathscr{C}$.

If $\mathscr{C} = (\mathscr{C}_\bullet, b, B)$ is a ϕ-parachain complex, then we have a ϕ-invariant subcomplex $\mathscr{C}^\phi = (\mathscr{C}^\phi_\bullet, b, B)$, where $\mathscr{C}^\phi_m$ is the ϕ-invariant submodule of $\mathscr{C}_m$. Here $\mathscr{C}^\phi$ is actually a mixed complex, since $bB + Bb = 1 - \phi^{-1} = 0$ on $\mathscr{C}^\phi_\bullet$. This allows us to form the mixed bicomplex $C^\flat(\Gamma, \mathscr{C}^\phi)$.

Bearing this in mind, let $\nu_\phi : C_\bullet(\Gamma) \to C_\bullet(\Gamma)$ be the $k\Gamma$-module map defined by

$$\nu_\phi(\psi_0, \ldots, \psi_m) = \frac{1}{r^{m+1}} \sum_{0 \leqslant j \leqslant m} \sum_{0 \leqslant \ell_j \leqslant r-1} (\phi^{\ell_0}\psi_0, \ldots, \phi^{\ell_m}\psi_m), \qquad \psi_j \in \Gamma. \tag{6.10}$$

Thus, ν_ϕ is the projection onto the ϕ-invariant submodule $((k\Gamma)^\phi)^{\otimes(m+1)} \subset (k\Gamma)^{\otimes(m+1)} = C_m(\Gamma)$. For $\phi = 1$ this is just the identity map.

We also let $\varepsilon : C_\bullet(\Gamma) \to C_\bullet(\Gamma)$ be the antisymmetrization map,

$$\varepsilon(\psi_0, \ldots, \psi_m) = \frac{1}{(m+1)!} \sum_{\sigma \in \mathfrak{S}_m} \left(\psi_{\sigma^{-1}(0)}, \ldots, \psi_{\sigma^{-1}(m)}\right), \qquad \psi_j \in \Gamma, \tag{6.11}$$

where $\mathfrak{S}_m$ is the group of permutations of $\{0, \ldots, m\}$.

As ν_ϕ and ε both are projections and commute with each other, the composition $\varepsilon\nu_\phi$ is a projection as well. It can be checked that $\varepsilon\nu_\phi : C^\phi_\bullet(\Gamma) \to C^\flat(\Gamma)$ is a map of parachain complexes. At the level of ordinary chains, the chain map $\varepsilon\nu_\phi : (C_\bullet(\Gamma), \partial) \to (C_\bullet(\Gamma), \partial)$ is chain homotopic to the identity (compare [**8**, **33**]). Since this is a projection, it yields a chain homotopy equivalence, which is a deformation retract onto its range. In addition, we observe, that given any ϕ-parachain complex, the map $(\varepsilon\nu_\phi) \otimes 1 : C_p(\Gamma) \otimes_\Gamma \mathscr{C}_q \to C_p(\Gamma) \otimes_\Gamma \mathscr{C}_q$ maps to $C_p(\Gamma) \otimes \mathscr{C}^\phi_q$, so that we get a parachain bicomplex map $(\varepsilon\nu_\phi) \otimes 1 : C^\phi_{\bullet,\bullet}(\Gamma, \mathscr{C}) \to C^\flat_{\bullet,\bullet}(\Gamma, \mathscr{C}^\phi)$. Therefore, by using Lemma 3.2 we obtain the following result.

PROPOSITION 6.12. *Suppose that ϕ is a central element of Γ of finite order.*

(1) *The S-map $\varepsilon\nu_\phi : C^\phi_\bullet(\Gamma)^\natural \to C^\flat_\bullet(\Gamma)^\natural$ has an S-homotopy inverse $(\varepsilon\nu_\phi)^\flat : C^\flat_\bullet(\Gamma)^\natural \to C^\phi_\bullet(\Gamma)^\natural$, which is a coextension of $\varepsilon\nu_\phi$.*

(2) *For any ϕ-parachain complex $\mathscr{C}$, the S-maps $(\varepsilon\nu_\phi)\otimes 1 : \text{Tot}_\bullet(C^\phi(\Gamma, \mathscr{C}))^\natural \to \text{Tot}_\bullet(C^\flat(\Gamma, \mathscr{C}^\phi))^\natural$ and $(\varepsilon\nu_\phi)^\flat \otimes 1 : \text{Tot}_\bullet(C^\flat(\Gamma, \mathscr{C}^\phi))^\natural \to \text{Tot}_\bullet(C^\phi(\Gamma, \mathscr{C}))^\natural$ are S-homotopy inverses.*

REMARK 6.13. If we take tensor products with k, then we get an S-homotopy equivalence of S-modules $\varepsilon\nu_\phi : C^\phi_\bullet(\Gamma, k)^\natural \to C^\flat_\bullet(\Gamma, k)^\natural$, where $C^\phi_\bullet(\Gamma, k)$ is the cyclic

k-module $(C_\bullet(\Gamma,k),d,s_\phi,t_\phi)$ and $C^\flat_\bullet(\Gamma,k)$ is the mixed complex $(C_\bullet(\Gamma,k),\partial,0)$. Here $C_p(\Gamma,k)=C_p(\Gamma)\otimes_\Gamma k$. The fact this map is a quasi-isomorphism was established by Marciniak [**33**] in case k is a field of characteristic 0. Marciniak's result gives an algebraic proof of Burghelea's isomorphisms (2.7) with coefficients in k.

Combinings Propositions 6.10, 6.11, and 6.12 together gives the following result.

THEOREM 6.14. *Let $\phi\in\Gamma$ have finite order, and suppose we are given a quasi-isomorphism of ϕ-parachain complexes $\alpha: C^\phi_\bullet(\mathcal{A})\to\mathscr{C}_\bullet$.*

(1) *The following S-maps are quasi-isomorphisms,*

$$\text{Tot}_\bullet\left(C^\flat(\Gamma_\phi,\mathscr{C}^\phi)\right)^\natural \xleftarrow{(\varepsilon\nu_\phi)\otimes\alpha} \text{Tot}_\bullet\left(C^\phi(\Gamma_\phi,\mathcal{A})\right)^\natural \underset{\text{AW}^\natural}{\overset{\sqcup\!\sqcup^\natural}{\rightleftarrows}} \text{Diag}_\bullet\left(C^\phi(\Gamma_\phi,\mathcal{A})\right)^\natural \xrightarrow{\mu_\phi} C_\bullet(\mathcal{A}_\Gamma)^\natural_{[\phi]}. \tag{6.12}$$

Moreover, if α has a quasi-inverse (resp., homotopy inverse) $\beta:\mathscr{C}_\bullet\to C^\phi_\bullet(\mathcal{A})$, then $(\varepsilon\nu_\phi)^\flat\otimes\beta$ is a quasi-inverse (resp., homotopy inverse) of $(\varepsilon\nu_\phi)\otimes\alpha$.

(3) *This yields isomorphisms,*

$$\text{HC}(\mathcal{A}_\Gamma)_{[\phi]}\simeq\text{HC}_\bullet(\text{Tot}(C^\flat(\Gamma_\phi,\mathscr{C}^\phi))),\qquad \text{HP}(\mathcal{A}_\Gamma)_{[\phi]}\simeq\text{HP}_\bullet(\text{Tot}(C^\flat(\Gamma_\phi,\mathscr{C}^\phi))).$$

As mentioned above, the mixed bicomplex $C^\flat(\Gamma_\phi,\mathscr{C}^\phi)$ gives rise to two triangular S-modules: the horizontal triangular S-module $C^\flat(\Gamma_\phi,\mathscr{C}^\phi)^{\bar\sigma}=C^\flat(\Gamma_\phi)^\natural\otimes_{\Gamma_\phi}\mathscr{C}^\phi$ and the vertical triangular S-module $C^\flat(\Gamma_\phi,\mathscr{C}^\phi)^\sigma=C^\flat(\Gamma_\phi)\otimes_{\Gamma_\phi}(\mathscr{C}^\phi)^\natural$. The total complexes of these triangular S-modules both agree with the cyclic complex $\text{Tot}(C^\flat(\Gamma_\phi,\mathscr{C}^\phi))^\natural$. The filtration by columns of $\text{Tot}(C^\flat(\Gamma_\phi,\mathscr{C}^\phi)^{\bar\sigma})$ gives rise to the spectral sequence (5.4). The filtration by rows of $\text{Tot}(C^\flat(\Gamma_\phi,\mathscr{C}^\phi)^\sigma)$ gives rise to the spectral sequence (5.5). Both spectral sequences converge to $\text{HC}_\bullet(\text{Tot}(C^\flat(\Gamma_\phi,\mathscr{C}^\phi)))$. As the B-differential of $C^\flat(\Gamma)$ is trivial, we observe that the filtration by rows of $\text{Tot}(C^\flat(\Gamma_\phi,\mathscr{C}^\phi)^\sigma)$ is a filtration of chain complexes, and so it gives rise to a 3rd spectral sequence converging to $\text{HC}_\bullet(\text{Tot}(C^\flat(\Gamma_\phi,\mathscr{C}^\phi)))$. Combining this with Theorem 6.14 we then arrive at the following result.

COROLLARY 6.15. *Let $\phi\in\Gamma$ have finite order, and suppose we are given a quasi-isomorphism of ϕ-parachain complexes $\alpha: C^\phi_\bullet(\mathcal{A})\to\mathscr{C}_\bullet$.*

(1) *The filtration by columns of* $\text{Tot}_\bullet(C^\flat(\Gamma_\phi)^\natural\otimes_{\Gamma_\phi}\mathscr{C}^\phi)$ *and the quasi-isomorphisms* (6.12) *give rise to the spectral sequence,*

$${}^{I}E^2_{p,q}=H_p(\Gamma_\phi,H_q(\mathscr{C}^\phi))\oplus H_{p-2}(\Gamma_\phi,H_q(\mathscr{C}^\phi))\oplus\cdots\Longrightarrow\text{HC}_{p+q}(\mathcal{A}_\Gamma)_{[\phi]}. \tag{6.13}$$

Here $H_\bullet(\mathscr{C}^\phi)$ is the ordinary homology of $\mathscr{C}^\phi$.

(2) *The filtrations by rows and columns of* $\text{Tot}_\bullet(C^\flat(\Gamma_\phi)\otimes_{\Gamma_\phi}(\mathscr{C}^\phi)^\natural)$ *and the quasi-isomorphisms* (6.12) *give rise to spectral sequences,*

$${}^{II}E^2_{p,q}=\text{HC}_p\left(H_q(\Gamma_\phi,\mathscr{C}^\phi)\right)\Longrightarrow\text{HC}_{p+q}(\mathcal{A}_\Gamma)_{[\phi]}, \tag{6.14}$$

$${}^{III}E^2_{p,q}=H_p(\Gamma_\phi,\text{HC}_q(\mathscr{C}^\phi))\Longrightarrow\text{HC}_{p+q}(\mathcal{A}_\Gamma)_{[\phi]}. \tag{6.15}$$

Here $H_q(\Gamma_\phi,\mathscr{C}^\phi)$ is the mixed complex $(H_q(\Gamma_\phi,\mathscr{C}^\phi_\bullet),b,B)$.

REMARK 6.16. The spectral sequence (6.14) is a refinement of the spectral sequence of Getzler-Jones [**24**, Theorem 4.5].

REMARK 6.17. When $\mathscr{C} = C^{\phi}(\mathcal{A})$ and $\alpha = \mathrm{id}$ the spectral sequence (6.15) is the spectral sequence of Feigin-Tsygan [**22**, Theorem 4.1.1] in the finite order case.

REMARK 6.18. The spectral sequence (6.13) seems to be new.

6.5. Infinite order case. In what follows we let $\phi \in \Gamma$ be central element of infinite order. We set $\overline{\Gamma} = \Gamma/\langle\phi\rangle$, where $\langle\phi\rangle$ is the subgroup generated by ϕ.

6.5.1. *From $C^{\phi}(\Gamma, \mathscr{C})$ to $C^{\sigma}(\overline{\Gamma}, \mathscr{C})$.* The canonical projection $\overline{\pi} : \Gamma \to \overline{\Gamma}$ yields a $(\Gamma, \overline{\Gamma})$-equivariant paracyclic module map $\overline{\pi} : C^{\phi}_{\bullet}(\Gamma) \to C_{\bullet}(\overline{\Gamma})$. Composing this map with the natural projection $\pi^{\natural} : C_{\bullet}(\overline{\Gamma})^{\natural} \to C_{\bullet}(\overline{\Gamma})$ and the antisymmetrization map $\varepsilon : C_{\bullet}(\overline{\Gamma}) \to C_{\bullet}(\overline{\Gamma})$ in (6.11), we get a $(\Gamma, \overline{\Gamma})$-equivariant chain map $\overline{\varepsilon}^{\natural} : C^{\phi}_{\bullet}(\Gamma)^{\natural} \to C_{\bullet}(\overline{\Gamma})$. If $\mathscr{M}$ is a ϕ-invariant $k\Gamma$-module, then the action of Γ on $\mathscr{M}$ descends to an action of $\overline{\Gamma}$, and so we get a chain map $\overline{\varepsilon}^{\natural} \otimes 1 : C^{\phi}_{\bullet}(\Gamma, \mathscr{M})^{\natural} \to C_{\bullet}(\overline{\Gamma}, \mathscr{M})$.

Let $C^{\phi}(\Gamma)^{\lambda} = (C^{\phi}_{\bullet}(\Gamma)^{\lambda}, \partial)$ be the Connes complex of $C^{\phi}(\Gamma)$, where $C^{\phi}_{\bullet}(\Gamma)^{\lambda} = C_{\bullet}(\Gamma)/\operatorname{ran}(1 - \tau_{\phi})$ with $\tau_{\phi} = (-1)^m t_{\phi}$ on $C_m(\Gamma)$. Note that ϕ acts like the identity on $C^{\phi}(\Gamma)^{\lambda}$, and so the action of Γ descends to an action of $\overline{\Gamma}$. The projection $\pi^{\lambda} : C^{\phi}_{\bullet}(\Gamma)^{\natural} \to C^{\phi}_{\bullet}(\Gamma)^{\lambda}$ is a chain map. Furthermore, the projection $\overline{\pi} : C_{\bullet}(\Gamma) \to C_{\bullet}(\overline{\Gamma})$ descends to a chain map $\overline{\pi} : C^{\phi}_{\bullet}(\Gamma)^{\lambda} \to C_{\bullet}(\overline{\Gamma})$ in such a way that $\overline{\varepsilon}^{\natural} = \varepsilon\overline{\pi}\pi^{\lambda}$.

It was shown by Marciniak [**33**] that the chain complex $C^{\phi}(\Gamma)^{\lambda}$ gives rise to a projective resolution of the trivial $k\overline{\Gamma}$-module k (see also [**8**]). It then follows that the chain map $\varepsilon\overline{\pi}$ has a chain homotopy inverse $\gamma : C_{\bullet}(\overline{\Gamma}) \to C^{\phi}_{\bullet}(\Gamma)^{\lambda}$. Moreover, results of Kassel [**30**] ensure that there is an explicit graded $k\Gamma$-module map $\iota : C_{\bullet}(\Gamma) \to C_{\bullet}(\Gamma)^{\natural}$ such that, for every ϕ-invariant $k\Gamma$-module $\mathscr{M}$, the k-module map $\iota \otimes 1 : C_{\bullet}(\Gamma, \mathscr{M}) \to C^{\phi}_{\bullet}(\Gamma, \mathscr{M})^{\natural}$ descends to a chain map $(\iota \otimes 1)^{\lambda} : C_{\bullet}(\Gamma, \mathscr{M})^{\lambda} \to C^{\phi}_{\bullet}(\Gamma, \mathscr{M})^{\natural}$ which is a homotopy inverse of $\pi^{\lambda} \otimes 1$. Therefore, we arrive at the following statement.

PROPOSITION 6.19. *For any ϕ-invariant $k\Gamma$-module $\mathscr{M}$, the chain maps $\overline{\varepsilon}^{\natural} \otimes 1 : C^{\phi}_{\bullet}(\Gamma, \mathscr{M})^{\natural} \to C_{\bullet}(\overline{\Gamma}, \mathscr{M})$ and $(\iota \otimes 1)^{\lambda}(\gamma \otimes 1) : C_{\bullet}(\overline{\Gamma}, \mathscr{M}) \to C^{\phi}_{\bullet}(\Gamma, \mathscr{M})^{\natural}$ are chain homotopy inverses.*

REMARK 6.20. For $\mathscr{M} = k$ the fact that $\overline{\varepsilon}^{\natural} \otimes 1 : C^{\phi}_{\bullet}(\Gamma, k)^{\natural} \to C_{\bullet}(\overline{\Gamma}, k)$ is a quasi-isomorphism was established by Marciniak [**33**] in case k is a field of characteristic 0. This yields an algebraic proof of Burghelea's isomorphism (2.11) with coefficients in k.

The periodicity operator $S : C^{\phi}_{\bullet}(\Gamma) \to C^{\phi}_{\bullet-2}(\Gamma)$ is closely related to the cap product with the Euler class $e_{\phi} \in H^2(\overline{\Gamma}, k)$ of the Abelian extension $1 \to \langle\phi\rangle \to \overline{\Gamma} \to \Gamma \to 1$ (*cf.* [**8**]). Let $u_{\phi} \in C^2(\overline{\Gamma}, k)$ be any 2-cocycle representing e_{ϕ}. The cap product with u_{ϕ} then gives rise to a chain map $u_{\phi} \frown - : H_{\bullet}(\overline{\Gamma}) \to H_{\bullet-2}(\overline{\Gamma})$.

LEMMA 6.21 (compare [**27**]). *There is an explicit $(\Gamma, \overline{\Gamma})$-equivariant map $h_{\phi} : C^{\phi}_{\bullet}(\Gamma)^{\natural} \to C_{\bullet-1}(\overline{\Gamma})$ such that $\overline{\varepsilon}^{\natural} S - (u_{\phi} \frown -)\overline{\varepsilon}^{\natural} = \partial h_{\phi} + h_{\phi}(\partial + B_{\phi} S)$.*

Lemma 6.21 implies that, for any ϕ-invariant $k\Gamma$-module, under the quasi-isomorphism provided by Proposition 6.19, the periodicity operator of $\mathrm{HC}_{\bullet}(C^{\phi}(\Gamma, \mathscr{M}))$ is given by the cap product with the Euler class $e_{\phi} \in H^2(\overline{\Gamma}, k)$.

As the cap product with u_{ϕ} is a chain map, we get an S-module $C^{\sigma}(\overline{\Gamma}) := (C_{\bullet}(\overline{\Gamma}), \partial, u_{\phi} \frown -)$. Given any ϕ-invariant mixed complex $\mathscr{C} = (\mathscr{C}_{\bullet}, b, B)$, we shall denote by $C^{\sigma}(\overline{\Gamma}, \mathscr{C})$ the triangular S-module given by the tensor product $C^{\sigma}(\overline{\Gamma}) \otimes_{\overline{\Gamma}}$

$\mathscr{C}$ as defined in Section 5. This gives rise to the total S-module $\operatorname{Tot}(C^{\sigma}(\overline{\Gamma},\mathscr{C})) = (\operatorname{Tot}_{\bullet}(C^{\sigma}(\overline{\Gamma},\mathscr{C})),\mathbf{d},u_{\phi}\frown -)$, where

(6.16)
$$\operatorname{Tot}_m\left(C^{\sigma}(\overline{\Gamma},\mathscr{C})\right) = \bigoplus_{p+q=m} C_p(\overline{\Gamma},\mathscr{C}_q), \qquad \mathbf{d} = \partial + (-1)^p b + (-1)B(u_{\phi}\frown -).$$

We then get a chain map $\theta : \operatorname{Tot}_{\bullet}(C^{\phi}(\Gamma,\mathscr{C}))^{\natural} \to \operatorname{Tot}_{\bullet}(C^{\sigma}(\overline{\Gamma},\mathscr{C}))$ by letting

$$\theta = \bar{\varepsilon}^{\natural}\otimes 1 + (-1)^{p-1}(1\otimes B)(h_{\phi}\otimes 1) \qquad \text{on } C_{p,q}(\Gamma,\mathscr{C}). \tag{6.17}$$

By using Proposition 6.19 and Lemma 6.21 we then obtain the following result.

PROPOSITION 6.22. *For any ϕ-invariant mixed complex $\mathscr{C}$, the chain map* (6.17) *is a quasi-isomorphism. Moreover, we have*

$$\theta S - (u_{\phi}\frown -)S = \mathbf{d}(h_{\phi}\otimes 1) + (h_{\phi}\otimes 1)\mathbf{d},$$

where we have also denoted by $\mathbf{d}$ *the differential of* $\operatorname{Tot}(C^{\phi}(\Gamma,\mathscr{C}))^{\natural}$.

6.5.2. *Good actions.* Given a ϕ-parachain complex $\mathscr{C} = (\mathscr{C}_{\bullet}, b, B)$, we can form its ϕ-coinvariant complex $\mathscr{C}_{\phi} = (\mathscr{C}_{\phi,\bullet}, b, B)$, where $\mathscr{C}_{\phi,m} = \mathscr{C}_m/\operatorname{ran}(\phi-1)$. Note that $\mathscr{C}_{\phi}$ is a ϕ-invariant mixed complex. Moreover, the canonical projection $\pi : \mathscr{C}_{\bullet} \to \mathscr{C}_{\phi,\bullet}$ is a parachain complex map.

DEFINITION 6.23. We shall say that a ϕ-parachain complex $\mathscr{C}$ is *good* when the canonical projection $\pi : \mathscr{C}_{\bullet} \to \mathscr{C}_{\phi,\bullet}$ is a quasi-isomorphism.

REMARK 6.24. It can be shown that $\mathscr{C}$ is a good ϕ-parachain complex if and only if it is quasi-isomorphic to a ϕ-invariant mixed complex (see [**43**]). For instance, $\mathscr{C}$ is good if we have a splitting $\mathscr{C}_{\bullet} = \mathscr{C}_{\bullet}^{\phi} \oplus \operatorname{ran}(\phi-1)$, where $\mathscr{C}^{\phi} = \ker(\phi-1)$, since in this case the inclusion of $\mathscr{C}_{\bullet}^{\phi}$ into $\mathscr{C}_{\bullet}$ is a quasi-isomorphism (see [**25**, Proposition 2.1]).

REMARK 6.25. We refer to [**26**, Example 3.10] for an example of ϕ-paracyclic module which does not give rise to a good ϕ-parachain complex.

DEFINITION 6.26. Let $\phi \in \Gamma$. We shall say that the action of ϕ on $\mathcal{A}$ is *good* when $C^{\phi}(\mathcal{A})$ is a good ϕ-parachain complex.

Combining Proposition 6.22 with Propositions 6.10– 6.11 gives the following result.

THEOREM 6.27. *Let $\phi \in \Gamma$ have infinite order, and set $\overline{\Gamma}_{\phi} = \Gamma_{\phi}/\langle\phi\rangle$. Assume that the action of ϕ on $\mathcal{A}$ is good, and let $\alpha : C_{\bullet}^{\phi}(\mathcal{A}) \to \mathscr{C}_{\bullet}$ be a parachain map and quasi-isomorphism, where $\mathscr{C}$ is a ϕ-invariant mixed complex.*

(1) *The following S-maps are quasi-isomorphisms,*

$$\operatorname{Tot}_{\bullet}(C^{\sigma}(\overline{\Gamma}_{\phi},\mathscr{C})) \xleftarrow{\theta(1\otimes\alpha)} \operatorname{Tot}_{\bullet}\left(C^{\phi}(\Gamma_{\phi},\mathcal{A})\right)^{\natural} \underset{\mathrm{AW}^{\natural}}{\overset{\sqcup\!\sqcup^{\natural}}{\rightleftarrows}} \operatorname{Diag}_{\bullet}\left(C^{\phi}(\Gamma_{\phi},\mathcal{A})\right)^{\natural} \xrightarrow{\mu_{\phi}} C_{\bullet}(\mathcal{A}_{\Gamma})^{\natural}_{[\phi]}. \tag{6.18}$$

(2) *Under these quasi-isomorphisms the periodicity operator of $\operatorname{HC}_{\bullet}(\mathcal{A})_{[\phi]}$ is given by the cap product,*

$$e_{\phi}\frown - : H_{\bullet}(\operatorname{Tot}(C^{\sigma}(\overline{\Gamma}_{\phi},\mathscr{C}))) \longrightarrow H_{\bullet-2}(\operatorname{Tot}(C^{\sigma}(\overline{\Gamma}_{\phi},\mathscr{C}))).$$

where $e_{\phi} \in H^2(\overline{\Gamma}_{\phi},k)$ is the Euler class of the extension $1 \to \langle\phi\rangle \to \Gamma_{\phi} \to \overline{\Gamma}_{\phi} \to 1$.

By combining Theorem 6.27 with Proposition 5.1 we obtain the following corollary.

COROLLARY 6.28. *Under the assumptions of Theorem* 6.27, *the quasi-isomorphisms* (6.18) *and the filtration by columns of* $\mathrm{Tot}_\bullet(C^\sigma(\overline{\Gamma}_\phi, \mathscr{C}))$ *give rise to a spectral sequence,*

$$E^2_{p,q} = H_p(\overline{\Gamma}_\phi, H_q(\mathscr{C})) \Longrightarrow \mathrm{HC}_{p+q}(\mathcal{A}_\Gamma)_{[\phi]}, \tag{6.19}$$

where $H_\bullet(\mathscr{C})$ *is the ordinary homology of* $\mathscr{C}$. *If the* b*-differential of* $\mathscr{C}$ *is zero, then* $E^2_{p,q} = H_p(\overline{\Gamma}_\phi, \mathscr{C}_q)$ *and the* E_2*-differential is* $(-1)^p B(u_\phi \frown -) : H_p(\overline{\Gamma}_\phi, \mathscr{C}_q) \longrightarrow H_{p-2}(\overline{\Gamma}_\phi, \mathscr{C}_{q+1})$.

REMARK 6.29. For $\mathscr{C} = C^\phi(\mathcal{A})$ and $\alpha = \mathrm{id}$ the spectral sequence (6.19) specializes to the spectral sequence of Feigin-Tsygan (Theorem 4.1.1 and Remark 4.1.2 of [**22**]).

The k-module of group cochains $C^\bullet(\overline{\Gamma}_\phi, k)$ is a differential graded ring under the cup product. The cap product yields an associative differential graded action of $C^\bullet(\overline{\Gamma}_\phi, k)$ on $C_\bullet(\overline{\Gamma}_\phi)$. By equivariance it extends to a graded bilinear map,

$$- \frown - : C^\bullet(\overline{\Gamma}_\phi) \times \mathrm{Tot}_\bullet\left(C^\sigma(\overline{\Gamma}_\phi, \mathscr{C})\right) \longrightarrow \mathrm{Tot}_\bullet\left(C^\sigma(\overline{\Gamma}_\phi, \mathscr{C})\right). \tag{6.20}$$

We obtain a graded differential action of $C^\bullet(\overline{\Gamma}_\phi, k)$ on $\mathrm{Tot}_\bullet(C^\sigma(\overline{\Gamma}_\phi, \mathscr{C}))$. Combining this with Theorem 6.27 we then arrive at the following corollary.

COROLLARY 6.30. *Under the assumptions of Theorem* 6.27, *the quasi-isomorphisms* (6.18) *and the cap product* (6.20) *give rise to a graded action of the cohomology ring* $H^\bullet(\overline{\Gamma}_\phi, k)$ *on* $\mathrm{HC}_\bullet(\mathcal{A}_\Gamma)_{[\phi]}$. *The periodicity operator* $\mathrm{HC}_\bullet(\mathcal{A}_\Gamma)_{[\phi]}$ *is given by the action of the Euler class* e_ϕ. *In particular,* $\mathrm{HP}(\mathcal{A}_\Gamma)_{[\phi]} = 0$ *whenever* e_ϕ *is nilpotent in* $H^\bullet(\overline{\Gamma}_\phi, k)$.

REMARK 6.31. Nistor [**37**] proved that, for any infinite order element $\phi \in \Gamma$, the cyclic homology $\mathrm{HC}_\bullet(\mathcal{A}_\Gamma)_{[\phi]}$ is a module over $H^\bullet(\overline{\Gamma}_\phi, k)$ and the action of e_ϕ agrees with the periodicity operator. Therefore, Corollary 6.30 provides us with a simple derivation of Nistor's result when the action of ϕ on $\mathcal{A}$ is good. In this case, we actually get a more precise result since the cap product (6.20) already provides us with a differential graded action at the level of chains.

6.6. Infinite order case (general actions). Let us now look at arbitrary actions of infinite order elements of Γ.

6.6.1. *Quasi-isomorphisms.* We have the following result.

THEOREM 6.32. *Let* $\phi \in \Gamma$ *have infinite order, and set* $\overline{\Gamma}_\phi = \Gamma_\phi/\langle\phi\rangle$. *In addition, let* $\alpha : C^\phi_\bullet(\mathcal{A}) \to \mathscr{C}_\bullet$ *be a quasi-isomorphism of* ϕ*-parachain complexes.*

(1) *The following* S*-maps are quasi-isomorphisms,*

$$\mathrm{Tot}_\bullet\left(C^\phi(\Gamma_\phi, \mathscr{C})\right)^\natural \xleftarrow{1\otimes\alpha} \mathrm{Tot}_\bullet\left(C^\phi(\Gamma_\phi, \mathcal{A})\right)^\natural \underset{\mathrm{AW}^\natural}{\overset{\sqcup^\natural}{\rightleftarrows}} \mathrm{Diag}_\bullet\left(C^\phi(\Gamma_\phi, \mathcal{A})\right)^\natural \xrightarrow{\mu_\phi} C_\bullet(\mathcal{A}_\Gamma)^\natural_{[\phi]}. \tag{6.21}$$

(2) *These quasi-isomorphisms and the filtration by columns of* $\mathrm{Tot}_\bullet(C^\phi(\Gamma_\phi)^\natural \otimes_{\Gamma_\phi} \mathscr{C})$ *give rise to a spectral sequence,*

$$E^2_{p,q} = H_p(\overline{\Gamma}_\phi, H_q(\mathscr{C})) \Longrightarrow \mathrm{HC}_{p+q}(\mathcal{A}_\Gamma)_{[\phi]}. \tag{6.22}$$

REMARK 6.33. As in Remark 6.29, the spectral sequence (6.22) gives back the spectral sequence of Feigin-Tsygan [**22**].

6.6.2. *$H^\bullet(\overline{\Gamma},k)$-module structure on* $\mathrm{HC}_\bullet(\mathcal{A}_\Gamma)_{[\phi]}$. The results of Nistor [**37**] alluded to in Remark 6.31 involve various deep homological algebra arguments. We shall now explain how to get Nistor's results as a consequence of the construction *at the level of chains* of a coproduct for paracyclic modules.

The ordinary Alexander-Whitney map yields a coassociative coproduct for simplicial modules. For instance, consider the diagonal map $\delta: C_\bullet(\Gamma,k)\rightarrow \mathrm{Diag}_\bullet(C(\Gamma,k)\otimes C(\Gamma,k))$ given by $\delta(\eta)=\eta\otimes\eta$, $\eta\in C_m(\Gamma,k)$. This is a map of simplicial modules. Composing it with the Alexander-Whitney map we obtain a coassociative graded coproduct $\Delta:=\mathrm{AW}\circ\delta: C_\bullet(\Gamma,k)\rightarrow \mathrm{Tot}_\bullet(C(\Gamma,k)\otimes C(\Gamma,k))$. By duality this gives the standard cup product on the group cohomology complex $(C^\bullet(\Gamma,k),\partial)$.

Suppose that ϕ is a central element of Γ. Let $\delta: C^\phi(\Gamma)\rightarrow \mathrm{Diag}_\bullet(C^\phi(\Gamma,k)\otimes C^\phi(\Gamma))$ be the graded $k\Gamma$-module map given by

$$\delta(\psi_0,\ldots,\psi_m)=[(\psi_0,\ldots,\psi_m)\otimes_\Gamma 1]\otimes(\psi_0,\ldots,\psi_m),\qquad \psi_j\in\Gamma.$$

Here the action of Γ on $C_\bullet(\Gamma,k)\otimes C_\bullet(\Gamma)$ is on the second factor. We obtain a map of parachain complexes, and so this gives rise to an S-map $C^\phi(\Gamma)^\natural\rightarrow \mathrm{Diag}(C^\phi(\Gamma,k)\otimes C^\phi(\Gamma))$. Composing this map with the paracyclic Alexander-Whitney map from Theorem 4.1 we then obtain an S-map $\mathrm{AW}^\natural\circ\delta: C^\phi(\Gamma)^\natural\rightarrow \mathrm{Tot}(C^\phi(\Gamma,k)\otimes C^\phi(\Gamma))^\natural$.

Let $C^\flat(\overline{\Gamma},k)$ be the mixed complex $(C_\bullet(\overline{\Gamma},k),\partial,0)$. As $\varepsilon\overline{\pi}B_\phi=0$ we get a parachain bicomplex map $(\varepsilon\overline{\pi})\otimes 1: C^\phi_\bullet(\Gamma,k)\otimes C^\phi_\bullet(\Gamma)\rightarrow C^\flat_\bullet(\overline{\Gamma},k)\otimes C^\phi_\bullet(\Gamma)$. This then gives rise to a parachain S-module map $(\varepsilon\overline{\pi})\otimes 1: \mathrm{Tot}(C^\phi(\Gamma,k)\otimes C^\phi(\Gamma))^\natural\rightarrow \mathrm{Tot}(C^\flat(\overline{\Gamma},k)\otimes C^\phi(\Gamma))^\natural$. As explained in Section 5, the para-S-module $\mathrm{Tot}(C^\flat(\overline{\Gamma},k)\otimes C^\phi(\Gamma))^\natural$ can be realized as the total para-S-module of some vertical triangular para-S-module. In the case of $\mathrm{Tot}(C^\flat(\overline{\Gamma},k)\otimes C^\phi(\Gamma))^\natural$ that triangular para-S-module is the tensor product $C(\overline{\Gamma},k)\otimes C^\phi(\Gamma)^\natural$ as defined in (5.1). We thus obtain a para-S-module map $(\varepsilon\overline{\pi})\otimes 1:\mathrm{Tot}_\bullet(C^\phi_\bullet(\Gamma,k)\otimes C^\phi(\Gamma))^\natural\rightarrow \mathrm{Tot}_\bullet(C^\flat(\overline{\Gamma},k)\otimes C^\phi(\Gamma)^\natural)$. Composing it with the S-map $\mathrm{AW}^\natural\circ\delta$ above we obtain a para-S-module map,

$$\Delta^\natural:=((\varepsilon\overline{\pi})\otimes 1)\,\mathrm{AW}^\natural\circ\delta: C^\phi(\Gamma)^\natural\longrightarrow \mathrm{Tot}_\bullet(C^\flat(\overline{\Gamma},k)\otimes C^\phi(\Gamma)^\natural).$$

Given any ϕ-parachain complex $\mathscr{C}$, by (5.2) and (5.3) we have $\mathrm{Tot}(C^\phi(\Gamma,\mathscr{C}))^\natural=\mathrm{Tot}(C^\phi(\Gamma)^\natural\otimes_\Gamma\mathscr{C})$ and $\mathrm{Tot}(C^\flat(\overline{\Gamma},k)\otimes\mathrm{Tot}(C^\phi(\Gamma,\mathscr{C}))^\natural)=\mathrm{Tot}(\mathrm{Tot}(C^\flat(\overline{\Gamma},k)\otimes C^\phi(\Gamma))^\natural\otimes_\Gamma\mathscr{C})$. Therefore, we get an S-module map,

$$\Delta^\natural\otimes 1:\mathrm{Tot}_\bullet\left(C^\phi(\Gamma,\mathscr{C})\right)^\natural\longrightarrow \mathrm{Tot}_\bullet\left(C^\flat(\overline{\Gamma},k)\otimes\mathrm{Tot}(C^\phi(\Gamma,\mathscr{C}))^\natural\right).$$

Combining this with the duality between $C_\bullet(\overline{\Gamma},k)$ and $C^\bullet(\overline{\Gamma},k)$ then provides us with a differential graded bilinear map,

$$\triangleright: C^\bullet(\overline{\Gamma},k)\times\mathrm{Tot}_\bullet\left(C^\phi(\Gamma,\mathscr{C})\right)^\natural\longrightarrow \mathrm{Tot}_\bullet\left(C^\phi(\Gamma,\mathscr{C})\right)^\natural. \tag{6.23}$$

More precisely, given any cochain $u\in C^p(\overline{\Gamma},k)$, $p\geqslant 0$, and chains $\eta\in C^\phi_\bullet(G)^\natural$ and $\xi\in\mathscr{C}_\bullet$, we have

$$u\triangleright(\eta\otimes_\Gamma\xi)=\left[(u\otimes 1)\Delta^\natural\eta\right]\otimes_\Gamma\xi\in C^\phi_{\bullet-p}(\Gamma)^\natural\otimes_\Gamma\mathscr{C}_\bullet.$$

PROPOSITION 6.34 (see [**42**]). *Let $\mathscr{C}$ be a ϕ-parachain complex. Then the bilinear map* (6.23) *descends to an associative graded action,*

$$\triangleright: H^\bullet(\overline{\Gamma},k)\times\mathrm{HC}_\bullet\left(\mathrm{Tot}(C^\phi(\Gamma,\mathscr{C}))\right)\longrightarrow \mathrm{HC}_\bullet\left(\mathrm{Tot}(C^\phi(\Gamma,\mathscr{C}))\right).$$

Moreover, the periodicity operator of $\mathrm{HC}_\bullet\big(\mathrm{Tot}(C^\phi(\Gamma,\mathscr{C}))\big)$ *is given by the action of the Euler class of the extension* $1 \to \langle\phi\rangle \to \overline{\Gamma} \to \Gamma \to 1$.

Specializing this to $\mathscr{C} = C^\phi(\mathcal{A})$ and using Proposition 6.11 gives the following result.

THEOREM 6.35 (compare [**37**]). *Let* $\phi \in \Gamma$ *have infinite order, and set* $\overline{\Gamma}_\phi = \Gamma_\phi/\langle\phi\rangle$.

(1) *The quasi-isomorphisms* (6.9) *and the bilinear map* (6.23) *for* $\mathscr{C} = C^\phi(\mathcal{A})$ *give rise to an action of the cohomology ring* $H^\bullet(\overline{\Gamma}_\phi, k)$ *on the cyclic homology* $\mathrm{HC}_\bullet(\mathcal{A}_\Gamma)_{[\phi]}$.
(2) *The periodicity operator of* $\mathrm{HC}_\bullet(\mathcal{A}_\Gamma)_{[\phi]}$ *is given by the action of the Euler class* $e_\phi \in H^2(\overline{\Gamma}_\phi, k)$ *of the extension* $1 \to \langle\phi\rangle \to \overline{\Gamma}_\phi \to \Gamma_\phi \to 1$. *In particular,* $\mathrm{HP}_\bullet(\mathcal{A}_\Gamma)_{[\phi]} = 0$ *whenever* e_ϕ *is nilpotent in* $H^\bullet(\Gamma_\phi, k)$.

7. Back to Group Actions on Manifolds

We shall now explain how the results of Section 6 enables us to obtain the results of Section 2 for group actions on manifold.

Throughout this section we let M be closed manifold and Γ a group acting on M by diffeomorphisms. We denote by $\mathcal{A}$ the algebra $C^\infty(M)$ equipped with its standard locally convex topology. We also keep on using the notation of Section 2 and Section 6.

7.1. Twisted CHKR theorem.

DEFINITION 7.1. We say that the action of $\phi \in \Gamma$ on M is *clean* if, for every $x_0 \in M^\phi$, the following conditions are satisfied:

(1) M^ϕ is a submanifold near x_0.
(2) $T_{x_0}M^\phi = \ker(1-\phi'(x_0))$ & $T_{x_0}M = T_{x_0}M^\phi \oplus \mathrm{ran}(1-\phi'(x_0))$.

REMARK 7.2. The action of ϕ is always clean if it preserves an affine connection on TM, e.g., if ϕ preserves a Riemannian metric. This always happens if ϕ has finite order.

Let $\phi \in \Gamma$ act cleanly on M. In this case the fixed-point set M^ϕ is a disjoint union of submanifolds of M parametrized by the values of the rank of $\phi'(x)$. As in Section 2 we define the de Rham complex $(\Omega^\bullet(M^\phi), d)$ as the direct sum of the de Rham complexes of the submanifold components of M^ϕ. Here $(\Omega^\bullet(M^\phi), 0, d)$ is a mixed complex and we have a map of parachain complexes $\alpha^\phi : (C^\phi_\bullet(\mathcal{A}), b) \to (\Omega^\bullet(M^\phi), 0)$ given by

$$\alpha^\phi(f^0 \otimes \cdots \otimes f^m) = \frac{1}{m!}\left(f^0 df^1 \wedge \cdots \wedge df^m\right)_{|M^\phi}, \qquad f^j \in C^\infty(M).$$

THEOREM 7.3 ([**7**]; see also [**39**]). *If* ϕ *acts cleanly, then* $\alpha^\phi : (C^\phi_\bullet(\mathcal{A}), b) \to (\Omega^\bullet(M^\phi), 0)$ *is a quasi-isomorphism of chain complexes.*

7.2. Finite order case. If $\phi \in \Gamma$, then ϕ acts cleanly on M, and so Theorem 7.3 holds. In this case the mixed complex $\mathrm{Tot}(C^\flat(\Gamma_\phi)\otimes_{\Gamma_\phi}\Omega^\bullet(M^\phi))^\natural$ is precisely the mixed-equivariant complex $C^{\Gamma_\phi}(M^\phi)$. Thus, by combining Theorem 6.14 and Theorem 7.3 we get the following result.

THEOREM 7.4. *Let $\phi \in \Gamma$ have finite order. Then the following S-maps are quasi-isomorphisms,*

$$C^{\Gamma_\phi}(M^\phi)^\natural \xleftarrow{(\varepsilon\nu_\phi)\otimes\alpha^\phi} \mathrm{Tot}_\bullet\left(C^\phi(\Gamma_\phi,\mathcal{A})\right)^\natural \underset{\mathrm{AW}^\natural}{\overset{\sqcup\!\sqcup^\natural}{\rightleftarrows}} \mathrm{Diag}_\bullet\left(C^\phi(\Gamma_\phi,\mathcal{A})\right)^\natural \xrightarrow{\mu_\phi} C_\bullet(\mathcal{A}_\Gamma)^\natural_{[\phi]}.$$

This allows us to obtain Theorem 2.10.

REMARK 7.5. If $\phi = 1$, then μ_1 is a isomorphism and has an explicit inverse (see [**42**]). Thus, we get an explicit S-map and quasi-isomorphism $[(\varepsilon\nu_1)\otimes \alpha^\phi]\,\mathrm{AW}^\natural\,\mu_1^{-1} : C_\bullet(\mathcal{A}_\Gamma)^\natural_{[1]} \to C^\Gamma_\bullet(M)^\natural$. This is a dual version of the characteristic map of Connes [**13**].

7.3. Infinite order case. Let $\phi \in \Gamma$ have infinite order and act cleanly. Theorem 7.3 then holds, and so $C^\phi(\mathcal{A})$ is a good ϕ-parachain complex (*cf.* Remark 6.24), and so the action of ϕ is good. In this case the S-module $\mathrm{Tot}_\bullet(C^\sigma(\overline{\Gamma}_\phi,\mathscr{C}))$ is just the S-module $(C^\sigma_\bullet(\overline{\Gamma}_\phi, M^\phi), \partial + (-1)^p(u_\phi \frown -)d, u_\phi \frown -)$ defined in (2.8)–(2.9). Therefore, by combining Theorem 6.27 and Theorem 7.3 we obtain the following result.

THEOREM 7.6. *Let $\phi \in \Gamma$ have infinite order and acts cleanly on M. The following S-maps are quasi-isomorphisms,*

$$C^\sigma_\bullet(\overline{\Gamma}_\phi, M^\phi) \xleftarrow{\theta(1\otimes\alpha)} \mathrm{Tot}_\bullet\left(C^\phi(\Gamma_\phi,\mathcal{A})\right)^\natural \underset{\mathrm{AW}^\natural}{\overset{\sqcup\!\sqcup^\natural}{\rightleftarrows}} \mathrm{Diag}_\bullet\left(C^\phi(\Gamma_\phi,\mathcal{A})\right)^\natural \xrightarrow{\mu_\phi} C_\bullet(\mathcal{A}_\Gamma)^\natural_{[\phi]}.$$

This allows us to get Theorem 2.15.

Finally, if $\phi \in \Gamma$ does not act cleanly on M we obtain Theorem 2.19 directly from Theorem 6.35.

References

[1] R. Akbarpour, M. Khalkhali: *Hopf algebra equivariant cyclic homology and cyclic homology of crossed product algebras.* J. Reine Angew. Math. **559** (2003), 137–152.

[2] P. Baum, A. Connes, *Chern character for discrete groups*, *A fête of topology*, Academic Press, Boston, 1988, pp. 163-232.

[3] A. Bauval, *Le Théorème d'Eilenberg-Zilber en homologie cyclique entière*, preprint, 1998, `arXiv:1611.08437`.

[4] R. Bott, *On some formulas for the characteristic classes of group actions*, Lecture Notes in Math., 652, Springer, Berlin, 1978, pp. 25–61.

[5] J. Brodzki, S. Dave, V. Nistor, *The periodic cyclic homology of crossed products of finite type algebras*, Adv. Math. **306** (2017), 494–523.

[6] R. Brown, *The twisted Eilenberg-Zilber theorem*, Simposio di Topologia (Messina, 1964), Edizioni Oderisi, Gubbio, 1965, pp. 33–37.

[7] J.-L. Brylinski, V. Nistor, *Cyclic cohomology of étale groupoids*, K-Theory **8** (1994), 341–365.

[8] D. Burghelea, *The cyclic homology of the group rings*, Comment. Math. Helv. **60** (1985), 354–365.

[9] D. Burghelea, *Cyclic homology and the algebraic K-theory of k-modules. I*, Contemp. Math., 55, Amer. Math. Soc. Providence, RI, 1986, pp. 89–115.

[10] A. Connes, *Spectral sequence and homology of currents for operator algebras*, Math. Forschungsinstitut Oberwolfach Tagungsbericht 42/81, Funktionalanalysis und C*-Algebren, 27.9.–3.10, 1981.

[11] A. Connes, *Cohomologie cyclique et foncteur* Ext^n, C. R. Acad. Sci. Paris Ser. I Math. **296** (1983), 953–958.

[12] A. Connes, *Noncommutative differential geometry*, Inst. Hautes Études Sci. Publ. Math. **62** (1985), 257–360.

[13] A. Connes, *Cyclic cohomology and the transverse fundamental class of a foliation*, Pitman Res. Notes in Math., 123, Longman, Harlow, 1986, pp. 52–144

[14] A. Connes, H. Moscovici, *Cyclic cohomology, the Novikov conjecture and hyperbolic groups*, Topology **29** (1990), no. 3, 345–388, 1990.

[15] A. Connes, H. Moscovici, *The local index formula in noncommutative geometry*, Geom. Funct. Anal. **5** (1995), 174–243.

[16] A. Connes, H. Moscovici, *Hopf algebras, cyclic cohomology and the transverse index theorem*, Commun. Math. Phys. **198** (1998), 199–246.

[17] A. Connes, H. Moscovici, *Differentiable cyclic cohomology Hopf algebraic structures in transverse geometry*, Monographie 38 de L'Enseignement Mathématique, pp. 217–256, 2001.

[18] M. Crainic, *Cyclic cohomology of étale Groupoids: the general case*, K-Theory **17** (1999), 319–362.

[19] J. Cuntz, D. Quillen, *Cyclic homology and nonsingularity*, J. Amer. Math. Soc. **8** (1995), 373–442.

[20] J.L. Dupont, *Simplicial De Rham cohomology and characteristic classes of flat bundles*, Topology **15** (1976), 233–245.

[21] I. Emmanouil, *On a class of groups satisfying Bassõ conjecture*, Invent. Math. **132** (1998) 307–330.

[22] B. Feigin, B. Tsygan, *Cyclic homology of algebras with quadratic relations, universal enveloping algebras and group algebras*, Lecture Notes in Math., 1289, Springer, 1987, pp. 210–239.

[23] E. Getzler, *The equivariant Chern character for noncompact Lie groups*, Adv. Math. **109** (1994), 88–107.

[24] E. Getzler, J.D.S. Jones, *The cyclic cohomology of crossed product algebras*, J. Reine Angew. Math. **445** (1993), 161–174.

[25] T. Hadfield, U. Krähmer, *Twisted homology of quantum $SL(2)$*, K-Theory **34** (2005), 327–360.

[26] T. Hadfield, U. Krähmer, *Braided homology for quantum groups*, J. K-Theory **4** (2009), 299–332.

[27] R. Ji, *Nilpotency of Connes' periodicity operator and the idempotent conjectures*, K-Theory **9** (1995), 59–76.

[28] J.D. Jones, C. Kassel, *Bivariant cyclic theory*, K-Theory **3** (1989) 339–365.

[29] C. Kassel, *Cyclic homology, comodules and mixed complexes*, J. Algebra **107** (1987), 195–216.

[30] C. Kassel, *Homologie cyclique, caractère de Chern et lemme de perturbation*, J. reine angew. Math. **408** (1990), 159–180.

[31] M. Khalkhali, B. Rangipour, *On the generalized cyclic Eilenberg-Zilber Theorem*, Canad. Math. Bull. Vol. **47** (1), (2004) 38–48.

[32] J.-L. Loday, *Cyclic homology*, Springer, Berlin, 1992.

[33] Z. Marciniak, *Cyclic homology of group rings*. Banach Center Publ., 18, PWN, Warsaw, 1986, pp. 210–239.

[34] H. Moscovici, *Geometric construction of Hopf cyclic characteristic classes*, Adv. Math. **274** (2015), 651–680.

[35] H. Moscovici, B. Rangipour, *Hopf cyclic cohomology and transverse characteristic classes*, Adv. Math. **227** (2011), 654–729.

[36] R. Nest, *Cyclic cohomology of crossed products with $\mathbb{Z}$*, J. Funct. Anal. **80** (1988), no. 2, 235–283.

[37] V. Nistor, *Group cohomology and the cyclic cohomology of crossed products*, Invent. Math. **99** (1990), 411–424.

[38] N. Neumaier, M.J. Pflaum, H.B. Posthuma, X. Tang, *Homology of formal deformations of proper ètale Lie groupoids*, J. Reine Angew. Math. **593** (2006) 117Ð168.

[39] M.J. Pflaum, H.B. Posthuma, X. Tang, *On the Hochschild homology of convolution algebras of proper Lie groupoids*, preprint `arXiv:2009.03216`, to appear in J. Noncommutat. Geom..

[40] R. Ponge, *The cyclic homology of crossed-product algebras, I*, C. R. Acad. Sci. Paris, sér. I, **355** (2017), 618–622.

[41] R. Ponge, *The cyclic homology of crossed-product algebras, II*, C. R. Acad. Sci. Paris, sér. I, **355** (2017), 623–627.

[42] R. Ponge, *Cyclic homology and group actions*, J. Geom. Phys. **123** (2018), 30–52.

[43] R. Ponge, *Periodicity and cyclic homology. Para-S-modules and perturbation lemmas.* J. Noncomm. Geom. **14** (2020), 1365–1444.

[44] R. Ponge, *The Paracyclic Eilenberg-Zilber theorem*, in preparation.

[45] R. Ponge, H. Wang, *Noncommutative geometry and conformal geometry. II, Connes-Chern character and the local equivariant index theorem.* J. Noncomm. Geom. **10** (2016), 307–378.
[46] B. Tsygan, *Homology of matrix Lie algebras over rings and Hochschild homology*, Uspekhi Math. Nawk. **38** (1983), 217–218.
[47] C. Voigt, *Equivariant periodic cyclic homology*, J. Inst. Math. Jussieu **6** (2007), 689–763.

SCHOOL OF MATHEMATICAL SCIENCES, SICHUAN UNIVERSITY, CHENGDU, PEOPLE'S REPUBLIC OF CHINA
Email address: ponge.math@icloud.com

Proceedings of Symposia in Pure Mathematics
Volume **105**, 2023
https://doi.org/10.1090/pspum/105/01911

Cyclic cocycles and quantized pairings in materials science

Emil Prodan

Abstract. The pairings between the cyclic cohomologies and the K-theories of separable C^*-algebras supply topological invariants that often relate to physical response coefficients of materials. Using three numerical simulations, we exemplify how some of these invariants survive throughout the full Sobolev domains of the cocycles. These interesting phenomena, which can be explained by index theorems derived from Alain Connes' quantized calculus, are now well understood in the independent electron picture. Here, we review recent developments addressing the dynamics of correlated many-fermions systems, obtained in collaboration with Bram Mesland. They supply a complete characterization of an algebra of relevant derivations over the C^*-algebra of canonical anti-commutation relations indexed by a generic discrete Delone lattice. It is argued here that these results already supply the means to generate interesting and relevant states over this algebra of derivations and to identify the cyclic cocycles corresponding to the transport coefficients of the many-fermion systems. The existing index theorems for the pairings of these cocycles, in the restrictive single fermion setting, are reviewed and updated with an emphasis on pushing the analysis on Sobolev domains. An assessment of possible generalizations to the many-body setting is given.

Contents

This work was supported by the U.S. National Science Foundation through the grants DMR-1823800 and CMMI-2131760.

1. Introduction, motivations and challenges

Cyclic cocycles and their pairings with the K-theories of underlining algebras of macroscopic physical observables can be often related to physical response coefficients quantifying the reactions of physical systems to external stimuli. For example, the cyclic cocycles generated from circle actions and invariant traces often relate to transport coefficients [**45**], which quantify the relation between the current-density of particle ensembles and macroscopic potential-gradients across samples. One immediate value of this connection to the physicists comes from the fact that disorder can be incorporated in the analysis with virtually no extra effort. For example, the dynamics of un-correlated electrons in a perfect crystal is governed by Hamiltonians drawn from the C^*-algebra $C^*(\mathbb{Z}^d)$ (its stabilization, to be more precise), where d is the space dimension of the sample. Including disorder amounts to promoting the group C^*-algebra to a crossed product algebra $C(\Xi) \rtimes \mathbb{Z}^d$, where Ξ is the space of disorder configurations.

Noncommutative Geometry [**23**] supplies a framework where physics applications can be developed and, most importantly, where difficult calculations can be completed, *e.g* by following existing general templates. One such example is the charge transport theory by Bellissard and his collaborators [**10, 11**], which was developed from first principles and culminated with expressions of the transport coefficients that rigorously incorporate the role of dissipation. In particular, with an expression for the Hall conductance at low temperatures and in the presence of disorder. Physicists often look at this expression through a left regular representation π_ξ of the algebra $C(\Xi) \rtimes \mathbb{Z}^2$ on $\ell^2(\mathbb{Z}^2)$, which up to a multiplicative factor reads as

$$\sigma_H = \lim_{V \to \mathbb{R}^2} \frac{1}{|V|} \sum_{x \in \mathbb{Z}^2 \cap V} \langle x | P_F [[X_1, P_F], [X_2, P_F]] | x \rangle, \tag{1.1}$$

where $P_F = \pi_\xi(p_F)$ is the Fermi projection corresponding to the Fermi energy E_F, i.e. the spectral projection of the Hamiltonian onto $[-\infty, E_F]$, and (X_1, X_2) is the position operator on $\ell^2(\mathbb{Z}^2)$. In the C^*-language, this expression reads

$$\sigma_H = \sum_{\lambda \in \mathcal{S}_2} (-1)^\lambda \mathcal{T}\big(p_F(\partial_{\lambda_1} p_F)(\partial_{\lambda_2} p_F)\big), \tag{1.2}$$

where $\mathcal{S}_N$ denotes the permutation group of N objects and $(\mathcal{T}, \partial)$ is the standard differential structure on $C(\Xi) \rtimes \mathbb{Z}^2$, with the trace $\mathcal{T}$ induced from a measure on Ξ. One immediately recognizes on the right side of (1.2) a pairing of a cyclic cocycle with p_F, hence one can understand on the spot the topological content of (1.2). However, this simplicity is deceiving because we are dealing here with elements from the weak closure of the C^*-algebra, unless the Fermi energy falls in a spectral gap. This is the central theme of the applications we will discuss here, which are first developed in a C^*-algebraic context and then pushed into Sobolev domains.

For us, expression (1.2) also supplied the key to a very efficient numerical algorithm to evaluate expressions like (1.1). The difficulty in the numerical evaluation of (1.1) comes from the incompatibility between the position operator and the periodic boundary conditions assumed in the computer simulations of finite-samples. The solution presented itself once we chose to work directly with the C^*-algebra rather than a representation [**43, 46**] and, to give a flavor of what the C^*-formalism

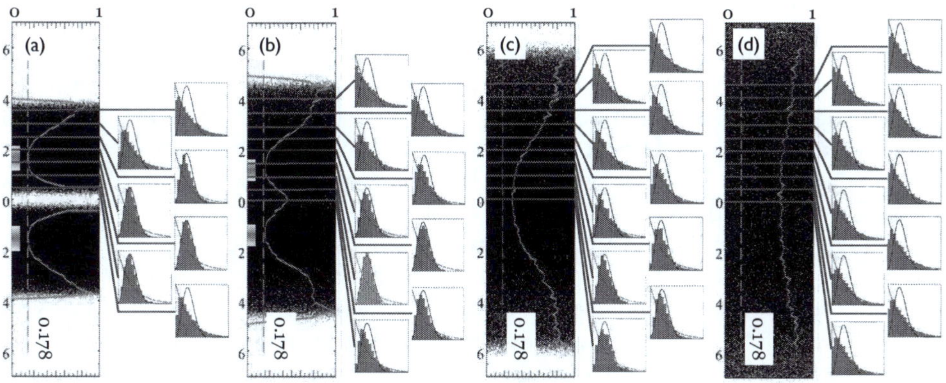

FIGURE 1.1. (Reproduced with permission from [**41**]) Spectrum of H_ξ from (1.4), sampled at many random configurations, together with statistical analysis indicating regions of localized and delocalized spectrum (see [**41**] for full details). From left to right, the four panels correspond to disorder strengths $W = 3$, 5, 8 and 11, respectively. Note the vanishing of the spectral gap, which occurs around $W = 4$.

offered, we present below the optimal solution, which consists of the replacement

$$[X_j, A] \to \sum_{z^{2L_j+1}=1} c_z z^{X_j} A z^{-X_j}, \quad c_z = \begin{cases} \frac{z^{L_j}}{1-z}, & z \neq 1 \\ 0, & z = 1 \end{cases} \tag{1.3}$$

where $2L_j + 1$ is the size of the sample in direction j. Under mild assumption on the Hamiltonians, the above recipe leads to numerical algorithms that converge exponentially fast to the thermodynamic limit [**46**]. This gave us the opportunity, for the first time, to get close enough to the topological phase transitions, mentioned next, and compute related critical exponents and other physical quantities of interest. These ideas crystallized around the year 2009 and the first published results can be found in [**40**]. After that, the workstations in my computational physics lab worked uninterrupted for years, on mapping the remarkable properties of the cyclic cocycle pairings. The best way to communicate what I mean by "remarkable" is to showcase a few examples.

Let us first examine the disordered Haldane model Hamiltonian [**26**]

$$H_\xi = \sum_{\langle x,y\rangle} |x\rangle\langle y| + 0.6\imath \sum_{\langle\langle x,y\rangle\rangle} \eta_x\big(|x\rangle\langle y| - |y\rangle\langle x|\big) + W\sum_x \xi_x |x\rangle\langle x|, \tag{1.4}$$

defined over the honeycomb lattice and where $\langle\cdot,\cdot\rangle/\langle\langle\cdot,\cdot\rangle\rangle$ indicate first/second near-neighboring sites and ξ_x are drawn randomly and independently from the interval $[-1/2, 1/2]$ (see [**41**] for the rest of the details). W will be referred to as the disorder strength. H_ξ is a left regular representation of an element from $M_2(\mathbb{C}) \otimes C\big([-\frac{1}{2},\frac{1}{2}]^{\mathbb{Z}^2}\big) \rtimes \mathbb{Z}^2$ and the phenomena exhibited here are generic, so the particular expression of this element is really not that important, though one needs to make sure that the Fermi projection falls into a non-trivial K-theoretic class when $W = 0$ (the constant 0.6 in Eq. (1.4) assures just that).

Fig. 1.1 reports the spectrum of H_ξ at various levels of disorder, evaluated for a large but finite sample. Let us point out that the spectral gap, still visible for $W = 3$, vanishes at larger disorder strengths. Fig. 1.1 also reports statistical analysis that detects the spatial character of the eigenvectors corresponding to the different parts of the spectrum. Specifically, at the highlighted points where the blue curves in Fig. 1.1 touch the value 0.178, the eigenvectors are uniformly spread throughout the sample, while for the rest of the spectrum the eigenvectors have a localized character. In an influential paper [**1**], Anderson and his collaborators gave strong evidence that, in one and two space dimensions, the eigenvectors of a disordered Hamiltonian should *all* be localized, unless extraordinary conditions are met. This is the same as saying that two dimensional materials are barred from conducting electricity. But according to the data from Fig. 1.1, the two dimensional model displays spectral regions where diffusive dynamics still occurs, even when the randomly fluctuating part is five time stronger than the non-fluctuating part!

The resilience against the localization predicted by Anderson is almost always related to topological aspects of the dynamics. In Fig. 1.2, we report the evaluation of the pairing (1.2) for the model (1.4), as function of the Fermi energy. The calculation is for disorder strength $W = 4$ where the spectral gap does not exist anymore, yet the pairing displays quantized values (the flat non-fluctuating values are quantized with more than 10 digits of precision!). It is clear from this data that something more than the ordinary quantization of the cocycle pairing is at work here. Indeed, the scaling analysis from panels (b) and (c) indicates that the quantization of the pairings holds for all Fermi energies except for the critical values where the de-localized spectral regions were detected in Fig. 1.1. So it seems that the change in the quantized value of the pairing and the existence of de-localized spectrum are interconnected. The phenomenon was observed in laboratories for real samples, *e.g.* in the classic integer quantum Hall experiments [**29**] but also in samples that are not subjected to magnetic fields [**19, 20**], as is the case for the model (1.4).

Perhaps a more convincing example is the following one dimensional model generated from the algebra $M_2(\mathbb{C}) \otimes C\big([-\frac{1}{2}, \frac{1}{2}]^{\mathbb{Z}}\big) \rtimes \mathbb{Z}$:

$$\begin{aligned} H_\xi = \sum_{x\in\mathbb{Z}} \big\{ \tfrac{1}{2} t(\tau_x \xi) \big[\sigma_+ \otimes |x\rangle\langle x+1| \\ + \sigma_- \otimes |x+1\rangle\langle x| \big] + m(\tau_x\xi)\sigma_2 \otimes |x\rangle\langle x| \big\}, \end{aligned} \tag{1.5}$$

where $t, m : [-\frac{1}{2}, \frac{1}{2}]^{\mathbb{Z}} \to \mathbb{R}$ are two continuous functions, τ is the standard shift on their domain and σ's are Pauli's matrices. We point out that the model has the key symmetry $\sigma_3 H_\xi \sigma_3 = -H_\xi$. We are going to solve the Schroedinger equation $H_\xi \psi = E\psi$ at $E = 0$,

$$t_x \psi^\alpha_{x-\alpha} + i\alpha m_x \psi^\alpha_x = 0 \;\Rightarrow\; \psi^\alpha_x = \prod_{j=1}^{x} (t_x/m_x)\, \psi^\alpha_0, \tag{1.6}$$

where $\alpha = \pm 1$ is the index for the two components of ψ. As observed in [**37**], we are in a rare situation where the Lyapunov exponent quantifying the asymptotic

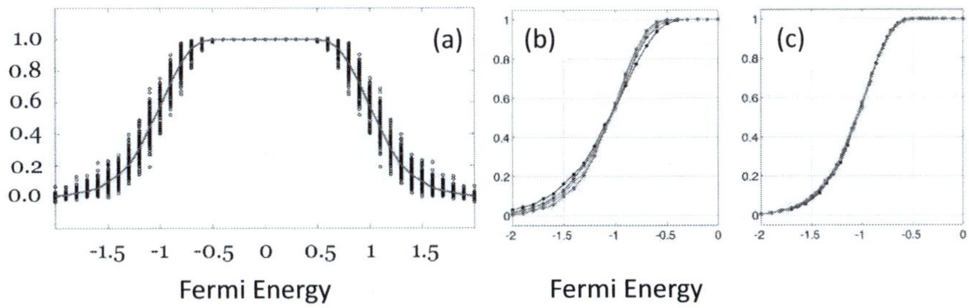

FIGURE 1.2. (Reproduced with permission from [**46**]) Numerical evaluation of the pairing (1.2) as function of E_F, using the algorithm outlined in the text and Hamiltonian (1.4) at disorder strength $W = 4$. Panel (a) shows results for individual random values ξ (the dots), as well as the average over 100 disorder configurations (red line). Panel (b) overlays the results obtained for different sample sizes and panel (c) confirms that the curves from panel (b) overlap almost perfectly when the horizontal axis is rescaled, proving the existence of a critical point (see [**46**] for full details).

spatial profile of the solution can be computed explicitly. Indeed

$$\Lambda := \max_{\alpha=\pm}\Big[-\lim_{x\to\infty}\tfrac{1}{x}\log|\psi_x^\alpha|\Big] = \Big|\lim_{x\to\infty}\tfrac{1}{x}\sum_{n=1}^{x}\Big(\ln|t(\tau_x\xi)| - \ln|m(\tau_x\xi)|\Big)\Big| \tag{1.7}$$

and Birkhoff's theorem can be applied at this point. For example, if we take $\mathbb{P}$ to be the normalized product measure over $[-\frac{1}{2},\frac{1}{2}]^{\mathbb{Z}}$ and $t(\{\xi_x\}) = 1 + W_1\,\xi_0$ as well as $m(\{\xi_x\}) = m + W_2\,\xi_0$, then

$$\Lambda = \left|\ln\left[\frac{|2+W_1|^{1/W_1+1/2}}{|2-W_1|^{1/W_1-1/2}}\frac{|2m-W_2|^{m/W_2-1/2}}{|2m+W_2|^{m/W_2+1/2}}\right]\right|. \tag{1.8}$$

The remarkable fact here is the existence of an entire manifold of parameters where Λ vanishes. This critical manifold is shown in Fig. 1.3 and, as one can see, it separates the parameter space (W_1, W_2, m) into two disjoint regions.[1] This critical manifold was recently detected in a laboratory using cold atom experiments [**34**], which was one of the highlights in the field for year 2018.

Due to its particular symmetry,

$$\frac{H_\xi}{|H_\xi|} = \begin{pmatrix} 0 & U_\xi^* \\ U_\xi & 0 \end{pmatrix}, \tag{1.9}$$

and the topology of the dynamics is encoded in the unitary element U_ξ. When a spectral gap is open around $E = 0$, U_ξ is the left regular representation of a unitary element $u \in C\big([-\frac{1}{2},\frac{1}{2}]^{\mathbb{Z}}\big) \rtimes \mathbb{Z}$ and we can evaluate the following odd pairing

$$\nu = \imath\,\mathcal{T}\big(u\partial u^{-1}\big). \tag{1.10}$$

As in the previous example, the numerical algorithm is not bound to the spectrally gapped cases and maps of this pairing are shown in Figs. 1.3(b,c) for entire sections

[1] It was verified numerically in [**37**] that Λ is finite for all parameters except for those belonging to the shown critical manifold.

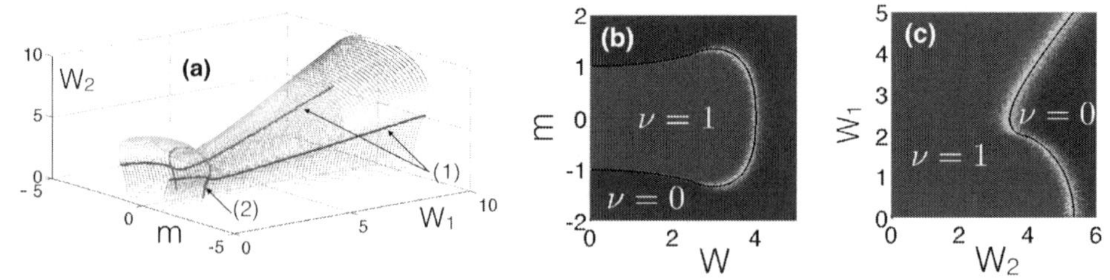

FIGURE 1.3. (Reproduced with permission from [**37**]) (a) The critical surface ($\Lambda = 0$) in the 3-dimensional phase space (W_1, W_2, m). (b,c) Maps of the pairing (1.10) for two sections of the phase space, defined by the constraints $W_2 = 2W_1 = W$ (b) and $m = 0.5$ (c). The analytic critical manifold is shown in these panels by the black line (see [**37**] for more details).

of the parameter space. They reveal again that the pairing remains quantized everywhere except at the critical manifold, where an abrupt change between the quantized values 1 and 0 is observed. Thus, we see again a quantization of the pairing that holds beyond the C^*-umbrella and a strong connection between the jumps in the quantized values and the emergence of delocalized spectrum.

The connection between the phenomena we just discussed and Noncommutative Geometry was established by Bellissard and his collaborators [**7**]. Let us denote the Fermi projection as P_ξ to emphasize its dependence on the random variable and note that the second moment $\sum_{x\in\mathbb{Z}^2} |x|^2 |\langle x|P_\xi|0\rangle|^2$ is a measure of the spatial spread of the kernel of this operator. If finite, it gives a length scale for the dynamics at energies near the Fermi level. For a non-disordered system, this quantity diverges as soon as the Fermi level dives into the spectrum, but so it does for a disordered system for ξ in a set of non-zero measure. The quantity that remains finite is the average [**3**][Th. 1.1]

$$l_F^2 = \int_\Xi d\mathbb{P}(\xi) \sum_{x\in\mathbb{Z}^2} |x|^2 |\langle x|P_\xi|0\rangle|^2 < \infty, \tag{1.11}$$

which in the C^*-language is simply $\mathcal{T}(|\nabla p|^2)$. What seems to happen in the simulations we just showed is that the parings of the cocycles remain quantized not only over the smooth sub-algebra of $C(\Xi) \rtimes \mathbb{Z}^2$, but also over a much larger Sobolev space. As we shall see in subsection 5.3, these Sobolev spaces can be regarded as the natural domains of the cyclic cococycles. But this, by no means, guaranties the observed quantization. For the latter, guided by Alaine Connes' quantized calculus [**23**][Ch. 4], Bellissard and collaborators established in [**7**] and index theorem

$$\operatorname{Ind} F_\xi = 2\pi\imath \sum_{\lambda\in\mathcal{S}_2} (-1)^\lambda \mathcal{T}\big(p_F(\partial_{\lambda_1} p_F)(\partial_{\lambda_2} p_F)\big) \tag{1.12}$$

with the equality holding $\mathbb{P}$-almost surely. Above, F_ξ comes from a Dirac Fredholm module and, in a nutshell, condition (1.11) assures the $(2+\epsilon)$-summability of the Fredholm module, but only $\mathbb{P}$-almost surely, and the right side of (1.12) is the evaluation of the Connes-Chern character, after the observation that the index is $\mathbb{P}$-almost surely independent of ξ, hence it can be averaged out. The index theorem explains the phenomenon seen in our first example because a change in the quantized value of the pairing can only happen when $l_F \to \infty$.

The evaluation of the Connes-Chern character in [**7**] shunts the Connes-Moscovici local index formula [**24**] by using a geometric identity due to Alain Connes [**22**]. Many people have found inspiration in different parts of this work by Alain that is celebrated in this volume. As strange as it may sound, Lemma 2 from [**22**][Sec. 9], where this geometric identity is proved, had a pivotal influence on my research. Indeed, sometime in 2012, my student Bryan Leung and myself just completed a calculation of the magneto-electric α_{EM} coefficient in three space dimensions and the result showed that, under a cyclic deformation of a model in some parameter space, the variation of α_{EM} per cycle is given by [**31**]

$$\Delta\alpha_{EM} = \Lambda_4 \sum_{\lambda \in \mathcal{S}_4} (-1)^\lambda \mathcal{T}\Big(p_F \prod_{j=1}^{4} \partial_{\lambda_j} p_F\Big) \tag{1.13}$$

and the right hand side is again a pairing of a cyclic cocycle. At that time, we were not aware of any index theorem that can be related to this pairing, although hints appeared in the literature (*e.g.* in a preprint later published in [**18**][Ch. 5]).[2] However, in order to push the index theorem into the Sobolev domain, a generalization of Alain's geometric identity was needed. And together with Jean Bellissard and Bryan Leung, we found the following generalization [**42**]:

$$\int_{\mathbb{R}^d} dw \, \mathrm{tr}_\gamma\Big(\gamma_0 \prod_{i=1}^{d} \gamma \cdot \big(\widehat{y_i - w} - \widehat{y_{i+1} - w}\big)\Big) = \Lambda_d \sum_{\lambda} (-1)^\lambda \prod_{i=1}^{d} (y_i)_{\lambda_i}, \tag{1.14}$$

where γ's are the Clifford matrices for even dimension d, $y_i \in \mathbb{R}^d$ and $\hat{x} = x/|x|$. Many readers will immediately see the connection between the left side of this identity and the Connes-Chern characters derived from a Dirac Fredholm module either over $\mathbb{R}^d$ or $\mathbb{Z}^d$, and between the right side and the pairing (1.13). An equivalent identity for odd dimensions was also derived in [**44**]. The index theorems derived from these identities proved half of the conjectured table of strong topological phases [**28, 48, 50, 51**] (*i.e.* those labeled by $\mathbb{Z}$), specifically, it confirmed that all the entries there are thermodynamically distinct because, when passing from one phase to another, one necessarily has to cross an experimentally detectable localization-delocalization phase transition.

The formalism faced a challenge in 2017, when two research groups observed similar topological phase transitions in amorphous metamaterials [**2, 36**]. These models can no longer be generated from a crossed product algebra and the whole program, with both its numerical and theoretical components, had to be upgraded to groupoid algebras (see section 3). Once it was understood that the differential calculus comes from an $\mathbb{R}^d$-action, the numerical algorithm was implemented almost without modifications and, in Fig. 1.4, we illustrate the outcome of a simulation [**14**]. It was performed for the amorphous pattern $\mathcal{L}$ shown in panel (a) subjected to a perpendicular uniform magnetic field. The dynamics was generated by the

[2]The index theorems, in their explicit form, were established later by Chris Bourne in his thesis [**13**], which also deals with the Clifford indices.

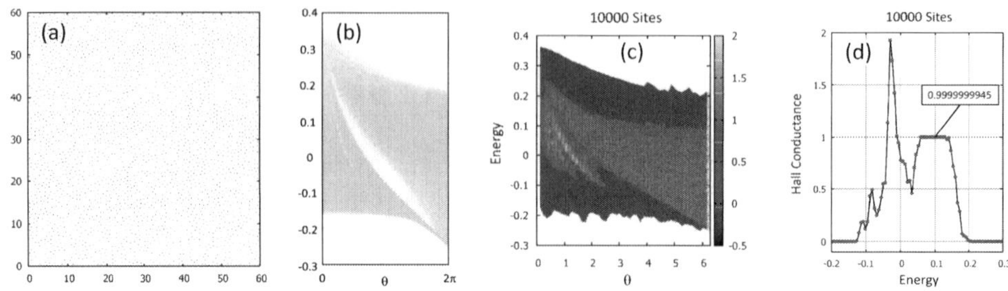

FIGURE 1.4. (Reproduced with permission from [**14**]) (a) Section of an amorphous Delone set containing 1000 sites. (b) The spectrum of the Hamiltonian (1.15) over the pattern from panel (a), rendered as function of the magnetic flux θ. (c) Map of the pairing (1.2) in the plane of the Fermi energy and magnetic flux, as evaluated for the Hamiltonian (1.15). (d) A section of the map from panel (c) cut at $\theta = 1.5$, confirming a high degree of quantization for the pairing.

Hamiltonian

$$H_{\mathcal{L}} : \ell^2(\mathcal{L}) \to \ell^2(\mathcal{L}), \quad H_{\mathcal{L}} = \sum_{x,x' \in \mathcal{L}} e^{\imath\theta x \wedge x'} e^{-3|x-x'|} |x\rangle\langle x'| \tag{1.15}$$

and panel (b) reports the spectrum of this Hamiltonian as function of the magnetic flux θ. The light regions seen in this plot are regions of low spectral density, as opposed to spectral gaps, hence we are again in a gap-less situation yet the pairing (1.2), reported in panels (c) and (d), displays again quantized values. The index theorems that are behind this phenomenon were derived in [**14, 15**] using the local index formula [**24**], as extended to the semifinite noncommutative geometries in [**18**]. In the present work, we undertook the challenge of proving the index theorems via a proper generalization of the identity (1.14) (see Lemma 5.19).

We are now coming to the greatest challenge of all, which is extending the formalism to correlated many-fermion systems. The present paper is geared towards this challenge and reports, in great part, on a contribution that was recently completed by my collaborator Bram Mesland and myself [**35**], luckily just in time for this celebratory volume. During this work, we first stepped back and re-assessed the groupoid algebras introduced by Bellissard and Kellendonk in the single fermion picture, [**6, 8, 27**], already mentioned above. These algebras were often referred to as the algebras of physical observables, but we took a different point of view and replaced the latter with the algebra $\mathbb{K}(\ell^2(\mathcal{L}))$ of compact operators. Then the Bellissard-Kellendonk groupoid algebra canonically associated with the pattern $\mathcal{L}$ can be seen as formalizing physically relevant derivations over the algebra $\mathbb{K}(\ell^2(\mathcal{L}))$. This apparently minor conceptual shift played an important role because it enabled us to identify more clearly the task when the context is upgraded to that of many-fermion systems: Repeat the program completed by Bellissard and Kellendonk, this time looking for algebras of derivations while replacing $\mathbb{K}(\ell^2(\mathcal{L}))$ by $\mathrm{CAR}(\ell^2(\mathcal{L}))$, the algebra of canonical anti-commutation relations over the Hilbert space of the lattice. My work with Mesland was restricted to derivations over $\mathrm{GICAR}(\ell^2(\mathcal{L}))$, the sub-algebra of gauge invariant elements, and the reader can find our conclusion reproduced here in subsection 4.2. As I will argue in this work, we are now in a position where we can generate interesting and relevant states over this algebra of physical derivations, as well as write down relevant cyclic cocycles and, more importantly, jump-start the numerical simulations of many-body transport coefficients. We are eagerly looking forward to seeing the outcomes of these simulations as well as to working on index theorems for the many-body context.

Unfortunately there was no room or time to survey all the work done in connection with Noncommutative Geometry and materials science, and I apologize to the reader for mostly showcasing works in which I was directly involved. My primary goal for this somewhat un-conventional introduction was to communicate to a broader audience certain aspects that I believe are really valued by the materials science community. As one can see, I did not focus at all on the classification of states, primarily because experimental scientists are more interested in learning about observable effects and much less in abstract labels and, secondly, because with meta-materials the possibilities are endless and cannot be captured by a table. Indeed, the algebra of physical observables can be engineered to be any AF-algebra. Lastly, this introduction and the paper itself left out half of the story, specifically, the so called bulk-boundary and bulk-defect correspondences. They refer to extremely robust spectral and dynamical effects that are observed along the interface between two samples with distinct bulk topological invariants (see [**47**] for a unifying reformulation using groupoid algebras). Here again, certain index theorems pushed to Sobolev domains play an important role in establishing the dynamical characteristics of these interfaces [**45**, Sec. 6.6].

The paper is organized as it follows. Section 2 covers classical concepts related to Delone sets and their many-body covers introduced in [**35**]. In particular, we review the Bellissard-Kellendonk groupoids canonically associated to a pattern and their natural generalization to the covers we just mentioned. Section 3 investigates the groupoid C^*-algebras and the bi-equivariant subalgebras relative to a specific 2-action of the symmetric group introduced in [**35**]. Additionally, this section introduces a differential calculus over the many-body groupoids and writes down cyclic cocycles that are relevant for the transport experiments. Section 4 introduces the algebra of physical derivations over the GICAR-algebra and summarizes the conclusions from [**35**], connecting this algebra and the many-body groupoid algebras introduced in section 3. The paper concludes with a new proof of the index theorems for the pairings of the cyclic cocycles mentioned above for the case $N = 1$, based on a generalization of the identity (1.14) that works on generic Delone sets. These can be found in section 5.

2. Spaces of Patterns and Associated Concepts

This section is about the subsets of the abelian topological group $\mathbb{R}^d$, their interactions with the group itself and the canonical algebraic and topological structures that emerge from these interactions. Given a Delone point set $\mathcal{L}_0$, one considers the topological space $\Omega_{\mathcal{L}_0}$ supplied by the closure of its $\mathbb{R}^d$-orbit in a specific topology. Everything discussed in this section is built on the resulting topological dynamical system $(\Omega_{\mathcal{L}_0}, \mathbb{R}^d)$. Specifically, the restriction of its transformation groupoid to a so called canonical transversal supplies the Bellissard-Kellendonk groupoid of the pattern [**6, 8, 27**]. Then blow-ups of this groupoid, induced by so called many-body covers of $\Omega_{\mathcal{L}_0}$, generate canonical many-body groupoids that will be later connected with the dynamics of a fermion gas populating the lattice $\mathcal{L}_0$.

2.1. The metric space of closed $\mathbb{R}^d$ subsets. The metric space $(\mathcal{K}(\mathbb{R}^d), \mathrm{d_H})$ of compact subsets of $\mathbb{R}^d$ equipped with the Hausdorff metric is well known and extensively used in pattern and image analysis (see *e.g.* [**5**]). However, when dealing with extended physical systems, as in the present case, one needs to consider the larger space of closed subsets of $\mathbb{R}^d$. One way to deal with these subsets is to promote them to compact subsets of the one-point compactification $\alpha\mathbb{R}^d$. We recall that, since $\mathbb{R}^d$ is proper, $\alpha\mathbb{R}^d$ is metrizable in an essentially canonical way. Hence, one can put forward the following:

DEFINITION 2.1 ([**25, 30**]). The space of patterns consists of the set $\mathcal{C}(\mathbb{R}^d)$ of closed sub-sets of $\mathbb{R}^d$ endowed with the metric

$$\mathrm{D}(\mathcal{L}, \mathcal{L}') = \bar{\mathrm{d}}_{\mathrm{H}}(\bar{\mathcal{L}}, \bar{\mathcal{L}}'), \quad \mathcal{L}, \mathcal{L}' \in \mathcal{C}(\mathbb{R}^d), \tag{2.1}$$

where $\bar{\mathcal{L}}$ and $\bar{\mathcal{L}}'$ are the closures of the embeddings of $\mathcal{L}$ and $\mathcal{L}'$ in the one-point compactification of $\mathbb{R}^d$ and $\bar{\mathrm{d}}_{\mathrm{H}}$ is the Hausdorff metric on this compactification.

The following statements supply the characterization we need for this space:

PROPOSITION 2.2 ([**30**]). *The space of patterns is bounded, compact and complete. Furthermore, there is a continuous action of $\mathbb{R}^d$ by translations, that is, a homomorphism between topological groups,*

$$\mathfrak{t} : \mathbb{R}^d \to \mathrm{Homeo}\big(\mathcal{C}(\mathbb{R}^d), D\big), \quad \mathfrak{t}_x(\Lambda) = \Lambda - x. \tag{2.2}$$

2.2. Delone sets and and their many-body covers. We will exclusively work with discrete subsets. Specifically:

DEFINITION 2.3. Let $\mathcal{L} \subset \mathbb{R}^d$ be discrete and fix $0 < r < R \in \mathbb{R}$. Then:

(1) $\mathcal{L}$ is r-uniformly discrete if $|B_r(x) \cap \mathcal{L}| \leq 1$ for all $x \in \mathbb{R}^d$.
(2) $\mathcal{L}$ is R-relatively dense if $|B_R(x) \cap \mathcal{L}| \geq 1$ for all $x \in \mathbb{R}^d$.

An r-uniform discrete and R-relatively dense set $\mathcal{L}$ is called an (r, R)-Delone set.

REMARK 2.4. Throughout, if S is a subset of a measure space, then $|S|$ denotes its measure. If S is discrete, then the measure is the counting measure and $|S|$ is the cardinal of S. Also, $B_s(x)$ denotes the open ball in $\mathbb{R}^d$ of radius s and centered at x. $\diamond$

PROPOSITION 2.5 ([**9**]). *The set $\mathrm{Del}^r_R(\mathbb{R}^d)$ of (r, R)-Delone sets in $\mathbb{R}^d$ is a compact subset of $(\mathcal{C}(\mathbb{R}^d), D)$.*

The Delone space $\mathrm{Del}^r_R(\mathbb{R}^d)$ accepts natural covers, which in [**35**] were dubbed many-body covers for reasons that will become apparent later. We describe them here.

PROPOSITION 2.6 ([**35**]). *Fix $r < R$ and consider the set $\widehat{\mathrm{Del}}^{(N)}_{(r,R)}(\mathbb{R}^d)$ of triples $(\chi_V, V, \mathcal{L})$, where $N \in \mathbb{N}^\times$ and:*

- $\mathcal{L} \in \mathrm{Del}^r_R(\mathbb{R}^d)$;
- *V is a compact subset of $\mathcal{L}$ with $|V| = N$;*
- *$\chi_V : \{1, \ldots, N\} \to V$ a bijection.*

Then $\widehat{\mathrm{Del}}^{(N)}_{(r,R)}(\mathbb{R}^d)$ *can be topologized such that*

$$(\chi_V, V, \mathcal{L}) \mapsto \mathfrak{a}_N(\chi_V, V, \mathcal{L}) := \mathcal{L} \tag{2.3}$$

becomes a connected (infinite) cover of $\mathrm{Del}^r_R(\mathbb{R}^d)$.

While we will not discuss the topology of $\widehat{\mathrm{Del}}^{(N)}_{(r,R)}(\mathbb{R}^d)$ explicitly, for orientation, we want to mention that this topology is such that the map $(\chi_V, V, \mathcal{L}) \to (V, \mathcal{L})$ is a $N!$ cover if the set of tuples $(V, \mathcal{L})$ is equipped with the topology inherited from $\mathcal{K}(\mathbb{R}^d) \times \mathrm{Del}^r_R(\mathbb{R}^d)$. We call this map the order cover while the map $\mathfrak{a}_N$ is referred to as the N-body cover, for reasons that will become apparent in the following sections. The hat in $\widehat{\mathrm{Del}}^{(N)}_{(r,R)}(\mathbb{R}^d)$ and other places is there to indicate that we are dealing with a cover of the familiar space $\mathrm{Del}^r_R(\mathbb{R}^d)$.

REMARK 2.7. The case $N = 1$ is special because χ_V is trivial and can be discarded. In this case, the topology on the tuples $(x, \mathcal{L})$, $x \in \mathcal{L}$, is the one inherited from the embedding in $\mathbb{R}^d \times \mathrm{Del}^r_R(\mathbb{R}^d)$. ◇

The points of $\widehat{\mathrm{Del}}^{(N)}_{(r,R)}(\mathbb{R}^d)$ will be denoted by greek letters like ξ, ζ, etc., and the triple corresponding to ξ will take the form $(\chi_\xi, V_\xi, \mathcal{L}_\xi)$. The following statement assures us that the $\mathbb{R}^d$ action extends over the many-body covers:

PROPOSITION 2.8. *The group action of* $\mathbb{R}^d$ *from Eq.* (2.2) *lifts to an action on* $\widehat{\mathrm{Del}}^{(N)}_{(r,R)}(\mathbb{R}^d)$ *by homeomorphisms*

$$\hat{\mathfrak{t}}_x(\xi) = \big(\mathfrak{t}_x \circ \chi_\xi, \mathfrak{t}_x(V_\xi), \mathfrak{t}_x(\mathcal{L}_\xi)\big). \tag{2.4}$$

There is a second group action on the many-body covers, which will play a central role in our framework:

PROPOSITION 2.9. *Let* $\mathcal{S}_N$ *be the group of permutations of* N *objects, which we identify here with the group of bijections from* $\{1, \ldots, N\}$ *to itself. Then* $\mathcal{S}_N$ *acts by homeomorphisms on* $\widehat{\mathrm{Del}}^{(N)}_{(r,R)}(\mathbb{R}^d)$ *via*

$$\Lambda_s(\xi) := \big(\chi_\xi \circ s^{-1}, V_\xi, \mathcal{L}_\xi\big), \quad s \in \mathcal{S}_N. \tag{2.5}$$

REMARK 2.10. The action of $\mathcal{S}_N$ described above coincide with the group of deck transformations for the order cover $(\chi_V, V, \mathcal{L}) \mapsto (V, \mathcal{L})$. ◇

2.3. Hulls, transversals and their many-body covers.

DEFINITION 2.11. Let $\mathcal{L}_0 \subset \mathbb{R}^d$ be a fixed Delone point set. The hull of $\mathcal{L}_0$ is the topological dynamical system $(\Omega_{\mathcal{L}_0}, \mathfrak{t}, \mathbb{R}^d)$, where

$$\Omega_{\mathcal{L}_0} = \overline{\{\mathfrak{t}_a(\mathcal{L}_0) = \mathcal{L}_0 - a, \ a \in \mathbb{R}^d\}} \subset \mathcal{C}(\mathbb{R}^d). \tag{2.6}$$

The above dynamical system is always transitive but may fail to be minimal. Indeed, $\Omega_{\mathcal{L}}$ may not coincide with $\Omega_{\mathcal{L}_0}$ for $\mathcal{L} \in \Omega_{\mathcal{L}_0}$, but it is always true that $\Omega_{\mathcal{L}} \subseteq \Omega_{\mathcal{L}_0}$ for any $\mathcal{L} \in \Omega_{\mathcal{L}_0}$.

REMARK 2.12. Proposition 2.5 assures us that all the subsets contained in $\Omega_{\mathcal{L}_0}$ are (r, R)-Delone point sets if $\mathcal{L}_0$ is one. Note, however, that the hull can be very well defined for just uniformly discrete patterns, a setting that is relevant for the so called bulk-defect correspondence (see [**47**]). ◇

As a closed subset of the space of Delone sets, $\Omega_{\mathcal{L}_0}$ accepts many-body covers:

DEFINITION 2.13. The N-body cover above $\Omega_{\mathcal{L}_0}$ is defined as

$$\widehat{\Omega}_{\mathcal{L}_0}^{(N)} := \mathfrak{a}_N^{-1}(\Omega_{\mathcal{L}_0}). \tag{2.7}$$

Proposition 2.8 assures us that $\big(\widehat{\Omega}_{\mathcal{L}_0}^{(N)}, \hat{\mathfrak{t}}, \mathbb{R}^d\big)$ is again a topological dynamical system.

DEFINITION 2.14. The canonical transversal of a hull $(\Omega_{\mathcal{L}_0}, \mathfrak{t}, \mathbb{R}^d)$ of a Delone set $\mathcal{L}_0$ is defined as

$$\Xi_{\mathcal{L}_0} = \{\mathcal{L} \in \Omega_{\mathcal{L}_0},\ 0 \in \mathcal{L}\}.$$

The transversal is a closed hence compact subspace of $\Omega_{\mathcal{L}_0}$.

As it was the case for the hull, $\Xi_{\mathcal{L}_0}$ accepts many-body covers, which are defined slightly different:

DEFINITION 2.15. The N-body cover of $\Xi_{\mathcal{L}_0}$ is defined as

$$\widehat{\Xi}_{\mathcal{L}_0}^{(N)} := \big\{\xi \in \widehat{\Omega}_{\mathcal{L}_0}^{(N)},\ \chi_\xi(1) = 0\big\}. \tag{2.8}$$

REMARK 2.16. The N-body cover $\widehat{\Xi}_{\mathcal{L}_0}^{(N)}$ is a closed subset of $\widehat{\mathrm{Del}}_{(r,R)}^{(N)}(\mathbb{R}^d)$. Except for the case $N = 1$, when the many-body covers reduce to the ordinary hulls and transversals, $\widehat{\Omega}_{\mathcal{L}_0}^{(N)}$ and $\widehat{\Xi}_{\mathcal{L}_0}^{(N)}$ are locally compact but not compact spaces. $\diamond$

2.4. Canonical groupoid of a point pattern. We review here the groupoid introduced by Bellissard and Kellendonk [**6**, **8**, **27**]. We start with a few general concepts:

DEFINITION 2.17 (Generalized equivalence relations, [**54**] p. 5). Let G be a group and X a set. Then $X \times G \times X$ can be given the structure of a groupoid by adopting the inversion function

$$(x, g, y)^{-1} = (y, g^{-1}, x), \tag{2.9}$$

and by declaring that two triples (x, g, y) and (w, h, z) are composible if $y = w$, in which case the product is

$$(x, g, y) \cdot (y, h, z) := (x, gh, z). \tag{2.10}$$

If X is a topological space and G is a topological group, then $X \times G \times X$ becomes a topological groupoid if endowed with the product topology.

DEFINITION 2.18 ([**54**] p.6). Let X be a G-space. Then the transformation groupoid of (X, G) is the topological sub-groupoid of the generalized equivalence relations introduced in Definition 2.17, consisting of the triples (x, g, y) with the entries obeying the constraint $x = g \cdot y$.

When applied to the dynamical system $(\Omega_{\mathcal{L}_0}, \mathfrak{t}, \mathbb{R}^d)$, this definition gives:

PROPOSITION 2.19. *The transformation groupoid $\widetilde{\mathcal{G}}_{\mathcal{L}_0}$ associated to the dynamical system $(\Omega_{\mathcal{L}_0}, \mathfrak{t}, \mathbb{R}^d)$ consists of triples*

$$\big(\mathfrak{t}_x(\mathcal{L}), x, \mathcal{L}\big) \in \Omega_{\mathcal{L}_0} \times \mathbb{R}^d \times \Omega_{\mathcal{L}_0} \tag{2.11}$$

endowed with the multiplication

$$\big(\mathfrak{t}_{x'+x}(\mathcal{L}), x', \mathfrak{t}_x(\mathcal{L})\big)\big(\mathfrak{t}_x(\mathcal{L}), x, \mathcal{L}\big) = \big(\mathfrak{t}_{x'+x}(\mathcal{L}), x' + x, \mathcal{L}\big) \tag{2.12}$$

and inversion

$$\big(\mathfrak{t}_x(\mathcal{L}), x, \mathcal{L}\big)^{-1} = \big(\mathcal{L}, -x, \mathfrak{t}_x(\mathcal{L})\big). \tag{2.13}$$

The topology of the transformation groupoid is that inherited from the product topology of $\Omega_{\mathcal{L}_0} \times \mathbb{R}^d \times \Omega_{\mathcal{L}_0}$.

The source and the range maps of the groupoid $\widetilde{\mathcal{G}}_{\mathcal{L}_0}$ act as

$$\tilde{\mathfrak{s}}\big(\mathfrak{t}_x(\mathcal{L}), x, \mathcal{L}\big) = (\mathcal{L}, 0, \mathcal{L}), \quad \tilde{\mathfrak{r}}\big(\mathfrak{t}_x(\mathcal{L}), x, \mathcal{L}\big) = \big(\mathfrak{t}_x(\mathcal{L}), 0, \mathfrak{t}_x(\mathcal{L})\big), \tag{2.14}$$

hence its space of units coincides with the hull $\Omega_{\mathcal{L}_0}$ of the pattern.

DEFINITION 2.20. The topological groupoid $\mathcal{G}_{\mathcal{L}_0}$ canonically associated to a Delone set $\mathcal{L}_0$ is the restriction of $\widetilde{\mathcal{G}}_{\mathcal{L}_0}$ to the transversal $\Xi_{\mathcal{L}_0} \subset \Omega_{\mathcal{L}_0}$:

$$\mathcal{G}_{\mathcal{L}_0} := \tilde{\mathfrak{s}}^{-1}(\Xi_{\mathcal{L}_0}) \cap \tilde{\mathfrak{r}}^{-1}(\Xi_{\mathcal{L}_0}). \tag{2.15}$$

Since $\Xi_{\mathcal{L}_0}$ is a closed sub-set, the restriction is indeed a topological groupoid.

This groupoid can be presented more explicitly as:

PROPOSITION 2.21 ([**6**, **8**, **27**]). *The topological groupoid canonically associated to a Delone set $\mathcal{L}_0$ consists of:*

1. *The set*

$$\mathcal{G}_{\mathcal{L}_0} = \big\{(x, \mathcal{L}),\ \mathcal{L} \in \Xi_{\mathcal{L}_0},\ x \in \mathcal{L}\big\} \subset \mathbb{R}^d \times \Xi_{\mathcal{L}_0}, \tag{2.16}$$

equipped with the inversion map

$$(x, \mathcal{L})^{-1} = \big(-x, \mathfrak{t}_x(\mathcal{L})\big) = \mathfrak{t}_x(0, \mathcal{L}). \tag{2.17}$$

2. *The subset of $\mathcal{G}_{\mathcal{L}_0} \times \mathcal{G}_{\mathcal{L}_0}$ of composable elements*

$$\Big\{\big((x', \mathcal{L}'), (x, \mathcal{L})\big) \in \mathcal{G}_{\mathcal{L}_0} \times \mathcal{G}_{\mathcal{L}_0},\ \mathcal{L}' = \mathfrak{t}_x(\mathcal{L})\Big\} \tag{2.18}$$

equipped with the composition

$$\big(x', \mathfrak{t}_x(\mathcal{L})\big) \cdot (x, \mathcal{L}) = (x' + x, \mathcal{L}). \tag{2.19}$$

The topology on $\mathcal{G}_{\mathcal{L}_0}$ is the relative topology inherited from $\mathbb{R}^d \times \Xi_{\mathcal{L}_0}$.

PROOF. From (2.14), one finds

$$\tilde{\mathfrak{s}}^{-1}(\mathcal{L}, 0, \mathcal{L}) = \big(\mathfrak{t}_x(\mathcal{L}), x, \mathcal{L}\big), \quad \tilde{\mathfrak{r}}^{-1}(\mathcal{L}, x, \mathcal{L}) = \big(\mathcal{L}, -x, \mathfrak{t}_x(\mathcal{L})\big), \tag{2.20}$$

hence $\big(\mathfrak{t}_x(\mathcal{L}), x, \mathcal{L}\big)$ belongs to the intersection (2.15) if and only if $\mathcal{L}$ and $\mathfrak{t}_x(\mathcal{L})$ both belong to $\Xi_{\mathcal{L}_0}$. This automatically constrains x to be on the lattice $\mathcal{L}$ and $\mathcal{G}_{\mathcal{L}_0}$ can then be presented as the sub-groupoid of $\widetilde{\mathcal{G}}_{\mathcal{L}_0}$ consisting of the triples

$$\big(\mathfrak{t}_x(\mathcal{L}), x, \mathcal{L}\big) \in \widetilde{\mathcal{G}}_{\mathcal{L}_0}, \quad \mathcal{L} \in \Xi_{\mathcal{L}_0} \subset \Omega_{\mathcal{L}_0}, \quad x \in \mathcal{L}, \tag{2.21}$$

endowed with the topology inherited from $\widetilde{G}_{\mathcal{L}_0}$, hence from $\Xi_{\mathcal{L}_0} \times \mathbb{R}^d \times \Xi_{\mathcal{L}_0}$. Lastly, we can drop the redundant first entry of the triples. □

REMARK 2.22. The algebraic structure of $\mathcal{G}_{\mathcal{L}_0}$ can be also described as coming from the equivalence relation on $\Xi_{\mathcal{L}_0}$

$$R_{\Xi_{\mathcal{L}_0}} = \big\{(\mathcal{L}, \mathcal{L}') \in \Xi_{\mathcal{L}_0} \times \Xi_{\mathcal{L}_0},\ \mathcal{L}' = \mathcal{L} - a \text{ for some } a \in \mathbb{R}^d\big\}.$$

However, the topology of $\mathcal{G}_{\mathcal{L}_0}$ is not the one inherited from $\Xi_{\mathcal{L}_0} \times \Xi_{\mathcal{L}_0}$. ◇

REMARK 2.23. Since $\Xi_{\mathcal{L}_0}$ is an abstract transversal for $(\Omega_{\mathcal{L}_0}, \mathbb{R}^d)$, the transformation groupoid $\widetilde{\mathcal{G}}_{\mathcal{L}_0}$ and $\mathcal{G}_{\mathcal{L}_0}$ are equivalent [**38**]. In particular, their corresponding C^*-algebras are Morita equivalent. ◇

REMARK 2.24. In a typical situation, the quantum resonators carry internal structures and, to account for them, the hull and the transversal are multiplied by a finite set. This complication can be avoided by duplicating the pattern such that the transversal of the new pattern becomes $\Xi_{\mathcal{L}_0} \times \{1, 2, \ldots\}$. This will be tacitly assumed in the following. ◇

2.5. Invariant measures. The topological dynamical system $(\Omega_{\mathcal{L}_0}, \mathfrak{t}, \mathbb{R}^d)$, like any other topological dynamical system, accepts invariant and ergodic measures. The measure with physical meaning, denoted here by $d\bar{\mathbb{P}}$, is fixed by $\mathcal{L}_0$ itself:

$$\int_{\Omega_{\mathcal{L}_0}} d\bar{\mathbb{P}}(\mathcal{L})\, \phi(\mathcal{L}) := \lim_{V \to \mathbb{R}^d} \frac{1}{|V|} \int_{V \subset \mathbb{R}^d} dy\, \phi\big(\mathfrak{t}_y(\mathcal{L}_0)\big). \tag{2.22}$$

The real interest, however, is in invariant and ergodic measures on the transversal $\Xi_{\mathcal{L}_0}$ relative to the groupoid action. We describe them briefly here and we start from the following reassuring statement:

PROPOSITION 2.25 ([**12, 21**]). *There is a one-to-one correspondence between measures on $\Omega_{\mathcal{L}_0}$ invariant under the $\mathbb{R}^d$-action and measures on the transversal $\Xi_{\mathcal{L}_0}$ invariant under the groupoid action.*

The transition between the two spaces is achieved as follows. One considers first the sub-set $\Xi_{\mathcal{L}_0}(s)$ of $\Omega_{\mathcal{L}_0}$ consisting of those $\mathcal{L}$'s with the property that there is a closed ball $\bar{B}_x(s)$ of radius s centered at one of the points x of $\mathcal{L}$ that contains the origin of $\mathbb{R}^d$. Since all $\mathcal{L} \in \Omega_{\mathcal{L}_0}$ belong to $\mathrm{Del}^r_R(\mathbb{R}^d)$, if we take $s < r/2$, then the point sets included in $\Xi_{\mathcal{L}_0}(s)$ have exactly one point such that $0 \in \bar{B}_s(x)$. Then we can define the following canonical map

$$\mathfrak{e} : \Xi_{\mathcal{L}_0}(s) \to \Xi_{\mathcal{L}_0}, \quad \mathfrak{e}(\mathcal{L}) = \mathfrak{t}_x(\mathcal{L}), \tag{2.23}$$

where $x \in \mathcal{L}$ is the unique point mentioned above and it is easy to check that $\mathfrak{e}$ respects the groupoid action

$$\mathfrak{t}_y\big(\mathfrak{e}(\mathcal{L})\big) = \mathfrak{e}\big(\mathfrak{t}_y(\mathcal{L})\big), \quad y \in \mathcal{L}. \tag{2.24}$$

Then the trace of $\bar{\mathbb{P}}$ on $\Xi_{\mathcal{L}_0}(s)$ can be pushed forward through $\mathfrak{e}$ to obtain the corresponding groupoid invariant measure $\mathbb{P} := \mathfrak{e}_* \bar{\mathbb{P}}$ on $\Xi_{\mathcal{L}_0}$. If $\bar{\mathbb{P}}$ was ergodic in the first place, then, for any measurable $\phi : \Xi_{\mathcal{L}_0} \to \mathbb{C}$,

$$\begin{aligned} \int_{\Xi_{\mathcal{L}_0}} d\mathbb{P}(\mathcal{L})\, \phi(\mathcal{L}) &:= \int_{\Xi_{\mathcal{L}_0}(s)} d\bar{\mathbb{P}}(\mathcal{L})\, (\phi \circ \mathfrak{e})(\mathcal{L}) \\ &= \lim_{V \to \mathbb{R}^d} \frac{1}{|V|} \int_V dy\, (\phi \circ \mathfrak{e})\big(\mathfrak{t}_y(\mathcal{L})\big) \\ &= \lim_{V \to \mathbb{R}^d} \frac{|\bar{B}_s(0)|}{|V|} \sum_{x \in \mathcal{L} \cap V} \phi(\mathfrak{t}_x \mathcal{L}), \end{aligned} \tag{2.25}$$

with the equality holding $\mathbb{P}$-almost surely. We normalize $\mathbb{P}$ such that $\mathbb{P}(\Xi_{\mathcal{L}_0}) = 1$ by dividing with the factor $|\bar{B}_s(0)|\, \mathrm{Dens}_{\mathcal{L}}$, where

$$\mathrm{Dens}_{\mathcal{L}} = \lim_{V \to \mathbb{R}^d} \frac{1}{|V|} \sum_{x \in \mathcal{L} \cap V} 1 \tag{2.26}$$

is the macroscopic density of the point set $\mathcal{L}$. This quantity is non-fluctuating, i.e. it coincides with $\mathrm{Dens}_{\mathcal{L}_0}$, $\mathbb{P}$-almost surely. In fact, we will fix the units such that $\mathrm{Dens}_{\mathcal{L}_0} = 1$. With this choice, the Borel measure on $\mathbb{R}^d$ defined by

$$|V| := \int_{\Omega_{\mathcal{L}_0}} \mathrm{d}\bar{\mathbb{P}}(\mathcal{L})\, |\mathcal{L} \cap V|, \tag{2.27}$$

coincides with the Haar measure of $\mathbb{R}^d$. This detail will become relevant for the proof of Lemma 5.19 containing an important geometric identity.

2.6. Canonical many-body groupoids. The groupoid canonically associated to a Delone pattern, as presented in Proposition 2.21, accepts the following generalizations over the many-body covers:

DEFINITION 2.26 ([**35**]). The groupoid associated to the dynamics of N fermions over a fixed Delone set $\mathcal{L}_0$ consists of:

1. The topological space

$$\widehat{\mathcal{G}}^{(N)}_{\mathcal{L}_0} = \big\{(\zeta, \xi) \in \widehat{\Omega}^{(N)}_{\mathcal{L}_0} \times \widehat{\Omega}^{(N)}_{\mathcal{L}_0},\ \mathcal{L}_\xi = \mathcal{L}_\zeta \in \Xi_{\mathcal{L}_0},\ \chi_\xi(1) = 0\big\}, \tag{2.28}$$

equipped with the topology inherited from $\widehat{\mathrm{Del}}^{(n)}_{(r,R)}(\mathbb{R}^d)^{\times 2}$ and the inversion map

$$(\zeta, \xi)^{-1} = \hat{\mathfrak{t}}_{\chi_\zeta(1)}(\xi, \zeta). \tag{2.29}$$

2. The set of composable elements

$$\Big\{\big(\hat{\mathfrak{t}}_{\chi_\zeta(1)}(\zeta', \zeta), (\zeta, \xi)\big) \in \widehat{\mathcal{G}}^{(N)}_{\mathcal{L}_0} \times \widehat{\mathcal{G}}^{(N)}_{\mathcal{L}_0}\Big\}, \tag{2.30}$$

equipped with the composition map

$$\hat{\mathfrak{t}}_{\chi_\zeta(1)}(\zeta', \zeta) \cdot (\zeta, \xi) := (\zeta', \xi). \tag{2.31}$$

REMARK 2.27. It is obvious that, for $N = 1$, the above groupoid reduces to the standard groupoid $\mathcal{G}_{\mathcal{L}_0}$ of the pattern. ◇

PROPOSITION 2.28 ([**35**]). *$\widehat{\mathcal{G}}^{(N)}_{\mathcal{L}_0}$ is a second-countable, Hausdorff étale groupoid.*

REMARK 2.29. Of course, the above statement is good news because an étale groupoid comes with a standard system of Haar measures, hence with a standard groupoid C^*-algebra, which is necessarily separable [**52**]. As such, the K-theories of these algebras are countable, a fact that gives us hopes about classifying the topological dynamical phases of N fermions hopping over a given lattice. ◇

There is an alternative characterization of the many-body groupoids, based on the following construction:

DEFINITION 2.30 ([**54**]). Let $\mathcal{G}$ be a locally compact Hausdorff groupoid with open range map. Suppose that Z is locally compact Hausdorff and that $f : Z \to \mathcal{G}^{(0)}$ is a continuous open map. Then

$$\mathcal{G}[Z] := \big\{(z, \gamma, w) \in Z \times \mathcal{G} \times Z : f(z) = r(\gamma) \text{ and } s(\gamma) = f(w)\big\} \tag{2.32}$$

is a topological groupoid when considered with the natural operations

$$(z, \gamma, w)(w, \eta, x) = (z, \gamma\eta, x) \text{ and } (z, \gamma, w)^{-1} = (w, \gamma^{-1}, z), \tag{2.33}$$

and the topology inherited from $Z \times \mathcal{G} \times Z$.

REMARK 2.31. The space of units for $\mathcal{G}[Z]$ is

$$\mathcal{G}[Z]^{(0)} = \{(z, f(z), z),\ z \in Z\},$$

hence it can be naturally identified with Z. For this reason, the map f is called a blow-up of the unit space and the groupoid $\mathcal{G}[Z]$ is referred to as the blow-up of $\mathcal{G}$ through f. ◇

Now, if $\hat{\mathfrak{s}}$ and $\hat{\mathfrak{r}}$ denote the source and the range maps of the groupoid $\widehat{\mathcal{G}}^{(N)}_{\mathcal{L}_0}$, respectively, then

$$\hat{\mathfrak{s}}(\zeta, \xi) = (\xi, \xi), \quad \hat{\mathfrak{r}}(\zeta, \xi) = \hat{\mathfrak{t}}_{\chi_\zeta(1)}(\zeta, \zeta). \tag{2.34}$$

As such, the space of units of the groupoid $\widehat{\mathcal{G}}^{(N)}_{\mathcal{L}_0}$ coincides with the N-body cover $\widehat{\Xi}^{(N)}_{\mathcal{L}_0}$ of the transversal, introduced in Eq. (2.8).

PROPOSITION 2.32 ([**35**]). *Consider the canonical groupoid $\mathcal{G}_{\mathcal{L}_0}$. Then*

$$\widehat{\Xi}^{N}_{\mathcal{L}_0} \ni \xi \mapsto \mathfrak{u}(\xi) := \mathcal{L}_\xi \in \Xi_{\mathcal{L}_0} \tag{2.35}$$

is a blow-up map for $\mathcal{G}_{\mathcal{L}_0}$ and the associated blow-up groupoid can be presented as

$$\begin{aligned} \widetilde{\mathcal{G}}^{(N)}_{\mathcal{L}_0} = \big\{\big(\zeta, (x, \mathcal{L}), \xi\big),\ (x, \mathcal{L}) \in \mathcal{G}_{\mathcal{L}_0},\ \xi \in \mathfrak{a}_N^{-1}(\mathcal{L}), \\ \zeta \in \mathfrak{a}_N^{-1}\big(\mathfrak{t}_x(\mathcal{L})\big),\ \chi_\xi(1) = \chi_\zeta(1) = 0\big\}. \end{aligned} \tag{2.36}$$

Furthermore, the map

$$\widehat{\mathcal{G}}^{(N)}_{\mathcal{L}_0} \ni (\zeta, \xi) \mapsto \big(\mathfrak{t}_{\chi_\zeta(1)}(\zeta), \big(\chi_\zeta(1), \mathcal{L}_\xi\big), \xi\big) \in \widetilde{\mathcal{G}}^{(N)}_{\mathcal{L}_0} \tag{2.37}$$

is an isomorphism of topological groupoids.

3. Canonical Groupoid C^*-Algebras and Associated Structures

The groupoids introduced in the previous section supply canonical C^*-algebras [**49**], which will be briefly reviewed here. The focus of the section, however, is on natural 2-actions of the permutation groups and identification of certain bi-equivariant sub-algebras that are directly related to the dynamics of many fermions. As we shall see, these sub-algebras can be endowed with differential structures induced by natural $\mathbb{R}^d$-actions and canonical semi-finite traces that are invariant against these actions. This enables us to define standard cyclic cocycles that are connected with the physical transport coefficients of many-fermion systems.

3.1. Groupoid C^*-algebras and their regular representations. It will be useful to spell out the algebraic operations of the reduced groupoid algebras. For this, we remind that, on the core $C_c\big(\widehat{\mathcal{G}}^{(N)}_{\mathcal{L}_0}\big)$ containing the compactly supported $\mathbb{C}$-valued continuous functions on $\widehat{\mathcal{G}}^{(N)}_{\mathcal{L}_0}$, the associative multiplication works as

$$(f_1 * f_2)(\zeta, \xi) := \sum_{(\zeta', \xi') \in \hat{\mathfrak{r}}^{-1}\left(\hat{\mathfrak{r}}(\zeta, \xi)\right)} f_1(\zeta', \xi') \cdot f_2\big((\zeta', \xi')^{-1}(\zeta, \xi)\big). \tag{3.1}$$

From (2.34), one finds

$$\hat{\mathfrak{r}}^{-1}\big(\hat{\mathfrak{r}}(\zeta, \xi)\big) = \big\{\hat{\mathfrak{t}}_{\chi_\eta(1)}(\zeta, \eta),\ \eta \in \mathfrak{a}_N^{-1}(\mathcal{L}_\xi)\big\} \tag{3.2}$$

and, since $\big(\hat{\mathfrak{t}}_{\chi_\eta(1)}(\zeta,\eta)\big)^{-1} = \hat{\mathfrak{t}}_{\chi_\zeta(1)}(\eta,\zeta)$, the multiplication can be written more explicitly as

$$(f_1 * f_2)(\zeta,\xi) = \sum_{\eta\in\mathfrak{a}_N^{-1}(\mathcal{L}_\xi)} f_1\big(\hat{\mathfrak{t}}_{\chi_\eta(1)}(\zeta,\eta)\big)\cdot f_2(\eta,\xi). \tag{3.3}$$

The $*$-operation is

$$f^*(\zeta,\xi) = f\big((\zeta,\xi)^{-1}\big)^* = f\big(\hat{\mathfrak{t}}_{\chi_\zeta(1)}(\xi,\zeta)\big)^*. \tag{3.4}$$

Now, if $f\in C_c\big(\widehat{\mathcal{G}}^{(N)}_{\mathcal{L}_0}\big)$ and $\varphi\in C_0\big(\widehat{\Xi}^{(N)}_{\mathcal{L}_0}\big)$, the C^*-algebra of $\mathbb{C}$-valued continuous and compactly supported functions over the space of units, then the product

$$(f\,\varphi)(\zeta,\xi) = f(\zeta,\xi)\cdot\varphi(\hat{\mathfrak{s}}(\zeta,\xi)) = f(\zeta,\xi)\cdot\varphi(\xi). \tag{3.5}$$

makes $C_c\big(\widehat{\mathcal{G}}^{(N)}_{\mathcal{L}_0}\big)$ into a right module over $C_0\big(\widehat{\Xi}^{(N)}_{\mathcal{L}_0}\big)$ and the restriction map

$$\rho : C_c\big(\widehat{\mathcal{G}}^{(N)}_{\mathcal{L}_0}\big)\to C_0\big(\widehat{\Xi}^{(N)}_{\mathcal{L}_0}\big),\quad (\rho f)(\xi) = f(\xi,\xi), \tag{3.6}$$

supplies an expectation. By composing with the evaluation maps

$$j_\eta : C_c\big(\widehat{\Xi}^{(N)}_{\mathcal{L}_0}\big)\to\mathbb{C},\quad j_\eta(\phi)=\phi(\eta),\quad \eta\in\widehat{\Xi}^{(N)}_{\mathcal{L}_0}, \tag{3.7}$$

one obtains a family of states $\rho_\eta = j_\eta\circ\rho$ on $C_r^*\big(\widehat{\mathcal{G}}^{(N)}_{\mathcal{L}_0}\big)$, indexed by the transversal $\widehat{\Xi}^{(N)}_{\mathcal{L}_0}$. The GNS-representations corresponding to these states supply the left-regular representations of $C_r^*\big(\widehat{\mathcal{G}}^{(N)}_{\mathcal{L}_0}\big)$. They are carried by the Hilbert spaces

$$\mathcal{H}_\eta := \ell^2\big(\hat{\mathfrak{s}}^{-1}(\eta)\big) = \ell^2\big(\mathfrak{a}_N^{-1}(\mathcal{L}_\eta)\big), \tag{3.8}$$

and take the explicit form

$$[\pi_\eta(f)\psi](\zeta) = \sum_{\xi\in\mathfrak{a}_N^{-1}(\mathcal{L}_\eta)} f\big(\hat{\mathfrak{t}}_{\chi_\xi(1)}(\zeta,\xi)\big)\cdot\psi(\xi). \tag{3.9}$$

We recall that these left regular representations supply the C^*-norm on $C_r^*\big(\widehat{\mathcal{G}}^{(N)}_{\mathcal{L}_0}\big)$,

$$\|f\| = \sup_{\eta\in\widehat{\Xi}^{(N)}_{\mathcal{L}_0}} \|\pi_\eta(f)\|. \tag{3.10}$$

These representations will be later compared with the Fock representation of the physical Hamiltonians (see section 4).

3.2. The 2-actions and the bi-equivariant groupoid algebras. There is a direct consequence of Proposition 2.32:

Corollary 3.1 ([**54**], Th. 2.52). *The groupoid C^*-algebras $C_r^*(\mathcal{G}_{\mathcal{L}_0})$ and $C_r^*\big(\widehat{\mathcal{G}}^{(N)}_{\mathcal{L}_0}\big)$ are Morita equivalent and, as a consequence, their K-theories coincide.*

The interest is, however, not entirely on $C_r^*\big(\widehat{\mathcal{G}}^{(N)}_{\mathcal{L}_0}\big)$, but rather on a bi-equivariant sub-algebra discussed next. We first recall that every topological groupoid $\mathcal{G}$ comes with a topological group of bisections $\mathcal{S}(\mathcal{G})$, consisting of continuous maps

$$\mathcal{S}(\mathcal{G}) := \big\{b : \mathcal{G}^{(0)}\to\mathcal{G} : \mathfrak{s}\circ b = \mathrm{Id},\quad \mathfrak{r}\circ b \text{ is a homeomorphism}\big\}$$

and a standard group structure. Above, $\mathcal{G}^{(0)}$ is the space of units, and $\mathfrak{s}$, $\mathfrak{r}$ are the source and range maps of the groupoid, respectively.

Definition 3.2 ([**35**]). Let Γ be a discrete group and $\mathcal{G}$ a locally compact Hausdorff groupoid. A 2-action of Γ on $\mathcal{G}$ is a group homomorphism $\alpha:\Gamma\to\mathcal{S}(\mathcal{G})$.

PROPOSITION 3.3 ([**35**]). *A 2-action of* Γ *on* $\mathcal{G}$ *automatically induces commuting left and right actions of* Γ *on* $\mathcal{G}$ *via:*

$$\gamma_1 \cdot g \cdot \gamma_2 := \alpha_{\gamma_1}\big(\mathfrak{r}(g)\big) \cdot g \cdot \alpha_{\gamma_2^{-1}}\big(\mathfrak{s}(g)\big)^{-1}. \tag{3.11}$$

PROPOSITION 3.4 ([**35**]). *Let* $s \in \mathcal{S}_N$ *be a permutation. The formula*

$$\alpha_s(\xi) := \big(\Lambda_s(\xi), \xi\big) \tag{3.12}$$

defines a homomorphism $\alpha : \mathcal{S}_N \to \mathcal{S}(\mathcal{G}_N)$ *and thus a 2-action of* $\mathcal{S}_N$ *on* $\widehat{\mathcal{G}}_{\mathcal{L}_0}^{(N)}$. *The induced left and right actions, denoted by* $s_1 \cdot (\xi, \zeta) \cdot s_2$, *for* $s_i \in \mathcal{S}_N$, *are given by*

$$s_1 \cdot (\zeta, \xi) \cdot s_2 = \hat{\mathfrak{t}}_{\chi_\xi \circ s_2(1)}\big(\Lambda_{s_1}(\zeta), \Lambda_{s_2}^{-1}(\xi)\big). \tag{3.13}$$

For a C^*-algebra $\mathcal{A}$, we denote by $UM(\mathcal{A})$ the unitary group of the multiplier algebra of $\mathcal{A}$. There is a similar notion of 2-action of a group on C^*-algebras:

DEFINITION 3.5 ([**35**]). Let Γ be a discrete group and $\mathcal{A}$ a C^*-algebra. A 2-action of Γ on $\mathcal{A}$ is a group homomorphism $u : \Gamma \to UM(\mathcal{A})$ from Γ to the unitary multipliers on $\mathcal{A}$.

REMARK 3.6. A 2-action of Γ on $\mathcal{A}$ induces left and right actions on $\mathcal{A}$ by setting

$$\gamma_1 \cdot a \cdot \gamma_2 := u_{\gamma_1} a u_{\gamma_2}^*, \tag{3.14}$$

for $\gamma_i \in \Gamma$ and $a \in \mathcal{A}$. $\diamond$

LEMMA 3.7 ([**35**]). *Given 2-actions of* Γ *on a* C^**-algebra* $\mathcal{A}$ *and groupoid* $\mathcal{G}$, *we write*

$$C_{c,\Gamma}(\mathcal{G}, \mathcal{A}) := \Big\{ f \in C_c(\mathcal{G}, \mathcal{A}) : f(\gamma_1 \cdot g \cdot \gamma_2) = \gamma_1 \cdot f(g) \cdot \gamma_2 \Big\} \tag{3.15}$$

for the space of Γ*-bi-equivariant and compactly supported maps from* $\mathcal{G}$ *to* $\mathcal{A}$. *Then the subspace* $C_{c,\Gamma}(\mathcal{G}, \mathcal{A})$ *is a* $*$*-subalgebra of* $C_c(\mathcal{G}, \mathcal{A})$.

DEFINITION 3.8 ([**35**]). Let Γ be a discrete group. Suppose that $\mathcal{G}$ is a locally compact Hausdorff étale groupoid and $\mathcal{A}$ is a C^*-algebra, both of which carry a 2-action by Γ. The Γ-bi-equivariant reduced $\mathcal{A}$-C^*-algebra $C^*_{r,\Gamma}(\mathcal{G}, \mathcal{A})$ of $\mathcal{G}$ is the closure of the image of $C_{c,\Gamma}(\mathcal{G}, \mathcal{A})$ inside $C^*_r(\mathcal{G}, \mathcal{A})$.

Applied to our context, this gives:

DEFINITION 3.9 ([**35**]). The bi-equivariant groupoid C^*-algebra associated with the dynamics of N fermions hopping on the Delone sets from the transversal of a $\mathcal{L}_0$ is the $\mathcal{S}_N$-bi-equivariant sub-algebra $C^*_{r,\mathcal{S}_N}(\widehat{\mathcal{G}}_{\mathcal{L}_0}^{(N)}, \mathbb{C})$, where the 2-action of $\mathcal{S}_N$ on $\mathbb{C}$ is simply $\mathcal{S}_N \ni s \mapsto (-1)^s \in UM(\mathbb{C})$.

The left regular representations of the bi-equivariant sub-algebra drop to representations on the Fock spaces:

PROPOSITION 3.10. *Let* $|\xi\rangle\!\rangle$ *denote the canonical basis of* $\ell^2\big(\mathfrak{a}_N^{-1}(\mathcal{L}_\eta)\big)$. *Then, since*

$$\langle\!\langle \Lambda_{s_1}(\zeta)|\pi_\eta(f)|\Lambda_{s_2}^{-1}(\xi)\rangle\!\rangle = f\big(s_1 \cdot \mathfrak{t}_{\chi_\xi(1)}(\zeta, \xi) \cdot s_2\big), \tag{3.16}$$

the sub-space $\mathcal{N}$ *spanned by*

$$|\xi\rangle\!\rangle - (-1)^s|\Lambda_s(\xi)\rangle\!\rangle, \quad \xi \in \mathfrak{a}_N^{-1}(\mathcal{L}_\xi), \quad s \in \mathcal{S}_N, \tag{3.17}$$

is contained in the kernel of the left regular representation π_η of $C^*_{r,\mathcal{S}_N}\big(\widehat{\mathcal{G}}^{(N)}_{\mathcal{L}_0},\mathbb{C}\big)$. As such, π_η drops to a representation on the quotient space $\ell^2\big(\mathfrak{a}_N^{-1}(\mathcal{L}_\eta)\big)/\mathcal{N}$, which coincides with the N-fermion sector of the Fock space over the lattice $\mathcal{L}_\eta$ (see section 4).

3.3. $\mathbb{R}^d$ actions, invariant semi-finite traces and cyclic cocycles. We will step back to the groupoid C^*-algebra $C^*_r\big(\widehat{\mathcal{G}}^{(N)}_{\mathcal{L}_0}\big)$ and introduce first a differential structure:

PROPOSITION 3.11. *$C^*_r\big(\widehat{\mathcal{G}}^{(N)}_{\mathcal{L}_0}\big)$ accepts a natural $\mathbb{R}^d$-action*

$$[k\cdot f](\zeta,\xi)=e^{\imath N\langle k, x_\zeta - x_\xi\rangle} f(\zeta,\xi),\quad k\in\mathbb{R}^d, \tag{3.18}$$

where x_η is the center of mass of the subset V_η for any $\eta\in\widehat{\mathrm{Del}}^{(N)}_{(r,R)}(\mathbb{R}^d)$ and $\langle,\rangle$ denotes the standard scalar product of $\mathbb{R}^d$.

PROOF. We need to show that the rule from Eq. (3.18) defines automorphisms. From (3.3), we have

$$[k\cdot(f_1 * f_2)](\zeta,\xi)=e^{\imath N\langle k, x_\zeta - x_\xi\rangle}\sum_{\eta\in\mathfrak{a}_N^{-1}(\mathcal{L}_\xi)} f_1\big(\mathfrak{t}_{\chi_\eta(1)}(\zeta,\eta)\big)\cdot f_2(\eta,\xi). \tag{3.19}$$

Then the statement follows from the observation that

$$x_\zeta - x_\xi = x_{\hat{\mathfrak{t}}_y(\zeta)} - x_{\hat{\mathfrak{t}}_y(\eta)} + x_\eta - x_\xi \tag{3.20}$$

holds for any $y\in\mathbb{R}^d$. □

REMARK 3.12. As we shall see in the next section, this action is related to familiar $U(1)$-gauge transformations on the CAR algebra over $\ell^2(\mathcal{L}_0)$. As such, they are relevant to the computation of the physical transport coefficients. ◇

DEFINITION 3.13. The $\mathbb{R}^d$-action supplies a set of d derivations,

$$(\partial_j f)(\zeta,\xi)=\imath N(x_\zeta - x_\xi)_j f(\zeta,\xi),\quad j=1,\dots,d, \tag{3.21}$$

and we will denote by $C^\infty\big(\mathcal{G}^{(N)}_{\mathcal{L}_0}\big)$ the Fréchet subalgebra of the smooth elements from $C^*_r\big(\widehat{\mathcal{G}}^{(N)}_{\mathcal{L}_0}\big)$.

REMARK 3.14. Since the $\mathbb{R}^d$-action τ commutes with the 2-action of the permutation group, these derivations drop on the sub-algebra $C^*_{r,\mathcal{S}_N}\big(\widehat{\mathcal{G}}^{(N)}_{\mathcal{L}_0},\mathbb{C}\big)$. ◇

We now turn our attention to a particular semi-finite trace. First, we endow the (discrete) fibers of the blow-up map (2.35) with the counting measure. The composition with a $\mathcal{G}_{\mathcal{L}_0}$-invariant measure on $\Xi_{\mathcal{L}_0}$ then supplies a $\widehat{\mathcal{G}}^{(N)}_{\mathcal{L}_0}$-invariant measure on $\widehat{\Xi}^{(N)}_{\mathcal{L}_0}$.

PROPOSITION 3.15. *Let $\mathbb{P}$ be a $\mathcal{G}_{\mathcal{L}_0}$-invariant measure on $\Xi_{\mathcal{L}_0}$. Then*

$$\widehat{\mathcal{T}}^{(N)}_{\mathcal{L}_0}(f):=\int_{\Xi_{\mathcal{L}_0}}\mathrm{d}\mathbb{P}(\mathcal{L})\sum_{\xi\in\mathfrak{u}^{-1}(\mathcal{L})} f(\xi,\xi) \tag{3.22}$$

is a positive linear map over the core algebra $C_c\big(\widehat{\mathcal{G}}^{(N)}_{\mathcal{L}_0}\big)$, whose kernel includes the commutator sub-space. As such, $\widehat{\mathcal{T}}^{(N)}_{\mathcal{L}_0}$ can be promoted to a semi-finite trace on the C^*-algebra of the N-body groupoid.

PROPOSITION 3.16. *If $\mathbb{P}$ is ergodic and f is in the domain of the trace, then $\mathbb{P}$-almost surely,*

$$\widehat{\mathcal{T}}_{\mathcal{L}_0}^{(N)}(f) = \lim_{V \to \mathbb{R}^d} \frac{1}{|V \cap \mathcal{L}|} \sum_{\xi \in \mathfrak{a}_N^{-1}(\mathcal{L})}^{\chi_\xi(1) \in V} \langle \xi | \pi_\eta(f) | \xi \rangle, \tag{3.23}$$

for any $\eta \in \mathfrak{u}^{-1}(\mathcal{L})$.

PROOF. We have from (2.25), that $\mathbb{P}$-almost surely,

$$\int_{\Xi_{\mathcal{L}_0}} d\mathbb{P}(\mathcal{L})\, \phi(\mathcal{L}) = \lim_{V \to \mathbb{R}^d} \frac{1}{|V \cap \mathcal{L}|} \sum_{x \in V \cap \mathcal{L}} \phi(\mathfrak{t}_x \mathcal{L}). \tag{3.24}$$

Then the statement follows from definition (3.22) of the trace and the expression (3.9) of the left regular representations. □

REMARK 3.17. As we shall see, $\widehat{\mathcal{T}}_{\mathcal{L}_0}^{(N)}$ essentially matches the trace per volume defined for the Fock representation of the physical Hamiltonians. ◇

Since the semi-finite trace is invariant against the $\mathbb{R}^d$-action supplying the derivations, the following expressions supply cyclic cocycles over $C^\infty(\widehat{\mathcal{G}}_{\mathcal{L}_0}^{(N)})$:

$$\Sigma_J(f_0, f_1, \ldots, f_{|J|}) := \Lambda_J \sum_{\lambda \in \mathcal{S}_J} (-1)^\lambda\, \widehat{\mathcal{T}}_{\mathcal{L}_0}^{(N)}\Big(f_0 \prod_{j \in J} \partial_{\lambda_j} f_j\Big), \tag{3.25}$$

where $J \subseteq \{1, 2, \ldots, d\}$ is a sub-set of indices. We defer the physical interpretation of these cyclic cocycles to the next section.

4. The algebra of many-body Hamiltonians

The first part of this section focuses on physical aspects, specifically on the dynamics of a gas of fermions populating a Delone set $\mathcal{L}$. The algebra of physical observables is that of canonical anticommutation relations (CAR) over the Hilbert space $\ell^2(\mathcal{L})$. The time evolutions of the physical observables, which can be experimentally mapped and quantified, are generated by particular derivations that are analyzed here. These derivations supply the core algebra of physical Hamiltonians and the main challenge is to complete this and related algebras to C^*- and pro-C^*-algebras. This was accomplished in [**35**] via the essential extensions of the groupoid C^*-algebras introduced in the previous section. This solution as well as other related aspects are discussed in the second part of the section.

4.1. The algebra of local observables. The algebra of local observables for fermions hopping over a Delone set $\mathcal{L}$ is the universal C^*-algebra $\mathrm{CAR}(\mathcal{L})$ of the anti-commutation relations over the Hilbert space $\ell^2(\mathcal{L})$ [**17**]:

$$a_x a_{x'} + a_{x'} a_x = 0, \quad a_x^* a_{x'} + a_{x'} a_x^* = \delta_{x,x'} \cdot 1, \quad x, x' \in \mathcal{L}. \tag{4.1}$$

This algebra accepts many useful presentations, but here we will be interested in the following symmetric presentation:

PROPOSITION 4.1. *For $\xi \in \widehat{\mathrm{Del}}_{(r,R)}^{(n)}(\mathbb{R}^d)$, let*

$$\mathfrak{a}(\xi) := a_{\chi_\xi(n)} \cdots a_{\chi_\xi(1)} \in \mathrm{CAR}(\mathcal{L}). \tag{4.2}$$

Then any element from CAR($\mathcal{L}$) *accepts a unique presentation as a convergent sum of the type*

$$A = \sum_{n,m\in\mathbb{N}} \tfrac{1}{\sqrt{n!m!}} \sum_{\zeta\in\mathfrak{a}_n^{-1}(\mathcal{L})} \sum_{\xi\in\mathfrak{a}_m^{-1}(\mathcal{L})} c(\zeta,\xi)\,\mathfrak{a}(\zeta)^*\mathfrak{a}(\xi), \tag{4.3}$$

where the coefficients are bi-equivariant in the sense

$$c\big(\Lambda_{s_1}(\zeta),\Lambda_{s_2}^{-1}(\xi)\big) = (-1)^{s_1}c(\zeta,\xi)(-1)^{s_2}, \quad s_1\in\mathcal{S}_n,\ s_2\in\mathcal{S}_m. \tag{4.4}$$

REMARK 4.2. The symmetric presentation puts all possible orderings of the generators on equal footing and this is desirable when working with generic Delone sets or when deformations of the underlying lattice are allowed. ◇

REMARK 4.3. The first sum in (4.3) includes the cases $n=0$ or $m=0$. In such situations, we use the convention that $\mathfrak{a}_0^{-1}(\mathcal{L})=\emptyset$ and $\mathfrak{a}(\emptyset)=1$, the unit of CAR($\mathcal{L}$). Therefore, $c(\emptyset,\emptyset)$ in (4.3) is the coefficient corresponding to the unit. ◇

The GICAR($\mathcal{L}$) subalgebra consists of the elements that are invariant against the $U(1)$-twist of the generators $a_x\mapsto\lambda a_x$, $\lambda\in\mathbb{S}^1$, hence the name of gauge invariant elements. The symmetric presentation of such elements takes the form

$$A = \sum_{n\in\mathbb{N}} \tfrac{1}{n!} \sum_{\zeta,\xi\in\mathfrak{a}_n^{-1}(\mathcal{L})} c(\zeta,\xi)\,\mathfrak{a}(\zeta)^*\mathfrak{a}(\xi), \tag{4.5}$$

The interest, however, is not in CAR($\mathcal{L}$) or GICAR($\mathcal{L}$), but in the dynamics of the local observables, i.e. in group homomorphisms $\alpha:\mathbb{R}\to\mathrm{Aut}\big(\mathrm{CAR}(\mathcal{L})\big)$ and their generators. The physical constraints on these generators will be described next.

4.2. The physical Hamiltonians. We will be dealing exclusively with dynamics that conserves the fermion number, hence with time evolutions that descend on the GICAR subalgebra. Specifically:

DEFINITION 4.4. A Galilean and gauge invariant Hamiltonian with finite interaction range is a correspondence

$$\mathrm{Del}_R^r(\mathbb{R}^d)\ni\mathcal{L}\mapsto H_{\mathcal{L}} = \sum_{n\in\mathbb{N}^\times} \tfrac{1}{n!} \sum_{\zeta,\xi\in\mathfrak{a}_n^{-1}(\mathcal{L})} h_n(\zeta,\xi)\,\mathfrak{a}^*(\zeta)\mathfrak{a}(\xi), \tag{4.6}$$

where the coefficients h are continuous $\mathbb{C}$-valued maps over the many-body covers obeying the constraint $h_n(\zeta,\xi)=\overline{h_n(\xi,\zeta)}$ and:

c1. h_n's are bi-equivariant w.r.t. $\mathcal{S}_n$;
c2. h_n's are equivariant w.r.t. shifts, $h_n\big(\hat{\mathfrak{t}}_x(\zeta,\xi)\big)=h_n(\zeta,\xi)$, $x\in\mathbb{R}^d$;
c3. h_n's vanish whenever the diameter of $V_\zeta\cup V_\xi$ exceeds a fixed value R_{i}.

REMARK 4.5. It is important to stress that *c2-3* are motivated by the experimental realities. For example, *c*2 is a consequence of the fact that the physical processes involved in the coupling of the quantum resonators are Galilean invariant. Also, *c*3 relates to the simple fact that a finite team of experimenters can only probe the couplings (encoded in the h_n's) between a finite number of sites. Any Hamiltonian that cannot be approximated by a finite range Hamiltonian is beyond the control of the experimental team. Continuity of the coefficients comes from a similar argument: The experimenters can only probe a finite number of lattices, hence they need to rely on extrapolations. As such, Hamiltonians with discontinuous coefficients are beyond their control. ◇

The following statement gives substance to the formal series in Eq. (4.6):

PROPOSITION 4.6. *Let $\{\mathcal{L}_k\}$ be a net of finite sets converging to $\mathcal{L}$ and let $H_{\mathcal{L}_k}$ be the truncation of $H_{\mathcal{L}}$ from Eq. (4.6) to an element of* $\mathrm{CAR}(\mathcal{L}_k)$. *Then the map*

$$\mathrm{ad}_{H_{\mathcal{L}}}(A) := \lim_{k\to\infty} \imath[A, H_{\mathcal{L}_k}], \quad A \in \mathcal{D}(\mathcal{L}) := \bigcup_k \mathrm{CAR}(\mathcal{L}_k),$$

is a derivation that leaves $\mathcal{D}(\mathcal{L})$ invariant. Furthermore,

$$\mathrm{ad}_{H_{\mathcal{L}}}(A)^* := \mathrm{ad}_{H_{\mathcal{L}}}(A^*), \quad \forall\ A \in \mathcal{D}(\mathcal{L}), \tag{4.7}$$

and $\mathrm{ad}_{H_{\mathcal{L}}}$ *is closable and in fact a pre-generator of a time evolution.*

REMARK 4.7. Obviously, we are dealing with outer derivations that are inner-limit [**16**]. Such derivations leave the ideals of GICAR($\mathcal{L}$) invariant, hence they descend on the quotients. This together with the fact that GICAR($\mathcal{L}$) is a solvable C^*-algebra were the essential ingredients for the characterization of these derivations supplied in [**35**] and summarized in Theorem 4.12. ◇

All derivations (4.7) leave their common domain invariant, hence they can be composed among themselves. This enables us to speak of:

DEFINITION 4.8 ([**35**]). The core algebra of physical derivations is the subalgebra $\dot{\Sigma}(\mathcal{L}) \subset \mathrm{End}(\mathcal{D}(\mathcal{L}))$ generated by derivations $\mathrm{ad}_{Q^n_{\mathcal{L}}}$ corresponding to

$$Q^n_{\mathcal{L}} = \tfrac{1}{n!} \sum_{\zeta,\xi\in\mathfrak{a}_n^{-1}(\mathcal{L})} q_n(\zeta,\xi)\mathfrak{a}^*(\zeta)\mathfrak{a}(\xi), \tag{4.8}$$

where q's are defined over the many-body covers and satisfy the constraints c1-3. Henceforth, an element of $\dot{\Sigma}(\mathcal{L})$ can be presented as a finite sum

$$\mathcal{Q} = \sum_{\{Q\}} c_{\{Q\}}\ \mathrm{ad}_{Q^{n_1}_{\mathcal{L}}} \circ \ldots \circ \mathrm{ad}_{Q^{n_k}_{\mathcal{L}}}, \quad c_{\{Q\}} \in \mathbb{C}. \tag{4.9}$$

The multiplication is given by the composition of linear maps over $\mathcal{D}(\mathcal{L})$.

REMARK 4.9. The linear maps from Eq. (4.9) can be also associated with formal sums as in Eq. (4.6), but the coefficients will no longer have finite range. The linear space spanned by $\mathrm{ad}_{Q^n_{\mathcal{L}}}$ is, however, invariant against the Lie bracket

$$(\mathrm{ad}_{Q^n_{\mathcal{L}}}, \mathrm{ad}_{Q^m_{\mathcal{L}}}) \mapsto \mathrm{ad}_{Q^n_{\mathcal{L}}} \circ \mathrm{ad}_{Q^m_{\mathcal{L}}} - \mathrm{ad}_{Q^m_{\mathcal{L}}} \circ \mathrm{ad}_{Q^n_{\mathcal{L}}}. \tag{4.10}$$

Indeed, the right side has again coefficients with finite range. Hence, $\dot{\Sigma}(\mathcal{L})$ is the associative envelope of this Lie algebra. ◇

The interest is in formalizing the whole algebra $\dot{\Sigma}(\mathcal{L})$ but also in its representations on the sectors with finite fermions of the Fock space.

DEFINITION 4.10. The vacuum state over CAR($\mathcal{L}$) is defined as

$$\omega(A) = c(\emptyset,\emptyset), \tag{4.11}$$

where $A \in \mathrm{CAR}(\mathcal{L})$ is assumed to be in its symmetric presentation (4.3). The corresponding GNS representation, also known as the Fock representation, will be denoted by π_ω and the equivalence class of $\mathfrak{a}(\xi)^*$ for this representation will be denoted by $|\xi\rangle := \mathfrak{a}(\xi)^* + \mathrm{Null}(\omega)$, $\xi \in \mathfrak{a}_n^{-1}(\mathcal{L})$ for some $n \in \mathbb{N}$. When restricted to GICAR subalgebra, the Fock representation breaks into a direct sum $\pi_\omega = \bigoplus_{N\in\mathbb{N}} \pi_\omega^N$, where π_ω^N is carried by the Hilbert space $\mathcal{F}_N^{(-)}$ linearly spanned by $|\xi\rangle$, $\xi \in \mathfrak{a}_N^{-1}(\mathcal{L})$.

The Fock representation extends over the core algebra $\dot{\Sigma}(\mathcal{L})$. More precisely:

PROPOSITION 4.11 ([**35**]). *The Fock representation supplies representations of $\dot{\Sigma}(\mathcal{L})$ by bounded operators on the N-fermion sectors $\mathcal{F}_N^{(-)}$ of the Fock space. Explicitly:*

1) *If $n > N$, then*

$$\pi_\eta^N(Q_{\mathcal{L}}^n) = 0.$$

2) *If $n = N$, then*

$$\pi_\eta^N(Q_{\mathcal{L}}^N) = \tfrac{1}{N!} \sum_{\zeta,\xi \in \mathfrak{a}_N^{-1}(\mathcal{L})} q_N\big(\mathfrak{t}_{\chi_\xi(1)}(\zeta,\xi)\big)|\zeta\rangle\langle\xi|. \tag{4.12}$$

3) *If $n < N$, then*

$$\pi_\eta^N(Q_{\mathcal{L}}^n) = \tfrac{1}{n!(N-n)!} \sum_{\zeta,\xi \in \mathfrak{a}_n^{-1}(\mathcal{L})} \ \sum_{\gamma \in \mathfrak{a}_{N-n}^{-1}(\mathcal{L})} q_n(\xi,\zeta)|\xi \vee \gamma\rangle\langle\zeta \vee \gamma|.$$

Above, $|\xi \vee \gamma\rangle$ is the equivalence class of $\mathfrak{a}(\xi)^*\mathfrak{a}(\gamma)^*$ in the GNS representation corresponding to ω.

We point to the similarity of the above representation in the case $n = N$ and the left regular representations (3.9) of the groupoid algebra over the N-body covers. There are two exceptional factors at work here, which enabled us in [**35**] to ultimately link the two representations. Firstly, $\mathcal{F}_N^{(-)}$ is isomorphic to the Hilbert space $\ell^2\big(\mathfrak{a}_N^{-1}(\mathcal{L})\big)/\mathcal{N}$ appearing in Proposition 3.10, with the isomorphism supplied by $|\xi\rangle\rangle + \mathcal{N} \mapsto |\xi\rangle$. Secondly, the sub-space of $\pi_\omega\big(\dot{\Sigma}(\mathcal{L})\big)$ spanned by $\pi_\omega(Q_{\mathcal{L}}^N)$, with $Q_{\mathcal{L}}^N$ as in (4.8), is an ideal of $\pi_\omega^N\big(\dot{\Sigma}(\mathcal{L})\big)$. We now can state the main conclusion of [**35**]:

THEOREM 4.12 ([**35**]). *Let $\mathcal{L} \in \Xi_{\mathcal{L}_0}$. Then:*

i) *All representations of the physical Hamiltonians on the finite sectors of the Fock space can be generated from a left-regular representation of an essential extension of the bi-equivariant groupoid algebras. More precisely, for any $N \in \mathbb{N}^\times$, there exists an embedding*

$$\pi_\omega^N\big(\dot{\Sigma}(\mathcal{L})\big) \rightarrowtail \pi_\eta\Big(\mathcal{M}\big(C^*_{r,\mathcal{S}_N}(\widehat{\mathcal{G}}_{\mathcal{L}_0}^{(N)},\mathbb{C})\big)\Big), \tag{4.13}$$

where π_η is the left regular representation corresponding to any $\eta \in \mathfrak{a}_N^{-1}(\mathcal{L})$ and $\mathcal{M}$ indicates the extension to multiplier algebra.

ii) *The algebra $\dot{\Sigma}(\mathcal{L})$ itself accepts a completion to a pro-C^*-algebra:*

$$\dot{\Sigma}(\mathcal{L}) \rightarrowtail \varprojlim \bigoplus_{n=1}^{N} \pi_\eta\Big(\mathcal{M}\big(C^*_{r,\mathcal{S}_n}(\widehat{\mathcal{G}}_{\mathcal{L}_0}^{(n)},\mathbb{C})\big)\Big) \otimes \pi_\eta\Big(\mathcal{M}\big(C^*_{r,\mathcal{S}_n}(\widehat{\mathcal{G}}_{\mathcal{L}_0}^{(n)},\mathbb{C})\big)\Big)^{\mathrm{op}}.$$

REMARK 4.13. Given the above, the pro-C^*-algebra

$$\Sigma(\mathcal{L}_0) := \varprojlim \bigoplus_{n=1}^{N} \mathcal{M}\big(C^*_{r,\mathcal{S}_n}(\widehat{\mathcal{G}}_{\mathcal{L}_0}^{(n)},\mathbb{C})\big) \otimes \mathcal{M}\big(C^*_{r,\mathcal{S}_n}(\widehat{\mathcal{G}}_{\mathcal{L}_0}^{(n)},\mathbb{C})\big)^{\mathrm{op}} \tag{4.14}$$

can be seen as the sought completion of the core algebra of Hamiltonians. For this reason, we will refer to $\Sigma(\mathcal{L}_0)$ as the algebra of physical Hamiltonians over the Delone set $\mathcal{L}_0$. $\diamond$

REMARK 4.14. There are direct physical interpretations of the essential ideal $C^*_{r,\mathcal{S}_N}\big(\widehat{\mathcal{G}}^{(N)}_{\mathcal{L}_0},\mathbb{C}\big)$ and of the corona of the multiplier algebra: Under a time evolution generated by Hamiltonians from the algebra $C^*_{r,\mathcal{S}_N}\big(\widehat{\mathcal{G}}^{(N)}_{\mathcal{L}_0},\mathbb{C}\big)$, the N fermions evolve in a single cluster while, for a dynamics generated by Hamiltonians from the corona of the multiplier algebra, the fermions can scatter into two or more clusters. ◇

4.3. Transport coefficients. The anticommutation relations (4.1) are invariant against the $U(1)$-gauge transformations $a_x \mapsto e^{\imath\langle k,x\rangle}a_x$, $k\in\mathbb{R}$. Then the universality of the CAR-algebra assures us that this transformation is in fact an automorphism of $\mathrm{CAR}(\mathcal{L})$. Any automorphism can be lifted to inner-limit derivations, in particular to the ones generating the core algebra $\dot\Sigma(\mathcal{L})$. It is straightforward to verify that these $U(1)$-gauge transformations coincide with the natural $\mathbb{R}^d$-action over $C^*_{r,\mathcal{S}_N}\big(\widehat{\mathcal{G}}^{(N)}_{\mathcal{L}_0},\mathbb{C}\big)$ introduced in Proposition 3.11.

To clarify the physical meaning of the $\mathbb{R}^d$-actions and of the associated derivations, we recall two important macroscopic physical observables, one corresponding to the fermion density $\mathfrak{n}$ and one to the fermion position $\mathfrak{X}$,

$$\mathfrak{n} = \sum_{x\in\mathcal{L}} a_x^* a_x, \quad \mathfrak{X} = \sum_{x\in\mathcal{L}} x\, a_x^* a_x, \tag{4.15}$$

which can be defined for any $\mathcal{L}$ from $\mathrm{Del}^r_R(\mathbb{R}^d)$. Note that $\mathfrak{n}$ belongs to the core algebra $\dot\Sigma(\mathcal{L})$, but $\mathfrak{X}$ does not. Yet, we have

$$\partial_j A = \imath[\mathfrak{X}_j, A], \quad A\in \mathrm{CAR}(\mathcal{L}). \tag{4.16}$$

We now consider $h\in C^\infty_{\mathcal{S}_N}\big(\widehat{\mathcal{G}}^{(N)}_{\mathcal{L}_0},\mathbb{C}\big)$ and the corresponding Hamiltonian

$$H_{\mathcal{L}} = \tfrac{1}{N!}\sum_{\zeta,\xi\in\mathfrak{a}_N^{-1}(\mathcal{L})} h\big(\mathfrak{t}_{\chi_\xi(1)}(\zeta,\xi)\big)\mathfrak{a}^*(\zeta)\mathfrak{a}(\xi) \tag{4.17}$$

on $\ell^2(\mathcal{L})$ with $\mathcal{L}\in\Xi_{\mathcal{L}_0}$. The macroscopic physical observables corresponding to the particle current-density vector of components is defined as

$$\mathfrak{J}_j := \frac{d\mathfrak{X}_j}{dt} = \imath[\mathfrak{X}_j, H_{\mathcal{L}}], \tag{4.18}$$

and we observe that

$$\pi^N_\omega(\mathfrak{J}_j) = \pi_\eta(\partial_j h), \quad j=1,\dots,d, \tag{4.19}$$

for any $\eta\in\mathfrak{u}^{-1}(\mathcal{L})$.

REMARK 4.15. Some of the above attributes must have confused the reader. Indeed, we declared in subsection 4.1 that the algebra of physical observables is $\mathrm{CAR}(\mathcal{L})$. The elements $\mathfrak{n}$ and $\mathfrak{J}$ do not belong to $\mathrm{CAR}(\mathcal{L})$ and this is why we used the terminology macroscopic physical observables. Their measurement requires entirely different experimental setups capable of probing the samples over large, macroscopic regions. Note that, in order to evaluate expected values for these macroscopic observables, one actually needs a state on the algebra $\dot\Sigma(\mathcal{L})$. There is, of course, a dynamics on this algebra, generated by the inner derivation $\imath[\cdot,H_{\mathcal{L}}]$. ◇

Now, let p be a spectral projection of h, both seen as elements of $C^*_{r,\mathcal{S}_N}\big(\widehat{\mathcal{G}}^{(N)}_{\mathcal{L}_0},\mathbb{C}\big)$. If p belongs to the domain of the trace, then $f\mapsto\widehat{\mathcal{T}}^{(N)}_{\mathcal{L}_0}(fp)$ defines a state over

$C^*_{r,\mathcal{S}_N}\big(\widehat{\mathcal{G}}^{(N)}_{\mathcal{L}_0},\mathbb{C}\big)$, which, like any other state, can be uniquely extended over the multiplier algebra, hence over $\pi^N_\omega\big(\dot{\Sigma}(\mathcal{L})\big)$. In particular, we can evaluate the expected value of the fermion density $\mathfrak{n}$ and the result will be a non-zero value. Hence,

$$\dot{\Sigma}(\mathcal{L}) \ni Q \mapsto \widehat{\mathfrak{T}}^{(N)}_{\mathcal{L}_0}\big(\pi^N_\omega(Q)\,\pi^N_\omega(p)\big) \tag{4.20}$$

generates a state at finite fermion density over the core algebra $\dot{\Sigma}(\mathcal{L})$, which is invariant against the dynamics. Above, $\widehat{\mathfrak{T}}^{(N)}_{\mathcal{L}_0}$ is evaluated via the relation (3.23). The state (4.20) corresponds to a uniform population of the energy spectrum covered by p, something that is routinely achieved in cold atom experiments, for example. Such states can be definitely achieved in electron systems if interest exists.

We can attach numerical invariants to these spectral states, by pairing the even cyclic cocycles (3.25) with the K-theoretic class of the spectral projection:

$$\mathrm{Ch}_J(p) = \Lambda_J \sum_{\lambda\in\mathcal{S}_J} (-1)^\lambda\, \widehat{\mathfrak{T}}^{(N)}_{\mathcal{L}_0}\Big(p\prod_{j=1}^{|J|}\partial_{\lambda_j}p\Big). \tag{4.21}$$

They have a direct connection with various transport coefficients, which are all computed from the current correlation functions

$$\langle \mathfrak{J}_{j_1}\dots\mathfrak{J}_{j_k}\rangle = \widehat{\mathfrak{T}}^{(N)}_{\mathcal{L}_0}\big((\partial_{j_1}h)\dots(\partial_{j_k}h)\big). \tag{4.22}$$

By repeating the arguments from [**7**, **11**], which are purely algebraic in nature, the cocycles with $|J|=2$ can be given the physical interpretation of the linear transport coefficient for the plane corresponding to the indices J. Furthermore, while we have not covered yet the presence of a magnetic field, we are confident that, once that is done, the calculation from Section 5.6 of [**45**] can be repeated and the cocycles with $|J|\geq 4$ can be identified with specific non-linear transport coefficients.

There is definitely a whole lot to be learned about the physical interpretation of the states (4.20). Various strategies for completing the thermodynamic limit $N\to\infty$ have been listed in [**35**]. In particular, it was pointed out that, if a Hamiltonian is generated as above at a fix value of N, then such Hamiltonian lands in the corona of the multiplier algebra of $C^*_{r,\mathcal{S}_N}\big(\widehat{\mathcal{G}}^{(N)}_{\mathcal{L}_0},\mathbb{C}\big)$, for larger N's. However, such Hamiltonian can be pulled into $C^*_{r,\mathcal{S}_N}\big(\widehat{\mathcal{G}}^{(N)}_{\mathcal{L}_0},\mathbb{C}\big)$ by sandwiching it between quasi-central approximations of the identity. These approximations can be crafted to control the fermion density and, essentially, generate finite-volume approximations that can be used to achieve thermodynamic limits with targeted fermion densities. Hence, the states (4.20) are expected to be quite accurate for N large enough. In this respect, we want to point out for the reader that the many-body Hamiltonians and associate states for various fractional Hall sequences constructed in [**39**] are very close in spirit with the prescription we just described.

REMARK 4.16. While the relation between the states (4.20) and the equilibrium thermodynamic states is definitely of great importance, we want to point to the reader that the focus in materials science is firmly shifting from equilibrium to dynamical states. The activities related to the latter consists precisely in identifying and classifying Hilbert modules that are invariant under the dynamics, with the interest coming from a search and discovery of new manifestations of the so called bulk-defect correspondence principle [**47**]. We believe that the framework supplied by the groupoid approach and the states (4.20) designed with finite number of fermions will be an abundant source of interesting new examples, as already hinted by the numerical experiments from [**32**]. $\diamond$

5. Index Theorems for the 1-Fermion Setting

At this moment, we are not able to produce a complete proof of the index theorem for the pairings of the cocycles (3.25) and the K-theory of the corresponding C^*-algebras for the cases $N > 1$. However, as a preparation, we reworked the proofs of the index theorems from [**14, 15**] for the case $N = 1$, by generalizing the geometric identities from periodic lattices [**42, 44**] to generic Delone sets. The parings (4.21) for generic fermion numbers are very close in spirit to the case $N = 1$ and, in our opinion, an index theorem is now within reach if we follow the template presented here. We recall that the challenge here is to validate the proofs in Sobolev regimes. We will showcase only the case of even dimensions because the case of odd dimensions is very similar.

Since we retreat to the case $N = 1$, hence to the groupoid $\mathcal{G}_{\mathcal{L}_0}$ introduced in subsection 2.6, we can simplify the notation a bit. Indeed, in this case, the left regular representations are indexed by $\mathcal{L} \in \Xi_{\mathcal{L}_0}$, hence they will be denoted by $\pi_{\mathcal{L}}$. The semi-finite trace $\widehat{\mathcal{T}}^{(N)}_{\mathcal{L}_0}$ becomes a continuous trace, which will be denoted by $\mathcal{T}$.

5.1. Dirac operators on Delone sets. We introduce here a class of Dirac operators over a point set $\mathcal{L} \in \Omega_{\mathcal{L}_0}$, with $\mathcal{L}_0$ fixed in $\mathrm{Del}^r_R(\mathbb{R}^d)$, and establish some of their basic properties. For this, let $\gamma_1, \ldots, \gamma_d$ be an irreducible representation of the complex Clifford algebra on $\mathbb{C}^n$, $n = 2^{\frac{d}{2}}$:

$$\gamma_i\gamma_j + \gamma_j\gamma_i = 2\delta_{ij}, \quad i, j = 1, \ldots, d. \tag{5.1}$$

When d is even, the Clifford algebra accepts the grading $\gamma_0 = -\imath^{-n}\gamma_1 \ldots \gamma_d$. We denote by tr_γ the ordinary trace over $\mathbb{C}^n$ and let

$$\mathcal{H}_{\mathcal{L}} = \mathbb{C}^n \otimes \ell^2(\mathcal{L}), \tag{5.2}$$

which is the Hilbert space which will carry the index theorems. We will also use the same symbol Tr for the ordinary trace over the Hilbert space $\mathcal{H}_{\mathcal{L}}$.

DEFINITION 5.1. For $\mathcal{L} \in \Omega_{\mathcal{L}_0}$, we call $\delta_{\mathcal{L}} : \mathcal{L} \to \mathbb{R}^d$ a set of acceptable shifts if the set $\{\delta_{\mathcal{L}}(x),\ x \in \mathcal{L}\}$ is bounded in $\mathbb{R}^d$ and $x - \delta_{\mathcal{L}}(x) \neq 0$ for all $x \in \mathcal{L}$.

DEFINITION 5.2. The Dirac operator $D_{\mathcal{L},\delta_{\mathcal{L}}}$ associated to $\mathcal{L} \in \Omega_{\mathcal{L}_0}$ and a set $\delta_{\mathcal{L}}$ of acceptable shifts is the unbounded self-adjoint operator

$$D_{\mathcal{L},\delta_{\mathcal{L}}} := \sum_{j=1}^{d} \sum_{x \in \mathcal{L}} \gamma_j \otimes \big(x - \delta_{\mathcal{L}}(x)\big)_j\, |x\rangle\langle x|. \tag{5.3}$$

The spectrum of the operator is

$$\mathrm{Spec}(D_{\mathcal{L},\delta_{\mathcal{L}}}) = \big\{ \pm |x - \delta_{\mathcal{L}}(x)|,\ x \in \mathcal{L} \big\} \tag{5.4}$$

and, by construction, $0 \notin \mathrm{Spec}(D_{\mathcal{L},\delta_{\mathcal{L}}})$. This will enable us to define the phase of the Dirac operator but the consideration of the shifts has a much deeper motivation, revealed in the proof of the main index theorem. Let us add that, under the natural unitary map between $\ell^2(\mathcal{L})$ and $\ell^2\big(\mathfrak{t}_y(\mathcal{L})\big)$, $y \in \mathbb{R}^d$, we have

$$D_{\mathcal{L},\delta_{\mathcal{L}}} \mapsto D_{\mathfrak{t}_y(\mathcal{L}),\mathfrak{t}_y\circ\delta_{\mathcal{L}}\circ\mathfrak{t}_y^{-1}}. \tag{5.5}$$

and note that $\mathfrak{t}_y \circ \delta_{\mathcal{L}} \circ \mathfrak{t}_y^{-1}$ is again an acceptable set of shifts for $\mathfrak{t}_y(\mathcal{L})$.

NOTATION 5.3. Given a metric space $(Y, |\cdot|)$ and $y \in Y$, we will consistently use the notation $\hat{y} := y/|y|$ if $|y| \neq 0$ and $\hat{y} := 0$ otherwise. Furthermore, we introduce the shorthand $\sum_{j=1}^{d} \gamma_j \otimes v_j = \gamma \cdot v$, for any d-tuple v from a linear space. ◇

DEFINITION 5.4. Let $\mathcal{L} \in \Omega_{\mathcal{L}_0}$ and $\delta_{\mathcal{L}}$ an acceptable set of shifts. Then, we define the phase of the corresponding Dirac operator as

$$\hat{D}_{\mathcal{L},\delta_{\mathcal{L}}} = D_{\mathcal{L},\delta_{\mathcal{L}}} / \sqrt{D^2_{\mathcal{L},\delta_{\mathcal{L}}}}. \tag{5.6}$$

Before stating the standard properties of the Dirac operators and their phases, we mention the following simple asymptotic estimate, which will come handy on several occasions:

PROPOSITION 5.5. *For any $x \in \mathbb{R}^d$ with $|x| = 1$ and $s \in \mathbb{R}$, we have*

$$\lim_{s \to \infty} s\big(\widehat{sx + y} - \widehat{sx}\big) = y - \tfrac{1}{2}\langle x, y\rangle\, x. \tag{5.7}$$

PROPOSITION 5.6. *Let $\mathcal{L} \in \Omega_{\mathcal{L}_0}$ and $\delta_{\mathcal{L}}$ and $\delta'_{\mathcal{L}}$ be any two acceptable sets of shifts. Then:*

i) $\hat{D}_{\mathcal{L},\delta_{\mathcal{L}}} - \hat{D}_{\mathcal{L},\delta'_{\mathcal{L}}}$ *is a compact operator.*
ii) $|D_{\mathcal{L},\delta_{\mathcal{L}}}|^{-d} - |D_{\mathcal{L},\delta'_{\mathcal{L}}}|^{-d}$ *is a trace class operator.*

PROOF. The statements follow from elementary estimates involving the asymptotic expression from Eq. (5.7) (see *e.g.* [**45**] or [**15**]). □

DEFINITION 5.7. Let $\mathcal{L} \in \Omega_{\mathcal{L}_0}$ and $B \in \mathbb{B}\big(\ell^2(\mathcal{L})\big)$, the algebra of bounded operators over $\ell^2(\mathcal{L})$. We say that B has finite range if there exists $M \in \mathbb{R}_+$ such that

$$\langle x|B|x'\rangle = 0, \quad \forall\ |x - x'| > M. \tag{5.8}$$

PROPOSITION 5.8. *Let $\mathcal{L} \in \Omega_{\mathcal{L}_0}$ and $A \in \mathbb{B}\big(\ell^2(\mathcal{L})\big)$ such that A is the limit in $\mathbb{B}\big(\ell^2(\mathcal{L})\big)$ of a sequence of operators with finite range. Then $[\hat{D}_{\mathcal{L},\delta_{\mathcal{L}}}, A]$ is a compact operator.*

PROOF. We consider first an operator B of finite range. Then,

$$\langle x|[\hat{D}_{\mathcal{L},\delta_{\mathcal{L}}}, B]|x'\rangle = \gamma \cdot \big(\widehat{x + \delta_{\mathcal{L}}(x)} - \widehat{x' + \delta_{\mathcal{L}}(x')}\big)\langle x|B|x'\rangle, \tag{5.9}$$

hence $[\hat{D}_{\mathcal{L},\delta_{\mathcal{L}}}, B]$ has finite range. Furthermore, using the asymptotic behavior described in Eq. (5.7), we can concluded that $\langle x|[\hat{D}_{\mathcal{L},\delta_{\mathcal{L}}}, B|x'\rangle|$ converges to zero as $|x| \to \infty$, uniformly in x' (because B has finite range). This assures us that $[\hat{D}_{\mathcal{L},\delta_{\mathcal{L}}}, B]$ is a compact operator. Now, if A is generic but obeys the stated assumptions, then $[\hat{D}_{\mathcal{L},\delta_{\mathcal{L}}}, A]$ can be approximated in norm by $[\hat{D}_{\mathcal{L},\delta_{\mathcal{L}}}, B_n]$, where B_n is a sequence in $\mathbb{B}\big(\ell^2(\mathcal{L})\big)$ of finite range operators. Thus $[\hat{D}_{\mathcal{L},\delta_{\mathcal{L}}}, A]$ is a limit of compact operators, hence compact. □

5.2. Dixmier trace. While singular traces are essential tools in the modern index theory [**24**], in the present approach, they will only be used to establish the summability of the Fredholm modules, following [**7**] and [**42**] as models. A comprehensive treatment of singular traces over $\mathbb{K}\big(\ell^2(\mathcal{L})\big)$ can be found in [**33**]. Here, we briefly recall that Dixmier trace involves the algebra $\mathbb{K}\big(\ell^2(\mathcal{L})\big)$ of compact operators over $\ell^2(\mathcal{L})$. Its unitization, as any other unital AF-algebra, can be pushed through Stratila and Voiculescu machinery [**53**], to construct its m.a.s.a (maximal abelian subalgebra), which is just $C(\alpha\mathcal{L})$, the algebra of continuous functions over the one-point compactification of $\mathcal{L}$. The associated subgroup Γ, as defined in [**53**], is the subgroup of homeomorphisms of $\alpha\mathcal{L}$ sending ∞ to itself. Furthermore, the conditional expectation P from $\mathbb{C}\cdot I\oplus\mathbb{K}(\ell^2(\mathcal{L}))$ to its m.a.s.a $C(\alpha\mathcal{L})$ is

$$\mathfrak{P}(cI+K)=G,\quad G(x)=c+\langle x|K|x\rangle. \tag{5.10}$$

Traces on the positive cone of $\mathbb{K}(\ell^2(\mathcal{L}))$ can be generated from Γ-invariant additive maps τ on the positive cone of $C(\alpha\mathcal{L})$, via the composition $\tau\circ\mathfrak{P}$. In fact, according to [**53**], all trace states on a unital AF-algebra come in this way.

Γ-invariant maps can be constructed from weighted partial sums, such as

$$s_N(f)=\frac{1}{C_N}\sum_{x\in\mathcal{L}}f(x)\mathfrak{c}_N(x),\quad f:\mathcal{L}\to\mathbb{R}_+, \tag{5.11}$$

where $\mathfrak{c}_N$ is the indicator function of the d-dimensional cube of size N and centered at the origin. The coefficients C_N can be essentially supplied by any monotone increasing function from $\mathbb{N}$ to $\mathbb{R}_+$. Such weighted sums supply maps $s(f):\mathbb{N}\to[0,\infty]$ and any ultrafilter ω on $\mathbb{N}$ supplies an ultrafilter $s(f)_*\omega$, the push forward ultrafilter, on the compact Hausdorff space $[0,\infty]$. Necessarily, $s(f)_*\omega$ has a unique limit in $[0,\infty]$ and, as such, one can define

$$\tau_\omega(f):=\lim s(f)_*\omega. \tag{5.12}$$

This map is additive and, furthermore, if ω is a free filter, then the limit in (5.12) coincides with $\lim_{N\to\infty}s_N(f)$, whenever the latter exists.

Definition 5.9. The Dixmier trace $\mathrm{Tr}_{\mathrm{Dix}}$ over $\mathbb{K}(\ell^2(\mathcal{L}))$ is defined as the map $\tau_\omega\circ P$ with τ_ω as above and with the choice

$$C_N=\ln\sum_{x\in\mathcal{L}}\mathfrak{c}_N(x) \tag{5.13}$$

in Eq. (5.11) and ω assumed to be a free ultrafilter.

Remark 5.10. We have included the above digression because we developed an interest in singular traces over generic AF-algebras and it seems to us that the machinery invented by Stratila and Voiculescu [**53**] supplies the right environment and the tools for studying such singular traces. $\diamond$

The statement involving the Dixmier trace, of great value for us, is:

Lemma 5.11. *If $g\in L^1(\Xi_{\mathcal{L}_0},\mathrm{d}\mathbb{P})$ and $\varphi:\mathcal{L}\to\mathbb{C}$ is uniformly bounded, then any positive operator B over $\ell^2(\mathcal{L})$ with diagonal*

$$\mathfrak{P}(B)(x)=\varphi(x)g(\mathfrak{t}_x(\mathcal{L}))|x+\delta_{\mathcal{L}}(x)|^{-d} \tag{5.14}$$

belongs to the domain of the Dixmier trace. In particular, it is trace class.

Proof. From [**14**][Lemma 6.1] (see also [**4**][Th. 1.2]), one knows that $G(x):=g(\mathfrak{t}_x(\mathcal{L}))|x+\delta_{\mathcal{L}}(x)|^{-d}$ is in the domain of τ_ω, independently of the choice of the ultrafilter ω, and, furthermore,

$$\tau_\omega(G)=|\mathbb{S}^{d-1}|^{-1}\int_{\Xi_{\mathcal{L}_0}}\mathrm{d}\mathbb{P}(\mathcal{L})\,g(\mathcal{L}). \tag{5.15}$$

The statement follows because the domain of τ_ω is an ideal in $C_b(\mathcal{L})$, the algebra of bounded functions over $\mathcal{L}$. $\square$

5.3. The Sobolev space. The L^s-spaces relative to the trace $\mathcal{T}$, denoted here by $L^s(\mathcal{G}_{\mathcal{L}_0},\mathcal{T})$, are defined as the completion of the convolution algebra $C_c(\mathcal{G}_{\mathcal{L}_0})$ under the norms

$$\|f\|_s = \mathcal{T}\big(|f|^s\big)^{\frac{1}{s}}. \tag{5.16}$$

The space $L^\infty(\mathcal{G}_{\mathcal{L}_0},\mathcal{T})$ is actually a von-Neumann algebra.

We need several simple facts about the latter:

PROPOSITION 5.12. *The following hold:*

i) $\phi(f) \in L^\infty(\mathcal{G}_{\mathcal{L}_0},\mathcal{T})$ *for any* $f \in C_r^*(\mathcal{G})$ *and Borel function* ϕ *over the real line.*

ii) $[\hat{D}_{\mathcal{L},\delta_{\mathcal{L}}},\pi_{\mathcal{L}}(f)]$ *is* $\mathbb{P}$*-almost surely a compact operator for* $f \in L^\infty(\mathcal{G}_{\mathcal{L}_0},\mathcal{T})$.

PROOF. i) $L^\infty(\mathcal{G}_{\mathcal{L}_0},\mathcal{T})$ is the bicommutant of $C_r^*(\mathcal{G}_{\mathcal{L}_0})$ hence it is stable for the Borel functional calculus. ii) First, note that $\pi_{\mathcal{L}}(f)$, $f \in L^\infty(\mathcal{G}_{\mathcal{L}_0},\mathcal{T})$, is $\mathbb{P}$-almost surely a bounded operator that can be approximated in norm by a sequence of finite range operators. Then Proposition 5.8 assures us that $[D_{\mathcal{L},\delta_{\mathcal{L}}},\pi_{\mathcal{L}}(f)]$ is indeed $\mathbb{P}$-almost surely a compact operator. □

The Sobolev space that we need for the index theorem is supplied by the completion of $C_c(\mathcal{G}_{\mathcal{L}_0})$ under the norm

$$\|f\|_W = \|f\|_\infty + \sum_{i=1}^{d} \|\partial_i f\|_d. \tag{5.17}$$

It will be denoted by $W(\mathcal{G}_{\mathcal{L}_0},\mathcal{T})$.

PROPOSITION 5.13. $W(\mathcal{G},\mathcal{T})$ *is a locally convex algebra.*

PROOF. $W(\mathcal{G},\mathcal{T})$ is already a locally convex topological vector space, hence we only need to verify that the product is jointly continuous. We have,

$$\|f_1 f_2\|_W = \|f_1 f_2\|_\infty + \sum_i \|(\partial_i f_1) f_2 + f_1(\partial_i f_2)\|_d \tag{5.18}$$

$$\leq \|f_1\|_\infty \|f_2\|_\infty + \sum_i \|(\partial_i f_1)\|_d \|f_2\|_\infty + \|f_1\|_\infty \|(\partial_i f_2)\|_d \tag{5.19}$$

$$\leq \Big(\|f_1\|_\infty + \sum_{i=1,d} \|\partial_i f_1\|_d\Big)\Big(\|f_2\|_\infty + \sum_{i=1,d} \|\partial_i f_2\|_d\Big), \tag{5.20}$$

where Hölder's inequality was used at places. Thus, $\|f_1 f_2\|_W \leq \|f_1\|_W \|f_2\|_W$ and the statement follows. □

The following statements re-assess the cocycles (3.25) in this new context:

PROPOSITION 5.14. *The relations*

$$W(\mathcal{G}_{\mathcal{L}_0},\mathcal{T})^{\times d} \ni (f_0, f_1, \ldots, f_d) \mapsto \mathcal{T}\big(f_0(\partial_1 f_1) \ldots (\partial_d f_d)\big). \tag{5.21}$$

defines a jointly continuous multi-linear map over $W(\mathcal{G}_{\mathcal{L}_0},\mathcal{T})$.

PROOF. We have

$$\Big|\mathcal{T}\big(f_0(\partial_1 f_1) \ldots (\partial_d f_d)\big) \leq \|f_0\|_\infty \|(\partial_1 f_1) \ldots (\partial_d f_d)\|_1 \tag{5.22}$$

$$\leq \|f_0\|_\infty \prod_{i=1}^{d} \|(\partial_i f_i)\|_d, \tag{5.23}$$

where Hölder's inequality was used at places. Therefore,

$$\big|\mathcal{T}\big(f_0(\partial_1 f_1)\,\ldots,(\partial_d f_d)\big)\big| \leq \|f_0\|_W \|f_1\|_W \cdots \|f_d\|_W, \tag{5.24}$$

and the statement follows. □

PROPOSITION 5.15. *Consider the cyclic and jointly continuous multi-linear map*

$$\Sigma_d(f_0, f_1, \ldots, f_d) = \Lambda_d \sum_{\sigma \in \mathcal{S}_d} (-1)^\sigma \mathcal{T}\big(f_0(\partial_1 f_{\sigma_1})\,\ldots(\partial_d f_{\sigma_d})\big) \tag{5.25}$$

over the Sobolev space. Then

$$|\Sigma_d(f_0, f_1, \ldots, f_d) - \Sigma_d(f_0', f_1', \ldots, f_d')| \leq \text{ct.} \sum_{i=0}^{d} \big(\|f_i - f_i'\|_W - \|f_i - f_i'\|_\infty\big), \tag{5.26}$$

where the constant in front is determined by the Sobolev norms of the entries.

PROOF. We have

$$\Sigma_d(f_0, f_1, \ldots, f_d) - \Sigma_d(f_0', f_1', \ldots, f_d') \tag{5.27}$$

$$= \sum_{k=0}^{d} \Sigma_d\big(f_0', \ldots, f_{k-1}', (f_k - f_k'), f_{k+1}, \ldots, f_d\big) \tag{5.28}$$

$$= \sum_{k=0}^{d} \Sigma_d\big(f_{k+1}, \ldots, f_d, f_0', \ldots, f_{k-1}', (f_k - f_k')\big). \tag{5.29}$$

Then the statement follows from the estimate in Eq. (5.22). □

REMARK 5.16. The estimate in Eq. (5.26) communicates that the value of the cocycle varies continuously under deformations inside $W(\mathcal{G}, \mathcal{T})$ that are continuous only w.r.t. the differential piece of the Sobolev norm, *i.e.* the second term in Eq. (5.17). This observation is crucial because the deformations occurring in the applications are never continuous w.r.t. the full Sobolev norm. Note that this special property is facilitated by the multi-linear map being cyclic. ◇

5.4. The Fredholm module and its quantized pairing. As in the previous sections and subsections, we will fix a Delone set $\mathcal{L}_0$. The C^*-algebra $C_r^*(\mathcal{G}_{\mathcal{L}_0})$ can be represented on $\mathcal{H}_\mathcal{L}$ by $1 \otimes \pi_\mathcal{L}$ and, to simplify the notation, we will denote this representation by the same symbols $\pi_\mathcal{L}$. Here $\mathcal{L}$ is assumed from $\Xi_{\mathcal{L}_0}$. Likewise, the natural actions of γ's on $\mathcal{H}_\mathcal{L}$ will be denoted by the same symbols. It is clear that $\pi_\mathcal{L}(f)\gamma_0 = \gamma_0 \pi_\mathcal{L}(f)$, for any $f \in C_r^*(\mathcal{G}_{\mathcal{L}_0})$. As for the Dirac operators, we will restrict the acceptable shifts to uniform shifts $\delta_\mathcal{L}(x) = w$, $w \in \mathbb{R}^d$ and simplify the notation $\hat{D}_{\mathcal{L},\delta_\mathcal{L}}$ to $\hat{D}_\mathcal{L}(w)$.

PROPOSITION 5.17. *The $\mathcal{L}$-dependent tuple*

$$\Big(W(\mathcal{G}_{\mathcal{L}_0}, \mathcal{T}), \pi_\mathcal{L} : M_n(\mathbb{C}) \otimes W(\mathcal{G}_{\mathcal{L}_0}, \mathcal{T}) \to \mathbb{B}(\mathcal{H}_\mathcal{L}), \hat{D}_\mathcal{L}(w), \gamma_0\Big)$$

is $\mathbb{P}$-almost surely a $d + \epsilon$-summable even Fredholm module.

PROOF. The tuple fulfills, $\mathbb{P}$-almost surely, the characteristics of an even Fredholm module, in part due to Proposition 5.12. Thus, it remains to prove the summability. We are going to evaluate the Dixmier trace of the positive operator

$$\text{tr}_\gamma\Big(\imath[\hat{D}_\mathcal{L}(w), \pi_\mathcal{L}(f)]^d\Big), \quad f \in W(\mathcal{G}_{\mathcal{L}_0}, \mathcal{T}).$$

After projecting on the diagonal part, using Eq. (5.9) as well as the asymptotic behavior from Eq. (5.7), we find

$$\mathrm{Tr}_{\mathrm{Dix}}\Big\{\mathrm{tr}_\gamma\left(\imath[\hat{D}_{\mathcal{L}}(w),\pi_{\mathcal{L}}(f)]^d\right)\Big\} = \tau_\omega(G), \tag{5.30}$$

where $G : \mathcal{L} \to \mathbb{R}_+$ is given by

$$G(x) = \mathrm{tr}_\gamma\Big(\langle 0|\prod_{i=1}^{d}\gamma(\hat{x})\cdot \imath[X,\pi_{\mathfrak{t}_x\mathcal{L}}(f)]|0\rangle\Big)|x+w|^{-d} \tag{5.31}$$

$$= \mathrm{tr}_\gamma\Big(\langle 0|\prod_{i=1}^{d}\gamma(\hat{x})\cdot \pi_{\mathfrak{t}_x\mathcal{L}}(\nabla f)]|0\rangle\Big)|x+w|^{-d}, \tag{5.32}$$

with $\gamma_i(\hat{x}) = \gamma_i - \hat{x}_i(\hat{x}\cdot\gamma)$. We can see that $G(x)$ is covered by Lemma 5.11, whenever $f \in W(\mathcal{G}_{\mathcal{L}_0},\mathcal{T})$. □

THEOREM 5.18. *Let $p \in W(\mathcal{G}_{\mathcal{L}_0},\mathcal{T})$ be a projection and let $\pi_{\mathcal{L}}^{\pm}$ be the restrictions of $\pi_{\mathcal{L}}$ representation on the sectors of the grading γ_0. Then, $\mathbb{P}$-almost surely, the operator*

$$\pi_{\mathcal{L}}^{-}(p)\hat{D}_{\mathcal{L}}(w)\pi_{\mathcal{L}}^{+}(p) \tag{5.33}$$

is in the Fredholm class. Furthermore, if the hull $\Omega_{\mathcal{L}_0}$ and its measure are invariant against the inversion operation $\mathcal{L} \mapsto -\mathcal{L}$, then, $\mathbb{P}$-almost surely,

$$\mathrm{Index}\left(\pi_{\mathcal{L}}^{-}(p)\hat{D}_{\mathcal{L}}(w)\pi_{\mathcal{L}}^{+}(p)\right) = \Lambda_d \sum_{\lambda\in\mathcal{S}_d}(-1)^{\lambda}\,\mathcal{T}\Big(p\prod_{i=1}^{d}\partial_{\lambda_i}p\Big), \tag{5.34}$$

where $\Lambda_d = \frac{(2\imath\pi)^{\frac{d}{2}}}{(d/2)!}$.

PROOF. Proposition 5.17 assures us that the operator (5.33) is indeed Fredholm, $\mathbb{P}$-almost surely. We observe first that the index in Eq. (5.34) is insensitive to the choice of $w \in \mathbb{R}^d$, as long as the latter supplies an acceptable set of shifts, which is the case almost surely with respect to the Lebesgue measure of $\mathbb{R}^d$. Next, we also observe that the index is $\mathbb{P}$-almost surely insensitive to the choice of $\mathcal{L}$ from $\Xi_{\mathcal{L}_0}$. To prove the claim, it is enough to demonstrate that the index is invariant against the groupoid action on $\Xi_{\mathcal{L}_0}$, specifically, against replacing $\mathcal{L}$ by $\mathfrak{t}_y(\mathcal{L})$, $y \in \mathcal{L}$. For this, we perform a unitary transformation from $\ell^2(\mathcal{L})$ to $\ell^2\big(\mathfrak{t}_y(x)\big)$ and use the statement from Eq. (5.5) to observe that the Fredholm operator transforms into

$$\pi_{\mathfrak{t}_y(\mathcal{L})}^{-}(p)\hat{D}_{\mathfrak{t}_y(\mathcal{L})}(y+w)\pi_{\mathfrak{t}_y(\mathcal{L})}^{+}(p). \tag{5.35}$$

According to Proposition 5.6, this is a compact perturbation of

$$\pi_{\mathfrak{t}_y(\mathcal{L})}^{-}(p)\hat{D}_{\mathfrak{t}_y(\mathcal{L})}(w)\pi_{\mathfrak{t}_y(\mathcal{L})}^{+}(p), \tag{5.36}$$

hence the index in Eq. (5.34) is indeed invariant against $\mathcal{L} \mapsto \mathfrak{t}_y(\mathcal{L})$.

We now start the computation of the index. Proposition 5.17 assures us that, $\mathbb{P}$-almost surely, the index is supplied by the pairing with the Connes-Chern character,

$$\mathrm{Ind}\big(\pi_{\mathcal{L}}^{-}(p)\hat{D}_{\mathcal{L}}(w)\pi_{\mathcal{L}}^{+}(p)\big) = \tfrac{1}{2}\,\mathrm{Tr}\big\{\gamma_0\hat{D}_{\mathcal{L}}(w)[\hat{D}_{\mathcal{L}}(w),\pi_{\mathcal{L}}(p)]^{d+1}\big\}. \tag{5.37}$$

We express the trace as the absolutely convergent sum

$$\sum_{x\in\mathcal{L}}\mathrm{tr}_\gamma\big(\gamma_0\langle x|\hat{D}_{\mathcal{L}}(w)[\hat{D}_{\mathcal{L}}(w),\pi_{\mathcal{L}}(p)]^{d+1}|x\rangle\big) \tag{5.38}$$

and observe that, by using the elementary identities,

$$\hat{D}_{\mathcal{L}}(w)[\hat{D}_{\mathcal{L}}(w),\pi_{\mathcal{L}}(p)] = -[\hat{D}_{\mathcal{L}}(w),\pi_{\mathcal{L}}(p)]\hat{D}_{\mathcal{L}}(w), \quad \hat{D}_{\mathcal{L}}(w)|x\rangle = |x\rangle\,(\gamma\cdot\widehat{x+w}),$$

as well as the cyclic property of tr_γ, the summand can be processed as

$$\begin{aligned}\mathrm{tr}\big(\gamma_0\langle x|\hat{D}_{\mathcal{L}}(w)\big(\hat{D}_{\mathcal{L}}(w)\,\pi_{\mathcal{L}}(p)-\pi_{\mathcal{L}}(p)\,\hat{D}_{\mathcal{L}}(w)\big)[\hat{D}_{\mathcal{L}}(w),\pi_{\mathcal{L}}(p)]^d|x\rangle\big)\\ = 2\,\mathrm{tr}\big(\gamma_0\langle x|\pi_{\mathcal{L}}(p)[\hat{D}_{\mathcal{L}}(w),\pi_{\mathcal{L}}(p)]^d|x\rangle\big).\end{aligned}$$

Therefore,

$$\mathrm{Ind}\big(\pi_{\mathcal{L}}^-(p)\hat{D}_{\mathcal{L}}(w)\pi_{\mathcal{L}}^+(p)\big) = \sum_{x\in\mathcal{L}}\mathrm{tr}_\gamma\big(\gamma_0\langle x|\pi_{\mathcal{L}}(p)[\hat{D}_{\mathcal{L}}(w),\pi_{\mathcal{L}}(p)]^d|x\rangle\big), \tag{5.39}$$

and note that the sum remains absolutely convergent. We now insert resolutions of the identity and write the right side as

$$\sum_{x\in\mathcal{L}}\sum_{x_i\in\mathcal{L}}\Big(\prod_{i=0}^{d}\langle x_i|\pi_{\mathcal{L}}(p)|x_{i+1}\rangle\Big)\mathrm{tr}_\gamma\Big(\gamma_0\prod_{i=1}^{d}\gamma\cdot\big(\widehat{x_i+w}-\widehat{x_{i+1}+w}\big)\Big), \tag{5.40}$$

where $x_0 = x_{d+1} = x$. Our next goal is to de-couple the sums over x and over x_i's by using the insensitivity to the choice of $\mathcal{L}$ or w. For this, it is convenient to introduce the new coordinates y_i that live on $\mathfrak{t}_x(\mathcal{L})$ and such that $x_i = x + y_i$. Then the right side of Eq. (5.40) becomes

$$\begin{aligned}&\sum_{x\in\mathcal{L}}\sum_{y_i\in\mathfrak{t}_x(\mathcal{L})}\Big(\prod_{i=0}^{d}\langle y_i|\pi_{\mathfrak{t}_x(\mathcal{L})}(p)|y_{i+1}\rangle\Big)\\ &\qquad\times\mathrm{tr}_\gamma\Big(\gamma_0\prod_{i=1}^{d}\gamma\cdot\big(\widehat{x+y_i+w}-\widehat{x+y_{i+1}+w}\big)\Big).\end{aligned} \tag{5.41}$$

where $y_0 = y_{d+1} = 0$. Consider a term in the above sum that corresponds to a particular $x\in\mathcal{L}$ and let $\mathcal{L}' = \mathfrak{t}_x(\mathcal{L})$. Note that $y := \mathfrak{t}_x(0) = -x$ belongs to $\mathcal{L}'$. Then such term can be completely written in term of the shifted lattice $\mathcal{L}'$, as

$$\sum_{y_i\in\mathcal{L}'}\Big(\prod_{i=0}^{d}\langle y_i|\pi_{\mathcal{L}'}(p)|y_{i+1}\rangle\Big)\mathrm{tr}_\gamma\Big(\gamma_0\prod_{i=1}^{d}\gamma\cdot\big(\widehat{y_i+w-y}-\widehat{y_{i+1}+w-y}\big). \tag{5.42}$$

Note also that $y-w$ belongs to $\mathfrak{t}_w(\mathcal{L}')$. Thus, if we let $\mathcal{L}'$ sample the entire $\Xi_{\mathcal{L}_0}$ and take into account the invariance of the index, we can write

$$\begin{aligned}\mathrm{Ind}\big(\pi_{\mathcal{L}}^-(p)\hat{D}_{\mathcal{L}}(w)\pi_{\mathcal{L}}^+(p)\big) &= \int_{\Xi_{\mathcal{L}_0}}\mathrm{d}\mathbb{P}(\mathcal{L}')\sum_{y_i\in\mathcal{L}'}\Big(\prod_{i=0}^{d}\langle y_i|\pi_{\mathcal{L}'}(p)|y_{i+1}\rangle\Big)\\ &\quad\times\sum_{y\in\mathfrak{t}_w(\mathcal{L}')}\mathrm{tr}_\gamma\Big(\gamma_0\prod_{i=1}^{d}\gamma\cdot\big(\widehat{y_i-y}-\widehat{y_{i+1}-y}\big)\Big).\end{aligned} \tag{5.43}$$

By giving values to w, we can sample the entire hull $\Omega_{\mathcal{L}_0}$ in the second line and, at this point, we indeed achieved the decoupling we mentioned at the beginning:

$$\begin{aligned}\operatorname{Ind}\big(\pi^-_{\mathcal{L}}(p)\hat{D}_{\mathcal{L}}(w)\pi^+_{\mathcal{L}}(p)\big) &= \int_{\Xi_{\mathcal{L}_0}} \mathrm{d}\mathbb{P}(\mathcal{L}') \sum_{y_i\in\mathcal{L}'} \Big(\prod_{i=0}^{d}\langle y_i|\pi_{\mathcal{L}'}(p)|y_{i+1}\rangle\Big) \\ &\times \int_{\Omega_{\mathcal{L}_0}} \mathrm{d}\bar{\mathbb{P}}(\mathcal{L}'') \sum_{y\in\mathcal{L}''} \operatorname{tr}_\gamma\Big(\gamma_0\prod_{i=1}^{d}\gamma\cdot\big(\widehat{y_i-y}-\widehat{y_{i+1}-y}\big)\Big).\end{aligned} \tag{5.44}$$

Then the statement follows from the geometric identity stated below. □

LEMMA 5.19. *Let $y_1,\dots,y_{d+1}$ be points of $\mathbb{R}^d$ with $y_{d+1}=0$. Then, if the hull $\Omega_{\mathcal{L}_0}$ and its measure are invariant against the inversion operation $\mathcal{L}\mapsto -\mathcal{L}$, the following identity holds:*

$$\int_{\Omega_{\mathcal{L}_0}} \mathrm{d}\bar{\mathbb{P}}(\mathcal{L}) \sum_{w\in\mathcal{L}} \operatorname{tr}_\gamma\Big(\gamma_0\prod_{i=1}^{d}\gamma\cdot\big(\widehat{y_i-w}-\widehat{y_{i+1}-w}\big)\Big) = \Lambda_d\sum_{\lambda}(-1)^{\lambda}\prod_{i=1}^{d}(y_i)_{\lambda_i}. \tag{5.45}$$

PROOF. As in [**42**], the proof relies on a geometric interpretation of the trace

$$\operatorname{tr}_\gamma\big(\gamma_0(\gamma\cdot y_1)\cdots(\gamma\cdot y_d)\big\} = \alpha_d\ \operatorname{Vol}\big[0,y_1,\dots,y_d\big],\quad \alpha_d=(2\imath)^{\frac{d}{2}}d!, \tag{5.46}$$

where $[y_0,y_1,\dots,y_d]$ is the simplex with vertices at y_0, y_1, …, y_d and Vol indicates the oriented volume of such simplex. Note that $y_0=0$ in Eq. 5.46. Expanding the left hand side of Eq. 5.45, we obtain

$$\alpha_d\int_{\Omega_{\mathcal{L}_0}} \mathrm{d}\bar{\mathbb{P}}(\mathcal{L}) \sum_{w\in\mathcal{L}}\sum_{j=1}^{d+1}(-1)^{j+1}\operatorname{Vol}\big[\mathbf{0},\widehat{y_1-w},\dots,\underline{\widehat{y_j-w}},\dots,\widehat{y_{d+1}-w}\big], \tag{5.47}$$

where the underline means the entry is omitted. It is convenient to translate the simplexes and move the first vertex to a proper place to work with the simplexes

$$\mathfrak{S}_j(w)=[w+\widehat{y_1-w},\dots,w,\dots,w+\widehat{y_{d+1}-w}], \tag{5.48}$$

where w is located at the j-th position. We will also denote by $\mathfrak{S}$ the simplex

$$\mathfrak{S}=[y_1,\dots,y_{d+1}], \tag{5.49}$$

where we recall that y_{d+1} coincides with the origin. To summarize,

$$\begin{aligned}\int_{\Omega_{\mathcal{L}_0}} \mathrm{d}\bar{\mathbb{P}}(\mathcal{L}) \sum_{w\in\mathcal{L}} \operatorname{tr}_\gamma\Big(\gamma_0\prod_{i=1}^{d}\gamma\cdot\big(\widehat{y_i-w}-\widehat{y_{i+1}-w}\big)\Big) \\ = \alpha_d\int_{\Omega_{\mathcal{L}_0}} \mathrm{d}\bar{\mathbb{P}}(\mathcal{L}) \sum_{w\in\mathcal{L}}\sum_{j=1}^{d+1}\operatorname{Vol}\{\mathfrak{S}_j(w)\}.\end{aligned} \tag{5.50}$$

The factor $(-1)^{j+1}$ disappeared because we changed the order of the vertices and note that, if w is located inside $\mathfrak{S}$, then the orientations of the simplexes $\mathfrak{S}_j(w)$ coincide with that of $\mathfrak{S}$, because the former can be continuously deformed into the latter while keeping the volume finite (see Fig. 5.1(a) for a demonstration).

Excepting w, the vertices of the simplexes $\mathfrak{S}_j(w)$ are all located on the unit sphere centered at w. As such, for any given simplex $\mathfrak{S}_j(w)$, the facets stemming from w define a sector of the unit ball $B_1(w)$. This sector will be denoted by $B_1^j(w)$. This construction is illustrated in Fig. 5.1(b) for the 2-dimensional case.

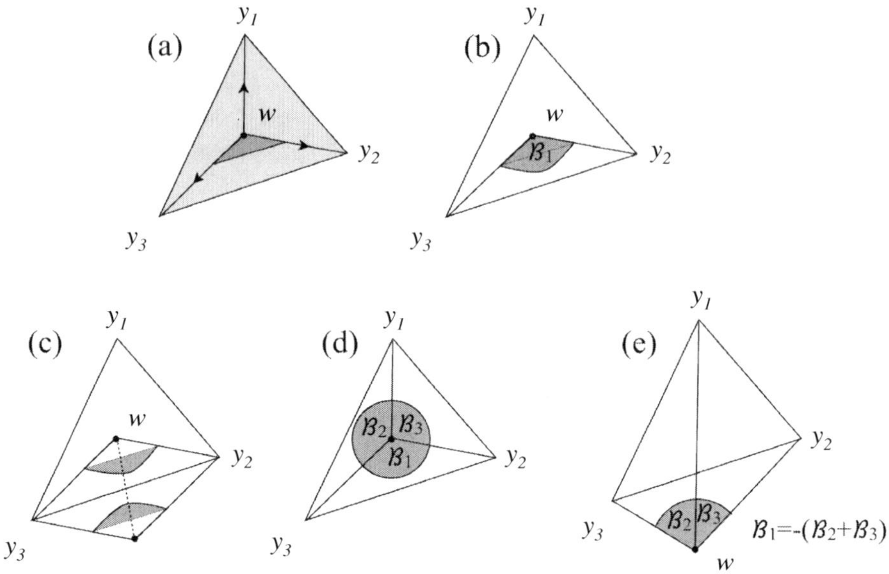

FIGURE 5.1. (a) Illustration of the simplexes $\mathfrak{S} = (y_1, y_2, y_3)$ (light gray) and $\mathfrak{S}_1(w) = (w, w + \widehat{y_2 - w}, w + \widehat{y_3 - w})$ (darker gray), together with the interpolation process that takes $\mathfrak{S}_1(w)$ into $\mathfrak{S}$. (b) Illustration of the ball sector $B_1^1(w)$ (labeled as $\mathcal{B}_1$ in the figure) corresponding to the simplex $\mathfrak{S}_1(w)$. (c) Illustration of the inversion operation of w relative to the center of the segment (y_2, y_3). The volume of $B_1^1(w) - \mathfrak{S}_1(w)$ (shaded in gray) changes sign under this operation. (d) The ball sectors $B_1^1(w)$, $B_1^2(w)$ and $B_1^3(w)$(denoted as $\mathcal{B}_j$ in the figure) have same orientation and they add up to the full unit disk when w is inside $\mathfrak{S}$. (e) When w is outside of $\mathfrak{S}$, the ball sector $B_1^1(w)$ has the opposite orientation of $B_1^2(w)$ and $B_1^3(w)$ and, as a consequence, they add up to zero.

The orientations of $B_1^j(w)$'s are considered to be the same as those of $\mathfrak{S}_j(w)$'s. Now, an important observation is that

$$\text{Vol}\{\mathfrak{S}_j(w)\} - \text{Vol}\{B_1^j(w)\} \sim |w|^{-(d+1)} \tag{5.51}$$

asymptotically as $|w| \to \infty$. Consequently, we can write

$$\begin{aligned}\int_{\Omega_{\mathcal{L}_0}} \mathrm{d}\bar{\mathbb{P}}(\mathcal{L}) \sum_{w\in\mathcal{L}} \sum_{j=1}^{d+1} \text{Vol}\{\mathfrak{S}_j(w)\} &= \int_{\Omega_{\mathcal{L}_0}} \mathrm{d}\bar{\mathbb{P}}(\mathcal{L}) \sum_{w\in\mathcal{L}} \sum_{j=1}^{d+1} \text{Vol}\{B_1^j(w)\} \\ &+ \sum_{j=1}^{d+1} \int_{\Omega_{\mathcal{L}_0}} \mathrm{d}\bar{\mathbb{P}}(\mathcal{L}) \sum_{w\in\mathcal{L}} \big(\text{Vol}\{\mathfrak{S}_j(w)\} - \text{Vol}\{B_1^j(w)\}\big),\end{aligned} \tag{5.52}$$

with the integrals in the second line being absolutely convergent. In fact,

$$\int_{\Omega_{\mathcal{L}_0}} \mathrm{d}\bar{\mathbb{P}}(\mathcal{L}) \sum_{w\in\mathcal{L}} \big(\text{Vol}\{\mathfrak{S}_j(w)\} - \text{Vol}\{B_1^j(w)\}\big) = 0 \tag{5.53}$$

for all $j = \overline{1, d+1}$, because, given our assumption, the integrand is odd under the inversion of w relative to the center of the facet $\{x_1, \ldots, \underline{x_j}, \ldots, x_{d+1}\}$ of the simplex $\mathfrak{S}$ and these two cases are sampled with equal weight. This fact is illustrated in Fig. 5.1(c) for the 2-dimensional case. As for the remaining term,

$$\sum_{j=1}^{d+1} \mathrm{Vol}\{B_1^j(w)\} = \begin{cases} \mathrm{Vol}\big(B_1(w)\big) & \text{if } w \text{ inside } \mathfrak{S}, \\ 0 & \text{if } w \text{ outside } \mathfrak{S}. \end{cases} \tag{5.54}$$

This is illustrated in Figs. 5.1(d-e) for the 2-dimensional case. Then the right side of Eq. (5.50) reduces to

$$\alpha_d \mathrm{Vol}\big(B_1(0)\big) \int_{\Omega_{\mathcal{L}_0}} \mathrm{d}\mathbb{P}(\mathcal{L})\ |\mathcal{L} \cap \mathfrak{S}| = \alpha_d \mathrm{Vol}\big(B_1(0)\big)|\mathfrak{S}|. \tag{5.55}$$

We write the volume of the simplex $\mathfrak{S}$ as a determinant, in which case

$$\alpha_d \mathrm{Vol}\big(B_1(0)\big)\mathrm{Vol}(\mathfrak{S}) = \frac{(2\imath\pi)^{\frac{d}{2}}}{(d/2)!} \mathrm{Det}\big(y_1, y_2, \ldots, y_d\big) \tag{5.56}$$

and the statement follows. □

5.5. Concluding remarks and outlook. The main difficulty in extending the above proof to the cases with $N > 1$ comes from the fact that the center of mass of the subsets live on a different, much finer lattice than $\mathcal{L}$. A more difficult task is to map the range of the pairings and this is where our efforts will go in the near future. For this, we need to develop new tools to compute the K-theories of the bi-equivariant subalgebra $C^*_{r,\mathcal{S}_N}(\widehat{\mathcal{G}}^{(N)}_{\mathcal{L}_0}, \mathbb{C})$. In the meantime, we plan to examine the problem numerically.

References

[1] E. Abrahams, P.W. Anderson, D. Licciardello, T. Ramakrishnan, *Scaling theory of localization: Absence of quantum diffusion in two dimensions*, Phys. Rev. Lett. 42, 673-676 (1979).

[2] A. Agarwala, V. B. Shenroy, *Topological insulators in amorphous systems*, Phys. Rev. Lett. **118**, 236402 (2017).

[3] Michael Aizenman, Alexander Elgart, Serguei Naboko, Jeffrey H. Schenker, and Gunter Stolz, *Moment analysis for localization in random Schrödinger operators*, Invent. Math. **163** (2006), no. 2, 343–413, DOI 10.1007/s00222-005-0463-y. MR2207021

[4] N. Azamov, E. Hekkelman, E. McDonald, F. Sukochev, and D. Zanin, *An application of singular traces to crystals and percolation*, J. Geom. Phys. **179** (2022), Paper No. 104608, 22, DOI 10.1016/j.geomphys.2022.104608. MR4448303

[5] Michael F. Barnsley, *Fractals everywhere*, 2nd ed., Academic Press Professional, Boston, MA, 1993. Revised with the assistance of and with a foreword by Hawley Rising, III. MR1231795

[6] Jean Bellissard, *K-theory of C*-algebras in solid state physics*, Statistical mechanics and field theory: mathematical aspects (Groningen, 1985), Lecture Notes in Phys., vol. 257, Springer, Berlin, 1986, pp. 99–156, DOI 10.1007/3-540-16777-3_74. MR862832

[7] J. Bellissard, A. van Elst, and H. Schulz-Baldes, *The noncommutative geometry of the quantum Hall effect*, J. Math. Phys. **35** (1994), no. 10, 5373–5451, DOI 10.1063/1.530758. Topology and physics. MR1295473

[8] J. Bellissard, *Gap labeling theorems for Schroedinger operators*, in: M. Waldschmidt, P. Moussa, J.-M. Luck, C. Itzykson (Eds.), From Number Theory to Physics, (Springer, Berlin, 1995).

[9] J. Bellissard, D. J. L. Herrmann, and M. Zarrouati, *Hulls of aperiodic solids and gap labeling theorems*, Directions in mathematical quasicrystals, CRM Monogr. Ser., vol. 13, Amer. Math. Soc., Providence, RI, 2000, pp. 207–258. MR1798994

[10] Jean Bellissard, *The noncommutative geometry of aperiodic solids*, Geometric and topological methods for quantum field theory (Villa de Leyva, 2001), World Sci. Publ., River Edge, NJ, 2003, pp. 86–156, DOI 10.1142/9789812705068_0002. MR2009996

[11] J. Bellissard, *Coherent and dissipative transport in aperiodic solids*, Lecture Notes in Physics **597**, 413–486 (2003).

[12] Jean Bellissard, Riccardo Benedetti, and Jean-Marc Gambaudo, *Spaces of tilings, finite telescopic approximations and gap-labeling*, Comm. Math. Phys. **261** (2006), no. 1, 1–41, DOI 10.1007/s00220-005-1445-z. MR2193205

[13] Chris Bourne, *Topological states of matter and noncommutative geometry*, Bull. Aust. Math. Soc. **94** (2016), no. 2, 349, DOI 10.1017/S000497271600037X. MR3568935

[14] Chris Bourne and Emil Prodan, *Non-commutative Chern numbers for generic aperiodic discrete systems*, J. Phys. A **51** (2018), no. 23, 235202, 38, DOI 10.1088/1751-8121/aac093. MR3810727

[15] C. Bourne and B. Mesland, *Index theory and topological phases of aperiodic lattices*, Ann. Henri Poincaré **20** (2019), no. 6, 1969–2038, DOI 10.1007/s00023-019-00764-9. MR3956166

[16] Ola Bratteli, *Derivations, dissipations and group actions on C^*-algebras*, Lecture Notes in Mathematics, vol. 1229, Springer-Verlag, Berlin, 1986, DOI 10.1007/BFb0098817. MR871870

[17] Ola Bratteli and Derek W. Robinson, *Operator algebras and quantum-statistical mechanics. II*, Texts and Monographs in Physics, Springer-Verlag, New York-Berlin, 1981. Equilibrium states. Models in quantum-statistical mechanics. MR611508

[18] A. L. Carey, V. Gayral, A. Rennie, and F. A. Sukochev, *Index theory for locally compact noncommutative geometries*, Mem. Amer. Math. Soc. **231** (2014), no. 1085, vi+130. MR3221983

[19] C.-Z. Chang, W. Zhao, J. Li, J. K. Jain, C. Liu, J. S. Moodera, M. H. W. Chan, *Observation of the quantum-anomalous-Hall insulator to Anderson insulator quantum phase transition and its scaling behavior*, Phys. Rev. Lett. **117**, 126802 (2016).

[20] J. G. Checkelsky, R. Yoshimi, A. Tsukazaki, K. S. Takahashi, Y. Kozuka, J. Falson, M. Kawasaki, Y. Tokura, *Trajectory of the anomalous Hall effect towards the quantized state in a ferromagnetic topological insulator*, Nat. Phys. **10**, 731–736 (2014).

[21] Alain Connes, *Sur la théorie non commutative de l'intégration* (French), Algèbres d'opérateurs (Sém., Les Plans-sur-Bex, 1978), Lecture Notes in Math., vol. 725, Springer, Berlin, 1979, pp. 19–143. MR548112

[22] Alain Connes, *Noncommutative differential geometry*, Inst. Hautes Études Sci. Publ. Math. **62** (1985), 257–360. MR823176

[23] Alain Connes, *Noncommutative geometry*, Academic Press, Inc., San Diego, CA, 1994. MR1303779

[24] A. Connes and H. Moscovici, *The local index formula in noncommutative geometry*, Geom. Funct. Anal. **5** (1995), no. 2, 174–243, DOI 10.1007/BF01895667. MR1334867

[25] Alan Forrest, John Hunton, and Johannes Kellendonk, *Topological invariants for projection method patterns*, Mem. Amer. Math. Soc. **159** (2002), no. 758, x+120, DOI 10.1090/memo/0758. MR1922206

[26] F. D. M. Haldane, *Model for a quantum Hall-effect without Landau levels: Condensed-matter realization of the parity anomaly*, Phys. Rev. Lett. **61**, 2015–2018 (1988)

[27] Johannes Kellendonk, *Noncommutative geometry of tilings and gap labelling*, Rev. Math. Phys. **7** (1995), no. 7, 1133–1180, DOI 10.1142/S0129055X95000426. MR1359991

[28] A. Kitaev, *Periodic table for topological insulators and superconductors*, in Advances in Theoretical Physics: Landau Memorial Conference, edited by V. Lebedev and M. Feigel'man, AIP Conf. Proc. No. 1134 (AIP, New York, 2009), p. 22.

[29] K. von Klitzing, G. Dorda, M. Pepper, *New method for high-accuracy determination of the fine-structure constant based on quantized Hall resistence*, Phys.Rev. Lett. **45**, 494–497 (1980).

[30] Daniel Lenz and Peter Stollmann, *Delone dynamical systems and associated random operators*, Operator algebras and mathematical physics (Constanţa, 2001), Theta, Bucharest, 2003, pp. 267–285. MR2018235

[31] Bryan Leung and Emil Prodan, *A non-commutative formula for the isotropic magnetoelectric response*, J. Phys. A **46** (2013), no. 8, 085205, 14, DOI 10.1088/1751-8113/46/8/085205. MR3030191

[32] Y. Liu, L. F. Santos, E. Prodan, *Topological gaps in quasi-periodic spin chains: A numerical and K-theoretic analysis*, Phys. Rev. B **105**, 035115 (2022).

[33] Steven Lord, Fedor Sukochev, and Dmitriy Zanin, *Singular traces*, De Gruyter Studies in Mathematics, vol. 46, De Gruyter, Berlin, 2013. Theory and applications. MR3099777

[34] E. J. Meier, F. A. An, A. Dauphin, M. Maffei, P. Massignan, T. L. Hughes, B. Gadway, *Observation of the topological Anderson insulator in disordered atomic wires*, Science *362*, 929-933 (2018).

[35] Bram Mesland and Emil Prodan, *A groupoid approach to interacting fermions*, Comm. Math. Phys. **394** (2022), no. 1, 143–213, DOI 10.1007/s00220-022-04397-8. MR4456123

[36] N. P. Mitchell, L. M. Nash, D. Hexner, A. Turner, W. T. M. Irvine, *Amorphous topological insulators constructed from random point sets*, Nat. Phys. **14**, 380–5 (2018).

[37] I. Mondragon-Shem, J. Song, Taylor L. Hughes, E. Prodan, *Topological criticality in the chiral-symmetric AIII class at strong disorder*, Phys. Rev. Lett. **113**, 046802 (2014).

[38] Paul S. Muhly, Jean N. Renault, and Dana P. Williams, *Equivalence and isomorphism for groupoid C^*-algebras*, J. Operator Theory **17** (1987), no. 1, 3–22. MR873460

[39] E. Prodan, F.D.M. Haldane, *Mapping the braiding properties of the Moore–Read state*, Phys. Rev. B **80**, 11512 (2009).

[40] E. Prodan, T.L. Hughes, B.A. Bernevig, *Entanglement spectrum of a disordered topological Chern insulator*, Phys. Rev. Lett. **105**, 115501 (2010).

[41] Emil Prodan, *Disordered topological insulators: a non-commutative geometry perspective*, J. Phys. A **44** (2011), no. 11, 113001, 50, DOI 10.1088/1751-8113/44/11/113001. MR2773856

[42] Emil Prodan, Bryan Leung, and Jean Bellissard, *The non-commutative nth-Chern number* $(n \geqslant 1)$, J. Phys. A **46** (2013), no. 48, 485202, 20, DOI 10.1088/1751-8113/46/48/485202. MR3138388

[43] Emil Prodan, *Quantum transport in disordered systems under magnetic fields: a study based on operator algebras*, Appl. Math. Res. Express. AMRX **2** (2013), 176–255, DOI 10.1093/amrx/abs017. MR3106880

[44] Emil Prodan and Hermann Schulz-Baldes, *Non-commutative odd Chern numbers and topological phases of disordered chiral systems*, J. Funct. Anal. **271** (2016), no. 5, 1150–1176, DOI 10.1016/j.jfa.2016.06.001. MR3522005

[45] Emil Prodan and Hermann Schulz-Baldes, *Bulk and boundary invariants for complex topological insulators: From K-theory to physics*, Mathematical Physics Studies, Springer, [Cham], 2016, DOI 10.1007/978-3-319-29351-6. MR3468838

[46] Emil Prodan, *A computational non-commutative geometry program for disordered topological insulators*, SpringerBriefs in Mathematical Physics, vol. 23, Springer, Cham, 2017, DOI 10.1007/978-3-319-55023-7. MR3618067

[47] Emil Prodan, *Topological lattice defects by groupoid methods and Kasparov's KK-theory*, J. Phys. A **54** (2021), no. 42, Paper No. 424001, 18. MR4318633

[48] X.-L. Qi, T. L. Hughes, S.-C. Zhang, *Topological field theory of time-reversal invariant insulators*, Phys. Rev. B **78**, 195424 (2008).

[49] Jean Renault, *A groupoid approach to C^*-algebras*, Lecture Notes in Mathematics, vol. 793, Springer, Berlin, 1980. MR584266

[50] S. Ryu, A. P. Schnyder, A. Furusaki, A. W. W. Ludwig, *Topological insulators and superconductors: Tenfold way and dimensional hierarchy*, New J. Phys. **12**, 065010 (2010).

[51] A. P. Schnyder, S. Ryu, A. Furusaki, A. W. W. Ludwig, *Classification of topological insulators and superconductors in three spatial dimensions*, Phys. Rev. B **78**, 195125 (2008).

[52] Aidan Sims, Gábor Szabó, and Dana Williams, *Operator algebras and dynamics: groupoids, crossed products, and Rokhlin dimension*, Advanced Courses in Mathematics. CRM Barcelona, Birkhäuser/Springer, Cham, [2020] ©2020. Lecture notes from the Advanced Course held at Centre de Recerca Matemàtica (CRM) Barcelona, March 13–17, 2017; Edited by Francesc Perera, DOI 10.1007/978-3-030-39713-5. MR4321941

[53] Şerban Strătilă and Dan Voiculescu, *Representations of AF-algebras and of the group $U(\infty)$*, Lecture Notes in Mathematics, Vol. 486, Springer-Verlag, Berlin-New York, 1975. MR0458188

[54] Dana P. Williams, *A tool kit for groupoid C^*-algebras*, Mathematical Surveys and Monographs, vol. 241, American Mathematical Society, Providence, RI, 2019, DOI 10.1016/j.physletb.2019.06.021. MR3969970

DEPARTMENT OF PHYSICS AND DEPARTMENT OF MATHEMATICAL SCIENCES, YESHIVA UNIVERSITY, NEW YORK, NEW YORK 10016

Email address: prodan@yu.edu

Proceedings of Symposia in Pure Mathematics
Volume **105**, 2023
https://doi.org/10.1090/pspum/105/01912

Periodic cyclic homology of crossed products

Michael Puschnigg

ABSTRACT. We discuss the cyclic homology of crossed product algebras from the Cuntz-Quillen point of view. The periodic cyclic homology of a crossed product algebra $A \rtimes G$ is described in terms of the G-action on periodic cyclic bicomplexes of crossed products of A by the cyclic subgroups of G .

1. Introduction

The work of Burghelea [**Bu**] and Nistor [**Ni**] on the cyclic homology of group rings and crossed product algebras, carried out soon after the invention of cyclic homology fourty years ago, constitutes an integral part of the theory. They observed that the cyclic chain complex of the crossed product of a complex algebra A by the action of an abstract group G decomposes as a direct sum of contributions labeled by the conjugacy classes of G and express the various corresponding summands of cyclic homology in terms of classical derived functors. The contribution of the unit class turns out to be isomorphic to the hyperhomology of G with values in the cyclic chain complex of A, and a similar description is given for the contributions of the conjugacy classes of torsion elements in G. Less explicit qualitative results are obtained for the contributions of the other conjugacy classes.

The 40th Birthday Conference of cyclic homology seems to be an appropriate occasion to review the work of Burghelea and Nistor from the point of view of Cuntz and Quillen [**CQ1**],[**CQ2**], which provides a great deal of conceptual insight into cyclic theory. Typical features of the Cuntz-Quillen approach are the strictly $\mathbb{Z}/2\mathbb{Z}$-graded setting, emphasizing periodic cyclic rather than ordinary cyclic or Hochschild theory, and the extensive use of universal algebras.

We recall the"derived functor analogy" of Cuntz-Quillen, which serves as a guiding principle for our work. Classical derived functors on the derived category of an abelian category are constructed by replacing an arbitrary chain complex by a well behaved quasi-isomorphic complex and by evaluating the functor to be derived on the latter. Cuntz and Quillen propose to replace a given algebra by a well behaved topologically nilpotent extension, and to study the given functor, in our case the periodic cyclic bicomplex, on the latter algebra.

2020 *Mathematics Subject Classification.* Primary 16E40, 18G35, 20J06.
Key words and phrases. Cyclic homology, crossed product, homological algebra.

In this spirit we introduce for a given set X a natural nilpotent extension $k\langle X\rangle$ of the ground field $k \supset \mathbb{Q}$. In particular, if a group G acts on X, then $k\langle X\rangle$ becomes a kG-algebra. For $X = G$, equipped with the translation action, and a kG-algebra A the G-algebra $A\langle G\rangle = A \otimes_k k\langle G\rangle$ is then a natural nilpotent extension of A which is free as a G-module. It serves as our "well behaved model" of A and will play a key role throughout this paper.

In order to formulate our results conveniently we introduce periodic cyclic bicomplexes $\widehat{CC}(A : R)$, given by the usual direct product of tensor powers of the given algebra A, but this time relative to a not necessarily commutative ground ring R.

The cyclic bicomplex $\widehat{CC}_*(A\langle G\rangle \rtimes G : k \rtimes G)_{[e]}$, i.e. the contribution of the conjugacy class of the unit to $\widehat{CC}_*(A\langle G\rangle \rtimes G : k \rtimes G)$, has then two canonical bases and can be written in two ways. On the one hand we find the bicomplex $\widehat{CC}_*(A \rtimes G)_{[e]}$, and on the other hand the space of G-coinvariants of the complex $\widehat{CC}_*(A\langle G\rangle)$. As the latter is a complex of free G-modules $HP_*(A \rtimes G)_{[e]}$ equals the hyperhomology of G with coefficients in $\widehat{CC}_*(A\langle G\rangle)$. Goodwillie's theorem on the invariance of periodic cyclic homology under nilpotent extensions allows to identify this with the hyperhomology of G with coefficients in $\widehat{CC}_*(A)$, recovering the result of Nistor [**Ni**]. A similar picture emerges for the contributions of the other conjugacy classes. For elements $v \in G$ of finite order we find a natural isomorphism

$$HP_*(A \rtimes G)_{[v]} \xrightarrow{\simeq} \widehat{\mathbb{H}}_*\left(Z_v, \widehat{CC}(A \rtimes v^{\mathbb{Z}} : k \rtimes v^{\mathbb{Z}})_{\{v\}}\right). \tag{1.1}$$

where Z_v denotes the centralizer of v in G. It follows that the elliptic part of the periodic cyclic homology of a crossed product, i.e. the contribution of the conjugacy classes of all torsion elements except the unit, is given by

$$HP_*(A \rtimes G)_{\mathrm{ell}} \simeq \widehat{\mathbb{H}}_*\left(G, \widehat{\bigoplus_{\substack{v\in G \\ 1<|v|<\infty}}} \widehat{CC}_*(A \rtimes v^{\mathbb{Z}} : k \rtimes v^{\mathbb{Z}})_{\{v\}}\right). \tag{1.2}$$

For conjugacy classes of elements $v \in G$ of infinite order we obtain partial results similar to those of Nistor [**Ni**].

Roughly speaking one could say that the periodic cyclic homology of a crossed product $A \rtimes G$ is determined by a family of cyclic bicomplexes of the crossed products of A by the cyclic subgroups of G and the G-action on this family.

Our approach not only provides quite explicit and rather simple formulas for the contributions of various conjugacy classes to the periodic cyclic (co)homology of a crossed product, but also extends nicely to a topological setting. The attempt of adapting Nistor's original approach to Banach crossed products was, on the contrary, a quite frustrating experience for the author and motivated the present research.

2. Preliminaries

2.1. Base change of cyclic complexes. Fix a field k of characteristic 0 and a unital associative k-algebra R.

DEFINITION 2.1. An R-algebra is a unitary R-bimodule A, together with an R-bimodule homomorphism

$$m : A \otimes_R A \to A, \tag{2.1}$$

which defines a (not necessarily unital) k-algebra structure on A. A homomorphism $\varphi : A \to B$ of R-algebras is a map which is simultaneously a homomorphism of R-bimodules and of k-algebras.

With this definition R-algebras form a category.

DEFINITION 2.2. Let A be an R-algebra. The graded R-bimodule

$$\Omega^*(A:R) = \bigoplus_{n=0}^{\infty} \Omega^n(A:R) \tag{2.2}$$

of **algebraic differential forms** of A over R is defined as

$$\begin{array}{rcl} \Omega^n(A:R) & \simeq & A^{\otimes_R^{n+1}} \oplus A^{\otimes_R^n} \\ a_0 da_1 \dots da_n & \leftrightarrow & a_0 \otimes a_1 \otimes \dots \otimes a_n \\ da_1 \dots da_n & \leftrightarrow & a_1 \otimes \dots \otimes a_n \end{array} \tag{2.3}$$

with R-bimodule-structure induced by the natural one on the tensor powers of A. The commutator quotient with respect to this bimodule structure is the graded k-vector space

$$\Omega^*(A:R)_\natural = \Omega^*(A:R)/[\Omega^*(A:R), R]. \tag{2.4}$$

Note that this differs from the notion of relative differential forms in [**CQ1**]. Under the canonical linear projection

$$\Omega^* A = \Omega^*(A:k)_\natural \to \Omega^*(A:R)_\natural \tag{2.5}$$

the well known Hochschild- and Connes-operators

$$b : \Omega^* A \to \Omega^{*-1} A, \quad B : \Omega^* A \to \Omega^{*+1} A \tag{2.6}$$

descend to linear operators

$$b : \Omega^*(A:R)_\natural \to \Omega^{*-1}(A:R)_\natural, \quad B : \Omega^*(A:R)_\natural \to \Omega^{*+1}(A:R)_\natural, \tag{2.7}$$

satisfying the usual identities

$$b^2 = B^2 = bB + Bb = 0. \tag{2.8}$$

DEFINITION 2.3.

i) The **cyclic bicomplex** of A over R is given by

$$CC_*(A:R) = (\Omega^*(A:R)_\natural,\ b+B). \tag{2.9}$$

This is a $\mathbb{Z}/2\mathbb{Z}$-graded naturally contractible chain complex.

ii) The **Hodge filtration** of the cyclic bicomplex is the descending filtration by the subcomplexes

$$\mathrm{Fil}^m_{\mathrm{Hodge}} CC_*(A:R) = \bigoplus_{n=m}^{\infty} \Omega^n(A:R)_\natural \oplus b(\Omega^m(A:R)_\natural),\ m \in \mathbb{N}, \tag{2.10}$$

generated by the algebraic differential forms of degree at least m.

iii) The associated graded complex is quasi-isomorphic to the Hochschild complex

$$C_*(A:R) = (\Omega^*(A:R)_\natural,\ b)\,. \tag{2.11}$$

iv) The **periodic cyclic bicomplex** of A over R is the completion of $CC_*(A:R)$ with respect to the topology defined by the Hodge filtration:

$$\widehat{CC}_*(A:R) = \left(\prod_{n=0}^{\infty} \Omega^n(A:R)_\natural,\ b+B\right). \tag{2.12}$$

This is a complete $\mathbb{Z}/2\mathbb{Z}$-graded chain complex, the grading being given by the parity of forms.

v) The **periodic cyclic homology** $HP_*(A:R)$ of A over R is the homology of the periodic cyclic bicomplex $\widehat{CC}_*(A:R)$. The **periodic cyclic cohomology** $HP^*(A:R)$ of A over R is the cohomology of the topologically dual complex of k-linear functionals on $\widehat{CC}_*(A:R)$ which are continuous (i.e., vanishing on $\mathrm{Fil}^m_{\mathrm{Hodge}}\widehat{CC}_*(A:R)$ for $m \gg 0$).

Every homomorphism $f: A \to B$ of R-algebras gives rise to a continuous morphism $\widehat{CC}(f): \widehat{CC}_*(A:R) \to \widehat{CC}_*(B:R)$ of periodic cyclic bicomplexes. So the periodic cyclic bicomplex defines a functor from the category of R-algebras to the category of $\mathbb{Z}/2\mathbb{Z}$-graded topologically complete chain complexes.
If $S \subset R$ is a unital subalgebra, the canonical projection $\Omega^*(A:S)_\natural \to \Omega^*(A:R)_\natural$ induces a morphism

$$\pi_{S,R}: \widehat{CC}_*(A:S) \longrightarrow \widehat{CC}_*(A:R) \tag{2.13}$$

of topologically complete chain complexes.

EXAMPLE 2.4. For unital A one has

$$\widehat{CC}(A:A) = \prod_{n=0}^{\infty} A/[A,A], \quad HP_0(A:A) = A/[A,A], \quad HP_1(A:A) = 0. \tag{2.14}$$

2.2. Auxiliary algebras. Let X be a set and let $k\langle X\rangle$ be the k-vector space over X. The projection

$$p: X \times X \to X,\ (x,y) \mapsto y \tag{2.15}$$

turns $k\langle X\rangle$ into an associative k-algebra with multiplication

$$m: k\langle X\rangle \otimes_k k\langle X\rangle = k\langle X \times X\rangle \xrightarrow{k\langle p\rangle} k\langle X\rangle. \tag{2.16}$$

If X consists of a single element the algebra thus obtained is canonically isomorphic to the ground field k. Every map of sets $f: X \to Y$ gives rise to an algebra homomorphism

$$k\langle f\rangle: k\langle X\rangle \to k\langle Y\rangle. \tag{2.17}$$

In particular, every action of a group G on X induces a G-action on $k\langle X\rangle$ by k-algebra automorphisms. The constant map to a point defines an augmentation homomorphism

$$\epsilon_X : k\langle X\rangle \to k. \tag{2.18}$$

The multiplication in $k\langle X\rangle$ is characterized by the identity

$$a \cdot b = \epsilon_X(a)b, \quad \forall a, b \in k\langle X\rangle. \tag{2.19}$$

In particular, the augmentation ideal $I_\epsilon = \mathrm{Ker}(\epsilon_X)$ satisfies

$$I_\epsilon^2 = I_\epsilon \cdot k\langle X\rangle = 0. \tag{2.20}$$

The tensor product of a unital k-algebra A and $k\langle X\rangle$ in the category of k-algebras is denoted by

$$A\langle X\rangle = A \otimes_k k\langle X\rangle. \tag{2.21}$$

The algebra extension

$$0 \longrightarrow I_\epsilon \otimes_k A \longrightarrow A\langle X\rangle \xrightarrow{\epsilon_A} A \longrightarrow 0 \tag{2.22}$$

is nilpotent by (2.20) and splits (non-canonically) as extension of k-algebras. If a group G acts on the k-algebra A and acts also freely on the set X, the diagonal action turns $A\langle X\rangle$ into a G-algebra whose underlying G-module is free.

3. Cyclic complexes of crossed products

3.1. Cyclic complexes attached to crossed products. Let G be a group and let A be a unital G-algebra over k. We let $k\langle G\rangle$ carry the G-action induced by left translation and equip $A\langle G\rangle$ with the diagonal G-action.

LEMMA 3.1. *Let $U \subset G$ be a subgroup and let $\sigma : U\backslash G \to G$ be a set theoretic section of the canonical projection $G \to U\backslash G$.*

i) *The linear map*

$$\begin{array}{rccc} \iota_\sigma : & (A \rtimes U)\langle U\backslash G\rangle & \to & A\langle G\rangle \rtimes U, \\ & (u_g a)\langle x\rangle & \mapsto & u_g(a\langle\sigma(x)\rangle) \end{array} \tag{3.1}$$

is a homomorphism of k-algebras.

ii) *The composition of chain maps*

$$\begin{array}{ccc} CC_*((A \rtimes U)\langle U\backslash G\rangle) & \xrightarrow{CC(\iota_\sigma)} & CC_*(A\langle G\rangle \rtimes U) \\ \| & & \downarrow \pi_{k,k\rtimes U} \\ CC_*((A \rtimes U)\langle U\backslash G\rangle) & \xrightarrow{\simeq} & CC_*(A\langle G\rangle \rtimes U : k \rtimes U) \end{array} \tag{3.2}$$

is an isomorphism of filtered bicomplexes (with respect to Hodge filtrations).

PROOF. i) The map ι_σ is an algebra homomorphism as

$$\begin{aligned} \iota_\sigma((u_g a)\langle x\rangle)\iota_\sigma((u_h b)\langle y\rangle) &= u_g a\langle\sigma(x)\rangle u_h b\langle\sigma(y)\rangle = u_{gh}h^{-1}(a\langle\sigma(x)\rangle)b\langle\sigma(y)\rangle \\ = u_{gh}h^{-1}(a)\langle h^{-1}\sigma(x)\rangle b\langle\sigma(y)\rangle &= u_{gh}h^{-1}(a)b\langle h^{-1}\sigma(x)\rangle\langle\sigma(y)\rangle = u_{gh}h^{-1}(a)b\langle\sigma(y)\rangle \\ &= \iota_\sigma((u_{gh}h^{-1}(a)b)\langle y\rangle) = \iota_\sigma((u_g a)\langle x\rangle \cdot (u_h b)\langle y\rangle). \end{aligned}$$

ii) Note that the k-linear map

$$(3.3)\qquad \begin{array}{ccc} kU\otimes_k A\langle U\backslash G\rangle \otimes_k kU & \longrightarrow & A\langle G\rangle \rtimes U \\ u_g\otimes a\langle x\rangle \otimes u_h & \mapsto & u_g a\langle \sigma(x)\rangle u_h = u_{gh}h^{-1}(a)\langle h^{-1}\sigma(x)\rangle \end{array}$$

is an isomorphism of kU-bimodules, so that $A\langle G\rangle \rtimes U$ becomes a free kU-bimodule with basis $A\langle U\backslash G\rangle$. Consequently

$$(A\langle G\rangle \rtimes U)^{\otimes^n_{kU}} \simeq (kU\otimes_k A\langle U\backslash G\rangle \otimes_k kU)^{\otimes^n_{kU}} \simeq (kU\otimes_k A\langle U\backslash G\rangle)^{\otimes^n_k}\otimes_k kU$$

and

$$(A\langle G\rangle \rtimes U)^{\otimes^n_{kU}}/[(A\langle G\rangle \rtimes U)^{\otimes^n_{kU}}, kU] \simeq (kU\otimes_k A\langle U\backslash G\rangle)^{\otimes^n_k} \simeq ((A\rtimes U)\langle U\backslash G\rangle)^{\otimes^n_k}$$

as k-vector spaces, which shows that the composition of the two chain maps is an isomorphism. The second assertion follows from the fact that all occuring maps preserve the degree of differential forms. □

In the sequel we denote by $\mathrm{Ad}(U)$ the set U with the adjoint U-action, and let $\mathrm{Vect}_k(\mathrm{Ad}(U))$ be the associated k-vector space with linear U-action.

LEMMA 3.2. *Let $i : A\langle G\rangle \to A\langle G\rangle \rtimes U$ be the canonical inclusion. The linear map*

$$(3.4)\qquad \begin{array}{ccc} \mathrm{Vect}_k(\mathrm{Ad}(U))\otimes_k \Omega^*(A\langle G\rangle) & \longrightarrow & \Omega^*(A\langle G\rangle \rtimes U) \\ u_g\otimes \omega & \mapsto & u_g\cdot i_*(\omega) \end{array}$$

fits into a commutative diagram

$$(3.5)\qquad \begin{array}{ccc} \mathrm{Vect}_k(\mathrm{Ad}(U))\otimes_k \Omega^*(A\langle G\rangle) & \longrightarrow & \Omega^*(A\langle G\rangle \rtimes U) \\ \downarrow & & \downarrow \\ (\mathrm{Vect}_k(\mathrm{Ad}(U))\otimes_k \Omega^*(A\langle G\rangle))_U & \xrightarrow{\simeq} & \Omega^*(A\langle G\rangle \rtimes U : k\rtimes U)_\natural \end{array}$$

whose lower horizontal arrow is an isomorphism. Here the left vertical arrow is the projection onto the space of coinvariants under the diagonal U-action and the right vertical arrow is given by the canonical projection of absolute onto relative algebraic differential forms. Every subgroup of G centralizing U acts naturally on this diagram.

PROOF. Let Φ be the composition of the upper horizontal and the right vertical map. We find for $h\in U$

$$\Phi(h(u_g\otimes a_0\langle g_0\rangle \otimes \ldots \otimes a_n\langle g_n\rangle)) = \Phi(u_{hgh^{-1}}\otimes h(a_0\langle g_0\rangle)\otimes \ldots \otimes h(a_n\langle g_n\rangle))$$

$$= u_h u_g u_{h^{-1}} h(a_0\langle g_0\rangle)\otimes \ldots \otimes h(a_n\langle g_n\rangle) = u_g u_{h^{-1}} h(a_0\langle g_0\rangle)\otimes \ldots \otimes h(a_n\langle g_n\rangle)u_h$$

$$= u_g u_{h^{-1}} h(a_0\langle g_0\rangle)u_h \otimes \ldots \otimes u_{h^{-1}} h(a_n\langle g_n\rangle)u_h = \Phi(u_g\otimes a_0\langle g_0\rangle \otimes \ldots \otimes a_n\langle g_n\rangle),$$

which shows that Φ is constant on U-orbits. Using the identification of $\Omega^*((A\langle G\rangle) \rtimes U : k\rtimes U)_\natural$ with $\Omega^*((A\rtimes G)\langle U\backslash G\rangle)$ of Lemma 3.1, one may construct a k-linear section σ' of Φ by putting

$$(3.6)\qquad \sigma'(u_{g_0}a_0\langle x_0\rangle \otimes \ldots \otimes u_{g_{n-1}}a_{n-1}\langle x_{n-1}\rangle \otimes u_{g_n}a_n\langle x_n\rangle) =$$

$$u_{g_0\ldots g_n}\otimes(g_1\ldots g_n)^{-1}(a_0\langle\sigma(x_0)\rangle)\otimes\ldots\otimes g_n^{-1}(a_{n-1}\langle\sigma(x_{n-1})\rangle)\otimes a_n\langle\sigma(x_n)\rangle.$$

The image of σ' intersects each U-orbit in exactly one element. This proves our assertion. □

3.2. The homogeneous decomposition. We recall the well known homogeneous decomposition of cyclic complexes of crossed product algebras [**Bu**],[**Ni**]. Let G be a group and denote by $[v]$ the conjugacy class of $v\in G$. For a G-algebra A over k let $\Omega^*(A\rtimes G)_{[v]}$ be the linear span of the differential forms

$$a_0u_{g_0}d(a_1u_{g_1})\ldots d(a_nu_{g_n})\in\Omega^*(A\rtimes G), g_0g_1\ldots g_n\in[v],$$

and

$$d(b_1u_{h_1})\ldots d(b_nu_{h_n})\in\Omega^*(A\rtimes G), h_1\ldots h_n\in[v].$$

Put

$$CC_*(A\rtimes G)_{[v]} = (\Omega^*(A\rtimes G)_{[v]}, b+B). \tag{3.7}$$

There are canonical decompositions

(3.8)

$$\Omega^*(A\rtimes G) = \bigoplus_{[v]}\Omega^*(A\rtimes G)_{[v]} \quad\text{and}\quad CC_*(A\rtimes G) = \bigoplus_{[v]}CC_*(A\rtimes G)_{[v]},$$

where the sum runs over the set of conjugacy classes of G. If $U\subset G$ is a subgroup, the homogeneous decompositions (3.8) descend under the canonical projections (2.13) to similar decompositions

$$\Omega^*(A\rtimes G:k\rtimes U)_\natural = \bigoplus_{[v]}\Omega^*(A\rtimes G:k\rtimes U)_{\natural\,[v]} \tag{3.9}$$

and

$$CC_*(A\rtimes G:k\rtimes U) = \bigoplus_{[v]}CC_*(A\rtimes G:k\rtimes U)_{[v]}. \tag{3.10}$$

The corresponding completions $\widehat{CC}_*(A\rtimes G)_{[v]}$ and $\widehat{CC}_*(A\rtimes G:k\rtimes U)_{[v]}$ are topological direct summands of $\widehat{CC}_*(A\rtimes G)$ and $\widehat{CC}_*(A\rtimes G:k\rtimes U)$, respectively, but the latter complexes are in general neither the direct sum nor the direct product of their factors in the homogeneous decomposition. The partition of the set of conjugacy classes of G into the singleton given by the conjugacy class of the unit, the union of the other conjugacy classes of elements of finite order, and the union of the conjugacy classes of elements of infinite order gives rise to a decomposition of the cyclic bicomplex, its completion, and the periodic cyclic (co)homology of a crossed product algebra into the direct sum of three factors, called their **homogeneous**, **elliptic** and **inhomogeneous** parts.

For $v\in G$ let Z_v be its centralizer in G. The cyclic subgroup U of G, generated by v, fits then into a central extension

$$1\longrightarrow U=v^{\mathbb{Z}}\longrightarrow Z_v\longrightarrow N_v\longrightarrow 1. \tag{3.11}$$

PROPOSITION 3.3. *The canonical chain map*

$$CC_*(A\langle G\rangle\rtimes U:k\rtimes U)\longrightarrow CC_*(A\langle G\rangle\rtimes G:k\rtimes G)\overset{(3.2)}{\simeq}CC_*(A\rtimes G)$$

induces an isomorphism

$$(CC_*(A\langle G\rangle \rtimes U : k \rtimes U)_{\{v\}})_{Z_v} \xrightarrow{\simeq} CC_*(A \rtimes G)_{[v]} \tag{3.12}$$

of filtered chain complexes (with respect to the Hodge filtrations).

PROOF. It is clear that the chain map preserves Hodge filtrations. Therefore it suffices to show that the underlying map of vector spaces is an isomorphism. By Lemma 3.2, applied to $U = v^{\mathbb{Z}}$, and (3.9) there is a canonical linear isomorphism

$$\Omega^*(A\langle G\rangle)_U \xrightarrow{\simeq} \Omega^*(A\langle G\rangle \rtimes U : k \rtimes U)_{\natural\{v\}} \tag{3.13}$$

By construction it commutes with the action of the centralizer Z_v, so that one obtains an isomorphism of the spaces of Z_v-coinvariants, which fits into the commutative diagram

$$\begin{array}{ccc} \Omega^*(A\langle G\rangle)_{Z_v} & \xrightarrow{\simeq} & (\Omega^*(A\langle G\rangle \rtimes U : k \rtimes U)_{\natural\{v\}})_{Z_v} \\ \downarrow & & \downarrow \\ (Vect_k([v]) \otimes_k \Omega^*(A\langle G\rangle))_G & \xrightarrow{\simeq} & \Omega^*(A\langle G\rangle \rtimes G : k \rtimes G)_{\natural[v]} \end{array} \tag{3.14}$$

The upper horizontal arrow is an isomorphism by (3.13), the lower one by Lemma 3.2, applied to $U = G$, and the left vertical arrow is an isomorphism because Z_v is the stabilizer of the element $v \in [v]$ under the transitive adjoint action of G on $[v]$. It results that the right vertical arrow is an isomorphism, which is our assertion. □

3.3. Hyperhomology. It is well known and easy to prove that quasi-isomorphic $\mathbb{Z}_+$-graded chain complexes of G-modules have isomorphic hyperhomology groups $\mathbb{H}_*(G, -)$. In general this is no longer true for $\mathbb{Z}/2\mathbb{Z}$-graded complexes. This forces us to add this technical section dealing with several cases where the previous result continues to hold.

We consider the category of $\mathbb{Z}/2\mathbb{Z}$-graded complexes of $k \rtimes G$-modules, equipped with an adic topology, defined by a decreasing chain of subcomplexes. The morphisms are given by continuous, G-equivariant chain maps. A morphism which becomes an isomorphism in the associated chain-homotopy category is called a continuous equivariant chain homotopy equivalence. A filtration defining the given topology will be called admissible. The completion of such a chain complex in the sense of general topology equals

$$\widehat{C}_* \simeq \lim_{\overleftarrow{n}} C_*/\mathrm{Fil}^n C_* \tag{3.15}$$

for any admissible filtration. The algebraic tensor product $C_* \otimes_k C'_*$ of two complexes is topologized by

$$\mathrm{Fil}_n(C_* \otimes_k C'_*) = \sum_{p+q=n} \mathrm{Fil}_p C_* \otimes_k \mathrm{Fil}_q C'_* \tag{3.16}$$

This is independent of the choice of the admissible filtrations on the individual factors. The completion $\widehat{C_* \otimes_k C'_*}$ of the tensor product $C_* \otimes_k C'_*$ contains in general the algebraic tensor product $\widehat{C_*} \otimes_k \widehat{C'_*}$ of the individual completions as dense subcomplex. The complex of G-coinvariants of C_* equals

$$(C_*)_G = C_* \otimes_{k \rtimes G} k \tag{3.17}$$

equipped with the filtration induced by the given one on C_* and the constant filtration on k. It is the largest quotient complex of C_* on which G acts trivially. The complex $(C_*)_G$ is complete if C_* is. The Hom-complex $L(C_*, C'_*)$ is the complex of all k-linear maps from C_* to C'_*, equipped with the differential

$$\partial_{L(C_*,C'_*)}(\varphi) = \partial_{C'_*} \circ \varphi - (-1)^{\deg(\varphi)} \varphi \circ \partial_{C_*}, \quad \varphi \in L(C_*, C'_*). \tag{3.18}$$

Its subcomplex of continuous linear maps is denoted by $\mathcal{L}(C_*, C'_*)$.
To a $\mathbb{Z}_+$-graded chain complex P_* one may associate the $\mathbb{Z}/2\mathbb{Z}$-graded chain complex

$$(P_*^{\pm}, \partial),\ P_{\text{ev}}^{\pm} = \bigoplus_{n=0}^{\infty} P_{2n},\ P_{\text{odd}}^{\pm} = \bigoplus_{n=0}^{\infty} P_{2n+1}, \tag{3.19}$$

topologized by the filtration

$$\text{Fil}^m P_*^{\pm} = \partial P_m \oplus \bigoplus_{n=m}^{\infty} P_n \tag{3.20}$$

Its completion equals

$$\widehat{P}_*^{\pm} = \prod_{n=0}^{\infty} P_n. \tag{3.21}$$

Definition 3.4. Let G be a group and let P_* be a projective resolution of the constant G-module k. Let C_* be a $\mathbb{Z}/2\mathbb{Z}$-graded chain complex of $k \rtimes G$-modules, topologized by a decreasing family of subcomplexes, and let $C^* = \mathcal{L}(C_*, k) = \mathcal{L}(\widehat{C}_*, k) = \widehat{C}^*$ be the topologically dual complex.

i) The **hyperhomology** of G with coefficients in C_* equals

$$\mathbb{H}_*(G, C_*) = H_*((P_*^{\pm} \otimes_k C_*)_G). \tag{3.22}$$

ii) The **continuous hyperhomology** of G with coefficients in $\widehat{C_*}$ equals

$$\widehat{\mathbb{H}}_*(G, \widehat{C}_*) = H_*((\widehat{P}_*^{\pm} \widehat{\otimes}_k \widehat{C}_*)_G). \tag{3.23}$$

iii) The **hypercohomology** of G with coefficients in C_* equals

$$\mathbb{H}^*(G, C_*) = H^*(L(P_*^{\pm} \otimes_k C_*, k)^G). \tag{3.24}$$

iv) The **continuous hypercohomology** of G with coefficients in $\widehat{C}^*$ equals

$$\widehat{\mathbb{H}}^*(G, \widehat{C}_*) = H^*(\mathcal{L}(\widehat{P}_*^{\pm} \widehat{\otimes}_k \widehat{C}_*, k)^G). \tag{3.25}$$

This is (up to canonical isomorphism) independent of the choice of resolution. The continuous hyper(co)homology depends only on the topology of $\widehat{C}_*$, but not on the particular choice of filtration used to define it. If there exists a resolution of finite length of the constant G-module k by finitely generated, projective kG-modules, then the natural maps

$$\mathbb{H}_*(G, \widehat{C}_*) \to \widehat{\mathbb{H}}_*(G, \widehat{C}_*) \quad \text{and} \quad \widehat{\mathbb{H}}^*(G, \widehat{C}_*) \to \mathbb{H}^*(G, \widehat{C}_*) \tag{3.26}$$

are isomorphisms.

Lemma 3.5. *Let A be an R-algebra and let G be a group acting on A by R-algebra automorphisms such that $\Omega^*(A : R)_\natural$ becomes a degree-wise free kG-module.*

Let P_* be a projective resolution of the constant G-module k. Then the natural chain map

$$\epsilon \otimes_k \mathrm{id} : (P_*^{\pm} \otimes CC_*(A : R))_G \longrightarrow (CC_*(A : R))_G \tag{3.27}$$

becomes a continuous chain-homotopy equivalence after completion. In particular

$$\widehat{\mathbb{H}}_*(G, \widehat{CC}_*(A : R)) \xrightarrow{\simeq} H_*(\widehat{CC}_*(A : R)_G). \tag{3.28}$$

PROOF. Let $h : \widetilde{P}_* \to \widetilde{P}_{*+1}, * \geqslant -1$, be a k-linear contracting chain homotopy of the augmented resolution $\widetilde{P}_*$ of the constant G-module k. By assumption $\Omega^n(A : R)_\natural$ is a free G-module i.e., $\Omega^n(A : R)_\natural \simeq kG \otimes_k V_n$ as kG-module for some k-vector space V_n and all $n \geqslant 0$. Let $\widetilde{h} : \widetilde{P}_*^{\pm} \otimes_k CC_*(A : R) \to \widetilde{P}_{*+1}^{\pm} \otimes_k CC_*(A : R)$ be the unique morphism of kG-modules satisfying $\widetilde{h}|_{\widetilde{P}_* \otimes_k V_*} = h \otimes_k \mathrm{id}$. The chain map

$$\varphi = \mathrm{id} - (\partial \circ \widetilde{h} + \widetilde{h} \circ \partial) : \widetilde{P}_*^{\pm} \otimes_k CC_*(A : R) \to \widetilde{P}_*^{\pm} \otimes_k CC_*(A : R)$$

satisfies

$$\varphi(\widetilde{P}_{\geqslant n} \otimes_k \Omega^{\geqslant m}(A : R)_\natural) \subset \widetilde{P}_{\geqslant n+1} \otimes_k \Omega^{\geqslant m-1}(A : R)_\natural. \tag{3.29}$$

So the infinite series

$$\nu = \sum_{j=0}^{\infty} \widetilde{h} \circ \varphi^j. \tag{3.30}$$

converges to a continuous G-equivariant contracting chain homotopy of the completion of $\widetilde{P}_*^{\pm} \otimes_k CC_*(A : R)$. It follows that the augmentation morphism

$$\epsilon \otimes_k \mathrm{id} : (P_*^{\pm} \otimes CC_*(A : R))_G \longrightarrow (CC_*(A : R))_G \tag{3.31}$$

becomes a chain homotopy equivalence after completion. Taking homology one obtains a natural isomorphism

$$\widehat{\mathbb{H}}_*(G, \widehat{CC}_*(A : R)) \xrightarrow{\simeq} H_*(\widehat{CC}_*(A : R)_G).$$

□

A similar proof establishes the following lemma for groups of finite cohomological dimension over k. As we do not want to make this restriction a more elaborate argument is necessary.

LEMMA 3.6. *Let A be a unital G-algebra over k and let $U \subset G$ be a normal subgroup. Let X be a G-set such that $\epsilon_A : A\langle X \rangle \to A$ has a U-equivariant k-algebra section. Then*

$$\epsilon_{A*} : \widehat{\mathbb{H}}_*(G, \widehat{CC}_*(A\langle X \rangle \rtimes U : kU)) \xrightarrow{\simeq} \widehat{\mathbb{H}}_*(G, \widehat{CC}_*(A \rtimes U : kU)). \tag{3.32}$$

is a natural isomorphism compatible with homogeneous decompositions.

PROOF. Let $\varphi : A \to A\langle X \rangle$ be a U-equivariant homomophism of k-algebras that splits the canonical epimorphism $\epsilon_A : A\langle X \rangle \to A$. These homomorphisms induce chain maps

$$(\epsilon_A \rtimes \mathrm{id})_* : \widehat{CC}(A\langle X \rangle \rtimes U : k \rtimes U) \longrightarrow \widehat{CC}(A \rtimes U : k \rtimes U)$$

and

$$(\varphi \rtimes \mathrm{id})_* : \widehat{CC}(A \rtimes U : k \rtimes U) \longrightarrow \widehat{CC}(A\langle X \rangle \rtimes U : k \rtimes U).$$

Moreover

$$s \mapsto (1 - s)(\varphi \circ \epsilon_A) \rtimes \mathrm{id} + s \cdot \mathrm{id},\ s \in [0, 1], \tag{3.33}$$

defines an affine homotopy of $k \rtimes U$-algebra homomorphisms between $(\varphi \circ \epsilon_A) \rtimes \mathrm{id} : A\langle X\rangle \rtimes U \to A\langle X\rangle \rtimes U$ and the identity. The corresponding chain homotopy of periodic cyclic complexes (see [Lo], page 117) between $\widehat{CC}((\varphi \circ \epsilon_A) \rtimes \mathrm{id})$ and the identity descends to a continuous linear Cartan homotopy operator

$$h_0 : \widehat{CC}_*(A\langle X\rangle \rtimes U : k \rtimes U) \to \widehat{CC}_{*+1}(A\langle X\rangle \rtimes U : k \rtimes U). \tag{3.34}$$

Note that

$$h_0((A\langle X\rangle \rtimes U)\,\Omega^n(A\langle X\rangle \rtimes U)) \subset \Omega^{n+1}(A\langle X\rangle \rtimes U) \tag{3.35}$$

by (2.19). We cannot use this operator directly if U is of infinite cohomological dimension because it does not preserve Hodge filtrations. To overcome this difficulty we use an auxiliary preliminary homotopy. Recall that the cone of a morphism $f : C_* \to C'_*$ of complexes is defined as the complex

$$\mathrm{Cone}(f)_* = C_* \oplus C'_{*+1}, \ \partial_{\mathrm{Cone}} = \begin{pmatrix} \partial_{C_*} & 0 \\ f & -\partial_{C'_*} \end{pmatrix}. \tag{3.36}$$

Put

$$h_1 = \begin{pmatrix} h_0 & \widehat{CC}(\varphi \rtimes \mathrm{id}) \\ 0 & 0 \end{pmatrix} : \mathrm{Cone}_*(\widehat{CC}(\epsilon_A \rtimes \mathrm{id})) \to \mathrm{Cone}_{*+1}(\widehat{CC}(\epsilon_A \rtimes \mathrm{id})). \tag{3.37}$$

This is a k-linear contracting homotopy of $\mathrm{Cone}_*(\widehat{CC}(\epsilon_A \rtimes \mathrm{id}))$.
Let $h_2 : \mathrm{Cone}_*(\widehat{CC}(\epsilon_A \rtimes \mathrm{id})) \to \mathrm{Cone}_{*+1}(\widehat{CC}(\epsilon_A \rtimes \mathrm{id}))$ be the k-linear map sending $\omega \in d\Omega^*(A\langle X\rangle \rtimes U)$ to $-d(\varphi(1_A)u_e)\omega$ and annihilating $(A\langle X\rangle \rtimes U)d\Omega^*(A\langle X\rangle \rtimes U)$, and sending $\omega' \in d\Omega^*(A \rtimes U)$ to $d(1_A u_e)\omega'$ and annihilating $(A \rtimes U)d\Omega^*(A \rtimes U)$. Let

$$\varphi_2 := \mathrm{id} - (h_2 \circ \partial_{\mathrm{Cone}(\widehat{CC}(\epsilon_A \rtimes \mathrm{id}))} + \partial_{\mathrm{Cone}(\widehat{CC}(\epsilon_A \rtimes \mathrm{id}))} \circ h_2) \tag{3.38}$$

and put

$$h_3 = h_1 \circ \varphi_2 + h_2 : \mathrm{Cone}_*(\widehat{CC}(\epsilon_A \rtimes \mathrm{id})) \to \mathrm{Cone}_{*+1}(\widehat{CC}(\epsilon_A \rtimes \mathrm{id})). \tag{3.39}$$

The operator h_3 is again a k-linear contracting homotopy of $\mathrm{Cone}_*(\widehat{CC}(\epsilon_A \rtimes \mathrm{id}))$. But in addition, it strictly increases the degree of algebraic differential forms.

Let P_* be a degree-wise free resolution i.e., $P_n \simeq kG \otimes_k W_n, n \geqslant 0$, of the constant G-module k (it always exists). We let

$$h : P_*^{\pm} \otimes_k \mathrm{Cone}_*(\widehat{CC}(\epsilon_A \rtimes \mathrm{id})) \to P_*^{\pm} \otimes_k \mathrm{Cone}_{*+1}(\widehat{CC}(\epsilon_A \rtimes \mathrm{id})) \tag{3.40}$$

be the unique G-equivariant linear operator such that $h_*|_{W_* \otimes_k \mathrm{Cone}} = \mathrm{id} \otimes_k h_3$ and put

$$\psi = \mathrm{id} - (\partial \circ h + h \circ \partial) : P_*^{\pm} \otimes_k \mathrm{Cone}(\widehat{CC}(\epsilon_A \rtimes \mathrm{id}))_* \to P_*^{\pm} \otimes_k \mathrm{Cone}(\widehat{CC}(\epsilon_A \rtimes \mathrm{id}))_*.$$

The operator $\sum_{n=0}^{\infty} h \circ \psi^n$ is then a well defined, continuous and G-equivariant contracting chain homotopy of $\widehat{P}_*^{\pm} \widehat{\otimes}_k \mathrm{Cone}_*(\widehat{CC}(\epsilon_A \rtimes \mathrm{id}))$. Consequently

$$\mathrm{id} \otimes_k \widehat{CC}(\epsilon_A \rtimes \mathrm{id}) : \widehat{P}_*^{\pm} \widehat{\otimes}_k \widehat{CC}_*(A\langle G\rangle \rtimes U : k \rtimes U) \to \widehat{P}_*^{\pm} \widehat{\otimes}_k \widehat{CC}_*(A \rtimes U : k \rtimes U) \tag{3.41}$$

is a continuous G-equivariant chain homotopy equivalence and

$$(\epsilon_A \rtimes \mathrm{id})_* : \widehat{\mathbb{H}}_*(G, \widehat{CC}_*(A\langle G\rangle \rtimes U : k \rtimes U)) \xrightarrow{\simeq} \widehat{\mathbb{H}}_*(G, \widehat{CC}_*(A \rtimes U : k \rtimes U))$$

is an isomorphism. □

4. Periodic cyclic homology of crossed products

THEOREM 4.1. *Let G be a group and let A be a G-algebra over k. Let $v \in G$ and let*

$$1 \longrightarrow U = v^{\mathbb{Z}} \longrightarrow Z_v \longrightarrow N_v \longrightarrow 1$$

be the associated central extension. Then there is a natural isomorphism

$$HP_*(A \rtimes G)_{[v]} \xrightarrow{\simeq} \widehat{\mathbb{H}}_* \left(N_v, \widehat{CC}_*(A\langle Z_v \rangle \rtimes U : k \rtimes U)_{\{v\}} \right). \tag{4.1}$$

PROOF. By Proposition 3.3 there is an isomorphism

$$\left(CC_*(A\langle G \rangle \rtimes U : k \rtimes U)_{\{v\}} \right)_{Z_v} \xrightarrow{\simeq} CC_*(A \rtimes G)_{[v]}$$

of filtered chain complexes. By definition (see Lemma 3.2) the restriction of the canonical Z_v-action on $\Omega^*(A\langle G \rangle \rtimes U : k \rtimes U)_\natural$ to U is trivial so that

$$HP_*(A \rtimes G)_{[v]} \xrightarrow{\simeq} H_*((\widehat{CC}_*(A\langle G \rangle \rtimes U : k \rtimes U)_{\{v\}})_{N_v}).$$

From (3.13) we learn that it is a degree-wise free $k \rtimes N_v$-module. So by Lemma 3.5 there is a natural isomorphism

$$\widehat{\mathbb{H}}_* \left(N_v, \widehat{CC}_*(A\langle G \rangle \rtimes U : k \rtimes U)_{\{v\}} \right) \xrightarrow{\simeq} H_*((\widehat{CC}_*(A\langle G \rangle \rtimes U : k \rtimes U)_{\{v\}})_{N_v}).$$

Lemma 3.6 may be applied to both projections in the diagram

$$A\langle G \rangle \leftarrow A\langle G \times Z_v \rangle \rightarrow A\langle Z_v \rangle$$

and obtains the natural quasi-isomorphism

$$\widehat{\mathbb{H}}_* \left(N_v, \widehat{CC}_*(A\langle Z_v \rangle \rtimes U : k \rtimes U)_{\{v\}} \right) \xrightarrow{\simeq} \widehat{\mathbb{H}}_* \left(N_v, \widehat{CC}_*(A\langle G \rangle \rtimes U : k \rtimes U)_{\{v\}} \right)$$

This concludes the proof of the theorem. □

4.1. The homogeneous part.

THEOREM 4.2. *Let G be a group and let A be a k-algebra on which G acts by automorphisms. There are natural isomorphisms*

$$HP_*(A \rtimes G)_{[e]} \xrightarrow{\simeq} \widehat{\mathbb{H}}_*(G, \widehat{CC}_*(A)), \tag{4.2}$$

and

$$HP^*(A \rtimes G)_{[e]} \xrightarrow{\simeq} \widehat{\mathbb{H}}^*(G, \widehat{CC}_*(A)), \tag{4.3}$$

which identify the homogeneous part of the periodic cyclic (co)homology of the crossed product $A \rtimes G$ with the continuous hyper(co)homology of G with coefficients in the periodic cyclic bicomplex of A.

PROOF. We apply Theorem 4.1 to the case $v = e$, $U = 1$, $N_v = G$ and obtain a natural isomorphism

$$HP_*(A \rtimes G)_{[e]} \xrightarrow{\simeq} \widehat{\mathbb{H}}_*(G, \widehat{CC}_*(A\langle G \rangle)).$$

Goodwillie's Theorem in the strengthened form of Lemma 3.6 shows that

$$\widehat{\mathbb{H}}_*(G, \widehat{CC}_*(A\langle G \rangle)) \xrightarrow{\simeq} \widehat{\mathbb{H}}_*(G, \widehat{CC}_*(A)),$$

is a natural isomorphism as well. A similar argument establishes (4.3). □

The **bivariant periodic cyclic cohomology** [**CQ2**] of a pair (A, B) of k-algebras is defined as the homology of the $\mathbb{Z}/2\mathbb{Z}$-graded chain complex

$$\widehat{CC}_*(A,B) = \mathcal{L}(\widehat{CC}_*(A), \widehat{CC}_*(B)). \tag{4.4}$$

We recall that the assumptions of the following two theorems are satisfied if some classifying space of G has the homotopy type of a finite CW-complex.

THEOREM 4.3. *Let G be a group such that the constant G-module k possesses a resolution of finite length by finitely generated free kG-modules. Let A be a G-algebra over k and let B be a k-algebra. Then there is a natural isomorphism*

$$HP_*(A \rtimes G, B)_{[e]} \xrightarrow{\simeq} \mathbb{H}^*(G, \widehat{CC}_*(A,B)) \tag{4.5}$$

identifying the homogeneous part of the bivariant periodic cyclic cohomology of the pair $(A \rtimes G, B)$ with the hypercohomology of G with coefficients in the bivariant periodic cyclic bicomplex of the pair (A, B).

PROOF. We learn from Proposition 3.3, applied to $v = e$, that there is a canonical isomorphism

$$(CC_*(A\langle G\rangle))_G \xrightarrow{\simeq} CC_*(A \rtimes G)_{[e]}$$

of chain complexes preserving Hodge filtrations. Let P_* be a resolution of finite length of the constant G-module k by finitely generated free kG-modules. As $\Omega^*(A\langle G\rangle)$ is degree-wise free as kG-module

$$(P_*^{\pm} \otimes_k CC_*(A\langle G\rangle))_G \longrightarrow (CC_*(A\langle G\rangle))_G$$

becomes a continuous chain-homotopy equivalence after completion by Lemma 3.5. Lemma 3.6, and in particular assertion (3.41), applied to $U = \{e\}$, $X = G$, and $\varphi : A \to A\langle G\rangle, a \mapsto a\langle e\rangle$, show that the left hand arrow in the natural diagram

$$(P_*^{\pm} \otimes_k CC_*(A))_G \longleftarrow (P_*^{\pm} \otimes_k CC_*(A\langle G\rangle))_G \longrightarrow CC_*(A \rtimes G)_{[e]} \tag{4.6}$$

becomes a continuous chain homotopy equivalence after completion, too.
There are natural isomorphisms of chain complexes

$$\mathrm{Hom}_{kG}\left(\widehat{P}_*^{\pm}, \mathcal{L}(\widehat{CC}_*(A), \widehat{CC}_*(B))\right) \xrightarrow{\simeq} \mathrm{Hom}_k\left(\widehat{P}_*^{\pm}, \mathcal{L}(\widehat{CC}_*(A), \widehat{CC}_*(B))\right)^G$$

$$\xrightarrow{\simeq} \mathcal{L}\left(\widehat{P}_*^{\pm}\widehat{\otimes}_k\widehat{CC}_*(A), \widehat{CC}_*(B)\right)^G \xrightarrow{\simeq} \mathcal{L}\left((\widehat{P}_*^{\pm}\widehat{\otimes}_k\widehat{CC}_*(A))_G, \widehat{CC}_*(B)\right).$$

Making use of (4.6) again and taking cohomology we arrive at the desired result. □

THEOREM 4.4. *Let G be a group such that the constant G-module k possesses a resolution of finite length by finitely generated free kG-modules. Let A be a k-algebra and let B be a G-algebra over k. Then there is a natural isomorphism*

$$HP_*(A, B \rtimes G)_{[e]} \xrightarrow{\simeq} \mathbb{H}_*(G, \widehat{CC}_*(A,B)) \tag{4.7}$$

identifying the homogeneous part of the bivariant periodic cyclic cohomology of the pair $(A, B \rtimes G)$ with the hyperhomology of G with coefficients in the bivariant periodic cyclic bicomplex of the pair (A, B).

PROOF. Let P_* be a resolution of finite length of the constant G-module k by finitely generated free kG-modules. So $P_* \simeq kG \otimes_k W_*$ as graded kG-modules for some finite dimensional graded k-vector space W_*. For a finite dimensional vector space V we denote by $\check{V}$ its dual. Let $\widehat{(P_*^\pm \otimes_k CC_*(B))_G}$ be the completion of $(P_*^\pm \otimes_k CC_*(B))_G$ with respect to the filtration topology. There are natural isomorphisms of linear spaces

$$\mathcal{L}\left(\widehat{CC}_*(A), \widehat{(P_*^\pm \otimes_k CC_*(B))_G}\right) \xrightarrow{\simeq} \mathcal{L}\left(\widehat{CC}_*(A), W_*^\pm \otimes_k \widehat{CC}_*(B)\right)$$

$$\xrightarrow{\simeq} \mathcal{L}\left(\widehat{CC}_*(A), \check{\check{W}}{}_*^\pm \otimes_k \widehat{CC}_*(B)\right) \xrightarrow{\simeq} \mathcal{L}\left(\widehat{CC}_*(A), \mathcal{L}(\check{W}_*^\pm, \widehat{CC}_*(B))\right)$$

$$\xrightarrow{\simeq} \mathcal{L}\left(\check{W}_*^\pm \otimes_k \widehat{CC}_*(A), \widehat{CC}_*(B)\right) \xrightarrow{\simeq} Hom_k\left(\check{W}_*^\pm, \mathcal{L}(\widehat{CC}_*(A), \widehat{CC}_*(B))\right)$$

$$\xrightarrow{\simeq} \check{\check{W}}{}_*^\pm \otimes_k \mathcal{L}(\widehat{CC}_*(A), \widehat{CC}_*(B)) \xrightarrow{\simeq} \left(P_*^\pm \otimes_k \mathcal{L}(\widehat{CC}_*(A), \widehat{CC}_*(B))\right)_G.$$

It is easily verified that this is a morphism of chain complexes. Making use of (4.6) again and taking homology, we prove our claim as before. □

4.2. The elliptic part.

THEOREM 4.5. *Let G be a group and let A be a k-algebra on which G acts by automorphisms. There is a natural isomorphism*

$$HP_*(A \rtimes G)_{\mathrm{ell}} \simeq \widehat{\mathbb{H}}_*\left(G, \widehat{\bigoplus_{\substack{v\in G\\ 1<|v|<\infty}}} \widehat{CC}_*(A \rtimes U : k \rtimes U)_{\{v\}}\right) \tag{4.8}$$

where the direct sum is labeled by the set of torsion elements in G and we denote by U the finite cyclic subgroup generated by the torsion element v. The subscripts $\{v\}$ and $[v]$ denote the conjugacy classes of v in U and G, respectively. The complex $\widehat{\bigoplus}_{\substack{v\in G\\ 1<|v|<\infty}} \widehat{CC}(A \rtimes U : k \rtimes U)_{\{v\}}$ is the completion of $\bigoplus_{\substack{v\in G\\ 1<|v|<\infty}} CC_*(A \rtimes U : k \rtimes U)_{\{v\}}$, equipped with the direct sum of the individual filtrations. For a single torsion element one obtains a natural isomorphism

$$HP_*(A \rtimes G)_{[v]} \simeq \widehat{\mathbb{H}}_*(Z_v, \widehat{CC}_*(A \rtimes U : k \rtimes U)_{\{v\}}) \tag{4.9}$$

where Z_v is the centralizer of v in G.

The theorem expresses the elliptic part of the periodic cyclic homology of the crossed product $A \rtimes G$ of A by G in terms of the hyperhomology of G with values in the completed direct sum of the periodic cyclic bicomplexes of the crossed products of A by the various finite cyclic subgroups of G.

PROOF. Let $v \in G$ be an element of finite order and let $U \subset G$ be the finite cyclic group generated by v. By Proposition 3.3 there is an isomorphism of filtered complexes

$$\left(CC_*(A\langle G\rangle \rtimes U : k \rtimes U)_{\{v\}}\right)_{Z_v} \xrightarrow{\simeq} CC_*(A \rtimes G)_{[v]}.$$

As k is of characteristic zero every projective $k \rtimes N_v$-module is projective as $k \rtimes Z_v$-module as well. Let P_* be a projective resolution of the G-module k. It is at the same time a projective resolution of the Z_v-module k. The finite central subgroup $U \subset Z_v$ acts trivially on the complex $CC_*(A\langle G\rangle \rtimes U : k \rtimes U)_{\{v\}}$ The underlying vector space $\Omega^*(A\langle G\rangle \rtimes U : k \rtimes U)_\natural$ is a degree-wise free N_v-module and thus projective as Z_v-module. Lemma 3.5 implies therefore that

$$\left(P_*^{\pm} \otimes_k CC_*(A\langle G\rangle \rtimes U : k \rtimes U)_{\{v\}}\right)_{Z_v} \longrightarrow \left(CC_*(A\langle G\rangle \rtimes U : k \rtimes U)_{\{v\}}\right)_{Z_v}$$

becomes a continuous chain homotopy equivalence after completion. Consider now the canonical homomorphism $\epsilon_A : A\langle G\rangle \longrightarrow A$. Its linear section

$$\varphi : A \longrightarrow A\langle G\rangle,\ a \mapsto \frac{1}{|U|}\sum_{h\in U} a\langle h\rangle$$

is actually a homomorphism of $k \rtimes U$-algebras. So one may deduce from Lemma 3.6 and (3.41) that

$$\left(P_*^{\pm} \otimes_k CC_*(A\langle G\rangle \rtimes U : k \rtimes U)_{\{v\}}\right)_{Z_v} \longrightarrow \left(P_*^{\pm} \otimes_k CC_*(A \rtimes U : k \rtimes U)_{\{v\}}\right)_{Z_v}$$

becomes a continuous chain homotopy equivalence after completion. Thus one derives from the previously constructed chain maps after completion and passage to homology the natural diagram of isomorphisms

$$\widehat{\mathbb{H}}_*(Z_v, \widehat{CC}_*(A \rtimes U : k \rtimes U)_{\{v\}}) \overset{\simeq}{\leftarrow} \widehat{\mathbb{H}}_*(Z_v, \widehat{CC}_*(A\langle G\rangle \rtimes U : k \rtimes U)_{\{v\}})$$
$$\overset{\simeq}{\rightarrow} HP_*(A \rtimes G)_{[v]},$$

which is our second claim. Our previous arguments show that there exists a natural chain map

$$\left(\bigoplus_{w\in[v]} P_*^{\pm} \otimes_k CC_*(A\langle G\rangle \rtimes w^{\mathbb{Z}} : k \rtimes w^{\mathbb{Z}})_{\{w\}}\right)_G \longrightarrow CC_*(A \rtimes G)_{[v]}$$

which becomes a continuous chain homotopy equivalence after completion. Passing to direct sums over all conjugacy classes of torsion elements we obtain a chain map

$$\left(\bigoplus_{\substack{v\in G\\ 1<|v|<\infty}} P_*^{\pm} \otimes_k CC_*(A\langle G\rangle \rtimes U : k \rtimes U)_{\{v\}}\right)_G \longrightarrow CC_*(A \rtimes G)_{\mathrm{ell}}$$

with similar properties because the filtration shifts are the same in all direct summands. Passing to completions and taking homology, we arrive at the first assertion of the theorem. □

There are corresponding results for the elliptic part of periodic cyclic cohomology and bivariant periodic cyclic homology of crossed products. The details are left to the reader.

4.3. The inhomogeneous part. In general it seems to be difficult to express the inhomogeneous part of the periodic cyclic homology of a crossed product in terms of simpler homological invariants. We consider here only a special case treated already by Nistor [**Ni**].
Recall that the extension

$$1 \longrightarrow U = v^{\mathbb{Z}} = \mathbb{Z} \longrightarrow Z_v \overset{\pi}{\longrightarrow} N_v \longrightarrow 1. \tag{4.10}$$

is classified up to isomorphism by its extension class

$$\alpha \in H^2(N_v, \mathbb{Z}). \tag{4.11}$$

If $\sigma : N_v \to Z_v$ is a set theoretic section of π, a homogeneous group cocycle representing α is given by

$$c_\alpha : (N_v)^3 \to \mathbb{Z}, \ (h_0, h_1, h_2) \mapsto \sigma(h_1^{-1}h_2)\sigma(h_0^{-1}h_2)^{-1}\sigma(h_0^{-1}h_1). \tag{4.12}$$

We follow the classical approch and begin with the calcuation of the inhomogeneous part of Hochschild homology. For an A-bimodule M let

$$H_*(C_*(M, A)) = HH_*(M, A) = Tor_*^{A\otimes_k A^{op}}(M, A) \tag{4.13}$$

be the Hochschild homology of (M, A). Let G be a group acting on A. For $v \in G$ let $A^{(v)}$ be the A-bimodule with underlying vector space A and bimodule structure

$$A \otimes_k A^{(v)} \otimes_k A \to A^{(v)}, \ a' \otimes a \otimes a'' \mapsto v^{-1}(a')aa''. \tag{4.14}$$

THEOREM 4.6. *Let G be a group and let A be a unital G-algebra. Let $v \in G$ be an element of infinite order. There is a natural isomorphism*

$$HH_*(A \rtimes G)_{[v]} \xrightarrow{\simeq} \mathbb{H}_*\left(N_v, C_*(A^{(v)}, A) \otimes^{\mathbb{L}}_{k\rtimes U} k\right). \tag{4.15}$$

PROOF. We proceed in several steps.
Step 1:
Lemma 3.1, Lemma 3.2 and Proposition 3.3 hold for Hochschild complexes as well as for cyclic complexes. So there is an natural isomorphism

$$HH_*(A \rtimes G)_{[v]} \xrightarrow{\simeq} H_*(C_*(A\langle G\rangle^{(v)}, A\langle G\rangle)_{Z_v}). \tag{4.16}$$

Step 2:
The Eilenberg-Zilber theorem in Hochschild theory [**Lo**] states that for unital k-algebras A, B, A-bimodules M and B-bimodules N there is a natural chain-homotopy equivalence

$$C_*(M \otimes_k N, A \otimes_k B) \xrightarrow{\Delta} C_*(M, A) \otimes_k C_*(N, B). \tag{4.17}$$

Step 3:
The algebra $k\langle X\rangle$ is not unital, but naturally H-unital [**Lo**] in the sense that

$$h' : C'_*(k\langle X\rangle) \to C'_*(k\langle X\rangle), \langle x_0\rangle \otimes \ldots \otimes \langle x_n\rangle \mapsto \langle x_0\rangle \otimes \ldots \otimes \langle x_n\rangle \otimes \langle x_n\rangle \tag{4.18}$$

is a contracting chain-homotopy of the b'-complex of $k\langle X\rangle$ which is natural with respect to maps $X \to Y$ of sets. It follows then from step 2 and excision in Hochschild homology [**Lo**] that there is a Z_v-equivariant chain homotopy equivalence

$$C_*((A\langle G\rangle)^{(v)}, A\langle G\rangle) \xrightarrow{\Delta} C_*(A^{(v)}, A) \otimes_k C_*(k\langle G\rangle).$$

(Note that $k\langle G\rangle^{(v)} = k\langle G\rangle$ by (2.19).) So one may pass to coinvariants and obtains a natural chain-homotopy equivalence

$$\Big(C_*((A\langle G\rangle)^{(v)}, A\langle G\rangle)\Big)_{Z_v} \xrightarrow{\Delta} \Big(C_*(A^{(v)}, A) \otimes_k C_*(k\langle G\rangle)\Big)_{Z_v}. \tag{4.19}$$

Step 4:
Observe that the natural G-equivariant chain map

$$\begin{array}{ccc} C_*^{\mathrm{Bar}}(G, k) & \hookrightarrow & (\Omega^*(k\langle G\rangle), b) \\ [g_0, \ldots, g_n] & \mapsto & \langle g_0\rangle d\langle g_1\rangle \ldots d\langle g_n\rangle \end{array} \tag{4.20}$$

is equivariantly linearly split with equivariantly contractible cokernel. In particular there is a natural chain-homotopy equivalence

$$(4.21) \qquad \left(C_*(A^{(v)}, A) \otimes_k C_*(k\langle G\rangle)\right)_{Z_v} \xrightarrow{\sim} \left(C_*(A^{(v)}, A) \otimes_k C_*^{Bar}(G, k)\right)_{Z_v}.$$

As the Bar-complex of a group is a free resolution of the constant G-module one deduces from (4.16), (4.19), and (4.21) a natural isomorphism

$$HH_*(A \rtimes G)_{[v]} \xrightarrow{\simeq} \mathbb{H}_*\left(Z_v, C_*(A^{(v)}, A)\right).$$

Step 5:
Observe finally the isomorphism

$$- \otimes^{\mathbb{L}}_{k \rtimes Z_v} k \simeq (- \otimes^{\mathbb{L}}_{k \rtimes U} k) \otimes^{\mathbb{L}}_{k \rtimes N_v} k$$

of derived functors from $\mathbb{D}_-(k \rtimes Z_v)$ to $\mathbb{D}_-(k)$, which implies

$$\mathbb{H}_*\left(Z_v, C_*(A^{(v)}, A)\right) \simeq H_*\left(C_*(A^{(v)}, A) \otimes^{\mathbb{L}}_{k \rtimes Z_v} k\right) \simeq$$

$$H_*\left((C_*(A^{(v)}, A) \otimes^{\mathbb{L}}_{k \rtimes U} k) \otimes^{\mathbb{L}}_{k \rtimes N_v} k\right) \simeq \mathbb{H}_*\left(N_v, C_*(A^{(v)}, A) \otimes^{\mathbb{L}}_{k \rtimes U} k\right).$$

The theorem is proved. □

PROPOSITION 4.7. *Let A be a G-algebra and let $v \in G$ be an element of infinite order. Then $HP_*(A \rtimes G)_{[v]}$ is a $H^*(N_v, k)$-module such that the extension class $\alpha \in H^2(N_v, k)$ of* (4.10) *acts as the identity.*

PROOF. Step 1:
A weak version of the Eilenberg-Zilber theorem in periodic cyclic homology [**Pu**], valid for non-unital algebras as well, states the existence of a natural chain map

$$(4.22) \qquad \Delta_{cyc} : \widehat{CC}(A \otimes_k B) \longrightarrow \widehat{CC}(A) \widehat{\otimes}_k \widehat{CC}(B)$$

for every pair A, B of k-algebras, where on the right hand side the completion of the algebraic tensor product of the periodic cyclic complexes of A and B occurs. In particular, if A and B are G-algebras, then Δ is G-equivariant equivariant. It is compatible with base change in the sense that it descends under the natural projections to a natural chain map

$$(4.23) \qquad \Delta_{cyc} : \widehat{CC}(A \otimes_k B : R \otimes_k S) \longrightarrow \widehat{CC}(A : R) \widehat{\otimes}_k \widehat{CC}(B : S).$$

This is a natural chain homotopy equivalence if A and B are unital.
Step 2:
In step 4 of the proof of the previous theorem we observed the close connection between the Bar-complex $C_*^{Bar}(G, k)$ of a group G with coefficients in k and the Hochschild complex of $k\langle G\rangle$. In fact, every alternating homogeneous group cocycle $c_\beta \in Z^n(G, k)$ defines a G-invariant periodic cyclic cocycle c'_β on $k\langle G\rangle$ by the formula
(4.24)

$$c'_\beta(\langle g_0\rangle d\langle g_1\rangle \ldots d\langle g_n\rangle) = c_\beta([g_0, \ldots, g_n]), \;\; c'_\beta(d\langle g_1\rangle \ldots d\langle g_n\rangle) = 0, \;\; g_0, \ldots, g_n \in G.$$

We apply this in the case $G = N_v$. By abuse of language we call the N_v-equivariant natural chain map

$$\cap_{c_\beta} : \widehat{CC}_*(A\langle Z_v\rangle \rtimes U : k \rtimes U)_{\{v\}} \stackrel{(\mathrm{id} \times \pi) \circ \mathrm{diag}}{\longrightarrow} \widehat{CC}_*(A\langle Z_v \times N_v\rangle \rtimes U : k \rtimes (U \times 1))_{\{(v,1)\}}$$

(4.25)

$$\stackrel{\Delta_{cyc}}{\longrightarrow} \widehat{CC}_*(A\langle Z_v\rangle \rtimes U : k \rtimes U)_{\{v\}}\widehat{\otimes}_k\widehat{CC}_*(k\langle N_v\rangle) \stackrel{\mathrm{id}\otimes c'_\beta}{\longrightarrow} \widehat{CC}_*(A\langle Z_v\rangle \rtimes U : k \rtimes U)_{\{v\}}$$

the "cap-product" with the cocycle c_β. Up to chain homotopy it only depends on the cohomology class of c_β. Via the identification (4.1) it turns $HP_*(A \rtimes G)_{[v]}$ into a module over the cohomology ring $H^*(N_v, k)$ of N_v with coefficients in k.
Step 3:
Let $j : U \to (k, +)$ be the homomorphism sending the generator $v \in U$ to 1. Let $\eta : \widehat{CC}(k\langle Z_v\rangle \rtimes U : kU)_{\{v\}} \to k$ be the N_v-invariant continuous k-linear functional given on one-forms by
(4.26)

$$\eta(u_v\langle g_0\rangle d\langle g_1\rangle) = \frac{1}{2}\left(j(g_0\sigma(g_0^{-1}g_1)g_1^{-1}) - j(g_1\sigma(g_1^{-1}g_0)g_0^{-1})\right), \quad \eta(u_v d\langle g_1\rangle) = 1$$

and vanishing on forms of other degrees. The N_v-equivariant continuous linear operator

$$\tilde{\eta} : \widehat{CC}_*(A\langle Z_v\rangle \rtimes U : k \rtimes U)_{\{v\}} \stackrel{\mathrm{diag}}{\longrightarrow} \widehat{CC}_*(A\langle Z_v \times Z_v\rangle \rtimes (U \times U) : k \rtimes (U \times U))_{\{(v,v)\}}$$

$$\stackrel{\Delta_{cyc}}{\longrightarrow} \widehat{CC}_*(A\langle Z_v\rangle \rtimes U : kU)_{\{v\}}\widehat{\otimes}_k\widehat{CC}_*(k\langle Z_v\rangle \rtimes U : kU)_{\{v\}}$$

$$\stackrel{\mathrm{id}\otimes\eta}{\longrightarrow} \widehat{CC}_{*+1}(A\langle Z_v\rangle \rtimes U : k \rtimes U)_{\{v\}} \tag{4.27}$$

defines then an N_v-equivariant chain homotopy between $\cap c_\alpha$ and the identity, where c_α is the cocycle (4.12) representing the extension class (4.11). □

THEOREM 4.8. *Let G be a group and let A be a k-algebra on which G acts by automorphisms. Let $v \in G$ be an element of infinite order that acts trivially on A and let $[v]$ be its conjugacy class. There is a natural isomorphism*

$$HP_*(A \rtimes G)_{[v]} \simeq \widehat{\mathbb{H}}_*(N_v, \widehat{CC}_*(A))[\alpha^{-1}] \tag{4.28}$$

where

$$\widehat{\mathbb{H}}_*(N_v, \widehat{CC}_*(A))[\alpha^{-1}] = \mathbb{R}\varprojlim_{\cap\alpha} \widehat{\mathbb{H}}_*(N_v, \widehat{CC}_*(A)) \tag{4.29}$$

fits into the natural exact sequence

$$0 \to \varprojlim_{\cap\alpha}{}^1(\widehat{\mathbb{H}}_{*-1}(N_v, \widehat{CC}_*(A)) \to \widehat{\mathbb{H}}_*(N_v, \widehat{CC}_*(A))[\alpha^{-1}]$$

$$\to \varprojlim_{\cap\alpha}(\widehat{\mathbb{H}}_*(N_v, \widehat{CC}_*(A)) \to 0. \tag{4.30}$$

Here the derived projective limit is taken over the iterated cap product

$$\cap\alpha : \widehat{\mathbb{H}}_*(N_v, \widehat{CC}_*(A)) \longrightarrow \widehat{\mathbb{H}}_*(N_v, \widehat{CC}_*(A)) \tag{4.31}$$

with the extension class of (4.10).

THEOREM 4.9. *Let G be a group and let A be a k-algebra on which G acts by automorphisms. Let $v \in G$ be an element of infinite order that acts trivially on A and let $[v]$ be its conjugacy class. There is a natural isomorphism*

$$HP^*(A \rtimes G)_{[v]} \simeq \widehat{\mathbb{H}}^*(N_v, \widehat{CC}_*(A))[\alpha^{-1}] \tag{4.32}$$

where

$$\widehat{\mathbb{H}}^*(N_v, \widehat{CC}_*(A))[\alpha^{-1}] = \varinjlim_{\cup\alpha} \widehat{\mathbb{H}}^*(N_v, \widehat{CC}_*(A)). \tag{4.33}$$

Here the direct limit is taken over the iterated cup product

$$\cup\alpha : \widehat{\mathbb{H}}^*(N_v, \widehat{CC}_*(A)) \longrightarrow \widehat{\mathbb{H}}^*(N_v, \widehat{CC}_*(A)) \tag{4.34}$$

with the extension class of (4.10).

PROOF. Step 1:
Recall the Eilenberg-Zilber theorem in its strong form [**Pu**] which states that the morphisms (4.21) and (4.22) are in fact natural chain-homotopy equivalences if A and B are unital. The chain map Δ_{cyc} preserves Hodge filtrations and induces on the associated graded Hochschild complexes the chain homotopy equivalence (4.17). As in Step 3 of the proof of Theorem 4.6 one may claim that this holds for the non-unital algebras $B = k\langle X\rangle$, X a set, as well. In particular, one obtains a natural, continuous, Z_v-equivariant chain homotopy equivalence

$$\widehat{CC}_*(A\langle Z_v\rangle \rtimes U : k \rtimes U)_{\{v\}} \xrightarrow{\sim} \widehat{CC}_*(A)\widehat{\otimes}_k\widehat{CC}_*(k\langle Z_v\rangle \rtimes U : k \rtimes U)_{\{v\}}. \tag{4.35}$$

if v acts trivially on A. It follows from Theorem 4.1 that in this case there is a natural isomorphism

$$HP_*(A \rtimes G)_{[v]} \xrightarrow{\simeq} \widehat{\mathbb{H}}_* \left(N_v, \widehat{CC}_*(A)\widehat{\otimes}_k\widehat{CC}_*(k\langle Z_v\rangle \rtimes U : k \rtimes U)_{\{v\}}\right). \tag{4.36}$$

Step 2:
Consider the composition

$$\pi_* : \widehat{CC}_*(k\langle Z_v\rangle \rtimes U : k \rtimes U)_{\{v\}} \to \widehat{CC}_*(k\langle N_v\rangle \rtimes U : k \rtimes U)_{\{v\}}$$

$$\xrightarrow{\Delta_{cyc}} \widehat{CC}_*(k\langle N_v\rangle)\widehat{\otimes}_k\widehat{CC}(k \rtimes U : k \rtimes U)_{\{v\}} \overset{(2.14)}{\simeq} \widehat{CC}_*(k\langle N_v\rangle) \tag{4.37}$$

This N_v-equivariant chain map preserves Hodge filtrations because the quasi-isomorphism

$$\widehat{CC}_*(k\langle N_v\rangle : k\langle N_v\rangle)_{\{v\}} \xrightarrow{\sim} k$$

vanishes on $\mathrm{Fil}^1_{\mathrm{Hodge}}\widehat{CC}_*(k\langle N_v\rangle : k\langle N_v\rangle)_{\{v\}}$. The N_v-equivariant continuous linear operator $\pi_* \circ \widetilde{\eta}$ defines then an N_v-equivariant chain homotopy between π_* and $\cap c_\alpha \circ \pi_*$ (see (4.27)), so that altogether we have constructed an N_v-equivariant continuous chain map

$$\Psi : \widehat{CC}_*(k\langle Z_v\rangle \rtimes U : kU)_{\{v\}} \longrightarrow \mathrm{Cone}(\cap c_\alpha - \mathrm{id} : \widehat{CC}_*(k\langle N_v\rangle) \to \widehat{CC}_*(k\langle N_v\rangle)). \tag{4.38}$$

Step 5:
Denote by $\widehat{CC}_*(k\langle N_v\rangle)[-2]$ the complex $\widehat{CC}_*(k\langle N_v\rangle)$ with the same grading, but the shifted filtration

$$\mathrm{Fil}^m_{\mathrm{Hodge}}\widehat{CC}_*(k\langle N_v\rangle)[-2] = \mathrm{Fil}^{m-2}_{\mathrm{Hodge}}\widehat{CC}_*(k\langle N_v\rangle).$$

This has the effect that $\cap c_\alpha : \widehat{CC}_*(k\langle N_v\rangle) \to \widehat{CC}_*(k\langle N_v\rangle)[-2]$ preserves filtrations while $\mathrm{id} : \widehat{CC}_*(k\langle N_v\rangle) \to \widehat{CC}_*(k\langle N_v\rangle)[-2]$ shifts them by $+2$. Consequently

$$\Psi : \widehat{CC}_*(k\langle Z_v\rangle \rtimes U : kU)_{\{v\}} \longrightarrow \mathrm{Cone}(\cap c_\alpha - \mathrm{id} : \widehat{CC}_*(k\langle N_v\rangle) \to \widehat{CC}_*(k\langle N_v\rangle)[-2])$$

preserves the given filtrations and one may pass to the associated graded situation. Up to natural quasi-isomorphism one finds

(4.39)
$$\mathrm{Gr}(\Psi): C_*(k\langle Z_v\rangle \rtimes U : kU)_{\{v\}} \longrightarrow \mathrm{Cone}(\cap c_\alpha : C_*(k\langle N_v\rangle) \to C_*(k\langle N_v\rangle)[-2])$$

where $\cap c_\alpha$ is the "classical" cap-product with the extension cocycle c_α.

Step 6:

We recall a classical result of Gysin, which states in our framework that

$$\left(C_*^{Bar}(Z_v, k)\right)_U \quad \to \quad C_*^{Bar}(N_v, k) \quad \xrightarrow{\cap\alpha} \quad C_*^{Bar}(N_v, k)[-2] \tag{4.40}$$

is a distinguished triangle in the chain-homotopy category of (bounded below) complexes of $k \rtimes N_v$-modules. By (4.20) and Lemma 3.2 the same holds for

$$C_*(k\langle Z_v\rangle \rtimes U : kU)_{\{v\}} \quad \to \quad C_*(k\langle N_v\rangle) \quad \xrightarrow{\cap\alpha} \quad C_*(k\langle N_v\rangle)[-2]. \tag{4.41}$$

This is a distinguished triangle of complexes of free N_v-modules, so that it remains distinguished after tensoring with $C_*(A)$. Passing to hyperhomology, one arrives at the vanishing result

$$\mathbb{H}_*\left(N_v, \mathrm{Cone}(\mathrm{id}_{C_*(A)} \otimes \mathrm{Gr}(\Psi))\right) = 0. \tag{4.42}$$

Iterated use of the five lemma yields then

$$\mathbb{H}_*(N_v, \mathrm{Cone}(\mathrm{id}_{\widehat{CC}_*(A)} \widehat{\otimes} \Psi)/\mathrm{Fil}^n \mathrm{Cone}(\mathrm{id}_{\widehat{CC}_*(A)} \widehat{\otimes} \Psi)) = 0$$

for $n \geqslant 0$ and finally

$$\widehat{\mathbb{H}}_*(N_v, \mathrm{Cone}(\mathrm{id}_{\widehat{CC}_*(A)} \widehat{\otimes} \Psi)) = 0 \tag{4.43}$$

by the lim^1-exact sequence.

Step 7:

The vanishing result (4.43), (4.36), and Lemma 3.6 imply a natural isomorphism

$$HP_*(A \rtimes G)_{[v]} \xrightarrow{\simeq} \widehat{\mathbb{H}}_*(N_v, \mathrm{Cone}(\cap c_\alpha - \mathrm{id} : \widehat{CC}_*(A) \to \widehat{CC}_*(A))). \tag{4.44}$$

This means that $HP_*(A \rtimes G)_{[v]}$ fits into an exact triangle

$$\begin{array}{ccc} \prod_{n=0}^{\infty} \mathbb{H}_n(N_v, \widehat{CC}_*(A)) & \xrightarrow{\cap\alpha - \mathrm{id}} & \prod_{n=0}^{\infty} \mathbb{H}_n(N_v, \widehat{CC}_*(A)) \\ \nwarrow & & \swarrow \delta \\ & HP_*(A \rtimes G)_{[v]} & \end{array} \tag{4.45}$$

which is equivalent to the claim

$$\widehat{\mathbb{H}}_*(N_v, \mathrm{Cone}(\cap c_\alpha - \mathrm{id})) \xrightarrow{\simeq} \mathbb{R}\underset{\overleftarrow{\cap\alpha}}{\lim} \widehat{\mathbb{H}}_*(N_v, \widehat{CC}_*(A)). \tag{4.46}$$

This establishes Theorem 4.8. The proof of Theorem 4.9 is left to the reader. □

REMARK 4.10. If one follows the same strategy in the general case one ends up instead at (4.45) at an exact triangle

(4.47)
$$\begin{array}{ccc} \widehat{\mathbb{H}}_*\left(N_v, \mathrm{Gr}_0(\widehat{CC}_*(A\langle Z_v\rangle \rtimes U : k \rtimes U)_{\{v\}})\right) & \xrightarrow{\delta} & \widehat{\mathbb{H}}_*\left(N_v, \mathrm{Gr}_1(\widehat{CC}_*(A\langle Z_v\rangle \rtimes U : k \rtimes U)_{\{v\}})\right) \\ \nwarrow & & \swarrow \delta \\ & HP_*(A \rtimes G)_{[v]} & \end{array}$$

Acknowledgment

I thank the referee for a very thorough reading of the manuscript, his suggestions for improvement, and for pointing out an error in the first version of the paper.

References

[Bu] Dan Burghelea, *The cyclic homology of the group rings*, Comment. Math. Helv. **60** (1985), no. 3, 354–365, DOI 10.1007/BF02567420. MR814144

[Co] Alain Connes, *Noncommutative differential geometry*, Inst. Hautes Études Sci. Publ. Math. **62** (1985), 257–360. MR823176

[CQ1] Joachim Cuntz and Daniel Quillen, *Algebra extensions and nonsingularity*, J. Amer. Math. Soc. **8** (1995), no. 2, 251–289, DOI 10.2307/2152819. MR1303029

[CQ2] Joachim Cuntz and Daniel Quillen, *Cyclic homology and nonsingularity*, J. Amer. Math. Soc. **8** (1995), no. 2, 373–442, DOI 10.2307/2152822. MR1303030

[KS] Masaki Kashiwara and Pierre Schapira, *Sheaves on manifolds*, Grundlehren der mathematischen Wissenschaften [Fundamental Principles of Mathematical Sciences], vol. 292, Springer-Verlag, Berlin, 1990. With a chapter in French by Christian Houzel, DOI 10.1007/978-3-662-02661-8. MR1074006

[Lo] Jean-Louis Loday, *Cyclic homology*, Grundlehren der mathematischen Wissenschaften [Fundamental Principles of Mathematical Sciences], vol. 301, Springer-Verlag, Berlin, 1992. Appendix E by María O. Ronco, DOI 10.1007/978-3-662-21739-9. MR1217970

[Ni] V. Nistor, *Group cohomology and the cyclic cohomology of crossed products*, Invent. Math. **99** (1990), no. 2, 411–424, DOI 10.1007/BF01234426. MR1031908

[Pu] Michael Puschnigg, *Explicit product structures in cyclic homology theories*, K-Theory **15** (1998), no. 4, 323–345, DOI 10.1023/A:1007748727044. MR1656222

Institut de Mathématiques de Marseille (I2M), UMR 7373 du CNRS, Université d'Aix-Marseille (AMU), France

Email address: michael.puschnigg@univ-amu.fr

Proceedings of Symposia in Pure Mathematics
Volume **105**, 2023
https://doi.org/10.1090/pspum/105/01913

Trace Expansions and Equivariant Traces on an Algebra of Fourier Integral Operators on $\mathbb{R}^n$

Anton Savin and Elmar Schrohe

ABSTRACT. We consider the operator algebra $\mathscr{A}$ on $\mathscr{S}(\mathbb{R}^n)$ generated by the Shubin type pseudodifferential operators, the Heisenberg-Weyl operators and the lifts of the unitary operators on $\mathbb{C}^n$ to metaplectic operators.

With the help of an auxiliary operator in the Shubin calculus, we find trace expansions for these operators in the spirit of Grubb and Seeley. Moreover, we can define a noncommutative residue generalizing that for the Shubin pseudodifferential operators and obtain a class of localized equivariant traces on the algebra.

CONTENTS

1. Introduction

On the space $\mathscr{S}(\mathbb{R}^n)$ of Schwartz functions on $\mathbb{R}^n$ we study the algebra $\mathscr{A}$ of all operators

$$D = \sum R_g T_w A, \tag{1.1}$$

where the sum is finite, the operators A are classical Shubin type pseudodifferential operators of integer order, the operators T_w, $w \in \mathbb{C}^n$, are Heisenberg-Weyl operators, acting on $u \in \mathscr{S}(\mathbb{R}^n)$ by

$$T_w u(x) = e^{ikx - iak/2} u(x-a), \quad w = a - ik,$$

2020 *Mathematics Subject Classification.* Primary 58J40, 58J42.

The second author gratefully acknowledges the support of Deutsche Forschungsgemeinschaft through grant SCHR 319/10-1. The first author is grateful for support to the Russian Foundation for Basic Research, project Nr. 21-51-12006.

and finally the operators R_g, $g \in \mathrm{Sp}(n) \cap \mathrm{O}(2n) \cong \mathrm{U}(n)$ are lifts of unitary operators on $\mathbb{C}^n \cong T^*\mathbb{R}^n$ to elements of the complex metaplectic group on $L^2(\mathbb{R}^n)$. More details will be given, below. We have denoted by $\mathrm{Sp}(n)$ and $\mathrm{O}(2n)$ the groups of symplectic and orthogonal matrices, respectively, on $T^*\mathbb{R}^n \cong \mathbb{R}^n \times \mathbb{R}^n$ and identified $T^*\mathbb{R}^n$ with $\mathbb{C}^n$ via $(x,p) \mapsto p - ix$.

The group $\mathrm{Mp}(n)$ of metaplectic operators on $L^2(\mathbb{R}^n)$ is generated by the operators $S_q = e^{i \operatorname{op}^{\mathrm{W}} q}$, where $\operatorname{op}^{\mathrm{W}} q$ is the Weyl quantization of a quadratic form $q = q(x,p)$ on $\mathbb{R}^n \times \mathbb{R}^n$. The operator S_q defines a symplectic map on $T^*\mathbb{R}^n$ by evaluating the Hamiltonian flow generated by the function q at time 1. In fact, this yields a surjection $\pi : \mathrm{Mp}(n) \to \mathrm{Sp}(n)$. It is well-known that this is a double covering. For more details on the metaplectic group we refer to Leray [**17**] and de Gosson [**6**] as well as to [**27**].

The complex metaplectic group $\mathrm{Mp}^c(n)$ is generated by the operators $S_{q,\phi} = e^{i(\operatorname{op}^{\mathrm{W}} q + \phi)}$ with q as before and $\phi \in \mathbb{R}$. As shown in [**27**, Proposition 1], the map

$$R : \mathrm{U}(n) \to \mathrm{Mp}^c(n), g \mapsto R_g = \pi^{-1}(g)\sqrt{\det g},$$

first defined for g close to the identity $g = I \in \mathrm{U}(n)$ with the choices $\pi^{-1}(I) = I$ and $\sqrt{\det I} = 1$, extends continuously to all of $\mathrm{U}(n)$. These are the operators used above. The metaplectic operators form a special class of Fourier integral operators, compatible with the Shubin pseudodifferential operators in the sense that, for a Shubin symbol a and a metaplectic operator S,

$$S^{-1} \operatorname{op}^{\mathrm{W}}(a) S = \operatorname{op}^{\mathrm{W}}(a \circ \pi(S)).$$

This algebra has been studied before in [**28**], where explicit formulae for the Chern-Connes character in the local index formula of Connes and Moscovici [**4**] were derived and applications to noncommutative tori and toric orbifolds were given. The subalgebra without the Heisenberg-Weyl operators has been used in [**27**] in the context of index theory.

The choice of this algebra is motivated by the quest for an index theory for algebras of quantized canonical transformations, as they were considered in [**25**], [**26**] by the authors and B. Sternin or in [**10**] by Gorokhovsky, de Kleijn and Nest; see also related work by Pflaum, Posthuma and Tang, [**22, 23**]. In [**25**] a Fredholm criterion has been obtained, in [**26**] localized analytic and algebraic indices were introduced and their equality established, while the authors of [**10**] and [**22**] derived an algebraic index theorem. Establishing concrete index expressions, however, seemed difficult. The great advantage of the present algebra is that it allows concrete computations – essentially due to the Mehler formula for the metaplectic operators – while encompassing nontrivial examples as those in [**28**].

In the present article, we are interested in trace expansions for operators in this algebra. We choose an elliptic auxiliary operator H of positive order in the Shubin calculus (a good choice would be the harmonic oscillator $H_0 = \frac{1}{2}(|x|^2 - \Delta)$), for which the resolvent $(H - \lambda)^{-1}$ exists for large λ in a sector about the negative real axis. Given an operator $D = R_g T_w A$ as above, the operator $D(H - \lambda)^{-K}$ will be of trace class for sufficiently large K. Moreover, the trace will be a holomorphic function of λ. We will show that, as $\lambda \to \infty$ in the sector, we have an expansion

$$\text{(1.2)} \qquad \operatorname{Tr}(R_g T_w A(H-\lambda)^{-K}) \sim \sum_{j=0}^{\infty} c_j(-\lambda)^{(2m+\operatorname{ord} A-j)/\operatorname{ord} H-K} + \sum_{j=0}^{\infty}(c'_j \ln(-\lambda)+c''_j)(-\lambda)^{-j-K}$$

with suitable coefficients c_j, c'_j, c''_j, $j = 0, 1, \ldots$. Here, $\operatorname{ord} A$ and $\operatorname{ord} H$ denote the orders of A and H, respectively, and m is the complex dimension of the fixed point set of g. In this expansion, the coefficients c_j and c'_j are 'local' in the sense that they can be determined from the terms in the asymptotic expansion of the symbols of D and H, respectively. They can therefore – at least in principle – be computed explicitly, while the c''_j are 'global'; they depend also on the residual part of the symbol and hence cannot be determined in general. This can be seen as a generalization of the corresponding result by Grubb and Seeley [**12**, Theorem 2.7].

The derivation of this expansion is the heart of the analysis. We obtain from it two other results. The first concerns the operator zeta function $\zeta_{R_g T_w A}(z) = \operatorname{Tr}(R_g T_w A H^{-z})$. Here, we additionally assume $(H-\lambda)^{-1}$ to exist in the closed sector, including $\lambda = 0$. The zeta function then is defined and holomorphic for $\operatorname{Re} z > (2n + \operatorname{ord} A)/\operatorname{ord} H$. Moreover, it has a meromorphic extension to the whole complex plane with at most simple poles. In fact, we have the pole structure

$$\text{(1.3)} \qquad \Gamma(z)\zeta_{R_g T_w A}(z) \sim \sum_{j=0}^{\infty} \frac{\tilde{c}_j}{z-(2m+\operatorname{ord} A-j)/\operatorname{ord} H} + \sum_{j=0}^{\infty}\Big(\frac{\tilde{c}'_j}{(z+j)^2}+\frac{\tilde{c}''_j}{z+j}\Big)$$

with suitable coefficients $\tilde{c}_j, \tilde{c}'_j$ and $\tilde{c}''_j$ related to those above by universal constants. The analysis of such operator zeta functions goes back to Seeley's classical article on complex powers of an elliptic operator [**29**], where $\operatorname{Tr}(H^{-z})$ was studied for an elliptic pseudodifferential operator H. Grubb and Seeley in [**13**] also showed how this expansion can be derived from the expansion of the resolvent trace under rather mild conditions.

Assuming, moreover, that the sector in which the resolvent is holomorphic is larger than the left half plane, we can study the 'heat trace' $\operatorname{Tr}\big(R_g T_w A e^{-tH}\big)$. It exists for all $t > 0$ and has an expansion

$$\text{(1.4)} \quad \operatorname{Tr}(R_g T_w A e^{-tH}) \sim \sum_{j=0}^{\infty} \tilde{c}_j t^{(j-2m-\operatorname{ord} A)/\operatorname{ord} H} + \sum_{j=0}^{\infty}(-\tilde{c}'_j \ln t + \tilde{c}''_j)t^j.$$

Note that the coefficients coincide with those in (1.3). Just as the result on operator zeta functions, this statement is a consequence of the resolvent expansion together with abstract arguments. Heat expansions have been studied by many authors. They provide links to index theory, see e.g. Gilkey's book [**9**] and the references therein, also [**19**] for more recent work. Via Tauberian theorems, they yield eigenvalue asymptotics and, for Dirac and Laplace type operators on manifolds, links to the geometry of the underlying space.

The coefficients c'_0 and $\tilde{c}'_0$ play a remarkable role in these expansions. For one thing, they are the same; moreover, after multiplication by $\operatorname{ord} H$ they are completely independent of the auxiliary operator H. In fact they allow us to define an analog of the noncommutative residue for our algebra. Restricted to the Shubin pseudodifferential operators it coincides with the residue introduced by Boggiatto and Nicola [**1**] for more general anisotropic pseudodifferential operators on $\mathbb{R}^n$.

We recall that the noncommutative residue for the algebra of classical pseudodifferential operators on a closed manifold, discovered independently by Wodzicki [**31**] and Guillemin [**14**] defines the unique trace – up to multiples – on this algebra (see e.g. [**8**] for a simple proof). The same is true for the residue of Boggiatto and Nicola. Meanwhile, analogs of this residue trace have been discovered in many contexts, see e.g. [**5, 18, 20, 24**].

Our algebra $\mathscr{A}$ is the algebraic twisted crossed product of the algebra Ψ of all Shubin type pseudodifferential operators with the group $\mathbb{C}^n \rtimes \mathrm{U}(n)$ via the representations of $\mathbb{C}^n$ by the operators T_w and of $\mathrm{U}(n)$ by the operators R_g. Given a discrete subgroup G of $\mathbb{C}^n \rtimes \mathrm{U}(n)$ and an element (w_0, g_0) of G, we define the residue trace of the operator $D = \sum R_g T_w A$ localized at the conjugacy class $\langle (w_0, g_0)\rangle$ by

$$\operatorname{res}_{\langle (w_0,g_0)\rangle} D = \operatorname{ord} H \sum_{(w,g)\in\langle (w_0,g_0)\rangle} c_0'(R_g T_w A) \tag{1.5}$$

and show that these are actually traces on the algebra $\mathscr{A}$.

The present article in this sense continues the work of Dave [**5**]. He defined corresponding equivariant traces for the algebra $\Psi_{\mathrm{cl}}(M) \rtimes \Gamma$ of operators on a closed manifold M, generated by the classical pseudodifferential operators and a finite group Γ of diffeomorphisms of M. He also computed the cyclic homology in terms of the de Rham cohomology of the fixed point manifolds S^*M^g, $g \in \Gamma$, and deduced that there are no other traces (noting that there are actually two independent traces, if $\dim M^g = 1$) – a question that remains to be explored in the present situation. In the same vein, Perrot [**21**] studied the local index theory for shift operators associated to non-proper and non-isometric actions of Lie groupoids; he also computed the residue of the corresponding operator zeta functions in zero.

It is shown in [**15**, Théorème 2.6.7] that the Shubin pseudodifferential operators are stable under conjugation with more general Fourier integral operators. One could therefore think of studying corresponding algebras $\Psi \rtimes G'$ for a group G' by (suitable) Fourier integral operators. However, it seems not clear how to obtain then formulae as concrete as those derived here.

Organisation of the article. We recall in Section 2 the essential facts about the operator classes. Section 3 contains the statements of the main results: Theorem 3.2 on the resolvent expansion, Theorem 3.3 on the structure of the singularities of the operator zeta function, Theorem 3.4 on the 'heat trace' expansion and Theorem 3.6 on the form of the noncommutative residue. Theorems 3.3 and 3.4 follow from Theorem 3.2 by abstract results in [**13**]. Section 4 contains a sketch of the proof of the resolvent expansion, while Section 5 treats the residue; full details will be given elsewhere. The appendix contains the essential material on the weakly parametric calculus of Grubb and Seeley.

2. The Operators

Shubin type pseudodifferential operators. A smooth function a on $\mathbb{R}^n \times \mathbb{R}^n$ is called a (Shubin type) pseudodifferential symbol of order $\mathfrak{a} \in \mathbb{R}$, provided that, for all multi-indices α, β, it satisfies the estimates

$$D_p^\alpha D_x^\beta a(x,p) \lesssim (1 + |x|^2 + |p|^2)^{(\mathfrak{a}-|\alpha|-|\beta|)/2}, \quad (x,p) \in \mathbb{R}^n \times \mathbb{R}^n.$$

From such a symbol, we can define the associated pseudodifferential operators $\operatorname{op} a$ and $\operatorname{op}^{\mathrm{W}} a$, respectively, by standard and Weyl quantization.

By $\Psi^{\mathfrak{a}}$ we denote the space of all Shubin type pseudodifferential operators of order $\mathfrak{a}$ and by Ψ the union over all $\mathfrak{a} \in \mathbb{R}$. An important example is the operator

$$H_0 = \frac{1}{2}(|x|^2 - \Delta) \in \Psi^2, \tag{2.1}$$

the so-called harmonic oscillator, which is selfadjoint and positive on $L^2(\mathbb{R}^n)$.

These operators act on the Sobolev spaces $\mathcal{H}^s(\mathbb{R}^n)$, $s \in \mathbb{R}$, consisting of all tempered distributions u such that $\operatorname{op}((1+|x|^2+|p|^2)^{s/2})u \in L^2(\mathbb{R}^n)$: In fact, $\Psi^{\mathfrak{a}} \hookrightarrow \mathscr{B}(\mathcal{H}^s(\mathbb{R}^n), \mathcal{H}^{s-\mathfrak{a}}(\mathbb{R}^n))$ for all $s, \mathfrak{a} \in \mathbb{R}$. In particular, zero order operators are bounded on $L^2(\mathbb{R}^n)$ and those of order $< -2n$ are of trace class.

We will work here mostly with classical symbols, i.e. symbols a of order $\mathfrak{a} \in \mathbb{Z}$ that have an asymptotic expansion

$$a \sim \sum_{j=0}^{\infty} a_{\mathfrak{a}-j}$$

with $a_{\mathfrak{a}-j}$ smooth and positively homogeneous in (x,p) of degree $\mathfrak{a}-j$ for $|(x,p)| \geq 1$. We call $a_{\mathfrak{a}}$ the principal symbol and write $\Psi^{\mathfrak{a}}_{\mathrm{cl}}$ and Ψ_{cl} for the classes of operators with classical symbols, respectively.

A Egorov type theorem holds: Given an element S in the metaplectic group $\mathrm{Mp}(n)$ and $A = \operatorname{op}^{\mathrm{W}} a$ a Weyl-quantized classical Shubin pseudodifferential operator with symbol a, then

$$S^{-1} \operatorname{op}^{\mathrm{W}}(a) S = \operatorname{op}^{\mathrm{W}}(a \circ \pi(S)), \tag{2.2}$$

where $\pi(S) \in \mathrm{Sp}(n)$ is the canonical transformation associated with S, see [**6**, Theorem 7.13]. This implies that conjugation by metaplectic operators leaves each of the classes $\Psi^{\mathfrak{a}}$ invariant, since every Shubin type pseudodifferential operator can be written in Weyl-quantized form. As the principal symbol of $\operatorname{op}^{\mathrm{W}}(a)$ coincides with that of $\operatorname{op}(a)$, we find the following relation for principal symbols:

$$\sigma(S^{-1}AS) = \sigma(A) \circ \pi(S). \tag{2.3}$$

The above operators need not be scalar, it is often important to also consider systems of operators over $\mathbb{R}^n$, see e.g. [**28**].

Heisenberg-Weyl operators. For $w = a - ik \in \mathbb{C}^n$, $a, k \in \mathbb{R}^n$, we define

$$T_w u(x) = e^{ikx - iak/2} u(x-a). \tag{2.4}$$

We note that, for $v, w \in \mathbb{C}^n$,

$$T_v T_w = e^{-i\operatorname{Im}(v,w)/2} T_{v+w}, \qquad \text{where } (v,w) = v\overline{w}, \tag{2.5}$$

so that $T_w^{-1} = T_{-w}$. Given a Shubin symbol $a = a(x,p)$ and $w = a - ik$,

$$T_w^{-1} \operatorname{op}(a) T_w = \operatorname{op}(a(x+a, p+k)). \tag{2.6}$$

In particular, conjugation by Heisenberg-Weyl operators leaves the spaces $\Psi^{\mathfrak{a}}_{\mathrm{cl}}$, $\mathfrak{a} \in \mathbb{Z}$, invariant and preserves the principal symbols.

Metaplectic operators. Since $\mathrm{U}(n)$ is generated by $\mathrm{O}(n)$ and $\mathrm{U}(1)$, see [**27**, Lemma 1], we can define a unitary representation $g \mapsto R_g$ of $\mathrm{U}(n)$ by complex metaplectic operators $R_g \in \mathrm{U}(L^2(\mathbb{R}^n))$ by the following two assignments:

(i) $R_g u(x) = u(g^{-1}x)$, if $g \in \mathrm{O}(n) \subset \mathrm{U}(n)$. Then R_g is a so-called shift operator;

(ii) $R_g u(x) = e^{i\varphi(1/2-H_1)}u(x)$, where $g = \mathrm{diag}(e^{i\varphi}, 1, \ldots, 1)$, and $H_1 = \frac{1}{2}(x_1^2 - \partial_{x_1}^2)$ is the one-dimensional harmonic oscillator. R_g is a so-called fractional Fourier transform with respect to x_1. According to Mehler's formula [**16**, Section 4] it is given by

$$\begin{aligned} R_g u(x) = &\sqrt{\frac{1 - i \operatorname{ctg}\varphi}{2\pi}} \\ &\times \int \exp\left(i \left((x_1^2 + y_1^2)\frac{\operatorname{ctg}\varphi}{2} - \frac{x_1 y_1}{\sin\varphi} \right) \right) u(y_1, x_2, \ldots, x_n) dy_1. \end{aligned}$$

Here we choose the square root so that the real part is positive.

In (ii) we may assume that $0 < \varphi < 2\pi$, $\varphi \neq \pi$, as the cases $\varphi = 0, \pi$ are covered by (i): For $\varphi = 0$, R_g is the identity, and for $\varphi = \pi$, it is the reflection in the first component. Moreover, for $g = (\exp(i\pi/2), 1, \ldots, 1)$, R_g coincides with the Fourier transform in the first variable, see the discussion on p. 427 in [**16**]. By Theorem 7.13 in [**6**]

$$R_g T_w R_g^{-1} = T_{gw}. \tag{2.7}$$

The operator algebra. In (1.1) we introduced the operators D given as finite sums

$$D = \sum_j R_{g_j} T_{w_j} A_j,$$

where $g_j \in \mathrm{U}(n)$, $w_j \in \mathbb{C}^n$, $A_j \in \Psi(\mathbb{R}^n)$, and the sum is finite. That these indeed form an algebra follows from the fact that

$$\begin{aligned} (R_{g_1} T_{w_1} A_1)(R_{g_2} T_{w_2} A_2) &= R_{g_1} R_{g_2} (R_{g_2}^{-1} T_{w_1} R_{g_2}) T_{w_2} (T_{w_2}^{-1} (R_{g_2}^{-1} A_1 R_{g_2}) T_{w_2}) A_2 \\ &= R_{g_1 g_2} T_{g_2^{-1} w_1} T_{w_2} A_3 A_2, \end{aligned}$$

where $A_3 = T_{w_2}^{-1}(R_{g_2}^{-1} A_1 R_{g_2}) T_{w_2}$ is a Shubin type operator of the same order as A_1 as we noticed in (2.6) and after (2.2), and $T_{g_2^{-1}w_1} T_{w_2} = e^{-i \operatorname{Im}\langle g_2^{-1} w_1, w_2 \rangle} T_{g_2^{-1} w_1 + w_2}$ is a multiple of a Heisenberg-Weyl operator.

3. Main Results

Let E be a (trivial) vector bundle over $\mathbb{R}^n$ and h a classical Shubin symbol of order $\mathfrak{m} > 0$ whose principal symbol is positive and scalar, i.e. a multiple of id_E, positively homogeneous of degree $\mathfrak{m}$ for $|x|^2 + |p|^2 \geq 1$. Denote by H the associated pseudodifferential operator acting on sections of E.

For $0 < \delta \leq \pi$ write S_δ for the sector

$$S_\delta = \{\lambda \in \mathbb{C} \setminus \{0\} \mid |\arg \lambda - \pi| < \delta\} \tag{3.1}$$

and $U_r(0)$ for the disk of radius $r > 0$ about $0 \in \mathbb{C}$. Standard pseudodifferential techniques show:

PROPOSITION 3.1. *For every $\delta < \pi$, $H - \lambda$ is invertible for all $\lambda \in S_\delta$ with $|\lambda|$ sufficiently large. The function $\lambda \mapsto (H-\lambda)^{-1}$ is meromorphic in $S_\delta \cup U_r(0)$ for suitably small $r > 0$. In particular, we may assume that $H - \lambda$ is invertible for all $\lambda \in S_\delta$ after replacing H by $H + c$ for suitable $c > 0$.*

We shall establish the following results:

THEOREM 3.2. *Fix $g \in \mathrm{U}(n)$, $w \in \mathbb{C}^n$, and a pseudodifferential operator $A \in \Psi$ of order* ord A. *Choose $K \in \mathbb{N}$ so large that $-K \operatorname{ord} H + \operatorname{ord} A < -2n$ and fix $\delta < \pi$. Then*

(a) *The operator $R_g T_w A(H-\lambda)^{-K}$ is of trace class for all λ in S_δ, $|\lambda|$ sufficiently large.*
(b) *The function $\lambda \mapsto \operatorname{Tr}(R_g T_w A (H-\lambda)^{-K})$ is meromorphic in S_δ and near $\lambda = 0$. It is $O(|\lambda|^{-\varepsilon})$ with suitable $\varepsilon > 0$ for $\lambda \to \infty$ in S_δ.*
(c) *As $\lambda \to \infty$ in the sector S_δ we have the expansion* (1.2).
(d) *If the fixed point set of the affine mapping $\mathbb{C}^n \to \mathbb{C}^n, v \mapsto gv + w$ is empty, then the trace is $O(\lambda^{-\infty})$.*
(e) *The coefficient c'_0 is independent of the choice of H (subject to the properties stated above) when multiplied by the factor* ord H.

A sketch of the proof of Theorem 3.2 will be given in Section 4, below. Statements (a) and (b) follow immediately from Proposition 3.1 and the fact that the embedding $\mathcal{H}^{2n+\varepsilon}(\mathbb{R}^n) \hookrightarrow L^2(\mathbb{R}^n)$ is trace class for every $\varepsilon > 0$. Deriving the expansion (1.2) in (c) is the crucial task. Statements (d) and (e) will be proven along the way of showing (c).

As a consequence, we obtain two more results. Theorems 3.3 and 3.4 follow from Theorem 3.2 with the help of Propositions 2.9 and 5.1 in [**13**].

THEOREM 3.3. *With the notation of Theorem 3.2 assume, moreover, that $H - \lambda$ is invertible for all $\lambda \in S_\delta \cup U_r(0)$ for some $r > 0$. Then*

(a) *The function $z \mapsto \zeta_{R_g T_w A}(z) := \operatorname{Tr}(R_g T_w A H^{-z})$ is defined and holomorphic in the half-plane $\{z \in \mathbb{C} : \operatorname{ord} H \operatorname{Re} z > 2n + \operatorname{ord} A\}$.*
(b) *$\zeta_{R_g T_w A}$ has a meromorphic continuation to $\mathbb{C}$ with at most simple poles in the points $(2m + \operatorname{ord} A - j)/\operatorname{ord} H$, where m is the complex dimension of the fixed point set of g. The function $\Gamma\zeta_{R_g T_w A}$ has the pole structure* (1.3).
(c) *The coefficients $\tilde{c}_j, \tilde{c}'_j$ and $\tilde{c}''_j$ in* (1.3) *are related to the coefficients c_j, c'_j and c''_j in* (1.2) *by universal constants. In particular, we have*

$$\operatorname{Res}_{z=0}\zeta_{R_g T_w A} = c'_0 = \tilde{c}'_0.$$

(d) *If the fixed point set of the affine mapping $\mathbb{C}^n \to \mathbb{C}^n, v \mapsto gv + w$ is empty, then $\zeta_{R_g T_w A}$ has no poles.*
(e) *$\zeta_{R_g T_w A}$ has rapid decay along vertical lines $z = c + it$, $t \in \mathbb{R}$:*

$$|\zeta_{R_g T_w A}(z)\Gamma(z)| \le C_N (1 + |z|)^{-N}, \quad \textit{for all } N \ge 0, \quad |\operatorname{Im} z| \ge 1, \tag{3.2}$$

uniformly for c in compact intervals.

THEOREM 3.4. *We use the notation in Theorem 3.2 and assume additionally that $H - \lambda$ is invertible for $\lambda \in \overline{S}_\delta$ for some $\delta \in]\pi/2, \pi]$. Then:*

(a) *The function $t \mapsto \operatorname{Tr}(R_g T_w A \exp(-tH))$ is defined for all $t > 0$.*
(b) *As $t \to 0^+$ it has the asymptotic expansion* (1.4) *with the same coefficients as for the zeta function in Theorem 3.3.*

The noncommutative residue. Let G be a subgroup of $\mathbb{C}^n \rtimes \mathrm{U}(n)$.

DEFINITION 3.5. For $w_0 \in \mathbb{C}^n$, $g_0 \in \mathrm{U}(n)$ such that $(w_0, g_0) \in G$ we define the noncommutative residue $\mathrm{res}_{\langle(w_0,g_0)\rangle}$ localized at the conjugacy class $\langle(w_0, g_0)\rangle$ in G by

$$\begin{aligned}\mathrm{res}_{\langle(w_0,g_0)\rangle} D &= \operatorname{ord} H \sum_{(w,g)\in\langle(w_0,g_0)\rangle} \mathrm{Res}_{z=0}\zeta_{A,g,w}(z)\\ &= \operatorname{ord} H \sum_{(w,g)\in\langle(w_0,g_0)\rangle} c_0'(R_gT_wA) = \operatorname{ord} H \sum_{(w,g)\in\langle(w_0,g_0)\rangle} \tilde{c}_0'(R_gT_wA)\end{aligned}$$

with the coefficients c_0' and $\tilde{c}_0'$ introduced in Theorem 3.2 and Theorem 3.3, respectively, and the order $\operatorname{ord} H$ of H. The expression is independent of H by Theorem 3.2(e).

While the definition in terms of the zeta residue or the coefficient in the heat trace expansion is more common, that in terms of the coefficient in the resolvent expansion is slightly more general, since we need less assumptions on the auxiliary operator to define it.

One can compute the residue explicitly. Consider the case of diagonal g:

$$g = \operatorname{diag}\Big(\underbrace{e^{i\varphi_1},\dots,e^{i\varphi_{m_1}}}_{m_1},\underbrace{i,\dots,i}_{m_2},\underbrace{-i,\dots,-i}_{m_3},\underbrace{-1,\dots,-1}_{m_4},\underbrace{1,\dots,1}_{m_5}\Big) \in \mathrm{U}(n), \tag{3.3}$$

where $\varphi_j \in \,]-\pi,\pi[\, \setminus \pi\mathbb{Z}/2$ and $m_5 = \dim_{\mathbb{C}}(\mathbb{C}^n)^g$. For notational convenience we will also write $\varphi_j = \pi/2$ for $j = m_1+1,\dots,m_1+m_2$ and $\varphi_j = -\pi/2$ for $j = m_1+m_2+1,\dots,m_1+m_2+m_3$.

THEOREM 3.6. *For $D = R_gT_wA$ with g as above, $w = a - ik \in \mathbb{C}^n$ and $A \in \Psi^{-2m_5}$ we find the explicit formula:*

$$\begin{aligned}\mathrm{res}_{\langle(w,g)\rangle} D &= (2\pi)^{-n+m_2+m_3} \prod_{j=1}^{m_1} \sqrt{1+i\operatorname{tg}\varphi_j}\\ &\quad\times \prod_{j=1}^{m_1+m_2+m_3} e^{\frac{i}{4}\operatorname{ctg}(\varphi_j/2)(k_j^2+a_j^2)} \int_{\mathbb{S}^{2m_5-1}} \operatorname{tr}_E a_{-2m_5}(\theta)\, dS,\end{aligned} \tag{3.4}$$

where we choose the square root in the right half plane, $\mathbb{S}^{2m_5-1}$ is the unit sphere in the $+1$-eigenspace of the matrix g, and a_{-2m_5} is the homogeneous component of degree $-2m_5$ in the asymptotic expansion of the symbol of A.

In particular, the residue vanishes for $A \in \Psi^{\mathfrak{a}}$ with $\mathfrak{a} < -2m_5$. For $\mathfrak{a} > -2m_5$, additional terms containing derivatives of components $a_{\mathfrak{a}-j}$, $\mathfrak{a} - j > -2m_5$ enter. For details see Section 5.

REMARK 3.7. For an arbitrary element $g \in \mathrm{U}(n)$, there exists a unitary $u \in \mathrm{U}(n)$ such that $g = ug_0u^{-1}$, where g_0 is diagonal and unitary. Then

$$\operatorname{Tr}(R_gT_wAH^{-z}) = \operatorname{Tr}(R_{g_0}T_{w'}A'(H')^{-z}),$$

where $A' = R_u^{-1}AR_u$ is a Shubin type operator by Egorov's theorem, $T_{w'} = R_u^{-1}T_wR_u$ with $w' = u^{-1}w$, and $H' = R_{u^{-1}}HR_u$ is an admissible auxiliary operator of the same order, so that Theorem 3.6 applies.

In case $H = H_0$ is the harmonic oscillator of (2.1), we even have $H' = H_0$, since H_0 commutes with R_u.

THEOREM 3.8. *For every choice of a conjugacy class $\langle(w_0, g_0)\rangle$, the functional $\mathrm{res}_{\langle(w_0,g_0)\rangle}$ is a trace.*

PROOF. We use a modification of an argument given by Dave [**5**]. By linearity it is sufficient to consider two elements $B_1 = R_{g_1}T_{w_1}A_1$ and $B_2 = R_{g_2}T_{w_2}A_2$. Equations (2.5) and (2.7) imply that

$$R_{g_1}T_{w_1}R_{g_2}T_{w_2} = R_{g_1}R_{g_2}T_{g_2^{-1}w_1}T_{w_2} = e^{-i\,\mathrm{Im}(g_2^{-1}w_1,w_2)}R_{g_1g_2}T_{g_2^{-1}w_1+w_2},$$

so we want to show that

$$0 = \mathrm{res}_{\langle(g_2^{-1}w_1+w_2,g_1g_2)\rangle}([B_1,B_2]) = 2\mathrm{Res}_{z=0}([B_1,B_2]H_0^{-z}).$$

We have chosen here H_0 as auxiliary operator; we know from Theorem 3.2(e) that the result does not depend on the choice.

For large $\mathrm{Re}(z)$, the operators $B_1B_2H_0^{-z}$, $B_2B_1H_0^{-z}$ and $B_1H_0^{-z}B_2$ are trace class operators on $L^2(\mathbb{R}^n)$ and therefore

$$\mathrm{Tr}([B_1,B_2]H_0^{-z}) = \mathrm{Tr}(B_1[B_2,H_0^{-z}]).$$

The fact that H_0 and hence H_0^{-z} commutes with R_{g_2} implies that

$$\begin{aligned}[B_2,H_0^{-z}] &= [R_{g_2}T_{w_2}A_2,H_0^{-z}] = R_{g_2}[T_{w_2}A_2,H_0^{-z}] + [R_{g_2},H_0^{-z}]T_{w_2}A_2\\ &= R_{g_2}T_{w_2}[A_2,H_0^{-z}] + R_{g_2}[T_{w_2},H_0^{-z}]A_2.\end{aligned}$$

Let us show that the contribution to the residue from both terms vanishes.

For the first we study the asymptotic expansion of the symbol of the commutator $[A_2, H_0^{-z}]$. The principal symbol of H_0^{-z} is proportional to $h(x,p)^{-z} = (|x|^2+|p|^2)^{-z}$, while the terms of degree $-2z-j$ in the asymptotic expansion are linear combinations of expressions of the form

$$(-z)(-z-1)\cdots(-z-k+1)d_{k,j}(x,p)h(x,p)^{-z-k}, \quad 1\le k\le 2j,$$

where $d_{k,j}$ is a polynomial in x and p of degree $2k-j$, see e.g. [**29**, Equation (25)]. Hence for arbitrary N, the commutator can be written in the form

$$[A_2,H_0^{-z}] = \mathrm{op}(q_N(z)+r_N(z)),$$

where

- $q_N(z)$ is a linear combination of terms of the form
$$z\cdots(z+k-1)\partial_x^\alpha\partial_p^\beta a(x,p)d(x,p)(|x|^2+|p|^2)^{-k-z}$$
with suitable derivatives of a, polynomials d in x and p and $k\ge 1$.
- $z\mapsto r_N(z)$ is a holomorphic family of symbols of order $-N$.

We can therefore write, for suitable N',

$$[A_2,H_0^{-z}] = \sum_{k=1}^{N'} z^kC_kH_0^{-z} + R_N(z).$$

with Shubin pseudodifferential operators C_k of order $\le \mathrm{ord}\,A_2$ and a family $z\mapsto R_N(z)$ of trace class operators that is holomorphic near $z=0$. Since $\mathrm{Tr}(BH^{-z})$ has at most a simple pole in $z=0$ for any operator B in the calculus, we conclude that the contribution from the terms under the summation is zero. On the other hand, if N is sufficiently large, then $\mathrm{Tr}(BR_N(z))$ will also be holomorphic in $z=0$ and therefore not contribute to the residue. This shows that the contribution from the first term is zero.

For the second, we recall that

$$H_0^{-z} = \frac{1}{2\pi i}\int_{\mathscr{C}} \lambda^{-z}(\lambda - H_0)^{-1}\,d\lambda,$$

where $\mathscr{C}$ is a contour surrounding the spectrum of H_0; specifically we can choose here $\mathscr{C} = \{re^{i\vartheta} : \infty > r \geq 0\} \cup \{re^{-i\vartheta} : 0 \leq r < \infty\}$ for some angle $\vartheta > 0$. The operator T_{w_2} is bounded on $L^2(\mathbb{R}^n)$ and

$$T_{w_2}^{-1}(\lambda - H_0)^{-1}T_{w_2} \quad = \quad (T_{w_2}^{-1}(\lambda - H_0)T_{w_2})^{-1} = (\lambda - T_{w_2}^{-1}H_0T_{w_2})^{-1},$$

so that $T_{w_2}^{-1}H_0^{-z}T_{w_2} = H_{w_2}^{-z}$ with $H_{w_2} = T_{w_2}^{-1}H_0T_{w_2}$. We conclude that

$$[T_{w_2}, H_0^{-z}] = T_{w_2}(H_0^{-z} - T_{w_2}^{-1}H_0^{-z}T_{w_2}) = T_{w_2}(H_0^{-z} - H_{w_2}^{-z}).$$

As a consequence,

$$\begin{aligned} \operatorname{Tr}(B_1R_{g_2}[T_{w_2}, H_0^{-z}]A_2) &= \operatorname{Tr}(A_2B_1R_{g_2}[T_{w_2}, H_0^{-z}]) \\ &= \operatorname{Tr}(A_2B_1R_{g_2}T_{w_2}H_0^{-z}) - \operatorname{Tr}(A_2B_1R_{g_2}T_{w_2}H_{w_2}^{-z}). \end{aligned}$$

Now the principal symbol of H_{w_2} coincides with that of H_0; it is therefore also an admissible auxiliary operator. Since we know that the residue is independent of the choice of the auxiliary operator, we see that the difference of the residues must vanish. □

4. The Resolvent Expansion

Recall that, by assumption, H is a classical Shubin type operator of order $\operatorname{ord} H = \mathfrak{m} \in \mathbb{Z}_{>0}$. Its principal symbol $h_{\mathfrak{m}} = h_{\mathfrak{m}}(x,p)$ is scalar and, moreover, strictly positive and homogeneous of degree $\mathfrak{m}$ in (x,p) for $|(x,p)| \geq 1$. We write $-\lambda = \mu^{\mathfrak{m}}$ for μ in the sector

$$S = \{z \in \mathbb{C} \setminus \{0\} : |\arg z| < \delta/\mathfrak{m}\}$$

so that $h_{\mathfrak{m}}(x,p) + \mu^{\mathfrak{m}}$ is invertible for all $\mu \in S$.

Given a Shubin pseudodifferential operator A of order $\mathfrak{a}$, choose $K \in \mathbb{N}$ so large that $\mathfrak{a} - K\mathfrak{m} < -2n$, so that the operators $A(H+\mu^{\mathfrak{m}})^{-K}$ and $R_gT_wA(H+\mu^{\mathfrak{m}})^{-K}$ are of trace class on $L^2(\mathbb{R}^n)$ for every μ in the sector S, $|\mu|$ sufficiently large.

Denote by $q(y,p;\mu)$ the (complete) symbol of $A(H+\mu^{\mathfrak{m}})^{-K}$ in y-form, i.e.

$$\Big(A(H+\mu^{\mathfrak{m}})^{-K}\Big)u(x) = (2\pi)^{-n}\int e^{i\langle x-y,p\rangle}q(y,p;\mu)u(y)dydp. \tag{4.1}$$

For $w = a - ik$ and a diagonal element g as in (3.3), consider the function

$$\mu \mapsto \operatorname{Tr}(R_gT_wA(H+\mu^{\mathfrak{m}})^{-K})$$

Our first step and the basis of the subsequent analysis is the proof of the following proposition.

PROPOSITION 4.1. *Up to terms that are $O(|\mu|^{-\infty})$ as $\mu \to \infty$ in S,*

$$\operatorname{Tr}(R_gT_wA(H+\mu^{\mathfrak{m}})^{-K}) = C_{\text{res}}\int e^{i\phi(u,v)}\operatorname{tr}_E q(B(u,v)-b_0, y,\eta;\mu)\,du\underline{dv}dyd\eta.$$

Here C_{res} is the constant in front of the integral in Theorem 3.6 above, tr_E is the trace in the vector bundle E,

$$(x_1,p_1,\dots x_{n-m_5},p_{n-m_5}) = B(u,v) - b_0, \quad u,v \in \mathbb{R}^{n-m_5}$$

is an affine linear change of coordinates.

As a mapping $(u_j, v_j) \mapsto (x_j, p_j)$, $j = 1, \dots, n - m_5$ *the matrix* B *is a diagonal block matrix with entries* $\frac{1}{\sqrt{2}}\begin{pmatrix} 1 & 1 \\ -1 & 1 \end{pmatrix}$ *for* $j \neq m_1 + 1, \dots, m_1 + m_2 + m_3$, *for the latter it is the identity. In particular,* B *is orthogonal. Moreover,*

$$b_0 = (u_{0,1}, v_{0,1}, \dots, u_{0,n-m_5}, v_{0,n-m_5})$$

with

$$(u_{0,j}, v_{0,j}) = \frac{1}{2}\left(a_j - k_j \operatorname{ctg}\frac{\varphi_j}{2}, a_j \operatorname{ctg}\frac{\varphi_j}{2} + k_j\right);$$

y *and* η *comprise the variables* $y = (x_{n-m_5+1}, \dots x_n)$, $\eta = (p_{n-m_5+1}, \dots p_n)$. *The notation '*$d\underline{v}$*' indicates that the integration is not over* $v_{m_1+1}, \dots v_{m_1+m_2+m_3}$; *for these variables,* q *is evaluated at the points* $v_j = \sin\varphi_j u_{0,j}$.

The phase ϕ *is given by*

$$\begin{aligned} \phi(u, v) &= -\sum_{j=1}^{m_1}(\lambda_j^- u_j^2 + \lambda_j^+ v_j^2) - \sum_{j=m_1+1}^{m_1+m_2} u_j^2 + \sum_{j=m_1+m_2+1}^{m_1+m_2+m_3} u_j^2 \\ &\quad - \sum_{j=m_1+m_2+m_3+1}^{n-m_5} (-u_j^2 + v_j^2) \end{aligned} \tag{4.2}$$

with $\lambda_j^{\pm}$ *defined in* (4.3), *below.*

That b_0 has no entries for $j = n - m_5 + 1, \dots, n$ is due to the fact that the expansion is $O(|\mu|^{-\infty})$ whenever $a_j \neq 0$ or $k_j \neq 0$ for some $j > n - m_5$; hence we assume $a_j = k_j = 0$ for all $j > n - m_5$.

Proof. The proof is lengthy; here is a sketch:

Step 1. A computation very similar to that for Equations (76) and (77) in the proof of [**28**, Theorem 5] with explicit constants shows that the trace is $O(\mu^{-\infty})$ unless $a_j = k_j = 0$ for $j > n - m_5$ and then

$$\begin{aligned} &\operatorname{Tr}(R_g T_w A(H + \mu^{\mathrm{m}})^{-K}) \\ &= (2\pi)^{-n+m_2+m_3} e^{-i\langle a,k\rangle/2} \prod_{j=1}^{m_1} \sqrt{1 + i \operatorname{tg}\varphi_j} \\ &\quad \times \int e^{i\phi_1(x,p')} \operatorname{tr}_E q(x, p', p''; \mu)_{|p_j''=\pm x_j - k_j}\, dp'dx. \end{aligned}$$

Here, the notation $q(x, p', p''; \mu)_{|p_j''=\pm x_j - k_j}$ indicates that we evaluate p_j'' at $x_j - k_j$ for $j = m_1 + 1, \dots, m_1 + m_2$ and at $-x_j - k_j$ for $j = m_1 + m_2 + 1, \dots, m_1 + m_2 + m_3$, and the phase ϕ_1 is given by

$$\begin{aligned} \phi_1(x, p) &= -\sum_{j=1}^{m_1} \frac{\operatorname{tg}\varphi_j}{2}\left(x_j^2 - 2x_j p_j\left(\frac{1 - \cos\varphi_j}{\sin\varphi_j}\right) + p_j^2 + 2x_j \frac{k_j}{\sin\varphi_j}\right. \\ &\quad \left. + 2p_j\left(\frac{a_j}{\operatorname{tg}\varphi_j} + k_j\right) + k_j^2\right) - \sum_{j=m_1+1}^{m_1+m_2+m_3} (x_j + a_j)\left(\frac{x_j}{\sin\varphi_j} - k_j\right) \\ &\quad - \sum_{j=m_1+m_2+m_3+1}^{n-m_5} (2x_j p_j + a_j p_j + k_j x_j), \end{aligned}$$

noting that

$$\begin{aligned} \operatorname{ctg}\varphi_j - \frac{1}{\sin\varphi_j\cos\varphi_j} &= -\operatorname{tg}\varphi_j \text{ and} \\ \frac{1}{\cos\varphi_j} - 1 &= \operatorname{tg}\varphi_j \frac{1-\cos\varphi_j}{\sin\varphi_j}. \end{aligned}$$

Step 2. We next make several changes of coordinates of the form

$$\begin{pmatrix} x_j \\ p_j \end{pmatrix} = B_j \begin{pmatrix} x_j \\ p_j \end{pmatrix} - \begin{pmatrix} u_{0,j} \\ v_{0,j} \end{pmatrix}, \quad j = 1, \dots, n - m_5.$$

For $j = 1, \dots, m_1$ we let $B_j = \frac{1}{\sqrt{2}}\begin{pmatrix} 1 & 1 \\ -1 & 1 \end{pmatrix}$,

$$\begin{aligned} \alpha_j &= -2\frac{1-\cos\varphi_j}{\sin\varphi_j}, \quad \beta_j = \frac{2k_j}{\sin\varphi_j}, \quad \gamma_j = \frac{2a_j}{\operatorname{tg}\varphi_j} + 2k_j, \text{ and} \\ u_{0,j} &= \frac{2\beta_j - \gamma_j\alpha_j}{4-\alpha_j^2} = \frac{a_j - k_j\operatorname{ctg}(\varphi_j/2)}{2}, \quad v_{0,j} = \frac{2\gamma_j - \beta_j\alpha_j}{4-\alpha_j^2} = \frac{a_j\operatorname{ctg}(\varphi_j/2) + k_j}{2}. \end{aligned}$$

For $j = m_1+1, \dots, m_1+m_2+m_3$, where $\varphi_j = \pm\frac{\pi}{2}$, we let $u_j = x_j + \frac{a_j - k_j\sin\varphi_j}{2}$ and $v_j = p_j + \frac{a_j\sin\varphi_j + k_j}{2}$.

Finally for $j = m_1+m_2-m_3+1, \dots, n-m_5$, we let B_j be as above, $u_{0,j} = a_j/2$ and $v_j = k_j/2$.

This leads to the expression for $\operatorname{Tr}(R_g T_w A(H+\mu^{\mathrm{m}})^{-K})$ in the statement for the phase (4.2) with

$$\lambda_j^\pm = \frac{\operatorname{tg}\varphi_j}{4}(2\pm\alpha_j) = \frac{1}{2}\frac{\sin\varphi_j \mp (1-\cos\varphi_j)}{\cos\varphi_j}, \; j = 1, \dots, m_1 \tag{4.3}$$

(note that $\lambda_j^\pm \neq 0$ for $j = 1, \dots, m_1$) and the constant C_{res}. □

A Simplification: Omitting the Heisenberg-Weyl Operators. If $a_j = k_j = 0$ for all j, the above formula becomes

$$\begin{aligned} \operatorname{Tr}(R_g A(H+\mu^{\mathrm{m}})^{-K}) &= (2\pi)^{-n+m_2+m_3}\prod_{j=1}^{m_1}\sqrt{1+i\operatorname{tg}\varphi_j} \\ &\times \int e^{i\phi(u,v)} \operatorname{tr}_E q(B\begin{pmatrix} u \\ v \end{pmatrix}, x'''', p''''; \mu)\, du dv dx'''' dp'''' \end{aligned}$$

with u, v subsuming the above coordinates $u', u'', u''', v', v'', v'''$, where v'' is evaluated at $v_j'' = \sin\varphi_j(u_j'' - \frac{a_j - k_j\sin\varphi_j}{2}) - k_j$ and

$$\begin{aligned} \phi &= -\sum_{j=1}^{m_1}(\lambda_j^- u_j^2 + \lambda_j^+ v_j^2) - \sum_{j=m_1+1}^{m_1+m_2} u_j^2 + \sum_{j=m_1+m_2+1}^{m_1+m_2+m_3} u_j^2 \\ &\quad - \sum_{j=m_1+m_2+m_3+1}^{n-m_5}(-u_j^2 + v_j^2). \end{aligned}$$

Sketch of the Proof of Theorem 3.2. We proceed in several steps.

Step 1. Notation. In order to keep the notation simple, we shall rewrite the expression found in Proposition 4.1, namely

$$\operatorname{Tr}(R_g T_w A(H+\mu^{\mathfrak{m}})^{-K}) = C_{\mathrm{res}} \int e^{i\phi(u,v)} \operatorname{tr}_E q(B\begin{pmatrix} u \\ v \end{pmatrix} - b_0, y, \eta; \mu)\, du d\underline{v} dy d\eta.$$

in the form

$$\operatorname{Tr}(R_g T_w A(H+\mu^{\mathfrak{m}})^{-K}) = C_{\mathrm{res}} \int_{\mathbb{R}^{n'}} e^{i\phi(w)} \tilde{q}(\tilde{B}w - \tilde{b}_0; \mu)\, dw. \tag{4.4}$$

Here,

- $w = (w_1, \ldots, w_{n'})$ where $n' = 2n - m_2 - m_3$ denotes the number of variables used in the integration,
- $\tilde{q}$ is $\operatorname{tr}_E q$ with the variables v_j, $j = m_1+1, \ldots, m_2+m_3$, evaluated at $v_j = \sin\varphi_j(u_j - \frac{a_j - k_j \sin\varphi_j}{2}) - k_j$ (recall that for these j, we have $\varphi_j = \pm\pi/2$)
- $\tilde{B} \in \mathrm{O}(n')$ and $\tilde{b}_0 \in \mathbb{R}^{n'}$ are made up from the corresponding components of B and b_0.
- $\phi(w) = \sum_{j=1}^{n'} \tilde{\lambda}_j w_j^2$, where the $\tilde{\lambda}_j$ are determined by the coefficients in the representation of ϕ, i.e. the $\lambda_j^{\pm}$, $j = 1, \ldots, m_1$, the coefficients -1 and $+1$ of u_j^2, $j = m_1+1, \ldots, m_1+m_2+m_3$, the coefficients $+1$ of u_j^2 and -1 of v_j^2 for $j = m_1+m_2+m_3+1, \ldots, 2n-m_5$, and where we complement the summation by letting $\tilde{\lambda}_j = 0$ for the variables corresponding to (u_j, v_j), $j > 2n - 2m_5$.

Step 2. Reduction of the integrand. Instead of considering the full symbol $\tilde{q}(\tilde{B}w - \tilde{b}_0; \mu)$ in (4.4), it will suffice to study the homogeneous components in the expansion of the symbol of $\tilde{q}$.

This is a consequence of the fact that, for $p \in S^{m,d}(\mathbb{R}^{n'}; S)$ and $N \in \mathbb{N}$,

$$p(w+b;\mu) \quad = \quad \sum_{|\alpha|<N} \frac{b^\alpha}{\alpha!} \partial_w^\alpha p(w;\mu) + r_N(w;\mu),$$

with r_N in $S^{m-N,d}(\mathbb{R}^{n'}; S)$ (see the appendix for notation).

Moreover, the residual terms have the right behavior, since for $r \in S^{-N,-K\mathfrak{m}}$ and $n_0 \in \mathbb{N}$ such that $N > 2n + n_0$, Theorem 6.2 implies that

$$\int e^{i\phi(w)} r(w;\mu)\, dw = \sum_{k=0}^{n_0-1} c_k \mu^{-K\mathfrak{m}-k} + O(\mu^{-K\mathfrak{m}-n_0}) \tag{4.5}$$

where the coefficients c_k are given by integrals of terms of the form (6.2).

These terms furnish the contributions to the coefficients c_j''. It follows from the parametrix construction in Theorem 6.3, below, that the c_k are zero unless k is a multiple of $\mathfrak{m}$, so that only integer powers of $-\lambda = \mu^{\mathfrak{m}}$ show up, see also Grubb and Seeley [**12**, p.501].

Step 3. Focusing on the homogeneous terms. It is actually sufficient to study expressions of the form

$$d(w)(h_{\mathfrak{m}}(w) + \mu^{\mathfrak{m}})^{-K-\ell}, \tag{4.6}$$

where $\ell \in \mathbb{N}_0$ and d is a Shubin symbol, which is homogeneous of some degree $\mathfrak{d} = \mathfrak{a} + \ell\mathfrak{m} - j$, $j \in \mathbb{N}_0$, for $|w| \geq 1$. This is a corollary of the following proposition:

PROPOSITION 4.2. (a) *For each $N \in \mathbb{N}$ the components of the complete right symbol $q = q(y, p; \mu)$ of $A(H + \mu^{\mathfrak{m}})^{-K}$ can be written as a finite sum of terms of the form*

$$d(y,p)(h_{\mathfrak{m}}(y,p) + \mu^{\mathfrak{m}})^{-K-\ell}, \tag{4.7}$$

where d is homogeneous in (y, p) of degree $\mathfrak{d} = \mathfrak{a} + \ell\mathfrak{m} - j$ for $j = 0, \ldots, N-1$, plus a remainder in $S^{-N,-K\mathfrak{m}} \cap S^{-N-K\mathfrak{m},0}$.

(b) *Each d is a polynomial in derivatives of the homogeneous components in the symbol expansions of a and h. If $\ell = 0$ in* (a)*, then d can only be one of the homogeneous components of a.*

(c) *The statement corresponding to* (a) *is also true for the symbol $\tilde{q}$. Here, d is additionally scalar, since we take the fiber trace in E.*

(d) *In the expansion of $\tilde{q}(Bw - \tilde{b}_0; \mu)$ into homogeneous components, the only possible coefficient functions of $(h_{\mathfrak{m}}(\tilde{B}w) + \mu^{\mathfrak{m}})^{-K}$ are of the form*

$$\frac{(-\tilde{b}_0)^\alpha}{\alpha!}\partial_w^\alpha \operatorname{tr}_E a_{\mathfrak{a}-j}(\tilde{B}w) \tag{4.8}$$

for suitable $j \in \mathbb{N}_0$ and a multi-index α.

So we are led to considering integrals of the form

$$\int_{\mathbb{R}^{n'}} e^{i\phi(w)} d(w)(h_{\mathfrak{m}}(w) + \mu^{\mathfrak{m}})^{-K-\ell}\, dw \tag{4.9}$$

with $\ell \in \mathbb{N}_0$ and d homogeneous of degree $\mathfrak{d}$ for $|w| \geq 1$.

Similarly as Grubb and Seeley in their analysis of weakly parametric symbols [**12**, Theorem 2.1], we split the integral over $\mathbb{R}^{n'}$ into the integrals over $\{|w| \leq 1\}$, $\{1 \leq |w| \leq |\mu|\}$, and $\{|w| \geq |\mu|\}$. It is not difficult to see that the integral over $|w| \leq 1$ produces powers $\mu^{-(K+\ell+j)\mathfrak{m}}$, $j \in \mathbb{N}_0$, using a binomial series expansion of the integrand. These terms also contribute to the terms $c_j''(-\lambda)^{-K-j}$ in (1.2), even though they are 'local'.

For the other two integrals we introduce polar coordinates and use the fact that ϕ is 2-homogeneous and d is homogeneous of degree $\mathfrak{d}$.

Step 4. The integral over $|w| \geq |\mu|$. Letting $w/|\mu| = v = r\theta$ we obtain

$$\begin{aligned}
&\int_{|w|\geq|\mu|} e^{i\phi(w)} d(w)(h_{\mathfrak{m}}(w) + \mu^{\mathfrak{m}})^{-K-\ell}\, dw \\
&\quad = |\mu|^{\mathfrak{d}-\mathfrak{m}(K+\ell)+n'} \int_{|v|\geq 1} e^{i\phi(v)|\mu|^2} d(v)(h_{\mathfrak{m}}(v) + (\mu/|\mu|)^{\mathfrak{m}})^{-K-\ell}\, dv \\
&\quad = |\mu|^{\mathfrak{d}-\mathfrak{m}(K+\ell)+n'} \int_1^\infty I(\hbar, r, \omega)_{|\hbar=(|\mu|^2 r^2)^{-1}} r^{\mathfrak{d}+n'-1}\, dr \qquad (4.10)
\end{aligned}$$

with

$$I(\hbar, r, \omega) = \int_{\mathbb{S}^{n'-1}} e^{i\phi(\theta)/\hbar} d(\theta)(h_{\mathfrak{m}}(r\theta) + \omega^{\mathfrak{m}})^{-K-\ell}\, dS(\theta).$$

The stationary phase approximation, see e.g. [**32**, Theorem 3.16] or [**7**], yields an expansion for this integral: Since $\phi(\theta) = \sum_{j=1}^{n'} \tilde{\lambda}_j \theta_j^2$, the stationary points on the sphere are the unit length eigenvectors of the matrix $\operatorname{diag}(\tilde{\lambda}_1, \ldots, \tilde{\lambda}_{n'})$. We denote by $\{\mu_l\}$ the set of *different* $\tilde{\lambda}_j$ and by $\varkappa_l$ the multiplicity of μ_l. To fix the notation let us assume that $\mu_1 = 0$ so that, by (3.3), $\varkappa_1 = 2\dim_{\mathbb{C}} \mathbb{C}^n_g$ is the real dimension of the fixed point set of $g \in \mathrm{U}(n)$. The set of stationary points therefore is the disjoint union of the spheres $\mathbb{S}^{\varkappa_l-1}$ in the eigenspaces for μ_l. We find, for $N \in \mathbb{N}$,

$$\begin{aligned} &I(\hbar, r, \omega) \\ (4.11) = &\sum_{J=0}^{N-1} \hbar^{J+\frac{n'-\varkappa_l}{2}} \sum_l e^{\frac{i}{\hbar}\mu_l} \int_{\mathbb{S}^{\varkappa_l-1}} A_{2J}(\theta', \partial_{\theta'}) \left(d(\theta)(h_\mathfrak{m}(r\theta)+\omega^\mathfrak{m})^{-K-\ell}\right) dS \\ &+ O\Big(\sum_l \hbar^{N+\frac{n'-\varkappa_l}{2}} \sum_{|\alpha|\le 2N+n'-\varkappa_l+1} \sup |\partial_{\theta'}^\alpha \left(d(\theta)(h_\mathfrak{m}(r\theta)+\omega^\mathfrak{m})^{-K-\ell}\right)|\Big) \end{aligned}$$

with differential operators A_{2J} of order $2J$ in the variables θ' transversal to the fixed point set. We insert this expression into the integral over r. The contribution of the remainder term turns out to be a low power of $|\mu|$ as desired. The terms in the expansion, however, give us expressions of the form

$$e^{i\mu_l|\mu|^2}|\mu|^{-(K+\ell)\mathfrak{m}-j} g_j(\mu/|\mu|) \tag{4.12}$$

for $j \in \mathbb{N}_0$ and smooth functions g_j. This is not quite what we want: There should be no exponential terms with $\mu_l \neq 0$ and, moreover, expressions in μ (and its powers), not in $|\mu|$ and $\mu/|\mu|$. We shall address this problem in Step 6.

Step 5. The integral over $1 \le |w| \le |\mu|$. We first expand $(h_\mathfrak{m} + \mu^\mathfrak{m})^{-K-\ell}$ into a finite number of powers of $h_\mathfrak{m}\mu^{-\mathfrak{m}}$ and a remainder term. Proceeding similarly as in Step 4, we have to treat terms of the form

$$\mu^{-(K+\ell)\mathfrak{m}-k\mathfrak{m}} \int_1^{|\mu|} \int_{\mathbb{S}^{n'-1}} e^{ir^2\phi(\theta)} d(\theta) h_\mathfrak{m}(\theta)^k \, dS \, r^{\mathfrak{d}+\mathfrak{m}k+n'-1} \, dr \tag{4.13}$$

and the corresponding integrals over the remainder terms. An application of the stationary phase approximation with $\hbar = r^{-2}$ for the integral over the sphere results in terms

$$\mu^{-(K+\ell)\mathfrak{m}-k\mathfrak{m}} c_{l,J} \sum_l \sum_{J=0}^{N-1} \int_1^{|\mu|} e^{ir^2\mu_l} r^{\mathfrak{d}+\mathfrak{m}k-2J+\varkappa_l-1} \, dr \tag{4.14}$$

with

$$c_{l,J} = \int_{\mathbb{S}^{\varkappa_l-1}} A_{2J,l}(\theta', \partial_{\theta'})(d(\theta)h_\mathfrak{m}(\theta)^k) \, dS \tag{4.15}$$

plus a remainder term that is easy to handle.

For $\mu_l \neq 0$ integration by parts in (4.14) yields, as in Step 4, terms of the form (4.12) above, but here – and only here – we also encounter logarithmic terms, namely for the eigenvalue $\mu_1 = 0$, when

$$\mathfrak{d} + \mathfrak{m}k - 2J + \varkappa_1 = 0. \tag{4.16}$$

Note from (4.14) that the μ-powers associated with log-terms are always integer multiples of $\mathfrak{m}$ as stated in (1.2).

Of particular interest is the coefficient of $\mu^{-K\mathfrak{m}} \ln|\mu|$, since this yields the noncommutative residue. It is immediate from (4.14) that it can only arise for $\ell = k = 0$. Together with Proposition 4.2, this gives us a wealth of information: By 4.2(b), the only terms with $\ell = 0$ in the parametrix construction yielding the symbol of q are of the form $a_{\mathfrak{a}-j}(h_\mathfrak{m} + \mu^\mathfrak{m})^{-K}$ for some term $a_{\mathfrak{a}-j}$ in the asymptotic expansion of the symbol of A. For $\operatorname{tr}_E q(\tilde{B}w - \tilde{b}_0; \mu)$, 4.2(d) shows that there may be contributions from the Taylor expansion, leading to terms

$(-\tilde{b}_0)^\alpha \partial_w^\alpha \operatorname{tr}_E(a_{\mathfrak{a}-j}(\tilde{B}w)(h_{\mathfrak{m}}(\tilde{B}w)+\mu^{\mathfrak{m}})^{-K})/\alpha!$. However, none of the derivatives must fall on $(h_{\mathfrak{m}}(\tilde{B}w)+\mu^{\mathfrak{m}})^{-K}$, for then we would have $\ell \neq 0$. Hence d does not contain factors involving terms $h_{\mathfrak{m}-j}$ from the asymptotic expansion of h or their derivatives. The fact that k equals 0 moreover implies that the integral (4.13), and hence the coefficient of $\mu^{-K}\ln|\mu|$, is independent of the choice of the auxiliary operator H. This proves statement (e) in Theorem 3.2: The fact that

$$\ln(-\lambda) = \ln(\mu^{\mathfrak{m}}) = \mathfrak{m} \ln \mu$$

causes a factor $1/\mathfrak{m} = 1/\operatorname{ord} H$ in the expansion with respect to $-\lambda$.

Finally, one has to treat the remainder terms in the expansion of $(h_{\mathfrak{m}}+\mu^{\mathfrak{m}})^{-K-\ell}$. The analysis is slightly more involved, but similar to the above, and also produces terms of the form (4.12).

Step 6. Correcting the expansion. We have an expansion of $\operatorname{Tr}(R_g T_w A(H+\mu^{\mathfrak{m}})^{-K})$ not only into powers of μ and $\ln\mu$, but also in powers of $|\mu|$ times factors of the form $e^{i|\mu|^2\mu_l} g_l(\mu/|\mu|)$, $\mu_l \neq 0$, or $h(\mu/|\mu|)$ that are not of the form stated in Theorem 3.2. This is analogous to a phenomenon already observed by Grubb and Seeley. They excluded this possibility in [**12**, Lemma 2.3]. An argument along similar lines shows that the expansion only contains the terms listed in (1.2). In particular, the terms involving factors $e^{i|\mu|^2\mu_l}$ for $\mu_l \neq 0$ necessarily add up to zero.

This last fact has also been established before in the context of the analysis of the zeta function $\operatorname{Tr}(R_g T_w A H^{-z})$ in [**28**, Theorem 5].

Step 7. Conclusion. As the terms involving the eigenvalues $\mu_l \neq 0$ all cancel and, moreover, only powers of μ and $\ln\mu$ occur, we see that we obtain the expansion (1.2): In fact, we saw already how the terms $(c_j'\ln(-\lambda)+c_j'')^{-K-j}$ arise. As for the terms $c_j(-\lambda)^{(2m+\mathfrak{a}-j)/\mathfrak{m}-K}$: they result from inserting the terms for $\mu_l = 0$ in the stationary phase expansion (e.g. in (4.11) with $\hbar = r^{-2}|\mu|^{-2}$) into the corresponding integral (4.10), noting that $\varkappa_1 = 2m$. This completes our sketch. □

5. The Noncommutative Residue

In Section 4 we saw that there is only one point, where logarithmic terms appear, namely in the analysis of the integral (4.13). We also observed that, as a consequence of Proposition 4.2, the only possible coefficient functions of $(h_{\mathfrak{m}}(\tilde{B}w)+\mu^{\mathfrak{m}})^{-K}$ in the expansion of $\tilde{q}(\tilde{B}w-\tilde{b}_0)$ are the functions from (4.8) for some choice of j and α. In addition, the application of the stationary phase expansion may add further derivatives $A_{2J,1}(\theta',\partial_{\theta'})$ of order $2J$ on $a_{\mathfrak{a}-j}$ as we saw in (4.14) and (4.15).

Next, Equation (4.16) implies that a nonzero contribution to $\mu^{-K}\ln|\mu|$ will only arise if the homogeneity degree satisfies

$$\mathfrak{a} - j - 2J + \varkappa_1 = 0, \tag{5.1}$$

with $\varkappa_1 = 2m_5$.

In order to understand the contribution to the noncommutative residue, we distinguish the cases where $\mathfrak{a} = \operatorname{ord} A = -2m_5$, $\mathfrak{a} < -2m_5$ or $\mathfrak{a} > -2m_5$.

(i) Let $\mathfrak{a} = -2m_5$. According to (5.1), the only possibility for obtaining a nontrivial contribution to the $\mu^{-K}\ln|\mu|$-term then is when $j=0$ and $\alpha = 0$ in (4.8) and, moreover, $J=0$. Setting $k=\ell=J=0$ and $l=1$ in (4.14) and (4.15) shows that the coefficient is

$$\int_{\mathbb{S}^{2m_5-1}} \operatorname{tr}_E a_{-2m_5}(\tilde{B}\theta)\,dS. \tag{5.2}$$

On the other hand, we know that $\tilde{B}$ is the identity on $(x_{n-m_5+1}, p_{n-m_5+1}, \dots x_n, p_n)$ and that, on $\mathbb{S}^{2m_5-1}$ the other variables are zero. Hence (5.2) reduces to

$$\int_{\mathbb{S}^{2m_5-1}} \operatorname{tr}_E a_{-2m_5}(\theta)\, dS \tag{5.3}$$

as asserted.

(ii) The same argument shows that the residue term vanishes if $\operatorname{ord} A < -2m_5$.

(iii) If $\operatorname{ord} A > -2m_5$, then we reconsider Equation (4.13). The standard application of the stationary phase expansion does not yield explicit expressions; therefore we take another approach. We know that there will be no contribution to the $\mu^{-K} \ln |\mu|$-term unless $k = \ell = 0$ in (4.13).

For $\varkappa_1 = 2m_5$ we write $\mathbb{S}^{2m_5-1} = \{\theta \in \mathbb{S}^{n'-1} : \theta' = (\theta_1, \dots, \theta_{n'-2m_5}) = 0\}$. Up to $O(\mu^{-\infty})$ contributions, the integral in (4.13) with $k = 0$, $l = 1$ and d from (4.8) then can be rewritten in the form

$$\int_1^{|\mu|} \int_{S^{2m_5-1}} \int_{[-\varepsilon,\varepsilon]^{n'-2m_5}} e^{ir^2\langle Q\theta',\theta'\rangle/2} d(\theta', \sigma\sqrt{1-|\theta'|^2})\, d\theta' dS(\sigma)\, r^{\eth+n'-1}\, dr, \tag{5.4}$$

where $Q = \operatorname{diag}(2\tilde{\lambda}_1, \dots, 2\tilde{\lambda}_{n'-2m_5})$ with $\tilde{\lambda}_j$ introduced after (4.4).

To the inner integral we apply the explicit version of the stationary phase expansion for quadratic exponentials with $\hbar = r^{-2}$, see [**32**, Theorem 3.13]:

$$\begin{aligned}&\int e^{\frac{i}{\hbar}\langle Q\theta',\theta'\rangle} d(\theta', \sigma\sqrt{1-|\theta'|^2})\, d\theta' \\ &\quad = \quad (2\pi\hbar)^{\frac{n'-2m_5}{2}} \frac{e^{i\frac{\pi}{4}\operatorname{sgn} Q}}{|\det Q|^{1/2}} \left(\sum_{J=0}^{N-1} \frac{\hbar^J}{J!} \left(\frac{\langle Q^{-1}D, D\rangle}{2i} \right)^J d(0,\sigma) + O(\hbar^N) \right),\end{aligned}$$

where $D = -i\partial/\partial\theta'$. This furnishes an expansion into decreasing powers of r, which we then insert into (5.4). Since integration over $[1, |\mu|]$ will only give a nontrivial contribution to the $\mu^{-K} \ln |\mu|$-term, if the total r-power is -1, only finitely many terms can contribute. For these we obtain an explicit formula.

6. Appendix

The parameter-dependent calculus of Grubb and Seeley. In order to make the text more accessible, we recall some results from [**12**, Section 1].

Let S be an open sector in $\mathbb{C}$ with vertex at the origin and let $p = p(x, \xi, \mu)$ be a smooth function on $\mathbb{R}^\nu \times \mathbb{R}^n \times \overline{S}$, which is additionally holomorphic for $\mu \in S$.

The symbol class $S^{m,0}(\mathbb{R}^\nu, \mathbb{R}^n; S)$ consists of all p for which

$$\partial_t^j p(\cdot, \cdot, 1/t) \in S^{m+j}(\mathbb{R}^\nu \times \mathbb{R}^n), \quad 1/t \in S, \tag{6.1}$$

while $S^{m,d}(\mathbb{R}^\nu, \mathbb{R}^n; S)$ consists of those p for which

$$\partial_t^j (t^d p(\cdot, \cdot, 1/t)) \in S^{m+j}(\mathbb{R}^\nu \times \mathbb{R}^n), \quad 1/t \in S,$$

with uniform estimates in S^{m+j} for $|t| \le 1$ and $\frac{1}{t}$ in closed subsectors of S.

Example 6.1. If $a(x, \xi)$ is homogeneous of degree $m > 0$ in ξ for $|\xi| \ge 1$ and smooth in (x, ξ) on $\mathbb{R}^\nu \times \mathbb{R}^n$, and if $a(x, \xi) + \mu^m$ is invertible for $(x, \xi, \mu) \in \mathbb{R}^\nu \times \mathbb{R}^n \times \overline{S}$, then the symbol p defined by

$$p(x, \xi, \mu) = (a(x, \xi) + \mu^m)^{-1}$$

is an element of $S^{0,-m}(\mathbb{R}^\nu, \mathbb{R}^n; S) \cap S^{-m,0}(\mathbb{R}^\nu, \mathbb{R}^n; S)$, see [**12**, Theorem 1.17].

THEOREM 6.2. (a) *For* $p_1 \in S^{m_1,d_1}$ *and* $p_2 \in S^{m_2,d_2}$ *we have* $p_1p_2 \in S^{m_1+m_2,d_1+d_2}$.
(b) *For* $p \in S^{m,d}$ *set*

$$p_{(d,k)}(x,\xi) = \frac{1}{k!}\partial_t^k(t^d p(x,\xi,\frac{1}{t}))_{|t=0}. \tag{6.2}$$

Then $p_{(d,k)} \in S^{m+k}$, *and for any* N,

$$p(x,\xi,\mu) - \sum_{k=0}^{N-1} \mu^{d-k} p_{(d,k)}(x,\xi) \in S^{m+N,d-N}. \tag{6.3}$$

THEOREM 6.3. *Let* $a \in S^m$, $m > 0$, *with principal symbol* a_m *which is homogeneous of degree* $m > 0$ *for* (x,ξ), $|\xi| \geq 1$ *and such that* $a_m(x,\xi) + \mu^m$ *is invertible for all* μ *in* S. *Then we find a symbol* $p = p(x,\xi,\mu)$ *such that*

$$p \circ (a + \mu^m) \sim 1 \sim (a+\mu^m) \circ p \mod S^{-\infty,-m}(\mathbb{R}^n,\mathbb{R}^n;S).$$

The symbol p *is an element of* $S^{-m,0} \cap S^{0,-m}$ *with an asymptotic expansion* $p \sim \sum_{j=0}^\infty p_{-m-j}$, *where* $p_{-m}(x,\xi,\mu) = (a_m(x,\xi)+\mu^m)^{-1} \in (S^{-m,0} \cap S^{0,-m})(\mathbb{R}^n,\mathbb{R}^n;S)$ *and* $p_{-m-j} \in (S^{-m-j,0} \cap S^{m-j,-2m})(\mathbb{R}^n,\mathbb{R}^n;S)$.

In fact, the parametrix is constructed by letting

$$\begin{aligned} p_{-m}(x,\xi,\mu) &= (a_m(x,\xi)+\mu^m)^{-1} \quad \text{and} \\ p_{-m-j}(x,\xi,\mu) &= -\sum_{l,k,\alpha} \frac{1}{\alpha!}(\partial_\xi^\alpha p_{-m-l})(\partial_x^\alpha a_{m-k})p_{-m}, \end{aligned}$$

where the sum is over all k, l, α with $l < j$ and $l + k + |\alpha| = j$. In fact, p_{-m-j} is a linear combination of terms $p_{-m}(\partial_\xi^{\alpha_1}\partial_x^{\beta_1} a_{m-k_1}) \dots p_{-m}(\partial_\xi^{\alpha_r}\partial_x^{\beta_r} a_{m-k_r})p_{-m}$, where, for $j > 0$, we have $r \geq 1$ and $\sum_{l=1}^r |\alpha_l| = \sum_{l=1}^r |\beta_l|$ and $\sum_{l=1}^r |\alpha_l| + \sum_{l=1}^r |k_l| = j$.

References

[1] Paolo Boggiatto and Fabio Nicola, *Non-commutative residues for anisotropic pseudo-differential operators in* $\mathbb{R}^n$, J. Funct. Anal. **203** (2003), no. 2, 305–320, DOI 10.1016/S0022-1236(03)00194-0. MR2003350

[2] Nicolas Bourbaki, *Functions of a real variable: Elementary theory*, Elements of Mathematics (Berlin), Springer-Verlag, Berlin, 2004. Translated from the 1976 French original [MR0580296] by Philip Spain, DOI 10.1007/978-3-642-59315-4. MR2013000

[3] Alain Connes, *Noncommutative geometry*, Academic Press, Inc., San Diego, CA, 1994. MR1303779

[4] A. Connes and H. Moscovici, *The local index formula in noncommutative geometry*, Geom. Funct. Anal. **5** (1995), no. 2, 174–243, DOI 10.1007/BF01895667. MR1334867

[5] Shantanu Dave, *An equivariant noncommutative residue*, J. Noncommut. Geom. **7** (2013), no. 3, 709–735, DOI 10.4171/JNCG/132. MR3108693

[6] Maurice de Gosson, *Symplectic geometry and quantum mechanics*, Operator Theory: Advances and Applications, vol. 166, Birkhäuser Verlag, Basel, 2006. Advances in Partial Differential Equations (Basel), DOI 10.1007/3-7643-7575-2. MR2241188

[7] M. V. Fedoryuk, *The stationary phase method and pseudodifferential operators*, Russian Math. Surveys, 6:1 (1971), 65–115

[8] Boris V. Fedosov, François Golse, Eric Leichtnam, and Elmar Schrohe, *The noncommutative residue for manifolds with boundary*, J. Funct. Anal. **142** (1996), no. 1, 1–31, DOI 10.1006/jfan.1996.0142. MR1419415

[9] Peter B. Gilkey, *Invariance theory, the heat equation, and the Atiyah-Singer index theorem*, 2nd ed., Studies in Advanced Mathematics, CRC Press, Boca Raton, FL, 1995. MR1396308

[10] Alexander Gorokhovsky, Niek de Kleijn, and Ryszard Nest, *Equivariant algebraic index theorem*, J. Inst. Math. Jussieu **20** (2021), no. 3, 929–955, DOI 10.1017/S1474748019000380. MR4260645

[11] Gerd Grubb and Elmar Schrohe, *Trace expansions and the noncommutative residue for manifolds with boundary*, J. Reine Angew. Math. **536** (2001), 167–207, DOI 10.1515/crll.2001.055. MR1837429

[12] Gerd Grubb and Robert T. Seeley, *Weakly parametric pseudodifferential operators and Atiyah-Patodi-Singer boundary problems*, Invent. Math. **121** (1995), no. 3, 481–529, DOI 10.1007/BF01884310. MR1353307

[13] Gerd Grubb and Robert T. Seeley, *Zeta and eta functions for Atiyah-Patodi-Singer operators*, J. Geom. Anal. **6** (1996), no. 1, 31–77, DOI 10.1007/BF02921566. MR1402386

[14] Victor Guillemin, *A new proof of Weyl's formula on the asymptotic distribution of eigenvalues*, Adv. in Math. **55** (1985), no. 2, 131–160, DOI 10.1016/0001-8708(85)90018-0. MR772612

[15] Bernard Helffer, *Théorie spectrale pour des opérateurs globalement elliptiques* (French), Astérisque, vol. 112, Société Mathématique de France, Paris, 1984. With an English summary. MR743094

[16] Lars Hörmander, *Symplectic classification of quadratic forms, and general Mehler formulas*, Math. Z. **219** (1995), no. 3, 413–449, DOI 10.1007/BF02572374. MR1339714

[17] Jean Leray, *Analyse lagrangienne et mécanique quantique* (French), Séminaire sur les Équations aux Dérivées Partielles (1976–1977), I, Exp. No. 1, Collège de France, Paris, 1977, pp. 303. MR0501198

[18] Matthias Lesch, *On the noncommutative residue for pseudodifferential operators with log-polyhomogeneous symbols*, Ann. Global Anal. Geom. **17** (1999), no. 2, 151–187, DOI 10.1023/A:1006504318696. MR1675408

[19] Matthias Lesch, Henri Moscovici, and Markus J. Pflaum, *Connes-Chern character for manifolds with boundary and eta cochains*, Mem. Amer. Math. Soc. **220** (2012), no. 1036, viii+92, DOI 10.1090/S0065-9266-2012-00656-3. MR3025890

[20] Cyril Lévy, Carolina Neira Jiménez, and Sylvie Paycha, *The canonical trace and the noncommutative residue on the noncommutative torus*, Trans. Amer. Math. Soc. **368** (2016), no. 2, 1051–1095, DOI 10.1090/tran/6369. MR3430358

[21] Denis Perrot, *Local index theory for operators associated with Lie groupoid actions*, J. Topol. Anal. **14** (2022), no. 2, 297–341, DOI 10.1142/S1793525321500059. MR4446909

[22] M. J. Pflaum, H. B. Posthuma, and X. Tang, *An algebraic index theorem for orbifolds*, Adv. Math. **210** (2007), no. 1, 83–121, DOI 10.1016/j.aim.2006.05.018. MR2298821

[23] M. J. Pflaum, H. Posthuma, and X. Tang, *Cyclic cocycles on deformation quantizations and higher index theorems*, Adv. Math. **223** (2010), no. 6, 1958–2021, DOI 10.1016/j.aim.2009.10.012. MR2601006

[24] Raphaël Ponge, *Noncommutative residue for Heisenberg manifolds. Applications in CR and contact geometry*, J. Funct. Anal. **252** (2007), no. 2, 399–463, DOI 10.1016/j.jfa.2007.07.001. MR2360923

[25] A. Savin, E. Schrohe, and B. Sternin, *Elliptic operators associated with groups of quantized canonical transformations*, Bull. Sci. Math. **155** (2019), 141–167, DOI 10.1016/j.bulsci.2019.01.010. MR3996327

[26] Anton Savin and Elmar Schrohe, *Analytic and algebraic indices of elliptic operators associated with discrete groups of quantized canonical transformations*, J. Funct. Anal. **278** (2020), no. 5, 108400, 45, DOI 10.1016/j.jfa.2019.108400. MR4046211

[27] Anton Savin and Elmar Schrohe, *An index formula for groups of isometric linear canonical transformations*, Doc. Math. **27** (2022), 983–1013. MR4452230
Adv. Math. 409 Part A, 108624, 2022

[28] Anton Savin and Elmar Schrohe, *Local index formulae on noncommutative orbifolds and equivariant zeta functions for the affine metaplectic group*, Adv. Math. **409** (2022), Paper No. 108624, DOI 10.1016/j.aim.2022.108624. MR4470610

[29] R. T. Seeley, *Complex powers of an elliptic operator*, Singular Integrals (Proc. Sympos. Pure Math., Chicago, Ill., 1966), Amer. Math. Soc., Providence, R.I., 1967, pp. 288–307. MR0237943

[30] M. A. Shubin, *Pseudodifferential operators and spectral theory*, Springer Series in Soviet Mathematics, Springer-Verlag, Berlin, 1987. Translated from the Russian by Stig I. Andersson, DOI 10.1007/978-3-642-96854-9. MR883081

[31] Mariusz Wodzicki, *Noncommutative residue. I. Fundamentals*, K-theory, arithmetic and geometry (Moscow, 1984), Lecture Notes in Math., vol. 1289, Springer, Berlin, 1987, pp. 320–399, DOI 10.1007/BFb0078372. MR923140

[32] Maciej Zworski, *Semiclassical analysis*, Graduate Studies in Mathematics, vol. 138, American Mathematical Society, Providence, RI, 2012, DOI 10.1090/gsm/138. MR2952218

Peoples' Friendship University of Russia (RUDN University), 6 Miklukho-Maklaya St, Moscow, 117198, Russian Federation
Email address: a.yu.savin@gmail.com

Leibniz University Hannover, Institute of Analysis, Welfengarten 1, 30167 Hannover, Germany
Email address: schrohe@math.uni-hannover.de

Proceedings of Symposia in Pure Mathematics
Volume **105**, 2023
https://doi.org/10.1090/pspum/105/01914

Cartan motion group and orbital integrals

Yanli Song and Xiang Tang

This paper is dedicated to the occasion of the 40th birthday of cyclic cohomology

ABSTRACT. In this short note, we study the variation of orbital integrals, as traces on the group algebra G, under the deformation groupoid. We show that orbital integrals are continuous under the deformation. And we prove that the pairing between orbital integrals and K-theory elements of $C_r^*(G)$ stays constant with respect to the deformation for regular group elements, but varies at singular elements.

1. Introduction

Connes introduced a beautiful construction of tangent groupoid [**11**] to present a groupoid proof of the Atiyah-Singer index theorem. The tangent groupoid provides a deep link between the Fredholm index of an elliptic operator and its local geometric information and has turned out to be extremely powerful in studying various generalized index problems, e.g. [**2, 3, 14–19, 30, 31**], etc.

The Connes tangent groupoid was applied [**6**] to study the K-theory of $C_r^*(G)$ and formulate a new approach to the Connes-Kasparov conjecture. More precisely, let G be a connected real reductive Lie group, K be a maximal compact subgroup of G. And let $\mathfrak{g}$ and $\mathfrak{k}$ be the Lie algebras of G and K, and θ be the Cartan involution on $\mathfrak{g}$. $\mathfrak{k}$ is the subspace of θ fixed points and $\mathfrak{p}$ is the -1 eigenspace of θ. The adjoint action of K on $\mathfrak{g}$ defines a K action on $\mathfrak{g}/\mathfrak{k}$, which is isomorphic to $\mathfrak{p}$. The Cartan motion group is the semidirect product $K \ltimes \mathfrak{p}$, where $\mathfrak{p}$ is viewed as an abelian group. The deformation groupoid $\mathcal{G}$ is a smooth family of Lie groups over the interval $[0,1]$,

$$\mathcal{G} := K \times \mathfrak{p} \times [0,1].$$

At $t = 0$, G_0 is the Cartan motion group $K \ltimes \mathfrak{p}$; at $t > 0$, G_t is isomorphic to G under the isomorphism $\varphi_t : K \times \mathfrak{p} \to G$ by

$$\varphi_t(k,v) = k \exp_G(tv),\ (k_1, v_1, t) \cdot_t (k_2, v_2, t) = \varphi_t^{-1}\big(\varphi_t(k_1, v_1) \cdot \varphi_t(k_2, v_2)\big).$$

The Connes-Kasparov isomorphism conjecture ([**8, 9, 29, 33**]) has the following simple but beautiful formulation.

The deformation groupoid $\mathcal{G}$ defines a natural isomorphism (c.f. Remark 4.2.*)*

$$K_\bullet\big(C_r^*(G_0)\big) \to K_\bullet\big(C_r^*(G_1)\big).$$

Key words and phrases. Cartan motion group, Mackey, orbital integrals.

The first author was supported in part by NSF Grants DMS-1800667 and DMS-1952557. The second author was supported in part by NSF Grants DMS-1800666 and DMS-1952551.

Higson [**25**] connected Mackey's approach to induced representation theory to the study of the Connes-Kasparov isomorphism conjecture. Afgoustidis [**1**] obtained a new proof of the Connes-Kasparov isomorphism conjecture via the above deformation groupoid using Vogan's theory.

In this article, we study variations of traces on $C_c^\infty(G)$ under the deformation groupoid. The traces we consider are from orbital integrals.

For an element $x \in K$, let G_x be the centralizer subgroup of G associated to x. The orbital integral associated to x is a trace τ_x on $C_c^\infty(G)$ defined by

$$\tau_x(f) := \int_{G/G_x} f(g\theta g^{-1})dg.$$

The orbital integral τ_k plays a fundamental role in Harish-Chandra's theory of Plancherel measure, e.g. [**24**].

Let $C_r^*(G)$ be the reduced group C^*-algebra of G. Harish-Chandra [**23**] showed that τ_x defines a trace on the Harish-Chandra Schwartz algebra $\mathcal{C}(G)$. Recently, Hochs and Wang [**26**] applied the orbital integrals to study the G-index of an invariant Dirac operator on a proper cocompact manifold. They proved a delocalized index theorem to detect interesting information of the Connes-Kasparov index map and the K-theory of the reduced group C^* algebra $C_r^*(G)$ when G has discrete series representations. For a general G, we [**32**] introduced a generalization of the orbital integral τ_x to a higher degree cyclic cocycle, called higher orbital integral, associated to a cuspidal parabolic subgroup P of G and extract the character information of limits of discrete series representations from the pairing between higher orbital integrals and $K_0(C_r^*(G))$, and with Hochs [**27**] we obtained a higher index theorem for the pairing between the higher orbital integrals and the G-index of an invariant Dirac operator.

In this paper, we focus on the classical orbital integrals τ_x. We start with observing the following continuity property of the orbital integral τ_x on the deformation groupoid $\mathcal{G}$.

THEOREM (Theorem 3.3). *Suppose that f is a smooth function on $\mathcal{G}$ with compact support and $x \in K$ is a regular element. Choose the Haar measure on G_t following Definition 3.2. Then*

$$\lim_{t\to 0}\int_{G_t} f(g\cdot x\cdot g^{-1}, t)\, d_t g = \int_K\int_{\mathfrak{p}} f\left(kxk^{-1}, w - Ad_{kxk^{-1}}w, 0\right)\, dk dw,$$

which by Lemma 2.5 is equal to $det_{\mathfrak{p}}(\mathrm{id} - x)$ times the orbital integral of $f(-,0)$ associated to x for the group G_0.

Next we study the pairing between orbital integrals τ_x and K_0 group and prove the following rigidity property of the index pairing.

THEOREM (Theorem 4.4). *Let E be an irreducible K representation and P_t^E an element of $K_0(C_r^*(G_t))$ associated to E in the image of the Connes-Kasparov (Baum-Connes) assembly map. The index pairing $\tau_x\left([P_t^E]\right) \in \mathbb{C}$ is independent of t. When $t = 0$, we have that*

$$\tau_x\left([P_0^E]\right) = (-1)^{\frac{\dim \mathfrak{p}}{2}} \cdot \frac{\sum_{w\in W_K}(-1)^w \cdot e^{w(\mu+\rho)}(x)}{\prod_{\alpha\in\Delta^+(\mathfrak{g},\mathfrak{t})}\left(e^{\frac{\alpha}{2}} - e^{\frac{-\alpha}{2}}\right)(x)}.$$

It is worth pointing out that it is crucial to consider a regular element x in the above theorem. In contrast, we show in Theorem 4.5 that the L^2-trace does vary with respect to the parameter t and vanishes at $t = 0$ with order $\dim(\mathfrak{p})$.

The above theorems suggest that it is possible to use the deformation groupoid to study the limit of the pairing between the higher cyclic cocycles introduced **[31]** and **[32]** on $K_\bullet(C^*_r(G))$. Such a study is expected to exhibit more interesting properties of the higher cyclic cocycles on the Harish-Chandra Schwartz algebra $\mathcal{C}(G)$. We plan to study this problem in the near future.

The article is organized as follows. In Section 2, we briefly introduce the Cartan motion group, its group C^*-algebra and Fourier transform; in Section 3, we investigate the variation of orbital integrals with respect to the deformation groupoid and prove Theorem 3.3; in Section 4, we study the variation of the pairing between orbital integrals and K_0 groups and prove Theorem 4.4 and 4.5.

2. Cartan Motion Group

Let K be a connected compact Lie group, and $\mathfrak{p}$ an even dimensional Eculidean space on which K acts, and T a maximal torus of K. For simplicity, we assume that the K-action on $\mathfrak{p}$ is spin so that we have the following diagram.

$$\begin{array}{ccc} & & \mathrm{Spin}(\mathfrak{p}) \\ & \nearrow & \downarrow \\ K & \longrightarrow & \mathrm{SO}(\mathfrak{p}) \end{array}$$

We point out that the results in this section do not really need to require the K-action on $\mathfrak{p}$ to be spin by working with exterior algebra. But the spin assumption is used crucially later in the study of K-theory of $C^*_r(G)$. Hence we have decided to work with the spin assumption throughout the article. Let $S_\mathfrak{p} = S^+_\mathfrak{p} \oplus S^-_\mathfrak{p}$ be the $\mathbb{Z}_2$-graded spin representation of $\mathrm{Cliff}(\mathfrak{p})$. In addition, we assume that the T-fixed part of $\mathfrak{p}$ is trivial, which is the case when $\mathfrak{g} = \mathfrak{k} \oplus \mathfrak{p}$ is the Lie algebra of a real reductive group satisfying the equal rank condition.

DEFINITION 2.1. The semidirect product $K \ltimes \mathfrak{p}$ is the group whose underlying set is the Cartesian product $K \times \mathfrak{p}$, equipped with the product law

$$(k_1, v_1) \cdot_0 (k_2, v_2) = \left(k_1 k_2, k_2^{-1} \cdot v_1 + v_2\right), \quad k_1, k_2 \in K, \quad \text{and} \quad v_1, v_2 \in \mathfrak{p},$$

which is called the *Cartan motion group*.

DEFINITION 2.2. The *reduced group C^*-algebra* of $K \ltimes \mathfrak{p}$, denoted by $C^*_r(K \ltimes \mathfrak{p})$, is the completion of the convolution algebra $C^\infty_c(K \ltimes \mathfrak{p})$ in the norm obtained from the left regular representation of $C^\infty_c(K \ltimes \mathfrak{p})$ as bounded convolution operators on the Hilbert space $L^2(K \ltimes \mathfrak{p})$.

Fix the Haar measure on K so that the volume of K equals one. Together with a K-invariant measure on vector space $\mathfrak{p}$, we obtain a Haar measure on $K \ltimes \mathfrak{p}$. If f is a smooth, compactly supported function on $K \ltimes \mathfrak{p}$, we define its *Fourier transform*, a smooth function from $\widehat{\mathfrak{p}}$ into the smooth functions on $K \times K$ by the formula

$$\widehat{f}(\varphi)(k_1, k_2) = \int_\mathfrak{p} f(k_1 k_2^{-1}, v)\varphi(k_2^{-1} \cdot v)\, dv.$$

Here $\varphi \in \widehat{\mathfrak{p}}, k_1, k_2 \in K, v \in \mathfrak{p}$ and e is the identity element in K.

For a fixed $\varphi \in \widehat{\mathfrak{p}}$, we shall think of $\widehat{f}(\varphi)$ as an integral kernel and hence as a compact operator on $L^2(K)$. The Fourier transform $\widehat{f}$ is therefore a function from $\widehat{\mathfrak{p}}$ into $\mathbb{K}\left(L^2(K)\right)$, where $\mathbb{K}\left(L^2(K)\right)$ denotes the space of compact operators on $L^2(K)$. The function is equivariant for the natural action of K on $\widehat{\mathfrak{p}}$ and the conjugacy action of K on $\mathbb{K}\left(L^2(K)\right)$ induced from the right regular representation of K on $L^2(K)$. We define

$$\begin{aligned}&C_0\left(\widehat{\mathfrak{p}}, \mathbb{K}\left(L^2(K)\right)\right)\\ &=\left\{\text{continuous functions from } \widehat{\mathfrak{p}} \text{ into } \mathbb{K}\left(L^2(K)\right), \text{ vanishing at infinity}\right\}\end{aligned}$$

and

$$C_0\left(\widehat{\mathfrak{p}}, \mathbb{K}\left(L^2(K)\right)\right)^K = \left\{K\text{-equivariant functions in } C_0\left(\widehat{\mathfrak{p}}, \mathbb{K}\left(L^2(K)\right)\right)\right\}.$$

The Fourier transform provides a concrete description of the (reduced) group C^*-algebra $C_r^*(K \ltimes \mathfrak{p})$.

THEOREM 2.3. *The Fourier transform $f \mapsto \widehat{f}$ extends to a C^*-algebra isomorphism*

$$C_r^*(K \ltimes \mathfrak{p}) \cong C_0\left(\widehat{\mathfrak{p}}, \mathbb{K}\left(L^2(K)\right)\right)^K$$

PROOF. See [**25**, Theorem 3.2]. □

We study the *orbital integrals* on Cartan motion group. Let $T \subseteq K$ be the maximal torus of K and $x \in T$ be a regular element. In particular, x gives an endomorphism from $\mathfrak{p}$ to itself such that $\det_{\mathfrak{p}}(\mathrm{id} - x) \neq 0$ as the T-fixed part of $\mathfrak{p}$ is assumed to be trivial. We write $\mathcal{O}_x$ the conjugacy class of $(x, 0) \in K \ltimes \mathfrak{p}$. Let dk be the Haar measure on K with total volume 1 and dv the measure on V associated to the Euclidean metric. And $dkdv$ determines a Haar measure on $K \ltimes \mathfrak{p}$. As $\mathrm{id} - x$ is invertible, the centralizer group of x is T. Choose the Haar measure on T so that the volume of T is 1. These choices decide a natural measure dm_x on $\mathcal{O}_x$.

DEFINITION 2.4. For any $f \in C_c(K \ltimes \mathfrak{p})$, we define the orbital integral

$$\tau_x(f) := \int_{(k,v) \in \mathcal{O}_x} f(k, v)\, dm_x.$$

LEMMA 2.5. *If x is regular in T, then the orbital integral associated to x for the Cartan motion group equals*

$$\tau_x(f) = \frac{1}{\det_{\mathfrak{p}}(\mathrm{id} - x)} \cdot \int_K \int_{\mathfrak{p}} f(k \cdot x \cdot k^{-1}, v)\, dkdv.$$

PROOF. For any $(k, v) \in K \ltimes \mathfrak{p}$, we have that

$$\begin{aligned}(k, v) \cdot_0 (x, 0) \cdot_0 (k, v)^{-1} &= (k, v) \cdot_0 (x, 0) \cdot_0 (k^{-1}, -k \cdot v)\\ &= (k \cdot x, x^{-1} \cdot v) \cdot_0 (k^{-1}, -k \cdot v)\\ &= (k \cdot x \cdot k^{-1}, kx^{-1} \cdot v - k \cdot v).\end{aligned}$$

The lemma follows immediately from change of variable

$$v \mapsto kx^{-1} \cdot v - k \cdot v = kx^{-1}(\mathrm{id} - x) \cdot v.$$

□

LEMMA 2.6. *The determinant function* $\det_{\mathfrak{p}}(\mathrm{id}-x)$ *has the following formula,*

$$\det_{\mathfrak{p}}(\mathrm{id}-x)=(-1)^{\frac{\dim\mathfrak{p}}{2}}\cdot\Big(\chi_{S_{\mathfrak{p}}^+}(x)-\chi_{S_{\mathfrak{p}}^-}(x)\Big)^2,$$

where $\chi_{S_{\mathfrak{p}}^{\pm}}$ *is the character of* $S_{\mathfrak{p}}^{\pm}$.

PROOF. Let $\mathfrak{t}$ be the Lie algebra of T and $\Delta^+(\mathfrak{p},\mathfrak{t})$ be a fixed set of positive roots for the T-action on $\mathfrak{p}$ that is compatible with the $\mathbb{Z}/2\mathbb{Z}$ grading on the spin representation. Suppose that $x=e^H$ with $H\in\mathfrak{t}$. We compute that

$$\begin{aligned}\det_{\mathfrak{p}}(\mathrm{id}-e^H)=&\prod_{\alpha\in\Delta^+(\mathfrak{p},\mathfrak{t})}\Big(1-e^{\alpha(H)}\Big)\cdot\Big(1-e^{-\alpha(H)}\Big)\\=&(-1)^{\frac{\dim\mathfrak{p}}{2}}\cdot\prod_{\alpha\in\Delta^+(\mathfrak{p},\mathfrak{t})}\Big(e^{\frac{\alpha(H)}{2}}-e^{-\frac{\alpha(H)}{2}}\Big)^2.\end{aligned}$$

□

PROPOSITION 2.7. *Let* E *be a finite dimensional* K*-representation and* x *be a regular element in* T. *Denote by* χ_E *the character of* E. *Suppose that* $g(z)$ *is a* K*-invariant Schwartz function on* $\widehat{\mathfrak{p}}$ *and* $f\in C_0(K\ltimes\mathfrak{p})$ *is the inverse Fourier transform of*

$$g(z)\cdot\mathrm{id}_E\in C_0\left(\widehat{\mathfrak{p}},\mathbb{K}\left(L^2(K)\right)\right)^K,\quad z\in\widehat{\mathfrak{p}},$$

where by the Peter-Weyl theorem id_E *is an element of* $\mathbb{K}\left(L^2(K)\right)$. *Then*

$$\tau_x(f)=(-1)^{\frac{\dim\mathfrak{p}}{2}}\cdot\frac{\chi_E(x)}{\Big(\chi_{S_{\mathfrak{p}}^+}(x)-\chi_{S_{\mathfrak{p}}^-}(x)\Big)^2}\cdot g(0).$$

PROOF. By Lemma 2.5,

$$\tau_x(f)=\frac{1}{\det_{\mathfrak{p}}(\mathrm{id}-x)}\cdot\int_K\int_{\mathfrak{p}}f(k\cdot x\cdot k^{-1},v)\,dkdv.$$

For any regular $z\in\widehat{\mathfrak{p}}$, it determines a $K\ltimes\mathfrak{p}$-representation on $L^2(K)$ given by

$$\left(\pi_z(x,v)s\right)(u)=e^{i\langle\mathrm{Ad}_u z,v\rangle}\cdot s(x^{-1}u),\quad u\in K,$$

where $s\in L^2(K)$. By the Fourier transform, we have that

$$\begin{aligned}f(k\cdot x\cdot k^{-1},v)=&\int_{\widehat{\mathfrak{p}}}g(z)\cdot\mathrm{tr}\left(\pi_z(kxk^{-1},v)\circ\mathrm{id}_E\right)\,dz\\=&\chi_E(x)\cdot\int_{\widehat{\mathfrak{p}}}\int_K g(z)\cdot e^{i\langle\mathrm{Ad}_u z,v\rangle}\,dudz\end{aligned}$$

Therefore,

$$\begin{aligned}\tau_x(f)=&\frac{\chi_E(x)}{\det_{\mathfrak{p}}(\mathrm{id}-x)}\cdot\int_K\int_{\mathfrak{p}}\int_{\widehat{\mathfrak{p}}}\int_K g(z)\cdot e^{i\langle\mathrm{Ad}_u z,v\rangle}\,dudzdkdv\\=&(-1)^{\frac{\dim\mathfrak{p}}{2}}\cdot\frac{\chi_V(x)}{\Big(\chi_{S_{\mathfrak{p}}^+}(x)-\chi_{S_{\mathfrak{p}}^-}(x)\Big)^2}\cdot\int_{\mathfrak{p}}\int_{\widehat{\mathfrak{p}}}g(z)\cdot e^{i\langle z,v\rangle}\,dzdv\\=&(-1)^{\frac{\dim\mathfrak{p}}{2}}\cdot\frac{\chi_V(x)}{\Big(\chi_{S_{\mathfrak{p}}^+}(x)-\chi_{S_{\mathfrak{p}}^-}(x)\Big)^2}\cdot g(0).\end{aligned}$$

□

3. Deformation to the Cartan motion group

In this section, we study the limit of the orbital integral under the deformation to the Cartan motion groups. Let G be a connected, linear, real reductive Lie group and K be its maximal compact subgroup. Let $\mathfrak{g}$ and $\mathfrak{k}$ be the Lie algebra of G and K, and $\mathfrak{g} = \mathfrak{k} \oplus \mathfrak{p}$ be the Cartan decomposition. Let dg be the standard Haar measure on G. Let $\mathfrak{a}$ be the maximal abelian sub-algebra in $\mathfrak{p}$ and $\mathfrak{a}^+$ a positive Weyl chamber for the roots of $(\mathfrak{g}, \mathfrak{a})$.

LEMMA 3.1. *The following identity holds*

$$\int_G f(g) dg = \int_K \int_{\mathfrak{a}^+} \int_K f\left(k \exp(v) k'\right) \cdot \prod_{\alpha \in \Delta(\mathfrak{p}, \mathfrak{a})} \left(\sinh(\alpha(v))\right)^{n_\alpha} \, dk dv dk',$$

where $\Delta(\mathfrak{p}, \mathfrak{a})$ is the set of restricted roots of $\mathfrak{a}$-action on $\mathfrak{p}$ and n_α is the multiplicity.

PROOF. See [**28**, Proposition 5. 28]. □

For every non-zero real number t, we define a group G_t by using the global diffeomorphism $\varphi_t \colon K \times \mathfrak{p} \to G$ given by

$$(k, v) \mapsto k \exp_G(tv).$$

We define a family of groups

$$\mathcal{G} = \begin{cases} G_t, & t > 0, \\ G_0 = K \ltimes \mathfrak{p}, & t = 0. \end{cases}$$

The bijection

$$K \times \mathfrak{p} \times [0, 1] \to \mathcal{G}$$

defined by the formula

$$(k, v, t) \mapsto \begin{cases} (k, v, 0), & t = 0, \\ (k \exp(tv), t)\,, & t > 0, \end{cases}$$

is a diffeomorphism. As a result, every smooth function f on $\mathcal{G}$ has the following form

$$\begin{cases} f(k, v, 0) = F(k, v, 0), \\ f(k \exp(tv), t) = F(k, v, t), \quad t > 0, \end{cases}$$

for some smooth function F on $K \times \mathfrak{p} \times [0, 1]$.

DEFINITION 3.2. For every positive real number t, we define the Haar measure $d_t g$ on G_t by the following formula

$$\int_{G_t} f(g, t) \, d_t g = \frac{1}{t^{\dim \mathfrak{p} - \dim \mathfrak{a}}} \int_K \int_{\mathfrak{a}^+} \int_K f\left(k \exp(tv) k', t\right) \cdot \prod_{\alpha \in \Delta(\mathfrak{p}, \mathfrak{a})} \left(\sinh(\alpha(tv))\right)^{n_\alpha} \, dk dv dk'.$$

THEOREM 3.3. *Suppose that f is a smooth function on $\mathcal{G}$ with compact support and $x \in K$ is a regular element. Then*

$$\lim_{t\to 0}\int_{G_t} f(g\cdot x\cdot g^{-1},t)\,d_tg = \int_K\int_{\mathfrak{p}} f\left(kxk^{-1}, w - Ad_{kxk^{-1}}w, 0\right)\,dkdw.$$

PROOF. By Definition 3.2 of Haar measure on G_t, we have that

$$\begin{aligned}
&\int_{G_t} f(g\cdot x\cdot g^{-1},t)\,d_tg\\
=&\int_K\int_{\mathfrak{a}^+}\int_K f\left(k'\exp(tv)kxk^{-1}\exp(-tv)k'^{-1},t\right)\\
&\qquad\cdot\prod_{\alpha\in\Delta(\mathfrak{p},\mathfrak{a})}\left(\frac{\sinh(\alpha(tv))}{t}\right)^{n_\alpha} dkdvdk'\\
=&\int_K\int_{\mathfrak{a}^+}\int_K f\left(\exp(t\,\mathrm{Ad}_{k'}v)k'kxk^{-1}\exp(-tv)k'^{-1},t\right)\\
&\qquad\cdot\prod_{\alpha\in\Delta(\mathfrak{p},\mathfrak{a})}\left(\frac{\sinh(\alpha(tv))}{t}\right)^{n_\alpha} dkdvdk'\\
=&\int_K\int_{\mathfrak{a}^+}\int_K f\left(\exp(t\,\mathrm{Ad}_{k'}v)\exp(-t\,\mathrm{Ad}_{k'kxk^{-1}k'^{-1}}\circ Ad_{k'}v)k'kxk^{-1}k'^{-1},t\right)\\
&\qquad\cdot\prod_{\alpha\in\Delta(\mathfrak{p},\mathfrak{a})}\left(\frac{\sinh(\alpha(tv))}{t}\right)^{n_\alpha} dkdvdk'\\
=&\int_K\int_{\mathfrak{a}^+}\int_K f\left(\exp(t\,\mathrm{Ad}_{k'}v)\exp(-t\,\mathrm{Ad}_{kxk^{-1}}\circ Ad_{k'}v)kxk^{-1},t\right)\\
&\qquad\cdot\prod_{\alpha\in\Delta(\mathfrak{p},\mathfrak{a})}\left(\frac{\sinh(\alpha(tv))}{t}\right)^{n_\alpha} dkdvdk',
\end{aligned}$$

where in the last equation we have changed $k'k$ to k. Let $w = \mathrm{Ad}_{k'}v \in \mathfrak{p}$. Then

$$\begin{aligned}
&\lim_{t\to 0} f\left(\exp(t\,\mathrm{Ad}_{k'}v)\exp(-t\,\mathrm{Ad}_{kxk^{-1}}\circ Ad_{k'}v)kxk^{-1},t\right)\\
=&f\left(kxk^{-1}, w - \mathrm{Ad}_{kxk^{-1}}w, 0\right).
\end{aligned}$$

Moreover, we have

$$\lim_{t\to 0}\frac{\sinh(\alpha(tv))}{t} = \alpha(v).$$

Hence, by the compact support assumption on f, we can commute the integral with the limit with respect to t and compute

$$\begin{aligned}
&\lim_{t\to 0}\int_{G_t} f(g\cdot x\cdot g^{-1},t)\,d_tg\\
=&\int_K\int_{\mathfrak{a}^+}\int_K f\left(kxk^{-1}, \mathrm{Ad}_{k'}v - \mathrm{Ad}_{kxk^{-1}}\circ\mathrm{Ad}_{k'}v, 0\right)\cdot\prod_{\alpha\in\Delta(\mathfrak{p},\mathfrak{a})}(\alpha(v))^{n_\alpha}\,dkdvdk'\\
=&\int_K\int_{\mathfrak{p}} f\left(kxk^{-1}, w - \mathrm{Ad}_{kxk^{-1}}w, 0\right)\,dwdk,
\end{aligned}$$

where the last equation follows from the fact that the term

$$\prod_{\alpha\in\Delta(\mathfrak{p},\mathfrak{a})}(\alpha(v))^{n_\alpha}$$

is the Jacobian of the map $\mathfrak{a}^+\times K\to\mathfrak{p}$. □

As we assume in Section 2 that the K action on $\mathfrak{p}$ is spin, G_t/K is equipped with a G_t invariant spin structure. Let $D_{G_t/K}$ be the associated Dirac operator on the homogeneous space $G_t/K\cong G/K$ for $t\geq 0$. Let κ_t^s be the smoothing kernel of the heat operator $e^{-sD^2_{G_t/K}}$. It is known [**5**, **21**] that κ_t^s gives a Harish-Chandra Schwartz function on G_t.

LEMMA 3.4. *We have the following*

$$\lim_{t\to 0}\kappa_t^s(k\exp(tv))=\kappa_0^s(k,v).$$

PROOF. It is proved in [**7**, Theorem 2.48] that the heat kernel on a compact manifold is smooth with respect to the parameter t, indexing a smooth family of Riemannian metrics. The same proof extends to G_t/K with analogous estimates. See [**20**, Lemma 2.2, Theorem 2.2]. □

We define a function $\Phi\colon\mathcal{G}\to\mathbb{C}$ by

$$\begin{cases}\Phi(k,v,0)=s^{-\dim G/2}\cdot\exp\left(\frac{-|v|^2}{16s}\right),\\[2ex] \Phi(k\exp(tv),t)=s^{-\dim G/2}\cdot\exp\left(\frac{-|v|^2}{16s}\right),\quad t>0.\end{cases}$$

By the estimate in [**10**], there exists a positive constant C such that, for all $t\in[0,1]$,

$$\|\kappa_t^s(g)\|\leq C\cdot\Phi(g),\quad g\in G_t. \tag{3.1}$$

By the same computation in Theorem 3.3, as $\Phi(g)$ is independent of t, the integral

$$\int_{G_t}f(g\cdot x\cdot g^{-1},t)\,d_tg$$

is uniformly bounded for all $t\in[0,1]$. Hence, the following theorem follows from the dominated convergence theorem:

THEOREM 3.5. *For any $s>0$,*

$$\lim_{t\to 0}\int_{G_t}\kappa_t^s(g\cdot x\cdot g^{-1},t)\,d_tg=\int_K\int_{\mathfrak{p}}\kappa_0^s\left(kxk^{-1},w-Ad_{kxk^{-1}}w,0\right)\,dkdw.$$

4. K-theory Pairing

We have defined the deformation groupoid $\mathcal{G}$, a family of Lie groups $\{G_t\}_{0\leq t\leq 1}$, in last section. We can consider the reduced C^*-algebra of each G_t for the corresponding Haar measure. The field

$$\{C_r^*(G_t)\}_{0\leq t\leq 1}$$

is then a continuous field of C^*-algebras [**25**, Section 6.2].

Let E be an irreducible K-representation and $S_{\mathfrak{p}} = S_{\mathfrak{p}}^{+} \oplus S_{\mathfrak{p}}^{-}$ be the $\mathbb{Z}_2$-graded spin module in Section 2. We consider a family of homogeneous spaces G_t/K and a family of Dirac operators $\{D_t^E\}_{0 \leq t \leq 1}$ acting on

$$\left[L^2(G_t) \otimes S_{\mathfrak{p}} \otimes E\right]^K.$$

We denote by $D_{t,\pm}^E$ its restriction to

$$\left[L^2(G_t) \otimes S_{\mathfrak{p}}^{\pm} \otimes E\right]^K.$$

We consider the following Connes-Moscovici projection [**13**]

$$P_t^E = \begin{pmatrix} e^{-D_{t,-}^E D_{t,+}^E} & e^{\frac{-D_{t,-}^E D_{t,+}^E}{2}} \cdot \frac{(1-e^{-D_{t,-}^E D_{t,+}^E})}{D_{t,-}^E D_{t,+}^E} \cdot D_{t,-}^E \\ e^{-\frac{D_{t,+}^E D_{t,-}^E}{2}} \cdot D_{t,+}^E & 1 - e^{-D_{t,+}^E D_{t,-}^E} \end{pmatrix}$$

which is an idempotent and

$$[P_t^E] - \begin{pmatrix} 0 & 0 \\ 0 & 1 \end{pmatrix}$$

defines a generator in $K\left(C_r^*(G_t)\right)$. The following is the well-known Connes-Kasparov conjecture proved by Lafforgue [**29**].

THEOREM 4.1 (Connes-Kasparov isomorphism). *If G is spin, then the map*

$$R(K) \to K^*(C_r^* G) \colon [E] \to [P_t^E] - \begin{pmatrix} 0 & 0 \\ 0 & 1 \end{pmatrix}$$

is an isomorphism, where $R(K)$ denotes the character ring of K.

REMARK 4.2. One can find different approaches to the Connes-Kasparov conjecture in [**1, 6, 9, 25, 33**]. Let us point out the Afgoustidis' proof is based on the study of Cartan motion group suggested in [**6, 25**]. To be more precise, we consider

$$\mathcal{C} \colon = C^*\text{-algebra of continuous sections of } C_r^*(G_t).$$

The evaluation maps at $t = 0$ and $t = 1$ induce C^*-algebra morphism from $\mathcal{C}$ to $C_r^*(G_0)$ and $C_r^*(G)$, respectively, and in turn induce two isomorphisms

$$\alpha_0 \colon K(\mathcal{C}) \to K(C_r^*(G_0)), \quad \alpha_1 \colon K(\mathcal{C}) \to K(C_r^*(G)).$$

On the other hand, it is easy to show that $K(C_r^*(G_0)) \cong R(K)$. Chabert-Echterhoff-Nest [**8**] generalized the above Theorem 4.1 to almost connected Lie groups.

For any regular element $x \in T$, the orbital integral τ_x is a trace on $\mathcal{C}(G_t)$,

$$\tau_x(f \star g) = \tau_x(g \star f),$$

where $\star$ denotes the convolution product on $\mathcal{C}(G)$. Thus, it induces a continuous (by Theorem 3.3) family of traces

$$\tau_x \colon K(C_r^*(G_t)) \to \mathbb{C}.$$

THEOREM 4.3. *Assume that G satisfies the equal rank condition. Let E be an irreducible K-representation with highest weight μ. For all $t > 0$, we have that*

$$\tau_x\left([P_t^E]\right) = (-1)^{\frac{\dim \mathfrak{p}}{2}} \cdot \frac{\sum_{w \in W_K} (-1)^w \cdot e^{w(\mu+\rho_K)}(x)}{\prod_{\alpha \in \Delta^+(\mathfrak{g},\mathfrak{t})} \left(e^{\frac{\alpha}{2}} - e^{\frac{-\alpha}{2}}\right)(x)},$$

which is the character for the (limit of) discrete series representation of G with Harish-Chandra parameter μ.

PROOF. See [**26**, Theorem 3.1]. □

THEOREM 4.4. *When $t = 0$, we have that*

$$\tau_x\left([P_0^E]\right) = (-1)^{\frac{\dim \mathfrak{p}}{2}} \cdot \frac{\sum_{w\in W_K}(-1)^w \cdot e^{w(\mu+\rho_K)}(x)}{\prod_{\alpha\in\Delta^+(\mathfrak{g},\mathfrak{t})}\left(e^{\frac{\alpha}{2}} - e^{\frac{-\alpha}{2}}\right)(x)}.$$

In particular, the index pairing

$$\tau_x\left([P_t^E]\right) \in \mathbb{C}.$$

is independent of t.

PROOF. Here we provide two different proofs.

Proof I: By Theorem 4.4, we know that

$$\tau_x\left([P_t^E]\right) = (-1)^{\frac{\dim \mathfrak{p}}{2}} \cdot \frac{\sum_{w\in W_K}(-1)^w \cdot e^{w(\mu+\rho_K)}(x)}{\prod_{\alpha\in\Delta^+(\mathfrak{g},\mathfrak{t})}\left(e^{\frac{\alpha}{2}} - e^{\frac{-\alpha}{2}}\right)(x)},$$

for all $t > 0$. Applying Theorem 3.5,

$$\lim_{t\to 0}\tau_x\left([P_t^E]\right) = \tau_x\left([P_0^E]\right).$$

Proof II. We compute $\tau_x\left([P_0^E]\right)$ directly. To be more precise, let $f^{\pm} \in C_0(K \ltimes \mathfrak{p})$ be the smoothing kernel of $e^{-D_{t,\mp}^E D_{t,\pm}^E}$. One can compute that the Fourier transforms

$$\widehat{f}_{\pm}(z) = e^{-|z|^2} \cdot \mathrm{id}_{S_{\mathfrak{p}}^{\pm}\otimes E} \in C_0\left(\widehat{\mathfrak{p}}, \mathbb{K}\left(L^2(K)\right)\right)^K, \quad z \in \widehat{\mathfrak{p}}$$

respectively. By Proposition 2.7, we have that

$$\begin{aligned}\tau_x(f^+ - f^-) &= \frac{\chi_{S_{\mathfrak{p}}^+\otimes E}(x) - \chi_{S_{\mathfrak{p}}^-\otimes E}(x)}{\left(\chi_{S_{\mathfrak{p}}^+}(x) - \chi_{S_{\mathfrak{p}}^-}(x)\right)^2} \\ &= \frac{\chi_E(x)}{\left(\chi_{S_{\mathfrak{p}}^+}(x) - \chi_{S_{\mathfrak{p}}^-}(x)\right)}.\end{aligned}$$

The theorem follows immediately from Weyl character formula for K. □

It is crucial that the group element x in Theorem 4.4 is a regular element. In the following we show that the pairing between the L^2-trace τ_e and $K_0(C_r^*(G))$ does change with respect to t.

THEOREM 4.5. *For the L^2-trace, we have that*

$$\tau_e\left([P_t^E]\right) = t^{\dim \mathfrak{p}} \cdot \tau_e\left([P^E]\right).$$

PROOF. We follow the approach in [**12**]. First fix a volume form

$$\omega \in \Lambda^{\dim \mathfrak{p}}\mathfrak{p}^*.$$

For any top degree form $\lambda \in \Lambda^{\dim \mathfrak{p}}\mathfrak{p}^*$, we define $\lambda([\mathfrak{p}]) \in \mathbb{C}$ by

$$\lambda = \lambda([\mathfrak{p}]) \cdot \omega.$$

Then we identify the highest weight μ of E as an element in $\mathfrak{g}^*$ and define

$$\beta_\mu \in \Lambda^2\mathfrak{p}^*$$

by the following

$$\beta_\mu(X,Y) = \mu\left([X,Y]_{\mathfrak{g}}\right), \quad X,Y \in \mathfrak{p}.$$

By [**12**, Proposition 7.1 A],

$$\begin{aligned}
&\tau_e\left([P^E]\right)\\
(4.1)\qquad &= \text{formal degree of the discrete series representation determined by } E\\
&= \frac{1}{n\cdot 2^n}\left(\Lambda^n\beta_\mu\right)([\mathfrak{p}]).
\end{aligned}$$

where $n = \frac{\dim \mathfrak{p}}{2}$.

The family of Lie algebras $\mathfrak{g}_t$ can be identified with $\mathfrak{g}$ by the following map: for any $X = X_{\mathfrak{k}} + X_{\mathfrak{p}} \in \mathfrak{k} \oplus \mathfrak{p}$, we define

$$X^t = X_{\mathfrak{k}} + \frac{1}{t}\cdot X_{\mathfrak{p}} \in \mathfrak{g}_t.$$

One can check that

$$(4.2)\qquad \begin{cases} [X,Y]_{\mathfrak{g}_t} = [X,Y]_{\mathfrak{g}} & X,Y \in \mathfrak{k},\\ [X,Y]_{\mathfrak{g}_t} = t^2\cdot[X,Y]_{\mathfrak{g}} & X,Y\in\mathfrak{p},\\ [X,Y]_{\mathfrak{g}_t} = [X,Y]_{\mathfrak{g}} & X\in\mathfrak{k}, Y\in\mathfrak{p}.\end{cases}$$

Applying the formula (4.1) to $\mathfrak{g}_t$, we obtain the formula of β^t_μ from (4.2)

$$\beta^t_\mu = t^2\cdot\beta_\mu.$$

Therefore, we conclude

$$\tau_e\left([P^E_t]\right) = t^{\dim\mathfrak{p}}\cdot\tau_e\left([P^E]\right).$$

□

Theorem 4.5 suggests that in general for a singular element x, $\tau_x([P^E_t])$ changes with respect to t and the vanishing order depends on x.

Acknowledgments

We would like to thank Alexandre Afgoustidis, Wushi Gold-ring, Nigel Higson, Yuri Kordyukov, Shiqi Liu, Ryszard Nest, Sanaz Pooya, and Sven Raum for inspiring discussions. Part of the research was carried out within the online Research Community on Representation Theory and Noncommutative Geometry sponsored by the American Institute of Mathematics.

References

[1] Alexandre Afgoustidis, *On the analogy between real reductive groups and Cartan motion groups: a proof of the Connes-Kasparov isomorphism*, J. Funct. Anal. **277** (2019), no. 7, 2237–2258, DOI 10.1016/j.jfa.2019.02.023. MR3989145

[2] Iakovos Androulidakis, Omar Mohsen, and Robert Yuncken, *The convolution algebra of Schwartz kernels along a singular foliation*, J. Operator Theory **85** (2021), no. 2, 475–503, DOI 10.7900/jot. MR4249560

[3] Iakovos Androulidakis, Omar Mohsen, and Robert Yuncken, *A pseudodifferential calculus for maximally hypoelliptic operators and the Helffer-Nourrigat conjecture* **2** (2022).`arXiv:2201.12060`

[4] Michael Atiyah and Wilfried Schmid, *A geometric construction of the discrete series for semisimple Lie groups*, Invent. Math. **42** (1977), 1–62, DOI 10.1007/BF01389783.

[5] Dan Barbasch and Henri Moscovici, *L^2-index and the Selberg trace formula*, J. Funct. Anal. **53** (1983), no. 2, 151–201, DOI 10.1016/0022-1236(83)90050-2.

[6] Paul Baum, Alain Connes, and Nigel Higson, *Classifying space for proper actions and K-theory of group C^*-algebras*, C^*-algebras: 1943–1993 (San Antonio, TX, 1993), Contemp. Math., vol. 167, Amer. Math. Soc., Providence, RI, 1994, pp. 240–291, DOI 10.1090/conm/167/1292018. MR1292018

[7] Nicole Berline, Ezra Getzler, and Michèle Vergne, *Heat kernels and Dirac operators*, Grundlehren Text Editions, Springer-Verlag, Berlin, 2004. Corrected reprint of the 1992 original. MR2273508

[8] Jérôme Chabert, Siegfried Echterhoff, and Ryszard Nest, *The Connes-Kasparov conjecture for almost connected groups and for linear p-adic groups*, Publ. Math. Inst. Hautes Études Sci. **97** (2003), 239–278, DOI 10.1007/s10240-003-0014-2.

[9] Pierre Clare, Nigel Higson, Yanli Song, and Xiang Tang, *On the Connes-Kasparov Isomorphism, I. The Reduced C^*-algebra of a Real Reductive Group and the K-theory of the Tempered Dual*, submitted.

[10] Siu Yuen Cheng, Peter Li, and Shing Tung Yau, *On the upper estimate of the heat kernel of a complete Riemannian manifold*, Amer. J. Math. **103** (1981), no. 5, 1021–1063, DOI 10.2307/2374257.

[11] Alain Connes, *Noncommutative geometry*, Academic Press, Inc., San Diego, CA, 1994. MR1303779

[12] Alain Connes and Henri Moscovici, *The L^2-index theorem for homogeneous spaces of Lie groups*, Ann. of Math. (2) **115** (1982), no. 2, 291–330, DOI 10.2307/1971393.

[13] Alain Connes and Henri Moscovici, *Cyclic cohomology, the Novikov conjecture and hyperbolic groups*, Topology **29** (1990), no. 3, 345–388, DOI 10.1016/0040-9383(90)90003-3. MR1066176

[14] A. Connes and G. Skandalis, *The longitudinal index theorem for foliations*, Publ. Res. Inst. Math. Sci. **20** (1984), no. 6, 1139–1183, DOI 10.2977/prims/1195180375. MR775126

[15] Claire Debord, Jean-Marie Lescure, and Victor Nistor, *Groupoids and an index theorem for conical pseudo-manifolds*, J. Reine Angew. Math. **628** (2009), 1–35, DOI 10.1515/CRELLE.2009.017. MR2503234

[16] Erik van Erp, *The Atiyah-Singer index formula for subelliptic operators on contact manifolds. Part I*, Ann. of Math. (2) **171** (2010), no. 3, 1647–1681, DOI 10.4007/annals.2010.171.1647. MR2680395

[17] Erik van Erp, *The Atiyah-Singer index formula for subelliptic operators on contact manifolds. Part II*, Ann. of Math. (2) **171** (2010), no. 3, 1683–1706, DOI 10.4007/annals.2010.171.1683. MR2680396

[18] Erik van Erp and Robert Yuncken, *On the tangent groupoid of a filtered manifold*, Bull. Lond. Math. Soc. **49** (2017), no. 6, 1000–1012, DOI 10.1112/blms.12096. MR3743484

[19] Erik van Erp and Robert Yuncken, *A groupoid approach to pseudodifferential calculi*, J. Reine Angew. Math. **756** (2019), 151–182, DOI 10.1515/crelle-2017-0035. MR4026451

[20] D. Gong and M. Rothenberg, *Analytic torsion forms for noncompact fiber bundles*, Preprint (1997), https://archive.mpim-bonn.mpg.de/id/eprint/1027/.

[21] Harish-Chandra, *Harmonic analysis on real reductive groups. I. The theory of the constant term*, J. Functional Analysis **19** (1975), 104–204, DOI 10.1016/0022-1236(75)90034-8.

[22] Harish-Chandra, *Harmonic analysis on real reductive groups. I. The theory of the constant term*, J. Functional Analysis **19** (1975), 104–204, DOI 10.1016/0022-1236(75)90034-8.

[23] Harish-Chandra, *Invariant distributions on Lie algebras*, Amer. J. Math. **86** (1964), 271–309, DOI 10.2307/2373165.

[24] Harish-Chandra, *Harmonic analysis on real reductive groups. III. The Maass-Selberg relations and the Plancherel formula*, Ann. of Math. (2) **104** (1976), no. 1, 117–201, DOI 10.2307/1971058.

[25] Nigel Higson, *The Mackey analogy and K-theory*, Group representations, ergodic theory, and mathematical physics: a tribute to George W. Mackey, Contemp. Math., vol. 449, Amer. Math. Soc., Providence, RI, 2008, pp. 149–172, DOI 10.1090/conm/449/08711. MR2391803

[26] Peter Hochs and Hang Wang, *Orbital integrals and K-theory classes*, Ann. K-Theory **4** (2019), no. 2, 185–209, DOI 10.2140/akt.2019.4.185.

[27] Peter Hochs, Yanli Song, and Xiang Tang, *An index theorem for higher orbital integrals*, Math. Ann. **382** (2022), no. 1-2, 169–202, DOI 10.1007/s00208-021-02233-3. MR4377301

[28] Anthony W. Knapp, *Representation theory of semisimple groups*, Princeton Mathematical Series, vol. 36, Princeton University Press, Princeton, NJ, 1986. An overview based on examples, DOI 10.1515/9781400883974.

[29] Vincent Lafforgue, *Une démonstration de la conjecture de Baum-Connes pour les groupes réductifs sur un corps p-adique et pour certains groupes discrets possédant la propriété (T)* (French, with English and French summaries), C. R. Acad. Sci. Paris Sér. I Math. **327** (1998), no. 5, 439–444, DOI 10.1016/S0764-4442(99)80019-6. MR1652538

[30] Bertrand Monthubert and François Pierrot, *Indice analytique et groupoïdes de Lie* (French, with English and French summaries), C. R. Acad. Sci. Paris Sér. I Math. **325** (1997), no. 2, 193–198, DOI 10.1016/S0764-4442(97)84598-3. MR1467076

[31] Markus J. Pflaum, Hessel Posthuma, and Xiang Tang, *The transverse index theorem for proper cocompact actions of Lie groupoids*, J. Differential Geom. **99** (2015), no. 3, 443–472. MR3316973

[32] Yanli Song and Xiang Tang, *Higher Orbit Integrals, Cyclic Cocycles, and K-theory of Reduced Group C^*-algebra,* `arXiv:1910.00175`.

[33] Antony Wassermann, *Une démonstration de la conjecture de Connes-Kasparov pour les groupes de Lie linéaires connexes réductifs* (French, with English summary), C. R. Acad. Sci. Paris Sér. I Math. **304** (1987), no. 18, 559–562. MR894996

Department of Mathematics and Statistics, Washington University, St. Louis, Missouri 63130

Email address: `yanlisong@wustl.edu`

Department of Mathematics and Statistics, Washington University, St. Louis, Missouri, 63130

Email address: `xtang@wustl.edu`

Proceedings of Symposia in Pure Mathematics
Volume **105**, 2023
https://doi.org/10.1090/pspum/105/01915

On noncommutative crystalline cohomology

Boris Tsygan

Abstract. We outline several constructions of a noncommutative version of crystalline cohomology for an associative algebra over a field of positive characteristic.

1. Introduction

In this paper we review several approaches to generalizing cohomology of schemes over a field of characteristic p to the case of noncommutative algebras or, more generally, differential graded categories.

Recall that the classical (commutative) theory uses several ideas. One starts from a lifting of an $\mathbb{F}_p$-algebra to a $\mathbb{Z}_p$-algebra. If a good lifting exists, one defines crystalline cohomology as the De Rham cohomology of the lifting and proves that the result does not depend on a lifting. If not, one has to use other methods [**24**], [**3**]. Alternatively one can define De Rham forms and cohomology extending the method by which Witt vectors are defined [**22**]. In a yet another approach, one uses topological/bornological methods in non-Archimedean geometry [**37**].

Noncommutative versions of all of the above have been advanced in recent years. In noncommutative geometry, the analog of De Rham cohomology is (periodic) cyclic homology [**8**], [**9**], [**10**], [**35**], [**38**], [**47**]. It is more suited in characteristic zero, and it had been long recognized that topological Hochschild and cyclic homology [**5**], [**6**], [**26**], [**39**] works better for arithmetics. However, some versions of the standard theory can be used to give some basic definitions, and sometimes they give an alternative way of looking at the invariants from the topological theory. Below we present four different approaches along these lines.

2. Hochschild-Witt homology

Here we follow Kaledin's works [**29**], [**30**], [**31**].

2.1. Noncommutative Witt vectors. For a $\mathbb{Z}$-module V, consider the action of the cyclic group C_{p^n} on $V^{\otimes p^n}$ by permutations. We denote the generator of C_{p^n} by σ and write

$$N = 1 + \sigma + \ldots + \sigma^{p^n - 1} \tag{2.1}$$

2020 *Mathematics Subject Classification.* Primary 16E40, 16S38.

Key words and phrases. Hochschild and cyclic homology, crystalline cohomology, De Rham cohomology.

We denote the image of N by Norms.

LEMMA 2.1. *Let x and y be two elements of V. Then*

$$(x+y)^p - x^p - y^p \in \text{Norms}$$

COROLLARY 2.2. *In an associative algebra A over $\mathbb{F}_p$, for any x and y in A*

$$(x+y)^p = x^p + y^p$$

in $A/[A,A]$.

LEMMA 2.3. *(Noncommutative Dwork's lemma). Let x and y be two elements of V. Then*

$$(x+py)^{p^n} - x^{p^n} \in p\text{Norms}$$

in $V^{\otimes p^n}$.

PROOF. Let V be a free $\mathbb{Z}$-module with a basis $\{x_j | j \in J\}$. We will denote a monomial in $V^{\otimes p^n}$ (with respect to this basis) by X. We say a monomial is primitive if it is not a pth power of another monomial. Any monomial is of the form

$$X = Y^{p^{n-k}}$$

where y is a primitive monomial in $V^{\otimes p^k}$. This happens if and only if the C_{p^n}-orbit of X is of order p^k.

Let M_k be a set of representatives of all primitive monomials in $V^{\otimes p^k}$ (two monomials are equivalent if one is obtained from the other by a permutation from C_{p^k}).

Let V be a free $\mathbb{Z}$-module with the basis $\{x, y\}$. Then

$$(x+y)^{p^n} = \sum_{k=0}^{n} \sum_{Y \in M_k} N_{p^k}(Y^{p^{n-k}}) \tag{2.2}$$

where

$$N_{p^k} = 1 + \sigma + \ldots + \sigma^{p^k - 1} \tag{2.3}$$

(in particular $N = N_{p^n}$). Now let y be divisible by p. Then, unless $k = 0$ and $Y = x$, $N_{p^k}(Y^{p^{n-k}})$ is in the image of $p^{p^{n-k}} N_{p^k}$. But the image of $p^{n-k+1} N_{p^k}$ is in the image of pN. Indeed, $n - k + 1 \leq p^{n-k}$. This proves Lemma 2.3. Lemma 2.1 follows from (2.2) when $n = 1$. □

DEFINITION 2.4. Let V be a free $\mathbb{Z}$-module. Put

$$W_n(V) = (V^{\otimes p^n})^{C_{p^n}} / \text{Norms}$$

In other words,

$$W_n(V) = \check{H}^0(C_{p^n}, V^{\otimes p^n})$$

(the Tate cohomology of degree zero). Put also

$$W'_n(V) = (V^{\otimes p^n})^{C_{p^n}} / p\text{Norms}$$

LEMMA 2.5.

$$W_n(V) = \bigoplus_{k=0}^{n-1} \bigoplus_{Y \in M_k} (\mathbb{Z}/p^{n-k}\mathbb{Z}) N_{p^k}(Y^{p^{n-k}})$$

$$W'_n(V) = \bigoplus_{k=0}^{n} \bigoplus_{Y \in M_k} (\mathbb{Z}/p^{n-k+1}\mathbb{Z}) N_{p^k}(Y^{p^{n-k}})$$

(Recall that M_k is a set of representatives of primitive monomials of length p^k up to cyclic permutation).

The proof is clear: one only has to compute $M^{C_{p^n}}/N(M)$ and $M^{C_{p^n}}/pN(M)$ for a C_{p^n}-module M induced from a trivial representation of C_{p^k}.

LEMMA 2.6. *Let f and g be two linear maps $V_1 \to V_2$ that differ modulo p. Then $f^{\otimes p^n}$ and $g^{\otimes p^n}$ define the same maps $W_n(V_1) \to W_n(V_2)$ and $W'_n(V_1) \to W'_n(V_2)$.*

This follows from noncommutative Dwork's lemma 2.3.

COROLLARY 2.7. *For a vector space E over $\mathbb{F}_p$ choose a free $\mathbb{Z}$-module $\widetilde{E}$ together with an isomorphism $\widetilde{E}/p\widetilde{E} \xrightarrow{\sim} E$. For any $n \geq 0$, $E \mapsto W_n(\widetilde{E})$ is a well-defined functor from vector spaces over $\mathbb{F}_p$ to modules over $\mathbb{Z}/p^n\mathbb{Z}$.*

We will denote this functor by the same symbol W_n, or $W_n(E)$ where E is a vector space over $\mathbb{F}_p$.

LEMMA 2.8. *There is a natural isomorphism*

$$W'_n(V) \xrightarrow{\sim} W_{n+1}(V)$$

PROOF. First observe that the two sides become isomorphic if one identifies the terms corresponding to the same primitive monomial Y in the decomposition from Lemma 2.5. It remains to see that this isomorphism is natural. We call a linear map $V \to (V^{\otimes p})^{C_p}$ standard if the induced map

$$V/pV \to (V^{\otimes p})^{C_p}/\text{Norms}$$

is the isomorphism sending each v to v^p. From Lemma 2.6 we see that any standard map defines the same map $W'_n(V) \to W_{n+1}(V)$. On the other hand, the map $x_j \mapsto x_j^p, j \in J$, induces precisely the isomorphism above. □

2.1.1. *Restriction and Verschiebung.*

DEFINITION 2.9. Define the natural transformation

$$R : W_{n+1}(V) \to W_n(V)$$

by

$$W_{n+1}(V) \xleftarrow{\sim} W'_n(V) \to W_n(V)$$

where the isomorphism on the left is from Lemma 2.8 and the map on the right is the obvious projection.

In terms of the decomposition from Lemma 2.5, R is the projection

$$(\mathbb{Z}/p^{n+1-k}\mathbb{Z}) N_{p^k}(Y^{p^{n-k}}) \to (\mathbb{Z}/p^{n-k}\mathbb{Z}) N_{p^k}(Y^{p^{n-k}})$$

for every primitive monomial Y. If Y is of length p^n then it maps to zero.

DEFINITION 2.10. Define the natural transformation

$$V : W_n(V^{\otimes p}) \to W_{n+1}(V)$$

by

$$N_p : ((V^{\otimes p})^{\otimes p^n})^{C_{p^n}} \xrightarrow{\sim} (V^{\otimes p^{n+1}})^{C_{p^n}} \to (V^{\otimes p^{n+1}})^{C_{p^{n+1}}}$$

(Recall that
$$N_p = 1 + \sigma + \ldots + \sigma^{p-1};$$
note that V takes norms to norms. Indeed, on the left hand side the norm is given by
$$N = 1 + \sigma^p + \ldots + \sigma^{p(p^n-1)};$$
therefore its composition with N_p is the norm on the right).

2.2. Trace functors. Following Kaledin, we define the trace functor from a monoidal category $(\mathcal{A}, \otimes)$ to a category $\mathcal{K}$ as a functor $\mathrm{Tr} : \mathcal{A} \to \mathcal{K}$ together with a natural transformation

$$\tau_{M,N} : \mathrm{Tr}(M \otimes N) \xrightarrow{\sim} \mathrm{Tr}(N \otimes M) \tag{2.4}$$

such that

$$\tau_{M \otimes N, L} \tau_{N \otimes L, M} \tau_{L \otimes M, N} = \mathrm{id}_{\mathrm{Tr}(L \otimes M \otimes N)} \tag{2.5}$$

and

$$\tau_{M,1} = \tau_{1,M} = \mathrm{id}_{\mathrm{Tr}(M)} \tag{2.6}$$

Given a trace functor Tr from k-modules to a category $\mathcal{K}$, Kaledin defines a cyclic object $\mathrm{Tr}^\natural(A)$ of $\mathcal{K}$ for any k-algebra A. Namely, we put

$$\mathrm{Tr}^\natural(A)[n] = \mathrm{Tr}(A^{\otimes(n+1)}) \tag{2.7}$$

The face maps $d_0, \ldots, d_{n-1}$ are induced by the ones on $A^{\otimes(n+1)}$ and so are the degeneracy maps. The action of the cyclic permutation is by $\tau_{A^{\otimes n}, A}$.

More generally, for a k-algebra A with an authomorphism α of order p one defines a p-cyclic object $\mathrm{Tr}^\natural(A, \alpha)$ of $\mathcal{K}$.

2.3. The construction.

LEMMA 2.11. *The cyclic permutation*
$$\sigma : (M \otimes N)^{\otimes p^n} \xrightarrow{\sim} (N \otimes M)^{\otimes p^n},$$
$$v_1 \otimes w_1 \otimes \ldots \otimes v_{p^n} \otimes w_{p^n} \mapsto w_1 \otimes v_{p^n} \otimes \ldots \otimes w_{p^n} \otimes v_1,$$
$v_i \in M,\ w_i \in N$, *turns* W_n *into a trace functor.*

Now for an $\mathbb{F}_p$-algebra A define the cyclic $\mathbb{Z}/p^n\mathbb{Z}$-module

$$W_n^\natural(A)[k] = W_n(A^{\otimes k+1}) \tag{2.8}$$

DEFINITION 2.12.
$$W_n\mathrm{HH}_\bullet(A) = \mathrm{HH}_\bullet(W_n^\natural(A));\ W_n\mathrm{HC}_\bullet(A) = \mathrm{HC}_\bullet(W_n^\natural(A))$$

The following theorems are from [**29**] and [**30**].

THEOREM 2.13. *For a finitely generated smooth commutative algebra over* $\mathbb{F}_p$ *there is a natural isomorphism*
$$W_n\mathrm{HH}_\bullet(A) \xrightarrow{\sim} W_n\Omega^\bullet_A$$
where the right hand side denotes De Rham - Witt forms of Deligne-Illusie [**22**]. *This isomorphism intertwines the cyclic differential B with the De Rham differential.*

THEOREM 2.14. *For any algebra over* $\mathbb{F}_p$ *there is a natural isomorphism*
$$W_n\mathrm{HH}_0(A) \xrightarrow{\sim} W_n^H(A)$$
where the right hand side denotes Hesselholt's generalized Witt vectors [**25**] .

3. The construction of Petrov and Vologodsky

In [**41**] (cf. also [**40**]), another approach to noncommutative crystalline cohomology is proposed. For an algebra (or a DG category) over $\mathbb{F}_p$, a homology theory $\mathrm{HC}^{\mathrm{crys}}_\bullet(A)$ is constructed. It has the following properties.

(1) When A admits a lifting $\widetilde{A}$ to an algebra over $\mathbb{Z}_p$ then

$$\mathrm{HC}^{\mathrm{crys}}_\bullet(A) \xrightarrow{\sim} \widehat{\mathrm{HC}}^{\mathrm{per}}_\bullet(\widetilde{A});$$

(2)

$$\mathrm{HC}^{\mathrm{crys}}_\bullet(A) \xrightarrow{\sim} \widehat{\mathrm{TP}}_\bullet(A)$$

(TP denotes periodic topological cyclic homology; in both cases the hat denotes the p-adic completion).

A few words about the construction in general. When we do not have a lifting, what we can do is compute the periodic cyclic homology *over* $\mathbb{F}_p$. We do this in the derived sense, i. e. replace A by a DG algebra flat over $\mathbb{Z}$ and then computing the p-adically completed standard complex over the DG resolution flat over $\mathbb{Z}$. Explicitly, this resolution is

$$R = (\mathbb{Z}[\xi], p\frac{\partial}{\partial \xi}) \tag{3.1}$$

where ξ is a free commutative variable of degree -1.

There are two problems with this. First, $\mathrm{HC}^{\mathrm{per}}_\bullet(R)$ is too big.

$$\widehat{\mathrm{CC}}^{\mathrm{per}}_\bullet(R) \xrightarrow{\sim} \widehat{\mathrm{CC}}^{\mathrm{per}}_\bullet(\mathbb{Z}[\xi]) \tag{3.2}$$

(with zero differential on the algebra in the right hand side). This may be computed directly, or deduced from section 4. The right hand side projects to $\mathbb{Z}_p$ (the projection induced by $\mathbb{Z}[\xi] \to \mathbb{Z}$, $\xi \mapsto 0$). So the solution would be: tensor over $\widehat{\mathrm{CC}}^{\mathrm{per}}_\bullet(\mathbb{Z}[\xi])$ by $\mathbb{Z}_p$.

But then the second problem arises: neither side of (3.2) is in any natural way a commutative algebra. In fact, for a commutative algebra A, while $\mathrm{CC}^{\mathrm{per}}_\bullet(A)$ is an A_∞ algebra, it is not clear why it is a homotopy commutative ($\mathbb{E}_\infty$-)algebra, and if yes why is there a *preferred* E_∞ structure. Actually the known A_∞ structure [**27**] is manifestly *not* commutative.

The problem is solved as follows. As it is well known, periodic cyclic homology of a cyclic module can be computed in terms of the projection of the module onto the quotient by a certain subcategory. The projection does have a symmetric monoidal structure, and (3.1) can be established at the level of this quotient.

REMARK 3.1. The subtle issue of the symmetric monoidal structure on (periodic) cyclic homology is also studied in [**36**].

4. Operations on (co)chains and noncommutative crystalline cohomology

The contents of this chapter are from [**38**].

DEFINITION 4.1. A lifting of an $\mathbb{F}_p$-algebra A is a torsion-free p-adically complete $\mathbb{Z}_p$-module $\widetilde{A}$ with a p-adically continuous bilinear (not necessarily associative) product together with a product-preserving isomorphism $\widetilde{A}/p\widetilde{A} \xrightarrow{\sim} A$. For two $\mathbb{F}_p$-algebras A and B, a morphism of liftings $\widetilde{A} \to \widetilde{B}$ is a p-adically continuous linear map whose reduction modulo p is a morphism of algebras.

THEOREM 4.2. *For any lifting $\widetilde{A}$ of an $\mathbb{F}_p$-algebra A there exists a complex $\widehat{\mathrm{CC}}_{\bullet}^{\mathrm{PER}}(\widetilde{A})$ such that:*

(1) *If the product on $\widetilde{A}$ is associative then there is a quasi-isomorphism*
$$\widehat{\mathrm{CC}}_{\bullet}^{\mathrm{PER}}(\widetilde{A}) \xrightarrow{\sim} \widehat{\mathrm{CC}}_{\bullet}^{\mathrm{per}}(\widetilde{A})$$
where the right hand side is the p-adic completion of the periodic cyclic complex of the algebra $\widetilde{A}$;

(2) *For any $n \geq 1$ and for any chain of morphisms of liftings*
$$\widetilde{A}_0 \xrightarrow{f_1} \widetilde{A}_1 \xrightarrow{f_2} \dots \xrightarrow{f_n} \widetilde{A}_n$$
there is a map
$$T(f_1, \dots, f_n) : \widehat{\mathrm{CC}}_{\bullet}^{\mathrm{PER}}(\widetilde{A}_0) \to \widehat{\mathrm{CC}}_{\bullet}^{\mathrm{PER}}(\widetilde{A}_n)$$
such that
$$[b + uB, T(f_1, \dots, f_{n+1})] = \sum_{j=2}^{n} (-1)^j T(f_1, \dots, \widehat{f_j} \dots, f_{n+1}) -$$
$$\sum_{j=2}^{n} (-1)^j T(f_j, \dots, f_{n+1}) T(f_1, \dots, f_j)$$

Here $b + uB$ denotes the differential in $\widehat{\mathrm{CC}}_{\bullet}^{\mathrm{PER}}(\widetilde{A})$ for any $\widetilde{A}$.

4.1. Hochschild and cyclic cohomology of DG coalgebras. We start with the Hochschild and cyclic homology of second kind introduced and studied in [**42**], [**43**], [**44**]. Note that the established generality for them is a *conilpotent* (curved) DG coalgebra which ours will automatically be. Let (C, d) be a DG coalgebra with counit. We define Hochschild and cyclic complexes of C in a way dual to what one does for algebras, with one nuance. Namely, we put

$$C_{II}^{\bullet}(C) = \bigoplus_{n \geq 0} C \otimes \overline{C}[-1]^{\otimes n} \tag{4.1}$$

with the differential $b + d$ and

$$\mathrm{CC}_{II}^{\bullet}(C) = (C_{II}^{\bullet}(C)[[u]], b + d + uB) \tag{4.2}$$

with the differential $b + d + vB$ Here $\overline{C}$ is the kernel of the counit and
$$b : C \otimes \overline{C}[-1]^{\otimes n} \to C \otimes \overline{C}[-1]^{\otimes (n+1)}; \ B : C \otimes \overline{C}^{\otimes n}[-1] \to C \otimes \overline{C}[-1]^{\otimes (n-1)}$$
are defined dually to the standard Hochschild and cyclic differentials for algebras.

An important feature of both is that they are *not* invariant with respect to quasi-isomorphisms of DG (co)algebras. This suits us well because we are going to consider the example $C = \mathrm{Bar}(A)$ where A is a DG algebra, and C is contractible when A has a unit.

THEOREM 4.3. *For an associative algebra A there are natural quasi-isomorphisms*
$$C_{II}^{\bullet}(\mathrm{Bar}(A)) \xrightarrow{\sim} C_{\bullet}(A); \ \mathrm{CC}_{II}^{\bullet}(\mathrm{Bar}(A)) \xrightarrow{\sim} \mathrm{CC}_{\bullet}^{-}(A)$$
where the right hand side is the standard Hochschild, resp. negative cyclic, complex of A.

This is proven in [**45**] and [**17**].

4.2. The multiplicative structure on cyclic cochains. We start by extending the cyclic Alexander-Whitney constructions from [**2**] and [**34**]. Given two associative algebras A_1 and A_2, there are two homotopy inverse maps ($C_\bullet$ denotes the Hochschild complex):

$$C_\bullet(A_1 \otimes A_2) \rightleftarrows C_\bullet(A_1) \otimes C_\bullet(A_2) \tag{4.3}$$

(i.e. the Alexander-Whitney the Eilenberg-Zilber morphisms). (The left hand side is the total complex of a bisimplicial Abelian group and the right hand side is the complex of the diagonal). The maps satisfy an associativity condition when we consider three algebras. In particular, the Hochschild complex of a bialgebra is a coalgebra. Dualizing this, we see that the Hochschild complex (4.1) of a bialgebra viewed as a coalgebra acquires an algebra structure.

The above can be extended to cyclic (co)chains. For any n coalgebras $C_1 \ldots, C_n$ there is a map

$$\mathrm{CC}_{II}(C_1) \otimes \ldots \otimes \mathrm{CC}_{II}(C_n) \to \mathrm{CC}_{II}(C_1 \otimes \ldots \otimes C_n) \tag{4.4}$$

satisfying the A_∞ relation. This implies that for any bialgebra H its cyclic complex $\mathrm{CC}_{II}(H)$ is an A_∞ algebra.

Therefore for any bialgebra we have an associated A_∞ algebra. It can be constructed explicitly when H is a cocommutative DG Hopf algebra.

4.2.1. *The DG algebra* $H \star \mathrm{Cobar}(\overline{H})$. To start with, there are two subalgebras of $\mathrm{CC}_{II}(H)$ for a bialgebra H. One is H itself (the degree zero cyclic cochains). The other is the subalgebra $k \otimes \overline{H}[-1]^{\otimes \bullet}$ which is identified with the cobar construction of the coalgebra $\overline{H}$. It turns out that the Hochschild complex of a Hopf algebra H is isomorphic to the cross product of the two. For the cyclic complex, the same is true up to an A_∞ quasi-isomorphism and with the addition of a cross term in the differential. Below are the details.

For a bialgebra H and an algebra A, an action of H on A is a linear map $H \otimes A \to A$, $x \otimes a \mapsto \rho(x)a$, such that

$$\rho(xy) = \rho(x)\rho(y);\ \rho(x)(ab) = \sum \rho(x^{(1)})(a)\rho(x^{(2)})(b)$$

If H is a Hopf algebra acting on A then one can define a cross product

$$H \star A = A \otimes H; (a \otimes x)(b \otimes y) = a\rho(x^{(1)})b \otimes Sx^{(2)}y \tag{4.5}$$

Let $A = \mathrm{Cobar}(\overline{H})$. Put

$$\rho(x)(x_1|\ldots|x_n) = \sum (x^{(1)}x_1 S(x^{(n+1)})|\ldots|x^{(n)}x_n S(x^{(2n)})) \tag{4.6}$$

where S is the antipode. The action commutes with the differential on $\mathrm{Cobar}(\overline{\mathrm{H}})$ (which we denote by b), and we get a DG algebra $H \star \mathrm{Cobar}(\overline{H})$.

REMARK 4.4. Note that the comultiplication on $\overline{H}$ is given by

$$\Delta x = \sum x^{(1)} \otimes x^{(2)} - 1 \otimes x - x \otimes 1$$

In other words, $H \star \mathrm{Cobar}(\overline{H})$ is the DG algebra generated by a subalgebra H and by elements (x), linear in $x \in \overline{H}[1]$, subject to

$$x \cdot (y) = \sum (x^{(1)} y S(x^{(2)})) \cdot x^{(3)};\ bx = 0;\ b(x) = \sum (x^{(1)})(x^{(2)}) \tag{4.7}$$

This DG algebra admits a derivation B determined by

$$Bx = 0,\ x \in H;\ B(x) = x,\ x \in \overline{H}[-1].$$

It is easy to see that B is well defined and commutes with b. Of course, if H is a DG Hopf algebra, then its differential d induces an extra differential on $H \star \mathrm{Cobar}(\overline{H})$.

PROPOSITION 4.5. *For a cocommutative DG Hopf algebra H,*
1) there is an isomorphism of DG algebras

$$C^{\bullet}_{\mathrm{II}}(H) \xrightarrow{\sim} (H \star \mathrm{Cobar}(\overline{H}), b + d);$$

2) there is a natural $k[[u]]$-linear (u)-adically continuous A_∞ isomorphism

$$\mathrm{CC}^{\bullet}_{\mathrm{II}}(H) \xrightarrow{\sim} ((H \star \mathrm{Cobar}(\overline{H}))[[u]], b + d + uB)$$

(An A_∞ isomorphism is an A_∞ morphism whose first term is invertible).

5. The action on the periodic cyclic complex

5.1. The A_∞ action of $\mathrm{CC}^{\bullet}_{\mathrm{II}}(U(\mathfrak{g}_A))$. For an associative algebra A, let $\mathfrak{g}_A$ denote the DG Lie algebra $C^{\bullet+1}(A, A)$ with the Gerstenhaber bracket which we identified with $\mathrm{Coder}(\mathrm{Bar}(A))$. We do not assume that the coderivations preserve the coaugmentation. In other words, the Abelian subalgebra $A[1]$ is contained in $\mathfrak{g}_A$. The latter defines the action of $U(\mathfrak{g}_A)$ on $\mathrm{Bar}(A)$ as well as linear maps

$$\mu_N : U(\mathfrak{g}_A)^{\otimes N} \otimes \mathrm{Bar}(A) \to \mathrm{Bar}(A) \tag{5.1}$$

(composition of the above action with the n-fold product on $U(\mathfrak{g}_A)$).

LEMMA 5.1. *The above are morphisms of DG coalgebras.*

PROOF. Clear. □

COROLLARY 5.2. *The compositions of*

$$\mathrm{CC}^{\bullet}_{\mathrm{II}}(U(\mathfrak{g}_A))^{\otimes N} \otimes \mathrm{CC}^{\bullet}_{\mathrm{II}}(\mathrm{Bar}(A)) \longrightarrow \mathrm{CC}^{\bullet}_{\mathrm{II}}(U(\mathfrak{g}_A)^{\otimes N} \otimes \mathrm{Bar}(A))$$

(cf. (4.4)) with the morphism induced by μ_N define on $\mathrm{CC}^{\bullet}_{\mathrm{II}}(\mathrm{Bar}(A))$ a structure of an A_∞ module over the A_∞ algebra $\mathrm{CC}^{\bullet}_{\mathrm{II}}(U(\mathfrak{g}_A))$.

REMARK 5.3. In classical calculus on a manifold M, the counterpart of the periodic cyclic complex is the De Rham complex $(\Omega^{\bullet}_M((u)), ud)$. If $\mathfrak{g}$ is the algebra of multivector fields then the DG Lie algebra $(\mathfrak{g}[[u]][\epsilon], u\frac{\partial}{\partial\epsilon})$ acts on this complex: an element $X + \epsilon Y$ acts by $L_X + \iota_Y$. In characteristic zero, the DGLA $(\mathfrak{g}_A[[u]][\epsilon], u\frac{\partial}{\partial\epsilon})$ does act on $\mathrm{CC}^{-}_{\bullet}(A)$ as on an L_∞ module. Cf. for example [**48**] and, for a new and different perspective, [**7**]. The A_∞ algebra

$$\mathrm{CC}^{\bullet}_{II}(U(\mathfrak{g}_A)) \xrightarrow{\sim} (U(\mathfrak{g}_A) \star \mathrm{Cobar}(\overline{U}(\mathfrak{g}_A))[[u]], b + \delta + uB),$$

while L_∞ quasi-isomorphic to the former in characteristic zero, seems to be better suited to working over $\mathbb{Z}$, and perhaps also in a more general categorical context.

5.2. The construction. Let $\mathfrak{a}$ be the graded Lie algebra with the basis consisting of one element R of degree two. We start by constructing

$$x \in \prod_{n=1}^{\infty} u^{-n} (U(\mathfrak{a}) \star_1 \mathrm{Cobar}(\overline{U}(\mathfrak{a})))^{2n} \tag{5.2}$$

satisfying

$$(\partial_{\mathrm{Cobar}} + uB)x + x^2 = -R \tag{5.3}$$

Here the graded component of degree n is spanned by elements of degree $2n$ with respect to R and of degree 0 with respect to u. We are looking for a solution of the form

$$x_F(R) = \sum_{n=1}^{\infty} x_n(R)(R^n) \text{ where } F(R,y) = \sum_{n=1}^{\infty} x_n(R)y^n \tag{5.4}$$

Equation (5.3) translates into the following two: first,

$$F(y_1+y_2) - F(y_1) - F(y_2) + F(y_1)F(y_2) = 0$$

which implies

$$x_n(R) = -\frac{1}{n!}f(R)^n$$

for some f, and second,

$$-u\sum_{n=1}^{\infty}\frac{1}{n!}f(R)^n R^n = -R.$$

This implies

$$f(R)R = \log(1+\frac{R}{u});\ 1-F(R,y) = \exp(\frac{y}{R}\log(1+\frac{R}{u})) = (1+\frac{R}{u})^{\frac{y}{R}};$$

we conclude that

$$F(R,y) = -\sum_{n=1}^{\infty}\frac{u^{-n}}{n!}y(y-R)\dots(y-(n-1)R) \tag{5.5}$$

A crucial observation for us is that the homogeneous part of $x_F(R)$ ((5.4)) of total degree n in R has the denominator $n!$.

Now assume that we have a chain of morphisms

$$\widetilde{A}_0 \xrightarrow{f_1} \widetilde{A}_1 \xrightarrow{f_2} \dots \xrightarrow{f_n} \widetilde{A}_n \tag{5.6}$$

Here $\widetilde{A}_j$ is a lifting of an algebra A_j and f_j are morphisms modulo p. Let

$$\mathcal{U}(m_0,\dots,m_n) = U(\mathrm{Lie}(m_0,\dots,m_n)) \star_1 \mathrm{Cobar}(\overline{U}(\mathrm{Lie}(m_0,\dots,m_n)) \tag{5.7}$$

where $\mathrm{Lie}(m_0,\dots,m_n)$ stands for the free Lie algebra with generators $m_0,\dots,m_n$ each of degree one. We claim that there is a pairing

$$\mathcal{U}(m_0,\dots,m_n) \otimes \widehat{\mathrm{CC}}_\bullet^{\mathrm{PER}}(\widetilde{A}_0) \to \widehat{\mathrm{CC}}_\bullet^{\mathrm{PER}}(\widetilde{A}_n) \tag{5.8}$$

Moreover: define the DG category as follows. An object is a lifting $\widetilde{A}$ of an $\mathbb{F}_p$-algebra A. A morphism from $\widetilde{A}_0$ to $\widetilde{A}_n$ is a pair of a chain $\{f_j\}$ (5.6) and an element of $\mathcal{U}(m_0,\dots,m_n)$. Composition is concatenation of chains together with the product

$$\mathcal{U}(m_0,\dots,m_n)\otimes\mathcal{U}(m_n,\dots,m_{n+k}) \to \mathcal{U}(m_0,\dots,m_{n+k}) \tag{5.9}$$

(embed the two algebras on the left into the one on the right, and multiply there). Then (5.8) extends to an A_∞ functor from this DG category to the category of complexes.

To see that, we need to refer to section 8, namely to 8.1. Define for any chain (5.6) a morphism of coalgebras

$$U(\mathrm{Lie}(m_0)) \otimes \dots \otimes U(\mathrm{Lie}(m_n)) \to \mathrm{Bar}(\mathbf{C}^\bullet(A_0, A_n)) \tag{5.10}$$

as follows:

$$m_0^{k_0}\otimes\ldots\otimes m_n^{k_n}\mapsto f_{1*}\widetilde{m}_0^{\bullet k_0}\bullet\ldots\bullet f_{n*}\widetilde{m}_{n-1}^{\bullet k_{n-1}}\bullet\widetilde{m}_n^{\bullet k_n} \tag{5.11}$$

Here $\widetilde{m}_j$ are cochains in $C^2(\widetilde{A}_j,\widetilde{A}_j)$ corresponding to the product. The above intertwines the $\bullet$ product(8.7) with the product in $U(\mathrm{Lie}(m_0))\otimes\ldots\otimes U(\mathrm{Lie}(m_{n+k}))$. Now follow (5.10) by

$$U(\mathrm{Lie}(m_0,\ldots,m_n))\to U(\mathrm{Lie}(m_0))\otimes\ldots\otimes U(\mathrm{Lie}(m_{n+k}))$$

and apply $\mathrm{CC}^{\bullet}_{II}$. We then get (5.8).

Next, let $\mathfrak{g}$ be the free graded Lie algebra over $k[[u]]$ generated by three elements λ of degree 1 and $\delta\lambda$, R of degree 2. Define a derivation δ of degree one by

$$\delta:\lambda\mapsto\delta\lambda\mapsto[R,\lambda];\ R\mapsto 0.$$

(For us, this is a subalgebra of $\mathrm{Lie}(m_0,m_1)[\frac{1}{2}][[u]]$ with $\lambda=m_1-m_0$, $R=m_0^2$, and $\delta=[m_0,\ ])$.

We assign weight one to λ and $\delta\lambda$, and weight two to R. Extend the weight to the algebra

$$\mathcal{U}=U(\mathfrak{g})\star_1\mathrm{Cobar}(\overline{U}(\mathfrak{g})) \tag{5.12}$$

multiplicatively. Denote by $\mathfrak{g}(n)$, $\mathcal{U}(n)$, etc. the span of all homogeneous elements of weight at least n. Let

$$\widehat{\mathcal{U}}=\prod_{k=0}^{\infty}\frac{u^{-k}k!}{n!}\mathcal{U}(k)[[u]] \tag{5.13}$$

For any $r\in\mathfrak{g}^2(1)$, define an element of

$$x_F(r)\in\widehat{\mathcal{U}} \tag{5.14}$$

by (5.4) with R replaced by r.

Consider the differential

$$\mu(t)=\delta+x(R)t-(-1)^l tx(R-\delta\lambda+\lambda^2) \tag{5.15}$$

for $t\in\mathcal{U}^l$.

We construct an invertible element t_{01} of degree zero in the completion $\widehat{U}$, such that

$$(\mu+\partial_{\mathrm{Cobar}}+uB)t_{01}+t_{01}\lambda=0 \tag{5.16}$$

We write $t_{01}=1+x_1+x_2+\ldots$ where x_k is in $\mathcal{U}(k)$. We find x_n recursively just as we did above, using the acyclicity of the differential induced by μ on $\mathcal{U}(n)/\mathcal{U}(n+1)$. For example, $x_1=-\frac{(\lambda)}{u}$.

It remains to prove that $\mathrm{gr}(\mathcal{U})$ is indeed acyclic. Let $\mathfrak{g}_0$ be the graded Lie algebra $\mathfrak{g}$ with the differential δ_0 defined by

$$\delta_0:\lambda\mapsto\delta\lambda\mapsto 0;\ R\mapsto 0.$$

Define the DGA

$$\mathcal{U}_0=U(\mathfrak{g}_0)\star_1\mathrm{Cobar}(\overline{U}(\mathfrak{g}_0)) \tag{5.17}$$

with the differential $\delta_0+\partial_{\mathrm{Cobar}}+uB$. Then

$$\mathrm{gr}(\mathcal{U})\xrightarrow{\sim}\oplus\frac{u^{-k}}{k!}\mathcal{U}_0(k)$$

Looking at the differential δ_0 as the leading term of a spectral sequence, we see that the above is quasi-isomorphic to itself with $\mathfrak{g}_0$ is replaced by $\mathfrak{a}_0$, the latter being the free graded Lie algebra generated by R of degree two (and weight one). Looking at uB as the leading term in a spectral sequence, we see that our complex is indeed acyclic.

More generally, let $\mathfrak{g}_n$ be the free graded algebra with generators λ_{0j} of degree one and $\delta\lambda_{0j}$ of degree two, $1 \le j \le n$, as well as R of degree two. Let the weight of $\delta\lambda_{0j}$ and λ_{0j} be one, and the weight of R be two. Define a derivation δ of degree one by

$$\delta : \lambda_{0j} \mapsto \delta\lambda_{0j} \mapsto [R, \lambda_{0j}]. \tag{5.18}$$

REMARK 5.4. Consider the graded Lie algebra generated by elements $m_0, \dots, m_n$ of degree one and R of degree two, subject to the relation $[m_0, m_0] = 2R$. Let $\delta = [m_0, \,]$. The span of all monomials of degree > 1 and of $m_0 - m_j$, $1 \le j \le n$, is a graded subalgebra stable under δ. It maps to $\mathfrak{g}_n$ via $m_0 - m_j \mapsto \lambda_{0j}$; $R \mapsto R$.

To finish the proofs, recall the A_∞ module structure given by Proposition 4.5 and Corollary 5.2. Together with the above, it gives an A_∞ functor to the category of complexes from the following DG category: objects are liftings of $\mathbb{F}_p$-algebras; morphisms are:

(1) morphisms of liftings (recall: they preserve the product only modulo p, and the products are associative only modulo p);
(2) in addition, every object has an endomorphism of degree one and square zero; and
(3) morphisms in (1) commute with the morphisms in (2).

The value of the A_∞ functor on an object $\widetilde{A}$ is the complex $\widehat{\mathrm{CC}}_\bullet^{\mathrm{PER}}(\widetilde{A})$.

6. Comparisons between different constructiions

Let us start by comparing sections 3 and 4. They obviously give the same result when A admits a lifting to an algebra over $\mathbb{Z}_p$. When this is not the case, the constructions diverge; we strongly expect them to give the same answer. Note that the general construction of section 3 is carried out in terms quite close to the methods of section 4. Clarifying these connections could be quite instructive.

Next we compare sections 2 and 4. We start by giving more details on the approach of section 2.

6.1. More on Frobenius under the trace. Let E be a finite dimensional space over $\mathbb{F}_p$. Corollary 2.2 provides a linear map

$$F : E \to (E^{\otimes p})_{C_p} \tag{6.1}$$

sending e to e^p for any e in E. Passing to dual spaces and taking the adjoint operator, we get

$$C : (E^{\otimes p})^{C_p} \to E \tag{6.2}$$

For a more general perfect field of scalars k, those maps are linear if the action on E is twisted by the Frobenius automorphism of k.

Explicitly: when a basis of E is chosen, F and C act on the corresponding basis of $E^{\otimes p^n}$ as follows: F sends a monomial to its pth power; C sends a monomial X

to Y if $X = Y^p$ for some Y; if there is no such Y then X is sent to zero. Applying this to $E^{\otimes p^n}$ instead of E, we get an inverse system

$$C : (E^{\otimes p^n})^{C_{p^n}} \to (E^{\otimes p^{n-1}})^{C_{p^{n-1}}} \tag{6.3}$$

LEMMA 6.1. *For a free $\mathbb{Z}$-module M there is a natural isomorphism*

$$W_{n+1}(M)/pW_{n+1}(M) \xrightarrow{\sim} ((M/pM)^{\otimes p^n})^{C_{p^n}}$$

which intertwines the restriction R with C as in (6.3).

PROOF. This was proven in [**31**], [**29**] and essentially follows from Lemma 2.5. □

6.1.1. *Recollection on cyclic objects.* Fix a unital commutative ring K. For a small category $\mathcal{C}$, a $\mathcal{C}$-module is by definition a functor $\mathcal{C} \to K-\mathrm{mod}$. For a functor $f : \mathcal{C}_1 \to \mathcal{C}_2$, we have the restriction functor

$$f^* : \mathcal{C}_2 - \mathrm{mod} \to \mathcal{C}_1 - \mathrm{mod} \tag{6.4}$$

There are two functors

$$f_!,\ f_* : \mathcal{C}_1 - \mathrm{mod} \to \mathcal{C}_2 - \mathrm{mod} \tag{6.5}$$

The functor $f_!$ is left adjoint and the functor f_* is right adjoint to f^*. A bit informally,

$$f_!M = \mathcal{C}_2 \times_{\mathcal{C}_1} M;\ f_*M = \mathrm{Hom}_{\mathcal{C}_1}(\mathcal{C}_2, M).$$

Recall the definition of the cyclic category Λ ([**11**]) and its generalizations Λ_p for any natural number p ([**16**], [**39**]). For $p = 1$, $\Lambda = \Lambda_p$. By definition, a cyclic K-module is a Λ^{op}-module; a p-cyclic K-module is a Λ_p^{op}-module. One has

$$\mathrm{Aut}_{\Lambda_p}([n]) \xrightarrow{\sim} C_{p(n+1)}$$

(the cyclic group). the subgroups C_p for all n form the center of Λ_p.

REMARK 6.2. One also defines the category Λ_∞ ([**14**], [**4**], [**16**]) with the same objects, with

$$\mathrm{Aut}_{\Lambda_\infty}([n]) \xrightarrow{\sim} \mathbb{Z}.$$

For any $p \geq 1$, Λ_p is the quotient (in an easy precise sense) of Λ_∞ over the subgroup $p(n+1)\mathbb{Z}$.

Let Δ be the category whose objects are $[n] = \{0, \ldots, n\}$ for all $n \geq 0$ and whose morphisms are order-preserving maps. By definition, simplicial K-modules are Δ^{op}-modules.There are inclusions $j_p : \Delta \to \Lambda_p$ for all p.

For any associative algebra A over K one defines the Λ^{op}-module $A^\natural$. More generally, for an algebra A together with an automorphism α of degree p one defines the Λ_p^{op}-module $A_\alpha^\natural$.One has

$$A_\alpha^\natural([n]) = A^{\otimes n+1} \tag{6.6}$$

Each group $\mathrm{Aut}_{\Lambda_p}([n])$ has a generator t_n that acts by

$$t_n(a_0 \otimes \ldots \otimes a_n) = \alpha(a_n) \otimes a_0 \otimes \ldots \otimes a_{n-1}$$

We put

$$\tau_n = t_n^{n+1} \tag{6.7}$$

For any small category $\mathcal{C}$ define

$$\mathrm{pr}_{\mathcal{C}} : \mathcal{C} \to *$$

to be the unique morphism to the category with one morphism. For a simplicial module M define

(6.8) $$\mathrm{HH}_\bullet(M) = \mathbb{L}_\bullet \mathrm{pr}_{\Delta !}(M)$$

For a p-cyclic module M define

(6.9) $$\mathrm{HC}_\bullet(M) = \mathbb{L}_\bullet \mathrm{pr}_{\Lambda_p !}(M);\ \mathrm{HH}_\bullet(M) = \mathrm{HH}_\bullet(j_p^* M)$$

For an algebra A with an automorphism α of degree p one has

(6.10) $$\mathrm{HH}_\bullet(A^\natural_\alpha) \xrightarrow{\sim} H_\bullet(A, {}_\alpha A)$$

(the Hochschild homology of A with coefficients in A on which A acts by multiplication on the right, and by multiplication twisted by α on the left). Also,

(6.11) $$\mathrm{HC}_\bullet(A^\natural_\alpha) \xrightarrow{\sim} \mathrm{HC}_\bullet(A, \alpha)$$

(the generalized cyclic homology from [**17**]). This is the generalized theorem of Connes [**11**].

6.2. Subdivisions. Here we follow [**28**]. There are two functors

(6.12) $$\pi_p, i_p : \Lambda_p \to \Lambda.$$

The functor π_p sends τ_n to 1. One has

(6.13) $$\pi_{p!}(M) = M_{C_p};\ \pi_{p*}(M) = M^{C_p}$$

where C_p is the center. To define i_p, start with any algebra A and pass to the algebra $A^{\otimes p}$ with the automorphism α acting by cyclic permutation. Now construct the p-cyclic module $(A^{\otimes p})^\sharp_\alpha$. We have

$$(A^{\otimes p})^\sharp_\alpha[n] = A^\sharp[p(n+1)-1]$$

and it is clear from the construction that the rest of the p-cyclic structure on $(A^{\otimes p})^\sharp_\alpha$. is constructed in terms of the cyclic structure on $A^\sharp$. In other words, there exists unique functor $i_p : \Lambda_p \to \Lambda$ such that

(6.14) $$(A^{\otimes p})^\sharp_\alpha = i_p^*(A^\sharp)$$

In particular

$$i_p([n]) = p(n+1) - 1$$

PROPOSITION 6.3. *There are natural isomorphisms*

$$\mathrm{HH}_\bullet(i_p^* M) \xrightarrow{\sim} \mathrm{HH}_\bullet(M);\ \mathrm{HC}_\bullet(i_p^* M) \xrightarrow{\sim} \mathrm{HC}_\bullet(M)$$

where M is any cyclic module.

6.2.1. *Frobenius under the trace and subdivisions.* Let A be an algebra over $\mathbb{F}_p$. Apply Lemma 6.1 and (6.3) to $E = A^{\otimes(n+1)}$. It is straightforward that the morphisms C and F are compatible with the cyclic structures and therefore define morphisms

(6.15) $$F : A^\natural \to \pi_{p!} i_p^* A^\natural$$

(6.16) $$C : \pi_{p*} i_p^* A^\natural \to A^\natural$$

There is an inverse system of morphisms of cyclic modules

(6.17) $$C : \pi_{p^n *} i_{p^n}^* A^\natural \to \pi_{p^{n-1} *} i_{p^{n-1}}^* A^\natural$$

and the diagram where the vertical maps are isomorphisms:

$$\begin{array}{ccc} W^{\natural}_{n+1}(A)/pW^{\natural}_{n+1}(A) & \xrightarrow{R} & W^{\natural}_{n}(A)/pW^{\natural}_{n}(A) \\ \downarrow & & \downarrow \\ \pi_{p^n *} i^*_{p^n} A^{\natural} & \xrightarrow{C} & \pi_{p^{n-1} *} i^*_{p^{n-1}} A^{\natural} \end{array} \tag{6.18}$$

To summarize: We observe that the periodic version of the Hochschild-Witt complex has some resemblance to the construction in section 4. Both use a lifting of A. Both, when reduced modulo p, become nonstandard (larger) complexes that compute periodic cyclic homology of A or something close to it. We hope that a common framework for operations on higher Hochschild chains and cochains, as discussed in the end of this article, will allow us to compare the two constructions directly.

7. Noncommutative dagger completions

In [**12**] and [**13**] Cortiñas, Cuntz, Meyer, Mukherjee, and Tamme introduce and study another approach to noncommutative crystalline cohomology, the one based on Monsky and Washnitzer's work [**37**]. It would be very interesting to compare this with another approaches outlined in this article.

8. What do DG categories form?

8.1. A category in cocategories. For every two algebras A and B and any two morphisms $f, g : A \to B$ we consider the Hochschild cochain complex $C^\bullet(A,\, {}_fB_g)$ where B is viewed as an A-bimodule on which A acts on the left via f and on the right via g. We use the picture

$$A \underset{g}{\overset{f}{\rightrightarrows}} B \tag{8.1}$$

to refer to such cochains. There is the cup product

$$C^\bullet(A,\, {}_fB_g) \otimes C^\bullet(A,\, {}_gB_h) \to C^\bullet(A,\, {}_fB_h).$$

In other words, given two cochains described by the picture

$$A \underset{h}{\overset{f}{\overset{g}{\Rrightarrow}}} B \tag{8.2}$$

one produces a cochain corresponding to

$$A \underset{h}{\overset{f}{\rightrightarrows}} B$$

Also, given a cochain φ in $C^\bullet(B,\, {}_{g_1}C_{g_2})$,

$$A \xrightarrow{f} B \underset{g_2}{\overset{g_1}{\rightrightarrows}} C \xrightarrow{h} D \tag{8.3}$$

one defines a cochain $h_*f^*\varphi$ in $C^\bullet(A, {}_{hg_1f}D_{hg_2f})$.

$$A \underset{hg_2f}{\overset{hg_1f}{\rightrightarrows}} D$$

Given two cochains as shown below, i.e. φ in $C^\bullet(A, {}_{f_1}B_{g_1})$ and ψ in $C^\bullet(B, {}_{f_2}C_{g_2})$,

$$A \underset{g_1}{\overset{f_1}{\rightrightarrows}} B \underset{g_2}{\overset{f_2}{\rightrightarrows}} C \tag{8.4}$$

there are two ways to produce a cochain n $C^\bullet(A, {}_{f_2f_1}C_{g_2g_1})$.

$$A \underset{g_2g_1}{\overset{f_2f_1}{\rightrightarrows}} C$$

One is $f_{2*}\varphi \cup g_1^*\psi$ and the other is $f_1^*\psi \cup g_{2*}\varphi$. There is a homotopy between the two, given by the brace operation $\psi\{\varphi\}$.

One may ask whether any two ways to compose the cup product, f_*, and f^* are essentially the same (up to homotopy). Below we outline an affirmative answer.

Let us first look at vertical compositions, i.e. at the cup product. It is associative, so we can define a DG category $\mathbf{C}^\bullet(A,B)$ whose objects are morphisms $A \xrightarrow{f} B$ and whose morphisms are $C^(A, {}_fB_g)$, with the cup product being the composition. Naively we could expect these to form a 2-category, in which case we would have a functor $\mathbf{C}^\bullet(A,B)\otimes\mathbf{C}^\bullet(B,C)\to\mathbf{C}^\bullet(A,C)$ satisfying the associativity condition. As we have seen, there are not one but two candidates for such a functor, and the associativity could be true up to homotopy at best.

In reality, algebras form a category not in categories but in cocategories. Namely, let

$$\mathbf{B}(A,B) = \mathrm{Bar}(\mathbf{C}^\bullet(A,B)) \tag{8.5}$$

be the bar construction. Recall that for a (small, conilpotent) DG category $\mathbf{C}$

$$\mathrm{Bar}(\mathbf{C})(f,g) = \bigoplus_{n=0}^{\infty} \bigoplus_{h_1,\ldots,h_n\in\mathrm{Ob}(\mathbf{C})} \mathbf{C}(f,h_1)[1]\otimes\ldots\otimes\mathbf{C}(h_n,g)[1] \tag{8.6}$$

The coproduct is the deconcatenation. The differential ∂_{Bar} is the usual bar differential plus the one induced by the one in $\mathbf{C}$. The term with $n = 0$ stands for $\mathbf{C}(f,g)[1]$.

Generalizing [**18**] and [**19**], we define a morphism of DG cocategories

$$\bullet : \mathrm{Bar}(\mathbf{C}^\bullet(A,B))\otimes\mathrm{Bar}(\mathbf{C}^\bullet(B,C))\to\mathrm{Bar}(\mathbf{C}^\bullet(A,C)) \tag{8.7}$$

which satisfies the associativity property (when four algebras A, B, C, D are chosen).

We can define $\mathrm{CC}^\bullet_{II}(\mathbf{B})$ for any DG cocategory $\mathbf{B}$. We get an A_∞ category whose objects are algebras and whose morphisms are $\mathrm{CC}^\bullet_{II}(\mathbf{B}(A,B))$. We also have an A_∞ module $A\mapsto\mathrm{CC}^\bullet_{II}(\mathrm{Bar}(A))$.

8.2. Category in categories. One can obtain a version of a two-category structure on the category of algebras when one applies to (8.7) the functor Cobar. Now to any two algebras one puts in correspondence the DG category

$$\mathcal{C}(A,B) = \mathrm{Cobar}(\mathrm{Bar}(\mathbf{C}^\bullet(A,B))) \tag{8.8}$$

which is quasi-isomorphic to $\mathbf{C}^\bullet(A,B)$. Those still do not form a strict two-category because Cobar is not compatible to the tensor product in the strictest possible way. Rather, Cobar is a lax monoidal functor which is enough to define a reasonable 2-category structure on algebras ([**38**], [**48**]).

REMARK 8.1. Note that, at least in the classical (not derived) sense, algebras form a 3-category, namely, a symmetric monoidal 2-category (the monoidal structure being the tensor product of algebras). The contents of 8.3 below address some aspects of this structure in the derived situation. The other very important aspect of this is the issue of the multiplicative structure on Hochschild and periodic cyclic chains (as we saw in section 3).

8.3. Higher Hochschild complexes. Now, in addition to cochains described by the "bigon" (8.1), let us consider more general $2k$-gons such as the one below (for k=2)

$$\begin{array}{ccc} A_1 & \xrightarrow{f_{11}} & B_1 \\ {\scriptstyle g_{12}}\downarrow & & \uparrow{\scriptstyle g_{21}} \\ B_2 & \xleftarrow[f_{22}]{} & A_2 \end{array} \tag{8.9}$$

Namely, for any algebras $A_j, B_j, 1 \le j \le k$, and for any morphisms $f_{jj} : A_j \to B_j$ and $g_{j,j+1} : A_j \to B_{j+1}$, $1 \le j \le k$ (the indices are added modulo k, we define the complex

$$C^\bullet(\otimes_{j=1}^k A_j, \ {}_{\otimes f_{jj}}(\otimes_{j=1}^k B_j)_{\otimes g_{j,j+1}}) \tag{8.10}$$

One can generalize the cup product (8.2); namely, for two cochains

$$\begin{array}{ccc} A_1 & \xrightarrow{f_{11}} & B_1 \\ {\scriptstyle g_{12}}\downarrow & & \updownarrow{\scriptstyle g_{21}} \\ B_2 & \xleftarrow[g_{22}=f_{22}]{f_{22}} & A_2 \\ {\scriptstyle f_{32}}\uparrow & & \downarrow{\scriptstyle f_{23}} \\ A_3 & \xrightarrow[g_{33}]{} & B_3 \end{array} \tag{8.11}$$

one produces a cochain

$$\begin{array}{ccc} A_1 & \xrightarrow{f_{11}} & B_1 \\ {\scriptstyle g_{12}}\downarrow & & \uparrow{\scriptstyle g_{21}} \\ B_2 & & A_2 \\ {\scriptstyle f_{32}}\uparrow & & \downarrow{\scriptstyle f_{23}} \\ A_3 & \xrightarrow[g_{33}]{} & B_3 \end{array}$$

Also, for a morphism $f : A'_j \to A_j$, resp. $B_j \to B'_j$, one defines f^*, resp. g_*.

Furthermore, one can generalize (8.4) and define the $\bullet$ product that takes two cochains as shown below

(8.12)

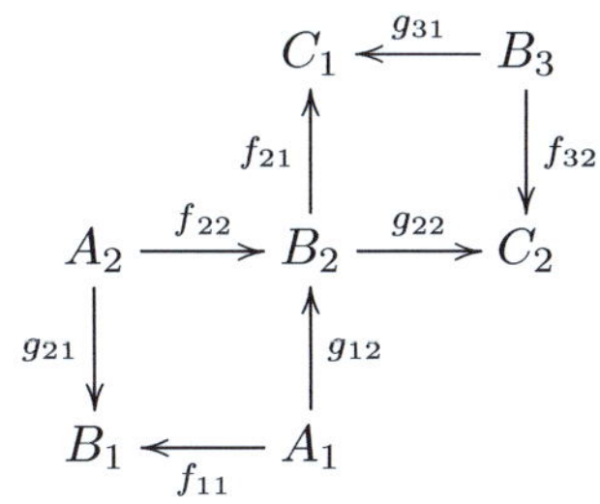

and produces a cochain described by

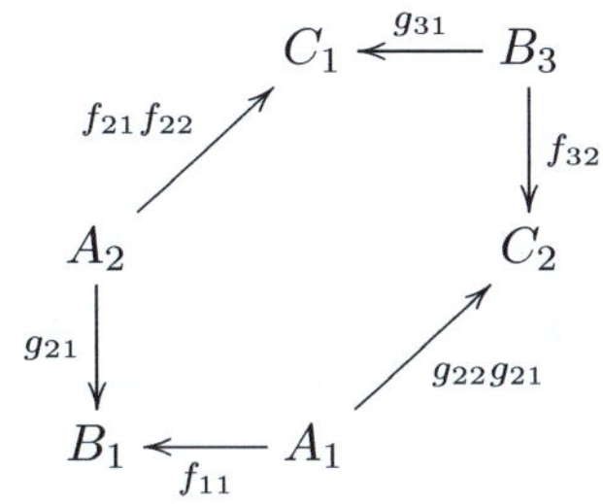

As in (8.4), this product is a homotopy between two different ways to compose the two cochains using the cup product and the operations of direct and inverse image. When one takes $A_j = B_j = A$ and $f_{jj} = g_{j,j+1} = \mathrm{id}_A$ for all j, one defines the Kontsevich-Vlassopoulos bracket on

$$\prod_{k=1}^{\infty} C^{\bullet+1}(A^{\otimes k}, {}_{\alpha}(A^{\otimes k}))^{C_k} \tag{8.13}$$

of degree -1 in k. As above, α is the cyclic permutation.

One would expect a generalization of the construction mentioned in 8.2. Namely, a strict structure would be as follows. We have defined a complex $C^\bullet(K)$ corresponding to a $2k$-gon K. For any picture which is a union of $2k$-gons such as (8.11) or (8.12),

$$K = \cup_{j=1}^{m} K_j \tag{8.14}$$

there should be an operation

$$\mathrm{Op}(K_1, \ldots, K_m) : \otimes_{i=1}^{m} C^\bullet(K_j) \to C^\bullet(K) \tag{8.15}$$

and an associativity condition for $\mathrm{Op}(K_1, \ldots, K_n)$ and $\mathrm{Op}(K_{j1}, \ldots, K_{j,nj})$ for every "double subdivision"

$$K = \cup_{i=1}^{m} K_j;\ K_j = \cup_{i=1}^{n_j} K_{j,i},\ 1 \le j \le m \tag{8.16}$$

More realistically, there should be a version of the notion of an operad (related to and generalizing Batanin's two-operads):

(1) a collection of complexes $\mathcal{O}(K; K_1, \ldots, K_m)$ for any subdivision (8.14);
(2) compositions

$$\mathcal{O}(K; \{K_j\}) \otimes \otimes_{j=1}^{m} \mathcal{O}(K_j; \{K_{ji}\}) \to \mathcal{O}(K; \{K_{ji}\}) \tag{8.17}$$

for any "double subdivision" (8.16);

(3) an associativity condition for any "triple subdivision"

$$K = \cup_{i=1}^{m} K_j;\ K_j = \cup_{i=1}^{n_j} K_{j,i},\ 1 \le j \le m;\ K_{j,i} = \cup_\ell K_{\ell,j,i} \tag{8.18}$$

An algebra over such a generalized operad will be

(1) A complex $C^\bullet(K)$ for each K;
(2) A morphism

$$\mathcal{O}(K;\{K_j\}) \otimes \otimes_{j=1}^{m} C^\bullet(K_j) \to C^\bullet(K)$$

for every subdivision (8.14) which is compatible with composition for any (8.16).

We expect the higher Hochschild complexes (8.10) to form an algebra over a generalized operad $\mathcal{O}$ which is homotopically constant, i.e. such that $\mathcal{O}(K;\{K_j\})$ are all weakly homotopy equivalent to the scalar ring k. This would generalize Tamarkin's theorem [**46**].

Furthermore, for a $2k$-gon $\{A_j, B_j, f_{jj}, g_{j,j+1}\}$ there is also the chain complex

$$C_\bullet(\otimes_{j=1}^{k} A_j,\ {}_{\otimes f_{jj}}(\otimes_{j=1}^{k} B_j)_{\otimes g_{j,j+1}}) \tag{8.19}$$

(In fact, when $k > 1$, there are also mixed chain-cochain complexes). The above should generalize to this situation, within the context of (generalized) multi-colored operads.

REMARK 8.2. This is not quite straightforward because chains have different functoriality properties. For example, given morphisms as in (8.3), at the level of cochains we get morphisms of complexes $C^\bullet(B,\ {}_{g_1}C_{g_2}) \to C^\bullet(A,\ {}_{hg_1f}D_{hg_2f})$ (as we saw earlier); but at the level of chains we get

$$C^\bullet(A,\ {}_{g_1f}C_{g_2f}) \to C^\bullet(B,\ {}_{hg_1}D_{hg_2})$$

In [**38**] we proved two results about the structure of chains and cochains in a two-categorical language. Firstly, we showed that the category in cocategories $\mathrm{Bar}(\mathbf{C}^\bullet(\_,\_))$ admits a trace functor up to homotopy (in a precise sense); this structure involves only the Hochschild chain complexes $\mathrm{TR}_A(f) = C_\bullet(A, {}_f A)$ for endomorphisms f of algebras. The structure on all $C_\bullet(A,\ {}_fB_g)$ is what we call a twisted tetramodule structure over $\mathrm{Bar}(\mathbf{C}^\bullet(\_,\_))$.

To summarize: we expect a rich but homotopically constant structure on higher Hochschild chains and cochains. When restricted to the case when all algebras are the same and all morphisms are identities, the structure stops being homotopically constant (not unlike passing from EG to BG for a group G). We then should recover a package consisting of the Kontsevich-Vlassopoulos Lie algebra structure on Hochschild cochains, the action(s) of this Lie algebra on Hochschild chains, and the subdivision isomorphism of Proposition 6.3 *(which produces the Frobenius in characteristic p).*

Acknowledgments

The author would like to thank B. Antieau, A. Beilinson, V. Drinfeld, D. Kaledin, M. Kontsevich, K. Magidson, R. Nest, A. Petrov, M. Rivera, N. Rozenblyum, D. Tamarkin, V. Vologodsky for fruitful discussions about various aspects of this work.

References

[1] M. A. Batanin, *The Eckmann-Hilton argument and higher operads*, Adv. Math. **217** (2008), no. 1, 334–385, DOI 10.1016/j.aim.2007.06.014. MR2365200

[2] A. Bauval, *Théorème d'Eilenberg-Zilber en homologie cyclique entière*, `arXiv:1611.08437` (1–25).

[3] Pierre Berthelot, *Cohomologie cristalline des schémas de caractéristique* $p > 0$ (French), Lecture Notes in Mathematics, Vol. 407, Springer-Verlag, Berlin-New York, 1974. MR0384804

[4] Jonathan Block, Ezra Getzler, and John D. S. Jones, *The cyclic homology of crossed product algebras. II. Topological algebras*, J. Reine Angew. Math. **466** (1995), 19–25. MR1353312

[5] M. Bökstedt, *Topological Hochschild homology*, Preprint, Universität Bielefeld (1985).

[6] M. Bökstedt, W. C. Hsiang, and I. Madsen, *The cyclotomic trace and algebraic K-theory of spaces*, Invent. Math. **111** (1993), no. 3, 465–539, DOI 10.1007/BF01231296. MR1202133

[7] C. Brav, N. Rozenblyum, *to appear.*

[8] A. Connes, *Non Commutative Differential Geometry, Part I, The Chern Character in K-Homology,* Preprint I.H.E.S. (1982).

[9] A. Connes, *Non Commutative Differential Geometry, Part II, De Rham Homology and Non Commutative Algebra*, Preprint I.H.E.S. (1983).

[10] Alain Connes, *Noncommutative differential geometry*, Inst. Hautes Études Sci. Publ. Math. **62** (1985), 257–360. MR823176

[11] Alain Connes, *Cohomologie cyclique et foncteurs* Ext^n (French, with English summary), C. R. Acad. Sci. Paris Sér. I Math. **296** (1983), no. 23, 953–958. MR777584

[12] Guillermo Cortiñas, Joachim Cuntz, Ralf Meyer, and Georg Tamme, *Nonarchimedean bornologies, cyclic homology and rigid cohomology*, Doc. Math. **23** (2018), 1197–1245. MR3874948

[13] Guillermo Cortiñas, Ralf Meyer, and Devarshi Mukherjee, *Nonarchimedean analytic cyclic homology*, Doc. Math. **25** (2020), 1353–1419. MR4164727

[14] W. G. Dwyer, M. J. Hopkins, and D. M. Kan, *The homotopy theory of cyclic sets*, Trans. Amer. Math. Soc. **291** (1985), no. 1, 281–289, DOI 10.2307/1999909. MR797060

[15] B. L. Feĭgin and B. L. Tsygan, *Additive K-theory and crystalline cohomology* (Russian), Funktsional. Anal. i Prilozhen. **19** (1985), no. 2, 52–62, 96. MR800920

[16] B. L. Feĭgin and B. L. Tsygan, *Additive K-theory*, K-theory, arithmetic and geometry (Moscow, 1984), Lecture Notes in Math., vol. 1289, Springer, Berlin, 1987, pp. 67–209, DOI 10.1007/BFb0078368. MR923136

[17] B. L. Feĭgin and B. L. Tsygan, *Cyclic homology of algebras with quadratic relations, universal enveloping algebras and group algebras*, K-theory, arithmetic and geometry (Moscow, 1984), Lecture Notes in Math., vol. 1289, Springer, Berlin, 1987, pp. 210–239, DOI 10.1007/BFb0078369. MR923137

[18] Murray Gerstenhaber and Alexander A. Voronov, *Homotopy G-algebras and moduli space operad*, Internat. Math. Res. Notices **3** (1995), 141–153, DOI 10.1155/S1073792895000110. MR1321701

[19] E. Getzler, J. D. S. Jones, Operads, homotopy algebra, and iterated integrals for double loop spaces, hep-th/9305013, 1–70 (1993).

[20] Ezra Getzler, *Cartan homotopy formulas and the Gauss-Manin connection in cyclic homology*, Quantum deformations of algebras and their representations (Ramat-Gan, 1991/1992; Rehovot, 1991/1992), Israel Math. Conf. Proc., vol. 7, Bar-Ilan Univ., Ramat Gan, 1993, pp. 65–78. MR1261901

[21] Ezra Getzler and John D. S. Jones, *A_∞-algebras and the cyclic bar complex*, Illinois J. Math. **34** (1990), no. 2, 256–283. MR1046565

[22] Luc Illusie, *Complexe de de Rham-Witt et cohomologie cristalline* (French), Ann. Sci. École Norm. Sup. (4) **12** (1979), no. 4, 501–661. MR565469

[23] Natalia Iyudu, Maxim Kontsevich, and Yannis Vlassopoulos, *Pre-Calabi-Yau algebras as noncommutative Poisson structures*, J. Algebra **567** (2021), 63–90, DOI 10.1016/j.jalgebra.2020.08.029. MR4156101

[24] A. Grothendieck, *Crystals and the de Rham cohomology of schemes*, Dix exposés sur la cohomologie des schémas, Adv. Stud. Pure Math., vol. 3, North-Holland, Amsterdam, 1968, pp. 306–358. Notes by I. Coates and O. Jussila. MR269663

[25] Lars Hesselholt, *Correction to: "Witt vectors of non-commutative rings and topological cyclic homology" [Acta Math.* **178** *(1997), no. 1, 109–141; MR1448712]*, Acta Math. **195** (2005), 55–60, DOI 10.1007/BF02588050. MR2233685

[26] Lars Hesselholt and Thomas Nikolaus, *Topological cyclic homology*, Handbook of homotopy theory, CRC Press/Chapman Hall Handb. Math. Ser., CRC Press, Boca Raton, FL, [2020] ©2020, pp. 619–656. MR4197995

[27] Christine E. Hood and John D. S. Jones, *Some algebraic properties of cyclic homology groups*, K-Theory **1** (1987), no. 4, 361–384, DOI 10.1007/BF00539623. MR920950

[28] D. B. Kaledin, *Cartier isomorphism for unital associative algebras*, Proc. Steklov Inst. Math. **290** (2015), no. 1, 35–51, DOI 10.1134/S0081543815060048. Published in Russian in Tr. Mat. Inst. Steklova **290** (2015), 43–60. MR3488779

[29] D. Kaledin, *Hochschild-Witt complex*, Adv. Math. **351** (2019), 33–95, DOI 10.1016/j.aim.2019.05.007. MR3949985

[30] D. Kaledin, *Witt vectors, commutative and non-commutative* (Russian, with Russian summary), Uspekhi Mat. Nauk **73** (2018), no. 1(439), 3–34, DOI 10.4213/rm9796; English transl., Russian Math. Surveys **73** (2018), no. 1, 1–30. MR3749617

[31] D. Kaledin, *Witt vectors as a polynomial functor*, Selecta Math. (N.S.) **24** (2018), no. 1, 359–402, DOI 10.1007/s00029-017-0365-z. MR3769733

[32] Masoud Khalkhali, *An approach to operations on cyclic homology*, J. Pure Appl. Algebra **107** (1996), no. 1, 47–59, DOI 10.1016/0022-4049(95)00026-7. MR1377655

[33] Masoud Khalkhali, *On Cartan homotopy formulas in cyclic homology*, Manuscripta Math. **94** (1997), no. 1, 111–132, DOI 10.1007/BF02677842. MR1468938

[34] M. Khalkhali and B. Rangipour, *On the generalized cyclic Eilenberg-Zilber theorem*, Canad. Math. Bull. **47** (2004), no. 1, 38–48, DOI 10.4153/CMB-2004-006-x. MR2032267

[35] Jean-Louis Loday, *Cyclic homology*, Grundlehren der mathematischen Wissenschaften [Fundamental Principles of Mathematical Sciences], vol. 301, Springer-Verlag, Berlin, 1992. Appendix E by María O. Ronco, DOI 10.1007/978-3-662-21739-9. MR1217970

[36] Tasos Moulinos, Marco Robalo, and Bertrand Toën, *A universal Hochschild-Kostant-Rosenberg theorem*, Geom. Topol. **26** (2022), no. 2, 777–874, DOI 10.2140/gt.2022.26.777. MR4444269

[37] P. Monsky and G. Washnitzer, *Formal cohomology. I*, Ann. of Math. (2) **88** (1968), 181–217, DOI 10.2307/1970571. MR248141

[38] R. Nest, B. Tsygan, *Cyclic Homology*, book in progress, available at: `https://sites.math.northwestern.edu/~tsygan/Part1.pdf`.

[39] Thomas Nikolaus and Peter Scholze, *On topological cyclic homology*, Acta Math. **221** (2018), no. 2, 203–409, DOI 10.4310/ACTA.2018.v221.n2.a1. MR3904731

[40] Alexander Petrov, Dmitry Vaintrob, and Vadim Vologodsky, *The Gauss-Manin connection on the periodic cyclic homology*, Selecta Math. (N.S.) **24** (2018), no. 1, 531–561, DOI 10.1007/s00029-018-0388-0. MR3769739

[41] A. Petrov, V. Vologodsky, *On the periodic topological cyclic homology of DG categories in characteristic p*, `arXiv:1912.03246`, 1–15 (2019).

[42] Alexander Polishchuk and Leonid Positselski, *Hochschild (co)homology of the second kind I*, Trans. Amer. Math. Soc. **364** (2012), no. 10, 5311–5368, DOI 10.1090/S0002-9947-2012-05667-4. MR2931331

[43] Leonid Positselski, *Homological algebra of semimodules and semicontramodules*, Instytut Matematyczny Polskiej Akademii Nauk. Monografie Matematyczne (New Series) [Mathematics Institute of the Polish Academy of Sciences. Mathematical Monographs (New Series)], vol. 70, Birkhäuser/Springer Basel AG, Basel, 2010. Semi-infinite homological algebra of associative algebraic structures; Appendix C in collaboration with Dmitriy Rumynin; Appendix D in collaboration with Sergey Arkhipov, DOI 10.1007/978-3-0346-0436-9. MR2723021

[44] Leonid Positselski, *Two kinds of derived categories, Koszul duality, and comodule-contramodule correspondence*, Mem. Amer. Math. Soc. **212** (2011), no. 996, vi+133, DOI 10.1090/S0065-9266-2010-00631-8. MR2830562

[45] Daniel Quillen, *Algebra cochains and cyclic cohomology*, Inst. Hautes Études Sci. Publ. Math. **68** (1988), 139–174 (1989). MR1001452

[46] Dmitry Tamarkin, *What do dg-categories form?*, Compos. Math. **143** (2007), no. 5, 1335–1358, DOI 10.1112/S0010437X07002771. MR2360318

[47] B. L. Tsygan, *Homology of matrix Lie algebras over rings and the Hochschild homology* (Russian), Uspekhi Mat. Nauk **38** (1983), no. 2(230), 217–218. MR695483
[48] Boris Tsygan, *Noncommutative calculus and operads*, Topics in noncommutative geometry, Clay Math. Proc., vol. 16, Amer. Math. Soc., Providence, RI, 2012, pp. 19–66. MR2986860

Department of Mathematics, Northwestern Universty, Evanston Illinois 60208
Email address: b-tsygan@northwestern.edu

Proceedings of Symposia in Pure Mathematics
Volume **105**, 2023
https://doi.org/10.1090/pspum/105/01916

Cyclic cocycles and one-loop corrections in the spectral action

Teun D.H. van Nuland and Walter D. van Suijlekom

ABSTRACT. We present an intelligible review of recent results concerning cyclic cocycles in the spectral action and one-loop quantization. We show that the spectral action, when perturbed by a gauge potential, can be written as a series of Chern–Simons actions and Yang–Mills actions of all orders. In the odd orders, generalized Chern–Simons forms are integrated against an odd (b, B)-cocycle, whereas, in the even orders, powers of the curvature are integrated against (b, B)-cocycles that are Hochschild cocycles as well. In both cases, the Hochschild cochains are derived from the Taylor series expansion of the spectral action $\mathrm{Tr}(f(D+V))$ in powers of $V = \pi_D(A)$, but unlike the Taylor expansion we expand in increasing order of the forms in A. We then analyze the perturbative quantization of the spectral action in noncommutative geometry and establish its one-loop renormalizability as a gauge theory. We show that the one-loop counterterms are of the same Chern–Simons–Yang–Mills form so that they can be safely subtracted from the spectral action. A crucial role will be played by the appropriate Ward identities, allowing for a fully spectral formulation of the quantum theory at one loop.

CONTENTS

1. Introduction

The spectral action [5, 6] is one of the key instruments in the applications of noncommutative geometry to particle physics. With inner fluctuations [12] of a noncommutative manifold playing the role of gauge potentials, the spectral action principle yields the corresponding Lagrangians. Indeed, the asymptotic behavior of the spectral action for small momenta leads to experimentally testable field theories, by interpreting the spectral action as a classical action and applying the usual renormalization group techniques. In particular, this provides the simplest way known to geometrically explain the dynamics and interactions of the gauge bosons and the Higgs boson in the Standard Model Lagrangian as an effective field theory [**7**] (see also the textbooks [**13, 34**]). More general noncommutative manifolds (spectral triples) can also be captured by the spectral action principle, leading to models beyond the standard model as well. As shown in [**15**], if one restricts to the scale-invariant part, one may naturally identify a Yang–Mills term and a Chern–Simons term to elegantly appear in the spectral action. From the perspective of quantum field theory, the appearance of these field-theoretic action functionals sparks hope that we might find a way to go beyond the classical framework provided by the spectral action principle. It is thus a natural question whether we can also field-theoretically describe the full spectral action, without resorting to the scale-invariant part.

Motivated by this, we study the spectral action when it is expanded in terms of inner fluctuations associated to an arbitrary noncommutative manifold, without resorting to heat-kernel techniques. Indeed, the latter are not always available and an understanding of the full spectral action could provide deeper insight into how gauge theories originate from noncommutative geometry. Let us now give a more precise description of our setup.

We let $(\mathcal{A}, \mathcal{H}, D)$ be an finitely summable spectral triple. If $f : \mathbb{R} \to \mathbb{C}$ is a suitably nice function we may define the spectral action [**6**]:

$$\mathrm{Tr}(f(D)).$$

An inner fluctuation, as explained in [**12**], is given by a Hermitian universal one-form

$$A = \sum_{j=1}^{n} a_j db_j \in \Omega^1(\mathcal{A}), \tag{1}$$

for elements $a_j, b_j \in \mathcal{A}$. The terminology 'fluctuation' comes from representing A on $\mathcal{H}$ as

$$V := \pi_D(A) = \sum_{j=1}^{n} a_j[D, b_j] \in \mathcal{B}(\mathcal{H})_{\mathrm{sa}}, \tag{2}$$

and fluctuating D to $D + V$ in the spectral action. The variation of the spectral action under the inner fluctuation is then given by

$$\mathrm{Tr}(f(D+V)) - \mathrm{Tr}(f(D)). \tag{3}$$

As spectral triples can be understood as noncommutative spin^c manifolds (see [**14**]) encoding the gauge fields as an inner structure, one could hope that perturbations of the spectral action could be understood in terms of noncommutative versions of geometrical, gauge theoretical concepts. Hence we would like to express (3) in terms of universal forms constructed from A. To express an action functional in terms of universal forms, one is naturally led to cyclic cohomology. As it turns out, hidden inside the spectral action we will identify an odd (b, B)-cocycle $(\tilde{\psi}_1, \tilde{\psi}_3, \ldots)$ and an even (b, B)-cocycle $(\phi_2, \phi_4, \ldots)$ for which $b\phi_{2k} = B\phi_{2k} = 0$, i.e., each Hochschild cochain ϕ_{2k} forms its own (b, B)-cocycle $(0, \ldots, 0, \phi_{2k}, 0, \ldots)$. On the other hand, the odd (b, B)-cocycle $(\tilde{\psi}_{2k+1})$ is truly infinite (in the sense of [**11**]).

The key result is that for suitable $f : \mathbb{R} \to \mathbb{C}$ we may expand

$$\mathrm{Tr}(f(D+V) - f(D)) = \sum_{k=1}^{\infty} \left(\int_{\psi_{2k-1}} \mathrm{cs}_{2k-1}(A) + \frac{1}{2k} \int_{\phi_{2k}} F^k \right), \tag{4}$$

in which the series converges absolutely. Here ψ_{2k-1} is a scalar multiple of $\tilde{\psi}_{2k-1}$, $F_t = tdA+t^2A^2$, so that $F = F_1$ is the curvature of A, and $\mathrm{cs}_{2k-1}(A) = \int_0^1 AF_t^{k-1}dt$ is a generalized noncommutative Chern–Simons form.

As already mentioned, a similar result was shown earlier to hold for the scale-invariant part $\zeta_D(0)$ of the spectral action. Indeed, Connes and Chamseddine [**15**] expressed the variation of the scale-invariant part in dimension ≤ 4 as

$$\zeta_{D+V}(0) - \zeta_D(0) = -\frac{1}{4}\int_{\tau_0} (dA + A^2) + \frac{1}{2}\int_{\psi}\left(AdA + \frac{2}{3}A^3\right),$$

for a certain Hochschild 4-cocycle τ_0 and cyclic 3-cocycle ψ.

It became clear in [**28**] that an extension of this result to the full spectral action is best done by using multiple operator integrals [**32**] instead of residues. It allows for stronger analytical results, and in particular allows to go beyond dimension 4. Moreover, for our analysis of the cocycle structure that appears in the full spectral action we take the Taylor series expansion as a starting point, and for working with such expansions multiple operator integrals provide the ideal tools, as shown by the strong results in [**1, 8, 31, 33**]. In [**28**] we pushed these results further still, by proving estimates and continuity properties for the multiple operator integral when the self-adjoint operator has an s-summable resolvent, thereby supplying the discussion here with a strong functional analytic foundation. This article will start with a review of the results of [**28**] without involving multiple operator integration techniques. Through the use of abstract brackets, we will investigate the interesting cyclic structure that exists within the spectral action, with all analytical details taking place under the hood.

We work out two interesting possibilities for application of our main result and the techniques used to obtain it. The first application is to index theory. One can show that the (b, B)-cocycles ϕ and ψ are *entire* in the sense of [**10**]. This makes it meaningful to analyze their pairing with K-theory, which we find to be trivial in Section 3.5.

The second application is to quantization. In Section 4, though evading analytical difficulties, we will take a first step towards the quantization of the spectral action within the framework of spectral triples. Using the asymptotic expansion proved in Theorem 3.9, and some basic quantum field theoretic techniques, we will propose a one-loop quantum effective spectral action and show that it satisfies a similar expansion formula, featuring in particular a new pair of (b, B)-cocycles.

Although the main aim of this paper is to give a simple review of the results of [**28**] and [**29**], some essential novelty is also provided. In order to connect to the quantization results of [**29**], the results of [**28**] are slightly generalized as well as put into context. Moreover, this paper gives a mathematically precise underpinning of the results presented in [**29**], which was geared towards a physics audience. We hope that the discussion presented here is clear to mathematicians with or without affinity to physics.

2. Taylor expansion of the spectral action

Consider a finitely summable spectral triple $(\mathcal{A}, \mathcal{H}, D)$ (in the sense that for some s the operator $(i - D)^{-s}$ is trace-class). Given the fluctuations of D to $D + V$ as explained in the introduction, we are interested in a Taylor expansion of the spectral action:

$$\begin{aligned}\operatorname{Tr}(f(D+V)-f(D)) &= \sum_{n=1}^{\infty}\frac{1}{n!}\frac{d^n}{dt^n}\operatorname{Tr}(f(D+tV))\big|_{t=0}\\ &= \sum_{n=1}^{\infty}\frac{1}{n}\langle V,\dots,V\rangle, \qquad (5)\end{aligned}$$

where $\langle V,\dots,V\rangle$ is a notation for ($1/(n-1)!$ times) the nth derivative of the spectral action, defined below, and dependent on f and D. Such an expansion exist under varying assumptions on f, D, and V, see for instance [**22, 27, 28, 31, 33**]. When we are interested in the inner fluctuations of the form $V = \pi_D(A)$ as in Equation (2), a convenient function class in which f should lie is given as in [**28**] by

$$\mathcal{E}_s^\gamma := \left\{ f \in C^\infty \;\middle|\; \begin{array}{l}\text{there exists } C_f \geq 1 \text{ s.t. } \|\widehat{(fu^m)^{(n)}}\|_1 \leq (C_f)^{n+1} n!^\gamma \\ \text{for all } m = 0,\dots,s \text{ and } n \in \mathbb{N}_0\end{array}\right\}, \qquad (6)$$

for $\gamma \in (0,1]$ a number, and s the summability of the pertinent spectral triple. Indeed, as shown in [**28**] if $f \in \mathcal{E}_s^\gamma$ we have good control over the expansion appearing on the right-hand side of (5).

For our present expository purposes, however, it is sufficient to assume that f' is compactly supported and analytic in a region of $\mathbb{C}$ containing a rectifiable curve Γ which surrounds the spectrum of D, and that $V_1,\dots,V_n$ are, say, trace class. In this case we have

$$\langle V_1,\dots,V_n\rangle = \frac{1}{2\pi i}\oint_\Gamma f'(z)\operatorname{Tr}\left(\prod_{j=1}^n V_j(z-D)^{-1}\right). \qquad (7)$$

A concrete expression can be also obtained in terms of divided differences of f. Indeed, for a self-adjoint operator D in $\mathcal{H}$ with compact resolvent, we let $\varphi_1, \varphi_2, \dots$ be an orthonormal basis of eigenvectors of D, with corresponding eigenvalues $\lambda_1, \lambda_2, \dots$. Recall Cauchy's integral formula for divided differences [**16**, Chapter I.1]:

$$g[x_0,\dots x_n] = \frac{1}{2\pi i}\oint \frac{g(z)}{(z-x_0)\cdots(z-x_n)}dz,$$

with the contour enclosing the points x_i. This then yields

$$\langle V,\dots,V\rangle = \frac{1}{n}\sum_{i_1,\dots,i_n\in\mathbb{N}} f'[\lambda_{i_1},\dots,\lambda_{i_n}]V_{i_1 i_2}\cdots V_{i_{n-1}i_n}V_{i_n i_1}. \qquad (8)$$

where $V_{kl} := \langle \varphi_k, V\varphi_l\rangle$ denote the matrix elements of V. This formula appears in [**22**, Corollary 3.6] and, in higher generality, in [**33**, Theorem 18]. The formula (8) gives a very concrete way to calculate derivatives of the spectral action, as well as to calculate the Taylor series of a perturbation of the spectral action.

For our algebraic results we only need two simple properties of the bracket $\langle\cdot\rangle$, stated in the following lemma.

LEMMA 2.1. *For $V_1,\dots,V_n \in \mathcal{B}(\mathcal{H})$ and $a \in \mathcal{A}$ we have*

(I) $\langle V_1,\dots,V_n\rangle = \langle V_n, V_1,\dots,V_{n-1}\rangle$,
(II) $\langle aV_1, V_2,\dots,V_n\rangle - \langle V_1,\dots,V_{n-1},V_n a\rangle = \langle V_1,\dots,V_n,[D,a]\rangle$.

PROOF. We will omit all analytical details and give a proof for finite-dimensional Hilbert spaces only. The full proof involving multiple operator integrals can be found in [**28**] (as Lemma 14).

In finite-dimensions we may use formula (7) for the bracket. Clearly (I) then follows directly from the tracial property. Note that the left-hand side of equality (II) comes down to the commutator of a with the resolvent $(z-D)^{-1}$, for which we have the equality

$$(z-D)^{-1}a - a(z-D)^{-1} = (z-D)^{-1}[D,a](z-D)^{-1}$$

This readily leads to the right-hand side in (II). □

3. Cyclic cocycles in the spectral action

We now generalize a little and consider a collection of functions $\prec\cdot\succ\colon \mathcal{B}(\mathcal{H})^{\times n} \to \mathbb{R}$, $n \in \mathbb{N}$, satisfing

(I) $\prec V_1,\ldots,V_n\succ = \prec V_n, V_1,\ldots,V_{n-1}\succ$,
(II) $\prec aV_1, V_2,\ldots,V_n\succ - \prec V_1,\ldots,V_{n-1},V_n a\succ = \prec V_1,\ldots,V_n,[D,a]\succ$

In view of Lemma 2.1 above, the brackets $\langle\cdot\rangle$ that appear in the Taylor expansion of the spectral action form a special case of these generalized brackets $\prec\cdot\succ$ —and of course form the key motivation for introducing them. However, such structures pop up in other places as well, for instance [**21, 26**], cf. [**24**, Proposition 3.2 and Remark 3.2]. In Section 4, we will introduce yet another instance of $\prec\cdot\succ$, in order to obtain one-loop corrections.

Therefore, in contrast to [**28**], the following discussion will involve the abstract bracket $\prec\cdot\succ$ instead of the explicit $\langle\cdot\rangle$.

3.1. Hochschild and cyclic cocycles. When the above brackets $\prec\cdot\succ$ are evaluated at one-forms $a[D,b]$ associated to a spectral triple, the relations (I) and (II) can be translated nicely in terms of the coboundary operators appearing in cyclic cohomology. This is very similar to the structure appearing in the context of index theory, see for instance [**18, 23**].

Let us start by recalling the definition of Hochschild cochains and the boundary operators b and B from [**9**].

DEFINITION 3.1. If $\mathcal{A}$ is an algebra, and $n \in \mathbb{N}_0$, we define the space of *Hochschild n-cochains*, denoted by $\mathcal{C}^n(\mathcal{A})$, as the space of $(n+1)$-linear functionals ϕ on $\mathcal{A}$ with the property that if $a_j = 1$ for some $j \geq 1$, then $\phi(a_0,\ldots,a_n) = 0$.

For such cochains we may use, as in [**11**], an integral notation on universal differential forms that is defined by linear extension of

$$\int_\phi a_0 da_1 \cdots da_n := \phi(a_0,a_1,\ldots,a_n).$$

DEFINITION 3.2. Define operators $b : \mathcal{C}^n(\mathcal{A}) \to \mathcal{C}^{n+1}(\mathcal{A})$ and $B : \mathcal{C}^{n+1}(\mathcal{A}) \to \mathcal{C}^n(\mathcal{A})$ by

$$\begin{aligned} b\phi(a_0, a_1, \ldots, a_{n+1}) &:= \sum_{j=0}^{n} (-1)^j \phi(a_0, \ldots, a_j a_{j+1}, \ldots, a_{n+1}) \\ &\quad + (-1)^{n+1} \phi(a_{n+1} a_0, a_1, \ldots, a_n), \\ B\phi(a_0, a_1, \ldots, a_n) &:= \sum_{j=0}^{n} (-1)^{nj} \phi(1, a_j, a_{j+1}, \ldots, a_{j-1}). \end{aligned}$$

Note that $B = \mathbf{A} B_0$ in terms of the operator $\mathbf{A}$ of cyclic anti-symmetrization and the operator defined by $B_0\phi(a_0, a_1, \ldots, a_n) = \phi(1, a_0, a_1, \ldots, a_n)$. Note that in integral notation we simply have

$$\int_{B_0\phi} a_0 da_1 \cdots da_n = \int_\phi da_0 da_1 \cdots da_n.$$

One may check that the pair (b, B) defines a double complex, *i.e.* $b^2 = 0$, $B^2 = 0$, and $bB + Bb = 0$. Hochschild cohomology now arises as the cohomology of the complex $(\mathcal{C}^n(\mathcal{A}), b)$. In contrast, we will be using *periodic cyclic cohomology*, which is defined as the cohomology of the totalization of the (b, B)-complex. That is to say,

$$\mathcal{C}^{\mathrm{ev}}(\mathcal{A}) = \bigoplus_k \mathcal{C}^{2k}(\mathcal{A}); \qquad \mathcal{C}^{\mathrm{odd}}(\mathcal{A}) = \bigoplus_k \mathcal{C}^{2k+1}(\mathcal{A}),$$

form a complex with differential $b + B$ and the cohomology of this complex is called periodic cyclic cohomology. We will also refer to a periodic cyclic cocycle as a cyclic cocycle or a (b, B)-cocycle. Explicitly, an odd (b, B)-cocycle is thus given by a sequence

$$(\phi_1, \phi_3, \phi_5, \ldots),$$

where $\phi_{2k+1} \in \mathcal{C}^{2k+1}(\mathcal{A})$ and

$$b\phi_{2k+1} + B\phi_{2k+3} = 0,$$

for all $k \geq 0$, and also $B\phi_1 = 0$. An analogous statement holds for even (b, B)-cocycles.

3.2. Cyclic cocycles associated to the brackets. In terms of the generic bracket $\prec \cdot \succ$ satisfying (I) and (II), we define the following Hochschild n-cochain:

$$\phi_n(a_0, \ldots, a_n) := \prec a_0[D, a_1], [D, a_2], \ldots, [D, a_n] \succ \qquad (a_0, \ldots, a_n \in \mathcal{A}). \tag{9}$$

We easily see that $B_0\phi_n$ is invariant under cyclic permutations, so that $B\phi_n = nB_0\phi_n$ for odd n and $B\phi_n = 0$ for even n. Also, $\phi_n(a_0, \ldots, a_n) = 0$ when $a_j = 1$ for some $j \geq 1$. We put $\phi_0 := 0$.

LEMMA 3.3. *We have $b\phi_n = \phi_{n+1}$ for odd n and we have $b\phi_n = 0$ for even n.*

PROOF. We only consider the case $n = 1$ while referring to [**28**, Lemma 17] for the proof of the general case. We combine the definition of the b-operator with Leibniz' rule for $[D, \cdot]$ to obtain:

$$\begin{aligned} \int_{b\phi_1} a_0 da_1 da_2 &= \prec a_0 a_1 [D, a_2] \succ - \prec a_0 [D, a_1 a_2] \succ + \prec a_2 a_0 [D, a_1] \succ \\ &= - \prec a_0 [D, a_1] a_2 \succ + \prec a_2 a_0 [D, a_1] \succ = \prec a_0 [D, a_1], [D, a_2] \succ \end{aligned}$$

where we used (II) for the last equality. □

LEMMA 3.4. *Let n be even. We have $bB_0\phi_n = 2\phi_n - B_0\phi_{n+1}$.*

PROOF. Again we only consider the first case $n = 2$ while referring to [**28**, Lemma 17] for the proof of the general case

$$\begin{aligned}\int_{bB_0\phi_2} a_0 da_1 da_2 &= \int_{B_0\phi_2} a_0 a_1 da_2 - \int_{B_0\phi_2} a_0 d(a_1a_2) + \int_{B_0\phi_2} a_2 a_0 da_1\\ &= \prec [D, a_0a_1], [D, a_2]\succ - \prec [D, a_0], [D, a_1a_2]\succ + \prec [D, a_2a_0], [D, a_1]\succ\\ &= \cdots = 2 \prec a_0[D, a_1], [D, a_2]\succ - \prec [D, a_0], [D, a_1], [D, a_2]\succ,\end{aligned}$$

combining Leibniz' rule with (I) and (II). □

Motivated by these results we define

$$\psi_{2k-1} := \phi_{2k-1} - \tfrac{1}{2}B_0\phi_{2k}, \tag{10}$$

so that

$$B\psi_{2k+1} = 2(2k+1)b\psi_{2k-1}.$$

We can rephrase this property in terms of the (b, B)-complex as follows.

PROPOSITION 3.5. *Let ϕ_n and ψ_{2k-1} be as defined above and set*

$$\tilde{\psi}_{2k-1} := (-1)^{k-1}\frac{(k-1)!}{(2k-1)!}\psi_{2k-1}.$$

(i) *The sequence (ϕ_{2k}) is a (b, B)-cocycle and each ϕ_{2k} defines an even Hochschild cocycle: $b\phi_{2k} = 0$.*
(ii) *The sequence $(\tilde{\psi}_{2k-1})$ is an odd (b, B)-cocycle.*

3.3. The brackets as noncommutative integrals. We will now describe how brackets $\prec V, \ldots, V\succ$ can be written as noncommutative integrals of certain universal differential forms defined in terms of $A = \sum a_j db_j \in \Omega^1(\mathcal{A})$, using only property (I) and (II).

At first order not much exciting happens and we simply have

$$\prec V\succ = \sum_j \prec a_j[D, b_j]\succ = \sum_j \int_{\phi_1} a_j db_j = \int_{\phi_1} A.$$

More interestingly, at second order we find using property (II) of the bracket that

$$\begin{aligned}\prec V, V\succ &= \sum_{j,k} \prec a_j[D, b_j], a_k[D, b_k]\succ\\ &= \sum_{j,k} \prec a_j[D, b_j]a_k, [D, b_k]\succ + \sum_{j,k} \prec a_j[D, b_j], [D, a_k], [D, b_k]\succ\\ &= \int_{\phi_2} A^2 + \int_{\phi_3} AdA.\end{aligned}$$

Continuing like this, while only using property (II) of the bracket we find

$$\begin{aligned}\prec V, V, V\succ &= \int_{\phi_3} A^3 + \int_{\phi_4} AdAA + \int_{\phi_5} AdAdA,\\ \prec V, V, V, V\succ &= \int_{\phi_4} A^4 + \int_{\phi_5} (A^3dA + AdAA^2) + \int_{\phi_6} AdAdAA + \int_{\phi_7} AdAdAdA.\end{aligned}$$

This implies that, at least when the infinite sum on the left-hand side makes sense:

$$\sum_n \frac{1}{n} \prec V, \dots, V \succ = \int_{\phi_1} A + \frac{1}{2} \int_{\phi_2} A^2 + \int_{\phi_3} \left(\frac{1}{2} AdA + \frac{1}{3} A^3\right) + \int_{\phi_4} \left(\frac{1}{3} AdAA + \frac{1}{4} A^4\right) + \dots,$$

where the dots indicate terms of degree 5 and higher. Using $\phi_{2k-1} = \psi_{2k-1} + \frac{1}{2} B_0 \phi_{2k}$, this becomes

$$\sum_n \frac{1}{n} \prec V, \dots, V \succ = \int_{\psi_1} A + \frac{1}{2} \int_{\phi_2} (A^2 + dA) + \int_{\psi_3} \left(\frac{1}{2} AdA + \frac{1}{3} A^3\right) + \frac{1}{4} \int_{\phi_4} \left(dAdA + \frac{2}{3}(dAA^2 + AdAA + A^2 dA) + A^4\right) + \dots.$$

Notice that, if ϕ_4 would be tracial, we would be able to identify the terms dAA^2, $AdAA$ and $A^2 dA$, and thus obtain the Yang–Mills form $F^2 = (dA + A^2)^2$, under the fourth integral. In the general case, however, cyclic permutations under $\int_\phi$ produce correction terms, of which one needs to keep track. Indeed, using [**28**, Corollary 24] we may re-order the integrands to yield

$$\sum_n \frac{1}{n} \prec V, \dots, V \succ = \int_{\psi_1} A + \int_{\phi_2} \tfrac{1}{2}(dA + A^2) + \int_{\psi_3} (\tfrac{1}{2} dAA + \tfrac{1}{3} A^3) + \tfrac{1}{4} \int_{\phi_4} (dA + A^2)^2 + \int_{\psi_5} (\tfrac{1}{3}(dA)^2 A + \tfrac{1}{2} dAA^3 + \tfrac{1}{5} A^5) + \tfrac{1}{6} \int_{\phi_6} (dA + A^2)^3 + \dots$$

where the dots indicate terms of degree 7 and higher. Writing $F = dA + A^2$ and $\mathrm{cs}_1(A) := A$, $\mathrm{cs}_3(A) := \frac{1}{2} dAA + \frac{1}{3} A^3$, etc., we can already discern our desired result in low orders.

As a preparation for the general result, we briefly recall from [**30**] the definition of Chern–Simons forms of arbitrary degree.

DEFINITION 3.6. The (universal) **Chern–Simons form** of degree $2k - 1$ is given for $A \in \Omega^1(\mathcal{A})$ by

$$\mathrm{cs}_{2k-1}(A) := \int_0^1 A (F_t)^{k-1} \, dt, \tag{11}$$

where $F_t = t dA + t^2 A^2$ is the curvature two-form of the (connection) one-form $A_t = tA$.

EXAMPLE 3.7. For the first three Chern–Simons forms one easily derives the following explicit expressions:

$$\mathrm{cs}_1(A) = A; \qquad \mathrm{cs}_3(A) = \frac{1}{2}\left(AdA + \frac{2}{3} A^3\right);$$

$$\mathrm{cs}_5(A) = \frac{1}{3}\left(A(dA)^2 + \frac{3}{4} AdAA^2 + \frac{3}{4} A^3 dA + \frac{3}{5} A^5\right).$$

3.4. Cyclic cocycles in the Taylor expansion of the spectral action. We now apply the above results to the brackets appearing in the Taylor expansion of the spectral action:

$$\mathrm{Tr}(f(D+V)-f(D)) = \sum_{n=1}^{\infty} \frac{1}{n}\langle V,\cdots,V\rangle.$$

In order to control the full Taylor expansion of the spectral action we naturally need a growth condition on the derivatives of the function f, and this is accomplished by considering the class $\mathcal{E}_s^\gamma$ defined in (6). The following result is [**28**, Theorem 27].

THEOREM 3.8. *Let $(\mathcal{A},\mathcal{H},D)$ be an s-summable spectral triple, and let $f \in \mathcal{E}_s^\gamma$ for $\gamma \in (0,1)$. There exist entire cyclic cocycles ϕ_{2k} and $\tilde{\psi}_{2k-1} = (-1)^{k-1}(k-1)!/(2k-1)!\psi_{2k-1}$ such that the spectral action fluctuated by $V = \pi_D(A) \in \Omega_D^1(\mathcal{A})_{\mathrm{sa}}$ can be written as*

$$\mathrm{Tr}(f(D+V)-f(D)) = \sum_{k=1}^{\infty}\left(\int_{\psi_{2k-1}} \mathrm{cs}_{2k-1}(A) + \frac{1}{2k}\int_{\phi_{2k}} F^k\right),$$

where the series converges absolutely.

Under less restrictive conditions on the function f we also have the following asymptotic version of this result [**28**, Proposition 28]

THEOREM 3.9. *Let $(\mathcal{A},\mathcal{H},D)$ be a spectral triple, and let $\prec\cdot\succ$ satisfy* (I) *and* (II), *with associated cyclic cocycles ϕ and $\tilde{\psi}$. For $A \in \Omega^1(\mathcal{A})$ and $V = \pi_D(A)$, we asymptotically have*

$$\sum_n \frac{1}{n}\prec V,\ldots,V\succ \sim \sum_{k=1}^{\infty}\left(\int_{\psi_{2k-1}} \mathrm{cs}_{2k-1}(A) + \frac{1}{2k}\int_{\phi_{2k}} F^k\right),$$

by which we mean that, for every $K \in \mathbb{N}$, there exist forms $\omega_l \in \Omega^l(\mathcal{A})$ for $l = K+1,\ldots,2K+1$ such that

$$\sum_{n=1}^{K} \frac{1}{n}\prec V,\ldots,V\succ - \sum_{k=1}^{K}\left(\int_{\psi_{2k-1}} \mathrm{cs}_{2k-1}(A) + \frac{1}{2k}\int_{\phi_{2k}} F^k\right) = \sum_{l=K+1}^{2K+1}\int_{\phi_l}\omega_l.$$

In particular, by taking $\prec\cdot\succ = \langle\cdot\rangle$, we obtain the following corollary.

COROLLARY 3.10. *For $f \in C^\infty$, and $V = \pi_D(A) \in \Omega_D^1(\mathcal{A})_{\mathrm{sa}}$ such that the Taylor expansion of the spectral action converges, we asymptotically have*

$$\begin{aligned}\mathrm{Tr}(f(D+V)-f(D)) &= \sum_{n=1}^{\infty}\frac{1}{n!}\frac{d^n}{dt^n}\mathrm{Tr}(f(D+tV))\Big|_{t=0}\\ &\sim \sum_{k=1}^{\infty}\left(\int_{\psi_{2k-1}} \mathrm{cs}_{2k-1}(A) + \frac{1}{2k}\int_{\phi_{2k}} F^k\right).\end{aligned}$$

3.5. Gauge invariance and the pairing with K-theory. Since the spectral action is a spectral invariant, it is in particular invariant under conjugation of D by a unitary $U \in \mathcal{A}$. More generally, in the presence of an inner fluctuation we find that the spectral action is invariant under the transformation

$$D+V \mapsto U(D+V)U^* = D+V^U; \qquad V^U = U[D,U^*]+UVU^*.$$

This transformation also holds at the level of the universal forms, with a gauge transformation of the form $A \mapsto A^U = UdU^* + UAU^*$. Let us analyze the behavior of the Chern–Simons and Yang–Mills terms appearing in Theorem 3.8 under this gauge transformation, and derive an interesting consequence for the pairing between the odd (b, B)-cocycle $\tilde{\psi}$ with the odd K-theory group of $\mathcal{A}$. As an easy consequence of the fact that ϕ_{2k} is a Hochschild cocycle, we have

LEMMA 3.11. *The Yang–Mills terms* $\int_{\phi_{2k}} F^k$ *with* $F = dA + A^2$ *are invariant under the gauge transformation* $A \mapsto A^U$ *for every* $k \geq 1$.

We are thus led to the conclusion that the sum of Chern–Simons forms is gauge invariant as well. Indeed, arguing as in [**15**], since both $\operatorname{Tr}(f(D+V))$ and the Yang–Mills terms are invariant under $V \mapsto V^U$, we find that, under the assumptions stated in Theorem 3.8:

$$\sum_{k=0}^{\infty} \int_{\psi_{2k+1}} \mathrm{cs}_{2k+1}(A^U) = \sum_{k=0}^{\infty} \int_{\psi_{2k+1}} \mathrm{cs}_{2k+1}(A).$$

Each individual Chern–Simons form behaves non-trivially under a gauge transformation. Nevertheless, it turns out that we can conclude, just as in [**15**], that the pairing of the whole (b, B)-cocycle with K-theory is trivial. Since the (b, B)-cocycle $\tilde{\psi}$ is given as an infinite sequence, we should first carefully study the analytical behavior of $\tilde{\psi}$. In fact, we should show that it is an *entire cyclic cocycle* in the sense of [**10**] (see also [**11**, Section IV.7.α]). It turns out [**28**, Lemma 36] that our assumptions on the growth of the derivatives of f ensure that the brackets define entire cyclic cocycles.

LEMMA 3.12. *Fix* $f \in \mathcal{E}_s^\gamma$ *for* $\gamma < 1$ *and equip* $\mathcal{A}$ *with the norm* $\|a\|_1 = \|a\| + \|[D, a]\|$. *Then, for any bounded subset* $\Sigma \subset \mathcal{A}$ *there exists* C_Σ *such that*

$$\left|\tilde{\psi}_{2k+1}(a_0, \ldots, a_{2k+1})\right| \leq \frac{C_\Sigma}{k!},$$

for all $a_j \in \Sigma$. *Hence,* ϕ *and* $\tilde{\psi}$ *are entire cyclic cocycles.*

We thus have the following interesting consequence of Theorem 3.8.

THEOREM 3.13. *Let* $f \in \mathcal{E}_s^\gamma$ *for* $\gamma < 1$. *Then the pairing of the odd entire cyclic cocycle* $\tilde{\psi}$ *with* $K_1(\mathcal{A})$ *is trivial, i.e.*

$$\langle U, \tilde{\psi} \rangle = (2\pi i)^{-1/2} \sum_{k=0}^{\infty} (-1)^k k! \tilde{\psi}_{2k+1}(U^*, U, \ldots, U^*, U) = 0$$

for all unitary $U \in \mathcal{A}$.

4. One-loop corrections to the spectral action

We now formulate a quantum version of the spectral action. To do this, we must first interpret the spectral action, expanded in terms of generalized Chern–Simons and Yang–Mills actions by Theorem 3.8, as a classical action, which leads us naturally to a noncommutative geometric notion of a vertex. Enhanced with a spectral gauge propagator derived from the formalism of random matrices (and in particular, random finite noncommutative geometries) this gives us a concept of one-loop counterterms and a proposal for a one-loop *quantum effective spectral action*, without leaving the spectral framework. We will show here that, at least in a finite-dimensional setting, these counterterms can again be written as Chern–Simons and Yang–Mills forms integrated over (quantum corrected) cyclic cocycles. We therefore discern a renormalization flow in the space of cyclic cocycles.

4.1. Conventions. We let $\varphi_1, \varphi_2, \ldots$ be an orthonormal basis of eigenvectors of D, with corresponding eigenvalues $\lambda_1, \lambda_2, \ldots$. For any $N \in \mathbb{N}$, we define

$$H_N := (M_N)_{\mathrm{sa}}, \quad M_N := \operatorname{span}\{|\varphi_i\rangle\langle\varphi_j| : i, j \in \{1, \ldots, N\}\},$$

and endow H_N with the Lebesgue measure on the coordinates $Q \mapsto \mathrm{Re}(Q_{ij})$ ($i \leq j$) and $Q \mapsto \mathrm{Im}(Q_{ij})$ ($i < j$). Here and in the following, $Q_{ij} := \langle \varphi_i, Q\varphi_j \rangle$ are the matrix elements of Q. For simplicity, we will assume that the perturbations $V_1, \ldots, V_n$ are in $\cup_K H_K$.

For us, a *Feynman diagram* is a finite multigraph with a number of marked vertices of degree 1 called external vertices, all other vertices being called internal vertices or, by abuse of terminology, vertices. An edge, sometimes called a propagator, is called external if it connects to an external vertex, and internal otherwise. The external vertices are simply places for the external edges to attach to, and are often left out of the discussion. An n-point diagram is a Feynman diagram with n external edges. A Feynman diagram is called one-particle-irreducible if any multigraph obtained by removing one of the internal edges is connected.

4.2. Diagrammatic expansion of the spectral action. Viewing the spectral action as a classical action, and following the background field method, the vertices of degree n in the corresponding quantum theory should correspond to n^{th}-order functional derivatives of the spectral action. However, in the paradigm of noncommutative geometry, a base manifold is absent, and functional derivatives do not exist in the local sense. Therefore, a more abstract notion of a vertex is needed. The brackets $\langle \cdot \rangle$ from (8) that power the expansion of the spectral action in Theorems 3.8 and 3.9 are by construction cyclic and multilinear extensions of the derivatives of the spectral action, and as such provide an appropriate notion of *noncommutative vertices*. We define a noncommutative vertex with $V_1, \ldots, V_n \in \cup_K H_K$ on the external edges by

(12)
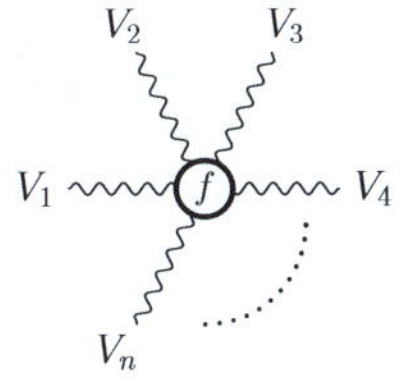

$$:= \langle V_1, \ldots, V_n \rangle.$$

In contrast to a normal vertex of a Feynman diagram, a noncommutative vertex is decorated with a cyclic order on the edges incident to it. By convention, the edges are attached clockwise with respect to this cyclic order. As such, with perturbations $V_1, \ldots, V_n$ decorating the external edges, the diagram (12) reflects the cyclicity of the bracket: $\langle V_1, \ldots, V_n \rangle = \langle V_n, V_1, \ldots, V_{n-1} \rangle$, the first property of Lemma 2.1. In order to diagramatically represent the second property of Lemma 2.1 as well, we introduce the following notation. Wherever a gauge edge meets a noncommutative vertex we can insert a dashed line decorated with an element $a \in \mathcal{A}$ before or after the gauge edge, with the following meaning:

With this notation, the equation

(13) $$\langle aV_1, \ldots, V_n\rangle - \langle V_1, \ldots, V_n a\rangle = \langle V_1, \ldots, V_n, [D, a]\rangle,$$

is represented as

(14) 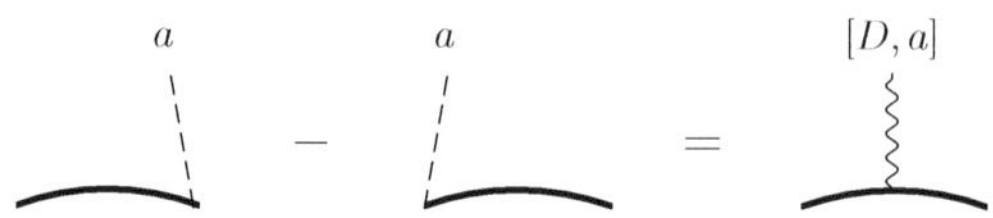

and is as such referred to as the *Ward identity.*

To illustrate, let us give the relevant lower order computations. The cyclic cocycles are expressed in terms of diagrams as

(15) $$\int_{\phi_n} a^0 da^1 \cdots da^n =$$ 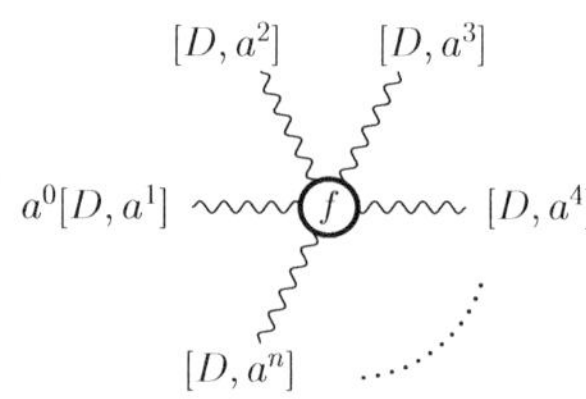

For one external edge we find, writing $A = \sum_j a_j db_j$ and suppressing summation over j,

(16) $$\langle V\rangle = \langle a_j[D, b_j]\rangle =$$ 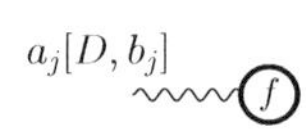

$$= \int_{\phi_1} A.$$

For two external edges, we apply the Ward identity (14) and derive

$\langle V, V\rangle =$ 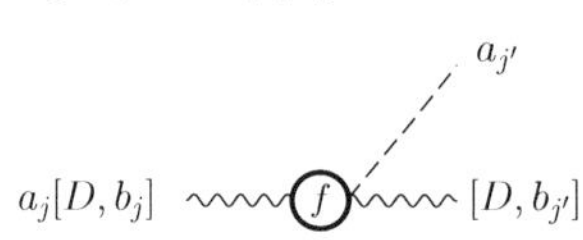

$=$ 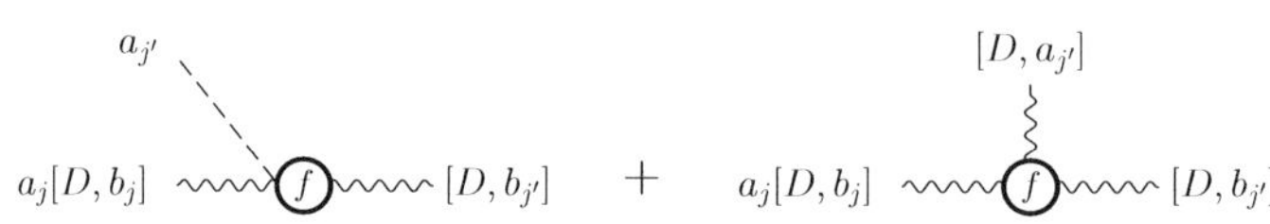

$$= \int_{\phi_2} A^2 + \int_{\phi_3} AdA.$$

4.2.1. *The propagator.* An important part of the quantization process introduced here is to find a mathematical formulation for the propagator. In other words, we need to introduce more general diagrams than the one-vertex diagram in (12), and assign each an amplitude. As usual in quantum field theory, the amplitudes depend on a cutoff N and are possibly divergent as $N \to \infty$.

What we will call a *noncommutative Feynman diagram* (or, for brevity, a diagram) is a Feynman diagram in which every internal vertex v is decorated with a cyclic order on the edges incident to v. These decorated vertices are what we call the noncommutative vertices, and are denoted as in (12). The edges of a diagram are always drawn as wavy lines. They are sometimes called gauge edges to distinguish them from any dashed lines in the diagram, which do not represent physical particles, but are simply notation. The *loop order* is defined to be $L := 1 - V + E$, where V is the amount of (noncommutative) vertices and E is the amount of internal edges. We also say the noncommutative Feynman diagram is L-loop, e.g., the noncommutative Feynman diagram in (12) is zero-loop. When the respective multigraph is planar, L corresponds to the number of internal faces. Following physics terminology, these faces are referred to as *loops*. As usual for Feynman diagrams, the external edges are marked, say by the numbers $1, \dots, n$.

Note that, by our definition, a noncommutative Feynman diagram is almost the same as a ribbon graph, the sole difference being that ribbons are sensitive to twisting, whereas our edges are not.

Each nontrivial noncommutative Feynman diagram will be assigned an *amplitude*, as follows. Here *nontrivial* means that every connected component contains at least one vertex with nonzero degree.

DEFINITION 4.1. Let $N \in \mathbb{N}$ and let $f \in C^\infty$ satisfy $f'[\lambda_i, \lambda_j] > 0$ for $i, j \leq N$. Given a nontrivial n-point noncommutative Feynman diagram G with external vertices marked by $1, \dots, n$, its **amplitude** at level $N \in \mathbb{N}$ on the gauge fields $V_1, \dots, V_n \in \cup_K H_K$ is denoted $\Gamma_N^G(V_1, \dots, V_n)$, and is defined recursively as follows. When G has precisely one vertex and the markings $1, \dots, n$ respect its cyclic order, we set $\Gamma_N^G(V_1, \dots, V_n) := \langle V_1, \dots, V_n \rangle$. Suppose the amplitudes of diagrams G_1 and G_2 with external edges $1, \dots, n$ and $n+1, \dots, m$ are defined. Then to the disjoint union G of the diagrams we assign the amplitude

$$\Gamma_N^G(V_1, \dots, V_m) := \Gamma_N^{G_1}(V_1, \dots, V_n)\Gamma_N^{G_2}(V_{n+1}, \dots, V_m).$$

Suppose the amplitude of a diagram G is defined. Then, for any two distinct numbers $i, j \in \{1, \dots, n\}$, let G' be the diagram obtained from G by connecting the two external edges i and j by a gauge edge (a propagator). We then define the amplitude of G' as

$$\Gamma_N^{G'}(V_1, \dots, \widehat{V_i}, \dots, \widehat{V_j}, \dots, V_n) := -\frac{\int_{H_N} \Gamma_N^G(V_1, \dots, \overset{i}{Q}, \dots, \overset{j}{Q}, \dots, V_n) e^{-\frac{1}{2}\langle Q, Q \rangle} dQ}{\int_{H_N} e^{-\frac{1}{2}\langle Q, Q \rangle} dQ}.$$

Well-definedness is a straightforward consequence of Fubini's theorem. Note that, in general, Γ_N^G is not cyclic in its arguments, as was the case in (12).

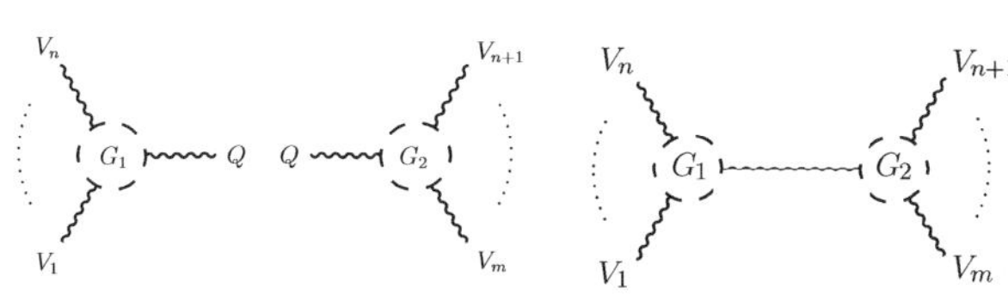

FIGURE 1. Constructing the propagator.

The assumption that $f'[\lambda_i, \lambda_j] > 0$ for $i, j \leq N$ can be accomplished by allowing f to be unbounded, and replacing the spectral action

$$\mathrm{Tr}(f(D))$$

with the regularized version

$$\mathrm{Tr}(f_N(D))$$

where $f_N := f\Phi_N$ for a sequence of bump functions Φ_N ($N \in \mathbb{N}$) that are 1 on $\{\lambda_k : k \leq N\}$. As quantization takes place on the finite level (for a finite N), it is natural to also regularize the classical action before we quantize. Because we can now easily require

$$f_N'[\lambda_k, \lambda_l] = f'[\lambda_k, \lambda_l] > 0,$$

for all $k, l \leq N$, Definition 4.1 makes sense and can be studied by Gaussian integration as in [**4**, Section 2].

4.3. Loop corrections to the spectral action. To obtain the propagator, we have chosen the approach of random noncommutative geometries (as done in [**2, 25**], see [**3, 19**] for computer simulations) in the sense that the integrated space in Definition 4.1 is the whole of H_N. Other approaches are conceivable by replacing H_N by a subspace of gauge fields particular to the gauge theory under consideration (like $\Omega^1_D(\mathcal{A})_{\mathrm{sa}}$ for a finite spectral triple $(\mathcal{A}, \mathcal{H}, D)$) but this should also take into account gauge fixing, and will quickly become very involved. We expect to require sophisticated machinery to perform such an integration, similar to the machinery in [**17**].

In our case, the propagator becomes quite simple, and can be explicitly expressed by the following result.

LEMMA 4.2. *Let $f \in C^\infty$ satisfy $f'[\lambda_k, \lambda_l] > 0$ for $k, l \leq N$. For $k, l, m, n \in \{1, \ldots, N\}$, we have*

$$\frac{\int_{H_N} Q_{kl} Q_{mn} e^{-\frac{1}{2}\langle Q, Q\rangle} dQ}{\int_{H_N} e^{-\frac{1}{2}\langle Q, Q\rangle} dQ} = \delta_{kn}\delta_{lm} G_{kl},$$

in terms of $G_{kl} := \frac{1}{f'[\lambda_k, \lambda_l]}$.

PROOF. By (8) we have the finite sum

$$\langle Q, Q\rangle = \sum_{k,l} f'[\lambda_k, \lambda_l]\left((\mathrm{Re}(Q_{kl}))^2 + (\mathrm{Im}(Q_{kl}))^2\right),$$

for all $Q \in H_N$. Moreover, we have

$$\begin{aligned}
&\int_{H_N} Q_{kl} Q_{mn} e^{-\frac{1}{2}\langle Q, Q\rangle} dQ \\
&\quad = \int_{H_N} (\mathrm{Re}(Q_{kl})\mathrm{Re}(Q_{mn}) - \mathrm{Im}(Q_{kl})\mathrm{Im}(Q_{mn})) e^{-\frac{1}{2}\langle Q, Q\rangle} dQ \\
&\qquad + i \int_{H_N} (\mathrm{Re}(Q_{kl})\mathrm{Im}(Q_{mn}) + \mathrm{Im}(Q_{kl})\mathrm{Re}(Q_{mn})) e^{-\frac{1}{2}\langle Q, Q\rangle} dQ.
\end{aligned}$$

The second integral on the right-hand side vanishes because its integrand is an odd function in at least one of the coordinates of H_N. The same holds for the first integral whenever $\{k, l\} \neq \{m, n\}$. Otherwise, we use that $\mathrm{Re}(Q_{lk}) = \mathrm{Re}(Q_{kl})$ and $\mathrm{Im}(Q_{lk}) = -\mathrm{Im}(Q_{kl})$ and see that the two terms of the first integral cancel when $k = m$ and $l = n$. When $k = n \neq l = m$, we instead find that these terms give the same result when integrated. By using symmetry of the divided difference (i.e., $f'[x,y]=f'[y,x]$) and integrating out all trivial coordinates, we obtain

$$\frac{\int_{H_N} Q_{kl}Q_{mn}e^{-\frac{1}{2}\langle Q,Q\rangle}dQ}{\int_{H_N} e^{-\frac{1}{2}\langle Q,Q\rangle}dQ} = \delta_{kn}\delta_{lm}\frac{2\int_{\mathbb{R}}(\mathrm{Re}(Q_{kl}))^2 e^{-f'[\lambda_k,\lambda_l](\mathrm{Re}(Q_{kl}))^2}d\mathrm{Re}(Q_{kl})}{\int_{\mathbb{R}} e^{-f'[\lambda_k,\lambda_l](\mathrm{Re}(Q_{kl}))^2}d\mathrm{Re}(Q_{kl})},$$

a Gaussian integral that gives the G_{kl} required by the lemma. When $k=l=n=m$, the result follows similarly. □

The above lemma allows us to leave out all integrals from the subsequent computations. In place of those integrals, we use the following notation.

DEFINITION 4.3. We define, with slight abuse of notation,

$$\overbrace{Q_{kl}\quad Q_{mn}} := \delta_{kn}\delta_{lm}G_{kl},$$

and refer to G_{kl} as the *propagator*.

As an example and to fix terminology, we will now compute the amplitudes of the three most basic one-loop diagrams with two external edges. These are given in Figure 2. Using Lemma 4.2 and Definition 4.3, we find the amplitude for the first diagram to be

$$= \sum_{\substack{i,j,k,l,\\ m,n\le N}} f'[\lambda_i,\lambda_j,\lambda_k](V_1)_{ij}Q_{jk}Q_{ki}f'[\lambda_l,\lambda_m,\lambda_n](V_2)_{lm}Q_{mn}Q_{nl}$$

$$= \sum_{i,k\le N} f'[\lambda_i,\lambda_i,\lambda_k]f'[\lambda_i,\lambda_k,\lambda_k](V_1)_{ii}(V_2)_{kk}(G_{ik})^2. \tag{17}$$

As V_1 and V_2 are assumed of finite rank, the above expression converges as $N\to\infty$. To see this explicitly, let K be such that $V_1,V_2\in H_K$, and let G be the diagram on the left-hand side of (17). We then obtain

$$\lim_{N\to\infty}\Gamma_N^G(V_1,V_2) = \sum_{i,k\le K} f'[\lambda_i,\lambda_i,\lambda_k]f'[\lambda_i,\lambda_k,\lambda_k](V_1)_{ii}(V_2)_{kk}(G_{ik})^2, \tag{18}$$

a finite number. In general we can say that if all summed indices of an amplitude occur in a matrix element of any of the perturbations (e.g., $(V_1)_{ii}$ and $(V_2)_{kk}$) then the amplitude remains finite even when the size N of the random matrices Q is sent to ∞. In physics terminology, the first diagram in Figure 2 is *irrelevant*, and can be disregarded for renormalization purposes.

We then turn to the second diagram in Figure 2, and compute

$$= \sum_{\substack{i,j,k,l,\\ m,n\le N}} f'[\lambda_i,\lambda_j,\lambda_k](V_1)_{ij}Q_{jk}Q_{ki}f'[\lambda_l,\lambda_m,\lambda_n](V_2)_{lm}Q_{mn}Q_{nl}$$

$$= \sum_{i,j,k\le N} (f'[\lambda_i,\lambda_j,\lambda_k])^2(V_1)_{ij}(V_2)_{ji}G_{ik}G_{kj}. \tag{19}$$

This diagram is planar, and the indices i,j,k correspond to regions in the plane, assuming the external edges are regarded to stretch out to infinity. The index k corresponds to the region within the loop, and is called a *running loop index*. As the index k is not restricted by V_1 and V_2 as in (17), we find that in general the amplitude (19) diverges as $N\to\infty$. In physical terms, this is a *relevant* diagram.

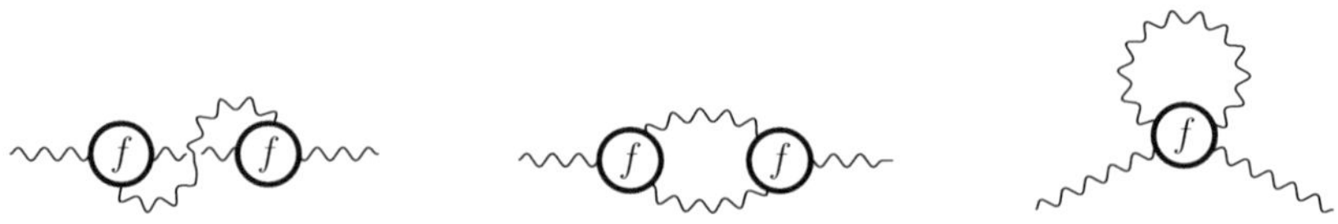

FIGURE 2. Two-point diagrams with one loop. The first one is irrelevant, the second and third are relevant.

The amplitude of the final diagram becomes

$$\begin{aligned} &= - \sum_{i,j,k,l \leq N} f'[\lambda_i, \lambda_j, \lambda_k, \lambda_l](V_1)_{ij} \overline{Q_{jk} Q}_{kl} (V_2)_{li} \\ (20) \qquad &= - \sum_{i,j,k \leq N} f'[\lambda_i, \lambda_j, \lambda_j, \lambda_k](V_1)_{ij}(V_2)_{ji} G_{jk}. \end{aligned}$$

Again, this amplitude contains a running loop index and is therefore potentially divergent in the limit $N \to \infty$.

4.3.1. *One-loop counterterms to the spectral action.* Because we are interested in the behavior of the one-loop quantum effective spectral action as $N \to \infty$, we wish to consider only one-loop noncommutative Feynman diagrams whose amplitudes involve a running loop index. For example, the final two diagrams in Figure 2, but not the first.

As dictated by the background field method, in order to obtain a quantum effective action we should further restrict to one-particle-irreducible diagrams whose vertices have degree ≥ 3.

Let us fix a one-loop one-particle-irreducible diagram G in which all vertices have degree ≥ 3, and investigate whether the amplitude of G contains a running loop index. Fix a noncommutative vertex v in G. The vertex v will have precisely two incident edges that belong to the loop of the diagram, and at least one external edge. Each index associated with v is associated specifically with two incident edges of v. If one of these edges is external, the index will not run, because it will be fixed by the gauge field attached. A running index can only occur if the two incident loop edges of v succeed one another, and the index is placed in between them. The latter of these two loop edges will attach to another noncommutative vertex, w, and the possibly running index will also be associated with the succeeding edge in w, which also has to be a loop edge if the index is to run. This process may continue throughout the loop until we end up at the original vertex v. By this argument, the amplitude of G will contain a running loop index if and only if G can be drawn in the plane with all noncommutative vertices oriented clockwise and all external edges extending outside the loop.

The wonderful conclusion is that the external edges of the relevant diagrams obtain a natural cyclic order. This presents us with a natural one-loop quantization of the bracket $\langle \cdot \rangle$, and thus with a natural proposal for the one-loop quantization of the spectral action.

DEFINITION 4.4. Let $N \in \mathbb{N}$ and let $f \in C^\infty$ satisfy $f'[\lambda_i, \lambda_j] > 0$ for $i, j \leq N$. We define

$$\langle\!\langle V_1, \ldots, V_n \rangle\!\rangle_N^{1L} := \sum_G \Gamma_N^G(V_1, \ldots, V_n),$$

where the sum is over all planar one-loop one-particle-irreducible n-point noncommutative Feynman diagrams G with clockwise vertices of degree ≥ 3 and external edges outside the loop and marked cyclically. The **one-loop quantum effective spectral action** is defined to be the formal series

$$\sum_{n=1}^{\infty} \frac{1}{n} \langle\!\langle V, \ldots, V \rangle\!\rangle_N^{1L}.$$

Directly from the definition of $\langle\!\langle \cdot \rangle\!\rangle_N^{1L}$, we see that

$$\langle\!\langle V_2, \ldots, V_n, V_1 \rangle\!\rangle_N^{1L} = \langle\!\langle V_1, \ldots, V_n \rangle\!\rangle_N^{1L}.$$

In other words, the property (I) holds for the bracket $\prec \cdot \succ = \langle\!\langle \cdot \rangle\!\rangle_N^{1L}$. In the next subsection we will show that (II) holds as well.

4.3.2. *Ward identity for the gauge propagator.* In addition to the Ward identity (14) for the noncommutative vertex, we claim that we also have the following Ward identity for the gauge edge:

(21)
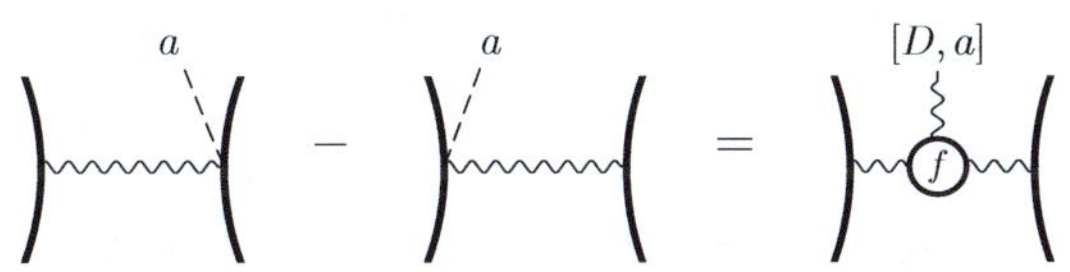

Indeed, the left-hand side yields terms

$$\sum_{m \leq N} \left(Q_{ik} Q_{lm} a_{mn} - a_{im} Q_{mk} Q_{ln} \right) = \sum_{m \leq N} \left(G_{ik} \delta_{im} \delta_{kl} a_{mn} - G_{ln} \delta_{mn} \delta_{kl} a_{im} \right)$$
$$= (G_{ik} - G_{nk}) \delta_{kl} a_{in},$$

for arbitrary values of i, k, l, and n determined by the rest of the diagram. The right-hand side, by the defining property of the divided difference, and because every internal edge adds a minus sign, yields the terms

$$- \sum_{p,q,r \leq N} Q_{ik} f'[\lambda_p, \lambda_q, \lambda_r] Q_{pq} [D, a]_{qr} Q_{rp} Q_{ln}$$
$$= - \sum_{p,q,r \leq N} f'[\lambda_p, \lambda_q, \lambda_r](\lambda_q - \lambda_r) a_{qr} G_{ik} \delta_{iq} \delta_{kp} G_{rp} \delta_{rn} \delta_{pl}$$
$$= \left(f'[\lambda_k, \lambda_n] - f'[\lambda_i, \lambda_k] \right) G_{ik} G_{nk} \delta_{kl} a_{in}.$$

Because $G_{kl} = 1/f'[\lambda_k, \lambda_l]$ (see Lemma 4.2) the two expressions coincide for every value of i, k, l, and n, thereby allowing us to apply the rule (21) whenever it comes up as part of a diagram. For example, by combining (21) with (14), we have

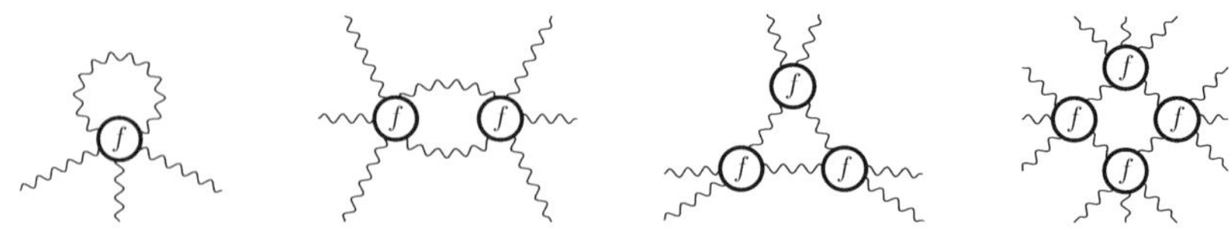

FIGURE 3. Relevant one-loop n-point functions with increasing number of vertices.

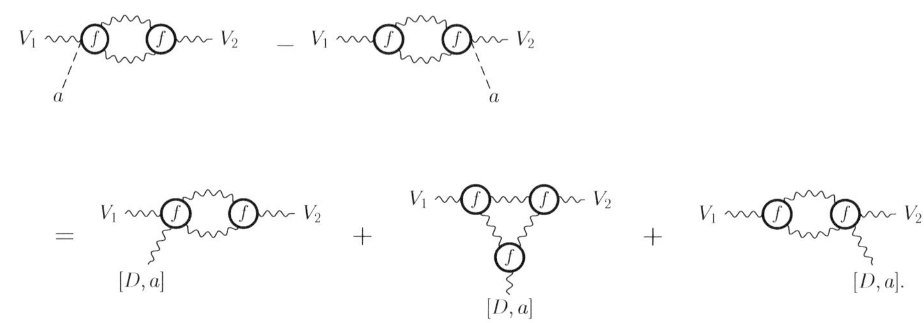

The Ward identity for the gauge propagator, in combination with the Ward identity for the fermion propagator (14) allows us to derive the so-called *quantum Ward identity*:

$$\langle\!\langle aV_1, \ldots, V_n \rangle\!\rangle_N^{1L} - \langle\!\langle V_1, \ldots, V_n a \rangle\!\rangle_N^{1L} = \langle\!\langle [D, a], V_1, \ldots, V_n \rangle\!\rangle_N^{1L}.$$

We derived this identity diagrammatically in [**29**] for low orders; below we give a general derivation. The quantum Ward identity, in combination with the obvious cyclicity, shows that $\langle\!\langle \cdot \rangle\!\rangle_N^{1L}$ is a special case of the generic bracket $\prec \cdot \succ$ satisfying property (I) and (II) on page 517, and hence allows us to apply Proposition 3.5 and Theorem 3.9. We thus obtain our final result: an expansion of the one-loop quantum effective action in terms of cyclic cocycles.

THEOREM 4.5. *There exist (b, B)-cocycles ϕ^N and $\tilde{\psi}^N$ (namely, those defined by taking $\prec \cdot \succ = \langle\!\langle \cdot \rangle\!\rangle_N^{1L}$ in (9) and (10)) for which the one-loop quantum effective spectral action can be expanded as*

$$\sum_{n=1}^{\infty} \frac{1}{n} \langle\!\langle V, \ldots, V \rangle\!\rangle_N^{1L} \sim \sum_{k=1}^{\infty} \left(\int_{\psi_{2k-1}^N} \mathrm{cs}_{2k-1}(A) + \frac{1}{2k} \int_{\phi_{2k}^N} F^k \right).$$

As before, $\tilde{\psi}_{2k-1}^N = (-1)^{k-1} \frac{(k-1)!}{(2k-1)!} \psi_{2k-1}^N$.

PROOF. Applying Definition 4.4, and combining two sums, we obtain

$$\langle\!\langle aV_1, \ldots, V_n \rangle\!\rangle_N^{1L} - \langle\!\langle V_1, \ldots, V_n a \rangle\!\rangle_N^{1L} = \sum_G \left(\Gamma_N^G(aV_1, \ldots, V_n) - \Gamma_N^G(V_1, \ldots, V_n a) \right),$$

where the sum is over all *relevant* diagrams G, by which we mean the planar one-loop one-particle-irreducible n-point noncommutative Feynman diagrams G with clockwise vertices of degree ≥ 3 and external edges outside the loop and marked cyclically. Let G be a relevant diagram marked $1, \ldots, n$. We let $I(G)$ denote the set of diagrams one can obtain from G by inserting a single gauge edge at any of the places one visits when walking along the outside of the diagram from the external edge n to the external edge 1. To be precise, if the edges n and 1 attach to the same noncommutative vertex v, we set

$$I(G) := \{G'\},$$

where G' is the diagram obtained from G by inserting an external edge marked $n+1$ at v between the edges marked n and 1. If the edges n and 1 attach to different vertices v and w, respectively, then the edge e succeeding the edge marked n on v necessarily attaches to w, preceding the edge marked 1. In this case, we set

$$I(G) := \{G_n, G_e, G_1\},$$

where G_n is obtained from G by inserting an external edge marked $n+1$ at v between n and e, G_e is obtained from G by inserting a noncommutative vertex v_0 along e and inserting an external edge marked $n+1$ along the outside of v_0, and G_1 is obtained from G by inserting an external edge marked $n+1$ at w between e and 1. By construction of $I(G)$, we find

$$\langle\!\langle aV_1, \dots, V_n \rangle\!\rangle_N^{1L} - \langle\!\langle V_1, \dots, V_n a \rangle\!\rangle_N^{1L} = \sum_G \sum_{G' \in I(G)} \Gamma_N^{G'}(V_1, \dots, V_n, [D, a]).$$

The sum over G and G' yields all relevant $n+1$-point diagrams, and, moreover, any relevant $n+1$-point diagram with labels $V_1, \dots, V_n, [D, a]$ is obtained in a unique manner from an insertion of an external edge in an n-point diagram, as described above. We are therefore left precisely with

$$\langle\!\langle aV_1, \dots, V_n \rangle\!\rangle_N^{1L} - \langle\!\langle V_1, \dots, V_n a \rangle\!\rangle_N^{1L} = \langle\!\langle V_1, \dots, V_n, [D, a] \rangle\!\rangle_N^{1L}.$$

In combination with cyclicity, $\langle\!\langle V_1, \dots, V_n \rangle\!\rangle_N^{1L} = \langle\!\langle V_n, V_1, \dots, V_{n-1} \rangle\!\rangle_N^{1L}$, this identity allows us to apply Proposition 3.5 and Theorem 3.9. We thus arrive at the conclusion of the theorem. □

We conclude that the passage to the one-loop renormalized spectral action can be realized by a transformation in the space of cyclic cocycles, sending $\phi \mapsto \phi + \phi^N$ and $\psi \mapsto \psi + \psi^N$. One could say the theory is therefore one-loop renormalizable in a generalized sense, allowing for infinitely many counterterms, as in [**20**]. Most notably, we have stayed within the spectral paradigm of noncommutative geometry.

References

[1] N. A. Azamov, A. L. Carey, P. G. Dodds, and F. A. Sukochev, *Operator integrals, spectral shift, and spectral flow*, Canad. J. Math. **61** (2009), no. 2, 241–263, DOI 10.4153/CJM-2009-012-0. MR2504014

[2] S. Azarfar and M. Khalkhali, *Random finite noncommutative geometries and topological recursion*, arXiv:1906.09362.

[3] J. W. Barrett and L. Glaser, *Monte Carlo simulations of random non-commutative geometries*, J. Phys. A **49** (2016), no. 24, 245001, 27, DOI 10.1088/1751-8113/49/24/245001. MR3512080

[4] D. Bessis, C. Itzykson, and J. B. Zuber, *Quantum field theory techniques in graphical enumeration*, Adv. in Appl. Math. **1** (1980), no. 2, 109–157, DOI 10.1016/0196-8858(80)90008-1. MR603127

[5] A. H. Chamseddine and A. Connes, *Universal formula for noncommutative geometry actions: unification of gravity and the standard model*, Phys. Rev. Lett. **77** (1996), no. 24, 4868–4871, DOI 10.1103/PhysRevLett.77.4868. MR1419931

[6] A. H. Chamseddine and A. Connes, *The spectral action principle*, Comm. Math. Phys. **186** (1997), no. 3, 731–750, DOI 10.1007/s002200050126. MR1463819

[7] A. H. Chamseddine, A. Connes, and M. Marcolli, *Gravity and the standard model with neutrino mixing*, Adv. Theor. Math. Phys. **11** (2007), no. 6, 991–1089. MR2368941
[8] A. Chattopadhyay and A. Skripka, *Trace formulas for relative Schatten class perturbations*, J. Funct. Anal. **274** (2018), no. 12, 3377–3410, DOI 10.1016/j.jfa.2018.02.019. MR3787595
[9] A. Connes, *Noncommutative differential geometry*, Inst. Hautes Études Sci. Publ. Math. **62** (1985), 257–360. MR823176
[10] A. Connes, *Entire cyclic cohomology of Banach algebras and characters of θ-summable Fredholm modules*, K-Theory **1** (1988), no. 6, 519–548, DOI 10.1007/BF00533785. MR953915
[11] A. Connes, *Noncommutative geometry*, Academic Press, Inc., San Diego, CA, 1994. MR1303779
[12] A. Connes, *Gravity coupled with matter and the foundation of non-commutative geometry*, Comm. Math. Phys. **182** (1996), no. 1, 155–176. MR1441908
[13] A. Connes and M. Marcolli, *Noncommutative geometry, quantum fields and motives*, American Mathematical Society Colloquium Publications, vol. 55, American Mathematical Society, Providence, RI; Hindustan Book Agency, New Delhi, 2008, DOI 10.1090/coll/055. MR2371808
[14] A. Connes, *On the spectral characterization of manifolds*, J. Noncommut. Geom. **7** (2013), no. 1, 1–82, DOI 10.4171/JNCG/108. MR3032810
[15] A. Connes and A. H. Chamseddine, *Inner fluctuations of the spectral action*, J. Geom. Phys. **57** (2006), no. 1, 1–21, DOI 10.1016/j.geomphys.2006.08.003. MR2265456
[16] W. F. Donoghue Jr., *Monotone matrix functions and analytic continuation*, Die Grundlehren der mathematischen Wissenschaften, Band 207, Springer-Verlag, New York-Heidelberg, 1974. MR0486556
[17] B. Eynard and A. P. Ferrer, *2-matrix versus complex matrix model, integrals over the unitary group as triangular integrals*, Comm. Math. Phys. **264** (2006), no. 1, 115–144, DOI 10.1007/s00220-006-1541-8. MR2212218
[18] E. Getzler and A. Szenes, *On the Chern character of a theta-summable Fredholm module*, J. Funct. Anal. **84** (1989), no. 2, 343–357, DOI 10.1016/0022-1236(89)90102-X. MR1001465
[19] L. Glaser and A. B. Stern, *Understanding truncated non-commutative geometries through computer simulations*, J. Math. Phys. **61** (2020), no. 3, 033507, 11, DOI 10.1063/1.5131864. MR4076626
[20] J. Gomis and S. Weinberg, *Are nonrenormalizable gauge theories renormalizable?*, Nuclear Phys. B **469** (1996), no. 3, 473–487, DOI 10.1016/0550-3213(96)00132-0. MR1403745
[21] H. Grosse, A. Sako, and R. Wulkenhaar, *The Φ^3_4 and Φ^3_6 matricial QFT models have reflection positive two-point function*, Nuclear Phys. B **926** (2018), 20–48, DOI 10.1016/j.nuclphysb.2017.10.022. MR3760253
[22] F. Hansen, *Trace functions as Laplace transforms*, J. Math. Phys. **47** (2006), no. 4, 043504, 11, DOI 10.1063/1.2186925. MR2226341
[23] N. Higson, *The residue index theorem of Connes and Moscovici*, Surveys in noncommutative geometry, Clay Math. Proc., vol. 6, Amer. Math. Soc., Providence, RI, 2006, pp. 71–126. MR2277669
[24] A. Hock. Matrix Field Theory. PhD thesis, WWU Münster (2020).
[25] M. Khalkhali and N. Pagliaroli, *Phase transition in random noncommutative geometries*, J. Phys. A **54** (2021), no. 3, Paper No. 035202, 14, DOI 10.1088/1751-8121/abd190. MR4209133
[26] Y. Liu, *Cyclic Structure behind Modular Gaussian Curvature*, `arXiv:2201.08730v1`
[27] T. D. H. van Nuland and A. Skripka, *Spectral shift for relative schatten class perturbations*, J. Spectr. Theor., to appear..
[28] T. D. H. van Nuland and W. D. van Suijlekom, *Cyclic cocycles in the spectral action*, J. Noncommut. Geom., to appear `arXiv:2104.09899`.
[29] T. D. H. van Nuland and W. D. van Suijlekom, *One-loop corrections to the spectral action*, J. High Energy Phys. **5** (2022), Paper No. 078, 14, DOI 10.1007/jhep05(2022)078. MR4430238
[30] D. Quillen, *Chern-Simons forms and cyclic cohomology*, The interface of mathematics and particle physics (Oxford, 1988), Inst. Math. Appl. Conf. Ser. New Ser., vol. 24, Oxford Univ. Press, New York, 1990, pp. 117–134. MR1103135
[31] A. Skripka, *Asymptotic expansions for trace functionals*, J. Funct. Anal. **266** (2014), no. 5, 2845–2866, DOI 10.1016/j.jfa.2013.12.021. MR3158710

[32] A. Skripka and A. Tomskova, *Multilinear operator integrals*, Lecture Notes in Mathematics, vol. 2250, Springer, Cham, [2019] ©2019. Theory and applications, DOI 10.1007/978-3-030-32406-3. MR3971571

[33] W. D. van Suijlekom, *Perturbations and operator trace functions*, J. Funct. Anal. **260** (2011), no. 8, 2483–2496, DOI 10.1016/j.jfa.2010.12.012. MR2772379

[34] W. D. van Suijlekom, *Noncommutative geometry and particle physics*, Mathematical Physics Studies, Springer, Dordrecht, 2015, DOI 10.1007/978-94-017-9162-5. MR3237670

Institute for Mathematics, Astrophysics and Particle Physics, Radboud University, Heyendaalseweg 135, 6525 AJ Nijmegen, The Netherlands; and School of Mathematics and Statistics, University of New South Wales, Kensington, New South Wales, 2052, Australia

Email address: t.van_nuland@unsw.edu.au

Institute for Mathematics, Astrophysics and Particle Physics, Radboud University, Heyendaalseweg 135, 6525 AJ Nijmegen, The Netherlands

Email address: waltervs@math.ru.nl

Proceedings of Symposia in Pure Mathematics
Volume **105**, 2023
https://doi.org/10.1090/pspum/105/01917

l^1-higher index, l^1-higher rho invariant and cyclic cohomology

Jinmin Wang, Zhizhang Xie, and Guoliang Yu

Dedicate to the 40th anniversary of cyclic cohomology

Abstract. In this paper, we study l^1-higher index theory and its pairing with cyclic cohomology for both closed manifolds and compact manifolds with boundary. We first give a sufficient geometric condition for the vanishing of the l^1-higher indices of Dirac-type operators on closed manifolds. This leads us to define an l^1-version of higher rho invariants. We prove a product formula for these l^1-higher rho invariants. A main novelty of our product formula is that it works in the general Banach algebra setting, in particular, the l^1-setting.

On compact spin manifolds with boundary, we also give a sufficient geometric condition for Dirac operators to have well-defined l^1-higher indices. More precisely, we show that, on a compact spin manifold M with boundary equipped with a Riemannian metric which has product structure near the boundary, if the scalar curvature on the boundary is sufficiently large, then the l^1-higher index of its Dirac operator D_M is well-defined and lies in the K-theory of the l^1-algebra of the fundamental group. As an immediate corollary, we see that if the Bost conjecture holds for the fundamental group of M, then the C^*-algebraic higher index of D_M lies in the image of the Baum-Connes assembly map.

By pairing the above K-theoretic l^1-index results with cyclic cocycles, we prove an l^1-version of the higher Atiyah-Patodi-Singer index theorem for manifolds with boundary. A key ingredient of its proof is the product formula for l^1-higher rho invariants mentioned above.

1. Introduction

The higher index for elliptic differential operators on closed manifolds is far-reaching generalization of the classical Fredholm index for Fredholm operators. Usually, the higher index an elliptic operators on a closed manifold takes values in the K-theory of the reduced or maximal group C^*-algebra of the fundamental group of the manifold.

Suppose (X, g) is a closed spin manifold equipped with a Riemannian metric g. If the scalar curvature κ of g is positive, then it follows from the Lichnerowicz formula that the Dirac operator D on X is invertible, which implies that the higher

2020 *Mathematics Subject Classification.* Primary 19K56, 19D55.

Key words and phrases. ℓ^1-higher index, product formula for ℓ^1-higher rho invariant, higher APS index theorem, Baum–Connes conjecture, Bost conjecture.

The first author is partially supported by NSF 1952693. The second author is partially supported by NSF 1800737 and 1952693. The third author is partially supported by NSF 2000082 and the Simons Fellows Program.

index $\mathrm{Ind}^\Gamma(D)$ of the Dirac operator D vanishes in $K_*(C_r^*(\Gamma))$ and $K_*(C_{\max}^*(\Gamma))$. Here $\Gamma = \pi_1(X)$ is the fundamental group of X, and $C_r^*(\Gamma)$ (resp. $C_{\max}^*(\Gamma)$) is the reduced (resp. maximal) group C^*-algebra of Γ. This vanishing result has some interesting geometric consequence. For example, it implies the non-existence of positive scalar metrics on aspherical manifolds whose fundamental groups satisfy the strong Novikov conjecture, cf. [**27**, **28**].

A standard construction of higher indices shows that the higher index of an elliptic operator on a closed manifold can always be represented by K-theory elements with finite propagation (cf. Definition 2.2). In particular, each such representative also lies in the K-theory $K_*(l^1(\Gamma))$ of the l^1-algebra of Γ. Let us denote the l^1-higher index class of an elliptic operator D by $\mathrm{Ind}_1^\Gamma(D)$. We will review the construction of these l^1-higher indices in Section 2.2. Unlike the C^*-algebraic setting, due to the lack of positivity in l^1-algebras, it remains an open question if the Lichnerowicz-type vanishing result still holds in the l^1-setting.

QUESTION 1.1. *Assume that (X, g) is a closed spin Riemannian manifold equipped with a Riemannian metric g whose scalar curvature is positive. Then does the l^1-higher index $\mathrm{Ind}_1^\Gamma\Gamma(D)$ of the Dirac operator D on X vanish in $K_*(l^1(\Gamma))$?*

Our first main theorem of the current paper is a partial answer to the above question. Namely, we show that, if the scalar curvature of g is sufficiently large in an appropriate sense, then the l^1-higher index $\mathrm{Ind}_1^\Gamma\Gamma(D)$ vanishes in $K_*(l^1(\Gamma))$. Before we state the theorem, let us fix some notation. Let $\widetilde{X}$ be the universal cover of X and $\mathcal{F}$ be a fundamental domain of the Γ-action on $\widetilde{X}$. Fix a finite symmetric generating set S of Γ. Let ℓ be the length function on Γ induced by S. We define

$$\tau = \liminf_{\ell(\gamma)\to\infty} \sup_{x\in\mathcal{F}} \frac{\mathrm{dist}(x, \gamma x)}{\ell(\gamma)} \tag{1.1}$$

and

$$K_\Gamma = \inf\{K : \exists C > 0 \text{ s.t. } \#\{\gamma \in \Gamma : \ell(\gamma) \leqslant n\} \leqslant Ce^{Kn}\}. \tag{1.2}$$

THEOREM 1.2. *Let (X, g) be an n-dimensional closed spin Riemannian manifold. Denote the fundamental group $\pi_1(X)$ of X by Γ. If the scalar curvature κ of g satisfies that*

$$\inf_{x\in X} \kappa(x) > \frac{16K_\Gamma^2}{\tau^2}, \tag{1.3}$$

then the l^1-higher index $\mathrm{Ind}_1^\Gamma(D)$ of the Dirac operator D on X vanishes in $K_n(l^1(\Gamma))$.

The numbers K_Γ and τ may depend on the choice of generating set, but Theorem 1.2 holds as long as the inequality (1.3) is satisfied for one choice of generating set. In particular, if the group Γ has sub-exponential growth, i.e., $K_\Gamma = 0$, then Theorem 1.2 holds as long as the scalar curvature is positive everywhere.

Recall that in the C^*-algebraic setting, the vanishing of the higher index of an elliptic operator together with a specific trivialization naturally induces a C^*-algebraic secondary invariant called higher rho invariant. These secondary invariants have various interesting applications in geometry and topology. For example, a positive scalar curvature metric on a closed spin manifold naturally defines a higher rho invariant associated to the given metric. The higher rho invariants for two positive scalar curvature metrics are the same if they are homotopic along a path of positive scalar curvature metric. Consequently, we can use higher rho invariants to distinguish different connected components of the space of positive scalar curvature metrics on a given closed spin manifold. In the l^1-setting, under the condition that the scalar curvature is sufficiently large (1.3), we introduce an l^1-higher rho invariant, denoted by $\rho_1(g)$, that lies in the K-theory of some geometric Banach algebra $\mathcal{B}_{L,0}(\widetilde{X})^\Gamma$, cf. Definition 2.16. This l^1-higher rho invariant also remains unchanged for a continuous path of metrics, provided each metric has sufficiently large scalar curvature. For C^*-algebraic higher rho invariant, there is a product formula (cf. [**24**] [**29**] [**30**]), which is important for geometric applications. In this paper, we prove a product formula for l^1-higher rho invariants. More precisely, we have the following theorem. Denote by $C_L^*(\mathbb{R})$ the localization algebra of the real line and $\mathrm{Ind}_L(D_\mathbb{R}) \in K_1(C_L^*(\mathbb{R}))$ the local index of the Dirac operator on $\mathbb{R}$, cf. [**26**].

THEOREM 1.3. *Assume that (X, g) is an m-dimensional closed spin Riemannian manifold such that the scalar curvature function κ of g satisfies the inequality in line* (1.3). *Then under the K-theory product map*

$$(1.4)\qquad {}_-\otimes_- : K_*(\mathcal{B}_{L,0}(\widetilde{X})^\Gamma) \otimes K_1(C_L^*(\mathbb{R})) \to K_{*+1}(\mathcal{B}_{L,0}(\widetilde{X} \times \mathbb{R})^\Gamma),$$

we have

$$(1.5)\qquad \rho_1(g) \otimes \mathrm{Ind}_L(D_\mathbb{R}) = \rho_1(g + dt^2) \in K_{m+1}(\mathcal{B}_{L,0}(\widetilde{X} \times \mathbb{R})^\Gamma),$$

where $\mathcal{B}_{L,0}(\widetilde{X})^\Gamma$ is a certain geometric Banach algebra (cf. Definition 2.16), and $g + dt^2$ is the product metric on $X \times \mathbb{R}$ induced by g and the standard Euclidean metric on $\mathbb{R}$.

When applied to the C^*-algebraic setting, our approach also proves the product formula for C^*-algebraic higher rho invariants. However, a main novelty of our approach is that it applies to the general Banach algebra setting, while the existing approaches in the C^*-algebra setting (cf. [**24**] [**29**] [**30**]) do not seem to generalize to the Banach algebra setting.

Now we turn to the case of manifolds with boundary. Given a compact spin manifold M with boundary, if M is equipped with a Riemannian metric which has product structure near the boundary and the scalar curvature is positive on the boundary, then the Dirac operator D on M defines a higher index $\mathrm{Ind}^\Gamma(D)$ which lies in the K-theory of the reduced group C^*-algebra of the fundamental group $\Gamma = \pi_1(M)$ of M, cf. Section 3. It is an open question whether such a higher index lies in the image of the Baum-Connes assembly map. Let us briefly recall the definition of the Baum-Connes assembly map. Denote by $\underline{E}\Gamma$ the classifying space of Γ for proper actions, and $K_*^\Gamma(\underline{E}\Gamma)$ its equivariant K-homology group. Each element of $K_*^\Gamma(\underline{E}\Gamma)$ can be represented by a Γ-equivariant Dirac operator (possibly twisted by an auxiliary vector bundle) on a complete *spin*^{c} manifold (without boundary) that is equipped with an isometric, proper and cocompact Γ-action. The Baum-Connes assembly map

$$\mu\colon K_*^\Gamma(\underline{E}\Gamma) \longrightarrow K_*(C_r^*(\Gamma))$$

is defined by mapping each twisted Γ-equivariant Dirac operator to its associated higher index.

The Baum–Connes conjecture claims that the assembly map is an isomorphism. The Baum-Connes conjecture has been verified for a large class of groups, including groups with Haagerup property [**8**] and hyperbolic groups [**11, 15**]. In general, the conjecture is still wide open. In fact, the following more special question is still wide open.

QUESTION 1.4. *Given a compact spin manifold with boundary whose Riemannian metric has product structure and positive scalar curvature near the boundary, does the higher index of its Dirac operator lie in the image of the Baum-Connes assembly map?*

A positive answer to the above question can be used to compute certain secondary index theoretic invariants, such as delocalized eta invariants and higher rho invariants associated to positive scalar curvature metrics on the boundary of a spin manifold, cf. [**22**, **25**].

Although we are mainly interested in Question 1.4 in the C^*-algebraic setting, the main tools we use in this paper for tackling such a question are in fact from the l^1-setting. Let us briefly explain how the l^1-setting and the C^*-algebraic setting are related. Recall that the algebra of l^1-functions on Γ is a Banach algebra and a dense subalgebra of $C_r^*(\Gamma)$. There is an analogue of the Baum-Connes conjecture in the l^1-setting, called the Bost conjecture, which states that

$$\mu_1\colon K_*^\Gamma(\underline{E}\Gamma) \longrightarrow K_*(l^1(\Gamma)) \tag{1.6}$$

is an isomorphism. Here we have used that fact the standard construction of higher indices of twisted Dirac operators on *spin*^{c} Γ-manifolds (without boundary) implies that these higher indices can be represented by elements with finite propagation (cf. Definition 2.2), hence in particular lie in $K_*(l^1(\Gamma))$. Consequently, the Baum-Connes assembly map $\mu\colon K_*^\Gamma(\underline{E}\Gamma) \to K_*(C_r^*(\Gamma))$ factors through the Bost assembly map $\mu_1\colon K_*^\Gamma(\underline{E}\Gamma) \to K_*(l^1(\Gamma))$ as follows:

$$K_*^\Gamma(\underline{E}\Gamma) \to K_*(l^1(\Gamma)) \to K_*(C_r^*(\Gamma)),$$

where the second homomorphism is induced by the natural inclusion

$$l^1(\Gamma) \hookrightarrow C_r^*(\Gamma).$$

One advantage of the l^1-setting is that the Bost conjecture has been proved for a larger class of groups, which in particular includes all lattices in reductive Lie groups [**11**], while it is not yet known whether or not the Baum-Connes conjecture holds for $\mathrm{SL}_3(\mathbb{Z})$. On the other hand, the Baum-Connes conjecture is more directly applicable to geometry and topology. In some sense, a main strategy of this paper is to work in a geometric setup where both conjectures intersect, that is, a geometric setup where methods from both the C^*-algebraic setting and the l^1-setting apply. In particular, our next theorem states that if the scalar curvature on the boundary of a spin manifold is sufficiently large (in the sense of the inequality from line (1.3)), then the higher index of the associated Dirac operator, which is a prior an element in $K_*(C_r^*(\Gamma))$, actually lies in $K_*(l^1(\Gamma))$.

Before we state the theorem, let us fix some notation. Let $\widetilde{M}$ be the universal cover of M and $\mathcal{F}_\partial$ be a fundamental domain of the Γ-action on $\partial\widetilde{M}$. Fix a finite symmetric generating set S of Γ. Let ℓ be the length function on Γ induced by S. We define

$$\tau_\partial = \liminf_{\ell(\gamma)\to\infty} \sup_{x\in\mathcal{F}_\partial} \frac{\mathrm{dist}(x,\gamma x)}{\ell(\gamma)} \tag{1.7}$$

and K_Γ as in line (1.2).

THEOREM 1.5. *Let M be an n-dimensional compact spin manifold with boundary. Suppose M is equipped with a Riemannian metric g that has product structure* *and positive scalar curvature near the boundary. Denote the fundamental group* $\pi_1(M)$ *of* M *by* Γ. *If the scalar curvature* κ *on* ∂M *satisfies that*

$$\inf_{x\in\partial M} \kappa(x) > \frac{16K_\Gamma^2}{\tau^2}, \tag{1.8}$$

then the C^**-algebraic higher index* $\mathrm{Ind}^\Gamma(D)$ *of the Dirac operator* D *on* M *admits a natural preimage* $\mathrm{Ind}_1^\Gamma(D)$ *in* $K_n(l^1(\Gamma))$ *under the homomorphism*

$$K_n(l^1(\Gamma)) \to K_n(C_r^*(\Gamma))$$

induced by the inclusion $l^1(\Gamma) \hookrightarrow C_r^*(\Gamma)$.

As before, the numbers K_Γ and τ_∂ may depend on the generating set, but Theorem 1.2 holds as long as the inequality (1.8) is satisfied for one choice of generating set. In particular, if the group Γ has sub-exponential growth, i.e., $K_\Gamma = 0$, then Theorem 1.5 holds as long as the scalar curvature on the boundary is positive. We point out that the Dirac operator on a general spin manifold with boundary does *not* automatically define an index class in $K_n(C_r^*(\Gamma))$. One usually requires the operator D to be invertible near the boundary in order for the index of D to lie in $K_n(C_r^*(\Gamma))$. In the above theorem, the positivity assumption on the scalar curvature of ∂M implies that D is invertible near ∂M, hence allows us define an index class of D in $K_n(C_r^*(\Gamma))$. The main point of the above theorem is that, if in addition the scalar curvature on ∂M is sufficiently large in the sense of the inequality (1.8), then we have a stronger consequence that the index of D in fact lies in $K_n(l^1(\Gamma))$.

As an application of Theorem 1.5, we have the following theorem, which gives a partial positive answer to Question 1.4 in the geometric setup of Theorem 1.5.

THEOREM 1.6. *Let* M *be an* n*-dimensional compact spin manifold with boundary* ∂M. *Suppose* M *is equipped with a Riemannian metric* g *that has product structure near the boundary such that the scalar curvature* κ *of* g *on* ∂M *satisfies that*

$$\inf_{x\in\partial M} \kappa(x) > \frac{16K_\Gamma^2}{\tau_\partial^2}.$$

If the Bost conjecture holds for $\Gamma = \pi_1(M)$, *then the* C^**-algebraic higher index* $\mathrm{Ind}^\Gamma(D)$ *of the Dirac operator* D *on* M *lies in the image of the Baum–Connes assembly map.*

Furthermore, by applying the techniques we develop for proving Theorem 1.5, we shall prove an l^1-version of the higher Atiyah-Patodi-Singer index formula. Roughly speaking, this l^1-higher Atiyah-Patodi-Singer index formula states that, for a given compact spin manifold M with boundary such that its Dirac operator is invertible on the boundary, the pairing between the higher index of the Dirac operator and a cyclic cohomology class φ of $\mathbb{C}\Gamma$ is equal to the sum of the integral of a local expression on M and a higher eta invariant from the boundary. In particular, if φ is a delocalized cyclic n-cocycle of $\mathbb{C}\Gamma$ (cf. Definition 5.1), then the local term in such a pairing vanishes, thus the index pairing is equal to a higher eta invariant from the boundary in this case.

To state our l^1-higher Atiyah-Patodi-Singer index theorem, we shall need the following notion of exponential growth rate for cyclic cocycles of Γ. We say a cyclic

n-cocycle φ of Γ has at most exponential growth with respect to a given length function ℓ on Γ if there exist $K > 0$ and $C > 0$ such that

$$|\varphi(\gamma_0, \gamma_1, \cdots, \gamma_n)| \leqslant Ce^{K(\ell(\gamma_0)+\ell(\gamma_1)+\cdots+\ell(\gamma_n))}, \ \forall \gamma_i \in \Gamma.$$

In this case, the infimum of such constants K is called the exponential growth rate of φ and will be denoted by K_φ from now on.

THEOREM 1.7. *Let M be a compact spin manifold with boundary ∂M and D the Dirac operator on M. Suppose φ is a delocalized cyclic cocycle of $\Gamma = \pi_1(M)$ that has exponential growth rate K_φ. If M is equipped with a Riemannian metric g that has product structure near the boundary such that the scalar curvature κ of g on ∂M satisfies that*

$$\inf_{x\in\partial M} \kappa(x) > \frac{16(K_\Gamma + K_\varphi)^2}{\tau_\partial^2} \tag{1.9}$$

then we have

$$\mathrm{ch}_\varphi(\mathrm{Ind}^\Gamma(D)) = -\frac{1}{2}\eta_\varphi(D_\partial), \tag{1.10}$$

where $\mathrm{ch}_\varphi(\mathrm{Ind}^\Gamma(D))$ is the pairing between φ and the Connes-Chern character of the higher index $\mathrm{Ind}^\Gamma(D)$, and $\eta_\varphi(D_\partial)$ is the higher eta invariant (with respect to φ) of the Dirac operator D_∂ on ∂M.

Let us recall the definitions of $\mathrm{ch}_\varphi(\mathrm{Ind}^\Gamma(D))$ and $\eta_\varphi(D_\partial)$ (cf. [**4**]). We will only give the formulas in the case where $\dim(M)$ is even. The odd dimensional case is similar. For simplicity, let us write $p = \mathrm{Ind}^\Gamma(D)$, a representative with finite support on Γ. If φ is a delocalized cyclic $2m$-cocycle, then we have

$$\mathrm{ch}_\varphi(\mathrm{Ind}^\Gamma(D)) := \frac{(2m)!}{m!}\varphi\#\mathrm{tr}(p^{\otimes 2m+1}) \tag{1.11}$$

and

$$\eta_\varphi(\widetilde{D}_\partial) := \frac{m!}{\pi i}\int_0^\infty \varphi\#\mathrm{tr}(\dot{u}_s u_s^{-1} \otimes (((u_s - 1) \otimes (u_s^{-1} - 1)))^{\otimes m})ds, \tag{1.12}$$

where $\dot{u}_s$ is the derivative of u_s with respect to $s \in (0, \infty)$ and u_s is defined as follows. Let us denote the universal covering space of M by $\widetilde{M}$. The boundary of $\widetilde{M}$ is a $\pi_1(M)$-covering space of ∂M. Let $\widetilde{D}_\partial$ be the lift of the Dirac operator D_∂ on ∂M to this covering space of ∂M. We define

$$u_s := e^{2\pi i\cdot f(s^{-1}\widetilde{D}_\partial)} \text{ with } f(x) = \frac{1}{\sqrt{\pi}}\int_{-\infty}^x e^{-y^2}dy.$$

It was proved in [**4**, Theorem 3.24] that the integral formula for the higher eta invariant $\eta_\varphi(D_\partial)$ in line (1.12) converges absolutely, provided that the scalar curvature of ∂M is sufficiently large, i.e., satisfying the condition in line (1.9). In proving Theorem 1.7, we will also show that the formula in (1.11) (which is a summation of infinitely many terms) absolutely converges, provided that the condition in line (1.9) holds. A key ingredient for the proof of Theorem 1.7 is Theorem 1.3—the product formula for l^1-higher rho invariants.

Theorem 1.7 improves the higher Atiyah-Patodi-Singer index formula in [**3**, Theorem 3.36]. The index formula (1.10) in Theorem 1.7 can be viewed as the pairing of the K-theoretic higher Atiyah-Patodi-Singer index formula (as in [**16, 24**]) with cyclic cohomology. The main difficulties for proving (1.10) are to justify the convergence of the formulas for $\mathrm{ch}_\varphi(\mathrm{Ind}^\Gamma(D))$ and $\eta_\varphi(D_\partial)$, and to establish the equality while working with K-theory of Banach algebras (instead of C^*-algebras). Furthermore, although we have only stated Theorem 1.7 for delocalized cyclic cocycles, the same result actually holds for all cyclic cocycles with at most exponential growth except there is an extra local term on the right hand side (cf. Section 5.1).

The paper is organized as follows. In Section 2, we discuss the l^1-higher index theory for closed manifolds. We prove a vanishing theorem for the l^1-higher index of Dirac operator when the given scalar curvature is sufficient large. Furthermore, we introduce an l^1-version of higher rho invariants and prove a product formula of these secondary invariants in the l^1-setting. In Section 3, we review the construction of C^*-algebraic higher index for Dirac operators on compact spin manifolds whose boundary has positive scalar curvature. In Section 4, we construct l^1-higher indices of Dirac operators on compact spin manifolds, provided the scalar curvature on the boundary is sufficiently large. We prove a generalized version of Theorem 1.5 (cf. Theorem 4.1 for the details). In Section Section 5, we apply the techniques in Section 4 to prove an l^1-higher Atiyah-Patodi-Singer index theorem (Theorem 1.7).

2. l^1-higher index theory for closed spin manifold

In this section, we review the construction of the l^1-higher index of the Dirac operator and prove a vanishing theorem for the index. As an application, we define a higher rho invariant and prove a product formula.

2.1. Geometric C^*-algebras and their Banach analogues. In this section, we review the construction of Roe algebras and their analogue in the l^1-setting.

Let X be a closed spin Riemannian manifold and $S(X)$ the spinor bundle over X. Denote by Γ the fundamental group of X. Let $\widetilde{X}$ be the universal cover of X and $S(\widetilde{X})$ be the lift of the spinor bundle.

Set $H = L^2(S(\widetilde{X}))$, the Hilbert space of all square-integrable sections of $S(\widetilde{X})$. Let $B(H)^\Gamma$ be the collection of all Γ-equivariant bounded operator on H. We choose a precompact fundamental domain $\mathcal{F}$ of the Γ-action on $\widetilde{X}$ and denote by ψ the characteristic function of $\mathcal{F}$.

Let $\|\cdot\|_{\mathrm{op}}$ be the operator norm on $B(H)$. For any $T \in B(H)^\Gamma$ and $\gamma \in \Gamma$, we define

$$T_\gamma := \psi \circ T \circ \gamma\psi.$$

DEFINITION 2.1. We define $\mathcal{B}(H)^\Gamma$ to be the following subspace of linear operators

$$\mathcal{B}(H)^\Gamma := \{T \in B(H)^\Gamma : \sum_{\gamma\in\Gamma} \|T_\gamma\|_{\mathrm{op}} < \infty\}$$

equipped with the norm

$$\|T\|_1 = \sum_{\gamma\in\Gamma} \|T_\gamma\|_{\mathrm{op}}. \tag{2.1}$$

It is easy to verify that $(\mathcal{B}(H)^\Gamma, \|\cdot\|_1)$ is a Banach $*$-algebra. Moreover, the definition of $\mathcal{B}(H)^\Gamma$ is independent of the choice of the fundamental domain $\mathcal{F}$.

DEFINITION 2.2.

(1) We say that $T \in B(H)^\Gamma$ has *finite propagation* if there exists $d > 0$ such that $\chi_A \circ T \circ \chi_B = 0$ for any two Borel sets A and B with $\mathrm{dist}(A, B) > d$. Here for example χ_A is the characteristic function of A. The infimum of such d is called the *propagation* of T.
(2) $T \in B(H)^\Gamma$ is called *locally compact* if $\chi_A \circ T$ and $T \circ \chi_A$ are compact for any precompact Borel set A.

DEFINITION 2.3. Let $\mathbb{C}(\widetilde{X})^\Gamma$ be the collection of operators $T \in B(H)^\Gamma$ such that T has finite propagation and is locally compact. Let $C^*(\widetilde{X})^\Gamma$ (resp. $\mathcal{B}(\widetilde{X})^\Gamma$) be the completions of $\mathbb{C}(\widetilde{X})^\Gamma$ with respect to the operator norm $\|\cdot\|_{\mathrm{op}}$ (resp. the l^1-norm $\|\cdot\|_1$ given in line (2.1)).

The map $T \mapsto \sum_{\gamma\in\Gamma} T_\gamma \gamma$ induces isomorphisms

$$C^*(\widetilde{X})^\Gamma \cong \mathcal{K} \otimes C_r^*(\Gamma) \text{ and } \mathcal{B}(\widetilde{X})^\Gamma \cong \mathcal{K} \otimes l^1(\Gamma)$$

where $\mathcal{K}$ is the algebra of compact operators and the norm on $\mathcal{K} \otimes l^1(\Gamma)$ is given by the l^1-norm

$$\Big\|\sum_{\gamma\in\Gamma} a_\gamma \gamma\Big\| := \sum_{\gamma\in\Gamma} \|a_\gamma\|_{op}, \; a_\gamma \in \mathcal{K},$$

which coincides with the maximal tensor product norm.

It is a well-known fact that K-theory of C^*-algebras is stable, that is,

$$K_*(A \otimes \mathcal{K}) = K_*(A)$$

for any C^*-algebra A. The following lemma shows that K-theory is also stable for l^1-algebras.[1]

PROPOSITION 2.4. *Let p be a rank one projection in $\mathcal{K}$. The following map*

$$\iota \colon l^1(\Gamma) \to \mathcal{K} \otimes l^1(\Gamma), \; a \mapsto p \otimes a$$

induces an isomorphism at the level of K-theory.

PROOF. Given an element $\sum_{\gamma\in\Gamma} a_\gamma\gamma \in \mathcal{K} \otimes l^1(\Gamma)$, for $\varepsilon > 0$, there exist a finite subset $F \subset \Gamma$ and finite dimensional matrices $a'_\gamma \in M_n(\mathbb{C})$ for each $\gamma \in F$ such that

$$\Big\|\sum_{\gamma\in\Gamma} a_\gamma\gamma - \sum_{\gamma\in F} a'_\gamma\gamma\Big\|_1 = \sum_{\gamma\in F} \|a_\gamma - a'_\gamma\|_{op} + \sum_{\gamma\notin F} \|a_\gamma\|_{op} < \varepsilon.$$

Thus $\mathcal{K} \otimes l^1(\Gamma)$ is the direct limit of the following direct system:

$$l^1(\Gamma) \xrightarrow{\iota_1} M_1(\mathbb{C}) \otimes l^1(\Gamma) \xrightarrow{\iota_2} M_2(\mathbb{C}) \otimes l^1(\Gamma) \xrightarrow{\iota_3} \cdots$$

where the map

$$\iota_n \colon M_n(\mathbb{C}) \otimes l^1(\Gamma) \to M_{n+1}(\mathbb{C}) \otimes l^1(\Gamma),$$

is given by

$$\sum_{\gamma\in\Gamma} b_\gamma\gamma \mapsto \sum_{\gamma\in\Gamma} \begin{pmatrix} b_\gamma & \\ & 0 \end{pmatrix} \gamma,$$

and the norm on $M_n(\mathbb{C}) \otimes l^1(\Gamma)$ is given by

$$\Big\|\sum_{\gamma\in\Gamma} b_\gamma\gamma\Big\|_1 = \sum_{\gamma\in\Gamma} \|b_\gamma\|_{\mathrm{op}}.$$

[1] It remains an open question whether K-theory is stable for general Banach algebras. More precisely, for any Banach algebra B, is there a suitable topological tensor product $B \otimes \mathcal{K}$ such that $K_*(B \otimes \mathcal{K}) \cong K_*(B)$?

Clearly, the map ι_n induces an isomorphism at the level of K-theory, hence follows the lemma.

□

Now we fix a symmetric generating set of Γ and denote by ℓ the corresponding length function. We will need the following weighted l^1-completions of $\mathbb{C}\Gamma$.

DEFINITION 2.5. Suppose K is a non-negative real number.

(1) Let $l^1_K(\Gamma)$ be the completion of $\mathbb{C}\Gamma$ with respect to the following norm

$$\|\sum_{\gamma\in\Gamma} a_\gamma\gamma\|_{1,K} := \sum_{\gamma\in\Gamma} e^{K\ell(\gamma)}\|a_\gamma\|, \tag{2.2}$$

for all finite sums $\sum_{\gamma\in\Gamma} a_\gamma\gamma \in \mathbb{C}\Gamma$.

(2) Similarly, define $\mathcal{B}(\widetilde{X})^\Gamma_K$ to be the completion of $\mathbb{C}(\widetilde{X})^\Gamma$ with respect to the following norm

$$\|T\|_{1,K} := \sum_{\gamma\in\Gamma} e^{K\ell(\gamma)}\|T_\gamma\|_{\mathrm{op}},$$

for all $T = \sum_{\gamma\in\Gamma} T_\gamma\gamma \in \mathbb{C}(\widetilde{X})^\Gamma$.

It is clear that $l^1_K(\Gamma)$ and $\mathcal{B}(\widetilde{X})^\Gamma_K$ are Banach algebras and $\mathcal{B}(\widetilde{X})^\Gamma_K$ is isomorphic to the maximal tensor product $l^1_K(\Gamma)\otimes\mathcal{K}$.

2.2. l^1-higher index of Dirac operators. Now we recall the definition of the higher index in the l^1-setting.

DEFINITION 2.6. A continuous function $F\colon \mathbb{R}\to[-1,1]$ is called a normalizing function if

(1) F is an odd function, that is, $F(-t) = -F(t)$,
(2) $\lim_{x\to\pm\infty} F(x) = \pm 1$,
(3) the Fourier transform of F' is supported on $[-N_F, N_F]$ for some $N_F > 0$.

Note that, in this case, the Fourier transform of F is also supported on $[-N_F, N_F]$ as a tempered distribution. Throughout the paper, we use the following choice of formula for the Fourier transform

$$\hat{f}(\xi) = \int f(x)e^{-ix\xi}dx, \tag{2.3}$$

and its inverse Fourier transform is given by

$$f(x) = \frac{1}{2\pi}\int \hat{f}(\xi)e^{ix\xi}dx. \tag{2.4}$$

Let $\widetilde{D}$ be the associated Dirac operator on the universal cover $\widetilde{X}$. By the Fourier inverse transform formula (2.4), we see that the operator $F(\widetilde{D})$ has propagation no more than N_F, since the wave operator $e^{i\xi\widetilde{D}}$ has propagation $\leqslant |\xi|$. Therefore, we have

$$\|F(\widetilde{D})\|_{1,K} < \infty$$

for any $K \geqslant 0$.

When X is even dimensional, the spinor bundle S on X admits a natural $\mathbb{Z}_2$-grading and the Dirac operator D is an odd operator. As F is an odd function,

$F(\widetilde{D})$ is an odd operator with respect to the $\mathbb{Z}_2$-grading on the spinor bundle $\widetilde{S}$ of $\widetilde{X}$, that is,

$$F(\widetilde{D}) = \begin{pmatrix} 0 & F(\widetilde{D})^- \\ F(\widetilde{D})^+ & 0 \end{pmatrix}. \tag{2.5}$$

Let us write $U = F(\widetilde{D})^+$ and $V = F(\widetilde{D})^-$, and define an invertible element

$$W_{F(\widetilde{D})} := \begin{pmatrix} 1 & U \\ 0 & 1 \end{pmatrix} \begin{pmatrix} 1 & 0 \\ -V & 1 \end{pmatrix} \begin{pmatrix} 1 & U \\ 0 & 1 \end{pmatrix} \begin{pmatrix} 0 & -1 \\ 1 & 0 \end{pmatrix} \tag{2.6}$$

This allows us to define the following idempotent

$$\begin{aligned} p_{F(\widetilde{D})} &= W_{F(\widetilde{D})} \begin{pmatrix} 1 & 0 \\ 0 & 0 \end{pmatrix} W_{F(\widetilde{D})}^{-1} \\ &= \begin{pmatrix} 1-(1-UV)^2 & (2-UV)U(1-VU) \\ V(1-UV) & (1-VU)^2 \end{pmatrix}. \end{aligned} \tag{2.7}$$

It is easy to see that $p_{F(\widetilde{D})} - \left(\begin{smallmatrix} 1 & 0 \\ 0 & 0 \end{smallmatrix}\right)$ is locally compact hence lies in $\mathcal{B}(\widetilde{X})_K^\Gamma$.

Definition 2.7. If $\dim X$ is even, then the l^1-higher index of $\widetilde{D}$ is defined to be

$$\operatorname{Ind}_{1,K}^\Gamma(D) := [p_{F(\widetilde{D})}] - [\left(\begin{smallmatrix} 1 & 0 \\ 0 & 0 \end{smallmatrix}\right)] \in K_0(\mathcal{B}(\widetilde{X})_K^\Gamma).$$

When X is odd-dimensional, we see that $e^{2\pi i \frac{F(\widetilde{D})+1}{2}}$ is locally compact and lies in the unitalization of $\mathcal{B}(\widetilde{X})_K^\Gamma$, as $\|\cdot\|_{1,K}$ is an algebraic norm.

Definition 2.8. If $\dim X$ is odd, then the l^1-higher index of $\widetilde{D}$ is defined to be

$$\operatorname{Ind}_{1,K}^\Gamma(D) := [e^{2\pi i \frac{F(\widetilde{D})+1}{2}}] \in K_1(\mathcal{B}(\widetilde{X})_K^\Gamma).$$

Obviously, the definition of the l^1-higher index is independent of the choice of the normalizing function. In fact, any function F in Definition 2.6 with the condition (3) replaced by that

$$\|F(\widetilde{D})\|_{1,K} < \infty$$

also defines the same index class in $K_*(\mathcal{B}(\widetilde{X})_K^\Gamma)$.

2.3. l^1-norm inequalities. In this subsection, we prove some l^1-norm estimates of various operators.

We first fix some notations. Choose a finite symmetric generating set S of Γ. Let ℓ be the length function on Γ induced by S. We define

$$K_\Gamma = \inf\{K : \exists C > 0 \text{ s.t. } \#\{\gamma \in \Gamma : \ell(\gamma) \leqslant n\} \leqslant Ce^{Kn}\} \tag{2.8}$$

and

$$\tau = \liminf_{\ell(\gamma)\to\infty} \sup_{x\in\mathcal{F}} \frac{\operatorname{dist}(x, \gamma x)}{\ell(\gamma)}, \tag{2.9}$$

where $\mathcal{F}$ is a fundamental domain of the Γ-action on $\widetilde{X}$.

For convenience, instead of the definitions of K_Γ and τ in line (2.8) and (2.9), we assume that

$$\#\{\gamma \in \Gamma : \ell(\gamma) \leqslant n\} \leqslant Ce^{K_\Gamma n}, \ \forall n \geqslant 0 \tag{2.10}$$

and

$$\text{dist}(x, \gamma x) > \tau\ell(\gamma) - C_0,\ \forall \gamma \in \Gamma,\ \forall x \in \mathcal{F} \tag{2.11}$$

with some $C > 0$ and $C_0 > 0$. We recall that ψ is the characteristic function of $\mathcal{F}$. We denote the diameter of $\mathcal{F}$ by diam$\mathcal{F}$.

LEMMA 2.9. *Given $f \in C_0(\mathbb{R})$, suppose its Fourier transform $\hat{f} \in L^1(\mathbb{R})$. Then for any $\mu > 1$, we have*

$$\|f(\widetilde{D})_\gamma\|_{\text{op}} \leqslant \frac{1}{2\pi}\int_{|\xi| > \frac{\tau\ell(\gamma)}{\mu}} |\hat{f}(\xi)| d\xi$$

for all $\gamma \in \Gamma$ satisfying

$$\ell(\gamma) > \mathfrak{N}_\mu := \frac{\sqrt{\mu}(C_0 + \text{diam}\mathcal{F})}{\tau(\sqrt{\mu} - 1)}, \tag{2.12}$$

where $f(\widetilde{D})_\gamma = \psi \circ f(\widetilde{D}) \circ \gamma\psi$.

PROOF. Fix $\mu > 1$. For each $\gamma \in \Gamma$ such that $\ell(\gamma) > \mathfrak{N}_\mu$, let χ be a smooth function on $\mathbb{R}$ that vanishes on the interval $[-\frac{\tau\ell(\gamma)}{\mu}, \frac{\tau\ell(\gamma)}{\mu}]$ and equals 1 on $(-\infty, -\frac{\tau\ell(\gamma)}{\sqrt{\mu}})$ and $(\frac{\tau\ell(\gamma)}{\sqrt{\mu}}, +\infty)$.

By the Fourier inverse transform formula, we have

$$f(\widetilde{D}) = \frac{1}{2\pi}\int \hat{f}(\xi) e^{i\xi\widetilde{D}} d\xi.$$

Let us define

$$g(\widetilde{D}) = \frac{1}{2\pi}\int \chi(\xi)\hat{f}(\xi) e^{i\xi\widetilde{D}} d\xi.$$

Since the wave operator $e^{i\xi\widetilde{D}_c}$ has propagation $\leqslant |\xi|$, it follows that $f(\widetilde{D}_c) - g(\widetilde{D}_c)$ has propagation $\leqslant \frac{\tau\ell(\gamma)}{\sqrt{\mu}}$. Therefore, from line (2.11), we have $f(\widetilde{D})_\gamma - g(\widetilde{D})_\gamma = 0$ in this case. We conclude that

$$\|f(\widetilde{D})_\gamma\|_{\text{op}} = \|g(\widetilde{D})_\gamma\|_{\text{op}} \leqslant \|g(\widetilde{D})\|_{\text{op}} \leqslant \frac{1}{2\pi}\int_{|\xi| > \frac{\tau\ell(\gamma)}{\mu}} |\hat{f}(\xi)| d\xi.$$

□

DEFINITION 2.10. Let f be a Schwartz function on $\mathbb{R}$. We say f has Gaussian decay with rate C if there exists $A > 0$ such that

$$|f(x)| \leqslant Ae^{-Cx^2}$$

for all $x \in \mathbb{R}$.

For example, the Gaussian function e^{-x^2} as well as its Fourier transform has Gaussian decay.

LEMMA 2.11. *Given a Schwartz function f on $\mathbb{R}$, suppose both f and its Fourier transform $\hat{f}$ have Gaussian decay with rates C and $\widehat{C}$ respectively. Let σ be the infimum of the spectrum of $|\widetilde{D}|$. If there exists a constant $K \geqslant 0$ such that*

$$\sigma > \frac{K_\Gamma + K}{\tau\sqrt{C\widehat{C}}},$$

then there exist $\varepsilon > 0$ and $\lambda > 0$ such that

$$\|f(t\widetilde{D})\|_{1,K} \leqslant \lambda \cdot e^{-\varepsilon t^2}$$

where $\|\cdot\|_{1,K}$ *is the weighted* l^1*-norm given in Definition* 2.5.

PROOF. By assumption, there exist positive constants A_0 and A_1 such that

$$|f(x)| \leqslant A_0 e^{-Cx^2} \text{ and } |\hat{f}(\xi)| \leqslant A_1 e^{-\widehat{C}\xi^2}.$$

Applying Lemma 2.9, it follows that for any $\mu > 1$, there exists $A_2 > 0$ such that

$$\|f(t\widetilde{D})_\gamma\|_{\mathrm{op}} \leqslant A_2 e^{-\widehat{C}\frac{\tau^2\ell(\gamma)^2}{t^2\mu^2}}$$

for all $\gamma \in \Gamma$ with $\ell(\gamma) \geqslant \mathfrak{N}_\mu$, where $\mathfrak{N}_\mu$ is the number given in line (2.12). On the other hand, since σ is the spectral gap of $\widetilde{D}$ at zero, we have

$$\|f(t\widetilde{D})_\gamma\|_{\mathrm{op}} \leqslant \|f(t\widetilde{D})\|_{\mathrm{op}} \leqslant A_0 e^{-Ct^2\sigma^2}.$$

By assumption that $\sigma > \frac{K_\Gamma + K}{\tau\sqrt{C\widehat{C}}}$, we have

$$\frac{C\sigma^2}{K_\Gamma + K} > \frac{(K_\Gamma + K)\mu^2}{\widehat{C}\tau^2}$$

for some $\mu > 1$. Fix such a μ for the rest of the proof. There exists a positive number s such that

$$\frac{C\sigma^2}{(K_\Gamma + K)s} > 1 > \frac{(K_\Gamma + K)s}{\widehat{C}\tau^2\mu^{-2}s^2}.$$

It follows that there exists $\varepsilon > 0$ such that

$$C\sigma^2 - (K_\Gamma + K)s > \varepsilon > 0 \text{ and } \frac{\widehat{C}\tau^2 s^2}{\mu^2} - (K_\Gamma + K)s > \varepsilon > 0.$$

We conclude that

$$\begin{aligned}
\|f(tD)\|_{1,K} = & \sum_{\ell(g) \leqslant st^2 + \mathfrak{N}_\mu} e^{K\ell(\gamma)} \|f(tD_c)_\gamma\|_{\mathrm{op}} + \sum_{\ell(g) > st^2 + \mathfrak{N}_\mu} e^{K\ell(\gamma)} \|f(tD_c)_\gamma\|_{\mathrm{op}} \\
\leqslant & A_0 e^{-Ct^2\sigma^2} e^{(K_\Gamma + K)(st^2 + \mathfrak{N}_\mu)} + A_2 \sum_{n > st^2 + \mathfrak{N}_\mu} e^{-\widehat{C}\frac{\tau^2 n^2}{t^2\mu^2}} e^{(K_\Gamma + K)\cdot n} \\
\leqslant & A_0 e^{(K_\Gamma + K)\mathfrak{N}_\mu} e^{-(C\sigma^2 - K_\Gamma s - Ks)t^2} + A_2 \sum_{n > st^2 + \mathfrak{N}_\mu} e^{-(\widehat{C}\frac{\tau^2 s}{\mu^2} - K_\Gamma - K)\cdot n} \\
\leqslant & A_3 e^{-\varepsilon t^2} + A_4 e^{-\varepsilon t^2}
\end{aligned}$$

for some $A_3 > 0$ and $A_4 > 0$. The lemma follows by setting $\lambda = A_3 + A_4$. □

2.4. A vanishing theorem for l^1-higher index. In this subsection, we show that the l^1-higher index of an elliptic differential operator $\widetilde{D}$ vanishes if the spectral gap of $\widetilde{D}$ at zero is sufficiently large.

THEOREM 2.12. *For any* $K \geqslant 0$, *if*

$$|\widetilde{D}| > \frac{2(K_\Gamma + K)}{\tau}, \tag{2.13}$$

then $\mathrm{Ind}^\Gamma_{1,K}(D) = 0$ *in* $K_*(\mathcal{B}(\widetilde{X})^\Gamma_K)$.

For example, by the Lichnerowicz formula, line (2.13) holds when the scalar curvature function $\kappa(x)$ on M satisfying that there exists $K \geqslant 0$ such that

$$\kappa(x) > \frac{16(K_\Gamma + K)^2}{\tau^2} \text{ for all } x \in X.$$

In particular, if $K = 0$ and the group Γ has sub-exponential growth (i.e. $K_\Gamma = 0$), then line (2.13) holds as long as $\widetilde{D}$ is invertible.

PROOF OF THEOREM 2.12. Let sgn be the sign function

$$\operatorname{sgn}(x) = \begin{cases} 1 & \text{if } x > 0, \\ 0 & \text{if } x = 0, \\ -1 & \text{if } x < 0. \end{cases}$$

We will prove in Lemma 2.13 and Lemma 2.15 that under the assumption (2.13), $\|\operatorname{sgn}(\widetilde{D})\|_{1,K}$ is finite, and there exists a family of normalizing function $\{G_t\}$ parameterized by $t \in [1, +\infty)$ such that $\|G_t(\widetilde{D})\|_{1,K}$ is uniformly bounded and

$$\lim_{t\to\infty} \|G_t(\widetilde{D}) - \operatorname{sgn}(\widetilde{D})\|_{1,K} = 0.$$

Since $\operatorname{sgn}(\widetilde{D})^2 = 1$, this shows that $\operatorname{Ind}^\Gamma_{1,K}(D)$ is trivial. □

We consider the normalizing function

$$G(x) = \frac{2}{\sqrt{\pi}} \int_{-\infty}^{x} e^{-y^2}\, dy - 1 \tag{2.14}$$

and set $G_t(x) = G(tx)$.

LEMMA 2.13. *For each $t > 0$, $\|G_t(\widetilde{D})\|_{1,K}$ is finite.*

PROOF. Without loss of generality, let us assume $t = 1$.
The distributional Fourier transform of G is given by

$$\widehat{G}(\xi) = \frac{2i}{\xi} e^{-\frac{\xi^2}{4}} - 2\pi\delta(\xi), \tag{2.15}$$

where δ is the delta function concentrated at the origin. Let χ be a cut-off function such that $\chi(\xi) = 1$ if $|\xi| < 1$, $\chi(\xi) = 0$ if $|\xi| > 2$ and $|\chi(\xi)| \leqslant 1$ for all $\xi \in \mathbb{R}$. We write G as a sum of two functions $G = G^{(1)} + G^{(2)}$, where the Fourier transforms of $G^{(1)}$ and $G^{(2)}$ are $\widehat{G}^{(1)} = \chi \cdot \widehat{G}$ and $\widehat{G}^{(2)} = (1-\chi)\cdot \widehat{G}$ respectively.

As $\widehat{G}^{(1)}$ has compact support, it follows from the Fourier inverse transform formula (cf. line (2.4)) that $G^{(1)}(\widetilde{D})$ has finite propagation. Therefore $\|G^{(1)}(\widetilde{D})\|_{1,K}$ is finite. On the other hand, $\widehat{G}^{(2)}$ is a Schwartz function with Gaussian decay. By Lemma 2.9, there exists $\varepsilon > 0$ such that

$$\|G^{(2)}(\widetilde{D})_\gamma\|_{\mathrm{op}} \leqslant e^{-\varepsilon \ell(\gamma)^2}$$

as long as $\ell(\gamma)$ is sufficiently large. It follows that $\|G^{(2)}(\widetilde{D})\|_{1,K}$ is also finite. This finishes the proof. □

REMARK 2.14. The proof of Lemma 2.13 above does not require the invertibility of $\widetilde{D}$.

Note that, under the assumption that $\widetilde{D}$ is invertible, it is clear that

$$G_t(\widetilde{D}) \to \operatorname{sgn}(\widetilde{D}) \text{ in the operator norm, as } t \to \infty.$$

The following lemma shows that the above convergence still holds with respect to the $\|\cdot\|_{1,K}$ norm, provided the spectral gap of $\widetilde{D}$ at zero is sufficiently large.

LEMMA 2.15. *If $|\widetilde{D}| > \frac{2(K_\Gamma+K)}{\tau}$, then $\|\operatorname{sgn}(\widetilde{D})\|_{1,K}$ is finite. Furthermore, there exist $C > 0$ and $\varepsilon > 0$ such that for any $t \geqslant 1$,*

$$\|G_t(\widetilde{D}) - \operatorname{sgn}(\widetilde{D})\|_{1,K} \leqslant Ce^{-\varepsilon t^2}. \tag{2.16}$$

PROOF. It follows from the proof of Lemma 2.13 that $G_t(\widetilde{D})$ is a differentiable in t with respect to the $\|\cdot\|_{1,K}$ norm. We have

$$\frac{d}{dt}G_t(\widetilde{D}_c) = \frac{2}{\sqrt{\pi}}\widetilde{D}_c e^{-t^2\widetilde{D}_c^2} = \frac{1}{t}h(t\widetilde{D}_c), \tag{2.17}$$

where $h(x) = \frac{2}{\sqrt{\pi}}xe^{-x^2}$. Note that h and $\hat{h}$ have Gaussian decay with rates 1 and 1/4 respectively. By Lemma 2.11, there exist positive constants C_1 and $\varepsilon > 0$ such that

$$\|h(t\widetilde{D})\| < C_1 e^{-\varepsilon t^2}. \tag{2.18}$$

We conclude that

$$\begin{aligned}\|G_t(\widetilde{D}) - G_T(\widetilde{D})\|_{1,K} &\leqslant \int_t^T \|\frac{d}{dt}G_t(\widetilde{D})\|_{1,K} \leqslant \int_t^T \frac{1}{s}C_1 e^{-\varepsilon s^2}ds \\ &\leqslant Ce^{-\varepsilon t^2}\end{aligned}$$

for some fixed $C > 0$ and for all $t < T$. The lemma follows by letting T go to infinity. □

2.5. l^1-higher rho invariant. In this subsection, we introduce l^1-higher rho invariants for Dirac operators on complete spin manifolds whose scalar curvature is sufficiently large.

DEFINITION 2.16. We denote by $\mathcal{B}_L(\widetilde{X})_K^\Gamma$ the completion of the collection of maps $f\colon [0,\infty) \to \mathbb{C}(\widetilde{X})^\Gamma$ that satisfy

(1) f is uniformly bounded and uniformly continuous in t with respect to $\|\cdot\|_{1,K}$, cf., Definition 2.5,
(2) the propagation of f goes to zero as t goes to infinity,

with respect to the norm $\|\cdot\|_{1,K}$ given by

$$\|f\|_{1,K} := \sup_{s\geqslant 0}\|f(s)\|_{1,K}.$$

Furthermore, we define

$$\mathcal{B}_{L,0}(\widetilde{X})_K^\Gamma = \{f \in \mathcal{B}_L(\widetilde{X})_K^\Gamma : f(0) = 0\}.$$

Let us first prove the following technical lemma, which gives a rather general sufficient condition for producing elements in the Banach localization algebra $\mathcal{B}_L(\widetilde{X})_K^\Gamma$.

LEMMA 2.17. *Let $f \in C_0(\mathbb{R})$ be a continuous function on $\mathbb{R}$ vanishing at infinity. Let χ be a smooth cut-off function on $\mathbb{R}$ such that $\chi(\xi) = 0$ on $[-1,1]$, $\chi = 1$ for $|\xi| \geqslant 2$ and $|\chi(\xi)| \leqslant 1$ for all $\xi \in \mathbb{R}$. Denote $\chi_N(\xi) = \chi(\xi/N)$ for $N > 0$. If there exist positive numbers N, C_1, and C_2 such that $\chi_N \cdot f \in C_0(\mathbb{R})$ and*

$$|\chi_N(\xi)\hat{f}(\xi)| \leqslant C_1 e^{-C_2|\xi|},$$

then there exists $s_0 \geqslant 0$ such that the path

$$h(s) = f\left(\tfrac{\widetilde{D}}{s+s_0}\right) \text{ with } s \in [0,\infty)$$

defines an element in $\mathcal{B}_L(\widetilde{X})^\Gamma_K$. *Moreover, the norm of* h *only depends on* N, C_1, C_2 *and the geometric data of* X.

PROOF. For any $\alpha > N$, we define

$$g_\alpha(x) = \frac{1}{2\pi}\int \chi(\alpha^{-1}\xi)\hat{f}(\xi)e^{ix\xi}d\xi.$$

Set $f_s(x) = f(x/s)$ and $g_{\alpha,s}(x) = g_\alpha(x/s)$. Since

$$|g_\alpha(x)| \leqslant \frac{1}{2\pi}\int_{|\xi|>\alpha} C_1 e^{-C_2|\xi|}d\xi \leqslant \frac{C_1}{\pi C_2}e^{-C_2\alpha},$$

we have

$$\|g_{\alpha,s}(\widetilde{D})_\gamma\|_{\mathrm{op}} \leqslant \|g_{\alpha,s}(\widetilde{D})\|_{\mathrm{op}} \leqslant \frac{C_1}{\pi C_2}e^{-C_2\alpha}. \tag{2.19}$$

On the other hand, the Fourier transform of $g_{\alpha,s}$ is given by

$$\hat{g}_{\alpha,s}(\xi) = s\cdot\hat{g}_\alpha(s\xi) = \chi(\alpha^{-1}s\xi)\hat{f}(s\xi).$$

If $s \geqslant 1$, then $|\hat{g}_{\alpha,s}(\xi)| \leqslant C_1 e^{-C_2 s|\xi|}$. By Lemma 2.9, there exist $C_1' > 0$ and $C_2' > 0$ independent of α, s such that

$$\|g_{\alpha,s}(\widetilde{D})_\gamma\|_{\mathrm{op}} \leqslant C_1' e^{-C_2' s\ell(\gamma)}. \tag{2.20}$$

By combining line (2.19) and (2.20) together, we have for any $s \geqslant 1$,

$$\|g_{\alpha,s}(\widetilde{D})_\gamma\|_{\mathrm{op}} \leqslant Ce^{-\frac{C_2\alpha}{2}-\frac{C_2' s}{2}\ell(\gamma)},$$

where $C = \sqrt{C_1C_1'}$. Therefore, if we choose s_0 large enough such that

$$e^{(K+K_\Gamma)-s_0C_2'/2} \leqslant \frac{1}{2},$$

then for any $s \geqslant s_0$,

$$\begin{aligned}\|g_{\alpha,s}(\widetilde{D}_X)\|_{1,K} &\leqslant \sum_{\gamma\in\Gamma} e^{K\ell(\gamma)}Ce^{-\frac{C_2\alpha}{2}-\frac{C_2' s}{2}\ell(\gamma)} \\ &\leqslant Ce^{-\frac{C_2\alpha}{2}}\sum_{n=0}^{\infty} e^{-\frac{C_2' s}{2}n+(K_\Gamma+K)n} \leqslant 2Ce^{-\frac{C_2\alpha}{2}}.\end{aligned}$$

Hence if $s \geqslant s_0$, then

$$\|f_s(\widetilde{D}) - (f_s(\widetilde{D}) - g_{\alpha,s}(\widetilde{D}))\|_{1,K} = \|g_{\alpha,s}(\widetilde{D})\|_{1,K} \leqslant 2Ce^{-\frac{C_2\alpha}{2}}.$$

Now observe that the operator $(f_s(\widetilde{D}) - g_{\alpha,s}(\widetilde{D}))$ has propagation $\leqslant 2\alpha/s$. In particular, the path

$$s \mapsto T_\alpha(s) = f_s(\widetilde{D}) - g_{\alpha,s}(\widetilde{D})$$

with $s \in [0,\infty)$ is a uniformly bounded and uniformly continuous with respect to the $\|\cdot\|_{1,K}$-norm and the propagation of $T_\alpha(s)$ goes to 0, as $s \to \infty$. We conclude that the element

$$h(s) = f\big(\tfrac{\widetilde{D}}{s+s_0}\big)$$

lies in $\mathcal{B}_K(\widetilde{X})^\Gamma$ for each $s \in [0,\infty)$, and the path h can be approximated uniformly by paths such as T_α by letting $\alpha \to \infty$. This shows that h defines an element in $\mathcal{B}_L(\widetilde{X})^\Gamma_K$, hence finishes the proof. □

The following theorem gives a large collection of functions f in $C_0(\mathbb{R})$ that satisfy the condition of Lemma 2.17.

THEOREM 2.18 ([**19**, Theorem IX.14]). *Assume that $f \in C_0(\mathbb{R})$ admits an analytic continuation to $\{z \in \mathbb{C} : |\mathrm{Im} z| < a\}$ for some $a > 0$. Let $\varphi(x) = f(x + ib)$ for some fixed $b \in \mathbb{R}$. Suppose there is $M > 0$ such that $\varphi \in L^1(\mathbb{R})$ and $\|\varphi\| \leqslant M$ for all $|b| < 2a/3$. Then $\hat{f}$ is a bounded continuous function on $\mathbb{R}$. Furthermore, there exists $C > 0$ that only depends on M and a such that*

$$|\hat{f}(\xi)| \leqslant Ce^{-\frac{a|\xi|}{3}}$$

for all $\xi \in \mathbb{R}$.

Given a closed spin Riemannian manifold X with metric g, we assume that the scalar curvature function $\kappa(x)$ of g satisfying that there exists $K \geqslant 0$ such that

$$\kappa(x) > \frac{16(K_\Gamma + K)^2}{\tau^2} \quad \text{for all } x \in X, \tag{2.21}$$

where K_Γ and τ are given in line (2.10) and (2.11) respectively.

Assume $\dim X$ is odd for the moment. Let us define

$$u_s = \begin{cases} \exp(2\pi i F(s^{-1}\widetilde{D})) & s > 0, \\ 1 & s = 0, \end{cases} \tag{2.22}$$

where $F(x) := \frac{G(x)+1}{2}$ and G is the function defined in line (2.14).

LEMMA 2.19. *The path $\{u_s\}_{s\in[0,\infty)}$ given in line (2.22) is an invertible element in $(\mathcal{B}_{L,0}(\widetilde{X})^\Gamma_K)^+$, the unitization of $\mathcal{B}_{L,0}(\widetilde{X})^\Gamma_K$.*

PROOF. Let sgn be the sign function and H the Heaviside step function $\frac{\mathrm{sgn}+1}{2}$. By the invertibility of $\widetilde{D}$, the element $H(\widetilde{D})$ is an idempotent. Therefore, we have

$$\exp(2\pi i \cdot H(\widetilde{D}_X)) = 1.$$

Hence

$$u_s = \exp(2\pi i F(s^{-1}\widetilde{D}) - 2\pi i H(s^{-1}\widetilde{D})).$$

The function $2\pi i(F - H)$ satisfies the condition in Lemma 2.17. It follows that there exists $s_0 \geqslant 0$ such that the path

$$\{2\pi i F(s^{-1}\widetilde{D}_X) - 2\pi i H(s^{-1}\widetilde{D}_X)\}_{s\in[s_0,\infty)}$$

defines an element in $\mathcal{B}_L(\widetilde{X})^\Gamma_K$. On the other hand, it follows from Lemma 2.13 and Lemma 2.15 that u_s is a continuous map from $[0, s_0]$ to $\mathcal{B}(\widetilde{X})^\Gamma_K$. Therefore the path $\{u_s\}_{s\in[0,\infty)}$ defines an element in $(\mathcal{B}_{L,0}(\widetilde{X})^\Gamma_K)^+$. Similarly, the path

$$u_s^{-1} = \begin{cases} \exp(-2\pi i F(s^{-1}\widetilde{D})) & s > 0, \\ 1 & s = 0, \end{cases} \tag{2.23}$$

lies in $(\mathcal{B}_{L,0}(\widetilde{X})^\Gamma_K)^+$ and is the inverse of $\{u_s\}_{s\in[0,\infty)}$. □

Now we are ready to define the l^1-higher rho invariant.

DEFINITION 2.20. Let (X, g) be a closed spin Riemannian manifold and $\widetilde{X}$ the universal cover of X. Suppose the scalar curvature of g satisfies the inequality in line (2.21).

(1) If $\dim X$ is odd, then we define the l^1-higher rho invariant $\rho_{1,K}(g)$ to be the class $[u] \in K_1(\mathcal{B}_{L,0}(\widetilde{X})^\Gamma_K)$, where $u = \{u_s\}_{s\in[0,\infty)}$.
(2) If $\dim X$ is even, then we define the l^1-higher rho invariant $\rho_{1,K}(g)$ to be the class $[p] - [(\begin{smallmatrix}1&0\\0&0\end{smallmatrix})] \in K_0(\mathcal{B}_{L,0}(\widetilde{X})^\Gamma_K)$, where $p = \{p_s\}_{s\in[0,\infty)}$ and

$$p_s = \begin{cases} p_{G(s^{-1}\widetilde{D})} & s > 0, \\ (\begin{smallmatrix}1&0\\0&0\end{smallmatrix}) & s = 0, \end{cases} \tag{2.24}$$

with the formula for $p_{G(s^{-1}\widetilde{D})}$ given in line (2.7).

Observe that, if there exits a continuous path of metrics each of which satisfies the condition (2.21), then these metrics define the same l^1-higher rho invariant.

2.6. A product formula for l^1-higher rho invariant. In this subsection, we prove a product formula for the l^1-higher rho invariant. Our method of proof is inspired by Higson–Roe [**9, 10**] and Weinberger–Xie–Yu [**23**].

As in the previous subsection, assume that (X, g) is a closed spin Riemannian manifold with scalar curvature function $\kappa(x)$ of g satisfying that there exists $K \geqslant 0$ such that

$$\kappa(x) > \frac{16(K_\Gamma + K)^2}{\tau^2} \text{ for all } x \in X,$$

where K_Γ and τ are given in line (2.10) and (2.11) respectively. We consider the Riemannian manifold $X\times\mathbb{R}$ with the product metric dt^2+g, whose scalar curvature is still uniformly bounded below by a positive number. Similar to Definition 2.3, Definition 2.16 and Definition 2.20, we define the geometric Banach algebras [2] $\mathcal{B}(\widetilde{X}\times\mathbb{R})^\Gamma_K$, $\mathcal{B}_L(\widetilde{X}\times\mathbb{R})^\Gamma_K$ and $\mathcal{B}_{L,0}(\widetilde{X}\times\mathbb{R})^\Gamma_K$ as well as the higher rho invariant

$$\rho_{1,K}(dt^2 + g) \in K_*(\mathcal{B}_{L,0}(\widetilde{X}\times\mathbb{R})^\Gamma_K).$$

Note that, although the Γ-action on $\widetilde{X}\times\mathbb{R}$ is no longer cocompact, these definitions still make sense due to the following observations.

(1) The l^1-norms of the operators in $\mathbb{C}(\widetilde{X}\times\mathbb{R})^\Gamma$ are defined by using the fundamental domain $\mathcal{F}\times\mathbb{R}$, where $\mathcal{F}$ is a precompact fundamental domain for the Γ-action on $\widetilde{X}$, where $\Gamma = \pi_1(X)$. Compare with Definition 2.1 and Definition 2.5.
(2) With respect to the given length function ℓ on Γ, we still have

$$\operatorname{dist}((x,t),(\gamma x,t')) > \tau\ell(\gamma) - C_0,\ \forall\gamma\in\Gamma,\ \forall x\in\mathcal{F},\ \forall t,t'\in\mathbb{R} \tag{2.25}$$

with the same constants C_0 and τ as in line (2.11).

Let $C^*_L(\mathbb{R})$ be the C^*-algebraic localization algebra of the metric space $\mathbb{R}$, and $\mathrm{Ind}_L(D_\mathbb{R}) \in K_1(C^*_L(\mathbb{R}))$ the local index class of the Dirac operator $D_\mathbb{R} = i\frac{d}{dx}$, cf. [**26**]. Then there is a natural product map

$$\mathcal{B}_{L,0}(\widetilde{X})^\Gamma_K \otimes_{\max} C^*_L(\mathbb{R}) \to \mathcal{B}_{L,0}(\widetilde{X}\times\mathbb{R})^\Gamma_K,$$

---

[2]Since the cocompactness fails for the Γ-action on $\widetilde{X}\times\mathbb{R}$, the Banach algebra $\mathcal{B}(\widetilde{X}\times\mathbb{R})^\Gamma_K$ is no longer isomorphic to $\mathcal{K}\otimes l^1_K(\Gamma)$.

which induces a homomorphism[3] at the level of K-theory

$$\_\otimes \mathrm{Ind}_L(D_{\mathbb{R}})\colon K_*(\mathcal{B}_{L,0}(\widetilde{X})_K^\Gamma) \to K_{*+1}(\mathcal{B}_{L,0}(\widetilde{X}\times\mathbb{R})_K^\Gamma). \tag{2.26}$$

THEOREM 2.21. *Assume that (X, g) is an m-dimensional closed spin Riemannian manifold with scalar curvature function $\kappa(x)$ of g satisfying that there exists $K \geqslant 0$ such that*

$$\kappa(x) > \frac{16(K_\Gamma + K)}{\tau} \quad \textit{for all } x \in X.$$

Then under the product map in line (2.26), we have

$$\rho_{1,K}(g) \otimes \mathrm{Ind}_L(D_{\mathbb{R}}) = \rho_{1,K}(g + dt^2) \in K_{m+1}(\mathcal{B}_{L,0}(\widetilde{X}\times\mathbb{R})_K^\Gamma).$$

Before we prove the theorem, let us introduce a different (but equivalent) representative of the l^1-higher rho invariant $\rho_{1,K}(g)$. First, we assume that $\dim X$ is odd. Set

$$v_s = \begin{cases} (\widetilde{D}+si)(\widetilde{D}-si)^{-1} & \text{if } s > 0, \\ 1 & \text{if } s = 0. \end{cases} \tag{2.27}$$

LEMMA 2.22. *For any $K \geqslant 0$, if $\dim X$ is odd and the scalar curvature κ on X satisfies that*

$$\inf_{x\in X} \kappa(x) > \frac{16(K_\Gamma + K)^2}{\tau^2},$$

then $v = \{v_s\}_{s\in[0,\infty)}$ in line (2.27) is an invertible element in $(\mathcal{B}_{L,0}(\widetilde{X})_K^\Gamma)^+$. Furthermore, we have

$$[v] = \rho_{1,K}(g) \in K_1(\mathcal{B}_{L,0}(\widetilde{X})_K^\Gamma).$$

PROOF. By Lemma 2.11, we have

$$\|e^{-y(s^2\widetilde{D}^2+1)}\|_{1,K} \leqslant Ce^{-y(\varepsilon s^2+1)}$$

for some $C > 0$ and $\varepsilon > 0$. It follows that the integral

$$\frac{1}{s^2\widetilde{D}^2+1} = -\int_0^\infty e^{-y(s^2\widetilde{D}^2+1)}dy \tag{2.28}$$

absolutely converges with respect to the $\|\cdot\|_{1,K}$-norm, hence

$$\frac{1}{s^2\widetilde{D}^2+1} \in \mathcal{B}(\widetilde{X})_K^\Gamma$$

and

$$\left\|\frac{1}{s^2\widetilde{D}^2+1}\right\|_{1,K} \leqslant \frac{C}{s^2\varepsilon^2+1} \tag{2.29}$$

for all $s > 0$. Similarly, the integral

$$\frac{1}{|\widetilde{D}|} = \frac{1}{\sqrt{\pi}}\int_{-\infty}^{+\infty} e^{-s^2\widetilde{D}^2}ds \tag{2.30}$$

absolutely converges with respect to the $\|\cdot\|_{1,K}$-norm. It follows that

$$\frac{1}{|\widetilde{D}|} \in \mathcal{B}(\widetilde{X})_K^\Gamma,$$

[3] In fact, a standard Elienberg swindle argument shows that the map

$$\_\otimes \mathrm{Ind}_L(D_{\mathbb{R}})\colon K_*(\mathcal{B}_{L,0}(\widetilde{X})_K^\Gamma) \to K_{*+1}(\mathcal{B}_{L,0}(\widetilde{X}\times\mathbb{R})_K^\Gamma)$$

is an isomorphism.

which implies that

$$\frac{1}{\widetilde{D}} = \frac{\operatorname{sgn}(\widetilde{D})}{|\widetilde{D}|} \in \mathcal{B}(\widetilde{X})_K^\Gamma$$

since by Lemma 2.15 we have $\operatorname{sgn}(\widetilde{D}) \in \mathcal{B}(\widetilde{X})_K^\Gamma$.

Let $\sigma(\widetilde{D}^{-1})$ be the spectrum of $\widetilde{D}^{-1}$ in $(\mathcal{B}(\widetilde{X})_K^\Gamma)^+$. For each $s \in \mathbb{R}\backslash\{0\}$, the element

$$(\widetilde{D}^{-1} - si)(\widetilde{D}^{-1} + si) = \frac{1 + s^2\widetilde{D}^2}{\widetilde{D}^2}$$

is invertible in $(\mathcal{B}(\widetilde{X})_K^\Gamma)^+$, with its inverse given by

$$\frac{\widetilde{D}^2}{1 + s^2\widetilde{D}^2} = \frac{1}{s^2}\Big(1 - \frac{1}{1 + s^2\widetilde{D}^2}\Big) \in (\mathcal{B}(\widetilde{X})_K^\Gamma)^+.$$

It follows that $(\widetilde{D}^{-1} - si)$ is invertible in $(\mathcal{B}(\widetilde{X})_K^\Gamma)^+$ for each $s \in \mathbb{R}\backslash\{0\}$. Equivalently, $si \notin \sigma(\widetilde{D}^{-1})$ for all $s \in \mathbb{R}\backslash\{0\}$.

Therefore, for any $s \in \mathbb{R}\backslash\{0\}$, the Cayley transform

$$\frac{\widetilde{D} + si}{\widetilde{D} - si} = 1 + \frac{2si}{\widetilde{D} - si}$$

is an invertible element in $(\mathcal{B}(\widetilde{X})_K^\Gamma)^+$. Furthermore, we have

$$\lim_{s\to 0}\Big\|\frac{\widetilde{D} + si}{\widetilde{D} - si} - 1\Big\|_{1,K} = 0,$$

Thus for any $T > 0$, the map $s \mapsto v_s$ is a continuous map from $[0, T]$ to $(\mathcal{B}(\widetilde{X})_K^\Gamma)^+$.

Set $f(x) = \frac{1}{x-i}$. We have

$$\hat{f}(\xi) = -ie^{-\xi}\chi_{(0,\infty)}(\xi),$$

where $\chi_{(0,\infty)}$ is the characteristic function of the interval $(0, \infty)$. Note that $\hat{f}$ decays exponentially as $|\xi| \to \infty$. Therefore by Lemma 2.17, there exists some $s_0 > 0$ such that the path $s \mapsto f_s(\widetilde{D})$ with $s \in [s_0, \infty)$ is an element of $\mathcal{B}_L(\widetilde{X})_K^\Gamma$. To summarize, we have proved that the path $\{v_s\}_{s\in[0,\infty)}$ is an invertible element in $(\mathcal{B}_{L,0}(\widetilde{X})_K^\Gamma)^+$.

Recall that the l^1-higher rho invariant $\rho_{1,K}(\widetilde{D})$ is defined to be K-theory class represented by $u = \{u_s\}_{s\in[0,\infty)} \in (\mathcal{B}_{L,0}(\widetilde{X})_K^\Gamma)^+$ in line (2.22). We shall show that

$$[v] = [u] \in K_1(\mathcal{B}_{L,0}(\widetilde{X})_K^\Gamma).$$

Note that for any fixed $s > 0$, we have the following identity in $(C^*(\widetilde{X})^\Gamma)^+$:

$$\begin{aligned} v_s &= \exp\Big(\int_0^s \frac{2i\widetilde{D}}{\widetilde{D}^2 + t^2}dt\Big) \\ &= \exp\big(i\arctan(s^{-1}\widetilde{D}) - \pi i \operatorname{sgn}(s^{-1}\widetilde{D})\big). \end{aligned}$$

Set

$$a_s = \int_0^s \frac{2i\widetilde{D}}{\widetilde{D}^2 + t^2}dt = i\arctan(s^{-1}\widetilde{D}) - \pi i \operatorname{sgn}(s^{-1}\widetilde{D}).$$

By the discussion above, we see that

$$t \mapsto \frac{2i\widetilde{D}}{\widetilde{D}^2 + t^2}$$

is a continuous path from $[0,s]$ to $\mathcal{B}(\widetilde{X})^\Gamma_K$. It follows the integral

$$\int_0^s \frac{2i\widetilde{D}}{\widetilde{D}^2+t^2}dt$$

absolutely converges in $\mathcal{B}(\widetilde{X})^\Gamma_K$. Therefore for any $T>0$, the map

$$a\colon [0,T]\to \mathcal{B}(\widetilde{X})^\Gamma_K \text{ by } a(s)=a_s$$

is a continuous map. On the other hand, the function

$$i\arctan(x)-\pi i\,\mathrm{sgn}(x)$$

satisfies the assumption of Lemma 2.17, which in particular implies there exists $s_0'>0$ such that $\{a_s\}_{s\in[s_0',\infty)}\in \mathcal{B}_L(\widetilde{X})^\Gamma_K$. Therefore, the path $\{a_s\}_{s\in[0,\infty)}$ is an element of $\mathcal{B}_{L,0}(\widetilde{X})^\Gamma_K$.

Recall that

$$u_s=\exp(2\pi i F(s^{-1}\widetilde{D}_X)-2\pi i H(s^{-1}\widetilde{D}_X)),$$

cf. the proof of Lemma 2.19. Now the following homotopy of invertible elements

$$\lambda\mapsto \exp\Big(\lambda a_s+(1-\lambda)\big(2\pi i F(s^{-1}\widetilde{D})-2\pi i H(\widetilde{D})\big)\Big),\ \lambda\in[0,1] \tag{2.31}$$

connects $v=\{v_s\}_{s\in[0,\infty)}$ and $u=\{u_s\}_{s\in[0,\infty)}$ in $(\mathcal{B}_{L,0}(\widetilde{X})^\Gamma_K)^+$. This finishes the proof. □

To prove the product formula for l^1-higher rho invariants, we shall give an alternative but equivalent description of l^1-higher rho invariants in terms of Clifford-linear Dirac operators. Recall the $\mathbb{C}\ell_n$-Dirac bundle $\mathfrak{S}(X)$ over X:

$$\mathfrak{S}(X)=P_{\mathrm{Spin}}(X)\times_\lambda \mathbb{C}\ell_n \tag{2.32}$$

where $\lambda\colon \mathrm{Spin}_n\to \mathrm{End}(\mathbb{C}\ell_n)$ is the representation given by left multiplication. Equip $\mathfrak{S}(X)$ with the canonical Riemannian connection determined by the presentation $\lambda\colon P_{\mathrm{Spin}}(X)\to \mathrm{End}(\mathbb{C}\ell_n)$. Lift all geometric data to $\widetilde{X}$ and denote by $\mathfrak{S}(\widetilde{X})$ the associated the $\mathbb{C}\ell_n$-Dirac bundle on $\widetilde{X}$. The decomposition $\mathbb{C}\ell_n=\mathbb{C}\ell_n^0\oplus\mathbb{C}\ell_n^1$ gives rise to a parallel decomposition

$$\mathfrak{S}(\widetilde{X})=\mathfrak{S}(\widetilde{X})^0\oplus\mathfrak{S}(\widetilde{X})^1.$$

Consider the (Clifford) volume element

$$\omega=i^{\frac{\dim X+1}{2}}e_1e_2\cdots e_n,$$

where $\{e_1,\cdots,e_n\}$ is an oriented local orthonormal frame of the tangent space. Let us denote by E the Clifford multiplication by ω.

The Clifford-linear Dirac operator $\not{D}$ on $\widetilde{X}$ is defined to be the $\mathbb{Z}_2$-graded Dirac operator acting on $\mathfrak{S}(\widetilde{X})$. In fact, when $\dim X$ is odd, it is equivalent to consider the following (ungraded) Dirac-type operator

$$\not{D}E\colon C_c^\infty(\widetilde{X},\mathfrak{S}(\widetilde{X})^0)\to C_c^\infty(\widetilde{X},\mathfrak{S}(\widetilde{X})^0)$$

Note that $\not{D}E$ is self-adjoint in this case, since $\not{D}$ and E commute when $\dim X$ is odd. With the natural identification $\mathbb{C}\ell_n^0\cong\mathbb{C}\ell_{n-1}$, the operator $\not{D}E$ becomes $\mathbb{C}\ell_{n-1}$-linear Dirac operator. From now on, if $\dim X$ is odd, we shall write $\not{D}$ to denote the operator

$$\not{D}\colon C_c^\infty(\widetilde{X},\mathfrak{S}(\widetilde{X})^0)\to C_c^\infty(\widetilde{X},\mathfrak{S}(\widetilde{X})^1),$$

unless otherwise specified.

The l^1-higher rho invariant of the spin Dirac operator $\widetilde{D}$ on $\widetilde{X}$ coincides with the l^1-higher rho invariant of the Clifford-linear Dirac operator $\not{D}$. If $\dim X$ is odd, it follows from Lemma 2.22 that the l^1-higher rho invariant of $\not{D}$ can be represented by

$$[\{(\not{D}E+si)(\not{D}E-si)^{-1}\}_{s\in[0,\infty)}] = [\{(\not{D}+siE)(\not{D}-siE)^{-1}\}_{s\in[0,\infty)}]$$

in $K_1(\mathcal{B}_{L,0}(\widetilde{X})^\Gamma_K)$. An alternative description of the l^1-higher rho invariant for the eve dimensional case is given in Lemma 2.24 below.

DEFINITION 2.23. Let $\mathcal{D}(\widetilde{X})^\Gamma_K$ be the $\|\cdot\|_{1,K}$-completion of operators in $\mathcal{B}(H)^\Gamma$ with finite propagation and *pseudo-local*, i.e., $\chi_A \circ T - T \circ \chi_A$ is compact for any precompact Borel set A. Similarly, we also define $\mathcal{D}_L(\widetilde{X})^\Gamma_K$ and $\mathcal{D}_{L,0}(\widetilde{X})^\Gamma_K$, which are the obvious analogues of $\mathcal{B}_L(\widetilde{X})^\Gamma_K$ and $\mathcal{B}_{L,0}(\widetilde{X})^\Gamma_K$.

Obviously, $\mathcal{B}(\widetilde{X})^\Gamma_K$ (resp. $\mathcal{B}_{L,0}(\widetilde{X})^\Gamma_K$) is a closed two-sided ideal of $\mathcal{D}(\widetilde{X})^\Gamma_K$ (resp. $\mathcal{D}_{L,0}(\widetilde{X})^\Gamma_K$).

LEMMA 2.24. *For a given $K \geqslant 0$, suppose* $\dim X$ *is even and the scalar curvature κ on X satisfies that*

$$\inf_{x\in X} \kappa(x) > \frac{16(K_\Gamma + K)^2}{\tau^2}.$$

Let us denote

$$P_{s,+} := \varphi(\not{D} + sE) \text{ and } P_{s,-} := \varphi(\not{D} - sE)$$

where

$$\varphi(x) = \begin{cases} 1, & x > 0, \\ 0, & x \leqslant 0. \end{cases}$$

Then the elements $\{P_{s,+}\}_{s\in[0,\infty)}$ and $\{P_{s,-}\}_{s\in[0,\infty)}$ lie in $(\mathcal{D}_L(\widetilde{X})^\Gamma_K)^+$, and the element $\{P_{s,+} - P_{s,-}\}_{s\in[0,\infty)}$ lies in $\mathcal{B}_{L,0}(\widetilde{X})^\Gamma_K$, Furthermore, we have

$$\rho_{1,K}(g) = [P_{s,+}] - [P_{s,-}]$$

in $K_0(\mathcal{B}_{L,0}(\widetilde{X})^\Gamma_K)$.

PROOF. Note that $P_{s,+} = \varphi(\not{D} + sE) = \varphi(s^{-1}\not{D} + E)$ for any $s > 0$. Recall the function G in line (2.14) given by

$$G(x) = \frac{2}{\sqrt{\pi}} \int_{-\infty}^{x} e^{-y^2} dy - 1.$$

By Lemma 2.13, Lemma 2.15 and Lemma 2.22, we have for any $r \geqslant 1$, the element $\{G(r(s^{-1}\not{D} + E))\}_{s\in[0,\infty)}$ lies in $\mathcal{D}_L(\widetilde{X})^\Gamma_K$, and

$$\begin{aligned} &\|G(r(s^{-1}\not{D} + E)) - \operatorname{sgn}(s^{-1}\not{D} + E)\|_{1,K} \\ \leqslant &\int_r^\infty \|(s^{-1}\not{D} + E)e^{-s^{-1}t^2\not{D}^2 - t^2}\| dt \\ = &\int_r^\infty \frac{1}{t} C_1 e^{-\frac{t^2\varepsilon}{s}} e^{-t^2} dt \leqslant C_2 \int_r^\infty e^{-t^2} dt, \end{aligned}$$

where sgn is the sign function and C_1, C_2, ε are positive constants independent of r and s. By varying r, we see that the element $\{\operatorname{sgn}(s^{-1}\not{D} + E)\}_{s\in[0,\infty)}$ lies in $(\mathcal{D}_L(\widetilde{X})^\Gamma_K)^+$. Since $P_{s,+} = \frac{\operatorname{sgn}(s^{-1}\not{D}+E)+1}{2}$, it follows that $\{P_{s,+}\}_{s\in[0,\infty)}$ lies in $(\mathcal{D}_L(\widetilde{X})^\Gamma_K)^+$. Similarly, we also have $(s\mapsto P_+(\not{D}-sE))\in \mathcal{D}_L(\widetilde{X})^\Gamma_K$. It follows from [**9**, Lemma 5.8] that the difference

$$\{P_{s,+}-P_{s,-}\}_{s\in[0,\infty)}$$

lies in $\mathcal{B}_{L,0}(\widetilde{X})^\Gamma_K$.

Now we shall adapt the proof of [**10**, Theorem 5.5] to show that

$$\rho_{1,K}(g) = [P_{s,+}] - [P_{s,-}].$$

First, if we decompose $\mathfrak{S}(\widetilde{X}) = \mathfrak{S}(\widetilde{X})^+ \oplus \mathfrak{S}(\widetilde{X})^-$ with respect to the grading operator E, then the operator E and $\not{D}$ become

$$E = \begin{pmatrix} 1 & 0 \\ 0 & -1 \end{pmatrix} \text{ and } \not{D} = \begin{pmatrix} 0 & \not{D}^- \\ \not{D}^+ & 0 \end{pmatrix}.$$

For each $s>0$, let $\chi_s\colon \mathbb{R}\to[-1,1]$ be the normalizing function given by

$$\chi_s(x) = \frac{x}{\sqrt{x^2+s^2}}.$$

With the decomposition $\mathfrak{S}(\widetilde{X}) = \mathfrak{S}(\widetilde{X})^+ \oplus \mathfrak{S}(\widetilde{X})^-$, we have

$$\chi_s(\not{D}) = \begin{pmatrix} 0 & V_s \\ U_s & 0 \end{pmatrix}$$

where $V_s = \not{D}^-(\not{D}^+\not{D}^- + s^2)^{-1/2}$ and $U_s = \not{D}^+(\not{D}^-\not{D}^+ + s^2)^{-1/2}$. By the explicit construction of $\rho_{1,K}(g)$ (cf. formula (2.7)), we have

$$\begin{aligned}\rho_{1,K}(g) &= \left[\begin{pmatrix} 1-(1-U_sV_s)^2 & (2-U_sV_s)U_s(1-V_sU_s) \\ V_s(1-U_sV_s) & (1-V_sU_s)^2 \end{pmatrix}\right] - \left[\begin{pmatrix} 1 & 0 \\ 0 & 0 \end{pmatrix}\right] \\ &= \left[\begin{pmatrix} 1-\left(\frac{s^2}{\not{D}^+\not{D}^-+s^2}\right)^2 & \frac{s^4\not{D}^+}{(\not{D}^-\not{D}^++s^2)^{5/2}} + \frac{s^2\not{D}^+}{(\not{D}^-\not{D}^++s^2)^{3/2}} \\ \frac{s^2\not{D}^-}{(\not{D}^+\not{D}^-+s^2)^{3/2}} & \left(\frac{s^2}{\not{D}^-\not{D}^++s^2}\right)^2 \end{pmatrix}\right] - \left[\begin{pmatrix} 1 & 0 \\ 0 & 0 \end{pmatrix}\right]\end{aligned}$$

Note that the idempotent

$$\begin{aligned}&\begin{pmatrix} 1-(1-U_sV_s)^2 & (2-U_sV_s)U_s(1-V_sU_s) \\ V_s(1-U_sV_s) & (1-V_sU_s)^2 \end{pmatrix} \\ &= \begin{pmatrix} 1-\left(\frac{s^2}{\not{D}^+\not{D}^-+s^2}\right)^2 & \frac{s^4\not{D}^+}{(\not{D}^-\not{D}^++s^2)^{5/2}} + \frac{s^2\not{D}^+}{(\not{D}^-\not{D}^++s^2)^{3/2}} \\ \frac{s^2\not{D}^-}{(\not{D}^+\not{D}^-+s^2)^{3/2}} & \left(\frac{s^2}{\not{D}^-\not{D}^++s^2}\right)^2 \end{pmatrix}\end{aligned}$$

is homotopic to the following projection

$$\begin{pmatrix} 1-\frac{s^2}{\not{D}^+\not{D}^-+s^2} & \frac{s\not{D}^+}{\not{D}^-\not{D}^++s^2} \\ \frac{s\not{D}^-}{\not{D}^+\not{D}^-+s^2} & \frac{s^2}{\not{D}^-\not{D}^++s^2} \end{pmatrix}$$

through the path of idempotents

$$\begin{pmatrix} 1-\left(\frac{s^2}{\not{D}^+\not{D}^-+s^2}\right)^{2-a} & \frac{(\not{D}^-\not{D}^++s^2)^{2-a}-s^{2(2-a)}}{\not{D}^-\not{D}^+(\not{D}^-\not{D}^++s^2)^{1-a}}\frac{s^{2-a}\not{D}^+}{(\not{D}^-\not{D}^++s^2)^{(3-a)/2}} \\ \frac{s^{2-a}\not{D}^-}{(\not{D}^+\not{D}^-+s^2)^{(3-a)/2}} & \left(\frac{s^2}{\not{D}^-\not{D}^++s^2}\right)^{2-a} \end{pmatrix}.$$

Here we have implicitly used the fact that

$$\left(\frac{s^2}{\not{D}^2+s^2}\right)^b \text{ lies in } (\mathcal{B}_{L,0}(\widetilde{X})^\Gamma_K)^+$$

for each $b>0$. In fact, similar to the proof of Lemma 2.22, for any $b>0$, we have

$$\frac{1}{(s^{-2}\not{D}^2+1)^b}=\frac{1}{C_b}\int_0^\infty e^{-t^{\frac{1}{b}}(s^{-2}\not{D}^2+1)}dt,$$

where

$$C_b=\int_0^\infty e^{-t^{\frac{1}{b}}}dt.$$

By Lemma 2.11, we have

$$\|e^{-y\widetilde{D}^2}\|_{1,K}\leqslant Ce^{-y\varepsilon}$$

for some $C>0$ and $\varepsilon>0$. Therefore,

$$\left\|\frac{1}{(s^{-2}\not{D}^2+1)^b}\right\|_{1,K}\leqslant\frac{1}{C_b}\int_0^\infty Ce^{-t^{\frac{1}{b}}(s^{-2}\varepsilon+1)}dt=\frac{C}{(s^{-2}\varepsilon+1)^b}\leqslant C. \tag{2.33}$$

Set $S_1=E$ and $S_2=\{\operatorname{sgn}(\not{D}+sE)\}_{s\in[0,\infty)}$, which are elements in $\mathcal{D}_L(\widetilde{X})^\Gamma_K$ whose squares are equal to 1. A straightforward calculation shows that

$$\frac{-S_2S_1S_2+1}{2}=\begin{pmatrix} 1-\frac{s^2}{\not{D}^+\not{D}^-+s^2} & \frac{s\not{D}^+}{\not{D}^-\not{D}^++s^2} \\ \frac{s\not{D}^-}{\not{D}^+\not{D}^-+s^2} & \frac{s^2}{\not{D}^-\not{D}^++s^2} \end{pmatrix}.$$

From the above discussion, we conclude that

$$\rho_{1,K}(g)=[\frac{-S_2S_1S_2+1}{2}]-[\frac{S_1+1}{2}].$$

On the other hand, using the fact that $\not{D}$ and E anticommute, we have that

$$[P_{s,+}]-[P_{s,-}]=[\frac{-S_1S_2S_1+1}{2}]-[\frac{S_2+1}{2}].$$

Now set $U=S_1S_2$, which is obviously invertible.

CLAIM. The spectrum of U excludes $\{-a : a\in\mathbb{R}, a\geqslant 0\}$.

In fact, for any $a>0$, we have

$$U+a=S_1S_2+aS_1^2=S_1(aS_1+S_2).$$

It suffices to show that $(aS_1+S_2)^2$ is invertible. In fact,

$$\begin{aligned}(S_2+aS_1)^2=&\Big(s\mapsto 1+a^2+\frac{2as}{\sqrt{D^2+s^2}}\Big)\\ =&\Big(s\mapsto 2a(\frac{1+a^2}{2a}+\frac{1}{\sqrt{s^{-2}D^2+1}})\Big)\end{aligned}$$

When $a > 0$, we have $\frac{1+a^2}{2a} > 1$.

Let r be the spectral radius of $(s \mapsto \frac{1}{s^{-2}D^2+1})$. By the spectral radius theorem, we have

$$r = \lim_{n\to\infty} \left\| \frac{1}{(s^{-2}\not{D}^2+1)^{\frac{n}{2}}} \right\|_{1,K}^{\frac{1}{n}} \leqslant \lim_{n\to\infty} \sqrt[n]{C} = 1.$$

Therefore, we see that $S_2 + aS_1$ is invertible for any $a > 0$. Our claim follows.

Since the spectrum of U excludes $\{-a : a \in \mathbb{R}, a \geqslant 0\}$, $\sqrt{z}$ is a holomorphic function on the spectrum of U. Thus there is an invertible element $V \in \mathcal{D}_L(\widetilde{X})_K^\Gamma$ such that $V^2 = U$.

Note that

$$S_1U = U^{-1}S_1, \ S_2U = U^{-1}S_2.$$

As $S_1^2 = S_2^2 = 1$, it follows from the holomorphic functional calculus that

$$S_1V = V^{-1}S_1, \ S_2V = V^{-1}S_2.$$

Set $W = VS_2$. Then we have

$$W^{-1}S_1W = S_2V^{-1}S_1VS_2 = S_2S_1V^2S_2 = S_2$$

and similarly $W^{-1}S_2W = S_1$. Therefore

$$[P_{s,+}] - [P_{s,-}] = [W^{-1}\frac{-S_1S_2S_1+1}{2}W] - [W^{-1}\frac{S_2+1}{2}W] = \rho_{1,K}(g).$$

This finishes the proof. □

Now we are ready to prove Theorem 2.21.

PROOF OF THEOREM 2.21. Let us first prove the even dimensional case, that is, when $n = \dim X$ is even. Recall that, the (Clifford) volume element on $\widetilde{X}$ is given by

$$\omega = i^{\frac{n}{2}} e_1 e_2 \cdots e_n,$$

where $\{e_1, \cdots, e_n\}$ is an oriented local orthonormal frame of the tangent space of $\widetilde{X}$. The corresponding volume element on $\widetilde{X} \times \mathbb{R}$ is given by

$$\omega_{\widetilde{X}\times\mathbb{R}} = i^{\frac{n}{2}+1} e_1 e_2 \cdots e_n e_{\mathbb{R}},$$

where $e_{\mathbb{R}}$ corresponds to the Clifford multiplication by the unit vector of $\mathbb{R}$. Let us denote by E the operator determined by multiplication of ω. Recall the natural isomorphism:

$$\mathbb{C}\ell_n \cong \mathbb{C}\ell_{n+1}^0$$

by sending e_j to $e_{n+1}e_j$ for all $1 \leq j \leq n$. In particular, this induces natural identifications

$$\mathfrak{S}(\widetilde{X}) \cong \mathfrak{S}(\widetilde{X} \times \mathbb{R})^0$$

and

$$\mathfrak{S}(\widetilde{X} \times \mathbb{R})^1 \xrightarrow{e_{\mathbb{R}}} \mathfrak{S}(\widetilde{X} \times \mathbb{R})^0 \cong \mathfrak{S}(\widetilde{X}),$$

Under these identifications, the operators

$$1 \widehat{\otimes} D_{\mathbb{R}} + \not{D} \widehat{\otimes} 1 \pm E \widehat{\otimes} e_{\mathbb{R}} \colon C_c^\infty(\widetilde{X} \times \mathbb{R}, \mathfrak{S}(\widetilde{X} \times \mathbb{R})^0) \to C_c^\infty(\widetilde{X} \times \mathbb{R}, \mathfrak{S}(\widetilde{X} \times \mathbb{R})^1)$$

become

$$1 \otimes \frac{d}{dt} + \not{D} \otimes 1 \pm E \otimes 1 \colon C_c^\infty(\widetilde{X} \times \mathbb{R}, \mathfrak{S}(\widetilde{X})) \to C_c^\infty(\widetilde{X} \times \mathbb{R}, \mathfrak{S}(\widetilde{X})).$$

In particular, the l^1-higher rho invariant $\rho_{1,K}(g+dt^2)$ is represented by

$$
\begin{aligned}
&\Big[(1 \mathbin{\widehat{\otimes}} D_{\mathbb{R}} + \not{D} \mathbin{\widehat{\otimes}} 1 + isE \mathbin{\widehat{\otimes}} ie_{\mathbb{R}})(1 \mathbin{\widehat{\otimes}} D_{\mathbb{R}} + \not{D} \mathbin{\widehat{\otimes}} 1 - isE \mathbin{\widehat{\otimes}} ie_{\mathbb{R}})^{-1}\Big] \\
=&\Big[(s^{-1} \otimes \frac{d}{dt} + s^{-1}\not{D} \otimes 1 - E \otimes 1)(s^{-1} \otimes \frac{d}{dt} + s^{-1}\not{D} \otimes 1 + E \otimes 1)^{-1}\Big] \\
=&\Big[(1 \otimes iD_s + s^{-1}\not{D} \otimes 1 - E \otimes 1)(1 \otimes iD_s + s^{-1}\not{D} \otimes 1 + E \otimes 1)^{-1}\Big]
\end{aligned}
$$

in $K_1(\mathcal{B}_{L,0}(\widetilde{X} \times \mathbb{R})^{\Gamma}_K)$, where we have

$$D_s := (1+s)^{-1}\frac{1}{i}\frac{d}{dt}.$$

Note that the path

$$\frac{1 \otimes iD_s + s^{-1}\not{D} \otimes 1 - E \otimes 1}{1 \otimes iD_s + s^{-1}\not{D} \otimes 1 + E \otimes 1} = 1 - \frac{2E \otimes 1}{1 \otimes iD_s + (s^{-1}\not{D} + E) \otimes 1}$$

is homotopic to the path

$$1 - \frac{2E(s^{-2}\not{D}^2 + 1)^{-1/2} \otimes 1}{1 \otimes iD_s + (s^{-1}\not{D} + E)(s^{-2}\not{D}^2 + 1)^{-1/2} \otimes 1}$$

through the family of paths

$$
\begin{aligned}
&\frac{1 \otimes iD_s + (s^{-1}\not{D} - E)(s^{-2}\not{D}^2 + 1)^{-r/2} \otimes 1}{1 \otimes iD_s + (s^{-1}\not{D} + E)(s^{-2}\not{D}^2 + 1)^{-r/2} \otimes 1} \\
=&1 - \frac{2E(s^{-2}\not{D}^2 + 1)^{-r/2} \otimes 1}{1 \otimes iD_s + (s^{-1}\not{D} + E)(s^{-2}\not{D}^2 + 1)^{-r/2} \otimes 1}
\end{aligned}
$$

where $s \in [0,\infty)$ and $r \in [0,1]$. Let us denote

$$
b_r(s) = \begin{cases} \dfrac{2E(s^{-2}\not{D}^2 + 1)^{-r/2} \otimes 1}{1 \otimes iD_s + (s^{-1}\not{D} + E)(s^{-2}\not{D}^2 + 1)^{-r/2} \otimes 1} & \text{if } s > 0, \\ 0 & \text{if } s = 0. \end{cases}
$$

CLAIM. For each $r \in [0,1]$, the path

$$\{b_r(s)\}_{s \in [0,\infty)} \text{ lies in } \mathcal{B}_{L,0}(\widetilde{X} \times \mathbb{R})^{\Gamma}_K.$$

Moreover, the family $\{b_r\}_{r \in [0,1]}$ is continuous in r.

Recall that

$$(s^{-1}\not{D} \pm E)^2 = s^{-2}\not{D}^2 + 1.$$

Therefore, there exists a constant $\delta > 0$ such that

$$\begin{aligned}
&\frac{2E(s^{-2}\not{D}^2+1)^{-r/2}\otimes 1}{1\otimes iD_s+(s^{-1}\not{D}+E)(s^{-2}\not{D}^2+1)^{-r/2}\otimes 1}\\
=&\frac{\big(2E(s^{-2}\not{D}^2+1)^{-r/2}\otimes 1\big)\cdot\big(-1\otimes iD_s+(s^{-1}\not{D}+E)(s^{-2}\not{D}^2+1)^{-r/2}\otimes 1\big)}{1\otimes D_s^2+(s^{-2}\not{D}^2+1)^{1-r}\otimes 1}\\
=&\frac{\big(2E(s^{-2}\not{D}^2+1)^{-r/2}\otimes 1\big)\cdot\big(-1\otimes iD_s+(s^{-1}\not{D}+E)(s^{-2}\not{D}^2+1)^{-r/2}\otimes 1\big)}{1\otimes (D_s^2+\delta)+((s^{-2}\not{D}^2+1)^{1-r}-\delta)\otimes 1}\\
=&\frac{-2E(s^{-2}\not{D}^2+1)^{-r/2}\otimes iD_s(D_s^2+\delta)^{-1}+2E(s^{-1}\not{D}+E)(s^{-2}\not{D}^2+1)^{-r}\otimes(D_s^2+\delta)^{-1}}{1\otimes 1+((s^{-2}\not{D}^2+1)^{1-r}-\delta)\otimes(D_s^2+\delta)^{-1}},
\end{aligned}$$

where the last term lies in $\mathcal{B}_{L,0}(\widetilde{X}\times\mathbb{R})_K^{\Gamma}$ by estimates similar to those in the proof of Lemma 2.24 (cf. line (2.33)). Furthermore, the same estimates also show that the family $\{b_r\}_{r\in[0,1]}$ is continuous in $r\in[0,1]$.

Let $P_{s,+}$ be the positive projection of $s^{-1}\not{D}+E$, that is, $P_{s,+}=\varphi(s^{-1}\not{D}+E)$, where

$$\varphi(x)=\begin{cases}1, & \text{if } x\geqslant 0,\\ 0, & \text{if } x<0.\end{cases}$$

Let $Q_{s,+}=1-P_{s,+}$. Similarly, we define $P_{s,-}=\varphi(s^{-1}\not{D}-E)$ and $Q_{s,-}=1-P_{s,-}$. Then we have

$$\begin{aligned}
b_1(s)&=1-\frac{2E(s^{-2}\not{D}^2+1)^{-1/2}\otimes 1}{1\otimes iD_s+(s^{-1}\not{D}+E)(s^{-2}\not{D}^2+1)^{-1/2}\otimes 1}\\
&=\frac{1\otimes iD_s+(s^{-1}\not{D}-E)(s^{-2}\not{D}^2+1)^{-1/2}\otimes 1}{1\otimes iD_s+(s^{-1}\not{D}+E)(s^{-2}\not{D}^2+1)^{-1/2}\otimes 1}\\
&=\frac{1\otimes iD_s+(P_{s,-}-Q_{s,-})\otimes 1}{1\otimes iD_s+(P_{s,+}-Q_{s,+})\otimes 1}\\
&=\frac{P_{s,-}\otimes(iD_s+1)+Q_{s,-}\otimes(iD_s-1)}{P_{s,+}\otimes(iD_s+1)+Q_{s,+}\otimes(iD_s-1)}
\end{aligned}$$

A straightforward calculation shows that

$$\begin{aligned}
&\big(P_{s,+}\otimes(iD_s+1)+Q_{s,+}\otimes(iD_s-1)\big)^{-1}\\
=&P_{s,+}\otimes(iD_s+1)^{-1}+Q_{s,+}\otimes(iD_s-1)^{-1}.
\end{aligned}$$

It follows that

$$\begin{aligned}
&\frac{P_{s,-}\otimes(iD_s+1)+Q_{s,-}\otimes(iD_s-1)}{P_{s,+}\otimes(iD_s+1)+Q_{s,+}\otimes(iD_s-1)}\\
=&P_{s,-}P_{s,+}\otimes 1+P_{s,-}Q_{s,+}\otimes(iD_s+1)(iD_s-1)^{-1}\\
&\quad+Q_{s,-}P_{s,+}\otimes(iD_s-1)(iD_s+1)^{-1}+Q_{s,-}Q_{s,+}\otimes 1\\
=&\big(P_{s,-}\otimes(iD_s+1)(iD_s-1)^{-1}+(1-P_{s,-})\otimes 1\big)\\
&\quad\cdot\big(P_{s,+}\otimes(iD_s-1)(iD_s+1)^{-1}+(1-P_{s,+})\otimes 1\big)
\end{aligned}$$

whose K-theory class is precisely the K-theory product

$$([P_{s,+}] - [P_{s,-}]) \otimes [(D_s + i)(D_s - i)^{-1}] = \rho_{1,K}(g) \otimes \mathrm{Ind}_L(D_{\mathbb{R}}).$$

Now let us prove the odd dimensional case, that is, when $n = \dim X$ is odd. Let $g + dx^2 + dy^2$ be the product metric on $X \times \mathbb{R}^2$ determined by the metric g on X.

CLAIM. The l^1-higher rho invariant $\rho_{1,K}(g+dx^2+dy^2)$ associated to the metric $g + dx^2 + dy^2$ satisfies the following product formula:

$$\rho_{1,K}(g + dx^2 + dy^2) = \rho_{1,K}(g) \otimes \mathrm{Ind}_L(D_{\mathbb{R}^2}) \tag{2.34}$$

in $K_1(\mathcal{B}_{L,0}(\widetilde{X} \times \mathbb{R}^2)^\Gamma_K)$, where $\otimes$ stands for the K-theory product

$$\_\otimes\_ : K_1(\mathcal{B}_{L,0}(\widetilde{X})^\Gamma_K) \otimes K_0(C^*_L(\mathbb{R}^2)) \to K_1(\mathcal{B}_{L,0}(\widetilde{X} \times \mathbb{R}^2)^\Gamma_K).$$

The proof of this claim is similar to the even dimensional case (cf. [**23**, Proposition D.3]), and we shall omit the details. Now by the product formula from the even dimensional case, we have

$$\rho_{1,K}(g + dx^2 + dy^2) = \rho_{1,K}(g + dx^2) \otimes \mathrm{Ind}_L(D_{\mathbb{R}}).$$

Recall that we also have the following product formula for the Dirac operators in the C^*-algebraic setting:

$$\mathrm{Ind}_L(D_{\mathbb{R}^2}) = \mathrm{Ind}_L(D_{\mathbb{R}}) \otimes \mathrm{Ind}_L(D_{\mathbb{R}}).$$

These, together with the formula (2.34), imply that

$$\rho_{1,K}(g + dx^2) = \rho_{1,K}(g) \otimes \mathrm{Ind}_L(D_{\mathbb{R}})$$

for the case where $\dim X$ is odd. This finishes the proof. □

3. Higher index on manifolds with boundary

In this section, we review the construction of higher indices of Dirac operators on spin manifolds with boundary whose metrics are assumed to have product structure near the boundary and also have positive scalar curvature near the boundary. This definition was originally introduced in [**2**, **20**] and is more detailed stated in [**24**].

GEOMETRIC SETUP 3.1. Let M be a compact spin manifold with boundary ∂M. Assume that the metric g on M has product structure near the boundary and we denote by g_∂ the restriction of g to ∂M. Let M_∞ be the manifold obtained by attaching an infinite cylinder to M, that is, $M_\infty = M \cup_{\partial M} \partial M \times [0, \infty)$. Since g has product structure near the boundary, g extends canonically to a Riemannian metric g_∞ on M_∞, where $g_\infty = g_\partial + dt^2$ on the cylindrical part.

Now let $\widetilde{M}_\infty$ be the universal cover of M_∞ and Γ the fundamental group of M_∞, which is also the fundamental group of M. If we denote the universal cover of M by $\widetilde{M}$, then we have $\widetilde{M}_\infty = \widetilde{M} \cup_{\partial\widetilde{M}} \partial\widetilde{M} \times [0, \infty)$. Let $S(\widetilde{M}_\infty)$ be the spinor bundle on $\widetilde{M}_\infty$. Endow $\widetilde{M}_\infty$ with the metric $\widetilde{g}_\infty$ which is the lift of the metric g_∞ on M_∞. The fundamental group Γ acts isometrically on $\widetilde{M}_\infty$ and $S(\widetilde{M}_\infty)$. Let $\widetilde{D}$ be the associated Dirac operator on $\widetilde{M}_\infty$. Note that $\widetilde{D}$ is Γ-equivariant, that is,

$$\widetilde{D}(\gamma f) = \gamma \widetilde{D} f, \forall \gamma \in \Gamma,$$

where f is any compactly support smooth section of $S(\widetilde{M}_\infty)$.

For $0 \leqslant s < \infty$, we denote $M_s = M \cup_{\partial M} \partial M \times [0, s] \subset M_\infty$ and $\widetilde{M}_s = \widetilde{M} \cup_{\partial \widetilde{M}} \partial \widetilde{M} \times [0, s] \subset \widetilde{M}_\infty$.

Using the Hilbert space $L^2(\widetilde{M}_\infty, S(\widetilde{M}_\infty))$, the Hilbert space of all square-integrable sections of $S(\widetilde{M}_\infty)$, we define the C^*-algebra $C^*(\widetilde{M}_\infty)^\Gamma$ as in Definition 2.3.

DEFINITION 3.2. We say an operator $T \in B(H)^\Gamma$ is supported on $\widetilde{M}_s$ if

$$\chi_{\partial \widetilde{M} \times [s, +\infty)} \circ T \text{ and } T \circ \chi_{\partial \widetilde{M} \times [s, +\infty)}$$

are both zero. We define $C^*(\widetilde{M}_s)^\Gamma$ to be the operators in $C^*(\widetilde{M}_\infty)^\Gamma$ supported on $\widetilde{M}_s$.

For any $0 < s < \infty$, we have

$$C^*(\widetilde{M}_s)^\Gamma \cong \mathcal{K} \otimes C_r^*(\Gamma).$$

3.1. Even dimensional case. Let us assume the same notation as in the Geometric Setup 3.1. Recall that when $\dim M$ is even, the spinor bundle on M_∞ admits a natural $\mathbb{Z}_2$-grading and the Dirac operator D on M_∞ is an odd operator with respect to this $\mathbb{Z}_2$-grading.

Recall the definition of normalizing function given in Definition 2.6. Given a normalizing function F, we obtain $p_{F(\widetilde{D})}$ as in line (2.7).

The difference from the definition of the higher index for closed manifold in Section 2.2 is the following cut-off function.

DEFINITION 3.3. Let Ψ be a smooth cut-off function on $\widetilde{M}_\infty$ such that $|\Psi(x)| \leqslant 1$, $\Psi(x) = 1$ for all $x \in \widetilde{M}_0 = \widetilde{M}$ and $\Psi(x) = 0$ on $\partial \widetilde{M} \times [1, +\infty)$. Also, we choose Ψ so that on the subspace $\partial \widetilde{M} \times [0, 1]$, it only depends on the parameter $t \in [0, 1]$. We furthermore define Ψ_T on $\widetilde{M}_\infty$ such that $\Psi_T = 1$ on $\widetilde{M}_T$ and $\Psi_T(x, t) = \Psi(x, t - T)$ for $(x, t) \in \partial \widetilde{M} \times [T, +\infty)$.

We define

$$q_{F(\widetilde{D}),T} = \left(p_{F(\widetilde{D})} - \left(\begin{smallmatrix} 1 & 0 \\ 0 & 0 \end{smallmatrix}\right)\right)\Psi_T + \left(\begin{smallmatrix} 1 & 0 \\ 0 & 0 \end{smallmatrix}\right). \tag{3.1}$$

Clearly, $q_{F(\widetilde{D}),T}$ lies in $M_2((C^*(\widetilde{M}_{T+5N_F})^\Gamma)^+)$, where $(C^*(\widetilde{M}_{T+5N_F})^\Gamma)^+$ is the unitalization of $C^*(\widetilde{M}_{T+5N_F})^\Gamma$.

LEMMA 3.4. *With the same setup as above, let $\widetilde{D}$ be the Dirac operator on $\widetilde{M}_\infty$. Then there exist $T > 0$ and a smooth normalizing function F whose Fourier transform is compactly supported such that*

$$\|q_{F(\widetilde{D}),T}^2 - q_{F(\widetilde{D}),T}\|_{\mathrm{op}} < 1/2. \tag{3.2}$$

PROOF. Let φ be a smooth normalizing function whose Fourier transform is compactly supported. For each $t > 0$, we define a normalizing function φ_t by setting $\varphi_t(x) = \varphi(tx)$ for all $x \in \mathbb{R}$. Let $\widetilde{D}_c$ be the Dirac operator on $\partial \widetilde{M} \times \mathbb{R}$, where the metric $g_{\partial \widetilde{M} \times \mathbb{R}}$ is given by $g_\partial + dt^2$. By the Lichnerowicz formula

$$\widetilde{D}_c^2 = \nabla^* \nabla + \frac{\kappa}{4}$$

where κ is scalar curvature of the metric $g_{\partial\widetilde{M}\times\mathbb{R}}$, we see that $\widetilde{D}_c$ is an invertible operator, since the metric g_∂ has positive scalar curvature,

It follows from the Fourier inverse transform formula (cf. line (2.4)) that $\varphi_t(\widetilde{D})$ has propagation $\leqslant N_\varphi \cdot t$. Moreover, the operator

$$p_{\varphi_t(\widetilde{D})} - q_{\varphi_t(\widetilde{D}),5N_\varphi t} = \big(p_{\varphi_t(\widetilde{D})} - \left(\begin{smallmatrix}1&0\\0&0\end{smallmatrix}\right)\big)(1-\Psi_{5tN_\varphi})$$

on $\widetilde{M}_\infty$ coincides with the operator

$$\big(p_{\varphi_t(\widetilde{D}_c)} - \left(\begin{smallmatrix}1&0\\0&0\end{smallmatrix}\right)\big)(1-\Psi_{5tN_\varphi})$$

on $\partial\widetilde{M}\times\mathbb{R}$ under the canonical identification between the subspace $\partial\widetilde{M}\times\mathbb{R}_{\geqslant 0}$ of $\partial\widetilde{M}\times\mathbb{R}$ and the cylindrical part $\partial\widetilde{M}\times\mathbb{R}_{\geqslant 0}$ of $\widetilde{M}_\infty$. Since $\widetilde{D}_c$ is invertible, we have

$$\begin{aligned}&\lim_{t\to\infty}\max\{\|1-\varphi_t(\widetilde{D}_c)^-\varphi_t(\widetilde{D}_c)^+\|_{\mathrm{op}}, \|1-\varphi_t(\widetilde{D}_c)^+\varphi_t(\widetilde{D}_c)^-\|_{\mathrm{op}}\}\\ &=\lim_{t\to\infty}\|\varphi_t(\widetilde{D}_c)^2-1\|_{\mathrm{op}}=0.\end{aligned}$$

Since $\|\varphi_t(\widetilde{D}_c)^\pm\|_{\mathrm{op}}\leqslant 1$ for all $t>0$, we have

$$\lim_{t\to\infty}\|p_{\varphi_t(\widetilde{D})} - q_{\varphi_t(\widetilde{D}),5tN_\varphi}\|_{\mathrm{op}} = 0. \tag{3.3}$$

Moreover, we have $\|p_{\varphi_t(\widetilde{D})}\|_{\mathrm{op}}\leqslant 64$ for all $t>0$ and $p^2_{\varphi_t(\widetilde{D})} = p_{\varphi_t(\widetilde{D})}$. It follows that

$$\lim_{t\to\infty}\|q^2_{\varphi_t(\widetilde{D}),5N_\varphi t} - q_{\varphi_t(\widetilde{D}),5tN_\varphi}\|_{\mathrm{op}} = 0. \tag{3.4}$$

Now the proof is finished by setting $F=\varphi_t$ and $T=5tN_\varphi$ for some sufficiently large $t\gg 0$. □

Let Θ be the function on the complex plane defined by

$$\Theta(z)=\begin{cases}0 & \mathrm{Re}\, z\leqslant 1/2,\\ 1 & \mathrm{Re}\, z>1/2.\end{cases}$$

Let $q_{F(\widetilde{D}),T}$ be the operator constructed in the above lemma. It follows that Θ is holomorphic on the spectrum of $q_{F(\widetilde{D}),T}$. Therefore $\Theta(q_{F(\widetilde{D}),T})$ is an idempotent in $M_2((C^*(\widetilde{M}_{T+5N_F})^\Gamma)^+)$ and $\Theta(q_{F(\widetilde{D}),T}) - \left(\begin{smallmatrix}1&0\\0&0\end{smallmatrix}\right)$ lies in $M_2(C^*(\widetilde{M}_{T+5N_F})^\Gamma)$.

Definition 3.5. The higher index $\mathrm{Ind}^\Gamma(D)$ of $\widetilde{D}$ on $\widetilde{M}_\infty$ is defined to be

$$[\Theta(q_{F(\widetilde{D}),T})] - [\left(\begin{smallmatrix}1&0\\0&0\end{smallmatrix}\right)] \in K_0(C^*(\widetilde{M}_{T+5N_F})^\Gamma)\cong K_0(C^*_r(\Gamma)). \tag{3.5}$$

The higher index $\mathrm{Ind}^\Gamma(D)$ is independent of the choice of T and F, as long as the inequality (3.2) in Lemma 3.4 is satisfied.

3.2. Odd dimensional case. Let $\dim M$ be odd and $\widetilde{D}$ the associated Dirac operator on $\widetilde{M}_\infty$.

Consider the function $P=(F+1)/2$ where F is a normalizing function. More precisely, we assume P is a continuous function on $\mathbb{R}$ satisfying

(P1) $\lim_{x\to-\infty}P(x)=0$ and $\lim_{x\to+\infty}P(x)=1$;
(P2) and the Fourier transform of P' is supported on the interval $[-N_P,N_P]$ for some $N_P>0$.

Again, it follows the Fourier inverse transform formula (cf. line (2.4)) that $P(\widetilde{D})$ has propagation no more than N_P.

We define

$$
\begin{aligned}
f_n(x) &= \sum_{k=1}^{n} \frac{(2\pi i)^k}{k!} x^k - \Big(\sum_{k=1}^{n} \frac{(2\pi i)^k}{k!}\Big) x \\
&= \Big(\sum_{k=1}^{n} \frac{(2\pi i)^k}{k!}\Big)(x^2 - x) + \sum_{k=1}^{n} \frac{(2\pi i)^k}{k!} \sum_{j=0}^{k-2} x^j (x^2 - x),
\end{aligned}
\tag{3.6}
$$

which satisfies

$$
e^{2\pi i x} - (1 + f_n(x)) = \sum_{k=n+1}^{\infty} \frac{(2\pi i)^k}{k!} x^k + \Big(\sum_{k=1}^{n} \frac{(2\pi i)^k}{k!}\Big) x. \tag{3.7}
$$

Note that $f_n(P(\widetilde{D}))$ is locally compact and has propagation $\leqslant n \cdot N_P$.

LEMMA 3.6. *With the same notation as above, there exist $n > 0$, $T > 0$ and a continuous function P satisfying conditions* (P1)-(P2) *above such that*

$$
\| e^{2\pi i P(\widetilde{D})} - \big(1 + f_n(P(\widetilde{D}))\Psi_T\big) \|_{\mathrm{op}} < 1. \tag{3.8}
$$

where Ψ_T is a compactly supported smooth function on $\widetilde{M}_\infty$ as Definition 3.3.

PROOF. Let φ be a smooth function on $\mathbb{R}$ satisfying conditions (P1)-(P2). Define the function φ_t by setting $\varphi_t(x) = \varphi(tx)$.

By the Fourier inverse transform formula (cf. line (2.4)), $\varphi_t(\widetilde{D})$ has propagation $\leqslant tN_\varphi$. Hence $f_n(\varphi_t(\widetilde{D}))$ has propagation $\leqslant ntN_\varphi$. Moreover, the operator $f_n(\varphi_t(\widetilde{D}))(1 - \Psi_{ntN_\varphi})$ on $\widetilde{M}_\infty$ coincides with the operator $f_n(\varphi_t(\widetilde{D}_{\mathrm{c}}))(1 - \Psi_{ntN_\varphi})$ on $\partial\widetilde{M} \times \mathbb{R}$ under the canonical identification between the subspace $\partial\widetilde{M} \times \mathbb{R}_{\geqslant 0}$ of $\partial\widetilde{M} \times \mathbb{R}$ and the cylindrical part $\partial\widetilde{M} \times \mathbb{R}_{\geqslant 0}$ of $\widetilde{M}_\infty$.

Since $\|\varphi_t(\widetilde{D})\|_{\mathrm{op}} \leqslant 1$ and $\|\varphi_t(\widetilde{D}_{\mathrm{c}})\|_{\mathrm{op}} \leqslant 1$, we have for any $t > 0$

$$
\| e^{2\pi i \varphi_t(\widetilde{D})} - \big(1 + f_n(\varphi_t(\widetilde{D}))\big) \|_{\mathrm{op}} < \frac{1}{3} \text{ and } \| e^{2\pi i \varphi_t(\widetilde{D}_{\mathrm{c}})} - \big(1 + f_n(\varphi_t(\widetilde{D}_{\mathrm{c}}))\big) \|_{\mathrm{op}} < \frac{1}{3}
$$

when n is sufficiently large.

As $\widetilde{D}_{\mathrm{c}}$ is invertible, we also have

$$
\| e^{2\pi \varphi_t(\widetilde{D}_{\mathrm{c}})} - 1 \|_{\mathrm{op}} < \frac{1}{3}.
$$

as long as t is sufficiently large. Therefore, when both n and t are sufficiently large, we have

$$
\begin{aligned}
&\| e^{2\pi i \varphi_t(\widetilde{D})} - \big(1 + f_n(\varphi_t(\widetilde{D}))\Psi_{ntN_\varphi}\big) \|_{\mathrm{op}} \\
<&\| e^{2\pi i \varphi_t(\widetilde{D})} - \big(1 + f_n(\varphi_t(\widetilde{D}))\big) \|_{\mathrm{op}} + \| f_n(\varphi_t(\widetilde{D}))(1 - \Psi_{ntN_\varphi}) \|_{\mathrm{op}} \\
=&\| e^{2\pi i \varphi_t(\widetilde{D})} - 1 - f_n(\varphi_t(\widetilde{D})) \|_{\mathrm{op}} + \| f_n(\varphi_t(\widetilde{D}_{\mathrm{c}}))(1 - \Psi_{ntN_\varphi}) \|_{\mathrm{op}} \\
<&\| e^{2\pi i \varphi_t(\widetilde{D})} - 1 - f_n(\varphi_t(\widetilde{D})) \|_{\mathrm{op}} \\
&\quad + \| 1 + f_n(\varphi_t(\widetilde{D}_{\mathrm{c}})) - e^{2\pi i \varphi_t(\widetilde{D}_{\mathrm{c}})} \|_{\mathrm{op}} + \| e^{2\pi i \varphi_t(\widetilde{D}_{\mathrm{c}})} - 1 \|_{\mathrm{op}} < 1.
\end{aligned}
$$

Now the proof is finished by setting $P = \varphi_t$ and $T = nN_\varphi \cdot t$ for some sufficiently large n and t. □

Now let (P, n, T) be the triple chosen as in Lemma 3.6 above so that the inequality (3.8) is satisfied. Note that $\|e^{2\pi i P(\widetilde{D})}\|_{\mathrm{op}} = 1$. It follows that $1 + f_n(P(\widetilde{D}))\Psi_T$ is invertible in $(C^*(\widetilde{M}_\infty)^\Gamma)^+$. Moreover, $1 + f_n(P(\widetilde{D}))\Psi_T$ lies in $(C^*(\widetilde{M}_{T+nN_P})^\Gamma)^+$ by construction, hence is also invertible in $(C^*(\widetilde{M}_{T+nN_P})^\Gamma)^+$, since $(C^*(\widetilde{M}_{T+nN_P})^\Gamma)^+$ is a C^*-subalgebra of $(C^*(\widetilde{M}_\infty)^\Gamma)^+$.

DEFINITION 3.7. The higher index $\mathrm{Ind}^\Gamma(D)$ of $\widetilde{D}$ is defined to be

$$[1 + f_n(P(\widetilde{D}))\Psi_T] \in K_1(C^*(\widetilde{M}_{T+nN_P})^\Gamma) \cong K_1(C_r^*(\Gamma)).$$

The higher index $\mathrm{Ind}^\Gamma(D)$ is independent of the choice of (P, n, T), as long as the inequality (3.8) in Lemma 3.6 is satisfied.

4. Sufficiently large spectral gap on the boundary and l^1-index of Dirac operator

In this section, we construct the l^1-indices of Dirac operators on spin manifolds whose scalar curvatures are sufficiently large on their boundaries. As a consequence, we prove Theorem 4.1, which is in fact a slightly improved version of Theorem 1.5.

Let us assume the same notation as in the Geometric Setup 3.1. Let $\mathcal{F}_\partial$ be a fundamental domain of the Γ-action on $\partial\widetilde{M}$. Fix a finite symmetric generating set S of Γ. Let ℓ be the length function on Γ induced by S. We define

$$\tau_\partial = \liminf_{\ell(\gamma)\to\infty} \sup_{x\in\mathcal{F}_\partial} \frac{\mathrm{dist}(x, \gamma x)}{\ell(\gamma)} \tag{4.1}$$

and

$$K_\Gamma = \inf\{K : \exists C > 0 \text{ s.t. } \#\{\gamma \in \Gamma : \ell(\gamma) \leqslant n\} \leqslant Ce^{Kn}\}. \tag{4.2}$$

THEOREM 4.1. *With the same notation as in the Geometric Setup 3.1, if there exists a constant $K \geqslant 0$ such that the scalar curvature κ of the metric g_∂ on ∂M satisfies*

$$\inf_{x\in\partial M} \kappa(x) > \frac{16(K_\Gamma + K)^2}{\tau_\partial^2}, \tag{4.3}$$

then the higher index $\mathrm{Ind}^\Gamma(D)$ of the Dirac operator $\widetilde{D}$ on $\widetilde{M}_\infty$ admits a natural preimage $\mathrm{Ind}^\Gamma_{1,K}(D)$ in $K_n(l^1_K(\Gamma))$ under the inclusion

$$l^1_K(\Gamma) \hookrightarrow C_r^*(\Gamma).$$

where the Banach algebra $l^1_K(\Gamma)$ is defined in Definition 2.5.

By the Lichnerowicz formula

$$\widetilde{D}_{\mathrm{c}}^2 = \nabla^*\nabla + \frac{\kappa}{4},$$

the lower bound on κ in line (4.3) implies that

$$|\widetilde{D}_{\mathrm{c}}| > \frac{2(K_\Gamma + K)}{\tau_\partial},$$

where $\widetilde{D}_{\mathrm{c}}$ is the Dirac operator on $\partial\widetilde{M} \times \mathbb{R}$ equipped with the metric $\widetilde{g}_\partial + dt^2$.

As in Definition 2.3 and Definition 2.5, we define the Banach algebras $\mathcal{B}(\widetilde{M}_\infty)^\Gamma_K$. Although the Γ-action on $\widetilde{M}_\infty$ is not cocompact, the definitions make sense due to the following observations.

(1) To define the l^1-norm on $\mathbb{C}(\widetilde{X}\times\mathbb{R})^\Gamma$, we should use a fundamental domain $\mathcal{F}_\infty \subset \widetilde{M}_\infty$ such that
- $\mathcal{F}_\infty \cap \widetilde{M}$ is a precompact fundamental domain of the Γ-action on $\widetilde{M}$;
- $\mathcal{F}_\infty \cap (\partial\widetilde{M}\times[0,+\infty)) = \mathcal{F}_\partial \times [0,+\infty) \subset \widetilde{M}_\infty$, where $\mathcal{F}_\partial$ is a precompact fundamental domain of the Γ-action on $\partial\widetilde{M}$.

(2) Given a length function on Γ, we still have

$$\operatorname{dist}(x,\gamma x) > \tau'\ell(\gamma) - C_0', \ \forall \gamma\in\Gamma, \ \forall x \in \mathcal{F}_\infty \tag{4.4}$$

for some positive constants C_0 and τ.

Similarly to Definition 3.2, we define $\mathcal{B}(\widetilde{M}_s)_K^\Gamma$ to be the operators in $\mathcal{B}(\widetilde{M}_\infty)_K^\Gamma$ supported on $\widetilde{M}_s$. For any $0 < s < \infty$, we have

$$\mathcal{B}(\widetilde{M}_s)_K^\Gamma \cong \mathcal{K}\otimes l_K^1(\Gamma).$$

4.1. Even dimensional case. Let us prove Theorem 4.1 in the case where $\dim M$ is even.

We assume the same notation as in the Geometric setup 3.1. In particular, $\widetilde{D}$ is the Dirac operator on $\widetilde{M}_\infty$ and $\widetilde{D}_{\mathrm{c}}$ is the Dirac operator on the cylinder $\partial\widetilde{M}\times\mathbb{R}$.

Let $q_{F(\widetilde{D}),T}$ be the element defined in line (3.1), where T is some positive number and F is a smooth normalizing function with F' a Schwartz function. In particular, $\|q_{F(\widetilde{D}),T}\|_{1,K}$ is finite, since the Fourier transform of F has compact support.

Similar to the construction of the higher index $\operatorname{Ind}^\Gamma(D)\in K_0(C_r^*(\Gamma))$ as in Definition 3.5. The proof of Theorem 4.1 for the even dimensional case can be reduced to the following claim, which is an l^1-analogue of Lemma 3.4.

Claim 4.2. If $|\widetilde{D}_{\mathrm{c}}| > \frac{2(K_\Gamma+K)}{\tau_\partial}$, then there exist a positive number T and a smooth normalizing function F whose Fourier transform is compactly supported such that

$$\|q^2_{F(\widetilde{D}),T} - q_{F(\widetilde{D}),T}\|_{1,K} < 1/2,$$

where $\|\cdot\|_{1,K}$ is the weighted l^1-norm given in Definition 2.5.

Proof of Claim 4.2. Let χ be a compactly supported smooth even function on $\mathbb{R}$ such that $\chi(x)=1$ if $|x|<1$, $\chi(x)=0$ if $|x|>2$ and $|\chi(x)|\leqslant 1$ for all $x\in\mathbb{R}$. Write $\chi_s(x)=\chi(x/s)$.

Let G be the normalizing function given in line (2.14). The derivative of G is $G'(x)=\frac{2}{\sqrt{\pi}}e^{-x^2}$. Set $G_t(x)=G(tx)$. Recall that in Lemma 2.13 and Lemma 2.15, we have seen that

(1) $\forall t\geqslant 1$, $\|G(t\widetilde{D})\|_{1,K}$ and $\|G(t\widetilde{D}_c)\|_{1,K}$ are uniformly bounded in t.
(2) $\|\operatorname{sgn}(\widetilde{D}_c)\|_{1,K}<\infty$.
(3) $\lim_{t\to\infty}\|G_t(\widetilde{D}_c)-\operatorname{sgn}(\widetilde{D}_c)\|_{1,K}=0$.

It follows that

$$\lim_{t\to\infty}\|G_t(\widetilde{D}_c)^2-1\|=0 \tag{4.5}$$

as $\operatorname{sgn}(\widetilde{D}_c)^2=1$.

We define

$$F_{a,t}(x)=\int_0^{xt}(G' * \widehat{\chi}_{at})(y)dy-1, \tag{4.6}$$

where a is a fixed positive number satisfying $a > \frac{4(K_\Gamma+K)}{\tau_\partial}$. Note that the distributional Fourier transform of $F_{a,t}$ is supported on $[-2at^2, 2at^2]$.

Let Ψ_T be the cut-off function on $\widetilde{M}_\infty$ which vanishes on $\partial\widetilde{M} \times [T, \infty)$ given in Definition 3.3. Let $p_{F_{a,t}(\widetilde{D})}$ and $q_{F_{a,t}(\widetilde{D}),T}$ be the elements as defined in line (2.7) and (3.1) respectively. We have

$$\begin{aligned}
&q^2_{F_{a,t}(\widetilde{D}),T} - q_{F_{a,t}(\widetilde{D}),T} = q^2_{F_{a,t}(\widetilde{D}),T} - q_{F_{a,t}(\widetilde{D}),T} - p^2_{F_{a,t}(\widetilde{D})} + p_{F_{a,t}(\widetilde{D})} \\
=& - q_{F_{a,t}(\widetilde{D}),T}(p_{F_{a,t}(\widetilde{D})} - q_{F_{a,t}(\widetilde{D}),T}) - (p_{F_{a,t}(\widetilde{D})} - q_{F_{a,t}(\widetilde{D}),T})p_{F_{a,t}(\widetilde{D})} \\
&+ (p_{F_{a,t}(\widetilde{D})} - q_{F_{a,t}(\widetilde{D}),T}) \\
=& - q_{F_{a,t}(\widetilde{D}),T}\big(p_{F_{a,t}(\widetilde{D})} - \left(\begin{smallmatrix}1&0\\0&0\end{smallmatrix}\right)\big)(1-\Psi_T) \\
&- \big(p_{F_{a,t}(\widetilde{D})} - \left(\begin{smallmatrix}1&0\\0&0\end{smallmatrix}\right)\big)(1-\Psi_T)\cdot p_{F_{a,t}(\widetilde{D})} + \big(p_{F_{a,t}(\widetilde{D})} - \left(\begin{smallmatrix}1&0\\0&0\end{smallmatrix}\right)\big)(1-\Psi_T)
\end{aligned}$$

The propagations of $q_{F_{a,t}(\widetilde{D}),T}$ and $p_{F_{a,t}(\widetilde{D})}$ are no more than $12at^2$. Therefore, if we set $T = 24at^2$, then the operator

$$q^2_{F_{a,t}(\widetilde{D}),24at^2} - q_{F_{a,t}(\widetilde{D}),24at^2}$$

on $\widetilde{M}_\infty$ coincides with the operator

$$q^2_{F_{a,t}(\widetilde{D}_c),24at^2} - q_{F_{a,t}(\widetilde{D}_c),24at^2}$$

on $\partial\widetilde{M} \times \mathbb{R}$ under the canonical identification between the subspace $\partial\widetilde{M} \times \mathbb{R}_{\geqslant 0}$ of $\partial\widetilde{M} \times \mathbb{R}$ and the cylindrical part $\partial\widetilde{M} \times \mathbb{R}_{\geqslant 0}$ of $\widetilde{M}_\infty$.

Note that $F_t - G_t$ is a Schwartz function and its Fourier transform is

$$\widehat{F}_t - \widehat{G}_t(\xi) = \widehat{G}_t(\xi)\cdot(1 - \chi_{at^2}(\xi)) = \frac{1}{t}\widehat{G}(t^{-1}\xi)\big(1 - \chi(\frac{\xi}{at^2})\big). \tag{4.7}$$

Since $\widehat{G}$ has Gaussian decay (away from the origin) with rate $1/4$, the function $\widehat{F}_t - \widehat{G}_t$ has Gaussian decay on the whole real line with rate $1/4$. Therefore, by Lemma 2.9, for any $\mu > 1$, there exists $C_1 > 0$ such that

$$\|F_{a,t}(\widetilde{D}_c)_\gamma - G_t(\widetilde{D}_c)_\gamma\|_{\mathrm{op}} \leqslant C_1 e^{-\frac{\tau_\partial^2 \ell(\gamma)^2}{4\mu^2 t^2}}$$

if $\ell(\gamma)$ is sufficiently large. Moreover, since the function $1 - \chi(\frac{\xi}{at^2})$ is supported on $\{\xi \in \mathbb{R} : |\xi| \geqslant at^2\}$, there exists $C_2 > 0$ such that

$$\begin{aligned}
\|F_{a,t}(\widetilde{D}_c)_\gamma - G_t(\widetilde{D}_c)_\gamma\|_{\mathrm{op}} &\leqslant \|F_{a,t}(\widetilde{D}_c) - G_t(\widetilde{D}_c)\|_{\mathrm{op}} \\
&\leqslant \frac{1}{2\pi}\int_{|\xi|\geqslant at^2} |\frac{1}{t}\widehat{G}(t^{-1}\xi)|d\xi \leqslant C_2 e^{-\frac{a^2t^2}{4}}
\end{aligned} \tag{4.8}$$

Similar to the proof of Lemma 2.11, since we have $a > 4(K_\Gamma + K)/\tau_\partial$, there exist positive constants C and δ such that for any $t > 0$,

$$\|F_{a,t}(\widetilde{D}_c) - G_t(\widetilde{D}_c)\|_{1,K} \leqslant Ce^{-\delta t^2}. \tag{4.9}$$

Therefore the norm $\|F_{a,t}(\widetilde{D}_c)\|_{1,K}$ is uniformly bounded for all $t \geqslant 1$. By the definition of $p_{F_{a,t}(\widetilde{D}_c)}$ (cf. line (2.7)), we see that the norm $\|p_{F_{a,t}(\widetilde{D}_c)}\|_{1,K}$ is also uniformly bounded for all $t \geqslant 1$, which in turn implies that the norm $\|q_{F_{a,t}(\widetilde{D}_c),T}\|_{1,K}$ is also uniformly bounded for all $t \geqslant 1$ and all $T \geqslant 0$, since the operator given by multiplication by Ψ_T has zero propagation and $\|\Psi_T\|_{op} \leqslant 1$ holds for all $T \geqslant 0$.

Furthermore, we have

$$\begin{aligned}&\|F_{a,t}(\widetilde{D}_c)^2-1\|_{1,K}\\ &\leqslant \|F_{a,t}(\widetilde{D}_c)-G_t(\widetilde{D}_c)\|_{1,K}\cdot\|F_{a,t}(\widetilde{D}_c)+G_t(\widetilde{D}_c)\|_{1,K}+\|G_t(\widetilde{D}_c)^2-1\|_{1,K},\end{aligned}$$

where the right hand side goes to zero as t goes to infinity by line (4.9) and (4.5). It follows that

$$\lim_{t\to\infty}\|p_{F_{a,t}(\widetilde{D}_c)}-\left(\begin{smallmatrix}1&0\\0&0\end{smallmatrix}\right)\|_{1,K}=0. \tag{4.10}$$

On the other hand, we have

$$\begin{aligned}&q^2_{F_{a,t}(\widetilde{D}_c),24at^2}-q_{F_{a,t}(\widetilde{D}_c),24at^2}\\ &=-\,q_{F_{a,t}(\widetilde{D}_c),24at^2}\big(p_{F_{a,t}(\widetilde{D}_c)}-\left(\begin{smallmatrix}1&0\\0&0\end{smallmatrix}\right)\big)(1-\Psi_{24at^2})\\ &\quad-\big(p_{F_{a,t}(\widetilde{D}_c)}-\left(\begin{smallmatrix}1&0\\0&0\end{smallmatrix}\right)\big)(1-\Psi_{24at^2})\cdot p_{F_{a,t}(\widetilde{D}_c)}\\ &\quad+\big(p_{F_{a,t}(\widetilde{D}_c)}-\left(\begin{smallmatrix}1&0\\0&0\end{smallmatrix}\right)\big)(1-\Psi_{24at^2}).\end{aligned}$$

Applying line (4.10), we have

$$\lim_{t\to\infty}\|q^2_{F_{a,t}(\widetilde{D}_c),24at^2}-q_{F_{a,t}(\widetilde{D}_c),24at^2}\|_{1,K}=0.$$

Hence the proof is finished by setting $T=24at^2$ and $F=F_{a,t}$ for some sufficiently large $t\gg 0$. □

4.2. Odd dimensional case. Let us now turn to the proof of Theorem 4.1 for the case where $\dim M$ is odd. Again, assume the same notation as in the Geometric setup 3.1. In particular, $\widetilde{D}$ is the Dirac operator on $\widetilde{M}_\infty$ and $\widetilde{D}_c$ is the Dirac operator on the cylinder $\partial\widetilde{M}\times\mathbb{R}$.

PROOF OF THEOREM 4.1 FOR THE ODD DIMENSIONAL CASE. Similar to the construction of the higher index $\mathrm{Ind}^\Gamma(D)\in K_1(C_r^*(\Gamma))$ as in Definition 3.7. The proof of Theorem 4.1 for the odd dimensional case can be reduced to the following claim, which is an l^1-analogue of Lemma 3.6.

CLAIM 4.3. If $|\widetilde{D}_c|>\frac{2(K_\Gamma+K)}{r_\partial}$, then there exist a positive number T, a positive integer n, and a smooth function P satisfying condition (P1)-(P2) in Section 3.2 such that $\|1+f_n(\pm P(\widetilde{D}))\Psi_T\|_{1,K}<\infty$. Furthermore, we have both

$$\|(1+f_n(P(\widetilde{D}))\Psi_T)(1+f_n(-P(\widetilde{D}))\Psi_T)-1\|_{1,K}<1,$$

and

$$\|(1+f_n(-P(\widetilde{D}))\Psi_T)(1+f_n(P(\widetilde{D}))\Psi_T)-1\|_{1,K}<1,$$

where Ψ_T is the cut-off function on $\widetilde{M}_\infty$ which vanishes on $\partial\widetilde{M}\times[T,\infty)$ given in Definition 3.3 and f_n is the polynomial given in line (3.6).

Let us assume we have verified the claim for the moment. We conclude that the element $1+f_n(P(\widetilde{D}))\Psi_T$ admits both left and right inverses in $(\mathcal{B}(\widetilde{M}_{T+nN_P})^\Gamma_K)^+$, hence is invertible in $(\mathcal{B}(\widetilde{M}_{T+nN_P})^\Gamma_K)^+$, where $(\mathcal{B}(\widetilde{M}_{T+nN_P})^\Gamma_K)^+$ the unitization of $\mathcal{B}(\widetilde{M}_{T+nN_P})^\Gamma_K$ (cf. Definition 2.5). It follows that the higher index $\mathrm{Ind}^\Gamma(\widetilde{D})$ represented by this invertible element lies in $K_1(\mathcal{B}(\widetilde{M}_{T+nN_P})^\Gamma_K)\cong K_1(l^1_K(\Gamma))$, which proves Theorem 4.1 for the odd dimensional case.

We remark that an l^1-analogue of the inequality from line (3.8) does *not* imply that $(1+f_n(P(\widetilde{D}))\Psi_T)$ is invertible, since the weighted l^1-norm of $e^{2\pi i P(\widetilde{D})}$ may be much greater than 1. This is the reason for trying to prove the two inequalities in Claim 4.3 instead. □

Now let us prove Claim 4.3, hence complete the proof of Theorem 4.1 for the odd dimensional case.

PROOF OF CLAIM 4.3. Let G_t and $F_{a,t}$ be the functions given in line (2.14) and (4.6) respectively. Set

$$P_{a,t} = \frac{F_{a,t}+1}{2} \text{ and } Q_t = \frac{G_t+1}{2},$$

where a is a fixed positive number satisfying $a > \frac{4(K_\Gamma+K)}{\tau\partial}$. By line (4.9), we see that there exists $A > 0$ such that $\|P_{a,t}(\widetilde{D}_c)\|_{1,K} \leqslant A$ and $\|Q_t(\widetilde{D}_c)\|_{1,K} \leqslant A$ for all $t \geqslant 1$. It follows from line (3.7) that

$$\|e^{2\pi i P_{a,t}(\widetilde{D}_c)}\|_{1,K} \leqslant e^{2\pi A}, \; \|e^{2\pi i Q_t(\widetilde{D}_c)}\|_{1,K} \leqslant e^{2\pi A},$$

$$\|1 + f_n(P_{a,t}(\widetilde{D}_c))\|_{1,K} \leqslant e^{2\pi A} + e^{2\pi} A,$$

and

$$\|e^{2\pi i P_{a,t}(\widetilde{D}_c)} - (1 + f_n(P_{a,t}(\widetilde{D}_c)))\|_{1,K} \leqslant \frac{e^{2\pi A} + e^{2\pi} A}{(n+1)!}. \tag{4.11}$$

By the inequality from line (4.9), we see that

$$\|P_{a,t}(\widetilde{D}_c) - Q_t(\widetilde{D}_c)\|_{1,K} \to 0,$$

as t goes to infinity. Therefore we have

$$\begin{aligned} &\|e^{2\pi i P_{a,t}(\widetilde{D}_c)} - e^{2\pi i Q_t(\widetilde{D}_c)}\|_{1,K} \\ \leqslant&\|e^{2\pi i Q_t(\widetilde{D}_c)}\|_{1,K}\|e^{2\pi i(P_{a,t}(\widetilde{D}_c)-Q_t(\widetilde{D}_c))} - 1\|_{1,K} \\ \leqslant&e^{2\pi A}(e^{2\pi\|P_{a,t}(\widetilde{D}_c)-Q_t(\widetilde{D}_c)\|_{1,K}} - 1), \end{aligned} \tag{4.12}$$

which goes to zero as t goes to infinity.

Let H be the Heaviside step function $H(x) = \frac{\operatorname{sgn}(x)+1}{2}$. Since $\widetilde{D}_c$ is invertible, we have that $e^{2\pi i H(\widetilde{D}_c)} = 1$. By Lemma 2.15, we have

$$\begin{aligned} \|e^{2\pi i Q_t(\widetilde{D}_c)} - 1\|_{1,K} =&\|e^{2\pi i(Q_t(\widetilde{D}_c)-H(\widetilde{D}_c))} - 1\|_{1,K} \\ \leqslant&e^{2\pi\|Q_t(\widetilde{D}_c)-H(\widetilde{D}_c)\|_{1,K}} - 1. \end{aligned} \tag{4.13}$$

which goes to zero as t goes to infinity.

By combining the inequalities from line (4.11), (4.12) and (4.13), we see there exist $t_c > 0$ and $n_c > 0$ such that if $n \geqslant n_c$ and $t \geqslant t_c$, then

$$(1 + e^{2\pi A} + e^{2\pi} A)\|f_n(P_{a,t}(\widetilde{D}_c))\|_{1,K} < 1/6. \tag{4.14}$$

The same inequality also holds if we replace $P_{a,t}(\widetilde{D}_c)$ by $-P_{a,t}(\widetilde{D}_c)$. Furthermore, since the operator given by the multiplication of Ψ_T has zero propagation and $\|\Psi_T\|_{op} \leqslant 1$ for all $T \geqslant 0$, it follows that

$$(1 + e^{2\pi A} + e^{2\pi} A)\|f_n(P_{a,t}(\widetilde{D}_c))\Psi_T\|_{1,K} < 1/6. \tag{4.15}$$

for all $T \geqslant 0$. We conclude that

$$\|(1 + f_n(P_{a,t}(\widetilde{D}_c)))(1 + f_n(-P_{a,t}(\widetilde{D}_c))) - 1\|_{1,K} < 1/3,$$

and

$$\|(1+f_n(P_{a,t}(\widetilde{D}_c))\Psi_T)(1+f_n(-P_{a,t}(\widetilde{D}_c))\Psi_T)-1\|_{1,K}<1/3,$$

for all $n \geqslant n_c$ and $t \geqslant t_c$, which implies that

$$\begin{aligned}\|\big(1+f_n(P_{a,t}(\widetilde{D}_c))\big)\big(1+f_n(-P_{a,t}(\widetilde{D}_c))\big)\\ -\big(1+f_n(P_{a,t}(\widetilde{D}_c))\Psi_T\big)\big(1+f_n(-P_{a,t}(\widetilde{D}_c))\Psi_T\big)\|_{1,K}<2/3.\end{aligned} \tag{4.16}$$

for all $n \geqslant n_c$ and $t \geqslant t_c$.

Now we turn to the Dirac operator $\widetilde{D}$ on $\widetilde{M}_\infty$. Note that $\widetilde{D}$ is not invertible in general, and the norm $\|P_{a,t}(\widetilde{D})\|_{1,K}$ may not uniformly bounded for $t \geqslant 1$ in general. On the other hand, by construction, $P_{a,t}(\widetilde{D})$ has propagation $\leqslant 2at^2$, and $\|P_{a,t}(\widetilde{D})\|_{\mathrm{op}} \leqslant 1$ for all $t>0$. Therefore by the definition of $\|\cdot\|_{1,K}$ in Definition 2.5, there exist $C>0$ and $\lambda>0$ such that

$$\|P_{a,t}(\widetilde{D})\|_{1,K}=\sum_{\gamma\in\Gamma}e^{K\ell(\gamma)}\|P_{a,t}(\widetilde{D})_\gamma\|_{op}\leqslant Ce^{\lambda(K+K_\Gamma)at^2},$$

cf. line (4.4). Let us denote $A_t=Ce^{\lambda(K+K_\Gamma)at^2}$. We have

$$\|e^{2\pi iP_{a,t}(\widetilde{D})}\|_{1,K}\leqslant e^{2\pi A_t},\ \|1+f_n(P_{a,t}(\widetilde{D}))\|_{1,K}\leqslant e^{2\pi A_t}+e^{2\pi}A_t,$$

and

$$\|e^{2\pi iP_{a,t}(\widetilde{D})}-\big(1+f_n(P_{a,t}(\widetilde{D}))\big)\|_{1,K}\leqslant\frac{e^{2\pi A_t}+e^{2\pi}A_t}{(n+1)!}. \tag{4.17}$$

The same inequalities hold if we replace $P_{a,t}(\widetilde{D})$ by $-P_{a,t}(\widetilde{D})$. Note that, for the positive numbers n_c and t_c from above, there exists a positive integer n_0 such that $n_0 \geqslant n_c$ and

$$\frac{(e^{2\pi A_{t_c}}+e^{2\pi}A_{t_c})^2}{(n_0+1)!}<1/3. \tag{4.18}$$

In particular, it follows that

$$\|(1+f_{n_0}(P_{a,t_c}(\widetilde{D})))(1+f_{n_0}(-P_{a,t_c}(\widetilde{D})))-1\|_{1,K}<1/3, \tag{4.19}$$

since $e^{2\pi iP_{a,t_c}(\widetilde{D})}e^{-2\pi iP_{a,t_c}(\widetilde{D})}=1$.

Note that

$$\begin{aligned}&(1+f_{n_0}(P_{a,t_c}(\widetilde{D})))(1+f_{n_0}(-P_{a,t_c}(\widetilde{D})))\\ &-(1+f_{n_0}(P_{a,t_c}(\widetilde{D}))\Psi_T)(1+f_{n_0}(-P_{a,t_c}(\widetilde{D}))\Psi_T)\\ =&f_{n_0}(P_{a,t_c}(\widetilde{D}))(1-\Psi_T)+f_{n_0}(-P_{a,t_c}(\widetilde{D}))(1-\Psi_T)\\ &+f_{n_0}(P_{a,t_c}(\widetilde{D}))(1-\Psi_T)f_{n_0}(-P_{a,t_c}(\widetilde{D}))\Psi_T\\ &+f_{n_0}(P_{a,t_c}(\widetilde{D}))f_{n_0}(-P_{a,t_c}(\widetilde{D}))(1-\Psi_T).\end{aligned}$$

Since the propagation of $f_{n_0}(\pm P_{a,t_c}(\widetilde{D}))$ is no more than $2n_0at_c^2$, if we choose $T=4n_0at_c^2$, then the operator

$$\begin{aligned}&(1+f_{n_0}(P_{a,t_c}(\widetilde{D})))(1+f_{n_0}(-P_{a,t_c}(\widetilde{D})))\\ &-(1+f_{n_0}(P_{a,t_c}(\widetilde{D}))\Psi_{4n_0at_c^2})(1+f_{n_0}(-P_{a,t_c}(\widetilde{D}))\Psi_{4n_0at_c^2})\end{aligned}$$

on $\widetilde{M}_\infty$ is identified with

$$(1+f_{n_0}(P_{a,t_c}(\widetilde{D}_c)))(1+f_{n_0}(-P_{a,t_c}(\widetilde{D}_c)))$$
$$-(1+f_{n_0}(P_{a,t_c}(\widetilde{D}_c))\Psi_{4n_0at_c^2})(1+f_{n_0}(-P_{a,t_c}(\widetilde{D}_c))\Psi_{4n_0at_c^2})$$

on $\partial\widetilde{M}\times\mathbb{R}$ under the canonical identification between the subspace $\partial\widetilde{M}\times\mathbb{R}_{\geqslant 0}$ of $\partial\widetilde{M}\times\mathbb{R}$ and the cylindrical part $\partial\widetilde{M}\times\mathbb{R}_{\geqslant 0}$ of $\widetilde{M}_\infty$. Thus it follows from line (4.16) that

$$\begin{aligned}&\|(1+f_{n_0}(P_{a,t_c}(\widetilde{D})))(1+f_{n_0}(-P_{a,t_c}(\widetilde{D})))\\ &\quad-(1+f_{n_0}(P_{a,t_c}(\widetilde{D}))\Psi_{4n_0at_c^2})(1+f_{n_0}(-P_{a,t_c}(\widetilde{D}))\Psi_{4n_0at_c^2})\|_{1,K}<2/3\end{aligned}\tag{4.20}$$

Together with line (4.19), we have proved

$$\|(1+f_n(P(\widetilde{D}))\Psi_T)(1+f_n(-P(\widetilde{D}))\Psi_T)-1\|_{1,K}<1,$$

by setting $P=P_{a,t_c}, n=n_0$ and $T=4n_0at_c^2$. The other inequality

$$\|(1+f_n(-P(\widetilde{D}))\Psi_T)(1+f_n(P(\widetilde{D}))\Psi_T)-1\|_{1,K}<1,$$

can be proved in the exact same way in this case. This finishes the proof. □

5. An l^1-Atiyah-Patodi-Singer higher index formula

In this section, we prove an l^1-version of the Atiyah-Patodi-Singer higher index theorem.

Let us first recall the following definition of cyclic cocycles (cf. [**5**, part II][**6**, chapter 3, §1])

DEFINITION 5.1. Let Γ be a discrete group.

(1) We say a map

$$\varphi\colon\Gamma^{n+1}=\underbrace{\Gamma\times\cdots\times\Gamma}_{(n+1)\text{ copies}}\to\mathbb{C}$$

is a cyclic n-cocyle on Γ if φ is cyclic, that is,

$$\varphi(\gamma_n,\gamma_0,\cdots,\gamma_{n-1})=(-1)^n\varphi(\gamma_0,\gamma_1,\cdots,\gamma_n),\ \forall\gamma_i\in\Gamma,$$

and φ is closed, that is,

$$\begin{aligned}(b\varphi)(\gamma_0,\cdots,\gamma_{n+1}):=&\sum_{i=0}^{n}(-1)^i\varphi(\cdots,\gamma_i\gamma_{i+1},\cdots)\\ &+(-1)^{n+1}\varphi(\gamma_{n+1}\gamma_0,\gamma_1,\cdots,\gamma_n)=0,\end{aligned}$$

for $\forall\gamma_i\in\Gamma$.

(2) A cyclic n-cocycle φ of Γ is said to be delocalized if

$$\varphi(\gamma_0,\gamma_1,\cdots,\gamma_n)=0$$

for all γ_i such that $\gamma_0\gamma_1\cdots\gamma_n=I$ where I is the identity element of Γ.

Let us fix a length function ℓ on Γ. We have the following notion of exponential growth for maps on Γ^{n+1}.

DEFINITION 5.2. A map $\varphi\colon \Gamma^{n+1} \to \mathbb{C}$ is said to have at most exponential growth with respect to the length function ℓ on Γ if there are positive numbers C and K_φ such that

$$|\varphi(\gamma_0, \gamma_1, \cdots, \gamma_n)| \leqslant Ce^{K_\varphi(\ell(\gamma_0)+\ell(\gamma_1)+\cdots+\ell(\gamma_n))}, \ \forall \gamma_i \in \Gamma.$$

If φ has exponential growth rate K_φ, then it extends to a bounded linear functional on $l^1_{K_\varphi}(\Gamma)^{\otimes(n+1)}$, where the maximal tensor product is used. Let $\mathcal{S}$ be the algebra of trace class operators on a Hilbert space. The map φ induces the following bounded linear functional

$$\varphi\#\mathrm{tr}\colon \mathcal{S} \otimes l^1_{K_\varphi}(\Gamma)^{\otimes(n+1)} \to \mathbb{C}$$

by setting

$$\varphi\#\mathrm{tr}(a) = \sum_{\gamma_0,\cdots,\gamma_n\in\Gamma} \mathrm{tr}(a_{\gamma_0,\cdots,\gamma_n})\varphi(\gamma_0, \cdots, \gamma_n),$$

for all

$$a = \sum_{\gamma_0,\cdots,\gamma_n\in\Gamma} a_{\gamma_0,\cdots,\gamma_n}\gamma_0 \otimes \cdots \otimes \gamma_n \in \mathcal{S} \otimes l^1_{K_\varphi}(\Gamma)^{\otimes(n+1)}.$$

Now let us recall the definition of the delocalized higher eta invariant (cf. [**4**, section 3]). Let X be a complete Riemannian spin manifold. Let Γ be a finitely represented group and $\widetilde{X}$ a normal Γ-cover of X.

DEFINITION 5.3. Suppose the Dirac operator $\widetilde{D}_X$ on $\widetilde{X}$ is invertible.

(1) If $\dim X$ is odd and φ is a delocalized cyclic $(2m)$-cocycle of Γ, we define the delocalized higher eta invariant of $\widetilde{D}_X$ at φ to be

$$\eta_\varphi(\widetilde{D}_X) := \frac{m!}{\pi i}\int_0^\infty \varphi\#\mathrm{tr}(\dot{u}_s u_s^{-1} \otimes (((u_s - 1) \otimes (u_s^{-1} - 1)))^{\otimes m})ds, \tag{5.1}$$

where $\{u_s\}_{s\in[0,\infty)}$ is the path of invertible elements given in line (2.22).

(2) If $\dim X$ is even and φ is a delocalized cyclic $(2m+1)$-cocycle, we define the delocalized higher eta invariant of $\widetilde{D}_X$ at φ to be

$$\eta_\varphi(\widetilde{D}_X) := \frac{(2m)!}{m!\pi i}\int_0^\infty \varphi\#\mathrm{tr}([\dot{p}_s, p_s] \otimes (p_s - \left(\begin{smallmatrix}1&0\\0&0\end{smallmatrix}\right))^{\otimes(2m+1)})ds, \tag{5.2}$$

where p_s is given in line (2.24).

In [**4**, Section 3.3] the authors showed the integral of $\eta_\varphi(\widetilde{D}_X)$ converges absolutely, under the conditions that the delocalized cyclic cocycle φ has at most exponential growth with growth rate K_φ and

$$|\widetilde{D}_X| > \frac{4(K_\Gamma + K_\varphi)}{\tau}.$$

In general, it is an open question whether the integral converges.

Recall that $l^1_{K_\varphi}(\Gamma) \otimes \mathcal{S}$ is a smooth dense subalgebra of $l^1_{K_\varphi}(\Gamma) \otimes \mathcal{K}$, that is, $l^1_{K_\varphi}(\Gamma)\otimes\mathcal{S}$ is dense and closed under holomorphic functional calculus in $l^1_{K_\varphi}(\Gamma)\otimes\mathcal{K}$. In particular, we have

$$K_*(l^1_{K_\varphi}(\Gamma) \otimes \mathcal{K}) \cong K_*(l^1_{K_\varphi}(\Gamma) \otimes \mathcal{S}).$$

Denote by $(l^1_{K_\varphi}(\Gamma) \otimes \mathcal{S})^+$ the unitization of $l^1_{K_\varphi}(\Gamma) \otimes \mathcal{S}$. For a delocalized n-cocycle φ with exponential growth, we extend $\varphi\#\mathrm{tr}$ from $(l^1_{K_\varphi}(\Gamma) \otimes \mathcal{S})^{\otimes(n+1)}$ to

$((l^1_{K_\varphi}(\Gamma)\otimes\mathcal{S})^+)^{\otimes(n+1)}$ by setting its value to be zero when encountering the extra identity.

It follows that for each cyclic $(2m)$-cocycle φ with exponential growth rate K_φ, the Connes-Chern character map

$$\mathrm{ch}_\varphi(p) := \frac{(2m)!}{m!}\varphi\#\mathrm{tr}(p^{\otimes 2m+1}) \tag{5.3}$$

for idempotents $p\in M_j(\mathbb{C})\otimes(l^1_{K_\varphi}(\Gamma)\otimes\mathcal{S})^+$, defines a homomorphism

$$\mathrm{ch}_\varphi\colon K_0(l^1_{K_\varphi}(\Gamma))\to\mathbb{C}.$$

Here $(l^1_{K_\varphi}(\Gamma)\otimes\mathcal{S})^+$ is the unitization of $l^1_{K_\varphi}(\Gamma)\otimes\mathcal{S}$. If φ is a cyclic $(2m+1)$-cocycle φ with exponential growth rate K_φ, then the Connes-Chern character map

$$\mathrm{ch}_\varphi\colon K_1(l^1_{K_\varphi}(\Gamma)\otimes\mathcal{K})\to\mathbb{C}$$

is given by

$$\mathrm{ch}_\varphi(u) := \frac{1}{m!}\varphi\#\mathrm{tr}((u^{-1}\otimes u)^{\otimes m}), \tag{5.4}$$

for invertible elements u in $M_j(\mathbb{C})\otimes(l^1_{K_\varphi}(\Gamma)\otimes\mathcal{S})^+$.

We have the following l^1-version of the Atiyah-Patodi-Singer higher index theorem.

THEOREM 5.4. *Assume the same notation from the geometric setup* 3.1. *Let φ be a delocalized cyclic cocycle that has exponential growth rate K_φ. If the scalar curvature k on ∂M satisfies that*

$$\inf_{x\in\partial M} k(x) > \frac{16(K_\Gamma+K_\varphi)^2}{\tau_\partial^2},$$

then we have

$$\mathrm{ch}_\varphi(\mathrm{Ind}^\Gamma_{1,K_\varphi}(D)) = -\frac{1}{2}\eta_\varphi(\widetilde{D}_\partial).$$

PROOF. We prove the case where φ is an even-degree cocycle and $\dim\partial M$ is odd. The case where φ is an odd-degree cocycle and $\dim\partial M$ is even can be proved in the same way.

As shown in [**4**, Section 6], the formula in line (5.1) extends to a homomorphism

$$\bar\eta_\varphi\colon K_1(\mathcal{B}_{L,0}(\partial\widetilde{M})^\Gamma_{K_\varphi})\to\mathbb{C}$$

in the sense that (cf. [**4**, Theorem 6.1(b)])

$$\eta_\varphi(\widetilde{D}_\partial) = \bar\eta_\varphi(\rho_{1,K_\varphi}(g_\partial)), \tag{5.5}$$

where $\rho_{1,K_\varphi}(g_\partial)$ is the l^1-higher rho invariant of $(\partial M, g_\partial)$ given in Definition 2.20. More precisely, the same proof of [**4**, Theorem 6.1(b)] can be used to prove the equality (5.5) above.

Consider the following short exact sequence

$$0\longrightarrow\mathcal{B}_{L,0}(\partial\widetilde{M})^\Gamma_{K_\varphi}\longrightarrow\mathcal{B}_L(\partial\widetilde{M})^\Gamma_{K_\varphi}\longrightarrow\mathcal{B}(\partial\widetilde{M})^\Gamma_{K_\varphi}\longrightarrow 0.$$

We denote the boundary map in the associated K-theory long exact sequence by

$$\partial\colon K_*(\mathcal{B}(\partial\widetilde{M})^\Gamma_{K_\varphi})\cong K_*(l^1_{K_\varphi}(\Gamma))\longrightarrow K_{*+1}(\mathcal{B}_{L,0}(\partial\widetilde{M})^\Gamma_{K_\varphi}).$$

By [**4**, Proposition 7.2], we have that

$$\mathrm{ch}_\varphi(\mathrm{Ind}^\Gamma_{1,K_\varphi}(D)) = -\frac{1}{2}\bar{\eta}_\varphi\big(\partial(\mathrm{Ind}^\Gamma_{1,K_\varphi}(D))\big). \tag{5.6}$$

Let Y be a subspace of $\partial\widetilde{M} \times \mathbb{R}$. We define $\mathcal{B}_{L,0}(Y, \partial\widetilde{M} \times \mathbb{R})^\Gamma_{K_\varphi}$ to be the completion of the collection of functions $f\colon [0,\infty) \to \mathbb{C}(\partial\widetilde{M} \times \mathbb{R})^\Gamma_{K_\varphi}$ satisfying

(1) $\|f(s)\|_{1,K_\varphi}$ is uniformly bounded and uniformly continuous in s,
(2) the propagation of $f(s)$ goes to zero as s goes to infinity,
(3) $f(0) = 0$,
(4) and there exists $\lambda > 0$ such that for any $s \geqslant 0$, $f(s)\cdot\chi_{Y,\lambda} = 0$, where $\chi_{Y,\lambda}$ is the characteristic function on the λ-neighborhood of Y.

with respect to the norm $\|f\|_{1,K_\varphi} := \sup_{s\geqslant 0}\|f(s)\|_{1,K_\varphi}$. Note that for each subspace Y of $\partial\widetilde{M} \times \mathbb{R}$, the Banach algebra $\mathcal{B}_{L,0}(Y, \partial\widetilde{M} \times \mathbb{R})^\Gamma_{K_\varphi}$ is a closed two-sided ideal of $\mathcal{B}_{L,0}(\partial\widetilde{M} \times \mathbb{R})^\Gamma_{K_\varphi}$.

Observe that

$$\mathcal{B}_{L,0}(\partial\widetilde{M} \times \mathbb{R})^\Gamma_{K_\varphi} = \mathcal{B}_{L,0}(\partial\widetilde{M} \times \mathbb{R}_+, \partial\widetilde{M} \times \mathbb{R})^\Gamma_{K_\varphi} + \mathcal{B}_{L,0}(\partial\widetilde{M} \times \mathbb{R}_+, \partial\widetilde{M} \times \mathbb{R})^\Gamma_{K_\varphi}$$

and

$$\mathcal{B}_{L,0}(\partial\widetilde{M}, \partial\widetilde{M} \times \mathbb{R})^\Gamma_{K_\varphi} = \mathcal{B}_{L,0}(\partial\widetilde{M} \times \mathbb{R}_+, \partial\widetilde{M} \times \mathbb{R})^\Gamma_{K_\varphi} \cap \mathcal{B}_{L,0}(\partial\widetilde{M} \times \mathbb{R}_+, \partial\widetilde{M} \times \mathbb{R})^\Gamma_{K_\varphi}.$$

Therefore, we have the following Mayer-Vietoris sequence in K-theory:

$$\begin{array}{ccccc}
K_0(\mathcal{B}_{L,0}(\partial\widetilde{M})^\Gamma_{K_\varphi}) & \longrightarrow & \begin{array}{c} K_0(\mathcal{B}_{L,0}(\partial\widetilde{M} \times \mathbb{R}_+, \partial\widetilde{M} \times \mathbb{R})^\Gamma_{K_\varphi}) \\ \oplus \\ K_0(\mathcal{B}_{L,0}(\partial\widetilde{M} \times \mathbb{R}_-, \partial\widetilde{M} \times \mathbb{R})^\Gamma_{K_\varphi}) \end{array} & \to & K_0(\mathcal{B}_{L,0}(\partial\widetilde{M} \times \mathbb{R})^\Gamma_{K_\varphi}) \\
\Big\uparrow & & & & \Big\downarrow{\scriptstyle \partial_{MV}} \\
K_1(\mathcal{B}_{L,0}(\partial\widetilde{M} \times \mathbb{R})^\Gamma_{K_\varphi}) & \leftarrow & \begin{array}{c} K_1(\mathcal{B}_{L,0}(\partial\widetilde{M} \times \mathbb{R}_+, \partial\widetilde{M} \times \mathbb{R})^\Gamma_{K_\varphi}) \\ \oplus \\ K_1(\mathcal{B}_{L,0}(\partial\widetilde{M} \times \mathbb{R}_-, \partial\widetilde{M} \times \mathbb{R})^\Gamma_{K_\varphi}) \end{array} & \longleftarrow & K_1(\mathcal{B}_{L,0}(\partial\widetilde{M})^\Gamma_{K_\varphi})
\end{array} \tag{5.7}$$

where we have used the identification

$$K_*(\mathcal{B}_{L,0}(\partial\widetilde{M}, \partial\widetilde{M} \times \mathbb{R})^\Gamma_{K_\varphi}) \cong K_*(\mathcal{B}_{L,0}(\partial\widetilde{M})^\Gamma_{K_\varphi}).$$

By the standard constructions of the boundary maps ∂ and ∂_{MV}, we have

$$\partial\big(\mathrm{Ind}^\Gamma_{1,K_\varphi}(D)\big) = \partial_{MV}\big(\rho_{1,K_\varphi}(g_\partial + dt^2)\big) \in K_1(\mathcal{B}_{L,0}(\partial\widetilde{M})^\Gamma_{K_\varphi}), \tag{5.8}$$

where $\rho_{1,K_\varphi}(g_\partial + dt^2)$ is the l^1-higher rho invariant on $\partial\widetilde{M} \times \mathbb{R}$ (cf. Definition 2.20) and ∂_{MV} is the boundary map

$$\partial_{MV}\colon K_0(\mathcal{B}_{L,0}(\partial\widetilde{M} \times \mathbb{R})^\Gamma_{K_\varphi}) \longrightarrow K_1(\mathcal{B}_{L,0}(\partial\widetilde{M})^\Gamma_{K_\varphi})$$

from the Mayer-Vietoris sequence above. By the standard construction of the boundary map ∂_{MV}, we also have the following commuting diagram

$$\begin{array}{ccc}
K_1(\mathcal{B}_{L,0}(\partial\widetilde{M} \times \mathbb{R})^\Gamma_{K_\varphi}) & \xrightarrow{\partial_{MV}} & K_0(\mathcal{B}_{L,0}(\partial\widetilde{M})^\Gamma_{K_\varphi}) \\
{\scriptstyle \otimes\mathrm{Ind}_L(D_{\mathbb{R}})}\Big\uparrow & \diagup\!\!\diagup & \\
K_0(\mathcal{B}_{L,0}(\partial\widetilde{M})^\Gamma_{K_\varphi}) & &
\end{array}$$

where $\mathrm{Ind}_L(D_{\mathbb{R}})$ is the local higher index of the Dirac operator $D_{\mathbb{R}} = \frac{1}{i}\frac{d}{dx}$ on $\mathbb{R}$ (cf. [**26**]). On the other hand, the product formula from Theorem 2.21 implies that

$$\partial_{MV}\big(\rho_{1,K_\varphi}(g_\partial + dt^2)\big) = \rho_{1,K_\varphi}(g_\partial) \in K_1(\mathcal{B}_{L,0}(\partial\widetilde{M})^\Gamma_{K_\varphi}). \tag{5.9}$$

Hence we have

$$\begin{aligned}\mathrm{ch}_\varphi(\mathrm{Ind}^\Gamma_{1,K_\varphi}(D)) &= -\frac{1}{2}\bar{\eta}_\varphi(\partial(\mathrm{Ind}^\Gamma_{1,K_\varphi}(\widetilde{D}))) = -\frac{1}{2}\bar{\eta}_\varphi(\partial_{MV}\big(\rho_{1,K_\varphi}(g_\partial + dt^2)\big))\\ &= -\frac{1}{2}\bar{\eta}_\varphi(\rho_{1,K_\varphi}(g_\partial)) = -\frac{1}{2}\eta_\varphi(\widetilde{D}_\partial)\end{aligned}$$

This finishes the proof. □

5.1. An alternative proof of Theorem 5.4. As pointed out in the introduction, Theorem 5.4 is an strengthening of the higher Atiyah-Patodi-Singer index formula in [**3**, Theorem 3.36]. Since [**3**, Theorem 3.36] was stated and proved using the language of b-calculus. In this subsection, for the convenience of the reader, we give an alternative proof of Theorem 5.4 using the language of b-calculus.

We retain the same notation from the geometric setup 3.1. For brevity, let us assume the dimension of ∂M is odd. The other case is completely similar.

Let us briefly review Lott's noncommutative differential higher eta invariant. We shall follow closely the notation in Lott's paper [**13**]. For any $K > 0$, let $l^1_K(\Gamma)$ be the weighted l^1-algebra of Γ from Definition 2.5. The universal graded differential algebra of $l^1_K(\Gamma)$ is defined to be

$$\Omega_*(l^1_K(\Gamma)) = \bigoplus_{j=0}^{\infty} \Omega_k(l^1_K(\Gamma))$$

where as a vector space, $\Omega_j(l^1_K(\Gamma)) = l^1_K(\Gamma) \otimes (l^1_K(\Gamma)/\mathbb{C})^{\otimes j}$. As $l^1_K(\Gamma)$ is a Banach algebra, we shall consider the Banach completion of $\Omega_*(l^1_K(\Gamma))$, which will still be denoted by $\Omega_*(l^1_K(\Gamma))$.

Let $S(\partial M)$ be the spinor bundle on ∂M. We denote the corresponding $l^1_K(\Gamma)$-vector bundle by $\mathfrak{S} = (\partial\widetilde{M} \times_\Gamma l^1_{K_\varphi}(\Gamma)) \otimes S(\partial M)$ and the space of smooth sections by $C^\infty(\partial M; \mathfrak{S})$. Now suppose ψ is a smooth function on $\partial\widetilde{M}$ with compact support such that

$$\sum_{\gamma\in\Gamma} \gamma\psi = 1.$$

Then we have a superconnection $\nabla\colon C^\infty(\partial M; \mathfrak{S}) \to C^\infty(\partial M; \mathfrak{S} \otimes_{l^1_{K_\varphi}(\Gamma)} \Omega_1(l^1_{K_\varphi}(\Gamma)))$ given by

$$\nabla(f) = \sum_{\gamma\in\Gamma} (\psi \cdot \gamma f) \otimes_{l^1_{K_\varphi}(\Gamma)} d\gamma.$$

See [**18**] for more details of the superconnection formalism.

Definition 5.5 ([**13**, Section 4.4 & 4.6]). Lott's noncommutative differential higher eta invariant $\widetilde{\eta}(\widetilde{D}_\partial)$ is defined by the formula

$$\widetilde{\eta}(\widetilde{D}_\partial) = \int_0^\infty \mathrm{STR}(\widetilde{D}_\partial\, e^{-(t\widetilde{D}_\partial + \nabla)^2})dt,$$

where STR is the corresponding supertrace, cf. [**13**, Proposition 22].

Combining Quillen's superconnection formalism with Melrose's b-calculus formalism [**14**], for each $t > 0$, one defines the b-Connes-Chern character of $\widetilde{D}$ on $\widetilde{M}_\infty$ to be

$$b\text{-Ch}_t(\widetilde{D}) = b\text{-STR}(e^{-(t\widetilde{D}+\nabla)^2}) \in \Omega_*(l^1_K(\Gamma)),$$

where b-STR is the corresponding b-supertrace in the b-calculus setting. See for example [**12**] for more details.

Recall that every cyclic n-cocycle φ of Γ can be decomposed as $\varphi = \varphi_e + \varphi_d$, where φ_d is the delocalized part of φ, i.e.,

$$\varphi_d(\gamma_0, \gamma_1, \cdots, \gamma_n) = \begin{cases} \varphi(\gamma_0, \gamma_1, \cdots, \gamma_n) & \text{if } \gamma_0\gamma_1\cdots\gamma_n \neq e, \\ 0 & \text{otherwise.} \end{cases}$$

The following theorem is an strengthening of the higher Atiyah-Patodi-Singer index formula in [**3**, Theorem 3.36].

THEOREM 5.6. *Assume that φ is a cyclic cocycle of Γ with exponential growth rate K_φ. If the scalar curvature k on ∂M satisfies that*

$$\inf_{x\in\partial M} k(x) > \frac{16(K_\Gamma + K_\varphi)^2}{\tau_\partial^2}, \tag{5.10}$$

then

$$\mathrm{ch}_\varphi(\mathrm{Ind}^\Gamma_{1,K_\varphi}(D)) = \langle \varphi_e, \int_M \hat{A}\wedge\omega\rangle - \frac{1}{2}\langle\varphi, \widetilde{\eta}(\widetilde{D}_\partial)\rangle, \tag{5.11}$$

*where $\hat{A}$ is the associated $\hat{A}$-form on M and ω is an element in $\Omega^*(M)\otimes\Omega_*(l^1_{K_\varphi}(\Gamma))$ (cf. [**12**, Theorem 13.6]) and φ_e (resp. φ_d) is the localized (reps. delocalized) part of φ. Consequently, if both Γ and φ have sub-exponential growth, then the equality in line* (5.11) *holds as long as $\widetilde{D}_\partial$ is invertible.*

PROOF. Under the assumptions of the theorem, we have the following equality

$$\begin{aligned} &b\text{-Ch}_{t_1}(\widetilde{D}) - b\text{-Ch}_{t_0}(\widetilde{D}) \\ &= -\frac{1}{2}\int_{t_0}^{t_1} \mathrm{STR}(\widetilde{D}_\partial e^{-(s\widetilde{D}_\partial+\nabla)^2})ds + d\int_{t_0}^{t_1} b\text{-STR}(\widetilde{D}e^{-(\nabla+s\widetilde{D})^2})ds \end{aligned} \tag{5.12}$$

holds in $\Omega_*(l^1_{K_\varphi}(\Gamma))$, where $d\colon \Omega_*(l^1_{K_\varphi}(\Gamma)) \to \Omega_{*+1}(l^1_{K_\varphi}(\Gamma))$ is the differential on $\Omega_*(l^1_{K_\varphi}(\Gamma))$, cf. [**12**, Proposition 14.2]. By pairing both sides of (5.12) with φ, we have

$$\langle\varphi, b\text{-Ch}_{t_1}(\widetilde{D})\rangle - \langle\varphi, b\text{-Ch}_{t_0}(\widetilde{D})\rangle = -\frac{1}{2}\int_{t_0}^{t_1}\langle\varphi, \mathrm{STR}(\widetilde{D}_\partial e^{-(s\widetilde{D}_\partial+\nabla)^2})\rangle ds.$$

By [**3**, Theorem 3.36], we have

$$\lim_{t\to\infty}\langle\varphi, b\text{-Ch}_t(\widetilde{D}))\rangle = \langle\varphi_e, \int_W \hat{A}\wedge\omega\rangle - \frac{1}{2}\langle\varphi, \widetilde{\eta}(\widetilde{D}_\partial)\rangle. \tag{5.13}$$

Now by Theorem 4.1, the higher index $\mathrm{Ind}^\Gamma(\widetilde{D})$ of $\widetilde{D}$ lies in $K_0(l^1_{K_\varphi}(\Gamma))$. By applying the techniques developed by Bismut [**1**] (cf.[4] [**12**, Section 12 & 14]), one shows that

$$\lim_{t\to\infty}\langle\varphi, b\text{-Ch}_t(\widetilde{D}))\rangle = \mathrm{ch}_\varphi(\mathrm{Ind}^\Gamma(\widetilde{D})).$$

[4]The construction of the higher index $\mathrm{Ind}^\Gamma(\widetilde{D})$ in Section 4 can be thought as the K-theoretic equivalent of taking b-trace.

This finishes the theorem. □

Let $\widetilde{\eta}(\widetilde{D}_\partial)$ be Lott's noncommutative differential higher eta invariant from above. Clearly, we have

$$\langle \varphi, \widetilde{\eta}(\widetilde{D}_\partial)\rangle = \langle \varphi_e, \widetilde{\eta}(\widetilde{D}_\partial)\rangle + \langle \varphi_d, \widetilde{\eta}(\widetilde{D}_\partial)\rangle$$

In the following, we shall prove under the same assumptions as in Theorem 5.6 that

$$\langle \varphi_d, \widetilde{\eta}(\widetilde{D}_\partial)\rangle = \eta_{\varphi_d}(\widetilde{D}_\partial), \tag{5.14}$$

where $\eta_{\varphi_d}(\widetilde{D}_\partial)$ is the delocalized higher eta invariant of $\widetilde{D}_\partial$ at φ_d given in Definition 5.3. Consequently, for delocalized cyclic cocycles, Theorem 5.6 recovers Theorem 5.4.

The equality (5.14) for identifying two different formulas of the delocalized higher eta invariant was first established by the authors in [**4**, Section 8], under the assumption that Γ has polynomial growth and φ is a delocalized cyclic cocycle with polynomial growth.[5] It was stated as an open question whether the equality holds in general. The following proposition answers this question positively for all groups, but under the condition that the spectral gap of $\widetilde{D}_\partial$ is sufficiently large, i.e. the equality in line (5.10) holds.

In order to make the discussion more transparent and also consistent with that from [**4**, Section 8], we shall work with the periodic version of Lott's noncommutative differential higher eta invariant.

DEFINITION 5.7 ([**13**, Section 4.4 & 4.6]). For each $\beta > 0$, we define $\widetilde{\eta}(\widetilde{D}_\partial, \beta)$ by the following integral

$$\widetilde{\eta}(\widetilde{D}_\partial, \beta) = \beta^{1/2}\int_0^\infty \mathrm{STR}(\widetilde{D}_\partial e^{-\beta(t\widetilde{D}_\partial+\nabla)^2})dt.$$

Then the periodic version of Lott's noncommutative differential higher eta invariant of $\widetilde{D}_\partial$ is defined to be

$$\widetilde{\eta}^{\mathrm{per}}(\widetilde{D}_\partial) = \int_0^\infty e^{-\beta}\widetilde{\eta}(\widetilde{D}, \beta^2)d\beta.$$

In [**4**, Section 8], using the Laplace transform, the authors showed that of $\langle \varphi_d, \widetilde{\eta}(\widetilde{D}_\partial)\rangle$ is *formally* equal to the following integral

$$\sqrt{\pi}\cdot\frac{m!}{\pi i}\int_0^\infty \varphi\#\mathrm{tr}\big(\dot{v}_s v_s^{-1}\otimes\big((v_s-1)\otimes(v_s^{-1}-1)\big)^{\otimes m}\big)ds, \tag{5.15}$$

where

$$v_s = \begin{cases} (\widetilde{D}+si)(\widetilde{D}-si)^{-1} & \text{if } s>0, \\ 1 & \text{if } s=0. \end{cases} \tag{5.16}$$

Now we are ready to prove the equality (5.14).

PROPOSITION 5.8. *Under the same assumptions as in Theorem 5.6, we have*

$$\langle \varphi_d, \widetilde{\eta}^{\mathrm{per}}(\widetilde{D}_\partial)\rangle = \sqrt{\pi}\cdot\eta_{\varphi_d}(\widetilde{D}_\partial).$$

[5]Here under the condition that Γ has polynomial growth, assuming φ to have polynomial growth does not impose any serious restrictions. Indeed, if Γ has polynomial growth, then every cyclic cohomology class of Γ has a representative that has polynomial growth and furthermore any two such representatives differ by a cyclic coboundary with polynomial growth.

PROOF. By the estimates from Lemma 2.17, we see that the integrand in line (5.15) is finite for each $t > 0$ and the integral converges for large t (cf. [**4**, Lemma 3.27]). In Lemma 2.22, under the large scalar curvature assumption in line (5.10), we prove that $\{v_s\}_{s\in[0,+\infty)}$ defines an invertible element in $(\mathcal{B}_{L,0}(\partial\widetilde{M})^{\Gamma}_{K_\varphi})^+$. Consequently, it follows that the integrand in line (5.15) is continuous for all $t \in [0,\infty)$, hence the integral also converges for small t. This proves that $\langle \varphi_d, \widetilde{\eta}^{\mathrm{per}}(\widetilde{D}_\partial)\rangle$ is equal to the integral in line (5.15).

It remains to verify that the integral in line (5.15) is equal to $\eta_{\varphi_d}(\widetilde{D}_\partial)$ from Definition 5.3. This can be done by applying a standard transgression formula as follows. Let $w_{\lambda,s}$, $\lambda \in [0,1]$ and $s \in [0,\infty)$, be the path of invertible elements connecting v_s and u_s from line (2.31). A straightforward calculation shows that

$$\partial_s\left(\varphi\#\mathrm{tr}(w^{-1}\partial_\lambda w \otimes (w\otimes w^{-1})^{\otimes m})\right) = \partial_\lambda\left(\varphi\#\mathrm{tr}(w^{-1}\partial_s w \otimes (w\otimes w^{-1})^{\otimes m})\right).$$

It follows that

$$\int_0^T \varphi\#\mathrm{tr}(\dot{v}v^{-1}\otimes(v\otimes v^{-1})^{\otimes m})ds - \int_0^T \varphi\#\mathrm{tr}(\dot{u}u^{-1}\otimes(u\otimes u^{-1})^{\otimes m})ds$$

$$= \int_0^1 \widetilde{\varphi\#\mathrm{tr}}(w^{-1}\partial_\lambda w\otimes(w\otimes w^{-1})^{\otimes m})\Big|_{s=0}^{s=T} d\lambda \tag{5.17}$$

By the l^1_K-norm estimates above (see also Lemma 2.17), we have

$$\varphi\#\mathrm{tr}(w^{-1}\partial_\lambda w\otimes(w\otimes w^{-1})^{\otimes m}) \to 0,$$

as $s\to\infty$. Also, note that $w_{\lambda,0}\equiv 1$ for all $\lambda\in[0,1]$. This implies that the right hand side of the equality (5.17) goes to zero as T goes to infinity. This proves that the integral in line (5.15) is equal to $\eta_{\varphi_d}(\widetilde{D}_\partial)$ from Definition 5.3, hence completes the proof.

□

References

[1] Jean-Michel Bismut, *The Atiyah-Singer index theorem for families of Dirac operators: two heat equation proofs*, Invent. Math. **83** (1986), no. 1, 91–151, DOI 10.1007/BF01388755. MR813584

[2] Ulrich Bunke, *A K-theoretic relative index theorem and Callias-type Dirac operators*, Math. Ann. **303** (1995), no. 2, 241–279, DOI 10.1007/BF01460989. MR1348799

[3] Xiaoman Chen, Hongzhi Liu, Hang Wang, and Guoliang Yu, *Higher rho invariant and delocalized eta invariant at infinity*, arXiv:2004.00486, 2020.

[4] Xiaoman Chen, Jinmin Wang, Zhizhang Xie, and Guoliang Yu. *Delocalized eta invariants, cyclic cohomology and higher rho invariants*, arXiv:1901.02378, 2019.

[5] Alain Connes, *Noncommutative differential geometry*, Inst. Hautes Études Sci. Publ. Math. **62** (1985), 257–360. MR823176

[6] Alain Connes, *Noncommutative geometry*, Academic Press, Inc., San Diego, CA, 1994. MR1303779

[7] Alain Connes and Henri Moscovici, *Cyclic cohomology, the Novikov conjecture and hyperbolic groups*, Topology **29** (1990), no. 3, 345–388, DOI 10.1016/0040-9383(90)90003-3. MR1066176

[8] Nigel Higson and Gennadi Kasparov, *E-theory and KK-theory for groups which act properly and isometrically on Hilbert space*, Invent. Math. **144** (2001), no. 1, 23–74, DOI 10.1007/s002220000118. MR1821144

[9] Nigel Higson and John Roe, *Mapping surgery to analysis. I. Analytic signatures*, K-Theory **33** (2005), no. 4, 277–299, DOI 10.1007/s10977-005-1561-8. MR2220522

[10] Nigel Higson and John Roe, *Mapping surgery to analysis. II. Geometric signatures*, K-Theory **33** (2005), no. 4, 301–324, DOI 10.1007/s10977-005-1559-2. MR2220523

[11] Vincent Lafforgue, *K-théorie bivariante pour les algèbres de Banach et conjecture de Baum-Connes* (French), Invent. Math. **149** (2002), no. 1, 1–95, DOI 10.1007/s002220200213. MR1914617

[12] Eric Leichtnam and Paolo Piazza, *The b-pseudodifferential calculus on Galois coverings and a higher Atiyah-Patodi-Singer index theorem* (English, with English and French summaries), Mém. Soc. Math. Fr. (N.S.) **68** (1997), iv+121. MR1488084

[13] John Lott, *Higher eta-invariants*, K-Theory **6** (1992), no. 3, 191–233, DOI 10.1007/BF00961464. MR1189276

[14] Richard B. Melrose, *The Atiyah-Patodi-Singer index theorem*, Research Notes in Mathematics, vol. 4, A K Peters, Ltd., Wellesley, MA, 1993, DOI 10.1016/0377-0257(93)80040-i. MR1348401

[15] Igor Mineyev and Guoliang Yu, *The Baum-Connes conjecture for hyperbolic groups*, Invent. Math. **149** (2002), no. 1, 97–122, DOI 10.1007/s002220200214. MR1914618

[16] Paolo Piazza and Thomas Schick, *Rho-classes, index theory and Stolz' positive scalar curvature sequence*, J. Topol. **7** (2014), no. 4, 965–1004, DOI 10.1112/jtopol/jtt048. MR3286895

[17] Paolo Piazza, Thomas Schick, and Vito Felice Zenobi, *Mapping analytic surgery to homology, higher rho numbers and metrics of positive scalar curvature*, `arXiv:1905.11861`, 2019.

[18] Daniel Quillen, *Superconnections and the Chern character*, Topology **24** (1985), no. 1, 89–95, DOI 10.1016/0040-9383(85)90047-3. MR790678

[19] Michael Reed and Barry Simon, *Methods of modern mathematical physics. II. Fourier analysis, self-adjointness*, Academic Press [Harcourt Brace Jovanovich, Publishers], New York-London, 1975. MR0493420

[20] John Roe, *Coarse cohomology and index theory on complete Riemannian manifolds*, Mem. Amer. Math. Soc. **104** (1993), no. 497, x+90, DOI 10.1090/memo/0497. MR1147350

[21] Charlotte Wahl, *The Atiyah-Patodi-Singer index theorem for Dirac operators over C^*-algebras*, Asian J. Math. **17** (2013), no. 2, 265–319, DOI 10.4310/AJM.2013.v17.n2.a2. MR3078932

[22] Jinmin Wang, Zhizhang Xie, and Guoliang Yu, *Approximations of delocalized eta invariants by their finite analogues*, `arXiv:2003.03401`, 2020.

[23] Shmuel Weinberger, Zhizhang Xie, and Guoliang Yu, *Additivity of higher rho invariants and nonrigidity of topological manifolds*, Comm. Pure Appl. Math. **74** (2021), no. 1, 3–113, DOI 10.1002/cpa.21962. MR4178180

[24] Zhizhang Xie and Guoliang Yu, *Positive scalar curvature, higher rho invariants and localization algebras*, Adv. Math. **262** (2014), 823–866, DOI 10.1016/j.aim.2014.06.001. MR3228443

[25] Zhizhang Xie and Guoliang Yu, *Delocalized eta invariants, algebraicity, and K-theory of group C^*-algebras*, Int. Math. Res. Not. IMRN **15** (2021), 11731–11766, DOI 10.1093/imrn/rnz170. MR4294131

[26] Guoliang Yu, *Localization algebras and the coarse Baum-Connes conjecture*, K-Theory **11** (1997), no. 4, 307–318, DOI 10.1023/a:1007766031161. MR1451759

[27] Guoliang Yu, *The Novikov conjecture for groups with finite asymptotic dimension*, Ann. of Math. (2) **147** (1998), no. 2, 325–355, DOI 10.2307/121011. MR1626745

[28] Guoliang Yu, *The coarse Baum-Connes conjecture for spaces which admit a uniform embedding into Hilbert space*, Invent. Math. **139** (2000), no. 1, 201–240, DOI 10.1007/s002229900032. MR1728880

[29] Rudolf Zeidler, *Positive scalar curvature and product formulas for secondary index invariants*, J. Topol. **9** (2016), no. 3, 687–724, DOI 10.1112/jtopol/jtw005. MR3551834

[30] Vito Felice Zenobi, *Mapping the surgery exact sequence for topological manifolds to analysis*, J. Topol. Anal. **9** (2017), no. 2, 329–361, DOI 10.1142/S179352531750011X. MR3622237

Department of Mathematics, Texas A&M University, College Station, Texas, 77843
Email address: `jinmin@tamu.edu`

Department of Mathematics, Texas A&M University, College Station, Texas, 77843
Email address: `xie@math.tamu.edu`

Department of Mathematics, Texas A&M University, College Station, Texas, 77843
Email address: `guoliangyu@math.tamu.edu`

SELECTED PUBLISHED TITLES IN THIS SERIES

105 **A. Connes, C. Consani, B. I. Dundas, M. Khalkhali, and H. Moscovici, Editors,** Cyclic Cohomology at 40: Achievements and Future Prospects, 2023

104 **A. Kechris, N. Makarov, D. Ramakrishnan, and X. Zhu, Editors,** Nine Mathematical Challenges, 2021

103 **Sergey Novikov, Igor Krichever, Oleg Ogievetsky, and Senya Shlosman, Editors,** Integrability, Quantization, and Geometry, 2021

102 **David T. Gay and Weiwei Wu, Editors,** Breadth in Contemporary Topology, 2019

101 **Avraham Aizenbud, Dmitry Gourevitch, David Kazhdan, and Erez M. Lapid, Editors,** Representations of Reductive Groups, 2019

100 **Chiu-Chu Melissa Liu and Motohico Mulase, Editors,** Topological Recursion and its Influence in Analysis, Geometry, and Topology, 2018

99 **Vicente Muñoz, Ivan Smith, and Richard P. Thomas, Editors,** Modern Geometry, 2018

98 **Amir-Kian Kashani-Poor, Ruben Minasian, Nikita Nekrasov, and Boris Pioline, Editors,** String-Math 2016, 2018

97 **Tommaso de Fernex, Brendan Hassett, Mircea Mustaţă, Martin Olsson, Mihnea Popa, and Richard Thomas, Editors,** Algebraic Geometry: Salt Lake City 2015 (Parts 1 and 2), 2018

96 **Si Li, Bong H. Lian, Wei Song, and Shing-Tung Yau, Editors,** String-Math 2015, 2017

95 **Izzet Coskun, Tommaso de Fernex, and Angela Gibney, Editors,** Surveys on Recent Developments in Algebraic Geometry, 2017

94 **Mahir Bilen Can, Editor,** Algebraic Groups: Structure and Actions, 2017

93 **Vincent Bouchard, Charles Doran, Stefan Méndez-Diez, and Callum Quigley, Editors,** String-Math 2014, 2016

92 **Kailash C. Misra, Daniel K. Nakano, and Brian J. Parshall, Editors,** Lie Algebras, Lie Superalgebras, Vertex Algebras and Related Topics, 2016

91 **V. Sidoravicius and S. Smirnov, Editors,** Probability and Statistical Physics in St. Petersburg, 2016

90 **Ron Donagi, Sheldon Katz, Albrecht Klemm, and David R. Morrison, Editors,** String-Math 2012, 2015

89 **D. Dolgopyat, Y. Pesin, M. Pollicott, and L. Stoyanov, Editors,** Hyperbolic Dynamics, Fluctuations and Large Deviations, 2015

88 **Ron Donagi, Michael R. Douglas, Ljudmila Kamenova, and Martin Rocek, Editors,** String-Math 2013, 2014

87 **Helge Holden, Barry Simon, and Gerald Teschl, Editors,** Spectral Analysis, Differential Equations and Mathematical Physics: A Festschrift in Honor of Fritz Gesztesy's 60th Birthday, 2013

86 **Kailash C. Misra, Daniel K. Nakano, and Brian J. Parshall, Editors,** Recent Developments in Lie Algebras, Groups and Representation Theory, 2012

85 **Jonathan Block, Jacques Distler, Ron Donagi, and Eric Sharpe, Editors,** String-Math 2011, 2012

84 **Alex H. Barnett, Carolyn S. Gordon, Peter A. Perry, and Alejandro Uribe, Editors,** Spectral Geometry, 2012

83 **Hisham Sati and Urs Schreiber, Editors,** Mathematical Foundations of Quantum Field Theory and Perturbative String Theory, 2011

82 **Michael Usher, Editor,** Low-dimensional and Symplectic Topology, 2011

For a complete list of titles in this series, visit the
AMS Bookstore at **www.ams.org/bookstore/pspumseries/**.